Physical Chemistry Problems and Solutions

Satoshi Hirayama · Masahiro Yamamoto

Physical Chemistry Problems and Solutions

Atoms, Molecules and Thermodynamics

Springer

Satoshi Hirayama
Kyoto Institute of Technology
Kyoto, Japan

Masahiro Yamamoto
Konan University
Kobe, Japan

ISBN 978-981-97-0164-3 ISBN 978-981-97-0165-0 (eBook)
https://doi.org/10.1007/978-981-97-0165-0

This Springer imprint is published by the registered company Springer Nature Singapore Pte Ltd.
The registered company address is: 152 Beach Road, #21-01/04 Gateway East, Singapore 189721, Singapore

$$\hat{l}_x + i\,\hat{l}_y = \hbar\, e^{i\varphi} \left[\frac{\partial}{\partial \vartheta} + i \cot\vartheta \frac{\partial}{\partial \varphi} \right]$$

$$k(t) = 4\pi R_{AB} D_{AB} N_A \left\{ 1 + \frac{R_{AB}}{(\pi D_{AB} t)^{1/2}} \right\}$$

Preface

When we started our research life as chemistry students, we were fascinated by the way the chemists, who were called giants of physical chemistry, solved previously unsolved chemical phenomena one after another as if they were freely manipulating their palms, and we aspired to become researchers of physical chemistry in the future. The term "Physical Chemist" sounded very comfortable and cool to us at the time.

It has already been more than a couple of decades since we started studying physical chemistry. Even though the subject matter has become more sophisticated and the research methods unimaginably advanced, physical chemistry itself, which forms the theoretical basis for understanding chemical phenomena, has not changed significantly. In fact, it can be said that the fundamentals have changed little.

On the other hand, computer software for analyzing theoretical equations and experimental results and the vast amount of scientific information available on the Internet have changed significantly our approach to solve scientific problems. As we often use them in this book, the computational software Mathematica and MITOCW, one of the free public classes, have greatly contributed to the understanding of not only physical chemistry but also other natural sciences. In addition to many attractive live lectures, MITOCW also offers demonstration experiments, problems, and answers, allowing students to enjoy classes at one of the world's top universities without leaving their own living places. It is a world away from the old days when one thought that studying abroad or somewhere at scientifically established places was the only way to come into contact with the world of science in the front. In this book, we have incorporated these elements as necessary.

This book covers seven areas of physical chemistry and is divided practically into two volumes, the first volume covering problems on atoms, molecules and thermodynamics and the second volume those on distributions, reactions and structures. This volume is the first one. We have provided a large number of problems, almost a thousand in total, all of which we considered useful for understanding the fundamentals of physical chemistry. All of them were solved and checked by ourselves. For this, we were able to provide each problem consistently with a very detailed solution beginning from an introductory to advanced approach with extensive explanations appropriately annotated when necessary.

The problems and accompanied solutions are categorized into seven areas as mentioned above but the arrangement of the problems within each area is not necessarily ordered. Consequently, it is not necessary to start reading from the beginning of the first volume. You can pick up any problem in any area of your interest. The reason for this is that the actual "questions" to be solved are not always prepared in order in the real world, and we thought it would be better to see how to solve the questions that one unexpectedly encounters, so that one could gain a deeper understanding of physical chemistry by doing so. Instead of putting problems in order, however, we have provided the extensive index so that one can locate one's own questions in the index. We hope that one will make full use of the index.

Unfortunately, unlike research, various educational efforts and attempts are not necessarily easily passed on. It would be a great pleasure for us, therefore, if the approaches to the problems attempted in this book, including, dare we say, various errors we may have made, could be of use not only to those who want to learn physical chemistry comprehensively but also those who are trying hard to teach it to students as a basic tool for understanding various kinds of chemical phenomena.

Kobe, Japan
Autumn 2024

Satoshi Hirayama
Masahiro Yamamoto

Acknowledgments

The completion of this book has been made possible not only by the questions, discussions, and ideas exchanged with many fellow researchers, students, research students and post-doctoral fellows throughout our long careers in education and research in the field of physical chemistry, but also by their warm advice, encouragement and support. We do not list up their names specifically here, but we would like to express our deepest gratitude to them. This book is dedicated to them and to our dear families.

Finally, we would like to express our deepest gratitude to Dr. Shinichi Koizumi of Springer Nature, the publisher. Without his precise advice and long-term support, this publication would not have been possible.

Contents

About the Authors

Satoshi Hirayama

1964 Graduated with a degree in Chemistry, Faculty of Science, Kyoto University
1972 Doctor of Science, Kyoto University
1983 Professor, Kyoto Institute of Technology
2006 Professor Emeritus, Kyoto Institute of Technology
During these periods, he worked as a visiting professor or researcher at the Department of Chemistry, University of Southampton (Scholar of the British Council), Southampton, UK; Department of Chemistry, University of Melbourne, Australia; Department of Applied Physics, University of Strathclyde, Scotland; Department of Chemistry, University of Saskatchewan, Canada; Visiting Professor at the Okazaki Institute for Molecular Science. (Visiting Professor at Okazaki Institute for Molecular Science (2 years))
2009–2015 Part-time lecturer, Faculty of Science and Technology, Konan University
Currently in charge of academic counseling, Department of Functional Molecular Chemistry, Faculty of Science and Technology, Konan University

Masahiro Yamamoto

1983 Graduated from Department of Industrial Chemistry, Faculty of Engineering, Kyoto University
1985 Master Degree, Department of Industrial Chemistry, Graduate School of Engineering, Kyoto University
1985 Research Associate, Institute of Atomic Energy, Kyoto University
1991 Doctor of Engineering, Kyoto University
1993 Visiting researcher, Ames Laboratory, USDOE, Department of Physics and Astronomy, Iowa State University, Ames, Iowa, USA (1 Year)
1996 Assistant Professor, Institute of Advanced Energy, Kyoto University
1999 Associate Professor, Department of Energy and Hydrocarbon Chemistry, Graduate School of Engineering, Kyoto University
2009 Professor, Department of Chemistry, Konan University

Part I
Problems

I-1 Electronic Structures and Properties of Atoms and Simple Molecules

Part of Problems

Problem 1-1

An atomic ion that has only one electron is called a hydrogen-like atom. Its electronic energy is determined by the principal quantum number alone, and the energy difference between the levels where the principal quantum number is specified by n and m is given by the following equation.

$$E_n - E_m = -A\left(\frac{1}{n^2} - \frac{1}{m^2}\right) \qquad (1)$$

(a) The ionization energy of H is 1312 kJ mol^{-1}. Find the value of A in eq(1).

(b) Calculate the ionization energy of Li^{2+} based on the value obtained above.

(c) The above equation also helped one to determine the atomic number of an element from the observed wavelength of the X-rays emitted by the element (*Moseley*'s equation, discovery of new elements (see the reference given in the solution)). When the X-ray (K_α) wavelength of an element is 3.35 Å, what is the atomic number of this element?

Problem 1-2

Answer the following questions about the periodic table given below.

(a) Classify all elements in the periodic table into s, p, d, and f elements according to the type of atomic orbital occupied by the outermost shell electrons. Answer the question by giving a schematic representation of the electronic configuration such as $\cdot\cdot ns^2np^6\cdots$.

(b) Show the electronic configuration of the elements of sodium, copper, and antimony.

「Periodic Table」

1	H																	He
2	Li	Be											B	C	N	O	F	Ne
3	Na	Mg											Al	Si	P	S	Cl	Ar
4	K	Ca	Sc	Ti	V	Cr	Mn	Fe	Co	Ni	Cu	Zn	Ga	Ge	As	Se	Br	Kr
5	Rb	Sr	Y	Zr	Nb	Mo	Tc	Ru	Rh	Pd	Ag	Cd	In	Sn	Sb	Te	I	Xe
6	Cs	Ba	La*	Hf	Ta	W	Re	Os	Ir	Pt	Au	Hg	Tl	Pb	Bi	Po	At	Rn
7	Fr	Ra	Ac	Rf														

Lanthanides	La	Ce	Pr	Nd	Pm	Sm	Eu	Gd	Tb	Dy	Ho	Er	Tm	Yb	Lu
Actinides	Ac	Th	Pa	U	Np	Pu	Am	Cm	Bk	Cf	Es	Fm	Md	No	Lr

Problem 1-3

What are the electron configurations of Ca ($Z = 20$), Fe ($Z = 26$), Zr ($Z = 40$), and Ag ($Z = 47$)?

S. Hirayama and M. Yamamoto, *Physical Chemistry Problems and Solutions*,
https://doi.org/10.1007/978-981-97-0165-0_1

Problem 1-4

When a particle with mass m is confined in a well-shaped potential as shown below, the particle in the lowest energy state has a wavelength of 2 a as a *de Broglie* wave (*Louis de Broglie*, Nobel Prize in Physics, 1929).

(a) From this, find the allowed energy states of this particle.

(b)The idea of *de Broglie* waves alone does not tell us correctly where the particle is located or what is the probability of its existence. When the same particle is treated properly by quantum mechanics, then you can obtain the probability of finding the particle at any location, e.g., at $0.25a$. Calculate the probability of finding the particle of the lowest energy in the range of $0.24a$ to $0.25a$.

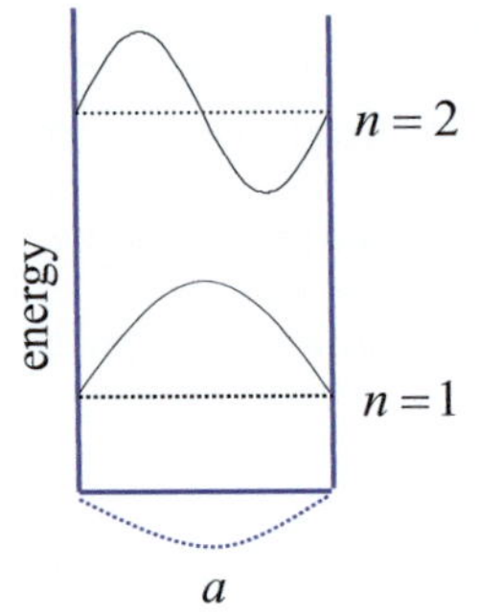

Fig.1 One-dimensional well-shaped potential and *de Broglie* wave.

Problem 1-5

(a) All of the crystals of NaI, KCl and LiF assume the ionic crystal structure of a type of rock salt known for NaCl. Tell which of those has the smallest ionic distance between a neighboring anion and cation based on the values of their ionic radii.

(b) It is assumed that the neighboring anion and cation are in contact with each other and that they form an ionic crystal of the type of sodium chloride. When the ionic radii of an anion and a cation are given by r_-, r_+, respectively, then obtain the smallest value for the ration of r_+/r_- possible for such ionic crystal.

Problem 1-6

Describe how the diatomic molecules CN, N_2, NO and O_2 are chemically bonded in terms of each own relevant molecular orbitals (*LCAO MO*: linear combination of atomic orbitals).

Problem 1-7

Nitric acid ion NO_3^- or butadiene $(CH_2=CH-CH=CH_2)$ normally assumes a planar structure. Explain why they are planar using their hybrid orbitals centered at the N-atom or each of the C-atoms.

Problem 1-8

(a) What kinds of MOs are constructed for diatomic molecules by using atomic orbitals up to 2p-orbitals?

(b) Based on the results of (a), answer which of N_2, NO, O_2, C_2,F_2 and CN become(s) more stable than the corresponding neutral molecule when i) one electron is added to form an anion or when ii) one electron is removed to form a cation.

Problem 1-9

For the following homo-nuclear diatomic molecules of H_2, N_2, O_2 and Cl_2, every component atom is assumed to be electronically in the ground state. Then list up all of conceivable electronic states for each of those diatomic molecules in terms of "spectral term".

Problem 1-10

For the hetero-nuclear diatomic molecules of HCl (H: ^2S, Cl: ^2P) and CO (C: ^3P, O: ^3P), give all of the possible molecular electronic states to be formed from the component atoms in the ground electronic states. Answer them in terms of the spectral term such as $^3\Sigma_u$, $^1\Pi_g$, etc.

Problem 1-11

Referring to Fig.2 given below,

(a) Answer the magnitude of the ionization energy of hydrogen atom.

(b) Answer the relation between the energy level of a hydrogen-like atom and the principal quantum number n.

(c) Answer the relation between the energy of the observed emission in the *Balmer* series and the principal quantum number.

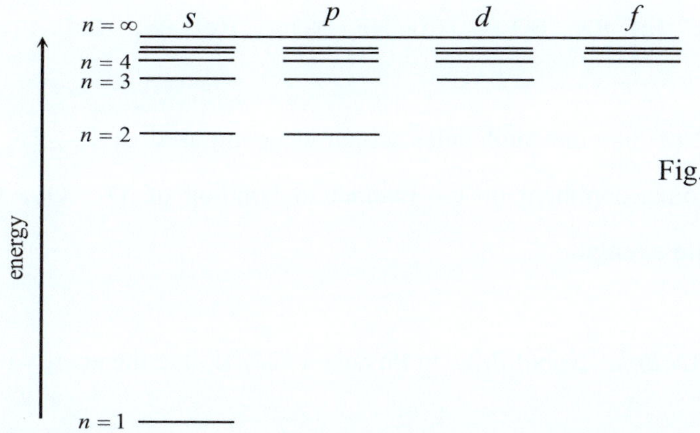

Fig.2 *Grotorian diagram* for hydrogen-like atoms. Principal quantum number n and energy level. Note that the energy level depends on n alone but not on the type of subshell.

Problem 1-12

Find the lowest ionization energy for each of Li, Be^+, B^{2+} and C^{3+} in an appropriate data book and plot its square root against the squared values of the nuclear charge as well as the effective nuclear charge of the corresponding atom or ion. How can you rationalize the obtained result ? (Find the theoretical value of the slope).

Problem 1-13

KCl forms an ionic crystal of the type of rock salt in a similar manner as does NaCl. Assuming that its *Madelung* constant is 1.75 and that the energy of sublimation for K (solid) → K(vapor) is $21.5\,\text{kcal}\cdot\text{mol}^{-1}$, answer the following questions.

(a) The electronic structures of the ions which construct the ionic crystal.

(b) The heat of the reaction of $\text{K (solid)} + \frac{1}{2}\text{Cl}_2\text{ (gas)} \rightarrow \text{K}^+\text{ (gas)} + \text{Cl}^-\text{ (gas)}$.

(c) The heat of the reaction of K^+ (gas) + Cl^- (gas)→KCl (solid).

(d) The heat of the reaction of K (solid) + $\frac{1}{2}Cl_2$ (gas)→ KCl (solid).

(e) The decreasing order of the lattice energies for the crystals of LiF, KI, NaF, KCl and NaCl.

The lattice energy is the energy required to dissociate the constituent ions of an ionic crystal to bring them into the gas phase. Necessary values are: ionization energy of atom in the gas phase, dissociation energy of diatomic molecule, electron affinity of gaseous atom, ionic distance between constituent ions in the crystal (for this purpose use the values for the ionic radii given below).

Table Ionic radii for alkaline and halide ions (Å= 0.1 nm)

Li^+	Na^+	K^+	Rb^+	Cs^+	F^-	Cl^-	Br^-	I^-
0.60	0.98	1.33	1.48	1.67	1.33	1.81	1.96	2.19

Problem 1-14

Describe the molecular orbitals (MO) of diatomic molecules which are composed of 1*s*-, 2*s*- and 2*p*-atomic orbitals of relevant component atoms with explanations for the type and nature of composed MO's. Based on your explanations, predict which one in each pair of molecules given in (a) through to (e) is more stable than the other.

(a) F_2, F_2^+; (b) NO, NO^-; (c) BN, BO; (d) NF, NO; (e) Be_2, Be_2^+.

Problem 1-15

Based on the molecular orbitals (MO) of diatomic molecules which are composed of 1*s*-, 2*s*- and 2*p*-atomic orbitals of relevant component atoms, comment on the manner of bonding of O_2, O_2^+, O_2^-, N_2 and N_2^+ in the increasing order of the bond strength.

Problem 1-16

(a) Employing two of *Bohr*'s hypotheses (*N.Bohr*: Nobel Prize in Physics 1922), derive the energy levels of a hydrogen atom.

(b) Based on the results obtained above in (a), derive the relations between the ionic radii as well as the ionization energies of hydrogen-like atoms and their atomic numbers.

(c) Answer the electron configuration for ${}_{11}Na^+$, ${}_{12}Mg^{2+}$ and ${}_{13}Al^{3+}$.

(d) Estimate the relative values of the ionic radius for Na^+, Mg^{2+} and Al^{3+} based on not only the fact that the observed first ionization potential for Na^+, Mg^{2+}, Al^{3+} are 47.29, 80.12 and 119.96 eV, respectively but also on the results obtained above in from (a) to (c),

Problem 1-17

The direction of a hybrid orbital, say a direction favorable for bonding, varies with the extent of mixing of each component orbital for hybridization. When the two atomic orbitals given below are mixed together on hybridization with the ratio of $\frac{1}{\sqrt{6}}$ to $\frac{1}{\sqrt{2}}$ as in $(\frac{1}{\sqrt{6}}p_x - \frac{1}{\sqrt{2}}p_y)$, then what does the shape of the resulting hybrid orbital look like? Answer it by drawing a rough shape of the hybrid orbital.

$$p_x = R\sin\vartheta\cos\varphi$$
$$p_y = R\sin\vartheta\sin\varphi \qquad (R = \text{constant})$$

Problem 1-18

Answer four different compounds in which at least one chemical bond can be described in terms of sp^2-hybridization.

Problem 1-19

The radial parts of 1s- and 2s-orbitals, $R_{1s}(r)$ and $R_{2s}(r)$, are given by the following wave functions.

$$R_{1s}(r) = 2\left(\frac{1}{a_0}\right)^{3/2} e^{-\frac{r}{a_0}}$$
$$(a_0 : \textit{Bohr}\ \text{radius})$$
$$R_{2s}(r) = \frac{1}{2\sqrt{2}}\left(\frac{1}{a_0}\right)^{3/2}\left(2-\frac{r}{a_0}\right)e^{-\frac{r}{2a_0}}$$

If the 'radius of an atomic orbital' is regarded as one equal to the distance at which the probability of finding the orbital electron takes a maximum value, then how many times is the radius of 2s-orbial bigger than that of 1s-orbital?

Problem 1-20

Based on the restriction imposed on the allowed combination of quantum numbers, show that the numbers of electrons which can be accommodated in K, L, M and N atomic shells are 2, 8, 18 and 32, respectively. These numbers are called magic numbers.

Problem 1-21

Let us assume that a unit cell of metal Ti is cubic as is shown in Fig.3 below. The atomic weight of Ti is 47.9 and the density of metal Ti is 4.55 $\text{kg}\cdot\text{dm}^{-3}$. Calculate the shortest distance between two Ti atoms in the unit cell.

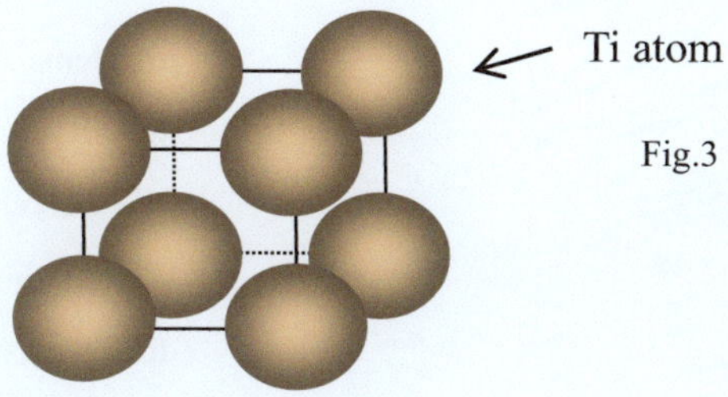

Fig.3 Assumed crystal structure of metal Ti (cubic lattice). Actual crystal structure of Ti is a close packed hexagonal one.

Problem 1-22

(a) How many electrons can be accommodated in 3d-orbitals?

(b) Refer to the figure in the preceding problem. It is seen that the electronic energy of 3d-orbitals start to decrease rapidly at around Z=2. Explain why this happens.

(c) Confirm whether the n+l rule* is satisfied at Z=50 (position marked by a vertical green line in the figure given in the preceding problem) for the order of the energy level of each atomic orbital.

* $n+l$ rule is one of the rules consisting of *Aufbau Principle* which is used to predict the electronic configuration of an atom. It tells us that the order of the energy levels of atomic orbitals is in the increasing order of the magnitude of $n+l$, where n is the principal quantum number and l is the azimuthal quantum

number. When a positive integer of n is given, l can take integer values of $1,\ 2,\ 3,\ \cdots$ up to $n-1$.

The atomic orbitals are specified by angular momentum of $l=0,\ 1,\ 2,\ 3,\ \cdots$ and they are usually named as *s*-, *p*-, *d*-, *f*-orbitals, respectively. In case of hydrogen, the electronic energy levels depend solely on the principal quantum number n alone, but for all the other atoms including helium, the energy levels come to depend on l as well. $n+l$ rule is a rule of thumb which tells us (1) that the energy of the level specified by n and l varies in parallel with the magnitude of $n+l$ and (2) that if $n+l$ is the same value for two levels, then the level with the smaller value of n is considered to be located lower than the other. Of course, it is not always applicable to any case. See also the solution to the problem.

Problem 1-23

The electronic energies of atomic orbitals of many electron atoms are plotted against the atomic number Z in Fig.4.

(a) The energy of 1*s*-orbital is seen to decrease smoothly with Z. Answer how its energy is related to Z by using the equation derived from *Bohr*'s hypothesis.

(b) For *d*- and *f*-orbitals, their orbital energies remain almost constant against Z up to certain atomic numbers and after that, they decrease rapidly. Explain how this happens.

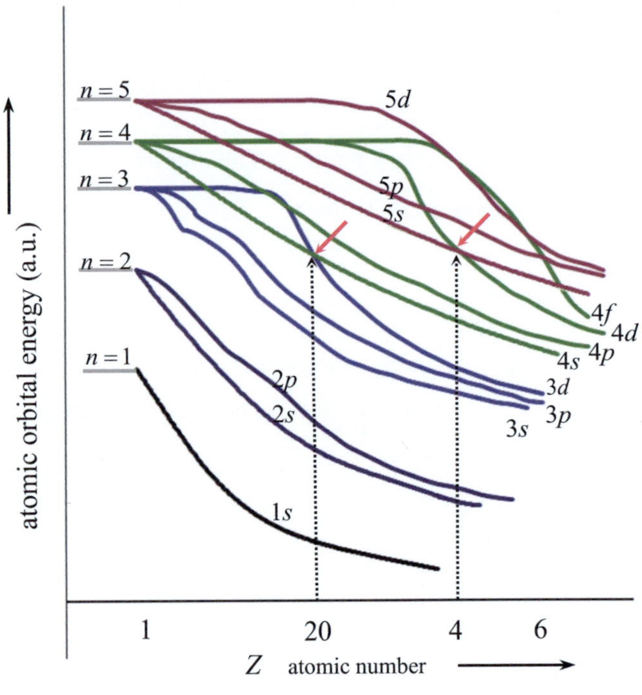

Fig.4 Orbital energies of atomic orbitals plotted against the atomic number Z. Note the scale of either axis is not linear. See **problem 1-11** for *Grotorian* diagram. (See *B.H. Mahan, University Chemistry* 2nd ed. ch.10 published by Addison Wesley in 1969 for an elementary discussion on the atomic orbital energy.) Fig.4 is modified after Fig.10.22 in *University Chemistry*. See **solution 1-22** for the unit of the vertical axis.

Problem 1-24

The magnetic quantum number (m_l) of *d*-orbital can assume five numbers of $2,\ 1,\ 0,\ -1,\ -2$. Corresponding to these numbers, there are five independent atomic orbitals. In many cases, however, these five orbitals are not specified by the magnetic quantum numbers m_l themselves but are indicated by using Cartesian coordinate x, y, z, like d_{z^2}, d_{xy}, d_{yz}, d_{zx}, $d_{x^2-y^2}$. Explain how these subscripts are related to the angular dependence of the atomic wavefunctions of *d*-orbitals. See **problem 1-117** to know why m_l is called "magnetic quantum number".

[hint] Consider the relation between the subscripts of *p*-orbitals and their angular shapes.

Problem 1-25

(a) Explain the molecular orbitals which are composed of 1*s* and 2*p*-atomic orbitals for homo-nuclear diatomic molecules.

(b) Express the chemical bonds of H_2, H_2^+, He_2^+, He_2 in terms of molecular orbitals considered in (a) and explain which molecule or ion corresponds to which potential curve shown in Fig.5 on the right from the viewpoint of the energy levels of the molecular orbitals (bonding or anti-bonding) used for its bonding.

(c) Estimate the bonding distance for H_2^+.

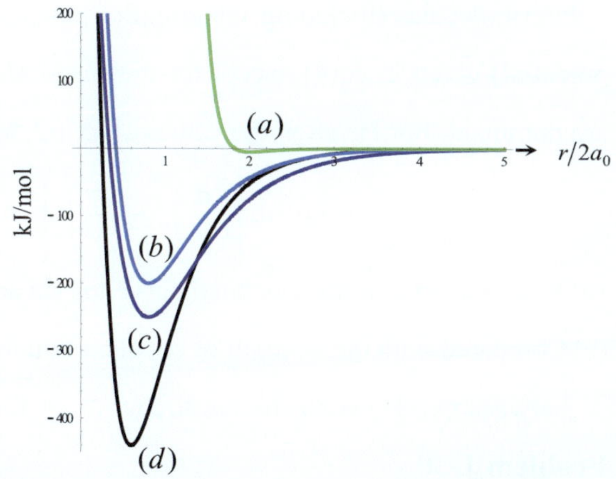

Fig.5 Potential energy curves for hydrogen or helium diatomic molecules.

Problem 1-26

The wave function $\varphi_n(x)$ given by eq(2) is for a particle with a mass *m* in a one-dimensional box of a length *a* and satisfies eq(3).

$$\varphi_n(x) = \left(\frac{2}{a}\right)^{1/2} \sin\left(\frac{n\pi x}{a}\right), \quad n: \text{positive integer} \qquad (2)$$

$$-\frac{h^2}{8m\pi^2}\frac{d^2}{dx^2}\varphi_n(x) = E_n\varphi_n(x) \qquad (3)$$

(a) Obtain the eigenvalues of energy E_n by substituting eq(2) into eq(3).

(b) Find a value of *x* where the probability of finding the particle takes a maximum for each of the states with the quantum numbers of $n = 1$, 2 and 3.

(c) For each of the states with the quantum numbers of $n = 1$, 2 and 3, obtain the *De Broglie* wavelength.

Problem 1-27

For one of the *d*-orbitals, its dependence on the polar coordinates of ϑ and φ is given by the function below. Draw its approximate shape in an *x*, *y*, *z*-space with explanations.

$$\left(\frac{15}{4\pi}\right)^{1/2} \sin\vartheta \cos\vartheta \cos\varphi$$

Problem 1-28

Based on the *De Broglie* relation, calculate the ratio of the lowest energies of electrons circulating separately along the circles A and B (no electric charge at neither center).

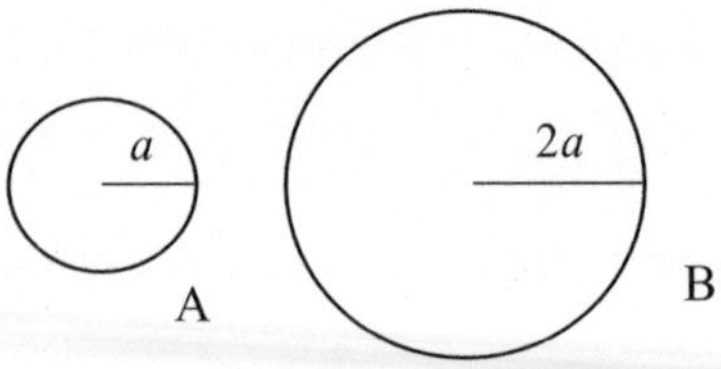

Fig.6 Two circular orbitals along which an electron can move freely.

Problem 1-29

Intermolecular (including interatomic) potentials can be approximated by *Lennard-Jones* potential (12-6 potential) given in eq(4) except for the cases where some chemical boning or special intermolecular forces are dominant. For He atom the values of $\sigma = 2.56\times10^{-10}\,\mathrm{m},\ \varepsilon = 84.7$ are known.

$$u(r) = 4\varepsilon\left[\left(\frac{\sigma}{r}\right)^{12} - \left(\frac{\sigma}{r}\right)^{6}\right] \qquad (4)$$

(a) Draw the interatomic potential curve for He and obtain the depth of the potential.

(b) Compared with the strength of covalent bonding $\mathrm{H-H}$ of H_2, what is a fraction of the attractive potential between two He atoms to that strength? Cf. Fig.5 in **problem 1-25**. .

Problem 1-30

Lennard-Jones potential function is beautiful in its functional form. However, the first term representing repulsive force cannot describe properly the real repulsive force. According to quantum mechanical calculation of the repulsive force, the repulsive force is known to be described by an exponential function of the interatomic (intermolecular) distance. Consequently, the potential function of eq(5), for which the first term of *Lennard-Jones* potential function is replaced by the exponential function, is called as exp-6 potential and is more widely used to treat interatomic and or inter molecular interactions.

$$u(r) = 4\varepsilon\left\langle \exp\left(-\frac{r-b}{a}\right) - \left(\frac{\sigma}{r}\right)^{6}\right\rangle \qquad (5)$$

(a) Determine the values of parameters a and b in eq(5) so that the potential energy not only takes a minimum value of $-\varepsilon$ but also it becomes zero at $r = \sigma$.

(b) Draw the potential curve by using the parameter values determined in (a) and describe in what points eq(5) is characteristic and superior as compared with 12-6 *Lennard-Jones* potential function given in **problem 1-29**.

Problem 1-31

An example of the data obtained by *Millikan*'s famous "*Oil Droplet*" experiments (*R.A.Millikan*, Nobel Prize in Physics in 1923) on charges of an oil droplet are as follows.

4.32×10^{-9} esu	2.88×10^{-9} esu	9.60×10^{-10} esu
3.84×10^{-9} esu	1.44×10^{-9} esu	

On the basis of these data alone, obtain the magnitude of the elementary charge e and from that value determine *Avogadro*'s number. Note that $1\ \mathrm{Coulomb} = 3.0\times10^{9}\,\mathrm{esu}$, where esu is the electrostatic unit. The elementary charge e is a fundamental unit of charge and all charges are expressed just by multiples of it.

The absolute value of the charge of an electron or a proton is equal to *e*.

Problem 1-32

Based on *Rutherford*'s *experiments on the scattering of* α *-particles* (*E.Rutherford*, Nobel Prize in chemistry in 1908), estimate the radius of an atomic nucleus whose atomic number is supposed to be 29. The velocity of the α -particle is $\upsilon = 1.8\times10^7\,\mathrm{m\cdot s^{-1}}$ and its mass is $m = 6.68\times10^{-27}\,\mathrm{kg}$.

Problem 1-33

The first ionization energies of the elements with the atomic number from 1 to 36 are plotted in Fig.7 below.

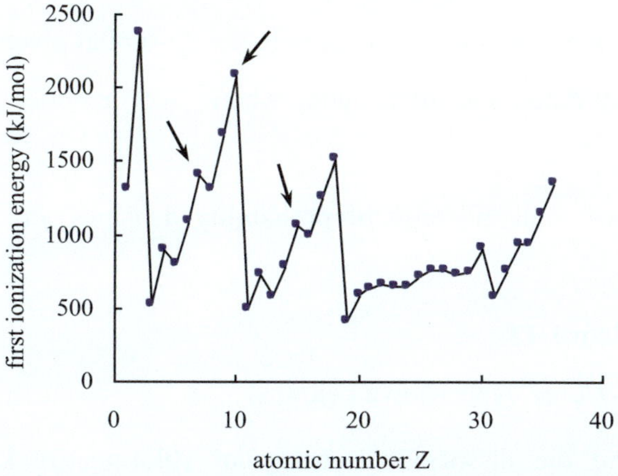

Fig.7 First ionization energy vs. atomic number.

(a) What are the elements marked by the arrows in Fig.7? Answer their names and their electronic configurations.

(b) Explain the reason why the first ionization energies vary against atomic number as shown in Fig.7 on the basis of the electronic configuration of an element. For instance, give a reason why the ionization energy of N atom is larger than that of O atom.

(c) Confirm the value of the first ionization energy of H atom by calculating it theoretically. If necessary, use the following values.

$$e = 1.602\times10^{-19}\,\mathrm{C},\quad h = 6.626\times10^{-34}\,\mathrm{J\cdot s},\quad m = 0.9109\times10^{-30}\,\mathrm{kg},\quad \varepsilon_0 = 8.854\times10^{-12}\,\mathrm{C^2\cdot N^{-1}\cdot m^{-2}}$$

(d) Obtain the effective nuclear charge (Z_{eff}) of Li.

(e) What is the element whose atomic number is 26 ? Answer it together with its electronic configuration.

Problem 1-34

Calculate the first ionization energy of He atom in the ground state by estimating its maximum and minimum values.

Problem 1-35

How many electronic states of an atom are there for which the principal quantum number is *n* and *Pauli*'s exclusion principle is satisfied. (*W.Pauli*, Nobel Prize in Physics in 1945.)

Problem 1-36

The 106th element Unh (Sg, *Seaborgium*) is a transition element and has the same electronic configuration for the outermost subshells as the 74th element W does. Answer the electronic configuration of Unh. (*G.T.Seaborg*, Nobel Prize in Chemistry in 1951.)

Problem 1-37

The atomic orbitals are generally expressed by eq(6) and by putting $n = 2,\ l = 1$ and $m = 0$, one of the 2*p*-orbitals of hydrogen atom is given by eqs(7) and (8), where a_0 is *Bohr* radius.

$$\psi(r,\vartheta,\varphi) = R_{nl}(r)Y_{lm}(\vartheta,\varphi) \qquad (6)$$

$$Y_{10} = \left(\frac{3}{4\pi}\right)^{1/2}\cos\vartheta \qquad (7),\quad R_{21}(r) = \frac{1}{4(6)^{1/2}}\left(\frac{1}{a_0}\right)^{3/2}\left(\frac{2r}{a_0}\right)\exp\left(-\frac{r}{2a_0}\right) \qquad (8)$$

(a) Obtain the probability density function $\rho(r)$ for finding the electron in the 2*p*-orbital given above in the range of $[r,\ r+dr]$ from the nucleus and then determine the distance where $\rho(r)$ assumes its maximum value and the probability density there.

(b) Using the same 2*p*-orbital examined above, find the ratio of probability densities $\rho(a_0)/\rho(4a_0)$ at $r = a_0$ and $r = 4a_0$.

Problem 1-38 a problem similar to **problem 1-13.**

Answer the following questions about *Born-Haber* cycle shown below.

(a) The heat of sublimation of sodium and the dissociation energy of chlorine are known to be 108.8 $\text{kJ}\cdot\text{mol}^{-1}$ and 241.8 $\text{kJ}\cdot\text{mol}^{-1}$, respectively. Employ $M = 1.7$ for the *Madelung* constant and assume that the closest distance between the cation and the anion is $r_0 = r_+ + r_- = 2.83$ Å. If necessary, find values required for the calculation in some data book. Then, calculate the energies associated with the processes from (1) to (3) in the cycle shown below.

(b) From the values obtained in (a), calculate the entalpy of formation ΔH_f° for the reaction of .

$$\text{Na}(s) + \frac{1}{2}\text{Cl}_2(g) \longrightarrow \text{NaCl}(s)\,.$$

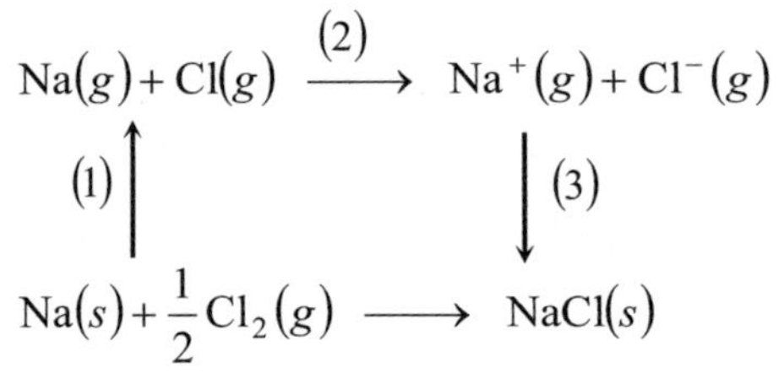

Problem 1-39

Assume that a radius of a nucleus of an atom is enlarged to a ping-pong ball of a diameter of 3 cm. Then how large would become the atom in its diameter? The radius of the nucleus R_N in the unit of m is given by the relation of $R_N = 1.33\times10^{-15}\times\sqrt[3]{A}$ in which *A* is the mass number.

Problem 1-40

The ionization potential of hydrogen (H) is 313.5 $\text{kcal}\cdot\text{mol}^{-1}$. On the other hand, that of potassium (K) is 100.1 $\text{kcal}\cdot\text{mol}^{-1}$. Estimate an effective nuclear charge which works on the outermost electron of K.

Problem 1-41a

Write down the electronic configuration of each of O_2^+, O_2, O_2^- and O_2^{2-}, and on its basis answer the bond order and the magnetic property about whether it is paramagnetic or diamagnetic for each molecule. The bond order is defined by $\text{b.o.} = \frac{1}{2}(n - n^*)$ in which n is the total number of electrons occupying bonding orbitals and n^* is the total number of electrons occupying antibonding orbitals.

Problem 1-41b

Which of the diatomic molecules given below are paramagnetic? Answer them by using their molecular orbitals.

Li_2, B_2, C_2, O_2, CO, NO, CN

Problem 1-42

Obtain the energy of the second lowest quantized level of a particle of a mass m moving in a plane (two-dimensional box) shown below in Fig.8. Also, answer how the probability of finding the particle at a certain place in the plane distributes.

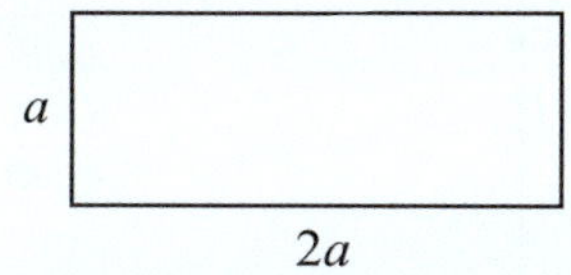

Fig.8 A plane (two-dimensional box) in which a particle with mass m can move freely

"hint" Since the motion of the particle along x- or y-direction, and hence the probability of finding it at x- or y-position is independent of each coordinate, the result obtained for a particle in a one-dimensional box is applicable.

Problem 1-43

Obtain the energy of a quantized state and its wave function of a particle in a 3-dimentional box (a cube of an edge of the length a shown below in Fig.9) and draw a gross shape of the probability distribution function. Discuss about analogy to be found between the obtained probability distribution and the shapes of atomic orbitals such as s-, p-, d-orbitals and so on.

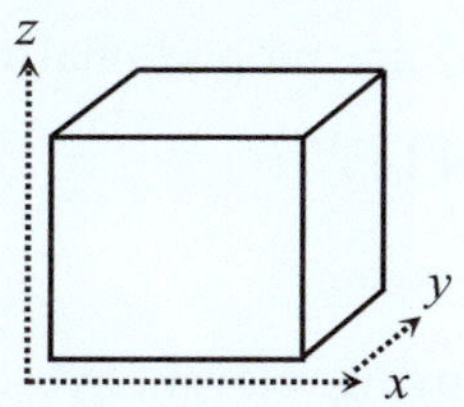

Fig.9 A particle in a three-dimensional box (a cube of an edge of a).

(As to the quantized motion of a particle in a sphere, see **problem 1-110**.)

Problem 1-44a

In order to indicate how large is an atom (or ion), the atomic radius (or ionic radius) defined in a certain way can be a measure. If we define the size of the atom in terms of the wave function, then the values defined in (1) and (2) below may be employed.

(1) A value of radial distance r where the probability density of the outer most electron takes a maximum.

(2) A value of radial distance r within which the outer most electron exists more than or equal to 95 %.

Obtain the atomic radii of Na atom based on the two different definitions given above.

Problem 1-44b

O^{2-}, Na^{+}, Ne, Mg^{2+} and F^{-} are an atom or ions all of which have the same number of electrons and the ions are called as isoelectronic ions. Arrange these ions and the atom in the increasing order of the ionic radii (atomic radius for Ne). See the preceding problem about the concept of the size of the atom.

Problem 1-45

For CO molecule、the relative positions of the energy levels of the atomic orbitals, hybrid orbitals and molecular orbitals are given below in Fig.10 together with their correlations. Referring to this diagram, describe the chemical bonding of CO.

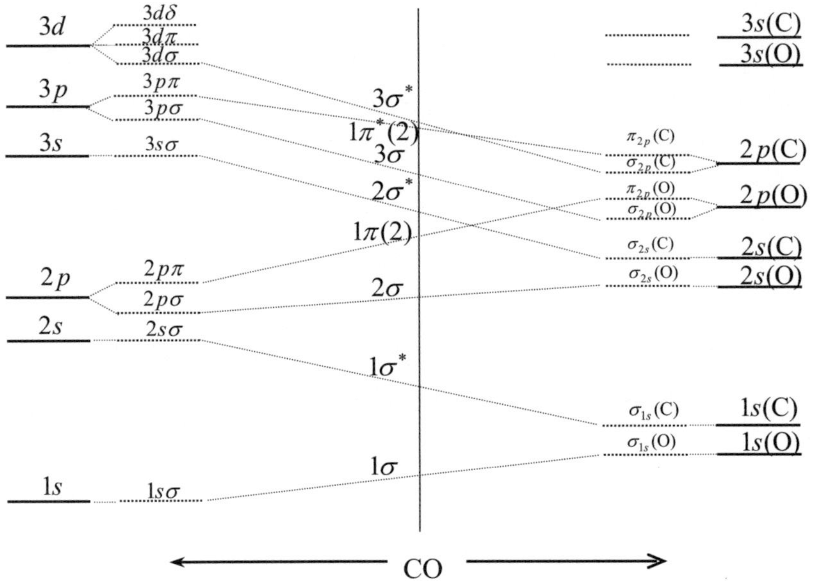

Fig.10 Energy level diagram of the atomic orbitals, hybrid orbitals and molecular orbitals of CO molecule.

Problem 1-46

For Li^{2+}, first three lines of *Lyman* series is observed at 740747, 877924 and 925933 cm^{-1}, Confirm that the energy levels of this hydrogen-like atom (ion) are proportional to $\frac{1}{n^2}$ where n is the principal quantum number and then obtain the ionization energy of Li^{2+}.

Problem 1-47

Explain the chemical bonding of H_2O in detail employing relevant hybrid and molecular orbitals.

Problem 1-48

Express the energy of the sp^3-hybrid orbitals employed in the preceding problem in terms of the energies of s-orbital ε_s and p-orbitals ε_p.

Problem 1-49

Explain the chemical bondings of the following compounds in terms of either hybrid orbitals or molecular orbitals.

(1) SO_4^{2-}, (2) NH_4^{+}, (3) NO_3^{-}, (4) CO_3^{2-}, (5) PO_4^{3-}, (6) CO_2, (7) butadiene ($CH_2=CH-CH=CH_2$)

Problem 1-50

Compared with the stable ions of $\mathbf{CO_3^{2-}}$ and $\mathbf{NO_3^-}$, $\mathbf{CO_4^{4-}}$ or $\mathbf{NO_4^{3-}}$ is not a stably existing ion. Why so? Give it reasons.

Problem 1-51

When the molecular orbitals are employed, the electronic configuration of CO is given by $1\sigma_{1s}{}^2 1\sigma_{1s}{}^{*\,2} 2\sigma_{2s}{}^2 2\sigma_{2s}{}^{*\,2} 1\pi_{2pa}^2 1\pi_{2pb}^2 3\sigma_{2p}{}^2$ as is explained in **solution 1-45**. On the other hand, when *Lewis*' dot electron expression is used, the *Lewis* structure which satisfies the octet rule is expressed as :C:::O: . Can the former electronic configuration based on the molecular orbitals of CO explain this *Lewis* structure?

Problem 1-52

(1) Show the electronic configurations of K and Ar atoms. (2) Ionization energy of K is known to be $419\ \mathrm{kJ \cdot mol^{-1}}$ What will be the effective nuclear charge of K ? (3) Using the result obtained in (2), estimate the electron affinity of Ar. The ionization energy of H atom is $1312\ \mathrm{kJ \cdot mol^{-1}}$.

Problem 1-53

The first ionization energy of Ne atom is $497.2\ \mathrm{kcal \cdot mol^{-1}}$. Using this value, estimate the electron affinity of F in a similar manner as is employed in the preceding problem.

Problem 1-54

There is a cube whose edge is *a* Å long. A cation with a positive unit charge is placed at each of four corners and an anion with a negative unit charge is placed at each of four unoccupied corners. In this manner, every corner of the cube is occupied either by one of the four cations or four anions. Draw an arrangement of the ions for which the total coulombic energy is lowest and obtain the energy for that arrangement.

Problem 1-55

Various molecular shapes of one-centered polyatomic molecules can be predicted by the **VSEPR** (*Valence Shell Electron-Pair Repulsion*) model as well as *Walsh* rule. Knowledge required for the former is,

(a) the number of valence shell electrons (electrons in the outermost shell) and the number of lone-pairs,

(b) *Steric Number* (SN): which is defined by "number of lone-pairs" + "number of bonds",

(c) a table or a list which gives the relation between a molecular shape and the numbers defined in (a) and (b).

Based on the **VSEPR** model, answer the molecular shapes of the following compounds.

(1) CO_2, (2) BF_3, (3) NH_3, (4) XeF_4, (5) SF_4, (6) PCl_6, (7) SF_6, (8) H_2O, (9) NO_2

It should be noted that the elements in the first and second periods satisfy the octet rule (i.e., the outermost electronic configuration of Ne type) but elements in the other periods do not always satisfy it.

Problem 1-56

Based on the **VSEPR** model, answer the molecular structures of the following compound ions (one-centered polyatomic ions).

(1) $CO_3{}^{-2}$, (2) $NH_4{}^{+1}$, (3) $NO_3{}^{-1}$, (4) $SO_4{}^{-2}$, (5) $PO_4{}^{-3}$, (6) $MnO_4{}^{-1}$, (7) $ClO_4{}^{-1}$

Problem 1-57

If we assume there exist the compounds whose chemical formulae are expressed by XeO_4 and FBr_3, then what molecular shapes they are expected to assume? Answer the question by using the **VSEPR** model.

Table Molecular structures of the molecules of $\mathbf{AX_n}$ (n=2-7) type predicted by the **VSEPR** model.

SN	Fundamental molecular structure and Number of lone-pairs			
	0	**1**	**2**	**3**
2	**X—A—X** linear			
3	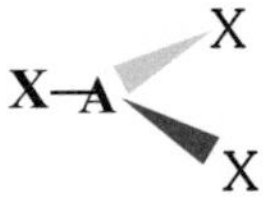 triangular plane	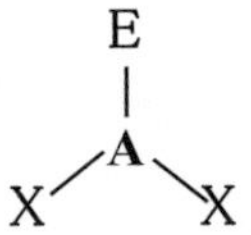 bent structure (in plane)		
4	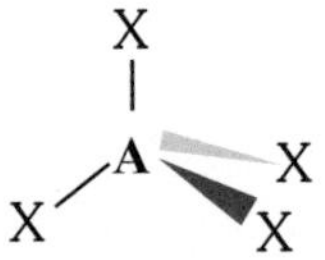 tetrahedral	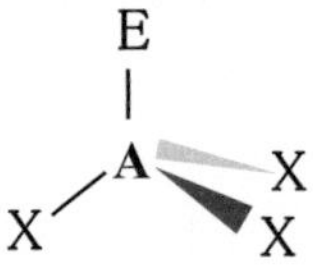 triangular pyramid	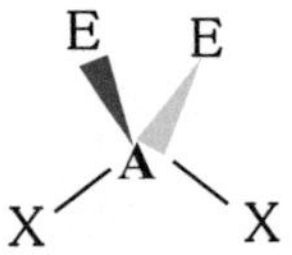 bent structure (perpendicular)	
5	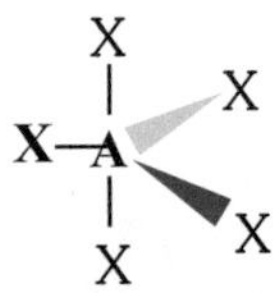 triangular bipyramid	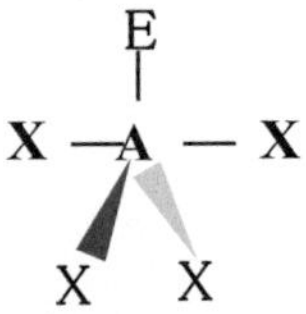 seesaw- shape	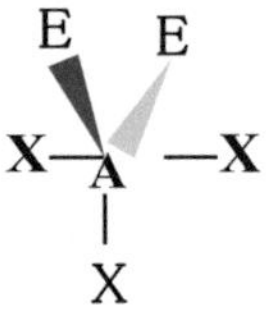 T-shape	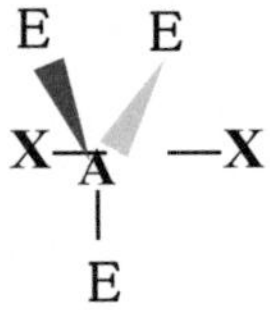 linear
6	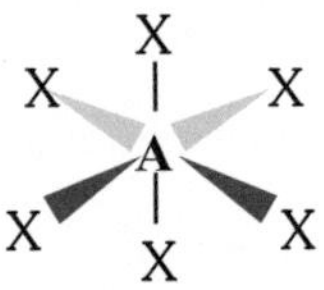 octahedral	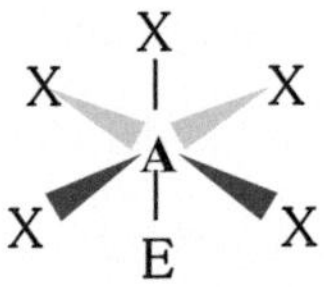 square pyramid	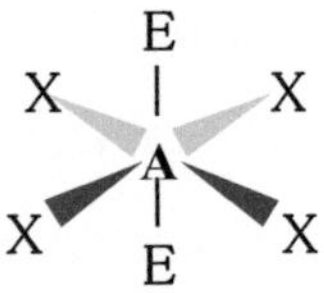 square	
7	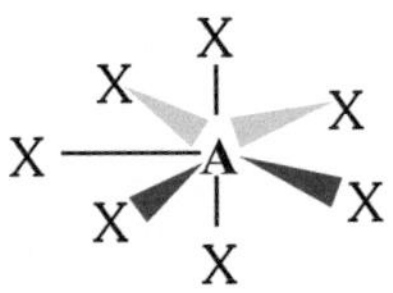 pentagonal bipyramid	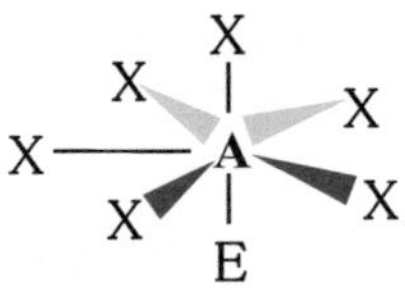 pentagonal pyramid		

"note" **X** stands for a chemical bond and **E** for a lone-pair. Multiple bonds between two atoms are counted as one bond. Valence electrons are illustrated for a few molecules by using dots of different colors not only to distinguish bond electron pairs from lone-pair electrons but also to clarify to which atom they originally belong.

Problem 1-58

Answer the following questions about the chemical bonds of CO_2 and NO_2, in which the atomic orbitals which participate in bond formation are considered to be 2*s*- and 2*p*- orbitals of C or N and 2*p*-orbitals of O.

(a) What kinds of σ -and π -bond are available for these molecules? Illustrate those bonds by clarifying the atomic orbitals involved.

(b) The diagram given below in Fig.11 shows how the energies of the molecular orbitals of a tri-atomic molecule of the type of AX_2 vary against the bond angle of $\angle$X - A - X. Describe what molecular shapes are predicted for CO_2 and NO_2 from this energy level diagram. It is assumed that the atomic orbitals consisting the molecular orbitals in the diagram are 2*s*- and 2*p*-orbitals of A and 2*p*-orbitals of X. Refer to *Walsh's* rule in **solution 1-49**.

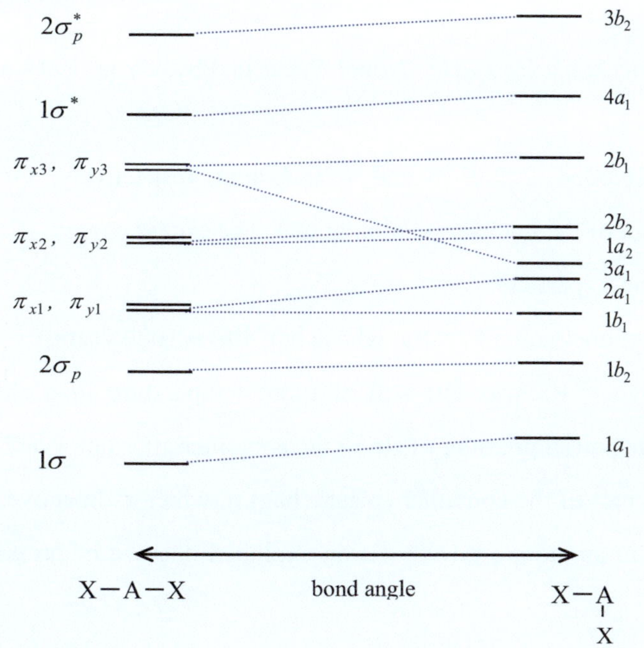

Fig.11 Molecular orbitals of triatomic molecules of the type of AX_2 and their energy levels. Atomic orbitals used to construct the molecular orbitals are,

A : 2*s*- and 2*p*-orbitals

X : 2*p*-orbitals.

Problem 1-59a

For which molecules or ions given below, *Walsh's* rule and the **VSEPR** model show a good agreement in predicting a molecular structure?

CO_2, CO_2^+, NO_2^+, NO_2, NO_2^-

Problem 1-59b

Water molecule is considered to have two equivalent lone-pairs capable of the hydrogen bonding based on which unique properties of water can be explained. On the other hand, in the energy correlation diagram drawn for a water molecule (see **solution 1-49**) only one lone-pair orbital ($2p_x$) is considered. What consideration is necessary to make the predictions by *Walsh's* rule and the **VSEPR** model consistent.

Problem 1-60

With respect to the three types of diatomic molecules A_2, B_2 and AB, the bond energies (eV) are known as $E_{\text{A-A}}, E_{\text{B-B}}$ and E_{AB}, respectively and the electronegativities of atom A and B as χ_A and χ_B, respectively.

(a) Explain what is electronegativity.

(b) By putting A as hydrogen atom and B as chlorine atom, calculate a value of χ_{Cl} from the given values of $\chi_\text{H} - 2.2$, $E_{\text{H-H}} = 4.48$ and $E_{\text{Cl-Cl}} = 2.48$.

(c) The ionic percentages % of A—B bond is given by the following equation.

$$16 \times |\chi_\text{A} - \chi_\text{B}| + 3.5 \times |\chi_\text{A} - \chi_\text{B}|^2 \qquad (9)$$

Calculate the ionic % of HCl bond and from its value find the dipole moment μ of HCl whose bond distance is known as 1.27 Å.

Problem 1-61a

(a) Fill an appropriate number in each [ ] to complete the α-degradation nuclear reaction shown below, where charges are omitted.

$${}^{226}_{88}\text{Ra} \rightarrow {}^{[\ ]}_{[\ ]}\text{Rn} + {}^{4}_{[\ ]}\text{He}$$

Radium (Ra) was discovered by Madam *Curie et.al.* (*M.Curie*, Nobel Prize in Physics in 1903 and in Chemistry in 1911)

(b)The masses of ^{226}Ra, ${}^{[\ ]}$Rn and ^{4}He are 226.0254, 222.0176 and 4.0026 amu, respectively. Find the velocity of the α-particle emitted as a product of the nuclear reaction above. The excess energy, if any, is totally converted into the kinetic energy of the α-particle.

(c) Find the number of α-particles emitted out of one mole of ^{226}Ra whose half-life is 1600 years.

[**advanced problem**] One gram of radioactive ^{226}Ra was left still at room temperature in a shielded container of a volume of 1 m^3. What will be the partial pressure of Rn in the container after one year? It is freshly remembered as a carless accident that a part of the container (plastic bag) in which radioactive waste which had been collected and shielded tightly in an atomic power station exploded because of pressurized gases produced by nuclear reaction(s).

Problem 1-61b

There are four disintegration processes of radioactive materials (atoms) as are given below.

(1) emission of neutron,

(2) emission of α-particle (He^{2+}),

(3) emission of β-particle (electron),

(4) emission of γ-ray (photon).

Every process is harmful to living things and special caution is needed depending on the nature of emission. In spite of this, radioactivity is expressed by Bq for any kind of radiation, which indicates the number of disintegration per unit time. Let us take processes (3) and (4) as examples to understand what is the expression of radiation by Bq.

(a) Naturally existing ^{40}K is a radioactive element and it emits β-ray (electron) to generate a stable isotope of ^{40}Ca as is shown by eq(10). Natural abundance of the radioactive ^{40}K is 0.0117%.

$$^{40}\mathrm{K} \xrightarrow{t_{1/2}=12.8\times10^{9}\,\mathrm{y}} {}^{40}\mathrm{Ca} + \beta(\mathrm{e}) \qquad (10)$$

Find the amount of radiation emitted in the unit of Bq from one kilogram of natural potassium.

(b) ^{137}Cs is a fission product of uranium and stabilizes via processes expressed by eqs(11) and (12). $^{137\mathrm{m}}$Ba in eq(11) is an energetically meta-stable isotope and decays accompanied by γ-radiation to a stable ^{137}Ba with a half-life of 2.55 min. By assuming that the branching ratio of eq(11) to eq(12) is 90 to 10, obtain the amount of radiation emitted from 1.0g of ^{137}Cs in the unit of Bq.

$$^{137}\mathrm{Cs} \xrightarrow{t_{1/2}=30.1\mathrm{y}} {}^{137\mathrm{m}}\mathrm{Ba} + \mathrm{e} \xrightarrow{t_{1/2}=2.55\mathrm{min}} {}^{137}\mathrm{Ba} + \gamma \qquad (11)$$

$$^{137}\mathrm{Cs} \longrightarrow {}^{137}\mathrm{Ba} + \mathrm{e} \qquad (12)$$

"note" The unit of Bq is the number of times of disintegration per second, i.e., [number /s]. Therefore, if we can specify a radioisotope then we can find its concentration in the target material from the measured amount of radiation from it. On the other hand, another frequently used unit of Sv measures the intensity of radiation to which a human or some other object is exposed. It is just like the unit to measure the intensity of light. Therefore, contrary to Bq, a value of Sv cannot tell the amount of radioactive isotopes to be found in the target material. The unit of Sv is determined on the basis of the extent to which living things are damaged by radiation.

Problem 1-62

(a) Calculate the *de Broglie* wavelength of the electron orbiting on the 1*s*-orbital of hydrogen atom (*L.V.de Broglie*, Nobel Prize in Physics in 1929) and find the relation, if any, between the orbital radius and the *de Broglie* wavelength of the electron (see **problem 1-16**).

(b) When the hydrogen atom absorbs a photon and the electron transfers just form the lowest energy level to the second lowest level, find the wavelength of this photon.

Problem 1-63

The charge distribution of fulvene in the ground state is shown below in Fig.12. Assuming that the bond distance of any C – C bond is 140 pm and that the five-membered ring is a regular pentagon, find the direction and magnitude of the dipole moment of this molecule.

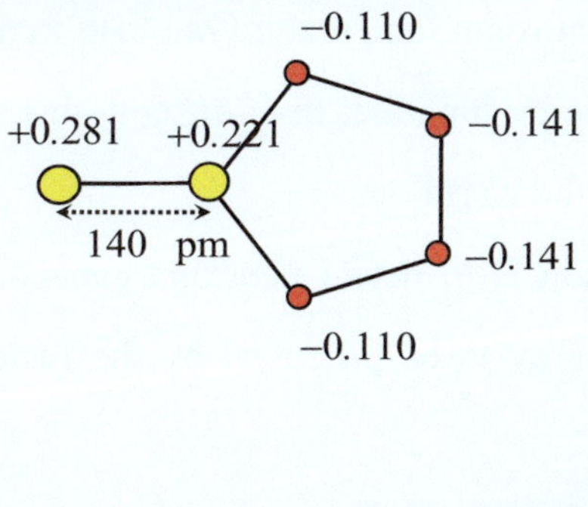

Fig.12 Charge distribution of fulvene in the ground state

Problem 1-64

(a) Describe what is the dipole moment by defining it.

(b) The dipole moment of H_2O molecule is 1.84 D. If the bond angle of the molecule is assumed to be $\angle H-O-H = 120^\circ$, how large is the dipole moment of the H – O bond?

(c) The dipole moment of the H_2O_2 molecule is 3.19 D. If the bond angle of the molecule is assumed to be $\angle H-O-O = 90^\circ$, what molecular structure of H_2O_2 is expected. Answer the question based on the result obtained in (b).

Problem 1-65

The dipole moment is generally recognized as one of the physical properties characteristic to a given molecule. Then, how can we define it for the molecules which are not electronically neutral, such as NO^- or NO^+ ?

Problem 1-66

The dipole moment of 1,2-dichloroethane CH_2ClCH_2Cl measured is 1.27 D at 25 °C and 1.57 D at 315 °C. What is speculated about a molecular structure of this molecule from this result?

Problem 1-67

Describe the chemical bonding in each of the following many atom ions, CO_3^{2-}, NO_3^-, SO_4^{2-} and PO_4^{3-}, in terms of hybrid orbitals. Explain also why the charges on those atomic groups are -2, -1, -2 and -3, respectively.

Problem 1-68

For many years, controlled nuclear fusion has risen our expectation for an eternal energy source under the slogan "Artificial Sun on the Earth!". One of such fusion reactions is given below and is called D-D reaction.

(a) The diameter of a nucleus is given by $R_0\sqrt[3]{A}$, where A is its mass number and $R_0 = 1.33\times10^{-15}$ m.
Find the coulomb energy required to bring two charged deuterium nuclei $^2D^+$'s to the distance at which the two nuclei contact. Use the following values for dielectric constant in vacuum: $4\pi\varepsilon_0 = 1.113 \times10^{-10}\,J\cdot C^2\cdot m^{-1}$ and for elementary charge: $e = 1.602\times10^{-19}$ C.

(b) Find how many times is larger the energy to be produced via the nuclear fusion given above than the coulomb energy obtained in (a). Use the following values. $^2D = 2.01474$, $^3He = 3.01698$, $^1n = 1.00898$.

(c) March 23rd in 1989 , Professors *Fleishmann* and *Pons* announced that they had succeeded in achieving the nuclear fusion given above not only at room temperature but also in a "test tube" and that they had been able to get more energy than the energy inputted. In an instant, this surprising news caused one of the greatest sensations in history all over the world.

One of the methods to prove the fusion is to detect neutrons generated by the fusion (see **"note"** below). By assuming that the whole energy to be produced by the fusion is converted to the kinetic energy of a single neutron, find its velocity in $\mathrm{m \cdot s^{-1}}$.

"note" The two professors *Fleishmann* and *Pons* failed to present any experimental facts to prove the occurrence of the nuclear fusion they claimed. They announced later that they caught neutrons emitted but, unfortunately, it turned out that their measurements were in error. A big fuss involving not only scientists but also mass media, economists, politicians, etc, came to the end shortly. Anyway, this was memorable incident which confronts us with questions about scientific ambition and ethics. The reference given below is worth reading.

[reference] *Frank Close "Too Hot to Handle—The Race for the Cold Fusion"*, Penguin Books.

[fifth wheel] They worked mainly in the chemistry department of Southampton University. Ten years before their breaking announcement, S.H. was spending his research life as a visiting scientist at the same department, though his research field was quite different from theirs and had not a word with them. At teatimes of 10 AM and 3 PM, however, he often saw Prof. *Fleishmann* in the common room of the department. S.H. still remember his face well and for this the incident above is indeed unforgettable for him.

Problem 1-69

How to decide the size of an atom is a very interesting issue. Let us imagine a hydrogen atom but if we assume that only an electron is replaced by a similarly charged particle but with a mass 207 times larger than the electron, then we have an atom-like par ticle called muonic hydrogen. Find, in this case, how many times smaller or larger the radius of the muonic particle is as compared with that of the hydrogen. Answer it by using *Bohr*'s hypothesis and equations. (Of course, wave equations can be used.)

Problem 1-70

The motion of an electron in a one-dimensional box is a good approximation to understand the behavior of π-electrons in the conjugated double-bond system of chain molecules. We assume that the length a of the box is $2.0\,\mathrm{nm}$ (equivalent to one unit of $\mathrm{C=C-C}$) and that the light absorption accompanied by the transition from the energy level $n=10$ to $n=11$ is observed. Find the wavelength of the absorbed light. The transition energy is given by $h\nu$ and is related to the wavelength by $h\nu = hc/\lambda$.

Problem 1-71

The motion of a π-electron in a chain-like or ring-like polyene can be treated approximately by the free electron model. We write here the number of carbon atoms in a polyene as $2N$ and the length of a $\mathrm{C=C-1}$ unit of a chain as d, as is shown below. Then, the length of the chain-like polyene is given by Nd.

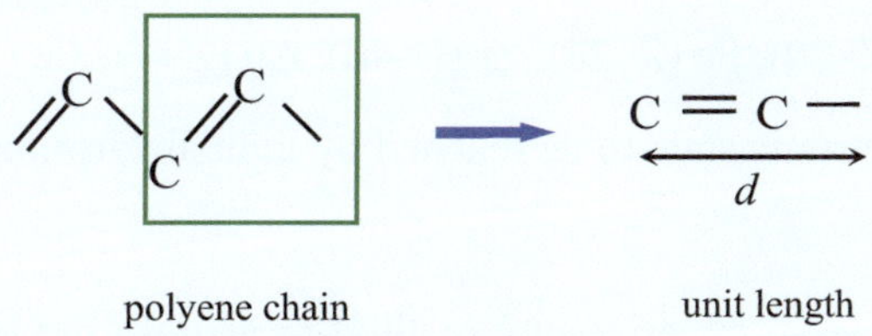

Fig.13 One unit of a polyene chain.

Use the one-dimensional free electron model to find the relation between the length of the molecule and the absorption wavelength. It is assumed that $a = Nd$.

Problem 1-72

The dipole moment μ of HCl is $3.60 \times 10^{-30}\,\mathrm{C \cdot m}$. When two HCl molecules are separated (center to center) by 1.0 nm, answer the following questions.

(a) Find the most stable orientation in which the dipole-dipole interaction energy takes the largest negative value.

(b) Calculate the dipole-dipole interaction energy per mole for that orientation found in (a).

Problem 1-73 [advanced problem]

Find the dipole-dipole interaction energy for the orientation of the two dipoles shown below in Fig.14. Both dipoles are in the same plane. Refer to the method employed in the **solution** to **problem 1-72.**

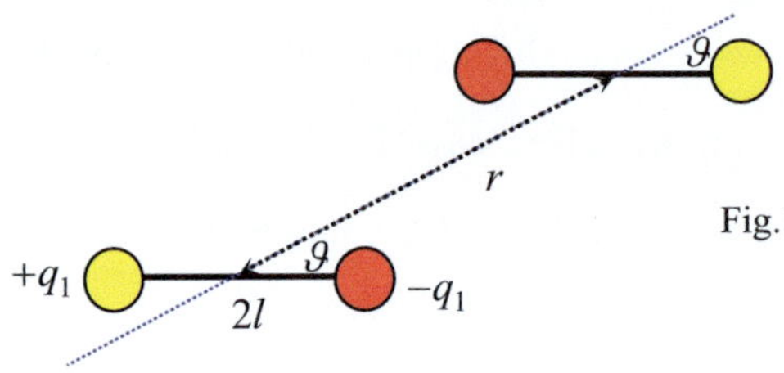

Fig.14 Parallel orientation of two dipoles at the inter molecular distance *r*.

Problem 1-74

The electronic configuration of Na atom in the ground state is given by [Ne] $3s^1$. One of the electronically excited states of the Na atom in which the outermost electron is excited is expressed by [Ne] $3p^1$. When the interaction between the spin angular momentum and the magnetic moment generated by the orbiting motion of that electron is taken into account, the electronic state expressed by the same electronic configuration is split into several states (fine structures). For each electronic configuration of the Na atom given above, list up the electronic states split according to the *Russell-Saunders'* rule.

Problem 1-75

Is it possible to excite Na atom directly from its ground state in the electronic configuration of $[\mathrm{Ne}]3s^1$ to the excited state in the electronic configuration of $[\mathrm{Ne}]3p^1$ by the light of an appropriate wavelength? Answer it by using the selection rules of electronic transition with respect to the parity (even or odd) of states.

Problem 1-76

(a) All of the following atoms are electronically in the ground states. Find parities (even or odd) of their electronic states.

H (Z=1), N (Z=7), S (Z=16), Ca (Z=20), Fe (Z=26), Zr (Z=40), Ag (Z=47, Eu (Z=63)

(b) Based on the parity, find at least one electronic state to be reached by light absorption for each atom.

Problem 1-77a

XPS (X-ray photoelectron spectroscopy) is an experimental method to acquire the ionization energies of inner shell electrons mainly of the atoms in compounds from the measurements of kinetic energies ($E_{kinetic}$)

of electrons ejected upon irradiation of high-energy X-ray ($E_X = h\nu_X$). This method is widely employed for the identification of atoms or for the study of states of chemical bonds on the spot.

The energy of the employed X-ray E_X is related to the ionization energy *I.P.* and the kinetic energy of the ejected electron by the following equation (*Koopman*'s theorem).

$$E_X = I.P. + E_{kinetic}$$

The XPS measurements on iron and its oxides vapor-deposited on the surface of a crystal of magnesium oxide yielded three peaks at 701, 708 and 711 eV as the ionization energy.

(a) Explain to which peak of the observed band group an inner-shell electron of the Fe atom can be ascribed.

(b) If it is assumed that each of the three bands can be assigned to one of the atoms in different oxidation states, i.e., Fe(0), Fe(II) and Fe(III), then try their assignments with explanations.

Problem 1-77b

XPS methods enable us to investigate states of chemical bonds of organic compounds by measuring the ionization energies of some specified atoms in the compounds. For instance, ethoxy flurومethyl ketone $CF_3 \cdot CO \cdot OC_2H_5$ yields clearly separated four peaks with intervals of 2 to 3eV which are ascribed to the ionizations of 1*s*-electrons of four carbon atoms in different bonding states.

(a) Find the oxidation number of each carbon atom of ethoxy fluoromethyl ketone.

(b) Arrange the ionization energies of the four carbon atoms of ethoxy fluromethyl ketone in the order of increasing energy.

Problem 1-78

The motion of an electron moving at the constant velocity along the *Bohr* orbit specified by the quantum number *n* can be regarded as a circular electric current. Use *Ampere*'s law to find that the magnetic moment of the electron circuiting in the orbit for *n*=1 is given by $eh/4\pi m_e$, which is equal to the *Bohr* magneton.

Problem 1-79

Recent development in scientific technology is amazingly fast and even to detect a single molecule in solution is not too difficult (single molecule detection). When a droplet of a radius of 10 μm is prepared from a liquid solution of a certain compound, find out how low the concentration of the solution should be so as for only a single molecule to exist in the prepared droplet.

Problem 1-80

Describe the manner of chemical bonding for Cl_2 and Cl_2^+ in terms of the molecular orbital (MO). Examine whether Cl_2^- can exist as a stable ion or not.

Problem 1-81

Answer the following questions about the hybrid orbitals.

The wave function of the 2*s*-atomic orbital is expressed by ϕ_s and the three mutually orthogonal wavefunctions of the $2p$-orbitals by ϕ_x, ϕ_y and ϕ_z. The eigenenergies of these wave functions are E_s for $2s$-orbital and E_p for $2p$-orbitals.

(a) Draw approximate shapes of ϕ_s and ϕ_x.

(b) One of the sp -hybrid orbitals is given by $\varphi_{sp} = \frac{1}{2}(\phi_s + \sqrt{3}\phi_x)$. Draw an approximate shape of φ_{sp}.

(c) Express the energy (eigenenergy) of φ_{sp} in terms of E_s and E_p.

(d) A hybrid orbital composed of two $2p$-orbitas is given by $\varphi_{pp} = \frac{1}{\sqrt{2}}(\phi_x + \phi_y)$. Draw an approximate shape of φ_{pp}.

(e) How many independent hybrid orbitals are obtained upon the sp^3 -hybridization?

(f) Two of the orbitals to be obtained in (e) may be expressed as follows. Show that φ_a and φ_b given by eq(13) are mutually orthogonal.

$$\varphi_a = \frac{1}{2}(\phi_s + \phi_x + \phi_y + \phi_z)\,, \quad \varphi_b = \frac{1}{2}(\phi_s - \phi_x - \phi_y + \phi_z) \qquad (13)$$

(g) After knowing the energies of the hybrid orbitals such as φ_a, φ_b, describe a role of hybridization in forming chemical bonds.

Problem 1-82

By employing the simple *Hückel* approximation, find the energy of the π -bonding of ethylene ($CH_2 = CH_2$). Also, find a value for the resonance integral β, if it is assumed that the bonding energies of a double bond of $-C=C-$ and a sing bond of $-C-C-$ are 167 and 88 $kcal \cdot mol^{-1}$, respectively.

Problem 1-83

A carbon frame of butadiene is shown in Fig.15 below. Based on the simple *Hückel* approximation, the molecular orbitals for the π -bonds of butadiene are obtained and the <u>signs</u> of the coefficients of the atomic orbitals of the molecular orbital of the lowest energy are just like those given on the left below.

Fig.15 Schematic representation of π-orbitals of butadiene.

(a) Following the above pictorial presentation of the molecular orbital for the lowest level, draw the other three molecular orbitals by clarifying the sign of each p-atomic orbital.

(b) Find the minimum energy required upon photo-excitation from the ground state of butadiene to the lowest excited state. Answer it in terms of the resonance integral β.

(c) Based on the results obtained above (a) and *Woodward-Hoffmann's* law (*R.Hoffmann* and *K.Fukui,* Nobel Prize in Chemistry in 1981, *R.B.Woodward*, Nobel Prize in Chemistry in 1965 for his work on the synthesis of vitamin B_{12}), predict with explanations the structures of products produced from the compound shown in Fig.16 on the right upon its photo-reaction or thermal reaction.

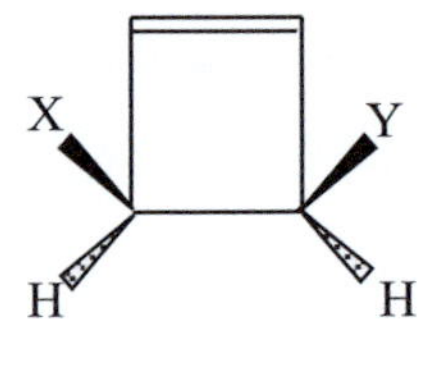

Fig.16 1,4-substituted cyclobutene.

Problem 1-84

Draw approximate shapes of the π -orbitals of benzene by using the relation between the eigenfunctions and the nodal planes.

Problem 1-85

Employing the simple *Hückel* approximation, obtain the molecular orbitals (LCAO MO) of π -bonds and their energies for (a) butadiene and (b) benzene.

Problem 1-86

Employing the simple *Hückel* approximation, find the energies of the molecular orbitals (LCAO MO) of π bonds of styrene (structure on the right). What can be said about the result from a comparison with benzene?

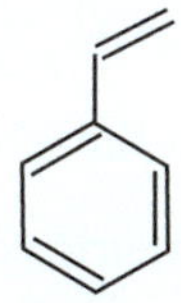

Fig.17 π-bonds of styrene.

Problem 1-87

Write three main resonance structures of a hetero-diatomic molecule CO. Explain how largely each of the three contributes to the bonding of CO. (*L.Pauling*, *Nature of the Chemical Bond,* p.194.)

Problem 1-88

Show that the oxidation numbers of the elements of Cr and Mn are +6 and +7, respectively from the electronic configurations and ionization energies of these atoms and, then, explain the chemical bond structures of $K_2Cr_2O_7$ and $KMnO_4$.

Problem 1-89

Find the oxidation number of each atom of every one of the molecules of H_2, O_2, H_2O, H_2O_2, $FeSO_4$, $K_2Cr_2O_4$, Fe_2O_3, NO, NO_2, NO_3^-, SO_2, SO_3 and H_2SO_4.

Problem 1-90

(a) For the following oxide ions CO_3^{2-}, NO_3^-, SO_4^{2-} and PO_4^{3-}, show that the charge number of each ion can be derived from the oxidation number of the central atom as is given as the superscript for each ion.

(b) Find the charge number n of the ion SiO_4^{n-} from the oxidation number of the Si atom.

Problem 1-91

A part of the chemical structure of benzoic acid expressed by using electron dots is shown below. Find the oxidation number of every atom (all of C, O and H) of the molecule. It should be noticed that every atom of the same kind need not have the same oxidation number.

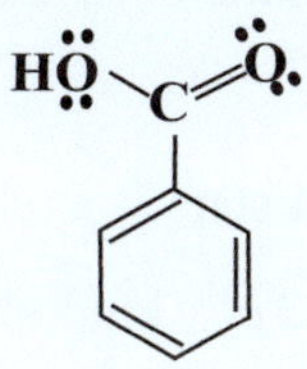

Fig.18 A part of the chemical structure of benzoic acid with lone-pair electrons (dots).

Problem 1-92

A reagent $K_2Cr_2O_7$ oxidizes H_2S via reaction shown below. For any atom or ion which undergoes a change in the oxidation number, describe explicitly how its oxidation number changes. In other words, answer what are oxidized or reduced.

$$3H_2S \quad + \quad Cr_2O_7^{2-} \quad + \quad 8H^+ \quad \rightarrow \quad 3S \quad + \quad 2Cr^{3+} \quad + \quad 7H_2O$$

Problem 1-93

Answer the following questions about the reaction given below.

$$5(COOH)_2 + (\text{ i })KMnO_4 + (\text{ ii })H_2SO_4 \rightarrow$$

$$(\text{ iii })CO_2 + 2MnSO_4 + 8H_2O + K_2SO_4$$

(a) Complete the above reaction by filling the parentheses (i), (ii) and (iii) with appropriate stoichiometric coefficients.

(b) Find a change in the oxidation number brought about by the reaction for every atom participating in the reaction. Even for the same element, if it has multiple oxidation numbers, answer all of them.

(c) From the results obtained in (b), specify what are (is) oxidized and what are (is) reduced.

Problem 1-94

The first ionization energy of an atom varies with the atomic number Z as is shown in Fig.19 on the right. Can you expect how the second and third ionization energies of the atom vary with its atomic number Z?

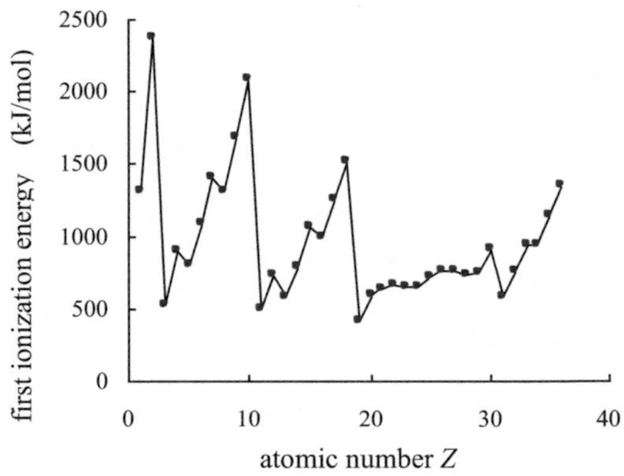

Fig.19 First ionization energy of an atom plotted against Z.

Problem 1-95a

For the atoms whose electronic configurations are expressed by (a) nd^2 or (b) np^1nd^1, find all possible spectral terms in each case.

Problem 1-95b

Based on the electronic configurations of the Ti、Fe and Co atoms in the ground states, find the lowest electronic state of each atom in terms of the spectral term. Refer to **"note 1"** in **solution 1-95a**.

Problem 1-96

(a) The electronic configuration of the Na atom in the ground state is $1s^2 2s^2 2p^6 3s^1$. Find the spectral terms for the ground state.

(b) When the outermost electron of the Na atom in the ground state is excited to a $3p$-orbital, the electronic configuration of this excited state is given by $1s^2 2s^2 2p^6 3p^1$. Find all possible spectral terms for this state.

(c) Discuss about a possibility of the electronic transition of $1s^2 2s^2 2p^6 3s^1 \rightarrow 1s^2 2s^2 2p^6 3p^1$ by photo-excitation based on the selection rules for the transition between the electronic states expressed by spectral terms.

(d) Show that the transition moment for the transition discussed in (c) is not zero.

Problem 1-97

Elements of F, Cl, Br and I in the group VII (IUPAC notation: group 17) have the electronic configuration expressed by ns^2np^5.

(a) Show that the spectral terms for that electronic configuration is given by ${}^2P_{1/2}$, ${}^2P_{3/2}$ and that these two states are energetically split with ${}^2P_{3/2}$ being located lower than ${}^2P_{1/2}$.

(b) The interval between the split spectral terms varies from the F atom to I atom as is shown in Table below. Find the reason for such variation.

halogen atoms	$\lvert E({}^2P_{1/2}) - E({}^2P_{3/2})\rvert$ / cm^{-1}
F	404
Cl	880
Br	3685
I	7600

Problem 1-98a

The element Eu widely used as a red fluorescent substance exists normally as a trivalent cation ${}_{63}Eu^{3+}$.

(a) Answer the electronic configuration of ${}_{63}Eu^{3+}$ in the ground state and list all of the spectral terms possible for the ground state. What is the spectral term for the lowest energy state?

(b) One of the electronically excited states of ${}_{63}Eu^{3+}$ is reached by promoting one of the *f*-orbital electrons from the state of $m_l = -2$ and $s = 1/2$ to the state specified by $m_l = -3$ and $s = -1/2$. Answer the spectral term of the state reached by this transition.

Problem 1-98b

Rare earth metal ions and lasers.

(a) Based on the electronic configuration of the rare earth element Nd which is used for NdYAG lasers, answer the electronic configuration of ${}_{60}Nd^{3+}$.

(b) Find the spectral terms of ${}_{60}Nd^{3+}$ in the ground state.

(c) The electronic transition used for the oscillation of NdYAG lasers is ${}^4F_{3/2} \rightarrow {}^4I_{11/2}$. Confirm that ${}^4F_{3/2}$ exists among the spectral terms which correspond to one electron excitation.

Problem 1-99

Ionization of an electron in the inner shell near the nucleus is considered not to be influenced significantly by shielding due to the electrons existing in outer shells. Find the ionization energy of 2*s* electron of the K atom in both units of eV and wavelength.

Problem 1-100

At a certain spot, the radioactive radiation of the strength of 100 Sv was observed. How can you identify the radioactive species responsible for that radiation? For instance, what knowledge do you need to ascribe the radiation to the radioactive ^{137}Cs ?

Problem 1-101

The coulomb force working between the two point charges of q_1 and q_2, which are separated by a distance R in vacuum, is defined by eq(14).

$$F = \frac{q_1 \times q_2}{4\pi\varepsilon_0 r^2} \qquad (14)$$

Using eq(14), show that the coulomb force working between the point charge of q_1 and the homogeneously charged sphere of radius a and charge density ρ is given by eq(15) (see Fig.20(a) below). *Gauss'* theorem should not be used for derivation here. The distance between the point charge and the center of the charged sphere is R and the total charge of the sphere is Q.

$$F = \frac{q_1 \times Q}{4\pi\varepsilon_0 R^2}, \qquad Q = \rho \times \frac{4\pi}{3} a^3 \qquad (15)$$

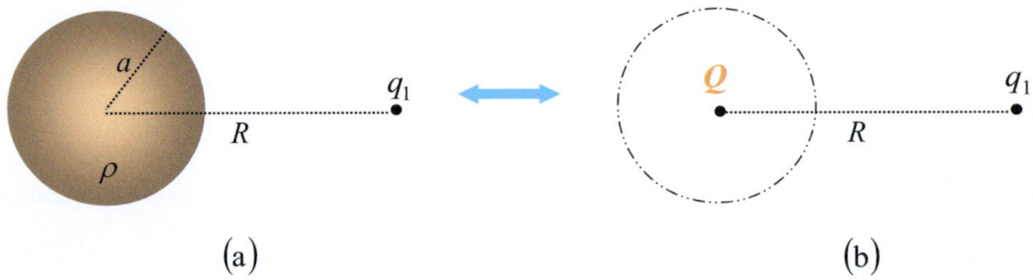

Fig.20 A point charge of q_1 and a homogeneously charged sphere of radius a and charge density ρ .

Eq(15) indicates that the coulomb force between two charged bodies (molecules) can be replaced by the equivalent coulomb force between two point charges. In that case, the amount of the point charge is equal to the total charge of the charged body (figure (b) above)

Problem 1-102

The azimuthal quantum number l of the d-orbitals is 2 and the magnetic quantum number m_l can take 2, 1, 0, −1, −2. The d-orbitals usually employed in chemistry are d_{xy}, d_{yz}, d_{zx}, $d_{x^2-y^2}$ and d_{z^2} but not all of them are necessarily eigenfunctions of the operator $\hat{l}_z$ for the magnetic quantum number, though d_{z^2}-orbital is the eigenfunction with the eigenvalue of zero. Show that d_{xy}-orbital is not the eigenfunction of $\hat{l}_z$. The operator $\hat{l}_z$ is given by $-i\dfrac{\partial}{\partial\varphi}$ for simplicity.

Problem1-103

Show that the function of $d_{-2} = 3d_{x^2-y^2} - i \times 3d_{xy}$ formed by a linear combination of $3d_{x^2-y^2}$ and $3d_{xy}$ is an eigenfunction of the operator of $\hat{l}_z$ (see **"note"** in preceding solution) with the eigenvalue of−2. Draw its approximate shape.

Problem 1-104

Calculate the transition moments of the radiative transitions of $1s \leftarrow 2p$ and $1s \leftarrow 3p$ for the hydrogen atom and comment on the results obtained.

Problem 1-105

Solid NaCl forms a crystal called a rock salt (rock salt structure) and its unit cubic lattice is shown on the right in Fig.21. Either of the Na atom or Cl atom occupies each corner of the cube alternatively as is shown in the figure schematically.

Show first that if we know the atomic distance *d* between Na and Cl, from the density ρ and the formula weight *M*, we can calculate *Avogadro*' s number and then find the number N_A calculated from the data of $d = 2.81 \times 10^{-10}$ m, $\rho = 2.17\ \mathrm{g \cdot cm^{-3}}$ and $M = 58.44\ \mathrm{g \cdot mol^{-1}}$ known for NaCl.

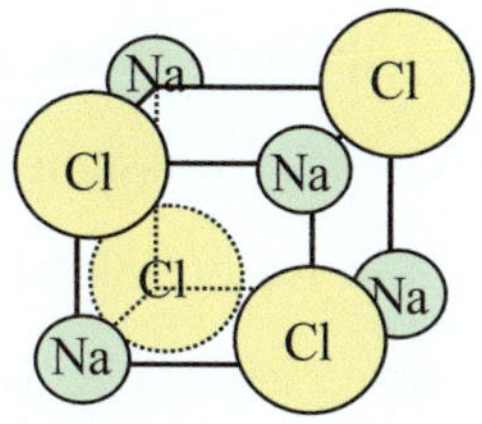

Fig.21 An illustration of a unit cubic lattice of rock salt.

Problem 1-106

By combining three of the mass of an electron m_e, the unit charge *e* and the Planck constant *h* in an appropriate manner, create a constant whose dimension is length. Show that the obtained constant represents "something quantum mechanical" well-known.

Problem 1-107

Owing to the discovery of new elements, specific names have been given to the atoms up to the atomic number of 120 apart from their general names. Based on the *Bohr* model, the velocity of the 1*s* electron is given as a function of the atomic number Z. What do you expect is the largest atomic number possible for the model? See also **solution 1-36**.

Problem 1-108

To describe the motion of atomic electrons quantum mechanically, there exist three kinds of quantum numbers, i.e., principal quantum number *n*, azimuthal quantum number *l* and magnetic quantum number m_l. The principal quantum number takes any positive integer and when it takes *n*, then *l* can take $l = n-1,\ n-2, \cdots,\ 0$, and for a given *l*, the allowed values for m_l are $m_l = l,\ l-1 \cdots,\ 0, \cdots, -l+1,\ -l$, and hence $2l+1$ different values in total. Needless to say, these are indispensable knowledge in chemistry and for more quantitative discussions, we need to understand well the wave functions and their shapes specified by those quantum numbers.

However, without solving *Schrödinger*'s equation or relying on the knowledge mentioned above, the number of allowed quantum states or their approximate shapes (phase) can easily be understood by the simple way introduced below.

For this, we only use the fact that "any state of motion we can know quantum mechanically is the state called eigenstate and different eigenstates are orthogonal to each other," from which we can derive the number corresponding to the three quantum numbers and draw approximate shapes of the eigenstates reflecting the eigenfunctions of *Schrödinger*'s equation. More specifically, it is sufficient to think about how many ways there exist to divide a space into two parts so as for one part to be orthogonal to the other (see "**note**"). One such example is given below.

In the case of three-dimensional central motion (one- or two-dimensional case can be treated in a similar way), the most stable state, in other word, lowest energy state is expressed by a positive homogeneous space. Of course, we assume there is only one state of this kind.

(a) How many simplest ways exist of dividing the 3-dimensional homogeneous space schematically shown on the right in Fig.22 (if the space is spherically symmetric, then a sphere) into two parts so as for one part to be orthogonal not only to the other but also to the undivided space.

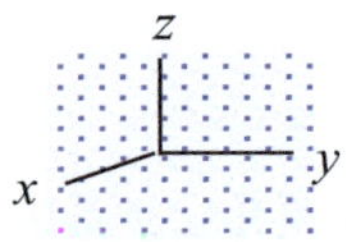

Fig.22 Three-dimensional homogeneous space.

(b) Let us assume the space to be divided is a sphere which represents a 1*s*-orbital. How many and what kind of atomic orbitals are obtained by dividing the sphere into two parts by a plane or a surface.

(c) Show that by dividing the space obtained in (a) or (b) further so as for the divided spaces to be orthogonal to each other, the spaces (phases) that correspond to 3*s*-, 3*p*- and 3*d*-orbitals are obtained. How about *f*-orbitals?

"**note**" What is meant here by "orthogonal" is that the integral of the superposed two parts of the space over the whole space becomes zero.

Problem 1-109

One-dimensional arrangement of ion crystal of NaCl is shown in Fig.23 below. By comparing the whole electrostatic energy of the one-dimensional crystal with that of one pair of Na^+Cl^-, find the *Madelung* constant for the one-dimensional ion crystal.

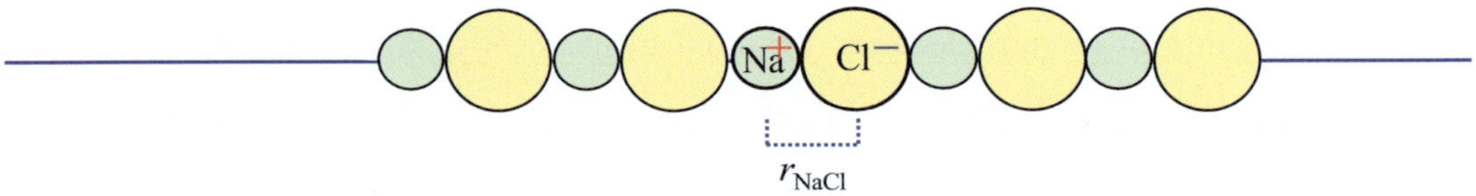

Fig.23 One-dimensional arrangement of ion crystal of NaCl.

Problem 1-110

The motion of a particle with mass *m* kept in a sphere of radius *a* is described by the wave functions which satisfy eq(16).

$$-\frac{\hbar^2}{2m}\frac{1}{r^2}\frac{d}{dr}\left(r^2\frac{d}{dr}\right)\psi(r)+V(r)\psi(r)=E\psi(r) \qquad (16)$$

As the boundary conditions for the equation, if $V(r)=0$ for $r<a$ and $V(r)=\infty$ for $\geq a$, are employed, one of the solutions is given by eq(17).

(a) Show that eq(17) satisfies the wave equation above. *A* is a proportional constant.

$$\psi(r)=A\cdot\frac{1}{r}\sin(\alpha\cdot r) \qquad (17)$$

vacuum in the inclining magnetic field applied perpendicular to the beam was split into two beams (see the figure below). We denote the mass of the Ag atom as $m_{\rm Ag}$, the magnetic moment inherent to the electron as μ_e and the velocity of the Ag atom in the beam as υ_z. The slope of the magnetic field is given by $(\partial B/\partial x)_{y,z}$ and the length of the magnet which provides the inclining magnetic field is *L*. Assuming that the velocity is approximated by the mean velocity $\sqrt{2\varepsilon/m_{\rm Ag}}$, where ε is the mean kinetic energy of the Ag atom, answer the following problems.

(a) Derive the equation which gives the interval *x* of the split beams at the exit end of the magnet (*Stern-Gerlach*'s experiment). Ag atoms fly along *z*-axis.

(b) When the temperature of the furnace used to generate Ag atomic vapor is 1000 K and the length of the magnet is 50 cm, how large the slope of the magnetic field must be in order to make the interval *x* larger than 1.0 mm? Fig.25 illustrates *Stern-Gerlach*'s experiment.

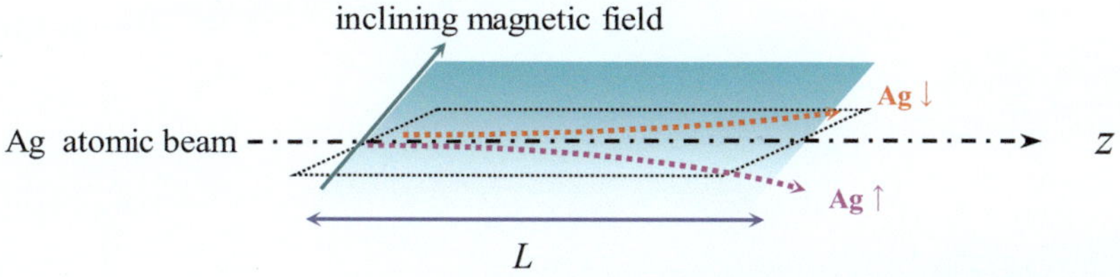

Fig.25 Illustration of *Stern-Gerlach*'s experiment.

Problem 1-117

Even when the total spin angular momentum is zero, i.e., $S = 0$, as far as the total angular momentum *L* is not zero, the orbiting motion of electrons has the magnetic moment μ_z depending on the magnetic quantum number *M*. Therefore, under the magnetic field of B_z, the electrons of an atom with different values of *M* come to have different energies given by $-\mu_z \cdot B_z$ with μ_z given by eq(24) (*P.Zeeman*, Nobel Prize in Physics in 1902). This is the reason why m_l and *M* are called magnetic quantum number.

$$\mu_z = -M \times \frac{eh}{4\pi m_e}, \qquad M = \sum_{i=1}^{\text{electrons}} m_{l_i} \qquad (24)$$

(a) Find how many different electronic states result due to the splitting when the atoms of the electronic states expressed by the spectral terms of ${}^1\mathrm{D}_2$ and ${}^1\mathrm{P}_1$ are put in a magnetic field. Calculate the maximum interval of the splitting under the magnetic field of the strength of 10 T.

(b) The selection rule for the magnetic quantum number *M* upon the electronic transition is given by eq(25).

$$\Delta M = 0, \ \pm 1 \qquad (25)$$

Assuming that the transition of ${}^1\mathrm{D}_2 \rightarrow {}^1\mathrm{P}_1$ under the magnetic field is an allowed transition ($\Delta L = \pm 1$), find how many split bands are observed.

Problem 1-118a

Select diamagnetic (not paramagnetic) substances among those given below and give each of them the reason. All substances are electronically in the ground state and are in the gas phase.

H, Li^+, B, Ne, Ti, F^-, H_2, O_2, N_2, HCl, CO_2, NO_2, NH_3, CH_4, $CH_3\cdot$

Problem 1-118b

According to the *Langevin-Debye* theory, the (atomic) diamagnetic susceptibility (volume) χ is given by eq(24). All units are in cgs. The minus sign indicates to be diamagnetic. N is the number of atoms per volume and the summation is carried out over all electrons in one atom (see **"note"** in the solution).

$$\chi = -\frac{Ne^2}{6m_e c^2} \sum_{i=1}^{\text{all electrons}} \overline{r_i^2} \qquad (26)$$

(a) Find the dimension of the diamagnetic susceptibility χ.

(b) Find the molar diamagnetic susceptibility of He ($1s^2$), assuming that $\overline{r_1^2} = \overline{r_2^2}$ = square of *Bohr* radius and the effective atomic number $Z_{eff} = 1.70$.

I-2 Gas-state Molecular Kinetics and State Equations

Part of Problems

Problem 2-1a

Consider a mixed gas consisted of methane (CH_4) 320 mg、argon 175 mg and neon 225 mg. The partial pressure of the neon at 300 K is 15.2 kPa. Find the volume and the total pressure of this mixed gas at 300 K by assuming the mixture behaves as an ideal gas.

Problem 2-1b

Air is substantially the binary mixture of oxygen and nitrogen whose volume ratio is 1:4. Find the apparent molecular weight of the air. It turned out that the air contained carbon dioxide at the rate of 1.0 % by volume, though the volume ratio of the oxygen to the nitrogen remained the same. Then what will be the apparent molecular weight? In any case, assume the gases behave ideally.

Problem 2-2

The density of mercury is $13.595 \times 10^3\,\mathrm{kg \cdot m^{-3}}$. Find the atmospheric pressure in Pa equivalent to the height of 0.76 m of the mercury when the acceleration of gravity is $9.807\ \mathrm{m \cdot s^{-2}}$. Instead of mercury, when water of the density of $1.000 \times 10^3\,\mathrm{kg \cdot m^{-3}}$ is employed, what will be the height of the water equivalent to that pressure?

Problem 2-3a

The parameters of a and b for the *van der Waals* equation of state of CO_2 are $3.592\ \mathrm{dm^6 \cdot atm \cdot mol^{-2}}$ and $4.267 \times 10^{-2}\mathrm{dm^3 \cdot mol^{-1}}$, respectively (*J.D. van der Waals*, Nobel Prize in Physics in 1910). The *P*-*V* relation of the CO_2 gas at 285 K is shown in the figure below. When the gas is compressed at this temperature, the liquefaction starts at the pressure of P_A atm (indicated by *A* in the figure). It is assumed that the *P*-*V* relation can be described by the *van der Waals* equation until the gas liquefies.

(a) Find the molar volume of CO_2 at the point *A* in the figure.

(b) When all the gas is liquefied at the point *B*, the volume of the liquid is $84.23 \times 10^{-3}\,\mathrm{dm^3 \cdot mol^{-1}}$. Between

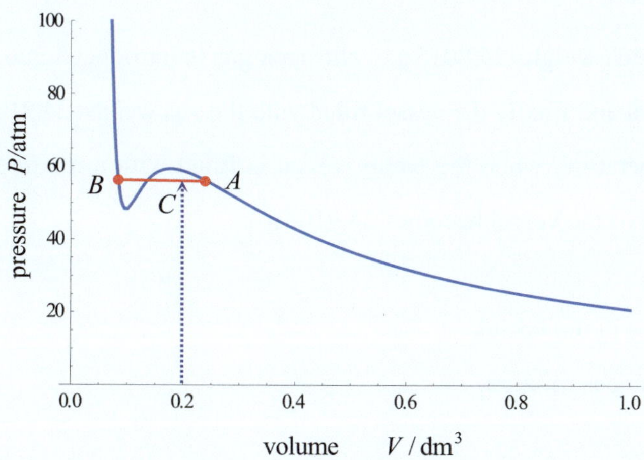

Fig.1 *P*-*V* curve of CO_2 at T=285 K expressed by the *van der Waals* equation of state.

S. Hirayama and M. Yamamoto, *Physical Chemistry Problems and Solutions*,
https://doi.org/10.1007/978-981-97-0165-0_2

the points A and B the gas and liquid coexist and the volume of the mixture at the point C is $V = 200 \times 10^{-3}\,\mathrm{dm}^3$. Then, find the ratio of the coexisting gas to liquid in the ratio of their mole numbers.

Problem 2-3b

As in the preceding problem, we consider the parameters of a and b for the *van der Waals* equation of state of CO_2 are given by $3.592\ \mathrm{dm}^6 \cdot \mathrm{atm} \cdot \mathrm{mol}^{-2}$ and $4.267 \times 10^{-2}\,\mathrm{dm}^3 \cdot \mathrm{mol}^{-1}$, respectively. The figure given below shows the P-V behavior of CO_2 at 273 K. When the gas is compressed along the P-V curve isothermally, the gas starts to liquefy at a certain pressure. Find this pressure by numerical calculation.

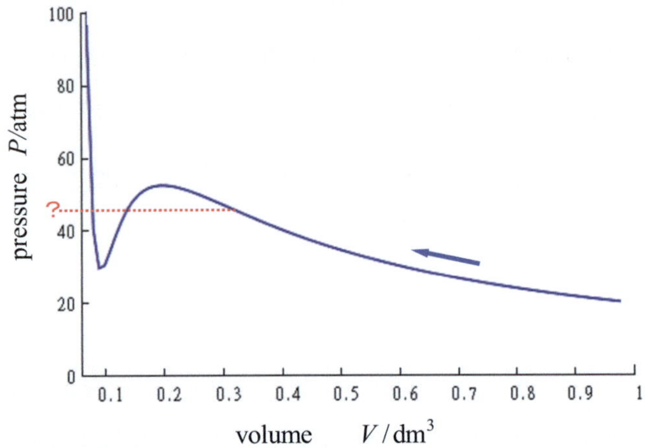

Fig.2 P-V curve expressed by the *van der Waals* equation of state of CO_2 at T=273 K.

Problem 2-4

Prove that the isothermal liquefaction line drawn along the equilibrium between the gas and its liquid on the P-V curve given by the *van der Waals* equation of state should fulfill the *Maxwell's* rule (*Maxwell*'s equal area construction rule). (How to draw *Maxwell*'s tie line.)

Problem 2-5

Find a collision number per molecule per second for the oxygen molecules at 25° C and 1 atm. Assume the collision diameter of the oxygen molecule is 3.57×10^{-10} m.

"note" The one-dimensional velocity υ distribution function of a gaseous molecule is given by , where M is $(2M/\pi RT)^{1/2} \exp(-M\upsilon^2/2RT)$ the molecular weight of the gas molecule.

Problem 2-6

There is an evacuated vessel made of glass which weighs 24.8015 g. Nitrogen gas is introduced into the vessel at 298.15 K up to the pressure of 101.32 kPa and finally the vessel filled with the gas weighs 24.9100 g. Under the same conditions of pressure and temperature, when the empty vessel is filled with a mixture of methane (CH_4) and ethane (C_2H_6) the total weight of the vessel becomes 24.9100 g.

(a) Find the capacity volume of this glass vessel.

(b) Calculate the number of the nitrogen molecules in the vessel.

(c) Find the molar ratio of methane to ethane in the gas mixture.

Problem 2-7

Answer the following questions about the ideal gas. Classify the following items (A) to (H) into each of three groups which i) increases with temperature, ii) decreases with increasing temperature, and iii) is invariant, all with explanations. In any case the gas is kept at a constant volume.

(A) mean free path, (B) mean square velocity, (C) heat capacity, (D) viscosity coefficient, (E) the number of molecules with velocities less than the mean velocity $\bar{u}$, (F) pressure of the gas, (G) collision radius, and (H) sound velocity in the gas.

Problem 2-8

Among the equations of state from i) to iii) given below, which one is able to describe the existence of a critical state? B and C are appropriate constants. For the equation of state you have selected, obtain the critical pressure, temperature and volume. Also, derive its corresponding equation of state by using the reduced pressure, temperature and volume.

$$\text{i)}\ PV = RT,\quad \text{ii)}\ PV = RT(1+BP),\quad \text{iii)}\ P = (RT/V) + (B/V^2) + (C/V^3)$$

Problem 2-9

As far as the temperature and volume are the same for ideal gases of the same number of moles, the pressure exerted by an ideal gas on a vessel wall does not depend on the kind of gases. Discuss this fact from a viewpoint of the kinetics of gaseous molecules. ("hint" Show that the "mass" of a gas molecule does not appear in the expression of the pressure the gas molecules exert on the wall.)

Problem 2-10

a) The distribution function of the one-dimensional velocity u is given by $A \cdot \exp(-mu^2/2k_BT)$, where A is a constant and m is the mass of one gaseous molecule. Find the mean velocity $\bar{u}$ and the mean square velocity $\overline{u^2}$.

b) The vapor pressure of Ge (mp. 985.5 °C) at 1000 °C is obtained by measuring the amount of substance effusing from a small aperture of a *Knudsen Cell* illustrated below. When the diameter of the aperture was 1.0 mm, the total effluent amount in two hours at the same temperature was 4.3×10^{-2} g. Find the vapor

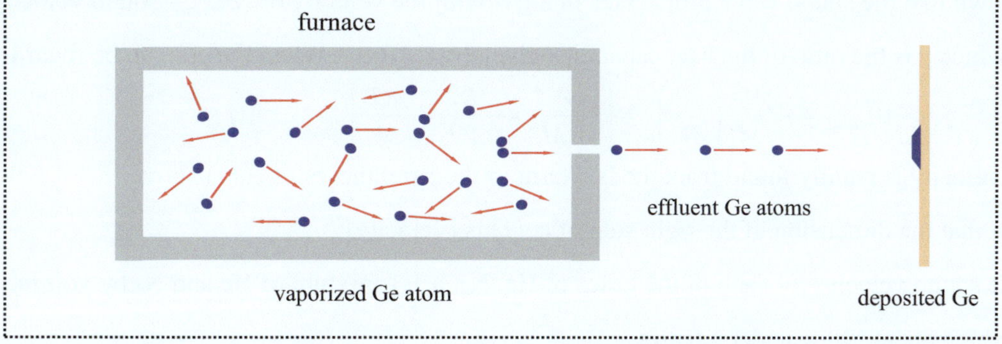

Fig.3 Illustration of the *Knudsen Cell*. The cell is placed in a vacuum chamber.

pressure of Ge at this temperature from this result by using the result obtained in a).

The atomic weight of Ge is 72.5 and use $1.38 \times 10^{-23}\,\mathrm{J \cdot K^{-1}}$ for the *Boltzmann* constant.

Problem 2-11

The *van der Waals* equation of state expressed in the form of PV/RT can be expanded in powers of $1/V$ into the virial equation of state. In this case, express the virial coefficients of B and C in terms of the parameters a and b associated with the *van der Waals* equation and temperature T. Also, from this expansion find the *Boyle* temperature T_B (temperature at which an approximation of $PV/RT = 1$ is reasonably applicable).

Problem 2-12

In the experiment to measure a vapor pressure of naphthalene ($C_{10}H_8$), air at STP (273.15 K, 101.32 kPa) was flowed at the rate of $1.000\ \mathrm{dm^3 \cdot min^{-1}}$ over crystals of naphthalene placed in a glass tube. The air flow carried sublimed naphthalene away out of the tube. Exactly in an hour, the amount of 0.0475 g of naphthalene was lost. Assuming that the air flow was saturated with naphthalene, find the vapor pressure of naphthalene at this temperature. The naphthalene vapor is assumed to behave like an ideal gas.

Problem 2-13

The distribution function $p(c)$ of the three-dimensional velocity c of a gas (mass: m) is as is given below.

$$p(c)dc = 4\pi\left(\frac{m}{2\pi k_B T}\right)^{3/2} \exp\left(-\frac{mc^2}{2k_B T}\right)c^2 dc$$

The molecular diameters of H_2, O_2 and CO_2 are known to be 218, 296 and 460 pm, respectively.

(a) For these three molecules, find the ratios of their mean velocities at each of 300 K and 600 K to the corresponding mean velocity of H_2.

(b) For these three molecules, find the ratios of the self-collision numbers per unit time at 300 K and 100 kPa to that of H_2.

(c) Under the same conditions as (b), find the ratios of the mean free-paths to that of H_2.

Problem 2-14

It is known that the sound wave propagates in a gas with the velocity of υ_{sound} (sound velocity) given by eq(1), in which γ is the ratio of the heat capacities given by C_P/C_V. When the gas can be regarded as ideal,

$$\upsilon_{sound}^2 = -\frac{\gamma V^2}{M}\left(\frac{\partial P}{\partial V}\right)_T = \left(-\frac{\gamma V^2}{M} \times -\frac{RT}{V^2} = \frac{\gamma RT}{M}\right) \qquad (1)$$

the sound velocity is readily found from the last term in the parentheses of eq(1) above.

(a) Confirm that the dimension of the right side of eq(1) is [velocity]^2.

(b) Find the sound velocity for each of the gases of He, N_2, a 1:1 mixture of He and N_2 by volume and the air, all of which are at 298 K under 1.0 atm.

(c) Compare the sound velocities found for He and N_2 in (a) above with the mean velocities of these gases under the same condition.

(d) It is assumed above in (a) to (c) that the sound velocity is independent of the frequency of a sound wave. Find the frequency of the sound propagating in He whose wavelength becomes equal to the mean free path of the He atom.

(e) The sound velocity varies with temperature. Show that the sound velocity in the air at t°C under the atmospheric pressure can be expressed by the equation below.

$$v_{sound} = 331.45 + 0.607 \times t^\circ\mathrm{C}$$

Problem 2-15

A sound source (vibrator) is often surrounded by a resonator to amplify the sound. One such resonator is illustrated in Fig.4. The inside of the resonator is normally filled with the air. However, when the air is intentionally replaced with He gas at 1 atm, how is the sound from the resonator audible to our ears? Give your answer with the reason.

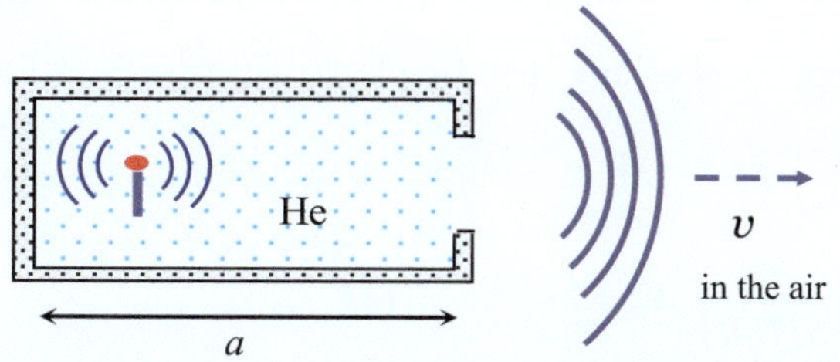

Fig.4 Illustration of a sound resonator

Problem 2-16

The virial equation of state can be expanded into the power series of P instead of the power series of $1/V$ as is expressed by eq(2).

$$\frac{PV}{RT} = 1 + B'(T)P + C'(T)P^2 + D'(T)P^3 + \cdots \qquad (2)$$

When the *van der Waals* equation of state is expanded in the same form as eq(2), how are the virial coefficients B' and C' expressed in terms of the *van der Waals* parameters a and b and temperature T?

Problem 2-17

The parameters of a and b of the *van der Waals* equation of state for carbon dioxide and ammonia are given by ($a = 3.640 / \mathrm{dm^6 \cdot atm \cdot mol^{-2}}$, $b = 4.267 \times 10^{-2} / \mathrm{dm^3 \cdot mol^{-1}}$) and ($a = 4.225/ \mathrm{dm^6 \cdot atm \cdot mol^{-2}}$, $b = 3.707 \times 10^{-2} / \mathrm{dm^3 \cdot mol^{-1}}$), respectively. Employing the virial expansion of the *van der Waals* equation of state given in the solutionr to the preceding problem (Solution 1), plot the compression factor $z = PV/RT$ against pressure in the range [0, 20 atm] at $T = 250,\ 300,\ 500,\ 1000$ K for carbon dioxide and ammonia. What can be concluded about the nature of these gases from a comparison of those plots?

Problem 2-18

Prove that the *van der Waals* equation of state satisfies the following relation.

$$z_c = \frac{P_c V_c}{RT_c} = \frac{3}{8}$$

Here, P_c, V_c and T_c are the critical pressure, critical volume and critical temperature, respectively. Therefore, z_c is the compression factor at the critical point.

Problem 2-19

Although the *van der Waals* equation of state (eq(3)) contains only two adjustable parameters, the equation is often employed under various circumstances as the one which can describe the behaviors of real gases quite well. However, the values for the parameters of a and b vary from gas to gas. Because of this, we tend to overlook the true nature which divides real gases from the ideal gas. Just as is the equation of state for the ideal gas irrelevant to the property inherent to each ideal gas (such as its mass), the *van der Waals* equation of state is expected to be rearranged into a sort of the universal equation which is independent of the properties inherent to an individual real gas.

Show that the *van der Waals* equation of state (eq(3)) can be rewritten as eq(6) by employing the reduced variables P_R,, V_R and T_R in eq(3) which are defined by the critical parameters given in eq(4) and are called the reduced pressure, reduced volume and reduced temperature, respectively. For this, eq(6) is called the reduced equation of state and is independent of any specific physical properties of a real gas.

$$P = \frac{RT}{(V-b)} - \frac{a}{V^2} \qquad (3)$$

$$P_c = \frac{a}{27b^2}, \quad T_c = \frac{8a}{27bR}, \quad V_c = 3b \qquad (4)$$

$$P_R = P/P_c, \quad V_R = V/V_c, \quad T_R = T/T_c \qquad (5)$$

$$P_R = \frac{8T_R}{(3V_R - 1)} - \frac{3}{{V_R}^2} \qquad (6) \qquad \text{(reduced equation of state)}$$

Problem 2-20

The reduced *van der Waals* equation of state is meaningful only when the reduced parameters satisfy certain conditions.

$$P_R = \frac{8T_R}{(3V_R - 1)} - \frac{3}{{V_R}^2} \qquad \text{(reduced equation of state, eq(6) in } \textbf{poblem 2-19}\text{)}$$

The condition of $V_R > 1/3$ is directly derived from the reduced equation above. Then, answer what condition(s) is imposed on T_R. Also, find in what temperature range is the reduced equation usable for carbon dioxide whose critical temperature is 304.2 K.

Problem 2-21

(a) In the solution to the **problem 2-18**, the critical compression factor z_c for the *van der Waals* gas is found as $z_c = 3/8 = 0.375$. As is seen in the table given below, the values for z_c are close to 0.3 for a number of real gases. Give the reason for that by comparing those values in the table one another.

(b) The polar gases are characterized by smaller values for z_c as compared with the other non-polar gases. However, as far as they are approximately treated as the *van der Waals* gases, the value for z_c is fixed at the constant value and cannot be adjusted even when the values for the parameter a and b are varied. Consider how the equation of state should be modified so as to yield a smaller value for z_c.

Table Critical constants of gases.

molecule	P_c/atm	V_c/cm^{3}mol^{-1}	T_c/K	z_c	T_B/K
He	2.26	57.76	5.21	0.305	22.64
Ar	48.0	75.25	150.7	0.292	411.5
H_2	12.8	64.99	33.23	0.305	110.0
N_2	33.54	90.10	123.6	0.292	327.2
O_2	50.14	78.0	154.8	0.308	405.9
HCl	81.5	81.0	324.7	0.248	——
NH_3	111.3	72.5	405.5	0.242	——
H_2O	218.3	55.3	647.4	0.227	——
CO_2	72.85	94.0	304.2	0.274	714.8
CH_4	45.6	98.7	190.6	0.288	510.0

Problem 2-22

For the O_2 molecule at 300 K, find (a) the one-dimensional mean velocity and (b) the three-dimensional mean velocity. Explain why the latter is larger than the former.

Problem 2-23

(a) Find the minimum velocities required for the H_2 and O_2 molecules to escape from the gravitational potential of the earth.

(b) Find the fractions of the molecules whose velocities are greater than the minimum velocities found above at room temperature.

(c) Is it possible to explain from the results obtained above why the currently existing amount of the H_2 molecule in the atmosphere is extremely small?

The gravitational potential energy for the one mole of H_2 is shown in Fig.5 given below.

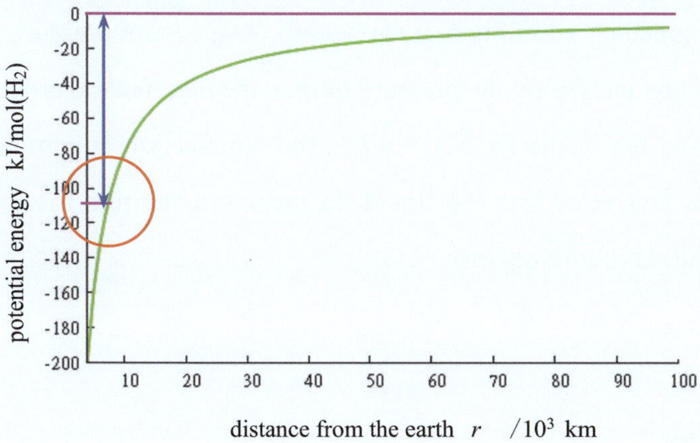

The origin is taken at the center of the earth and the potential energy is put equal to zero at the infinite distance from the origin. The blue arrow indicates the potential energy difference from the surface of the earth.

Fig.5 Gravitational potential energy of H_2 (1 mole) plotted against the distance from the surface of the earth.

Problem 2-24

The *van der Waals* equation of state can be rewritten as eq(7) to express the compression factor

$z = PV/RT$ or as eq(8) to express P itself. From these equations, z can be plotted as a function of P at a constant temperature by taking z along the y-axis and P along the x-axis (parametric plot), either of which is expressed as a function of V.

$$z = \frac{PV}{RT} = \frac{V}{(V-b)} - \frac{a}{RTV} \qquad (7)$$

$$P = \frac{RT}{(V-b)} - \frac{a}{V^2} \qquad (8)$$

Plot the compression factor z against P for CO_2 at several temperatures and compare the results with those (z *vs.* P plot) expected from the first and second virial approximations. The *van der Waals* parameters of a and b for CO_2 are $3.592\ \mathrm{dm^6 \cdot atm \cdot mol^{-2}}$ and $4.267 \times 10^{-2}\mathrm{dm^3 \cdot mol^{-1}}$, respectively.

Problem 2-25

For real gases, explain what kind of point the critical point on the curve of the compression factor z plotted against P ($z = PV/RT$ *vs.* P plot) is generally characterized as.

Problem 2-26

The virial equation of state is expressed in two different ways, i.e., eq(9) and eq(10) (see **problems 2-11** and **2-16**).

$$\frac{PV}{RT} = 1 + \frac{B(T)}{V} + \frac{C(T)}{V^2} + \frac{D(T)}{V^3} + \cdots \qquad (9)$$

$$\frac{PV}{RT} = 1 + B'(T)P + C'(T)P^2 + D'(T)P^3 + \cdots \qquad (10)$$

Since $P \propto 1/V$ at low pressures, the two equations seem to be equivalent though they look different in their expressions. However, the two equations are decisively different in the prerequisite for deriving the equation. Explain how different they are in an essential point.

Problem 2-27

When a gas is compressed at a constant temperature, the volume of the gas decreases and the pressure increases along the P-V isotherm (a part of the green curve in Fig.6 given below). At a certain pressure (the blue dot in the figure) the liquefaction begins. Upon increasing the pressure further, the liquefaction proceeds at the constant pressure (liquid-gas coexisting region in the figure). When the liquefaction is completed (the yellow dot in the figure), all of the gas is converted into the liquid. In order to compress the liquid further, much higher pressure than before is required (liquid region).

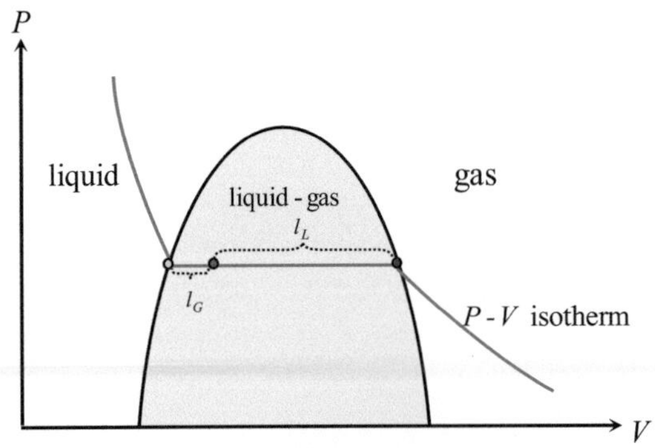

Fig.6 P-V isotherm of gas and liquid in equilibrium.

Now, let us consider the liquid and gas coexisting in equilibrium at the point indicated by the red dot in the figure. The distances from this point to the yellow and blue dots are denoted by l_G and l_L, respectively. Then, show that the ratio of the gas to the liquid in mole numbers is given by $l_L : l_G$ (lever rule).

Problem 2-28

Show that the mean translational kinetic energy is independent of the mass of a molecule and is equal to $3k_BT/2$ per molecule when it is calculated based on the three-dimensional velocity distribution function.

Problem 2-29

(a) The viscosity coefficient $\eta\,(\mathrm{kg \cdot m^{-1} \cdot s^{-1}})$ of a gas is expressed in terms of the mean free path λ, three-dimensional mean velocity $\bar{c}$, the diameter of the molecule d and its mass m.

$$\eta = \frac{1}{2} m\lambda\bar{c} \times \frac{N}{V} = \frac{m\bar{c}}{2\sqrt{2}\pi d^2}$$

Calculate the viscosity coefficient of the air at STP, assuming its apparent molecular weight is 28.8 (see **problem 2-15**) and its hypothetical molecular diameter is 3.0×10^{-10} m.

(b) The viscous friction for a spherical object with a radius r moving with a velocity υ in a media of viscosity η is given by $6\pi\eta r\upsilon$. Apply this formula to find the viscosity resistance for a ball with the radius of 1.5×10^{-1} m moving in the air with the velocity of 10 m.

(c) A ball falling down in the gravitational field of the earth with acceleration eventually starts to fall with a constant velocity because of the viscous resistance of the air. Find the velocity of the falling ball whose specific gravity is 1.5.

Problem 2-30

Show that the total number of collisions Z_{wall} per second with which the gaseous molecules in a chamber collide with its wall is given by the following equation, in which *S*, *P*, *N* and $\bar{c}$ denote, the surface area of the wall, the pressure of the gas, the number of the gaseous molecules and the mean velocity of the molecule at the temperature *T*, respectively. The collision number per unit time and area is denoted using a small letter as z_{wall}.

$$z_{wall} = \frac{Z_{wall}}{S} = \frac{1}{4}\bar{c} \times \frac{N}{V} = \frac{P}{(2\pi mk_BT)^{1/2}}$$

Problem 2-31

The wall of a chamber containing gaseous oxygen is consisted of the metal atoms whose atomic radius is 2.0×10^{-10} m. Employing the formula derived in **problem 2-30**, find the minimum pressure of the oxygen at which the collision number of the oxygen molecules with the single metal atom becomes less than one per second at 300 K. Assume the collision radius of the oxygen is 1.5×10^{-10} m.

Problem 2-32

Show that the relative mean velocity $\bar{c}_{rel}$ of the gaseous molecules of the same kind which obey the *Maxwell-Boltzmann* distribution law is given by $\bar{c}_{rel} = \sqrt{2} \times \bar{c}$, where $\bar{c}$ is the three-dimensional mean velocity.

Problem 2-33

Using the *Maxwell-Boltzmann* distribution function, answer the following questions about the oxygen molecule kept at 300 K.

(a) Find the formula as well as the value giving the velocity (speed) c_{max} at which the distribution function takes a maximum value.

(b) Find not only the value but also the formula for the mean velocity $\bar{c}$.

(c) Find the value and the formula for the square root of mean square velocity $\sqrt{\overline{c^2}}$.

(d) Find the fraction of the molecules whose velocities are in the range $390 - 400\ \mathrm{m \cdot s^{-1}}$.

Problem 2-34

One mole of gaseous oxygen was introduced into a vessel of a volume of $10\,\mathrm{dm}^3$ at 300 K and at the pressure *P*. However, it turned out that there was a hole of the area $1.0\,\mathrm{mm}^2$ in the wall of the vessel and that the gas was leaking out. Find how long it will take for the pressure to drop to one half of the initial pressure. Also, if the area of the hole is reduced to one tenth of the original one, find how many times longer it will take for the same initial pressure to be halved by leakage.

Problem 2-35

The parameter *b* in the *van der Waals* equation of state is associated with the excluded volume, which is defined as the minimum volume within which molecules cannot approach each other since the molecules are not just points with no volume but have finite volumes. Referring to Fig.7 given below, find the excluded volume of the molecules with the radius *R*.

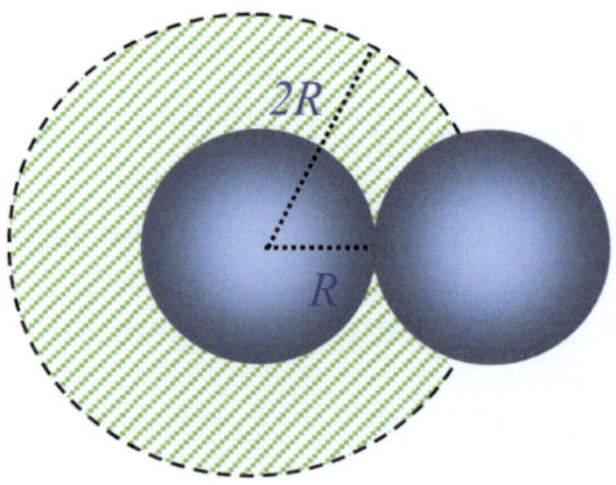

Fig.7 Illustration of the excluded volume for two identical spherical molecules.

Problem 2-36

Describe how the excluded volume is expressed in the case of a gas mixture of different molecules A and B. The molecular radii and the numbers of A and B are denoted R_1, R_2 and N_1, N_2, respectively. (Important examples of excluded volume or area are given in Chapters 3, 5 and 7.)

[just for a break] The following question is not necessarily associated with the concept of excluded volume in the *van der Waals* equation of state, however, it often happens that an object of a finite volume moving in a container can never occupy a certain part of the container. Let us see such an example as an arithmetic problem.

Find the area that a circle with the radius 1 cannot occupy when it moves around freely within the right triangle shown below in Fig.7'a.

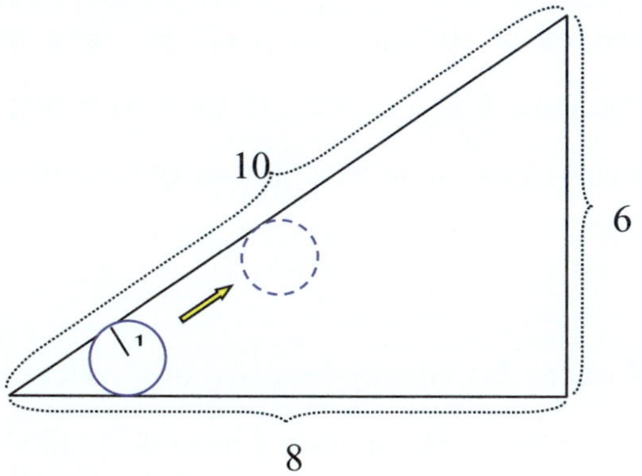

Fig.7'a A right triangle in which a circle of the radius 1 move freely.

(The figure is not drawn to the correct size.)

Example solution

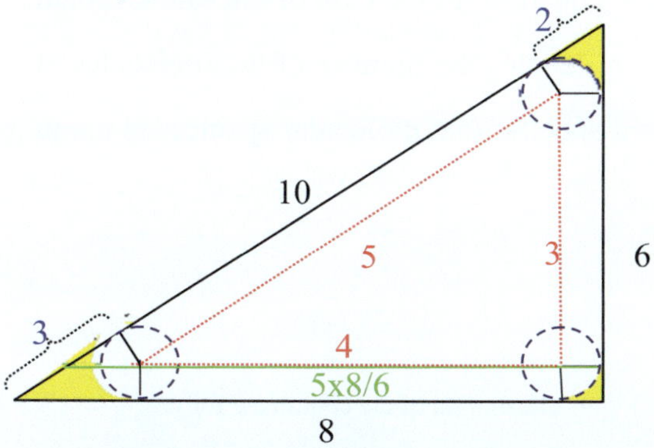

length of the green lline: $(6-1)\times\dfrac{8}{6}$

length of the base of the red triangle:

$$(6-1)\times\frac{8}{6}-1-1\times\frac{10}{6}=\frac{40-6-10}{6}=4$$

areas colored yellow: $1\times(1+2+3)=6$

area the circle of radius1 cannot occupy: $6-3.14\times(1)^2=2.86$

Fig.7'b How to work out the area of the right triangle which the circle of radius 1 cannot occupy.

Problem 2-37

There are two kinds of gases 1 and 2 each of which obeys the *van der Waals* equation of state. We consider here CO_2 (gas 1) and NH_3 (gas 2) taking them as an example. Their *van der Waals* parameters a and b are given below.

$$a_1=3.592\ \text{dm}^6\cdot\text{atm}\cdot\text{mol}^{-2},\ b_1=4.267\times10^{-2}\text{dm}^3\cdot\text{mol}^{-1}$$

$$a_2=4.248\ \text{dm}^6\cdot\text{atm}\cdot\text{mol}^{-2},\ b_2=3.785\times10^{-2}\text{dm}^3\cdot\text{mol}^{-1}$$

In Fig.8 are plotted the pressures calculated from their *van der Waals* equations of state and the ideal gas law ($n=1$) as well at $T=310\,\text{K}$ against V. Consider what is the most reasonable *van der Waals* equation of state to express the gas mixture composed of one mole of each gas of CO_2 and NH_3.

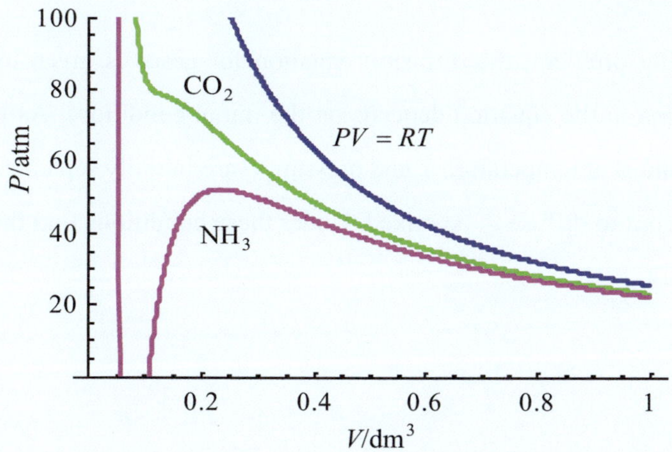

Fig.8 P plotted against V for two *van der Waals* gases and an ideal gas at T=310 K.

Problem 2-38

The parameters a and b of the *van der Waals* equation of state are $a = 4.248\ \text{dm}^6 \cdot \text{atm} \cdot \text{mol}^{-2}$ and $b = 3.785 \times 10^{-2}\,\text{dm}^3 \cdot \text{mol}^{-1}$, respectively for gaseous ammonia. It is interesting to know what percent of the molar volume of this gas at 300 K corresponds to the molar volume of an ideal gas under the same condition. Find the values at 1 atm and 5 atm.

Problem 2-39a

As is illustrated in Fig.9 shown below, a number of molecules initially localized at a certain place (for instance at $x = 0$) will eventually spread out with time, but how? The spreading process is called diffusion and the question is answered quantitatively by solving the diffusion equation. In the case of one-dimensional diffusion, the diffusion equation is given by eq(11), in which $N(x,t)$ denotes the number of the molecules at position x and at time t, and $D/\text{m}^2 \cdot \text{s}^{-1}$ is the diffusion coefficient specific to a molecular species. In many cases D takes a value in the range of $10^{-5}/\text{m}^2 \cdot \text{s}^{-1}$.

$$\frac{\partial N(x,t)}{\partial t} = D\frac{\partial^2 N(x,t)}{\partial x^2} \qquad (11)$$

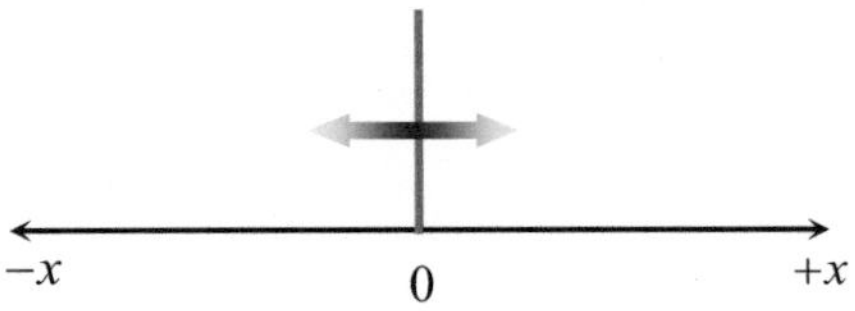

Fig.9 Illustration of the coordinate for the one-dimensional diffusion.

(a) Confirm that eq(12) given below is a solution of eq(11) first and then plot $N(x,t)$ at $t = 10$ s, 100 s and 900 s as a function of x for $D = 10^{-5}/\text{m}^2 \cdot \text{s}^{-1}$.

$$N(x,t) = \frac{N_0}{2\sqrt{\pi D t}} e^{-\frac{x^2}{4Dt}} \qquad (12)$$

(b) How far the molecules have spread out in time t is known by finding the value for the square root of the mean square of a distance of x a molecule has moved in that time. Find a formula to express $\sqrt{\overline{x^2}}$ by employing eq(12).

(c) Find how many seconds it will take for a molecule to reach $\sqrt{\overline{x^2}} = 0.1\,\text{m}$ and $1.0\,\text{m}$ by diffusion in the two cases of $D = 10^{-5}/\text{m}^2\text{s}^{-1}$ and $10^{-9}/\text{m}^2\text{s}^{-1}$ (see **problem 2-41**).

Problem 2-39b

As is known from the solution to the preceding problem, the diffusion equation for gases is given by eq(13) with D_A as a diffusion coefficient. A solution to the equation depends on the initial conditions. As is illustrated in Fig.10 given below, two gases of A and B at temperature T and pressure P are initially separated by a wall. Upon removing the wall the two gases start to diffuse to be mixed. Under these conditions find the solution of eq(13) and plot it as a function of x and t.

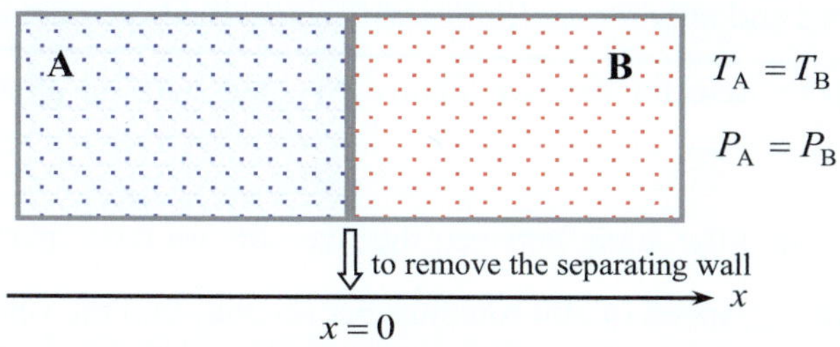

$$\frac{\partial C_A}{\partial t} = D_A\left(\frac{\partial^2 C_A}{\partial x^2}\right) \quad (13)$$

Fig.10 Illustration of mixing of two gases A and B initially contained in each chamber separated by a removable separating wall indicated by a thick vertical line.

Problem 2-40

The distance a molecule can travel per unit time, which is treated in the previous problems, can also be considered as follows. We denote here the distance a molecule can move without any collision with the other molecules by λ (mean free path) and then the (average) time required for this is given by $\Delta t = \lambda/\bar{u}$, where $\bar{u}$ is the mean velocity. In the one-dimensional case, the molecule can move in Δt by a distance of λ or $-\lambda$ with an equal probability of $1/2$.

(a) In the one-dimensional case, where can you find the molecule which is initially located at $x = 0$ ($t = 0$) after $3\Delta t$? Answer its positions together with its probabilities.

(b) Drive a general equation to predict where and with what probability the molecule is found after the time $n\Delta t$ has elapsed. Such a problem as this is called "one-dimensional random walk".

(c) Under what conditions does the result derived in (b) above lead to the same form as eq(12) given in the previous **problem 2-39a**?

Problem 2-41

Gaseous oxygen can be regarded as an ideal gas. Find values for the diffusion coefficient of the oxygen at 1 atm and 10 atm, both at 298 K. The collision cross section σ of the oxygen is assumed to be $0.40\times10^{-18}\,\mathrm{m}^2$.

Problem 2-42

We can sense a smell whatever it is aromatic or bad quite sensitively from far away. Then we can ask how long it takes for flavoring molecules to spread a certain distance so that we can smell them. Answer to this question referring to the previous **problems 2-39a** and **2-40**.

Problem 2-43

The *Maxwell-Boltzmann* distribution which universally governs the velocities of gaseous atoms or molecules was experimentally verified by *P.Kusch* (Nobel Prize in Physics in 1955) and his coworkers in 1950's by measuring the velocity distribution of molecules directly. A basic idea of their method is illustrated in Fig.11 given below. Molecules vaporized in the source (or furnace) flow out through an orifice in the source and pass through a "slit" in a linear flow of molecules like a molecular beam and reach an apparatus installed for selecting velocity (velocity selector). The selector consists of several discs placed at regular intervals. Each disc has a slit so that a part of molecules in the molecular beam can pass through the disc. The velocity selector itself is rotated at a constant speed and only the molecules with a specified constant velocity can pass through the discs of the rotating velocity selector and are finally detected with an appropriate apparatus.

(a) An interval of two neighboring discs is 1.0 cm and the angle between the two slits on those discs is 30 degrees. When the velocity selector is rotated at the speed of 300 rotations per second, find the velocity of the molecules (atoms) which can pass through the selector to be detected.

(b) The speed of rotation is increased to 400 rotations per second. Then find how many times the amount of molecules detected will increase or decrease. Take the K atom at 350 K as an example.

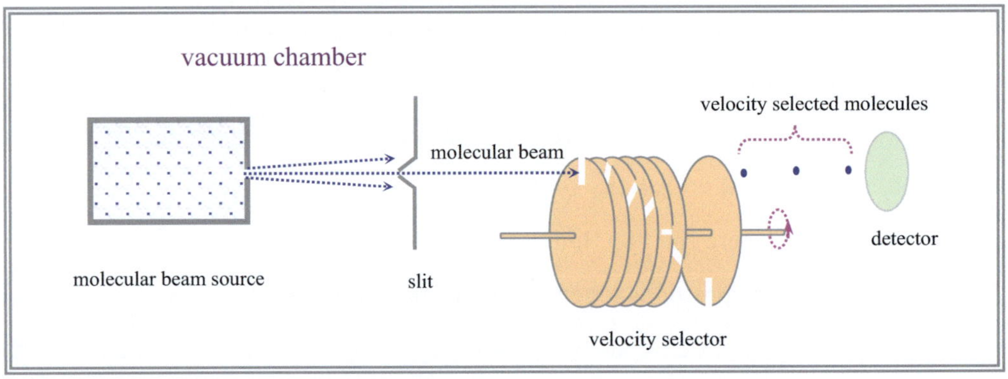

Fig.11 Illustration of the method for measuring molecular velocities and their distributions employed by *P.Kusch*.

Problem 2-44

The thermal conduction by gases can be treated in terms of the transfer of kinetic energies. Show the equation of thermal conductivity derived on the basis of the molecular dynamics (velocity distribution) takes the same form as the equation of diffusion and is given by the following equation.

$$\frac{\partial T}{\partial t} = \frac{\kappa}{\rho C_V}\left(\frac{\partial^2 T}{\partial x^2}\right) = D\left(\frac{\partial^2 T}{\partial x^2}\right)$$ D: diffusion coefficient, ρ:density, κ: coefficient of thermal conductivity

Problem 2-45

As is evident from the solution to the preceding problem, the equation of thermal conduction for the ideal gas is given by the following equation with D as a diffusion coefficient.

$$\frac{\partial T}{\partial t} = D\left(\frac{\partial^2 T}{\partial x^2}\right)$$ D: diffusion coefficient

A solution to this equation varies largely depending on the initial and boundary conditions. We discuss here a heating process of gases which we often encounter when the problems associated with the first and second laws of thermodynamic are solved. As is shown in the figure Fig.12 given below, when a gas is in contact with a solid heat reservoir, the gas eventually reaches the same temperature as the reservoir's temperature. In equilibrium thermodynamics it is of no concern how the gas reaches a constant temperature. But how is that

Table 1 Vapor pressure and temp.

T	P/ mmHg
247.9	10
258.7	20
278.36	60
304.61	200
341.74	760
366	1520
404.7	3800
439.6	7600
482.4	15200

Source: *Dortmund Data Bank*

Table 2 Density of the liquid against temp.

T	ρ / kg/m^3	T	ρ / kg/m^3
223	723.3	298.15	655.8
233	713	303	650.9
243	705.3	308.15	645.2
253	695.2	313	640.9
263	690.2	323	631.9
273	677.8	333	621.9
283	666.2		
288.15	663.7		
293	659.3		

Handbook of Chemistry:Pure Chemistry

Problem 2-50

The variation of atmospheric pressure with the altitude can be predicted by eq(15) called barometric equation.

$$P = P_0 e^{-Mgh/RT} \qquad (15)$$

P_0 is the atmospheric pressure at the sea level (zero altitude). M is the averaged molecular weight of the air, h is the altitude and g is the gravitational acceleration. Assuming $P_0 = 1$ atm, find the pressures at the altitudes of 10^3 m and 5×10^3 m. What are the shortcomings in this type of calculation?

Problem 2-51a

The *Poiseuille* equation modified so as to be applicable for the measurement of gas viscosity is given by the equation below. The left side of the equation is the volume of a gas flowing out through a pipe per unit time at the pressure of $P_\circ$. On the right side, l is the length of the pipe, R_0 is the radius of the pipe, and P_1 and P_2 are the pressures of the gas at the entrance and the exit of the pipe, respectively.

$$\frac{dV^g}{dt} = \frac{\pi(P^2{}_1 - P_2{}^2)R_0{}^4}{16 l \eta P_\circ}$$

Now the viscosity of the air was measured at 298 K, employing the pipe of 1.00×10^{-3} m in diameter and 1.00 m in length. A volume of 9.02×10^{-5} m^3 of the air flew out in 1.00×10^2 s. The pressure at the inlet was 765 Torr and the one at the outlet was 760 Torr. The volume of the air was measured at 760 Torr.

Find the viscosity of the air at 298 K from these data.

Problem 2-51b

One of the methods to measure the relative viscosity of a gas is the evacuation method. It utilizes the fact that the lower is the viscosity of a gas, the faster the high vacuum the gas can reach by pumping it out of a chamber by a vacuum pump. The viscosity of the gas is denoted η, the initial pressure is P_0 and the pressure at the time t after the start of pumping is P. Then these variables satisfy the following relation which is derived from the *Poiseuille* equation. Here, α is a constant dependent on the pumping system used.

$$\frac{1}{P}=\frac{\alpha}{\eta}t+\frac{1}{P_0}$$

By plotting $1/P$ against t, a value for η is found from the slope of the plot. The slopes were obtained at 298 K for the air and Ar as $2.50\ \mathrm{s}^{-1}\cdot\mathrm{bar}^{-1}$ and $2.02\ \mathrm{s}^{-1}\cdot\mathrm{bar}^{-1}$, respectively. Based on these results, find the viscosity of Ar if the viscosity of the air is taken as 1.81×10^{-2} cP.

Problem 2-52

Considering that the air is composed of nitrogen and oxygen in the mole ratio of 4 to 1, find the values of the mean free path for the nitrogen and the oxygen in the one atmospheric air at 298 K. The molecular diameters of nitrogen and oxygen are 0.373 and 0.353 nm, respectively and both molecules can be treated as hard spheres. If possible, find a value of the diffusion coefficient for each molecule.

Problem 2-53

For a gas mixture ($T,\ N_1,\ N_2,\ V$) composed of two different molecules 1 and 2, find a value of the mean free path for each component molecule along the method employed for a pure gas. From the obtained values, find the value of the diffusion coefficient for each molecule in the gas mixture, denoted D_{12} for the molecule 1 and D_{21} for the molecule 2, and show that $D_{1,2}\neq D_{2,1}$. Based on this finding discuss how the diffusion coefficient of the gas mixture is to be defined.

Problem 2-54a

One mole of oxygen is introduced into a vessel at 298 K under the pressure of 1 atm. Find the number of the oxygen molecules which collide with a unit area of 1 cm^2 of the vessel wall per second.

Problem 2-54b

Under the same conditions as in the preceding problem, but if it is assumed that once the oxygen molecules collide with the wall, they are all adsorbed on the wall surface, then find how long it takes for the wall to be covered thoroughly with oxygen molecules (one layer). Here, put the collision cross section of oxygen molecules as $\sigma=0.40\ \mathrm{nm}^2$.

Problem 2-55

The mean free path of an oxygen molecule in the standard state is 6.24×10^{-8} m. Find probabilities that the oxygen molecule can fly distances of two, three and ten times the mean free path without any collision with the other molecules.

Problem 2-56

The *Maxwell-Boltzmann* distribution law which concerns molecular velocities in the gas phase can be derived expediently in a few ways.

(a) Derive the *Maxwell-Boltzmann* distribution law based on the distance distribution law of the random walk.

(b) The velocity distribution must be isotropic in space. Derive the *Maxwell-Boltzmann* distribution law based on this condition alone.

Problem 2-57

There are several types of the equation of state to describe real gases. Among those which are characterized by two parameters, the representative ones are (i) *van der Waals*, (ii) *Dieterici* and (iii) *Berthelot* equations of state and they are given below (for $n=1$).

$$P=\frac{RT}{V-b}-\frac{a}{V^2} \quad \text{(i)}, \qquad P=\frac{RTe^{-\frac{a}{RTV}}}{V-b} \quad \text{(ii)}, \qquad P=\frac{RT}{V-b}-\frac{a}{TV^2} \quad \text{(iii)}$$

(a) Equations (ii) and (iii) are introduced to modify the *van der Waals* equation of state so as for them to describe real gases better than does the *van der Waals* equation of state. Answer for each equation what are essential improvements to meet that purpose.

(b) Show that the *Dieterici* equation of state approaches to the *van der Waals* equation of state at higher temperatures.

(c) In those three equations the same symbols a and b are employed for the two parameters. Answer the differences in the meaning of the parameters a and b in each equation by showing P-V curve for each equation.

Problem 2-58

The *Redlich-Kwong* equation of state is characterized by two parameters as is seen from eq(16) given below.

$$P=\frac{RT}{V-b}-\frac{a}{T^{1/2}V(V+b)} \qquad (16)$$

(a) Eq(16) is written for one mole of a pure gas. Give the equation which holds for n mole of the gas.

(b) Eq(16) is considered as an equation modified so as to improve the *Berthelot* equation of state. Explain in what points *Redlich* and *Kwong* attempted to improve the equation.

Problem 2-59

As is illustrated in Fig.6 below, sounds propagate in media as waves of condensation and rarefaction (longitudinal waves). A velocity of the wave, generally called sound velocity, is found by solving simultaneously (i) the eqeuation of motion and (ii) the equation of continuity (conservation of matter) which hold in a small space cut out of the media in which the wave is propagating as is shown by the dotted square in Fig.13.

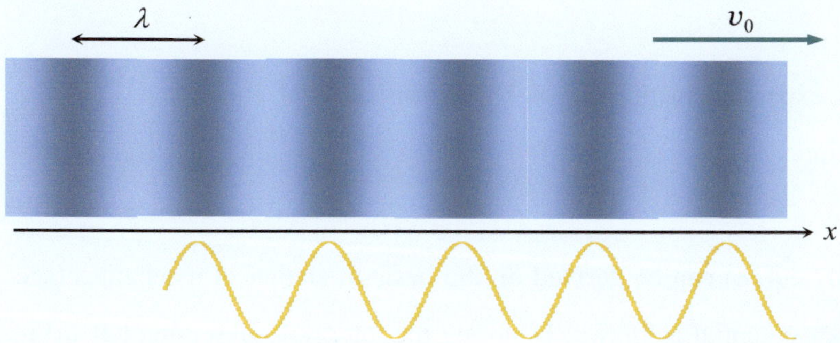

Fig.13 Sound wave propagating in a medium along x-axis with a velocity of v_0.

A part of Fig.13 is expanded and is shown in Figs.13a and b. Let us consider a small rectangle shown by the dotted line in Fig.13a or b, whose area of *yz*-plane is denoted A. Fig.13a helps to work out the equation of motion driven by the pressure difference and Fig.7b the equation of continuity (conservation of matter) upon mass transport.

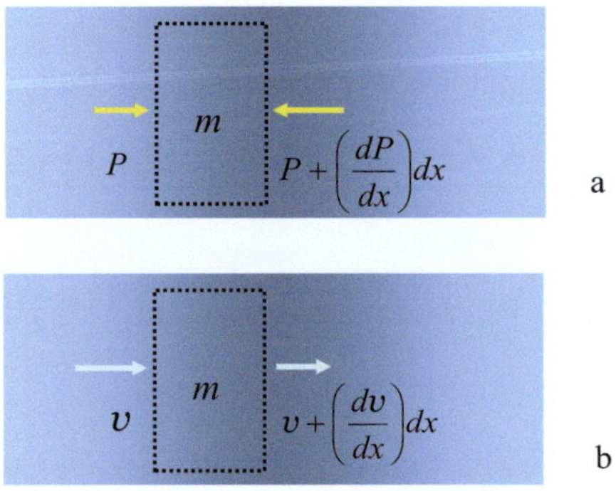

Figs.13 a and b Expanded view of a part of Fig.13.

(a) Referring to Fig.13 and equations given there, derive the equation of motion for the mass of the rectangle and the equation of continuity with respect to the rectangle.

(b) From the equations obtained in (a), show that the equation and its solution for the density of the medium denoted $\rho(x,t)$ are as given below (υ_0 : sound velocity, λ : wave.length).

$$\frac{\partial^2\rho}{\partial t^2}=\upsilon_0^2\frac{\partial^2\rho}{\partial x^2}=0\ ,\quad \rho\ (x,t)=\rho_0\left\{1+\phi_0\sin\frac{2\pi}{\lambda}(x-\upsilon_0 t)\right\}$$

(c) Show that the sound velocity υ_0 is given by the equation below, where $\gamma=C_P/C_V$ $\gamma=C_P/C_V$.

$$\upsilon_0=\sqrt{\gamma\left(\frac{\partial P}{\partial\rho}\right)_T}=\sqrt{-\gamma\left(\frac{\partial P}{\partial V}\right)_T\frac{V}{\rho}}$$

When the medium can be approximated as an ideal gas, the above equation is rewritten as below with M as the molecular weight of the gas. It is seen that the velocity depends on neither pressure nor number of moles.

$$\upsilon_0=\sqrt{\frac{\gamma RT}{M}}$$

Problem 2-60a

The *Dieterici* equation of state is given by eq(17).

$$P=\frac{RTe^{-\frac{a}{RTV}}}{(V-b)} \qquad (17)$$

Employing the reduced variables such as $T_R=T/T_c$, etc., derive the reduced equation of state of eq(17). Any equation of state described by using only two parameters can, in principle, be expressed in the form of $P_R=f(T_R,V_R)$.

Problem 2-60b

The *Dieterici* equation of state, when expanded in powers of $1/V$, forms a virial equation. Find the second coefficient $B(T)$ of this virial equation.

Problem 2-61

The virial equation of state expressed by eq(18) can be derived on the basis of statistical mechanics (see Chapter 5), provided that the intermolecular potential is known. Denoting the intermolecular potential $u(r)$, the second virial coefficient $B(T)$ is given by eq(19).

$$\frac{PV}{RT} = 1 + B(T)\cdot\frac{1}{V} + C(T)\cdot\frac{1}{V^2} + \cdots \qquad (18)$$

$$B(T) = -2\pi N_A \int_0^\infty [e^{-u(r)/k_BT} - 1] r^2 dr \qquad (19)$$

When the *Lennard-Jones* potential given by eq(20) is employed for the intermolecular potential, eq(19) can be rewritten as eq(21).

$$u(r) = 4\varepsilon\left[\left(\frac{\sigma}{r}\right)^{12} - \left(\frac{\sigma}{r}\right)^{6}\right] \qquad (20)$$

$$B(T) = -2\pi N_A \int_0^\infty \exp\left\{-\frac{4\varepsilon}{k_BT}\left[\left(\frac{\sigma}{r}\right)^{12} - \left(\frac{\sigma}{r}\right)^{6}\right] - 1\right\} r^2 dr \qquad (21)$$

If the values for ε and σ specific to the molecule under consideration are known, eq(21) can be evaluated.

(a) The values of (ε/k_B, σ) for the following gases are He (10.22, 256), N_2 (95.1, 370) and CO_2 (189, 449). Plot $B(T)$ against T for each gas listed above. Package software such as *Mathematica* is useful for this purpose.

(b) Find *Boyle* temperature defined by $B(T) = 0$ for each gas mentioned above.

(c) Compare the second virial coefficient thus obtained with that of the *van der Waals* equation of state and comment on that.

Problem 2-62

Initially a gas of carbon dioxide (CO_2) is at the pressure of 1 atm at 300 K. Find a % change in volume when the temperature is varied by 1K, assuming that the gas obeys the equation of state of (i) $PV = nRT$, (ii) $P(V - nb) = nRT$, (iii) *van der Waals* or (iv) *Dieterici*. Use the following values for the parameters: $a =$ 3.64 $\text{atm}\cdot\text{L}^2\cdot\text{mol}^{-2}$ and $b = 4.267$ $10^{-2}\text{L}\cdot\text{mol}^{-1}$. for (iii) and $a =$ 4.63 $\text{atm}\cdot\text{L}^2\cdot\text{mol}^{-2}$ and $b = 4.63$ $10^{-2}\text{L}\cdot\text{mol}^{-1}$ for (iv). What is the change in volume when the pressure is kept at 10 atm instead of 1 atm in each case?

Problem 2-63

Questions raised here are about the relation between the parameters *a* and *b* of the *van der Waals* equation of state and the mean intermolecular distance. For this the values of $a = 1.36$ $\text{atm}\cdot\text{L}^2\cdot\text{mol}^{-2}$ and $b = 3.22\times10^{-2}\,\text{L}\cdot\text{mol}^{-1}$ are employed as the representative ones.

(a) Find the mean intermolecular distance for an ideal gas at the standard state.

(b) When the pressure of the gas is increased to 10 atm, how largely does the mean intermolecular distance change? When the excluded volume given above is expressed by a cube of the same volume, find how many times the mean intermolecular distance at 10 atm is larger than the length of the edge of the cube.

(c) Show first that the mean energy due to the intermolecular force of the *van der Waals* gas at *P*, *V*, *T* per one mole is given by a/V, and then find the volume of the gas when that value becomes of the order of one tenth of the mean dipole-dipole interaction energy of $0.6\ \text{kJ}\cdot\text{mol}^{-1}$.

(d) Find the mean intermolecular distance for the volume found in (c).

Problem 2-64

The pressure exhibited by a gas was derived in **problem 2-9** on the basis of the kinetic theory of the gaseous molecules. We clarified the reason there why the equation of state for an ideal gas does not contain any molecular property specific to the molecular species of a gas. A basic idea there was that the force (pressure) exerted on the wall of a vessel containing the gas arises from a time-averaged change in the momentum of an individual molecule upon the collision to the wall.

Then, along the same line as the basic idea given above, explain why the pressure exerted by a *van der Waals* gas in which the attractive intermolecular force is operating is smaller than that of the ideal gas at the same temperature.

Problem 2-65

The molar volumes of gaseous Ar and ethane are measured at $T = 247$ K and $T = 500$ K, by varying the pressure so as for the reduced temperature to be kept at $T_R = 1.64$ for each gas. The obtained results are summarized in the table given below.

P/bar	$\overline{V}$ / L · mol^{-1}		P/bar	$\overline{V}$ / L · mol^{-1}		P/bar	$\overline{V}$ / L · mol^{-1}	
	Ar	ethane		Ar	ethane		Ar	ethane
0.5	83.076	40.506	100	0.3734	0.1785	350	0.1175	0.05626
2.0	20.723	10.106	120	0.3068	0.1462	400	0.1085	0.05219
10	4.105	1.999	160	0.2265	0.1076	450	0.1019	0.04919
20	2.028	0.9857	200	0.1819	0.0853	500	0.09676	0.04687
40	0.9907	0.4795	240	0.1548	0.07348	600	0.08937	0.04348
60	0.6461	0.3114	300	0.1303	0.06208	700	0.08421	0.04108
80	0.475	0.2279	350	0.1175	0.05626			

Calculate the value for the compression factor z (or reduced compression factor z_R) and plot them against the reduced pressure P_R. Confirm whether or not the reduced equation of state is satisfied for these two gases. Use the following values. Ar: $V_C = 0.0753$ L · mol^{-1}, $P_C = 49.288$ bar. ethane: $V_C = 0.1480$ L · mol^{-1}, $P_C = 48.714$ bar. (Modified from *D.A.McQuarrie and J.D.Simon, Physical Chemistry* problem 16-31.)

Problem 2-66

Effusion: *Graham*'s law

An ideal gas initially kept at P, V and T in a vessel was effused out through an orifice in the vessel wall. The total amount of the gas flowing out is calculated from the cross section of the orifice and the collision number onto it. The collision number is given by eq(22) where N is the number of molecules of the gas, the mean three-dimensional velocity is $\bar{c}$ and the collision cross section is denoted S_0 (see **problem 2-30**).

$$Z = \frac{1}{2}\bar{u} \times S_0 \times \frac{N}{V} = \frac{1}{4}\bar{c} \times S_0 \times \frac{N}{V}$$

$$= \frac{1}{4}\left(\frac{8k_BT}{\pi m}\right)^{1/2} \times S_0 \times \frac{P}{k_BT} = \left(\frac{1}{2\pi m k_BT}\right)^{1/2} \times P \times S_0 \qquad (22)$$

(a) Gases of He and N_2 are introduced into the same vessel as above each at 1 atm. Find the ratio of the numbers of the molecules effused per unit time. Assume that the amount of gases having effused is small enough to maintain the pressure in the vessel unchanged.

(b) Applying the kinetic theory of gaseous molecules to the effusion, the diameter of the orifice must be smaller than the mean free path λ (see **solution 2-52**) of the gas under discussion. Find how small the diameter of the orifice must be in the case of (a).

Problem 2-67

The effusion of a gas through an orifice of a vessel is assumed to obey the *Graham* law (see the preceding problem). The initial pressure of the gas in the vessel is P_0 and the pressure of the gas eventually drops due to the effusion. Find a rate equation to describe the pressure variation with time. Based on the derived equation, work out the time required for the pressure to drop to one half of the initial pressure of P_0 for the He gas at 300 K. The volume of the vessel is $1.0\,\text{dm}^3$ and no gas flows into the vessel during the effusion. The radius of the orifice is known to be $1\,\mu\text{m}$.

Problem 2-68

A rubber balloon was inflated with He gas and after an inlet part of the balloon was tightly closed, it was tied with a string to a hook so as for it not to fly away. Initially the volume of the balloon was 3 L but it got smaller gradually and became hard floating and finally fell to the ground. The cause for this is clearly the partial escape of He gas from the balloon. Modeling the escape of the gas as the effusion of He through a small hole in the balloon, estimate the time after which the inflated and tightly closed balloon comes to fall to the ground. The given conditions are that the empty balloon weighs 2 g, the apparent molecular weight of the air is 28.8, the radius of the small hole is $1\,\mu\text{m}$ and the balloon is inflated at 300 K under the atmospheric pressure of 1 atm.

Problem 2-69

(a) Isolation of gases: Based on the *Graham* law, gases of different molecular weights can be isolated. When a mixture of oxygen and nitrogen by the molar ratio of one to one is flowed through an orifice into a vacuum chamber of a large capacity, find the molar ratio of the gases in the chamber.

(b) The gas mixture in the chamber obtained in (a) is compressed appropriately and the collected gas mixture is again flowed through the orifice into a vacuum chamber. Such a cycle can be repeated many times. Then, find how many cycles are required at least to achieve the molar ratio of 1 (O_2) to 100 (N_2) of the mixture in the chamber.

(c) Radioactive ^{235}U is an element contained only ca. 1% in the compounds of the isotopic ^{238}U element.

One of the methods to concentrate ^{235}U is to employ the effusion (diffusion) method discussed in (b). For this purpose, gaseous fluorine compounds of uranium UF_6 are used. The Initial molar ratio of $^{235}UF_6$ to $^{238}UF_6$ is 1 to 100. In order to achieve the molar ratio of 10 to 1, find how many cycles of effusion-compression should be repeated.

Problem 2-70

There are two containers A and B of a volume V in which either the pure gas 1 or 2 is filled. The two containers are in contact with each other but are isolated initially. Eventually, through a small hole in their separating wall either gas starts to flows in the other container obeying the *Graham* law (see **problem 2-66**). The initial pressures of the gases in the containers kept at temperature T are P_1^0 and P_2^0.

(a) When the rate constant for the flowing out or in of the gas 1 is given by $k_1 = a/\sqrt{M_1}$, where a is a constant and M_1 is the molecular weight of the gas 1, find the rate constant for the flowing out or in of the gas 2.

(b) Accompanying the flowing out or in of the gases, the pressures in the containers A and B are expected to vary with time. Give a rate equation to describe a change in the pressure with time.

(c) Both containers are kept at 300 K. Initially, the container A is filled with N_2 at 2.0 atm and B with He at 1.0 atm. Find a change in the pressure with time in each container brought about by the effusion through the small hole and plot the result against time for each container.

Problem 2-71

By applying the laws of the conservations of mass, energy and momentum to the fluid (gas or liquid) as is done for the rigid body (for viscous fluid, see **problem 2-51**), *Bernoulli*'s equation of motion is derived (see the solution to this problem). When the fluid is assumed incompressible, *Bernoulli*'s equation is expressed as eq(23). The subscripts 1 and 2 specify the observation points of the fluid and P, ρ and υ indicate the pressure, density and velocity of the fluid at a relevant observation point, respectively.

$$P_1 + \frac{1}{2}\rho\upsilon_1^2 = P_2 + \frac{1}{2}\rho\upsilon_2^2 \qquad (23)$$

A *Pitot* tube (*Henri Pitot*, 1728) is an instrument employed to measure the velocities of flying objects such as airplanes from a difference in the pressures between the atmosphere and the gas flowing in the tube.

(a) Eq(23) implies that the pressure of the gas becomes the lower as the gas flows in the thinner tube. Explain what this means if the gas can be regarded as an ideal gas?

(b) While an airplane was flying in the atmosphere whose pressure was 0.5, the pressure difference a *Pitot* tube indicated was 0.10 atm. Find the speed of the airplane at that time.

Problem 2-72

One-dimensional diffusion equation in a fluid under an external force field is expressed by eq(24), where D is the diffusion coefficient of the fluid and u is the translational velocity driven by the external force, either of which is assumed constant, and $\rho(t,\ x)$ is the density of particles at the time t and at the position x.

$$\frac{\partial \rho(t,\ x)}{\partial t} = u\frac{\partial \rho(t,\ x)}{\partial x} + D\frac{\partial^2 \rho(t,\ x)}{\partial x^2} \qquad (24)$$

(a) Show that a solution for eq(24) is given by eq(25).

$$\rho(t,\ x) = \frac{N_0}{\sqrt{4\pi Dt}}\exp\left\{-\frac{(x+ut)^2}{4Dt}\right\} \qquad (25)$$

(a) Taking the x-axis along the vertical direction, eq(24) can express the diffusion under the gravitational force field. Find the solution when the diffusion has reached an equilibrium, in other words, the solution $\rho(x)$ which satisfies the left side of eq(26). Given conditions are $x \geq 0$ and $\rho(0) = \rho_0$.

$$\frac{\partial \rho(t,\ x)}{\partial x} = 0 \ \Rightarrow\ u\frac{\partial \rho(x)}{\partial x} + D\frac{\partial^2 \rho(x)}{\partial x^2} = 0 \qquad (26)$$

Problem 2-73

Brownian motion and *Avogadro*'s number (1). *Perrin*'s experiments (*J.B.Perrin*、Nobel Prize in Physics in 1926).

For particles of a uniform diameter and mass which were dispersed in water at constant temperature and pressure in a column vertically placed, *Perrin* measured the number of particles in equilibrium as a function of the height of h from the bottom of the column and obtained the results such as those shown in the table below. The radius and density of the particle were $r = 0.212\ \mu\text{m}$ and $d = 1.206\ \text{g}\cdot\text{cm}^{-3}$, respectively. Assume that the density of water was $1.0\,\text{g}\cdot\text{cm}^{-3}$.

height (μm)	5	35	65	95
number	100	47	22.6	12

Find the *Avogadro*'s number from these data assuming that the density of water was $1.0\,\text{g}\cdot\text{cm}^{-3}$.

Problem 2-74

Brownian motion and *Avogadro*'s number (2). Calculation from diffusion coefficient.

We now know in **solution 2-72** that the distribution function of the molecules or particles driven by an external force field such as the gravitational force filed is given by eq(27), when the diffusion process in a fluid has reached a stationary state (equilibrium). The force is considered to work along the z-axis and u is the translational velocity driven by the external force

$$\rho(z) = \rho_0 \exp\left(-\frac{u}{D}z\right) \qquad (27)$$

Under a constant force of f, the potential energy of the particle at the position z is given by $-f \times z$.

(a) Find the mean potential energy averaged over the distribution of particles.

(b) Show that under the gravitational force field, if it is assumed that the mean potential energy found for the particles above is equal to k_BT, just as is for molecules, then the particle distribution function is given by eq(28) in which g is the gravitational acceleration.

$$\frac{N(z)}{N_0} = \exp\left(-\frac{mgz}{k_BT}\right) = \exp\left(-\frac{N_A \times mgz}{RT}\right) \qquad (28)$$

Also, show that this equation is the barometric equation employed in **problem 2-50**.

(c) A particle driven by a constant external force f in a viscous fluid eventually comes to move in the fluid at a constant velocity u given by eq(29). The proportional constant μ is called mobility (easiness to move).

$$u = \mu \times f \qquad \mu: \text{mobility} \qquad (29)$$

The mobility of the particle of a radius of a moving in the fluid of the viscosity of η is known to be expressed by eq(30).

$$\mu = \frac{1}{6\pi\eta a} \qquad (30)$$

Then, show that the diffusion coefficient D of the particle is given by eq(31) which is called *Einstein*'s equation (*A.Einstein*、Nobel Prize in Physics in 1921).

$$D = \frac{RT}{6\pi\eta a N_A} \qquad (31)$$

(d) The migration distance of macroscopic particles by diffusion is given by eq(32). Show that *Avogadro*'s number can be worked out from the measurements of this migration distance.

$$\overline{x^2} = 2Dt \qquad (32)$$

Problem 2-75

The force F which works on a sphere of a radius a moving in a fluid of viscosity η at a constant velocity υ is a viscosity resistant force and is given by eq(33). Derive this equation based on the *Stokes* equation (see eq(312) in the solution).

$$F = 6\pi\eta a\upsilon \qquad (33)$$

Problem 2-76

The second virial coefficient is given by eq(34) in which $u(r)$ is the intermolecular potential (see **solution 5-98** for the detailed explanation).

$$B(T) = -2\pi N_A \int_0^\infty [e^{-u(r)/k_BT} - 1] r^2 dr \qquad (34)$$

It is assumed here that the intermolecular potential can be approximated by a square-well potential as shown on the right.

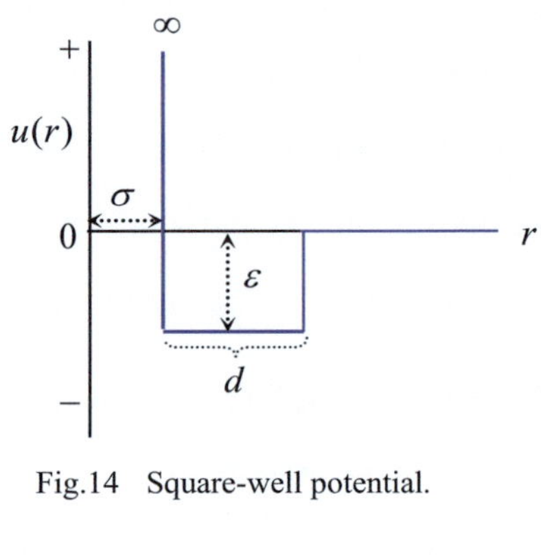

Fig.14 Square-well potential.

(a) For this potential, evaluate the integral of eq(34) and express $B(T)$ in terms of σ, d and ε.

(b) The second virial coefficient obtained by expanding the *van der Waals* equation of state in the powers of $1/V$ is given by eq(35) as has already been discussed in this chapter.

$$B(T) = b - a/RT \qquad (35)$$

Express the parameters a and b in terms of σ, d and ε and then discuss the molecular theoretical messages the parameters a and b can convey.

Problem 2-77

The virial theorem (virial theorem of *Clausius*) is a general law about the time-average over the dynamical properties of a system consisting of many particles. As a special case, it has a great significance in deriving equations of state of gases.

Representing the position vector of the *i*-th gas molecule ($i = 1, \cdots, N$) by $\boldsymbol{r}_i = (x_i, y_i, z_i)$ and the force vector working on the molecule by $\boldsymbol{F}_i = (X_i, Y_i, Z_i)$, the virial theorem predicts that the relation of eq(36) holds for the time-averages (indicated by over lines) of the kinetic energies, forces and positions taken over all the molecules (see the solution for its proof).

$$\overline{T} = \sum_i \overline{T_i} = -\frac{1}{2}\sum_i \overline{\boldsymbol{F}_i \cdot \boldsymbol{r}_i} = -\frac{1}{2}\sum_i (\overline{X_i \cdot x_i} + \overline{Y_i \cdot y_i} + \overline{Z_i \cdot z_i}) \qquad (36)$$

Show that the equation of state for the ideal gas is derived from eq(36) for *N* molecules occupying a cuboid ($a \times b \times c$) at the temperature *T* under the pressure *P* when the intermolecular forces are neglected.

Problem 2-78

Since the intermolecular distance is short in fluid, the space in which fluid molecules can move freely becomes much narrower as compared with those for gaseous molecules. This space is called "free volume". How to estimate the free volume depends on how the molecular level structure of the fluid is described.

Estimate the free volume in the following simplest case. Molecules are spheres of the same radius of *a*. The intermolecular distance is *R* (mean). It is assumed that a molecule can move freely when it is inside a spherical cage of radius of *R*.

Problem 2-79 [advanced problem]

Employing the *Bernoulli* equation (see solution **2-71**), derive the *Graham* law (see **solutions 2-66** and **2-69**). Solution to this problem is not provided.

I-3 Thermodynamics

Part of Problems

Problem 3-1

(a) Calculate the work done to the external by an ideal gas which is expanded from 5 dm^3 to 50 dm^3 by keeping it at a constant pressure of 100 kPa.

(b) One mole of the ideal gas at the temperature of 273 K under the pressure of 100 kPa is expanded isothermally to 50 dm^3. Find the following values associated with this process: heat energy the gas receives; work done by the gas; internal energy change $\Delta U(g)$; enthalpy change $\Delta H(g)$ entropy change $\Delta S(g)$. The symbol (g) indicates that a relevant value is for the gas. Also, positive values are for those gained by the system (gas) and negative values indicate those lost by the system (gas in the present case).

(c) Under the same conditions as (b), find the total entropy change given by the sum of the entropy changes for the gas and the surroundings.

Problem 3-2

There is a thermodynamic heat engine working between a hot heat reservoir at 150 °C and a cold heat reservoir at 30 °C.

(a) Find a maximum efficiency of this heat engine.

(b) When a heat energy of 100 kJ is taken out from the hot heat reservoir and a part of it is converted into a work, find a maximum work to be gained.

(c) Under the same conditions as (b), find an entropy change in each heat reservoir.

Problem 3-3

(a) Find a value of ΔG_r° (standard free energy change of the reaction) for the reaction given by eq(1) using the thermodynamic values given in the table below.

$$NH_3(g) + CO(g) \rightleftharpoons HCN(g) + H_2O(g) \qquad (1)$$

(b) Give the expression for the equilibrium constant K_P defined by a partial pressure of each gas in eq(1). Answer the relation between K_P and ΔG_r°, and then find a value for the equilibrium constant.

(c) At the time the above reaction started, the partial pressures of the gases in a vessel were 100 kPa both for NH_3 and CO and were 1 kPa both for HCN and H_2O, and the temperature was 50 °C.

Find a value for the free energy change of the reaction ΔG for this reaction at the time when the reaction started. Predict to which side the reaction (1) will proceed based on the value found for ΔG.

S. Hirayama and M. Yamamoto, *Physical Chemistry Problems and Solutions*,
https://doi.org/10.1007/978-981-97-0165-0_3

Table Thermodynamic values for simple reactions in the gas phase

	ΔH_f° / kJ · mol^{-1}	ΔG_f°/kJ · mol^{-1}	S_m°/J · K^{-1} · mol^{-1}
$H_2O(g)$	-241.82	-228.57	188.83
$H_2O(l)$	-285.83	-237.13	69.91
$CO(g)$	-110.53	-137.17	197.67
$NO(g)$	90.25	86.55	210.76
$CO_2(g)$	-393.51	-394.36	213.74
$NO_2(g)$	33.18	51.31	240.06
$N_2O_4(g)$	9.16	97.89	304.29
$SO_2(g)$	-296.83	-300.19	248.22
$NH_3(g)$	294.1	328.1	238.97
$HCl(g)$	-92.31	-95.30	186.91
$HCN(g)$	135.1	124.7	201.78

ΔH_f° : standard molr enthalpy of formation

ΔG_f° : standard molar free energy of formation

S_m° : standard molar entropy

(extracted from the table in *Atkins*)

Problem 3-4

Find the enthalpies of reaction at 298 K and 400 K for the following reaction. Use the values for the

$$2CO_2 + 2NH_3 \rightleftarrows 2H_2O + 2NO + C_2H_2$$

standard enthalpy of formation and relevant data for the temperature dependence of heat capacity known for each of the reactants and products.

Problem 3-5

(a) Find the entropy change ΔS in bringing 1 g of ice at 270 K into 1 g of water at 300 K under the following conditions. The molar heat capacities for ice and water are $C_P(ice) = 37.5\ \mathrm{J \cdot K^{-1} mol^{-1}}$ and $C_P(water) = 75.3\ \mathrm{J \cdot K^{-1} mol^{-1}}$, respectively. The heat of fusion of ice is given by $\Delta H_f(273\ \mathrm{K}) = 6010\ \mathrm{J \cdot mol^{-1}}$.

(b) In a thermally shielded container, 10 g of ice at 270 K and 100 g of water at 370 K were mixed and the temperature eventually reached a constant value. Find the temperature reached at that time and calculate the entropy change associated with this mixing as well.

Problem 3-6

Find the enthalpy of formation ΔH_f of $C_2H_5NH_2(g)$ from bond energies assignable to the chemical bonds of the compound. Search some appropriate data tables for the bond energies required for calculations.

Problem 3-7

In order to understand a relation between the equilibrium constant and the free energy deeper, some essential points useful are listed in the following as questions. A system under consideration is consisted of ideal gases.

(a) The *Gibbs* free energy of a pure substance is a function of temperature T and pressure P. When a gas of 1 mole at the standard pressure of P° is compressed or expanded to P at a constant temperature, find a change of the *Gibbs* free energy. For this, derive the relation of $(\partial G/\partial P)_T = V$ first, and then integrate it to find $\Delta G = G(P,T) - G(P^\circ, T)$.

(b) When the chemical reaction expressed as $a\,\mathrm{A} + b\,\mathrm{B} \rightleftarrows c\,\mathrm{C} + d\,\mathrm{D}$ has reached an equilibrium, find the relation that connects the free energies which the chemical species A, B, C and D have in the equilibrium.

(c) Based on the relation obtained above in (b), derive the equilibrium constant expressed in terms of the partial pressures of the reactants and products in the equilibrium.

(d) How is related the equilibrium constant expressed in terms of the partial pressure to that expressed in terms of the concentrations? Derive the equation relating these two constants.

(e) Referring to the considerations in (d), express $(\partial \ln K_C/\partial T)_P$ in terms of thermodynamic variables.

(f) Express $(\partial \ln K_P/\partial P)_T$ as well in terms of thermodynamic variables. P is the total pressure of the gases.

Problem 3-8a

For a solution of acetone (A) and chloroform (C) at 50 °C, the equilibrium vapor pressures P(kPa) above solutions are obtained as given in the table below as a function of the mole fractions of acetone in the solution phase X_A^l and in the gas phase X_A^g. Under the assumption that the gases of these compounds behave ideally, work out the values for the activity and activity coefficient for both compounds and plot them as a function of X_A^l. Italicized l sands for solution (or liquid) and g for gas. (Modified from *W.J.Moore Basic Physical Chemistry*).

Table Mole fractions and vapor pressures for solutions of acetone (A) and chloroform (C).

X_A^l	0	0.100	0.200	0.300	0.400	0.500	0.600	0.700	0.800	0.900	1.000
X_A^g	0	0.071	0.165	0.279	0.408	0.550	0.684	0.789	0.890	0.955	1.000
P	69.5	66.0	63.2	61.7	61.3	62.5	65.2	68.1	72.0	76.8	81.6

Problem 3-8b

For the mixed solutions consisting of acetone (A) and a substance (Z), equilibrium vapor pressures of

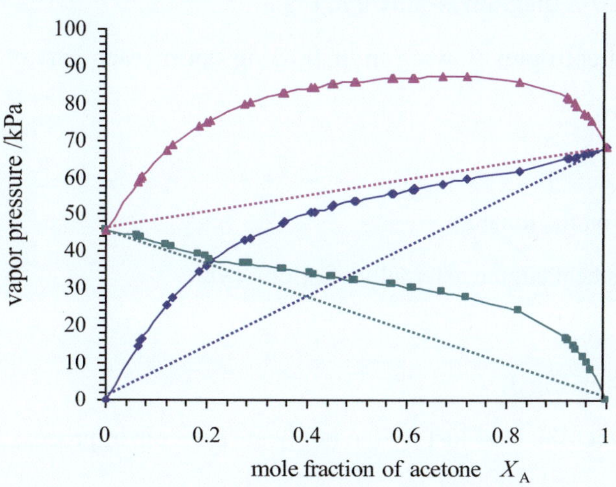

Fig.1 Vapor pressures and mole fractions for mixed solutions of acetone and substance Z.

Convex deviation from linearity is recognized. In some range the total vapor pressures are larger than either of the vapor pressures of the pure substances.

Zawidzki, Z.Physikal.Chem., **35**,129 (1900).

the solutions and the partial pressures of the component Z at a given temperature are plotted against the mole fraction of acetone as is shown in Fig.1 above (*P-X* diagram).

From this figure, what can be speculated about properties of the substance Z ?

Problem 3-9

The solid carbon dioxide CO_2 of 1 kg is placed in a steel chamber of a volume of $0.1\ m^3$ equipped with a pressure gauge and a valve. The surrounding temperature is $25\,^{\circ}C$. Find the pressure in the chamber when all of the solid CO_2 is vaporized. After that, the valve is opened so as for the pressure to drop to the atmospheric pressure (1 atm). Find the work done to the air by the compressed gas upon its expanding to the atmospheric pressure. If the same gas is reversibly expanded at the constant temperature until for it to reach the pressure of 1 atm, how large is the work done by the gas to the surroundings?

Problem 3-10

For a solution composed of two pure liquid substances A and B, when the concentration of A is small, generally *Henry*'s law holds for A and *Raoult* 's law holds for B.

(a) Show that the above statement is always approved by looking at qualitatively how the molecules A and B are co-existing in a dilute solution of A (vice versa for a dilute solution of B).

(b) Show the same thing as stated above but being based on the *Gibbs-Duhem* relation between the chemical potentials of A and B.

Problem 3-11a

Let us consider a heat engine called a *Carnot* cycle. Briefly speaking, the engine itself is composed of two heat reserviors, one at high and the other at low temperatures and a working gas in a container with a frictionless piston. The whole cycle consisting of four processes is shown in Fig.2 as a *P-V* diagram. The first process (1) is a reversible expansion of the gas at the constant high temperature T_h. The second process (2) is an adiabatic expansion of the gas. The third process (3) is a reversible compression of the gas at the constant low temperature T_l. The fourth and final process (4) is an adiabatic compression of the gas. After going through these four processes, the working gas is brought back completely to the same original state. The process 1 depicted as a *T-S* diagram is shown in Fig.3.

Paying attention to the relation between heats and entropies flowing in or flowing out of each part of the engine in each process, answer the following questions.

About the *P-V* diagram.

(a)Represent the work done by this system in process 1 on the diagram.

(b)Express on the diagram the work energy done by this heat engine externally after one cycle.

About the *T-S* diagram.

(c)Write the *T-S* relations for the processes from 2 to 4 in Fig.3.

(d) Based on the completed Fig.2 (see (c) above), depict the heat lost by the hot heat reservoir in process 1 in Fig.3.

(e) The same as (d), but the heat received by the cold heat reservoir. In process 3

(f) Define the work efficiency of this heat engine based on the results obtained in (b), (d) and (e) above.

(g) For the isothermal processes shown in Fig.3, the relation of P and V of the working gas is given by $PV = RT$. Then, find how they are related in the adiabatic processes of 2 and 4 (see **problem 3-26**).

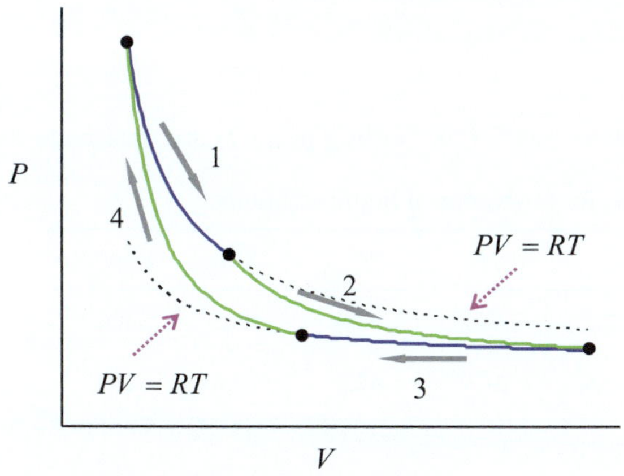

Fig.2 *P-V* diagram for a *Carnot* cycle. Either a starting or ending point of each process is indicated by ●.

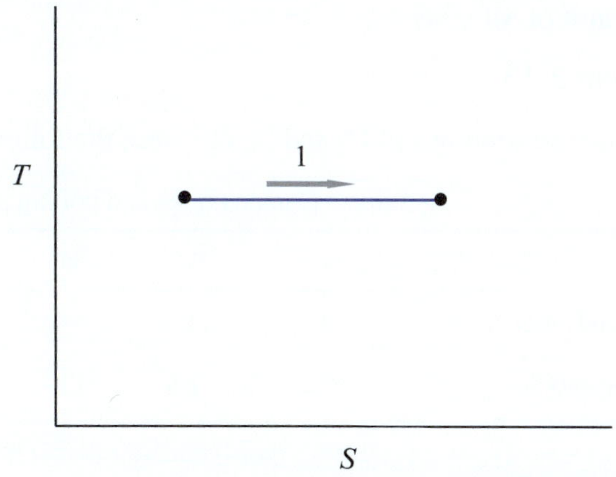

Fig.3 A part of *T-S* diagram for the *Carnot* cycle (left). This figure is not completed (see Fig.2).

Problem 3-11b

Find the work efficiency of the *Carnot* cycle presented in the previous problem in terms of the temperatures of the two heat reservoirs. Assume that the working gas is an ideal gas. The temperatures of the hot and cold heat reservoirs are given by T_h and T_l, respectively.

Problem 3-12

For a pure substance (for instance, water), the conditions for its liquid-vapor equilibrium are derived by knowing a variation of *Gibb*'s function of the substance as a function of *T* and *P*. We consider here how the vapor pressure of a certain pure substance varies with temperature in the following manners.

(a) When the temperature T and pressure P of a pure substance change by the amounts of dT and dP, respectively, find a change in its *Gibbs* free energy dG in terms of dT and dP.

(b) For a gas (g) and its liquid (l) in equilibrium with the gas, show that the temperature and pressure varies according to the relation given by $dP/dT = \Delta S_{vap}/\Delta V$ under the equilibrium by employing the relation obtained above for the variation of G with T and P for each of the gas and the liquid. Here, $\Delta S_{vap} = (S^g - S^l)$ is the entropy change upon vaporization and $\Delta V = (V^g - V^l)$ is the volume change upon vaporization.

(c) The *Clausius-Clapeyron* equation is given by $d\ln P/dT = \Delta H_{vap}/R^2T$. Show that this equation is derived from the results in (b). ΔH_{vap} is the latent heat of vaporization. Assume that the gas in the vapor phase behaves ideally and its molar volume is far greater than that of the liquid.

(d) Express $\ln P/P^\circ$ as a function of T by integrating the *Clausius-Clapeyron* equation under the assumption that ΔH_{vap} is independent of temperature. P° is the standard pressure and, in the present case, it is taken as 1 atm.

(e) From the relation between the altitude and the atmospheric pressure (see **problem 2-50**), it is known that at the top of Mt. *Everest*, the atmospheric pressure drops to 0.34 atm. Water start to boil when its vapor pressure becomes equal to the atmospheric pressure. Find at what temperature the water boils at the summit of Mt. *Everest*.

Problem 3-13

For mixed solutions of O_2 and N_2 at 1 atm, the following values for the boiling point T_b are obtained.

Table Compositions and boiling points for O_2-N_2 mixed liquid solutions.

T_b /K	77.3	78	79	80	82	84	86	88	90.1
O_2(liquid) mol%	0	8.1	21.6	33.4	52.2	66.2	77.8	88.5	100
O_2(gas) mol%	0	2.2	6.8	12	23.6	36.9	52.2	69.6	100

(a) Draw a *T-X* diagram and find the temperature range where the gas and the liquid coexist.

(b) Find the range of mole % of O_2 in which at least a liquid phase exists at 83 K.

(c) If the 80 % (% in mole numbers) of the mixed solution composed of 20 mole % of O_2 is removed from the solution by distillation, then find the composition of the remaining solution at T_b (modified from *Walter J. Moore, Basic Physical Chemistry*).

"note" As is seen in the table, the boiling temperature of pure N_2 is lower than that of O_2 by 13 degrees in K. Therefore, when a long empty test tube is dipped in a *Dewar* vessel filled with liquid nitrogen as is shown in Fig.4 on the right, the oxygen from the air cooled down to the temperature of liquid nitrogen starts to be liquefied first. After ca.10 min an enough volume of liquid oxygen (faintly blue-colored liquid) is obtained, which can be used for a demonstration to show the oxygen is essential in burning. A remarkable point of the demonstration is to prove that a stuff condensed from the air in front of students helps various materials even iron wires burn vigorously.

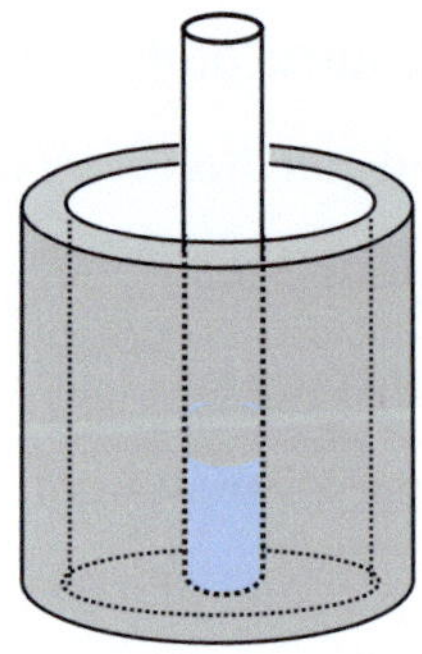

Fig.4 Air liquefied in a long tube inserted into a Dewar vessel.

Problem 3-14

For aqueous solutions of sucrose ($C_{12}H_{22}O_{11}, M = 342$), the vapor pressures in equilibrium with the solution at 323 K were measured at several mole fractions of sucrose. The results are summarized in the table given below (adopted from *Walter J. Moore, Basic Physical Chemistry*).

(a) Confirm whether or not these solutions satisfy *Raoult*'s law.

(b) Find the values for the activity and its coefficient of water at the mole fractions of 0.9864, 0.9559 and 0.9098 (indicated by parentheses in the table).

(c) The molar *Gibbs* free energy of pure water at this temperature, denoted G^0, is known to be $-231.7 \mathrm{kJ \cdot mol^{-1}}$. Find the values for the chemical potential of water at each mole fraction given in (b).

(d) At the mole fraction of $X_{\mathrm{H_2O}} = 0.9098$, the activity of sucrose was 0.2190. Find the molar *Gibbs* free energy of this solution. Here, the free energy of formation of sucrose (solid) ΔG_f° is given by $-1549\ \mathrm{kJ \cdot mol^{-1}}$ and its solubility in water is 72 (g /water 100g).

Table Saturated water vapor pressures over aqueous solutions of sucrose.

$X_{\mathrm{H_2O}}$ *	vapor pressure /kPa	activity $a_{\mathrm{H_2O}}$	activity coef. $\gamma_{\mathrm{H_2O}}$
1.000	12.333		
0.9940	12.258		
0.9864	12.252	( )	( )
0.9826	12.085		
0.9762	11.959		
0.9665	11.863		
0.9940	12.258		
0.9559	11.688	( )	( )
0.9439	11.468		
0.9323	11.153		
0.9098	10.801	( )	( )
0.8911	10.539		

* mole fraction of water.

Problem 3-15

Fig.5 shows a gas (G) - liquid (L) phase diagram for a certain two-component mixture. The blue line is a dew line and the green straight line is a vaporization line. In the pressure range under the dew line, the mixture can exist as a gas alone and the range above the vaporization line, the mixture can exist as a liquid. Initially, a mixture of 0.4 mole fraction is in the gas state and with increasing the pressure along the red broken line, it starts to be liquefied at the intersection with the blue line. With increasing the pressure further, the liquefaction proceeds and when the pressure reaches the one at the intersection with the green line after passing through the gas-liquid coexisting region, the mixture can exist only as a liquid.

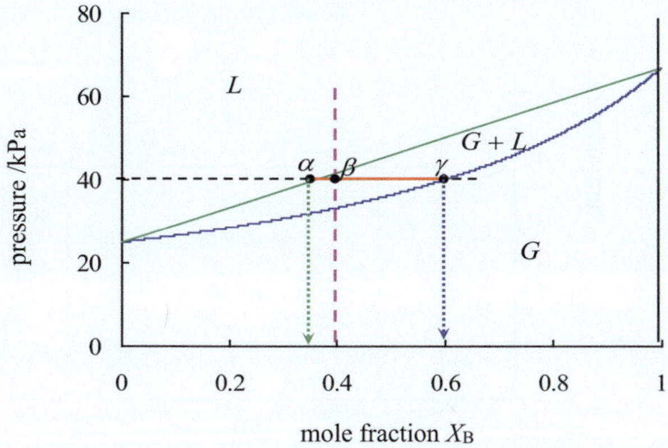

Fig.5 Gas(G)-liquid(L) phase diagram for two-component mixtures.

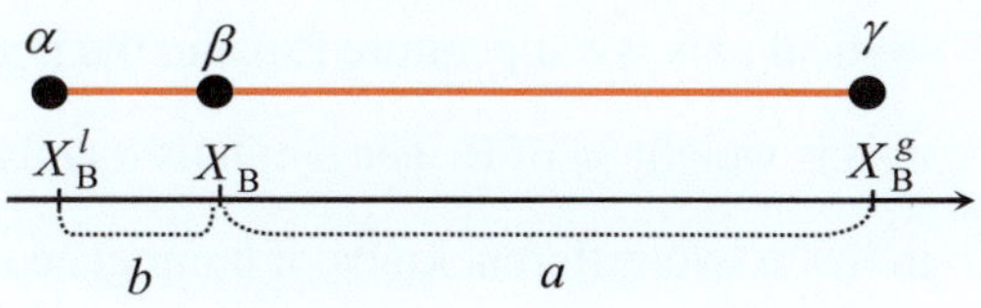

Fig.6 Enlargement of the part indicated by the orange line in Fig.5.

We need to know what relation exists between the amounts (mole numbers) of the gas and the liquid in their coexisting region. For instance, the mixture of the mole fraction of 0.4 exists at the pressure of 40 kPa (point β) as a gas mixture and a solution as is seen in Fig.5.

Show that the ratio of the amount of coexisting gas to that of liquid (solution) at this pressure is given by the ratio of [the number of moles of the gas : the number of moles of solution $= b:a$], i.e., [lever rule], by referring to the expanded figure given in Fig.6.

Problem 3-16a

Consider solutions composed of two kinds of pure compounds A and B. The boiling points of A and B are 414.0 K and 336.5 K, and the values for the latent heat of vaporization are $16.0\ \text{kJ}\cdot\text{mol}^{-1}$ and $13.0\ \text{kJ}\cdot\text{mol}^{-1}$, respectively. The solutions are regarded as ideal ones for which *Raoult*'s law is satisfied. Also, *Clausius-Clapeyron*'s equation is applicable to describe the variation of the vapor pressure of each pure substance with temperature.

Draw a *T-X* (temperature vs. mole fraction) diagram for this system of mixed solutions.

Problem 3-16b

As to the mixed solutions considered in the preceding problem, find the mole fraction of the solution whose boiling point becomes equal to 370.0 K under 1 atm.

Problem 3-16c

Based on the *T-X* diagram given in **solution 3-16a**, describe a method of obtaining component A or B as pure as possible from their mixed solutions.

Problem 3-17a

Consider an ideal solution composed of pure substance A and B. The vapor pressures of pure A and B at 298 K are $P_A^\circ = 25$ kPa and $P_B^\circ = 67$ kPa, respectively. Draw the equilibrium lines between the liquid and gas-liquid coexisting phases as well as between the gas and gas-liquid coexisting phases, i.e., a *P-X* diagram at this temperature.

Problem 3-17b

The figure shown on the right is a phase diagram for solutions composed of pure liquids A and B. The vertical axis is temperature (K) and the horizontal axis is weight % of B. The symbols α and β indicate two different kinds of homogeneous solutions. At the point P where the temperature is T_1 and the weight % is w_P, the two phases, whose components are specified by the weight % at the points M and L, coexist.

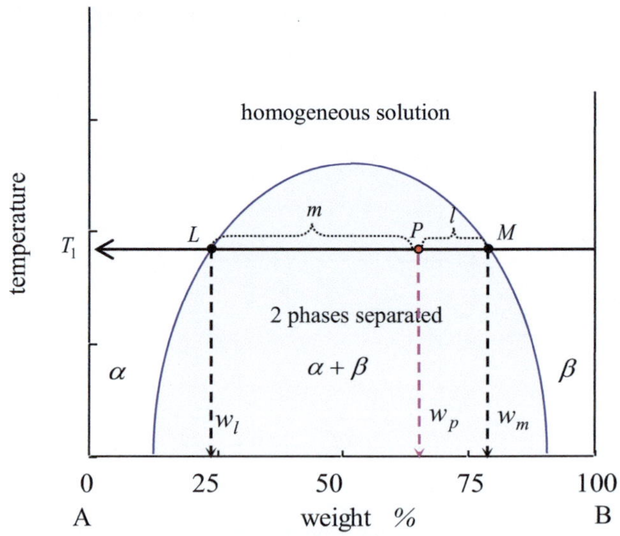

Fig.7 A phase diagram for solutions composed of pure liquids A and B.

Show that the ratio of the amounts (weights) of the solutions in the two phases is given by the lever rule.

Problem 3-17c

There are two kinds of pure liquids A and B which only mix each other homogeneously in limited ranges of temperature and pressure. Draw a plausible phase diagram for the mixtures of A and B as a *T-X* diagram. Describe how the *T-X* diagram changes when the pressure is varied.

Problem 3-17d

Phase equilibriums due to the phase separation, such as gas-liquid, liquid-liquid, liquid-solid and solid-solid are frequently encountered. In general, however, there does not exist any gas-gas equilibrium (except for equilibriums due to chemical reactions). Explain why this is the case in thermodynamic terms.

Problem 3-18

Using the *Legendre* transformation, derive the *Helmholtz* free energy $F \equiv (A)$ from the internal energy *U*. $F \equiv (A)$ thus introduced is a thermodynamic function specified normally by the thermodynamic variables *T* and *V*.

Problem 3-19a

The vapor pressures of pure benzene (A) and pure toluene (B) at room temperature (300 K) are 67 Torr ($P_A^\bullet$) and 25 Torr ($P_B^\bullet$), respectively. Assuming that mixed solutions of these solvents can be treated as ideal solutions, answer the following questions. 760 Torr is equivalent to 101 kPa.

(a) Denoting the mole fraction of benzene (A) X_A, find the vapor pressure *P* of the mixed solution in equilibrium with the solution (the pressure of the gas which emerges from the solution for the first time) as a function of X_A.

(b) Calculate the vapor pressure (the pressure of the gas which emerges from the solution for the first time) at $X_A = 0.2$.

(c) A and B are mixed so as for the mole fraction to become $X_A = 0.3$ (fraction for the total amount) and the vapor pressure is kept at the same one as found in (b) at room temperature. Find in what phase the mixture can exist. If it exists in two phases, then find the ratio of the amounts in the two phases.

Problem 3-19b

The system under consideration is the same as the one treated in the preceding problem but the questions are asked from a different point of view.

The vapor pressures of benzene (A) and toluene (B) at room temperature (300K) are 67 Torr and 25 Torr, respectively. It is assumed that mixed solutions of these compounds can be treated as ideal solutions.

Answer the following questions, referring to the figure given in **problem 3-15**.

(a) Under the conditions that only the liquid solution can exist, benzene (A) and toluene (B) were mixed together so as for the mole fraction to become $X_A = 0.2$. Find the free energy change ΔG_{mix} due to mixing.

(b) When the pressure over the solution mixed at the mole fraction given above was reduced, a gas phase appeared for the first time at a certain pressure. Find the pressure and the mole fraction of A in the gas phase at this point.

(c) The vapor pressure was reduced further until all the solution was transformed into the gas phase. Find the vapor pressure at this point.

(d) How much has the free energy of the gas in (c) changed as compared with that before mixing the two solvents?

Problem 3-20a

The density of a mixture composed of mixable pure substances A and B generally varies with the mixing ratio by volume. Explain how it varies with the volume ratio by providing an approximate shape of a curve representing the relation between the density and the mixing ratio for any of conceivable cases.

Problem 3-20b

Consider a mixture composed of two mixable pure substances A and B. It is assumed that the reciprocal of the density of the mixture, $1/\rho_{mix}$, is given by eq(2) as a function of the mole fraction of X_B.

$$\frac{1}{\rho_{mix}} = aX_B^2 + bX_B + \frac{1}{\rho_A}, \qquad \frac{1}{\rho_B} - \frac{1}{\rho_A} = a + b \qquad (2)$$

(a) Find a function which expresses the partial molar volume for each of A and B in the mixture.

(b) Show that the *Gibbs-Duhem* relation holds for the partial molar volumes found for A and B in (a).

Problem 3-20c

The specific volumes $\overline{V}_{spc}$ $(\mathrm{cm^3 \cdot g^{-1}})$ of mixtures of ethyl iodide $(\mathrm{A: CH_3CH_2I},\ M_A = 156.0)$ and ethyl acetate $(\mathrm{B: CH_3COOCH_2CH_3},\ M_B = 88.10)$ were measured as a function of the mixing mass ratio of $x_A = m_A/(m_A + m_B)$ and found as are listed in the table below.

(a) Find the molar volumes of ethyl iodide and ethyl acetate.

(b) Find the partial molar volume of each compound at the mixing ratio of 0.5974.

Modified problem based on 17-7 in *K.S.Pitzer and L.Brewer, Thermodynamics*, 2nd Ed., 1961.

x_A	$\overline{V}_{spc}\ /\ \mathrm{cm^3 \cdot g^{-1}}$	x_A	$\overline{V}_{spc}\ /\ \mathrm{cm^3 \cdot g^{-1}}$
0.0000	1.15866	0.7450	0.70040
0.19082	1.04358	0.82792	0.64814
0.35007	0.94654	0.89093	0.60749
0.49517	0.85737	0.9497	0.56927
0.59741	0.79377	1.0000	0.53622
0.68529	0.73880		

Problem 3-21

Answer the following questions about the partial molar volume.

(a) What is the definition of the partial molar volume?

(b) The figure given below shows the partial molar volumes of mixed solutions of ethanol and water plotted against the mole fractions of ethanol. The subscripts e and w denote ethanol and water, respectively.

i) What are the molar volumes of pure ethanol and water?

ii) What is the volume of the solution prepared by mixing $90\ \text{cm}^3$ of water and $110\ \text{cm}^3$ of ethanol?

(c) The partial molar volume is just one of the generally defined partial molar quantities and, hence, it satisfies the *Gibbs-Duhem* relation. Explain what this relation means employing the figure shown below and confirm quantitatively whether or not the relation is fulfilled at each point of the mole fractions of $a = 0.0$, $b = 0.08$, $c = 0.40$, $d = 0.60$ and $e = 1.0$. (See **problems 3-20b** and **3-123.**)

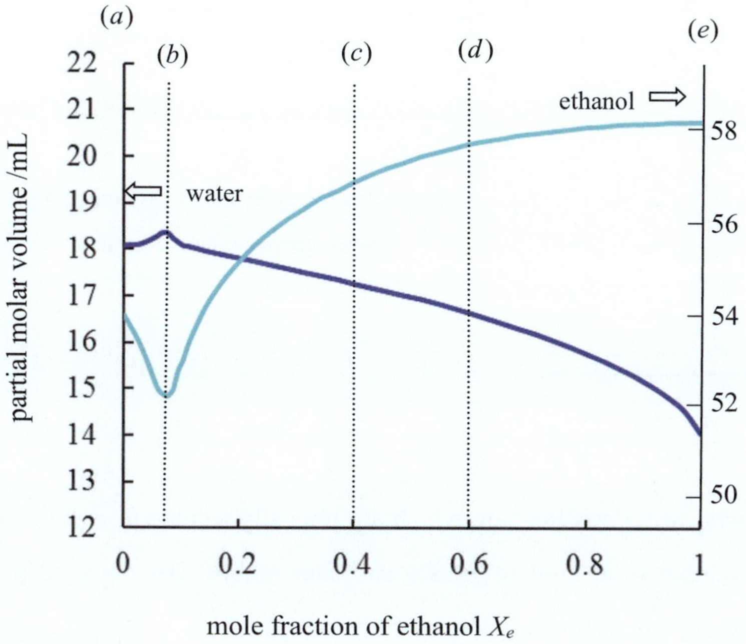

Fig.8 Partial molar volumes and mole fractions of ethanol-water mixed solutions. Partial molar volumes of ethanol are along the vertical axis on the right and those of water along the axis on the left. (The curves in the figure are newly drawn for this book after *Atkins, Physical Chemistry, 6th* edition.) See also the figure 1.2 for the partial molar volumes given by *Progpgenue and Defey* in *Chemical Thermodynamics* I (Japanese version).

Problem 3-22

The vapor pressures of some pure liquids A and B at room temperature are 100 Torr and 25 Torr, respectively. Assuming mixed solutions of these two liquid can be treated as ideal ones, answer the following questions. If necessary use the relation of $760\ \text{Torr} = 101\ \text{kPa}$.

(a) Denoting the mole fractions of A and B in a solution by X_A and X_B, respectively, find the vapor pressure P_v of the solution at room temperature (the pressure at which the vapor phase appears for the first time) as a function of X_A.

(b) At the same temperature as above, a gas mixture of the mole fraction of X_A^g is compressed to the pressure P_d until the liquid phase appears for the first time. Find P_d as a function of X_A^g.

(c) When the mixture is kept under the total pressure of 75Torr at the same temperature as above, find the range of X_A in which the gas and liquid can coexist in equilibrium.

Problem 3-23

It is important to understand how the entropy of an ideal gas varies by changing its volume or pressure.

(a) Express $\left(\dfrac{\partial S}{\partial V}\right)_T$ by using $\left(\dfrac{\partial S}{\partial P}\right)_T$.

(b) When the volume of an ideal gas is changed by three-times larger, i.e., $V \to 3V$, or the pressure of the same gas is reduced to one-third, i.e., $P \to P/3$, show that the entropy changes in both cases are the same by using the relation found in (a).

Problem 3-24

The figure given below shows how the molar free energies of a pure gas and its liquid vary with temperature at a constant pressure over a comparatively narrow range of temperature.

(a) Which one of the solid and broken lines represents the free energy of the gas?

(b) Draw a similar diagram as the figure given below for a pure liquid and its solid, indicating which one is for the solid.

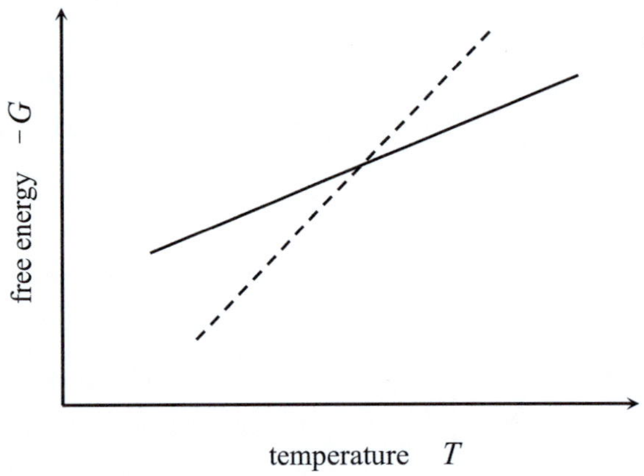

Fig.9 Phase equilibrium and free energy of a pure substance (conceptual diagram).

Problem 3-25

Three kinds of pure liquids of acetic acid, water and butanol, which are mutually soluble to some extent, are mixed at various mixing ratios to examine what kind of phases can exist stably. The obtained phase diagram is shown below in Fig.10. The mixing ratio is given in terms of the mole fraction.

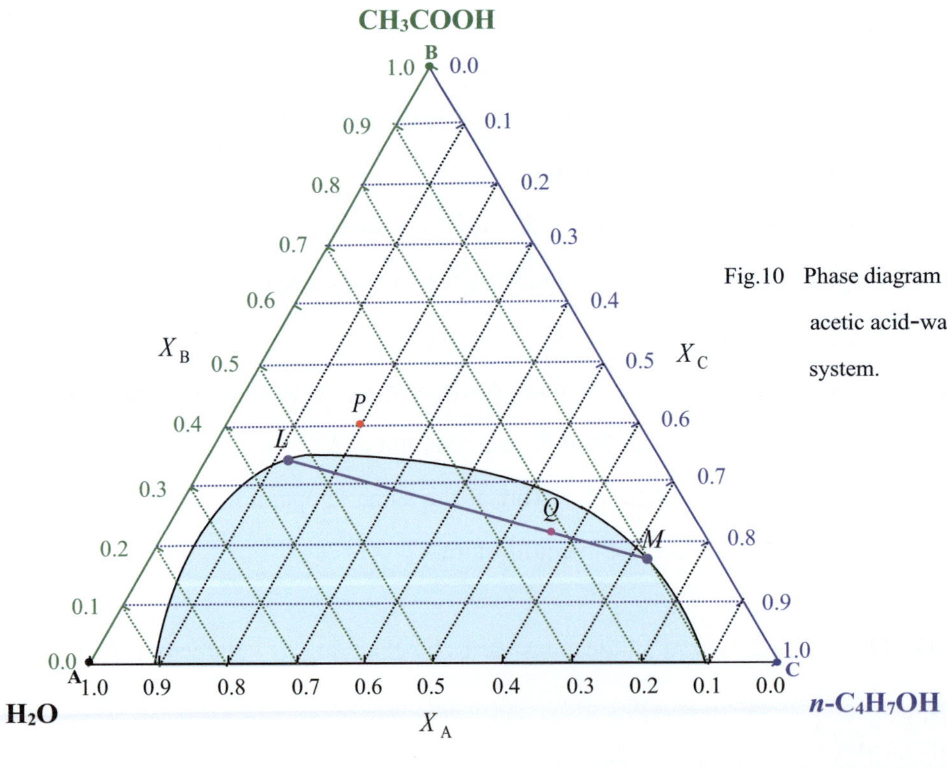

Fig.10 Phase diagram showing mutual solubility of acetic acid-water-butanol three-component system.

Answer the following questions referring to the given phase diagram.

(a) What kind of state do the points A, B and C in the phase diagram indicate?

(b) Find the mole ratio of water to butanol at the point *P* in the phase diagram.

(c) List up all combinations of the two solvents among acetic acid, water and butanol, i) which mix each other only in a limited range of mixing ratio, and ii) which mix in any mixing ratio.

(d) How many liquid phases exist at the point *Q* in the phase diagram?

(e) The segment *LM* is called *tie line*. Find the composition(s) of the solution(s) at the point *Q* on the tie line. If there exist more than one solution phases, answer the composition for each solution.

(f) Examine whether or not three kinds of pure solvents can form three phases based on the phase rule. If possible, draw an expected phase diagram.

Problem 3-26

The relation of *T* and *P* or *T* and *V* obeyed for a reversible adiabatic compression or expansion of an ideal gas is obtained by expressing first the partial derivative of $\left(\frac{\partial T}{\partial P}\right)_S$ or $\left(\frac{\partial T}{\partial V}\right)_S$ in terms of *P*, *V*, *T* and some known quantities such as the molar heat capacity C_P^m or C_V^m and then integrating the obtained expressions. Based on this method, show that eq(3) is satisfied for the adiabatic process expressed by $(T_i,\ V_i) \rightarrow (T,\ V)$. From this equation, find the temperature and the pressure of the gas when its volume is

$$T/T_i = (V/V_i)^{-\frac{R}{C_V^m}} \qquad (3)$$

adiabatically reduced to one half of the initial volume. For an alternative solution, see **problem 3-42**.

Problem 3-27

Not only for the most fundamental eight thermodynamic variables but also for the chemical potential and partial molar volume, classify them as an extensive variable, an intensive variable or neither of these.

Problem 3-28

An ideal gas of the volume *V* at the pressure *P* and the temperature *T* is compressed reversibly at the constant temperature to one half of the initial volume $(V/2)$.

Under these conditions, find the following values:

(a) the work done to the gas from the outside, (b) the amount of heat lost by the gas, (c) the amount of change in the internal energy of the gas, (d) the amount of change in the entropy of the gas.

Problem 3-29

(a) Express the differential dG of the *Gibbs* free energy in terms of the differentials of the independent variable *T* and *P*.

(b) From the relation found in (a), calculate a change in the *Gibbs* free energy associated with the process given in the preceding **problem 3-28**.

(c) From the result obtained in (b), discuss whether the isothermal compression of the ideal gas occurs naturally or not.

Problem 3-30

The values for the heat of combustion ΔH_c of C(s), H_2(g) and C_6H_6(l) are as given below from reaction (1) to (3), in which s, g and l indicate the solid, gas and liquid states, respectively.

$$C(s)+O_2(g)=CO_2(g) \qquad (1),\quad \Delta H_c(1)=-94.0\times 4.18\ \text{kJ}\cdot\text{mol}^{-1}$$

$$H_2(g)+\frac{1}{2}O_2(g)=H_2O(l) \qquad (2),\quad \Delta H_c(2)=-68.3\times 4.18\ \text{kJ}\cdot\text{mol}^{-1}$$

$$C_6H_6(l)+\frac{15}{2}O_2(g)=6CO_2(g)+3H_2O(l) \qquad (3),\quad \Delta H_c(3)=-783.4\times 4.18\ \text{kJ}\cdot\text{mol}^{-1}$$

(a) Give the equation of the formation reaction to yield benzene from simple substances (made of a single element).

(b) Find the heat of formation of benzene.

Problem 3-31

Show that the following equations hold for the ideal gas.

(a) $\left(\dfrac{\partial U}{\partial V}\right)_T=0$, $\left(\dfrac{\partial U}{\partial P}\right)_T=0$

(b) $\left(\dfrac{\partial U}{\partial T}\right)_V=C_V$, C_V: heat capacity at constant volume.

Problem 3-32

When one mole of an ideal gas of the volume of V_1 at the temperature T_1 is brought to the volume V_2 and the temperature T_2, find the entropy gained or lost by the gas. Notice that as far as the initial and final states are fixed, entropy change does not depend on how the temperature and pressure are varied since the entropy is a quantity of state.

Problem 3-33

We can consider a few examples of problems to which the results obtained in the previous **problems 31** and **32** can be applied in answering them. For instance, one mole of a mono-atomic ideal gas initially at 25°C under the pressure of 10 atm was expanded adiabatically against the external pressure of 1 atm until its volume became two times of the initial volume.

Find the following quantities the gas gained by this process; q, w, ΔU, ΔH, ΔS and ΔT. Put the positive sign to the quantities that the gas gained and the negative sign to those lost by the gas.

Comment on the value found for ΔS.

Problem 3-34

By using $\left(\dfrac{\partial S}{\partial V}\right)_T$, find a change in the entropy for the gas which obeys the *van der Waals* equation of state when its volume is changed two times larger by the reversible isothermal expansion. Compare the result with the entropy change expected for the ideal gas whose volume increases two times by a similar reversible isothermal expansion and give explanations for a difference found if any.

Problem 3-35

(a) Confirm that any arbitrarily chosen three thermodynamic variables *x*, *y* and *z* (corresponding to three of $U,\ S,\ P,\ V,\ T,\ H,\ A,\ G,\ \mu$) satisfy *Euler*'s chain relation given below.

$$\left(\frac{\partial y}{\partial x}\right)_z\left(\frac{\partial x}{\partial z}\right)_y\left(\frac{\partial z}{\partial y}\right)_x=-1$$

(b) Replacing *x*, *y* and *z* with *P*, *V* and *T*, show that *Euler*'s chain relation is satisfied for the ideal gas and the *van der Waals* gas as well.

Problem 3-36

The derivative expressing the adiabatic compression-expansion plays an important role in investigating the velocity of the sound in a gas. The sound velocity υ in the gas(es) is known to be given by eq(4). The density of the gas is denoted ρ.

$$\upsilon=\sqrt{\left(\frac{\partial P}{\partial \rho}\right)_S} \qquad (4)$$

Assuming the air behaves ideally, calculate the sound velocity in the air. See **problem 2-59**.

Problem 3-37

(a) If the temperature dependence of the heat of vaporization ΔH_{vap} is known for a substance, then we can tell how the vapor pressure of the substance varies with the temperature. When the enthalpy of vaporization is given by $\Delta H_{vap}(T)=H_0+a\cdot(T-T_0)$ where H_0, T_0 and *a* are the constants for the substance, find the vapor pressure of this substance as a function of temperature.

(b) The boiling point of a certain liquid is T_b and its molar enthalpy of vaporization is given by ΔH^m_{vap}. If $\Delta H^m_{vap}/T_b$ takes a constant value over a range of temperature, then show that the vapor pressure curve for such a liquid assumes an identical shape against T/T_b.

(c) The molar enthalpy of vaporization for water is given by $\Delta H^m_{vap}=40.7\ \mathrm{kJ\cdot mol^{-1}}$. Draw a curve for the vapor pressure of water over the temperature range $0^\circ\mathrm{C}-100^\circ\mathrm{C}$.

Problem 3-38a

The thermodynamic function *G* is an extensive variable which can be expressed in terms of the variables of *T*, *P* and the mole number n_i of the *i*-th component. The partial molar quantity μ_i (chemical potential) is defined by eq(5). On the other hand, *G* is *Euler*'s homogeneous function of the 1st-order as is expressed by eq(6).

$$\mu_i=\left(\frac{\partial G}{\partial n_i}\right)_{T,P,n_j,j\neq i} \qquad (5)$$

$$G(T,P,\lambda n_1,\lambda n_2,\cdots\lambda n_l)=\lambda\cdot G(T,P,n_1,n_2,\cdots n_l) \qquad (6)$$

Employing these relations, prove that *G* is expressed by eq(7).

$$G=\sum_{i=1}^{l}n_i\mu_i \qquad (7)$$

Problem 3-38b

The thermodynamic function *F* is an extensive variable which can be expressed in terms of the variables of *T*, *V* and the mole number n_i of the *i*-th component. The partial molar quantity ϕ_i is defined by eq(8). On the other hand, *F* is *Euler*'s homogeneous function of the 1st-order as is expressed by eq(9).

$$\phi_i = \left(\frac{\partial F}{\partial n_i}\right)_{T,V,n_j,j\neq i} \qquad (8)$$

$$F(T,\lambda V,\lambda n_1,\lambda n_2,\cdots\lambda n_l) = \lambda\cdot F(T,V,n_1,n_2,\cdots n_l) \qquad (9)$$

Employing these relations, prove that *F* is expressed by eq(10).

$$F + PV = \sum_{i=1}^{l} n_i\phi_i \qquad (10)$$

Problem 3-39

The latent heat ΔH_l associated with a phase transition is given by a function of *P* and *T*. Its temperature dependence $\left(\frac{d\,\Delta H_l}{dT}\right)_e$ is given by eq(11), which is called *Planck*'s equation.

$$\left(\frac{d\,\Delta H_l}{dT}\right)H + \frac{\Delta H_l}{T_e} - \Delta H_l\left(\frac{\partial \ln\Delta V_e}{\partial T}\right)_P \qquad (11)$$

Derive this equation. The subscript *e* indicates that the relevant values are those in equilibrium and that the differentiation is carried out by keeping the equilibrium. This problem is useful to understand the differentiation carried out by specifying the path along which the variables are changed.

Problem 3-40

The two lines representing two different adiabatic processes drawn in a *P*-*V* diagram of a *Carnot* cycle do not cross each other. Prove this from a viewpoint of the efficiency of work.

Problem 3-41a

Show that the chemical potential (free energy) of an ideal gas at *P* and *T* is given by the following equation, in which P° is the pressure at the standard state.

$$\mu(P,T) = \mu(P^\circ,T) + RT\ln\frac{P}{P^\circ}$$

Problem 3-41b

The chemical potential of a non-ideal gas is different from that of the ideal gas (see the preceding problem). However, when the equation of state is known for the non-ideal gas, its chemical potential can be found in a similar manner as employed for the ideal gas. Find the functional form of the chemical potential for a gas which obeys the *van der Waals* equation of state.

Problem 3-41c

The chemical potential $\mu(P,T)$ of a real gas is expressed by modeling it after the functional form of the chemical potential of the ideal gas given below.

$$\mu_{ideal}(P,T) = \mu_{ideal}(P^\circ,T) + RT\ln P/P^\circ$$

Hence, $\mu(P,T)$ is written as eq(12) by introducing the fugacity f whose meaning becomes clear later.

$$\mu(P,T) = \mu(f^\circ,T) + RT\ln\frac{f}{f^\circ} \qquad (12)$$

f° is the fugacity at the standard state and usually it is put equal to the pressure P° at the standard state of the gas. Also, f is expressed by eq(13) using the pressure P of the gas and the coefficient $\gamma(P,T)$, which is pressure and temperature dependent and is called fugacity coefficient.

$$f = \gamma \cdot P \qquad (13)$$

Summarizing these, the chemical potential of the real gas is expressed by eq(14), and as is evident from its form, any deviation of the real gas from the ideal behavior is ascribed to $\gamma(P,T)$ entirely.

$$\mu(P,T) = \mu(P^\circ,T) + RT\ln\frac{\gamma \cdot P}{P^\circ} \qquad (14)$$

Find the fugacity coefficient of a gas which obeys the *van der Waals* equation of state, using the result obtained in the preceding problem (see also **problem 3-41f**).

Problem 3-41d

The fugacity coefficient γ varies from gas to gas and is not unity in usual. However, just as equations of state for many gases are expressed by a single reduced equation of state in terms of the reduced variables, the fugacity coefficients of various gases can also be expressed by a single equation if the reduced variables are employed.

Express the fugacity coefficient γ of the *van der Waals* gases using the reduced variables P_R, V_R and T_R.

Problem 3-41e

Find the fugacity coefficient γ of CO_2 at $T_R = 373/304.2 = 1.23$ and $V_R = 20.0$, using the result obtained for the preceding problem. If necessary, use the following critical constants $T_c = 304.2$ and $V_c = 92.4\times10^{-6}\text{m}^3$.

Problem 3-41f

Give a reason why the fugacity f is regarded as a property of a gas to measure its escaping ability, the ability of keeping gaseous molecules far away from one another as much as possible.

Problem 3-42a

Show that the following relations hold for the adiabatic process of the ideal gas, and derive the relations of

$$dU = C_V dT = -PdV \qquad (15), \qquad dH = C_P dT = VdP \qquad (16)$$

of $(T,\ V)$ and $(T,\ P)$ for the adiabatic process of the ideal gas from these eqs(15) and (16).

Problem 3-42b

Employing eq(15) given in the preceding problem, find an equation to give the temperature when the ideal gas at the initial state of $(P_1,\ V_1,\ T_1)$ is converted to $(P_2,\ V_2)$ via reversible adiabatic process.

Problem 3-43a

The osmotic pressure Π is derived from the balance of the chemical potentials of a solvent molecule A in two phases and in the case of an ideal solution it is given by eq(17), in which n_B, m_B and M_B are the number of moles of a solute molecule B, the mass of the solute and the molecular weight of B, respectively.

$$\Pi = \frac{n_B RT}{V} = \frac{RT}{V} \cdot \frac{m_B}{M_B} = \frac{RT}{M_B} \cdot \frac{m_B}{V} \qquad (17)$$

However, especially for polymer molecules of a large molecular weight, eq(17) is not satisfactory and a correct molecular weight cannot be accessed by eq(17) (see **problem 3-43b**). For that, it is well known that when an approximate eq(18) which takes the terms of higher order with respect to m_B/V into account is employed, better results are obtained.

$$\Pi = \frac{RT}{M_B} \cdot \frac{m_B}{V} \left\{ 1 + B(T)\frac{m_B}{V} + C(T)\left(\frac{m_B}{V}\right)^2 + \cdots \right\} \qquad (18)$$

This way of modification is analogous to that of correcting the deviation of the real gas from the ideal behavior. When the first order approximation is sufficient in eq(18), eq(19) is obtained and since the term is linearly $\Pi / \left(\frac{m_B}{V}\right)$ correlated to $\frac{m_B}{V}$ in this equation, the molecular weight M_B is found from the intercept of eq(19).

$$\Pi / \left(\frac{m_B}{V}\right) \approx \frac{RT}{M_B} \left\{ 1 + B(T)\frac{m_B}{V} \right\} \qquad (19)$$

Derive eq(19) from the virial equation of state for a real gas.

Problem 3-43b

In order to find the molecular weight of bovine serum albumin (BSA), the osmotic pressure of a 0.5 M NaCl aqueous solution of BSA was measured at several concentrations of BSA at 273K. The obtained values are listed in the table below.

BSA/g·cm^{-3}	0.020	0.030	0.040	0.060
Π/Pa	820	1360	1900	3230

Find the molecular weight (molar weight) of BSA.

Problem 3-44

How the phase equilibrium is affected by varying pressure or temperature has been considered so far in terms of the *Gibbs* free energy. However, show that *Clausius-Clapeyron*'s equation can be derived from the consideration of the *Helmholtz* free energy change upon the gas-liquid phase transition.

Problem 3-45a

One mole of the gas which obeys the equation of state of $P(V-b) = RT$ is reversibly compressed at 300 K from the pressure of 100 kPa to 400 kPa. Find the work done to the gas with $b = 4.27 \times 10^{-2}\,\mathrm{L \cdot mol^{-1}}$ upon that process. Compare the result with the work done to an ideal gas which performs a similar compression.

Problem 3-45b

A similar problem as the preceding one but the gas obeys the *van der Waals* equation of state. Find the work done to the one mole of this gas when it is compressed under the same conditions as in the preceding problem. The *van der Waals* parameters are $a = 3.64\,\mathrm{atm \cdot L^2 \cdot mol^{-2}}$ and $b = 4.27 \times 10^{-2}\,\mathrm{L \cdot mol^{-1}}$.

Problem 3-46a

One of the equations called the thermodynamic inequality is given by eq(20).

$$\left(\frac{\partial P}{\partial V}\right)_T < 0 \qquad (20)$$

Show that eq(20) holds for the *van der Waals* gas as well as for the ideal gas.

Problem 3-46b

The equation given below is called the thermodynamic inequality. Intuitively, it is easy to understand what

$$\left(\frac{\partial P}{\partial V}\right)_T < 0 \qquad (20) \qquad \text{(See the preceding problem.)}$$

is implied by the inequality. However, to derive the inequality is another thing and requires a bit of knowledge of thermodynamics. Referring to, for instance, to *Statisitical Physics* by *Landau and Lifshitz*, §21, try to prove the inequality.

Problem 3-47

There are two partition-separated containers 1 and 2 whose volumes are V_1 and V_2, respectively and they are filled with a gas of the same kind but at different pressures of $P_1 = \frac{Nk_BT}{V_1}$ and $P_2 = \frac{Nk_BT}{V_2}$. N is the number of molecules. Find an entropy change when these gases are mixed together in the containers isothermally by removing a partition between the two containers.

Problem 3-48

The two partition-separated containers under the same conditions as given in the preceding problem are connected so as to mix the gases. Find the maximum work gained by mixing the gases in each case of (a) isothermal reversible process and (b) adiabatic reversible process. If the gases cannot be mixed, then give a reason for that.

Problem 3-49a

When learning thermodynamics for the first time, many of you may feel bewildered and incomprehensible because you encounter various types of partial differential coefficients. Although it is often said that there is no royal road to learning, there are a few basic approaches in order to comprehend a basic structure of thermodynamics more easily. For instance, based on the fact that the thermodynamic variables are quantities of state, we have already learned in this chapter that (1) new thermodynamic functions such as enthalpy or free energy can be introduced by means of the *Legendre* transformation and (2) *Maxwell*' s relations are readily derived. Now, by solving the problem here on how to express *Joule-Thomson*'s coefficient as a function of P, V and T *alone*, we will learn a simple but generally effective method to express complicated partial differentials by relevant functions of P, V and T (sometimes containing other thermodynamic variables as well). We hope that this will make thermodynamics a little easier to understand.

Joule-Thomson's coefficient is defined by $\mu = \left(\frac{\partial T}{\partial P}\right)_H$. (Conventionally, the coefficient is denoted as μ, but it should not be confused with the chemical potential.) It indicates how the temperature of a gas varies when the gas is expanded or compressed isenthalpically (keeping H invariant). It is obvious, however, μ

cannot be calculated directly from the equation of state for the gas. The point is, therefore, to find out a general way to express μ as a function containing *P*, *V* and *T* alone, all of which are measurable.

First of all, the thermodynamic variable *H*, which is kept constant during the process, is considered to be a function of *P* and *T* (usually two variables). Then the total differential of *H* is given by eq(21).

$$dH = \left[\quad \right] dP + \left(\frac{\partial H}{\partial T} \right)_P dT \qquad (21)$$

Since "isenthalpically" means that $dH = [\quad]$, eq(22) is derived from eq(21). C_P is the specific heat at a constant pressure.

$$\left(\frac{\partial T}{\partial P} \right)_H = -\frac{\left[\qquad \right]}{\left(\frac{\partial H}{\partial T} \right)_P} = -\frac{\left[\qquad \right]}{C_P} \qquad (22)$$

On the other hand, *dH* can be expressed by eq(23) as has already been learned in the *Legendre* transformation (see **II-2-1** (5)).

$$dH = TdS + VdP \qquad (23)$$

And then, after dividing both sides of eq(23) by *dP* at constant temperature, *Maxwell*'s relation is used to obtain eq(24).

$$\left(\frac{\partial H}{\partial P} \right)_T = T \left(\frac{\partial S}{\partial P} \right)_T + V = T \times \left[\quad \right] + V \qquad (24)$$

Consequently, from eqs(22) and (24) *Joule-Thomson*'s coefficient μ is expressed in terms of the variables of *P*, *V* and *T* alone as is intended.

(a) Fill the brackets [ ] in the above explanatory description with appropriate equations, values or constants.

(b) Find *Joule-Thomson*'s coefficient μ for an ideal gas.

Problem 3-49b

Thomson's coefficient μ_T is defined by $\mu_T = \left(\frac{\partial T}{\partial V} \right)_U$. Following the method given in the preceding problem, express μ_T as a function of *P*, *V* and *T*.

Problem 3-49c

When *Joule-Thomson*'s coefficient is positive, the temperature of a gas expanded at constant enthalpy is lowered. In fact, many gases can be liquefied practically by repeating *Joule-Thomson*'s process. It should be noted, however, that the coefficient is not positive unless the temperature of a gas is below a certain temperature and it varies from gas to gas.

(a) Find *Joule-Thomson*'s coefficient for a *van der Waals* gas.

(b) The *van der Waals* constants of hydrogen molecule is given by $a = 24.7 \times 10^{-3} \mathrm{m}^6 \cdot \mathrm{Pa} \cdot \mathrm{mol}^{-2}$ and $b = 26.6 \times 10^{-6} \mathrm{m}^3 \cdot \mathrm{mol}^{-1}$. In order to liquefy a hydrogen gas compressed to 2 MPa by *Joule-Thomson*'s process, find a maximum temperature at least below which the gas can be liquefied.

Problem 3-50

The latent heat ΔH_l of a phase transition is given by a function of *P* and *T*. Its temperature dependence is given by the following equation (*Planck*'s equation). In order to derive this equation starting from the total

$$\left(\frac{d\,\Delta H_l}{dT}\right)_e = \Delta C_P + \frac{\Delta H_l}{T_e} - \Delta H_l\left(\frac{\partial \ln\Delta V_e}{\partial T}\right)_P \qquad (25)$$

differential of H (eq(26) below), complete the answer given below by filling the brackets of the equations given there (see **problem 3-39**). The present problem is useful to understand the partial differential along the route specified.

[answer]

Starting from eq(26),

$$dH = \left(\frac{\partial H}{\partial T}\right)_P dT + \left(\frac{\partial H}{\partial P}\right)_T dP \qquad (26)$$

eq(27) is derived.

$$\left(\frac{d\,\Delta H_l}{dT}\right)_e = \left(\frac{d\,\Delta H_l}{dT}\right)_P + \left(\frac{d\,\Delta H_l}{dP}\right)_T [(a)] \qquad (27)$$

The subscript e indicates that the corresponding differential is evaluated along the route on which the equilibrium is maintained.

On the other hand, eq(28) is given by the definition of the enthalpy H.

$$dH = -TdS + VdP \qquad (28)$$

Dividing both sides of eq(28) by dP at a constant temperature, *Maxwell*'s relation is employed to derive eq(29) .

$$\left(\frac{\partial H}{\partial P}\right)_T = -T\left(\frac{\partial S}{\partial P}\right)_T + V = -T[(b)] + V \qquad (29)$$

By applying eq(29) to the latent heat ΔH_l, eq(30) is obtained.

$$\left(\frac{\partial \Delta H_l}{\partial P}\right)_T = -T\left(\frac{\partial \Delta V_e}{\partial T}\right)_P + [(c)] \qquad (30)$$

Then, by substituting *Clausius-Clapeyron*'s equation into eq(27), it is rewritten as eq(31).

$$\left(\frac{d\,\Delta H_l}{dT}\right)_e = [(d)] + \left\{-T\left(\frac{\partial \Delta V_e}{\partial T}\right)_P + \Delta V_e\right\}\frac{\Delta H_l}{T\,\Delta V_e} \qquad (31)$$

After rearranging eq(31), eq(32) is obtained, which is *Planck*'s equation required to derive.

$$\left(\frac{d\,\Delta H_l}{dT}\right)_e = \Delta C_P + \frac{\Delta H_l}{T} - \frac{\Delta H_l}{\Delta V_e}\left(\frac{\partial \Delta V_e}{\partial T}\right)_P = \Delta C_P + \frac{\Delta H_l}{T} - \Delta H_l\left(\frac{\partial \ln\Delta V_e}{\partial T}\right)_P \qquad (32)$$

Problem 3-51

Before you answer the following questions, you may refer to the solution to **problem 3-50**.

(a) Express $\left(\frac{\partial P}{\partial T}\right)_S$ in terms of P, T, V and C_P, though not necessary to use them all.

(b) Examine what relations exist among P, V and T for the isentropic processes expressed as $\left(\frac{\partial P}{\partial V}\right)_S$ and $\left(\frac{\partial V}{\partial T}\right)_S$.

Problem 3-52a

There is an open container consisted of two compartments separated by a semipermeable membrane. In one compartment a pure solvent A is filled and in the other compartment a solution of B dissolved in the solvent A is filled. Solvent A molecules alone pass through the semipermeable membrane freely. To stop the spontaneous flow of A across the semi-permeable membrane into the solution so as for the system to remain staionary, the external pressure is applied to the solution. Based on the conditions the chemical potentials of the solvent and solution should satisfy in the equilibrium, derive an equation which gives the osmotic pressure the solute B generates (derivation of colligative properties). See **problem 4-87** for the reverse osmosis membrane.

Problem 3-52b

In connection with the preceding problem, prove that when two kinds of non-volatile solutes B and C are dissolved, the osmotic pressure of the solution is additive with respect to each concentration of the solutes as is given by (additivity). Here, c_i denotes the molar concentration of the i-th species.

$$\Pi = RT(c_\mathrm{B} + c_\mathrm{C}) \qquad (33). \qquad \text{For } n \text{ kinds of solute, } \Pi = RT \times \sum_{i=1}^{n} c_i .$$

Problem 3-53a

Form the conditions that chemical potentials should satisfy in an equilibrium, find an equation which gives a rise in the boiling point of the solution prepared by dissolving a small amount of non-volatile solute B in pure solvent A (derivation of colligative properties).

Problem 3-53b

When substances A and B are both volatile, discuss whether the boiling point of the solution prepared by mixing A and B is higher or lower as compared with those of the pure solvents.

Problem 3-54

Form the conditions that chemical potentials should satisfy in an equilibrium, derive an equation which gives a depression in the freezing point of the solution (derivation of colligative properties).

Problem 3-55

Show that eq(34) is satisfied for a *van der Waals* gas.

$$\left(\frac{\partial U}{\partial V}\right)_T = \frac{a}{V^2} \qquad (34)$$

The quantity defined by the left side of eq(34) is called internal pressure. For the ideal gas the internal pressure is zero. Taking a *van der Waals* gas as an example, examine the cause of the non-zero internal pressure. Describe also how the internal pressure is expressed when the gas follows either the *Dieterici* or the virial equation of state.

Problem 3-56a

Verify that the *Joule-Thomson* coefficient $\mu_{JT} = \left(\frac{\partial T}{\partial P}\right)_H$, the specific heat at constant pressure

$C_P = \left(\frac{\partial H}{\partial T}\right)_P$ and the isothermal *Joule-Thomson* coefficient $\mu_T^T = \left(\frac{\partial H}{\partial P}\right)_T = \left(\frac{\partial T}{\partial V}\right)_U$ are related one another by eq(35). The symbol μ is employed here but it has nothing to do with the chemical potential.

$$\mu_{JT} \times C_P = -\mu_T^T \qquad (35)$$

Practically, μ_{JT} is obtained from the measurements of C_P and the *Thomson* coefficient μ_T^T both of which are easier to measure directly as compared with μ_{JT}.

Problem 3-56b

The *Joule-Thomson* coefficient μ_{JT} of N_2 around at $0^\circ C$ under the pressure of 1.0 atm is known as $0.26\ K \cdot atm^{-1}$. Then find the temperature of the N_2 gas when the gas at $0^\circ C$ under 1 atm is expanded to the pressure of 0.2 atm employing the *Joule-Thomson* process. Find also the temperature when the same gas is expanded adiabatically to 0.2 atm.

Problem 3-57

Temperature variation of chemical potential.

Show that the chemical potential of a gas (taking Ar as an example for instance) varies linearly with temperature within a certain range of temperature. See **problem 5-34** for the theoretical calculation of the chemical potential of a gas based on the partition function.

Problem 3-58

An ideal solution composed of two substances A and B with mole fractions of X_A and X_B, respectively, has the chemical potentials μ_A for A and μ_B for B given by the following equations,

$$\mu_A = \mu_A^\circ + RT \ln X_A \qquad (36), \qquad \mu_B = \mu_B^\circ + RT \ln X_B \qquad (37)$$

where μ_A° and μ_B° represent the molar free energies of the pure substances A and B, respectively.

(a) Find a free energy change ΔG_{mix} upon mixing A and B to prepare the ideal solution specified above.

(b) Find the entropy of mixing ΔS_{mix} using the result obtained in (a).

(c) Find the volume change ΔV_{mix} upon mixing.

Problem 3-59

A non-ideal solution composed of two substances A and B with mole fractions are X_A and X_B, respectively, has the chemical potentials μ_A for A and μ_B for B given by the following equations,

$$\mu_A = \mu_A^\circ + RT \ln a_A = \mu_A^\circ + RT \ln \gamma_A X_A \qquad (38)$$

$$\mu_B = \mu_B^\circ + RT \ln a_B = \mu_B^\circ + RT \ln \gamma_B X_B \qquad (39)$$

where a_A and b_B are the corresponding activities, and γ_A and γ_B are their activity coefficients. The molar free energies of the pure substances A and B are denoted μ_A° and μ_B°, the same as in the preceding problem.

(a) Find a free energy change ΔG_{mix} upon mixing A and B to prepare the non-ideal solution specified above.

(b) Find the entropy of mixing ΔS_{mix} using the result obtained in (a).

(c) The enthalpy of mixing ΔH_{mix} is not zero for non-ideal solutions. Explain a role of ΔH_{mix} to play in determining ΔS_{mix}.

(d) Find the volume change ΔV_{mix} upon mixing.

Problem 3-60

Among the three states of solid, liquid and gas of a pure substance A, the molar free energies (μ_A°) of two states arbitrarily chosen exhibit temperature dependences as are shown in the figure below by straight lines (i) and (ii).

(a) Express $\left(\frac{d\mu_A^\circ}{dT}\right)_P$, which corresponds to the slopes of the straight lines shown in the figure in terms of thermodynamic variables.

(b) Answer which state of the pure substance is represented by the lines (i) and (ii) based on the result obtained in (a). What does the temperature T_e at the intersection represent?

(c) T_e shown in the figure varies with pressure. Find a function to describe $\frac{dT_e}{dP}$ for a case of one specific equilibrium in your mind.

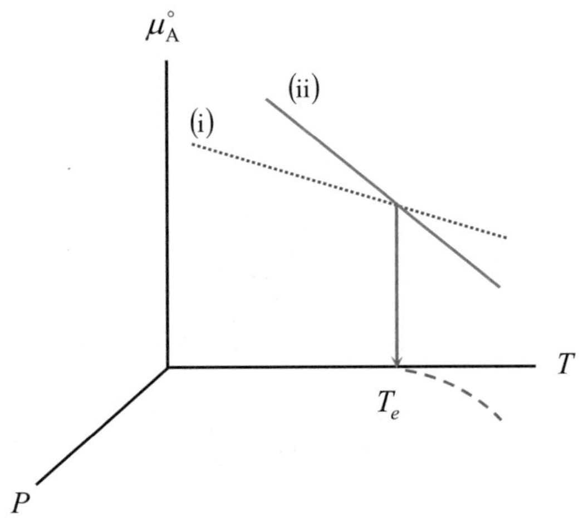

Fig.11 A schematic diagram to show the *P*-*T* dependence of the phase equilibrium of a pure substance and its chemical potentials (see **problem 3-57**)

Problem 3-61

For carbon dioxide CO_2 and water H_2O, their formation reactions and molar heats of formation are known as shown below.

$$C(graphite) + O_2 = CO_2 \qquad \Delta H = -393.51\ \text{kJ}\cdot\text{mol}^{-1} \qquad (40)$$

$$H_2 + \frac{1}{2}O_2 = H_2O(gas), \qquad \Delta H = -241.83\ \text{kJ}\cdot\text{mol}^{-1} \qquad (41)$$

Also, for ethanol, its combustion reaction and molar heat of combustion are given as below.

$$C_2H_5OH + 3O_2 = 2CO_2 + 3H_2O(gas), \qquad \Delta H = -1366.4\ \text{kJ}\cdot\text{mol}^{-1} \qquad (42)$$

Find the molar heat of formation of ethanol employing these relations and the values given.

Problem 3-62

In an appropriate text book of physical chemistry or a data book, find the values of $-(G^\circ - H_0^\circ)/T$ and ΔH_0° for each substance participating in the dehydrogenation reaction of ethane shown below,

$$C_2H_6 \rightleftarrows C_2H_4 + H_2$$

and then calculate not only the values of ΔG° for the reaction above at 298 and 500 K but also the equilibrium constant K_P at these temperatures. The superscript $^\circ$ refers to the standard state.

Problem 3-63

For a substance A in a solution, its chemical potential is μ_A, partial molar entropy is $\bar{S}_A$ and partial molar volume is $\bar{V}_A$. Both of $\bar{S}_A$ and $\bar{V}_A$ are defined based on the concept of the partial molar quantity. Show that those three quantities are related by the following equation.

$$d\mu_A = -\bar{S}_A dT + \bar{V}_A dP \qquad (43)$$

Problem 3-64

The *Gibbs* free energy of an *n*-components system consisting of n_1, n_2,⋯moles is given by G and the chemical potential of the *i*-th component, which is the partial molar free energy, is given by μ_i. Show that G is given by the equation $G = \sum_{i=1}^{n} n_i \mu_i$. More generally, show that any quantitative thermodynamic variable Λ is expressed by the equation $\Lambda = \sum_{i=1}^{n} n_i \lambda_i$ where λ_i is the corresponding partial molar quantity of Λ for the *i*-th component.

Problem 3-65

For the chemical reaction given below, referring to an appropriate data table (e.g. the appended table A.1 in *Basic Physical Chemistry* by *Walter J. Moore*), answer the following questions.

$$2Fe + 3H_2O \rightarrow Fe_2O_3 + 3H_2$$

(a) Find the enthalpy of the reaction $\Delta H_r^\circ(298\ \mathrm{K})$ at 298 K.

(b) Assuming that the specific heat of each substance in the reaction is constant against temperature, find the enthalpy of the reaction $\Delta H_r^\circ(400\ \mathrm{K})$ at 400 K.

Problem 3-66a

Diamond and graphite are in chemical equilibrium under certain pressure and temperature.

$$\mathrm{C}(\mathit{graphite}) \rightleftarrows \mathrm{C}(\mathit{diamond})$$

The equilibrium pressure at 298 K is known to be $P_0 = 1.51\times10^9\ \mathrm{Pa}$. Also at 298 K, the densities of diamond ρ_d and graphite ρ_g are $3.513\times10^3\ \mathrm{kg\cdot m^{-3}}$ and $2.260\times10^3\ \mathrm{kg\cdot m^{-3}}$, and the entropies S_d and S_g are $2.439\ \mathrm{J\cdot K^{-1}mol^{-1}}$ and $5.694\ \mathrm{J\cdot K^{-1}mol^{-1}}$, respectively. The heat of formation of diamond ΔH_f is $1900\ \mathrm{J\cdot mol^{-1}}$. This quantity is also the latent heat of phase transition $\Delta H_t(\mathit{graphite} \leftrightarrow \mathit{diamond})$.

Find the equilibrium pressure at 1300 K by assuming that those values given above are independent of temperature (adapted and modified from *Basic Physical Chemistry* by *Walter J. Moore*).

Problem 3-66b

Find the equilibrium constant of $[\mathrm{C}(\mathit{graphite})]/[\mathrm{C}(\mathit{diamond})]$ for the system given in the previous problem at 298 and 1300 K. Discuss the obtained result.

Problem 3-67

If a partial differential of any of *P*, *V* or *T* for an adiabatic process, i.e., a process at constant entropy, is known as a function of *P*, *V* and *T*, then the *P-V-T* relation along the adiabatic process can be found. In the case of the ideal gas, the relation is easily found starting from eq(44).

$$dU = C_V dT \,, \quad dH = C_P dT \qquad (44)$$

Express each of $\left(\dfrac{\partial T}{\partial P}\right)_S$, $\left(\dfrac{\partial V}{\partial P}\right)_S$ and $\left(\dfrac{\partial T}{\partial V}\right)_S$ as a function of *P*, *V* and *T*. It may include C_V and C_P as constants (see **problem 3-51**).

Problem 3-68

Generally, the temperature dependence of the equilibrium constant *K* is given by *van't Hoff*'s equation (*van't Hoff,* the first Nobel Prize in Chemistry in 1901) given below in which ΔH_r° is the standard enthalpy of the reaction.

$$\mathrm{A} \underset{}{\overset{K}{\rightleftarrows}} \mathrm{B} \qquad \Delta H_r^\circ \,, \qquad \left(\frac{\partial \ln K}{\partial T}\right)_P = \frac{\Delta H_r^\circ}{RT^2} \qquad (45)$$

Show that this equation is consistent with the principle of *Le Chatelier* with respect to the direction toward which the equilibrium shifts on increasing or decreasing temperature.

Problem 3-69

In thermodynamics, the chemical potential is generally defined by using the *Gibbs* free energy *G* as below.

$$\mu_i = \left(\frac{\partial G}{\partial n_i}\right)_{T,P,n_{j,j\neq i}} \qquad (46)$$

However, it can also be defined by employing the *Helmholtz* free energy $F(\equiv A)$.

Show that the following relation is satisfied.

$$\mu_i = \left(\frac{\partial G}{\partial n_i}\right)_{T,P,n_{j,j\neq i}} = \left(\frac{\partial F}{\partial n_i}\right)_{T,V,n_{j,j\neq i}} \qquad (47)$$

"note" It is straightforward to employ $F(\equiv A)$ when the chemical potential is defined by employing partition functions. This is because the partition functions contain the volume *V* of a system explicitly as a variable (for instance, through energy levels). See **I-5** and **II-2-2**.

Problem 3-70

One mole of superheated water at 383 K under 1 atm (as possible) is evaporated reversibly at the constant temperature. Find the entropy changes of the water (system) and its surroundings (outside world) associated with that process. Here, the latent heat of vaporization at 373 K is $47.3\ \mathrm{kJ \cdot mol^{-1}}$, the specific heat of the water is $75.4\ \mathrm{J \cdot mol^{-1}}$ and the one for the vapor is given by the function of $C_P = 30.36 + 9.61 \times 10^{-3} T + 11.8 \times 10^{-7} T^2 (\mathrm{J \cdot mol^{-1}})$. The specific heat of the water is assumed invariant with temperature.

Problem 3-71

The vapor pressure of water was measured at several temperatures and the values listed in the table below were obtained. Find a value for the heat of vaporization of water from these results assuming it is independent of temperature.

T/K	283	288	293	298
P/Torr	9.21	12.79	17.53	23.76

Problem 3-72a

The solubility of gas A into liquid B is found based on *Henry*'s law. Denoting the mole fraction of A as X_A, the pressure P_A of the gas A is related to X_A by eq(48), which is called *Henry*'s law. κ_A^H is a proportional constant specific to an individual combination of A and B and is named *Henry*'s law constant.

$$P_A = \kappa_A^H X_A \qquad (48)$$

(a) The *Henry*'s law constant of gaseous oxygen above water at room temperature is 4.4×10^4 bar. Find the solubility (mole/kg) of the oxygen at 1 atm into the water.

(b) The *Henry*'s constant of gaseous carbon dioxide above water is 1.25×10^6 Torr. When a cap of a bottle containing 500 ml of water saturated with gaseous carbon dioxide under the pressure of 2 atm is opened suddenly, how many litters of the gas are released into the air under the ambient temperature and pressure?

Problem 3-72b

Assuming that the *Henry*'s law constants of gases above blood are the same as those above pure water, when a diver who is diving 30 m under the sea with his or her lung fully filled with the air suddenly comes to the surface in emergency, find how many liters of nitrogen gas per 1 L of blood is released. The *Henry*'s law constant of nitrogen is 6.51×10^7 Torr.

"note" Diver's disease (decompression sickness) and *Henry*'s law constant

The *Henry*'s law constant of nitrogen is about two times larger than that of oxygen but the partial pressure of the former in the air is four times larger than that of the latter. As a result, the amount of nitrogen dissolved in blood under high pressure is two times larger. When the person diving deep under water is brought back to normal pressure suddenly, the over-saturated nitrogen gas is rapidly released as bubbles in blood vessels, which causes serious blood circulatory disorder or damages various tissues. Diver's disease means a sickness caused temporarily or chronically by blood circulatory disorder. A way to reduce the sickness is to use a pressurized air gas relatively enriched by oxygen or a pressurized gas mixture of oxygen and helium whose *Henry*'s law constant is larger than that of nitrogen.

Problem 3-73

The solubility of gaseous SO_2 into water was measured at three different temperatures and obtained values are listed in the table below. X_{SO_2} is the mole fraction of SO_2.

(a) Find values for the *Henry*'s law constant of SO_2 above water at the three temperatures.

(b) Explain the temperature dependence of the *Henry*'s law constant found for SO_2 from a molecular theoretical point of view.

$SO_2 / X_{SO_2} \times 10^2$	P / atm			$SO_2 / X_{SO_2} \times 10^2$	P / atm		
	10° C	20° C	30° C		10° C	20° C	30° C
27.4	0.624	0.918	—	0.843	0.0104	0.0186	0.0259
20.7	0.459	0.68	0.906	0.562	0.00605	0.0112	0.015
13.9	0.297	0.442	0.595	0.422	0.00408	0.00763	0.0107
6.98	0.138	0.212	0.284	0.281	0.00230	0.00421	0.00618
4.20	0.0776	0.121	0.165	0.141	0.000987	0.00158	0.00224
2.80	0.0487	0.0776	0.104	0.0562	0.000395	0.000658	0.00789
1.96	0.0311	0.0613	0.0684				
1.40	0.0205	0.0342	0.0474				

Problem 3-74

(a) Show that a relation between the vapor pressure P of a liquid and its temperature T is given by the following equation (eq(49)) (*Clausius-Clapeyron*'s equation) where ΔH_{vap} is a heat of vaporization regarded as constant. Assume that the vapor gas behaves ideally.

$$d\ \ln P / dT = \Delta H_{vap} / RT^2 \qquad (49)$$

(b) The boiling point and the heat of vaporization of octanol (A) and n-propanol (B) are known as given below (both at 1 atm).

$$\text{A}: (468\ \text{K},\ 52.3\ \text{kJ} \cdot \text{mol}^{-1}) \text{ and } \text{B}: (371\ \text{K},\ 40.1\ \text{kJ} \cdot \text{mol}^{-1}).$$

Find the mixing ratio of the two solvents so as to make the solution boil at 400 K. Assume the solution of any component ratio is ideal.

(c) For the mixtures, find the range of mole fraction where the gas and solution phases can co-exist at 400 K.

Problem 3-75a

(a) For a certain gas, its equation of state is given by a function of $P(V - nb) = nRT$ in which b is an excluded volume. Derive functions which describe the fugacity and fugacity coefficient of this gas (see **problem 3-41c**).

(b) Putting $b = 4.5 \times 10^{-2}\,\text{dm}^3 \cdot \text{mol}^{-1}$, find values for the fugacity coefficient of this gas at 300 K under the pressures of 5, 10 and 20 atm.

Problem 3-75b

(a) The equation of state for a certain gas is given by $\left(P + \frac{n^2 a}{V^2}\right)(V - b) = nRT$. Here a is a constant associated with the intermolecular force. Derive functions which describe the fugacity and fugacity coefficient of this gas (see **problem 3-41c**).

(b) Putting $a = 3.50 \times 10^{-1}\,\text{m}^6\text{Pa} \cdot \text{mol}^{-2}$, find values for the fugacity coefficient of this gas at 300 K under the pressures of 5, 10 and 20 atm.

Problem 3-76

Draw the *P-V* curves of the *van der Waals* equation of state $\left(P + \frac{a}{V^2}\right)(V - b) = RT$ for the given values of

$a = 3.50 \times 10^{-1} m^6 Pa \cdot mol^{-2}$ and $b = 4.50 \times 10^{-5} m^3 mol^{-1}$ at the temperatures of 240, 250, 260, 280, 290 and 300 K.

Problem 3-77

Using a mono-atomic ideal gas as a working substance, a cyclic process consisting of the three steps of (1) isovolumetric heating $\rightarrow$ (2) adiabatic expansion $\rightarrow$ (3) isobaric compression is considered and is shown in Fig.12 below. Based on this, answer the following questions.

(a) Find the work and the heat the working substance gives to or gain from the surroundings for each step.

(b) Find the work efficiency of this cyclic process.

(c) Confirm that the change of the internal energy of the working substance after one cycle is zero.

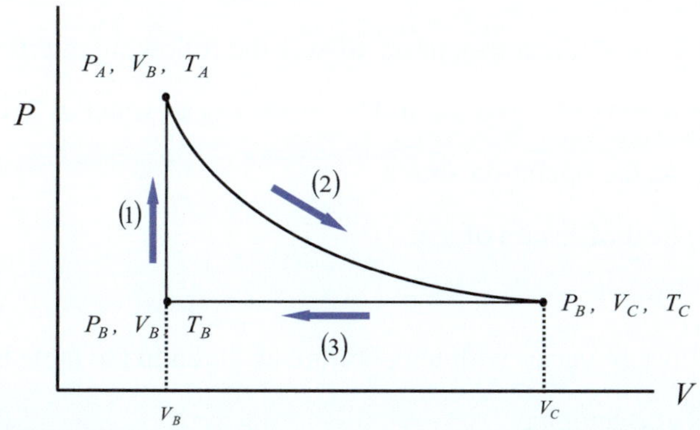

Fig.12 Cyclic process consisted of three steps for a mono-atomic gas as a working substance. (1): isometric heating, (2): adiabatic expansion, (3): isobaric compression.

Problem 3-78

Show that the work efficiency of the three-step cyclic process discussed in the preceding problem is poorer as compared with that of the *Carnot* cycle.

Problem 3-79

Find the efficiency of work for the cyclic process consisting of the three reversible steps of isovolumetric heating (1) $\rightarrow$ isothermal expansion (2) $\rightarrow$ isobaric compression (3). The conditions of *T*, *P* and *V* are as given in Fig.13 below.

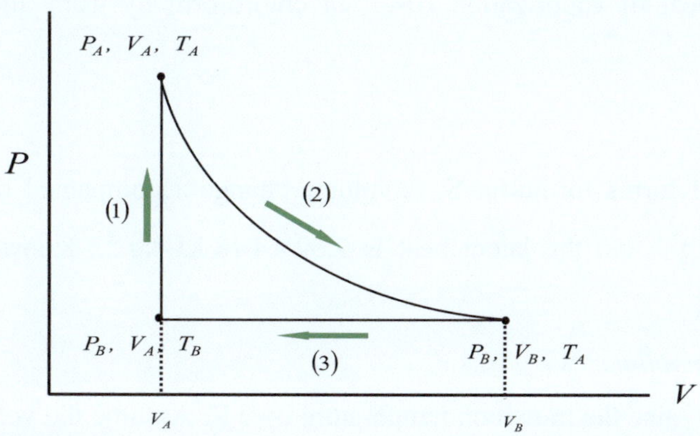

Fig.13 Cyclic process consisting of three reversible steps. The figure looks like the one given in **problem 3-77** but notice step (2) is not the adiabatic but isothermal one.

Problem 3-80

Assuming that the working substance is an ideal gas, find the efficiency of work for the *Stirling* cycle consisting of two isothermal and two isovolumetric steps. The *T-V* and *P-V* diagrams of each step of the cycle are as shown in the figures given below.

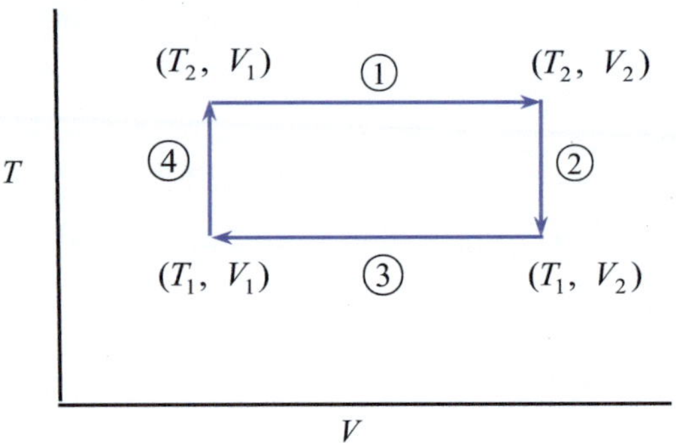

Fig.14 *T-V* diagram of *Stirling* cycle.

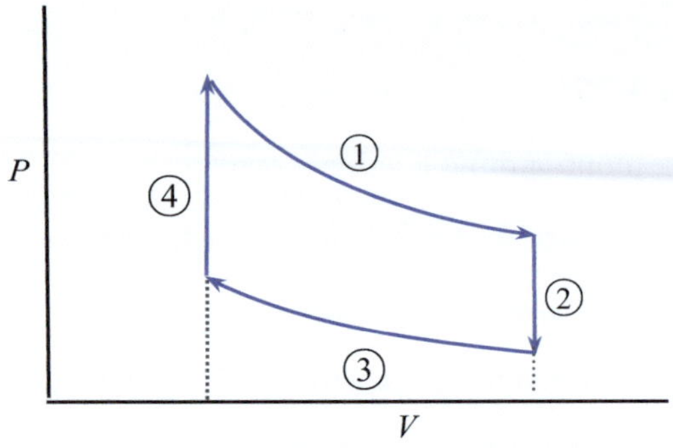

Fig.15 *P-V* diagram of *Stirling* cycle.

Problem 3-81

There is a refrigerator working in a room at 25 °C. For this refrigerator, answer the following questions.

(a) Find a minimum work W_{min} required in order to make 1 kg of ice at 0 °C from 1kg of water at 0 °C.

(b) Find the amount of heat emitted into the room on the operation above.

Use a value of $79.8 \times 4.18 \times 10^3 \text{J} \cdot \text{kg}^{-1}$ for the heat of fusion of ice.

Problem 3-82

Vapor pressure of chloroform in liquid-gas equilibrium varies with temperature as given in the table below.

Table Equilibrium vapor pressures of chloroform.

vapor pressure *P*/Torr	50	150	200	500	600	800
temperature *T*/K	272	293.4	298.8	322.3	327.1	335.5

(a) The vapor pressure *P* is most satisfactorily expressed by a function of temperature *T* given by eq(50).

$$P = a + \frac{b}{T} + c \times \ln T \qquad (50)$$

Employing the non-linear least squares fitting provided by the free software gnuplot, find the best-fitted values for the parameter *a*, *b* and *c*.

(b) Find the boiling point T_b and the heat of vaporization ΔH°_{vap} of chloroform by using the best-fitted function determined in (a).

Problem 3-83

There exist several varieties of crystal forms for sulfur S. A volume change accompanied by the phase transition given below is $1.26 \times 10^{-5} \text{m}^3\text{kg}^{-1}$ and the latent heat is $2.52 \times 4.18 \text{ kJ} \cdot \text{kg}^{-1}$. Knowing that the transition temperature is 368.6 K at 1 atm,

$$\textit{rhombic}\ \text{S} \rightarrow \textit{monoclinic}\ \text{S}$$

find the pressure to be applied in order to raise the transition temperature by 3 K. Assume the volume change as well as the latent heat remains constant. Can the transition temperature be lowered?

Problem 3-84

The volume (cubic) expansion coefficient α of metal Al at 298 K is given by $\alpha \equiv \frac{1}{V}\left(\frac{\partial V}{\partial T}\right)_P = 2.21 \times$

$10^{-5}\,\mathrm{K}^{-1}$, the compressibility coefficient by $\beta = -\frac{1}{V}\left(\frac{\partial V}{\partial P}\right)_T = 1.32\times10^{-6}\,\mathrm{bar}^{-1}$, and the density of Al is $\rho = 2.70\times10^3\,\mathrm{kg\cdot m^{-3}}$. Find the difference of the heat capacities $C_P - C_V$ for meta Al based on these data (adapted and modified from *Clyde R.Metz, Physical Chemistry*). What is generally concluded for the heat capacity of solid from the results obtained?

Problem 3-85

There was a gas mixture of hydrogen and oxygen in the ratio of 2:1 by volume. At 298 K and at the pressure of 1 atm, 1.0 dm^3 of the mixture was introduced into a vessel. When the mixture was suddenly compressed to 1/10 of the initial volume adiabatically, it exploded. If it had not exploded, what would be the temperature and pressure of the mixture? Assume the heat capacities of the gases are the same and the mixture behaves ideally.

Problem 3-86

Estimate the resonance energy for the *Kekulé* structure of toluene ($C_7H_8(l)$) using the thermodynamic values and bond energies given below for toluene.

reaction energies at the standard state (/mol)

$C_7H_8(l) + 9O_2(g) \rightarrow 7CO_2(g) + 4H_2O(l)$ $\quad \Delta H^\circ_{comb.} = -3910\,\mathrm{kJ}$ (51) (heat of combustion)

$C_7H_8(l) \rightarrow C_7H_8$ $\quad \Delta H^\circ_{vap.} = 28.1\,\mathrm{kJ}$ (52) (heat of vaporization)

$H_2(g) \rightarrow 2H(g)$ $\quad \Delta H^\circ_{diss.} = 436\,\mathrm{kJ}$ (53) (heat of dissociation)

$C(s) \rightarrow C(g)$ $\quad \Delta H^\circ_{sub.} = 715\,\mathrm{kJ}$ (54) (heat of sublimation)

$H_2(g) + \frac{1}{2}O_2(g) \rightarrow H_2O(l)$ $\quad \Delta H^\circ_{comb.} = -286\,\mathrm{kJ}$ (55) (heat of combustion)

$C(s) + O_2(g) \rightarrow CO_2(g)$ $\quad \Delta H^\circ_{comb.} = -394\,\mathrm{kJ}$ (56) (heat of combustion)

bond energies (/mol)

C−H: $E_{bond} = 413\,\mathrm{kJ}$, C−C: $E_{bond} = 346\,\mathrm{kJ}$, C=C: $E_{bond} = 610\,\mathrm{kJ}$.

Problem 3-87

The heat (heat of dissolution) produced when 1 mole of sulfuric acid is diluted with n moles of water is given in the table below. From this data, find the amount of heat produced when $1.0\,\mathrm{dm}^3$ of the aqueous sulfuric acid solution of the concentration of $1.0\,\mathrm{mol\cdot dm^{-3}}$ is diluted ten times with water. Also, obtain how many degrees the temperature after dilution has changed compared to the beginning. Assume that the specific heat of the solution remains constant at 1.0 and that the volume of the solution is additive.

water n mol	5	100	400	1600
heat of dissolution ($\mathrm{kJ\cdot mol^{-1}}$)	57.3	73.6	75.7	79.4

Problem 3-88

We consider obtaining metal Cd by reducing CdS with H_2. Is it possible to obtain metal Cd if the reduction reaction is carried out at 1100 °C? Answer the question based on eqs(57) and (58) given below together with the relevant free energy changes at $1100\,^{\circ}\mathrm{C}$.

$$\mathrm{H_2}(g)+\frac{1}{2}\mathrm{S_2}(g) \rightarrow \mathrm{H_2S}(g) \qquad \Delta G_1 = -49.10\ \mathrm{kJ\cdot mol^{-1}} \tag{57}$$

$$\mathrm{Cd}(g)+\frac{1}{2}\mathrm{S_2}(g) \rightarrow \mathrm{CdS}(g) \qquad \Delta G_2 = -127.7\ \mathrm{kJ\cdot mol^{-1}} \tag{58}$$

Problem 3-89

The internal energy U of the ideal gas is known to be given by a function of temperature alone, as is derived from the equation of state and its change is expressed as $dU = C_V dT$ with C_V constant. Then, find a functional form to express the internal energy of a gas which obeys the *van der Waals* equation of state. Similarly, find also a function for the entropy of the gas.

Problem 3-90

The solubility of picric acid in benzene varies with temperature as is given in the table below. Based on these data, find a value for the heat of fusion of picric acid. The molecular weight and the melting point are 78 and 278.5 K for benzene and 229 and 394.8 K for picric acid.

T /K	278	283	288	293	298	308
g/100g (benzene)	3.70	5.37	7.29	9.56	12.66	21.38

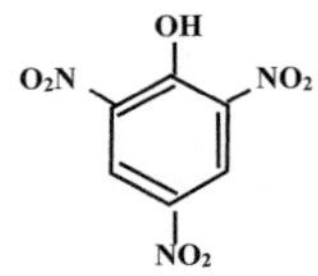

Fig.16 Picric acid.

Problem 3-91

The melting point of alloy composed of lead and antimony (Pb-Sb) varies with the composition (mass %) as given in the table below.

melting point /K	904	823	773	673	573	473	600
Pb(w/w%)	0	30	44	66	80	94	100

(a) Based on these data, draw a phase diagram for this alloy and explain each phase you draw.

(b) Based on the phase diagram obtained, find the amount of Sb to be deposited, when 20 kg of the alloy containing Pb 35w% is melted once and then cooled down to 637 K.

(c) Find the temperature to be lowered to in order to obtain the maximum amount of pure Sb by cooling the melt of the alloy given in (b). Answer the amount of Sb to be obtained as well.

Problem 3-92

For the ideal gas, the heat capacities at constant volume and at constant pressure are given by the following equations. Derive these relations by using *Jacobians*.

$$C_V = \left(\frac{\partial U}{\partial T}\right)_V = \left(\frac{\partial U}{\partial T}\right)_P = \left(\frac{\partial U}{\partial T}\right)_S \tag{59}$$

$$C_P = \left(\frac{\partial H}{\partial T}\right)_P = \left(\frac{\partial H}{\partial T}\right)_V = \left(\frac{\partial H}{\partial T}\right)_S \tag{60}$$

The above relations explicitly indicate that when the heat capacity is defined, a due attention should be payed to the meaning of condition of "constant volume" or "constant pressue" and the resulting effect.

Problem 3-93

One of the *Maxwell* relations is given by eq(61). Prove eq(61) for the ideal gas by expressing the left and right sides of eq(61) individually in terms of the variables *P, V* and *T* alone.

$$\left(\frac{\partial V}{\partial T}\right)_S = -\left(\frac{\partial S}{\partial P}\right)_V \qquad (61)$$

Problem 3-94

There is a solid substance whose molar specific heat is given by the following equation with constants given below.

$$C_V = A + 2BT \qquad (62)$$

$$A = 9.6\,\mathrm{J\cdot mol^{-1}\cdot K^{-1}}\ , \qquad B = 0.24\,\mathrm{J\cdot mol^{-1}\cdot K^{-2}}$$

Two blocks of the same substance, one at 400 K and the other at 200 K, are made contact thermally and are brought to the thermal equilibrium under otherwise isolated conditions. Each block is made of 1 mole of the substance.

(a) Find the temperature when the thermal equilibrium is established. It is assumed there is no volume change accompanied by heat transfer.

(b) Find the total entropy change.

Problem 3-95

Employing the same substance as given in the preceding problem, find the maximum work to be obtained when the two blocks (1 mole each), one at 400 K and the other at 200 K initially, are thermally contacted and are brought to thermal equilibrium finally. What is the final temperature?

Problem 3-96

(a) Show that the difference between the specific heat at constant pressure $C_P = (\partial H/\partial T)_P$ and the one at constant volume $C_V = (\partial U/\partial T)_V$ is given by the following equation (eq(63)),

$$C_P - C_V = \frac{TV\alpha^2}{\kappa_T}(= R) \qquad (63),\ \text{in which}\ \alpha = \frac{1}{V}\left(\frac{\partial V}{\partial T}\right)_P \ \text{and}\ \kappa_T = -\frac{1}{V}\left(\frac{\partial V}{\partial P}\right)_T,$$

where α is a thermal expansion coefficient and κ_T is a compression coefficient. The equality given in the parenthesis is for the ideal gas (see **"note"**).

(b) Show that for liquids or solids an approximate relation of $C_P \approx C_V$ is valid.

"note" An alternative way of proving the equation is to employ the relations of $C_V = T(\partial S/\partial T)_V$ and $C_P = T(\partial S/\partial T)_P$.

Problem 3-97

The heat capacity of pure solid of Na_2SO_4 was measured at very low temperatures and the data listed in the table below were obtained.

Table Temperature dependence of the heat capacity of Na_2SO_4 at constant pressure.*

T /K	13.74	16.25	20.43	27.73	41.11	52.72	68.15	82.96	95.71
C_P /cal·K^{-1}	0.171	0.286	0.626	1.615	4.346	7.032	10.48	13.28	15.33

*Data taken from *K.S.Pitzer and L.Brewer, Thermodynamics, 2nd Ed.,McGraw-Hill Book Company,* p.127, 1961.

Find the absolute entropy (third-law entropy) of Na_2SO_4 at 100 K based on these data.

Problem 3-98

Thermally insulated two containers 1 and 2 with a volume of 1.0 L each are connected through a valve. Before opening the valve, both containers are filled with He gas, one (1) at 20 °C under 0.5 atm and the other (2) at 80 °C under 1.0 atm. After opening the valve fully, both gas has reached the equilibrium at the same temperature and pressure. Assuming that the He behaves as a mono-atomic ideal gas, find the temperature and pressure reached.

Problem 3-99

A phase diagram (*P-T* figure) of water near at the triple point is shown in Fig.17 below. The equilibrium pressure for either gas-liquid or gas-solid is approximately expressed by eq(64) and eq(65), respectively.

$$\ln P = 2.303 \times 20.9586 - 4.0843 \times \ln T - 2.303 \times \frac{2825.4}{T} \qquad (64)$$

$$\ln P = 2.303 \times 11.5041 - 0.4001 \times \ln T - 2.303 \times \frac{2694.2}{T} \qquad (65)$$

The *P-T* relation for the liquid-solid equilibrium is expressed by a straight line passing through the triple point. From these given conditions and Fig.17, find values for the three kinds of latent heats, i.e., the heats of sublimation, condensation and solidification of water.

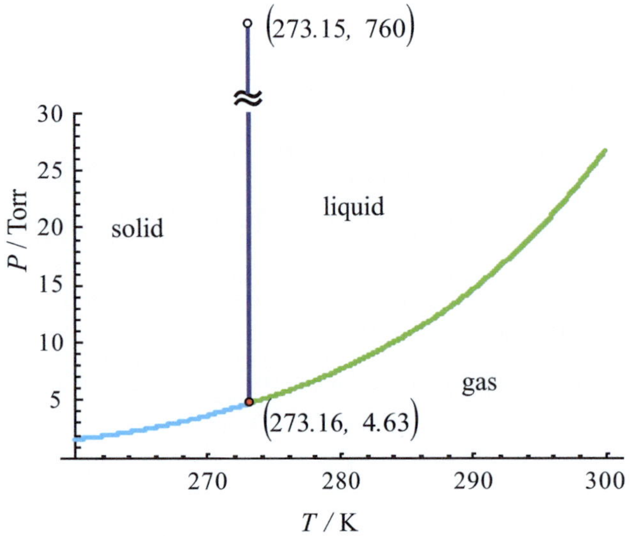

Fig.17 Phase diagram of water near the triple point.
red full circle: triple point,
while full circle: freezing point,
green line: gas-liquid equilibrium line, eq(64),
blue line: gas-solid equilibrium line, eq(65).

Problem 3-100

van der Waals constants of O_2 are given by $a = 1.378\ \text{atm}\cdot\text{dm}^6\cdot\text{mol}^{-2}$ and $b = 3.183\times10^{-2}\ \text{dm}^3\cdot\text{mol}^{-1}$.

(a) Find the boiling point of O_2 at 1 atm from its *van der Waals* equation of state.

(b) Calculate the heat of liquefaction of O_2 at that temperature.

Problem 3-101

(a) Consider a heat reservoir of a constant capacity at the temperature T_h whose heat capacity is given by C_V. Show that the maximum work W_{max} this hot reservoir is capable of doing to the surroundings at the temperature T_0 is given by eq(66) below.

$$W_{\max} = -\int_{T_h}^{T_0} C_V \left(\frac{T - T_0}{T} \right) \cdot dT \qquad (66)$$

(b) Consider 10 kg of water heated to 98 °C. The temperature of the air outside (surroundings) is 5 °C. Find the maximum work this hot water is able to do to the air outside. Use $18 \times 4.2\ \mathrm{J \cdot mol^{-1} K^{-1}}$ as the molar specific heat of water at constant volume.

(c) How many meters an object of 1 kg can be lifted by using the maximum work to be obtained above?

Problem 3-102

In order to keep a sample under current research at very low temperatures, the heat that keeps flowing into the sample must be removed constantly with an appropriate machine. Consider the sample initially kept at 0.001K and assume that the heat of 1J must be removed per second in order to keep the sample constantly at this temperature. When an ideal freezing machine which operates between at room temperature (298 K) and at 0.001K is employed, find the minimum work (J/s) needed to maintain that low temperature. What electric power in watts is this work converted to?

Problem 3-103

(a) Show that the inequality given below (eq(67)) is satisfied by using one of the inequalities in thermodynamics.

$$\frac{1}{\alpha} \cdot \left(\frac{\partial T}{\partial P} \right)_S > 0 \qquad (67), \qquad \text{where } \alpha = \frac{1}{V} \cdot \left(\frac{\partial V}{\partial T} \right)_P .$$

(b) Explain how the above inequality is satisfied from a viewpoint of *Le Chatelier*'s principle.

Problem 3-104

0.025 mol of a non-ideal pure gas is in a container of a volume of 1L at 100 K under the pressure of 0.2 atm. The container is connected to a vacuum container of a volume of 0.01L through a valve initially closed. When the valve is opened and let the gas expand freely into the vacuum container, find the accompanied enthalpy change of the gas, for which the following values are known: $C_V = 32\ \mathrm{J \cdot mol^{-1} K^{-1}}$, $\kappa_T = 0.3\ \mathrm{atm^{-1}}$ (compressibility) and $\alpha = 0.002\ \mathrm{K^{-1}}$ (expansion coefficient).

Problem 3-105

Starting with the relation of $(\partial U / \partial V)_S = -P$, answer what conditions must be satisfied for the thermodynamic function to derive the equation of state for an ideal gas $PV = nRT$. Based on your answer, show that the specific heat at constant volume need not be a constant, and that the equation of state of an ideal gas can be obtained even if it is an arbitrary function of temperature.

Problem 3-106

A heat engine consisting of two adiabatic processes and two isovolumetric processes with an ideal gas as a working substance is called *Otto* cycle. Its schematic cycle is illustrated in the figure below (Fig.18). Based on this *P-V* diagram, find the efficiency of work of the *Otto* cycle.

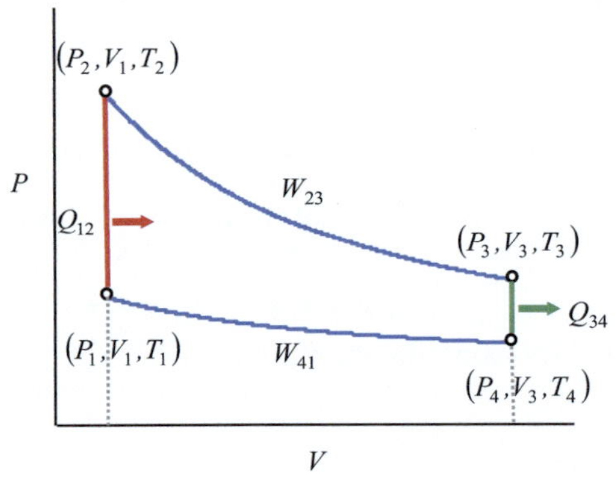

Fig.18 *Otto* cycle consisting of two adiabatic processes and two isovolumetric processes.

Problem 3-107

From a theoretical point of view, there exists a way of thinking of the fundamental equation for thermodynamics, in which the entropy S or the internal energy U is expressed as a function of the other extensive variables (U or S, V and N : number of particles) and S or U is defined specifically as a quantity of state for the system under consideration.

When the entropy is given by a function of U, V and N as eq(68) with $\alpha\,[\mathrm{J\cdot K^{-1}}]$ and $\beta\,[\mathrm{J\cdot m^3}]$ as constants, show that the system represents the ideal gas and that the equation of state of the ideal gas is derived from eq(68).

$$S(U,V,N) = \alpha N \ln \frac{UV}{\beta N^2} \qquad (68)$$

If eq(68) can be modified so as to yield the heat capacity at constant volume C_V equal to $3R/2$, answer how it is to be modified. (see **problem 5-68a** for *Sackur-Tetrode*'s equation).

Problem 3-108

The thermal expansion coefficient of a substance is usually measured unde1r a constant pressure and is defined by eq(69). In the case of gases, this thermal expansion coefficient is significantly different from that defined by eq(70) for the adiabatic process.

$$\alpha_{(P)} = \frac{1}{V}\left(\frac{\partial V}{\partial T}\right)_P \qquad (69), \qquad \alpha_{(S)} = \frac{1}{V}\left(\frac{\partial V}{\partial T}\right)_S \qquad (70)$$

(a) Find the expressions for $\alpha_{(P)}$ and $\alpha_{(S)}$ for the ideal gas.

(b) When the molar specific heat at constant pressure C_P ($\mathrm{J\cdot mol^{-1}\cdot K^{-1}}$) is given by eq(71) for the ideal gas (see **problem 3-105**), find the function to describe $\alpha_{(S)}$.

$$C_P = (6.26 + 2.746 \times T - 0.770 \times T^2) \times 4.184 \qquad (71)$$

Problem 3-109a

The compressibility of a substance is usually measured at constant temperature and is defined by eq(72).

Similarly as is asked in the preceding problem, examine how the compressibility defined by eq(73) under the adiabatic condition differs from that given by eq(72) in the case of a gas.

$$\kappa_T = -\frac{1}{V}\left(\frac{\partial V}{\partial P}\right)_T \quad (72)\ , \qquad \kappa_S = -\frac{1}{V}\left(\frac{\partial V}{\partial P}\right)_S \quad (73)$$

Problem 3-109b

The adiabatic compressibility κ_S and the isothermal compressibility κ_T are defined by the first and second equation of eq(74), respectively.

$$\kappa_S = -\frac{1}{V}\left(\frac{\partial V}{\partial P}\right)_S, \quad \kappa_T = -\frac{1}{V}\left(\frac{\partial V}{\partial P}\right)_T \quad (74)$$

Show that the ratio of these compressibilities is given by the ratio of the specific heat at constant volume to the one at constant pressure as is given by eq(75).

$$\frac{\kappa_S}{\kappa_T} = \frac{C_V}{C_P} \quad (75)$$

Problem 3-110

According to the third law of thermodynamics, the entropy of a substance approaches to zero when the temperature approaches to the absolute zero K. Namely, $S \to 0$ when $T \to 0$. Based on this fact, show that when $T \to 0$, the heat capacity at constant volume C_V approaches to zero. Does the heat capacity at constant pressure C_P also approach to zero?

Problem 3-111

For the phase equilibrium between two phases A and B, · *Clausius-Clapeyron*'s equation holds between the equilibrium pressure P_{eq} and the equilibrium temperature T_{eq}. This equation is derived by finding how the *Gibbs* energy varies with T and P along the phase equilibrium line. Changes in the entropy and volume given by eq(76) and eq(77) accompanied by the phase transition are not zero for the so called "first-order phase transition".

$$\Delta S_{A\to B} = S_B - S_A \quad (76)\ , \qquad \Delta V_{A\to B} = V_B - V_A \quad (77)$$

However, for the phase transition called " second-order transition", they both become zero. For that, a change in the *Gibbs* energy given by eq(78) is not sufficient to find out the relation between P_{eq} and T_{eq} as is usually performed in deriving *Clausius-Clapeyron*'s equation.

$$d(\Delta G_{A\to B})_{eq} = \Delta S_{A\to B} \times dT_{eq} + \Delta V_{A\to B} \times dP_{eq} = (S_B - S_A)dT_{eq} + (V_B - V_A)dP_{eq} = 0 \quad (78)$$

By expanding $d(\Delta G_{A\to B})_{eq}$ to the second orders of dP_{eq} and dT_{eq} (higher order approximation), derive the equation of dP_{eq}/dT_{eq} called <u>*Ehrenfest*'s equation</u>.

Problem 3-112a

A *P-T* diagram illustrating the phase equilibriums of a pure substance among gas (*G*)-liquid (*L*)-solid (*S*) phases are shown in the figure on the right. Present the same phase diagrams as a diagram in the *T-V* plane specifying each phase and its region.

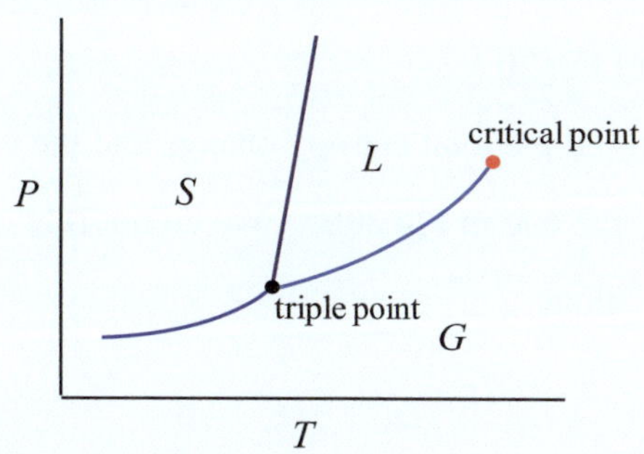

Fig.19 Phase diagram of a pure substance exhibiting a triple point.

Problem 3-112b

As is indicated by the full black circle in Fig.61 in the solution to the preceding problem, the three phases coexist at the triple point. When the sum of the volumes of the substance in these three phases at the triple point is V_t, then answer how the volume of the substance in each phase is expressed in terms of V_t. The total amount of the substance is 1.0 mole.

Problem 3-113

When the thermal expansion coefficient α is given by eq(79) (for instance for an ideal gas),

$$\alpha \equiv \frac{1}{V}\left(\frac{\partial V}{\partial T}\right)_P = \frac{1}{T} \qquad (79)$$

show that the specific heat at constant pressure C_P does not depend on pressure $\left(\frac{\partial C_P}{\partial P}\right)_T = 0$.

Problem 3-114

Under a constant temperature, the entropy change with volume is given by one of *Maxwell*'s relations (eq(80)).

$$(\partial S/\partial V)_T = (\partial P/\partial T)_V \qquad (80)$$

Based on this equation, show that the entropy of mixing is given by eq(81) for a mixed gas prepared by mixing two ideal gases A and B by the volumes of V_A and V_B, respectively at the constant temperature and pressure. X_A and X_B denote the mole fractions of A and B, respectively and n is the mole number of the mixed gas.

$$\Delta S_{mix} = -nR(X_A \ln X_A + X_B \ln X_B) \qquad (81)$$

Problem 3-115

(a) Two ideal gases of the same kind are mixed in the mole ratio of 1 to 1 at the constant temperature and pressure. Show that the entropy of mixing is zero in this case.

(b) If the gases are mixed not by the mole ratio of 1 to 1 but in the mole ratio of 1 to n, how should the argument given above be modified in order to draw the same conclusion as in (a), that is, the entropy of mixing is zero?

Problem 3-116

As is illustrated in Fig.20 below, two vessels 1 nad 2 of the volume of V_1 and of V_2 are immersed in a thermostat kept at constant T and they are filled with an ideal gas of the same kind under the condition of (P_1, V_1, n_1) and of (P_2, V_2, n_2), for the vessels 1 and 2, respectively. By opening a valve put in between the vessels, let the gases mix freely until a stationary state is reached at the final pressure of P_f as is illustrated in **B** of Fig.20.

(a)Find the total entropy change and the work done by the gases upon the change from the state in Fig.20A to the one in Fig.20B. We do not rely on the fact that the internal energy of the ideal gas is a function of T alone.

(b) Show that the entropy change accompanied by this process is positive.

(c) Answer whether this process is reversible or irreversible with explanation.

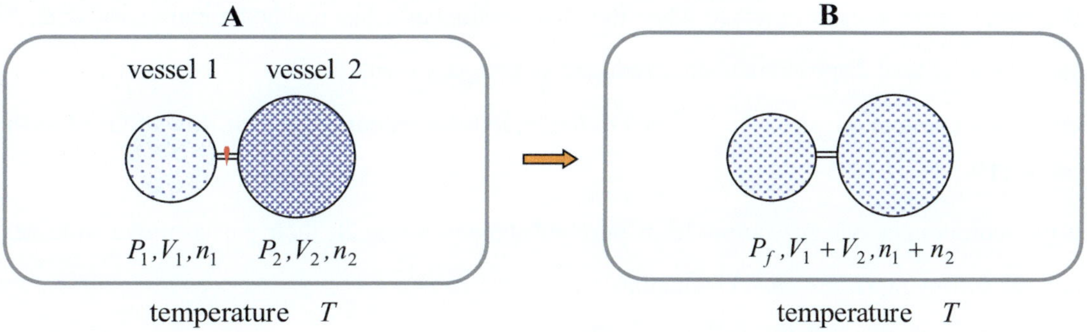

Fig.20 Illustration of mixing of two gases, initially kept at different pressures and volumes as shown in **A**, through isothermal process, reaching finally the stationary state as shown in **B**.

Problem 3-117

There are two kinds of phase transition. One is those seen for the gas-liquid, gas-solid, solid-solid (maltitude shapes), etc., and is named first-order transition. The other is recognized for the order-disorder transition of alloys or for the transition in ferroelectric or ferromagnetic substances and is named second-order transition.

Features of the second-order transition are that at the transition point, the thermodynamic values of *H*, *S*, *V* in each phase of I and II do change continuously upon phase transition, making a sharp contrast to the first-order phase transition in which these thermodynamic values change discontinuously upon phase transition. Noting these features for the second-order phase transition between phase I and II, find how the transition point varies when the transition temperature is varied by dT by deriving a function to describe dP/dT (see **problem 3-111**). Needless to add, it is well known for the first-order transition that *Clausius-Clapeyron*'s equation is satisfied.

Problem 3-118

The equation of heat conductance along the temperature gradient can be expressed by eq(82) in the case of one-dimensional (x-direction) transfer. Here, λ is the thermal conductivity, Q is the amount of heat transferred per unit time per unit area.

$$Q = -\lambda \frac{dT}{dx} \qquad (82)$$

(a) Let us consider the heat conductance through a window glass. Find the amount of heat disposed outside per unit time through the window under the following conditions;

thermal conductivity of the window glass : $\lambda = 0.95\,\mathrm{J \cdot s^{-1} \cdot m^{-1} \cdot K^{-1}}$,

temperature outside: $5\,^\circ\mathrm{C}$, room temperature: $20\,^\circ\mathrm{C}$

the area of the window: $0.90 \times 0.70\,\mathrm{m^2}$, thickness of the window: $3\,\mathrm{mm}$.

(b) In order to reduce the heat loss, the window glass is replaced with a double glazed window consisting of two window glasses of 2 mm thickness separated by an air layer of 2 mm in thickness. Taking the thermal conductivity of the air layer as $\lambda = 0.025\ \mathrm{J \cdot s^{-1} \cdot m^{-1} \cdot K^{-1}}$ and assuming that eq(82) is valid to each layer, find the temperature at each interface when the heat conductance has reached a stationary state. Find also to what extent the heat disposed outside is reduced as compared with (a).

(c) When the thickness of the air layer is tripled to 6 mm, to what extent is the heat flow reduced further?

Problem 3-119

For a two-component system composed of pure substances A and B, their partial molar volumes $\bar{\mathrm{v}}_{\mathrm{A}}$ and $\bar{\mathrm{v}}_{\mathrm{B}}$ satisfy the following *Gibbs-Duhem* equation.

$$n_{\mathrm{A}} d\bar{\mathrm{v}}_{\mathrm{A}} + n_{\mathrm{B}} d\bar{\mathrm{v}}_{\mathrm{B}} = 0 \tag{83}$$

Show that the above *Gibbs-Duhem* relation is really satisfied by obtaining $d\bar{\mathrm{v}}_{\mathrm{A}}$ and $d\bar{\mathrm{v}}_{\mathrm{B}}$ directly from the functions defining $\bar{\mathrm{v}}_{\mathrm{A}}$ and $\bar{\mathrm{v}}_{\mathrm{B}}$.

Problem 3-120

Theoretically and thermodynamically defined efficiency of work for a solar cell.

According to *Planck*'s law of black body radiation, the distribution of quantum states of wave number $\bar{\nu}$ is given by eq(84). Therefore, the energy distribution is expressed by eq(84'), which is derived by multiplying eq(84) by $h\nu\,(= hc\bar{\nu})$ (see **problem 5-67a**).

$$n(\bar{\nu}) = 8\pi\bar{\nu}^2 \cdot \frac{1}{e^{hc\bar{\nu}/k_B T} - 1} \tag{84}$$

$$\varepsilon(\bar{\nu}) = 8\pi hc\bar{\nu}^3 \cdot \frac{1}{e^{hc\bar{\nu}/k_B T} - 1} \tag{84'}$$

(a) If all of the radiation in the range of wave number of $15000 - 30000\ \mathrm{cm^{-1}}$ emitted from the sun can be converted into electric current by means of a solar cell, then find what percentage of the total radiation energy from the sun light is converted. Use eq(84') to answer the question.

(b) Is it possible to improve the conversion efficiency significantly by extending the range of the effective wave number usable for the solar cell?

(c) Supposing that there exists an ideal heat engine which works between the sun and the earth, find its efficiency of work.

Problem 3-121

As has been seen in many cases, a quantity of heat δq flowing in or out of a system is not the quantity conserved upon a change in the state of the system. However, the quantity derived by dividing it by *T*, i.e., $\delta q/T$ is a total differential conserved against any thermodynamical change. Prove this based on the first law of thermodynamics $\delta q = dU + pdV$. Such a role as *T* plays is called "integrating factor".

Problem 3-122

Concerning the *Carnot* cycle (see **3-11b**), find a change in the *Gibbs* free energy for each step of the cycle. Then, confirm that the maximum work for the cycle is equal to the sum of the changes in the *Gibbs* free energy for the isothermal reversible processes of the cycle and answer why it is called free energy.

Problem 3-123

(a) The *Gibbs* free energy is *Euler*'s homogeneous function of the first-order. Show that the chemical potential of the chemical species i, which is given by eq(85), is *Euler*'s homogeneous function of the zeroth-order. Intensive variable in thermodynamics are the zeroth-order function.

$$\mu_i(T, P, n_1, n_2, \cdots n_l) = \left(\frac{\partial G}{\partial n_i}\right)_{T,P,n_j,j\neq i} \qquad (85)$$

(b) Although the chemical potential is an intensive variable, it depends on components of chemical species. As is easily imagined, just by changing the concentration of the species i a little, the chemical potentials of all the other species vary. Therefore, the left side of eq(86) cannot be equated to zero.

$$\left(\frac{\partial \mu_i}{\partial n_i}\right)_{T,P,n_j,j\neq i} \neq 0 \qquad (86)$$

Show that when the mole number of each chemical species varies eq(87) is satisfied.

$$\sum_{i=1}^{l} n_i \left(\frac{\partial \mu_i}{\partial n_1}\right)_{T,P,n_2,n_3,\cdots,n_l} = 0 \qquad (87)$$

(c) Let us assume that the chemical potentials of a regular solution composed of components A and B are given by eq(88) with μ_{A}^{non} for A and μ_{B}^{non} for B, in which α is a constant (superscripts *non* standing for *non*-ideal). Then, show that the chemical potentials for this regular solution satisfy the relation given by eq(87) above.

$$\mu_{\mathrm{A}}^{non} = \mu_{\mathrm{A}}^{\circ} + RT\ln X_{\mathrm{A}} + \alpha \cdot X_{\mathrm{B}}^2 , \qquad \mu_{\mathrm{B}}^{non} = \mu_{\mathrm{B}}^{\circ} + RT\ln X_{\mathrm{B}} + \alpha \cdot X_{\mathrm{A}}^2 \qquad (88)$$

Problem 3-124

One of the well-known thermodynamic inequalities is $\left(\frac{\partial P}{\partial V}\right)_T < 0$. However, eq(89) holds for the critical state of a pure substance, as is confirmed for the *van der Waals* equation of state.

$$\left(\frac{\partial P}{\partial V}\right)_T = 0, \left(\frac{\partial^2 P}{\partial V^2}\right)_T = 0 \qquad (89)$$

Therefore, to check whether the critical state is a stable one or not, it is necessary to know the signs of differential terms of the order higher than the second one. In fact, for it to be a stable one, eq(90) must hold.

$$\left(\frac{\partial^3 P}{\partial V^3}\right)_T < 0 \qquad (90)$$

For the *van der Waals* equation of state, show that eq(90) holds at the critical point.

Problem 3-125

When pure liquid substances 1 and 2 are mixed, whether they form a stable homogeneous phase or separate into two phases can be predicted from the sign of the higher-order terms that affect the free energy change upon mixing. Denoting the relevant mole fractions of the mixture as X_1 and X_2, and chemical potentials as μ_1 and μ_2, the first derivative of each chemical potential derived by differentiating the chemical potential by the counterpart mole fraction (second derivative of the free energy) are expressed as are given in eq(91).

$$\mu_{12}=\left(\frac{\partial \mu_1}{\partial X_2}\right)_{T,P}, \quad \mu_{21}=\left(\frac{\partial \mu_2}{\partial X_1}\right)_{T,P} \qquad (91)$$

Practically, we can determine the state of the mixture knowing if the derivative of each chemical potential defined in eq(91) satisfy the following inequality.

$$\mu_{12}=\mu_{21}<0 \qquad (92)$$

(a) Show that eq (92) always holds for ideal solutions (not necessarily two components) whose chemical potentials are given by $\mu_1=\mu_1^\circ+RT\ln X_1$ and $\mu_2=\mu_2^\circ+RT\ln X_2$ as a function of mole fraction.

(b) In a two-component regular solution, the respective chemical potentials are represented by eq(93).

$$\left.\begin{aligned}\mu_1&=\mu_1^\circ+RT\ln X_1+\alpha X_2^2\\ \mu_2&=\mu_2^\circ+RT\ln X_2+\alpha X_1^2\end{aligned}\right\} \qquad (93)$$

When pure substances 1 and 2 are mixed with mole fractions X_1 and X_2, respectively, represent the region as a *T-X* diagram where the solution phase does not remain homogeneous stably but separates to form two solution phases.

Problem 3-126

Three substances *o*-, *m*-, and *p*-dinitrobenzene, which form homogeneous solutions above a certain temperature at ambient pressure, have melting points of 390.1, 363.0, and 446.7 K and heats of fusion of 16.33, 17.60, and 14.00 $\mathrm{kJ\cdot mol^{-1}}$, respectively It is assumed that the *Clausius-Clapeyron* equation derived for the liquid-vapor equilibrium holds formally for the dissolution equilibrium of each substance into one of the others (melting point depression).

(a) Draw *T-X* diagram (phase diagram) for the mutual solubility of *o*- and *p*-dinitrobenzene and find the temperature and compositions at the eutectic point.

(b) Based on the idea developed in (a), explain how the eutectic point and corresponding compositions of the three substances when all of them coexist in solid-liquid equilibrium can be predicted.

Problem 3-127

The following problems are concerned with the expression called *Clausius*' inequality, which is important in thermodynamics to consider reversible and irreversible processes. The objective is to deepen your understanding of irreversible, quasi-static, and reversible processes. Note that the *Clausius*' inequality is expressed as in eq (94) and eq (94') (see **"note 1"** in **solution 3-127**), but it is applicable only to cyclic closed processes. T is the temperature of the heat source associated with the process and δq is the thermal energy entering or leaving the heat source, and its sign (positive or negative) is determined with reference to the system performing the cyclic process. The notation $\oint$ represents the integral along the cyclic path when an infinite number of intervening heat sources are involved in the cycle. Equality indicates that the cyclic process is reversible, while inequality indicates that a part or whole of the cyclic process is irreversible.

$$\sum_{i=1}^{n}\frac{q_i}{T_i}\le 0 \qquad (94), \qquad \oint\frac{\delta q}{T}\le 0 \qquad (94')$$

(a) Consider two objects with temperatures T_1 and T_2, respectively, with T_1 higher than T_2 as is illustrated in

Fig.21. When these two objects are simply brought into thermal contact without any intervention, heat flows irreversibly from the hotter object to the cooler object, but can you show that this is an irreversible process on the basis of the *Clausius*' inequality? Assume that the heat capacity of the object is sufficiently large that the temperature of the object does not change as heat flows in or out.

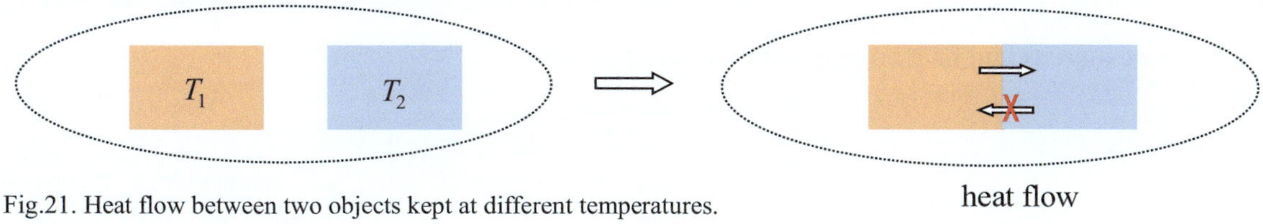

Fig.21. Heat flow between two objects kept at different temperatures.

(b) Consider a *Carnot* cycle in which an engine (an ideal gas in a piston-cylinder) that converts heat into work by performing two isothermal and two adiabatic processes (see **solution 3-11a**) is interposed between two heat sources, and thermal energy is obtained from the hotter T_1 heat source and a part of it is converted into work energy, while the other is discarded as heat to the colder heat source. A schematic diagram is shown in Fig.22. In a reversible *Carnot* cycle, what is the relationship between the amount of heat entering and leaving the heat engine upon one cycle and the temperatures of the heat sources?

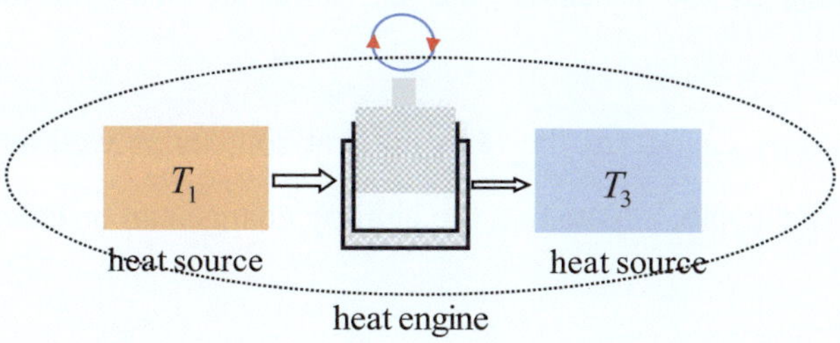

Fig.22. High (T_1) and low temperature (T_3) heat sources and a reversibly operating heat engine.

(c) Consider a heat engine in which heat transfer from or to a heat source occurs irreversibly. The heat engine receives heat at temperature T_1' lower than the temperature T_1 of the high-temperature heat source, and after reaching the temperature T_3' through isothermal and adiabatic expansion processes (see (b)), the heat engine again irreversibly releases heat to the low-temperature heat source at T_3. In other words, as illustrated in Fig.23 below, when the heat transfer processes to and from the heat engine are irreversible, what relationship can be established between the heat energy flowing in and out the heat engine and the temperatures of the external heat sources. Answer whether *Clausius*' inequality holds for this case.

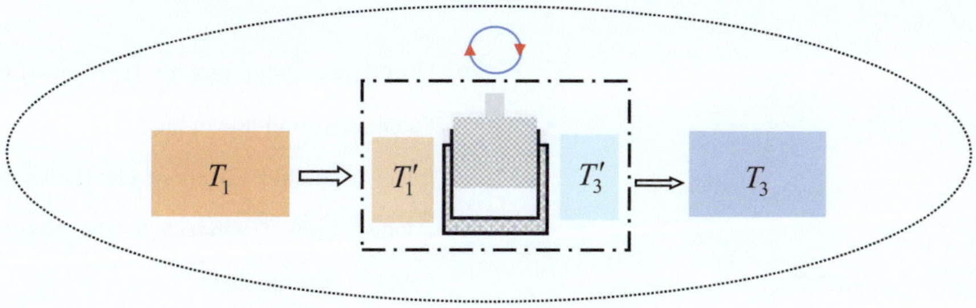

Fig.23. Heat sources and heat engine. There exists temperature diff- erence between external heat sources and heat engine working between them.

(d) Assume that the heat transfer between the heat engine shown in Fig.24 and either of the heat sources is reversible upon performing the *Carnot* cycle. However, the isothermal expansion and compression processes related to work energy are assumed to be irreversible (processes ① and ③ in Fig.25). For simplicity, we also assume that the adiabatic processes (② and ④) of the cycle are reversible. Given such a circulatory process, what is the relationship among the amounts of heat entering and leaving the external heat sources and heat engine and their temperatures?

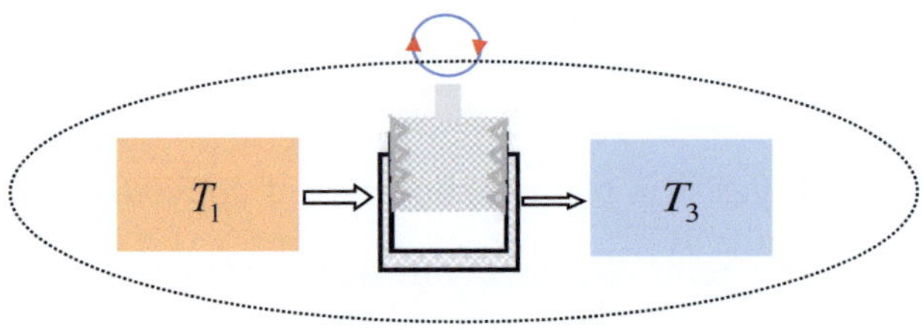

Fig. 24. *Carnot* cycle with irreversible processes.

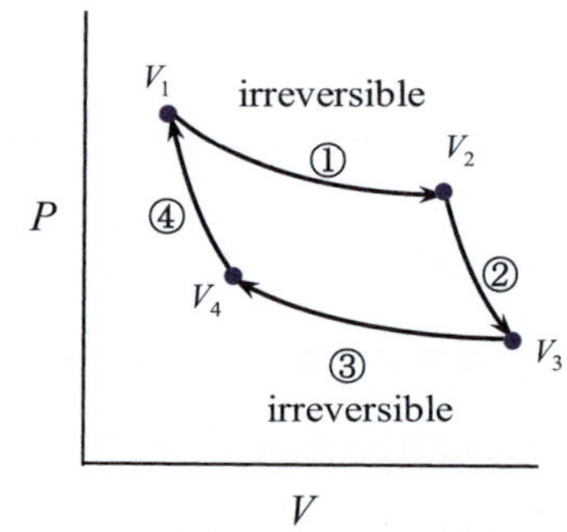

Fig. 25. Four processes of the *Carnot* cycle.

Irreversible processes are different from the paths represented by the solid lines shown in Fig.25, but the starting (initial) and ending (final) points of the processes are the same as those for the reversible processes.

(e) For each case of (b)-(d), answer the entropy change of the relevant heat sources as well as the entropy change of the relevant heat engine after one cycle. Note that the entropy change can be calculated by eq (95) only for reversible processes.

$$\Delta S = \int_{initial}^{final} \frac{\delta q}{T} \qquad (95)$$

Solution 3-128

Find the work efficiency of the reversible cyclic process of a heat engine whose cycle is shown by the closed solid curve in blue in Fig.26. Here, this cyclic process is tangent to each process of the reversible *Carnot* cycle working between a high-temperature heat source at T_1 and a low-temperature heat source at T_3 at four points as is shown by the dotted line in Fig.26.

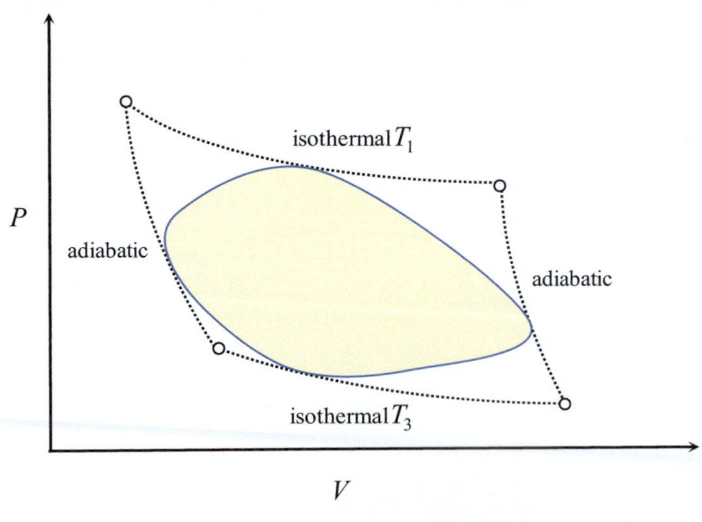

Fig.26. Reversible cyclic process represented by a circular solid line in blue.
The reversible *Carnot* cycle shown by dotted lines is tangent to the circular cycle at four points.

I-4 Equilibrium in Solutions and Interfaces
Part of Problems

Problem 4-1

The dissociation constants of sulfuric acid are given as $K_1 = 10^2$ (first dissociation) and $K_2 = 2 \times 10^{-2}$ (second dissociation). Find how many moles of undissociated H_2SO_4 exist in 100 mL of a 0.1 N aqueous sulfuric acid solution.

Problem 4-2

An aqueous solution containing a certain amount of a salt, which is composed of a weak acid and its strong conjugate base, is known to exhibits a *buffer action* against pH($= -\log_{10}[H^+]$) change. The following problems are related with this buffer action.

(a) Find pH of the aqueous solution containing acetic acid of 0.1 mol/L and sodium acetate of 0.1 mol/L. Assume the dissociation constant of acetic acid K_a is 1.8×10^{-5}.

(b) When the mixing ratio of acetic acid to sodium acetate is varied from 0.1 to 5.0, how does the pH value vary against the mixing ratio? The concentration of acetic acid is kept constant at 1.0 mol/L.

(c) Explain how the buffer action is related to the dissociation constant of the acid (or base) and the concentration of its conjugate salt.

Problem 4-3

(a) When 10 mL of an aqueous solution of the normality of 0.1N of strong acid HCl is titrated with an aqueous solution of the normality of 0.1N of strong base NaOH, answer how the pH value of the solution being titrated varies with the titrant volume of NaOH by drawing precisely the relation between pH and titrant volume (titration curve).

(b) When 10 mL of an aqueous solution of the normality of 0.1N of weak acid CH_3COOH is titrated with an aqueous solution of the normality of 0.1N of strong base NaOH, answer how the pH value of the solution being titrated varies with the titrant volume of NaOH by drawing precisely the relation between pH and titrant volume (titration curve). Use the value of $K = 1.8 \times 10^{-5}$ for the dissociation constant of acetic acid.

Problem 4-4

Highly pure solid of Na_2CO_3 is specified as the primary standard substance on the basis of JIS (Japanese Industrial Standards). When 10 mL of an aqueous solution of Na_2CO_3 of the concentration of $0.1\,\mathrm{mol \cdot L^{-1}}$ is titrated by an aqueous solution of HCl of the concentration of 0.1N, find how the pH value of the solution being titrated varies with the amount of the added HCl solution x mL by drawing an expected titration curve.

Problem 4-5

With regard to the preceding problem, find approximate functions to describe the pH values at the first and second equivalence points in terms of the corresponding dissociation constants of K_1 and K_2. Also, evaluate

S. Hirayama and M. Yamamoto, *Physical Chemistry Problems and Solutions*,
https://doi.org/10.1007/978-981-97-0165-0_4

pH values for $K_1 = 4.3 \times 10^{-7}$ and $K_2 = 5.6 \times 10^{-10}$ at the two equivalence points.

Problem 4-6

Consider a mixed solution prepared from 10 mL of $0.1\,\text{mol} \cdot \text{L}^{-1}$ NaOH aqueous solution and 10 mL of $0.1\,\text{mol} \cdot \text{L}^{-1}$ Na_2CO_3 aqueous solution. When this mixed solution is titrated with a 0.1N HCL aqueous solution (titrant), find how the pH value of the solution being titrated varies with the volume x mL of the titrant added by deriving a function to give the pH value. Draw the function against x. A method to determine the concentration of each individual component in solution by means of the titration is called differential titration (in the present case, NaOH and Na_2CO_3).

Problem 4-7

To 100 mL of a buffer solution containing 0.1 mol/L acetic acid and 0.1 mol/L sodium acetate, 1.0 mL of 0.1 N HCl solution was mistakenly added. Then what would be the pH value of the solution thus obtained?

Problem 4-8

Blood of a healthy person is kept at a constant pH due to several systems working as buffer solution. Their greatest action is ascribed to a system composed of carbonic acid and salt of hydrogen carbonic acid (H_2CO_3/HCO_3^-) in the blood, whose pH is in the range $\text{pH} = 7.35 - 7.45$.

(a) Taking the dissociation constant of carbonic acid as $K_a = 4.3 \times 10^{-7}$, find an appropriate range of the ratio of carbonic acid to salt of hydrogen carbonic acid in order to keep blood normal with respect to pH.

(b) Capacity of a buffer action in general can be estimated by knowing to what extent the pH value varies when an extra amount of acid or base is added to the solution under examination. When 0.1mL of 0.1N acid aqueous solution was added to 1.0 ml of blood of a healthy person, the pH value of the blood decreased by 0.1. Answer in what range the concentration of carbonic acid in the blood will be predicted.

Problem 4-9

A pH meter is an apparatus to find the concentration of hydrogen ion in solution by measuring an electromotive force between a reference electrode and a hydrogen electrode (see **problems 4-50** and **4-51** for further details). Find the electromotive force when the concentration of hydrogen ion is $5.0 \times 10^{-5}\,\text{mol} \cdot \text{L}^{-1}$. If the activity coefficient of hydrogen ion is not unity but is 0.9854, then how large will be the electromotive force and what will be the pH value?

Problem 4-10

(a) Hydrochloric acid (HCl) is fully ionized in aqueous solutions. Find the pH value of an aqueous solution of HCl at the concentration of 10^{-8} M. Assume the activity coefficient of H^+ in the solution is unity.

(b) The dissociation constant of acetic acid (CH_3COOH) is $K_a = 1.8 \times 10^{-5}$. Find the pH value of an aqueous solution of CH_3COOH at the concentration of 10^{-8} M. Assume the activity coefficient of H^+ in the solution is unity.

Problem 4-11

(a) The solubility product K_{sp} of silver chloride AgCl in aqueous solution is known as 1.6×10^{-10} at 25°C under 1.0 atm. Assuming that the density of the solution is unity, find the saturated concentration of aqueous AgCl solution. Its solubility produce is expressed as $K_{sp}=[Ag^{+}]\times[Cl^{-}]$ (see **problem 4-13**).

(b) The solubility product K_{sp} of calomel Hg_2Cl_2 (mercurous chloride) is given by 2.04×10^{-18}. Find the saturated concentration of aqueous Hg_2Cl_2 solution.

(c) How many grams of AgCl or Hg_2Cl_2 can be dissolved to the maximum in 1.0 L of an aqueous solution of sodium chloride of the concentration of $0.01\ \text{mol}\cdot\text{L}^{-1}$ at 25°C under 1.0 atm? Assume that the density of the solution is unity and is invariant upon dissolution.

Problem 4-12a

Obtain the solubility products of four silver halides, AgF, AgCl, AgB and AgI, from their standard reduction potentials, which are usually found in a data section of a standard text book of physical chemistry.

Problem 4-12b

The solubility products of four silver halides, AgF, AgCl, AgB and AgI, are known to be 0.44, 1.6×10^{-10}, 4.7×10^{-13} and 7.9×10^{-17}, respectively. Find the saturated concentration in aqueous solution for each salt and explain why AgF alone is practically soluble in water as compared with the other three salts. What do you consider about the solubilities of $AgNO_3$ and Ag_2SO_4 in water?

Problem 4-13

Silver chloride is a salt hard to be dissolved in water. Its dissolution equilibrium in water is expressed by eq(1). Find the solubility product of AgCl in water $K_{sp}=[Ag^{+}]\cdot[Cl^{-}]$ at 298 K from the following two approaches.

$$AgCl(s) \rightleftarrows Ag^{+}(aq)+Cl^{-}(aq) \qquad (1)$$

(a) Break down the reaction in eq(1) into two or three half-cell electrode reactions first and then use their standard electromotive forces.

(b) Break down the reaction in eq(1) into two or three elementary reactions first and then use their standard free energies of formation.

Problem 4-14

Concentrations of halide ions (Cl^{-}, Br^{-} or I^{-}) in water can be measured by the precipitation titration. When 10 mL of $0.10\ \text{mol}\cdot\text{L}^{-1}$ NaCl aqueous solution is titrated with $0.10\ \text{mol}\cdot\text{L}^{-1}$ $AgNO_3$ aqueous solution, the concentrations of Cl^{-} and Ag^{+} in the solution being titrated vary with the amount x of the $AgNO_3$ solution added. Find these concentrations as a function of x using the solubility product K_{sp} of AgCl given as 1.6×10^{-10}. Based on the results obtained, describe how the equivalence point is detected experimentally.

Problem 4-15

When 20 mL of an aqueous solution containing $0.05\ \text{mol}\cdot\text{L}^{-1}$ of NaCl and $0.05\ \text{mol}\cdot\text{L}^{-1}$ of NaBr is titrated with a $0.10\ \text{mol}\cdot\text{L}^{-1}$ $AgNO_3$ aqueous solution (titrant), the concentrations of Cl^- and Ag^+ in the solution being titrated vary with the titrant amount x mL. Find these ion concentrations as a function of x. The values for the solubility product K_{sp} of AgCl and AgBr are 1.6×10^{-10} and 4.7×10^{-13}, respectively.

Problem 4-16

20 mL of an aqueous solution containing $0.05\ \text{mol}\cdot\text{L}^{-1}$ of NaCl and $0.05\ \text{mol}\cdot\text{L}^{-1}$ of NaBr was titrated with a $0.10\ \text{mol}\cdot\text{L}^{-1}$ $AgNO_3$ aqueous solution. When exactly 10 mL of the $AgNO_3$ solution was added, find how many grams of solid AgCl and AgBr were precipitated in the solution. For AgCl and AgBr, the solubility products are 1.6×10^{-10} and 4.7×10^{-13}, and the formula weights are 58.45 and 102.9, respectively.

Problem 4-17

(a) For the depression of the freezing point ΔT_m of an aqueous solution of NaCl, the values given below are experimentally obtained. Assuming that the aqueous solution of NaCl behaves ideally, find the freezing point of the solution at each concentration listed in the table. The cryoscopic constant K_f in the present case is 1.853.

m / mol · kg^{-1}	0.01	0.02	0.05	0.1	0.20	0.5
ΔT_m / K	0.0361	0.0714	0.1758	0.3470	0.6850	1.677

(b) Sea water in a certain district starts to freeze at -0.5°C. How many grams of salts, converted to NaCl, are contained in this sea water? Assume its relative density is 1.0.

Problem 4-18

Based on *Debye-Hücke*'s limiting law, find the values for the mean activity coefficients $\gamma_\pm$ of aqueous solutions of KCl and $CaCl_2$ both at the three concentrations of 0.001, 0.01 and $0.05\ \text{mol}\cdot\text{L}^{-1}$ at room temperature. The density and the specific dielectric constant of any solution are assumed to be $d=1.0\times10^3\,\text{kg}\cdot\text{m}^{-3}$ and $\varepsilon_r=80.3$, respectively.

Problem 4-19

The activity of the ions in an aqueous solution of a 1:1 strong electrolyte MX with a unit charge for each ion is given by eq(2) under the assumption of *Debye-Hückel*'s limiting law.

$$a_{M^+}=[M^+]\times\gamma_\pm\,,\quad \gamma_\pm=e^{-1.172\sqrt{[M^+]}}\,,\quad Z_+=-Z_-=1\,,\quad I=[M^+]/\text{mol}\cdot\text{kg}^{-1} \qquad (2)$$

(a) Find the activity of the water a_{H_2O} in that solution based on the *Gibbs-Duhem* relation and express it as a function of the ionic strength I.

(b) Based on the function obtained in (a), find a function which describes the depression of the freezing point of the aqueous solution, evaluate from which the freezing points at the concentrations of 0.01, 0.02, 0.05, 0.1, 0.2 and $0.5\ \text{mol}\cdot\text{kg}^{-1}$, and compare the results with the experimentally observed values cited in **problem 4-17**.

Problem 4-20

Employing *Debye-Hückel*'s limiting law as well as the extended *Debye-Hückel* theory, find functions to express the relation between the freezing point depression and the concentration of a 1:1 strong electrolyte MX in water and compare the result with the experimental values obtained for NaCl solutions given in **problem 4-17**.

Problem 4-21

Strictly speaking, even the solubility product K_{sp} of an almost insoluble 1:1 salt MX must be defined by using the activity of the ions. Show that there exists a relation of $S = K_{sp}^{1/2} e^{1.172 S^{1/2}}$ between the solubility S ($= [M^+] = [X^-] (mol \cdot kg^{-1})$) and the solubility product.

Problem 4-22

The dissociation constant K_a of acetic acid is known as 1.75×10^{-5} at $25\,^\circ C$. Find the degree of dissociation of an aqueous solution of acetic acid at this temperature at the concentration of $1.0 \times 10^{-3}\,M$ by assuming that *Debye-Hückel*'s limiting law is applicable. Show also how largely the obtained result deviates from that obtained when ion-ion interactions are neglected.

Problem 4-23

What are the important points to be taken into account when you predict the freezing point of a dilute vinegar (dilute aqueous solution of acetic acid) very accurately? List them up.

Problem 4-24

Evaluate the equilibrium constant K of the disproportionation reaction shown below in eq(3) at 298 K in the two ways based on: (a) standard electrode potential and (b) standard free energy of formation.

$$2Cu^+(aq) \rightleftarrows Cu(s) + Cu^{2+}(aq) \quad \text{: disproportionation reaction} \qquad (3)$$

Problem 4-25

Silver chloride AgCl is an almost insoluble salt and its solubility product $K_{sp} = [Ag^+] \cdot [Cl^-]$ is 1.7×10^{-10} at $25\,^\circ C$. In order to increase the solubility (concentration), the temperature of the solution was raised to $80\,^\circ C$. Estimate how many times the solubility increased as compared with that at $25\,^\circ C$. How about a solubility increase expected for rather soluble $AgNO_3$ under the same temperature conditions?

Problem 4-26

(a) Although benzene is usually considered insoluble in water, it actually solves in water a little and vice versa. Show that the mutual solubility of benzene vs. water assumes a constant value at a constant temperature and pressure.

(b) When benzoic acid is dissolved in a two-phase separated solution of benzene and water, the acid is dispersed into the two phases and it dissociates partially into ions in a water-rich phase (eq(4)) and, on the other hand, in a benzene-rich phase, it forms partially an associated dimmer as is expressed by eq(5).

$$C_6H_5COOH \rightleftarrows C_6H_5COO^- + H^+ \qquad (4) \qquad \text{degree of dissociation}: \alpha = 0.170$$

$$2C_6H_5COOH \rightleftarrows (C_6H_5COOH)_2 \qquad (5) \qquad \text{association constant}: K_{assoc} = ?$$

The amounts of benzoic acid in each of the water and benzene phases were measured experimentally and were found as 0.0214 g/100 mL and 0.0724 g/100 mL, respectively. Employing 0.705 for the distribution coefficient ($[C_6H_5COOH](water)/[C_6H_5COOH](benzene)$), find a value for the association constant K_{assoc} of benzoic acid in the benzene-rich phase.

Problem 4-27

Based on the phase rule of *Gibbs*, answer the number of degrees of freedom for the following equilibrium systems first and then answer what are to be kept constant in order to make the number of degrees of freedom zero.

(a) Water and its vapor in equilibrium under the vapor pressure of 1 atm.

(b) Water and its vapor in equilibrium under the atmospheric pressure of 1 atm.

(c) Equilibrium between an aqueous solution of a volatile pure substance and the vapor phase above the solution.

(d) Equilibrium between an aqueous solution of a non-volatile pure substance and the vapor phase above the solution.

(e) Chemical equilibrium in water: $A + B \rightleftarrows C$, in which A, B and C are different non-volatile chemical compounds.

(f) Formation reaction of ammonium $N_2 + 3H_2 \rightleftarrows 2NH_3$ in the presence of water.

Problem 4-28

The dissolution of a nitrogen gas into water obeys *Henry*'s law and the *Henry*'s constant κ^H of nitrogen at room temperature is given as $6.80 \times 10^7\,\mathrm{Torr}^{-1}$. Assuming that the dissolution of nitrogen into human blood is the same as that into water and that the specific density of the blood is 1.0, answer the following questions.

(a) Find the molar concentration of nitrogen dissolved in the blood under the atmospheric pressure (1.0 atm).

(b) Assuming that the total amount of blood of an adult is 5L, calculate the amount of nitrogen gas dissolved in the blood in the unit of L at the standard state.

(c) An adult person who had dived under water where the water pressure reached 10 atm and had worked there for a while had to rise suddenly to the surface for emergency. How many liters of nitrogen gas are predicted to be released into the blood of that person in total?

(d) The total amount of nitrogen gas found in (c) exceeds the minimum above which gas babbles formed in the blood block vessels so as to hinder the blood circulation (diver's disease). Point out important things in order to maintain healthy breathing while diving deep under water and to avoid such disease.

Problem 4-29

The dissolution of gaseous ammonium into water, as is shown in Fig.1 below, does not follow *Henry*'s law which predicts a linear relationship between the pressure of the gas P_{NH_3} above the solution and the mole fraction X_{NH_3} of the solute in the solution.

(a) Propose a mechanism by which the non-linearity such as shown in Fig.1 can be interpreted.

(b) The solubility of gaseous ammonium increases with decreasing temperature similarly as compared with other gases (see **problem 3-79**) which obey *Henry*'s law. Give the reason for this in the manner consistent with the mechanism proposed in (a).

	P_{NH_3} / atm	
X_{NH_3}	10°C	20°C
0.458	1.3	1.91
0.425	1.03	1.54
0.388	0.790	1.24
0.346	0.297	0.396
0.241	0.250	0.392
0.209	0.190	0.298
0.175	0.136	0.218
0.137	0.092	0.150
0.0957	0.055	0.0916
0.0735	0.0393	0.0658
0.0502	0.0251	0.0417
0.0406	0.0212	0.0328
0.0308	0.0149	0.023
0.0258	___	0.0197
0.0207	___	0.0158

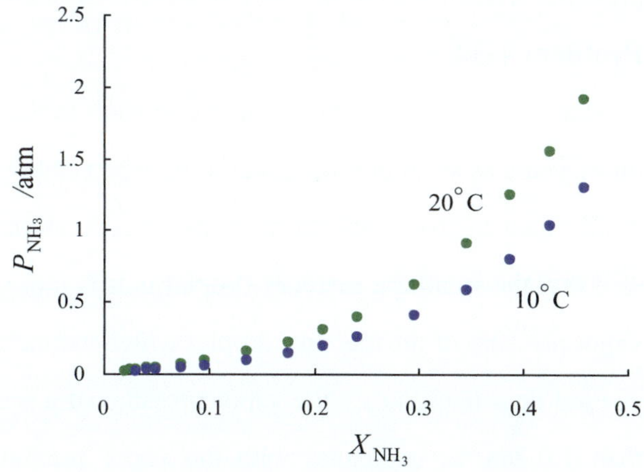

Fig.1 *P-X* plot of the data given in the table on the right.

Problem 4-30

It is well known that small droplets tend to become smaller spheres due to surface tension by reducing their surface areas as small as possible. Macroscopic balance of pressure applied on or in liquid is illustrated in Fig.2. Now let us consider a water droplet of a radius of *a* under the external pressure of P_0 as is shown in Fig.3. Under such circumstances, the pressure to which the water molecules in the droplet exposed is obviously different from P_0.

Denoting the pressure inside the droplet *P*, find a function to express the pressure *P* in terms of the radius *a* and the surface tension γ of water. Based on the found function, obtain the pressure inside a water droplet of the radius of 1.0 μm formed under the atmospheric pressure. Use a value of $\gamma = 72.0\times10^{-3}\,\text{N}\cdot\text{m}^{-1}$ for the surface tension of water.

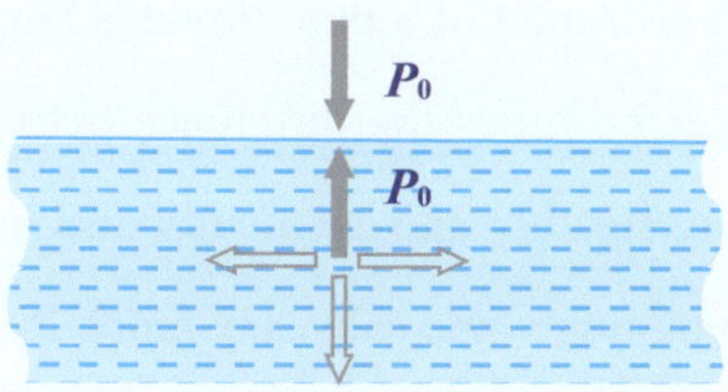

Fig.2 Pressure exerted on and in the liquid whose own weight is neglected.

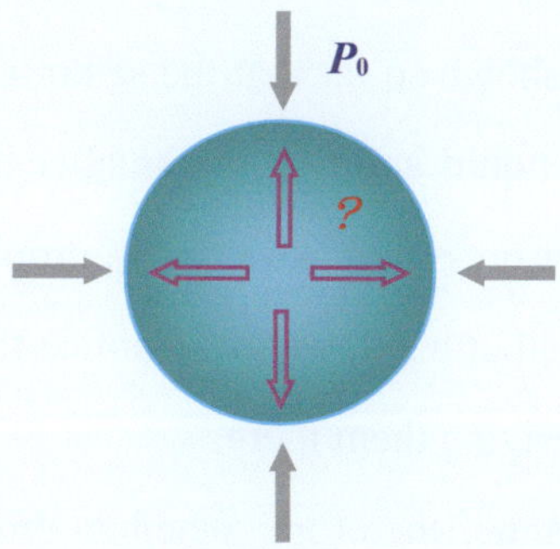

Fig.3 Pressure balance in a droplet whose surface tension cannot be neglected.

Problem 4-31

Equilibrium vapor pressures of liquids are varied by applying an external pressure. Consider a liquid whose equilibrium vapor pressure at temperature T is P_0. By applying an extra external pressure ΔP on the liquid, the equilibrium vapor pressure has changed to P'. Find a function to express the ratio of P'/P_0, assuming that the vapor gas behaves ideally and the volume of the liquid does not change with the applied pressure. Also, interpret the result obtained based on the principle of *Le Chatelier*.

Problem 4-32

Mercury is a liquid whose surface tension is the largest among liquids. In the past, it had been used for many years as an important material indispensable for a wide range of experimental works, though it is now totally banned from using it as commercial materials because of its seriously poisonous nature. It is often said that the smaller a mercury droplet is it is more poisonous. One of the reasons for this is related to that vapor pressure of the mercury droplet which increases with decreasing its size.

Find how many times the vapor pressure of a mercury droplet increases when its diameter becomes less than 1.0 μm as compared with the vapor pressure of bulk liquid mercury at room temperature under atmospheric pressure. Use $\gamma = 472 \times 10^{-3}\,\mathrm{N \cdot m^{-1}}$ for the surface tension of mercury and $\rho = 13.6\ \mathrm{g \cdot cm^{-3}}$ for its density.

Problem 4-33a

As is illustrated in Fig.4 on the right, an inert gas at the pressure P is blown into water and small bubbles of radius a are formed in water. Putting the saturated vapor pressure of water equal to P_0, find how the vapor pressure P' of water inside the bubble is related to the radius a of the bubble and the surface tension γ of water.

The result will show that $P' < P_0$. Explain in thermodynamic words why the vapor pressure inside the bubble is smaller than that above the bulk water.

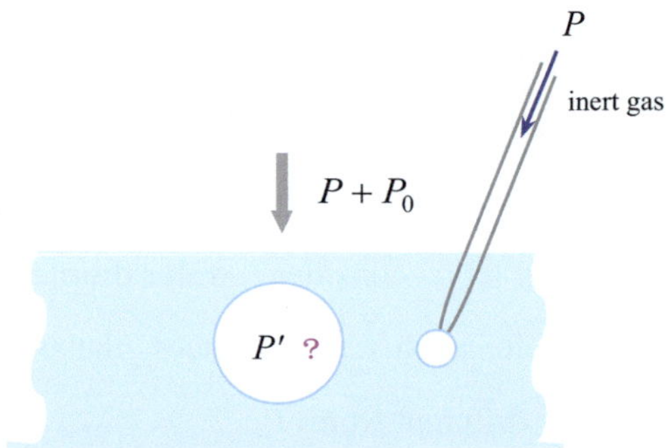

Fig.4 Gas-bubbling to form small babbles in water.

Problem 4-33b

It has been shown so far that the saturated vapor pressure over a droplet of a pure liquid is larger than that over the bulk liquid and that the smaller the droplet is the larger becomes the difference between the two saturated vapor pressures. As a result, droplets usually cannot stay steadily and eventually disappear just like mists. On the other hand, we often notice that water droplets containing some salts stay much longer or even get larger on leaving them there.

Let us examine, therefore, whether droplets of aqueous salt solution are formed stably or not in the following order by taking a salt solution of $0.02\ \mathrm{mol \cdot L^{-1}}$ at $25^\circ\mathrm{C}$ as an example.

(a) Find the saturated vapor pressure of this salt solution. The vapor pressure of bulk water at $25^\circ C$ is 60 Torr.

(b) Find the radius of a droplet of the salt solution which has the saturated vapor pressure same as that of the bulk water at room temperature.

(c) Explain why droplets of salt solutions are usually slow in disappearing or even grow larger on being left.

Problem 4-34

The saturated vapor pressure P_0 $(/\,\text{mmHg})$ of water at temperature T is given by eq(6) and it increases with temperature as is shown in Fig.5 on the right.

$$\log_{10}P_0 = 20.9586 - 4.0843 \times \log_{10}T - 2825/T \qquad (6)$$

(a) Find the radius of the smallest bubble to be formed stably existing at 300 K.

(b) How does the radius of the smallest bubble vary with temperature?

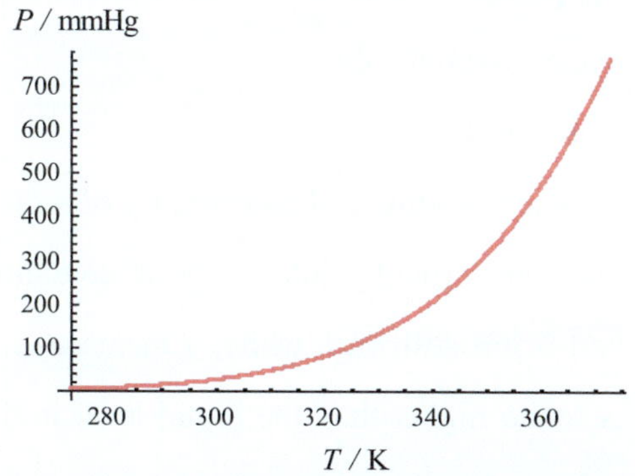

Fig.5 Variation of saturated vapor pressure of water with temperature.

(*E.A.Moelwyn-Hughes, Physical Chemistry,* 2nd revised ed., p.699, 936, 1965.)

Problem 4-35

Consider a system consisted of a pure liquid and a single bubble staying in the liquid. A change in the *Gibbs* free energy of the system under the conditions of constant temperature and pressure is expressed by eq(7), in which μ_l and μ_g are the chemical potentials of the liquid and the gas inside the bubble, n_l and n_g are the moles of the molecules consisting the liquid and the gas, and A is the surface area of the bubble.

$$dG_{T,P} = \mu_g dn_g + \mu_l dn_l + \gamma d\text{A} = 0 \qquad (7)$$

Based on this equation, find a function to describe the vapor pressure inside the bubble and then show that, from the function found, there exists the limiting smallest radius for the bubble which can stay in the liquid stably at temperature T.

Problem 4-36

So far we have seen a few cases in which micro particles whose fraction of the surface energy accounting for the total energy is largely increased exhibit various properties, for instance vapor pressures of droplets, which are significantly different from those specific to the corresponding bulks. The solubility of a solid is another example. It is worth knowing how the solubility of a pure solid substance in a solvent varies with its size when it is smashed into finer pieces. Denoting the solubility of the substance of conventional size by S_{bulk} and the solubility of its micro crystals approximated as spheres by S_{micro}, answer how the ratio of S_{bulk}/S_{micro} varies with a radius a of the micro crystals. Assume that the surface tension γ at the interface between the liquid and the particle is independent of the concentration. The solution for **problem 4-35** is helpful.

Problem 4-37

Using the results derived in the preceding problem, find how many times the saturated solubility of NaCl, when dissolved after broken into micro cubes of an edge of $1.0\,\mu m$, increases as compared with that known for a normal salt of NaCl ($356\ g\cdot kg^{-1}$). Use $72.0\,mN\cdot m^{-1}$ for the surface tension of water, $\rho = 2.17\ g\cdot cm^{-3}$ for the density of NaCl and 58.44 for the formula weight of NaCl.

In practice, however, it is considered difficult to increase the saturated solubility by making the solid finer. Give a reason for this.

Problem 4-38

When an empty glass capillary of a radius of r cm is inserted vertically into a vessel containing liquid (L) and is left still for a while, a meniscus in the capillary is found higher than the liquid level in the vessel by h cm (see Fig.6 on the right). At the interface where two different phases contact, the surface tension is working. The surface tensions at the interfaces liquid (L)-glass (S), air (V)-liquid (L) and glass (S)-air (V) are denoted γ_{LS}, γ_{VL} and γ_{SV}, respectively. The contact angle between the liquid and glass surfaces is denoted ϑ.

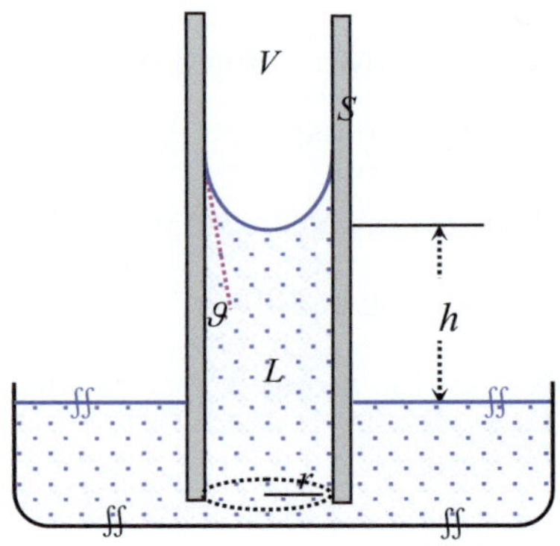

Fig.6 Illustration of capillary rise.

(a) How is the height h by which the liquid level in the capillary has risen expressed in terms of the surface tension, the contact angle and the density of the liquid? Assume that the density of the air is negligible.

(b) When the liquid is pure water, find how high the water level in the capillary of the inner radius of 0.1 mm rises by using $\gamma_{VL} = 72.0\ mN\cdot m^{-1}$ for the water-air surface tension and putting the contact angle equal to 10°. Estimate an error in the height if there exists an uncertainty of $\pm 2^\circ$ in the measured contact angle.

(c) When sea water is used instead of pure water, is the liquid level in the capillary higher or lower as compared with that for pure water?

Problem 4-39a

As is illustrated in Fig.7 below, we consider an interface between the substances A and O (for instance, water and air), on which a minute amount of the substance C (for instance, an oil lighter than water) is dropped. Then three interfaces are formed. Putting the surface tensions of these three interfaces A-O, O-B and B-A are equal to $\gamma_{AO} = 72.5$, $\gamma_{OB} = 30.0$ and $\gamma_{BA} = 52.1$ ($mN\cdot cm^{-1}$), respectively, answer which of the following shapes the substance B takes on the interface A-O. The gravitational force working on B is neglected.

(i) It spreads on the surface of A as a thin film.

(ii) It forms a floating droplet like a convex lens.

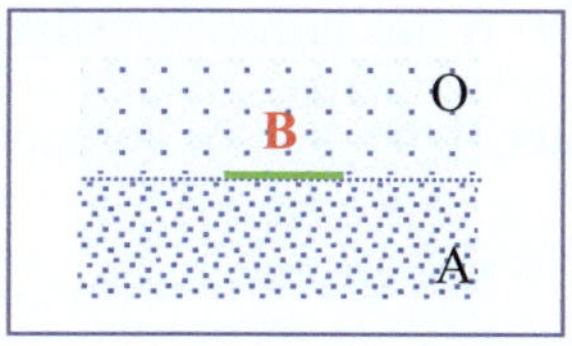

Fig.7 Membrane or lens formation in the interface of two substances.

Problem 4-39b

For the compounds listed in the table below, the values for the surface tension are measured as given there. Using these values, predict which form of a lens or a liquid film these substances assume on water surface. The surface tension of the interface of water (A) and air (O) is $\gamma_{AO} = 72.8\ \mathrm{mN \cdot m^{-1}}$.

substance B	γ_{OB}	γ_{BA}
n-octane	21.8	50.8
n-octanol	27.5	8.5
n-hexadecane	30.0	52.1
benzene	28.9	35.0
oleic acid	32.5	15.7

in $\mathrm{mN \cdot m^{-1}}$ units.

Problem 4-40

The surface tension γ of an aqueous solution of phenol was measured at 303 K as a function of the concentration of phenol and the results are listed in the table below. Here, c_2 is the concentration of phenol in the units of weight %. Base on these results, find the surface excess concentration of phenol at the solution concentration of 0.100 % (modified from *Walter J.Moore*, *Physical Chemistry,* 4th ed.).

c_2 wt %	0.024	0.047	0.118	0.471
γ / $\mathrm{mN \cdot m^{-1}}$	72.6	72.2	71.3	66.5

Problem 4-41

The surface tension of an aqueous solution of salt (NaCl) at 20 °C varies with temperature as shown in the table given below.

Table Concentration dependence of surface tension of NaCl aqueous solution.

c / $\mathrm{mol \cdot L^{-1}}$	0.496	0.963	1.922	2.829	4.523
γ / $\mathrm{mN \cdot m^{-1}}$	73.75	74.39	76.05	77.65	80.95

Data from *Theory and Problems of Physical Chemistry, C.R.Metz, Schaum's Outline Series, McGraw.*

(a) Based on these data, find the surface (excess) concentration of NaCl at each concentration given in the table.

(b) The effective thickness of the interface is defined by $|\text{[surface concentration]}|/\text{[bulk concnetration]}$. Find the thickness at each concentration.

Problem 4-42

Surface tensions of dilute aqueous solutions of organic compounds are found to be expressed satisfactorily by eq(8), where γ is the surface tension of the solution, γ_W is the surface tension of pure water and X_2 is the mole fraction of the organic compound.

$$\frac{\gamma}{\gamma_W} = 1 - 0.178 \times \ln\left(1 + \frac{X_2}{\alpha}\right) \qquad (8)$$

For the aqueous solution of *n*-valeric acid, $\alpha = 1.7 \times 10^{-4}$ is obtained. Using that value, find the surface (excess) concentration of *n*-valeric acid at 25 °C. What can be speculated from the obtained value of the surface excess concentration about the orientation of *n*-valeric acid in the interface?

n-valeric acid (pentanoic acid): $CH_3(CH_2)_3COOH$

Problem 4-43

In the close vicinity of the interface of the two phases I and II composed of such as multi-component liquid-liquid or liquid-gas phases, the distribution of each component varies from the phase I to the phase II continuously. When the interface is defined at the position where the surface excess concentration of the main component (component 1) becomes zero and, on the other hand, the surface (excess) concentration of the component 2 is defined by the equation of $\Gamma_2 = -(\partial\gamma/\partial\mu_2)_T/RT$, the latter concentration can take various values ranging from negative to positive, as is seen in **problems** from **4-40** to **4-42**.

Based on the profiles of the concentration distributions of the components 1 and 2, explain the meaning of the positive or negative sign of the surface (excess) concentration of each component.

Problem 4-44

The surface tension of the interface between the gas and liquid phases of a mixed solution varies as a function of the concentration of the mixed solution. Employing a mixed solution of water and ethanol as an example, we will try to find the surface (excess) concentration of ethanol (component 2 with mole fraction X_2) from the concentration dependence of the surface tension γ of the mixed solution. In the

X_2	γ / mN · m^{-1}	P_2 / mmHg
1.0	21.93	59.02
0.50	25.43	36.73
0.30	27.60	31.05
0.20	29.97	26.79
0.15	32.20	23.55
0.12	34.42	20.70
0.10	36.72	18.03
0.04	47.86	8.09
0.02	55.57	4.00
0	71.97	0

Partially adapted and modified from *K.S.Pitzer and L.Brewer, Thermodynamics*, p. 477.

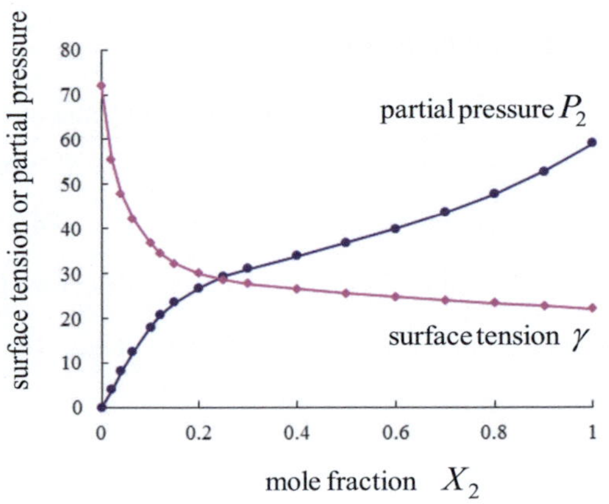

Fig.8 Surface tension or partial pressure of ethanol as a function of its mole fraction.
The data points more than those given in the table are plotted.

table given above, the values for the surface tension measured at several concentrations are listed. The values for the vapor pressure (P_2) of ethanol are also given for your reference. See Fig.8 above.

Using the values given in the table, answer the following questions.

(a) Explain the fact that the surface tension decreases with increasing concentration of ethanol from a view point of molecular interactions.

(b) Find the surface (excess) concentration of ethanol at the mole fraction of 0.1. From the bulk concentration at this mole fraction and the found surface concentration, find a volume and an area occupied by a molecule, respectively. What can be inferred from those values found?

(c) The surface concentration of ethanol decreases with increasing its mole fraction. Why is this?

Problem 4-45

(a) Show that when gaseous molecules are adsorbed on a solid surface, the surface pressure f of the adsorbed molecules is given by eq(9) based on the *Gibbs* adsorption isotherm. Here, P is the pressure of the gas, υ is the volume (converted to the gaseous state) of the molecules adsorbed on the surface area S and V_m is the molar volume of the gas. The gas is assumed to behave ideally.

$$df = -\frac{RT\upsilon}{V_m S} d\ln P \qquad (9)$$

(b) When f is expressed in some particular form, *Langmuir*'s adsorption isotherm (eq(10)) can be derived from eq(9). Find the function which satisfies that condition. Here, the fraction of surface coverage is given by $\vartheta = \upsilon/\upsilon_{max}$ and b is a constant.

$$\frac{1}{\vartheta} = 1 + \frac{1}{bP} \qquad (10)$$

Problem 4-46

For a monomolecular membrane of egg white albumin (protein), the data given in the table below are obtained. From these data, find the molecular weight of albumin as well as the surface area per molecule (modified from *Walter J.Moore*, *Physical Chemistry,* 4th ed., and *H.Bull*, *J.Amer.Chem.Soc.*, **67**, 4 (1945)).

surface pressure π / $\mathrm{mN \cdot m^{-1}}$	0.07	0.11	0.18	0.2	0.26	0.33	0.38
surface area A_s / $\mathrm{m^2 \cdot mg^{-1}}$	2.00	1.64	1.50	1.45	1.38	1.36	1.32

Problem 4-47

In the electrode reactions, it is essential to understand the fundamental electron transfer processes in detail. As an example, we consider the tunneling of an electron through a rectangular potential barrier whose height is V_0 and width is a, as is illustrated in Fig.9 on the right.

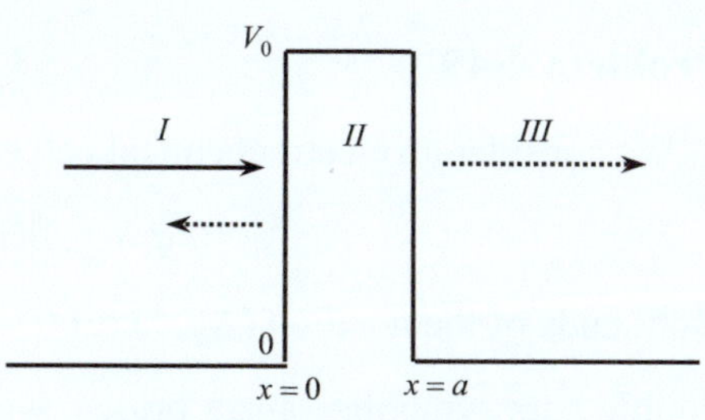

Fig.9 Tunneling of an electron through a rectangular potential barrier.

Find the function which gives the transmission probability

(coefficient) or the reflection probability (coefficient) when an electron collides with the potential barrier with the kinetic energy less than the potential height, i.e., $P^2/2m_e < V_0$. The wave function $\varphi(x)$ of the incident electron is expressed by $\varphi(x) = exp(-ikx)$ with $k = \sqrt{2mE}/\hbar$.

Also, calculate the transmission probability in the particular case of $E = 0.1$ eV, $V_0 = 1.0$ eV, $a = 0.20$ nm and $m = m_e = 9.109 \times 10^{-31}$ kg.

Problem 4-48

The transmission probability (coefficient) of a rectangular potential barrier with the height of V_0 is found approximately to be given by the following eq(16) as is derived in the solution to the previous problem.

$$\approx \exp\left(-\frac{2a\sqrt{2m(V_0 - E)}}{\hbar}\right) \quad (16) \quad \text{(reposted)}$$

In general, the potential barrier is not expressed by a rectangular function but the potential V is given as a function of the coordinate x. However, we can consider that the potential $V(x)$ is composed of a number of rectangles with different heights as is illustrated in the figure on the right and that the transmission probability through the potential $V(x)$ is given by the product of the transmission probability through each rectangular potential thus divided. See Fig.10.

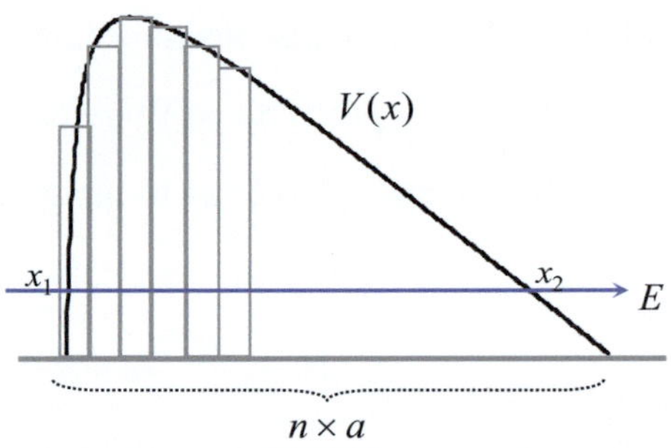

Fig.10 A potential barrier given by $V(x)$ which can be decomposed into a large number of rectangular potentials.

(a) Show that the transmission probability through the potential $V(x)$ is given by eq (11), in which x_1 and x_2 are the coordinates which satisfy the equation $(V(x) - E) = 0$.

$$\exp\left(-\int_{x_1}^{x_2} \frac{2\sqrt{2m(V(x) - E)}}{\hbar} dx\right) \quad (11)$$

(b) When $V(x)$ is given by eq(12), find the transmission probability for the electron with $E = 6\,\text{eV}$ to pass through the potential barrier. The potential employed here is of a magnitude close to an electric potential commonly found at an electrode interface.

$$V(x) = V_0 - Fx - \frac{e}{4\pi\varepsilon_0} \cdot \frac{1}{4x} = 10.0 - 5.0x - \frac{1.44}{4x} \quad (\text{V}^{-1}) \quad (12)$$

$$V_0 = 10.0\,\text{V}, \qquad F = 5.0 \times 10^7\,\text{V} \cdot \text{cm}^{-1}, \qquad \frac{e}{4\pi\varepsilon_0} = 1.44\,\text{J} \cdot \text{C}^{-1} \cdot \text{nm}$$

Problem 4-49

We consider an electrochemical cell expressed by a standard notation as shown below.

$$\text{Ag}|\text{Ag}^+(aq, a_{\text{Ag}^+})\|\text{HCl}(aq, a_{\text{Cl}^-})|\text{AgCl}(s)|\text{Ag}$$

Both ends of the notation (Ag, here) represent electrodes, the vertical bars indicate interfaces, and the vertical double bars symbolize a salt bridge, which is essential to connect two half-cells. Each part separated by the salt bridge ($\|$) is called a half-cell. The standard state reduction potentials of the half-cells of

$Ag|Ag^{+}(aq,a_{Ag^{+}})$ and $AgCl(s)|Ag$ are $E^{\circ}=0.7991$ V and $E^{\circ}=-0.2192$ V, respectively.

(a) Write the electrode reaction for each half-cell.

(b) Based on each half-cell reaction, answer the chemical reactions to occur in the electrochemical cell.

(c) From the standard state reduction potentials for the half-cells, find a value for the solubility constant for the dissolution reaction $AgCl \rightleftarrows Ag^{+}+Cl^{-}$.

Problem 4-50

We consider two electrodes, one of which is a hydrogen electrode (half-cell) written as $Pt(s)|H_2(g, f=1 \text{ atm})|HCl(aq,a_{H^+})$ by the standard notation and illustrated in Fig.11, and the other is a silver–silver chloride electrode (half-cell) written similarly as $Ag(s)|AgCl(s)|HCl(aq,a_{Cl^-})$ and is illustrated in Fig.12. The electrode reduction reaction at each electrode is given either by eq(13) or eq(14). (Each reverse reaction is the electrode oxidation reaction.) When these electrode reactions have reached equilibrium, the electrodes are at constant electric potentials of E_1 and E_2, respectively (see Figs.11 and12).

$$H^{+}+e \rightarrow \frac{1}{2}H_2 \quad (13) \qquad AgCl+e \rightarrow Ag+Cl^{-} \quad (14)$$

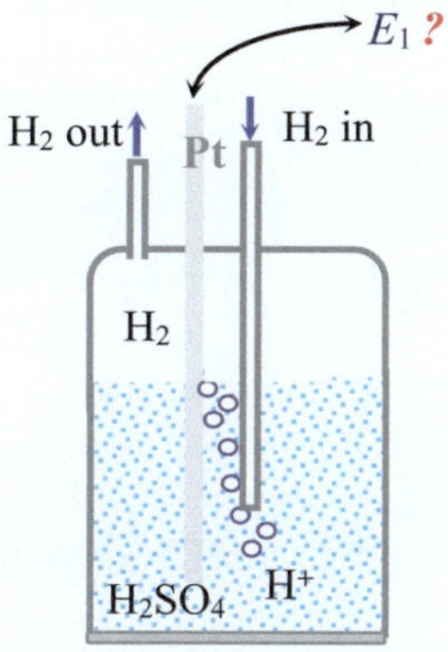

Fig.11 Illustration of H_2 electrode.

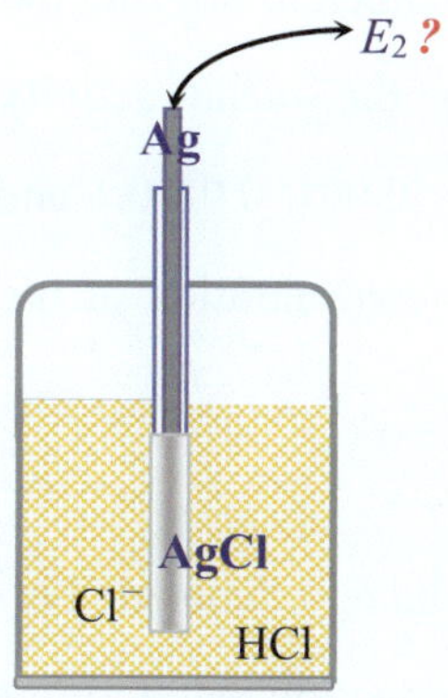

Fig.12 Illustration of Ag-AgCl electrode.

(a) Express the electric potentials of E_1 and E_2 in terms of the chemical potentials for the substances involved in the relevant electrode reactions.

(b) Find a chemical reaction by eliminating the electrons from eqs(13) and (14) and express the free energy change of the reaction ΔG_r in terms of the chemical potentials of the relevant substances.

(c) Show that ΔG_r found in (b) can be expressed as below by using the electric potentials of the half-cells E_1 and E_2 determined in (a). F is the *Faraday* constant.

$$\Delta G_r = -F(E_2 - E_1)$$

(d) Explain why it is easy to determine the potential difference of two half electrochemical cells chosen arbitrarily, though it is difficult to determine unambiguously the electric potential of a half-cell individually.

(e) The standard state electrical potential of a hydrogen electrode is set at zero. How can each of the chemical potentials of hydrogen gas, hydrogen ion and electron in the electrode be made consistent to this value set at zero? Based on this definition, the standard electric potential of any arbitrary half-cell is determined unambiguously.

Problem 4-51

Consider an electrochemical cell denoted $\mathrm{Pt(s)}|\mathrm{H_2}(g, f = 1\ \mathrm{atm})||\mathrm{HCl}(aq, m)|\mathrm{AgCl}(s)|\mathrm{Ag}$. Its electromotive force was measured at $25^\circ\mathrm{C}$ by varying the concentration of hydrochloric acid and the data obtained are given in the table below. Base on these data, answer the following questions.

m / mol · kg^{-1}	ΔE(V)	m / mol · kg^{-1}	ΔE(V)	m / mol · kg^{-1}	ΔE(V)
0.00100	0.57915	0.0200	0.43024	0.500	0.27231
0.00200	0.54425	0.0500	0.38588	1.000	0.23328
0.00500	0.49846	0.1000	0.35241	1.500	0.20719
0.0100	0.46417	0.200	0.31847	2.000	0.18631

(a) Find a function to describe the electromotive force of this cell. Answer it by using the fugacity of hydrogen and the concentration of hydrochloric acid m / mol · kg^{-1}.

(b) Using the function obtained in (a), find the value of the standard electromotive force ΔE° of this cell.

(c) Find the values for the mean activity coefficient $\gamma_\pm$ of the hydrochloric acid at the following concentrations of $m = 0.001,\ 0.01,\ 0.1$ and 1.0 / mol · kg^{-1}.

(d) Find the value of the concentration of the hydrochloric acid at which the activity of H^+ or Cl^- becomes equal to 1.0.

Problem 4-52

We consider a storage battery denoted $\mathrm{Pb}(s)|\mathrm{PbSO_4}(s)|\mathrm{H_2SO_4}(aq, m)|\mathrm{PbO_2}(s)|\mathrm{Pb}(s)$ by using the standard notation. The electrode reduction reaction to occur at the right-hand side electrode (Pb(s)) is given by eq(15) and its standard reduction potential is $E_1^\circ = 1.685$ V.

$$\frac{1}{2}\mathrm{PbO_2}(s) + 2\mathrm{H}^+ + \frac{1}{2}\mathrm{SO_4^{2-}} + e \rightarrow \frac{1}{2}\mathrm{PbSO_4}(s) + \mathrm{H_2O} \quad (15)$$

The reaction seen at the left-side electrode (Pb(s)) is given by eq(16) and its standard reduction potential is $E_2^\circ = -0.356$ V.

$$\frac{1}{2}\mathrm{PbSO_4}(s) + e \rightarrow \frac{1}{2}\mathrm{Pb}(s) + \frac{1}{2}\mathrm{SO_4^{2-}} \quad (16)$$

At the concentration of 1.0 mol · kg^{-1} of sulfuric acid, the electromotive force ΔE was measured by varying the temperature. It was found that ΔE varies with temperature as is given by eq(17), in which t is the temperature in Celsius ($^\circ\mathrm{C}$).

$$\Delta E = 1.917 + 56.1\times10^{-6}t + 108\times10^{-8}t^2 \quad (17)$$

(a) Answer which half-cell of this battery becomes a positive (or negative) electrode. Express the electrode reactions by a single net reaction and find a value for the standard electromotive force of this reaction.

(b) For that reaction, find the values of ΔG, ΔH and ΔS at $25^\circ\mathrm{C}$.

(c) A light bulb of 2V and 5W was connected to the battery and it was turned on for one hour. Find the mass of Pb consumed.

(d) To what extent has the electromotive force of the battery dropped after lighting the bulb for one hour?

Problem 4-53

The *Edison* storage battery is represented by the following standard notation.

$$\mathrm{Fe}(s)|\mathrm{FeO}(s)|\mathrm{KOH}(aq,a)|\mathrm{Ni_2O_3}(s)|\mathrm{NiO}(s)|\mathrm{Ni}(s)$$

Its two half-cell reactions are given by eqs(18) and (19).

$$\mathrm{Ni_2O_3}(s)+\mathrm{H_2O}(l)+2e \rightleftarrows \mathrm{NiO}(s)+2\mathrm{OH^-}(aq) \qquad E_1^\circ = 0.40\ \mathrm{V} \qquad (18)$$

$$\mathrm{FeO}(s)+\mathrm{H_2O}(l)+2e \rightleftarrows \mathrm{Fe}(s)+2\mathrm{OH^-}(aq) \qquad E_2^\circ = -0.87\ \mathrm{V} \qquad (19)$$

(a) Express the reactions in the battery by a single net reaction.

(b) How does the electric potential of the battery depend on the activity of KOH?

(c) From the net reaction found in (a), calculate the amount of the electrical energy for the work to be obtained per 1.0 kg of the working substance (Fe) of the battery.

Problem 4-54

The *Daniell* cell is expressed as $\mathrm{Zn}(s)|\mathrm{ZnSO_4}(aq,1\mathrm{M})||\mathrm{CuSO_4}(aq,1\mathrm{M})|\mathrm{Cu}(s)$.

Its standard electromotive force ΔE° depends on temperature and its dependence is known to be given by eq(20), in which t is the temperature in Celsius.

$$\Delta E^\circ = 1.1028 - 0.641\times10^{-3}t + 0.72\times10^{-5}t^2 \qquad (20)$$

(a) Express the reaction in the cell by a single net reaction.

(b) Using eq(19), find the values of ΔG°, ΔH° and ΔS° for the reaction in the cell at 25 °C.

(c) How is the solution of the cell to be prepared in order to achieve the electromotive force higher than the standard electromotive force ΔE°?

Problem 4-55

The *Volta* cell is a simple primary battery. Its electrodes are consisted of a zinc plate and a copper plate immersed separately in an acid aqueous solution. Thus, the cell is expressed by the standard notation as $\mathrm{Zn}|\mathrm{H_2SO_4}(aq)|\mathrm{Cu}$ (*cf.* secondary battery).

(a) Answer the electrode reactions seen at the positive and negative electrodes of this cell.

(b) What is the maximum electromotive force expected for this cell?

(c) Is it possible to turn on a LED in red or a wolfram miniature bulb with this cell?

Problem 4-56

The basic chemical reaction seen in a fuel cell which utilizes hydrogen as fuel is a combustion reaction of hydrogen gas expressed by eq(21).

$$\mathrm{H_2}(g)+\frac{1}{2}\mathrm{O_2}(g) \rightarrow \mathrm{H_2O}(l) \qquad (21)$$

(a) Find the standard electromotive force of the fuel cell of this sort using the standard free energy of formation of water.

(b) Express first the combustion reaction given by eq(21) by using the oxidation-reduction reactions of hydrogen, oxygen and water and then compare the value of the electromotive force found in (a) with that to be calculated from the standard electric potentials associated with those oxidation-reduction reactions;

(c) The heat of combustion of hydrogen is given by $\Delta H_f^\circ = -286\ \mathrm{kJ \cdot mol^{-1}}$. In comparison with the heat engine which utilizes heat energy produced by combustion, the efficiency of the fuel cell is considered higher. Give a reason for this. Defining the theoretical efficiency by $\left|\Delta G_f^\circ / \Delta H_f^\circ\right|$, answer which one has the higher efficiency, the heat engine ideally working between 1000 K and room temperature or the hydrogen fuel cell discussed here.

(d) When methanol is employed as fuel instead of hydrogen, how large is the theoretical efficiency?

Problem 4-57

Dry cells are generally divided into the two types; (a) non-rechargeable alkaline cell (primary cell) and (b) rechargeable cell (secondary cell) such as nickel-cadmium (NiCd) and lithium ion cells.

(a) As an example of a primary cell, we consider a zinc-manganese (ZnMn) cell (used to be called *Leclanche* cell, 1878) whose electrode reactions are described by eqs(22) and (23).

$$\text{(A electrode): } \mathrm{Zn(s) + 2OH^-}(aq) \rightarrow \mathrm{ZnO(s) + H_2O} + 2e \qquad \Delta E_1^\circ = 1.1\ \mathrm{V} \qquad (22)$$

$$\text{(B electrode): } \mathrm{2MnO_2(s) + H_2O}(l) + 2e \rightarrow \mathrm{Mn_2O_3(s) + 2OH^-}(aq) \qquad \Delta E_2^\circ = -0.76\ \mathrm{V} \qquad (23)$$

What are the overall reaction and the standard electromotive force of this cell? Which of A or B electrode works as a positive electrode?

(b) As an example of a secondary cell, historically speaking, NiCd cell was well known. However, because of the high toxicity of Cd, it is now replaced by hydrogen occluding alloy and its electrode reactions are typically described by eqs(24) and (25). What are the overall reaction and the standard electromotive force of this cell? Which of A or B electrode is a positive electrode? Why can this cell work as a rechargeable cell?

$$\text{(A electrode): } \mathrm{NiO(OH)(s) + H_2O}(l) + e \rightarrow \mathrm{Ni(OH)_2(s) + OH^-}(aq) \qquad \Delta E_3^\circ = -0.83\ \mathrm{V} \qquad (24)$$

$$\text{(B electrode): } \mathrm{MH(s) + OH}(aq) \rightarrow \mathrm{M + H_2O}(l) + e \qquad \Delta E_4^\circ = 0.52\ \mathrm{V} \qquad (25)$$

Problem 4-58

Values for the ion equivalent conductivity λ are given for several ions in the table below. Employing these values, solve the following problems.

	λ		λ		λ
H^+	34.96	OH^-	19.91	$CH_3CO_2^-$	4.09
Na^+	5.01	Cl^-	7.64	CO_3^{2-}	13.86
Ca^{2+}	11.90	Br^-	7.81	SO_4^{2-}	16.00

units: $\mathrm{mS \cdot m^2 \cdot mol^{-1}}$

(a) Find the conductivity of each of the following three aqueous solutions: 0.100 N hydrochloric acid HCl, 0.100 N sodium hydroxide NaOH and 0.100 M calcium chloride $CaCl_2$.

(b) When 25.0 mL of the 0.100 N HCL aqueous solution is titrated with a 0.100 N NaOH aqueous solution, find an equation which describes the conductivity of the solution being titrated as a function of the volume of the titrant added. Plot the conductivity against the volume added based on the equation found.

Problem 4-59

To know the molecular weight and the degree of polymerization of a sample of polystyrene with some degree of polymerization, the sample was dissolved in pure toluene and the osmotic pressure of the toluene solution was measured at 25 °C at a few concentrations. The obtained values are listed in the table below. Find the molecular weight and the degree of polymerization of this polystyrene. The density of any toluene solution examined can be put equal to 860 g/L.

concentration (g/L)	1.65	2.97	4.80	7.66
osmotic pressure*	0.71	1.35	2.32	4.09

*(height of liquid column /cm)

Problem 4-60

The molecular weight found in the preceding problem is the one averaged over a distribution of molecular weights. We denote the degree of polymerization i and the mean degree of polymerization $\bar{i}$. When the distribution p_i of the degree of polymerization i is given by a normal distribution function of eq(26), show that the molecular weight acquired from the measurements of the osmotic pressure is the mean molecular weight defined by eq(27).

$$p_i = A \cdot \exp\left\{-\frac{(i-\bar{i})^2}{\sigma^2}\right\} \qquad (26)$$

$$\overline{M} = \sum_i p_i M_i\,, \qquad M_i = i \times M_1\,, \qquad M_1:\ \text{molecular weight of monomer} \qquad (27)$$

Problem 4-61

In a melt consisted of components CaF_2 and $CaCl_2$, either component dissolves in the other completely in any mixing ratio. Solid samples of CaF_2 and $CaCl_2$ mixed at several mole fractions (X) were melted once completely and then gradually cooled down to measure the temperatures (t °C) at which the solidification starts and completes. The obtained data are listed in the table below.

(a) Based on these data, draw a t–X diagram and answer in what state or states the mixed substances can exist stably in each area, if separated one another, of the diagram.

(b) At the mole fraction of $X_{CaCl_2} = 0.5$, when the melted mixture is cooled down slowly from 1000 °C to 600 °C, how does a cooling curve look like? Answer the question by drawing its rough shape.

X_{CaCl_2}	start temp.	completion temp.	X_{CaCl_2}	start temp.	completion temp.
0.0	1300	1300	0.6	735	644
0.3	1050	737	0.7	700	644
0.4	950	737	0.8	644	644
0.5	820	737	0.9	710	644
0.58	737	644	1.0	774	774

Problem 4-62

In an apparatus to measure the transport numbers of ions some amount of a HCl aqueous solution at a known concentration was introduced and a constant electric current was passed through for a fixed duration between the two platinum electrodes immersed in the solution. An anode section of the apparatus was analyzed for electrolyte content. The initial amount of Cl^- was 0.117g but after the passage of the current it was found for its amount to have decreased to 0.163 g. The total amount of the current was measured with a silver coulometer connected to the apparatus in series and was found that 0.2508 g of silver metal were deposited after the passage of the current.

(a) Based on these results, find the transport numbers of H^+ and Cl^-.

(b) The molar conductivities of H^+ and Cl^- are known to be 34.96 and 7.64 $S \cdot cm^2 \cdot mol^{-1}$, respectively. Comment on this in comparison with the values found in (a).

Problem 4-63

The molar conductivity Λ_m of an aqueous solution of 0.1 M NaF was measured and found as 83.5 $S \cdot cm^2 \cdot mol^{-1}$. However, NaF does not fully dissociate into ions at this concentration. Only its 92% are dissociated. Putting the transport number of Na^+ equal to 0.45, find the mobility for both of Na^+ and F^-.

Problem 4-64

The molar conductivity of hydrogen ion is known to be 349.5 $S \cdot cm^2 \cdot mol^{-1}$ at $25\,^\circ C$. Employing a value of 8.99×10^{-3} P for the viscosity of water at this temperature, find a value for the mobility of the hydrogen ion and estimate its *Stokes*' radius.

Problem 4-65

To an aqueous solution of 0.1 M sodium acetate (CH_3COONa), a drop of a solution of phenolphthalein is added at $25\,^\circ C$ as an indicator (In). Assuming that the dissociation constant of acetic acid is $K_a = 1.8 \times 10^{-5}$ and that the dissociation constant of phenolphthalein is $K_{In} = 3.16 \times 10^{-10}$, find what percentage of the indicator is in a visible-color-presenting form in this solution.

Problem 4-66

Substances named amphoteric electrolyte (ampholyte) such as amino acids or proteins are in acid-base or association-dissociation equilibria in solution and, hence, exist usually in various chemical forms in aqueous solution. At a specific concentration of hydrogen ion, most of the amphoteric electrolytes exist in a form of zwitter-ion, which has a positive and a negative charge separated in the same molecule. The concentration of hydrogen ion at this point expressed by its pH value is called isoelectric point. Since the zwitter-ion is electronically neutral, the existence of the isoelectric point is one of the useful properties to identify an amphoteric electrolyte (e.g., by electrophoresis).

Employing appropriate equilibrium constants of the acid-base equilibria of an amphoteric electrolyte, find an equation to predict a pH value at its isoelectronic point.

Problem 4-67

Consider a mono-dispersed polymer dissolved in an organic solvent. The osmotic pressure of the solution was measured at 30 °C at several concentrations and the obtained values are summarized in the table below. Find the molecular weight of the polymer from these data.

Table Concentration dependence of the osmotic pressure of a polymer solution.

concentration c / mg · cm^{-3}	1.33	2.10	4.52	7.18	9.87
osmotic pressure Π / N · m^{-2}	30	51	132	246	390

Problem 4-68

The molar conductivity Λ_m was measured for aqueous solutions of monochloroacetic acid at several concentrations and the obtained values are given in the table below. It is assumed that mono-chloroacetic acid is in the dissociation equilibrium in aqueous solution expressed by eq(28)

$$CH_2ClCOOH \rightleftarrows CH_2ClCOO^- + H^+ \qquad (28).$$

(a) Assuming that the law of the independent migration of ions is valid, express the dissociation constant K_a in terms of molar conductivity Λ_m.

(b) Based on the data in the table, find values for the dissociation constant and the molar conductivity at the infinite dilution Λ_0.

(c) Compare the value of Λ_0 obtained in (b) with that to be calculated from the ion molar conductivity.

Table Concentration dependence of the molar conductivity of monochloroacetic acid solutions.

$1/[CH_2ClCOOH]_0$ (M^{-1})	16	32	64	128	256	512	1024
Λ_m (S · cm^2 · mol^{-1})	53.1	72.4	96.8	127.7	164	205.8	249.2

(S · cm^2 · mol^{-1} = 0.1 mS · m^2 · mol^{-1})

Problem 4-69

The concentration dependence of the molar conductivity is largely different between strong and weak electrolytes. The molar conductivity of a given electrolyte is found to depend on the concentration c as is shown in the table below. From this behavior, tell whether the electrolyte is a strong or weak electrolyte (modified from problem 27, p35 in *Thomas Engel & Philip Reid, Physical Chemistry,* 2004).

Table Concentration dependence of molar conductivity.

c (mol · L^{-1})	5.0×10^{-3}	1.0×10^{-2}	5.0×10^{-3}	1.0×10^{-2}	2.0×10^{-2}	5.0×10^{-2}	1.0×10^{-1}
Λ_m (S · cm^2 · mol^{-1})	124.5	123.7	120.7	118.5	115.8	111.1	106.7

Problem 4-70

In order to obtain the molecular weight (mass) of ribosome, the sedimentation rate of the ribosome was measured, and the following data were obtained:

diffusion coefficient: $D = 1.52\times10^{-7}\,cm^2\cdot s^{-1}$, partial molar volume: $\overline{V} = 0.61\ cm^3\cdot g^{-1}$,

density of the solution: $\rho = 1\ g\cdot cm^{-3}$ and sedimentation constant: $S = 82.6\times10^{-13}\,s$.

Find the molecular weight of the ribosome from these data (adapted from *Hans Kuhn and Horst-Dieter Försterling, Principles of Physical Chemistry, John Wiley & Sons, Ltd.,* p.789, 2000).

Problem 4-71

Amino acids, generally written as $NH_2CHRCOOH$ in chemical formula, have two functional groups of amino and carboxyl groups within a molecule. Because of this, they exhibit two-step acid-base equilibria with hydrogen ion H^+ in aqueous solution as are expressed by eqs(29) and (30).

$$H_3N^+-\overset{\overset{R}{|}}{C}H-CO_2H \rightleftarrows H_3N^+-\overset{\overset{R}{|}}{C}H-CO_2^- + H^+, \qquad K_{a_1} \qquad (29)$$

$$H_3N^+-\overset{\overset{R}{|}}{C}H-CO_2^- \rightleftarrows H_2N-\overset{\overset{R}{|}}{C}H-CO_2^- + H^+, \qquad K_{a_2} \qquad (30)$$

Denoting an amino acid with one carboxyl group AH, there exist three kinds of ionic species originating from the amino acid, H^+AH, H^+A^- and A^-. As H^+A^- has both of a negative and a positive ion within the molecule, it is called zwitter ion.

(a) 10.0 mL of a 0.1 N HCl aqueous solution containing 0.1M of asparagine ($C_4H_8N_2O_4$) is pipetted out and is titrated with a 0.1N NaOH aqueous solution. Draw a titration curve which shows the value of pH of the solution being titrated against the volume of the 0.1N NaOH aqueous solution added. The dissociation constants of asparagine are given by $pK_{a_1} = 2.02$ and $pK_{a_2} = 8.84$.

(b) Find a pH value at the isoelectric point where H^+AH and A^- exist in the solution in the same amount and also the concentration of the zwitter ion.

Problem 4-72

A few kinds of amyloses with different molecular weights were separately dissolved in a 0.33M potassium chloride aqueous solution and the intrinsic viscosity $[\eta_{in}]$ $(m^3 \cdot kg^{-1})$ of each solution was measured. The obtained data are summarized in the table below.

(a) Denoting the molecular weight *M*, fit the data to the equation $[\eta_{in}] = K \cdot M^{\alpha}$ (*Mark-Houwink-Sakurada*'s equation) and find the constants *K* and α characteristic to the amylose. The definition of the intrinsic viscosity is $[\eta_{in}] = \lim_{c \to 0} (\eta/\eta_0 - 1)/c$.

(b) For a given amylose, the intrinsic viscosity was found as $250\,m^3 \cdot kg^{-1}$. Find its molecular weight and estimate the chain length of this amylose.

Table Molecular weight and intrinsic viscosity of amyloses.

M.W. M	270	552	847	1050	1350	2220
viscosity η_{in}	60	90	115	126	152	173

Problem 4-73

A known amount of acetic acid is dissolved in a two-phase separated liquid consisting of benzene and water which do not mix each other. The acetic acid is dispersed into each phase and the system is eventually in equilibrium. The concentrations of acetic acid (M) in the two phases are measured for several different amounts of acetic acid dissolved and are summarized in the table below. In benzene, two acetic acid molecules are considered to associate partially to from a dimmer in equilibrium with the monomer molecules.

Under these conditions, find values for the coefficient K_1 for the dispersion of acetic acid into the two phases, the association constant K_{assoc} of acetic acid in benzene and its degree of association α.

Table Dispersion of acetic acid (M) into two-phase separated benzene and water.

benzene phase (M)	0.00303	0.00776	0.01551	0.0248	0.03586
water phase (M)	0.2291	0.4328	0.6661	0.8855	1.09

Problem 4-74

We consider here how the salt concentration of a sea water is related with the depth of the sea where the sea water is sampled. Since all of the potential energy, temperature, pressure, etc., vary in some ways with the depth of sea, it is an interesting issue of physical chemistry whether or not we can predict a kind of relationship between the depth of sea and the salt concentration.

Assuming that the sea water is an ideal solution, answer the following questions.

(a) Predict how the saturated concentration of a salt varies with the temperature if the temperature of sea water is assumed to vary with the depth in a known way.

(b) Contrary to (a), if the temperature of sea water is independent of the depth, how does the saturated salt concentration vary with the depth?

For your reference, two graphs are given below in Fig.13 which show the correlations of temperature-depth and salinity (salt concentration)-depth measured in a certain sea area (courtesy from Prof. *H.Obata*).

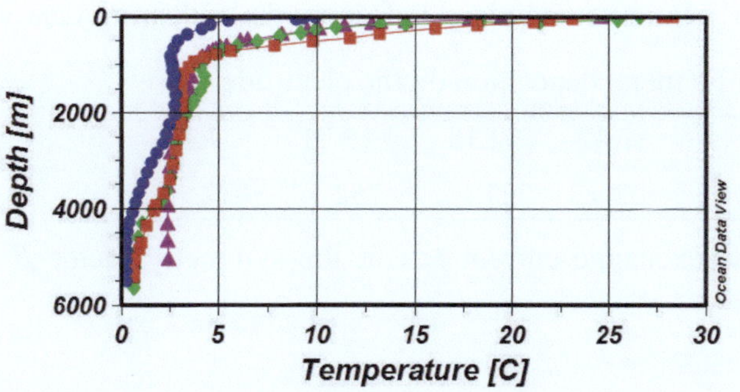

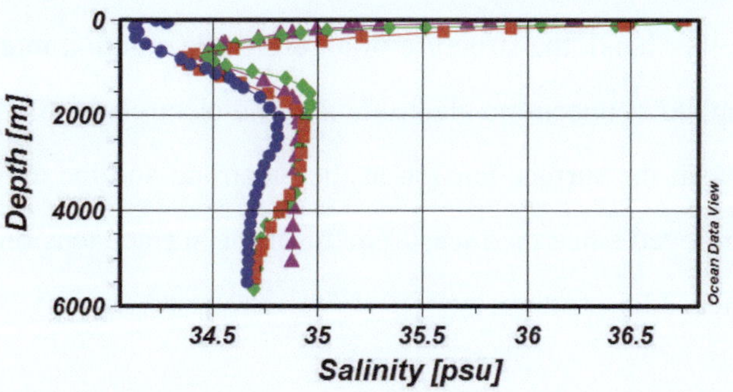

Fig.13 *H.Obata*, et al., Deep-Sea Research II, 54, 38 (2007). Courtesy of the authors.

Problem 4-75

Consider two aqueous solutions A and B, each containing an electrolyte, and the two solutions are separated by an ion-permeable membrane as is illustrated in Fig.14 given below. Between the solutions A and B there arises a stationary electric potential via membrane (membrane potential), which is different from the equilibrium potential.

Fig.14 Two solutions (A and B) separated by an ion-permiable membrane.

(a) Find the stationary potential described above when KCl is used as the electrolyte and the concentration of the solution A is 1.0 M and that of B is 0.10 M. Assume that the mobilities of K^+ and Cl^- ions are $u_{K^+} = 7.62$ and $u_{Cl^-} = 7.91$ in the units of $10^{-8} m^2 s^{-1} V^{-1}$, respectively.

(b) How largely does the separation potential change when the electrolyte is replaced by NaCl? Assume that the mobility of Na^+ is $u_{Na^+} = 5.19$ in the unit of $10^{-8} m^2 s^{-1} V^{-1}$.

Problem 4-76

For an electrode oxidation-reduction reaction expressed as $A + e \rightleftarrows A^-$, the overpotential η and reduction electric current i are measured and the values obtained are given in the table below. From the *Tafel* plot of the data in the table (η vs. $\ln i$ plot), find the exchange current density (i_0) and symmetry factor β of this electrode reaction.

overvoltage /V	0	0.001	0.005	0.01	0.02	0.03	0.05	0.1	0.2
i /A·cm^{-2}	0	0.04	0.195	0.39	0.8	1.23	2.27	6.89	49.38

Problem 4-77

Upon the electrolysis of an aqueous solution containing Fe^{2+} ions, the following data were obtained for the electrolysis potential and the rate of Fe metal deposition on the electrode.

rate of deposition / pmol · s^{-1}	1.47	2.18	3.11	7.26
electrolysis pot. E /V	-702	-727	-752	-812

From these data, find values for the exchange current i_0 and the symmetry factor β of the electrode reaction of $Fe^{2+} + 2e \rightleftarrows Fe$.

Problem 4-78

It is obvious that the surface tension inherent to an interface generally varies depending on the components and structures of the interface. On the other hand, the structure of an electrode-solution interface is expected to change with the electric potential applied between the electrode and the electrolyte solution. Therefore, it is important to clarify the relation between the surface tension at the electrode and the applied voltage. For this purpose, a mercury electrode is employed since the measurement of the surface tension arising from the interface between liquid-liquid (Hg) is feasible.

(a) The surface tension γ of the interface of a mercury electrode immersed in an aqueous solution of 1.0 N CsCl was measured at 298 K as a function of the applied voltage E_V standardized against a reference electrode and the obtained data are summarized in Table 1 below (electrocapillary curve). The charge density q_M in the surface of the mercury electrode is given by *Lippmann*'s equation (see the solution to this problem).

$$\left(\frac{\partial \gamma}{\partial E_V}\right)_{\mu_i,T,P} = -q_M \qquad (31) \qquad (\textit{Lippmann}\text{'s equation})$$

According to this equation, find the applied voltage at which the charge density becomes zero (electrocapillary maximum (e.c.m.))

(b) Find the surface charge density of the mercury electrode at the applied voltage of -200 mV.

How large is the capacitance of an electrified double-layer formed by the electrode and solution at this voltage? How would it change at -800 mV?

(c) At a constant applied voltage, the surface tension was measured by varying the concentration of CsCl and the values given in Table 2 were obtained. Find the surface excess concentration of Cl^- at this voltage from the data given in Table 2 (see **problem 4-41**, **4-45**).

Table 1 Surface tension γ and applied voltage E_V.

E_V / mV	γ/dyne · cm^{-1}	E_V / mV	γ/dyne · cm^{-1}
0	345	-700	419.9
-100	376.4	-800	414
-200	397.1	-900	405.6
-300	410.2	-1000	395.1
-400	418.8	-1100	382.9
-500	422.6	-1200	369.2
-600	422.9	-1300	356.6

Table 2 Surface tension γ and salt conc.*

mol/L	γ/dyne · cm^{-1}
3	390
0.3	398.9
0.1	402.9
0.03	406.4
0.01	410.8

*-1.0V vs. calomel electrode.

"note" Data adapted from *J.O'M.Bockris and A.K.N.Reddy, Modern Electrochemistry,* vol.2, table7.5, 7.6.

Problem 4-79

Acetic acid in aqueous solution is in equilibrium expressed as eq(32) with a degree of dissociation denoted α.

$$\begin{array}{ccccc} CH_3COOH & \rightleftarrows & CH_3COO^- & + & H^+ \\ C_0(1-\alpha) & & C_0\alpha & & C_0\alpha \end{array} \qquad (32)$$

The limiting molar conductivities (equivalent conductivity) of acetate ion and proton are denoted λ_0^- and λ_0^+, respectively.

(a) Find an equation to give the molar conductivity of the acetic acid aqueous solution at a molar concentration C_0. It is assumed that the molar conductivity of each ion, denoted λ^- and λ^+, follows *Kohlrauch*'s law.

(b) Putting $\lambda_0^- = 4.09$ and $\lambda_0^+ = 34.96$ in the units of $mS \cdot m^2 \cdot mol^{-1}$, draw a curve to show how the molar conductivity of the solution varies with the concentration of acetic acid whose dissociation constant is given by $K_a = 1.8 \times 10^{-5}$.

(c) Show whether or not the molar conductivity derived according to *Kohlrauch*'s law satisfies *Ostwald*'s dilution law, when the concentration of acetic acid is low. (*W.Ostwald,* Nobel Prize in Chemistry in 1909.)

Problem 4-80

The multi-step equilibrium of a poly-dentate coordination compound is considered here taking a four-dentate copper-ammonium coordination compound as an example. Its multi-step equilibrium is composed of the following four steps given by the equations from (33) to (36).

$$Cu^{2+} + NH_3 \rightleftarrows Cu^{2+}(NH_3) \qquad K_1 \qquad (33)$$

$$Cu^{2+}(NH_3) + NH_3 \rightleftarrows Cu^{2+}(NH_3)_2 \qquad K_2 \qquad (34)$$

$$Cu^{2+}(NH_3)_2 + NH_3 \rightleftarrows Cu^{2+}(NH_3)_3 \qquad K_3 \qquad (35)$$

$$Cu^{2+}(NH_3)_3 + NH_3 \rightleftarrows Cu^{2+}(NH_3)_4 \qquad K_4 \qquad (36)$$

When the total concentration of the copper ion is C_0, by putting the concentration of free ammonium x, find the concentrations of the other species given in the above scheme including from mono- to four-ammonium coordinated complexes as a function of the ammonium concentration x. Display the results to show visually how they vary with x.

Problem 4-81

Find the maximum work obtainable or the minimum work required upon preparing a saturated solution from a pure solvent and a pure solute to be dissolved. The solution is assumed ideal.

Problem 4-82

Consider a system composed of a solvent and a solute. Find the minimum energy of work required to prepare a solution of the concentration of $C'(>C)$ from the solution of the concentration of C by removing a part of its solvent from the solution. Any solution is assumed to behave ideally.

Problem 4-83

(a) Consider a tabular electric double layer composed of an electrode with a positive surface charge density σ and a layer negatively charged (*Helmholtz* layer with negative ions arrayed) and separated from the electrode by a distance d. The double layer is in a medium of a specific dielectric constant ε_r.

Under these conditions, find the strength of an electric field as well as the potential within the double layer as a function of the distance from the electrode based on *Gauss*' law.

(b) Positive and negative ions in an electrolyte solution are not distributed homogeneously in space but are considered distributing in the solution according to the *Boltzmann* distribution. Applying the *Debye-Hückel*'s *limiting law* (see **solution 4-17**) to know a distribution of ions in an interfacial area of the electrode, find how the electric field and potential in the vicinity of the electrode vary with a distance from the electrode by solving *Poisson*'s relevant equation.

Problem 4-84

Consider ions which exist inside and outside of a cell. They are separated by a flat cell membrane with a thickness of 10^{-8} m whose specific dielectric constant is 8.

(a) Both sides of the cell membrane can be charged, each oppositely, and can work as a capacitor. Find a value for the capacitance of the cell per $1.0\,\text{cm}^2$ of the cell membrane.

(b) When the electrical potential difference between the two faces of the cell membrane is 0.1 V, how large is the energy to be stored in a $1.0\ \text{cm}^2$ area of the membrane?

(c) Find the strength of the electric field inside the cell membrane. Find also the surface charge density per unit area of $1.0\ \text{cm}^2$ on both sides of the membrane.

(d) Find a minimum energy required for a single hydrogen ion to pass through the membrane. Assume that the membrane is permeable and there exists no steric hindrance in passing through the membrane.

(e) Instead of the surface charges on the membrane, in order to achieve the electrical potential difference of 0.1V between the inside and outside of the membrane by means of the concentration gap, how many times the concentration of hydrogen ion must be changed between the inside and outside of the membrane?

Problem 4-85

S. Arrhenius, who actively worked in science from the late of the 19th century to the beginning of the 20th century, is a Nobel Prize laureate well known for his theory of electrolytic dissociation or work on the concept of the activation energy of rate processes now widely used as the *Arrhenius* equation. He did not remain simply as an active physical chemist but worked as an astronomer as well and examined extensively the relation between the ratio of the component gases in the atmosphere such as water or carbon dioxide and the temperature of the earth surface based on the fundamental equation derived from the thermal equilibrium established between the earth and the atmosphere as a function of the season and latitude of an area on the earth. As a result, he found out that if the concentration of carbon dioxide is doubled, then the temperature on the surface of the earth will rise by ca. 5°C and published the obtained results in a scientific paper (see **"note"1**). His pioneering work seems to have foreseen our current most urgent issue of "*green house effect*" (see **"note"2**).

In accordance with his paper, we consider below how he could manage to draw the above-mentioned conclusion by extracting a few essential points as exercises in physical chemistry.

The whole system we consider is divided into four sub-systems of [sun] – [cosmic space (vacuum)] – [atmosphere] – [earth surface] and an equilibrium to be established between [atmosphere] and [earth surface] by means of the radiation energy from the sun received by the earth and the energy lost from the earth surface as heat through the atmosphere is examined.

The radiation energy (black body radiation) from a body at the absolute temperature T is governed by

Stefan-Boltzmann's law and is proportional to T^4. Its proportional constant is denoted γ, the temperature at the surface of the earth T, the temperature of the atmosphere θ, the reflectance at the interface between the earth and the air $(1-\nu)$ (*albedo*) (see **"note"**), the absorption coefficient of the air (mainly water and carbon dioxide) β_a, which is put equal to the emission coefficient β_e. Then, the energy transferred to the air from the earth surface by radiation is given by eq(37) (see problem (a)).

$$W = \nu \times \beta_a \times \gamma(T^4 - \theta^4) \qquad (37)$$

The radiation density of the sun light is given by A and the absorption probability by the air by α, then the radiation energy absorbed by the air is given by $\alpha \times \mathrm{A}$ and, taking the reflectance into account, $\nu \times (1-\alpha) \times \mathrm{A}$ is absorbed by the earth. Apart from the radiation energy, the energy transport by thermal conduction and convection must be considered in the air or on the earth and they are expressed by M and N, respectively. Then, noticing the radiation energy from the atmosphere at the temperature θ is given by $\beta_e \times \gamma\theta^4$, eq(38) is derived from the energy balance in the atmosphere.

$$\beta_e\gamma\theta^4 = W + \alpha\mathrm{A} + \mathrm{M} = \nu\beta_a\gamma(T^4 - \theta^4) + \alpha\mathrm{A} + \mathrm{M} \qquad (38)$$

Similarly, eq(39) is satisfied at the earth surface. Here, $(1-\beta_a)\nu\gamma T^4$ represents the energy dissipated from the earth surface and discarded into the cosmic space without being absorbed in the air. The equation on the right side is the energy the earth surface gains from the radiation of the sun as well as the energy gained by the thermal conductance and convection.

$$W + (1-\beta_a)\gamma\nu T^4 = \nu\beta_a\gamma(T^4 - \theta^4) + (1-\beta_a)\gamma\nu T^4 = (1-\alpha)\nu\mathrm{A} + \mathrm{N} \qquad (39)$$

By putting $\beta_a = \beta_e \equiv \beta$, the elimination of θ^4 from eqs(38) and (39) in order to solve for T^4 leads to eq(40).

$$T^4 = \frac{\mathrm{A}(1+(1-\alpha)\nu) + \mathrm{N}(1+\nu^{-1}) + \mathrm{M}}{\gamma(1+(1-\beta)\nu)} \qquad (40)$$

Eq(39) can be expressed in a simpler form as eq(39') with C as an appropriate constant. This equation shows us that the temperature at the earth surface will vary if the absorption coefficient of the air β as well as the reflectance at the earth surface ν changes (see problems (b) and (c)).

$$T^4 = \frac{C}{(1+(1-\beta)\nu)} \qquad (40') \qquad C: \text{ constant independent of } \beta$$

(a) Among the molecules found in the air, it is mainly H_2O or CO_2 that absorbs or emits the radiation in the infrared and far infrared regions and their rotational and vibrational modes are responsible for the absorption and emission of energy. Give a reason why the absorption and emission coefficients for their transitions can be put equal, i.e., $\beta_e = \beta_a$. (For greenhouse gases other than CO_2, see **"note 2"**.)

Based on eq(40'),

(b) Show that even when the compositions of the air remain the same, i.e., same β, the rate of the temperature rise at the earth surface is the larger, the less the earth surface is covered with snow and ice. In his paper it is described as "*The last formula shows that the temperature of the earth augments with* β *and the more rapidly the greater* ν *is.* (*The last formula* mentioned here is eq(40')).

(c) When the concentration of carbon dioxide is doubled, the absorption coefficient changes from 0.6 to 0.7 (see "**note 3**"). Assuming that $v = 0.90$ and the temperature at the earth surface is 288 K, predict how largely the temperature will rise if the concentration of CO_2 is doubled (see also "**note 4**").

(d) The absorption coefficient of H_2O is larger than that of CO_2. Furthermore, the former exists in the air in much larger quantity than the latter. Why is H_2O not treated as a primary cause in the global warming issue in spite of these facts?

"**note 1**" *Svante Arrhenius, Philosophical Magazine and Journal of Science,* S.5, SVol.41, 237-276 (1896). Available on the internet.

"**note 2**" *Al Gore, An Inconvenient Truth, Rodale,* 2006.

"**note 3**" To estimate values for β, which varies with the concentrations of carbon dioxide and steam, *Arrhenius* tabulated the values classified according to the season and the latitude of an area based on a huge amount of data he collected.

"**note 4**" *S.Manabe*, Nobel Prize in Physics in 2021 for his work on physical modelling of earth's climate, quantifying variability and reliably predicting global warming.

Problem 4-86

Antifreeze used for automobiles is mainly composed of a mixture of water and ethylene glycol and a small amount of chemicals for rust-proof and others. For the sake of simplicity, the freezing point of a mixture of water and ethylene glycol alone is considered here. The molecular weight of ethylene glycol $(\mathrm{CH_2OH})_2$ is 62.07, its freezing point is $-13.0\,^\circ\mathrm{C}$ and the latent heat of fusion is 2.778 kcal/mol. Ethylene glycol is, as is seen from its having two hydroxyl groups, a highly hydrophilic (moisture absorbing) substance. (*Techniques of Chemistry, vol.II, Organic Solvents–Physical Properties and Methods of Purification 3rd edition*).

Assuming the mixture is an ideal solution, answer the following problems.

(a) Find the freezing point of a mixture obtained by mixing 100 g of ethylene glycol with 40 g of water. The cryoscopic constant (coefficient of freezing point depression) of pure ethylene glycol is given by K_f^{E} and that of pure water is by K_f^{W}.

(b) Employing 103.26 K as the cryoscopic constant of water, estimate a value of K_f^{E} for ethylene glycol and compare the obtained value with the one to be calculated on the basis of the latent heat of fusion and the freezing point of pure ethylene glycol.

Problem 4-87

Water is indispensable for living things though, it is well known that sea water as it is cannot be taken in as water. For this reason, to convert a part of sea water into fresh water is extremely important technology and various methods are already put into practice. We consider here a technique of extracting fresh water from sea water by means of osmosis membrane, being based on the thermodynamic principle.

(a) When the salinity of the seawater collected in a certain sea is 3.3 wt %, being converted to the concentration of NaCl, how large is the osmotic pressure of this seawater against pure water? The density of the seawater is assumed 1.0.

(b) When a hydrostatic pressure of 50 atm is applied on the seawater separated by a semipermeable osmosis membrane from fresh water under 1 atm and the applied pressure is equilibrated with the osmotic pressure of the seawater, how largely will the concentration of the seawater (converted to the concentration of NaCl) be changed?

(c) Based on the result in (b), describe how fresh water can be obtained from the seawater by applying hydrostatic high pressure on the seawater. Find a volume of fresh water to be obtained from 1L of 3.3 wt% seawater by the above method. Instead of NaCl, if a part of 3.3 wt% of the seawater is ascribed to $MgCl_2$, then will the volume of fresh water to be obtained increase or decrease?

(d) The technique of obtaining fresh water using osmosis membranes is called reverse osmosis method. Explain why it is called "reverse" osmosis method.

Problem 4-88

Citric acid (H_3Cit) is a carboxyl acid and is contained ubiquitously in citrus fruits as is implied by its name. Hence, from its safeness as foods, it is used as a cleaning agent to remove alkaline stains at home. In addition to its acidic property, citric acid reacts with iron hydroxide in iron rust to form water-soluble complex as is shown in the following equations and, therefore, is used as household rust remover as well.

The solubility product of iron hydroxide is given by eq(41) and the equilibrium constant for the complex formation with citric acid is given by eq(42).

$$Fe(OH)_3(s) \rightleftarrows Fe^{3+} + 3OH^- \qquad (41) \qquad K_{sp} = 2.6\times10^{-39}$$

$$Fe^{3+} + Cit^{3-} \rightleftarrows FeCit \qquad (42) \qquad K_f = 6.3\times10^{11}$$

$$H_3Cit \rightleftarrows Cit^{3-} + 3H^+ \qquad (43) \qquad K_a = ?$$

On the other hand, the acid-base dissociation of citric acid consists of three steps as are given by each of the reactions from eq (43a) to eq(43c).

$$H_3Cit \rightleftarrows H_2Cit^- + H^+ \qquad (43a) \qquad K_{a_1} = 7.1\times10^{-4}$$

$$H_2Cit^- \rightleftarrows HCit^{2-} + H^+ \qquad (43b) \qquad K_{a_2} = 1.7\times10^{-5}$$

$$HCit^{2-} \rightleftarrows Cit^{3-} + H^+ \qquad (43c) \qquad K_{a_3} = 4.1\times10^{-7}$$

(a) Find the equilibrium constant K_a for the whole dissociation of citric acid given by eq(43).

(b) Find the equilibrium constant K_4 of eq(44) which corresponds to the complex formation and expresses the dissolution of iron rust.

$$Fe(OH)_3(s) + H_3Cit \rightleftarrows FeCit + 3H_2O \qquad (44) \qquad K_4$$

(c) How many grams of iron rust, being weighed as $Fe(OH)_3$, can be dissolved in 1L of 0.1 mol/L citric aqueous solution?

Problem 4-89

For an ideal solution, the free energy change of mixing (two kinds of solvent) ΔG_{ex} is given by eq(45) per mole of the solution.

$$\Delta G_{ex}^{ide} = RT\langle X_A \ln X_A + X_B \ln X_B \rangle \qquad (45)$$

A deviation from the ideal behavior can be expressed by a correction made to eq(45). One of the simplest ways is to add a correction term as is given by eq(46), in which a constant α introduced is assumed to be independent of temperature and pressure.

$$\Delta G_{ex}^{non} = RT\langle X_A \ln X_A + X_B \ln X_B \rangle + \alpha \cdot X_A X_B \qquad (46)$$

Concerning non-ideal solutions for which the free energy change of mixing is given by eq(46), answer the following problems.

(a) Express the free energy G of a solution obtained by mixing two kinds of solvent, denoted A and B, together in quantities of n_A and n_B moles, respectively. The chemical potentials of the pure solvents A and B are given by μ_A° and μ_A°, respectively.

(b) From the result obtained in (a), find the chemical potential of each component of the mixed solution.

(c) The vapor pressures of the pure solvents A and B are P_A° and P_B°, respectively at a given temperature. For each component, find the vapor pressure above the mixed solution as a function of the mole fraction in solution.

(d) For a non-ideal two-component solution whose free energy of mixing is give by eq(46), find the entropy of mixing, the enthalpy of mixing and the volume change upon mixing.

(e)The solutions characterized by the thermodynamic properties found in (d) are called regular solutions. Define the regular solution in your own words

Problem 4-90

Vapor pressures of benzene-carbon tetrachloride mixtures at 373.15 K are found as given in the table below.

Table

mole fraction of CCl_4 (solution)	0	0.1398	0.2378	0.3735	0.4919	0.6201	0.7585	0.8718	1
mole fraction of CCl_4 (gas)	0	0.1703	0.2774	0.4159	0.5295	0.6475	0.7739	0.8783	1
vapor pressure (P / kPa)	24.36	25.36	25.96	26.67	27.20	27.66	28.05	28.26	28.44

(a) Find a best-fitted curve which can describe the relation between the vapor pressure and the mole fraction of the solution.

(b) Judging from a shape of the best-fitted curve, is it possible to regard the present solution as a kind of regular solution?

Problem 4-91

In the **problem 3-8a**, the relation between the vapor pressure and the mole fraction of a mixed solution is examined for the two systems of (i) chloroform-acetone mixed solution and (ii) carbon disulfide-acetone mixed solution.

Can either or both of these mixed solutions be regarded as regular solutions?

Part II
Useful Information

II Useful Information

II-1 Units

II-1-1 Universal Constants and their Units

Constant	Symbol	Value*	Unit
atomic mass unit	amu	1.661×10^{-27}	kg
Avogadro's number	N_A	6.022×10^{23}	mol^{-1} (6.02214076×10^{23})**
Boltzmann constant	k_B	1.381×10^{-23}	$J\cdot K^{-1}$ (1.380649×10^{-23})**
		0.6951	$cm^{-1}\cdot K^{-1}$
electron rest mass	m_e	9.109×10^{-31}	kg
proton rest mass	m_p	1.673×10^{-27}	kg
neutron rest mass	m_n	1.675×10^{-27}	kg
Bohr radius	a_0 $(=\varepsilon_0 h^2/\pi m_e e^2)$	5.292×10^{-11}	m
gas constant	$R=N_A k_B$	8.315	$J\cdot K^{-1}\cdot mol^{-1}$
		0.08315	$dm^3\cdot bar\cdot K^{-1}\cdot mol^{-1}$
		0.08206	$dm^3\cdot atm\cdot K^{-1}\cdot mol^{-1}$
Faraday constant	$F=N_A e$	9.649×10^4	$C\cdot mol^{-1}$
vacuum permittivity	ε_0	8.854×10^{-12}	$C^2\cdot J^{-1}\cdot m^{-1}$
vacuum permeability	μ_0	$4\pi\times10^{-7}$	$J\cdot s^2\cdot C^{-2}\cdot m^{-1}$ $(T^2\cdot J^{-1}\cdot m^3)$
Planck constant	h	6.626×10^{-34}	$J\cdot s$ (6.62607015×10^{-34})**
elementary charge	e	1.602×10^{-19}	C ($1.602176634\times10^{-19}$)**
speed of light	c	2.998×10^8	$m\cdot s^{-1}$ (2.99792458×10^8)**
gravitational acceleration	g	9.807	$m\cdot s^{-2}$

*All values are given in four digits but the significant figures vary from constant to constant.

The **CODATA 2018 recommended values taken from **TABLE XXX** in *E.Tiesinga et al., Rev.Mod.Phys.*, 93, 025010-025063 (2021). These values are now adopted as exact constants based on quantum theory and are not to be determined experimentally. See **TABLE XXX** for other constants.

II-1-2 System of Units

Although the units employed throughout in this book are those of SI units (International System of Units), until a while ago the system of cgs units (cm, g, s) had been basically used in most of technical books and chemistry papers. Because of this, these two systems are compared in the table below for units most frequently used.

system	length	mass	time	charge	pressure	energy	force
SI	m	kg	s	C	Pa	J(*joule*)	N
cgs	cm	g	s	esu	atm	erg	dyne

S. Hirayama and M. Yamamoto, *Physical Chemistry Problems and Solutions*,
https://doi.org/10.1007/978-981-97-0165-0_5

$1\mathrm{C} = c(\text{dimensionless}) \times 10^{-1}\,\mathrm{esu}$,

$1e = 1.602 \times 10^{-19}\,\mathrm{C} = 1.602 \times 10^{-19} \times 10^{-1} \times 2.998 \times 10^{10} = 4.803 \times 10^{-10}\,\mathrm{esu}$,

$1\mathrm{J} = 10^7\,\mathrm{erg}$, $1\,\mathrm{dyne} = \mathrm{g{\times}cm/s^2} = 10^{-5}\,\mathrm{N}$, $1\mathrm{W} = 1\mathrm{J}\cdot\mathrm{s}^{-1}$.

II-1-3 Conversion of Units

1-3-1 energy units

	joule (J)	*calorie*(cal.)	eV	cm^{-1}	Hz
joule (J)	1	0.2390	$6.242{\times}10^{18}$	$5.034{\times}10^{22}$	$1.509{\times}10^{33}$
calorie(cal.)	4.184	1			
eV	$1.602{\times}10^{-19}$		1	8066	$2.418{\times}10^{14}$
cm^{-1}	$1.986{\times}10^{-23}$		$1.204{\times}10^{-4}$	1	$2.998{\times}10^{10}$
Hz	$6.626{\times}10^{-34}$		$4.136{\times}10^{-15}$	$3.334{\times}10^{-11}$	1

1-3-2 pressure units

	Pa*	bar	atm	Torr	psi#
Pa*	1	10^{-5}	$9.869{\times}10^{-6}$	$7.501{\times}10^{-3}$	
bar	10^5	1	0.9869	750.1	
atm	$1.013{\times}10^5$	1.013	1	760	
Torr	133.3	$133.3{\times}10^{-3}$	$1.316{\times}10^{-3}$	1	
psi	$6.895{\times}10^3$	$6.895{\times}10^{-2}$			

*$1\,\mathrm{Pa} = 1\,\mathrm{N}\cdot\mathrm{m}^{-2}$, $1\,\mathrm{kg}\cdot\mathrm{m}^{-1}\mathrm{s}^{-2}$. # psi : pounds per square inch.

1-3-3 other useful units

name	symbol	unit
angstrom	Å	$10^{-8}\,\mathrm{cm} = 10^{-10}\,\mathrm{m}$
poise	P	$0.1\ \mathrm{kg}\cdot\mathrm{m}^{-1}\cdot\mathrm{s}^{-1}(=1\mathrm{g}\cdot\mathrm{cm}^{-1}\cdot\mathrm{s}^{-1})$ $1\ \mathrm{cP} = 10^{-3}\,\mathrm{kg}\cdot\mathrm{m}^{-1}\cdot\mathrm{s}^{-1}$
Bohr magneton	μ_B	$9.274 \times 10^{-24}\,\mathrm{J}\cdot\mathrm{T}^{-1}$ $(\mu_B = e\hbar/2m_e)$
nuclear magneton*	μ_N	$5.051 \times 10^{-27}\,\mathrm{J}\cdot\mathrm{T}^{-1}$ $(\mu_N = e\hbar/2m_p)$
Debye	D	$3.3356 \times 10^{-30}\,\mathrm{C}\cdot\mathrm{m}$ (dipole moment unit)
Rydberg constant	R_H	$109677.581\ \mathrm{cm}^{-1}$ $e^4 m_e/8\varepsilon_0^2 h^3 c\ (\mathrm{cm}^{-1})$, $e^4 m_e/8\varepsilon_0^2 h^2\,(\mathrm{J})$
Hartree	E_h	$4.360 \times 10^{-18}\,\mathrm{J}$ $e^2/4\pi\varepsilon_0 a_0$
volt	V	J/C $1\,\mathrm{V} = 1\,\mathrm{J/C}$
ampere	A	C / s
ohm	Ω	V/A $1\,\Omega = 1\,\mathrm{V/A}$
siemens	S	Ω^{-1} $1\,\mathrm{S} = 1/1\,\Omega$
tesla	T	$\mathrm{N/C}\cdot\mathrm{m}\cdot\mathrm{s}^{-1} = \mathrm{N/A}\cdot\mathrm{m}$

gauss	Γ	$10^{-4}\,\mathrm{T}$ $(1\,\Gamma = 1\mathrm{Oe}$ exerts 1dyne on a unit magnetic pole.)
atomic mass unit	amu	$1.661\times10^{-27}\,\mathrm{kg} = {}^{12}\mathrm{C}/12/N_A = 1.00\times10^{-3}/6.022\times10^{23}$
dalton	Da	$1\,\mathrm{g/mol}$ (used to express molar masses of polymers)

* The magnitude of the nuclear magneton is approximately 1/1800 of the *Bohr* magneton.

II-2 Important Formulae Frequently Used in Physical Chemistry

II-2-1 Summary of Basic Formulae in Thermodynamics

(1) thermodynamic functions

extensive variables

U: internal energy, V: volume, S: entropy, H: enthalpy, A: *Helmholtz* free energy,

G: *Gibbs* free energy

intensive variables

T: temperature, P: pressure, μ : chemical potential, E(V): electric potential (electromotive force)

thermodynamic function	symbol (variables)	defining eq.[*1]	total differential	*Maxwell*'s relation[*2]
internal energy	$U(S,V)$		$dU = TdS - PdV$	$\left(\frac{\partial T}{\partial V}\right)_S = -\left(\frac{\partial P}{\partial S}\right)_V$
enthalpy	$H(S,P)$	$H = U + PV$	$dH = TdS + VdP$	$\left(\frac{\partial T}{\partial P}\right)_S = \left(\frac{\partial V}{\partial S}\right)_P$
Helmholtz free energy	$A(T,V)$ $(\equiv F(T,V))$	$A = U - TS$ $(\equiv F = U - TS)$	$dA = -SdT - PdV$	$\left(\frac{\partial S}{\partial V}\right)_T = \left(\frac{\partial P}{\partial T}\right)_V$
Gibbs free energy	$G(T,P)$	$G = H - TS$	$dG = -SdT + VdP$	$\left(\frac{\partial S}{\partial P}\right)_T = -\left(\frac{\partial V}{\partial T}\right)_P$

*[1] See section (5) below for derivation of the thermodynamic functions based on *Legendre* transformation.

*[2] Derivable straightforwardly from the fact that the total differentials of thermodynamic functions are exact (perfect) differentials.

(2) first law of thermodynamics

$dU = \delta q + \delta w$ $\quad$ δq : thermal energy, δw : work energy

$dU = \delta q - PdV$ $\quad$ d signifies a minute change of a state quantity

$dU = TdS - PdV$ $\quad$ δ indicates a minute change of a quantity usually other than state quantityies

(3) second law of thermodynamics

$$dS = \frac{\delta q_{rev}}{T}$$ *rev*: reversible process

$$\Delta S_{AB} = \int_A^B \frac{\delta q_{rev}}{T}$$

Instead of Δ, *Δ* is employed throughout the book to indicate the change of quantity value *in oder to distinguish it from Laplacian* Δ which is given by $\nabla \cdot \nabla$.

$$\Delta S_{irrev} > \Delta S_{rev} = \int \frac{\delta q_{rev}}{T}$$ *irrev*: irreversible process (*Clausius* inequality)

(4) third law of thermodynamics

$S(T = 0\,\text{K}) = 0$ (The entropy of a pure substance which forms a perfect crystal at 0 K is zero at 0 K.)

S. Hirayama and M. Yamamoto, *Physical Chemistry Problems and Solutions*,
https://doi.org/10.1007/978-981-97-0165-0_6

(5) derivation of thermodynamic functions by means of *Legendre* transformation

From the first and second laws of thermodynamics, the internal energy of a system, when expressed in the total differential, is given by the following equation.

$$dU = TdS - PdV$$

This relation indicates that the internal energy *U* is a function of *S* and *V*. All of the thermodynamic functions such as *H*, *A* and *G* other than those which appear in the above equation can readily be derived by means of *Legendre* transformation, starting with the above equation as are shown in the following.

(i) introduction of enthalpy $H = U + PV$ (variable exchange $V \Rightarrow P$, *Legendre* transformation)

$$dU + \underline{d(PV)} = TdS - PdV + \underline{d(PV)} \;\Rightarrow\; d(U + PV) = TdS - PdV + PdV + VdP = TdS + VdP$$

Putting $U + PV \equiv H$, a new thermodynamic function *H* is introduced and since its total differential is given by $dH = TdS + VdP$, *H* is a function of *S* and *P* and is expressed as $H(S,P)$. In short, *Legendre* transformation is, as is indicated by the underlined part, to exchange the conjugate variables (*P* and *V* in the present case) by adding or subtracting the total differential of their products

(ii) introduction of *Gibbs* free energy $G = H + PV$ (simultaneous variable exchanges $V,S \Rightarrow P,T$, *Legendre* transformation, see the underlined parts)

$$dU + \underline{d(PV)} - \underline{d(TS)} = TdS - PdV + \underline{d(PV)} - \underline{d(TS)}$$

$$\Rightarrow\; d(U + PV - TS) = TdS - PdV + PdV + VdP - TdS - SdT = -SdT + VdP$$

Putting $U + PV - TS \equiv H - TS \equiv G$, a new thermodynamic function *G* is introduced and as is obvious from its total differential given by $dG = -SdT + VdP$, *G* is a function of *T* and *P* and is expressed as $G(T,P)$.

"note" Each pair of *V* and *P*, *T* and *S*, and μ and *n* is called a pair of thermodynamic variables in a conjugative relation. It should be noted that the dimension of the product of any conjugative variables such as *PV* or *TS* is energy. By keeping either of the conjugative variables at a constant value, a change in the state of a system can be specified.

(6) *Euler*'s chain relation

Putting three different thermodynamic variables *x*, *y and z*, their partial derivatives are related to one another through *Euler*'s chain relation given below:

$$\left(\frac{\partial x}{\partial y}\right)_z \left(\frac{\partial y}{\partial z}\right)_x \left(\frac{\partial z}{\partial x}\right)_y = -1 \text{ (chain relation), for instance, } \left(\frac{\partial T}{\partial P}\right)_S \left(\frac{\partial P}{\partial S}\right)_T \left(\frac{\partial S}{\partial T}\right)_P = -1.$$

II-2-2 Summary of Important Formulae Used in Chemical Statistical Thermodynamics

(1) number of complexions *W* of mutually distinguishable particles (microcanonical ensembles)

$$W = \frac{N! \times g_1^{n_1} g_2^{n_2} \cdots g_l^{n_l}}{n_1! n_2! \cdots n_l!}, \qquad g_i \text{: degeneracy of state } i.$$

(2) *Stirling*'s approximate formula

$$\ln N! = N \ln N - N, \qquad N >> 1$$

(3) definition of molecular partition function *p.f.* (microcanonical ensembles)

$$p.f. = \sum_i g_i \exp(-\varepsilon_i / k_B T), \qquad \varepsilon_i, g_i: \text{energy and degeneracy of state } i, \text{ respectively.}$$

(4) distribution giving most probable number of complexions W_{max}

$$n_i = \frac{N \times g_i \exp(-\varepsilon_i / k_B T)}{p.f.}$$

(5) W_{max} and partition function $p.f.$ (mutually distinguishable particles)

$$\ln W_{max} = N \times \ln p.f. + \frac{E}{k_B T}$$

(6) W_{max} and partition function $p.f.$ (mutually indistinguishable particles)

$$\ln W_{max} = N \times (\ln \frac{p.f.}{N} + 1) + \frac{E}{k_B T}$$

(7) relation between (heat) energy change δQ and W_{max}

$$\delta \ln W_{max} = \frac{\delta Q}{k_B T}$$

(8) *Boltzmann* formula for entropy

$$k_B \ln W_{max} = S \qquad \text{(mutually distinguishable particles)}$$

$$k_B \ln \frac{W_{max}}{N!} = S \qquad \text{(mutually indistinguishable particles)}$$

(9) thermodynamic functions and (molecular) partition functions $p.f.$

$$E = Nk_B T^2 \left(\frac{\partial \ln p.f.}{\partial T} \right)_V, \qquad P = Nk_B T \left(\frac{\partial \ln p.f.}{\partial V} \right)_T$$

$$S = Nk_B \ln p.f. + E/T \quad \text{(solid)}, \qquad S = Nk_B (\ln \frac{p.f.}{N} + 1) + E/T \quad \text{(gas)}$$

$$A (\equiv F) = -Nk_B T \ln p.f. \quad \text{(solid)}, \qquad A = -Nk_B T (\ln \frac{p.f.}{N} + 1) \quad \text{(gas)}$$

$$G = -Nk_B T \ln p.f. + PV \quad \text{(solid)}, \qquad G = -Nk_B T (\ln \frac{p.f.}{N} + 1) + PV \quad \text{(gas)}$$

(10) canonical partition function $P.F.$ and molecular partition function $p.f.$ for distinguishable particles

$$P.F. = (p.f.)^N$$

(11) *Sackur-Tetrode* equation

$$S = Nk_B \left(\ln \frac{(2\pi)^{\frac{3}{2}} k_B^{\frac{5}{2}}}{h^3} + \frac{5}{2} + \frac{5}{2} \ln T - \ln P + \frac{3}{2} \ln m \right)$$

(12) grand canonical partition function $G.P.F. (\equiv \Xi)$ and molecular partition function

$$G.P.F. = \sum_N P.F. \times \exp(\mu N / k_B T) = \sum_N \{(p.f.) \times \exp(\mu / k_B T)\}^N = \frac{1}{1 - (p.f.) \times \exp(\mu / k_B T)}$$

μ: chemical potential

(13) *G.P.F.* and thermodynamic functions

$$PV = k_B T \ln G.P.F.$$

$$N = k_B T \left(\frac{\partial \ln G.P.F.}{\partial \mu} \right)_{V,T}$$

$$S = k_B T \left(\frac{\partial \ln G.P.F.}{\partial T} \right)_{V,\mu}$$

(14) definition of chemical potential (*Helmholtz* free energy $F(\equiv A)$ is convenient for this purpose.)

$$\mu = \left(\frac{\partial F(T,V,N)}{\partial N} \right)_{T,V} = \left(\frac{\partial G(T,P,N)}{\partial N} \right)_{T,P}$$

(15) chemical potentials and partition functions (ideal systems)

$$\mu_{\mathrm{A}}(solid) = -k_B T \ln p.f._{\cdot\mathrm{A}}(solid)$$ Upright subscript A signifies substance A.

$$\mu_{\mathrm{A}}(gas) = -k_B T \ln \left(\frac{p.f._{\cdot\mathrm{A}}(gas)}{N_{\mathrm{A}}} \right)$$ N_{A} is the number of particles of substance A.

(16) equilibrium constant K and partition function for a general chemical equilibrium

$$a\mathrm{A} + b\mathrm{B} \rightleftarrows c\mathrm{C} + d\mathrm{D}$$

$$K(T,V,N) = \frac{N_{\mathrm{C}}^c \times N_{\mathrm{D}}^d}{N_{\mathrm{A}}^a \times N_{\mathrm{B}}^b} = \frac{(p.f._{\cdot\mathrm{C}})^c \times (p.f._{\cdot\mathrm{D}})^d}{(p.f._{\cdot\mathrm{A}})^a \times (p.f._{\cdot\mathrm{B}})^b}$$

Here, K is given by the ratio of the numbers of particles involved in the equilibrium.

(17) rate constant k_r of the reaction $\mathrm{A} + \mathrm{B} \rightleftarrows \mathrm{X}^{\neq} \xrightarrow{k_r}$ Product **and partition functions**

$$k_r = \frac{k_B T}{h} \frac{p.f._{\cdot\mathrm{X}^{\neq}}}{p.f._{\cdot\mathrm{A}} \times p.f._{\cdot\mathrm{B}}} \exp(-\varepsilon_a / k_B T)$$

$$= Z_{col} \exp(-\varepsilon_0 / k_B T) \times \left[\frac{(p.f._{\cdot v})^5}{(p.f._{\cdot r})^5} \right], \quad Z_{col} = N_A \left(\frac{8 k_B T}{\pi \mu} \right)^{1/2} \pi r_{\mathrm{AB}}^2 = \frac{k_B T}{h} \left[\frac{(p.f._{\cdot t}^{\mathrm{X}^{\neq}})^3 (p.f._{\cdot r}^{\mathrm{X}^{\neq}})^2}{(p.f._{\cdot t}^{\mathrm{A}})^3 (p.f._{\cdot t}^{\mathrm{B}})^3} \right]$$

(18) rate constant expression in terms of thermodynamic functions

$$k_r = \frac{k_B T}{h} \cdot e^{\frac{\Delta S^{0\neq'}}{R}} e^{-\frac{\Delta H^{0\neq'}}{RT}} = \frac{k_B T}{h} \cdot e^{\frac{\Delta S^{0\neq'}}{R}} e^{-\frac{E_a - RT}{RT}} = \frac{k_B T}{h} \cdot e \cdot e^{\frac{\Delta S^{0\neq'}}{R}} e^{-\frac{E_a}{RT}},$$ E_a: activation energy

"note" Relation between activation energy and activation thermodynamic quantities.

II-2-3 Molecular Motions, Energy Levels and Partition Functions

(1) translational motions (particle of a mass *m* in a cuboid $a \times b \times c$), energy levels and partition functions

$$\varepsilon_x = \frac{n_x^2 h^2}{8ma^2}, \quad \varepsilon_y = \frac{n_y^2 h^2}{8mb^2}, \quad \varepsilon_z = \frac{n_z^2 h^2}{8mc^2}$$ $(n_x, n_y, n_z = 1,2,3\cdots)$, n_x, n_y, n_z : quantum numbers

$$\varepsilon_{total} = \varepsilon_x + \varepsilon_y + \varepsilon_z$$

$$p.f._{\cdot}^{x} = \sum_{n_x=1}^{\infty} e^{-\frac{1}{k_B T} \cdot \frac{n_x^2 h^2}{8ma^2}} = \sqrt{\frac{2\pi m k_B T}{h^2}} \cdot a$$

$$p.f.^{total} = \sum_{n_x,n_y,n_z=1}^{\infty} e^{-\frac{\varepsilon_{total}}{k_BT}} = p.f.^x \times p.f.^y \times p.f.^z = \left(\frac{2\pi mk_BT}{h^2}\right)^{\frac{3}{2}} V, \qquad V = a\times b\times c$$

(2) rotational motions, energy levels and partition functions

(a) diatomic molecules (or linear molecules)

$$\varepsilon_{rot} = \frac{J(J+1)}{8\pi^2 I}, \quad (J = 0,1,2,3\cdots) \quad J: \text{rotational quantum number, } I: \text{moment of inertia}$$

$$p.f._{rot} = \sum_{J=0}^{\infty} g_J e^{-\frac{\varepsilon_{rot}}{k_BT}} = \sum_{J=0}^{\infty}(2J+1)e^{-\frac{BJ(J+1)}{k_BT}} \approx \int_0^{\infty}(2J+1)e^{-\frac{BJ(J+1)}{k_BT}} dJ = \int_0^{\infty} e^{-\frac{B}{k_BT}\cdot x} dx = \frac{k_BT}{B}$$

$$g_J = (2J+1): \text{ degeneracy of energy level,} \quad B \equiv \frac{h^2}{8\pi^2 I}$$

Taking a symmetry number σ ($\sigma = 2$ for homonuclear diatomic molecules) into account, a correct rotational partition function is written as

$$p.f._{rot} = \frac{k_BT}{\sigma B} = \frac{8\pi^2 Ik_BT}{\sigma h^2} \quad \text{high temperature approximation for } T > \Theta_{rot}$$

$$\text{characteristic rotational temperature: } \Theta_{rot} = \frac{B}{k_B}.$$

(b) three dimensional (3D) rotational partition function (asymmetric top molecules)

$$p.f._{rot}^{total} = \frac{\sqrt{\pi}(8\pi^2 k_BT)^{\frac{3}{2}}(I_X I_Y I_Z)^{\frac{1}{2}}}{\sigma h^3} \quad I_X, I_Y, I_Z: \text{ moments of inertia around principal axes } X, Y \text{ and } Z, \text{ respectively.}$$

$$2D: \quad p.f._{rot}^{2D} = \frac{1}{\sigma}\times\frac{8\pi^2 Ik_BT}{h^2}$$

$$3D: \quad p.f._{rot}^{3D} = \frac{\sqrt{\pi}}{\sigma}\times\left(\frac{8\pi^2 I_X k_BT}{h^2}\right)^{\frac{1}{2}}\times\left(\frac{8\pi^2 I_Y k_BT}{h^2}\right)^{\frac{1}{2}}\times\left(\frac{8\pi^2 I_Z k_BT}{h^2}\right)^{\frac{1}{2}} \text{(as if product of 1}D\text{ } p.f.\text{'s)}$$

(3) vibrational energy levels and vibrational partition function

$$\varepsilon_{vib} = \left(n+\frac{1}{2}\right)h\nu, \qquad \nu = \frac{1}{2\pi}\sqrt{\frac{\kappa}{\mu}}, \qquad \frac{1}{\mu} = \frac{1}{m_1}+\frac{1}{m_2}, \qquad n: \text{ vibrational quantum number, } \kappa: \text{ force constant.}$$

$$p.f._{vib} = \sum_{n=0}^{\infty} e^{-\frac{h\nu}{2k_BT}} e^{-\left(\frac{h\nu}{k_BT}\right)\cdot n} = e^{-\frac{h\nu}{2k_BT}} \sum_{n=0}^{\infty} e^{-\left(\frac{h\nu}{k_BT}\right)\cdot n} = e^{-\frac{h\nu}{2k_BT}} \times \frac{1}{1-e^{-\left(\frac{h\nu}{k_BT}\right)\cdot n}}$$

$$\approx \frac{k_BT}{h\nu} \quad \text{for } h\nu/k_BT << 1,$$

$$\approx e^{-\frac{h\nu}{2k_BT}} \quad \text{for } h\nu/k_BT >> 1.$$

The vibrational energy is often measured from the zero point energy $\left(=\frac{1}{2}h\nu\right)$ and in that case the partition function is given by the equation below.

$$p.f._{vib} = \frac{1}{1-e^{-\left(\frac{h\nu}{k_BT}\right)}}$$

II-3 Linear and Non-linear Least Squares Methods

II-3-1 Errors Associated with the Fitting Parameters Employed in the Linear Least Squares Method — First Order Reaction Rate Constant as an Example

Needless to say, integrated reaction rate equations and many other experimental equations are described by a linear function expressed as eq(1), so that the most likely straight line which passes through n measured points

$$y = a + bx \qquad (1)$$

$(x_i,\ y_i)$ is searched for by means of the linear least squares method from which the best-fitted slope a and the intercept b in eq(1) are determined. Naturally, the parameters a and b thus determined contain errors associated not only with the errors in the measured values themselves but also the number of data employed. (Some of you may think that the values for a and b obtained in eight digits by employing software for the least-squares fitting do not contain errors but they are totally wrong about this.)

For this, it is necessary to mention the errors associated with the values of a and b in reporting these values in scientific reports or papers. How to find these errors is given in the following.

(i) To start with, how to find the values of a and b in eq(1) is described.

Even when the same fitting equation such as eq(1) is employed, the result will be different depending on what is minimized. We consider here the difference$\{\,y_i - (a+bx)\,\}$between the value of $a+bx_i$ calculated at x_i and the value of y_i measured at x_i and sum up its squared value over all points of measurement (with equal weight for simplicity). The total sum obtained thus is denoted S (eq(2)). The straight line which makes S minimum is considered to be the best one in reproducing the measured values.

$$S = \sum_{i=1}^{n} \{y_i - (a+bx)\}^2 \qquad (2)$$

Since S is a function of a and b, a and b must satisfy eq(3) for S to be minimized against them.

$$\left(\frac{\partial S}{\partial a}\right)_b = 0, \qquad \left(\frac{\partial S}{\partial b}\right)_a = 0 \qquad (3)$$

From these equations, eqs(4) and (5) are obtained.

$$\left(\frac{\partial S}{\partial a}\right)_b = -2\sum_{i=1}^{n}\{y_i - (a+bx_i)\} = 0 \quad \Rightarrow \quad a\times n + b\sum_{i=1}^{n} x_i = \sum_{i=1}^{n} y_i \qquad (4)$$

$$\left(\frac{\partial S}{\partial b}\right)_a = -2\sum_{i=1}^{n}\{x_i y_i - (a+bx_i)x_i\} = 0 \quad \Rightarrow \quad a\sum_{i=1}^{n} x_i + b\sum_{i=1}^{n} x_{ii}^2 = \sum_{i=1}^{n} x_i y_i \qquad (5)$$

The obtained equations can be regarded as the simultaneous equations for a and b and the solutions are found as eqs(6) and (7), respectively.

$$a = \frac{1}{n}\left(\sum_{i=1}^{n} y_i - b\sum_{i=1}^{n} x_i\right) = \bar{y} - b\bar{x}, \qquad \bar{x} = \frac{1}{n}\sum_{i=1}^{n} x_i, \quad \bar{y} = \frac{1}{n}\sum_{i=1}^{n} y_i \qquad (6)$$

S. Hirayama and M. Yamamoto, *Physical Chemistry Problems and Solutions*,
https://doi.org/10.1007/978-981-97-0165-0_7

$$b=\frac{\left(\sum_{i=1}^{n}x_iy_i-\frac{1}{n}\sum_{i=1}^{n}x_i\times\sum_{i=1}^{n}y_i\right)}{\left\{\sum_{i=1}^{n}x_i^2-\frac{1}{n}\left(\sum_{i=1}^{n}x_i\right)^2\right\}} \qquad (7) \qquad (=(13))$$

From eq(7), the best-fitted value of b is obtained and substituting it into eq(6), the best-fitted value of a is found as is given by eq(8).

$$a=\frac{1}{n}\frac{\sum_{i=1}^{n}y_i\sum_{i=1}^{n}x_i^2-\sum_{i=1}^{n}x_i\sum_{i=1}^{n}x_iy_i}{\sum_{i=1}^{n}x_i^2-\frac{1}{n}\left(\sum_{i=1}^{n}x_i\right)^2}=\frac{\bar{y}\sum_{i=1}^{n}x_i^2-\bar{x}\sum_{i=1}^{n}x_iy_i}{\sum_{i=1}^{n}x_i^2-\frac{1}{n}\left(\sum_{i=1}^{n}x_i\right)^2} \qquad (8)$$

(ii) errors associated with the best-fitted values of a and b

Since the measured values of x_i and y_i contain errors in themselves, some errors must be propagated to the best-fitted values of b and a given by eqs(7) and (8). There are a number of ways to estimate these errors, but we will derive here the equation which gives the errors contained in a and b on the basis of the propagation of errors.

We consider a function $f(x_1,x_2,\cdots,x_n)$ consisting of variables $x_1,x_2,x_3,\cdots,x_n$ in general. The total differential of f can be written as

$$df=\left(\frac{\partial f}{\partial x_1}\right)dx_1+\left(\frac{\partial f}{\partial x_2}\right)dx_2+\left(\frac{\partial f}{\partial x_3}\right)dx_3+\cdots+\left(\frac{\partial f}{\partial x_n}\right)dx_n$$

and, hence, the error of f originating from each individual of the error in x_i,, denoted dx_i, is given by df and its squared value is expressed as below.

$$(df)^2=\sum_{i=1}^{n}\left(\frac{\partial f}{\partial x_i}\right)^2dx_i^2+\sum_{i,j=1,i\neq j}^{n}\left(\frac{\partial f}{\partial x_i}\right)\left(\frac{\partial f}{\partial x_j}\right)dx_idx_j$$

The second sum in the above equation can be neglected because it will approach to zero when the product term of dx_i and dx_j which can take randomly positive or negative value is summed over all i and j ($i\neq j$). Consequently, $(df)^2$ can be approximated by the following equation.

$$(df)^2\approx\sum_{i=1}^{n}\left(\frac{\partial f}{\partial x_i}\right)^2dx_i^2$$

Then we can simply apply this relation to eqs(7) and (8). Denoting the errors in a and b by Δa and Δb they are found as given below after careful calculations.

$$(\Delta b)^2\equiv\sigma_b^2=\frac{\left(\sum_{i=1}^{n}x_i\Delta y_i\right)^2-\left(\sum_{i=1}^{n}x_i\sum_{i=1}^{n}\Delta y_i\right)^2/n^2}{\left\{\sum_{i=1}^{n}x_i^2-\left(\sum_{i=1}^{n}x_i\right)^2/n\right\}^2}=\frac{\sum_{i=1}^{n}x_i^2\Delta y_i^2-\left(\sum_{i=1}^{n}x_i\right)^2\left(\sum_{i=1}^{n}\Delta y_i\right)^2/n^2}{\left\{\sum_{i=1}^{n}x_i^2-\left(\sum_{i=1}^{n}x_i\right)^2/n\right\}^2}$$

$$=\frac{\sum_{i=1}^{n}x_i^2\sigma^2-\left(\sum_{i=1}^{n}x_i\right)^2n\sigma^2/n^2}{\left\{\sum_{i=1}^{n}x_i^2-\left(\sum_{i=1}^{n}x_i\right)^2/n\right\}^2}=\frac{\left\{\sum_{i=1}^{n}x_i^2-\left(\sum_{i=1}^{n}x_i\right)^2/n\right\}\sigma^2}{\left\{\sum_{i=1}^{n}x_i^2-\left(\sum_{i=1}^{n}x_i\right)^2/n\right\}^2}=\frac{\sigma^2}{\sum_{i=1}^{n}x_i^2-\left(\sum_{i=1}^{n}x_i\right)^2/n}$$

The error of b squared, therefore, is expressed as the last equation, i.e., eq(9) below.

$$\sigma_b^2 = \frac{\sigma^2}{\sum_{i=1}^{n} x_i^2 - \left(\sum_{i=1}^{n} x_i\right)^2 / n} \qquad (9)$$

In a similar manner, σ_a^2 is found as is given below by eq(10).

$$(\Delta a)^2 \equiv \sigma_a^2 = \frac{1}{n^2} \frac{\left(\sum_{i=1}^{n} \Delta y_i \sum_{i=1}^{n} x_i^2\right)^2 - \left(\sum_{i=1}^{n} x_i \sum_{i=1}^{n} x_i \Delta y_i\right)^2}{\left\{\sum_{i=1}^{n} x_i^2 - \left(\sum_{i=1}^{n} x_i\right)^2 / n\right\}^2} = \frac{1}{n^2} \frac{\left(\sum_{i=1}^{n} \Delta y_i\right)^2 \left(\sum_{i=1}^{n} x_i^2\right)^2 - \left(\sum_{i=1}^{n} x_i\right)^2 \left(\sum_{i=1}^{n} x_i \Delta y_i\right)^2}{\left\{\sum_{i=1}^{n} x_i^2 - \left(\sum_{i=1}^{n} x_i\right)^2 / n\right\}^2}$$

$$= \frac{1}{n^2} \frac{n\sigma^2 \left(\sum_{i=1}^{n} x_i^2\right)^2 - \left(\sum_{i=1}^{n} x_i\right)^2 \sum_{i=1}^{n} x_i^2 \sigma^2}{\left\{\sum_{i=1}^{n} x_i^2 - \left(\sum_{i=1}^{n} x_i\right)^2 / n\right\}^2} = \frac{1}{n^2} \frac{n\sigma^2 \sum_{i=1}^{n} x_i^2 \left(\sum_{i=1}^{n} x_i^2 - \left(\sum_{i=1}^{n} x_i\right)^2 / n\right)}{\left\{\sum_{i=1}^{n} x_i^2 - \left(\sum_{i=1}^{n} x_i\right)^2 / n\right\}^2}$$

$$\sigma_a^2 = \frac{1}{n} \frac{\sigma^2 \sum_{i=1}^{n} x_i^2}{\sum_{i=1}^{n} x_i^2 - \left(\sum_{i=1}^{n} x_i\right)^2 / n} \qquad (10)$$

Here, we have used the following relations.

$$\sum_{i,j=1, i\neq j}^{n} \Delta y_i \cdot \Delta y_j \approx 0, \quad \sum_{i,j=1, i\neq j}^{n} x_i \Delta y_i \cdot x_j \Delta y_j \approx 0, \quad (\Delta y_i)^2 \approx (\Delta y_j)^2 \approx \sigma^2 = \frac{S}{n-2}$$

The values of a and b determined by the least-squares method are considered to be associated with the errors of σ_a and σ_b, respectively, so that the result acquired by this method should be presented as below.

$a \pm \sigma_a$, $b \mp \sigma_b$, (The double sings correspond since the errors associated with a and b are correlated.)

Taking these errors into account, it becomes self-evident how many figures of a and b are significant.

(iii) relation between correlation coefficient R and errors

Software programs of the least-squares method often may not provide errors associated with the fitting parameter a and b but only output the best-fitted values of a and b together with the correlation coefficient squared R^2 of the best-fitted straight line (see the section of gnuplot which does provide these errors). Because of this, it is shown here that the values of σ_a and σ_b can be calculated readily from the value of the correlation coefficient R.

To start with, several values calculable from the data to be fitted are defined below.

$$S_{xx} = \sum_{i=1}^{n} (x_i - \bar{x})^2 = \sum_{i=1}^{n} x_i^2 - \left(\sum_{i=1}^{n} x_i\right)^2 / n \qquad (11a)$$

$$S_{yy} = \sum_{i=1}^{n} (y_i - \bar{y})^2 = \sum_{i=1}^{n} y_i^2 - \left(\sum_{i=1}^{n} y_i\right)^2 / n \qquad (11b)$$

$$S_{xy} = \sum_{i=1}^{n} (x_i - \bar{x})(y_i - \bar{y}) = \sum_{i=1}^{n} x_i y_j - \left(\sum_{i=1}^{n} x_i\right)\left(\sum_{i=1}^{n} y_i\right) / n \qquad (11c)$$

$$S = \frac{S_{yy} \cdot S_{xx} - S_{xy}{}^2}{S_{xx}} \geq 0 \qquad (11d)$$

When these defined values are employed, the slope and the intercept of the best-fitted straight line and the dispersion can be expressed as the following equations from (12) to (14).

$$\sigma = \sqrt{S/(n-2)} = \sqrt{\left(S_{yy} - \frac{S_{xy}{}^2}{S_{xx}}\right)/(n-2)} \qquad (12)$$

$$b = \frac{\left(\sum_{i=1}^{n} x_i y_i - \frac{1}{n}\sum_{i=1}^{n} x_i \times \sum_{i=1}^{n} y_i\right)}{\left\{\sum_{i=1}^{n} x_i^2 - \frac{1}{n}\left(\sum_{i=1}^{n} x_i\right)^2\right\}} = \frac{S_{xy}}{S_{xx}} \qquad (13)$$

$$\sigma_b = \frac{\sigma}{\sqrt{S_{xx}}} = \sqrt{\frac{S_{yy} \cdot S_{xx} - S_{xy}{}^2}{(n-2)S_{xx}{}^2}} \qquad (14)$$

Now, defining the correlation coefficient R by eq(15),

$$R = \frac{S_{xy}}{\sqrt{S_{yy} \cdot S_{xx}}} \leq 1 \qquad (15)$$

S_{yy} can be expressed as $S_{yy} = \frac{S_{xy}{}^2}{S_{xx} \cdot R^2}$. Substituting this into eq(14) and rewriting it, eq(16) is derived.

$$\sigma_b = \sqrt{\frac{S_{xy}^2 / R^2 - S_{xy}{}^2}{(n-2)S_{xx}{}^2}} = \sqrt{\frac{(1/R^2 - 1)}{n-2}} \times |b| \qquad (16)$$

Since the relation between σ_a and R is not obtained, σ_a has to be calculated directly from eq(10) or from eq(6), giving eq(17) as an estimate of the error in a.

$$\sigma_a = \sigma_b \times \bar{x} = \sqrt{\frac{(1/R^2 - 1)}{n-2}} \times |b| \times \bar{x} \qquad (17)$$

When any particular values are found through calculations based on experimental data, we have to pay due attention to the errors accompanied by the calculations. In this book, a number of practical examples are provided as problems, so it is expected for ones to deepen their understanding of the error estimate upon data analysis by solving those problems.

"note" The meaning of the correlation coefficient is clear from its definition, that is, $\sum_{i=1}^{n}(x_i - \bar{x}) \cdot (y_i - \bar{y})$ represents the correlation between the deviations of x_i and y_i from their mean values given by $x_i - \bar{x}$ and $y_i - \bar{y}$, respectively. This value is divided by the autocorrelation coefficients of x_i and y_i given by the square roots of $\sum_{i=1}^{n}(x_i - \bar{x})^2$ and $\sum_{i=1}^{n}(y_i - \bar{y})^2$, respectively for the purpose of normalization. When the correlation coefficient R is unity, there will exist any error neither in a nor b, as is obvious from eqs(16) and (17). Refer to the exercise given below.

[exercise 1] Show that $R^2 \leq 1$.

[exercise 2] In the case of a linear plot of two variables, it is not necessarily prefixed which variable (dependent variable) is plotted along the vertical axis but can be arbitrarily chosen. Therefore, when the same data are plotted in two ways by swapping the vertical and horizontal axes and both are analyzed by the linear least-squares method, we will find a different slope (b or b') for each of the best-fitted straight lines.

Show that there exists the following relation between those slopes.

$$b \times b' = R^2 \leq 1 \qquad (18)$$

[reference] *Philip R.Bevington, Data Reduction and Error Analysis for the Physical Science,* McGraw-Hill Book Company, 1969.

II-3-2 Deviation from True Value

When the standard deviation is calculated, the residual squared obtained for each sample is summed up over the total number of samples n and then divided not by n but by $n-1$. The reason for this is briefly given in the following.

The sum of the residual squared which measures the deviation of the value x_i for the sample i from the true value x_0 is given by eq(1). However, the true value x_0 is not known.

$$S_{true} = \sum_{i=1}^{n}(x_i - x_0)^2 = \sum_{i=1}^{n} x_i^2 - 2x_0 \sum_{i=1}^{n} x_i + nx_0^2$$

$$= \sum_{i=1}^{n} x_i^2 - 2nx_0\bar{x} + nx_0^2 \qquad (1)$$

On the other hand, the sum of the residual squared calculated similarly but based on the deviation of x_i from the mean value $\bar{x}$ is expressed as eq(2).

$$S_{mean} = \sum_{i=1}^{n}(x_i - \bar{x})^2 = \sum_{i=1}^{n} x_i^2 - 2\bar{x} \sum_{i=1}^{n} x_i + n\bar{x}^2$$

$$= \sum_{i=1}^{n} x_i^2 - 2n\bar{x}^2 + n\bar{x}^2$$

$$= \sum_{i=1}^{n} x_i^2 - n\bar{x}^2 \qquad (2)$$

The difference between eq(1) and eq(2) is given by eq(3). The important thing about eq(3) is that this differ-

$$S_{true} - S_{mean} = n(\bar{x}^2 - 2\bar{x}x_0 + x_0^2) = n(\bar{x} - x_0)^2 \geq 0 \qquad (3)$$

ence is always greater than zero because of $\bar{x} \neq x_0$.

Evaluating the difference $(\bar{x} - x_0)$ by eq(4), eq(5) is finally derived via following calculations.

$$(\bar{x} - x_0)^2 = \left(\frac{1}{n}\sum_{i=1}^{n} x_i - x_0\right)^2$$

$$= \frac{1}{n^2}\left\{\sum_{i=1}^{n}(x_i - x_0)\right\}^2 \approx \frac{1}{n^2}\sum_{i=1}^{n}(x_i - x_0)^2 = \frac{S_{true}}{n^2}, \quad \text{or } |\bar{x} - x_0| = \frac{\sqrt{S_{true}}}{n} \qquad (4)$$

$$S_{true} - S_{mean} = \frac{S_{true}}{n} \Rightarrow \left(1 - \frac{1}{n}\right)S_{true} = S_{mean} \Rightarrow \frac{S_{true}}{n} = \frac{S_{mean}}{n-1}$$

$$\sigma_{true} = \sqrt{\frac{S_{true}}{n}} = \sqrt{\frac{S_{mean}}{n-1}} \qquad (5) \quad Q.E.D.$$

II-3-3 Calculation of Standard Deviation of Best-Fitted Value − Liner Least-Squares Method calculation with the statistical function LINEST installed in Excel

The equation to be fitted is expressed as $y = a + bx$. The point is how to calculate the standard deviations Δa and Δb associated with the values of a and b which are determined by the linear least-squares method using LINEST.

Procedure

1. Input the data to be fitted into an Excel spreadsheet. ⇒ procedure 1 in the table given below.

   Sample data are those obtained for a temperature-dependent rate constant k and to be analyzed by *Arrhenius* plot. We put here $y = \ln k$ and $x = 10^3/T$.

2. Select any blank part of 2 rows × 2 columns on the Excel spreadsheet where the data are input (⇒ a light green cell area in the sample display).

3. Select [insert] from menu bars and then click on [functions]. Select [statistics] from the [category of functions] and among [function names] choose [LINEST].

4. [LINEST] input spreadsheet will appear. In the present example, select (D2;D6) for the known y values and (C2;C6) for the known x values on the spreadsheet. Also, enter TRUE into each of input boxes for constant and correction. Now. [LINEST] is ready to be executed, but, for this, we have to press Enter while pressing both of Shift and Ctrl keys simultaneously. If you fail and press just Enter alone, then you will obtain only the value of a. ⇒ procedure 4 in the table.

5. The obtained values will be displayed in the chosen cell area of the sheet. To what they correspond are as indicated in the sheet provided below.

6. We have chosen the cell area of 2 rows × 2 columns above for illustration, but if we choose a cell area of 5 rows × 2 columns and repeat the procedures 3 and 4, then we will be able to obtain the maximum number of the statistical values available by LINEST, as are shown in the grey cell area in the sheet.

7. For the liner least-squares method frequently practiced in physical chemistry, the display of the cell area of 2 rows × 2 columns would be sufficient.

8. If the standard plotting tool of Excel is used for a linear plotting of data, the best-fitted straight line found by the linear least-squares method is instantaneously returned but only the values of a, b and correlation coefficient squared R^2 are displayed. This is shown in the figure below. However, it should be noted that the errors associated with the fitting parameters a and b can be calculated from the value of R^2 as has already been seen.

9. With [LINEST], the errors associated with the fitting parameters are automatically found but the details of their calculations are as has been given in **II-3-1**.

Table Excel spreadsheet (what are shown in a real sheet are the cell areas filled with colors in the table).

	A	B	C	D	E	F	G
1	T(K)	k(s^{-1})	$1/T$x10^3	ln$k$			
2	523	0.018	1.912046	-4.01738			
3	541	0.15	1.848429	-1.89712	}	procedure 1	
4	560	0.6	1.785714	-0.51083			
5	576	1.6	1.736111	0.470004			
6	593	9.5	1.686341	2.251292			
7			x	y			
8		mean of x	1.793728				
9							
10			procedure 2, 3 ↓			procedure 6	
						LINEST()	
			LINEST()	*			
			procedure 4 ↓				
						procedure 4 ↓	
		a ⇒	-26.4469	46.69775	⇐ b	-26.4469	46.69775
		Δa ⇒	1.541439	2.767663	⇐ Δb	1.541439	2.767663
					R^2 ⇒	0.989912	0.275341
						294.3724	3
						22.31718	0.227438

* Do not make a mistake about where to input the values of x and y.

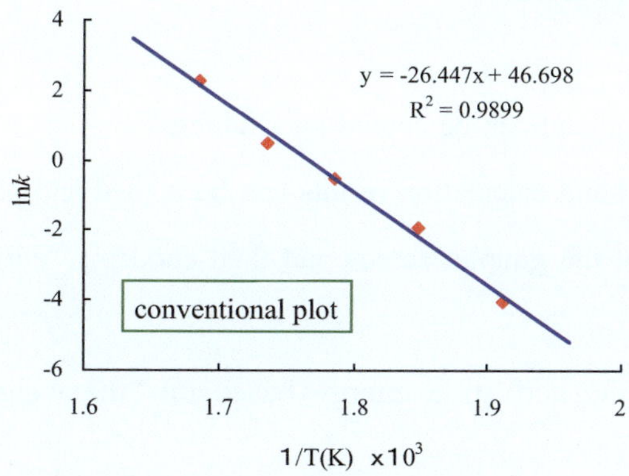

Fig.1 A plot of linear least-squares fitting by Excel.
The best-fitted function and the correlation coefficient squared are shown but the standard deviations Δa and Δb of a and b, respectively are not displayed, though these can be calculated from R^2.

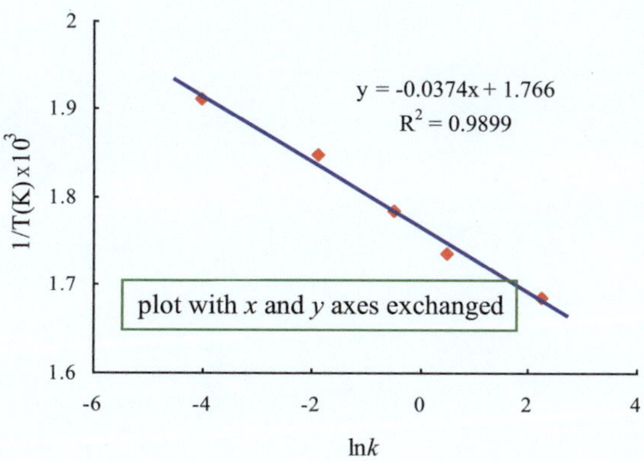

Fig.2 A plot of the same data in Fig.1 analyzed by the linear least-squares method after the x and y axes being exchanged.
The product of the two slopes (Figs.1 and 2) must be equal to the correlation coefficient squared, i.e.,

$$(-26.447)\times(-0.0374)\approx R^2=0.9899.$$

The correlation coefficients in Figs.1 and 2 are naturally the same.

II-3-4 (Weighted) Nonlinear Least Squares Method by Gnuplot ([nu:plot])

Gnuplot : simple manual

Gnuplot readily available on the internet is a very useful software upon fitting experimental data to various types of theoretical function. Only minimum necessary commands required in such practices are given here but you will be able to find more details (free manuals for instance) of gnuplot easily on the internet.

gnuplot>plot "data file"

data file such as "AE7.dat"

gnuplot>f(x)=a*exp(b/x)

f(x): arbitrary function to be fitted

gnuplot>fit f(x) "AE7.dat" using 1:2:3 via a,b

data "AE7.dat" is fitted to f(x) with weighting.

gnuplot>plot[xrange] f(x), "AE7.dat" with y error bars

[xrange] such as [400:800]

best-fitted f(x) and data "AE7.dat" are plotted with vertical error bars.

[preparation of data file]

The data to be used in gnuplot is created in Excel first and then the prepared data file is copied and pasted to Word as a text file. The pasted data file is converted to unformatted data and then after adding a filename extension dat. to it, it is saved in a gnuplot subfile [binary], for instance, like AE7.dat in the case of Word 97-2003. On Windows 7 or later, the filename extension can be left as txt.

[record of calculation results]

The results displayed as figures on a monitor by "plot" command can be copied by clicking "clipboardcopy" and can be pasted into Word for future use. The program used and calculation results can be copied and pasted into Word by clicking on the mark in the upper left corner of the gnuplot screen and then choosing "edit" ⇒ "range".

In the following, one practical example of obtaining the activation energy based on "the weighted nonlinear least-squares fitting" using gnuplot is given.

Data AE7.dat

T(K)	$k(s^{-1})$	Δk
417.9	112000	8250
480.7	260000	11700
520.1	396000	10710
633.2	938000	33720
666.8	1150000	27600
710.3	1610000	67620
737.9	1845000	72280

If the data AE7.dat prepared above can be read correctly by gnuplot, the “plot” command as indicated below in yellow followed by “enter” will bring up Fig.3 on the screen. (The colors and marks of the data points in the graph are default setting. They can be changed quite freely, though its details are not given here.)

◎gnuplot> plot "AE7.dat"＋enter

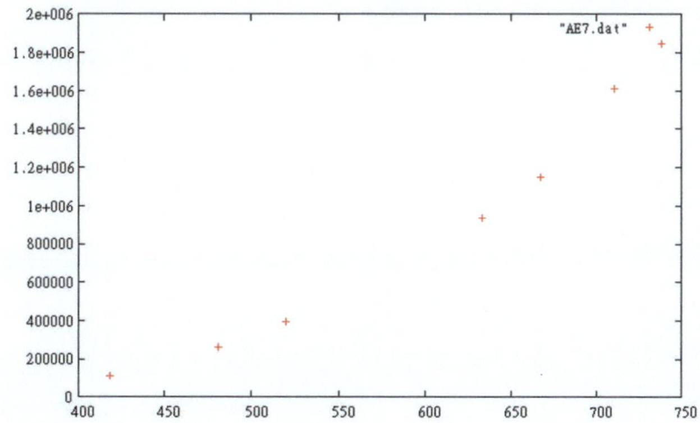

Fig.3 Default display on the monitor when data AE7 are plotted by using gnuplot.

◎gnuplot> plot "AE7.dat" with yerrorbars＋enter

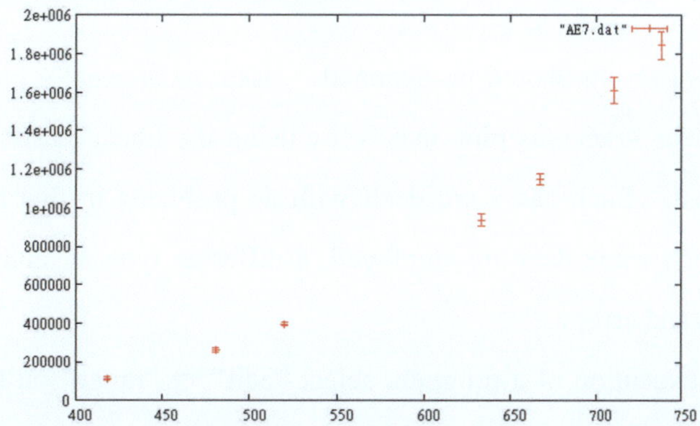

Fig.4 Default display on the monitor when data AE7 are plotted with errors associated with the rate constant *k* by using gnuplot.

The data with vertical error bars are displayed (Fig.4).

◎gnuplot> f(x)=a*exp(b/x)

gnuplot> fit f(x) "AE7.dat" using 1:2:3 via a,b＋enter

After 47 iterative calculations, the following final results are displayed on the screen.

Asymptotic Standard Error

===

a	= 6.21846e+007	+/- 6.111e+006	(9.827%)
b	= -2637.45	+/- 58.45	(2.216%)

correlation matrix of the fit parameters:

	a	b
a	1.000	
b	-0.987	1.000

===

◎gnuplot> plot f(x), "AE7.dat" with yerrorbars + enter

The data points with vertical error bars and the best-fitted curve is displayed on the screen (see Fig.5).

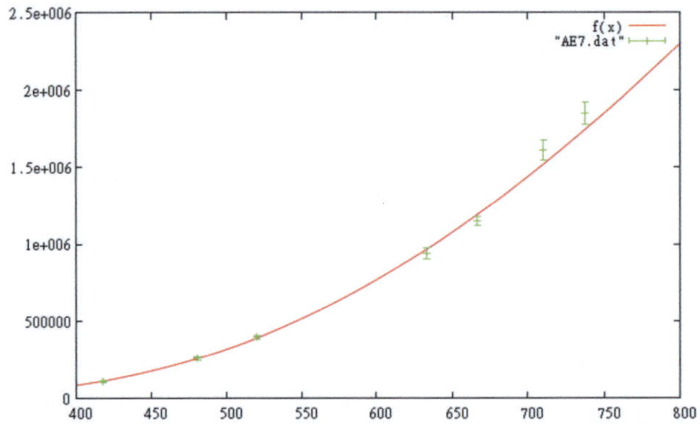

Fig.5 *Arrhenius* function best fitted by using gnuplot. to the data AE7 with errors associated with the rate constant k (see error bars)

Without the weights, different results would be obtained. Also, as is usually practiced, when the same data is analyzed by using a linear *Arrhenius* plot, that is, by using the linear least-squares method, another different result will be obtained. Such cases are dealt with as problems in this book. What should be noted here is that even when the same data are employed, a different type of analysis will yield different results on the best-fitted values and errors.

To record the results of the execution of a program, select "edit" ⇒ "range" on the screen of gnuplot to specify a necessary range and then select "copy". The specified range can be pasted into a suitable file (see above).

II-4 Useful Mathematical Expressions

1. Hyperbolic Functions and Trigonometric Functions

Hyperbolic functions

$$\sinh z = z + \frac{1}{3!}z^3 + \frac{1}{5!}z^5 + \cdots\cdots = \frac{e^z - e^{-z}}{2} = -i\sin(iz)$$

$$\cosh z = 1 + \frac{1}{2!}z^2 + \frac{1}{4!}z^4 + \cdots\cdots = \frac{e^z + e^{-z}}{2} = \cos(iz)$$

$$\coth z = \frac{e^z + e^{-z}}{e^z - e^{-z}}, \qquad \tanh z = \frac{e^z - e^{-z}}{e^z + e^{-z}}$$

Expressions of trigonometric functions in terms of exponentials

(*Euler*'s equation)

$$\left.\begin{aligned} e^{i\phi} &= \cos\phi + i\sin\phi \\ e^{-i\phi} &= \cos\phi - i\sin\phi \end{aligned}\right\} \Rightarrow \begin{aligned} \cos\phi &= \frac{1}{2}(e^{i\phi} + e^{-i\phi}) \\ \sin\phi &= \frac{1}{2i}(e^{i\phi} - e^{-i\phi}) \end{aligned}$$

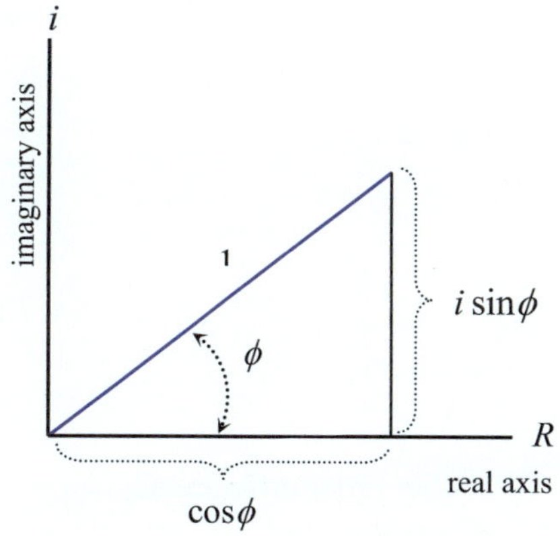

Fig.6 Imaginary coordinates.

Addition theorem (readily derived from exponential expressions)

$$\begin{aligned} e^{-i(\phi+\theta)} &= \cos(\phi+\theta) - i\sin(\phi+\theta) \\ &= e^{-i\phi} \times e^{-i\theta} \\ &= (\cos\phi - i\sin\phi) \times (\cos\theta - i\sin\theta) \\ &= \cos\phi\cos\theta - \sin\phi\sin\theta - i(\sin\phi\cos\theta + \cos\phi\sin\theta) \end{aligned}$$

$$\begin{aligned} \cos(\phi \pm \theta) &= \cos\phi\cos\theta \mp \sin\phi\sin\theta \\ \sin(\phi \pm \theta) &= \sin\phi\cos\theta \pm \cos\phi\sin\theta \end{aligned} \Rightarrow \begin{aligned} \cos^2\theta &= \frac{1}{2}(1 + \cos 2\theta) \\ \sin^2\theta &= \frac{1}{2}(1 - \cos 2\theta) \end{aligned}$$

double angle formulae

2. Differential of Functions

$$\frac{d\sin ax}{dx} = a\cos ax, \qquad \frac{d\cos ax}{dx} = -a\sin ax, \qquad \frac{d\tan ax}{dx} = \frac{a}{\cos^2 ax}$$

$$\frac{de^{ax}}{dx} = ae^{ax}, \qquad \frac{dx^n}{dx} = nx^{n-1}$$

$$\frac{d\ln x}{dx} = \frac{1}{x}, \qquad \frac{d}{dx}\int_a^x tf(t)dt = xf(x)$$

$$\frac{d}{dx}F(g(x)) \overset{y=g(x)}{=} \frac{d}{dx}F(y) = \frac{d}{dy}F(y) \cdot \frac{dy}{dx} = F'(y) \cdot g'(x)$$

S. Hirayama and M. Yamamoto, *Physical Chemistry Problems and Solutions*,
https://doi.org/10.1007/978-981-97-0165-0_8

Differential of inverse functions (for example, Arctan $ax \equiv \tan^{-1} ax$)

$$\frac{d}{dx}F^{-1}(x) = \frac{1}{F'(y)} \qquad y = \tan^{-1}ax \;\Rightarrow\; ax = \tan y \;\Rightarrow\; \frac{dx}{dy} = \frac{1}{a}\frac{d\tan y}{dy} = \frac{1}{a}\frac{1}{\cos^2 y}$$

$$\frac{dy}{dx} = \frac{d\tan^{-1}ax}{dx} = \frac{1}{\dfrac{dx}{dy}} = \frac{1}{\dfrac{1}{a}\dfrac{d\tan y}{dy}} = a\cos^2 y = \frac{a}{1+a^2x^2}$$

or
$$\frac{d\tan^{-1}ax}{dx} = \frac{d\tan^{-1}ax}{dax}\cdot\frac{dax}{dx} = \frac{a}{1+a^2x^2}$$

$$\left[\; y = F^{-1}(x) \;\Rightarrow\; x = F(y) \;\Rightarrow\; \frac{dy}{dx}\cdot\frac{dx}{dy} = 1 \;\Rightarrow\; \frac{dy}{dx} = \frac{1}{\dfrac{dx}{dy}} \;\right]$$

Higher order differentials

$$\frac{d^n}{dx^n}(f\cdot g) = \frac{d^{n-1}}{dx^{n-1}}(f'\cdot g + f\cdot g') = \frac{d^{n-2}}{dx^{n-2}}(f''\cdot g + 2f'\cdot g' + f\cdot g'')$$
$$= \frac{d^{n-3}}{dx^{n-3}}(f^{(3)}\cdot g + 3f^{(2)}\cdot g' + 3f'\cdot g^{(2)} + f\cdot g^{(3)})$$
$$= f^{(n)}\cdot g + C_1^n f^{(n-1)}\cdot g' + \cdots C_j^n f^{(n-j)}\cdot g^{(j)} + \cdots + f\cdot g^{(n)}$$

3. **Approximate Formulae**

$$\frac{1}{1\pm x} = 1 \mp x + x^2 \mp x^3 + \cdots\cdots$$

$$\sqrt{\frac{1}{1\pm x}} = 1 \mp \frac{1}{2}x + \frac{3}{8}x^2 \mp \frac{5}{16}x^3 + \frac{35}{32}x^4 + \cdots\cdots$$

$$\ln(1\pm x) = \pm x - \frac{x^2}{2} \pm \frac{x^3}{3} + \cdots\cdots$$

$$f(a+x) = f(a) + f'(a)x + \frac{f''(a)}{2!}x^2 + \frac{f'''(a)}{3!}x^3 + \cdots\cdots \qquad (\textit{Tayler}\ \text{expansion})$$

$$e^x = 1 + \frac{x}{1!} + \frac{x^2}{2!} + \cdots + \frac{x^n}{n!} + \cdots\cdots$$

$$e^{i\vartheta} = 1 + \frac{i\vartheta}{1!} + \frac{(i\vartheta)^2}{2!} + \frac{(i\vartheta)^3}{3!} + \frac{(i\vartheta)^4}{4!} + \frac{(i\vartheta)^5}{5!}$$
$$= \left(1 - \frac{\vartheta^2}{2!} + \frac{\vartheta^4}{4!} - \cdots\right) + i\left(\frac{\vartheta}{1!} - \frac{\vartheta^3}{3!} + \frac{\vartheta^5}{5!} - \cdots\right) \cdots\cdots$$
$$= \cos\vartheta + i\sin\vartheta$$

$$\cos\vartheta = \left(1 - \frac{\vartheta^2}{2!} + \frac{\vartheta^4}{4!} - \cdots\right) \qquad \sin\vartheta = \left(\frac{\vartheta}{1!} - \frac{\vartheta^3}{3!} + \frac{\vartheta^5}{5!} - \cdots\right)$$

4. **Constant** e (base of natural logarithm)

The constant e is an irrational number defined either by eq(1) or eq(2).

$$e = 1 + \frac{1}{1!} + \frac{1}{2!} + \frac{1}{3!} + \frac{1}{4!} + \frac{1}{5!} + \cdots\cdots = 2.71828 \qquad (1)$$

$$e = \lim_{n\to\infty}\left(1+\frac{1}{n}\right)^n = 2.71828 \qquad (2)$$

An approximate formula of e^x is given in the section 3.

5. Proof of $\dfrac{d\ \ln x}{dx} = \dfrac{1}{x}$

Utilizing eq(2) given in the section 4, we find the differential of $\ln x$.

$$\frac{d\ \ln x}{dx} = \lim_{h\to 0}\frac{1}{h}\{\ln(x+h) - \ln x\}$$

$$= \lim_{h\to 0}\frac{1}{h}\ln\frac{x+h}{x} = \lim_{h\to 0}\frac{1}{h}\ln\left(1+\frac{h}{x}\right)$$

$$= \lim_{h\to 0}\ln\left(1+\frac{1}{\frac{x}{h}}\right)^{\frac{1}{h}} = \lim_{h\to 0}\frac{1}{x}\ln\left(1+\frac{1}{\frac{x}{h}}\right)^{\frac{x}{h}}$$

$$= \frac{1}{x}\lim_{x/h\to\infty}\ln\left(1+\frac{1}{\frac{x}{h}}\right)^{\frac{x}{h}} = \frac{1}{x}\ln e$$

$$= \frac{1}{x} \qquad (Q.E.D.)$$

(e.g.) differential of a logarithmic function

$$\frac{d}{dx}\ln\left(\frac{1+x}{1-x}\right) = \frac{d\ \ln(1+x)}{dx} - \frac{d\ \ln(1-x)}{dx}$$

$$= \frac{1}{1+x}\frac{d(1+x)}{dx} - \frac{1}{1-x}\frac{d(1-x)}{dx}$$

$$= \frac{1}{1+x} + \frac{1}{1-x}$$

$$= \frac{2}{(1-x^2)}$$

[exercise] Confirm the following two equations: $\dfrac{de^{ax}}{dx} = ae^{ax}$ and $\dfrac{da^x}{dx} = a^x \ln a$. Also, show that the differential equation of $d\ \ln x/dx = 1/x$ can be readily verified by using the fact that the inverse function of e^x is $\ln x$.

6. Integrated Values (Definite Integrals)

Provided $a > 0$ and n is a positive integer.

$$\int_0^\infty x^n e^{-ax}dx = \frac{n!}{a^{n+1}}$$

$$\int_{-\infty}^\infty e^{-ax^2}dx = \sqrt{\frac{\pi}{a}}, \qquad \int_0^\infty xe^{-ax^2}dx = \frac{1}{2a}, \qquad \int_{-\infty}^\infty x^2e^{-ax^2}dx = \frac{1}{2a}\sqrt{\frac{\pi}{a}}, \qquad \int_0^\infty x^3e^{-ax^2}dx = \frac{1}{2a^2},$$

$$\int_0^\infty x^{2n}e^{-ax^2}dx = \frac{1\cdot 3\cdot 5\cdots(2n-1)}{2^{n+1}a^n}\sqrt{\frac{\pi}{a}},$$

$$\int_0^\infty x^{2n+1}e^{-ax^2}dx = \frac{n!}{2a^{n+1}},$$

$$\int_{-\infty}^{\infty}\frac{1}{x^2+a^2}dx=\frac{\pi}{a},\qquad \int_{0}^{x}\frac{a}{x^2+a^2}dx=\tan^{-1}\frac{x}{a}$$

$$erf(x)=\frac{2}{\sqrt{\pi}}\int_0^x e^{-t^2}dt\approx 1-\frac{1}{\sqrt{\pi}}\frac{e^{-x^2}}{x},\qquad erfc(x)=\frac{2}{\sqrt{\pi}}\int_x^{\infty}e^{-t^2}dt=1-erf(x),\quad erf(e)\approx 1.0$$

$$\int_0^{\pi}\sin^2 mx\,dx=\int_0^{\pi}\cos^2 mx\,dx=\frac{\pi}{2}$$

$$\int_0^a \sin\left(\frac{n\pi x}{a}\right)\times\sin\left(\frac{m\pi x}{a}\right)dx=\int_0^a\cos\left(\frac{n\pi x}{a}\right)\times\cos\left(\frac{m\pi x}{a}\right)dx=\frac{a}{2}\delta_{nm}\qquad \delta_{nm}\begin{cases}=1, & n=m\\ =0, & n\neq m\end{cases}$$

$$f(a)=\frac{i}{2\pi}\int_C\frac{f(z)}{z-a}dz\qquad (Cauchy\text{'s integral theorem})\Rightarrow\ \frac{d^n}{da^n}f(a)=\frac{i}{2\pi}\int_C\frac{f(z)}{(z-a)^{n+1}}dz$$

[exercise] Find the value of the integral $\int_0^{\infty}x^{1/2}e^{-ax}dx\quad (a>0)$.

7. Differential Equations and How to Solve them

(a) Solution of linear first-order differential equations by means of integrating factor (see **6-27**)
(b) Separation of variables

} not dealt with here.

(c) Simultaneous linear first-order differential equations

There exist various ways of solving the following two simultaneous first-order differential equations given by eqs(1) and (2). Some of them are dealt with below.

$$\frac{dx}{dt}=a_1x+b_1y \qquad (1)$$

$$\frac{dy}{dt}=a_2x+b_2y \qquad (2)$$

(i) substitution method

Rewriting eq(1) as $y=\frac{x'}{b_1}-\frac{a_1x}{b_1}$, it is substituted into eq(2). The obtained eq(3) is the second-order ordinary

$$\frac{dy}{dt}=\frac{x''}{b_1}-\frac{a_1x'}{b_1}=a_2x+b_2\left(\frac{x'}{b_1}-\frac{a_1x}{b_1}\right)\quad\Rightarrow\quad x''-(a_1+b_2)x'+(a_1b_2-a_2b_1)x=0 \qquad (3)$$

differential equation, into which $e^{\lambda t}$ is substituted as its particular solution. Then a quadratic equation with respect to λ is obtained as is shown in eq(4) (see *Laplace* transform in **II-4-14**).

$$\lambda^2-(a_1+b_2)\lambda+(a_1b_2-a_2b_1)=0 \qquad (4)$$

Denoting the two roots of eq(4) λ_1 and λ_2, two particular solutions of eq(3) are given by $e^{\lambda_1 t}$ and $e^{\lambda_2 t}$, so that eqs(5) and (6) are obtained as the general solutions of the two simultaneous first-order differential equations.

$$x(t)=c_1e^{\lambda_1 t}+c_2e^{\lambda_{21}t} \qquad (5)$$

$$y(t)=d_1e^{\lambda_1 t}+d_2e^{\lambda_{21}t} \qquad (6)$$

Here, c_1, c_2, d_1 and d_2 are the coefficients to be determined from given initial conditions.

(ex.) Solve the following simultaneous equations for the initial conditions of $x(0)=6,\ y(0)=3$, $x'(0)=-12$ and $y'(0)=0$.

$$\frac{dx}{dt}=-3x+2y$$
$$\frac{dy}{dt}=x-2y$$

(solution) Substituting the corresponding values into eq(4), the two roots of the quadratic equation are found as $\lambda_1=-4$ and $\lambda_2=-1$, so that *x* and *y* can be expressed as follows:

$$x(t)=c_1e^{-4t}+c_2e^{-1t} \qquad (5')$$
$$y(t)=d_1e^{-4t}+d_2e^{-1t} \qquad (6')$$

From the initial conditions at $t=0$ (if *x* and *y* are considered to specify the position of a particle, then from the components of its initial position and velocity) given by eqs($5''$) and ($6''$),

$$6=c_1+c_2 \qquad (5'')$$
$$3=d_1+d_2 \qquad (6'')$$

eqs(8) and (9) are derived.

$$\left(\frac{dx}{dt}\right)_0=-\lambda_1c_1-\lambda_2c_2=4c_1+c_2=-12 \qquad (8)$$
$$\left(\frac{dy}{dt}\right)_0=-\lambda_1d_1-\lambda_2d_2=4d_1+d_2=0 \qquad (9)$$

Solving these four simultaneous equations, $c_1=-6,\ c_2=12,\ d_1=-1$ and $d_2=4$ are found for the coefficients of eqs($5'$) and ($6'$), so that the solutions of the two simultaneous first-order differential equations are given by eqs(10) and (11).

$$x(t)=-6e^{-4t}+12e^{-1t} \qquad (10)$$
$$y(t)=-1e^{-4t}+4e^{-1t} \qquad (11)$$

(ii) <u>solution with the use of matrix and determinant</u>

By putting $dx/dt=x'$ and $dy/dt=y'$, eqs(1) and (2) are expressed as eq(12) with the use of a matrix.

$$\begin{pmatrix}x'\\y'\end{pmatrix}=\begin{pmatrix}a_1 & b_1\\a_2 & b_2\end{pmatrix}\cdot\begin{pmatrix}x\\y\end{pmatrix} \Rightarrow \begin{pmatrix}x'\\y'\end{pmatrix}=(A)\cdot\begin{pmatrix}x\\y\end{pmatrix},\quad (A)\equiv\begin{pmatrix}a_1 & b_1\\a_2 & b_2\end{pmatrix} \qquad (12)$$

Denoting two independent linear combinations of *x* and *y* as *u* and *v*, they are expressed as the left side of eq(13) in a matrix form and its inverse transformation is expressed as the right side of eq(13).

$$\begin{pmatrix}u\\v\end{pmatrix}=(C)\cdot\begin{pmatrix}x\\y\end{pmatrix} \Rightarrow \begin{pmatrix}x\\y\end{pmatrix}=(C)^{-1}\begin{pmatrix}u\\v\end{pmatrix} \qquad (13)$$

The differentiation of the right side of eq(13) with time is expressed as eq(14).

$$\begin{pmatrix}x'\\y'\end{pmatrix}=(C)^{-1}\begin{pmatrix}u'\\v'\end{pmatrix} \qquad (14)$$

From eqs(12), (13) and (14), eq(15) is derived.

$$\begin{pmatrix}x'\\y'\end{pmatrix}=(C)^{-1}\begin{pmatrix}u'\\v'\end{pmatrix}=(A)\cdot\begin{pmatrix}x\\y\end{pmatrix}=(A)\cdot(C)^{-1}\begin{pmatrix}u\\v\end{pmatrix} \Rightarrow \begin{pmatrix}u'\\v'\end{pmatrix}=(C)\cdot(A)\cdot(C)^{-1}=\begin{pmatrix}u\\v\end{pmatrix} \qquad (15)$$

By finding out a matrix (C) and its inverse matrix $(C)^{-1}$ which diagonalize the matrix A, they are expressed as eq(16).

$$(C)\cdot(A)\cdot(C)^{-1}=\begin{pmatrix}\lambda_1 & 0\\ 0 & \lambda_2\end{pmatrix} \qquad (16)$$

Then, the differential equations of eq(15) can be expressed as two linear first-order differential equations which are readily solvable. It is seen from eq(13) that the final solutions are given by eq(17).

$$\begin{pmatrix}u\\ v\end{pmatrix}=\begin{pmatrix}e^{\lambda_1 t}\\ e^{\lambda_2 t}\end{pmatrix} \quad \Rightarrow \quad \begin{pmatrix}x\\ y\end{pmatrix}=(C)^{-1}\begin{pmatrix}e^{\lambda_1 t}\\ e^{\lambda_2 t}\end{pmatrix} \qquad (17)$$

You are recommended to refer a text book on matrix and determinant for a practical procedure of diagonalizing matrixes, but in order to solve the simultaneous differential equations in eq(12), it is sufficient to solve eq(18) to find the eigenvalues of λ_1 and λ_2.

$$\begin{vmatrix}a_1-\lambda & b_1\\ a_2 & b_2-\lambda\end{vmatrix}=0 \qquad (18)$$

(ex.) Solve the simultaneous differential equations dealt with in (i) based on the method of (ii) given above.

(solution) Substituting the given values into eq(18), the right side of eq(19) is obtained when expanded.

$$\begin{vmatrix}-3-\lambda & 2\\ 1 & -2-\lambda\end{vmatrix}=0 \quad \Rightarrow \quad (3+\lambda)(2+\lambda)-2=0 \qquad (19)$$

Solving eq(19), $\lambda_1=-4,\ \lambda_2=-1$ are found as the eigenvalues of the determinant, so that x and y are expressed in a matrix form as eq(20), in which the coefficients ($(C)^{-1}$) are determined from the initial conditions.

$$\begin{pmatrix}x\\ y\end{pmatrix}=(C)^{-1}\begin{pmatrix}e^{-4t}\\ e^{-1t}\end{pmatrix} \qquad (20)$$

"note" Even for n-dimensional simultaneous differential equations, as far as they are of the first-order, the $n\times n$ determinant obtained by generalizing eq(18) can be solved to find particular solutions and, hence, general solutions of the n-dimensional simultaneous differential equations of this type can be expressed in terms of the linear combinations of these particular solutions. We often encounter this type of differential equations in chemical kinetics.

(d) Solution of autonomous differential equation

See **problems** from **6-46** to **6-48** for the method of critical point and a solution of a logistic equation. $\frac{dy}{dt}=ay-by^2-h$.

(e) Simple partial differential equations – variable separation method

(ex.) $\frac{1}{v^2}\frac{\partial^2 u(x,t)}{\partial t^2}=\frac{\partial^2 u(x,t)}{\partial x^2}$, $u(x,t)$: amplitude of wave,
solvable by putting $u(x,t)=\mathrm{X}(x)\cdot\mathrm{Y}(t)$.

8. Thermodynamics and Differential

In thermodynamics, differential values of thermodynamic quantities are indispensable. Taking the internal energy $U(T,V)$ (or you may express it as $U(S,V)$ or $U(S,P)$), its total differential can be written as eq(1).

$$dU = \left(\frac{\partial U}{\partial T}\right)_V dT + \left(\frac{\partial U}{\partial V}\right)_T dV \qquad (1)$$

Let us calculate the total differential employing a simple function as an example. Here, z is a function of x and y and is given by eq(2).

$$z(x,y) = x^2 + 3xy + y^2 \qquad (2)$$

The differential value dz is a variation in z when the variables x and y change infinitesimally to $x+dx$ and $y+dy$, respectively, so that it is expressed as eq(3).

$$dz = z(x+dx, y+dy) - z(x,y) \qquad (3)$$

This variation is calculated for the function picked up as an example as given below.

$$\begin{aligned} dz &= (x+dx)^2 + 3(x+dx)(y+dy) + (y+dy)^2 - x^2 - 3xy + y^2 \\ &= 2xdx + (dx)^2 + 3xdy + 3ydx + 3dxdy + 2ydy + (dy)^2 \\ &\approx (2x+3y)dx + (3x+2y)dy \end{aligned} \qquad (4)$$

In the last form of eq(4), the second-order terms of $(dx)^2$, $(dy)^2$ and $dxdy$ with respect to small values of dx and dy are neglected.

The general way of deriving the result of eq(4) is as follows. By adding zero to eq(3), that is adding $-z(x+dx,y)+z(x+dx,y)$, the obtained equation is rewritten as eq(5).

$$\begin{aligned} dz &= z(x+dx, y+dy) - z(x+dx, y) + z(x+dx, y) - z(x,y) \\ &= \{z(x+dx, y+dy) - z(x+dx, y)\} + \{z(x+dx, y) - z(x,y)\} \\ &= \left\{\frac{z(x+dx, y+dy) - z(x+dx, y)}{dy}\right\}dy + \left\{\frac{z(x+dx, y) - z(x,y)}{dx}\right\}dx \\ &= \left\{\frac{z(x+dx, y) - z(x,y)}{dx}\right\}dx + \left\{\frac{z(x+dx, y+dy) - z(x+dx, y)}{dy}\right\}dy \end{aligned} \qquad (5)$$

The first term of the last equality of eq(5) is the amount of change of z when x is varied infinitesimally while keeping y constant and, on the other hand, the second term is the amount of change of z when y is varied while keeping x constant. Denoting these dz_y and dz_x, respectively, eq(5) can be expressed as eq(6).

$$\begin{aligned} dz &= \left\{\frac{z(x+dx, y) - z(x,y)}{dx}\right\}dx + \left\{\frac{z(x+dx, y+dy) - z(x+dx, y)}{dy}\right\}dy \\ &= \frac{dz_y}{dx}dx + \frac{dz_x}{dy}dy \end{aligned} \qquad (6)$$

Each of the coefficients of dx and dy in eq(6) is called partial differential in the sense that these differential coefficients are calculated by keeping the other variable(s) constant and they are generally written as eq(7).

$$dz = \left(\frac{\partial z}{dx}\right)_y dx + \left(\frac{\partial z}{dy}\right)_x dy \qquad (7)$$

Consequently, the total differential dz is given by eq(8), which is of course identical with the previously given eq(4).

$$\left(\frac{\partial z}{dx}\right)_y = 2x+3y, \quad \left(\frac{\partial z}{dy}\right)_x = 3x+2y \quad \Rightarrow \quad dz = (2x+3y)dx + (3x+2y)dy \qquad (8)$$

The procedures of the equations from (3) to (7) applied successively to the internal energy $U(T,V)$ are shown below.

$$\begin{aligned} dU &= U(T+dT, V+dV) - U(T,V) \\ &= U(T+dT,V+dV) - U(T+dT,V) + U(T+dT,V) - U(T,V) \\ &= \langle U(T+dT,V+dV) - U(T+dT,V)\rangle + \langle U(T+dT,V) - U(T,V)\rangle \\ &= (dU)_T + (dU)_V \qquad (9) \end{aligned}$$

Note that $(dU)_T$ and $(dU)_V$ are defined by the equations given below.

$$\langle U(T+dT,V+dV) - U(T+dT,V)\rangle \approx \langle U(T,V+dV) - U(T,V)\rangle = (dU)_T$$

$$\langle U(T+dT,V) - U(T+dT,V)\rangle = (dU)_V$$

Eq(9) is rewritten in the general form of partial differential as eq(10).

$$\begin{aligned} dU &= (dU)_T + (dU)_V \\ &= \frac{(dU)_T}{dV} \times dV + \frac{(dU)_V}{dT} \times dT \\ &= \left(\frac{\partial U}{\partial V}\right)_T dV + \left(\frac{\partial U}{\partial T}\right)_V dT \qquad (10) \end{aligned}$$

If you master the procedures employed in deriving eq(10), it will be very useful in handling the thermodynamic functions and you could be a little free of thermodynamic allergy.

[exercise] Show that *Euler*'s chain relation $\left(\frac{\partial P}{\partial V}\right)_T \left(\frac{\partial V}{\partial T}\right)_P \left(\frac{\partial T}{\partial P}\right)_V = -1$ holds. Also, employing the equation of state $PV = nRT$ of the ideal gas, confirm that this chain relation is really satisfied.

[clue] The following relations are useful for proving the relation.

$$dP = \left(\frac{\partial P}{\partial V}\right)_T dV + \left(\frac{\partial P}{\partial T}\right)_V dT \quad \Rightarrow \quad 0 = \left(\frac{\partial P}{\partial V}\right)_T (dV)_P + \left(\frac{\partial P}{\partial T}\right)_V (dT)_P$$

9. Polar Coordinate Expressions of Differential Operators ∇ and ∇^2

The rectangular coordinates x, y and z are expressed in terms of the polar coordinates r, ϑ and φ as eq(1).

$$\left.\begin{aligned} x &= r\sin\vartheta\cos\varphi \\ y &= r\sin\vartheta\sin\varphi \\ z &= r\cos\vartheta \end{aligned}\right\} \qquad (1)$$

Regarding each of x, y and z as a function of r, ϑ and φ, the total differentials of x, y and z are found as follows (see the section **8**).

$$\left.\begin{aligned} dx &= \sin\vartheta\cos\varphi\cdot dr + r\cos\vartheta\cos\varphi\cdot d\vartheta - r\sin\vartheta\sin\varphi\cdot d\varphi \\ dy &= \sin\vartheta\sin\varphi\cdot dr + r\cos\vartheta\sin\varphi\cdot d\vartheta + r\sin\vartheta\cos\varphi\cdot d\varphi \\ dz &= \cos\vartheta\cdot dr - r\sin\vartheta\cdot d\vartheta \end{aligned}\right\} \qquad (2)$$

Eq(2) is expressed using a matrix as eq(3).

$$\begin{pmatrix} dx \\ dy \\ dz \end{pmatrix} = \begin{pmatrix} \sin\vartheta\cos\varphi & r\cos\vartheta\cos\varphi & -r\sin\vartheta\sin\varphi \\ \sin\vartheta\sin\varphi & r\cos\vartheta\sin\varphi & r\sin\vartheta\cos\varphi \\ \cos\vartheta & -r\sin\vartheta & 0 \end{pmatrix} \begin{pmatrix} dr \\ d\vartheta \\ d\varphi \end{pmatrix} \equiv (A)\cdot \begin{pmatrix} dr \\ d\vartheta \\ d\varphi \end{pmatrix} \quad (3)$$

From eq(3), using the inverse matrix $(A)^{-1}$ of the matrix (A), dr, $d\vartheta$ and $d\varphi$ are expressed in terms of dx, dy and dz as eq(4).

$$\begin{pmatrix} dr \\ d\vartheta \\ d\varphi \end{pmatrix} = (A)^{-1} \begin{pmatrix} dx \\ dy \\ dz \end{pmatrix}, \qquad (A)^{-1}(A) = (E), \qquad (E): \text{unit matrix} \quad (4)$$

The inverse matrix here can be expressed as eq(5) using the determinant $|A| = r^2\sin\vartheta$ and the cofactor (adjugate) matrix ${}_{adj}A$ of the matrix A.

$$(A)^{-1} = \frac{{}_{adj}A}{|A|} = \frac{1}{r^2\sin\vartheta} \begin{pmatrix} r^2\sin^2\vartheta\cos\varphi & r^2\sin^2\vartheta\sin\varphi & r^2\sin\vartheta\cos\vartheta \\ r\sin\vartheta\cos\vartheta\cos\varphi & r\sin\vartheta\cos\vartheta\sin\varphi & -r\sin^2\vartheta \\ -r\sin\varphi & r\cos\varphi & 0 \end{pmatrix} \quad (5)$$

check for $(A)^{-1}(A) = (E)$

$$\begin{pmatrix} r^2\sin^2\vartheta\cos\varphi & r^2\sin^2\vartheta\sin\varphi & r^2\sin\vartheta\cos\vartheta \\ r\sin\vartheta\cos\vartheta\cos\varphi & r\sin\vartheta\cos\vartheta\sin\varphi & -r\sin^2\vartheta \\ -r\sin\varphi & r\cos\varphi & 0 \end{pmatrix} \cdot \begin{pmatrix} \sin\vartheta\cos\varphi & r\cos\vartheta\cos\varphi & -r\sin\vartheta\sin\varphi \\ \sin\vartheta\sin\varphi & r\cos\vartheta\sin\varphi & r\sin\vartheta\cos\varphi \\ \cos\vartheta & -r\sin\vartheta & 0 \end{pmatrix}$$

$$= \begin{pmatrix} r^2\sin^3\vartheta\cos^2\varphi + r^2\sin^3\vartheta\sin^2\varphi + r^2\sin\vartheta\cos^2\vartheta & r^2\sin^2\vartheta\cos\vartheta\cos^2\varphi + r^2\sin^2\vartheta\cos\vartheta\sin^2\varphi - r^2\sin\vartheta\cos\vartheta & 0 \\ r\sin^2\vartheta\cos\vartheta\cos^2\varphi + r\sin^2\vartheta\cos\vartheta\sin^2\varphi - r\sin^2\vartheta\cos\vartheta & r^2\sin\vartheta\cos^2\vartheta\cos^2\varphi + r^2\sin\vartheta\cos^2\vartheta\sin^2\varphi + r^2\sin^3\vartheta & 0 \\ -r\sin\vartheta\sin\varphi\cos\varphi + r\sin\vartheta\sin\varphi\cos\varphi & -r\cos\vartheta\sin\varphi\cos\varphi + r\cos\vartheta\sin\varphi\cos\varphi & r^2\sin\vartheta \end{pmatrix}$$

$$= \begin{pmatrix} r^2\sin\vartheta & 0 & 0 \\ 0 & r^2\sin\vartheta & 0 \\ 0 & 0 & r^2\sin\vartheta \end{pmatrix}$$

Substituting eq(5) into eq(4), eq(6) is obtained.

$$\begin{pmatrix} dr \\ d\vartheta \\ d\varphi \end{pmatrix} = \frac{1}{r^2\sin\vartheta} \begin{pmatrix} r^2\sin^2\vartheta\cos\varphi & r^2\sin^2\vartheta\sin\varphi & r^2\sin\vartheta\cos\vartheta \\ r\sin\vartheta\cos\vartheta\cos\varphi & r\sin\vartheta\cos\vartheta\sin\varphi & -r\sin^2\vartheta \\ -r\sin\varphi & r\cos\varphi & 0 \end{pmatrix} \begin{pmatrix} dx \\ dy \\ dz \end{pmatrix} \quad (6)$$

[exercise] Confirm the relation $(A)^{-1}(A) = (E) = \begin{pmatrix} 1 & 0 & 0 \\ 0 & 1 & 0 \\ 0 & 0 & 1 \end{pmatrix}$ from eqs(3) and (5).

Now, $\dfrac{\partial}{\partial x}$ can be expressed in terms of r, ϑ and φ as eq(7), but remarking here that $\dfrac{\partial}{\partial x} = \left(\dfrac{\partial}{\partial x}\right)_{y,z}$.

$$\frac{\partial}{\partial x} = \frac{\partial}{\partial\vartheta}\cdot\frac{\partial\vartheta}{\partial x} + \frac{\partial}{\partial\varphi}\cdot\frac{\partial\varphi}{\partial x} + \frac{\partial}{\partial r}\cdot\frac{\partial r}{\partial x} \quad \left(= \frac{\partial\vartheta}{\partial x}\cdot\frac{\partial}{\partial\vartheta} + \frac{\partial\varphi}{\partial x}\cdot\frac{\partial}{\partial\varphi} + \frac{\partial r}{\partial x}\cdot\frac{\partial}{\partial r} \right) \quad (7)$$

$\dfrac{\partial\vartheta}{\partial x}$ must be put $\dfrac{\partial\vartheta}{\partial x} = \left(\dfrac{\partial\vartheta}{\partial x}\right)_{y,z}$. In such a case as eq(7) where the differentiation is expressed by transforming variables, it is essential to make it clear which variables are kept constant for each of differentiations. For this purpose, eq(6) which is calculated by using the inverse matrix is required. Each of the components of eq(6) can be expressed as follows from eq(8) to eq(8″).

$$d\vartheta = \frac{\cos\vartheta\cos\varphi}{r}dx + \frac{\cos\vartheta\sin\varphi}{r}dy - \frac{\sin\vartheta}{r}dz \qquad (8)$$

$$dr = \sin\vartheta\cos\varphi \cdot dx + \sin\vartheta\sin\varphi \cdot dy + \cos\vartheta \cdot dz \qquad (8')$$

$$d\varphi = -\frac{\sin\varphi}{r\sin\vartheta}dx + \frac{\cos\varphi}{r\sin\vartheta}dy \qquad (8'')$$

From eq(8), eq(9) is readily derived. In a similar manner, $\left(\frac{\partial r}{\partial x}\right)_{y,z}$, $\left(\frac{\partial \varphi}{\partial x}\right)_{y,z}$, …, are obtained as are shown in eqs(9′) and (9″).

$$\left.\begin{aligned} \left(\frac{\partial\vartheta}{\partial x}\right)_{y,z} &= \frac{\cos\vartheta\cos\varphi}{r} \\ \left(\frac{\partial\vartheta}{\partial y}\right)_{z,x} &= \frac{\cos\vartheta\sin\varphi}{r} \\ \left(\frac{\partial\vartheta}{\partial z}\right)_{x,y} &= -\frac{\sin\vartheta}{r} \end{aligned}\right\} (9) \qquad \left.\begin{aligned} \left(\frac{\partial r}{\partial x}\right)_{y,z} &= \sin\vartheta\cos\varphi \\ \left(\frac{\partial r}{\partial y}\right)_{z,x} &= \sin\vartheta\sin\varphi \\ \left(\frac{\partial r}{\partial z}\right)_{x,y} &= \cos\vartheta \end{aligned}\right\} (9') \qquad \left.\begin{aligned} \left(\frac{\partial\varphi}{\partial x}\right)_{y,z} &= -\frac{\sin\varphi}{r\sin\vartheta} \\ \left(\frac{\partial\varphi}{\partial y}\right)_{z,x} &= \frac{\cos\varphi}{r\sin\vartheta} \\ \left(\frac{\partial\varphi}{\partial z}\right)_{x,y} &= 0 \end{aligned}\right\} (9'')$$

Employing these results, eq(10) is obtained from eq(7) and eqs(11) and (12) are similarly obtained from the equations corresponding to eq(7).

$$\begin{aligned} \frac{\partial}{\partial x} &= \frac{\partial}{\partial\vartheta}\cdot\frac{\partial\vartheta}{\partial x} + \frac{\partial}{\partial\varphi}\cdot\frac{\partial\varphi}{\partial x} + \frac{\partial}{\partial r}\cdot\frac{\partial r}{\partial x} \\ &= \frac{\cos\vartheta\cos\varphi}{r}\frac{\partial}{\partial\vartheta} - \frac{\sin\varphi}{r\sin\vartheta}\frac{\partial}{\partial\varphi} + \sin\vartheta\cos\varphi\frac{\partial}{\partial r} \end{aligned} \qquad (10)$$

$$\begin{aligned} \frac{\partial}{\partial y} &= \frac{\partial}{\partial\vartheta}\cdot\frac{\partial\vartheta}{\partial y} + \frac{\partial}{\partial\varphi}\cdot\frac{\partial\varphi}{\partial y} + \frac{\partial}{\partial r}\cdot\frac{\partial r}{\partial y} \\ &= \frac{\cos\vartheta\sin\varphi}{r}\frac{\partial}{\partial\vartheta} + \frac{\cos\varphi}{r\sin\vartheta}\frac{\partial}{\partial\varphi} + \sin\vartheta\sin\varphi\frac{\partial}{\partial r} \end{aligned} \qquad (11)$$

$$\begin{aligned} \frac{\partial}{\partial z} &= \frac{\partial}{\partial\vartheta}\cdot\frac{\partial\vartheta}{\partial z} + \frac{\partial}{\partial\varphi}\cdot\frac{\partial\varphi}{\partial z} + \frac{\partial}{\partial r}\cdot\frac{\partial r}{\partial z} \\ &= -\frac{\sin\vartheta}{r}\frac{\partial}{\partial\vartheta} + \cos\vartheta\frac{\partial}{\partial r} \end{aligned} \qquad (12)$$

Based on these equations, the calculations of the second-order partial differentials with respect to x, y and z can be carried out as follows, though a bit lengthy and tedious.

$$\begin{aligned} \frac{\partial}{\partial x}\cdot\frac{\partial}{\partial x} &= \left(\frac{\cos\vartheta\cos\varphi}{r}\frac{\partial}{\partial\vartheta} - \frac{\sin\varphi}{r\sin\vartheta}\frac{\partial}{\partial\varphi} + \sin\vartheta\cos\varphi\frac{\partial}{\partial r}\right)\times\left(\frac{\cos\vartheta\cos\varphi}{r}\frac{\partial}{\partial\vartheta} - \frac{\sin\varphi}{r\sin\vartheta}\frac{\partial}{\partial\varphi} + \sin\vartheta\cos\varphi\frac{\partial}{\partial r}\right) \\ &= (A+B+C)\times(A+B+C) \\ &= A\cdot A + A\cdot B + A\cdot C + B\cdot A + B\cdot B + B\cdot C + C\cdot A + C\cdot B + C\cdot C \end{aligned}$$

Here, A, B and C represent the relevant operators including differentials. More specifically, the products of operators are expressed as follows.

$$A \cdot A = \frac{\cos\vartheta\cos\varphi}{r}\frac{\partial}{\partial\vartheta}\left(\frac{\cos\vartheta\cos\varphi}{r}\frac{\partial}{\partial\vartheta}\right) = \frac{\cos^2\vartheta\cos^2\varphi}{r^2}\frac{\partial^2}{\partial\vartheta^2} - \frac{\sin\vartheta\cos\vartheta\cos^2\varphi}{r^2}\frac{\partial}{\partial\vartheta}$$

$$A \cdot B = \frac{\cos\vartheta\cos\varphi}{r}\frac{\partial}{\partial\vartheta}\left(-\frac{\sin\varphi}{r\sin\vartheta}\frac{\partial}{\partial\varphi}\right) = -\frac{\cos\vartheta\sin\varphi\cos\varphi}{r^2 sin\vartheta}\frac{\partial^2}{\partial\vartheta\partial\varphi} + \frac{\cos^2\vartheta\sin\varphi\cos\varphi}{r^2\sin^2\vartheta}\frac{\partial}{\partial\varphi}$$

$$A \cdot C = \frac{\cos\vartheta\cos\varphi}{r}\frac{\partial}{\partial\vartheta}\left(\sin\vartheta\cos\varphi\frac{\partial}{\partial r}\right) = \frac{\sin\vartheta\cos\vartheta\cos^2\varphi}{r}\frac{\partial^2}{\partial\vartheta\partial r} + \frac{\cos^2\vartheta\cos^2\varphi}{r}\frac{\partial}{\partial r}$$

$$B \cdot A = -\frac{\sin\varphi}{r\sin\vartheta}\frac{\partial}{\partial\varphi}\left(\frac{\cos\vartheta\cos\varphi}{r}\frac{\partial}{\partial\vartheta}\right) = -\frac{\cos\vartheta\sin\varphi\cos\varphi}{r^2\sin\vartheta}\frac{\partial^2}{\partial\varphi\partial\vartheta} + \frac{\cos\vartheta\sin^2\varphi}{r^2\sin\vartheta}\frac{\partial}{\partial\vartheta}$$

$$B \cdot B = -\frac{\sin\varphi}{r\sin\vartheta}\frac{\partial}{\partial\varphi}\left(-\frac{\sin\varphi}{r\sin\vartheta}\frac{\partial}{\partial\varphi}\right) = \frac{\sin^2\varphi}{r^2\sin^2\vartheta}\frac{\partial^2}{\partial\varphi^2} + \frac{\sin\varphi\cos\varphi}{r^2\sin^2\vartheta}\frac{\partial}{\partial\varphi}$$

$$B \cdot C = -\frac{\sin\varphi}{r\sin\vartheta}\frac{\partial}{\partial\varphi}\left(\sin\vartheta\cos\varphi\frac{\partial}{\partial r}\right) = -\frac{\sin\varphi\cos\varphi}{r}\frac{\partial^2}{\partial\varphi\partial r} + \frac{\sin^2\varphi}{r}\frac{\partial}{\partial r}$$

$$C \cdot A = \sin\vartheta\cos\varphi\frac{\partial}{\partial r}\left(\frac{\cos\vartheta\cos\varphi}{r}\frac{\partial}{\partial\vartheta}\right) = \frac{\sin\vartheta\cos\vartheta\cos^2\varphi}{r}\frac{\partial^2}{\partial r\partial\vartheta} - \frac{\sin\vartheta\cos\vartheta\cos^2\varphi}{r^2}\frac{\partial}{\partial\vartheta}$$

$$C \cdot B = \sin\vartheta\cos\varphi\frac{\partial}{\partial r}\left(-\frac{\sin\varphi}{r\sin\vartheta}\frac{\partial}{\partial\varphi}\right) = -\frac{\sin\varphi\cos\varphi}{r}\frac{\partial^2}{\partial r\partial\varphi} + \frac{\sin\varphi\cos\varphi}{r^2}\frac{\partial}{\partial\varphi}$$

$$C \cdot C = \sin\vartheta\cos\varphi\frac{\partial}{\partial r}\left(\sin\vartheta\cos\varphi\frac{\partial}{\partial r}\right) = \sin^2\vartheta\cos^2\varphi\frac{\partial^2}{\partial r^2}$$

Similarly from eq(11),

$$\begin{aligned}\frac{\partial}{\partial y}\cdot\frac{\partial}{\partial y} &= \left(\frac{\cos\vartheta\sin\varphi}{r}\frac{\partial}{\partial\vartheta} + \frac{\cos\varphi}{r\sin\vartheta}\frac{\partial}{\partial\varphi} + \sin\vartheta\sin\varphi\frac{\partial}{\partial r}\right)\times\left(\frac{\cos\vartheta\sin\varphi}{r}\frac{\partial}{\partial\vartheta} + \frac{\cos\varphi}{r\sin\vartheta}\frac{\partial}{\partial\varphi} + \sin\vartheta\sin\varphi\frac{\partial}{\partial r}\right)\\ &= (A' + B' + C')\times(A' + B' + C')\\ &= A'\cdot A' + A'\cdot B' + A'\cdot C' + B'\cdot A' + B'\cdot B' + B'\cdot C' + C'\cdot A' + C'\cdot B' + C'\cdot C'\end{aligned}$$

$$A' \cdot A' = \frac{\cos\vartheta\sin\varphi}{r}\frac{\partial}{\partial\vartheta}\left(\frac{\cos\vartheta\sin\varphi}{r}\frac{\partial}{\partial\vartheta}\right) = \frac{\cos^2\vartheta\sin^2\varphi}{r^2}\frac{\partial^2}{\partial\vartheta^2} - \frac{\sin\vartheta\cos\vartheta\sin^2\varphi}{r^2}\frac{\partial}{\partial\vartheta}$$

$$A' \cdot B' = \frac{\cos\vartheta\sin\varphi}{r}\frac{\partial}{\partial\vartheta}\left(\frac{\cos\varphi}{r\sin\vartheta}\frac{\partial}{\partial\varphi}\right) = \frac{\cos\vartheta\sin\varphi\cos\varphi}{r^2\sin\vartheta}\frac{\partial^2}{\partial\vartheta\partial\varphi} - \frac{\cos^2\vartheta\sin\varphi\cos\varphi}{r^2\sin^2\vartheta}\frac{\partial}{\partial\varphi}$$

$$A' \cdot C' = \frac{\cos\vartheta\sin\varphi}{r}\frac{\partial}{\partial\vartheta}\left(\sin\vartheta\sin\varphi\frac{\partial}{\partial r}\right) = \frac{\sin\vartheta\cos\vartheta\sin^2\varphi}{r}\frac{\partial^2}{\partial\vartheta\partial r} + \frac{\cos^2\vartheta\sin^2\varphi}{r}\frac{\partial}{\partial r}$$

$$B' \cdot A' = \frac{\cos\varphi}{r\sin\vartheta}\frac{\partial}{\partial\varphi}\left(\frac{\cos\vartheta\sin\varphi}{r}\frac{\partial}{\partial\vartheta}\right) = \frac{\cos\vartheta\sin\varphi\cos\varphi}{r^2\sin\vartheta}\frac{\partial^2}{\partial\varphi\partial\vartheta} + \frac{\cos\vartheta\cos^2\varphi}{r^2\sin\vartheta}\frac{\partial}{\partial\vartheta}$$

$$B' \cdot B' = \frac{\cos\varphi}{r\sin\vartheta}\frac{\partial}{\partial\varphi}\left(\frac{\cos\varphi}{r\sin\vartheta}\frac{\partial}{\partial\varphi}\right) = \frac{\cos^2\varphi}{r^2\sin^2\vartheta}\frac{\partial^2}{\partial\varphi^2} - \frac{\sin\varphi\cos\varphi}{r^2\sin^2\vartheta}\frac{\partial}{\partial\varphi}$$

$$B' \cdot C' = \frac{\cos\varphi}{r\sin\vartheta}\frac{\partial}{\partial\varphi}\left(\sin\vartheta\sin\varphi\frac{\partial}{\partial r}\right) = \frac{\sin\varphi\cos\varphi}{r}\frac{\partial^2}{\partial\varphi\partial r} + \frac{\cos^2\varphi}{r}\frac{\partial}{\partial r}$$

$$C' \cdot A' = \sin\vartheta\sin\varphi\frac{\partial}{\partial r}\left(\frac{\cos\vartheta\sin\varphi}{r}\frac{\partial}{\partial\vartheta}\right) = \frac{\sin\vartheta\cos\vartheta\sin^2\varphi}{r}\frac{\partial^2}{\partial r\partial\vartheta} - \frac{\sin\vartheta\cos\vartheta\sin^2\varphi}{r^2}\frac{\partial}{\partial\vartheta}$$

$$C' \cdot B' = \sin\vartheta\sin\varphi\frac{\partial}{\partial r}\left(\frac{\cos\varphi}{r\sin\vartheta}\frac{\partial}{\partial\varphi}\right) = \frac{\sin\varphi\cos\varphi}{r}\frac{\partial^2}{\partial r\partial\varphi} - \frac{\sin\varphi\cos\varphi}{r^2}\frac{\partial}{\partial\varphi}$$

$$C' \cdot C' = \sin\vartheta\sin\varphi\frac{\partial}{\partial r}\left(\sin\vartheta\sin\varphi\frac{\partial}{\partial r}\right) = \sin^2\vartheta\sin^2\varphi\frac{\partial^2}{\partial r^2}$$

and finally from eq(12), the second-order partial differential with respect to z is calculated as follows.

$$\begin{aligned}\frac{\partial}{\partial z}\cdot\frac{\partial}{\partial z} &= \left(-\frac{\sin\vartheta}{r}\frac{\partial}{\partial\vartheta} + \cos\vartheta\frac{\partial}{\partial r}\right)\times\left(-\frac{\sin\vartheta}{r}\frac{\partial}{\partial\vartheta} + \cos\vartheta\frac{\partial}{\partial r}\right)\\ &= \left(A'' + C''\right)\times\left(A'' + C''\right)\\ &= A''\cdot A'' + A''\cdot C'' + C''\cdot A'' + C''\cdot C''\end{aligned}$$

$$A''\cdot A'' = \left(-\frac{\sin\vartheta}{r}\frac{\partial}{\partial\vartheta}\right)\left(-\frac{\sin\vartheta}{r}\frac{\partial}{\partial\vartheta}\right) = \frac{\sin^2\vartheta}{r^2}\frac{\partial^2}{\partial\vartheta^2} + \frac{\sin\vartheta\cos\vartheta}{r^2}\frac{\partial}{\partial\vartheta}$$

$$A''\cdot C'' = \left(-\frac{\sin\vartheta}{r}\frac{\partial}{\partial\vartheta}\right)\left(\cos\vartheta\frac{\partial}{\partial r}\right) = -\frac{\sin\vartheta\cos\vartheta}{r}\frac{\partial^2}{\partial\vartheta\partial r} + \frac{\sin^2\vartheta}{r}\frac{\partial}{\partial r}$$

$$C''\cdot A'' = \left(\cos\vartheta\frac{\partial}{\partial r}\right)\left(-\frac{\sin\vartheta}{r}\frac{\partial}{\partial\vartheta}\right) = -\frac{\sin\vartheta\cos\vartheta}{r}\frac{\partial^2}{\partial r\partial\vartheta} + \frac{\sin\vartheta\cos\vartheta}{r^2}\frac{\partial}{\partial\vartheta}$$

$$C''\cdot C'' = \left(\cos\vartheta\frac{\partial}{\partial r}\right)\left(\cos\vartheta\frac{\partial}{\partial r}\right) = \cos^2\vartheta\frac{\partial^2}{\partial r^2}$$

In summary, putting the results of the calculations carried out in the three steps together and they are arranged according to the type of the partial differential such as $\frac{\partial^2}{\partial\vartheta^2}$, $\frac{\partial}{\partial\vartheta}$, $\cdots$, paying attention to the relations of $\frac{\partial^2}{\partial\vartheta\partial\varphi} = \frac{\partial^2}{\partial\varphi\partial\vartheta}$, $\frac{\partial^2}{\partial\vartheta\partial r} = \frac{\partial^2}{\partial r\partial\vartheta}$, $\frac{\partial^2}{\partial r\partial\varphi} = \frac{\partial^2}{\partial\varphi\partial r}$ as follows. The left side of each equation is the result of arrangement

$$\frac{1}{r^2}\frac{\partial^2}{\partial\vartheta^2} = \frac{\cos^2\vartheta\cos^2\varphi}{r^2}\frac{\partial^2}{\partial\vartheta^2} + \frac{\cos^2\vartheta\sin^2\varphi}{r^2}\frac{\partial^2}{\partial\vartheta^2} + \frac{\sin^2\vartheta}{r^2}\frac{\partial^2}{\partial\vartheta^2}$$

$$\begin{aligned}\frac{\cos\vartheta}{r^2\sin\vartheta}\frac{\partial}{\partial\vartheta} = &-\frac{\sin\vartheta\cos\vartheta\cos^2\varphi}{r^2}\frac{\partial}{\partial\vartheta} + \frac{\cos\vartheta\sin^2\varphi}{r^2\sin\vartheta}\frac{\partial}{\partial\vartheta} - \frac{\sin\vartheta\cos\vartheta\cos^2\varphi}{r^2}\frac{\partial}{\partial\vartheta}\\ &-\frac{\sin\vartheta\cos\vartheta\sin^2\varphi}{r^2}\frac{\partial}{\partial\vartheta} + \frac{\cos\vartheta\cos^2\varphi}{r^2\sin\vartheta}\frac{\partial}{\partial\vartheta} - \frac{\sin\vartheta\cos\vartheta\sin^2\varphi}{r^2}\frac{\partial}{\partial\vartheta}\\ &+\frac{\sin\vartheta\cos\vartheta}{r^2}\frac{\partial}{\partial\vartheta} + \frac{\sin\vartheta\cos\vartheta}{r^2}\frac{\partial}{\partial\vartheta}\end{aligned}$$

$$\frac{\partial^2}{\partial r^2} = \sin^2\vartheta\cos^2\varphi\frac{\partial^2}{\partial r^2} + \sin^2\vartheta\sin^2\varphi\frac{\partial^2}{\partial r^2} + \cos^2\vartheta\frac{\partial^2}{\partial r^2}$$

$$\frac{2}{r}\frac{\partial}{\partial r} = \frac{\cos^2\vartheta\cos^2\varphi}{r}\frac{\partial}{\partial r} + \frac{\sin^2\varphi}{r}\frac{\partial}{\partial r} + \frac{\cos^2\vartheta\sin^2\varphi}{r}\frac{\partial}{\partial r} + \frac{\cos^2\varphi}{r}\frac{\partial}{\partial r} + \frac{\sin^2\vartheta}{r}\frac{\partial}{\partial r}$$

$$\frac{1}{r^2\sin^2\vartheta}\frac{\partial^2}{\partial\varphi^2} = \frac{\sin^2\varphi}{r^2\sin^2\vartheta}\frac{\partial^2}{\partial\varphi^2} + \frac{\cos^2\varphi}{r^2\sin^2\vartheta}\frac{\partial^2}{\partial\varphi^2}$$

$$\begin{aligned}\frac{\partial}{\partial\varphi} &= \frac{\cos^2\vartheta\sin\varphi\cos\varphi}{r^2\sin^2\vartheta}\frac{\partial}{\partial\varphi} + \frac{\sin\varphi\cos\varphi}{r^2\sin^2\vartheta}\frac{\partial}{\partial\varphi} + \frac{\sin\varphi\cos\varphi}{r^2} \\ &\quad - \frac{\cos^2\vartheta\sin\varphi\cos\varphi}{r^2\sin^2\vartheta}\frac{\partial}{\partial\varphi} - \frac{\sin\varphi\cos\varphi}{r^2}\frac{\partial}{\partial\varphi} - \frac{\sin\varphi\cos\varphi}{r^2\sin^2\vartheta}\frac{\partial}{\partial\varphi} \\ &= 0\end{aligned}$$

$$\begin{aligned}\frac{\partial^2}{\partial\varphi\partial\vartheta} &= -\frac{\cos\vartheta\sin\varphi\cos\varphi}{r^2\sin\vartheta}\frac{\partial^2}{\partial\varphi\partial\vartheta} - \frac{\cos\vartheta\sin\varphi\cos\varphi}{r^2\sin\vartheta}\frac{\partial^2}{\partial\varphi\partial\vartheta} \\ &\quad + \frac{\cos\vartheta\sin\varphi\cos\varphi}{r^2\sin\vartheta}\frac{\partial^2}{\partial\varphi\partial\vartheta} + \frac{\cos\vartheta\sin\varphi\cos\varphi}{r^2\sin\vartheta}\frac{\partial^2}{\partial\varphi\partial\vartheta} \\ &= 0\end{aligned}$$

$$\begin{aligned}\frac{\partial^2}{\partial\vartheta\partial r} &= \frac{\sin\vartheta\cos\vartheta\cos^2\varphi}{r}\frac{\partial^2}{\partial\vartheta\partial r} + \frac{\sin\vartheta\cos\vartheta\cos^2\varphi}{r}\frac{\partial^2}{\partial\vartheta\partial r} + \frac{\sin\vartheta\cos\vartheta\sin^2\varphi}{r}\frac{\partial^2}{\partial\vartheta\partial r} \\ &\quad + \frac{\sin\vartheta\cos\vartheta\sin^2\varphi}{r}\frac{\partial^2}{\partial\vartheta\partial r} - \frac{\sin\vartheta\cos\vartheta}{r}\frac{\partial^2}{\partial\vartheta\partial r} - \frac{\sin\vartheta\cos\vartheta}{r}\frac{\partial^2}{\partial\vartheta\partial r} \\ &= 0\end{aligned}$$

$$\begin{aligned}\frac{\partial^2}{\partial r\partial\varphi} &= -\frac{\sin\varphi\cos\varphi}{r}\frac{\partial^2}{\partial\varphi\partial r} - \frac{\sin\varphi\cos\varphi}{r}\frac{\partial^2}{\partial r\partial\varphi} + \frac{\sin\varphi\cos\varphi}{r}\frac{\partial^2}{\partial r\partial\varphi} + \frac{\sin\varphi\cos\varphi}{r}\frac{\partial^2}{\partial r\partial\varphi} \\ &= 0\end{aligned}$$

Summarizing all of these results obtained so far, the well-known polar coordinate expression of the *Laplacian* shown in eq(13) is derived.

$$\begin{aligned}\nabla^2 &= \frac{\partial^2}{\partial x^2} + \frac{\partial^2}{\partial y^2} + \frac{\partial^2}{\partial z^2} \\ &= \frac{1}{r^2}\frac{\partial}{\partial r}\left(r^2\frac{\partial}{\partial r}\right) + \frac{1}{r^2}\left\{\frac{1}{\sin\vartheta}\frac{\partial}{\partial\vartheta}\left(\sin\vartheta\frac{\partial}{\partial\vartheta}\right) + \frac{1}{\sin^2\vartheta}\frac{\partial^2}{\partial\varphi^2}\right\} \qquad (13)\end{aligned}$$

If the system being treated is spherically symmetrical, then only the first term of eq(13) remains as is given by eq(14).

$$\nabla^2 = \frac{1}{r^2}\frac{\partial}{\partial r}\left(r^2\frac{\partial}{\partial r}\right) \qquad (14)$$ (for instance, see diffusion-controlled reactions)

By the way, eq(13) is derivable far more simply as compared with the method described above by using the polar coordinate expressions for the operators of the angular momenta (section **11** and see **problem 7-42a** as well).

"note" Operator *Laplacian*: $\nabla^2 = (grad \cdot grad)$, operator nabla: $\nabla \equiv grad \equiv \boldsymbol{i}\frac{\partial}{\partial x} + \boldsymbol{j}\frac{\partial}{\partial y} + \boldsymbol{k}\frac{\partial}{\partial z}$

"remark" Based on eq(2),

$$\begin{aligned} dx &= \sin\vartheta\cos\varphi\cdot dr + r\cos\vartheta\cos\varphi\cdot d\vartheta - r\sin\vartheta\sin\varphi\cdot d\varphi \\ dy &= \sin\vartheta\sin\varphi\cdot dr + r\cos\vartheta\sin\varphi\cdot d\vartheta + r\sin\vartheta\cos\varphi\cdot d\varphi \qquad (2) \\ dz &= \cos\vartheta\cdot dr - r\sin\vartheta\cdot d\vartheta \end{aligned}$$

if you carelessly derive the relations such as

$$\frac{\partial \vartheta}{\partial x} = \frac{1}{r\cos\vartheta\cos\varphi}, \quad \frac{\partial \varphi}{\partial x} = -\frac{1}{r\sin\vartheta\sin\varphi}, \quad \frac{\partial r}{\partial x} = \frac{1}{\sin\vartheta\cos\varphi}, \text{ etc.,}$$

and calculate $\frac{\partial}{\partial x}, \frac{\partial^2}{\partial x^2}, \cdots,$ then you will be led to an erroneous result. This is because the partial differentiation is being carried out without correct recognition of which variable(s) should be kept constant upon differentiation.

[exercise] Show that the polar coordinate expression of $\frac{\partial^2}{\partial x^2} + \frac{\partial^2}{\partial y^2}$ is given by $\frac{1}{r^2}\frac{\partial^2}{\partial \vartheta^2}$ for a two-dimensional system when the radial coordinate r is kept constant. (rotation in a plane)

10. Polar Coordinate Expression of Volume Element

When the variables x, y and z of the orthogonal coordinates are transformed into the variables r, ϑ and φ of the polar coordinates, the volume element in the integral expressed in the orthogonal coordinates are transformed into the volume element in the polar coordinates through the determinant called *Jacobian* as is defined by eq(1) whose determinant value is given by eq(2) .

$$dxdydz = \begin{vmatrix} \frac{\partial x}{\partial r} & \frac{\partial x}{\partial \vartheta} & \frac{\partial x}{\partial \varphi} \\ \frac{\partial y}{\partial r} & \frac{\partial y}{\partial \vartheta} & \frac{\partial y}{\partial \varphi} \\ \frac{\partial z}{\partial r} & \frac{\partial z}{\partial \vartheta} & \frac{\partial z}{\partial \varphi} \end{vmatrix} \cdot drd\vartheta d\varphi \qquad (1)$$

$$= \begin{vmatrix} \sin\vartheta\cos\varphi & \sin\vartheta\sin\varphi & \cos\vartheta \\ r\cos\vartheta\cos\varphi & r\cos\vartheta\sin\varphi & -r\sin\vartheta \\ -r\sin\vartheta\sin\varphi & r\sin\vartheta\cos\varphi & 0 \end{vmatrix} \cdot drd\vartheta d\varphi$$

$$= r^2\sin\vartheta\, drd\vartheta d\varphi \qquad (2)$$

Let us consider the significance of this transformation from a different angle. The transformation of the relevant variables are given by eq(3) already described in the section **9.**

$$\begin{pmatrix} dx \\ dy \\ dz \end{pmatrix} = \begin{pmatrix} \sin\vartheta\cos\varphi & \sin\vartheta\sin\varphi & \cos\vartheta \\ r\cos\vartheta\cos\varphi & r\cos\vartheta\sin\varphi & -r\sin\vartheta \\ -r\sin\vartheta\sin\varphi & r\sin\vartheta\cos\varphi & 0 \end{pmatrix} \begin{pmatrix} dr \\ d\vartheta \\ d\varphi \end{pmatrix} \equiv (A) \begin{pmatrix} dr \\ d\vartheta \\ d\varphi \end{pmatrix} \qquad (3)$$

Rewriting the left side of eq(3) using a unit matrix, eq(4) which is expressed as eq(4′) is obtained.

$$\begin{pmatrix} 1 & 0 & 0 \\ 0 & 1 & 0 \\ 0 & 0 & 1 \end{pmatrix} \cdot \begin{pmatrix} dx \\ dy \\ dz \end{pmatrix} = \begin{pmatrix} \sin\vartheta\cos\varphi & \sin\vartheta\sin\varphi & \cos\vartheta \\ r\cos\vartheta\cos\varphi & r\cos\vartheta\sin\varphi & -r\sin\vartheta \\ -r\sin\vartheta\sin\varphi & r\sin\vartheta\cos\varphi & 0 \end{pmatrix} \cdot \begin{pmatrix} dr \\ d\vartheta \\ d\varphi \end{pmatrix} \qquad (4)$$

$$(E) \cdot \begin{pmatrix} dx \\ dy \\ dz \end{pmatrix} \Rightarrow (A) \cdot \begin{pmatrix} dr \\ d\vartheta \\ d\varphi \end{pmatrix} \qquad (4')$$

Here we note a very important property of a determinant which states “the absolute value of the determinant represents a volume of the space specified by its coordinate components (a plane in the two-dimensional case)”.

Utilizing this property of the determinant, eq(4) (or eq(4′)) is seen to indicates that upon the transformation of the coordinates from $(x,\ y,\ z)$ to $(r,\ \vartheta,\ \varphi)$ the volume element in the former coordinate system is converted into the volume element in the latter in the ratio of the absolute value of $|E|$ to that of $|A|$. Since the values of the determinants are $|E|=1$ and $|A|=r^2\sin\vartheta$, the relation given by eq(2) is readily obtainable.

This relation will be intuitively understood by looking at an illustrative figure of the volume element in the polar coordinates shown in Fig.7. In the orthogonal coordinates, the volume of a cube whose center is at $(x,\ y,\ z)$ and enclosed by edges $(dx,\ dy,\ dz)$ is given by $dxdydz$. On the other hand, taking the polar coordinates $(r,\ \vartheta,\ \varphi)$ which correspond to $(x,\ y,\ z)$ at a base, when the coordinates are varied by the amounts of $(dr,\ d\vartheta,\ d\varphi)$, the volume of the space enclosed by the increments associated with these variations is obviously given by $rd\vartheta \times dr \times r\sin\vartheta d\varphi = r^2\sin\vartheta \cdot drd\vartheta d\varphi$, as is seen from Fig.7.

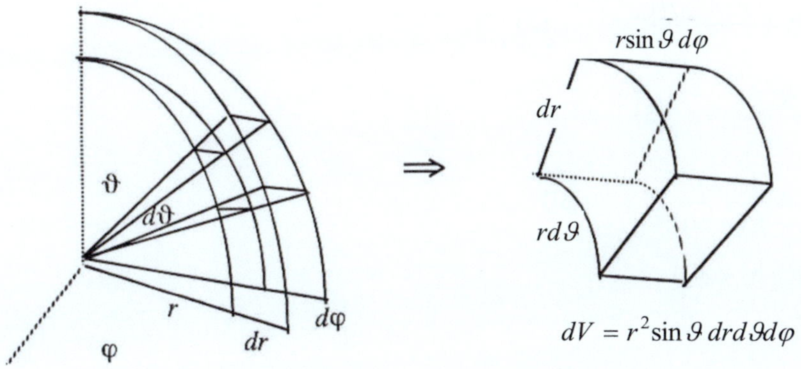

Fig.7 Polar coordinates and volume element.

In the case of the two dimensions, the area element in the polar coordination expression is given by $rd\vartheta dr$.

11. Ellipsoidal Coordinates and Volume Element.

To describe the coordinates of diatomic molecules, it is really convenient to take the origin of the coordinates at the center of each nucleus of A and B atoms. As is shown in Fig.8, the coordinates of a point X in the space can be expressed in terms of the three variables, the distance AX denoted r_A, the distance BX denoted r_B and the angle of rotation ϕ around the A-B (bond) axis. The variables r_A and r_B are practically transformed into λ and μ as are given in eq(2) in which R is the distance A-B between the two origins. To specify the point X, therefore, the three variables λ, μ and ϕ can be employed.

$$\left.\begin{aligned} &\lambda = \frac{r_A + r_B}{R} \\ &\mu = \frac{r_A - r_B}{R} \\ &\phi\text{: angle of rotation around } z \text{ axis} \\ &\quad\text{(same as polar coordinate)} \end{aligned}\right\} \quad (1)$$

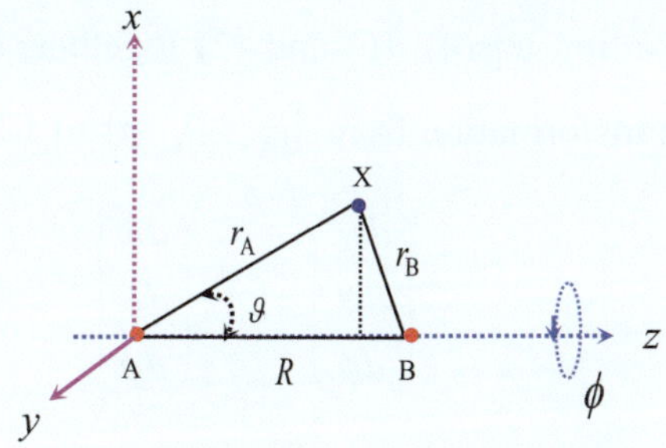

Fig.8 Ellipsoidal and polar coordinates.

The coordinates λ, μ and ϕ thus defined are called ellipsoidal coordinates. For the sake of comparison, the polar coordinates (r_A, ϑ, ϕ) with the origin taken at the center of A is also shown in Fig.8.

Since eq (1) represents ellipses and hyperbolas, any point in the space can be specified by the intersection of these two curves and the angle of rotation ϕ. An example of the ellipsoidal (elliptical) coordinates is given in Fig.9.

The relation between the ellipsoidal and polar coordinates is clarified by knowing the relation between the variables (λ, μ, ϕ) and (r_A, ϑ, $\phi = \varphi$). The variables r_A and r_B are given by eq(2) and ϑ is expressed by eq(3′).

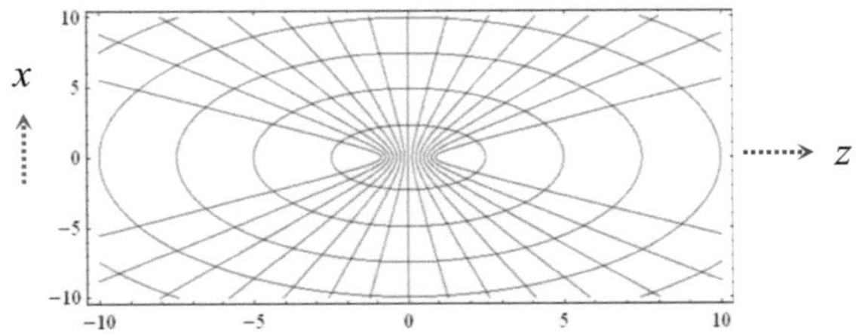

Fig.9 A grid of ellipsoidal coordinates.

$$r_A = \frac{R(\lambda+\mu)}{2}, \quad r_B = \frac{R(\lambda-\mu)}{2} \qquad (2)$$

$$\cos\vartheta = \frac{r_A{}^2 - r_B{}^2 + R^2}{2Rr_A} \qquad (3)$$

$$= \frac{R^2 + \lambda\mu R^2}{R^2(\lambda+\mu)} = \frac{1+\lambda\mu}{(\lambda+\mu)} \qquad (3')$$

The differential of r_A in eq(2) is given by eq(4).

$$dr_A = \frac{R}{2}d\lambda + \frac{R}{2}d\mu \qquad (4)$$

Also, eq(5) derived from the differentiation of eq(3′) can be rewritten with respect to $d\vartheta$ as eq(6).

$$d\cos\vartheta = -\sin\vartheta d\vartheta$$

$$= \frac{+\mu d\lambda + \lambda d\mu}{(\lambda+\mu)} - \frac{(1+\lambda\mu)(d\lambda+d\mu)}{(\lambda+\mu)^2}$$

$$= \frac{\mu(\lambda+\mu)d\lambda + \lambda(\lambda+\mu)d\mu - (1+\lambda\mu)d\lambda - (1+\lambda\mu)d\mu}{(\lambda+\mu)^2}$$

$$= \frac{(\mu^2-1)d\lambda + (\lambda^2-1)d\mu}{(\lambda+\mu)^2} \qquad (5)$$

Since ϕ is not transformed, eq(7) holds.

$$d\vartheta = -\frac{1}{\sin\vartheta}\left\{\frac{\mu^2-1}{(\lambda+\mu)^2}d\lambda + \frac{\lambda^2-1}{(\lambda+\mu)^2}d\mu\right\} \qquad (6)$$

$$d\phi = d\phi \qquad (7)$$

Putting eqs(4), (6) and (7) together, the transformation of the differentials associated with the coordinate transformation from $(r_A,\ \vartheta,\ \phi)$ to $(\lambda,\ \mu,\ \phi)$ can be expressed in a matrix form as eq(8).

$$\begin{pmatrix} dr_A \\ d\vartheta \\ d\phi \end{pmatrix} = (A)\cdot\begin{pmatrix} d\lambda \\ d\mu \\ d\phi \end{pmatrix}$$

$$= \begin{pmatrix} \frac{R}{2} & \frac{R}{2} & 0 \\ -\frac{1}{\sin\vartheta}\frac{\mu^2-1}{(\lambda+\mu)^2} & -\frac{1}{\sin\vartheta}\frac{\lambda^2-1}{(\lambda+\mu)^2} & 0 \\ 0 & 0 & 1 \end{pmatrix} \cdot \begin{pmatrix} d\lambda \\ d\mu \\ d\phi \end{pmatrix} \qquad (8)$$

The absolute value of the determinant $|A|$ is given by eq(9), so that the product of the variable differentials in each coordinate system is related by eq(10).

$$\|A\| = \frac{R(\lambda^2-\mu^2)}{2\sin\vartheta(\lambda+\mu)^2}, \quad \left(\because \lambda \geq \mu,\ 0 \leq \vartheta \leq \pi \right) \qquad (9)$$

$$dr_A d\vartheta d\phi = \frac{R(\lambda^2-\mu^2)}{2\sin\vartheta(\lambda+\mu)^2} d\lambda d\mu d\phi \qquad (10)$$

Summarizing all these, it is seen that changes in the volume element accompanied by the coordinate transformation of $(x,\ y,\ z) \Rightarrow (r_A,\ \vartheta,\ \phi) \Rightarrow (\lambda,\ \mu,\ \phi)$ are related by eq(11).

$$d\tau \equiv dxdydz = r_A^2\sin\vartheta \cdot dr_A d\vartheta d\phi = \frac{R(\lambda^2-\mu^2)}{2\sin\vartheta(\lambda+\mu)^2} \times r_A^2 \sin\vartheta \cdot d\lambda d\mu d\phi$$

$$= \frac{R^2(\lambda+\mu)^2}{4} \frac{R(\lambda^2-\mu^2)}{2(\lambda+\mu)^2} d\lambda d\mu d\phi$$

$$= \frac{R^3}{8}(\lambda^2-\mu^2) \cdot d\lambda d\mu d\phi \qquad (11)$$

Thus, the volume element and the range of each variable of the ellipsoidal coordinates are as follows:

$$d\tau = \frac{R^3}{8}(\lambda^2-\mu^2) \cdot d\lambda d\mu d\phi, \qquad 1 \leq \lambda < \infty, \qquad -1 \leq \mu \leq 1, \qquad \lambda \geq \mu, \qquad 0 \leq \phi \leq 2\pi.$$

[exercise] Calculate the volume of the spheroid shown below whose long and short radii are a and b, respectively by using the ellipsoidal coordinates.

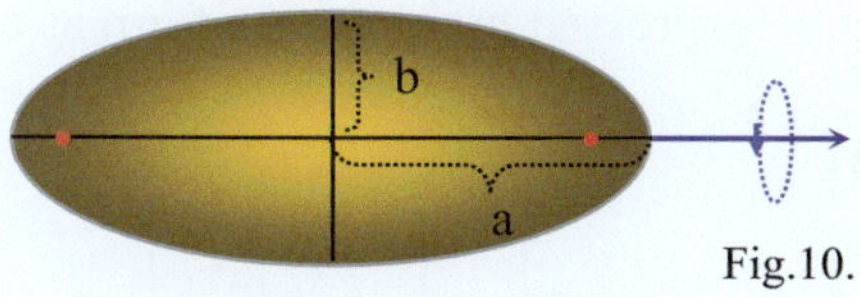

Fig.10. Long(a) and short (b) radii of a spheroid.

[solution]

The volume of a spheroid is easily calculated by integrating the volume element expressed in the ellipsoidal coordinates over the appropriate ranges of the relevant variables. The practical calculations are shown below. The result will be given by $\frac{4\pi a b^2}{3}$ as is well known for the volume of a spheroid.

$$V = \int_1^{2a/R}\int_{-1}^{1}\int_0^{2\pi} \frac{R^3}{8}(\lambda^2-\mu^2) \cdot d\lambda d\mu d\phi = 2\pi \times \frac{R^3}{8}\int_1^{2a/R}\left[\mu\lambda^2 - \frac{1}{3}\mu^3\right]_{-1}^{1} d\lambda$$

$$= \frac{\pi R^3}{4}\int_1^{2a/R}\left(2\lambda^2 - \frac{2}{3}\right)d\lambda = \frac{\pi R^3}{4}\left[\frac{2}{3}\lambda^3 - \frac{2}{3}\lambda\right]_1^{2a/R}$$

$$= \frac{\pi R^3}{4} \times \frac{2}{3}\left\{\left(\frac{2a}{R}\right)^3 - 1 - \frac{2a}{R} + 1\right\}$$

$$=\frac{\pi a R^2}{3}\frac{4a^2-R^2}{R^2}=\frac{4\pi ab^2}{3}\qquad\left\{\because b^2+\left(\frac{R}{2}\right)^2=a^2\right\}\tag{12}$$

12. Polar Coordinate Representations of Angular Momentum Operators

We denote the rectangular coordinate components x, y and z of the angular momentum operator by $\hat{l}_x$, $\hat{l}_y$ and $\hat{l}_z$, then they can be expressed as eq(1).

$$\left.\begin{array}{l}\hbar\hat{l}_x=y\hat{p}_z-z\hat{p}_y,\ \ \hbar\hat{l}_y=z\hat{p}_x-x\hat{p}_z,\ \ \hbar\hat{l}_z=x\hat{p}_y-y\hat{p}_x\\ \hat{p}_x=-i\hbar\dfrac{\partial}{\partial x},\ \ \hat{p}_y=-i\hbar\dfrac{\partial}{\partial y},\ \ \hat{p}_z=-i\hbar\dfrac{\partial}{\partial z}\end{array}\right\}\tag{1}$$

Their polar coordinate representations are found as eq(2) by using eq(1) above and the equations from (10) to (12) given in section **9**.

$$\left.\begin{array}{l}\hat{l}_x+i\hat{l}_y=e^{i\varphi}\left(\dfrac{\partial}{\partial\vartheta}+i\cot\vartheta\dfrac{\partial}{\partial\varphi}\right)\\ \hat{l}_x-i\hat{l}_y=e^{-i\varphi}\left(-\dfrac{\partial}{\partial\vartheta}+i\cot\vartheta\dfrac{\partial}{\partial\varphi}\right)\\ \hat{l}_z=-i\dfrac{\partial}{\partial\varphi}\end{array}\right\}\tag{2}$$

(For simplicity, the operators on the left side, $\hat{l}_x$, $\hat{l}_y$ and $\hat{l}_z$, are made dimensionless by multiplying them by $\hbar$. To make them the true angular momentum operators whose dimensions are [ML^2s^{-1}], the right side of eq(2) should be multiplied by $\hbar$ ([Js^{-1}]=[ML^2s^{-1}]), where M corresponds to mass and L to length.)

The operator $\hat{l}_x^{\,2}+\hat{l}_y^{\,2}+\hat{l}_z^{\,2}$ in the polar coordinate representation is seen to be readily obtainable from eq(2),

13. *Cramer*'s Formula : How to Solve Simultaneous Linear Equations (see section 7, c-(ii))

$$\begin{array}{l}a_{11}x_1+a_{12}x_2+\cdots+a_{1n}x_n=b_1\\ a_{21}x_1+a_{22}x_2+\cdots+a_{1n}x_n=b_2\\ \cdot\ \cdot\ \cdot\ \cdot\ \cdot\ \cdot\ \cdot\ \cdot\ \cdot\ \cdot\ \cdot\ \cdot\ \cdot\\ \cdot\ \cdot\ \cdot\ \cdot\ \cdot\ \cdot\ \cdot\ \cdot\ \cdot\ \cdot\ \cdot\ \cdot\ \cdot\\ a_{n1}x_1+a_{n2}x_2+\cdots+a_{nn}x_n=b_n\end{array}\tag{1}$$

A simultaneous linear equations usually written as eq(1) can be expressed in the matrix form as eq(1′)

$$(A)\cdot\begin{pmatrix}x_1\\x_2\\\cdot\\\cdot\\x_n\end{pmatrix}=\begin{pmatrix}a_{11}&a_{12}&\cdot&\cdot&a_{1n}\\a_{21}&a_{22}&\cdot&\cdot&a_{2n}\\\cdot&\cdot&\cdot&\cdot&\cdot\\\cdot&\cdot&\cdot&\cdot&\cdot\\a_{n1}&a_{n2}&\cdot&\cdot&a_{nn}\end{pmatrix}\cdot\begin{pmatrix}x_1\\x_2\\\cdot\\\cdot\\x_n\end{pmatrix}=\begin{pmatrix}b_1\\b_2\\\cdot\\\cdot\\b_n\end{pmatrix}\quad\Rightarrow\quad\begin{pmatrix}x_1\\x_2\\\cdot\\\cdot\\x_n\end{pmatrix}=(A)^{-1}\begin{pmatrix}b_1\\b_2\\\cdot\\\cdot\\b_n\end{pmatrix}\tag{1′}$$

and so, to find solutions, we only need to obtain the inverse matrix $(A)^{-1}$, but more simply, provided that the j-th column $\begin{pmatrix}a_{1j}\\a_{2j}\\\cdot\\\cdot\\a_{nj}\end{pmatrix}$ of the matrix (A) is replaced by the column $\begin{pmatrix}b_1\\b_2\\\cdot\\\cdot\\b_n\end{pmatrix}$ and the obtained matrix is expressed by (A_j), then the solutions are given by eq(3) in terms of the determinants of (A) and (A_j) denoted $|A|$ and $|A_j|$, respectively. Eq(3) is called *Cramer*'s formula.

$$x_1 = \frac{|A_1|}{|A|}, \quad x_2 = \frac{|A_2|}{|A|}, \quad \cdots x_n = \frac{|A_n|}{|A|} \qquad (3)$$

For example, the solutions of the following linear simultaneous equations are straightforwardly found as below because of this formula.

$$\begin{aligned} -3x + 2y + 4z &= 9 \\ 5x - 3y + 6z &= 12 \\ -4x + 5y - 5z &= 3 \end{aligned} \quad , \quad \text{solutions:} \quad x = \frac{297}{99} = 3, \quad y = \frac{495}{99} = 5, \quad z = \frac{198}{99} = 2.$$

14. *Laplace* Transformation and Its Inverse Transformation

The *Laplace* transformation (L) of a function $f(t)$ is defined by eq(1) and its inverse transformation (L^{-1}) is

$$Lf(t) \Rightarrow F(s) = \int_0^{\infty} f(t)e^{-st}dt \qquad (1)$$

expressed as eq(2).

$$L^{-1}F(s) = f(t) \qquad (2)$$

The results of the *Laplace* transformation (right side) performed on some important functions (left side) in physical chemistry are summarized in the table below. The inverse transformation is from right to left side.

In this book, the *Laplace* transformation is actually employed in solving the equation of diffusion of diffusion-controlled reactions and in examining the chemical relaxation by the phase shift method. It is also used to solve ordinary differential equations.

[exercise] Solve the following second-order differential equations using the *Laplace* transformation. The initial conditions are given by $y(0) = 1$ and $y'(0) = 0$.

$$\begin{aligned} &y'' - 9y = 0 \\ &y'' - 4y' + 3y = 0 \\ &y'' + 6y' + 5y = 0 \\ &y'' + 2y' + y = \exp(-t) \end{aligned}$$

$$Lf'(t) = sF(s) - f(0+)$$
$$Ltf(t) = F'(s)$$
$$L1 = 1/s$$
$$Lt = 1/s^2$$
$$Lt^{-1/2} = \pi^{1/2}s^{-1/2}$$
$$Le^{at} = 1/(s-a)$$
$$L\frac{1}{\sqrt{\pi t}}e^{-k^2/4t} = \frac{1}{\sqrt{s}}e^{-k\sqrt{s}}$$
$$Lt^{-3/2}e^{-k^2/4t} = \frac{2\sqrt{\pi}}{k}e^{-k\sqrt{s}}$$
$$L\cos\omega t = s/(s^2+\omega^2)$$
$$L\sin\omega t = \omega/(s^2+\omega^2)$$
$$L\cosh\omega t = s/(s^2-\omega^2)$$
$$L\sinh\omega t = \omega/(s^2-\omega^2)$$
$$Lt^b e^{at} = \frac{\Gamma(b+1)}{(s-a)^{b+1}}$$
$$L(1-e^{-t})t^{-1} = \ln\left(1+\frac{1}{s}\right)$$
$$L\,erfc\left(\frac{a}{\sqrt{4t}}\right) = \frac{\exp(-a\sqrt{s})}{s}$$

15. *Fourier* Transformation and Its Inverse Transformation

$$F(\omega) = \int_{-\infty}^{\infty} f(t)\cdot e^{-i\omega t}dt: \text{ definition}$$

$$\text{(ex.)} \quad \int_{-\infty}^{\infty} e^{-kt}e^{-i\omega t}dt = \frac{k}{k^2+\omega^2} - i\frac{\omega}{k^2+\omega^2} \quad (Lorentzian)$$

$$f(t) = \frac{1}{2\pi}\int_{-\infty}^{\infty} F(\omega)\cdot e^{i\omega t}d\omega: \text{ inverse transformation}$$

$$\text{(ex.)} \quad \frac{1}{2\pi}\int_{-\infty}^{\infty}\left(\frac{k}{k^2+\omega^2} - i\frac{\omega}{k^2+\omega^2}\right)\cdot e^{i\omega t}d\omega = e^{-kt} \quad \text{(exponentail decay)}$$

Molecules decaying single exponentially yield a spectrum of *Lorentzian* shape in frequency distribution. See **7-117a,b** for coherence spectrum.

16. Convolution and Deconvolution

The concept of convolution is very important in understanding various natural phenomena. This is because when we observe a particular natural phenomenon, the result having been obtained through our observation is nothing more than a convolution of the true figure of the phenomenon. This situation will be explained in the following taking the two cases of (i) spectral bandwidth and (ii) fluorescence decay curve as examples.

We denote a physical quantity observable as a function of x by $F(x)$ and a response function of the observation by $L(x)$. Let us assume that $D(x)$ is the true figure of the phenomenon being observed and totally independent of the manner of observation. These three functions are related one another through the integral of eq(1) which is called "convolution".

$$F(x) \equiv L(x) \otimes D(x) = \int_0^t L(x')D(x-x')dx' \qquad (1)$$

Let us examine the following two examples to comprehend the meaning of eq(1).

(i) spectral bandwidth (resolution)

We now consider the case in which the slit width is given by the δ-function ($L(x)$) and the intrinsic spectrum (true figure) is represented by a rectangular function ($D(x)$) as are illustrated in Fig.11.

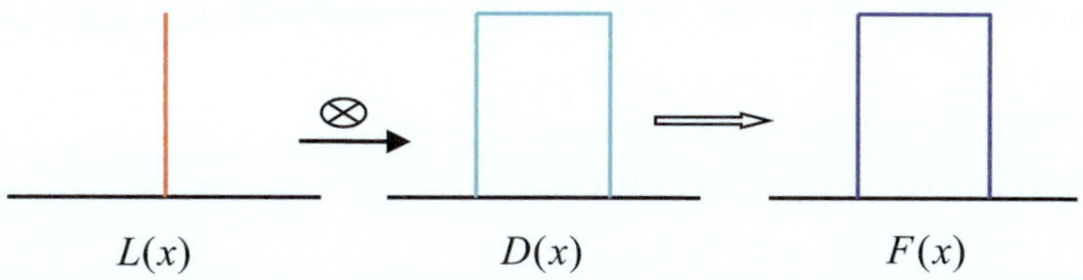

Fig.11 Conceptual diagram of convolution upon spectral measurement–infinitesimally narrow slit width.

In this case, the observed spectrum ($F(x)$) reflects exactly the true rectangular spectrum. However, when a slit of a finite width is employed instead of the infinitesimally narrow slit width, the spectrum observed is broadened and yields a spectrum of triangle shape in the case of a rectangular finite slit width as is illustrated in Fig.12. The observed spectrum ($F(x)$) is far from the true spectrum ($D(x)$).

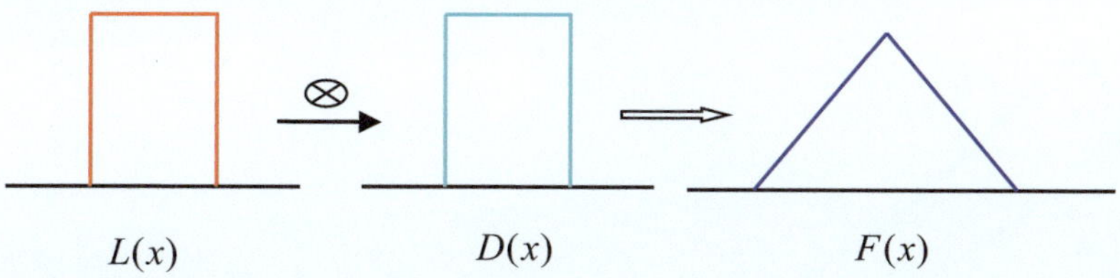

Fig. 12 Conceptual diagram of convolution upon spectral measurement–rectangular finite slit width.

As is obvious from this illustration, it is very important to realize correctly the relation between the slit width and the spectral resolution in the spectral measurements and then to consider how we can work out the true spectrum from the observed one (deconvolution).

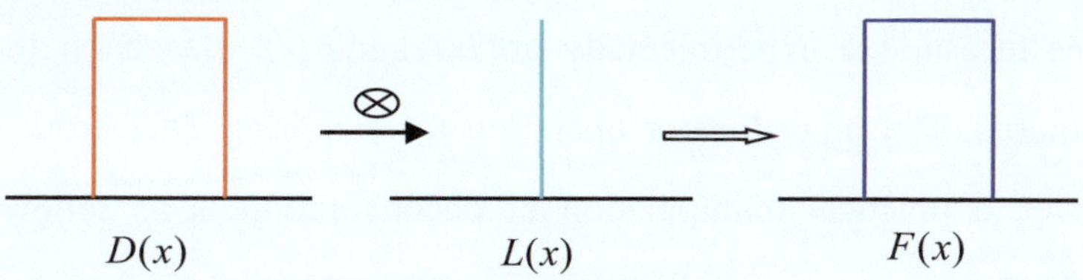

Fig.13 Conceptual diagram of convolution upon spectral measurement – symmetrical property of convolution.

"note" As is shown below, eq(1) holds even when $D(x)$ and $L(x)$ are interchanged (eq($1'$)). Putting

$x - x' = y$, the relations of $x' = x - y$ and $dx' = -dy$ are derived, from which the range of integration of eq(1) is found to change from $x' = 0$, $x' = x$ to $y = x$, $y = 0$, yielding eq(1′). Fig.13 illustrates this interchangeable property of convolution integral.

$$F(x) \equiv L(x) \otimes D(x) = \int_0^x L(x')D(x - x')dx'$$

$$= -\int_x^0 L(x - y)D(y)dy = \int_0^x L(x - y)D(y)dy$$

$$= \int_0^x D(y)L(x - y)dy = D(x) \otimes L(x) \qquad (1')$$

(ii) fluorescence decay curve

The time profile of an excitation light pulse is denoted $L(t)$ and the true fluorescence decay function is denoted $D(t)$. Then, the fluorescence decay $F(t)$ of the molecules excited with this light pulse is expressed by the convolution given by eq(2).

$$F(t) \equiv L(t) \otimes D(t) = \int_0^t L(t')D(t - t')dt' \qquad (2)$$

For instance, when the true fluorescence decay is describe by a function $D(t) = F_0 \exp(-kt)$ and the excitation light pulse by a function $L(t) = At^3 \exp(-Bt)$, the fluorescence decay curve $F(t)$ to be observed experimentally is given by the convolution expressed as eq(3).

$$F(t) = \int_0^t At'^3 \exp(-Bt')\exp - k(t - t')dt' \qquad (3)$$

Each function involved in the convolution of eq(3) is drawn by *Mathematica* and is shown in Fig.14 below.

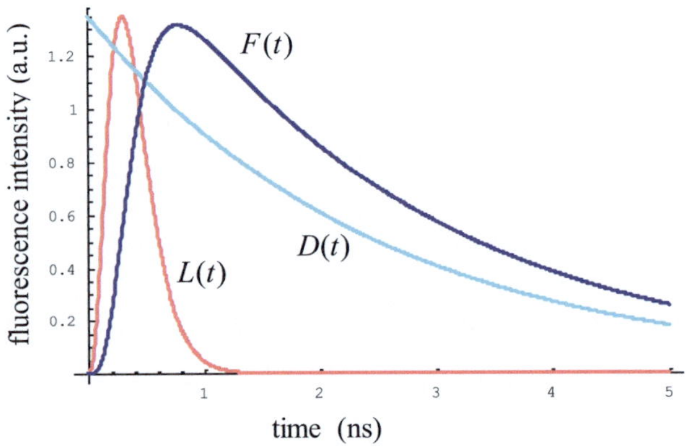

Fig.14 Fluorescence decay and convolution.

Usually, the quantities measurable experimentally are $L(t)$ and $F(t)$. Based on the information on these quantities, $D(t)$ is acquired. The procedure or operation for acquiring $D(t)$ from $L(t)$ and $F(t)$ is called deconvolution. However, a practical manipulation of deconvolution is as follows. Starting with a trial function $D_{test}(t)$ for $D(t)$, an approximate function $F_{app}(t)$ is obtained by the convolution of $D_{test}(t)$ with $L(t)$ and then is compared with $F(t)$. If $F_{app}(t)$ is not close enough to $F(t)$, then another new trial function for $D(t)$ is chosen so as to generate another approximate $F_{app}(t)$. This procedure is repeated until a newly chosen trial function for $D(t)$ yields an approximate $F_{app}(t)$ close enough to the observed $F(t)$. Then the last trial function $D_{test}(t)$ which has yielded a best fitting with $F(t)$ is regarded as $D(t)$. Therefore, the deconvolution in this case is just repetition of convolution procedures.

17. Determination of Fluorescence Lifetimes by the Moment Method

Generally, the i-th moment of a function $f(t)$ is defined by eq(1) and μ_i is called the i-th moment of the function $f(t)$.

$$\mu_0 = \int_0^\infty f(t)dt \equiv 1 \text{ (normalized to 1)},\ \mu_1 = \int_0^\infty tf(t)dt,\ \mu_2 = \int_0^\infty t^2 f(t)dt,\ \cdots\ \mu_i = \int_0^\infty t^i f(t)dt \tag{1}$$

As is obvious from the above definition, the i-th moment is the time average of t^i with the weight of $f(t)$.

In the following, we will see how to utilize the moments taking the fluorescence decay curve discussed in the preceding section as an example.

Putting $f(t) = D(t) = F_0 \exp(-kt)$ and $F_0 = k$, the first-order (primary) moment is calculated as eq(2).

$$\mu_1 = \int_0^\infty tf(t)dt = F_0 \int_0^\infty t\exp(-kt)dt = F_0 / k^2 = 1/k = \tau \tag{2}$$

Namely, the average of t (first-order moment) is equal to the mean lifetime.

When the fluorescence decay curve is given as a result of convolution with the excitation light pulse, the situation is slightly more complicated. The first-order moment of $F(t)$ is found as eq(3).

$$\begin{aligned}\mu_1 &= \int_0^\infty tF(t)dt = \left[\frac{1}{2}t^2 F(t)\right]_0^\infty - \int_0^\infty \frac{1}{2}t^2 F'(t)dt \\ &= -\frac{1}{2}\int_0^\infty t^2 L(t)dt + \frac{1}{2}k\int_0^\infty t^2 F(t)dt \end{aligned} \tag{3}$$

Note that the following relation is used (eq(3) in **16.**).

$$\begin{aligned}F'(t) &= L(t) - k\int_0^t At'^3 \exp(-Bt')\exp - k(t-t')dt' \\ &= L(t) - kF(t)\end{aligned} \tag{4}$$

We define the moments of the excitation light pulse by the following equations.

$$m_0 = \int_0^\infty L(t)dt \equiv 1,\ m_1 = \int_0^\infty tL(t)dt,\ m_2 = \int_0^\infty t^2 L(t)dt,\ \cdots,\ m_n = \int_0^\infty t^n L(t)dt$$

Rewriting eq(4) according to this definition, eq(5) which expresses the relation between the fluorescence lifetime ($\tau = 1/k$) and the moments is obtained.

$$2\mu_1 = -m_1 + \frac{1}{\tau}\mu_2 \quad \Rightarrow \quad \tau = \frac{\mu_2}{2\mu_1 + m_2},\ k = \frac{1}{\tau} \tag{5}$$

18. Special Functions

(i) *Hermite* polynomials of the n-th order

general formula of *Hermite* polynomials

$$H_n(x) = (2x)^n - \frac{n\cdot(n-1)}{1!}(2x)^{n-2} + \frac{n\cdot(n-1)\cdot(n-2)\cdot(n-3)}{2!}(2x)^{n-4} - \cdots \tag{1}$$

$$= (-1)^n e^{x^2} \frac{d^n}{dx^n} e^{-x^2} \tag{1'}$$

Hermite polynomials and differential equations

Putting $y = H_n(x)$, y is found to be a particular solution of the differential equation given by eq(2).

$$y'' - 2xy' + 2ny = 0 \qquad (2)$$

Also, by putting $z = e^{-x^2/2} H_n(x)$, z is found to be a particular solution (eigenfunction for a harmonic oscillator) of the differential equation given by eq(3).

$$z'' + (1 - x^2 + 2n)z = 0 \qquad (3)$$

Hermite polynomials of 0th to 5th order

$$H_0(x) = 1, \quad H_1(x) = 2x, \quad H_2(x) = 4x^2 - 2, \quad H_3(x) = 8x^3 - 12x,$$
$$H_4(x) = 16x^4 - 48x^2 + 12, \quad H_5(x) = 32x^5 - 160x^3 + 120x$$

Hermite polynomials and recurrence formulae

$$H_n^{'}(x) = 2nH_{n-1}(x), \qquad H_{n+1}(x) = 2xH_n(x) - 2nH_{n-1}(x) \qquad (4)$$

[exercise]

Confirm that $z = e^{-x^2/2} H_2(x)$ satisfies eq(3).

(ii) *Legendre* polynomials and associated *Legendre* polynomials

Legendre polynomial of the *l*-th order is given by eq(5) or eq(5′). The function is often associated with the angular momentum, so a letter l is employed to specify the order of the function.

$$P_l(x) = \frac{(2l)!}{2^l (l!)^2} \left\{ x^l - \frac{l \cdot (l-1)}{2 \cdot (2l-1)} x^{l-2} + \frac{l \cdot (l-1) \cdot (l-2) \cdot (l-3)}{2 \cdot 4 \cdot (2l-1) \cdot (2l-3)} x^{l-4} - \cdots \right\} \qquad (5)$$

$$= \frac{1}{2^l l!} \frac{d^l}{dx^l} (x^2 - 1)^l \qquad (5') \qquad (\textit{Rodrigues'} \text{ formula})$$

recurrence formulae of *Legendre* polynomials

$$(l+1)P_{l+1}(x) - (2l+1)xP_l(x) + lP_{l-1}(x) = 0 \qquad (6)$$
$$P'_{l+1}(x) - xP'_l = (l+1)P_l(x) \qquad (7)$$
$$xP'_l(x) - P'_{l-1}(x) = l \cdot P_l(x) \qquad (8)$$
$$P'_{l+1}(x) - P'_{l-1}(x) = (2l+1)P_l(x) \qquad (9)$$
$$(x^2 - 1)P'_l(x) = l \cdot \{xP_l(x) - P_{l-1}(x)\} \qquad (10)$$

Legendre polynomials and differential equations

Putting $y = P_l(x)$, y is found to be a special solution of the differential equation given by eq(11).

$$(1 - x^2)y'' - 2xy' + l \cdot (l+1)y = 0 \qquad (11)$$

Also, putting $x = \cos\vartheta$, the function $P_l(\cos\vartheta)$ satisfies the differential equation given by eq(11′).

$$x = \cos\vartheta \Rightarrow P_l(\cos\vartheta)$$

$$\frac{1}{\sin\vartheta} \frac{\partial}{\partial\vartheta} \left(\sin\vartheta \frac{\partial P_l}{\partial\vartheta} \right) + l \cdot (l+1)P_l = 0 \qquad (11')$$

associated *Legendre* polynomials and differential equations

Defining associated *Legendre* polynomials by eq(12), the function $P_l^m(\cos\vartheta)$ defined thus satisfies eq(13).

$$P_l^m(\cos\vartheta) = \sin^m\vartheta \frac{\partial^m}{\partial\cos\vartheta^m} P_l(\cos\vartheta) \qquad (12)$$

$$\frac{1}{\sin\vartheta}\frac{\partial}{\partial\vartheta}\left(\sin\vartheta\frac{\partial P_l^m}{\partial\vartheta}\right) + \left\{l\cdot(l+1) - \frac{m^2}{\sin^2\vartheta}\right\}P_l^m = 0 \qquad (13)$$

a few examples of *Legendre* and associated *Legendre* polynomials (not normalized)

$$P_0(x) = 1, \quad P_1(\cos\vartheta) = \cos\vartheta, \quad P_2(\cos\vartheta) = \frac{3}{2}\left(\cos^2\vartheta - \frac{1}{3}\right), \quad P_3(\cos\vartheta) = \frac{5}{2}\left(\cos^3\vartheta - \frac{3}{5}\cos\vartheta\right),$$

$$P_1^1(\cos\vartheta) = \sin\vartheta, \quad P_2^1(\cos\vartheta) = 3\sin\vartheta\cos\vartheta, \quad P_2^2(\cos\vartheta) = 3\sin^2\vartheta.$$

[exercise]

(a) Show that $P_2^2(\cos\vartheta)$ satisfies eq(13).

(b) Confirm that the recurrence formula given by eq(6) holds by using *Rodrigues*' formula of eq($5'$).

(See **[hint]** given below. This task may require a little ingenuity but will be a good practice in differentiation worth trying.)

(c) Confirm $y = P_n(x)$ satisfies eq(11) using the recurrence formulae from eq(7) to eq(10).

[hint] on answering the question (b)

Show that the following relations hold and then eliminate unnecessary terms if any.

$$P_{l+1}(x) = xP_l(x) + \frac{1}{2^l(l-1)!}\frac{d^{l-1}}{dx^{l-1}}(x^2-1)^l, \quad xP_l(x) = \frac{l+1}{2^l l!}\frac{d^{l-1}}{dx^{l-1}}(x^2-1)^l + P_{l-1}(x)$$

(iii) *Laguerre*'s polynomials, associated polynomials and differential equations

definition:

$$X = L_n(x) = (-1)^n\left\{x^n - \frac{n^2}{1!}x^{n-1} + \frac{n^2(n-1)^2}{2!}x^{n-2} + \cdots + (-1)^n n!\right\} \qquad (14)$$

$$= e^x\frac{d^n}{dx^n}x^n e^{-x} \qquad (14')$$

associated polynomials and differential equations:

$$xX'' + (1-x)X' + nX = 0 \qquad (15)$$

$$Y = L_n^k(x) \equiv \frac{d^k}{dx^k}L_n(x)$$

$$xY'' + (k+1-x)Y' + (n-k)Y = 0 \qquad (16)$$

$$Z = e^{-x/2}x^{(k-1)/2}L_n^k(x)$$

$$xZ'' + 2Z' + \left(n - \frac{k-1}{2} - \frac{x}{4} - \frac{k^2-1}{4x}\right)Z = 0 \qquad (17)$$

(iv) definition of Gamma function ($\Gamma(z)$) and some values

$$\Gamma(z)=\int_{0}^{\infty} e^{-t} t^{z-1} dt \qquad (18)$$

$$\Gamma(z)\Gamma(1-z)=\frac{\pi}{\sin \pi z}$$

$$\Gamma\left(\frac{1}{2}\right)=\sqrt{\pi}, \qquad \Gamma\left(\frac{3}{2}\right)=\frac{1}{2}\sqrt{\pi}, \qquad \Gamma\left(\frac{5}{2}\right)=\frac{3}{4}\sqrt{\pi}$$

(v) *Dirac*'s δ-function

properties of δ-function:

$$\left.\begin{array}{ll} \delta(x-a)=0, & x \neq a \\ \delta(x-a)=\infty, & x=a \\ \int_{-\infty}^{\infty} f(x)\delta(x-a)\cdot dx = f(a) & \end{array}\right\} \qquad (19)$$

an example of δ-function:

$$\delta(x-a)=\frac{1}{2\pi}\int_{-\infty}^{\infty} e^{i(x-a)t} dt \qquad (20)$$

See also § 15 (δ-function) in "*The Principles of Quantum Mechanics*" by *P.A.M.Dirac.*

19. Definition of *Jacobian* and Some Useful Relations

See **"note"** in **solution 3-35**.

20. Tensors

The second rank tensor encountered in physical chemistry is expressed by a 3-by-3 matrix and has nine components as are shown in eq(1).

$$\begin{pmatrix} A_{11} & A_{12} & A_{13} \\ A_{21} & A_{22} & A_{23} \\ A_{31} & A_{32} & A_{33} \end{pmatrix}, \qquad \begin{pmatrix} A_{xx} & A_{xy} & A_{xz} \\ A_{yx} & A_{yy} & A_{yz} \\ A_{zx} & A_{zx} & A_{zz} \end{pmatrix} \qquad (1)$$

Among others, we will find tensors of moment of inertia, polarization (dielectric or magnetic) tensors, stress tensors, etc. If a given tensor is a symmetrical matrix ($A_{ij}=A_{ji}$ $(i \neq j)$), the principal axes of the moment of inertia or polarization axes can be found by diagonalizing the matrix with appropriate coordinate transformation. The tensors of inertia and polarization play an important role in spectroscopy. The problem to find the principal axes of a molecule employing the tensors of inertia is given in **7-118**. The tensors are characterized by their ranks. For instance, the zeroth rank tensor A is a scalar A, the first rank tensor expressed by $(A_1, A_2, A_3, \cdots)$ represents the components of a vector $(A_1, A_2, A_3, \cdots)$. You are suggested to refer to a relevant text book on mathematics for a general discussion on tensors.

Part III
Worked Solutions

III-1 Electronic Structures and Properties of Atoms and Simple Molecules
Part of Worked Solutions

Solution 1-1

(a) Ionization energy (*I.P.*) is the energy required to bring an electron in the lowest energy state of the atomic orbital ($m=1$) to the orbital of the highest energy ($n=\infty$). Therefore, by substituting $m=1$ and $n=\infty$ ($E_\infty=0$) into the given equation (eq(1) given in **problem 1-1**), the value of $A=1312\ \mathrm{kJ\cdot mol^{-1}}$ is obtained.

Ans. $A=1312\ \mathrm{kJ\cdot mol^{-1}}$

(b) The atomic number Z of Li is 3 and Li^{2+} has only one electron in the 1*s*-orbital. For this, Li^{2+} is a hydrogen-like atom and can be treated as such. In other words, the energy levels of the ion are obtained from those for the hydrogen atom just by increasing the nuclear charge Ze from $1e$ to $3e$. Therefore, ionization energy of Li^{2+} is given by multiplying that of the hydrogen atom by $Z^2=9$.

Ans. 9 times 0f $1312\mathrm{kJ\cdot mol^{-1}}$

(c) K_α line is the *X*-ray line observed upon the transition of an electron in *L* shell ($n=2$) to the *K* shell ($m=1$). Therefore, when calculated as follows by substituting those principal quantum numbers into the given equation (eq(1) in **problem 1-1**), the value for A' per mole is obtained as $A' = 4.77\times10^8\mathrm{J\cdot mol^{-1}}$.

$$E_2-E_1=h\nu=h\cdot\frac{c}{\lambda}=6.63\times10^{-34}\times\frac{3.0\times10^8}{3.35\times10^{-10}}$$

$$=-A'\left(\frac{1}{2^2}-\frac{1}{1^2}\right)=A'\times0.75 \qquad (2)$$

Since the obtained value is proportional to the effective nuclear charge squared ($Z_{eff}^2e^2$), the value divided by the value of $A=1312\mathrm{kJ\cdot mol^{-1}}$ for a hydrogen atom yields a value for Z_{eff}^2. And hence its square root $\sqrt{A'/A}=\sqrt{363}=19.05$ is equal to Z_{eff}. Taking the shielding effect (1 from *K* electron shell) into account, the atomic number of the atom under examination is obtained as 19 + 1 = 20.

H.Moseley See *Nature* 46, 33 (1915) for his brief carrier and scientific achievement.

Ans. *Z*= 20 corresponding to Ca atom

"note" The calculated ionization energies of hydrogen-like atoms exactly match the measured values. It is useful to confirm this by referring to some data book or reference on the ionization energies of atoms. Even for non-hydrogen-like atoms, their ionization energies can be estimated approximately by the following equation (eq3)),

$$I.P.=A\times(Z_{eff}{}^2/n^2) \qquad (3)$$

in which Z_{eff} is the effective nuclear charge in the unit of e. There is a rule of thumb to estimate Z_{eff}. Refer to the **solution 1-11** and *University Chemistry* by *Maian*.

Solution 1-2

(a) The numbers of electrons that can be accommodated in the atomic orbital are schematically shown for

S. Hirayama and M. Yamamoto, *Physical Chemistry Problems and Solutions*,
https://doi.org/10.1007/978-981-97-0165-0_9

oxygen atom ($_8O$) in Fig.1, which illustrates how 8 electrons of oxygen atom occupy the relevant atomic orbitals in paired- or unpaired-manner. In one square, maximum two electrons with paired spins can be accommodated. See the note below for the general rule ($n+l$ rule, etc.) which can tell us how electrons will fill relevant atomic orbitals in the ascending order of energy.

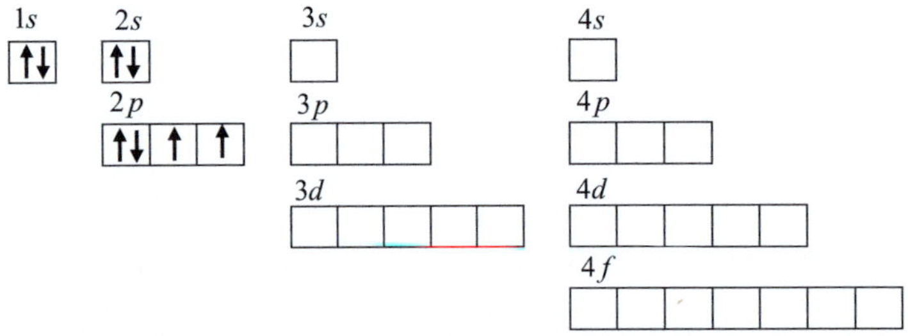

Fig.1 Schematic representation of atomic orbitals to show electronic configurations.

Accordingly, the atomic orbitals of each element are occupied by the number of electrons equal to the atomic number. As a result, each element in the periodic table can be characterized by the type of the highest occupied atomic orbital such as s, p, d, f and each element is classified into such as $2p$-, $3d$-element, etc., as is indicated by thick arrow lines of different color.

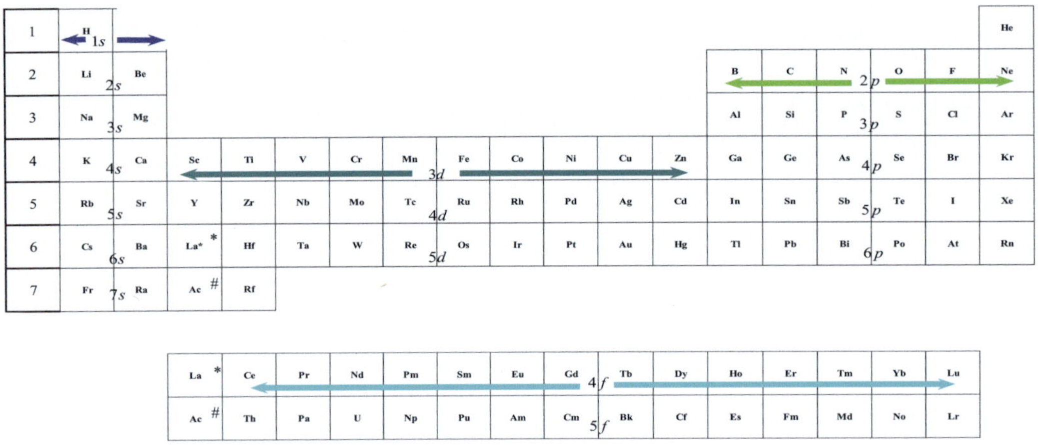

Fig.2. Periodic table to show the classification of elements based on their electronic configurations.

Ans. See text and Figs.1, 2

(b) Using the $n+l$ rule (see below), the same number of electrons as the atomic number are filled into relevant atomic orbitals as are given in Ans.

Ans. $\mathrm{Na}(1s^2 2s^2 2p^6 3s^1)$, $\mathrm{Cu}(1s^2 2s^2 2p^6 3s^2 3p^6 4s^2 3d^9)$, $\mathrm{Sb}(1s^2 2s^2 2p^6 3s^2 3p^6 4s^2 3d^{10} 4p^6 5s^2 4d^{10} 5p^3)$

"note" $n+l$ rule

When we place a required number of electrons in each atomic orbital, there are a number of rules concerning how to arrange the electrons into each orbital. A set of the rules is collectively called *Aufbau Principle*. Its first and most important one is $n+l$ rule where n is the principal quantum number and l is the azimuthal quantum number. Firstly, the energy level can be specified by n and l and is in the increasing order of their sum $n+l$. When the value for $n+l$ is the same for some orbitals, then the orbital which has the smaller quantum number n is located lower than the others. Secondly, for the orbital such as *p*-, *d*-, *f*-orbitals in which more than two electrons can be accommodated, the electrons assume parallel spin configuration as many as possible, i.e., *Hund's Rule*. Fig.1 illustrates the case for oxygen. Two electrons out of four electrons accommodated in the *p*-orbital are located in parallel to each other.

You can remember $n+l$ rule by using a picture shown in Fig.3. From the top, atomic orbitals are arranged in lines where the orbitals with a same principal quantum number are put in the same line. The atomic orbitals which have the same $n+l$ values are connected by a slant arrow line. The order of the energy is readily known by starting from the top (1*s*) and then following a dashed blue line and then when it meets another slant arrow line, you follow the arrow until you reach the head of the arrow. Repeating a similar procedure several times you will end up at the 8*s* orbital in Fig.3.

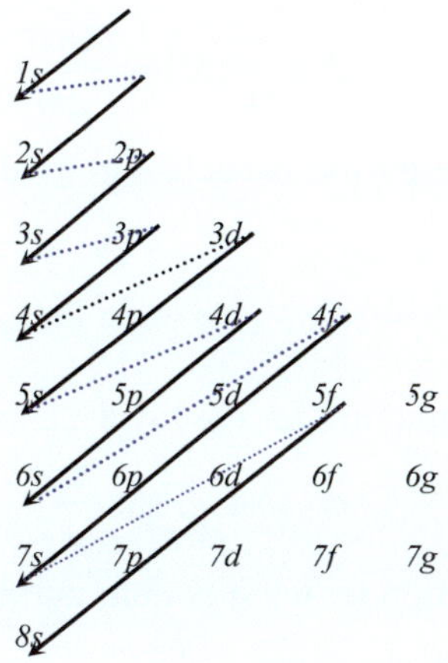

Fig.3 The order of the energy levels of atomic orbitals ($n+l$ rule).

When the orbitals from 1*s* to 8*s* are fully occupied in accordance with Fig.3, the total number of electrons becomes 120. Then what will be a name for the atom with atomic number 120 ? Since the general rule for the nomenclature is established, you can answer the name even for the atom with the atomic number of 1000, no matter how it exists or not. (see **problem 1-107**).

Solution 1-3

You can answer the question by using the order of energy levels of atomic orbitals and *Aufbau Principle* (supra) as follows.

$_{20}\mathrm{Ca}\left(1s^2 2s^2 2p^6 3s^2 3p^6 4s^2\right)$

$_{26}\mathrm{Fe}\left(1s^2 2s^2 2p^6 3s^2 3p^6 4s^2 3d^6\right)$

$_{40}\mathrm{Zr}\left(1s^2 2s^2 2p^6 3s^2 3p^6 4s^2 3d^{10} 4p^6 5s^2 4d^2\right)$

$_{47}\mathrm{Ag}\left(1s^2 2s^2 2p^6 3s^2 3p^6 4s^2 3d^{10} 4p^6 5s^1 4d^{10}\right)$

It should be noted that for a *d*-orbital (and *f*-orbital as well), when it is just half-filled or when it is completely filled, the resulting electronic configuration d^5 or d^{10} is particularly stable. As a result, the electron configuration of Ag is $[\mathrm{Kr}]5s^1 4d^{10}$ but not $[\mathrm{Kr}]5s^2 4d^9$. The element Zr came to be well known through the nuclear power plant accident in *Fukushima*. Because of its excellent heat and wear resistance, Zr has been used as a coating material for nuclear fuel.

Ans. See text.

"note" Through various photos of the real elements, in particular rare elements, given in " *Elements A Visual Exploration of Every Known Element in the Universe" by Theodor Grey* (2009), we can learn their familiar applications, though unfamiliar in daily life.

Solution 1-4

(a) The relationship between the wavelength of the *de Broglie* wave and the length of a one-dimensional box is given by eq(4), where υ is the velocity of the particle and n is the quantum number.

$$\lambda = \frac{h}{p} = \frac{h}{m\upsilon} = \frac{2a}{n} \qquad (n = 1, 2, 3, \cdots) \qquad (4)$$

Since the energy of the particle in a box in the absence of external filed is kinetic energy alone, it is given by the left side of eq (5).

$$E = \frac{1}{2}m\upsilon^2 = \frac{(m\upsilon)^2}{2m} = \frac{1}{2m} \times \left(\frac{h}{2a}\right)^2 = \frac{h^2}{8ma^2} \qquad (5)$$

The lowest energy the particle can assume is obtained as $h^2/8ma^2$ by inserting $m\upsilon = h/2a$ from eq(4) into eq(5).

Ans. the lowest energy: $h^2/8ma^2$

(b) The wave function for the lowest energy level $\varphi_1(x)$ is given by eq(6).

$$\varphi_1(x) = \sqrt{\frac{2}{a}}\sin\frac{\pi x}{a} \qquad (6)$$

The wave function with a quantum number of n and its energy are given by eq(7).

$$\varphi_n(x) = \sqrt{\frac{2}{a}}\sin\frac{n\pi x}{a}, \qquad E_n = \frac{n^2h^2}{8ma^2} \qquad (7)$$

For the lowest energy level, a probability of finding the particle in a given range is given by eq(8).

$$\int_{0.24a}^{0.26a} \varphi_1(x)^2 dx = \frac{2}{a}\int_{0.24a}^{0.26a}\left(\sin\frac{\pi x}{a}\right)^2 dx = 0.02 \qquad (8)$$

Ans. existence probability: 0.02

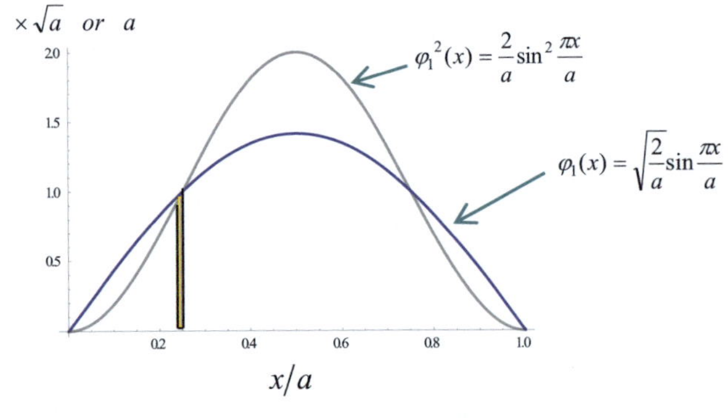

Fig.4 Wave function and probability distribution function for a particle in a one-dimensional box in the lowest energy level.

In Fig.4 above, eq (6) and its function squared ($\varphi_1^2(x)$) is shown together. It should be noted that these two functions look similar in shape but are notably different in their slopes at $x = 0$ and $x = a$, i.e.,

$$\left(\frac{\partial \varphi_n}{\partial x}\right)_{x=0,a} \neq 0 \ \text{but} \ \left(\frac{\partial {\varphi_n}^2}{\partial x}\right)_{x=0,a} = 0\,.$$

An approximate value of the integral of eq(8) can be obtained by multiplying the "width ($= 0.02a$)" by "median" as follows.

$$\frac{2}{a}\times\left(\sin\frac{\pi\cdot 0.25a}{a}\right)^2\times 0.02a = 0.04\times\sin^2(0.25\pi) = 0.02$$

Solution 1-5

(a) Values of the ionic radius in the unit of Å for alkali cations and halogen anions are listed in the table.

Table Ionic Radii for the Alkali Cations and Halogen Anions (1Å = 0.1 nm)

Li^+	Na^+	K^+	Rb^+	Cs^+	F^-	Cl^-	Br^-	I^-
0.60	0.98	1.33	1.48	1.67	1.33	1.81	1.96	2.19

For the crystal of LiF, the anion and cation in the nearest neighbor are in contact to each other as is illustrated in Fig.5 on the right, the ionic distance (dotted line) will be $0.60 + 1.32 = 1.92$ Å. Then the relation $1.92\times\sqrt{2} > 2\times 1.32$ holds and, hence, the nearest anions shown by white circles are not in contact to each other as is seen by the broken line.

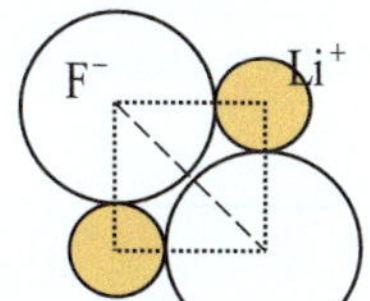

Fig.5 Two opposite ions in contact to each other in a cubic crystal.

Ans. LiF and its ionic distance is 1.92 Å.

(b) It is readily seen from the example given above that when two anions come to contact each other, the radius of the anion can assume the greatest value. Therefore, the relation of eq(9) holds

$$\sqrt{2}(r_+ + r_-) \ge 2r_- \qquad (9)$$

and both side of eq(9) is divided by r_- to yield eq(10).

$$\sqrt{2}(1 + r_+/r_-) \ge 2 \qquad (10)$$

Eq(10) can be rewritten as $r_+/r_- \ge \sqrt{2} - 1 = 0.414$. Therefore, the minimum value for the ratio r_+/r_- is 0.414 and the ratio thus obtained is called as limiting radius ratio.

Ans. limiting radius ratio: $r_+/r_- = 0.414$

Solution 1-6

The molecular orbitals for homonuclear diatomic molecules and the heteronuclear diatomic molecules which can be treated as the homonuclear diatomic molecules are classified based on their component atomic orbitals as follows;

① σ_{1s}, σ_{1s}^* orbitals composed of a linear combination of each $1s$-orbital ($1s - 1s$).

② σ_{2s}, σ_{2s}^* orbitals composed of a linear combination of each $2s$-orbital ($2s - 2s$).

③ σ_{2p_z}, $\sigma_{2p_z}^*$ orbitals composed of a linear combination of each $2p_z$-orbital whose axis is along the bond axis ($2p_z - 2p_z$).

④ π_{2p_y}, $\pi_{2p_y}^*$ orbitals composed of a linear combination of each $2p_x$ whose axis is perpendicular to the bond axis ($2p_x - 2p_x$).

⑤ π_{2p_y}, $\pi_{2p_y}^*$ orbitals composed of a linear combination of each $2p_x$ whose axis is perpendicular to the bond axis ($2p_x - 2p_x$).

The anti-bonding orbitals are marked by an upper-case*. The energy of each orbital generally increases in the order of

$$\sigma_{1s} < \sigma_{1s}^* < \sigma_{2s} < \sigma_{2s}^* < \sigma_{2p_z} < \pi_{2p_x} = \pi_{2p_y} < \pi_{2p_x}^* = \pi_{2p_x}^* < \sigma_{2p_z}^* \text{ (cf. "note").}$$

Based on this order, by filling each orbital with two electrons up to the total number of electrons belonging to a molecule, its molecular orbitals are constructed.

Taking oxygen molecule as an example, the energy levels of the component atomic orbitals (on both sides) and the constructed molecular orbitals (in the middle) are shown schematically in Fig.6. Each electron is indicated by either an up-arrow or a down-arrow. The last two electrons fill $\pi_{2p_y}^*$ orbitals but these orbitals are degenerate and hence the two electrons come to fill them separately with parallel spins in accordance with *Hund*'s rule.

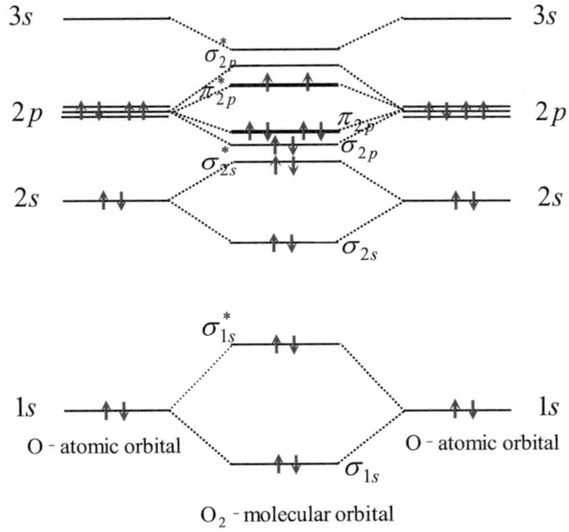

Fig.6 Atomic and molecular orbitals illustrate for a homo-nuclear diatomic molecule of oxygen. Both of π_{2p} and π_{2p}^* orbitals are doubly degenerated. σ_{2p} is for σ_{2p_z}.

For the diatomic molecule of CN, the total number of electrons are 13, which are distributed as $\sigma_{1s}{}^2\sigma_{1s}^{*2}\sigma_{2s}{}^2\sigma_{2s}^{*2}\sigma_{2p_z}{}^2\pi_{2p_x}^2\pi_{2p_y}^1$, resulting in one σ -bond and one and a half π -bonds.

In a similar way, the MO's for the other molecules are obtained as given in **Ans.**

Ans.

CN: $\sigma_{1s}{}^2\sigma_{1s}^{*2}\sigma_{2s}{}^2\sigma_{2s}^{*2}\sigma_{2p_z}{}^2\pi_{2p_x}^2\pi_{2p_y}^1$ resulting in σ -bond and one and a half π -bonds,

N_2: $\sigma_{1s}{}^2\sigma_{1s}^{*2}\sigma_{2s}{}^2\sigma_{2s}^{*2}\sigma_{2p_z}{}^2\pi_{2p_x}^2\pi_{2p_y}^2$ resulting in σ -bond and two π -bonds,

NO: $\sigma_{1s}{}^2\sigma_{1s}^{*2}\sigma_{2s}{}^2\sigma_{2s}^{*2}\sigma_{2p_z}{}^2\pi_{2p_x}^2\pi_{2p_y}^2\pi_{2p_x}^{*1}$ resulting in σ -bond and one and a half π -bonds,

O_2: $\sigma_{1s}{}^2\sigma_{1s}^{*2}\sigma_{2s}{}^2\sigma_{2s}^{*2}\sigma_{2p_z}{}^2\pi_{2p_x}^2\pi_{2p_y}^2\pi_{2p_x}^{*1}\pi_{2p_x}^{*1}$ resulting in σ -bond and one π -bonds with two unpaired electrons. Its spin quantum number is 1 and the state is called as triplet state. This electronic configuration is the most stable one and is denoted as ${}^3\Sigma_g^-$. (cf."note" in **solution 1-9**)

"**note**" For a little more precise discussions, the order of electronic energy levels of diatomic molecular orbitals is different for home-nuclear diatomic molecules constructed of H to N from for those constructed of O to F atoms. For the first group, σ_{2p} -orbital is located higher than π_{2p} -orbital which may lead to in a different electronic configuration from those given in Ans. Discussion here is only to comprehend a basic idea of MO's and electronic configurations. See *T.Engel and P.Reid, Physical Chemistry,* p.532, Person Benjamin Cummings, 2004.

Solution 1-7

NO_3^-: The electronic configuration of O atom is $1s^2 2s^2 2p^4$ and that of N atom is $1s^2 2s^2 2p^3$. The total number of electrons of NO_3^- is 32 but the valence bond electrons are 24. As is shown in Fig.7(a), there are three bonds between O and N atoms, and, hence, it is most appropriate to prepare sp^2-hybrid orbitals both at the O and N atoms to form three bonds of σ-type between the N and O atoms by using the three sp^2-hybrid orbitals of N and one of the three sp^2-hybrid orbitals of each O atom. The three bonds formed thus are on the same plane and are filled with two paired electrons each, 6 in total. The remaining two sp^2-hybrid orbitals on each O atom are filled with two paired electrons, again 6 in total, but these electrons are not shared with the N atom and do not contribute to bonding. Such pair of electrons is called lone pair. $2p_z$-orbitals of oxygen and nitrogen are perpendicular to the plane expanded by the three σ-bonds and do not contribute to σ-bonding but a set of $2p_z$-orbitals, one from nitrogen atom and one from each oxygen atom, forms a set of four π-orbitals. One of them is a bonding orbital filled with two electrons and two of them are non-bonding orbitals filled with four electrons. Summarizing the above argument, there are 6 electrons of the three σ-bonds, 12 electrons of the 6 non-bonding orbitals and 6 electrons in the 6π-orbitals (for further detail, see **7-39**). For this, neither O nor N atom satisfies Octet Rule but the three of N-O bonds are totally identical with no unpaired electron.

Butadiene: Each of four carbon atoms provides sp^2-hybrid orbitals and all of them are arranged to lie on the same plane (molecular plane) in order to form σ-bonds not only between C-H atoms but also between C-C atoms. Four $2p_z$-orbitals which are perpendicular to the molecular plane do not participate sp^2-hybridization can form π-orbitals in which four electrons are accommodated as is illustrated in Fig.7(b).

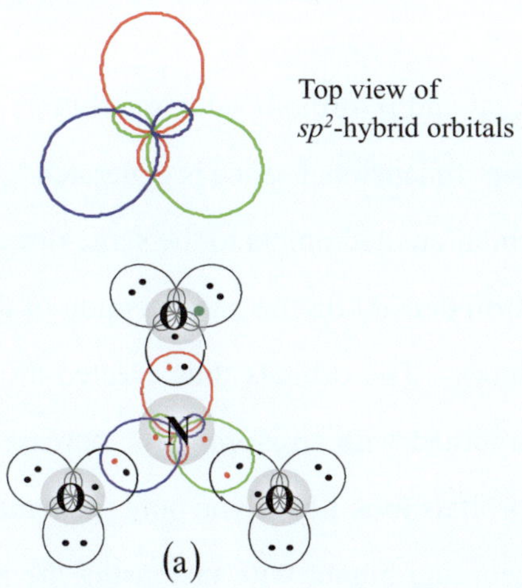

(a)

Fig.7 (a) sp^2-hybridization and chemical bonding of NO_3^-. indicates $2p_z$-orbital. Black dots and red dots are constructed from four $2p_z$-orbitals. respectively. The green dot indicates an electron accepted to form an anion.

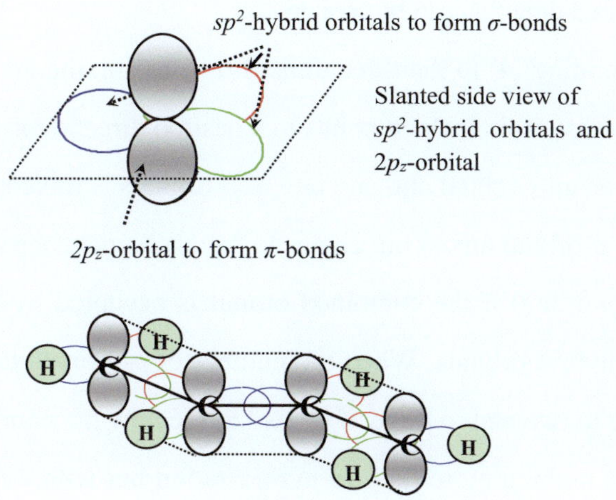

(b)

Fig.7(b) sp^2-hybridization and chemical bonding of butadiene. Dotted lines schematically indicate π-bonding constructed from four $2p_z$-orbitals.

The chemical bonding of NO_3^- is more simply given either by the structure on the left-hand or the resonance structures on the right-hand side given below. Any structure of the latter satisfies octet rule.

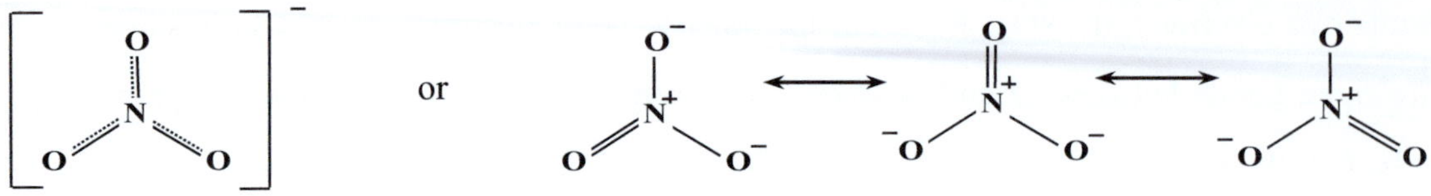

Fig.8 Resonance structures of NO_3^-.

Ans. See text.

How hybrid orbitals are constructed and used.

We limit the discussion here to the hybrid orbitals formed of *s*-and *p*-orbitals. In order to understand 'hybridization' correctly it will be most appropriate to discuss the matter in the following order.

(1) An *s*-orbital is represented by a spherically symmetric wave function.

(2) p_x-, p_y- and p_z-orbitals behave just like as unit vector s of x, y, and z, respectively.

(3) Because of the property of *p*-orbitals described above, when p_x- and p_y-orbitals are mixed by one to one, another *p*-orbital directed along the line making the angle of 45° against the *y*-axis is formed as shown in Fig.9 below.

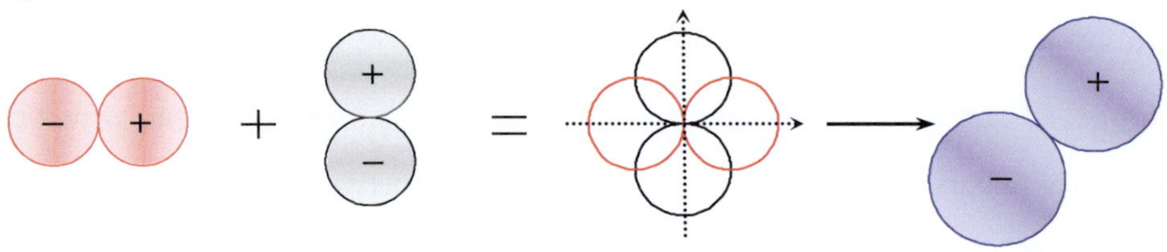

Fig.9 Hybridization of two orthogonal 2*p*-orbitals.

(4) Based on (3), by mixing p_x- and p_y-orbitals with an arbitrary weight, any *p*-orbital which points to a desired direction can be created.

(5) Extending (4) to the three dimensions, by mixing all of p_x-, p_y- and p_z-orbitals with an arbitrary weigh, a new *p*-orbital which points to a desired direction in the three-dimensional space is generated.

(6) When an *s*-orbital and any one of *p*-orbitals is mixed, the orbital created points to the same direction as the *p*-orbital mixed but comes to have a bigger robe of electron density on the one direction of the axis of *p*-orbital if the combined orbital is occupied by electron(s). The orbitals thus created are called *sp*-hybrid orbitals. When the number of *p*-orbitals to be combined with an *s*-orbital is increased from one to two and to three, it is easy to see how the combined orbitals look like if you note the remark (5). It is nothing more than *sp*-hybridization but with a robe becoming bigger with increasing the number of *p*-orbitals mixed. When one *s*- and two *p*-orbitals are combined, the resulting orbitals are called sp^2-hybrid orbitals, and when three, sp^3-hybrid orbitals.

(7) Using (5) and (6) described above, we can readily answer the question about the shape of sp^3-hybrid orbitals and C-H bonding of methane. For this, we consider a cube with the length of 2 whose coordinate of the center is assigned as (0, 0, 0). at which the C atom of methane is located. From a

symmetric point of view for tetragonal methane, we can chose four apexes for example as is shown in Fig.10 (a). In accord with the numbers given for the coordinate of any apex thus taken, we mix (or combine) three *p*-orbitals to yield a new *p*-orbital directed toward to the apex chosen. After this we can mix an *s*-orbital with any one of four newly generated *p*-orbitals, resulting four hybrid orbitals as is illustrated in Fig.10 (b). It should be noted that four independent hybrid orbitals are generated from four components but not one single sp^3-hybrid orbital which looks like a tetrapod. All of four independent orbitals shown in Fig.10 (b) are superposed at the center of the cube as is shown in Fig.10 (c) to prepare σ-bonding orbitals of methane. It should be emphasized again that sp^3-hybridization yields just one of the four orbitals all of which look like one of *sp*-hybrid orbitals but not yield an orbital resembling a tetrapod pointing four different directions. For exact forms of the wave functions of hybrid orbitals, see Chap.7.

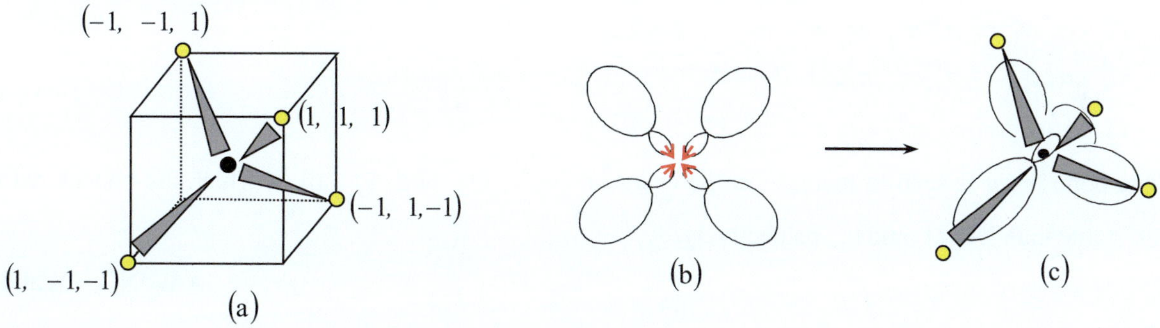

Fig.10 Various ways of representation of sp^3-hybrid orbitals.

(8) Including *d*-orbitals, there arise other types of hybridization such as sp^3d-hybridization and sp^3d^2-hybridization. However, the concept of hybridization itself is the same irrespective of the type of atomic orbitals employed, that is five independent hybrid orbitals are formed upon the sp^3d-hybridization and six upon the sp^3d^2-hybridization. However, because of the more complex angle dependence (ϑ, φ-dependence) of *d*-orbitals than that of *p*-orbitals, each of the resulting sp^3d- and sp^3d^2-hybrid orbitals exhibits more characteristic space dependence than does any of hybrid orbitals constructed of just *s*- and *p*-orbitals alone.

(9) When a pictorial shape of a hybrid orbital is drawn, the shape looks different whether a crude shape of its wave function or its probability distribution, i.e., the square of the wave function, is employed. In (3) and (6) above, image of wave functions of the constituent atomic orbitals are employed but in (7), an image of probability density distributions is in mind.

Solution 1-8

(a) See the solution to **problem 1-6**.

(b) The electronic configuration and bond order of each neutral diatomic molecule is modified upon the addition or removal of an electron as shown below. The modified electronic configuration is given in parentheses and the change in the bond order is indicated by an arrow.

(i) In the case of anion

N_2^- ($\sigma_{1s}^2\sigma_{1s}^{*2}\sigma_{2s}^2\sigma_{2s}^{*2}\sigma_{2p_z}^2\pi_{2p_x}^2\pi_{2p_y}^2\pi_{2p_x}^{*1}$) 3→2.5

NO^- ($\sigma_{1s}^2\sigma_{1s}^{*2}\sigma_{2s}^2\sigma_{2s}^{*2}\sigma_{2p_z}^2\pi_{2p_x}^2\pi_{2p_y}^2\pi_{2p_x}^{*1}\pi_{2p_y}^{*1}$) 2.5→2

O_2^- ($\sigma_{1s}^2\sigma_{1s}^{*2}\sigma_{2s}^2\sigma_{2s}^{*2}\sigma_{2p_z}^2\pi_{2p_x}^2\pi_{2p_y}^2\pi_{2p_x}^{*2}\pi_{2p_y}^{*1}$) 2→1.5

C_2^- ($\sigma_{1s}^2\sigma_{1s}^{*2}\sigma_{2s}^2\sigma_{2s}^{*2}\sigma_{2p_z}^2\pi_{2p_x}^2\pi_{2p_y}^1$) 2→2.5

F_2^- ($\sigma_{1s}^2\sigma_{1s}^{*2}\sigma_{2s}^2\sigma_{2s}^{*2}\sigma_{2p_z}^2\pi_{2p_x}^2\pi_{2p_y}^2\pi_{2p_x}^{*2}\pi_{2p_y}^{*2}\sigma_{3s}^1$) 1→1.5

The bond order is seen to increase in the cases of C_2^- and F_2^-, which are considered to be more stable than C_2 and F_2, respectively.

Ans. C_2 and F_2

(ii) In the case of cation

N_2^+ ($\sigma_{1s}^2\sigma_{1s}^{*2}\sigma_{2s}^2\sigma_{2s}^{*2}\sigma_{2p_z}^2\pi_{2p_x}^2\pi_{2p_y}^1$) 3→2.5

NO^+ ($\sigma_{1s}^2\sigma_{1s}^{*2}\sigma_{2s}^2\sigma_{2s}^{*2}\sigma_{2p_z}^2\pi_{2p_x}^2\pi_{2p_y}^2$) 2.5→3

O_2^+ ($\sigma_{1s}^2\sigma_{1s}^{*2}\sigma_{2s}^2\sigma_{2s}^{*2}\sigma_{2p_z}^2\pi_{2p_x}^2\pi_{2p_y}^2\pi_{2p_x}^{*1}$) 2→2.5

C_2^+ ($\sigma_{1s}^2\sigma_{1s}^{*2}\sigma_{2s}^2\sigma_{2s}^{*2}\sigma_{2p_z}^2\pi_{2p_x}^1$) 2→1.5

F_2^+ ($\sigma_{1s}^2\sigma_{1s}^{*2}\sigma_{2s}^2\sigma_{2s}^{*2}\sigma_{2p_z}^2\pi_{2p_x}^2\pi_{2p_y}^2\pi_{2p_x}^{*2}\pi_{2p_y}^{*1}$) 1→1.5

The bond order is seen to increase in the cases of NO^+, O_2^+ and F_2^+, all of which are considered to be more stable than NO, O_2 and F_2, respectively.

Ans. NO, O_2 and F_2

Solution 1-9

To answer the problem properly, we need the knowledge on the four terms of [angular momentum of electron], [spin angular momentum], [symmetry of improper rotation (rotation followed by reflection against a plane perpendicular to the rotation axis)], and [parity upon exchange of nuclei]. The [symmetry of improper rotation] is decomposed into the product of the two successive operations of [inversion at the center of an atom: $I = \pm 1$] and [rotation of 180° about the axis perpendicular to the plane of rotary reflection: $(-1)^l$].

(a) H_2 (H: ^2S (electronic state of atom), $D_{\infty h}$ (symmetry group of molecule))

Angular momenta of electrons: Since the total angular momentum L of the two electrons denoted by subscripts 1 and 2 is given by $L = l_1 + l_2 = 0 \Rightarrow \Sigma$, only the electronic states expressed as Σ are possible.

Symmetry of improper rotation: The product of the two symmetry operations is given as $(-1)^{l_1} I_1 \times (-1)^{l_2} I_2 = (-1)^{l_1+l_2} I_1 I_2 = 1 \times 1 \times 1 = 1$ and, hence, only the Σ^+ state is possible.

Spin angular momenta of electrons: The spin angular momenta can take two values of $S = s_1 + s_2 = 1$ and $S = s_1 + s_2 = 1 - 1 = 0$, yielding triplet and singlet state, respectively, i.e., $^3\Sigma^+$ and $^1\Sigma^+$ states.

Parity: The total electronic wave function including spin state for a molecule should be anti-symmetric upon the interchange of two electrons. The symmetry of the triplet state is symmetrical but the singlet state is antisymmetrical (ungerade) upon the interchange of electrons,

Summarizing all these points, it is concluded the electronic states for H_2 constructed of H in the state of ${}^2S$ are ${}^3\Sigma_u^+$ and ${}^1\Sigma_g^+$, the latter of which is for the ground state.

Ans. ${}^3\Sigma_u^+$ and ${}^1\Sigma_g^+$ (ground state)

(b) N_2 (N: ${}^4S(1s^2 2s^2 2p^3)$, $D_{\infty h}$)

Angular momenta of electrons: Since $L = l_1 + l_2 = 0$, only the states denoted by Σ are possible.

Symmetry of improper rotation: In a similar manner to the case of H_2, this symmetry is given by $(-1)^{l_1+l_2} I_1 I_2 = 1\times(-1)\times(-1) = 1$ and, hence, only the Σ^+ state is possible. Note that a p-orbital is antisymmetric against inversion.

Spin angular momenta of electrons: The spin angular momenta can take four values of $S = 3$, 2, 1 and 0, yielding heptad, quintet, triplet and singlet states, respectively. Therefore, the states of ${}^7\Sigma^+$, ${}^5\Sigma^+$, ${}^3\Sigma^+$ and ${}^1\Sigma^+$ are possible.

Parity: The total electronic wave function including spin state for a molecule should be anti-symmetric upon the interchange of two electrons. Heptad and triplet spin states are symmetric and quintet and singlet states are antisymmetric against spin exchange. Summarizing all these points, it is concluded the electronic states for N_2 constructed of N in the state of ${}^4S$ are ${}^7\Sigma_u^+$, ${}^5\Sigma_g^+$, ${}^3\Sigma_u^+$ and ${}^1\Sigma_g^+$.

Ans. ${}^7\Sigma_u^+$, ${}^5\Sigma_g^+$, ${}^3\Sigma_u^+$ and ${}^1\Sigma_g^+$

(c) Cl_2 (Cl: ${}^2P(\mathrm{Ne}3s^2 3p^5)$, $D_{\infty h}$)

Angular momenta of electrons: Since $L = l_1 + l_2 = 2$, $l_1 + l_2 - 1 = 1$, $l_1 + l_2 - 2 = 0$, the four states denoted by Δ, Π, $\Sigma^{\pm}$ are possible.

Symmetry of improper rotation: Based on the value of $(-1)^{l_1+l_2} I_1 I_2 = 1\times(-1)\times(-1) = 1$, when $m_1 = m_2 = 0$, only the Σ^+ state is possible. Note again that p-orbitals are antisymmetric against inversion.

Spin angular momenta of electrons: The spin angular momenta can take the two values of $S = 1$ and 0, yielding triplet and singlet states, respectively. Therefore, the states of ${}^{3,1}\Delta$, ${}^{3,1}\Pi$, $2\,{}^{3,1}\Sigma^+$ and ${}^{3,1}\Sigma^-$ are possible.

Parity: The total electronic wave function including spin state for a molecule should be antisymmetric upon the interchange of two electrons. Since Δ and Σ^+ states are symmetric, Σ^- states are antisymmetric and Π states can be symmetric or antisymmetric upon the interchange. Summarizing all these requirements, the electronic states for Cl_2 constructed of Cl in the state of ${}^2P$ are ${}^3\Delta_u$, ${}^1\Delta_g$, ${}^{3,1}\Pi_g$, ${}^{3,1}\Pi_u$, $2{}^3\Sigma_u^+$, $2{}^1\Sigma_g^+$, ${}^3\Sigma_g^-$ and ${}^1\Sigma_u^-$ states. Or these states are grouped into ${}^1\Delta_g$, ${}^1\Pi_g$, ${}^1\Pi_u$, $2{}^1\Sigma_g^+$ and ${}^1\Sigma_u^-$, and ${}^3\Delta_u$, ${}^3\Pi_g$, ${}^3\Pi_u$, $2{}^3\Sigma_u^+$ and ${}^3\Sigma_g^-$, based on the spin angular momentum.

Ans. ${}^1\Delta_g$, ${}^1\Pi_g$, ${}^1\Pi_u$, $2{}^1\Sigma_g^+$, ${}^1\Sigma_u^-$, ${}^3\Delta_u$, ${}^3\Pi_g$, ${}^3\Pi_u$, $2{}^3\Sigma_u^+$ and ${}^3\Sigma_g^-$

(d) O_2 (O: ${}^3P(\mathrm{He}2s^2 2p^4)$, $D_{\infty h}$)

Angular momenta of electrons: Same as for Cl_2.

Symmetry of improper rotation: Same as for Cl_2.

Parity : Same as for Cl_2.

Spin angular momenta of electrons: Since $S = s_1 + s_2 = 2$, 1, 0, resulting in quintet, triplet and singlet states. Referring to the above case (Cl_2), the possible electronic states are as given in the Ans. The ground state of O_2 is the triplet state $^3\Sigma_g^-$ and the first excited singlet state is $^1\Delta_g$.

Ans. $^1\Delta_g$, $^1\Pi_g$, $^1\Pi_u$, $2^1\Sigma_g^+$, $^1\Sigma_u^-$, $^3\Delta_u$, $^3\Pi_g$, $^3\Pi_u$, $2^3\Sigma_u^+$, $^3\Sigma_g^-$, $^5\Delta_g$, $^5\Pi_g$, $^5\Pi_u$, $2^5\Sigma_g^+$, $^5\Sigma_u^-$

"note 1" Detailed discussions on the electronic states of diatomic molecules are given (§75) in *Quantum Mechanics-non-relativistic theory*, by *L.D.Landau and E.M.Lifshitz.*

Diatomic molecular orbitals constructed from the linear combination of atomic orbitals can be described in terms of orbital angular momentum and symmetric property. The electronic states of diatomic molecules are therefore denoted as $^3\Sigma_u$, $^1\Pi_g$, $^1\Delta_g$, and so on. For instance, the ground electronic state of O_2 whose diatomic molecular orbitals are illustrated in Fig.6 in **solution 1-6** is denoted as $^3\Sigma_g^-$.

The rule of obtaining possible electronic states for a diatomic molecule is essentially to predict what kind of bonding is possible when two atoms in the ground electronic states are brought together to a distance of bonding. In this case, correspondence between the electronic states for atoms and a molecule is not one to one. Therefore, not just one but be a few electronic states may be expected for the molecule thus formed.

The rule is;

(1) Hetero-nuclear diatomic molecules ($C_{\infty v}$).

The angular momenta and spin angular momenta associated with the valence electrons of each atom are denoted by l_1, l_2 ($l_1 \geq l_2$) and s_1, s_2, respectively. The components of the angular momentum along the axis of bonding can assume the values of $m_1 = -l_1, -l_1+1, \cdots, l_1-1, l_1$, $m_2 = -l_2, -l_2+1, \cdots, l_2-1, l_2$ and hence the component of the orbital angular moment the molecule can take is $m_1 + m_2$. The total number of its combination is given by $(2l_1+1)\cdot(2l_2+1)$. Table 1 exemplifies all such combinations in the case of $l_1 = 3$ and $l_2 = 2$, yielding 35 ways of combination in total.

The electronic states of the molecule are distinguished in terms of the absolute magnitude of $\Lambda = |m_1 + m_2|$. Thus the components of 5 and -5 in Table 1 are considered to arise from the same electronic state. For this the electronic state is doubly degenerated. It is readily known from the table that the doubly degenerated electronic states characterized by the value of $\Lambda = 4$, 3, 2, 1 exist 2, 3, 4, 5 sets, respectively.

Table 1 Combination of orbital angular momenta.

m_2 \ m_1	3	2	1	0	-1	-2	-3
2	5	4	3	2	1	0	-1
1	4	3	2	1	0	-1	-2
0	3	2	1	0	-1	-2	-3
-1	2	1	0	-1	-2	-3	-4
-2	1	0	-1	-2	-3	-4	-5

Table 2 Combination of spin angular momenta.

s_2 \ s_1	$\frac{1}{2}$	$-\frac{1}{2}$
$\frac{1}{2}$	1	0
$-\frac{1}{2}$	0	1

(a)

s_2 \ s_1	1	0	−1
$\frac{1}{2}$	$\frac{3}{2}$	$\frac{1}{2}$	$-\frac{1}{2}$
$-\frac{1}{2}$	$\frac{1}{2}$	$-\frac{1}{2}$	$-\frac{3}{2}$

(b)

Two spin angular momenta s_1 and s_2, when combined, yield the total spin angular momenta S given by eq(11).

$$S = s_1 + s_2,\ s_1 + s_2 - 1,\ s_1 + s_2 - 2,\ \cdots,\ |s_1 - s_2| + 1,\ |s_1 - s_2| \qquad (11)$$

Table 2 illustrates how the components of the total summed spin angular momenta are obtained in the two cases of (a) ($s_1 = 1/2$, $s_2 = 1/2$) and (b) ($s_1 = 1$, $s_2 = 1/2$), where the sets of shadowed spin components correspond to triplet (a) and quartet states (b), respectively.

The electronic state of a relevant diatomic molecule can be described by a combination of the orbital angular momentum in the Table 1 and the spin angular momentum in the Table 2. Furthermore, when the angular momentum is zero we can specify the corresponding electronic states in more detail, as is described below.

The summed angular momentum becomes zero, i.e., $m_1 + m_2 = 0$ either in the case of $m_1 = -m_2 = m$ or $m_1 = m_2 = 0$, which correspond to 0 in Table 1. If the electronic wave function of each atom is represented by $\varphi^1_{m_1}$ and $\varphi^2_{m_2}$, respectively, then the wave function for the two atoms separated infinitely away can be given by $\varphi^1_{m_1} \times \varphi^2_{m_2}$, where superscripts 1 and 2 represent electron 1 of atom 1 and electron 2 of atom 2, respectively. There are two cases to be discussed.

When $m \neq 0$, i.e., $m_1 = -m_2 = m$. There are two independent molecular electronic wave functions, which are given by eqs(12) and (13).

$$\phi^+ = \varphi^1_m \varphi^2_{-m} + \varphi^1_{-m} \varphi^2_m \qquad (12)$$

$$\phi^- = \varphi^1_m \varphi^2_{-m} - \varphi^1_{-m} \varphi^2_m \qquad (13)$$

When the reflection against a plane containing the bond axis is performed to the wave functions of φ^1_m and φ^2_{-m}, they turn into φ^1_{-m} and φ^2_m, respectively, because the angular momentum is inverted upon such reflection. As a result of this, the function ϕ^+ of eq(12) is invariant but the function ϕ^- is inverted to $-\phi^-$ upon the reflection against the molecular plane. In terms of the operator of such reflection symbolized as R, the above situation is expressed by eq(14).

$$R\phi^+ = \phi^+, \qquad R\phi^- = -\phi^- \qquad (14)$$

For this it is derived that the Σ states where the component of the summed angular momentum is zero can be divided into Σ^+ and Σ^- states. Consequently, if $l_1 \geq l_2$, then there exist l_2 states of Σ^+ and the same number of Σ^- states as well.

When $m = 0$, the operation of reflection is divided into the two successive operations of 'inversion (I) at the center of an atom' and 'rotation by $180°$ against the axis vertical to the plane of the reflection'. When this successive operations are performed to φ^1_0, it changes a sign alone determined by $(-1)^{l_1} I_1$ for which I_1 takes a value of $+1$ or -1 depending on whether φ^1_0 is symmetric against the inversion at the center of the atom or antisymmetric, respectively. The same successive operations are performed on φ^2_0 as well and the two results are combined to yield the result of the reflection against the molecular plane. In conclusion, the resulting effect of the reflection is given by $(-1)^{l_1+l_2} I_1 I_2$ which takes either value of $+1$ or -1. As a result of this, the Σ states whose total number is given by $2l_2+1$ (five in Table 1) are classified into either Σ^+ or Σ^- states.

(2) Homo-nuclear diatomic molecules ($D_{\infty h}$)

How to construct the total angular momentum and spin momentum is the same as in the case of hetero-nuclear diatomic molecules described above. In the case of homo-nuclear diatomic molecules, however, an additional symmetric operation arises, i.e., the inversion at the center of the bond. This operation corresponds to the exchange of two atoms. The resulting effect of the operation is either symmetric (even) or antisymmetric (odd). We distinguish between the former and the latter by adding a subscript *g* or *u* to the state symbol, corresponding to German 'gerade' (even) and 'ungerade' (odd), respectively.

(a) Two atoms are in different electronic states (one atom can be an ion.).

All of the electronic states possible for the homo-nuclear atoms can be distinguished in terms of parity, i.e., *g* or *u* and hence the total number of electronic states possible in this case is just doubled.

(b) Two atoms are in the same electronic state.

Because of its length, the explanation is not provided here but it is found that the total number of the possible electronic states is the same as that for the hetero-nuclear diatomic molecules and each number of states is given as follows.

In summary, in terms of parity, the number of even states is N_g and odd states is N_u

When $\Lambda = |m_1 + m_2|$ is odd, then $N_g = N_u$.

When Λ is even and S is even as well, then $N_g = N_u + 1$.

When Λ is even but S is odd, then $N_g = N_u - 1$.

For the Σ states, it is necessary to distinguish between Σ^+ and Σ^- states and they are;

When S is odd, $N_g^+ = N_u^- + 1 = L + 1$.

When S is even, $N_g^+ = N_u^- + 1 = L + 1$.

"note 2" According to the group theory, the homo-nuclear diatomic molecules belong to the point group of $D_{\infty h}$ and the hetero-nuclear diatomic molecules to $C_{\infty v}$. The irreducible expressions for these point groups are $\Sigma_g^+, \Sigma_g^-, \Pi_g, \Delta_g, \cdots \Sigma_u^+, \Sigma_u^-, \Pi_u, \Delta_u, \cdots$ and $\Sigma^+, \Sigma^-, \Pi, \Delta, \Phi, \cdots$, respectively and the electronic states discussed above and classified in terms of the spectral terms such as Σ_u, Π_g are just the same as those irreducible expressions. See, for instance, *F.Alber Cotton, Chemical Applications of Group Theory*, Interscience Publishers, New York, 1964.

Solution 1-10

(a) HCl (H: ^2S, Cl: ^2P, $C_{\infty v}$)

Angular momentum: Since L can take two values of 1 and 0 ($L = l_1 + l_2 = 1,\ l_1 + l_2 - 1 = 0$), two kinds of electronic states Π and Σ are possible.

Symmetry of reflection: $(-1)^{l_1+l_2} I_1 I_2$ yields 1 and hence only Σ^+ is possible between Σ^+ or Σ^-. Note that s-orbital is symmetric against the inversion I but p-orbital is antisymmetric.

Spin angular momentum: Since S can take two values of 1 and 0 ($S = s_1 + s_2 = 1,\ s_1 + s_2 - 1 = 0$), both of triplet and singlet states are possible.

No further symmetry element to be considered exists for HCl. Putting the above requirements together, possible electronic states of HCl are $^3\Pi$, $^1\Pi$, $^3\Sigma^+$ and $^1\Sigma^+$ in spectral terms.

Ans. $^3\Pi$, $^1\Pi$, $^3\Sigma^+$ and $^1\Sigma^+$

(b) CO (C: ^3P, O: ^3P, $C_{\infty v}$)

Angular momentum: Since L can take three values of 2, 1 and 0 ($L = l_1 + l_2 = 2,\ l_1 + l_2 - 1 = 1,\ l_1 + l_2 - 2 = 0$), three kinds of electronic states $\Delta, \Pi, \Sigma^{\pm}$ are possible. Π state is described by either $\varphi^{\mathrm{C}}_{m_1}\varphi^{\mathrm{O}}_{m_2} = \varphi^{\mathrm{C}}_1\varphi^{\mathrm{O}}_0$ or $\varphi^{\mathrm{C}}_0\varphi^{\mathrm{O}}_1$.

Symmetry of reflection: Since $(-1)^{l_1+l_2} I_1 I_2$ given by $1\times(-1)\times(-1) = 1$ yields 1, Σ state to be derived from $m_1 = -m_2 = 0$ is Σ^+ alone. Note that p-orbital is antisymmetric against inversion I.

Spin angular momentum: Since S can take three values of 2, 1 and 0 ($S = s_1 + s_2 = 2,\ s_1 + s_2 - 1 = 1,\ s_1 + s_2 - 2 = 0$) quintet (5), triplet (3) and singlet (1) states are possible.

Taking degeneracy into account, there are $9 = 2 + 2\times 2 + 2 + 1$ different electronic states which agree with the number of states distinguishable on the basis of the components of angular momentum, i.e., $(2l_1 + 1)\cdot(2l_2 + 1) = 9$.

No further symmetry element to be considered for CO exists. Putting together the above requirements, possible electronic states of CO are $^{5,3,1}\Delta$, $2\,^{5,3,1}\Pi$, $2\,^{5,3,1}\Sigma^+$ and $^{5,3,1}\Sigma^-$ on the basis of the spectral terms.

Ans. $^{5,3,1}\Delta$, $2\,^{5,3,1}\Pi$, $2\,^{5,3,1}\Sigma^+$ and $^{5,3,1}\Sigma^-$

Solution 1-11

(a) The ionization energy of the hydrogen atom is the energy required for the electronic transition from the state specified by the principal quantum number $n = 1$ to that denoted by $n = \infty$, which is calculated to be 13.60 eV. Refer to eq(1) given in **problem 1-1** and the *Grotorian* diagram below.

Ans. 13.60 eV

(b) The energy level of the hydrogen-like atom is given by $E_n = -\dfrac{e^4 m_e}{8\varepsilon_0^2 h^2}\cdot\dfrac{Z^2}{n^2}$, where n is the principle quantum number, ε_0 is the dielectric constant in vacuum, Z is the atomic number and m_e is the mass of an electron.

Ans. See text (b).

(c) *Balmer* series is a group of the electronic transitions (emissions) from the energy levels specified by the principal quantum numbers of $l = 3,\ 4,\ 5,\ \cdots$ to the energy level of $n = 2$. Therefore, from the equation given in (b), the energy of the emission ($n \leftarrow l$) is given by $E_l - E_n = \dfrac{e^4 m_e Z^2}{8\varepsilon_0^2 h^2}\left(\dfrac{1}{2^2} - \dfrac{1}{l^2}\right)$.

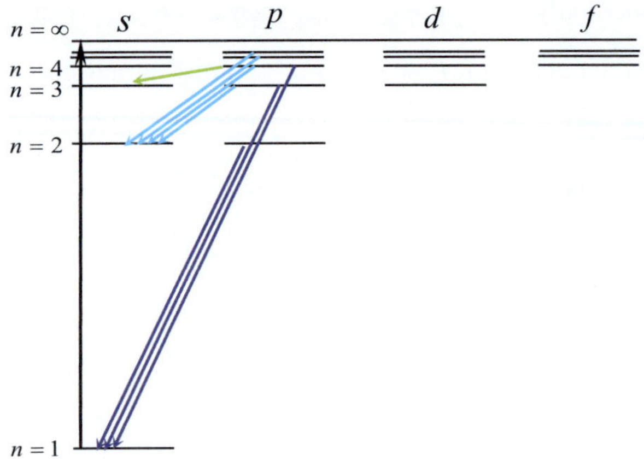

Fig. 11 *Grotorian* diagram of hydrogen atom. (——) ionization energy, (——) *Lyman* series (——) *Balmer* series、(——) *Paschen* series. The electronic energies of the hydrogen-like atoms depend only on the principal quantum number. Note that they do not depend on the magnitude of angular momentum. The electronic transition such as $s \leftrightarrow s$ or $p \leftrightarrow p$, etc., does not occur. See the selection rule for the electronic transition of an atom in Chapter 7.

Ans. See text (c).

Solution 1-12

In the following, the electronic configurations of the relevant atoms and ions before and after ionization are given together with their ionization energies.

$$\left.\begin{array}{lll} {}_3\mathrm{Li}(1s^2 2s^1) \rightarrow {}_3\mathrm{Li}(1s^2) + e, & 5.392\ \mathrm{eV} \\ {}_4\mathrm{Be}^{+}(1s^2 2s^1) \rightarrow {}_4\mathrm{Be}^{2+}(1s^2) + e, & 18.2\ 1\mathrm{eV} \\ {}_5\mathrm{B}^{2+}(1s^2 2s^1) \rightarrow {}_5\mathrm{B}^{3+}(1s^2) + e, & 37.93\ \mathrm{eV} \\ {}_6\mathrm{C}^{3+}(1s^2 2s^1) \rightarrow {}_6\mathrm{C}^{4+}(1s^2) + e, & 64.49\ \mathrm{eV} \end{array}\right\} \quad (15)$$

In the all cases above, an electron involved in the ionization from the outermost shell is the electron in 2s-orbital. The electron to be ionized is shielded from a nuclear charge effectively by two electrons occupying 1s-orbital and there is none other than these two electrons. For this the shielding effect is considered the same in the all cases above (see **"note"** below). In Fig.12 below the ionization energies given above are plotted against the squared values of $Z_{eff} = Z - \sigma$ employing three values of $\sigma = 2.0, 1.5$ and 0.

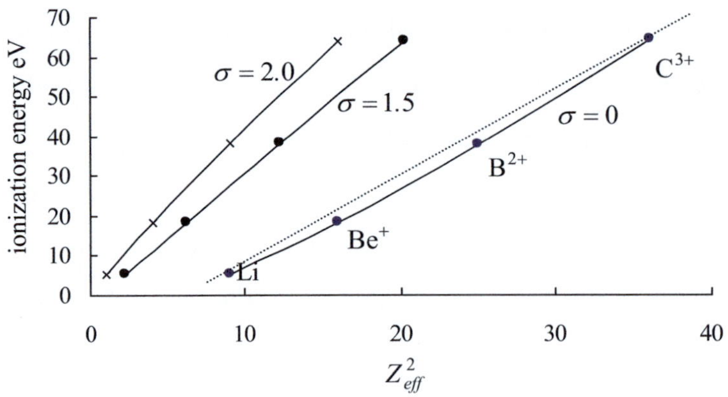

Fig.12 Ionization energies and shielding effect. The plot for $\sigma = 0$ is the one against Z^2 but yields less satisfactory relation as compared with that for $\sigma = 1.5(e)$.

It is seen from Fig.12 that if it is assumed that the 2s-electron is shielded by the two 1s-electrons by the amount of 1.5 charge units, then the value calculated for *I.P.* with $Z_{eff}^2 = (Z-\sigma)^2 = (Z-1.5)^2$ is very close to that observed.

Ans. See text.

"note 1" How to estimate a value of σ.

(1) Shell electrons are grouped as follows;

$1s$; $2s,2p$; $3s,3p$; $3d$; $4s,4p$; $4d$; $4f$; $5s,5p$;

The s and p shells of the same principal quantum number are classified to the same shell group but d,f shells belong to a different one.

(2) Shielding constant σ is given by the sum of the contribution from each of the above shell groups.

(a) The electrons in the shell groups locating outer than the shell of the electron to be ionized contribute null (0).

(b) Each of the electrons which belong to the same shell group as the electron to be ionized does contribute 0.35 but in the case of $1s$ electrons one contributes 0.30 to shield the other.

(c) When the electron to be ionized is located in either of s or p shell of the same group, the electrons in the nearest inner shell group contribute 0.85 and the electrons further inner shells contribute 1.00 per each.

(d) When the electron to be ionized is located either in d or f shell, all of the inner shell electrons contribute 1.00 per each.

For instance in the case of $C(1s^2 2s^2 2p^2)$, the effective nuclear charges for the electron in the relevant group is estimated as below:

$$1s: \quad Z_{eff} = 6 - 0.30 = 5.7$$

$$2s, 2p: \quad Z_{eff} = 6 - 3\times 0.35 - 2\times 0.85 = 3.25$$

[reference] *C.A.Coulson*,"*Valence*", Oxford University Press, p.40 (1961).

"note 2" For the ionization of bonding electrons of molecules see problems associated with PES (*Photoelectron Spectroscopy*) and XPS (*X-ray Photoelectron Spectroscopy*) given in Chapter 7 Molecular Structures and Photochemistry.

Solution 1-13

(a) Refer to **solution 1-2 .**

Ans. ${}_{19}K^+(1s^2 2s^2 2p^6 3s^2 3p^6)$, ${}_{17}Cl^-(1s^2 2s^2 2p^6 3s^2 3p^6)$

(b) The reaction under discussion is divided into several elementary reactions, each of which is schematically shown in Fig.13 given below together with the associated energies. The energy diagram as such is one of

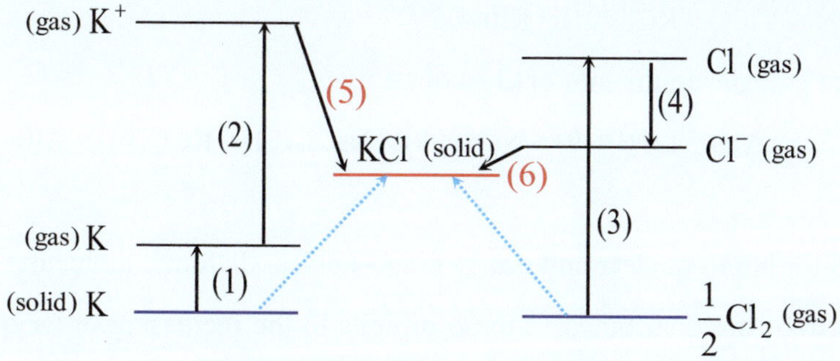

Fig.13 Schematic energy diagram for the reaction. Energy levels are only schematically drawn and hence do not correspond to the real values.

the figures representing *Born-Haber cycle*. Its advantage is it is easy to see pictorially how large is the total energy of the reaction to yield desired product(s) (red line) from reactants (blue lines).

The processes numbered in Fig.13 are as follow:

$$\text{(1) sublimation: } 21.5(\text{kcal}) \times 4.184 = 89.96\ \text{kJ} \cdot \text{mol}^{-1}$$

$$\text{(2) ionization: } 4.34(\text{eV}) \times 96.485 = 418.74\ \text{kJ} \cdot \text{mol}^{-1}$$

$$\text{(3) bond dissociation: } 239.2 \div 2 = 119.6\ \text{kJ} \cdot \text{mol}^{-1}$$

$$\text{(4) electron affinity: } 3.62(\text{eV}) \times 96.485 = 349.28\ \text{kJ} \cdot \text{mol}^{-1}$$

(b) The energy of the reaction of $K_{(solid)} + \frac{1}{2}Cl_{2\,(gas)} \rightarrow K^{+}{}_{(gas)} + Cl^{-}{}_{(gas)}$ can be obtained from Fig.13 as $(1)+(2)+(3)-(4) = 89.96 + 418.7 + 119.6 - 349.3 = 279\ \text{kJ} \cdot \text{mol}^{-1}$.

Ans. heat of reaction: $279\,\text{kJ} \cdot \text{mol}^{-1}$

(c) The energy of the reaction of $K^{+}{}_{(gas)} + Cl^{-}{}_{(gas)} \rightarrow KCl_{(solid)}$, i.e., the lattice energy U is calculated by inserting the values of M(*Madelung* constant) $= 1.75$, $z_{+} = z_{-} = 1$, $r_0 = r_{+} + r_{-} = 2.73 \times 10^{-10}\,\text{m}$, $4\pi\varepsilon_0 = 1.11 \times 10^{-10}\,\text{C}^2\text{J}^{-1}\text{m}^{-1}$ and $n = 10$ (n : a parametric constant associated with the repulsion potential between ions) into eq(16).

$$U = -\frac{N_A \cdot M \cdot z_{+} z_{-} e^2}{4\pi\varepsilon_0 r_0}\left(1 - \frac{1}{n}\right) \qquad (16)$$

For the reaction $K^{+}{}_{(gas)} + Cl^{-}{}_{(gas)} \rightarrow KCl_{(gas)}$ which occurs all in the gas phase, the energy for the reaction is equivalent to the energy required to bring the two ions located infinitely far apart to the bonding distance of r_0. Therefore, in that case the energy is given by the term of $N_A z_{+} z_{-} e^2 / 4\pi\varepsilon_0 r_0$ alone of eq(16), which is equal to $-510\,\text{kJ} \cdot \text{mol}^{-1}$. The parameters M and n in eq(16) is employed for estimating the coulombic potential of 3-dimensional ionic crystals.

Ans. $U = -803\ \text{kJ} \cdot \text{mol}^{-1}$

(d) The energy of the reaction of $K_{(solid)} + \frac{1}{2}Cl_{2\,(gas)} \rightarrow KCl_{\ (solid)}$ can be calculated by employing the value obtained in (c) as follows. The value in parentheses is the observed one.

$$(1)+(2)+(3)-(4)-(5)+(6)$$

$$= 89.96 + 418.7 + 119.6 - 349.3 - 803$$

$$= -524\ (-436.8)\ \text{kJ} \cdot \text{mol}^{-1}$$

Ans. $-524\,\text{kJ} \cdot \text{mol}^{-1}$

(e) For the ionic crystals of the same crystal form, their lattice energies become larger as their ionic distances get shorter. Thus they are in the decreasing order as follow.

$$\text{LiF}(1019) > \text{NaF}(909) > \text{NaCl}(771) > \text{KCl}(701) > \text{KI}(641)$$

The values in parentheses are observed lattice energies in the unit of $\text{kJ} \cdot \text{mol}^{-1}$.

Ans. $\text{LiF}(1019) > \text{NaF}(909) > \text{NaCl}(771) > \text{KCl}(701) > \text{KI}(641)$

Solution 1-14

Based on the molecular orbitals composed for homo-nuclear and nearly homo-nuclear diatomic molecules (see **solution 1-6**), required number of electrons are distributed to these orbitals in the increasing order of orbital energy. When the net number of electrons contributing to bonding, i.e., number of electrons in the bonding orbital minus the number of electrons in the antibonding orbitals, is the greater, the stronger the bond of a molecule under consideration is. To answer the question, the strength of bonding is qualitatively compared by using a mark of <.

(a) $F_2(\sigma_{1s}^2\sigma_{1s}^{*2}\sigma_{2s}^2\sigma_{2s}^{*2}\sigma_{2p_z}^2\pi_{2p_x}^2\pi_{2p_y}^2\pi_{2p_x}^{*2}\pi_{2p_y}^{*2}) < F_2^-(\sigma_{1s}^2\sigma_{1s}^{*2}\sigma_{2s}^2\sigma_{2s}^{*2}\sigma_{2p_z}^2\pi_{2p_x}^2\pi_{2p_y}^2\pi_{2p_x}^{*2}\pi_{2p_y}^{*2}\sigma_{3s}^1)$

(b) $NO(\sigma_{1s}^2\sigma_{1s}^{*2}\sigma_{2s}^2\sigma_{2s}^{*2}\sigma_{2p_z}^2\pi_{2p_x}^2\pi_{2p_y}^2\pi_{2p_x}^{*1}) < NO^+(\sigma_{1s}^2\sigma_{1s}^{*2}\sigma_{2s}^2\sigma_{2s}^{*2}\sigma_{2p_z}^2\pi_{2p_x}^2\pi_{2p_y}^2)$

(c) $BN(\sigma_{1s}^2\sigma_{1s}^{*2}\sigma_{2s}^2\sigma_{2s}^{*2}\sigma_{2p_z}^2\pi_{2p_x}^2\pi_{2p_y}^1) < BO(\sigma_{1s}^2\sigma_{1s}^{*2}\sigma_{2s}^2\sigma_{2s}^{*2}\sigma_{2p_z}^2\pi_{2p_x}^2\pi_{2p_y}^2)$

(d) $NF(\sigma_{1s}^2\sigma_{1s}^{*2}\sigma_{2s}^2\sigma_{2s}^{*2}\sigma_{2p_z}^2\pi_{2p_x}^2\pi_{2p_y}^2\pi_{2p_x}^{*2}) < NO(\sigma_{1s}^2\sigma_{1s}^{*2}\sigma_{2s}^2\sigma_{2s}^{*2}\sigma_{2p_z}^2\pi_{2p_x}^2\pi_{2p_y}^2\pi_{2p_x}^{*1})$

(e) $Be_2(\sigma_{1s}^2\sigma_{1s}^{*2}\sigma_{2s}^2\sigma_{2s}^{*2}) < Be_2^+(\sigma_{1s}^2\sigma_{1s}^{*2}\sigma_{2s}^2\sigma_{2s}^{*1})$

Ans. See text.

Solution 1-15

Based on the net number of bonding electrons, as is considered in the preceding problem, the order of the bonding strength is found as below;

$N_2(\sigma_{1s}^2\sigma_{1s}^{*2}\sigma_{2s}^2\sigma_{2s}^{*2}\sigma_{2p_z}^2\pi_{2p_x}^2\pi_{2p_y}^2) > N_2^+(\sigma_{1s}^2\sigma_{1s}^{*2}\sigma_{2s}^2\sigma_{2s}^{*2}\sigma_{2p_z}^2\pi_{2p_x}^2\pi_{2p_y}^1) \approx O_2^+(\sigma_{1s}^2\sigma_{1s}^{*2}\sigma_{2s}^2\sigma_{2s}^{*2}\sigma_{2p_z}^2\pi_{2p_x}^2\pi_{2p_y}^2\pi_{2p_x}^{*1}) >$
$O_2(\sigma_{1s}^2\sigma_{1s}^{*2}\sigma_{2s}^2\sigma_{2s}^{*2}\sigma_{2p_z}^2\pi_{2p_x}^2\pi_{2p_y}^2\pi_{2p_x}^{*1}\pi_{2p_y}^{*1}) > O_2^-(\sigma_{1s}^2\sigma_{1s}^{*2}\sigma_{2s}^2\sigma_{2s}^{*2}\sigma_{2p_z}^2\pi_{2p_x}^2\pi_{2p_y}^2\pi_{2p_x}^{*2}\pi_{2p_y}^{*1})$.

Since the orbital energies of $\pi_{2p_x}^*$ and $\pi_{2p_y}^*$ are almost equal, the two electrons which occupy these orbitals assume parallel orientation according to *Hund*'s rule, yielding a triplet state. Actually, the observed bonding energies of N_2^+ and O_2^+ are 6.34 eV and 6.48 eV, respectively. Incidentally, they are 7.37 eV and 5.08 eV for N_2 and O_2, respectively.

[reference] *G.Hergberg "Molecular Spectra and Molecular Structure 1. Spectra of Diatomic Molecules"*, D.van Nostrand Company,Inc., p343,1950.

Ans. $N_2 > N_2^+ \approx O_2^+ > O_2 > O_2^-$

Solution 1-16

(a) *Bohr*'s hypotheses consists of three premises and two of them are employed in the following to decide the orbital and energy of an electron of a hydrogen atom.

1. The centrifugal force working on the electron in motion is balanced with the coulombic force existent between the electron and the nucleus.
2. The angular momentum of the electron circulating around the nucleus is given by $n\times\left(\frac{h}{2\pi}\right)$ where $n = 1, 2, 3, \cdots$.

For the electron circulating along a circle of radius r with the velocity of υ,

hypothesis 1 leads us to eq(17) and

$$\frac{m_e\upsilon^2}{r} = \frac{e^2}{4\pi\varepsilon_0 r^2} \qquad (17)$$

hypothesis 2 to eq(18).

$$m_e v r = n\times\left(\frac{h}{2\pi}\right) \qquad (18)$$

From these two equations, v and r are determined uniquely as in eq(19)

$$r=\frac{\varepsilon_0 h^2 n^2}{\pi m_e e^2}, \qquad v=\frac{nh}{2\pi m_e}\cdot\frac{\pi m_e e^2}{\varepsilon_0 h^2 n^2}=\frac{e^2}{2\varepsilon_0 nh} \qquad (19)$$

and then the total energy E_n of the electron is given by eq(20).

$$E_n=\frac{1}{2}m_e v^2-\frac{e^2}{4\pi\varepsilon_0 r}=-\frac{m_e e^4}{8\varepsilon_0^2 h^2 n^2} \qquad (20)$$

Ans. $E_n=-\dfrac{m_e e^4}{8\varepsilon_0^2 h^2 n^2}$

(b) By replacing e^2 in the first part of eq(19) with Ze^2 and e^4 in eq(20) with Z^2e^4, radii and energy levels of an electron of a hydrogen-like atom can be obtained. Consequently, the first ionization potential can be calculated by putting it equal to the energy difference between the energy for $n=1$ and the one for $n=\infty$ as given in **Ans.**

Ans. $r=\dfrac{\varepsilon_0 h^2 n^2}{\pi m_e Z e^2}$, $I.P.=\dfrac{m_e Z^2 e^4}{8\varepsilon_0^2 h^2}$

(c) **Ans.** $_{11}Na^+(1s^2 2s^2 2p^6)$, $_{12}Mg^{2+}(1s^2 2s^2 2p^6)$, $_{13}Al^{3+}(1s^2 2s^2 2p^6)$

(d) When the effective nuclear charge of an atom is Z_{eff} (e), the effective nuclear charges of Ca^+ and Al^{3+} are considered to be given by $Z_{eff}+1$ and $Z_{eff}+2$, respectively, since those ions have the same electronic configuration as Na^+ does. The square root of the ionization potential of each ion under consideration should be proportional to its effective nuclear charge and hence eq(21) and (22) are obtained.

$$\sqrt{\frac{I.P.(Ca^{2+})}{I.P.(Na^+)}}=\frac{Z_{eff}+1}{Z_{eff}}=1.3016 \quad (21), \qquad \sqrt{\frac{I.P.(Al^{3+})}{I.P.(Ca^{2+})}}=\frac{Z_{eff}+2}{Z_{eff}+1}=1.2236 \quad (22)$$

Each of these two equations yields a different value for Z_{eff}, say Z_{eff_1} and Z_{eff_2}, as follows, so that their average value of 3.39 can be taken as the effective nuclear charge.

$$Z_{eff_1}=\frac{1}{0.3016}=3.32, \quad Z_{eff_2}=\frac{0.7746}{0.2236}=3.46 \quad \Rightarrow \quad Z_{eff}=\frac{Z_{eff_1}+Z_{eff_2}}{2}=3.39$$

Employing that value of 3.39 for the effective nuclear charge of Na^+ in the calculation of ionic radii explained in (b), the ratios of ionic radii of relevant ions are calculated as follow.

$$\frac{r(Ca^{2+})}{r(Na^+)}=\frac{Z_{eff}}{Z_{eff}+1}=\frac{3.39}{3.39+1}=0.772, \qquad \frac{r(Al^{3+})}{r(Na^+)}=\frac{Z_{eff}}{Z_{eff}+2}=\frac{3.39}{3.39+2}=0.692$$

Then, the ratios normalized to the value of $r(Na^+)$ are as given in **Ans.**

Ans. $r(Na^+):r(Ca^{2+}):r(Al^{3+})=1.0:0.772:0.692$

"note" The results given in (d) are valid only for the ions in the gas phase. The experimental values of ionic radius obtained in crystals are naturally quite different from those given in **Ans.**

Solution 1-17

The important point is the fact that as is treated in **problem 1-7** the hybridization of two different p-orbitals yields a new p-orbital which orients to a different direction from either of the two employed

p-orbitals depending on the magnitude of mixing. In fact, the function given by $1/\sqrt{6}\,p_x - 1/\sqrt{2}\,p_y$ is reduced to another p-type orbital rotated counterclockwise by 30 degrees from the negative y-axis as is readily seen from the composition of trigonometric functions given below in eq(23). For simplicity, the proportional constant for normalization is neglected there .

$$\begin{aligned}\frac{1}{\sqrt{6}}p_x - \frac{1}{\sqrt{2}}p_y &= \frac{1}{\sqrt{6}}\sin\vartheta\cos\varphi - \frac{1}{\sqrt{2}}\sin\vartheta\sin\varphi \\ &= \sqrt{\frac{2}{3}}\left(\frac{1}{2}\sin\vartheta\cos\varphi - \frac{\sqrt{3}}{2}\sin\vartheta\sin\varphi\right) \\ &= \sqrt{\frac{2}{3}}\sin\vartheta(\sin30^\circ\cos\varphi - \cos30^\circ\sin\varphi) \\ &= -\sqrt{\frac{2}{3}}\sin\vartheta\sin(\varphi - 30^\circ) \qquad (23)\end{aligned}$$

Generally speaking, any arbitrary combination of any two or three p-orbitals yields another p-type orbital with a different orientation and magnitude from either of the employed p-orbitals. In short this tells us that p-orbitals behave as if they were unit vectors in the three-dimensional space. Therefore, we can readily answer what kind of orbital is formed when not only two but also all three p-orbitals are combined. This is a basically important point in understanding what are the hybrid orbitals and what is their use.

In the above case, p_x and p_y can be represented in an $x-y$ plane by vectors along the x-axis and y-axis with the magnitudes of $1/\sqrt{6}$ and $-1/\sqrt{2}$, respectively, as shown below. The vectorial sum of these two vectors yields a vector shown by a red arrow which corresponds a hybrid orbital synthesized from the two p-orbitals given by $1/\sqrt{6}\,p_x - 1/\sqrt{2}\,p_y$.

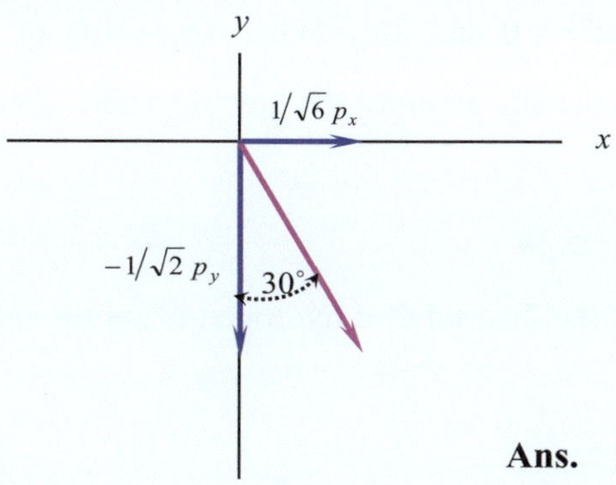

Fig.14 Vectorial representation of $p-p$ hybridization.

(——) component p-orbitals before hybridization

(——) p-orbital resulting from hybridization

Ans. See text. See **problem 1-7** for a shape of a p-orbital.

Solution 1-18

The σ-bonds which constitute molecular frames of $CH_2=CH_2$, $CH_3\cdot$, $AlCl_3$, NO_3^- and CO_3^{2-} are considered to be formed of sp^2-hybrid orbitals. Because of this, all these molecules assume a triangular shape. On the other hand, σ-bonds in NH_3 or PH_3 are considered not to be composed of sp^2-hybrid orbitals but are interpreted in terms of sp^3-hybrid orbitals. For the relation between the number of valence electrons and molecular shapes of one-centered polyatomic molecules such as NH_3 or PH_3, see **problem 1-55 (VSEPR theory)**.

Ans. See text.

Solution 1-19

The electron density function $\rho(r)$ is defined by using a radial wave function $R(r)$ as in eq(24) (**"note"**).

$$\rho(r)dr = r^2 R(r)^2 dr \qquad (24)$$

Consequently, the atomic radius of a relevant orbital is found by seeking a value of r at which the electron density function takes a maximum. By putting the first derivative of $\rho(r)$ equal to zero, $\rho'(r) = 0$, the 1s-orbital is found to take a maximum of the density at $r = a_0$ ($= Bohr$ radius) as shown in eq(25).

$$\rho'(r) = 4\left(\frac{1}{a_0}\right)^3 \left(r^2 e^{-\frac{2r}{a_0}}\right)' = 4\left(\frac{1}{a_0}\right)^3 \left(2re^{-\frac{2r}{a_0}} - r^2 \times \frac{2}{a_0} e^{-\frac{2r}{a_0}}\right) = 4\left(\frac{1}{a_0}\right)^3 \times 2re^{-\frac{2r}{a_0}}\left(1 - \frac{r}{a_0}\right) = 0 \qquad (25)$$

In a similar manner, the first derivative of $\rho(r)$ for the 2s-orbital is put equal to zero as in eq(26),

$$\rho'(r) = \frac{1}{8}\left(\frac{1}{a_0}\right)^3 \left\{\left(2r - \frac{r^2}{a_0}\right)^2 e^{-\frac{r}{a_0}}\right\}' = \frac{1}{8}\left(\frac{1}{a_0}\right)^3 \left\{2\left(2r - \frac{r^2}{a_0}\right)\left(2 - \frac{2r}{a_0}\right) - \frac{1}{a_0}\left(2r - \frac{r^2}{a_0}\right)^2\right\} e^{-\frac{r}{a_0}}$$

$$= \frac{1}{8}\left(\frac{1}{a_0}\right)^3 \times e^{-\frac{r}{a_0}} \left(2r - \frac{r^2}{a_0}\right)\left\{-\frac{1}{a_0}\left(2r - \frac{r^2}{a_0}\right) + 4\left(1 - \frac{r}{a_0}\right)\right\} = 0 \qquad (26)$$

which is simplified as eq(27).

$$r\left(2 - \frac{r}{a_0}\right)\left\{\left(\frac{r}{a_0}\right)^2 - 6\left(\frac{r}{a_0}\right) + 4\right\} = 0 \qquad (27)$$

Then the solutions of eq(27) are obtained as follows.

$$r = 0,\ (3-\sqrt{5})a_0,\ 2a_0,\ (3+\sqrt{5})a_0$$

In Fig.15 $\rho(r)$ is plotted against r for the 1s- and 2s-orbitals. It is found that $\rho(r)$ for the 2s-orbital takes maxima at $(3-\sqrt{5})a_0$ and $(3+\sqrt{5})a_0$ but minima at $r = 0$ and $2a_0$. When the value of r is compared between the 1s- and 2s-orbitals at which the electron density assumes the greatest value, their ratio is found as follows.

$$\frac{r_{max}(2s)}{r_{max}(1s)} = \frac{(3+\sqrt{5})a_0}{a_0} = (3+\sqrt{5}) = 5.24$$

From this it is concluded that the atomic radius of the 2s-orbital of hydrogen is approximately five times larger than that of the 1s-orbital.

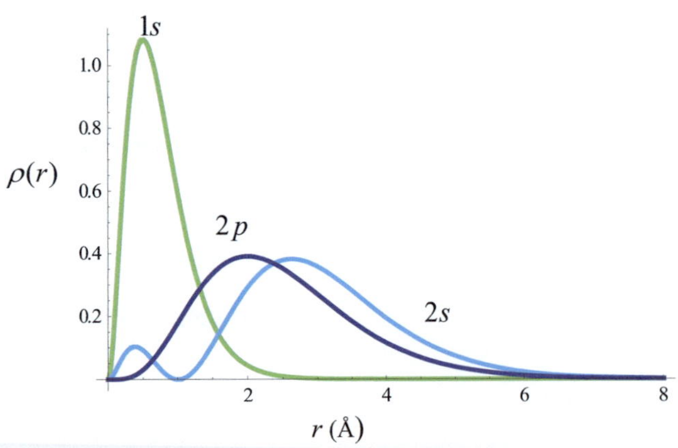

Fig.15 Electron density distributions for 1s-, 2s- and 2p-orbotals.

$\rho(r)$ is plotted for a_0 = 0.5 Å. It should be added that $\rho(r)$ for ns-orbital becomes zero at n-1 different positions except for r = 0 and for np-orbital it becomes zero at n-2 positions (relation between quantum number and number of nodal planes).

Ans. app. 5 times larger

"note" Volume element expressed in the polar coordinate is $d\tau = r^2 \sin\vartheta d\vartheta d\varphi dr$ and after its integration over ϑ and φ it is reduced to $d\tau = 4\pi r^2 dr$. For this it seems that a proportional constant 4π is dropped off in eqs(24)-(26), but the wave function used here is normalized to $(1/4\pi)^{1/2}$ and thus $\rho(r)$ shown in Fig.15 is normalized so as $\int_0^\infty \rho(r)dr = 1$.

Solution 1-20

K-shell is the shell for which the principal quantum number n is equal to 1. Since the spin angular momentum can take two values, the number of electrons accommodated in this shell is two and represented as $1s^2$. Thus, 2 for *K*-shell.

L-shell is for $n = 2$ and $l = 0$ and 1. The fully occupied *L*-shell is represented as $2s^2 2p^6$. Thus, 8 for *L*-shell .

M-shell is for $n = 3$ and $l = 0$, 1 and 2. The fully occupied *M*-shell is represented as $3s^2 3p^6 3d^{10}$. Thus, 18 for *M*-shell.

N-shell is for $n = 4$ and $l = 0$, 1, 2 and 3. The fully occupied *N*-shell is represented as $4s^2 4p^6 4d^{10} 4f^{14}$. Thus, 32 for *N*-shell.

There are $(2l+1)$ different atomic orbitals such as s, p, d, $f \cdots$ for an eigenstate specified by the azimuthal quantum number l, each of which can accommodate two electrons. The maximum number of electrons accommodated is therefore given by $2 \times (2l+1)$. See *Aufbau Principle* (**solution 1-2**) to know how to fill electrons in these orbitals one by one in order.

Ans. See text.

Solution 1-21

The volume of a cubic cell of edge length a is a^3. A Ti atom at each corner of this cube is shared by eight cubes. Therefore, only one eighth of each atom is regarded to belong to the unit cell under consideration. Such circumstance is illustrated in Fig.16 given below. Taking this into consideration, the value of

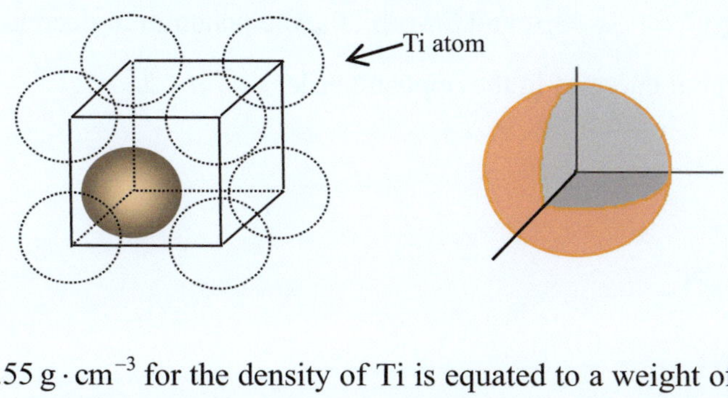

Fig.16 Density of unit cell.

An atom at each corner of a cube shown on the left of Fig.16 is seen to belong to the cube through only its one eighth as is illustrated on the right of Fig.16. Only gray-colored portion belongs to the cube. Thus the eight atoms together contribute just one atom to the cube.

$4.55\ \mathrm{g \cdot cm^{-3}}$ for the density of Ti is equated to a weight of one Ti atom divided by the volume of a unit cube a^3 as shown below. And a value for a is calculated from this relation to be 2.60 Å.

$$a^3 = \frac{47.9 \times 10^{-3}}{6.02 \times 10^{23} \times 4.55 \times 10^3} = \frac{4.79 \times 10^{-28}}{6.02 \times 4.55} \Rightarrow a = \left(\frac{4.79 \times 10^2}{6.02 \times 4.55}\right)^{1/3} \times 10^{-10} = 2.60\ \text{Å}$$

$$\rho = \frac{47.9\times10^{-3}(8\times1/8)}{a^3\times6.02\times10^{23}} = 4.55\times10^3(\text{kg}\cdot\text{m}^{-3}) \;\Rightarrow\; 4.55\,\text{g}\cdot\text{cm}^{-3}$$

Since the atomic radius of Ti is known to be 1.32 Å, it can be said from the result obtained above that two neighboring Ti atoms are in contact with each other.

Ans. 2.60 Å

Solution 1-22

(a) As is discussed in the solution to **problem 1-12**,

the orbital energy of $1s$-orbital is related to Z by eq(28),

$$E_{1s} = -\frac{m_e(Z-\sigma)^2 e^4}{8\varepsilon_0^2 h^2} \qquad (28)$$

where σ is the shielding constant and it takes a value of 0.30 for He and atoms with Z larger than 2. In addition, all of the outer shell electrons are assumed not to contribute in shielding a $1s$-orbital electron. (See "**note 1**" in **solution 1-12** for how to estimate a value for the shielding constant.)

Ans. eq(28)

(b) For an electron which occupies any of electronic orbitals with large principal quantum numbers, the other electron of the atom under discussion can be regarded as inner shell electrons which provide complete shielding and hence the shielding constant in eq(28) can be put equal to $\sigma = (Z-1)$. Because of this, for atoms with small atomic numbers, the orbital energies of outer shells such as d- and f-orbitals remain constant against Z for a while. When the atomic number increases further and electrons come to occupy the shells located outer the shell of the electron under consideration, then the inner shell orbital energy comes to decrease with Z since the outer shell electrons are not effective in shielding the electron in the inner shell under consideration. Under such circumstances, it is necessary to take it into account that the probability of finding an electron near the nucleus varies from orbital to orbital of sub-shells $(s, p, d, f, \cdots)$. We may say that the electron penetrates greater into near the nucleus when the probability is larger. When the negative orbital energy becomes the greater in its absolute value, the greater is the penetration effect. As is expected from the electron density distribution shown in Fig.17 for $3s$-, $3p$- and $3d$-orbitals, the penetration decreases in the order of $s > p > d > f$ leading to different orbital energies in the opposite order $s < p < d < f$.

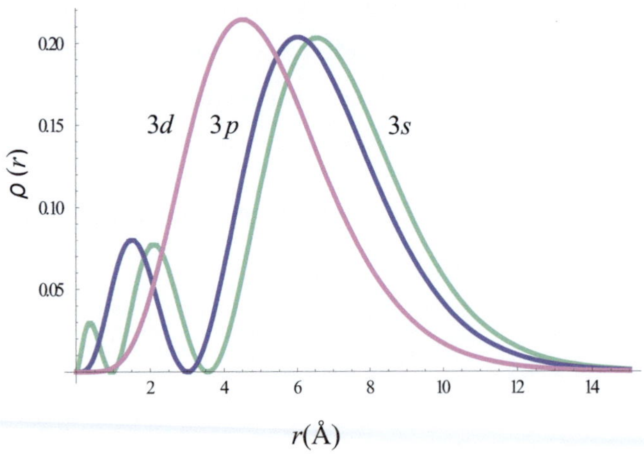

Fig.17 Radial density distributions for $3s$-, $3p$- and $3d$-orbitals of hydrogen atom.

The energy of 1*s*-orbital estimated from the ionization energy of 1*s*-orbital electron is plotted in against Z for atoms from Z=1 to 9. The curve in Fig.18 is a fitting curve drawn with $\sigma = 0.65$. The observed values are shown by blue dots for from H to Be.(from *Maian,* 2nd ed.) As is readily seen from Fig.19, the orbital energy of 1*s*-orbital can be given by eq(28).

The curve in Fig.19 looks different from those given in many text books (e.g., a figure in **solution 1-23**) The reason for this is that the *y*-axis employed in literature is not a linear scale but in common logarithmic scale. For comparison, the orbital energy expressed in log $_{10}$ scale is plotted against *Z* in Fig.19.

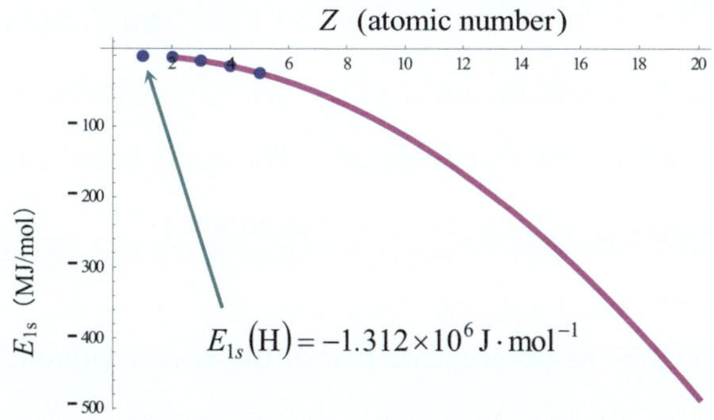

Fig.18 Orbital energy of 1*s*-electron for the atoms of *Z* from 1 to 9.
The red line represents eq(28). Note that the vertical scale is linear.

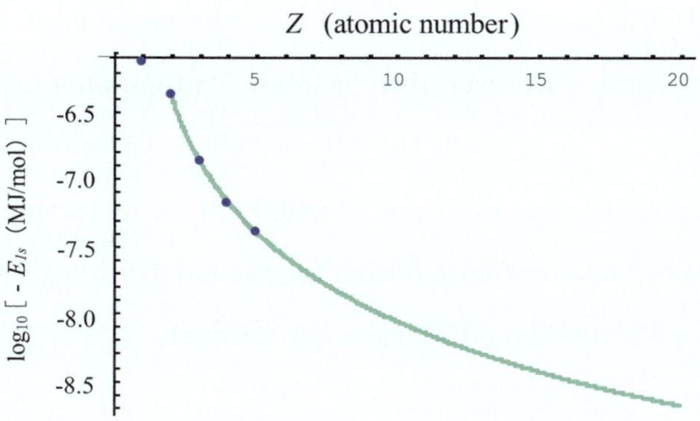

Fig.19 Logarithmic orbital energies of 1*s*-electron for the atoms of *Z* from 1 to 9.
The blue line represents eq(28) but the vertical scale is in log10 ($-E_{1s}$).

The vertical scale of Fig.10.22 given in *B.H.Maian*, *University Chemistry* 2nd ed. 1968 is a logarithmic scale but since no value is given there, we may misunderstand as if the orbital energy of 1*s*-electron decreases concavely but the energy expressed in linear scale does not. It decreases convexly as is shown in Fig.19.

"note 1" Calculation of orbital energy by SCF. In *Physical Chemistry* by *Berry, Rice & Ross* (p.176, Fig.5.2) orbital energies calculated by *Hartree-Fock* method is cited(*F.Herman and R.Skillman, Atomic Structure Calculations*:Prenti ce Hall, Englewood Cliffs, NB.J.,1963).

"note 2" Unit of energy. $1\,hartree\,(E_h) = 2\,rydbergs = 4.359814 \times 10^{-18}\,\mathrm{J} = 27.2116$ eV.

The orbital energies of inner shell electrons can be obtained experimentally by XPS(*X-ray photoelectron spectroscopy*). For a brief explanation of XPS see a text book by *Atkins* p.520 or by *T.Engel & P.Reid* p.500. When they are plotted against *Z*, a general trend is almost the same as that for I.P. used herewith.

Solution 1-23

(a) Since $l = 2$, the maximum number of electrons is $2 \times (2l + 1) = 10$.

Ans. 10 electrons

(b) The number of electrons accommodated before any 3*d*-orbital starts to be filled is $2+2+6+2+6=18$. On the basis of $n+l$ rule, in the case of $_{19}$K or $_{20}$Ca, before 3*d*-orbital of being filled, 4*s*-orbital comes to be occupied by one or two electrons first. Hereafter the 3*d*-orbital starts to be filled by electrons. However, the penetrating effect (discussed before) of the 3*d*-orbital is small and hence the effective nuclear charge increases monotonically with the atomic number. Therefore the electronic energy of the 3*d*-orbital decreases proportionally in accordance with $Z_{eff}^2 = (Z - [\text{constant}])^2$, exhibiting a behavior similar to the one shown in Fig.3 given in **solution 1-22.**

Ans. See text.

(c) The maximum number of electrons to be accommodated and the order of energy levels predicted by the $n+l$ rule are as follows. Up to 5*p*-orbitals, 54 electrons can be accommodated. The under bared parts with red lines show two orbitals with the order inverted according to the *n*+*l* rule (see **problem 1-2**).

$$1s^2 2s^2 2p^6 3s^2 3p^6 \underline{4s^2 3d^{10}} 4p^6 \underline{5s^2 4d^{10}} 5p^6$$

On the other hand, the orbital energies shown in Fig.20 below indicate that at the atomic number of 50 (marked by a green vertical line in Fig.20), the order of energy levels is in accord with the principal quantum number, that is, the energy level is higher the larger *n* is, and the inversion of the order based on the $n+l$ rule (indicated by sky blue under lines) is not recognized. However, the electronic configuration expected for the atom of Z=50 is $5p^2$ for the out most shell and the inner shells are fully occupied. Therefore, it does not matter whether the orbitals are filled in accordance with the $n+l$ rule or simply in the increasing order of *n* and the same electronic configuration results. For the atoms with greater atomic number, however, the order of energy levels is $5s$, $5p$, $5d < 4f$ and the $n+l$ rule should be taken into account.

$$1s < 2s < 2p < 3s < 3p < \underline{3d < 4s} < 4p < \underline{4d < 5s} < \underline{5p < 5d < 4f}$$

Ans. The $n+l$ rule is irrelevant for $3d-4s$ and $4d-5s$. See text and Fig.20.

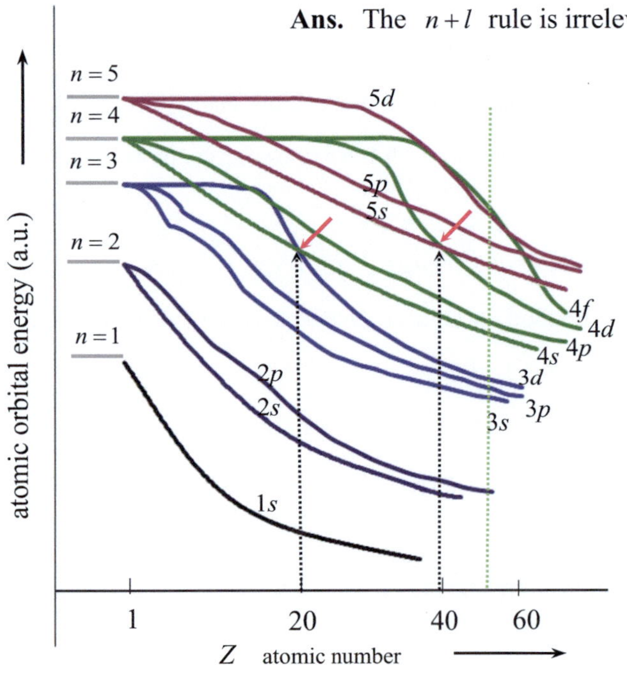

Fig. 20 Energy levels of atomic orbitals and $n+l$ rule. Level inversion is recognized at the atomic numbers indicated by (↙).
Fig.20 is modified after Fig.10.22 in *University Chemistry* 2nd ed. by *B.H. Maian*. See **solution 1-22** for the unit of the vertical axis.

Solution 1-24

Solution 1. Cartesian coordinates x, y, z can be expressed in terms of polar coordinates as in eq(29).

$$x = \sin\vartheta\cos\varphi,\quad y = \sin\vartheta\sin\varphi,\quad z = \cos\vartheta \qquad (29)$$

If we assume that the angular parts of the *d*-orbitals d_{z^2}, d_{xy}, d_{yz}, d_{zx}, $d_{x^2-y^2}$ are expressed in terms of combinations of x, y, z multiplied just in the same way as subscripts indicate, then their angular parts can be written as are given below (parts marked by dotted lines) in eq(30).

$$\left.\begin{aligned}
&d_{z^2} \propto \cos^2\vartheta \\
&d_{xy} \propto \sin\vartheta\cos\varphi \times \sin\vartheta\sin\varphi = \frac{1}{2}\sin^2\vartheta\sin 2\varphi \\
&d_{yz} \propto \sin\vartheta\sin\varphi \times \cos\vartheta = \frac{1}{2}\sin 2\vartheta\sin\varphi \\
&d_{zx} \propto \cos\vartheta \times \sin\vartheta\cos\varphi = \frac{1}{2}\sin 2\vartheta\cos\varphi \\
&d_{x^2-y^2} \propto \sin^2\vartheta\cos^2\varphi - \sin^2\vartheta\sin^2\varphi = \sin^2\vartheta(\cos^2\varphi - \sin^2\varphi) = \frac{1}{2}\sin^2\vartheta\cos 2\varphi
\end{aligned}\right\} \qquad (30)$$

On the other hand, the wave functions of 3*d*-orbitals in practical use are as given in Table 1.

Table 1 Five atomic wave functions of 3*d*-orbitals.

$3d_{z^2} = \dfrac{1}{81\sqrt{6\pi}}\left(\dfrac{Z}{a_0}\right)^{3/2}\sigma^2 e^{-\sigma/3}(3\cos^2\vartheta - 1)$	$3d_{xy} = \dfrac{1}{81\sqrt{2\pi}}\left(\dfrac{Z}{a_0}\right)^{3/2}\sigma^2 e^{-\sigma/3}\sin^2\vartheta\sin 2\varphi$
$3d_{yz} = \dfrac{\sqrt{2}}{81\sqrt{\pi}}\left(\dfrac{Z}{a_0}\right)^{3/2}\sigma^2 e^{-\sigma/3}\sin 2\vartheta\sin\varphi$	$3d_{zx} = \dfrac{\sqrt{2}}{81\sqrt{\pi}}\left(\dfrac{Z}{a_0}\right)^{3/2}\sigma^2 e^{-\sigma/3}\sin 2\vartheta\cos\varphi$
$3d_{x^2-y^2} = \dfrac{1}{81\sqrt{2\pi}}\left(\dfrac{Z}{a_0}\right)^{3/2}\sigma^2 e^{-\sigma/3}\sin^2\vartheta\cos 2\varphi$	$\sigma = Zr/a_0$, a_0 : *Bohr radius*

When the angular parts of d_{z^2}, d_{xy}, d_{yz}, d_{zx}, $d_{x^2-y^2}$ expressed in polar coordinates are compared with those given in Table 1, we see readily an exact matching between those two sets. In another word, the subscripts of atomic orbitals in Cartesian coordinates tells us how the relevant atomic orbitals depend on polar angles of ϑ and φ. On the other hand, the functions expressed with subscripts of the eigenvalues of magnetic angular momentum (m_l) d_2, d_1, d_0, d_{-1}, d_{-2} are exact eigenfunctions of the operator of $\hat{m}$. However, they are not appropriate in visualizing a shape of an atomic orbital since these functions are complex conjugate. It should also be noted that there exist no one to one correspondence between d_{z^2}, d_{xy}, d_{yz}, d_{zx}, $d_{x^2-y^2}$ and d_2, d_1, d_0, d_{-1}, d_{-2}. With regard this, it is often erroneously explained as if one of d_2, d_1, d_0, d_{-1}, d_{-2} behaves just like one of d_{z^2}, d_{xy}, d_{yz}, d_{zx}, $d_{x^2-y^2}$. For example, see the MIT OCW video lecture : *Chemistry, Principles of Chemistry Science* (2005), *Lecture 30, Crystal Field Theory* . The same is true for *p*-orbitals. None of p_x, p_y, p_z orbitals corresponds to one of the eigenfunctions of the operator of $\hat{m}$ whose eigenvalues are $1, 0, -1$ and their eigenfunctions are expressed as p_1, p_0 and p_{-1}, respectively. Instead, p_x, p_y, p_z can be expressed as appropriate linear combinations of p_1, p_0 and p_{-1}.

Solution 2. The wave functions of the principal quantum number $n=3$ have two nodal planes where the functions assume zero. As is readily seen in Table 1, the radial parts of d_{z^2}, d_{xy}, d_{yz}, d_{zx}, $d_{x^2-y^2}$ have the same functional form of $\sigma^2 e^{-\sigma/3}$ and hence they do not become zero at any radial distance, that is, they have no spherical nodal plane. For this reason, the angular part of any of *d*-orbitals should have two nodal planes. The suffixes *x, y, z* of the *d*-orbitals indicate where the nodal planes located. For example, the angular part of $3d_{xy}$-orbital is given by $\sin^2\vartheta \sin 2\varphi$ and the sign of the wave function $3d_{xy}$ is determined by $\sin 2\varphi$. As is seen from Fig.21 below, the function of $\sin 2\varphi$ changes its sign at $\varphi=0$, $\varphi=\pi/2$ and $\varphi=3\pi/2$. In other word, $3d_{xy}$-orbital has two nodal planes, one is a plane containing the *x*-axis and vertical to the page and the other is a plane vertical but containing the *y*-axis.

Similarly, in the case of $3d_{x^2-y^2}$-orbital, since its angular part is $\sin^2\vartheta \cos 2\varphi$, its sign is determined by $\cos 2\varphi$ and it changes sign against φ at $\varphi=\pi/4$, $\varphi=3\pi/4$ and $\varphi=5\pi/4$. That is, $3d_{x^2-y^2}$-orbital has two nodal planes, which are vertical to the page but contain either the line given by $x=y$ or $x=-y$. These two lines are related to the subscript of x^2-y^2 in the sense that the two lines are the solutions of $x^2-y^2=(x+y)\times(x-y)=0$.. If you remember the location of a nodal plane through a subscript as is explained here, it is easy to image a shape of any one of *d*-orbitals. See **solution 1-27** for three dimensional shapes of *d*-orbitals.

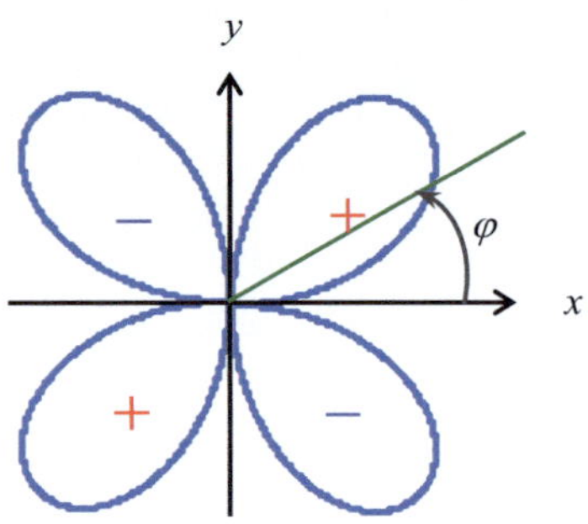

Fig.21 $3d_{xy}$-orbital and its nodal planes.

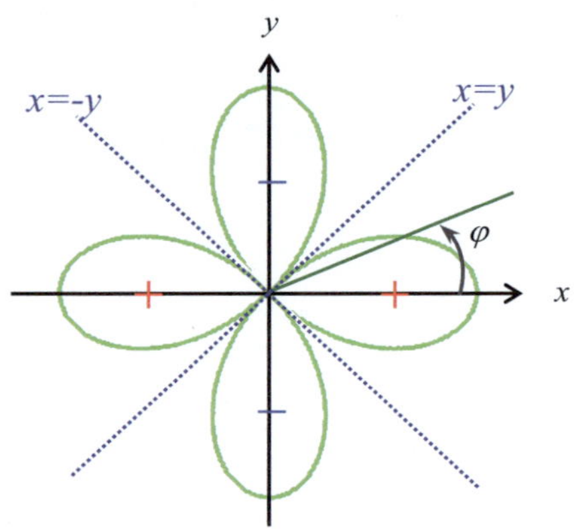

Fig.22 $3d_{x^2-y^2}$-orbital and its nodal planes.

Ans. See text.

Solution 1-25

(a) See **problem 1-6** and its solution.

(b) The electronic configurations of the diatomic molecules in the problem are as follows,

$$H_2^+(\sigma_{1s}^1),\ H_2(\sigma_{1s}^2),\ He_2^+(\sigma_{1s}^2\sigma_{1s}^{*1}),\ He_2(\sigma_{1s}^2\sigma_{1s}^{*2})$$

based on which it is easy to assign each of them to one of the potential curves given in Fig.23 below. $H_2^+(\sigma_{1s}^1)$ is considered more stable than $He_2^+(\sigma_{1s}^2\sigma_{1s}^{*1})$ because He_2^+ with the larger nuclear charge is expected to have the greater nuclear-nuclear repulsion energy as compared with H_2^+ (see **"note"** below).

Ans. See Fig.23 and text.

(c) The bond distance of H_2^+ is the one at whic its potential curve takes a minimum and is obtained from

Fig.23 as $r/2a_0 = 0.82$, yielding $r = 86\,\text{pm}$ for the bond length.

Ans. bond length: 86 pm

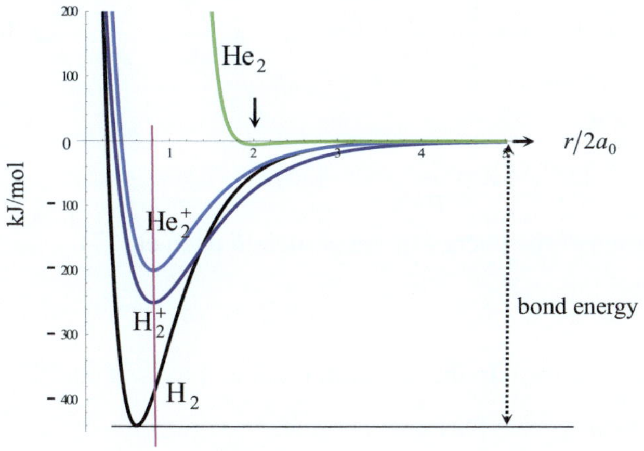

Fig.23 Potential energy curves for simple diatomic molecules and their bond distances.
The potential curves are drawn by using relevant *Morse* function. A small dip recognized for He_2 (marked by ↓) arises from *van der Waals* attraction between two He atoms.

"note"

Table Bond energies, bond lengths and bond orders for simple diatomic molecules.

	H_2^+	H_2	He_2^+	He_2	N_2	O_2^+	O_2	O_2^-	F_2	CO
bond energy ($kJ \cdot mol^{-1}$)	268	457	241	$<<1$	942	643	494	395	154	1071
bond length (pm)	106	74	108	~6000	110	112	121	135	141	113
bond order	1/2	1	1/2	0	3	5/2	2	3/2	1	

Solution 1-26

(a) The wave function of $\varphi_n(x) = \left(\dfrac{2}{a}\right)^{1/2} \sin\left(\dfrac{n\pi x}{a}\right)$ is inserted into the wave equation and is rewritten to give eq(31) below.

$$-\frac{h^2}{8m\pi^2}\frac{d^2}{dx^2}\varphi_n(x) = \left(\frac{h^2}{8m\pi^2}\right)\left(\frac{n\pi}{a}\right)^2\left(\frac{2}{a}\right)^{1/2}\sin\left(\frac{n\pi x}{a}\right) = \frac{n^2h^2}{8ma^2}\varphi_n(x) \qquad (31)$$

Therefore, from eq(3) given in the problem, the eigenenergy is obtained as $\dfrac{n^2h^2}{8ma^2}$.

Ans. $\dfrac{n^2h^2}{8ma^2}$

(b) The probability for finding the particle in the range $(x + dx,\ x)$, the probability of existence, is given by $\rho(x)dx$, where $\rho(x)$ is given by the squared wave function as in eq(32).

$$\rho(x)dx = \varphi_n^2(x)dx = \left(\frac{2}{a}\right)\sin^2\left(\frac{n\pi x}{a}\right)dx \qquad (32)$$

The function $\rho(x) = \varphi_n^2(x)$ takes maxima when its first and second derivatives satisfy eq(31) and eq(32) at x, respectively.

$$\frac{d\varphi_n^2(x)}{dx} = 2\times\left(\frac{2}{a}\right)\left(\frac{n\pi}{a}\right)\sin\left(\frac{n\pi x}{a}\right)\cos\left(\frac{n\pi x}{a}\right) = 0 \qquad (33)$$

$$\frac{d^2\varphi_n^2(x)}{dx^2} = 2\times\left(\frac{2}{a}\right)\left(\frac{n\pi}{a}\right)^2\left\{-\sin^2\left(\frac{n\pi x}{a}\right) + \cos^2\left(\frac{n\pi x}{a}\right)\right\} < 0 \qquad (34)$$

Notifying $\sin\left(\dfrac{n\pi x}{a}\right) \neq 0$, eqs(33) and (34) lead to $\cos\left(\dfrac{n\pi x}{a}\right) = 0$, which is satisfied at $n\pi x/a = \pi/2,\ 3\pi/2,$

$5\pi/2, \cdots$. Therefore, for $n=1$, $x=\frac{1}{2}a$; for $n=2$, $x=\frac{1}{4}a, \frac{3}{4}a$; for $n=3$, $x=\frac{1}{6}a, \frac{1}{2}a, \frac{5}{6}a$; and so on, are obtained.

Ans. $x=\frac{1}{2}a\ (n=1),\ x=\frac{1}{4}a, \frac{3}{4}a\ (n=2),\ x=\frac{1}{6}a, \frac{1}{2}a, \frac{5}{6}a\ (n=3)$

(c) From the equations given below,

$$E=\frac{1}{2}mv^2=\frac{(mv)^2}{2m}=\frac{p^2}{2m} \quad (35), \quad \lambda=\frac{h}{p} \quad (36)$$

the length of *de Broglie* wave λ is expressed in terms of the energy of the particle E as in eq(37).

$$E=\frac{h^2}{2m\lambda^2} \quad (37)$$

By putting the energy equal to its eigen energy, λ is related to the quantum number n as is in eq(38).

$$E=\frac{h^2}{2m\lambda^2}=\frac{n^2h^2}{8ma^2} \Rightarrow \lambda=2a/n \quad (38)$$

From this, the length of *de Broglie* wave λ for a relevant quantum number n is obtains as given in **Ans.**

Ans. For $n=1$, $\lambda=2a$: for $n=2$, $\lambda=a$; for $n=3$, $\lambda=2a/3$

"**note**" It can easily be shown that not only a particle moving along a circle (**problem 1-28**) but also an electron in an atomic orbital of a hydrogen-like atom the circumferences of their orbitals are equal to multiples of *de Broglie* wavelengths.

Solution 1-27

The function presented in the problem expresses the angular dependence of d_{zx}-orbital. Cartesian coordinates x, y, z are expressed in terms of polar coordinates, where r is fixed at $r=1$, as below.

$$x=\sin\vartheta\cos\varphi,\quad y=\sin\vartheta\sin\varphi,\quad z=\cos\vartheta \quad (39)$$

It is readily seen that the functional form of d_{zx} expressed in polar coordinates is given by multiplying z by x both of which are expressed in eq(39) (cf. **solution 1-24**).

In order to draw its approximate shape on the *zx*-plane, we set first $\varphi=0$ and then trace a point at the distance of $\sin 2\vartheta$ (eg. a black point in Fig.24) from the origin at (0, 0, 0) by varying ϑ. Since ϑ varies over the range of $[0, \pi]$, we can draw a shape on the right-hand side of the figure given in Fig.24 (cf. the figure given in **solution 1-24**). The distance in itself is an absolute value but distinction between negative or positive is essentially important and its sign is as is specified in Fig.24.

When a value of φ is increased from $-\pi/2$ to $\pi/2$, figures same in shape as given by a solid line but smaller by the factor of $\cos\varphi$ can be drawn. The same procedure is repeated in the range of $\varphi\ [\pi/2, 3\pi/2]$ to draw the figures on the left-hand of z-axis in Fig.24. The *d*-orbitals can be drawn directly on computer by using "polar plot" of *Mathematica* as are shown in Fig.25. It should be remarked that the shapes presented in Fig.25 do not show radial distance-dependence of the *d*-orbitals.

The electron density of d_{zx}-orbital which is given by $4\pi r^2 d_{zx}{}^2$ takes a maximum at the position marked by ● in Fig.24. In the case of hydrogen atom it is located at $r=9a_0$, $\vartheta=\pi/4$ and $\varphi=0$ (cf. Fig.21 in **solution 1-24**).

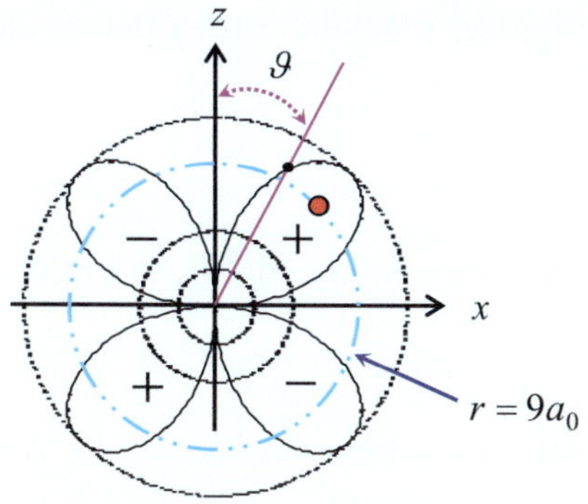

Fig.24 How to draw d_{zx}-orbital.

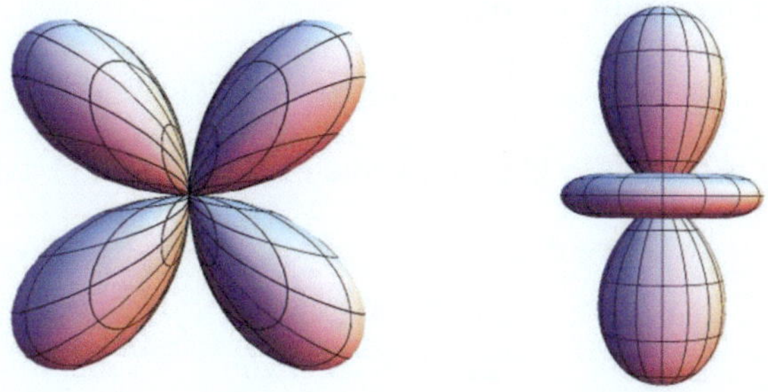

Fig.25 d_{zx}-orbital (left) and d_{z^2}-orbital (right).

Ans. See text.

"note" *Hiromichi.Niikura* et al. have observed experimentally for the first time shapes of atomic orbitals (*f*-orbitals) by using an atto-second laser system (*Science*, **356**,1150 (2017)) and achieved their visualization in momentum space (*Physical Review* A 106,063513 (2022)).

Solution 1-28

The total energy of the particle moving along each circle is kinetic energy alone and is given by eq(40),

$$E = \frac{p^2}{2m}, \qquad \lambda = \frac{h}{p} \qquad (40)$$

where p is its momentum and λ is its *de Broglie* wavelength ($m = m_e$ for an electron). Then the energy can be expressed in terms of λ as is given in eq(41).

$$E = \frac{h^2}{2m\lambda^2} \qquad (41)$$

Therefore, the energy takes the smallest value when *de Broglie* wavelength is the longest. In the cases of A and B, they are $\lambda = 2\pi a$ and $4\pi a$, respectively and hence the lowest energies are $\frac{h^2}{8\pi^2 ma^2}$ and $\frac{h^2}{32\pi^2 ma^2}$, respectively. From this, the ratio of the lowest energies is given by 4 : 1.

Ans. the ratio: 4 : 1

"note 1" Correctly, of course, relevant *Schrödinger* equation must be solved to obtain eigenenergies. However, as far as the eigenenergies are concerned, the same results are obtained. When the radius of a circle is *R*, the quantized energy is given by $E_n = n^2h^2/8\pi^2 mR^2$, where *n* is the quantum number which can take a value of $n = 0,\ 1,\ 2,\ 3,\cdots$. However, the eigenstate is doubly degenerated since the particle can rotate both clock wise and counterclockwise, i.e., corresponding to a positive and a negative angular momentum. Note that for $n = 0$, the *de Broglie* wavelength considered above becomes $\lambda = \infty$.

"note 2" The motion and the eigenenergy of an electron on a ring is often used to mimic the electronic states of aromatic hydrocarbons or porphyrins. Some such examples are given in Chapter 7.

Solution 1-29

(a) It is easy to see that *Lennard-Jones* potential of eq(42) (eq(4) given in the problem) is zero at $r = \sigma$.

$$u(r) = 4\varepsilon\left[\left(\frac{\sigma}{r}\right)^{12} - \left(\frac{\sigma}{r}\right)^{6}\right] = 0 \Rightarrow r = \sigma \qquad (42)$$

In order to determine where the potential takes a minimum, *Lennard-Jones* potential is rewritten as eq(43) by replacing r with $x \equiv \sigma/r$.

$$\frac{\sigma}{r} = x \;\Rightarrow\; u(x) = 4\varepsilon(x^{12} - x^{6}) \qquad (43)$$

Then it is easy to obtain a value of x where $u(x)$ takes a minimum. By solving eq(44), the value for x is obtained as in eq(45).

$$\frac{du(x)}{dr} = \frac{du(x)}{dx}\cdot\frac{dx}{dr} = 4\varepsilon(12x^{11} - 6x^{5})\times\left(-\frac{\sigma}{r^{2}}\right) = 0 \qquad (44)$$

$$(12x^{11} - 6x^{5}) = 6x^{5}(2x^{6} - 1) = 0 \;\Rightarrow\; x = 2^{-1/6} \qquad (45)$$

Inserting this value into eq(43), the minimum value of the potential is obtained as $-\varepsilon$ from eq(46). The value is just equal to a negative of the coefficient ε for the potential.

$$u(x = 2^{-1/6}) = 4\varepsilon\left\langle(2^{-1/6})^{12} - (2^{-1/6})^{6}\right\rangle = 4\varepsilon\left(\frac{1}{4} - \frac{1}{2}\right) = -\varepsilon \qquad (46)$$

From the values given for He, it is found that the potential takes the minimum value of $-84.7\,\mathrm{J\cdot mol^{-1}}$ at $r = \sigma \times 2^{1/6} = 2.56 \times 2^{1/6} = 2.87 \times 10^{-10}\,\mathrm{m}$. *Lennard-Jones* potential thus obtained for He is shown in Fig.26

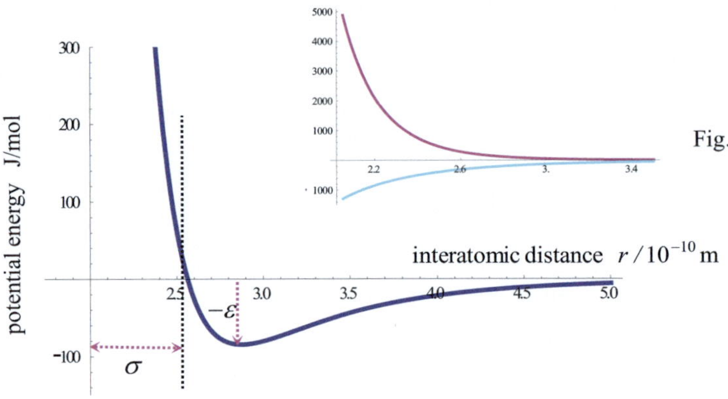

Fig.26 *Lennard-Jones* potential curve for He. The magnified potential curves in the inserted figure shows that the attractive second term (——) of the potential function dominates where r is large but where r is small the first term for repulsion (——) dominates.

Ans. the depth of the potential: $\varepsilon = 84.7\,\mathrm{J\cdot mol^{-1}}$

(b) Two atoms of He_2 do not form a covalent bond. When the two atoms are brought nearer, the second term of *Lennard-Jones* potential function is dominant and hence the attractive force proportional to $-1/r^{6}$ is operative but when brought further nearer the force between the two atoms turns to be repulsive, being proportional to $1/r^{12}$, via potential minimum. The maximum attractive potential energy between the two He atoms is $84.7\,\mathrm{J\cdot mol^{-1}}$, which is practically null compared with the bonding energy of $440\,\mathrm{kJ\cdot mol^{-1}}$ for hydrogen molecule. Even when compared with the thermal energy of $RT = 8.31\times 298\,\mathrm{J\cdot mol^{-1}}$ at ambient temperatures, its smallness is obvious.

Ans. See text. Ratio of energies: less than 2.0×10^{-4}

Solution 1-30

(a) Since the potential energy is zero at $r = \sigma$, $b = \sigma$ is obtained by putting $u(\sigma) = 0$ in eq(47).

$$u(r) = 4\varepsilon\left\langle \exp\left(-\frac{r-\sigma}{a}\right) - \left(\frac{\sigma}{r}\right)^{6}\right\rangle \qquad (47)$$

And then from the requirement that the minimum value of $u(r)$ is $-\varepsilon$, eq(48), the first derivative of $u(r)$, is equated to zero to obtain $r = r_{\min}$ at which eq(48) and eq(49) are satisfied.

$$\frac{du(r)}{dr}=4\varepsilon\left\langle -\frac{1}{a}\exp\left(-\frac{r-\sigma}{a}\right)+\frac{6}{\sigma}\left(\frac{\sigma}{r}\right)^7\right\rangle=0 \;\Rightarrow\; \exp\left(-\frac{r_{\min}-\sigma}{a}\right)=\frac{6a}{\sigma}\left(\frac{\sigma}{r_{\min}}\right)^7 \tag{48}$$

$$u(r_{\min})=4\varepsilon\left\langle \exp\left(-\frac{r_{\min}-\sigma}{a}\right)-\left(\frac{\sigma}{r_{\min}}\right)^6\right\rangle=-\varepsilon \;\Rightarrow\; \exp\left(-\frac{r_{\min}-\sigma}{a}\right)-\left(\frac{\sigma}{r_{\min}}\right)^6=-\frac{1}{4} \tag{49}$$

In order to solve eqs(48) and (49) for $r_{\min}$ and a, a bit of an idea is necessary. Firstly, the result of eq(48) (right-hand side) is inserted into eq(49) to get eq(50).

$$\frac{6a}{\sigma}\left(\frac{\sigma}{r_{\min}}\right)^7-\left(\frac{\sigma}{r_{\min}}\right)^6=-\frac{1}{4} \tag{50}$$

By putting $x=r_{\min}/\sigma$ and $\alpha=a/\sigma$, eqs(48)-(50) are rewritten as eqs$(48')-(50')$.

$$\exp\left(-\frac{x-1}{\alpha}\right)=6\alpha\left(\frac{1}{x}\right)^7 \tag{48'}$$

$$\exp\left(-\frac{x-1}{\alpha}\right)-\left(\frac{1}{x}\right)^6=-\frac{1}{4} \tag{49'}$$

$$6\alpha=\left(-\frac{1}{4}x^7+x\right) \tag{50'}$$

When eq$(50')$ is inserted into eq$(49')$, eq(51) is obtained.

$$\exp\left\langle -6(x-1)\Big/\left(-\frac{1}{4}x^7+x\right)\right\rangle=-\frac{1}{4}+\left(\frac{1}{x}\right)^6 \tag{51}$$

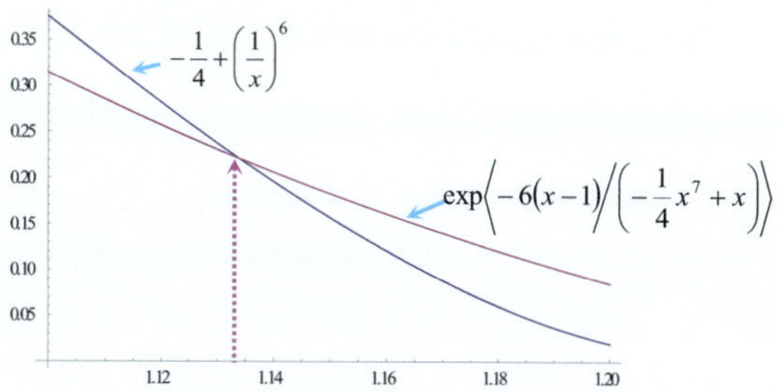

Fig.27 Graphical solution of eq(51).

An easy way to solve eq(51) is to plot the functions on both sides of eq(51) and to search for the point where the two functions cross each other as is illustrated in Fig.27. From Fig.27, the intersection is obtained as $x=1.134$ and then α is determined from eq$(50')$. From these values determined, the values for $r_{\min}$ and a are determined as follow.

$$x=1.134 \;\Rightarrow\; r_{\min}=1.134\times 2.56=2.90\times 10^{-10}\,\mathrm{m},\;\; a=\alpha\times\sigma=0.226$$

Ans. $b=\sigma$, $a=0.226\ (\times 10^{-10}\,\mathrm{m})$

(b) By equating $\sigma=2.56\ \times 10^{-10}\,\mathrm{m}$ together with the values obtained above, the potential function (red line) obtained is shown in Fig.28. For comparison, *Lennard-Jones* potential curve (blue line) is shown together. For both curves, not only the minimum value of potential but also the distance at which the potential becomes zero is put equal. Since the values of $r_{\min}$ are almost the same for both curves, the two potential

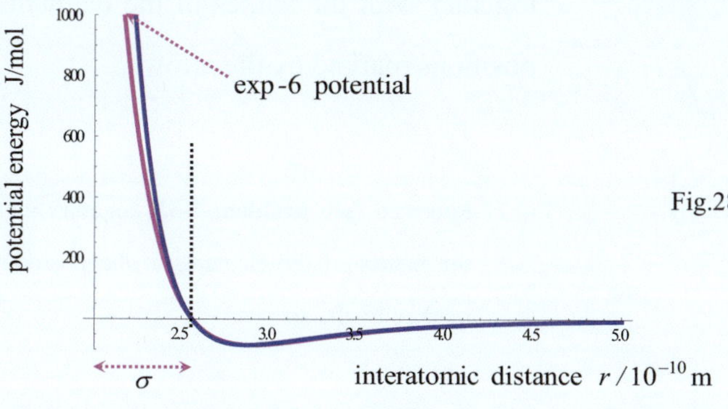

Fig.28 A comparison of *Lennard-Jones* potential (blue line) and exp-6 potential (red line).

curves behave almost the same where $r\geq\sigma$, but eq(47) is characterized by (i) the repulsive forces are largely different when $r<\sigma$, (ii) because of the number of parameters is larger by one, a shape of the potential curve can be more flexibly varied, and (iii) eq(47) treated in this problem is superior when applied to real gases in order to, for instance, obtain virial coefficients.

Solution 1-31

Since the *elementary charge* is the smallest unit of charge, any charges divided by the elementary charge will yield integral numbers. When the measured charges are divided by the smallest one, i.e., 9.60×10^{-10} esu, the following numbers are obtained.

$$\frac{1.44}{0.960}=1.5 \quad \frac{2.88}{0.960}=3.0 \quad \frac{3.84}{0.960}=4.0 \quad \frac{4.32}{0.960}=4.5$$

These numbers are not integer but if they are multiplied by 2, then all of them turn out to be integral numbers.

Therefore, it is concluded that a half of the smallest charge measured $9.6\times10^{-10}\,\text{esu}/2=4.8\times10^{-10}\,\text{esu}$ is the elementary charge. Also, *Avogadro*'s number N_A is obtained by dividing the charge for 1 mole of electrons by the elementary charge as follows.

$$e\times N_A=96500\ \text{C}=96500\times3\times10^9\,\text{esu} \ \Rightarrow\ N_A=\frac{96500\times3\times10^9}{4.8\times10^{-10}}=6.03\times10^{23}$$

Ans. $e=4.8\times10^{-10}\,\text{esu},\quad N_A=6.03\times10^{23}$

Solution 1-32

The incident α-particle is considered to have approached nearest to the nucleus when all of its kinetic energy has been converted into the repulsive potential energy between the nucleus and the particle. Therefore, eq(52) is obtained, where R is the nearest approachable distance from the (center of) nucleus.

$$\frac{1}{2}mv^2=\frac{e^2Z}{4\pi\varepsilon_0 R} \qquad (52)$$

This distance R is regarded as the maximum possible radius of the relevant nucleus. By substituting the given values and $e=1.602\times10^{-19}\,\text{C}$ into eq(52), R is determined to be $6.2\times10^{-15}\,\text{m}$ as below.

$$\frac{1}{2}\times6.68\times10^{-27}\times(1.8\times10^7)^2=\frac{29\times(1.602\times10^{-19})^2}{R}$$

Ans. nucleus radius: $R=6.2\times10^{-15}\,\text{m}$

Solution 1-33

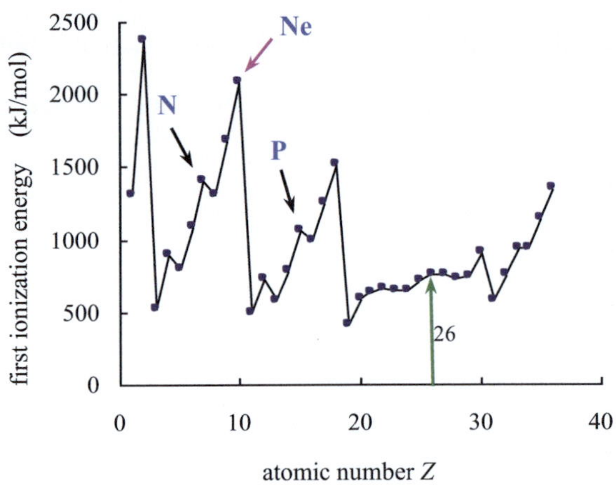

(a) Fig.7 of the first ionization energy vs. Z plot given in the problem is shown on the left again together with the names of the elements at the positions marked by the arrows.

Fig.7 Reposted (see **problem 1-33**) together with the names of the elements at the positions marked by the arrows.

Ans. From left to right in Fig.7, Nitrogen: $_7\text{N}\,(1s^2 2s^2 2p^3)$, Neon: $_{10}\text{Ne}\,(1s^2 2s^2 2p^6)$, Phosphorus: $_{15}\text{P}(1s^2 2s^2 2p^6 3s^2 3p^3)$.

(b) As is found for noble gases, when the atomic orbitals with the principal quantum number n are fully occupied, the electrons are in the most stable configuration. This situation is reflected by the appearance of outstanding peaks in Fig.7 of ionization energy vs.Z plot as is illustrated by the red arrow in the middle of the figure indicating Ne. The deep valleys which appear as if the sharp peaks were turned over represent the states in which the principal quantum number is increased by one from n to n+1 and the first electron to be added on this occasion occupies n+1 s-orbital with a higher energy than that of ns-orbital. As is easily seen, this electron is rather unstable and easily ionized.

N and P which are located on both sides of Ne and are indicated by the black arrows have three electrons in the relevant outmost p-orbitals and they assume the configuration so as to have the largest spin angular momentum according to *Hund*'s rule (See **"note"** in **solution 1-2**) as illustrated by | ↑ | ↑ | ↑ |. When the sub- shells such as p_x, p_y and p_z are all half-filled, the electrons in the subshells are stabilized significantly. Therefore, the electrons in those configurations are said less easily ionized. The ionization energies of the elements which have those configuration become larger than those of the elements which have one more electron because of the larger atomic number by one and whose electronic configurations are expressed by | ↑↓ | ↑ | ↑ | .

In summary, although both of regularity and irregularity are recognized in Fig.7 in which ionization energies are plotted against the atomic number, either feature can be well explained in terms of a periodic nature of electron configurations.

Ans. See text (b).

(c) The ionization energy (*I.P.*) of an atom is given by eq(53) using the effective nuclear charge Z_{eff} and the principal quantum n of the electron to be ionized.

$$I.P. = \frac{e^4 m_e}{8\varepsilon_0^2 h^2} \cdot \frac{Z_{eff}^{\ 2}}{n^2} \qquad (53)$$

By putting $Z_{eff} = 1$ and $n = 1$ for hydrogen and by substituting the other relevant values, the ionization potential of hydrogen is obtained as $1312\ \mathrm{kJ \cdot mol^{-1}}$, which is in an excellent agreement with a value of 1300

$$I.P. = \frac{(1.602\times10^{-19})^4 \cdot 9.109\times10^{-31}}{8\times(8.854\times10^{-12})^2 \cdot (6.626\times10^{-34})^2} = 1312\ \mathrm{kJ \cdot mol^{-1}}$$

$\mathrm{kJ \cdot mol^{-1}}$ to be read from Fig.7. (The theoretically calculated value is in perfect agreement with the experimental value.)

Ans. *I.P.*: $1312\ \mathrm{kJ \cdot mol^{-1}}$

(d) An approximate value of $550\,\text{kJ}\cdot\text{mol}^{-1}$ is read for the first ionization energy of Li from Fig.7. By putting $n=2$ in eq(53), a value for Z_{eff} of Li is obtained as 1.29 as is shown below.

$$I.P.(\text{Li}) = 550 = \frac{(1.602\times10^{-19})^4\cdot 9.109\times10^{-31}}{8\times(8.854\times10^{-12})^2\cdot(6.626\times10^{-34})^2}\cdot\frac{Z_{eff}^2}{2^2} \Rightarrow Z_{eff} = \sqrt{\frac{550\times4}{1312}} = 1.29$$

Ans. $Z_{eff} = 1.29$

(e) The 26th element is Fe and its electron configuration is $_{26}\text{Fe}(1s^2 2s^2 2p^6 3s^2 3p^6 4s^2 3d^6)$. Since in the configuration of $3d^5$ the d-orbitals are exactly half-filled, its configuration is relatively stable compared with those in which one electron less or more. Consequently, as is seen in Fig.7 reposted, the ionization energy does not linearly increase in the order of 24→25→26, but an increase seen upon 24→25 is greater than that for 25→26.

Ans. $_{26}\text{Fe}(1s^2 2s^2 2p^6 3s^2 3p^6 4s^2 3d^6)$

Solution 1-34

The electronic configuration of He atom in the ground state is $\text{He}(1s^2)$. Ionization is expressed by $\text{He}(1s^2)\rightarrow \text{He}^+(1s^1)+e$. The upper limiting value of the ionization energy is the energy required in order to remove an electron which is not shielded by the other electron at all and is calculated from eq(54) by equating $Z=2$.

$$E_{1s} = -\frac{m_e Z^2 e^4}{8\varepsilon_0^2 h^2} \qquad (54)$$

On the other hand, the lower limiting value is obtained when the electron is shielded by the other electron completely. Therefore equating $Z=1$ in eq(54), the energy which is totally equal to that of H atom is obtained. It is valuable to compare those values with that shown in Fig.7 of **1-33**.

Ans. upper limiting value: $5247\,\text{kJ}\cdot\text{mol}^{-1}$, lower limiting value: $1312\,\text{kJ}\cdot\text{mol}^{-1}$

Solution 1-35

Pauli's exclusion principle.

Quantum mechanical states that an electron can assume are represented by a set of quantum numbers allowed to the electron. The electron in an atom is characterized by the four quantum numbers, i.e., $n,\ l,\ m_l,\ m_s$. *Pauli*'s exclusion principle tells us that "in a system composed of two or more electrons, no two electrons can have the completely same set of values for these four individual quantum numbers at the same time". In the case of the principal quantum number of $n=3$, for instance, for the allowed azimuthal quantum number of $l=2,\ 1,\ 0$, the magnetic quantum number can take $m_{l=2}=-2,\ -1,\ 0,\ 1,\ 2$, $m_{l=1}=-1,\ 0,\ 1$ and $m_{l=0}=0$, respectively. For any combinations of l and m_l, two spin states of $m_s=\frac{1}{2}$ and $-\frac{1}{2}$ are allowed. Therefore, it is concluded that $(5+3+1)\times2=18$ sets of the combination of $l,\ m_l,\ m_s$ are possible and allowed.

Ans. number of allowed states: 18 states

Solution 1-36

The electronic configurations of W and Unh are given below. The order of filling electrons in relevant subshells are in accord with the $n+l$ rule (cf. **solution 1-2**). The periodic table given in **solution 1-2** can be referred to for finding subshells.

$$_{74}\mathrm{W}(1s^2 2s^2 2p^6 3s^2 3p^6 4s^2 3d^{10} 4p^6 5s^2 4d^{10} 5p^6 \underline{6s^2 4f^{14} 5d^4})$$

$$_{106}\mathrm{Unh}(1s^2 2s^2 2p^6 3s^2 3p^6 4s^2 3d^{10} 4p^6 5s^2 4d^{10} 5p^6 6s^2 4f^{14} 5d^{10} 6p^6 \underline{7s^2 5f^{14} 6d^4})$$

As is seen from a comparison of the under lined parts, two elements have the same outermost subshells filled by the same number of electrons in the same way except for the principal quantum number n.

Ans. See text.

"note" How to name new elements.

Almost all elements of the atomic numbers up to 112 have been named historically to commemorate the names of scientists who found them or the places where they were found. Except for some special cases, the elements with the atomic number of 109 or greater than that are to be called by their atomic numbers expressed in *Latin*. For instance, the 110th element is Unn because the initial of each digit of 110 is U(1)u(1)n(0). Similarly, the 118th element is Uuo, i.e., U(1)u(1)o(8) and the 888th element is Ooo. If you know a bit of Spanish, then this way of naming new elements is quite easy. See **problem 1-107** to speculate how large the atomic number can be for the element to exist practically. Since the Japanese scientists discovered and proved its existence experimentally, the 113th element was named "Nh: Nihonium" to commemorate Japan and Japanese scientists involved.

Solution 1-37

(a) For the wave function given by eq(55), the radial distribution density function $\rho(r)$ is given by eq(56) after integration over the solid angle.

$$\psi(r,\vartheta,\varphi) = R_{21}(r)Y_{10}(\vartheta,\varphi) \qquad (55)$$

$$\rho(r)dr = \int_0^{2\pi}\int_0^{\pi} |R_{21}(r)Y_{10}(\vartheta,\varphi)|^2 r^2 \sin\vartheta\, d\vartheta d\varphi dr = r^2 R_{21}(r)^2 dr \int_0^{2\pi}\int_0^{\pi} |Y_{10}(\vartheta,\varphi)|^2 \sin\vartheta\, d\vartheta d\varphi \qquad (56)$$

By putting $Z=1$ in eq(55), the corresponding wave functions are obtained as eqs(57) and (58) given below and after integration over the solid angle, eq(59) is obtained.

$$Y_{10} = \left(\frac{3}{4\pi}\right)^{1/2} \cos\vartheta \qquad (57), \qquad R_{21}(r) = \frac{1}{4(6)^{1/2}}\left(\frac{1}{a_0}\right)^{3/2}\left(\frac{2r}{a_0}\right)\exp\left(-\frac{r}{2a_0}\right) \qquad (58)$$

$$\int_0^{2\pi}\int_0^{\pi} |Y_{10}(\vartheta,\varphi)|^2 \sin\vartheta\, d\vartheta d\varphi = \int_0^{2\pi}\int_0^{\pi} \frac{3}{4\pi}\cos^2\vartheta \sin\vartheta\, d\vartheta d\varphi$$

$$= 2\pi \int_0^{\pi} \frac{3}{4\pi}\cos^2\vartheta \sin\vartheta\, d\vartheta = \frac{1}{2}[-\cos^3\vartheta]_0^{\pi} = 1 \qquad (59)$$

Consequently, the probability density function for the corresponding $2p$-orbital is obtained as eq(60),

$$\rho(r)dr = r^2 R_{21}(r)^2 dr = \frac{1}{24}\left(\frac{1}{a_0}\right)^5 r^4 \exp\left(-\frac{r}{a_0}\right)dr \qquad (60)$$

from which the distance where the density takes a maximum value is determined to be at $r = 4a_0$ from eq(61).

$$\frac{d\rho(r)}{dr} = \frac{1}{24}\left(\frac{1}{a_0}\right)^5 \left\{4 - \frac{r}{a_0}\right\} r^3 \exp\left(-\frac{r}{a_0}\right) = 0 \qquad (61)$$

The value of $\rho(r)$ at this distance is calculated to be $0.369/\text{Å}$ as shown below. Notice that the unit of the density is not per volume but per length.

$$\rho(4a_0) = \frac{1}{24}\left(\frac{1}{a_0}\right)^5 a_0^4 \exp\left(-\frac{4a_0}{a_0}\right) = 0.369/\text{Å}$$

Ans. eq(60), $r = 4a_0$, $\rho(4a_0) = 0.369/\text{Å}$

(b) The $1s$-orbital of hydrogen is well known to take its maximum value of the probability density at *Bohr* radius a_0. On the other hand, in the case of the $2p$-orbital the ratio of the probability densities at a_0 and $4a_0$ is calculated to be $\rho(a_0)/\rho(4a_0) = 0.078$, indicating the density at a_0 is much smaller than that at $4a_0$.

Ans. the ratio of the probability densities: 0.0785

Solution 1-38

(a) The energies associated with the processes (1)—(3) (see **problem 1-38**) are denoted as E_1、E_2 and E_3, respectively, and they are calculated as follows:

$$E_1 = 108.8 + \frac{1}{2} \times 241.8 = 229.7\ \text{kJ} \cdot \text{mol}^{-1}$$

$$E_2 = I.P.(\text{Na} \rightarrow \text{Na}^+) - E.A.(\text{Cl} + e^- \rightarrow \text{Cl}^-) = (5.139 - 3.615)\text{eV} \times 96.485 = 147.0\ \text{kJ} \cdot \text{mol}^{-1}$$

$$E_3 = -M \times \frac{N_A z_+ z_- e^2}{4\pi\varepsilon_\circ (r_+ + r_-)}\left(1 - \frac{1}{n}\right) = -791.9\ \text{kJ} \cdot \text{mol}^{-1}$$

For the calculation of E_3, the following values were employed where n is a parameter to estimate the lattice energy and is assumed to be 10.

$$M = 1.75,\quad r_+ + r_- = 1.16\text{Å} + 1.67\text{Å} = 2.83 \times 10^{-10}\,\text{m},\quad \varepsilon_0 = 8.85 \times 10^{-12}\,\text{J}^{-1}\text{C}^2\text{m}^{-1},$$

$$z_+ = 1,\quad z_- = 1,\quad e = 1.60 \times 10^{-19}\,\text{C},\quad n = 10$$

(b) A merit of employing *Born-Haber* cycle is based on the fact that a change in the total energy becomes zero after going through various processes to return to the initial state. When applied to the present problem, going through the processes from (1) to (4) clockwise results in a total energy change of zero, i.e.,

$$E_1 + E_1 + E_3 + E_4 = 0.$$

Since the enthalpy of formation under consideration is associated with the opposite process of (4), its value is obtained as a negative value of E_4 as below.

$$\Delta H_f^0 = -E_4 = E_1 + E_1 + E_3 = -415.2\ \text{kJ} \cdot \text{mol}^{-1}$$

From a data book on thermodynamics, a value of $\Delta H_f^0 = -411.15\ \text{kJ} \cdot \text{mol}^{-1}$ can be obtained, which is in a good agreement with the value calculated by employing *Born-Haber* cycle.

Ans. $-415.2\ \text{kJ} \cdot \text{mol}^{-1}$

Solution 1-39

For simplicity, a hydrogen atom is considered to answer the problem. By substituting $A = 1$ into the relation of $R_N = 1.33 \times 10^{-15} \times \sqrt[3]{A}$ (m), the value of $R_N(\text{H}) = 1.33 \times 10^{-15}$ m is obtained. On the other hand, when the *Bohr* radius of hydrogen which is 0.529×10^{-10} m is taken as a radius of a hydrogen atom, then the size of the atom is larger than that of the nucleus by a factor of $0.529 \times 10^{-10} / 1.33 \times 10^{-15} = 4 \times 10^4$. Thus, roughly speaking, when the nucleus is enlarged to a ball of 3 cm in diameter, then the hydrogen atom would become a big ball with a diameter of $3 \times 10^5\,\text{cm} = 3 \times 10^3\,\text{m} = 3\,\text{km}$.

Ans. 3 km

Solution 1-40

Referring to the solution to **problem 1-12**, the ionization potential of hydrogen is given by eq(62).

$$E_{I.P.} = \frac{m_e e^4}{8\varepsilon_0{}^2 h^2} = 313.5\,\text{kcal}\cdot\text{mol}^{-1} \qquad (62)$$

Substituting this value into eq(63), the shielding constant σ is calculated to be 16.74 as follows.

$$E_{I.P.}(\text{K}) = \frac{m_e (Z-\sigma)^2 e^4}{8\varepsilon_0{}^2 h^2} \times \frac{1}{n^2} \qquad (63)$$

$$= 313.5 \times (Z-\sigma)^2 \times \frac{1}{4^2} = 313.5 \times \frac{(19-\sigma)^2}{4^2} = 100.1$$

$$\frac{(19-\sigma)^2}{4^2} = \frac{100.1}{313.5} \quad \Rightarrow \quad \sigma = 19 - 4 \times \sqrt{\frac{100.1}{313.5}} = 19 - 2.26 = 16.74$$

It is, therefore, recognized that the outermost electron of $4s^1$ is shielded extensively by the inner subshell electrons. How to estimate a value for the shielding constant is described in detail in **"note 1"** in **solution 1-12.** Since the electronic configuration of K is given by $_{19}\text{K}(1s^2 2s^2 2p^6 3s^2 3p^6 4s^1)$, the shielding constant estimated by following the method (c) (given there and cited below again) is derived as $\sigma = 10 \times 1.0 + 0.85 \times 8 = 16.8$, which shows a good agreement with the value calculated above.

"note 1" Shielding constant σ : 2 (c) in **solution 1-12**. When the electron to be ionized is located in either of s or p shell of the same group, the electrons in the nearest inner shell group contribute 0.85 and the electrons further inner shells contribute 1.00 per each.

Ans. effective nuclear charge: $Z_{eff} = 2.26$

Solution 1-41a

For each molecule in the problem, the electronic configuration, bond order b.o. and spin angular momentum S are given below.

$$O_2^+ (\sigma_{1s}^2 \sigma_{1s}^{*2} \sigma_{2s}^2 \sigma_{2s}^{*2} \sigma_{2p_z}^2 \pi_{2p_x}^2 \pi_{2p_y}^2 \pi_{2p_x}^{*1}) \qquad \text{b.o.} = 2.5, \qquad S = \frac{1}{2}$$

$$O_2 (\sigma_{1s}^2 \sigma_{1s}^{*2} \sigma_{2s}^2 \sigma_{2s}^{*2} \sigma_{2p_z}^2 \pi_{2p_x}^2 \pi_{2p_y}^2 \pi_{2p_x}^{*1} \pi_{2p_y}^{*1}) \qquad \text{b.o.} = 2.0, \qquad S = 1$$

$$O_2^- (\sigma_{1s}^2 \sigma_{1s}^{*2} \sigma_{2s}^2 \sigma_{2s}^{*2} \sigma_{2p_z}^2 \pi_{2p_x}^2 \pi_{2p_y}^2 \pi_{2p_x}^{*2} \pi_{2p_y}^{*1}) \qquad \text{b.o.} = 1.5, \qquad S = \frac{1}{2}$$

$$O_2^{2-} (\sigma_{1s}^2 \sigma_{1s}^{*2} \sigma_{2s}^2 \sigma_{2s}^{*2} \sigma_{2p_z}^2 \pi_{2p_x}^2 \pi_{2p_y}^2 \pi_{2p_x}^{*2} \pi_{2p_y}^{*2}) \qquad \text{b.o.} = 1.0, \qquad S = 0$$

The definition of the bond order (b.o.) is given by the equation below, where n is the total number of elec-

$$\text{b.o.} = \frac{1}{2}(n - n^*) \qquad (64)$$

trons occupying bonding orbitals and n^* is the total number of electrons occupying antibonding orbitals. You can assume instead that b.o is 1.0 for a full-bonding but 0.5 for a half-bonding. Refer to **solution 1-6** and **1-14**. The magnetic property for these molecules is determined by their magnetic moments. Except for peroxide ion O_2^{2-} with $S = 0$, the other molecules with $S \neq 0$ are magnetic. This is one of the characteristic features of molecular oxygen and oxygen ions.

A video on the experiment to prove molecular oxygen to be really paramagnetic by using an electromagnet is available on internet (Access HP of *MITOCW* and click *Audio/Video Course* and select *Physics* in which the 21st lecture on *Electricity and Magnetism* delivered in 2002 is found. About 43 minutes later you will see the relevant experiment in the lecture). It is truly impressive to see a volume of liquid oxygen flowing out downward from a Dewar flask being attracted to the electromagnet instantaneously when it is turned on.

Ans. See text.

Solution 1-41b

The molecular orbitals of the homo-nuclear diatomic molecules constructed of the 1*s*-, 2*s*- and 2*p*-atomic orbitals change the order of the energy levels between a group of H_2–N_2 and that of O_2–F_2. How it changes is briefly illustrated in Fig.29. below. For the group of $H_2 - N_2$, σ_{2p}-orbital is located between π_{2p}- and π^*_{2p}-orbitals. On the other hand for the group of $O_2 - F_2$, σ_{2p}-orbital comes to be located below π_{2p}-orbital. Generally speaking, since the bonding energy of σ-bond is greater than that of π-bond, the order of the energy levels of the homo-nuclear diatomic molecules is simply like that given for F_2 on the right-hand side of Fig.29. However, for the group of $H_2 - N_2$, the correct order of the energy levels is considered like that given for Li_2 on the left-hand side of Fig29.

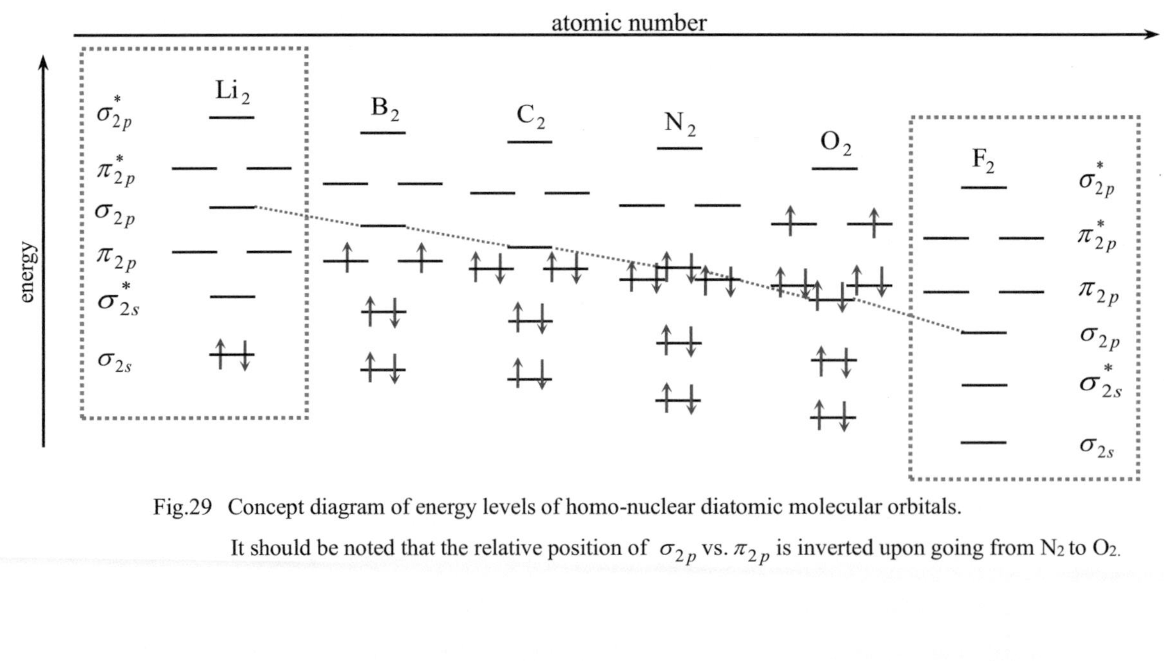

Fig.29 Concept diagram of energy levels of homo-nuclear diatomic molecular orbitals.
It should be noted that the relative position of σ_{2p} vs. π_{2p} is inverted upon going from N_2 to O_2.

It is sometimes critical which one of the two orders is employed for the discussion of chemical bonding of diatomic molecules. For instance, in the case of B_2 if the order on the left is relied on then B_2 is predicted to have two unpaired electrons and hence the molecule is expected to be paramagnetic. On the other hand, if we rely on the order on the right, then the molecule has no unpaired electron becoming diamagnetic. In the case of O_2, however, the same conclusion is drawn whichever order is relied on (cf. **solution 1-6**).

For those heteronuclear diatomic molecules whose component atomic numbers do not differ largely such as for CO, NO and CN, the above concept diagram can be employed for the discussion about their chemical bondings. Whichever order of the energy levels is used, the same number of lone-pair electrons is obtained for any of them. Paramagnetic molecules are given in the solution together with their spin angular momentums.

Ans. paramagnetic molecules: B_2 ($S=1$), O_2 ($S=1$), NO ($S=1/2$), CN ($S=1/2$)

"**note**" The reason why the energy levels for diatomic molecules come to be inverted.

The energy level of the molecular orbital of the same kind decreases by increasing the atomic number of a component atom. This is because the energy of the atomic orbital used decreases with increasing atomic number. However when σ_{2p}-orbital is constructed, it is necessary, as is trivial from their symmetric property, to employ not only $2p_z$-orbital but also $2s$-orbital in order to form sp-hybrid orbitals. At the same time the effect of overlap integral (**problem 7-38**) must be taken into account. A contribution of sp-hybrid orbital to σ-bonding would decrease with the atomic number. When the contribution of the sp-hybrid orbital to the bonding is completely neglected and the molecular orbitals composed of a single kind of atomic orbital alone are used, the order of the energy levels would look like that shown for F_2 on the right in Fig.29.

Solution 1-42

The energy of the particle in the plane (2-dimensonal box) $(2a \times a)$ is given by eq(65). Notice that motion along x-direction or y-direction is independent of each other. Therefore, the lowest energy is obtained when

$$E_{n_x} + E_{n_y} = \frac{n_x{}^2 h^2}{8m(2a)^2} + \frac{n_y{}^2 h^2}{8ma^2}, \qquad n_x, n_y = 1, 2, 3, \cdots \tag{65}$$

both quantum numbers are $n_x = n_y = 1$ and the second lowest one is obtained for $n_x = 2$ and $n_y = 1$.

Ans. second lowest energy: $h^2/4ma^2$

The wave function for the motion of the particle under consideration is given by the product of the wave functions one for the motion along x-direction and the other for y-direction, as is expressed by eq(66).

$$\varphi_{n_x,n_y}(x,y) = \sqrt{\frac{2}{2a}}\sin\left(\frac{n_x \pi x}{2a}\right) \times \sqrt{\frac{2}{a}}\sin\left(\frac{n_y \pi y}{a}\right) \tag{66}$$

$$\varphi_{n_x,n_y}{}^2(x,y) = \left|\sqrt{\frac{2}{2a}}\sin\left(\frac{n_x \pi x}{2a}\right) \times \sqrt{\frac{2}{a}}\sin\left(\frac{n_y \pi y}{a}\right)\right|^2 \tag{67}$$

Therefore, the probability of finding the particle at x and y is given by the wave function squared as is given

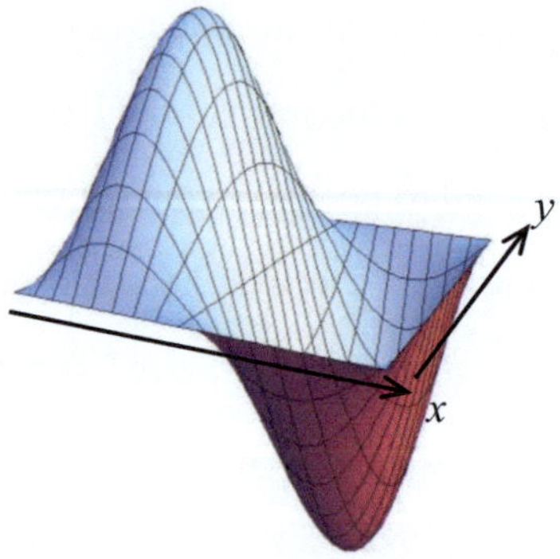

Fig.30 $\varphi_{2,1}(x, y)$.

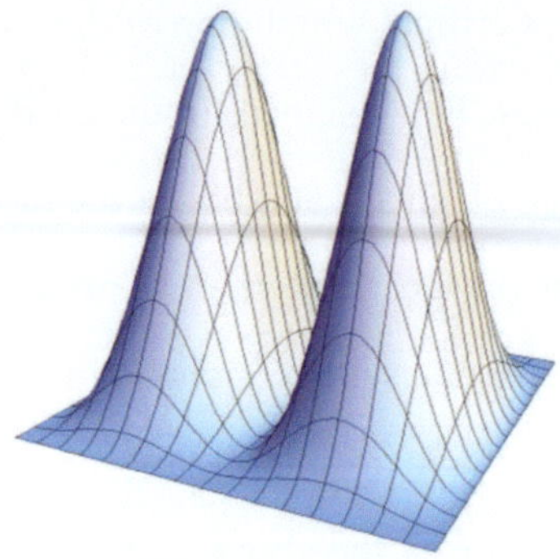

Fig.31 $\varphi_{2,1}{}^{2}(x, y)$.

by eq(67). The wave function and corresponding probability distribution function obtained by substituting $n_x = 2$ and $n_y = 1$ into eqs(66) and (67) are plotted in Figs.30 and 31, respectively.

Ans. Figs.30 and 31

Solution 1-43

Extending the discussion given in the preceding problem to the motion of a particle in a cube, the wave function for the particle is readily found to be given by the following equation.

$$\varphi_{n_x,n_y,n_z}(x,y,z) = \sqrt{\frac{2}{a}}\sin\left(\frac{n_x\pi x}{a}\right)\times\sqrt{\frac{2}{a}}\sin\left(\frac{n_y\pi y}{a}\right)\times\sqrt{\frac{2}{a}}\sin\left(\frac{n_z\pi z}{a}\right)$$

As compared with the two-dimensional motion, it is not easy to draw a precise shape of the probability distribution given by the square of the wave function above. However, its conceptual shapes can be drawn, for instance, for the lowest level ($n_x = 1,\ n_y = 1,\ n_z = 1$) and three second lowest levels ($n_x = 2,\ n_y = 1,\ n_z = 1$, etc.) as shown below. They are triply degenerate.

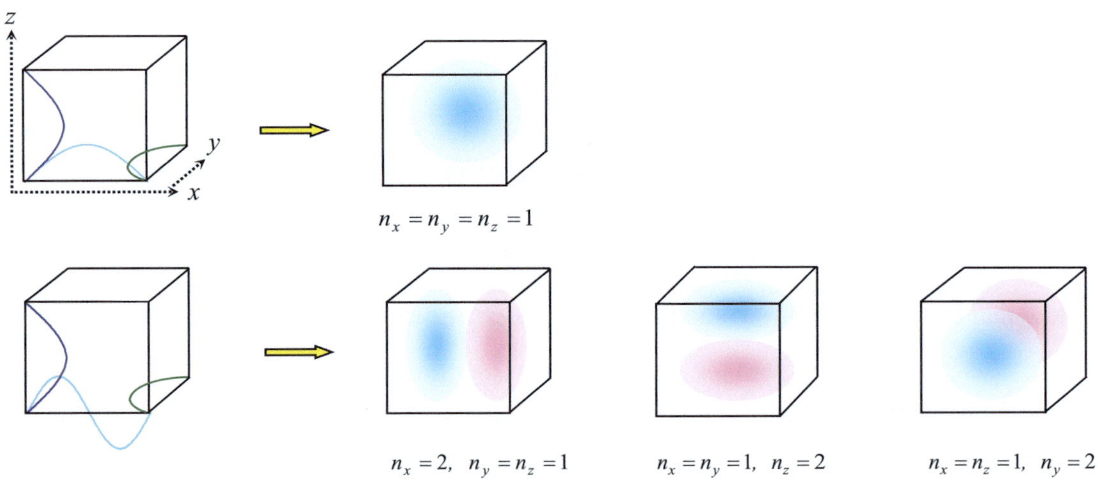

Fig.32 Schematic representation of probability distribution of a particle in a cube suggesting an *s*-orbital (t0p) and *p*-orbitals (bottom).

It is readily recognized that the probability distribution for the lowest energy level looks like an *s*-orbital and those for the second lowest levels look like one of *p*-orbitals. Furthermore, it is straightforward to see visually why three different and independent *p*-orbitals are existent. The analogy is not so good as for the *p*-orbitals but, putting two of n_x, n_y, n_z equal to 2 an the other equal to 1, three kinds of a crude probability distribution something like the figure shown below can be drawn, though they are only triply degenerate but not five-fold degenerate as the actual *d*-orbitals do. The analogy seen between the motions of a particle (can be an electron) in a cube and an electron under a coulombic potential is helpful to understand a shape of an atomic orbital or its degeneracy. See also **problem** and **its solution 1-108.**

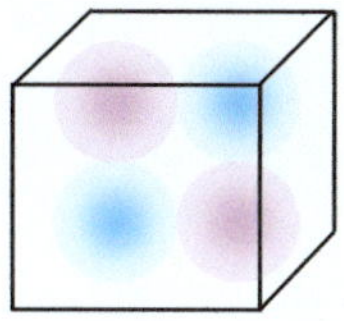

Fig.33 Probability distribution of a particle in a cube suggesting a *d*-orbital.

Ans. See text.

"note" The quantization of the kinetic energy discussed in this problem and solution is realized in cubic quantum dots. See also **Problem 110** and its **Solution.**

Solution 1-44a

(1) The electronic configuration of Na is given by $1s^2 2s^2 2p^6 3s^1$ and the outermost electron in 3*s*-orbital is completely shielded by the eight inner shell electrons. Thus the effective nuclear charge Z_{eff} is estimated as $Z_{eff} = 11 - \sum \sigma = 11 - 8 \times 0.85 - 2 \times 1 = 2.2.$ This value is adopted to insert the wave function of 3*s*-orbital as is in eq(68).

$$\varphi_{3s} = \frac{1}{81\sqrt{3\pi}} \left(\frac{Z_{eff}}{a_0} \right)^{3/2} \left(27 - 18 \times \frac{Z_{eff} r}{a_0} + 2 \times \left(\frac{Z_{eff} r}{a_0} \right)^2 \right) \cdot e^{-\frac{Z_{eff} r}{3a_0}} \tag{68}$$

Consequently, the radial probability density function is obtained as eq(69) below.

$$\rho(r) = 4\pi r^2 \varphi_{3s}{}^2 \propto r^2 \left(27 - 18 \times \frac{Z_{eff} r}{a_0} + 2 \times \left(\frac{Z_{eff} r}{a_0} \right)^2 \right)^2 e^{-\frac{2Z_{eff} r}{3a_0}} \tag{69}$$

By replacing r with $x = Z_{eff} r / a_0$, eq(69) can be rewritten as eq(70), for which a proportional constant is omitted.

$$\rho(x) \propto x^2 (27 - 18x + 2x^2)^2 e^{-\frac{2x}{3}} = \left\{ (27x - 18x^2 + 2x^3) \cdot e^{-\frac{x}{3}} \right\}^2 \tag{70}$$

The density function $\rho(x)$ takes maxima at x where eq(71) and, hence, eq(72) is satisfied.

$$\frac{d\rho(x)}{dx} \propto 2(27x - 18x^2 + 2x^3) \cdot e^{-\frac{x}{3}} \times \left\{ (27 - 36x + 6x^2) \cdot e^{-\frac{x}{3}} - \frac{1}{3}(27x - 18x^2 + 2x^3) \cdot e^{-\frac{x}{3}} \right\} = 0 \tag{71}$$

$$\frac{2}{3}(27x - 18x^2 + 2x^3) \times (81 - 108x + 18x^2 - 27x + 18x^2 - 2x^3) = 0$$

$$x(27 - 18x + 2x^2) \times (81 - 135x + 36x^2 - 2x^3) = 0 \tag{72}$$

Solving eq(72), the following six values of x, including x=0, are obtained.

$$x(27 - 18x + 2x^2) = 0 \ \Rightarrow \ x = 0, \ 3(3 \pm \sqrt{3})/2 = 1.90, \ 7.10$$

$$(81 - 135x + 36x^2 - 2x^3) = 0 \ \Rightarrow \ x = 0.74, \ 4.19, \ 13.07$$

In Fig.34 $\rho(x)$ is plotted against x together with the values of x where $\rho(x)$ takes maxima. For the estimation of the atomic radius of Na, the largest value of 13.07 of x can be adopted to yield 3.14 Å for r_{max} as follows, which can be regarded as the atomic radius.

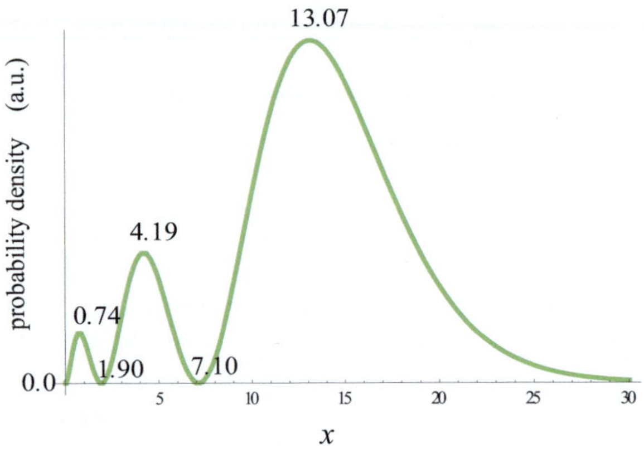

$$r_{max} = \frac{x \times a_0}{Z_{eff}} = \frac{13.07 \times 0.529}{2.2} = 3.14\,\text{Å}$$

Fig.34 $\rho(x)$ plotted against x indicating explicitly where it takes maxima.

Ans. atomic radius: 3.14 Å

(2) Employing $\rho(r)$ obtained in (1) above and integrating it over the range $[0,\ r_0]$, the value of r_0 which satisfies eq(73) is searched for.

$$\int_0^{r_0} \rho(r)dr = A^{-1}\int_0^{r_0} r^2\left(27 - 18\times\frac{Z_{eff}r}{a_0} + 2\times\left(\frac{Z_{eff}r}{a_0}\right)^2\right)^2 e^{-\frac{2Z_{eff}r}{3a_0}}\,dr \geq 0.95 \qquad (73)$$

A is a constant for normalization and is calculated from eq(74) after integration by putting Z_{eff}/a_0 = 2.2/0.529 acquired in (1) above.

$$A = \int_0^{\infty} r^2\left(27 - 18\times\frac{Z_{eff}r}{a_0} + 2\times\left(\frac{Z_{eff}r}{a_0}\right)^2\right)^2 e^{-\frac{2Z_{eff}r}{3a_0}}\,dr \qquad (74)$$

Using the value for A thus obtained, the value of r_0 which satisfies eq(73) is determined to be $r_0 = 5.14$ Å. To perform the integrations of eqs(73) and (74), the computer software *Mathematica* was employed.

In Fig.35 is plotted the radial distribution function of 3s-orbital of Na employing $Z_{eff} = 2.2$ (curve (a)). The position of r_0 is marked by a red dotted line. Compared with the atomic radius estimated in (1) above, which is marked by a black dotted line in Fig.35a, the radius evaluated on the basis of the premise (2) is far greater.

The curve (b) in blue is the radial distribution function of 3*s*-orbital of Na calculated with Z=11, i.e., without any consideration of shielding. When the curves (a) and (b) are compared, it is recognized that they look same in shape but their corresponding maxima are located at very different distances from the nucleus, indicating that the larger the shielding effect the outer the 3*s*-orbital expands. This tells us that any question about how large is an atom depends specifically on every type of phenomenon under discussion.

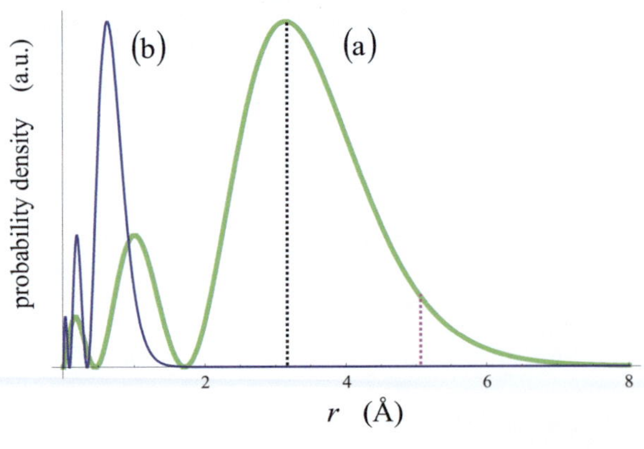

Fig.35 Radial probability distribution of 3*s*-orbital of Na.
(a) effective nuclear charge: $Z_{eff}e = 2.2e$,
(b) no shielding effect (Z=11),
black dotted line: atomic radius based on the premise (1),
red dotted line: atomic radius based on the premise (2), within which the probability of finding the electron is 0.95.

Ans. atomic radius: 5.14 Å

Solution 1-44b

Since the electronic configurations of the ions and atom under consideration are the same as far as the occupied atomic orbitals are concerned, the electron in the outermost subshell is attracted to the nucleus the stronger, the larger the nuclear charge is, making the ionic radius smaller. Therefore, the ions and atom in question can be arranged in the decreasing order of the ionic (atomic) radii as is given in **Ans.**

Ans. $O^{2-} > F^- > Ne > Na^+ > Mg^{2+}$

The ionic radii determined experimentally in solution are $O^{2-}(1.26) > F^-(1.19) > [Ne] > Na^+(1.16) > Mg^{2+}$ (0.86) (in the unit of $10^{-10}\,m = 1\,Å$), when compared for those whose coordination number n is 6.

Solution 1-45

When the chemical bonding of diatomic molecules are discussed, one of the bases is to examine the correlation between the symmetries and energy levels of the united atom and its far separated constituent atoms. As an example, a case of hetero-nuclear diatomic molecules is considered.

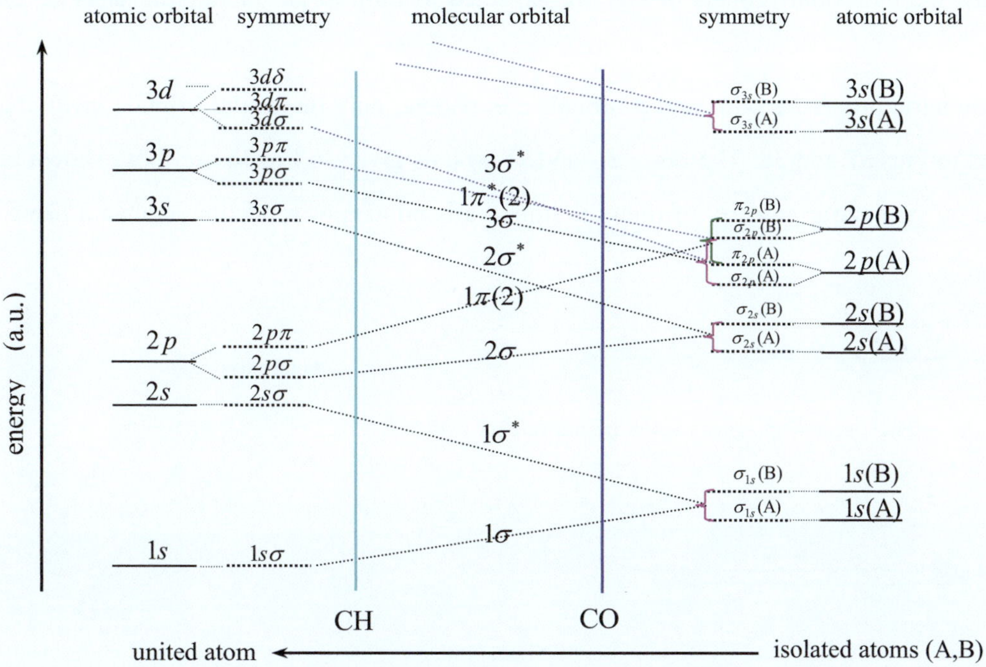

Fig.36 Concept of *united atom* and energy levels of molecular orbitals of a hetero-nuclear diatomic molecule. (Adapted from *G.Herzberg "Molecular Spectra and Molecular Structure 1. Spectra of Diatomic Molecules"*, D.van Nostrand Company,Inc., p328,1950.)

The atomic number of a united atom is larger than either that of A or that of B, and hence the energy levels of the atomic orbitals of the united atom are located lower than those for A and B. A mark { indicates a pair for linear combination (see **"note"**).

On the left side of Fig.36 are shown the energy levels of the united atom which is made of two atoms of A and B and hence it has the electrons and protons given by the sum of electrons of A and B and the sum of protons of A and B, respectively. On the right are shown the energy levels of the atomic orbitals of completely isolated atoms of A and B. This diagram is applicable to hetero-nuclear diatomic molecules.

Correlation lines which connect orbitals of the united and isolated atoms are drawn so as to connect those orbitals which have the same kind of symmetry against the rotation around the bond axis. Specifically, those of $\sigma \leftrightarrow \sigma$, $\pi \leftrightarrow \pi$, $\delta \leftrightarrow \delta$ are connected and any correlation line of the same symmetry type does not cross each other, for instance two $\sigma \leftrightarrow \sigma$ do not cross each other. A number in the parentheses, such as 2 of $1\pi(2)$ indicates that there are two orbitals of the same type.

The order of the energy levels of the molecular orbitals are determined depending on at which distance (inter atomic distance) the molecule under consideration is formed on the way moving from the isolated atoms to the united atom. In the case of CO, if it is assumed that the molecule is formed at the atomic distance indicated by the blue line in the diagram, 14 electrons in total are accommodated in the increasing order of energies into the orbitals as follows. Since the three bonding orbitals are full filled, it is predicted that CO is triply bonded (two π -bonds and one σ -bond).

$$1\sigma_{1s}^2 1\sigma_{1s}^{*\,2} 2\sigma_{2s}^2 2\sigma_{2s}^{*\,2} 1\pi_{2pa}^2 1\pi_{2pb}^2 3\sigma_{2p}^2$$

Incidentally, the molecular orbitals of CH are expected to form at an interatomic distance closer to the united atom.

Since symmetric properties of *s*- and *p*-orbitals are evident, only molecular orbitals involving *d*-orbitals are illustrated in Figs.37 and 38. The $d\sigma$-, $d\pi$- and $d\delta$-orbitals given in Fig.36 are those shown below. They are classified by symmetric property of rotation around a bond axis or reflection against a plane containing the bond axis.

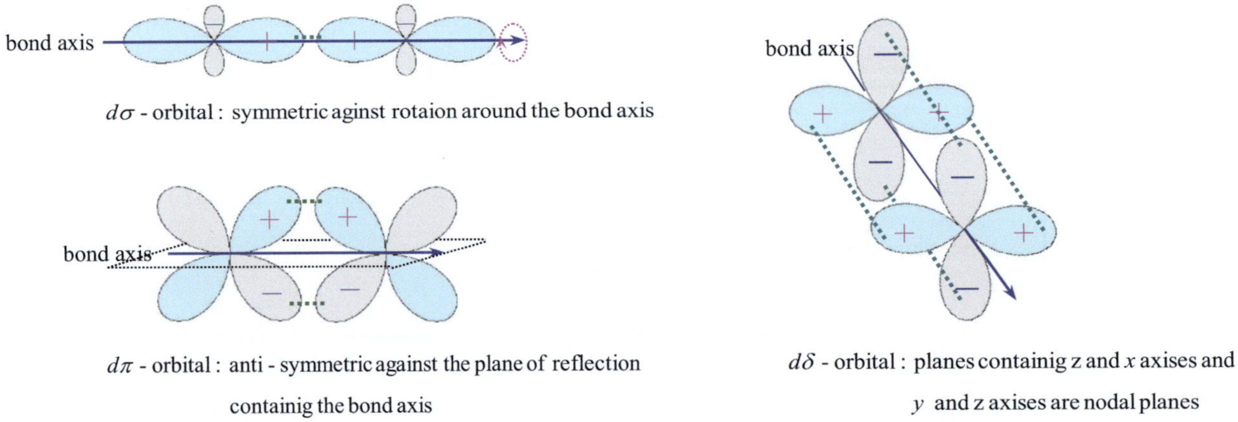

Fig.37 Symmetric property of *d*-orbitals and bonding molecular orbitals composed of the linear combination of d_A and d_B orbitals. The molecular orbitals thus formed are classified into one type of $d\sigma$-orbital, two types of $d\pi$-orbital and two types of $d\delta$-orbital. .

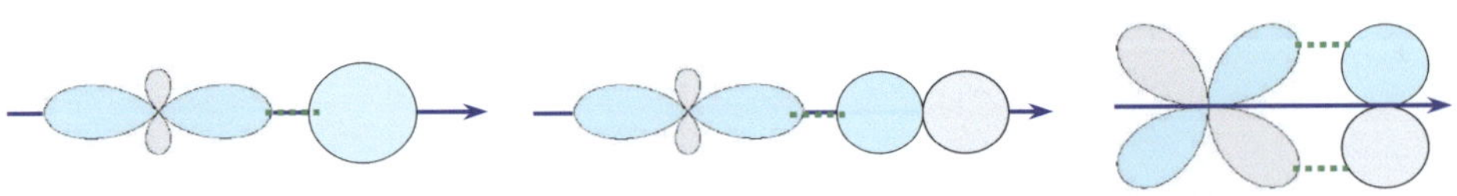

Fig.38 Molecular orbitals formed between a *d*-orbital and an *s*- or a *p*-orbital.

Owing to its symmetric property, *d*-orbital can also form molecular orbitals with *s*- and *p*-orbitals as is illustrated in Fig.38

Ans. See text.

"note 1" The reason why $2s$-orbital can correlate with σ^*-orbital is evident if it is noticed that a $2s$-orbital has a spherical nodal surface, that is, a double sphere as is represented by .

"note 2" Apart from the concept of united atoms, chemical bonding of diatomic molecules such as CO can be considered from a point of view of resonance or SALC molecular orbitals. For these, see **problem 1-41b, 1-51, 7-97** and their **solutions.**

Solution 1-46

Lyman series is the electronic transition from higher energy levels than K shell to K shell ($n=1$). The first three transitions correspond to the transitions of $1 \leftarrow 2$, $1 \leftarrow 3$ and $1 \leftarrow 4$. Employing the equation given in **solution 1-11**, the transition energy of $1 \leftarrow n$ is given by eq(75) and the energy in wave number by eq(76).

$$E_n - E_1 = \frac{m_e Z^2 e^4}{8\varepsilon_0 h^2}\left(\frac{1}{1^2} - \frac{1}{n^2}\right) \qquad (75)$$

$$\bar{\nu} = \frac{E_n - E_1}{hc} = \frac{m_e Z^2 e^4}{8\varepsilon_0 h^3 c}\left(\frac{1}{1^2} - \frac{1}{n^2}\right) = R \cdot \left(\frac{1}{1^2} - \frac{1}{n^2}\right) \qquad (76)$$

Therefore, it is enough to show that the given transition energies (cm^{-1}) satisfy eq(76). Also, since the ionization corresponds to the transition of $1 \rightarrow \infty$ and its energy is obtained from eq(76) by putting $n = \infty$, the proportional constant $R\,(\text{cm}^{-1})$ of eq(76) is found equal to the ionization energy.

Ans. Values for R of eq(76) remain constant for the observed three lines. The obtained constant $9.8766 \times 10^5\,\text{cm}^{-1}$ is equal to the ionization energy and exactly equal to 9 times (Z^2) of *Rydberg* constant.

Solution 1-47

The electronic configuration of O atom is expressed as $1s^2 2s^2 2p^4$. Six electrons in $2s$- and $2p$-orbitals are combined to prepare a sort of mixed orbitals beforehand in order to form stronger bonds with the $1s$-orbital of a hydrogen atom than those between s- or p-orbital of O alone and s-orbital. Such a mixed orbital is called hybrid orbital and in the present case sp^3-hybrid orbitals are most appropriate (see next problem). We have already mentioned about sp^3-hybrid orbitals in the **solution 1-7** and here we will express them in terms of wave functions. As has been explained before, each of these wave functions for $2p$-orbitals behaves like a unit vector along x-, y- or z-direction. Consequently it is straightforward to construct a wavefunction which is essentially p-orbital in its shape but directs toward one of the corners of a cube shown in Fig.39 by linearly combining three orbitals with coefficients given by the coordinate of the corner chosen, such as (-1, -1, 1). The four directions in Fig.39 are chosen so as to meet a tetrahedral and four wave functions thus created are given in eq(77) (not normalized). The function of ϕ_i and a p-orbital look alike. .

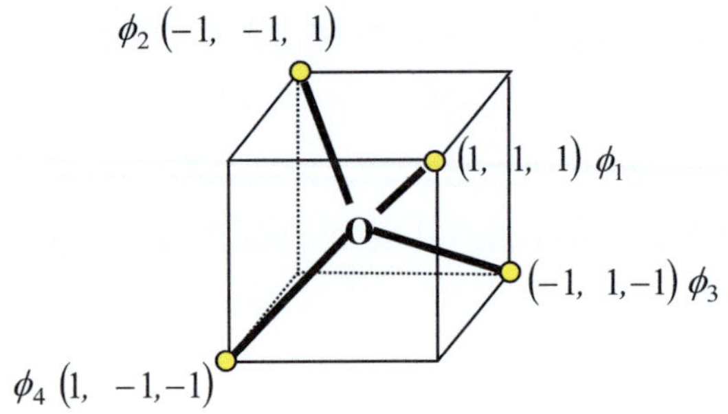

$$\left.\begin{aligned} \phi_1 &= p_x + p_y + p_z \\ \phi_2 &= -p_x - p_y + p_z \\ \phi_3 &= -p_x + p_y - p_z \\ \phi_4 &= p_x - p_y - p_z \end{aligned}\right\} \quad (77)$$

Fig.39 Relation between directions of p-hybrid orbitals and coordinates of corners of a cube.

By hybridizing any one of these p-orbitals given by ϕ_i and an s-orbital (correctly $2s$-orbital), four sp^3-hybrid orbitals h_1, h_2, h_3 and h_4 are formed as are given in eq(78), though the hybridization itself is nothing more than sp-hybridization. The reason why they are called as $sp^3$is evident because one s-orbital and three p-orbitals are involved. Conceptual shapes of the sp^3-orbitals are shown in Fig.40(a).

$$h_1 = \frac{1}{2}(s+\phi_1),\quad h_2 = \frac{1}{2}(s+\phi_2),\quad h_3 = \frac{1}{2}(s+\phi_3),\quad h_4 = \frac{1}{2}(s+\phi_4) \qquad (78)$$

In the case of H_2O, with O atom in the center of a cube, its four hybrid orbitals direct toward four corners of the cube and two of them are used to form two σ-bonds with $1s$-orbitals of two hydrogen atoms. The other two are filled with two electrons each, forming two lone-pair electrons. A schematic diagram of bond formation is given in Fig.40(b). Summarizing the explanations above, the configuration of bonding and non-bonding electrons of H_2O is as shown in Fig.41.

"note" Bonding of one-centered polyatomic molecules such as H_2O is explained by SALC molecular orbitals in a different way from the hybrid orbitals.

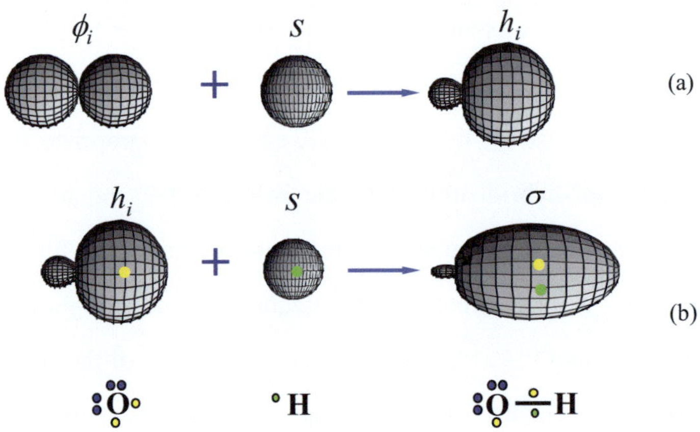

Fig.40 Formation of one of sp^3-hybrid orbital h_i (a) and σ-bond formation by linear combination of h_i and s. (a) is one-centered and (b) is two-centered hybridization.

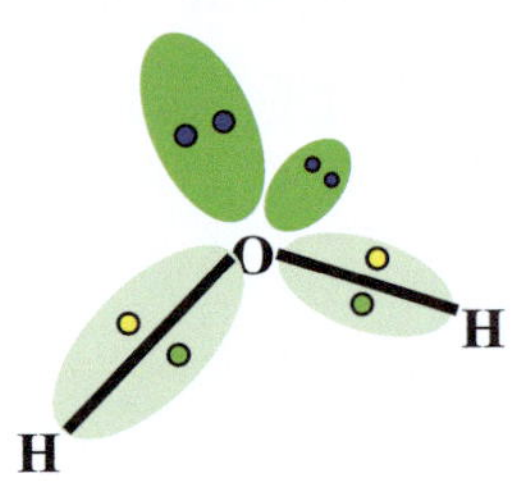

Fig.41 Chemical bonds and lone-pair electrons of H_2O.

orbitals shown in Fig.40(a),

orbitals shown in Fig.40(b).

Ans. See text.

Solution 1-48

The energy of one of the sp^3-orbitals, say h_1 in the preceding solution, is readily calculated as is given by eq(79), since all of the component orbitals used for hybridization are orthogonal to each other. Note that

$$\langle h_1|H|h_1\rangle = \frac{1}{4}(\varepsilon_s + 3\varepsilon_p) \qquad (79)$$

$\varepsilon_s = \langle s|H|s\rangle$ and $\varepsilon_p = \langle p|H|p\rangle$. For the other three orbitals the same orbital energy is obtained. Therefore, the sum of the orbital energies for the four hybrid orbitals is $\varepsilon_s + 3\varepsilon_p$, which is exactly the same as the sum of the energy of one *s*-orbital and those of the three *p*-orbitals. As is seen from this, the sum of orbital energies are invariant before and after hybridization. Namely, there is no gain in energy by hybridization, i.e., no stabilization by that. This is a natural consequence of the fact that atomic orbitals such as *s*-and *p*-orbitals are eigenfunctions of Hamiltonian of an atomic electron. Any linear combination of eigenfunctions does not yield any change in the eigenvalue of energy.

Now, let us examine the total energy of six electrons of O atom by comparing it before and after the hybridization. The total energy is $2\varepsilon_s + 4\varepsilon_p$ before the hybridization but after the hybridization it changes to $\frac{1}{4}(\varepsilon_s + 3\varepsilon_p)\times 6 = \frac{3}{2}(\varepsilon_s + 3\varepsilon_p)$. The difference amounts to $\frac{1}{2}(\varepsilon_p - \varepsilon_s) > 0$, which means that this amount of energy is required in order to put six electrons in the hybrid orbitals (see Fig42. below). The energy required is called "energy of promotion".

Then, why we consider "hybridization". This is because a chemical bond to be formed between a hybrid orbital of one atom and an atomic or a hybrid orbital of the other atom is much stronger than those to be formed between simple atomic orbitals such as pure *s*- or *p*-orbitals. The energy to be gained upon bond formation is usually far greater than the energy of promotion required for hybridization. In this sense, the concept of resonance energy and structures and the bond stabilization by hybridization are in common (see **problem 1-7**).

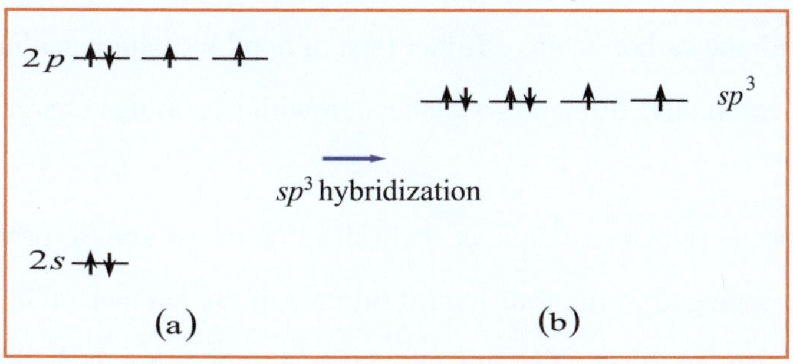

Fig.42 Electronic configuration (O atom) before and after hybridization.
The total energy of configuration (b) is larger than that of (a) by the amount of $\frac{1}{2}(\varepsilon_p - \varepsilon_s) > 0$, which is called "energy of promotion".

Ans. See text.

"note" See **problem 1-81** for the linear combinations of orthogonal complete functions and their eigenenergies.

Solution 1-49

(1) $\mathbf{SO_4^{2-}}$ The electronic configuration of S atom is $1s^2 2s^2 2p^6 3s^2 3p^4$ and hence sp^3-hybridization of 3*s*- and 3*p*-orbitals is applicable. Six electrons of S atom and two additional electrons ascribed to the anion (or as

if it were S^{2-}), eight electrons in total, are arranged pairwise in each of four hybrid orbitals and each of four oxygen atoms forms a bond with S atom by sharing these two electrons in a hybrid orbital, or in other words, by coordinating electrons so as to satisfy octet rule. Since these eight electrons cannot distinguish one another, four coordination bonds formed in this manner are identical. Because of this, SO_4^{2-} behaves as if it were a spherical ion. Furthermore, the reason why such an ion as SO_4^{3-} or SO_4^{1-} is not formed can be made clear.

(2) $\mathbf{NH_4^+}$ The electronic configuration of N atom is $1s^2 2s^2 2p^3$ and hence sp^3-hybridization of 2*s*- and 2*p*-orbitals is applicable. After taking one electron out of five valence-shell electrons of N atom to give it to a proton to prepare four hydrogen atoms, the remaining four electrons are arranged in the four hybrid orbitals and then each half-filled hybrid orbital is used to form a σ-bond with a hydrogen atom. Since the four bonds formed in this manner are identical, the ion must be spherical.

Alternatively, if we take NH_3 a base, three of hybrid orbitals can be used to form three σ_{NH}-bonds and the remaining hybrid orbital is used for a lone pair electrons, to which a proton coordinates. However, since four N-H bonds in NH_4^+ must be identical, the other three structures in each of which a proton is coordinated at a different hybrid orbital should be taken into account further for resonance.

(3) $\mathbf{NO_3^-}$ Again, the electronic configuration of N atom is $1s^2 2s^2 2p^3$ but instead of sp^3-hybridization of 2*s*- and 2*p*-orbitals, sp^2-hybridization should be applied. Five valence-shell electrons of N atom are arranged in the three hybrid orbitals separately in the manner of 2, 2 and 1 electron. Three oxygen atoms in the forms of O, O and O^- form three σ-bonds with the nitrogen atom so as to fulfill octet rule. This results in a compound of a charge of minus one with a planer structure. To consider the reason why sp^3-hybridization was not employed to predict a structure of NO_3^- is left to readers.

(4) $\mathbf{CO_3^{2-}}$ The electronic configuration of C atom is $1s^2 2s^2 2p^2$ and as is employed above sp^2-hybridization is applicable. Four electrons of C atom are arranged in the three hybrid orbitals in the manner of 2, 1 and 1 or 2, 2 and 0 electron. In the former, three oxygen atoms in the form of O, O^- and O^- and in the latter in the form of O, O and O^{2-} construct three σ-bonds with the carbon atom. Either type of bond formation can make all oxygen atoms satisfy the octet rule and the compound formed is a planer ion with two minus charges in either case.

(5) $\mathbf{PO_4^{3-}}$ The electronic configuration of P atom is $[Ne]3s^2 3p^5$ and sp^3-hybridization of 3*s*- and 3*p*-orbitals is applicable here. Five electrons of P atom are arranged in the four hybrid orbitals in the manner of 2, 1, 1 and 1 electron. Four oxygen atoms in the form of O, O^-, O^- and O^- construct four σ-bonds with the phosphorous atom so as to fulfill the octet rule. The resulting compound is a spherical (tetrahedral) ion with three minus charges

(6) $\mathbf{CO_2}$ The electronic configurations of C and O atoms are $1s^2 2s^2 2p^2$ and $1s^2 2s^2 2p^4$, respectively. A 2*s*- orbital of C atom prepares two *sp*-hybrid orbitals with $2p_z$-orbital which is as symmetric against the rotation around the bond axis as 2*s* does. $2p_z$-orbital of each oxygen atom constructs σ, σ^* bonds with one of

the two hybrid orbitals. It should be noticed here that the number of component atomic orbitals is four ($2s$- and $2p_z$-orbitals of C and $2p_z$-orbital of each O) and hence the number of the resulting molecular orbitals must be four (two σ- and two σ^*-orbitals).

The remaining $2p_x$- and $2p_y$-orbitals of each atom can construct π-bonds in zx- and yz-plane, respectively. There are three types of π-bonds or orbitals constructed from three atoms. They are a bonding, nonbonding and antibonding as is illustrated in Fig.43. (See **solutions 1-84** and **1-85** for molecular orbitals based on *Hückel* approximation.) Namely, two sets of three molecular orbitals each consisting three atomic orbitals are prepared.

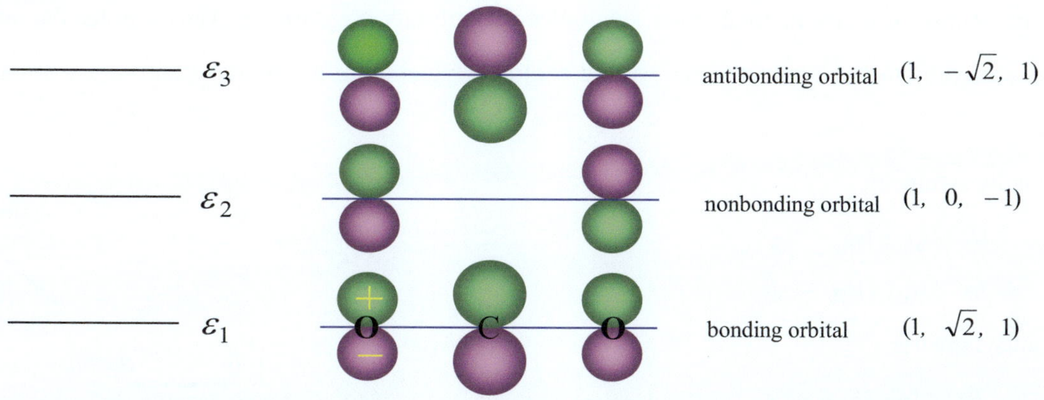

Fig.43 π-orbitals and their energy levels of CO_2.

Another set of three molecular orbitals is obtained by rotating those shown above by 90° around the bond axis.

Simple *Hückel* approximation yields the eigenenergies of $\varepsilon_1 = \alpha + \sqrt{2}\beta$, $\varepsilon_2 = \alpha$ and $\varepsilon_3 = \alpha - \sqrt{2}\beta$.

Values given in parentheses are coefficients of $2p$-oritals for the relevant molecular orbitals.

So far we have discussed chemical bonds on the basis of hybrid orbitals, but, alternatively, we can discuss the same issue on the basis of molecular orbitals which are delocalized over all constituent atoms of a molecule. In the case of CO_2 see **problem 1-58**.

At this point it is essentially important to understand why CO_2 is a linear molecule. Theories such as *Walsh* rule and **VSEPR** (*<u>V</u>alence <u>S</u>hell <u>E</u>lectron <u>P</u>air <u>R</u>epulsion*) (see **solution 1-55**) are available to predict shapes of tri-atomic molecules. *Walsh* rule is explained below by taking H_2O as an example and compared with what the molecular orbital treatment of H_2O predicts.

***Walsh* rule**

Based on their thorough studies, it is well known for one-centered poly-atomic molecules that there are close correlations between molecular shapes and sums of valence electrons. For a molecule of the type of AB_2, when the sum of its valence electrons is equal to or less than 16, the molecule assumes a linear shape. In the case of CO_2, since the sum is 16 (= 4+6+6), it is expected linear in shape as has been proved experimentally. On the other hand, the sum of the valence electrons of NO_2 is 17 and hence its molecular shape is expected to be non-linear. In fact, its bond angle is known to be ca. 134°.

linear or non-linear?

Fig.44 Molecular structures of NO_2 and NO_2^+.

The theoretical grounds of *Walsh* rule are made clear when the energies of molecular orbitals of a given molecule under examination are known about how they vary with bond angle(s). As an example of this, H_2O is examined here in detail.

Let us consider a linear molecule of $H-O-H$ first and assume that its molecular orbitals are consisted of the atomic orbitals $1s_{H_a}$ and $1s_{H_b}$of hydrogen a and b, respectively and $2s_O$, $2p_x$, $2p_y$ and $2p_z$ of oxygen, six atomic orbitals in total, then six molecular orbitals are formed. These molecular orbitals are classified according to the symmetry against the central oxygen atom as follows (coefficient of each atomic omitted).

totally symmetric

$1\sigma : 1s_{H_a} + 1s_{H_b} + 2s_O$

$1\sigma^* : 1s_{H_a} + 1s_{H_b} - 2s_O$

antisymmetric

$2\sigma_p : 1s_{H_a} - 1s_{H_b} - 2p_z$

$2\sigma_p^* : 1s_{H_a} - 1s_{H_b} + 2p_z$

antisymmetric against molecular plane

$2p_x$, $2p_y$

(to accommodate lone-pair electrons)

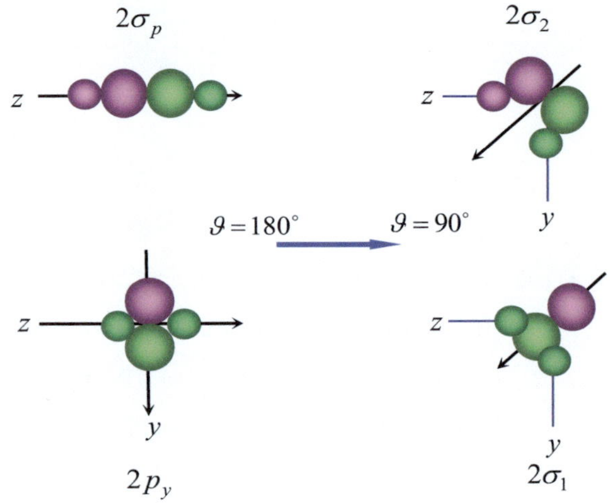

Fig.45 Molecular orbitals and bond angle.

Next, let us examine the molecular orbitals of a molecule bent at right angles $\begin{array}{l} O-H \\ | \\ H \end{array}$ and classify them according to the symmetry against the plane which is perpendicular to the molecular plane and dissects the bond angel of $H-O-H$. Six molecular orbitals are grouped as below.

totally symmetric

$1\sigma : 1s_{H_a} + 1s_{H_b} + 2s_O$

$1\sigma^* : 1s_{H_a} + 1s_{H_b} - 2s_O$

$2\sigma_1 : 1s_{H_a} + 1s_{H_b} + (2p_y - 2p_z)$

antisymmetric

$2\sigma_2 : -1s_{H_a} + 1s_{H_b} + (2p_y + 2p_z)$

$2\sigma_3 : 1s_{H_a} - 1s_{H_b} + (2p_y + 2p_z)$

antisymmetric against the molecular plane

$2p_x$ (to accommodate lone-pair electrons)

The molecular orbitals for the liner and right-angled molecules are correlated as is exemplified in Fig.45 for $2\sigma_p$ to $2\sigma_2$ and $2P_y$ to $2\sigma_1$. The orbital energies do vary with the bond angle but the symmetric property is conserved along variation of the bond angle. Taking it into account that none of energy levels of the same symmetry cross each other, the six energy levels of the linear and the right-angled molecules are correlated against the bond angle as is shown in Fig.46 (see **solution 1-59b**). When eight valence electrons are arranged in the molecular orbitals either of the linear or bent molecule to compare the corresponding total bond energies, it is readily seen from Fig.46 that the bent molecule is more stable. For this H_2O is concluded to be a bent molecule.

"note" Energies of molecular orbitals of a given molecule increases with increasing number of nodal plane. For instance, the orbitals discussed above are arranged in the increasing order of energy as shown in Fig.47.

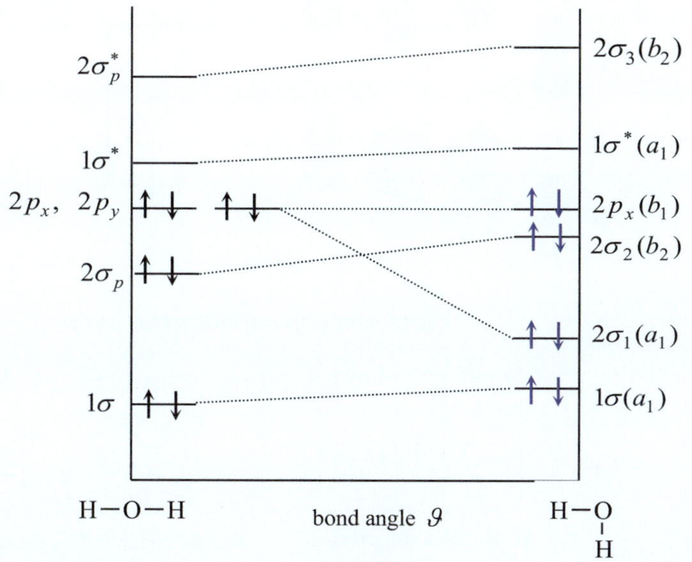

Fig.46 Energy level correlation diagram. Energy levels of molecular orbitals of H_2O and bond angle.

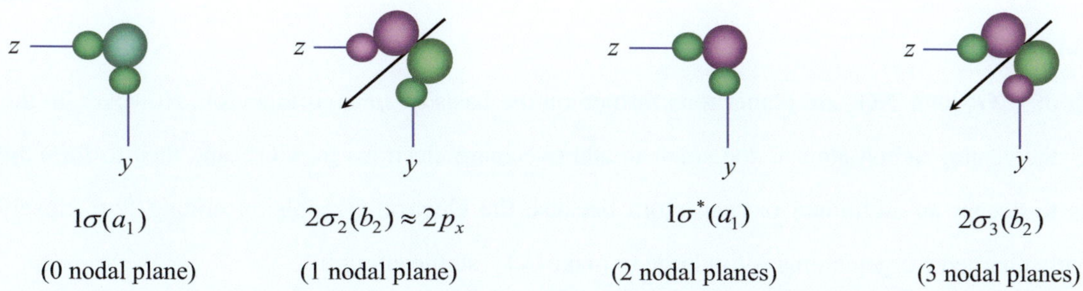

Fig.47 Increasing energies of molecular orbitals with increasing number of nodal plane.

(7) **Butadiene** As has been explained in **solution 1-7**, it is most appropriate to consider that the σ-bonds of C−C and C−H are composed of sp^2-hybrid orbitals. Therefore, a top view of the molecular frame of the σ-bonds in the molecular plane is just like either of those shown in Fig.48.

Four $2p_x$-orbitals perpendicular to the molecular plane are linearly combined to form four delocalized molecular orbitals. Their images are shown in Fig.49 together with the energy levels and their wave functions with correct coefficients, though not normalized to one Two of them are bonding and the other two are antibonding. Four electrons are accommodated in the bonding orbitals but it is clearly seen from Fig.49 that the double bond character is greater for the end C−C bonds than for the central C−C bond.

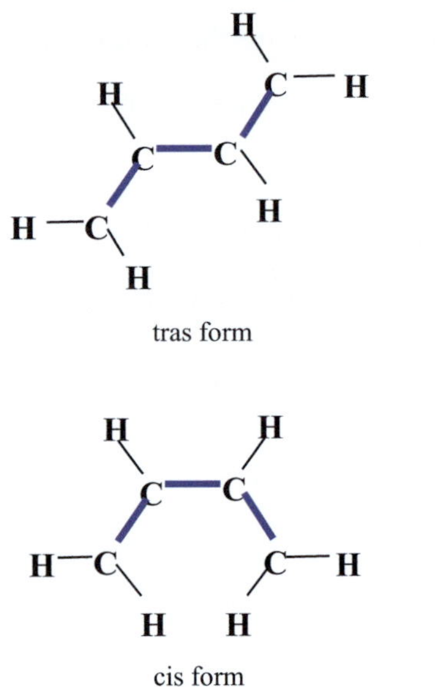

Fig.48 Molecular frame of butadiene. Eigenfunctions and eigenvalues of π-orbitals are exactly the same for both isomers in simple *Hückel* approximation.

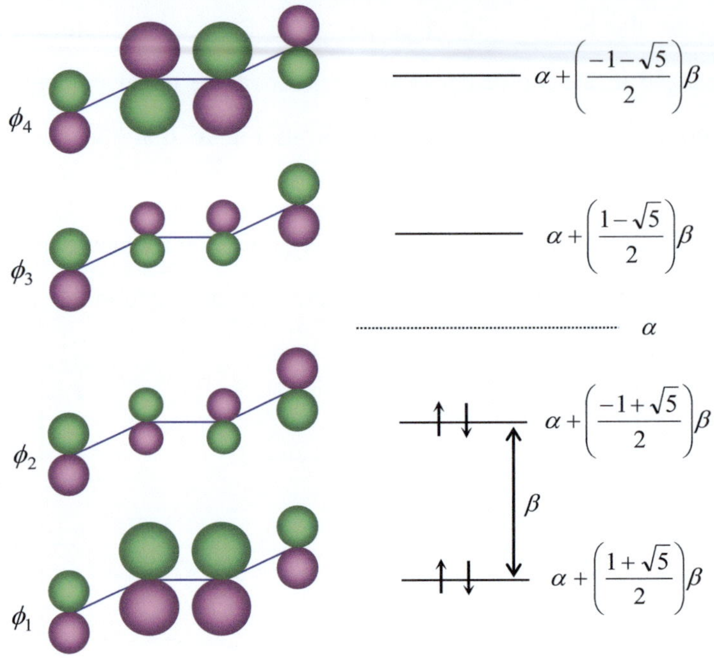

Fig.49 π-orbitals and their energy levels.

$$\phi_4 = p_1 - \left(\frac{1+\sqrt{5}}{2}\right)p_2 + \left(\frac{1+\sqrt{5}}{2}\right)p_3 - p_4, \quad \phi_3 = p_1 + \left(\frac{1-\sqrt{5}}{2}\right)p_2 + \left(\frac{1-\sqrt{5}}{2}\right)p_3 + p_4$$

$$\phi_2 = p_1 - \left(\frac{1-\sqrt{5}}{2}\right)p_2 + \left(\frac{1-\sqrt{5}}{2}\right)p_3 - p_4, \quad \phi_1 = p_1 + \left(\frac{1+\sqrt{5}}{2}\right)p_2 + \left(\frac{1+\sqrt{5}}{2}\right)p_3 + p_4$$

ϕ_1, ϕ_2 : bonding orbitals ϕ_3, ϕ_4 : antibonding obitals

Ans. See text (1)-(7).

Solution 1-50

Both of $\mathbf{CO_3^{2-}}$ and $\mathbf{NO_3^-}$ are planer ions formed on the basis of *sp*²-hybridization. However, in the case of $\mathbf{CO_4^{4-}}$, the central carbon atom is too small to add two more electrons to CO_3^{2-} and then to form *sp*³-hybrid orbitals to accept an additional oxygen atom because the electronic repulsion among four closely placed negatively charged oxygen atoms is too large to make CO_4^{4-} stable enough.

A similar situation holds for $\mathbf{NO_4^{3-}}$. Hypothetically, two more electrons are added to NO_3^- and then *sp*³-hybrid orbitals are formed to make four bonds with four oxygen atoms. Again, the central nitrogen atom is too small to make NO_4^{3-} stable enough because the electronic repulsion among four closely placed negatively charged oxygen atoms is too large. However, Si and P atoms, though they belong to the same group as C and N in the periodic table, respectively, are large enough to reduce the electronic repulsion among negatively charged oxygen atoms and ions such as $\mathbf{SiO_4^{4-}}$ and $\mathbf{PO_4^{3-}}$ come to exist under ordinal conditions. For the relation between the oxidation number of the central atom of an ion and its charge number, see **problem 1-90**.

Ans. See text.

Solution 1-51

The molecular orbitals employed are either bonding or antibonding and good enough to show that the C≡O bond is a triple bond, but none of which can be used to accommodate lone-pair electrons. Consequently, this type of molecular orbitals is not appropriate to explain the *Lewis* structure of CO.

Ans. Not applicable.

Then, what type of molecular orbitals is appropriate to meet the requirement in the problem. To form π-orbitals perpendicular to the molecular plane, nothing other than linear combinations of p-orbitals can be used but for σ-orbitals, instead of the linear combination of s-orbitals alone the linear combination of sp-hybrid orbitals can be employed in order to provide stronger σ-bonds as is illustrated below.

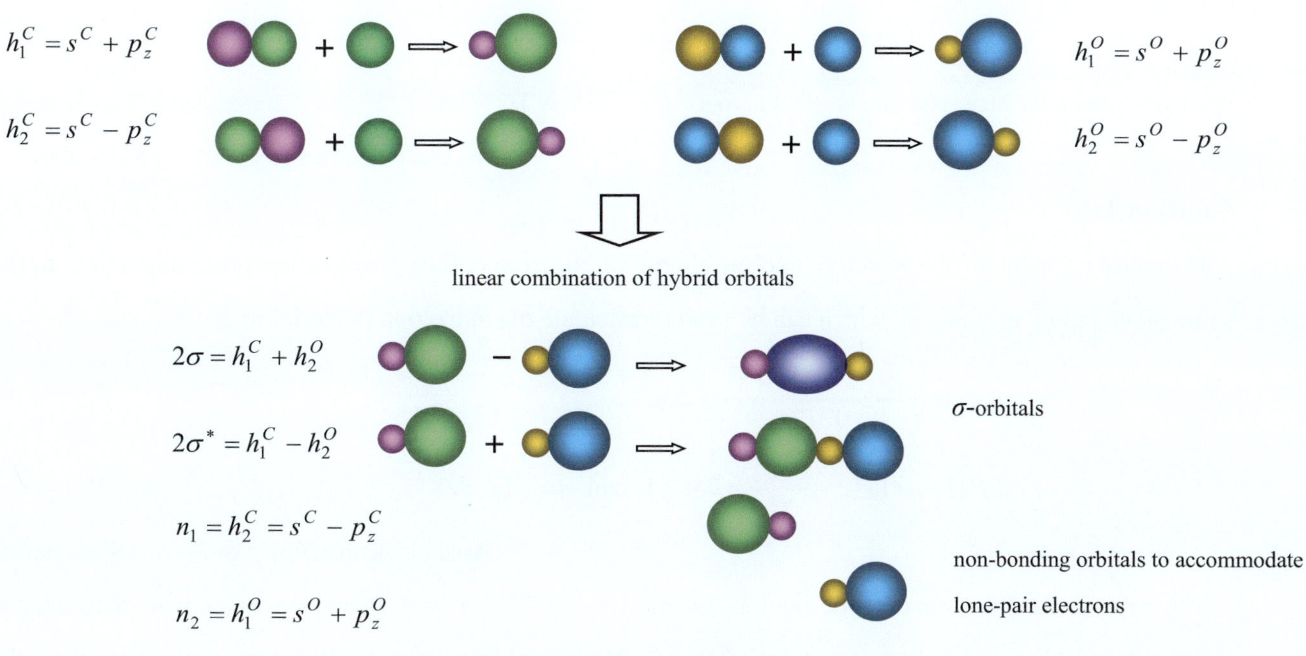

Fig.50 Preparation of hybrid orbitals of C and O.

Solution 1-52

(1) **Ans.** K: $(1s^2 2s^2 2p^6 3s^2 3p^6 4s^1)$, Ne: $(1s^2 2s^2 2p^6 3s^2 3p^6)$

(2) The ionization energy *I.P.* is given by eq(80). A value for constant A is determined by using the ionization

$$I.P. = A \times \frac{Z_{eff}^{\ 2} e^2}{n^2} \qquad (80)$$

energy of hydrogen atom as follows.

$$1312 \times 10^3 = A \times \frac{e^2}{1^2} \quad \Rightarrow \quad A = 1312\ \mathrm{kJ}/e^2$$

The value for A obtained thus can be used to evaluate the first ionization energy of K from eq(80). By substituting the given value of *I.P.* for K, the effective nuclear charge is obtained as below.

$$419 \times 10^3 = 1312 \times 10^3 \times \frac{Z_{eff}^{\ 2}}{4^2} \quad \Rightarrow \quad Z_{eff} = \sqrt{\frac{419}{1312}} \times 4 = 2.26$$

Ans. effective nuclear charge: 2.26

(3) The electron affinity of Ar can be regarded as the energy required for the reaction given by eq(81).

$$\mathrm{Ar} + \mathrm{e} \rightarrow \mathrm{Ar}^- \qquad (81)$$

On the other hand, the reverse reaction given in eq(82) can be regarded as the ionization reaction of Ar^-.

$$\mathrm{Ar}^- \rightarrow \mathrm{Ar} + \mathrm{e} \qquad (82)$$

Consequently, a negative value of the ionization energy of Ar^- can be used as an estimate for the electron affinity of Ar. Looking back at the result obtained in (1), the nuclear charge of +19 of K is found to be shielded by the amount of 16.74 (*e*). Since Ar^- has the same electronic configuration as K , though its nuclear charge is 18, the nuclear charge is expected to be shielded by the same amount as does K, yielding $Z_{eff} = 18 - 16.74 = 1.26$. Then eq(80) yields a value of $130\ \mathrm{kJ \cdot mol^{-1}}$ for the ionization potential of Ar^-. Its negative value is the estimated electron affinity of Ar.

$$I.P.(\mathrm{Ar}^-) = 1312 \times \frac{1.26^2}{4^2} = 130\ \mathrm{kJ \cdot mol^{-1}}$$

The experimentally observed one is $-35\ \mathrm{kJ \cdot mol^{-1}}\ (= -1.35\ \mathrm{eV})$.

Ans. electron affinity of Ar: $-130\ \mathrm{kJ \cdot mol^{-1}}$

Solution 1-53

Estimating a value of the effective nuclear charge of Ne from eq(80) given in the preceding solution, the value given below is obtained, which can be used to calculate the ionization potential of F^-.

$$Z_{eff} = \sqrt{\frac{497.2 \times 4.18 \times 2^2}{1312}} = 2.52$$

$$I.P.(\mathrm{F}^-) = 1312 \times \frac{1.52^2}{2^2} = 758\ \mathrm{kJ \cdot mol^{-1}} (= 7.56\ \mathrm{eV})$$

Ans. electron affinity of F: $-758\ \mathrm{kJ \cdot mol^{-1}}$

The experimentally observed one is $-332\ \mathrm{kJ \cdot mol^{-1}} (= -3.44\ \mathrm{eV})$ and again, the agreement between the estimated and observed values is not satisfactory. However, for a pair of elements next each other, the present method makes it possible to estimate a value of electron affinity for the element with the smaller atomic number than the other, when the ionization energy for the latter is known. In particular, it is useful in order to explain the periodicity found in the electron affinity of elements.

Solution 1-54

Coulomb energy due to4the coulomb force between two point charges of $q_1 e$ and $q_2 e$ is given by eq(83).

$$U_{12} = \frac{q_1 e \times q_2 e}{4\pi\varepsilon_0 r_{12}} \qquad (83)$$

When point charges of the same sign are placed far away from each other as much as possible and point changes of the opposite sing are placed near to each other as much as possible, the total coulomb energy attains the smallest value. Namely, the eight point charges arranged as shown in Fig.51 are most stable.

Coulomb energy between two oppositely charged points:

$$U_{+-} = \left(\frac{1}{4\pi\varepsilon_0}\right)\frac{-e \times e}{a \times 10^{-10}} \times 12 + \left(\frac{1}{4\pi\varepsilon_0}\right)\frac{-e \times e}{\sqrt{5}a \times 10^{-10}} \times 4 = -\left(\frac{e^2}{4\pi\varepsilon_0}\right)\frac{(12 + 4\sqrt{5}/5)}{a} \times 10^{10}$$

Coulomb energy between either two positively or negatively charged points:

$$U_{++}=U_{--}=\left(\frac{1}{4\pi\varepsilon_0}\right)\frac{e\times e}{\sqrt{2}a\times10^{-10}}\times6=\left(\frac{e^2}{4\pi\varepsilon_0}\right)\frac{3\sqrt{2}}{a}\times10^{10}$$

Consequently, total coulomb energy U_{total} is given by a sum of those energies as follows.

$$U_{total}=U_{+-}+U_{++}+U_{--}=-\left(\frac{e^2}{4\pi\varepsilon_0}\right)\left(\frac{(12+4\sqrt{5}/5)}{a}-\frac{6\sqrt{2}}{a}\right)\times10^{10}=-\left(\frac{e^2}{4\pi\varepsilon_0}\right)\times\frac{5.30}{a}\times10^{10}$$

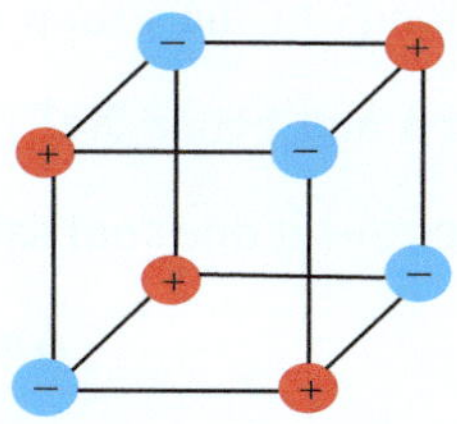

Fig.51 Most stable arrangement.

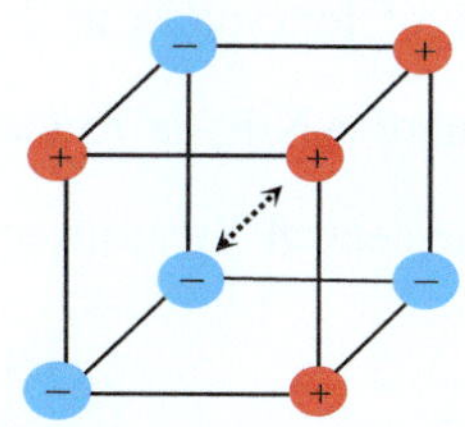

Fig.52 Oppositely charged pair points (←······→) exchanged.

When two oppositely charged points are swapped as is shown in Fig.52, the resulting coulomb energy is calculated as follows.

$$U_{total}=-\left(\frac{e^2}{4\pi\varepsilon_0}\right)\left(\frac{4-4\sqrt{2}}{a}\right)\times10^{10}=-\left(\frac{e^2}{4\pi\varepsilon_0}\right)\times\frac{-1.66}{a}\times10^{10}$$

The energy thus obtained is higher than that for the arrangement shown in Fig.51, indicating that the most stable (lowest energy) arrangement is the one shown in Fig.51.

Ans. Fig.51, $-\left(\frac{e^2}{4\pi\varepsilon_0}\right)\left(\frac{(12+4\sqrt{5}/5)}{a}-\frac{6\sqrt{2}}{a}\right)\times10^{10}$

Solution 1-55

The molecular structures or shapes of the compounds in question can be answered by using the VSEPR model (see below).

(1) **CO_2** The number of the valence electrons of C is 4. With respect to the central C atom, the number of bonds is 2 and the number of lone-pairs is 0. Therefore, CO_2 belongs to the group of AX_2 type with $SN=2$. From the table given below, a linear AX_2 is found to satisfy the condition and hence CO_2 is predicted to be a linear molecule. The bonding between C and O is multiple bonding, but the number of this multiple bond is counted just one. Both atoms satisfy the octet rule.

Ans. linear

(2) **BF_3** The number of the valence electrons of B is 3. With respect to the central B atom, the number of bonds is 3 and the number of lone-pairs is 0. Therefore, BF_3 belongs to the group of AX_3 type with $SN=3$. From the table given below, a planer triangle AX_3 is found to satisfy the condition and hence BF_3 is predicted to be a flat triangle molecule. The central B atom does not satisfy the octet rule.

Ans. flat triangle

(3) **NH_3** The number of the valence electrons of N is 5. With respect to the central N atom, the number of bonds is 3 and the number of lone-pairs is 1. Therefore, NH_3 belongs to the group of AX_3E type with

SN = 3. E stands for a lone pair. Similarly as above, a triangular pyramid AX_3E is found to satisfy the condition and hence $\mathbf{NH_3}$ is predicted to be a triangular pyramid molecule. The central N atom satisfies the octet rule.

Ans. triangular pyramid

(4) $\mathbf{XeF_4}$ The number of the valence electrons of Xe is 8. With respect to the central Xe atom, the number of bonds is 4 and the number of lone-pairs is 2. Therefore, XeF_4 belongs to the group of AX_4E_2 type with SN = 6. In the table, a square AX_4E_2 is found to satisfy the condition and hence XeF_4 is predicted to be a square molecule. Since the central Xe atom is surrounded by ten electrons it does not satisfies the octet rule.

Ans. square structure

(5) $\mathbf{SF_4}$ The number of the valence electrons of S is 6. With respect to the central S atom, the number of bonds is 4 and the number of lone-pairs is 1. Therefore, SF_4 belongs to the group of AX_4E type with SN = 5. From the table, a seesaw AX_4E is found to satisfy the condition and hence SF_4 is predicted to be a seesaw molecule. Although it may appear for the molecule to assume a tetrahedral structure judging from the number of the bonds formed, it does not. It is a seesaw-shaped molecule. The central S atom does not satisfy the octet rule.

Ans. seesaw structure

(6) $\mathbf{PCl_5}$ The number of the valence electrons of P is 5. With respect to the central P atom, the number of bonds is 5 and the number of lone-pairs is 0. Therefore, PCl_5 belongs to the group of AX_5 type with SN = 5. From the table, PCl_5 is expected to assume a triangular bipyramid structure. The central P atom does not satisfy the octet rule.

Ans. triangular bipyramid

(7) $\mathbf{SF_6}$ The number of the valence electrons of S is 6. Six F atoms share these six electrons with S to form six bonds, leaving no lone-pair electrons. Consequently, SF_6 is classified as the group of AX_6 with SN = 6 and is expected to assume an octahedral structure. Since the S atom has 12 electrons, it does not satisfy the octet rule.

Ans. octahedral structure

(8) $\mathbf{H_2O}$ The number of the valence electrons of O is 6. Two of the six electrons form <u>two</u> σ-bonds with the two hydrogen atoms and the remaining four electrons form <u>two</u> lone-pairs. Consequently, H_2O is classified as the group of AX_2E_2 with SN = 4 and is expected to assume a tetrahedral structure.

Ans. tetrahedral structure

(9) $\mathbf{NO_2}$ The number of the valence electrons of O is 5. The two O atoms tend to satisfy the octet rule by forming two double bonds with N to share four electrons each. With respect to the central N atom, therefore, the number of bonds is 2 but only one electron is left in a lone-pair (non-bonding) orbital. No category for this is found in the table. The VSEPR model cannot be applied to $\mathbf{NO_2}$. (See **1-49.**)

Ans. In contrast to $\mathbf{SO_2}$, **VSEPR** is not applicable to NO_2.

The **VSEPR** model has been developed by *Gillespie* and *Nyholm* to the present form. According to the model, the valence electrons of the central atom A of one-centered molecules of the type of AX_n (n = 2~7) are classified into two groups of "bond-electron pair" and "lone-pair". As is suggested by the name of the model, the model predicts that a molecule AX_n will assume a molecular configuration so as to make electronic repulsions among those classified electron pairs as small as possible. Although the model is criticized for lacking theoretical grounds, a molecular shape can be predicted correctly by the model for many of one-centered polyatomic molecules.

Specifically,

1) Based on *Lewis'* octet rule, electrons occupying the outermost shell of the central atom A are classified into "bond-electron pairs" and "lone-pairs".
2) The sum of the number of bond-electron pairs and the number of lone-pairs is called as ***Steric Number*** (SN). Bond-electron pairs associated with one multiple bond are counted as one.
   In equation, SN = **X** (number of bond-electron pairs) + **E** (number of lone-pairs).
3) The electron pairs tend to take configurations far away from each other as much as possible. The repulsive force between two electron pairs is considered to become weaker in the following order:
(lone-pair − lone-pair) > (lone-pair − bond-electron pair) > (bond-electron pair − bond-electron pair),
   i.e., the order in symbol **(E − E) > (E − X) > (X − X)**. Consequently, two lone-pairs tend to be most apart from each other.
4) For each number of SN from 2 to 7, the molecular shapes are classified according to the number of lone-pairs and the number of bond-electron pairs as given in the table on the next page.

The most important thing is how to classify the electrons of the molecule under consideration, though there are many molecules whose molecular shapes cannot be predicted correctly based on the VSEPR model. (See **problem 1-59**.)

Solution 1-56

The molecular structures of atomic group ions can also be predicted by using the **VSEPR** model. For an alternative way of considering the molecular structures of these molecules, see **solution 1-49** .

(1) $\mathbf{CO_3^{-2}}$ C atom normally has four valence electrons but two negative charges on this ion are assigned to the C atom making the number of valence electrons six. With respect to the central C atom, then the number of bonds is 3 (i.e., six valence electrons) and the number of lone-pairs is 0. Therefore, CO_3^{-2} belongs to the group of AX_3 type with SN = 3 and is found to be a triangular plane ion from the table given below after **solution 1-57**.

Ans. triangular plane structure

(2) $\mathbf{NH_4^{+1}}$ N atom normally has five valence electrons, but for this ion, all three electrons of the three H atoms are added to N to make the total valence electrons of N eight. With respect to the central N atom, then

the number of bonds becomes 4 (i.e., eight valence electrons) and the number of lone-pairs is 0. Therefore, NH_4^{+1} belongs to the group of AX_4 type with $SN = 4$ and is found to be a tetrahedral ion from the table given below (after **solution 1-57**).

Ans. tetrahedral structure

(3) **NO_3^{-1}** By assigning a negative charge of the ion to the N atom, the number of its valence electrons is considered 6. With respect to the central N atom, the number of bonds becomes 3 (i.e., six valence electrons) and the number of lone-pairs is 0. Therefore, NO_3^{-1} is classified as the group of AX_3 type with $SN = 3$ in the table given below and is expected to be a triangular plane ion.

Ans. triangular plane structure

(4) **SO_4^{-2}** Two negative charges of the ion are assigned to the S atom and the number of its valence electrons is considered 8. With respect to the central S atom, then the number of bonds becomes 4 (i.e., eight valence electrons) and the number of lone-pairs is 0. Therefore, SO_4^{-2} belongs to the group of AX_4 type with $SN = 4$ and is expected to be a tetrahedral ion from the table given below.

Ans. triangular plane structure

(5) **PO_4^{-3}**, (6) **MnO_4^{-1}**, (7) **ClO_4^{-1}** Similar as in the case (4) above, the number of the valence electrons becomes eight for any central atom of the ions from (6) to (8). With respect to any central atom, therefore, the number of bonds is 4 and the number of lone-pairs is 0. Then, all the ions are classified as the group of AX_4 type with $SN = 4$ in the table given below and are expected to be tetrahedral ion molecules.

Ans. tetrahedral structure for all from (6) to (8)

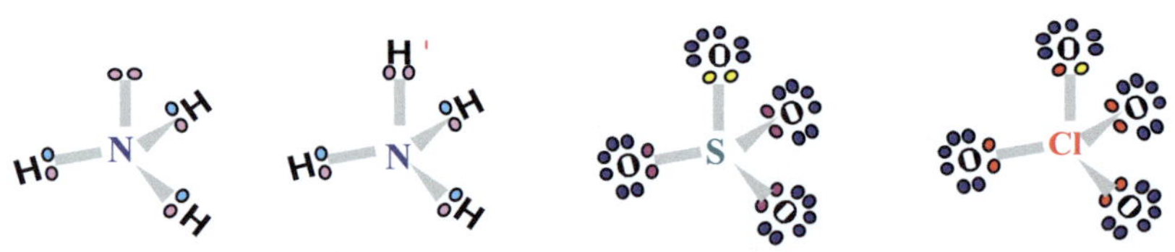

NH_3 NH_4^+ SO_4^{-2} ClO_4^-

Fig.53 Representation of lone-pair electrons and bond-electrons by dot electrons based on the octet rule.
- valence electrons of the central atom,
- added electrons,
- electron of H atom.

Solution 1-57

XeO_4 The number of the valence electrons of Xe is 8. Xe atom donates two electrons to each of the four O atoms so that each O atom can satisfy the octet rule. With respect to the central Xe atom, therefore, the number of bonds becomes 4 and the number of lone-pairs becomes 0. As a result, XeO_4 is classified into the group of AX_4 type with $SN = 3$. From the table given in the problem (or reposted below), XeO_4 is predicted to be a tetrahedral molecule. Since the central Xe atom is surrounded by eight electrons, it satisfies the octet rule. At a glance, **XeO_4** may appear to assume the same square structure as **XeF_4** but actually they are completely different in molecular structure.

Ans. tetrahedral structure

FBr_3 The number of the valence electrons of F is 7. The F atom is considered to donate one electron to each Br atom to make all three Br atoms satisfy the octet rule. In this case, the number of bonds is 3 and the

Table Molecular structures of the molecules of **AX_n** (n = 2-7) type predicted by the **VSEPR** model.

SN	Fundamental molecular structure and Number of lone-pairs			
	0	**1**	**2**	**3**
2	**X–A–X** linear			
3	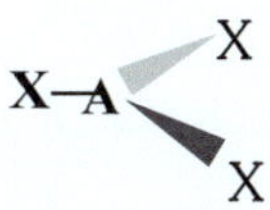 triangular plane	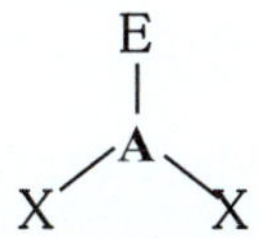 bent structure (in plane)		
4	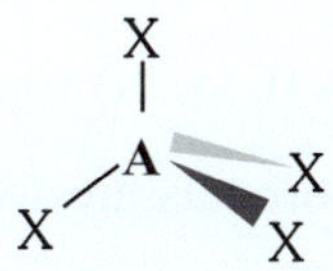 tetrahedral	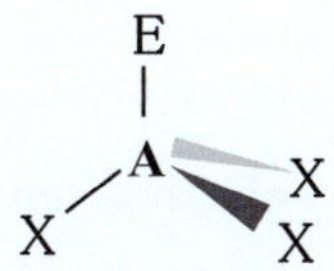 triangular pyramid	 bent structure (perpendicular)	
5	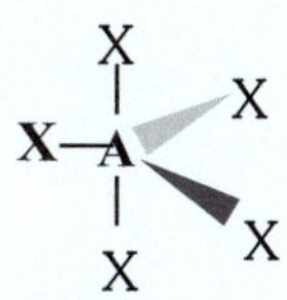 triangular bipyramid	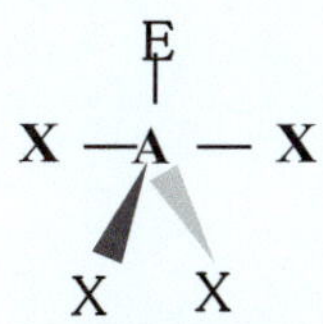 seesaw- shape	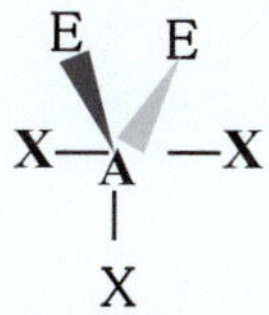 T-shape	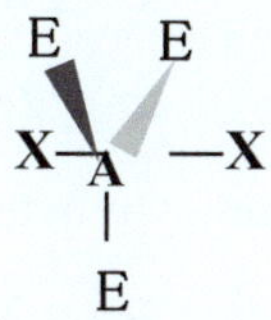 linear
6	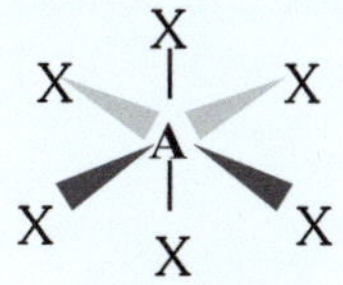 octahedral	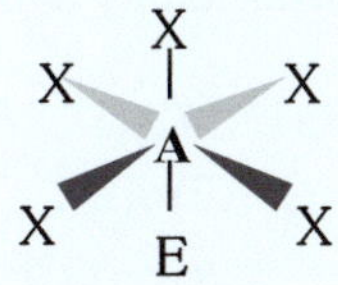 square pyramid	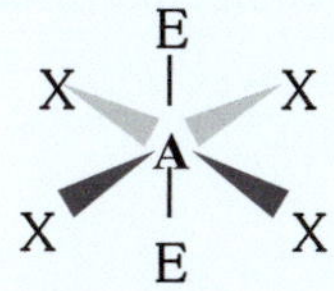 square	
7	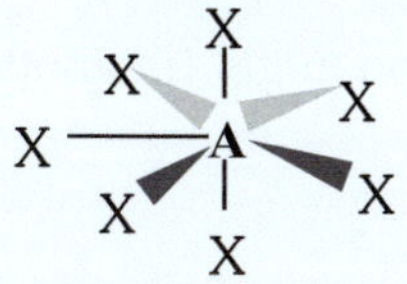 pentagonal bipyramid	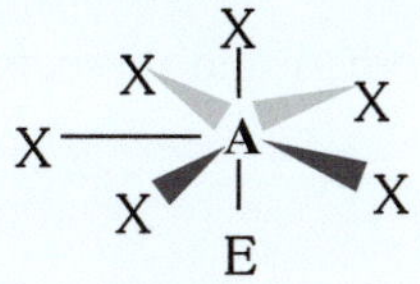 pentagonal pyramid		

"note" **X** stands for a chemical bond and **E** for a lone-pair. Multiple bonds between two atoms are counted as one bond. Valence electrons are illustrated for a few molecules by using dots of different colors not only to distinguish bond electron pairs from lone-pair electrons but also to clarify to which atom they originally belong.

The number of lone-pairs is 2 with respect to the central F atom. Therefore, FBr_3 is classified into the group of AX_3E_2 type with $SN = 5$. From the table given in the problem (reposted above), FBr_3 is predicted to be a T-shaped molecule. Then, the F atom is going to have 10 electrons in the outermost shell and does not satisfy the octet rule.

Ans. T-shaped structure

Solution 1-58

(a) To follow **problem 1-49**, CO_2 and NO_2 are examined here as examples in which *Walsh's* rule is applied to the molecules of the type of AX_2. We assume that the atomic orbitals involved in the bond formation are 2*s*- and 2*p*- orbitals of C or N atom and 2*p*-orbitals of O atom.

To start with, the molecular orbitals of a linear molecule of $O_a{-}A{-}O_b$ are formed by six atomic orbitals of two O atoms (three $2p_{O_a}$ and three $2p_{O_b}$) plus four atomic orbitals of atom A (C or N) (one $2s_A$ and three $2p_A$), in total, by ten atomic orbitals. In accordance with this, ten molecular orbitals are formed. They are classified according to the symmetry against the central atom A or against the molecular plane as follows.

totally symmetric

$$1\sigma : 2p_{O_a z} - 2p_{O_b z} + 2s_A$$

$$1\sigma^* : 2p_{O_a z} - 2p_{O_b z} - 2s_A$$

antisymmetric

$$2\sigma_p : 2p_{O_a z} - 2p_{O_b z} - 2p_{Az}$$

$$2\sigma_p^* : 2p_{O_a z} + 2p_{O_b z} + 2p_{Az}$$

antisymmetric against molecular plane (not classified by symmetry against the center)

$$\pi_{x1}: 2p_{O_a x} + 2p_{Ax} + 2p_{O_b x}, \qquad \pi_{y1}: 2p_{O_a y} + 2p_{Ay} + 2p_{O_b y}$$

$$\pi_{x2}: 2p_{O_a x} - 2p_{O_b x}, \qquad \pi_{y2}: 2p_{O_a y} - 2p_{O_b y}$$

$$\pi_{x3}: 2p_{O_a x} - 2p_{Ax} - 2p_{O_b x}, \qquad \pi_{y3}: 2p_{O_a y} - 2p_{Ax} - 2p_{O_b y}$$

As is seen, ten molecular orbitals (not normalized) in total are obtained.

Next, the molecular orbitals of a molecule bent at right angles 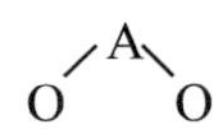 are classified by the symmetry against the plane which passes through the central atom A and dissects the bond angle (in another word, by the symmetry of C_{2v}).

symmetric

$$1\sigma : 2p_{O_a z} - 2p_{O_b x} + 2s_A$$

$$1\sigma^* : 2p_{O_a z} - 2p_{O_b x} - 2s_A$$

$$\pi_{y1}: 2p_{O_a y} + 2p_{Ay} + 2p_{O_b y}$$

π_{y3}: $2p_{\mathrm{O_a}y} - 2p_{\mathrm{A}x} - 2p_{\mathrm{O_b}y}$

π_{x1}: $2p_{\mathrm{O_a}x} + (2p_{\mathrm{A}x} - 2p_{\mathrm{A}x}) + 2p_{\mathrm{O_b}z}$

π_{x3}: $2p_{\mathrm{O_a}x} - (2p_{\mathrm{A}x} - 2p_{\mathrm{A}x}) - 2p_{\mathrm{O_b}z}$

antisymmetric

$2\sigma_p$: $2p_{\mathrm{O_a}z} - 2p_{\mathrm{O_b}x} - (2p_{\mathrm{A}z} + 2p_{\mathrm{A}x})$

$2\sigma_p^*$: $2p_{\mathrm{O_a}z} + 2p_{\mathrm{O_b}x} + (2p_{\mathrm{A}z} + 2p_{\mathrm{A}x})$

π_{y2} : $2p_{\mathrm{O_a}y} - 2p_{\mathrm{O_b}y}$

π_{x2} : $2p_{\mathrm{O_a}x} - 2p_{\mathrm{O_b}z}$

Again, ten orbitals in total are obtained. Those ten orbitals are classified into a_1, a_2, b_1 and b_2 in terms of the irreducible representations of C_{2v},. Except for σ^*- and $2\sigma_p^*$-orbitals, they are pictorially shown in Fig.54.

Ans. σ- and π-bonding orbitals: See text and Fig.54.

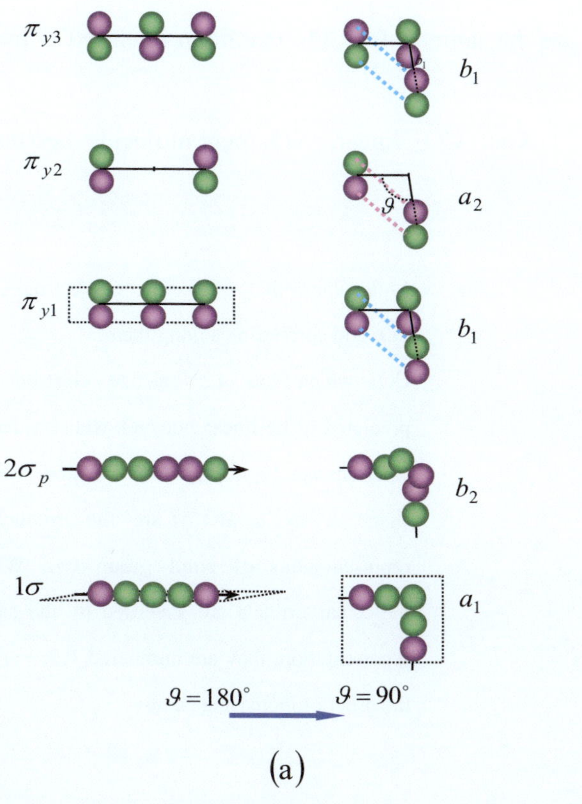

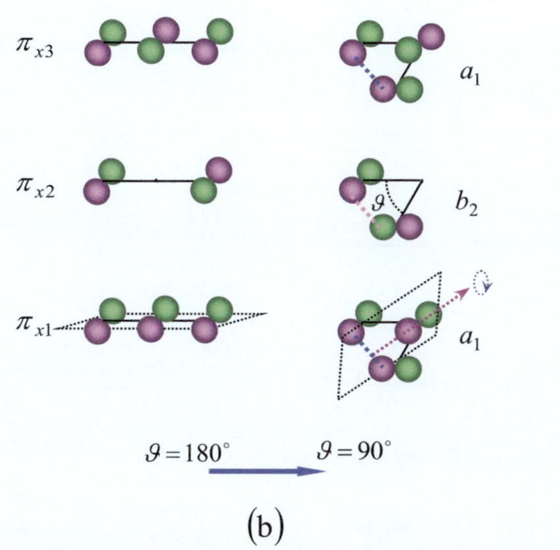

Fig.54 Molecular orbitals of linear and non-linear molecules of AO_2 type.

Variation of orbital energy against bond angle is visualized. For instance, in the case of $\pi_{x3} \rightarrow a_1$ the orbital energy decreases significantly because of an interaction favorable for bonding as is indicated by a blue dotted line.

(b) Thirdly, relative energy levels of two sets of ten orbitals are determined and then it is examined how each of ten orbitals of the linear molecule varies upon the variation of the bond angle from 180° to 90°. To do this, it is important to notice that the symmetry of a molecular orbital is conserved and any correlation line doe not cross each other if two orbitals are of the same symmetry (conservation of symmetry) upon the variation of the bond angle.

The energies of the molecular orbitals of the linear molecule are influenced greatly by changes in bond energies of O_a–A and O_b–A bonds and newly increasing interaction between O_a and O_b as the bond angle is decreased. For instance, in the case of a change upon $\pi_{x3} \rightarrow a_1$, a σ-bond like interaction between O_a *and* O_b newly arises as is shown by the dotted line in Fig.54(b). In addition, an antibonding character found in O_a-A and O_b-A is reduced in the non-linear molecule. Therefore, the energy of π_{x3} is considered to decrease greatly with decreasing the bond angel. Being similarly examined the other orbitals shown in Fig.54, though qualitatively, the energy level correlation diagram as shown in Fig.55 is obtained.

Based on the energy level correlation diagram thus obtained, 12 valence electrons are filled into the orbitals in the order of increasing energy and the total energies are compared for both shapes of molecule. For CO_2, the total energy is found lower for a linear shape and it is concluded that CO_2 is a linear molecule. In the case of CO_2^+, the number of the valence electrons is 11and, hence, a linear shape is more stable for this ion as well. On the other hand, NO_2 has 13 valence electrons and the 13th electron is assigned to either of π_{x3}- or $3a_1$-orbital as is indicated by red arrows but $3a_1$-orbital is located much lower than π_{x3}-orbital at 90°. As a result of this, NO_2 is concluded to be a bent molecule. However, since NO_2^+ has 12 valence electrons, one less than those of NO_2, the occupied molecular orbitals are the same as for CO_2, making a linear NO_2^+ more stable.

Ans. CO_2: linear, NO_2: bent molecule. See text.

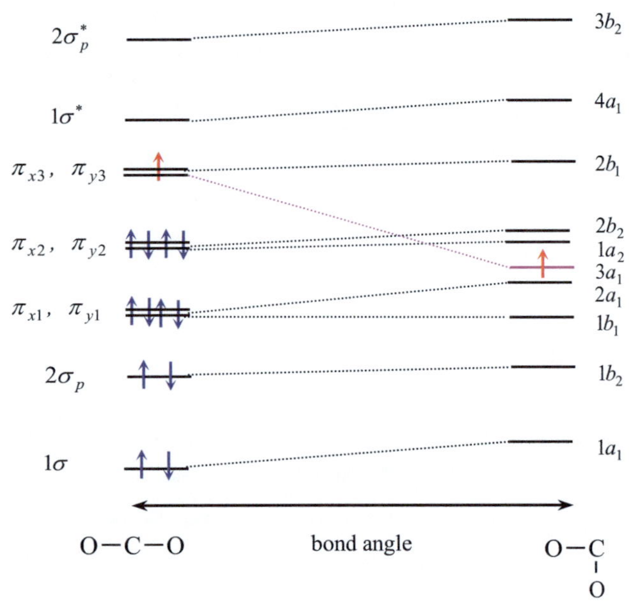

Fig.55 Molecular orbitals of the valence electrons of CO_2 and correlation among them. CO_2 which has 12 valence electrons is predicted to be linear but NO_2 whose valence electrons are 13 is expected to assume a bent shape. a_1, a_2, b_1 and b_2 are the irreducible representations of point group C_{2v}. When molecular orbitals are specified by the same representation, they are numbered 1, 2, ··· in the order of increasing energy.

Solution 1-59a

Walsh's rule predicts that CO_2, CO_2^+ and NO_2^+ are linear molecules and that NO_2 and NO_2^- are non-linear molecules. On the other hand, the **VSEPR** model tells that CO_2 and NO_2^+ are linear molecules and NO_2^- is a bent molecule. The model can be applied neither to CO_2^+ nor to NO_2. Hence, the rule and the model are consistent in prediction only for CO_2 and NO_2^+ (linear), and NO_2^- (bent).

Ans. CO_2 and NO_2^+ (linear), and NO_2^- (bent)

Solution 1-59b

It is essential to introduce hybrid orbitals in constructing molecular orbitals.

Accompanied by a variation of the bond angle ϑ from 180° to 90°, hybrid orbitals associated with the σ-bonds of H–O–H should be varied from *sp*- to sp^3-hybrid orbitals. Then, the energy level correlation diagram shown in Fig.57 is obtained. Based on this diagram, the examination on the molecular shape and its stability leads to a satisfactory agreement between predictions by *Walsh's* rule and those by the **VSEPR** model.

Molecular orbitals of H–O–H based on *sp*-hybrid orbitals. Classified by the symmetry of C_{2v}.

totally symmetric (a_1)

$$1\sigma\ :\ s_{H_a} + 1s_{H_b} + 2s_O \times 2 = 1s_{H_a} + 1s_{H_b} + (2s_O + 2p_Z) + (2s_O - 2p_Z)$$
$$= 1s_{H_a} + (2s_O + 2p_Z) + 1s_{H_b} + (2s_O - 2p_Z)$$

$$1\sigma^*\ :\ 1s_{H_a} + 1s_{H_b} - 2s_O \times 2 = 1s_{H_a} + 1s_{H_b} - (2s_O + 2p_Z) - (2s_O - 2p_Z)$$
$$= 1s_{H_a} - (2s_O + 2p_Z) + 1s_{H_b} - (2s_O - 2p_Z)$$

antisymmetric (b_2)

$$2\sigma_p\ :\ s_{H_a} - 1s_{H_b} - 2p_z \times 2 = 1s_{H_a} - 1s_{H_b} - (2s_O + 2p_Z) + (2s_O - 2p_Z)$$
$$= 1s_{H_a} - (2s_O + 2p_Z) - 1s_{H_b} + (2s_O - 2p_Z)$$

$$2\sigma_p^*\ :\ 1s_{H_a} - 1s_{H_b} + 2p_z \times 2 = 1s_{H_a} - 1s_{H_b} + (2s_O + 2p_Z) - (2s_O - 2p_Z)$$
$$= 1s_{H_a} + (2s_O + 2p_Z) - 1s_{H_b} - (2s_O - 2p_Z)$$

Besides those given above, there are two more 2*p*-orbitals.

$$2p_x(a_1),\ 2p_y(b_1)$$

As is clear from the rewritten formulae above (second line for each), those molecular orbitals are found to be essentially the same in formulae as the molecular orbitals constructed from *s*- and *p*-orbitals so far discussed. (See **solution 1-50.**)

Molecular orbitals of 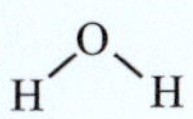 based on sp^3-hybrid orbitals. Classified by the symmetry of C_{2v}.

totally symmetric (a_1)

$$1\sigma\ :\ 1s_{H_a} + 1s_{H_b} + h_1 + h_2$$

$$1\sigma^*\ :\ 1s_{H_a} + 1s_{H_b} - h_1 - h_2$$

$$2\sigma_1\ :\ 1s_{H_a} + 1s_{H_b} - h_3 - h_4$$

$$2\sigma_1^*\ :\ 1s_{H_a} + 1s_{H_b} + h_3 + h_4$$

Besides those given above, there are two more non-bonding orbitals (n_1 and n_2) which cannot be classified into any irreducible representation of C_{2v}.

$$n_1,\ n_2\ :\quad h_3 + 1s_{H_a} + 1s_{H_b},\ \ h_4 + 1s_{H_a} + 1s_{H_b}$$

Here、$h_i (i = 1,2,3,4)$ indicates one of the sp^3-hybrid orbitals. (See **solution 1-47.**) The relation between the coordinates and the hybrid orbitals is shown in Fig.56. None of the hybrid orbitals can be described in terms of the irreducible representations of C_{2v}, however, by combining two of them, it is possible to construct molecular orbitals of appropriate irreducible representations as are given above.

From Fig.57, it is concluded that H_2O is not a linear molecule but a bent molecule with two lone-pairs.

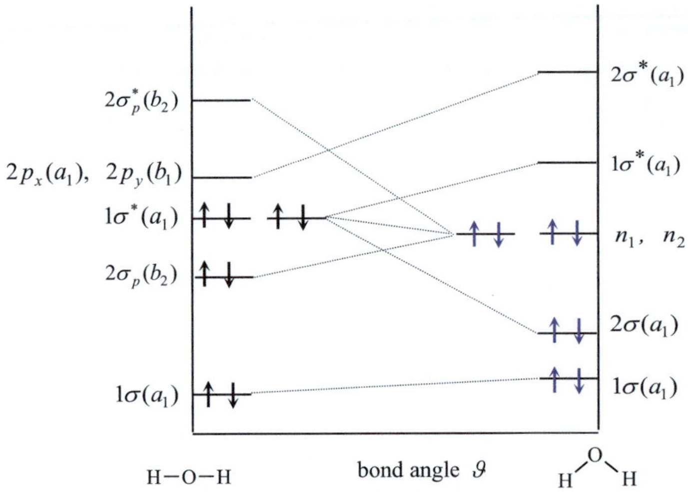

Fig.56 Energy level correlation with bond angle. It should be noted that there are two non-bonding orbitals (n_1 and n_2).

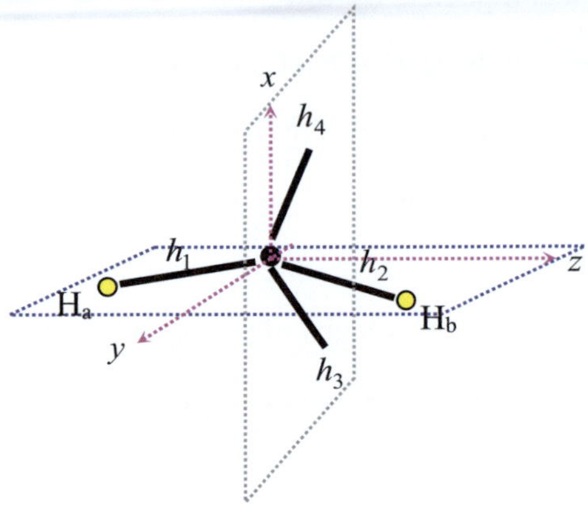

Fig.57 Components and their directions of sp^3-hybrid orbitals used to construct molecular orbitals of H_2O.

Ans. See text and Fig.57.

Solution 1-60

(a) Definition of electronegativity and its formula.

Electronegativity is a measure to know how strongly bonding electrons are attracted to either of two atoms of different nature. The idea was introduced by *Linus Pauling* for the first time (The author of *The Nature of the Chemical Bond*, Novel Prize in Chemistry in1954). Another measure is available which was defined by *Robert S.Mulliken* afterwards (Novel Prize in Chemistry in1966) on the basis of molecular orbital theory. They are defined by eq(84) and eq(85), respectively,

$$|\chi_A - \chi_B| = 0.102\left\{E_{A-B} - \frac{1}{2}(E_{A-A} + E_{B-B})\right\} \qquad (84)$$

$$\chi_{A,M} = \frac{1}{2}(I_A + EA_A) \qquad (85)$$

where I_A and EA_A are the ionization potential and the electron affinity of atom A.

Ans. Definition of electronegativity: See text.

(b) The given value for HCl is inserted into eq(85) to yield a value below for the electronegativity of Cl.

$$|2.2 - \chi_B| = 0.102\left\{428 - \frac{1}{2}(432 + 239)\right\} \Rightarrow \chi_B = 2.2 + 0.981 = 3.2$$

Ans. electronegativity of Cl: 3.2

(c) Ionic character of chemical bond and dipole moment.

According to *Pauling*, the ionic character (%) of the chemical bond is defined by eq(86) (eq(9) in the problem) in terms of elctronegativity.

$$16|\chi_A - \chi_B| + 3.5|\chi_A - \chi_B|^2 \qquad (86)$$

By substituting the value for χ_B obtained in (b) above into eq(86), a value of 19.5 is found and so the ionic per centage of the H-Cl bond is 19.5%. Most of you may have expected much higher ionic character of the H-Cl

$$16|2.2-3.2|+3.5|2.2-3.2|^2=19.5\%$$

bond but in fact a polar nature of this bond is found surprisingly low.

The dipole moment μ_{HCl} of HCl can be calculated from the apparent charge q on the H or Cl atom estimated from the ionic percentage of the HCl bond and the bond distance R as below, where D is *Debye* unit.

$$\begin{aligned}\mu_{\mathrm{HCl}} &= q\times R\\ &= 0.195\times e\times 1.27\times 10^{-10}/3.34\times 10^{-30}\\ &= 0.195\times 1.60\times 10^{-19}\times 1.27\times 10^{-10}/3.34\times 10^{-30}\\ &= 1.19\ \mathrm{D}\end{aligned}$$

Meanwhile, the electronegativity of Cl defined by *Mulliken* is found to be $\chi_{\mathrm{Cl},M}=8.29$ by substituting the values of $I_{\mathrm{Cl}}=12.97(\mathrm{eV})$ and $EA_{\mathrm{Cl}}=3.62(\mathrm{eV})$ into eq(86). Then from the equation which relates *Pauling's* value to that of *Mulliken* (**advanced problem** below) $\chi_{\mathrm{Cl}}=2.046\times\sqrt{\chi_{\mathrm{Cl},M}}-2.541$, a value of 3.35 is obtained. It is, therefore, found that there exists a comparatively good correlation between these two independently defined electronegativities.

Ans. ionic character: 19.5%, dipole moment: 1.19 D

[advanced problem] Employing the values of the ionization energy and the electronaffinity known for the elements in the second and third periods (E=C~F and Si~Cl), find the best fitted line which correlates *Pauling*'s χ_E with the root mean square of *Mulliken*'s electronegativity $\sqrt{\chi_{E,M}}$ mentioned above.

Solution 1-61a

(a) In the nuclear reactions, the mass number and the proton number of an atom are given as a superscript and a subscript, respectively. The sum of the mass number(s) for the reactant(s) on the left must be equal to that for the product(s) on the right. The same is the case for the sum of the proton numbers.

Ans. $^{226}_{88}\mathrm{Ra}\rightarrow\,^{[222]}_{[86]}\mathrm{Rn}+\,^{4}_{[2]}\mathrm{He}$

(b)The total mass defect Δm accompanied by the nuclear reaction is found as 0.0052 amu.

$$226.0254-(222.0176+4.0026)=0.0052\ \mathrm{amu}$$

The energy associated with this mass defect is assumed to be converted as a whole into the kinetic energy of the α -particle, yielding eq(87).

$$\Delta mc^2=\frac{1}{2}m_{\mathrm{He}}(v_{\mathrm{He}})^2 \qquad (87)$$

Then the velocity v_{He} of the α -particle is found as below.

$$v_{\mathrm{He}}=\left(\frac{2\Delta mc^2}{m_{\mathrm{He}}}\right)^{\frac{1}{2}}=\left(\frac{2\times 0.0052}{4.0026}\right)^{\frac{1}{2}}\times 2.998\times 10^8=1.53\times 10^7\,\mathrm{m\cdot s^{-1}}$$

Ans. particle velocity : $1.53\times 10^7\,\mathrm{m\cdot s^{-1}}$

(c) The rate equation of atomic disintegration is given by eq(88), where τ is the half-life.

$$\frac{d[N_{\mathrm{Ra}}]}{dt}=-\frac{\ln 2}{\tau}\times[N_{\mathrm{Ra}}] \qquad (88)$$

When τ is expressed in second, the number of the Ra atoms which disintegrate in second is given by the equation below and by equating $dt = 1\,\mathrm{s}$, the number is found as 8.27×10^{12}.

$$-d[N_{\mathrm{Ra}}] = \frac{\ln 2}{\tau}\times[N_{\mathrm{Ra}}]dt$$

$$= \frac{\ln 2}{1600\times365\times24\times60\times60}\times6.02\times10^{23} = 8.27\times10^{12}$$

Ans. the number of particles emitted: 8.27×10^{12}

Solution 1-61b

(a) The rate equation of disintegration of $^{40}\mathrm{K}$ is given by eq(89) with its rate constant expressed by $k_{\deg}$. Here, a general relation $t_{1/2} = \ln 2/k_{\deg}$ which holds between rate constant and half-life $t_{1/2}$ is employed to replace $k_{\deg}$ with $t_{1/2}$.

$$^{40}\mathrm{K} \xrightarrow{t_{1/2}=12.8\times10^{9}\,\mathrm{y}} {}^{40}\mathrm{Ca} + \beta(e) \qquad \text{(eq(10) in the problem)}$$

$$\frac{d^{40}\mathrm{K}}{dt} = -k_{\deg}\times{}^{40}\mathrm{K} = -\frac{\ln 2}{t_{1/2}}\times{}^{40}\mathrm{K} \qquad (89)$$

Eq(89) expresses nothing more than the number of atoms disintegrated per unit time and hence by substituting a value for the half-life in the unit of second, the number of disintegration per second is found as below.

$$\frac{\ln 2}{t_{1/2}}\times{}^{40}\mathrm{K} = \frac{\ln 2}{12.8\times10^{8}(\mathrm{y})\times365\times24\times60\times60(\mathrm{s/y})}\times\frac{1.0(\mathrm{g})\times0.0117(\%)\times10^{-2}}{39.10(\mathrm{g/mol})}\times6.02\times10^{23}(\mathrm{number/mol})$$

$$= 30.9 \ \ \mathrm{Bq} \ \ (= \mathrm{number/s})$$

For the sample in question a *Geiger* counter will respond about thirty one times a second.

Ans. number of disintegration of $^{40}\mathrm{K}/\mathrm{kg}$: 30.9 Bq

"note" In fact, the radioactive decay of $^{40}\mathrm{K}$ is consisted of two parallel disintegration processes and radiations of different kinds are emitted. However, since the half-life is common for any parallel processes, the number obtained above remains correct irrespective of single or parallel processes of radiation. (See (b) below.)

(b) In calculating the number of disintegration, it should be taken into account that two kinds of radiation are emitted in this case. Based on the radioactive decay processes of eqs(90) and (91), the rate equations can be written as eqs(4) and (5), where relevant rate constants are employed instead of half-life.

$$^{137}\mathbf{Cs} \xrightarrow{k_1} {}^{137\mathrm{m}}\mathbf{Ba} + \mathrm{e} \xrightarrow{k_3} {}^{137}\mathbf{Ba} + \gamma \qquad (90)$$

$$^{137}\mathbf{Cs} \xrightarrow{k_2} {}^{137}\mathbf{Ba} + \mathrm{e} \qquad (91)$$

$$\frac{d^{137}\mathbf{Cs}}{dt} = -(k_1 + k_2)\times{}^{137}\mathbf{Cs} \qquad (92)$$

$$\frac{d^{137\mathrm{m}}\mathbf{Ba}}{dt} = k_1\times{}^{137}\mathbf{Cs} - k_3\times{}^{137\mathrm{m}}\mathbf{Ba} \qquad (93)$$

As solutions of the rate equations, eqs(94) and (95) are obtained in which the initial conditions of $^{137}\mathbf{Cs}\,(t=0) = A_0$ and $^{137\mathrm{m}}\mathbf{Ba}\,(t=0)=0$ are employed. A_0 is the amount of $^{137}\mathbf{Cs}$ in the unit of number at $t=0$.

$$^{137}\mathbf{Cs} = A_0 e^{-(k_1+k_2)t} \qquad (94)$$

$$^{137\mathrm{m}}\mathbf{Ba} = \frac{k_1 A_0}{k_1+k_2-k_3}\left\{ e^{-k_3 t} - e^{-(k_1+k_2)t} \right\} \qquad (95)$$

From these equations, the number of disintegration per second is given by eq(96) and since $k_3 >> (k_1+k_2)$, its approximate value is given by eq(96').

$$(k_1+k_2)\times{}^{137}\mathbf{Cs} + k_3\times{}^{137\mathrm{m}}\mathbf{Ba} = (k_1+k_2)\times A_0 e^{-(k_1+k_2)t} + k_3 \times \frac{k_1 A_0}{k_1+k_2-k_3}\left\{ e^{-k_3 t} - e^{-(k_1+k_2)t} \right\} \qquad (96)$$

$$\approx (2k_1+k_2)\times A_0 e^{-(k_1+k_2)t} \qquad (96')$$

On the other hand, at $t=0$ eq (97) is derived form eq(96).

$$(k_1+k_2)\times{}^{137}\mathbf{Cs} + k_3\times{}^{137\mathrm{m}}\mathbf{Ba} = (k_1+k_2)\times A_0 \qquad (97)$$

From the relation between the rate constants and the half-life, the values for the rate constants are obtained and then inserted into eq(97) to yield the value of 3.21 TBq as the number of disintegration of Cs/g.

$$(k_1+k_2) = \frac{\ln 2}{30.1(\mathrm{y})\times 365\times 24\times 60\times 60(\mathrm{s/y})}$$

$$(k_1+k_2)\times A_0 = \frac{\ln 2}{30.1(\mathrm{y})\times 365\times 24\times 60\times 60(\mathrm{s/y})} \times \frac{1.0(\mathrm{g})}{137(\mathrm{g/mol})} \times 6.02\times 10^{23}(\mathrm{numbers/mol})$$

$$= 3.21\times 10^{12}\,\mathrm{Bq} = 3.21\,\mathrm{TBq}$$

However, this is not a correct way of calculation, because the amount of γ-radiation is not taken into account. To do correctly, it is evident that eq(96) or eq(96') should be employed. The amount of radiation to be emitted 0.1 days later is found as 6.098 TBq from eq(96), being approximately two times larger than that found above. Because of the long half-life of Cs, the amount of radiation hardly changes even after one year (5.960 TBq), only being a little more than 2% decrease in a year.

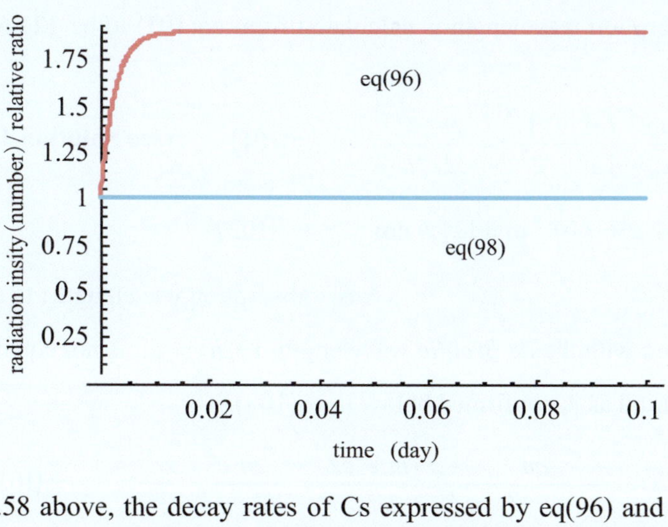

Fig.58 Time dependence (day) of the radiation intensity from ^{137}Cs in the unit of number/second. The intensity is normalized to that at t=0 (eq(98), 3.21 TBq).

In Fig.58 above, the decay rates of Cs expressed by eq(96) and eq(98) in the latter of which the process of γ-radiation is not taken into account are compared.

$$(k_1+k_2)\times{}^{137}\mathbf{Cs} + k_3\times{}^{137\mathrm{m}}\mathbf{Ba} = (k_1+k_2)\times A_0 e^{-(k_1+k_2)t} \qquad (98)$$

Ans. $6.10\times 10^{12}\,\mathrm{Bq} = 6.10\,\mathrm{TBq}$

The derivation of the rate equations associated with the present problem is discussed in detail in Chapter 6 (for instance, see **problem 6-37).** Therefore, only the simplest procedures to derive eq(95) are given below.

$$^{137m}\mathbf{Ba} = f(t)\times e^{-k_3 t} \qquad \text{(i)}$$

$$f'(t)\times e^{-k_3 t} - k_3 f(t)\times e^{-k_3 t} = k_1 A_0 e^{-(k_1+k_2)t} - k_3 f(t)\times e^{-k_3 t} \qquad \text{(ii)}$$

$$f'(t) = k_1 A_0 e^{-(k_1+k_2-k_3)t} \qquad \text{(iii)}$$

$$f(t) = -\frac{k_1 A_0}{k_1+k_2-k_3} e^{-(k_1+k_2-k_3)t} + C \qquad \text{(iv)}$$

$$f(0)=0 \;\Rightarrow\; C = \frac{k_1 A_0}{k_1+k_2-k_3} \qquad \text{(v)}$$

Solution1-62

(a) Employing the result obtained in the **problem 1-16**, the wavelength is calculated. As has already been explained, the radius of the orbital of the electron and its velocity are related by eq(99).

$$r = \frac{\varepsilon_0 h^2 n^2}{\pi m_e e^2},\quad \upsilon = \frac{nh}{2\pi m}\cdot\frac{\pi m_e e^2}{\varepsilon_0 h^2 n^2} = \frac{e^2}{2\varepsilon_0 hn} \qquad (99)$$

Substituting these relations into eq(100) which is the definition of the *de Broglie* wavelength, the relation between λ and the orbital radius r is derived as below, that is, the wavelength multiplied by integer n $(n\lambda)$ is equal to the circumference $2\pi r$ of the *ns*-orbital.

$$\lambda = \frac{h}{m_e \upsilon} \qquad (100)$$

$$= \frac{h}{m_e \times \dfrac{e^2}{2\varepsilon_0 hn}} = \frac{2\varepsilon_0 h^2 n}{m_e e^2} = \frac{2\pi r}{n}$$

Ans. radius of *ns*-orbital vs. *de Broglie* wavelength: $\lambda = 2\pi r / n$

(b) The electronic transition from the lowest to the second lowest energy level induced by the light absorption is due to the transition of $1s \rightarrow 2p$ according to the selection rule. For the hydrogen atom, the energies of the 2*s*- and 2*p*-orbitals are equal, so the absorption wavelength is calculated from eq(101) to be 121.6 nm as shown in eq(102).

$$h\nu = E(2p) - E(1s) = \frac{m_e e^4}{8\varepsilon_0^2 h^2}\left(\frac{1}{1^2} - \frac{1}{2^2}\right) = \frac{3m_e e^4}{32\varepsilon_0^2 h^2} \qquad (101)$$

(see **Solution 1-11**)

$$\lambda_{1s\rightarrow 2p} = \frac{c}{\nu} = \frac{32\varepsilon_0^2 h^3 c}{3m_e e^4} = 1.216\times 10^{-7}\,\text{m} = 121.6\ \text{nm} \qquad (102)$$

Ans. absorption wavelength: 121.6 nm

"note" Needless to say, the energy associated with the *de Broglie* wavelength, i.e. $h\nu = E$, is not equal to the total energy of the electron in the relevant orbital as is confirmed below in eq(103).

$$h\nu = h\times\frac{c}{\lambda} = h\times\frac{cn}{2\pi r} = h\times\frac{cn}{2\pi\times\dfrac{\varepsilon_0 h^2}{\pi m_e e^2}\cdot\dfrac{n^2}{Z}} = \boxed{\frac{m_e e^2 cZ}{2h\varepsilon_0 n} \neq -\frac{m_e e^4}{8\varepsilon_0^2 h^2}\cdot\frac{Z^2}{n^2}} \qquad (103)$$

Solution 1-63

When the center of the central carbon atom is taken as the origin (see **solution 1-65**), the component of the dipole moment along the y-axis is readily known to be zero from the picture shown below in Fig59. The component along the x-axis is calculated as follows in eq(104).

$$\mu_x = \left\langle 0.281\times(-140\times10^{-12}) + 2\left\{-0.110\times140\times10^{-12}\times\cos\left(\frac{3\pi}{10}\right)\right\} + 2\left\{-0.141\times\frac{140}{2}\times10^{-12}\times\tan\left(\frac{2\pi}{5}\right)\right\}\right\rangle$$

$$\times1.602\times10^{-19}$$

$$= -1.89\times10^{-29}\ \mathrm{C\cdot m} = -5.68\ \mathrm{D} \qquad (104)$$

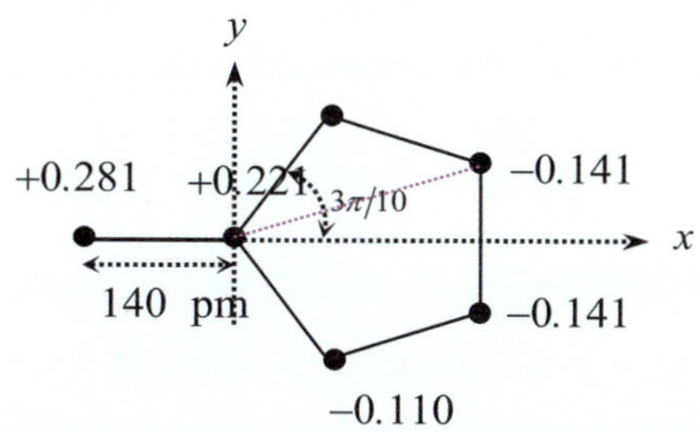

Fig.59 Coordinates to calculate the dipole moment of fluvene in the ground state.

Judging from the obtained value of –5.68D, the dipole moment of this molecule is said to be oriented along the negative direction of the x-axis defined in the figure above, that is, from the ring toward the substituent. To answer the question about the magnitude of a dipole moment, its absolute value, i.e. 5.68 D in the present case, is normally used. (See **solution 1-64**.)

Ans. from the ring toward the substituent, 5.68 D

Solution 1-64

(a) Definition of dipole moment.

When a positive and a negative point charge denoted by $+q$ and $-q$, respectively, are located with the distance of r apart, the dipole moment is defined as a three dimensional vector directing from the negative point charge toward the positive point charge and its magnitude is given by $q\times r$ (see **solution 1-65**). In terms of molecular terminology, q and r are of the magnitude of $e=1.602\times10^{-19}$ C and 10^{-10} m, respectively.

moment used to be defined by their product of $\mathrm{esu}\cdot\text{Å}$. We continue to use that historical unit. D (*Debye*) defined below is equivalent to $\mathrm{esu}\cdot\text{Å}$ and is now employed as the unit of the dipole moment ($\mathrm{C\cdot m}$).

$$1\ \mathrm{D} = 1\ \mathrm{esu}\cdot\text{Å} = \frac{e}{4.8}(\mathrm{C})\times10^{-10}(\mathrm{m}) = 3.33564\times10^{-30}\,\mathrm{C\cdot m}$$

(*P.Debye*, Nobel Prize in Chemistry in1936).

Ans. definition of dipole moment: See text.

(b) If we denote the dipole moment of one H – O by X, its contribution to the dipole moment of the H_2O molecule is given by $\mathrm{X}\times\cos60^\circ$. Therefore, the total dipole moment of the molecule is twice of this and is equal to 1.84D, i.e., $2\times\mathrm{X}\times\cos60^\circ = 1.84\ \mathrm{D}$. From this, X is found equal to 1.84 D.

Ans. dipole moment of H – O bond: 1.84 D

(c) The most plausible molecular structure of H_2O_2 is the one shown below (*Hertzberg*, vol.2, p.301).

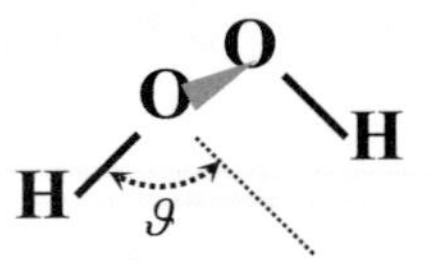

Fig.60 Most plausible structure of H_2O_2.

The angle ϑ spanned by the two $H-O$ bonds seems flexible and if it is fixed, the molecular structure is uniquely determined as far as the bond angle $\angle H-O-O$ is fixed as well. If the angle is taken equal to $\angle H-O-O=90^\circ$, then the dipole moment of H_2O_2 is given by $1.84\ \mathrm{D}\times(\cos\vartheta/2)\times 2=3.19\ \mathrm{D}$, from which the angle ϑ is calculated to be $\vartheta=59.8^\circ$ via $\cos\vartheta/2=0.8668$.

Ans. A twisted structure with the bond angle of 59.8° spanned by the two $H-O$ bonds (point group C_2)

Solution 1-65

General definition of dipole moment.

In general, the dipole moment is defined by a point charge distributed at the distance $\boldsymbol{r}_i$ (or charge density distribution $\rho(\boldsymbol{r}_i)$) measured from an arbitrarily taken coordinate origin by eq(105). It should be noted here that both of $\boldsymbol{\mu}$ and $\boldsymbol{r}_i$ are vectors.

$$\boldsymbol{\mu}=\sum_i q_i\times\boldsymbol{r}_i \qquad (105) \qquad (\boldsymbol{\mu}=\int\rho(\boldsymbol{r}_i)\times\boldsymbol{r}_i d\tau)$$

The sum of charges is given by eq(106) and for electronically neutral molecules the sum is zero.

$$Q=\sum_i q_i \qquad (106)$$

In such cases, it is readily proved that the magnitude of the dipole moment does not depend on where the coordinate origin is taken. To show this, let us take a new origin at the distance $\boldsymbol{R}$ from the old origin. Then the distance $\boldsymbol{r}_i'$ from the new origin to the point charge q_i is given by eq(107).

$$\boldsymbol{r}_i'=\boldsymbol{r}_i+\boldsymbol{R} \qquad (107)$$

The dipole moment $\boldsymbol{\mu}'$ defined on the basis of the new origin is given by the first equation of eq(105') according to the definition of the dipole moment (eq(105)), which can be rewritten as the last equation of eq(105').

$$\boldsymbol{\mu}'=\sum_i q_i\times\boldsymbol{r}_i'=\sum q_i\times(\boldsymbol{r}_i+\boldsymbol{R})=\sum q_i\times\boldsymbol{r}_i+\sum q_i\times\boldsymbol{R} \qquad (105')$$

In the case of a neutral molecule, the last term of eq(105') can be deleted because the sum of charges is zero. Therefore, eq(108) is derived which shows the magnitude of the dipole moment of a neutral molecule does not depend on where the coordinate origin is taken. Consequently, the dipole moment of the neutral molecule can be regarded as a physical property intrinsic to the molecule.

$$\boldsymbol{\mu}'=\boldsymbol{\mu} \qquad (108)$$

On the other hand, in the case of charged molecules the dipole moment defined according to eq(105) comes to depend on where the coordinate origin is taken. For the charged molecules, therefore, dipole moments even when defined should be carefully treated.

Ans. Cannot be defined as a meaningful physical property. See text.

"note" Direction of dipole moment

The direction of a dipole moment is evident from its definition by eq(105) and as is shown in Fig.61 below, is a vector directing from the negative charge toward the positive charge. Naturally, this is the direction which has long been employed in most text books of physics or physical chemistry.

Fig.61 Direction of a dipole moment.

Curiously enough, however, the exactly opposite direction has been practiced in organic chemistry. For instance, the dipole moment originating from bond polarization of CCl_4 is illustrated in many text books of organic chemistry as is shown in Fig.62. The reason for that is not clear, but inconsistency of this kind is not preferable in learning chemistry. A correct way of representing the dipole moment of CCl_4 is to write it as is shown in Fig.63 and a partial charge on the carbon atom should be written as $+4\delta$.

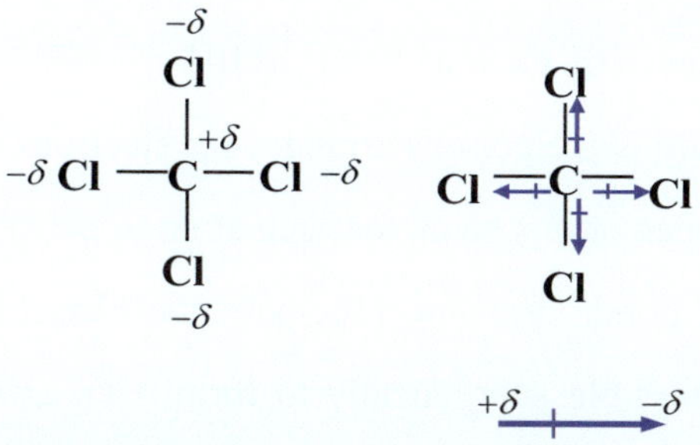

Fig.62 Representation of polarization and dipole moment actually used in organic chemistry.

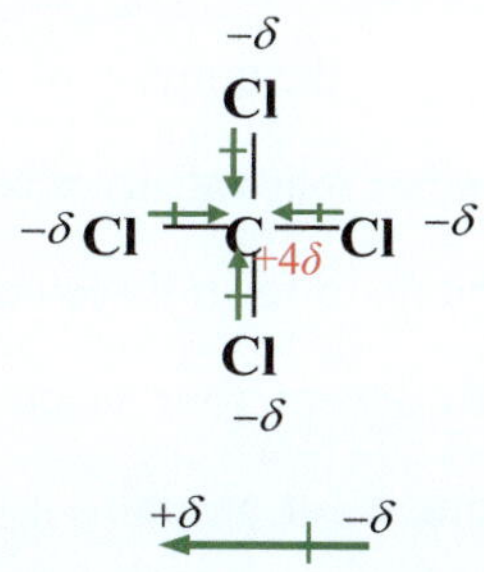

Fig.63 Recommended representation against Fig.62.

Solution 1-66

Fig.64 is a *Newman* diagram to show a steric configuration of 1,2-dichloroethane and Fig.65 illustrates three plausible stereoisomers of the compound.

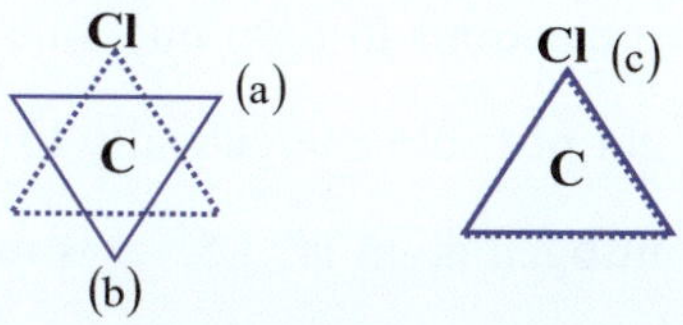

Fig.64 *Newman* diagram to show three kinds of steric configurations of 1,2-dichloroethane. Front positions of Cl are indicated by (a), (b) and (c).

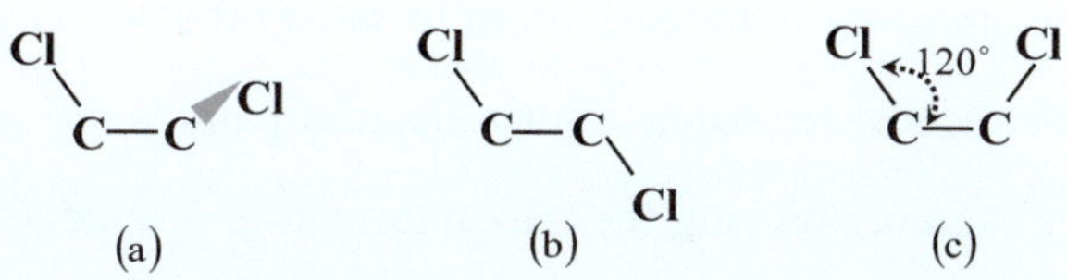

Fig.65 Three stereoisomers of 1,2-dichloroethane. For (b) and (c), all atoms are in the same plane.

From the measured dipole moments, the compound is considered to take a molecular configuration which makes the dipole moment larger at the higher temperatures. For the configuration shown in (b), the dipole moment is zero. Assuming the dipole moment for the configuration in (a) is 1.27 D, the dipole moment of the C – Cl bond μ_{C-Cl} is calculated to be 0.847 D as follows in eq(109).

$$\mu_{C-Cl} = 1.27/(2\times\cos30^\circ\times\cos30^\circ) = 1.27/\left(2\times\frac{3}{4}\right) = \frac{1.27\times2}{3} = 0.847\ \text{D} \qquad (109)$$

Employing this value, the dipole moment of the molecule which takes the configuration (c) is found as 1.47 D from the calculation below, which shows a reasonable agreement with the value measured at 315 °C.

$$2\times\mu_{\text{C-Cl}}\times\cos 30^\circ = 2\times 0.864\times\frac{\sqrt{3}}{2} = 1.47\ \text{D}$$

Consequently, the molecule is considered to come to assume the configuration in (c) at higher temperatures.

Ans. See text.

"note" The temperature dependence of the dielectric constant of 1,2-dichloroethane is positive, i.e., $(\partial\varepsilon/\partial T)_P > 0$, suggesting that its dipole moment becomes larger at higher temperatures. In contrast, the orientational dielectric constant normally decreases with temperature. (See **problem 5-40b.)**

Solution 1-67

Molecular structures of_CO_3^{2-}, NO_3^-, SO_4^{2-} and PO_4^{3-} can be predicted roughly to be either spherical (tetrahedral in the present case) or planer, by employing the sp^2- or sp^3-hybrid orbitals and applying the octet rule to each atom. When we write the oxidation number of the central atom of an ion as X and take that of oxygen equal to -2, the valence of the ion is given by eq (110).

$$\text{the number of oxygen atoms}\times(-2)+\text{X} = \text{valence of the ion} \qquad (110)$$

CO_3^{2-}: C atom has four valence electrons. C^{2-} formed by accepting two electrons arranges six electrons in its sp^2-hybrid orbitals to form three equivalent bonds with three oxygen atoms so as for each atom to satisfy the octet rule. (To accept four electrons to form three sp^3-hybrid bonds and one lone-pair, the electrostatic repulsion among electron pairs is too large. For this, it is not possible energetically to form CO_3^{4-}. Also, knowing the maximum value of oxidation number of the carbon atom is $+4$, then it is readily derived from eq(110) that the valence of the ion to be formed from one carbon atom and three oxygen atoms must be -2.)

NO_3^-: N atom has five valence electrons. N^- formed by accepting one electron comes to have six valence electrons and arranges six electrons in its sp^2-hybrid orbitals to form three equivalent bonds with three oxygen atoms so as for each atom to satisfy the octet rule. (To accept three electrons to form one lone-pair, the electrostatic repulsion among electron pairs is too large. For this, it is not possible energetically to form NO_4^{3-}. Again, knowing the maximum value of oxidation number of the nitrogen atom is $+5$, it is readily derived from eq(110) that the valence of the ion to be formed from one nitrogen atom and three oxygen atoms must be -1.)

SO_4^{2-}: S atom has six valence electrons. S^{2-} formed by accepting two electrons comes to have eight valence electrons and arranges eight electrons in its sp^3-hybrid orbitals to form four equivalent bonds with four oxygen atoms so as for each atom to satisfy the octet rule. (Knowing the maximum value of oxidation number of the sulfur atom is $+6$, then it is readily derived from eq(110) that the valence of the ion to be formed from one sulfur atom and four oxygen atoms must be -2.)

PO_4^{3-}: P atom has five valence electrons. P^{3-} formed by accepting three electrons comes to have eight valence electrons and arranges them in its sp^3-hybrid orbitals to form four equivalent bonds with four oxygen atoms so as for each atom to satisfy the octet rule. (It is evident from eq(110) that the valence of the ion to be formed from one phosphorus atom and four oxygen atoms must be -3.)

It is concluded from the explanations given above that in order for the ions under consideration to have equivalent bonds and for each atom to satisfy the octet rule, their valence should be $-2, -1, -2$ and -3, in the order of explanation.

Alternative explanations can be given by using the **VSEPR** model described in **solution 1-54** and we can answer immediately whether the ions under consideration assume a planer or a tetrahedral structure. However, it is indispensable to perform molecular orbital calculations to know the molecular structure of a molecule more precisely. The results of theoretical computer calculations on some popular molecules formed of five to six atoms are now freely available on the internet (e.g., http://www.colby.edu/chemistry/webmo/monitro.archive) and we can access to the precise values of bond angle, bond distance or dipole moment for these molecules. The dipole moments of the ions treated here are calculated to be zero Debye, which agrees with the prediction from the symmetry associated with the structures of the ions. As to the dipole moments of ions in general see **solution 1-65**.

Ans. See text (valence of many atom ion).

Solution 1-68

(a) The radius R of the nucleus of heavy hydrogen ^2H (=^{2}D) is found from the given formula as below.

$$R_0\sqrt[3]{A} = 1.33\times10^{-15}\times\sqrt[3]{2} = 1.68\times10^{-15}\,\text{m}$$

Therefore, the coulomb energy at the distance where the two nuclei $^2H^+$ contact each other is calculated to be $4.14\times10^{10}\,\text{J}\cdot\text{mol}^{-1}$ from the equation eq(111) given below.

$$\frac{e^2\times N_A}{4\pi\varepsilon_0\times 2R} = \frac{(1.602\times10^{-19})^2\times6.022\times10^{23}}{1.113\times10^{-10}\times2\times1.676\times10^{-15}} = 4.14\times10^{10}\,\text{J}\cdot\text{mol}^{-1} \qquad (111)$$

Ans. $4.14\times10^{10}\,\text{J}\cdot\text{mol}^{-1}$

(b) The mass defect Δm for the nuclear fusion given in the problem is calculated as below,

$$\Delta m = 2.01474\times2-(3.01698+1.00898) = 0.00352\ \text{amu}$$

and its energy is found to be $3.16\times10^{11}\,\text{J}\cdot\text{mol}^{-1}$, which is about eight times larger as compared with the

$$\Delta m\times c^2\times N_A = 0.00352\times1.661\times10^{-27}\times2.998\times10^8\times6.022\times10^{23} = 3.16\times10^{11}\,\text{J}\cdot\text{mol}^{-1}$$

coulomb energy obtained in (a), indicating that enormous energy is produced by nuclear fusions.

Ans. energy of nuclear fusion: $3.16\times10^{11}\,\text{J}\cdot\text{mol}^{-1}$, 7.6 times larger

"note" atomic mass unit (amu): 12 is assigned to the mass of one atom of the isotope ^{12}C and its one twelfth is defined as 1 amu. Therefore, its value is 1.6605655×10^{-27} kg/amu. (See **II-1-3-3**.)

(c) m_n stands for the mass of a neutron and υ its velocity. From the equation of $\frac{1}{2}m_n\upsilon^2 = \Delta m\times c^2$ which signifies the whole energy is converted into the kinetic energy of the emitted neutron, the velocity of the neutron is obtained as below.

$$\upsilon = \sqrt{\frac{2\Delta m}{m_n}}\times c = \sqrt{\frac{2\times0.00352}{1.009}}\times10^8 = 2.50\times10^7\,\text{m}\cdot\text{s}^{-1} \qquad (112)$$

Ans. velocity of neutron: $2.50\times10^7\,\text{m}\cdot\text{s}^{-1}$

Solution 1-69

The answer is given here by using the *Bohr* model. To make the model simpler, the nucleus is assumed to be at rest at the coordinate origin, To be precise, however, the relative motion of the electron against the nucleus should be described by using the center of gravity system. For this, the mass m_e of the electron which is used in eqs(17)-(20) in **solution 1-16** should be replaced by the reduced mass μ which is given by the formula below, where m_{H^+} stands for the mass of the proton. Therefore, it is necessary to replace m_e in eqs (17)-(20) in **solution 1-16** with μ in order to acquire more precise picture of the orbiting motion of the electron. $\frac{1}{\mu}=\frac{1}{m_e}+\frac{1}{m_{H^+}} \Rightarrow \mu=\frac{m_e m_{H^+}}{m_e+m_{H^+}}$

Similarly, in the case of deuterium (D), by replacing m_{H^+} with m_{D^+}, we could explain a slight difference observed in the electronic spectra between hydrogen and deuterium. (See advanced problem below.)

Now, in order to answer the question (c), m_e should be replaced with $207m_e$. For the hydrogen atom its reduced mass is found as

$$\mu_H=\frac{9.109\times10^{-31}\times1.673\times10^{-27}}{9.109\times10^{-31}+1.673\times10^{-27}}=9.104\times10^{-31}\,\text{kg}.$$

On the other hand, for the muonic hydrogen atom (μH) the reduced mass is found as

$$\mu_{\mu H}=\frac{207\times9.109\times10^{-31}\times1.673\times10^{-27}}{207\times9.109\times10^{-31}+1.673\times10^{-27}}=1.695\times10^{-28}\,\text{kg}.$$

These values thus obtained are separately substituted into eq(19) given in **solution 1-16** to calculate the ratio of the *Bohr* radii $a_0(\mu H)/a_0(H)$, which is given below in eq(113).

$$\frac{a_0(\mu H)}{a_0(H)}=\frac{\varepsilon_0 h^2}{\pi\mu_{\mu H}e^2}\Big/\frac{\varepsilon_0 h^2}{\pi\mu_H e^2}=\frac{\mu_H}{\mu_{\mu H}}=\frac{9.104\times10^{-31}}{1.695\times10^{-28}}=\frac{1}{186} \qquad (113)$$

Ans. ratio of the Bohr radii: $1/186$

[**advanced problem**] The wavelength of the electronic absorption band is expected to differ slightly from the H-atom to D-atom. Find the difference in cm^{-1} for the absorption bands located at the lowest energies among the *Lyman* series (see **solution 1-11**).

Solution 1-70

The energy level is given by $\varepsilon_n=\frac{n^2h^2}{8ma^2}$, where n is the quantum number. Therefore, the energy of the transition for $n=10\rightarrow n=11$ and the wavelength of a photon to be absorbed are calculated as below.

$$\varepsilon_{11}-\varepsilon_{10}=\frac{h^2}{8ma^2}(11^2-10^2)=\frac{(6.63\times10^{-34})^2\times21}{8\times9.11\times10^{-31}\times(2.0\times10^{-9})^2}=3.17\times10^{-19}\,\text{J}$$

$$\lambda=\frac{c}{\nu}=\frac{c}{(\varepsilon_{11}-\varepsilon_{10})/h}=\frac{3.0\times10^8\times6.63\times10^{-34}}{3.17\times10^{-19}}=6.27\times10^{-7}\,\text{m}=627\text{ nm}$$

Ans. wavelength of photon: 627 nm

Solution 1-71

There exists a relation between the length of a polyene and the absorption wavelength. For the polyene made of $2N$ carbon atoms and with the length of a $C=C-$ unit given by d, the length of a one-dimensional box can be expressed by $a=Nd$. Since the box is occupied by N free electrons (one electron from each carbon atom) and each energy level can accommodates two electrons, the quantum number of the highest occupied energy level becomes N. Therefore, the electronic transition of the lowest energy corresponds to $N \to N+1$ and the transition energy is given by eq(114) after substituting $a = Nd$ into $\varepsilon_n = \dfrac{n^2h^2}{8ma^2}$. From eq(114), the wavelength of the associated light absorption is derived readily as eq(115).

$$\varepsilon_{N+1} - \varepsilon_N = \frac{h^2}{8m_e(Nd)^2}\left\langle (N+1)^2 - N^2 \right\rangle = \frac{h^2}{4m_e Nd^2} \qquad (114)$$

$$\lambda = \frac{c}{\nu} = \frac{ch}{\varepsilon_{N+1} - \varepsilon_N} = \frac{4m_e cNd^2}{h} \propto N \qquad (115)$$

Ans. eq(115)

"note" The table below shows for some polyenes and polyacenes how the light absorption wavelength varies with the number of double bonds or the number of aromatic rings.

polyene	no. of C=C	wavelength (Å)	color	polyacene	no. of ring	wavelength (Å)	color
ethylene	1	1625		benzene	1	2690	
butadiene	2	2170		naphthalene	2	3110	
hexatriene	3	2680		anthracene	3	3750	
octatriene	4	3040		naphthacene	4	4750	orange
vitamin A	5	3280		pentacene	5	5800	violet
α-carotene	10	4450	yellow	hexacene	6	~6500	greenish blue
γ-carotene	11	4600	yellow	heptacene	7	~8300	dark green

[reference] *Yamaguchi, Hirayama, Taniguchi and Higashi, Basic Physical Chemistry* (in Japanese), Asakura Shoten p.68, 1989.

Solution 1-72

A dipole moment is represented by a three-dimensional vector. Therefore, a relative orientation of two identical dipole moments is more easily examined after decomposing the dipole into three-dimensional components. One of the two dipoles is assumed to be oriented along one of the three axes as is shown in Fig.66. The other dipole can take various orientations, but the orientations with different interaction energies are only those shown in Fig.67. That is, (a) parallel orientation, (b) orientation on the same line, and (c) orthogonal orientation.

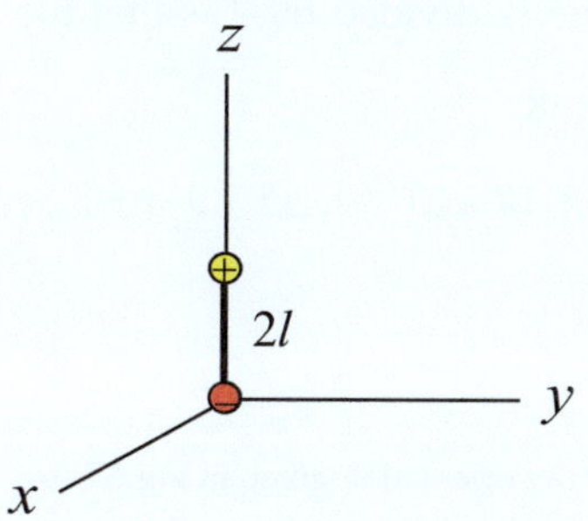

Fig.66 Orthogonal coordinates and a dipole oriented along z-axis.

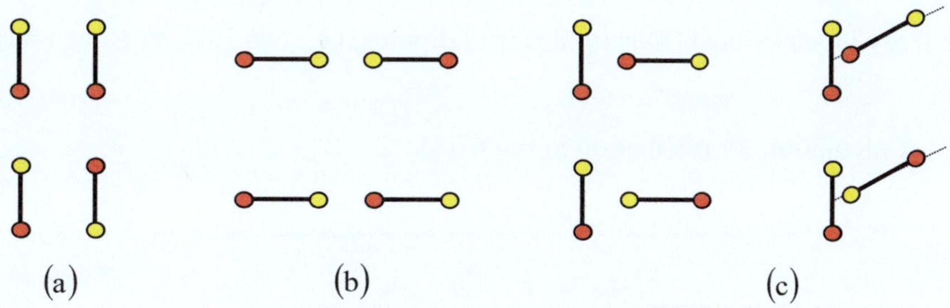

Fig.67 Relative orientations of two dipoles with different interaction energies.

For each orientation, the total of the coulomb energies between every two point charges chosen one from each dipole is nothing more than the dipole-dipole interaction energy. Some of such calculations are given below.

because of their symmetrical arrangements of positive and negative charges. For the orientations shown in (a) or (b), it is evident the one displayed in the bottom has the lowest (negative) energy. The exact orientations are shown in Figs.68 and 69, for which the interaction energies are calculated below. The dipole moment of HCl is $\mu = q_1 e \times 2l = 3.6\times10^{-30}\,\mathrm{C\cdot m}$.

Calculation for orientation in Fig.67(a).

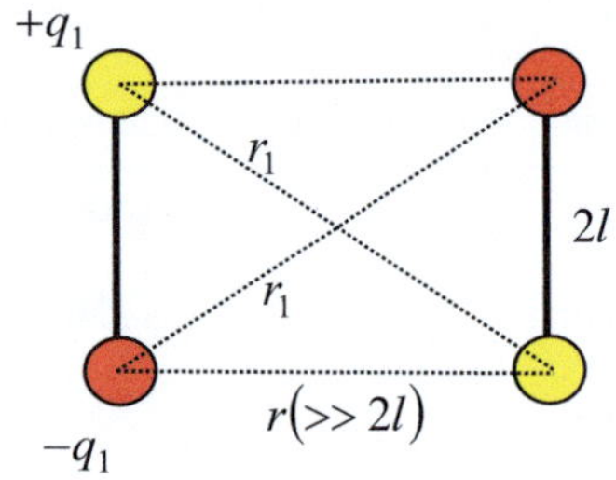

Fig.68 Details of the orientation given in Fig.67 (a). Intermolecular distance is r.

The total energy E_a of the coulomb interaction is given by eq(116).

$$E_a = \frac{1}{4\pi\varepsilon_0}\left(-\frac{2q_1^2e^2}{r}+\frac{2q_1^2e^2}{r_1}\right)=\frac{q_1^2e^2}{2\pi\varepsilon_0}\left(-\frac{1}{r}+\frac{1}{r_1}\right) \tag{116}$$

By substituting the approximate expression for $1/r_1$ given by eq(117) into eq(116),

$$\frac{1}{r_1}=\frac{1}{\sqrt{r^2+4l^2}}=\frac{1}{r}\cdot\frac{1}{\sqrt{\left(1+\frac{4l^2}{r^2}\right)}}\approx\frac{1}{r}\left(1-\frac{1}{2}\cdot\frac{4l^2}{r^2}\right) \tag{117}$$

E_a is found to be expressed by eq(118).

$$E_a = \frac{q_1^2e^2}{2\pi\varepsilon_0}\left\{-\frac{1}{r}+\frac{1}{r}\left(1-\frac{4l^2}{2r^2}\right)\right\}=-\frac{q_1^2e^2}{4\pi\varepsilon_0}\cdot\frac{4l^2}{r^3}=-\frac{\mu^2}{4\pi\varepsilon_0 r^3} \tag{118}$$

In the case of HCl, therefore, the interaction energy is calculated as follows. For 1 mole of HCl, multiplying

$$-\frac{\mu^2}{4\pi\varepsilon_0 r^3}=-\frac{(3.6\times10^{-30})^2}{1.11\times10^{-10}\times(1.0\times10^{-9})^3}=-1.17\times10^{-22}\,\mathrm{J}$$

this value by $N_A/2$ the dipole-dipole interaction energy is found to amount to $-1.17\times10^{-22}\times 6.02\times10^{23}/2=-35.1\ \mathrm{J\cdot mol^{-1}}$. It should be noted that the reason why the value is divided by 2 is that the energy given by eq(118) is for a pair of dipoles, i.e., two HCl, as is seen from Fig.68.

Ans. $1.17\times10^{-22}\,\mathrm{J}$, $35.1\ \mathrm{J\cdot mol^{-1}}$

Calculation for orientation in Fig.67(b).

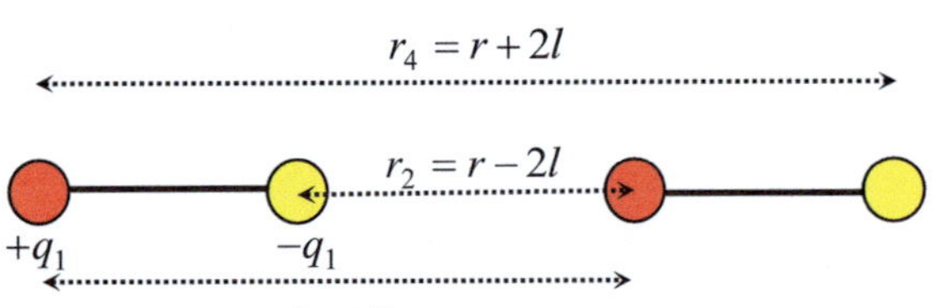

Fig.69 Details of the orientation given in Fig.67 (b). Intermolecular distance is r.

The total energy E_b of the coulomb interaction is given by eq(119).

$$E_b = -\frac{q_1^2 e^2}{4\pi\varepsilon_0}\left(\frac{1}{r_2} - \frac{2}{r_3} + \frac{1}{r_4}\right)$$

$$\approx -\frac{q_1^2 e^2}{4\pi\varepsilon_0}\left\{\frac{1}{r}\left(1 - \frac{2l}{r} + \frac{4l^2}{r^2}\right) - \frac{2}{r}\left(1 + \frac{2l}{r} + \frac{4l^2}{r^2}\right)\right\} = -\frac{q_1^2 e^2}{4\pi\varepsilon_0} \cdot \frac{8l^2}{r^3} = -\frac{\mu^2}{2\pi\varepsilon_0 r^3} \qquad (119)$$

For the same intermolecular distance r, it is found that the orientation (b) is two times more stable as compared with the orientation (a).

Ans. orientation in Fig.69 (same direction on a line) 2.34×10^{-22} J, 70.2 J · mol^{-1}

"note" The results given by eqs(67) and (68) are approximations for the dipole-dipole interaction energy under the condition of $r >> 2l$. In more precise approximation involving higher orders of l/r, energies for the interaction called dipole-quadrupole or quadrupole-quadrupole interaction are obtained. These terms in higher order approximation are called multi-pole expansion and are represented by *Legendre*'s polynomials. (See **II-4 -18.**)

Solution 1-73

As are indicated by blue arrows in the figure given below, each dipole is decomposed into two orthogonal components, one is on the line (dotted line) connecting the centers and the other orthogonal to the former.

1) The interaction energy E_1 along the line connecting the two molecular center is easily found by referring to the orientation (b) in the previous solution. Taking it into account that the component dipole moment is given by $\mu\cos\vartheta$ and the two components align oppositely, eq(120) is obtained for E_1.

$$E_1 = -\frac{1}{4\pi\varepsilon_0}\left(-\frac{q_1^2 e^2}{r_2} + \frac{2q_1^2 e^2}{r_3} - \frac{q_1^2 e^2}{r_4}\right) \approx \frac{2\mu^2\cos^2\vartheta}{4\pi\varepsilon_0 r^3} \qquad (120)$$

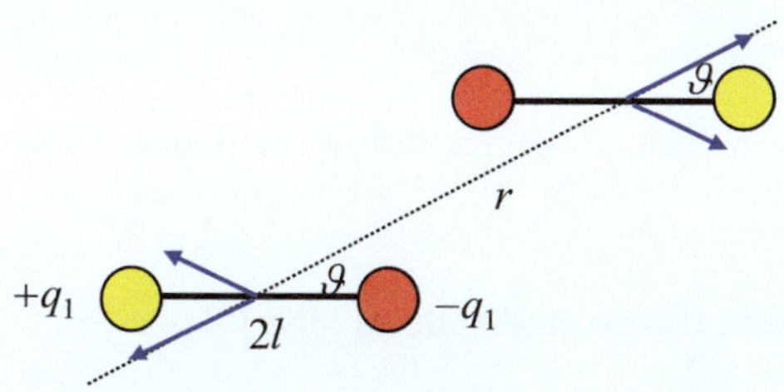

Fig.70 Anti-parallel orientation of two dipoles.

2) The interaction energy E_2 of the two components orthogonal to the line connecting the two molecular center is also readily found by referring to the orientation (a) in the previous solution and notifying the magnitude of the component dipole moment is $\mu \sin\vartheta$. The result is given by eq(121).

$$E_2 \approx -\frac{\mu^2\sin^2\vartheta}{4\pi\varepsilon_0 r^3} \qquad (121)$$

Summarizing the results above, the total interaction energy of the two dipoles in the parallel orientation (see above) is found to be given by eq(122). (See **"note1"** below.)

$$E_1 + E_2 = \frac{2\mu^2\cos^2\vartheta}{4\pi\varepsilon_0 r^3} - \frac{\mu^2\sin^2\vartheta}{4\pi\varepsilon_0 r^3} = -\frac{\mu^2}{4\pi\varepsilon_0 r^3}(\sin^2\vartheta - 2\cos^2\vartheta)$$

$$= -\frac{\mu^2}{4\pi\varepsilon_0 r^3}(1 - 3\cos^2\vartheta) \qquad (122)$$

Ans. $-\frac{\mu^2}{4\pi\varepsilon_0 r^3}(1 - 3\cos^2\vartheta)$

"note 1" The dipole-dipole interaction energy (potential) φ_{ab} of the two dipoles $\boldsymbol{\mu}_a$ and $\boldsymbol{\mu}_b$ separated by the distance of $\boldsymbol{r}_{ab}$ is generally given by eq(123). As is evident, the energy is independent of the manner to define the direction (positive or negative) of a dipole.

$$\varphi_{ab} = \frac{1}{4\pi\varepsilon_0 r_{ab}^3}\left\{(\boldsymbol{\mu}_a\cdot\boldsymbol{\mu}_b) - \frac{3(\boldsymbol{\mu}_a\cdot\boldsymbol{r}_{ab})(\boldsymbol{\mu}_b\cdot\boldsymbol{r}_{ab})}{r_{ab}^2}\right\} \qquad (123)$$

By employing eq(123), it is easy to confirm the results given in **solution 1-72** (eqs(118) and (119)). Also, the total energy to be found in the present problem is easily derived as eq(124) given below by using eq(123).

$$\begin{aligned} E_1 + E_2 &= \frac{1}{4\pi\varepsilon_0 r_{ab}^3}\left\{(\boldsymbol{\mu}_a\cdot\boldsymbol{\mu}_b) - \frac{3(\boldsymbol{\mu}_a\cdot\boldsymbol{r}_{ab})(\boldsymbol{\mu}_b\cdot\boldsymbol{r}_{ab})}{r_{ab}^2}\right\} \\ &= \frac{1}{4\pi\varepsilon_0 r^3}\left\{-\mu^2(\sin^2\vartheta + \cos^2\vartheta) - \frac{3\mu\cos(180^\circ - \vartheta)\cdot r \times \mu\cos\vartheta\cdot r}{r^2}\right\} \\ &= -\frac{\mu^2}{4\pi\varepsilon_0 r^3}(1 - 3\cos^2\vartheta) \qquad (124) \end{aligned}$$

The derivation of the general formula given by eq(123) can be achieved also by decomposing the dipole given as a function of polar coordinates (r, ϑ, φ) into the components expressed in Cartesian coordinates, as is shown in Fig.67 in **solution 1-72** and then by summing up the energies calculated for each component without missing any pair.

"note 2" The direction of a dipole is defined from a negative point charge to a positive one, i.e., $- \rightarrow +$, in the **"note"** in **solution 1-65**. Obviously, however, even when the direction of the dipole is defined in the opposite way, i.e., $+ \rightarrow -$, the interaction energies for the dipole-ion, dipole-dipole, etc., are invariant of the definition and have the same values.

Solution 1-74

How to describe the electronic states based on *Russell-Saunders'* rule is given below in detail. Before describing that the answers required are given here first.

Na atom in the ground state of $[\mathrm{Ne}]\ 3s^1$:

$$l = 0 \ \Rightarrow\ S$$

$$s = \frac{1}{2}$$

$$j = l + s = \frac{1}{2}$$

$$^{2s+1}l_j \ \Rightarrow\ {}^2S_{1/2}$$

Na atom in the ground state of $[\mathrm{Ne}]\ 3p^1$:

$$l = 1 \ \Rightarrow\ P$$

$$s = \frac{1}{2}$$

$$j = l + s = \frac{3}{2},\ \frac{1}{2}$$

$$^{2s+1}l_j \ \Rightarrow\ {}^2P_{3/2},\ {}^2P_{1/2}$$

Ans. ground state: ${}^2S_{1/2}$, excited state: ${}^2P_{3/2}$, ${}^2P_{1/2}$

"*Russell-Saunders'* Rule"

We have tried so far to deepen our understanding of the electronic configuration of atoms by solving various problems. When the electronic states of atoms are examined more closely, we come to know that their energies cannot be described by the electronic configuration alone even after taking the spin multiplicity into account. In other words, one electronic configuration does not correspond to solely one energy state alone. This is because (a) the electron has its intrinsic magnetic moment due to the spin angular momentum (spin quantum number s) and (b) the orbiting motion of the electron (angular momentum, azimuthal quantum number l) causes a magnetic moment which interacts with the spin magnetic moment leading to several different energy states. This interaction is called "*spin-orbit interaction*". The magnetic moments described in (a) and (b) are proportional to the spin quantum number and the azimuthal quantum number and are conventionally represented by vectors. Then, the energy of the spin-orbit interaction should be proportional to $(\boldsymbol{l} \cdot \boldsymbol{s})$. Writing the total angular momentum of one electron as $\boldsymbol{j}$, it is given by eq(125) as the sum of $\boldsymbol{s}$ and $\boldsymbol{l}$.

$$\boldsymbol{j} = \boldsymbol{l} + \boldsymbol{s} \qquad (125)$$

$$\begin{aligned}(\boldsymbol{j} \cdot \boldsymbol{j}) = \boldsymbol{j}^2 &= (\boldsymbol{l} + \boldsymbol{s}) \cdot (\boldsymbol{l} + \boldsymbol{s}) \\ &= (\boldsymbol{l} \cdot \boldsymbol{l}) + (\boldsymbol{s} \cdot \boldsymbol{s}) + 2(\boldsymbol{l} \cdot \boldsymbol{s}) \qquad \because \ (\boldsymbol{l} \cdot \boldsymbol{s}) = (\boldsymbol{s} \cdot \boldsymbol{l}) \\ &= \boldsymbol{l}^2 + \boldsymbol{s}^2 + 2(\boldsymbol{l} \cdot \boldsymbol{s}) \qquad (126)\end{aligned}$$

From eqs(125) and (126), eq(127) is derived.

$$2(\boldsymbol{l} \cdot \boldsymbol{s}) = \boldsymbol{j}^2 - (\boldsymbol{l}^2 + \boldsymbol{s}^2) \qquad (127)$$

Consequently, in order to find out the electronic states of different energies, it is sufficient to list up the terms (sets of $\boldsymbol{l}$ and $\boldsymbol{s}$) which yield different values for eq(127) on the right side.

Meanwhile, the quantum number j can take the values given by eq(128) . (See **"note"**.)

$$j = l + s, \ l + s - 1, \ l + s - 2, \cdots, \ |l - s| \qquad (128)$$

For instance,

for Na([Ne]$3s^1$), $l = 0$ and $s = \frac{1}{2}$ yield $j = \frac{1}{2}$, and

for Na*([Ne]$3p^1$), $l = 1$ and $s = \frac{1}{2}$ yield $j = \frac{3}{2}$ and $\frac{1}{2}$.

The electronic states described in terms of l, s and j are usually expressed by term symbols ${}^{2s+1}l_j$ defined below, but l in the term is replaced by a relevant capital letter in the alphabet given in the second line.

${}^{2s+1}l_j$	$l = 0,$	$l = 1,$	$l = 2,$	$l = 3,$	$l = 4,$	$\cdots$
term symbol	$S,$	$P,$	$D,$	$F,$	$G,$	$\cdots$

Following the definition above, the electronic states taken as examples can be denoted as Na([Ne]$3s^1$)$\equiv {}^2S_{1/2}$ and Na*([Ne]$3p^1$)$\equiv {}^2P_{3/2}$, ${}^2P_{1/2}$ as are given in **Ans.**

When two or more electrons are involved, the total orbital angular momentum L, total spin angular momentum S and the total angular momentum J are expressed by eq(129) as below. In the case of small

$$\begin{aligned}&L = l_1 + l_2, \ l_1 + l_2 - 1, \ l_1 + l_2 - 2, \ \cdots, \ |l_1 - l_2| \\ &S = s_1 + s_2, \ ls_1 + s_2 - 1, \ s_1 + s_2 - 2, \ \cdots, \ |s_1 - s_2| \qquad (129) \\ &J = L + S, \ L + S - 1, \ L + S - 2, \ \cdots, \ |L - S|\end{aligned}$$

spin-orbit interactions, the interaction energies can be ascribed to the $(\boldsymbol{L} \cdot \boldsymbol{S})$ coupling again

(*Russell-Saunders Coupling*). Therefore, as is in the case of one electron, resulting electronic states of different energy are described by term symbols of $^{2S+1}L_J$ using capital letters. Again, L is replaced by a corresponding capital letter given before.

For example, in the case of $Ca^*([Ar]3s^13p^1)$, the total angular momenta, total spin momenta, etc. are $L=1+0=1$, $S=\frac{1}{2}+\frac{1}{2}=1$, $\frac{1}{2}-\frac{1}{2}=0$, $J=2,\ 1,\ 0$, and, hence, the electronic states are expressed by term symbols as $Ca^*([Ar]3s^13p^1) \equiv {}^3P_2,\ {}^3P_1,\ {}^3P_0,\ {}^1P_1$. Apart from the spin states, it is known that there are four different electronic states for that electronic configuration.

"note" The operator for the angular momentum (or wave function) has the same property irrespective of the kind of angular momentum. Therefore, the quantum number for the synthesized angular momentum takes positive values decreasing by one successively as are found in eqs(128) and (129). Refer to **"note"** in **solution 1-9** for how to synthesize the angular momentum from the components of two individual angular momenta.

Solution 1-75

Since the problem is about the electronic transition of one electron involved, the selection rules are i) $\Delta s=0$ and ii) $\Delta l=\pm 1$. Therefore, the transition of $3s \to 3p$ is allowed. More precisely, however, the excited state of $[Ne]3p^1$ is energetically split because of ***LS*** coupling and the split excited states are expressed by the spectral terms of $^2P_{3/2}$ and $^2P_{1/2}$. The ground state is $^2S_{1/2}$. Therefore two types of electronic transition $^2S_{1/2} \to {}^2P_{3/2}$ and $^2S_{1/2} \to {}^2P_{1/2}$ are to be observed. For further detailed explanations refer to **solution 1-95**.

Ans. allowed transition

To judge correctly the transition under consideration is whether allowed or not on the basis of the calculation of the transition moment, it is necessary to examine whether the transition moment given by eq(130) is zero or non-zero,

$$\iiint_{r,\vartheta,\varphi} \psi_{3s}(r,\vartheta,\varphi)\alpha \cdot \begin{pmatrix} x \\ y \\ z \end{pmatrix} \cdot \psi_{3p}(r,\vartheta,\varphi)\alpha \cdot r^2 \sin\vartheta \, d\vartheta d\varphi dr \qquad (130)$$

in which α stands for the spin wave function. To make the calculation simpler, we choose $3p_z$ for $3p$ and calculate only the z-component of the transition moment. The wave functions associated with the transition are given by eq(131) (ground state) and eq(132) (excited state). The z-component of the transition dipole moment is given by eq(133) when it is expressed in the polar coordinates. Here, $\sigma = Zr/a_0 \ge 0$.

$$\psi_{3s}(r,\vartheta,\varphi) = \frac{1}{81\sqrt{3\pi}}\left(\frac{Z}{a_0}\right)^{3/2}(27-18\sigma+2\sigma^2)e^{-\sigma/3} \qquad (131)$$

$$\psi_{3p_z}(r,\vartheta,\varphi) = \frac{1}{81}\left(\frac{2}{\pi}\right)^{1/2}\left(\frac{Z}{a_0}\right)^{3/2}(6\sigma-\sigma^2)e^{-\sigma/3}\cos\vartheta \qquad (132)$$

$$z = r\cos\vartheta \qquad (133)$$

To tell whether the integral in eq(130) is zero or not, it is just enough to examine the parity of its integrand. As is clear from eqs(131) and (132), the parity of the wave functions can be examined against ϑ alone, since these wave functions cannot be inverted against non-negative σ or r to find parity. The part which is a function of ϑ in eq(130) is $\cos^2\vartheta\sin\vartheta$. Over the integral range of $[0,\ \pi]$, since $\cos\vartheta$ is an odd function and $\sin\vartheta$ is an even function, $\cos^2\vartheta\sin\vartheta$ is an even function. In fact, the integration of $\cos^2\vartheta\sin\vartheta$ over the range $[0,\ \pi]$ yields $2/3$ and is surely non-zero as is shown in eq(134).

$$\int_0^\pi \cos^2\vartheta\sin\vartheta\, d\vartheta = \left[-\frac{1}{3}\cos^3\vartheta\right]_0^\pi = \frac{2}{3} \neq 0 \qquad (134)$$

As is exemplified above, without carrying out complicated calculations we can easily tell whether the transition under consideration is allowed or prohibited by examining the parity of the wave functions involved in the transition. One of the merits of employing the atomic orbitals expressed in terms of the orthogonal coordinates such as p_x, p_y instead of quantum number m_l lies in their easiness to tell their parities. See the next problem and solution.

Solution 1-76

The parity is the property of the atoms or molecules which have a center of symmetry and when the coordinates of those atoms or molecules are inverted against the center (origin) their shapes remain totally invariant. The wave functions which describe their states, therefore, can change only the signs of the functions upon the inversion of the coordinates. When the sign remains the same the functions are named even (g for gerade in German) and when the sing changes named odd (u for ungerade).

The operation of inversion is equivalent to change the orthogonal coordinates in the manner $x \to -x$, $y \to -y$ and $z \to -z$ or the polar coordinates as $r \to r$, $\vartheta \to \pi - \vartheta$ and $\varphi \to \varphi + \pi$.

The wave function for the atomic orbital is written as $\Psi(r,\vartheta,\varphi)$ in the polar coordinates and is expressed by the product of the radial part $R_{n.l}(r)$ and the angular part $Y_m^l(\vartheta,\varphi)$ as is given by eq(135).

$$\Psi(r,\vartheta,\varphi) = R_{n.l}(r)\cdot Y_m^l(\vartheta,\varphi) \qquad (135)$$

$$Y_m^l(\vartheta,\varphi) = P_l^m(\cos\vartheta)\cdot e^{-im\varphi} \qquad (136)$$

Only the angular part changes its sign upon inversion. In fact, when the inversion is performed on $Y_m^l(\vartheta,\varphi)$,

$$\begin{aligned} Y_m^l(\pi-\vartheta,\varphi+\pi) &= P_l^m(\cos(\pi-\vartheta))\cdot e^{-im\varphi}\cdot e^{-im\pi} \qquad (137)\\ &= P_l^m(-\cos\vartheta)\cdot(-1)^m \times e^{-im\varphi}\\ &= (-1)^{l-m}\times P_l^m(\cos\vartheta)\cdot(-1)^m\cdot e^{-im\varphi}\\ &= (-1)^l P_l^m(\cos\vartheta)\cdot e^{-im\varphi} \qquad (138)\end{aligned}$$

eq(138) is derived via eq(137), in which the properties of the associated *Legendre* polynomials (see **II-4-18**) are utilized..

In conclusion, the atomic orbitals change their signs according to $(-1)^l$ upon the inversion of the coordinates. The electronic states of a many-electron atom are represented by the product of each wave function for each electron and, hence, their signs upon inversion are determined by the product of $(-1)^l$ over all electrons belonging to the atom as is given in eq(139).

$$\prod_i^{\text{all electrons}} (-1)^{l_i} = (-1)^{\sum_i l_i} \quad (139)$$

(a) Based on eq(139), the parities of the electronic states of the atoms listed in the problem are found as follows.

$H:(1s^1),\ \sum_i l_i = 0 \Rightarrow$ even

$N:(1s^2 2s^2 2p^3),\ \sum_i l_i = 1+1+1=3 \Rightarrow$ odd

$S:([Ne]3s^2 3p^4),\ \sum_i l_i = 1+1+1+1=4 \Rightarrow$ even

$Ca:([Ar]4s^2),\ \sum_i l_i = 0+0=0 \Rightarrow$ even

$Fe:([Ar]4s^2 3d^6),\ \sum_i l_i = 0+0+2\times 6=12 \Rightarrow$ even

$Zr:([Kr]5s^2 4d^2),\ \sum_i l_i = 0+0+2\times 2=4 \Rightarrow$ even

$Ag:([Kr]5s^1 4d^{10}),\ \sum_i l_i = 0+2\times 10=20 \Rightarrow$ even

$Eu:([Xe]6s^2 4f^7),\ \sum_i l_i = 0+0+3\times 7=21 \Rightarrow$ odd

Ans. H (g), N (u), S (g), Ca (g), Fe (g), Zr (g), Ag (g), Eu (u). g: even, u: odd

(b) As to the selection rule upon the transition due to the light absorption or emission, the allowed transition is between the odd and even states, i.e., g$\leftrightarrow$u, and the transitions between the states of the same parity, i.e., g$\leftrightarrow$g or u$\leftrightarrow$u are forbidden, This is easily derived because the parity of the dipole moment associated with the transition is always odd.

Consequently, an example of the transition for each atom is as follows.

$H:(1s^1 \to 2p^1) \Rightarrow g \to u$

$N:(1s^2 2s^2 2p^3 \to 1s^2 2s^2 2p^2 3s^1) \Rightarrow u \to g$

$S:([Ne]3s^2 3p^4 \to [Ne]3s^2 3p^3 3d^1) \Rightarrow g \to u$

$Ca:([Ar]4s^2 \to [Ar]4s^1 3d^1) \Rightarrow g \to g$ (forbidden)

$Fe:([Ar]4s^2 3d^6 \to [Ar]4s^2 3d^5 4p^1) \Rightarrow g \to u$

$Zr:([Kr]5s^2 4d^2 \to [Kr]5s^1 4d^2 5p^1) \Rightarrow g \to u$

$Ag:([Kr]5s^1 4d^{10} \to [Kr]5s^1 4d^{10} 5p^1) \Rightarrow g \to u$

$Eu:([Xe]6s^2 4f^7 \to [Xe]6s^1 4f^8) \Rightarrow u \to g$

Ans. $Ca:([Ar]4s^2 \to [Ar]4s^1 4p^1) \Rightarrow g \to u$

"note 1" Refer to the periodic table and electronic configurations (Fig.2) and the energy level diagram of atoms (Fig.3) given in **solution 1-2**.

"note 2" It is evident from eq(139) that the electronic transition of $d \leftrightarrow d$ is forbidden. The light absorptions arising from the ligand field splitting observed for many metal complexes are due to the forbidden transitions of $d \leftrightarrow d$ and, hence, the absorptions are not strong usually. (See Ch.7 Molecular structures and Photochemistry.)

Solution 1-77a

(a) The electronic configuration of $_{26}Fe$ in the ground state is expressed as $Fe:(1s^2 2s^2 2p^6 3s^2 3p^6 4s^2 3d^6)$. The ionization energies of inner shell electrons can be estimated from the effective nuclear charges of Fe by eq(140).

$$I.P. = A \times \frac{Z_{eff}^{\ 2} e^2}{n^2} \quad (140)$$

For this, the electrons of Fe are classified into the following groups: $1s$; $2s,2p$; $3s,3p$; $3d$; $4s,4p$; $4d$;

$4f$; and $5s,5p$. The most likely groups to explain the XPS spectrum are chosen and values for the effective nuclear charge $Z_{eff} = Z - \sigma$ are estimated as below. As to the estimation, refer to **solution 1-12.**

$$2s,2p\,;\ Z_{eff} = 26 - 0.85 \times 2 - 0.35 \times 7 = 21.85$$

$$3s,3p\,;\ Z_{eff} = 26 - 2.0 - 0.85 \times 8 - 0.35 \times 7 = 14.75$$

$$3d\,;\quad Z_{eff} = 26 - 2.0 - 8.0 - 0.85 \times 8 - 0.35 \times 5 = 7.45$$

From these values, the ionization energies for the electrons of the groups of $2s,2p$; $3s,3p$; can be estimated as below.

$$I.P.(\text{Fe}(2s,\ 2p)) \approx 1312 \times \frac{(21.85)^2}{2^2} = 156595\ \text{kJ/mol} = 1623\ \text{eV} \qquad 1\ \text{eV} = 1.602 \times 10^{-19}\ \text{J} = 96.485\ \text{kJ/mol}$$

$$I.P.(\text{Fe}(3s,\ 3p)) \approx 1312 \times \frac{(14.75)^2}{3^2} = 31715\ \text{kJ/mol} = 329\ \text{eV}$$

On the other hand, the values estimated similarly from Z_{eff} for the first and second ionization energy are as given below and in either case the values are smaller than the observed values given in the parentheses.

$$I.P.(\text{Fe}(4s)) \approx 1312 \times \frac{(2.55)^2}{4^2} = 533\,\text{kJ/mol}\,(761),\ \ I.P.(\text{Fe}^+(4s)) \approx 1312 \times \frac{(2.9)^2}{4^2} = 670\,\text{kJ/mol}\ \ (1560)$$

In contrast, the third ionization energy calculated in the same manner is more than two times greater than the observed one as seen below.

$$I.P.(\text{Fe}^{2+}(3d)) \approx 1312 \times \frac{(7.45)^2}{3^2} = 8091\,\text{kJ/mol}\,(2954)$$

Gathering these facts together, the ionization energies estimated for the inner shell electrons from Z_{eff} are considered to yield much larger values than the observed ones. Therefore, the ionization energies obtained from the XPS measurement can more reasonably be ascribed to those for $2s,2p$ electrons but not for $3s,3p$ electrons. Furthermore, since the order of the energies of 2*s* and 2*p* orbitals is considered as $E_{2s} < E_{2p}$ from Fig.17 in **solution 1-22**, the XPS bands observed are expected to originate from the ionization of the 2*p* electron which has the smaller ionization energy than 2*s* electron.

Ans. Due to the ionization of 2*p* electron.

(b) It is apparent from a comparison of the ionization energies ascribed to from the first through the third ones, the ionization energies of the outermost electrons vary significantly with a change in the quantum number and the effective nuclear charge brought about through a change in the oxidation state. As compared with this, the ionization of the inner shell electrons is not influenced much by a change in the oxidation state of elements. Approximately, therefore, the ionization energies are similar for Fe(0), Fe(II) and Fe(III). When looked at more closely, the positive net charge on the central atom or ion increases with the oxidation number and hence the attractive interaction between the atom or ion and the electron to be ejected will become stronger with increasing oxidation number. As a consequence of this, the work required to separate the electron completely away from the atom or ion will increases in the order of $\text{Fe}(0) < \text{Fe}(\text{II}) < \text{Fe}(\text{III})$, resulting in the same order of the ionization energy as well.

Ans. the order of ionization energy: $\text{Fe}(0) < \text{Fe}(\text{II}) < \text{Fe}(\text{III})$

Solution 1-77b

(a) The oxidation numbers of the carbon atoms (from 1 to 4) of the compound given below in (b) are allotted to +3, +3, −1, −2 from the left to the right. See **solution 1-88** for the method of determining the oxidation number.

Ans. See text.

(b) As far as the oxidation number is concerned, we cannot distinguish the carbon atom of the fluoro methyl group from that of the ketone but since the electronegativity of the F atom is larger than that of the O atom, the former carbon atom is more positively polarized than the latter, leading to the larger ionization energy for the former. Consequently, the ionization energy of the 1*s* electron decreases in the order given below.

$C^1 > C^2 > C^3 > C^4$ (for the numbering see the compound below)

The carbon atom of the methyl group is taken as a standard, then a decrease or an increase in the ionization energy from the standard value is considered to reflect the nature of chemical bonds and is called as chemical shift. (Chemical shift used in NMR is another thing.)

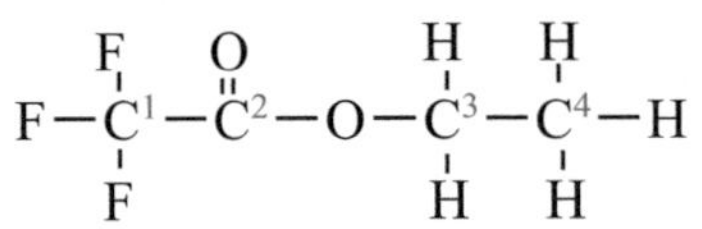

Fig. 71 Bond structure of ethoxy fluromethyl ketone.

Ans. Decreases in the order of $C^1 > C^2 > C^3 > C^4$.

Solution 1-78

The *Bohr* orbital of hydrogen atom is readily derived from the quantization of the angular momentum given by eq(141) and the balance of force within the scope of classical mechanics between the centrifugal force and the coulomb force given by eq(142).

$$m_e v r = n\left(\frac{h}{2\pi}\right) \qquad (141)$$

$$\frac{Ze^2}{4\pi\varepsilon_0 r^2} = \frac{m_e v^2}{r} \qquad (142)$$

Here we put the nuclear charge equal to *Ze* and carry out the calculation for the hydrogen-like atom. From eqs(141) and (142), the radius of the allowed circular orbit r_n is given by eq(143), where n is the quantum number and a_0 is the *Bohr* radius.

$$r_n = \frac{\varepsilon_0 h^2}{\pi m_e e^2}\cdot\frac{n^2}{Z} = a_0 \times \frac{n^2}{Z}, \qquad a_0 = \frac{\varepsilon_0 h^2}{\pi m_e e^2} = 5.29\times 10^{-11}\,\mathrm{m} \qquad (143)$$

Also、the velocity of the electron circuiting in the allowed orbit is directly obtained from eq(141) as eq(144).

$$v_n = \frac{n}{m_e r_n}\cdot\left(\frac{h}{2\pi}\right) = \frac{e^2 Z}{2\varepsilon_0 h n} \qquad (144)$$

The electric current I_n due to the circling motion of the electron (=C/s: converted to the number of electrons passed per sec) is given by eq(145). After knowing the magnitude of the current I_n and the area of the circle πr_n^2, the magnetic moment of the hydrogen-like atom is calculated by *Biot-Savart*'s law as in eq(146).

$$I_n = \frac{v_n \times e}{2\pi r_n} = \frac{enh}{4\pi^2 r_n^2 m_e} \qquad (145)$$

$$\mu_n = I_n \times \pi r_n^2 = \frac{enh}{4\pi^2 r_n^2 m_e} \times \pi r_n^2 = n \times \left(\frac{eh}{4\pi m_e}\right) = n \times \mu_B \qquad (146)$$

Here, the value of $\mu_B = \dfrac{eh}{4\pi m_e}$ is called the *Bohr* magneton. As is clear from eq(146), the magnitude of a magnet is quantized in the unit of μ_B. Its SI units are $\mathrm{C \cdot J \cdot s/kg = C \cdot kg \cdot (m \cdot s^{-1})^2 \cdot s/kg = C \cdot m^2 s^{-1}}$. In terms of the unit of the magnetic field strength $\mathrm{T / J \cdot A^{-1} \cdot m^{-2}}$, it is also expressed by the units of $\mathrm{J \cdot T^{-1}}$.

Ans. See text. $\mu_B = \dfrac{eh}{4\pi m_e}$

"note 1" The magnet moment was obtained here by using the *Bohr* orbit. However, it should be noted that the 1*s* electron is never circling in the *Bohr* orbit from the quantum mechanical point of view. The electric current due to angular motion of the electron is related to the magnetic moment only in the case when the magnetic quantum number m_l is not zero. More detailed mechanistic view of the motion of the electron is treated in **solutions 1-113** and **1-114.**

"note 2" The magnitude of the geomagnetic field (horizontal component) is 0.3×10^{-4} T on average.

"note 3" The nuclear magnetic moment is given by $\dfrac{eh}{4\pi m_p} = 5.051 \times 10^{-27}\,\mathrm{JT^{-1}}$ and is ca. 1/1800 of the *Bohr* magneton.

"note 4" *Biot-Savart*'s law: The law which gives the magnetic field at the distance $\boldsymbol{r}$ created by the current running through a segment of $d\boldsymbol{r}'$ is called "*Biot-Savart*'s law". Based on this law, the current I running through a closed curve and the magnetic field $B(\boldsymbol{r})$ at the distance $\boldsymbol{r}$ is related each other by eq(147), in which μ_0 is the magnetic permeability in vacuum and I is the current running through the closed curve C.

$$B(\boldsymbol{r}) = \frac{\mu_0 I}{4\pi} \oint_C \frac{(\boldsymbol{r} - \boldsymbol{r}')}{|\boldsymbol{r} - \boldsymbol{r}'|^3} d\boldsymbol{r}' \qquad (147)$$

When the closed curve C is taken as a circle of the radius R, then the magnetic field at the distance x just above from the center of the circle is given by eq(148) via eq(147) and in the limit of $x >> R$, it is given by the right hand side of eq(148). (**[advanced problem]** Show that eq(148) can be derived from eq(147).)

$$B(x) = \frac{\mu_0 I R^2}{2(R^2 + x^2)^{3/2}} \xrightarrow[x \to >> R]{} \frac{\mu_0 I R^2}{2x^3} \qquad (148)$$

Let's assume that the circle is replaced by a magnet and it is placed at the origin. By equating the magnetic field thus created equal to the one arising from the circle, the magnet is found to have the magnet moment of μ_M given by eq(149).

$$\mu_M = I\pi R^2 (= \text{current} \times \text{area of circle}) \qquad (149)$$

"note 5" Since the spin angular momentum of an electron is given by $(\pm) h/2\pi \cdot 1/2$, the magnetic moment of the spin is given by multiplying the momentum by a factor of e/m_e as in eq(150).

$$\frac{eh}{4\pi m_e} = \frac{h}{4\pi} \cdot \frac{e}{m_e} = \frac{1}{2} \cdot \frac{h}{2\pi} \times \frac{e}{m_e} = (\text{angular momentum}) \times \frac{e}{m_e} \qquad (150)$$

On the other hand, the magnetic moment of the electron due to its orbital angular momentum is given by multiplying it by $e/2m_e$, i.e., $(\text{angular momentum}) \times e/2m_e$.

Solution 1-79

If it is assumed that only one molecule exists in a droplet of the radius of 10 μm, then the number of molecules in 1 dm^3 is readily calculated as below by clarifying units associated with each value involved,

$$\frac{10^{-3}\,m^3(=1\ dm^3)}{\frac{4\pi}{3}\times(10\times10^{-6})^3 m^3\times 6.02\times10^{23}\,mol^{-1}}=3.97\times10^{-10}\,mol$$

and the result divided by the volume yields the concentration of the molecule in the droplet in the unit of M.

$$\frac{3.97\times10^{-10}\,mol}{1 dm^3}=3.97\times10^{-10}\,mol\cdot dm^{-3}=3.97\times10^{-10}\,M$$

Ans. below $3.97\times10^{-10}\,M$

Solution 1-80

Based on the electronic configuration (Cl: ${}^2P\,(\mathrm{Ne}3s^2 3p^5)$) of Cl, the molecular orbitals with which the seven valence electrons of the Cl atom are associated are depicted in Fig.72 in a similar manner to the case of F_2. Since the electronic configuration of Cl_2 in terms of the molecular orbitals is expressed as

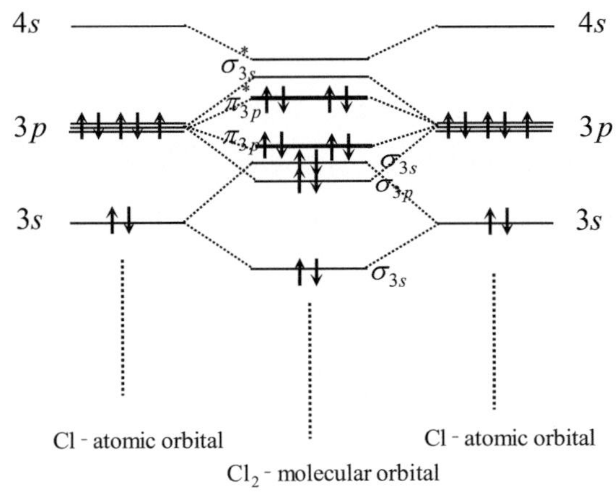

Fig.72 A part of energy levels and molecular orbitals constructed on the basis of the linear combinations of atomic orbitals.

$\sigma_{3s}<\sigma_{3p_z}<\sigma^*_{3s}<\pi_{3p_x}=\pi_{3p_y}<\pi^*_{3p_x}=\pi^*_{3p_x}<\sigma^*_{3p_z}$

$\sigma^2_{3s}\sigma^2_{3p}\sigma^{*2}_{3s}\pi^2_{3p_x}\pi^2_{3p_y}\pi^{*2}_{3p_x}\pi^{*2}_{3p_y}$, the chemical bond of Cl_2 is a sing bond of σ_{3p_z} type. Also, the order of the bond strength for Cl_2^+, Cl_2 and Cl_2^- is expected to be $Cl_2^+>Cl_2>Cl_2^-$ from the energy level diagram shown below.

Ans. the order of bond strength: $Cl_2^+>Cl_2>Cl_2^-$

Solution 1-81

The answers to (a), (b) and (d) are omitted. (See **solution 1-7.**)

(c) Expressing the Hamiltonian by $\hat{H}$ and eigenfunctions by ϕ_s, ϕ_x, ϕ_y and ϕ_z, then their eigenvalues are given by eq(151) and the eigenfunctions satisfy the normalization (eq(152)) and orthogonality (eq(153)) conditions.

$$\hat{H}\phi_s=E_s\phi_s,\ \hat{H}\phi_x=E_p\phi_x,\ \hat{H}\phi_y=E_p\phi_y,\ \hat{H}\phi_z=E_p\phi_z \qquad (151)$$

$$\int\phi_s^2 d\tau=1,\ \int\phi_i^2 d\tau=1,\ i=x,y,z \qquad (152)$$

$$\int\phi_s\cdot\phi_i d\tau=0,\ \int\phi_i\cdot\phi_j d\tau=0,\ i\neq j \qquad (153)$$

Also、from eqs(151) and (153), it is easy to show that these eigenfunctions satisfy eq(154).

$$\int\phi_s\hat{H}\phi_i d\tau=0,\ i=x,y,z,\ \int\phi_i\hat{H}\phi_j d\tau=0,\ i\neq j \qquad (154)$$

Since the wave functions employed here are real, the fact that the complex conjugate of any one of them is equal to itself, for instance $\phi_s^* = \phi_s$, is utilized throughout the calculations here.

The energy of φ_{sp} is calculated from eq(155),

$$E_{sp} = \int \varphi_{sp} \hat{H} \varphi_{sp} d\tau \qquad (155)$$

and by substituting the actual function for φ_{sp} and using eq(151) through eq(154), the energy is obtained as as eq(156) given below.

$$\int \frac{1}{2}(\phi_s + \sqrt{3}\phi_x)\hat{H}\frac{1}{2}(\phi_s + \sqrt{3}\phi_x)d\tau = \frac{1}{4}\times E_s + \frac{3}{4}\times E_p \qquad (156)$$

Ans. $\frac{1}{4}E_s + \frac{3}{4}E_p$

(e) Through the linear combination (hybridization) of <u>four independent component functions</u> (atomic orbitals), <u>four independent hybrid orbitals</u> are obtained.

Ans. four

(f) The orthogonality relation is confirmed by calculating directly the integrals in eqs(152) and (153) as below.

$$\int \varphi_a \cdot \varphi_b d\tau = \frac{1}{4}\int (\phi_s + \phi_x + \phi_y + \phi_z)\cdot(\phi_s - \phi_x - \phi_y + \phi_z)d\tau$$
$$= \frac{1}{4}(1-1-1+1) = 0 \qquad (157)$$

Ans. See text.

(g) The eigenenergies of the four hybrid orbitals of φ_a, φ_b, etc., can easily be calculated in a similar manner as in (c) above, yielding the same value of $\frac{1}{4}E_s + \frac{3}{4}E_p$ for any of them. Consequently, the sum of the energies of the four hybrid orbitals is equal to $E_s + 3E_p$, which is nothing more than the sum of the energy of one *s*-orbital and the energies of three *p*-orbitals. Namely, <u>it is of no value simply to create hybrid orbitals,</u> since <u>there is no gain in energy upon hybridization.</u> However, as is evident from the shape of the hybrid orbitals, they are far more suitable than *s*- or *p*-orbitals alone for bond formation. Actually, in the case of σ-bonds, for instance, the stronger bond is formed between one of the hybrid orbitals of C and 1*s*-orbital of H than simply between *s*- or one of *p*-orbitals of C and 1*s*-orbotal of H. Of course, it is needless to say that the symmetry of the hybrid orbitals employed should match the symmetry of the molecule whose bonding is under consideration. (See SALC: *<u>s</u>ymmetry <u>a</u>dapted <u>l</u>inear <u>c</u>ombination.*)

Ans. See text (g).

Solution 1-82

The *Hückel* approximation is generally employed so as to solve the determinant (secular equation) of a relevant matrix equation. Before following this, however, we present the problem here in order to understand basically what procedures are necessary to attain a molecular orbital and its energy. As to the explanation of the *Hückel* approximation, see **"note"** in **solution 1-86**.

The molecular plane of ethylene is taken as an *x-y* plane. The π-bond is constructed of the linear combination (LCAO) of the $2p_z$-orbitals which are not used for the sp^2-hybridization. The $2p_z$-orbital of each carbon atom of ethylene is denoted as ϕ_1 or ϕ_2. There are two independent linear combinations which are denoted by φ_1, φ_2 and are given by eq(158). They are called π -molecular orbitals from their shapes.

$$\left.\begin{aligned}\varphi_1 &= c_1\phi_1 + c_2\phi_2 \\ \varphi_2 &= c_1\phi_1 - c_2\phi_2\end{aligned}\right\} \qquad (158)$$

The atomic orbitals ϕ_1 and ϕ_2 are normalized to one and hence satisfy eq(159).

$$\int \phi_1\,\phi_1 d\tau \equiv \langle\phi_1|\phi_1\rangle = 1,\ \int \phi_2\phi_2 d\tau \equiv \langle\phi_2|\phi_2\rangle = 1 \qquad (159)$$

Also, in the simple *Hückel* approximation, the overlap integral S defined by eq(160) is put equal to zero (See **"note"**).

$$S = \int \phi_1\phi_2 d\tau = \int \phi_2\phi_1 d\tau \equiv \langle\phi_1|\phi_2\rangle = 0 \qquad (160)$$

The coefficients of c_1 and c_2 can be determined unambiguously from the conditions of orthogonality and normalization given by eq(161).

$$\int \varphi_i\varphi_j d\tau = \delta_{ij},\ \ i, j = 1 \text{ or } 2 \qquad (161)$$

The normalization condition yields $c_1^2 + c_2^2 = 1$, and the orthogonality condition leads to $c_1^2 - c_2^2 = 0$. From these the eq(162) is obtained.

$$c_1 = c_2 = \frac{\sqrt{2}}{2} \qquad (162)$$

On the other hand, the energies of the π-orbitals ε_1 and ε_2 are given by eq(163), in which the Hamiltonian $\hat{H}$ is different from that given in **problem 1-81** but it expresses the motion of the electron in the electric field generated by the carbon frame of ethylene molecule. α and β are named the coulomb integral and the resonance integral, respectively and both of them assume negative values.

$$\left.\begin{aligned}\varepsilon_1 &= \langle\varphi_1|\hat{H}|\varphi_1\rangle = \langle c_1\phi_1 + c_2\phi_2|\hat{H}|c_1\phi_1 + c_2\phi_2\rangle = c_1^2\alpha + c_2^2\alpha + 2c_1c_2\beta \\ \varepsilon_2 &= \langle\varphi_2|\hat{H}|\varphi_2\rangle = c_1^2\alpha + c_2^2\alpha - 2c_1c_2\beta\end{aligned}\right\} \qquad (163)$$

$$\langle\phi_1|\hat{H}|\phi_1\rangle = \langle\phi_2|\hat{H}|\phi_2\rangle = \alpha,\ \langle\phi_1|\hat{H}|\phi_2\rangle = \langle\phi_2|\hat{H}|\phi_1\rangle = \beta \qquad (164)$$

The energies of the π -orbitals are obtained by substituting the values determined for the coefficients c_1 and c_2 into eq(163) as eq(165).

$$\varepsilon_1 = \alpha + \beta,\ \ \varepsilon_2 = \alpha - \beta \qquad (165)$$

Since the energy of p_z-orbital is α, the bond energy of the π -bond becomes β per electron. Therefore, the π -bond energy of ethylene is given by 2β, which should be equal to the difference between the bond energies of the C-C double bond and the C-C single bond as assumed in the problem. For this, from the equation of $2\beta = -(167 - 88)\ \text{kcal}\cdot\text{mol}^{-1}$, β is found as $-39.5\ \text{kcal}\cdot\text{mol}^{-1}$.

Ans. the energies of the molecular orbitals: $\varepsilon_1 = \alpha + \beta$, $\varepsilon_2 = \alpha - \beta$; $\beta = -39.5\ \text{kcal}\cdot\text{mol}^{-1}$

"note" If it is assumed that the overlap integral in eq(160) is not zero, i.e., $S = <\phi_1|\phi_2> \neq 0$, then the energies are given by $\varepsilon_{1,2} = (\alpha \pm \beta)/(1 \pm S)$. As compared with the case of $S = 0$, it is found that the two carbon atoms tend to apart because of unstabilization due to repulsive interaction between the two atomic orbitals. See **problems 7-37** and **7-38**.

Solution 1-83

(a) In order to answer the problem, here we rely on the following two properties intrinsic to i) and ii). i) molecular orbitals constructed from the linear combinations of atomic orbitals and ii) the eigenfunctions and the eigenvalues (quantum numbers). (Of course, you can answer the problem using the LCAO MO method explained in **"note"** in **solution 1-86**.)

i) Since four atomic orbitals are employed, four independent molecular orbitals can be created. Four electrons are accommodated in these orbitals, two in the lowest and two in the second lowest orbital.

ii) There exists a general relation between the shape of an eigenfunction and its eigenenergy. The number of the points (nodes) or the planes (nodal planes) at or on which the eigenfunction takes zero against variables (location along bond in the case of molecular orbital) increases by one from zero with increasing the eigenenergy. When this general relation is applied to butadiene, its four molecular orbitals can be depicted roughly as shown in Fig.73 below. An analogy to the wavefunctions obtained for a particle (electron) in a

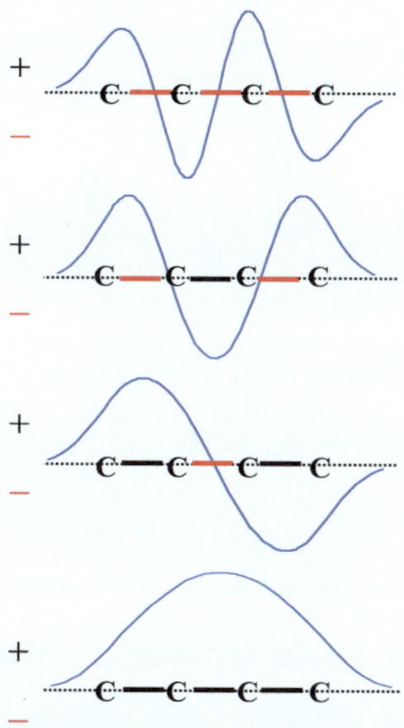

Fig.73 Molecular orbitals of butadiene depicted based on the intrinsic property of the eigenfunction.
As the number of nodes increases from 0 to 1, 2, 3, etc., the eigenenergy also increases. At the nodal points, the wavefunctions change their signs and, for this, the C – C bond there becomes antibonding. In the figure such bonds are indicated by red lines (——). As a whole, the bottom two molecular orbitals are bonding and the top two are antibonding.

one-dimensional box is obvious. The same idea can also be applied to the molecular orbitals of ring-shaped molecules. (See **problem 1-84**.)

Ans. The other three molecular orbitals (signs are distinguished by colors).

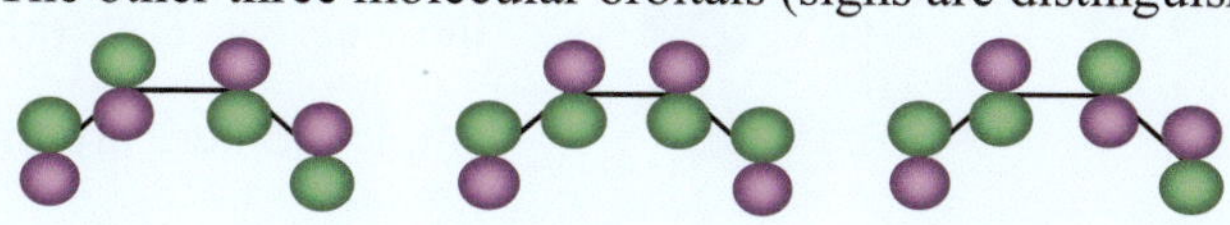

(b) See **solution 1-85**.

Ans. $(\sqrt{5}-1)\beta$

(c) According to the *Woodward-Hoffmann* law, the symmetry of the molecular orbital involved in the ring-opening or ring-closing should be conserved. Therefore, for the thermal ring-opening or ring-closing reactions which occur in the ground state, the symmetry of the highest occupied molecular orbital (HOMO) is conserved. On the other hand, since the corresponding photochemical reactions occur in the excited states, the symmetry of the lowest unoccupied molecular orbital (LUMO) must be conserved.

More specifically, the ring-opening reaction of cyclobutene which yields butadiene, whose highest occupied molecular π-orbital is anti-symmetric against the inversion at the central point as indicated in (a) above. Consequently, in the case of the thermal ring-opening, the split of the bond should lead the p-orbitals at the bond ends to the opposite rotation so as to form anti-symmetry p-orbitals. Namely, the directions of rotation are the same, either clockwise or counterclockwise. This type of rotation is called "*conrotatory*". On the other hand, in the case of the corresponding photochemical reactions, since the symmetry of LUMO is conserved, again as is evident from (a) above, the p-orbitals at both ends of the bond to be broken should orient to the same direction in order to conserve the symmetry during the reaction. Therefore, the rotations must occur in the opposite directions to each other. This type of rotation is called "*disrotatory*".

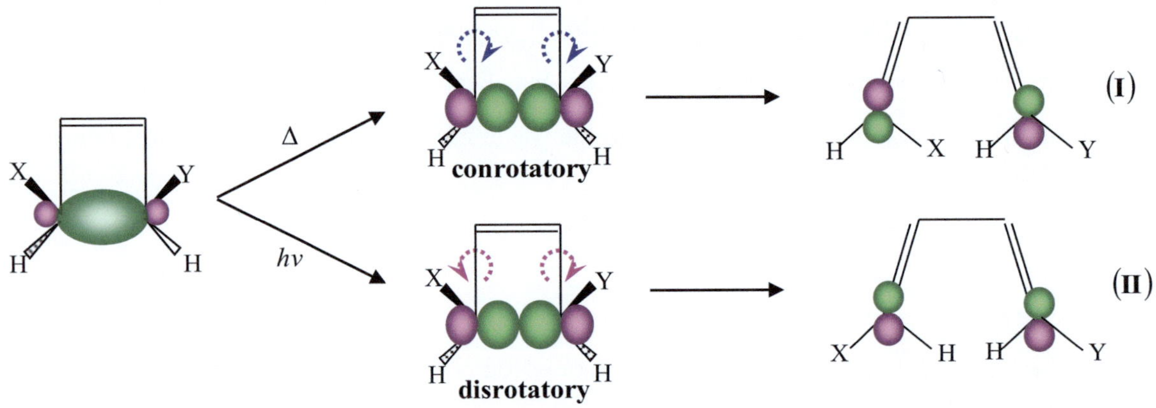

Fig.74 Illustration of symmetry conservation upon the thermal or photochemical ring opening reaction of cyclobutene.

Ans. product (**I**) for thermal reaction, product (**II**) for photo reaction

Solution 1-84

The answer is as is shown in the figure on the left. The signs of 2p-orbitals (plane views) are distinguished

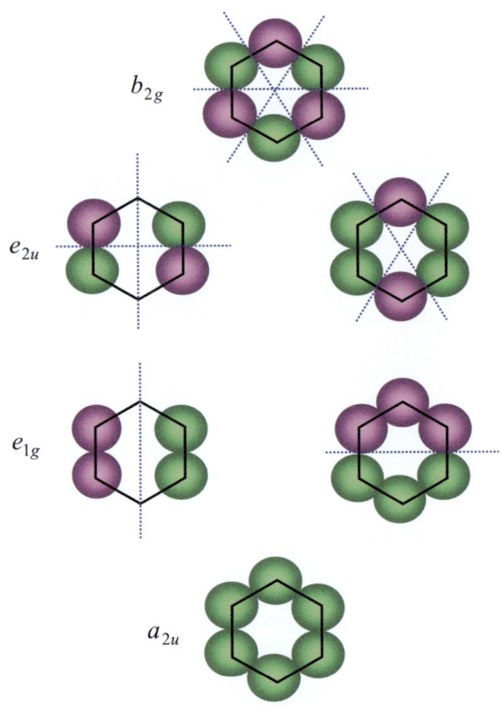

As the number of the nodal planes indicated by blue doted lines increases, the energy of the corresponding orbital increases. The orbitals indicated by e_{1g} and e_{2u} have the same number of nodal planes, 1 and 2, respectively and either of the e_{1g} or e_{2u} orbitals are degenerate. The symbols employed here are those of the irreducible representation for the point group D_{6h}. To represent the symmetry of molecular orbitals, small letters are used.

Fig.75 π-molecular orbitals of benzene and their nodal planes indicated by blue dotted lines.
The signs of 2p-orbitals are shown in different colors.

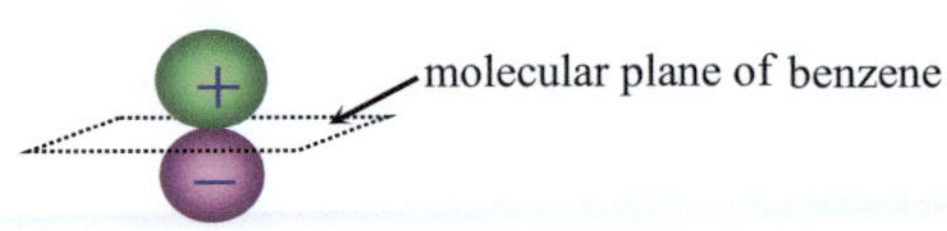

by using the different colors (see below). The orbital of the lowest energy (a_{2u}) has not a nodal plane perpendicular to the molecular plane.

Ans. See Fig.75.

Solution 1-85

The eigenenergy ε can be found by solving the secular equation (see **solution 1-86**).

(a) butadiene

The secular equation for butadiene is given by eq(166), in which α is the coulomb energy and β is the resonance energy.

$$\begin{vmatrix} x & 1 & 0 & 0 \\ 1 & x & 1 & 0 \\ 0 & 1 & x & 1 \\ 0 & 0 & 1 & x \end{vmatrix} = 0 \quad x = \frac{\alpha - \varepsilon}{\beta} \quad \Rightarrow \quad \varepsilon = \alpha - \beta x \qquad (166)$$

There exist various ways of solving the secular equation and taking eq(166) as an example, a few of them are introduced here.

(i) solution by expanding the equation (determinant)

$$\begin{vmatrix} x & 1 & 0 & 0 \\ 1 & x & 1 & 0 \\ 0 & 1 & x & 1 \\ 0 & 0 & 1 & x \end{vmatrix} = x\begin{vmatrix} x & 1 & 0 \\ 1 & x & 1 \\ 0 & 1 & x \end{vmatrix} - \begin{vmatrix} 1 & 0 & 0 \\ 1 & x & 1 \\ 0 & 1 & x \end{vmatrix} = x(x^3 - 2x) - (x^2 - 1) = x^4 - 3x^2 + 1 = 0 \qquad (167)$$

$$x^2 = \frac{3 \pm \sqrt{5}}{2} = \left(\frac{1 \pm \sqrt{5}}{2}\right)^2 \quad \Rightarrow \quad x = \frac{1+\sqrt{5}}{2},\ \frac{\sqrt{5}-1}{2},\ \frac{-\sqrt{5}+1}{2},\ -\frac{1+\sqrt{5}}{2} \qquad (167')$$

Ans. eigenenergies: $\alpha + \frac{1+\sqrt{5}}{2}\beta,\ \alpha + \frac{\sqrt{5}-1}{2}\beta,\ \alpha - \frac{\sqrt{5}-1}{2}\beta,\ \alpha - \frac{1+\sqrt{5}}{2}\beta$

The coefficients of the orbitals ($\varphi_i = c_{i1}\phi_1 + c_{i2}\phi_2 + c_{i3}\phi_3 + c_{i4}\phi_4$) are found by substituting the corresponding eigenenergy ε_i into eq(168) and by solving the resultant matrix equation. However, as is clear from eq(166),

$$\begin{pmatrix} x & 1 & 0 & 0 \\ 1 & x & 1 & 0 \\ 0 & 1 & x & 1 \\ 0 & 0 & 1 & x \end{pmatrix}\begin{pmatrix} c_{i1} \\ c_{i2} \\ c_{i3} \\ c_{i4} \end{pmatrix} = 0 \qquad (168)$$

the solutions of eq(168') are not unique. In order to decide them unambiguously, it is necessary to employ the normalization condition of $\sum_{j=1}^{4} c_{ij}{}^2 = 1$. We just calculate here the coefficients for the orbital of the lowest energy ($x = -(1+\sqrt{5}/2)$) of butadiene. By expanding eq(168'), the ratios of each coefficient, for instance, to

$$\begin{pmatrix} -(1+\sqrt{5})/2 & 1 & 0 & 0 \\ 1 & -(1+\sqrt{5})/2 & 1 & 0 \\ 0 & 1 & -(1+\sqrt{5})/2 & 1 \\ 0 & 0 & 1 & -(1+\sqrt{5})/2 \end{pmatrix}\begin{pmatrix} c_{11} \\ c_{12} \\ c_{13} \\ c_{14} \end{pmatrix} = 0 \qquad (168')$$

c_{11}, are found as below in eq(169).

$$\left.\begin{array}{l} -(1+\sqrt{5})/2\times c_{11}+c_{12}=0 \;\Rightarrow\; c_{12}=c_{11}(1+\sqrt{5})/2 \\ c_{11}-(1+\sqrt{5})/2\times c_{12}+c_{13}=0 \;\Rightarrow\; c_{13}=c_{11}\left(\left\{(1+\sqrt{5})/2\right\}^2-1\right)=c_{11}(1+\sqrt{5})/2 \\ c_{13}-(1+\sqrt{5})/2\times c_{14}=0 \;\Rightarrow\; c_{14}=c_{11}2\left(\left\{(1+\sqrt{5})/2\right\}^2-1\right)/(1+\sqrt{5})=c_{11} \end{array}\right\} \quad (169)$$

The ratios found in eq(169) are substituted into the normalization equation in order to solve for c_{11}, which is determined as in eq(170) via following equation.

$$\sum_{j=1}^{4} {c_{1j}}^2 = {c_{11}}^2\left\{1+\frac{(1+\sqrt{5})^2}{4}+\frac{(1+\sqrt{5})^2}{4}+1\right\}={c_{11}}^2\left(1+\frac{6+2\sqrt{5}}{2}+1\right)={c_{11}}^2(5+\sqrt{5})=1$$

$$c_{11}=\frac{1}{\sqrt{5+\sqrt{5}}} \qquad (170)$$

Each coefficient is finally determined from eq(169) by substituting eq(170).

Summarizing the calculation above, the normalized molecular orbital of the lowest energy is determined as in **Ans**. The coefficients of the other orbitals are found in a similar manner.

Ans. $\varphi_1=\left(\dfrac{1}{5+\sqrt{5}}\right)^{1/2}\left(\phi_1+\dfrac{1+\sqrt{5}}{2}\phi_2+\dfrac{1+\sqrt{5}}{2}\phi_3+\phi_4\right)$

(ii) utilization of mathematical software

Since an appropriate program to handle determinants is installed in most of mathematical software, it is straightforward to solve the secular equation to acquire molecular orbitals, if the system under consideration is not complicated. For instance, if we use the software of *Mathematica* to solve eq(166), then the eigenvalues calculated manually above are instantaneously found. At the same time, the eigenvectors, i.e., the coefficients c_{ij} of the molecular orbitals, are obtained. The components of the four eigenvectors corresponding to the four eigenenergies thus obtained are shown below. (See **solution 7-36**.)

$$\varphi_1:\left(1,\ \frac{1+\sqrt{5}}{2},\ \frac{1+\sqrt{5}}{2},\ 1\right),\ \varphi_2:\left(1,\ \frac{\sqrt{5}-1}{2},\ -\frac{\sqrt{5}-1}{2},\ -1\right),\ \varphi_3:\left(1,\ -\frac{\sqrt{5}-1}{2},\ -\frac{\sqrt{5}-1}{2},\ 1\right),\ \varphi_4:\left(1,\ -\frac{1+\sqrt{5}}{2},\ \frac{1+\sqrt{5}}{2},\ -1\right)$$

(b) benzene

The secular equation for benzene is given by eq(171).

$$\begin{vmatrix} x & 1 & 0 & 0 & 0 & 1 \\ 1 & x & 1 & 0 & 0 & 0 \\ 0 & 1 & x & 1 & 0 & 0 \\ 0 & 0 & 1 & x & 1 & 0 \\ 0 & 0 & 0 & 1 & x & 1 \\ 1 & 0 & 0 & 0 & 1 & x \end{vmatrix}=0 \quad x=\frac{\alpha-\varepsilon}{\beta} \quad \Rightarrow \quad \varepsilon=\alpha-\beta x \qquad (171)$$

(i) solution by expanding the equation (determinant)

(A)

$$\begin{vmatrix} x & 1 & 0 & 0 & 0 & 1 \\ 1 & x & 1 & 0 & 0 & 0 \\ 0 & 1 & x & 1 & 0 & 0 \\ 0 & 0 & 1 & x & 1 & 0 \\ 0 & 0 & 0 & 1 & x & 1 \\ 1 & 0 & 0 & 0 & 1 & x \end{vmatrix}=x\begin{vmatrix} x & 1 & 0 & 0 & 1 \\ 1 & x & 1 & 0 & 0 \\ 0 & 1 & x & 1 & 0 \\ 0 & 0 & 1 & x & 1 \\ 0 & 0 & 0 & 1 & x \end{vmatrix}-\begin{vmatrix} 1 & 0 & 0 & 0 & 1 \\ 1 & x & 1 & 0 & 0 \\ 0 & 1 & x & 1 & 0 \\ 0 & 0 & 1 & x & 1 \\ 0 & 0 & 0 & 1 & x \end{vmatrix}-\begin{vmatrix} 1 & 0 & 0 & 0 & 1 \\ x & 1 & 0 & 0 & 0 \\ 1 & x & 1 & 0 & 0 \\ 0 & 1 & x & 1 & 0 \\ 0 & 0 & 1 & x & 1 \end{vmatrix}$$

The first term on the right hand side is calculated to give eq(172) and from the second and third terms eq(173) is obtained.

$$x\begin{vmatrix} x & 1 & 0 & 0 & 1 \\ 1 & x & 1 & 0 & 0 \\ 0 & 1 & x & 1 & 0 \\ 0 & 0 & 1 & x & 1 \\ 0 & 0 & 0 & 1 & x \end{vmatrix} = xx\begin{vmatrix} x & 1 & 0 & 0 \\ 1 & x & 1 & 0 \\ 0 & 1 & x & 1 \\ 0 & 0 & 1 & x \end{vmatrix} - x\begin{vmatrix} 1 & 0 & 0 & 1 \\ 1 & x & 1 & 0 \\ 0 & 1 & x & 1 \\ 0 & 0 & 1 & x \end{vmatrix} = x^2(x^4 - 3x^2 + 1) - x\begin{vmatrix} x & 1 & 0 \\ 1 & x & 1 \\ 0 & 1 & x \end{vmatrix}$$

$$= x^2(x^4 - 3x^2 + 1) - x(x^3 - 2x) \qquad (172)$$

$$-\begin{vmatrix} 1 & 0 & 0 & 0 & 1 \\ 1 & x & 1 & 0 & 0 \\ 0 & 1 & x & 1 & 0 \\ 0 & 0 & 1 & x & 1 \\ 0 & 0 & 0 & 1 & x \end{vmatrix} - \begin{vmatrix} 1 & 0 & 0 & 0 & 1 \\ x & 1 & 0 & 0 & 0 \\ 1 & x & 1 & 0 & 0 \\ 0 & 1 & x & 1 & 0 \\ 0 & 0 & 1 & x & 1 \end{vmatrix} = -\begin{vmatrix} x & 1 & 0 & -1 \\ 1 & x & 1 & 0 \\ 0 & 1 & x & 1 \\ 0 & 0 & 1 & x \end{vmatrix} - \begin{vmatrix} 1 & 0 & 0 & -x \\ x & 1 & 0 & -1 \\ 1 & x & 1 & 0 \\ 0 & 1 & x & 1 \end{vmatrix}$$

$$= -x\begin{vmatrix} x & 1 & 0 \\ 1 & x & 1 \\ 0 & 1 & x \end{vmatrix} + \begin{vmatrix} 1 & 0 & -1 \\ 1 & x & 1 \\ 0 & 1 & x \end{vmatrix} - \begin{vmatrix} 1 & 0 & -1 \\ x & 1 & 0 \\ 1 & x & 1 \end{vmatrix} - x\begin{vmatrix} x & 1 & 0 \\ 1 & x & 1 \\ 0 & 1 & x \end{vmatrix}$$

$$= -x(x^3 - 2x) + (x^2 - 2) - (2 - x^2) - x(x^3 - 2x) \qquad (173)$$

By adding eq(172) to eq(173), eq(174) is obtained,

$$\begin{aligned} (172) + (173) &= x^2(x^4 - 3x^2 + 1) - x(x^3 - 2x) - x(x^3 - 2x) + (x^2 - 2) - (2 - x^2) - x(x^3 - 2x) \\ &= x^6 - 6x^4 + 9x^2 - 4 \\ &= x^2(x^4 - 6x^2 + 5) + 4(x^2 - 1) \\ &= x^2(x^2 - 1)\cdot(x^2 - 5) + 4(x^2 - 1) \\ &= (x^2 - 1)\cdot(x^4 - 5x^2 + 4) \\ &= (x^2 - 1)^2(x^2 - 4) = 0 \qquad (174) \end{aligned}$$

and as the solutions of eq(174), the values of $x = 2,\ 1,\ 1,\ -1,\ -1,\ -2$ are found, from which it is seen that the second and the third energy levels are doubly degenerate.

Ans. eigenenergies: $\alpha \pm 2\beta$, $\alpha \pm \beta$ (doubly degenerated)

(B) As an exercise for the calculation of determinants, a method of arranging a determinant so as to factor out just one cofactor is shown below (cofactor expansion).

Firs of all, we add all rows (from the second to the sixth) to the first row to find that all the elements of the first row of the rearranged determinant are equal to $x + 2$. Hence, the determinant can be rearranged as below by factoring out $x + 2$.

$$\begin{vmatrix} x & 1 & 0 & 0 & 0 & 1 \\ 1 & x & 1 & 0 & 0 & 0 \\ 0 & 1 & x & 1 & 0 & 0 \\ 0 & 0 & 1 & x & 1 & 0 \\ 0 & 0 & 0 & 1 & x & 1 \\ 1 & 0 & 0 & 0 & 1 & x \end{vmatrix} = \begin{vmatrix} x+2 & x+2 & x+2 & x+2 & x+2 & x+2 \\ 1 & x & 1 & 0 & 0 & 0 \\ 0 & 1 & x & 1 & 0 & 0 \\ 0 & 0 & 1 & x & 1 & 0 \\ 0 & 0 & 0 & 1 & x & 1 \\ 1 & 0 & 0 & 0 & 1 & x \end{vmatrix} = (x+2)\begin{vmatrix} 1 & 1 & 1 & 1 & 1 & 1 \\ 1 & x & 1 & 0 & 0 & 0 \\ 0 & 1 & x & 1 & 0 & 0 \\ 0 & 0 & 1 & x & 1 & 0 \\ 0 & 0 & 0 & 1 & x & 1 \\ 1 & 0 & 0 & 0 & 1 & x \end{vmatrix}$$

By subtracting the 1st column from each column of the determinant above, the resulting determinant is found to be expressed by the cofactor for the element at the 1st row and 1st column (1,1) as below. This cofactor is further rearranged by adding the 2nd row to the 5th row and then by subtracting the 2nd column from the 5th column, the determinant is found to be expressed by the cofactor of (5,5) as shown below.

In this way, the determinant is reduced step by step to the lower order by factoring out cofactors (see **"note"**) of the elements indicated by yellow marks to find that eq(171) is finally reduced to eq(175).

$$\begin{vmatrix} 1 & 1 & 1 & 1 & 1 & 1 \\ 1 & x & 1 & 0 & 0 & 0 \\ 0 & 1 & x & 1 & 0 & 0 \\ 0 & 0 & 1 & x & 1 & 0 \\ 0 & 0 & 0 & 1 & x & 1 \\ 1 & 0 & 0 & 0 & 1 & x \end{vmatrix} = \begin{vmatrix} x-1 & 0 & -1 & -1 & -1 \\ 1 & x & 1 & 0 & 0 \\ 0 & 1 & x & 1 & 0 \\ 0 & 0 & 1 & x & 1 \\ -1 & -1 & -1 & 0 & x-1 \end{vmatrix} = \begin{vmatrix} x-1 & 0 & -1 & -1 & -1 \\ 1 & x & 1 & 0 & 0 \\ 0 & 1 & x & 1 & 0 \\ 0 & 0 & 1 & x & 1 \\ 0 & x-1 & 0 & 0 & x-1 \end{vmatrix} = \begin{vmatrix} x-1 & 1 & -1 & -1 & -1 \\ 1 & x & 1 & 0 & 0 \\ 0 & 1 & x & 1 & 0 \\ 0 & -1 & 1 & x & 1 \\ 0 & 0 & 0 & 0 & x-1 \end{vmatrix}$$

$$= (x-1)\cdot\begin{vmatrix} x-1 & 1 & -1 & -1 \\ 1 & x & 1 & 0 \\ 0 & 1 & x & 1 \\ 0 & -1 & 1 & x \end{vmatrix} = (x-1)\cdot\begin{vmatrix} x-1 & 0 & -1 & -1 \\ 1 & x+1 & 1 & 0 \\ 0 & 1+x & x & 1 \\ 0 & 0 & 1 & x \end{vmatrix} = (x-1)\cdot\begin{vmatrix} x-1 & 0 & -1 & -1 \\ 1 & x+1 & 1 & 0 \\ -1 & 0 & x-1 & 1 \\ 0 & 0 & 1 & x \end{vmatrix}$$

$$= (x-1)\cdot(x+1)\cdot\begin{vmatrix} x-1 & -1 & -1 \\ -1 & x-1 & 1 \\ 0 & 1 & x \end{vmatrix} = (x-1)\cdot(x+1)\cdot\begin{vmatrix} x-2 & x-2 & 0 \\ -1 & x-1 & 1 \\ 0 & 1 & x \end{vmatrix} = (x-1)\cdot(x+1)\cdot\begin{vmatrix} x-2 & 0 & 0 \\ -1 & x & 1 \\ 0 & 1 & x \end{vmatrix}$$

$$= (x-1)\cdot(x+1)\cdot(x-2)\cdot\begin{vmatrix} x & 1 \\ 1 & x \end{vmatrix} = (x-1)^2(x+1)^2(x-2)$$

$$\text{eq}(171) = (x+2)\cdot(x-1)^2(x+1)^2(x-2) \qquad (175)$$

The eigenfuction (coefficients of c_{1i} $(i = 1,2,\cdot\cdot,6)$) for the eigenvalue of $x = -2$ is determined to satisfy the following equation(eq(176)).

$$\begin{pmatrix} -2 & 1 & 0 & 0 & 0 & 1 \\ 1 & -2 & 1 & 0 & 0 & 0 \\ 0 & 1 & -2 & 1 & 0 & 0 \\ 0 & 0 & 1 & -2 & 1 & 0 \\ 0 & 0 & 0 & 1 & -2 & 1 \\ 1 & 0 & 0 & 0 & 1 & -2 \end{pmatrix}\begin{pmatrix} c_{11} \\ c_{12} \\ c_{13} \\ c_{14} \\ c_{15} \\ c_{16} \end{pmatrix} = 0 \qquad (176)$$

One example of solving eq(176) is shown below. By expanding eq(176), the following equations are obtained.

$$-2c_{11} + c_{12} + c_{16} = 0 \;\Rightarrow\; c_{16} = 2c_{11} - c_{12} \qquad \text{(i)}$$

$$c_{11} - 2c_{12} + c_{13} = 0 \;\Rightarrow\; c_{13} = -c_{11} + 2c_{12} \qquad \text{(ii)}$$

$$c_{12} - 2c_{13} + c_{14} = 0 \;\Rightarrow\; c_{14} = -c_{12} + 2c_{13} = -c_{12} + 2(-c_{11} + 2c_{12}) = 3c_{12} - 2c_{11} \qquad \text{(iii)}$$

$$c_{13} - 2c_{14} + c_{15} = 0 \;\Rightarrow\; c_{15} = -c_{13} + 2c_{14} = c_{11} - 2c_{12} + 2(3c_{12} - 2c_{11}) = 4c_{12} - 3c_{11} \qquad \text{(iv)}$$

$$c_{14} - 2c_{15} + c_{16} = 0 \;\Rightarrow\; c_{16} = -c_{14} + 2c_{15} = -(3c_{12} - 2c_{11}) + 2(4c_{12} - 3c_{11}) = 5c_{12} - 4c_{11} \qquad \text{(v)}$$

$$c_{11} + c_{15} - 2c_{16} = 0 \qquad \text{(vi)}$$

From eq(i) and (v), eq(vii) are derived and substituting it into eqs(ii)-(iv), eq(viii) is obtained. The value of

$1/\sqrt{6}$ is found from the normalization condition.

$$5c_{12} - 4c_{11} = 2c_{11} - c_{12} \;\Rightarrow\; c_{12} = c_{11} \qquad \text{(vii)}$$

$$c_{12} = c_{13} = c_{14} = c_{15} = c_{16} = c_{11}(=1/\sqrt{6}) \qquad \text{(viii)}$$

The coefficients c_{2i} $(i = 1, 2, \cdot\cdot, 6)$ of the molecular orbital with the eigenvalue of $x = -1$ are determined so as to satisfy eq(177).

$$\begin{pmatrix} -1 & 1 & 0 & 0 & 0 & 1 \\ 1 & -1 & 1 & 0 & 0 & 0 \\ 0 & 1 & -1 & 1 & 0 & 0 \\ 0 & 0 & 1 & -1 & 1 & 0 \\ 0 & 0 & 0 & 1 & -1 & 1 \\ 1 & 0 & 0 & 0 & 1 & -1 \end{pmatrix} \begin{pmatrix} c_{21} \\ c_{22} \\ c_{23} \\ c_{24} \\ c_{25} \\ c_{26} \end{pmatrix} = 0 \qquad (177)$$

Expanding eq(177), the following equations are obtained.

$$-c_{21} + c_{22} + c_{26} = 0 \;\Rightarrow\; c_{26} = c_{21} - c_{22} \qquad \text{(i')}$$

$$c_{21} - c_{22} + c_{23} = 0 \;\Rightarrow\; c_{23} = -c_{21} + c_{22} \qquad \text{(ii')}$$

$$c_{22} - c_{23} + c_{24} = 0 \;\Rightarrow\; c_{24} = -c_{22} + c_{23} = -c_{22} + (-c_{21} + c_{22}) = -c_{21} \qquad \text{(iii')}$$

$$c_{23} - c_{24} + c_{25} = 0 \;\Rightarrow\; c_{25} = -c_{23} + c_{24} = c_{21} - c_{22} - c_{21} = -c_{22} \qquad \text{(iv')}$$

$$c_{14} - c_{15} + c_{16} = 0 \;\Rightarrow\; c_{26} = -c_{24} + c_{25} = c_{21} - c_{22} \qquad \text{(v')}$$

Here the two coefficients are rewritten as eq(vi'),

$$c_{21} = a, \;\; c_{22} = b \qquad \text{(vi')}$$

then from eqs(ii')-(v'), each coefficients can be expressed in terms of *a* and *b* as in eq(vii').

$$c_{26} = a - b, \;\; c_{23} = -a + b, \;\; c_{24} = -a, \;\; c_{25} = -b \qquad \text{(vii')}$$

The normalization condition for the coefficient given by eq(viii') is also expressed in terms of *a* and *b*.

$$\sum_{i=1}^{6} c_{2i}{}^{2} = a^2 + b^2 + (-a + b)^2 + a^2 + b^2 + (a - b)^2 = 4(a^2 - ab + b^2) = 1 \qquad \text{(viii')}$$

When a is put equal to $1/2$ as in eq(ix'), values for *b* are obtained as in eq(x').

$$a = \frac{1}{2} \qquad \text{(ix')}, \qquad -\frac{1}{2}b + b^2 = 0 \quad \Rightarrow \quad b = 0, \;\; b = \frac{1}{2} \qquad \text{(x')}$$

From these values of *a* and *b*, two sets of the coefficients c_{2i} $(i = 1, 2, \cdot\cdot, 6)$ are obtained as below.

$$\frac{1}{2}\times(1, \;0, \;-1, \;-1, \;0, \;1), \;\; \frac{1}{2}\times(1, \;1, \;0, \;-1, \;-1, \;0)$$

Therefore, there are two molecular orbitals for the second lowest energy level as are given by eq(178).

$$\varphi_2 = \frac{1}{2}\times(\phi_1 - \phi_3 - \phi_4 + \phi_6), \;\; \varphi_3 = \frac{1}{2}\times(\phi_1 + \phi_2 - \phi_4 - \phi_5) \qquad (178)$$

However, these two molecular orbitals are not mutually orthogonal as is readily confirmed by eq(179).

$$\langle\varphi_2|\varphi_3\rangle = \frac{1}{4}\langle(\phi_1 - \phi_3 - \phi_4 + \phi_6)|(\phi_1 + \phi_2 - \phi_4 - \phi_5)\rangle = \frac{1}{2} \neq 0 \qquad (179)$$

The two molecular orbitals which are orthogonal to each other can be constructed by linearly combining φ_2 and φ_3 as are given in eq(180).

$$\left.\begin{aligned} \varphi_2 + \varphi_3 &= \frac{1}{2}(2\phi_1 + \phi_2 - \phi_3 - 2\phi_4 - \phi_5 + \phi_6) \\ \varphi_2 - \varphi_3 &= \frac{1}{2}(-\phi_2 - \phi_3 + \phi_5 + \phi_6) \end{aligned}\right\} \qquad (180)$$

The sets of coefficients for the other three molecular orbitals are given below (not normalized).

$$(-1,\ 1,\ 0,\ -1,\ 1,\ 0),\ (-1,\ 0,\ 1,\ -1,\ 0,\ 1),\ (-1,\ 1,\ -1,\ 1,\ -1,\ 1)$$

The sets of coefficients for the molecular orbitals all of which are mutually orthogonal are as follows.

$$\begin{aligned}&(c_{j1},\ c_{j2},\ c_{j3},\ c_{j4},\ c_{j5},\ c_{j6} : j=1-6) =\\ &\quad (1,\ 1,\ 1,\ 1,\ 1,\ 1),\ (0,\ -1,\ -1,\ 0,\ 1,\ 1),\ (2,\ 1,\ -1,\ -2, -1,\ 1),\\ &\quad (2,\ -1,\ -1,\ 2,\ -1,\ -1),\ (0,\ 1,\ -1,\ 0,\ 1,\ -1),\ (-1,\ 1,\ -1,\ 1,\ -1,\ 1)\end{aligned} \tag{181}$$

Since the sign of a wavefunction is arbitrary by a factor of ± 1, each set of the coefficients in eq(181) can be multiplied by ± 1. For instance, $(0, -1, -1,\ 0,\ 1,\ 1)$ and $(0,\ 1,\ 1,\ 0, -1, -1)$ represent the same molecular orbital.

Ans. molecular orbitals (not normalized): eq(181)

(ii) utilization of mathematical software

The eigenvalues of $x = 2,\ 1,\ 1,\ -1,\ -1,\ -2$ are obtained in a second as well as the coefficients of the eigenvectors.

"note" A matrix of the n-th order is represented by $A = (a_{ij})$. The determinant of the matrix which is formed by removing the i-th row and j-th column is denoted by $\widetilde{d}_{ij}$. Then the cofactor of a_{ij} is defined by the equation below and written as $\widetilde{a}_{ij}$. The calculations above, the sign of $(-1)^{i+j}$ is neglected for simplicity since they are just for solving det$A = 0$

$$A = (a_{ij}),\quad \widetilde{a}_{ij} = (-1)^{i+j}\,\widetilde{d}_{ij}$$

Solution 1-86

(a) The secular equation to be solved is eq(182). It is formally divided into two parts, benzene part (yellow) and ethylene part (pale green), both of which are interacting each other through the elements marked pink. If the pink-marked elements are zero, then the secular equation is decomposed into two diagonal determinants and hence the eigenvalues are those for benzene plus those for ethylene.

$$\begin{vmatrix} x & 1 & 0 & 0 & 0 & 0 & 0 & 0\\ 1 & x & 1 & 0 & 0 & 0 & 0 & 0\\ 0 & 1 & x & 1 & 0 & 0 & 0 & 1\\ 0 & 0 & 1 & x & 1 & 0 & 0 & 0\\ 0 & 0 & 0 & 1 & x & 1 & 0 & 0\\ 0 & 0 & 0 & 0 & 1 & x & 1 & 0\\ 0 & 0 & 0 & 0 & 0 & 1 & x & 1\\ 0 & 0 & 1 & 0 & 0 & 0 & 1 & x \end{vmatrix} = 0 \quad x = \frac{\alpha - \varepsilon}{\beta} \quad \Rightarrow \quad \varepsilon = \alpha - \beta x \tag{182}$$

(i) Eq(182) is expanded into eq(183) and hence we just need to solve eq(183).

$$x^8 - 8x^6 + 19x^4 - 16x^2 + 4 = 0 \tag{183}$$

Some time before, it used to one of the basically important excises in physical chemistry to acquire the molecular orbitals and their energies of several compounds. Nowadays, however, in many cases pc can yield required answers in an instant and just to solve secular equations is not so important in itself as before.

(ii) Utilization of mathematical software.

Ans. 8 eigenenergies: $\varepsilon = \alpha \pm \beta,\ \alpha \pm \sqrt{2}\beta,\ \alpha \pm \frac{1}{2}\sqrt{2(5-\sqrt{17})}\beta,\ \alpha \pm \frac{1}{2}\sqrt{2(5+\sqrt{17})}\beta$

(b) The energy levels obtained for the molecular orbitals of the π -conjugated systems of ethylene, benzene and styrene are shown in Fig.76. As compared with those of benzene,

(i) The two doubly degenerated levels come to be undegenerated for styrene to yield eight single levels.

(ii) The lowest energy level is lowered.

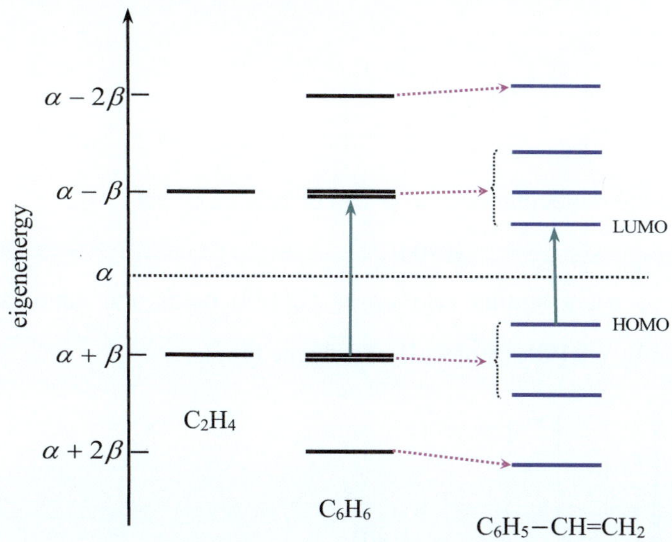

Fig.76 Energy levels of the molecular orbitals of the π-conjugated systems of ethylene, benzene and styrene.

As to the electronic transition due to the light absorption, see **solution 7-4**.

(iii) The total energy of the 8 electrons of styrene is smaller than the sum of the total energies of 2 electrons of ethylene and 6 of benzene as are calculated below. In other words, by substituting a vinyl group to benzene, the whole π-system comes to be stabilized more than benzene.

$$\varepsilon(\text{stylene}) = 2\left(\alpha + \frac{1}{2}\sqrt{2(5+\sqrt{17})}\beta\right) + 2\left(\alpha + \sqrt{2}\beta\right) + 2(\alpha + \beta) + 2\left(\alpha + \frac{1}{2}\sqrt{2(5-\sqrt{17})}\beta\right)$$
$$= 8\alpha + 10.42\beta \qquad (184)$$

$$\varepsilon(\text{ethylene}+\text{benzene}) = 2(\alpha + 2\beta) + 4(\alpha + \beta) + 2(\alpha + \beta) = 8\alpha + 10\beta \qquad (185)$$

(iv) The energy separation between HOMO and LUMO becomes narrower. In other words, the absorption wavelength shifts to longer wavelength. Also, since the molecular symmetry of styrene is lowered, the molecular extinction coefficient is expected to increase and at the same time the fine vibrational structures of the electronic absorption band will become obscure. It is worth comparing the two spectra measured for benzene and styrene in the gas phase. (*J.Phys.Chem.*,92, 4914 (1988).)

Ans. See (b) text.

"note" Simple *Hückel* Approximation

The molecular orbital φ of a molecule consisting of n atoms (of the same kind) are assumed to be expressed by the linear combinations of the atomic orbital ϕ_i of the i-th atom as in eq(186). Here, c_i is a proportional constant.

$$\varphi = \sum_{i=1}^{n} c_i \phi_i \qquad (186)$$

The Hamiltonian for that molecule is written as H and the eigenenergy as ε. Then, the molecular orbital φ satisfies eq(187).

$$H\varphi = \varepsilon\varphi \qquad (187)$$

When it is assumed that (i) the overlap integral between ϕ_i and ϕ_j is zero; (ii) the coulomb integral takes the same value of α for any component atomic orbital; (iii) for a pair of atoms in the nearest neighbor the resonance integral takes the same value of β. All these approximations are expressed by eq(188) and the approximation as a whole is called "Simple *Hückel* Approximation".

$$\left.\begin{array}{l} \langle\phi_i|\phi_j\rangle = \delta_{ij} \\ \langle\phi_i|H|\phi_i\rangle = \alpha \\ \langle\phi_i|H|\phi_{i+1}\rangle = \langle\phi_i|H|\phi_{i-1}\rangle = \beta \end{array}\right\} \qquad (188)$$

In this approximation, the atomic orbital ϕ_i and the molecular orbital φ can be related by eq(189).

$$\langle\phi_i|H|\varphi\rangle = \alpha c_i + \beta c_{i-1} + \beta c_{i+1} = \varepsilon c_i \qquad (189)$$

For each of all atomic orbitals $\phi_i(i=1,\ 2,\ 3,\cdots,\ n)$, a similar relation as eq(189) holds and all of such relations can be expressed in a matrix form given by eq(190) (molecule of polyene type).

$$\begin{pmatrix} \alpha & \beta & 0 & 0 & \cdot & 0 & 0 \\ \beta & \alpha & \beta & \cdot & & & 0 \\ 0 & \beta & \alpha & \beta & & & \cdot \\ 0 & \cdot & \beta & \cdot & & & \cdot \\ \cdot & & & & & \cdot & 0 \\ 0 & & & & \beta & \alpha & \beta \\ 0 & 0 & \cdot & \cdot & 0 & \beta & \alpha \end{pmatrix} \begin{pmatrix} c_1 \\ c_2 \\ \\ \\ \\ \\ c_n \end{pmatrix} = \varepsilon \begin{pmatrix} c_1 \\ c_2 \\ \\ \\ \\ \\ c_n \end{pmatrix} \qquad (190)$$

The right side of eq(190) is transferred to the left side and is rewritten as eq(191), in which $(\alpha-\varepsilon)/\beta$ is replaced by x.

$$\begin{pmatrix} \alpha-\varepsilon & \beta & 0 & 0 & \cdot & 0 & 0 \\ \beta & \alpha-\varepsilon & \beta & \cdot & & & 0 \\ 0 & \beta & \alpha-\varepsilon & \beta & & & \cdot \\ 0 & \cdot & \beta & \cdot & & & \cdot \\ \cdot & & & & \cdot & & 0 \\ 0 & & & & \beta & \alpha-\varepsilon & \beta \\ 0 & 0 & \cdot & \cdot & 0 & \beta & \alpha-\varepsilon \end{pmatrix} \begin{pmatrix} c_1 \\ c_2 \\ \\ \\ \\ \\ c_n \end{pmatrix} = 0 \Rightarrow \begin{pmatrix} x & 1 & 0 & 0 & \cdot & 0 & 0 \\ 1 & x & 1 & \cdot & & & 0 \\ 0 & 1 & x & 1 & & & \cdot \\ 0 & \cdot & 1 & \cdot & & & \cdot \\ \cdot & & & & & \cdot & 0 \\ 0 & & & & 1 & x & 1 \\ 0 & 0 & \cdot & \cdot & 0 & 1 & x \end{pmatrix} \begin{pmatrix} c_1 \\ c_2 \\ \\ \\ \\ \\ c_n \end{pmatrix} = 0 \qquad (191)$$

For eq(191) to be solved under the condition that at least one of $(c_1,\ c_2,\ c_3,\cdots,\ c_n)$ is not zero, the determinant of the n-th order must be zero as expressed by eq(192).

$$\begin{vmatrix} x & 1 & 0 & 0 & \cdot & 0 & 0 \\ 1 & x & 1 & \cdot & & & 0 \\ 0 & 1 & x & 1 & & & \cdot \\ 0 & \cdot & 1 & \cdot & & & \cdot \\ \cdot & & & & & \cdot & 0 \\ 0 & & & & 1 & x & 1 \\ 0 & 0 & \cdot & \cdot & 0 & 1 & x \end{vmatrix} = 0 \qquad (192)$$

Eq(192) is what is called secular equation in simple *Hückel* approximation. By solving the secular equation, a set of n molecular orbitals and n eigenenergies are obtained. When we wrote eq(187), we did not expect explicitly that the number of n molecular orbitals were obtained at a time. Of course, the number of n independent components (atomic orbitals) can create the number of n independent components by linear combination. Therefore, by solving eq(192), not only the molecular orbital in the ground state but also those in the excited states are simultaneously obtained. But why is that so?

Solution 1-87

Carbon monoxide CO has the largest bond energy among the diatomic molecules. The dipole moment is only 0.117 D and directs from O to C. The molecule is not paramagnetic. Taking these facts as well as that the number of valence electrons is four for C and 6 for O into account, possible limiting molecular structures which satisfy the octet rule are speculated as follows.

The molecular orbitals considered for diatomic molecules (see **solution 1-45**) suggest a triple bond formation for CO as is so for N_2 which has the same number of the valence electrons as CO. Since the molecule has no unpaired electron, the structure (a) is considered as one of the limiting structures. As an unpolarized structure expected from the small dipole moment, the structure (b) is conceivable. For a structure in which the dipole moment orients oppositely to that of (a), the structure (c) is possible.

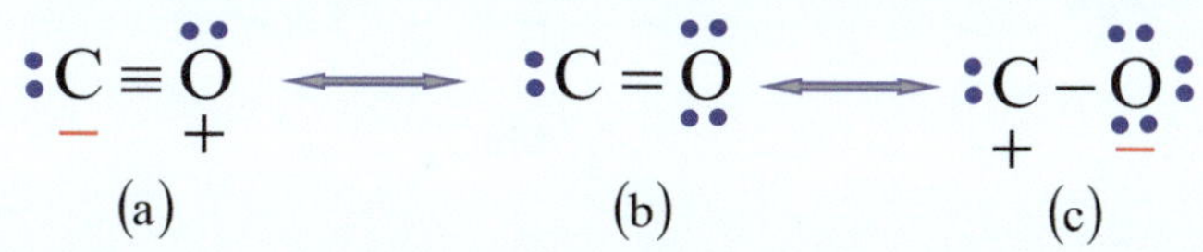

Fig.77 Three possible limiting molecular structures of CO which satisfy the octet rule.

Considering that the bond energy is larger than that of N_2 (triple bond), the structure (a) is most appropriate, and based on the electronegativities of C and O, the structure (c) is likely, and from the small value of the dipole moment, the structure (a) and (c) are expected to contribute comparably to the resonance. The structure (b) is the most ordinary structure jugging from its bond strength and unpolarized structure. Summarizing all these, it can be said that each of the structures (a), (b) and (c) contributes to the resonance to comparable extent.

Ans. See text and Fig.77.

Solution 1-88

The electronic configurations of Cr and Mn are Cr([Ar]$4s^2 4d^4$) and Mn([Ar]$4s^2 4d^5$), respectively. The oxidation number (o.n.) of K is $+1$ from the rule ① given below. Therefore the charges of the atomic groups in the problem are expressed by $Cr_2O_7^{-2}$ and MnO_4^{-1}. Most simply, the oxidation number of a target atom is obtained by putting the oxidation number of oxygen as -2 so as to make the sum of the oxidation numbers of the constituent atoms of the atomic group equal to the charges of the corresponding atomic group as exemplified below:

x stands for o.n. of Cr, $2x+(-2)\times 7=-2 \Rightarrow x=6,$ hence o.n. of Cr is $+6$,

x stands for o.n. of Mn, $x+(-2)\times 4=-1 \Rightarrow x=7,$ hence o.n. of Mn is $+7$.

These oxidation numbers are also obtained from the numbers of valence electrons. They are 6 and 7 for Cr and Mn, respectively. When they are most extensively oxidized and lose all valence electrons (to other atoms bonded with), their oxidation numbers become $+6$ and $+7$, respectively.

From a viewpoint of chemical structures, all the atoms constituting a compound are assumed to form all bonds so as to satisfy the octet rule. All of the bonding electrons are assigned either to each atom linked by a bond to the atom under consideration or the latter atom itself according to the rules given below.

The two compounds treated here are depicted in Fig.78 so as for every constituent atom to satisfy the octet rule. Blue dots indicate electrons from O atom and yellow dots are those from outside (K atom for instance). Green or violet dots are electrons from the valence electrons of Cr (6 electrons) or those of Mn (7 electrons). As is clear from Fig.78, every O atom satisfies the octet rule and no electrons are assigned to the central atom of Cr or Mn. Therefore, it is evident that the oxidation numbers of these atoms are $+6$ and $+7$, respectively.

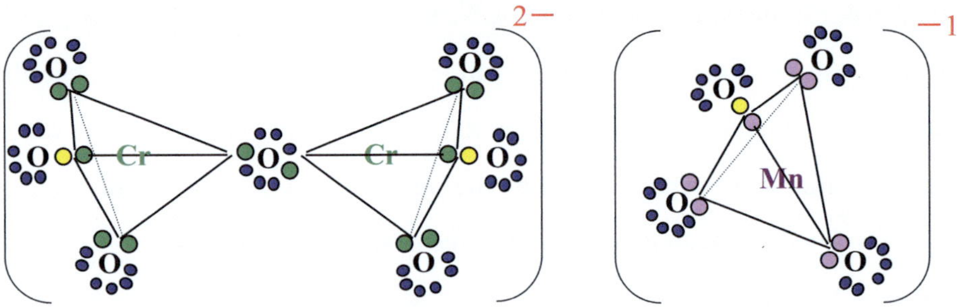

$Cr_2O_7^{-2}$ MnO_4^-

Fig.78 Chemical structure and oxidation number. Bonding electrons are classified by different colors.

Ans. See text.

Definition of oxidation number (o.n.).

In order to clarify which atoms in chemical reactions are oxidized or reduced, the concept of the oxidation number has been introduced to describe in and out of electrons between before and after chemical reactions. The oxidation number indicates that a certain atom of a molecule or an ion has less or more electrons compared with the number of the electrons the atom has in the isolated state and is expressed such as $+1$, $+2$, -1, -2 -2, etc.

Upon reaction (an increase in o.n. is Oxyidation, a decrease in o.n. is Reduction.)

Rules ①–⑦ in acquiring the oxidation number.

Rule ① The oxidation number of the O atom in compounds is put equal to -2. That of H, Na and K is $+1$ and that of F and Cl is -1.

Rule ② The oxidation number of the atom in a simple substance (composed of the same elements), such as O_2 and H_2 is put equal to 0.

Rule ③ The bond electrons of a bond between two atoms of the same element are shared and hence –no electron is given to or taken from one atom.

Rule ④ For a single ionic atom, the charge on the ion is regarded as equal to the oxidation number.

Rule ⑤ The oxidation number of a polyatomic ion is determined so as for the sum of the oxidation numbers over all the constituent atoms to become equal to the charge of the ion.

Rule ⑥ For an electronically neutral compound, the sum of the oxidation numbers over all the constituent atoms is zero.

Rule ⑦ For a covalent bond between hetero atoms (especially for organic compounds), the shared electrons are ascribed to the atom of the greater electronegativity.

Based on these rules, a few examples are given below to show how to decide the oxidation numbers.

(ex.1) The oxidation number of the O atom in H_2O.

From the structure of $H^+-O^{--}-H^+$, the oxidation number of O is found as -2.

(ex.2) The oxidation number of the O atom in H_2O_2.

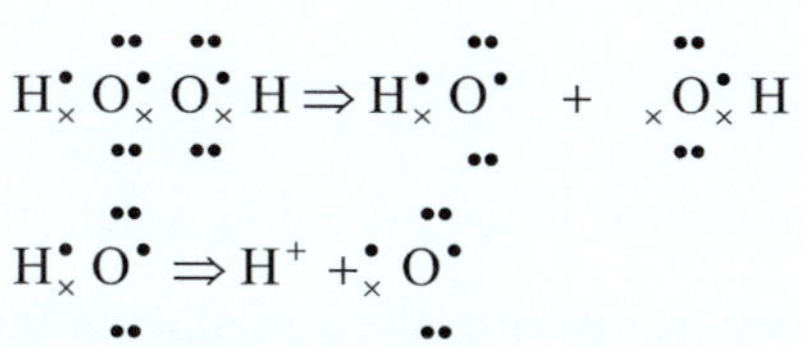

Fig.79 Arrangement of the outmost octet electrons of H_2O_2.

Since the bond electrons indicated by ● and × are assigned to the O atom as shown in the figure on the left, the number of electrons of the O atom amounts to 7 yielding the oxidation number of -1.

(ex.3) The oxidation numbers of the Cl atom in Cl^-, ClO^-, ClO_2^-, ClO_2, ClO_3^- and ClO_4^-.

They are -1, $+1$, $+3$, $+4$, $+5$, $+7$ in the above order.

(The calculations of the oxidation numbers are easy but to associate the results with their chemical structures is not an easy task.)

How to decide the oxidation numbers in ex.3 is explained in detail in the following. In the case of ClO_4^-, for instance, according to the rules ① and ⑤, the oxidation number of the Cl atom put equal to x should satisfy the equation of $x+(-2)\times 4=-1$, from which $x=+7$ is obtained. Based on the chemical structure of ClO_4^- which is assumed to be a tetrahedral, all the bond electrons are assigned to each atom as is shown in Fig.80. The Cl atom at the center of the molecule has not only its original 7 electrons marked in red but also 1 electron marked in yellow, totaling to 8 electrons, all of which are arrange in the four sp^3-hybrid orbitals with two in each. When the oxidation number of the Cl atom is considered, every two electrons shared with one of the four O atoms are assigned to the O atom. Consequently the Cl atom is regarded as having lost its original 7 electrons and its oxidation number becomes +7.

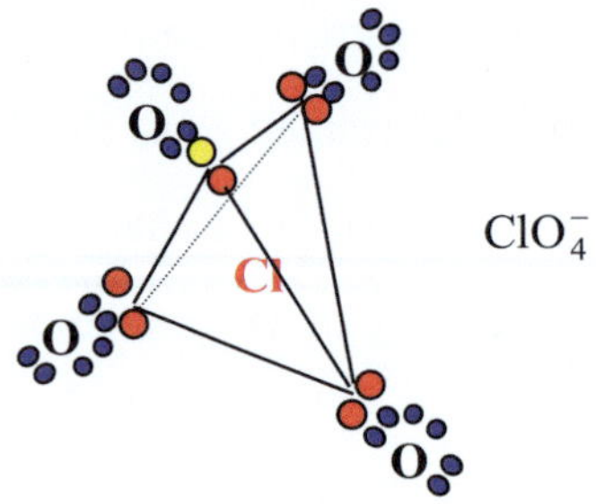

Fig.80 Chemical bond structure of ClO_4^- based on the octet rule.

For other ions in ex.3, their ion structures can be drawn as below by using the **VSERP** model. By assigning all bond electrons between the O and Cl atom to the O atom ($\chi_O > \chi_{Cl}$), the oxidation number of the Cl atom can easily be determined as given in parentheses.

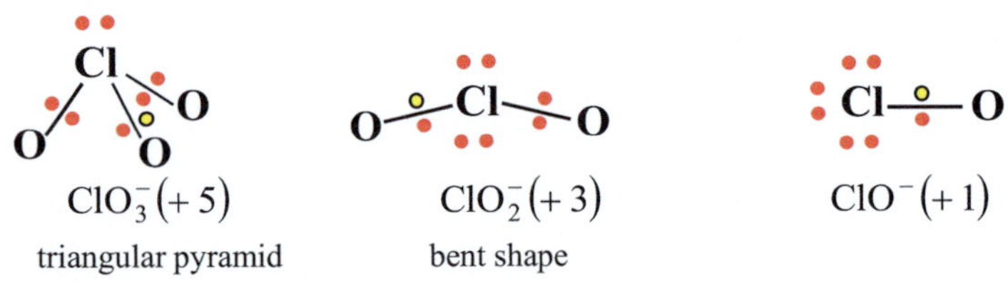

Fig.81 Chemical bond structures of chlorate, chlorite, and hypochlorite ions based on the octet rule.

Solution 1-89

Based on the definition of the oxidation number presented in the preceding solution, the oxidation number of each atom of all the compounds in the problem are as given in the parentheses in **Ans**. The sum of the oxidation numbers over all the constituent atoms of a compound is zero for neutral compounds and is equal to the charge number for ions.

Ans. $\mathbf{H_2}$ (H: 0), $\mathbf{O_2}$ (O: 0), $\mathbf{H_2O}$ (H: +1, O: −2), $\mathbf{H_2O_2}$ (H:+1, O: −1), $\mathbf{FeSO_4}$ (Fe: +2, S:+6, O: −2), $\mathbf{K_2Cr_2O_4}$ (K: +1, Cr: +3, O: −2), $\mathbf{Fe_2O_3}$ (Fe: +3, O: −2), **NO** (N: +2, O: −2), $\mathbf{NO_2}$ (N: +4, O: −2), $\mathbf{NO_3^-}$ (N: +5, O:−2), $\mathbf{SO_2}$ (S: +4, O: −2), $\mathbf{SO_3}$ (S: +6, O: −2), $\mathbf{H_2SO_4}$ (H: +1, S: +6, O: −2)

Solution 1-90

(a) The oxidation number of the central atom is put equal to $+l$, which is equal to the number of the valence electrons, and the number of the O atoms is m. Then the charge number of the oxide ion n is given by eq(193). (See **solution 1-88,** rule ⑤.)

$$+l + (-2)\times m = \text{charge number } n \qquad (193)$$

For the oxide ions, the central atoms are considered to be oxidized most extensively. Therefore, in the case of CO_3^{2-}, substituting $l=4$ and $m=3$ into eq(193), the charge number of the ion is obtained as equal to −2. Similarly, substituting $l=5$ and $m=3$ for NO_3^{n-}, $l=6$ and $m=4$ for SO_4^{n-}, and $l=5$ and $m=4$ for PO_4^{n-}, the charge number well known for each ion is recovered. In other words, the central atoms of oxide ions can be said normally in the most extensively oxidized states.

Ans. See text.

(b) By substituting $l=4$ and $m=4$ into eq(193) given above, $n=4$ is obtained for the charge number of SiO_4^{n-}.

Ans. $n=4$

Solution 1-91

For all of the chemical bonds of benzoic acid, the valence electrons are divided into three groups depending on to which atom the electrons are originally assigned by using different colors for dot electrons as shown in Fig.82.

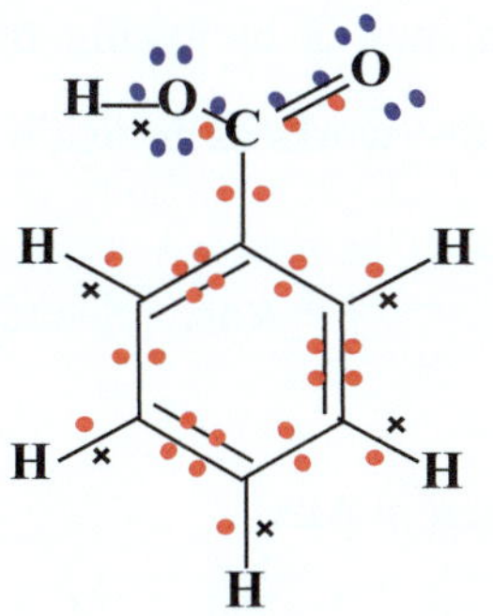

Fig.82 Chemical bonds and valence electrons of benzoic acid. All valence electrons are assigned to each atom. • C atom; • O atom; ×H atom.

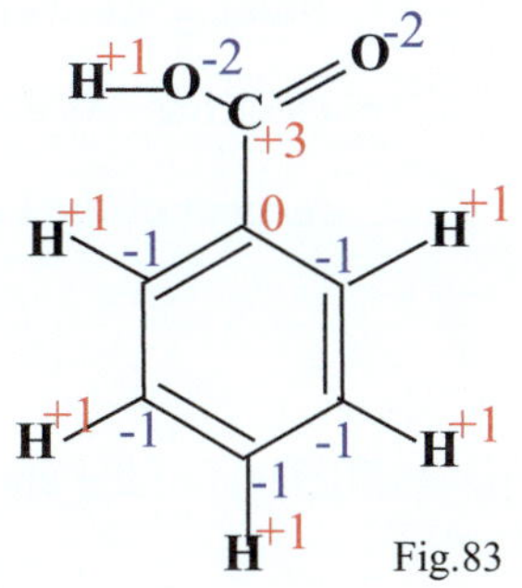

Fig.83 Oxidation numbers.

All the valence electrons shown in Fig.82 are assigned to relevant atoms according to the rules of determining the oxidation number (**solution 1-88**) as is explained in the following.

* The oxidation number of the O atom is -2. 8 electrons are assigned to any one of the O atoms.
* The oxidation number of any H atom is $+1$. None of electrons is assigned to the H atom.
* The oxidation number of the C atom of the carboxyl group is $+3$. One electron of the two valence electrons in the $C-C$ covalent bond is assigned to the C atom of the carboxyl group.
* The oxidation number of the C atom in the benzene ring at the substituted position is 0. Two electrons in a $C=C$ double bond, one electron in each of two $C-C$ single bond, in total 4 electrons are assigned to it.
* The oxidation number of the C atoms at the unsubstituted positions is -1. Two electrons in a $C=C$ double bond, one electron in a $C-C$ single bond, in total 3 electrons are assigned to them.

All these results are summarized in Fig.83. It is seen that the oxidation numbers of the carbon atoms are not the same but depend on the manner of bonding.

Ans. Fig.83

"note" Neglecting the manner of bonding, if the acetic acid $CH_3CO\text{-}OH$ in structural formula is written as $C_2H_4O_2$ in molecular formula and the oxidation number of -2 assigned to O and $+1$ assigned to H are substituted into the equation $2x+4-2\times 2=0$ in which x is the oxidation number of C, a value of x is obtained as 0 from the equation. This is, however, totally wrong since the oxidation number is closely related to the molecular bonding structure.

Solution 1-92

In the chemical reaction given below (eq(194)), there is no change in the oxidation number for neither H (H^+) atom nor O atom.

$$3H_2S + Cr_2O_7^{2-} + 8H^+ \rightarrow 3S + 2Cr^{3+} + 7H_2O \qquad (194)$$

With respect to the S and Cr atoms, their oxidation numbers given in the parentheses change before and after the reaction as explained below.

From S(− 2) to S(0) Since the oxidation number increases, the S atom itself is oxidized. Namely, it works as a reducing reagent of the valency 2 by losing two electrons.

From Cr(+ 6) to Cr(+ 3) Since the oxidation number decreases the Cr atom itself is reduced. Namely, it works as an oxidizing reagent of the valency 3 by taking three electrons. As a chemical, potassium dichromate $K_2Cr_2O_7$ works as an oxidizing reagent of the valency of $2 \times 3 = 6$.

Ans. See text.

Solution 1-93

(a) The stoichiometric coefficients of the chemical reaction in the problem are as given in **Ans**.

Ans. (i) 2, (ii) 3 and (iii) 10

(b) The oxidation numbers of the O, H and K atoms are -2, $+1$ and $+1$, respectively (premise). Assuming that these oxidation numbers do not change throughout the reaction, the fact that the sum of the oxidation numbers over the constituent atoms of a compound is zero for the neutral compound is used to find out a change in the oxidation number of each atom in the reactants and products of the given reaction. The obtained results are summarized below.

<u>reactants</u>	premise	results
$(COOH)_2$	$(CO^{-2}O^{-2}H^{+1})_2$	$(C^{+3}O^{-2}O^{-2}H^{+1})_2$
$KMnO_4$	$K^{+1}Mn(O^{-2})_4$	$K^{+1}Mn^{+7}(O^{-2})_4$
H_2SO_4	$(H^{+1})_2S(O^{-2})_4$	$(H^{+1})_2S^{+6}(O^{-2})_4$
<u>products</u>	premise	results
CO_2	$C(O^{-2})_2$	$C^{+4}(O^{-2})_2$
$Mn\ SO_4$	$Mn\ S^{+6}(O^{-2})_4$	$Mn^{+2}\ S^{+6}(O^{-2})_4$
H_2O	$(H^{+1})_2O^{-2}$	$(H^{+1})_2O^{-2}$
K_2SO_4	$(K^{+1})_2S^{+6}(O^{-2})_4$	$(K^{+1})_2S^{+6}(O^{-2})_4$

It is found from these results that the atoms (ions) which undergo a change in the oxidation number are C for $C^{+3} \rightarrow C^{+4}$ and Mn for $Mn^{+7} \rightarrow Mn^{+2}$. In a word of chemical reaction it is said that the carbon atom of oxalic acid is oxidized into carbon dioxide by the oxidizing reagent of potassium permanganate.

Ans. oxidized substance: carbon atom of oxalic acid, reduced substance: manganese

"note" Since the oxidation number is closely related to the manner of bonding, it is misleading to assign automatically the same number to the atoms of the same kind. In particular, it often happens that the atoms of the same kind in a compound have various oxidation numbers. Just to make sure, how to acquire the oxidation number of each atom of oxalic acid is illustrated below in Fig.85.

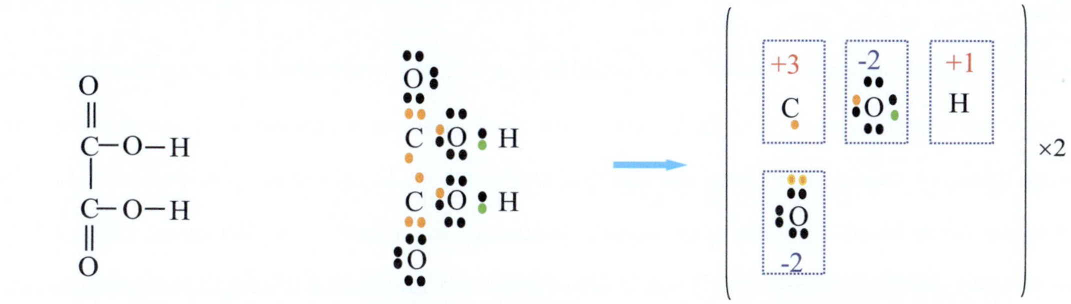

chemical bonds of oxalic acid

chemical bonds of oxalic acid expressed by dot electrons

assignments of electrons and oxidation numbers

Fig.85 How to assign the oxidation number of each component atom of oxalic acid.

Solution 1-94

When the first ionization energies are plotted against the atomic number, the second ionization energies against Z–1, and the third ionization energies against Z–2, Fig.86 shown below is obtained. As is clearly seen from the figure, not only all three different plots show peaks and dips at the same positions of the horizontal axis but also the details of the variation patterns of the plots are almost identical. This indicates that the ion produced by taking an electron away from the atom of the atomic number Z has the same electronic configuration as the atom of the atomic number Z–1 does, and upon two electrons being taken away, the ion produced is the same as the atom of the atomic number Z–2 in their electronic configurations (at least when Z<40), reflecting the characteristic structure of the principal shells and subshells of the atom.

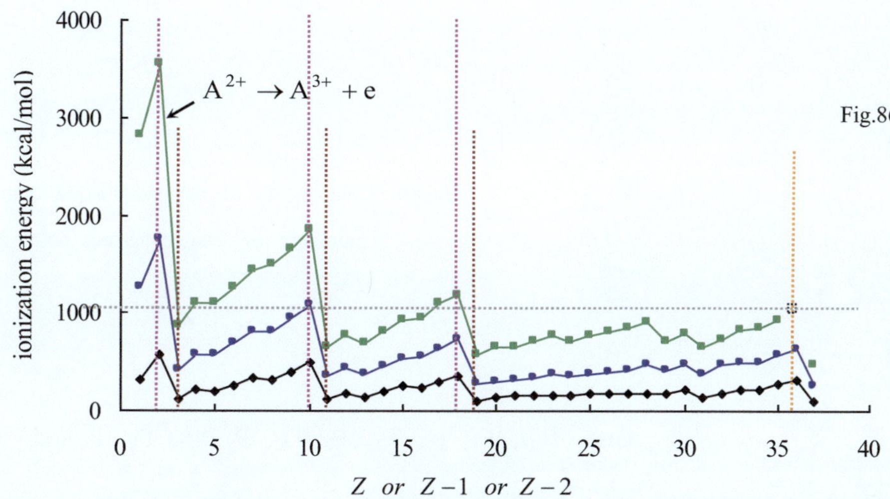

Fig.86 Periodicity of ionization energy. atom A (——), mono-ion A^+(——), di-ionA^{2+}(——). Horizontal axis is Z, Z–1 and Z–2 from top to bottom.

Therefore, when corresponding ions are compared, a behavior similar to the periodicity known for neutral atoms is expected. For example, the third ionization energy of the atom of Z=38 is predicted to be around 1050 kJ (marked by a small square) from the figure above.

Ans. See text.

Solution 1-95a

For two electrons, it is necessary to consider separately depending on whether they have the same set of the principal and azimuthal quantum numbers or either quantum number is different for them.

(a) Solution 1.

Since the azimuthal quantum number for *d*-orbitals is $l=2$, the synthesized angular momentum L for the two electrons can take the values of 4, 3, 2, 1 and 0 and their component momentums are as shown in Table 1 below using different colors. Therefore, the spectral terms are G, F, D, P and S, respectively (see **"note 1"** below). On the other hand, the total spin angular momenta are 1 and 0 but because of *Pauli*'s exclusion principle, the two electrons cannot take a same set of quantum numbers including spin quantum number. In the table below, therefore, the components of the synthesized angular momentum at the diagonal positions indicated by the broken red line cannot have the same spin component, that is, the two electrons can only have opposite spins resulting in the singlet spin state of $S=0$. As a result of this, the possible spectral terms are ^1G, $^{3,1}F$, ^1D, $^{3,1}P$ and ^1S. Furthermore, these spectral terms can be specified in more detail by using the total angular momenta $J=L+S$, leading to the 11 terms of $^1G_4$, $^3F_4$, $^3F_3$, $^3F_2$, $^1F_3$, $^1D_2$, $^3P_2$, $^3P_1$, $^3P_0$, $^1P_1$ and $^1S_0$. Consequently, the total number of the electronic states are found by summing it over the spectral term multiplied by the multiplicity of $(2J+1)$ as in eq(195) below, resulting in 55 states.

$$\sum_{i=1}^{11}(2J+1)_i = 9+9+7+5+7+5+5+3+1+3+1=55 \qquad (195) \qquad i \text{ indicates spectral term.}$$

The electronic states for the nd^2 configuration amount to 100 in total as shown below (see Fig.86),

$$25(\text{combinations of angular momentum } 5\times5)\times((3+1)(=\text{combinations of spin angular momentum: triplet+singlet}))=100.$$

Out of 100 states, only 55 states are possible states.

m_{l_2} \ m_{l_1}	2	1	0	-1	-2
2	4	3	2	1	0
1	3	2	1	0	-1
0	2	1	0	-1	-2
-1	1	0	-1	-2	-3
-2	0	-1	-2	-3	-4

Fig.86 Components of the synthesized angular momenta of two electrons both with $l=2$.

Ans. $^1G_4$, $^3F_4$, $^3F_3$, $^3F_2$, $^1F_3$, $^1D_2$, $^3P_2$, $^3P_1$, $^3P_0$, $^1P_1$ and $^1S_0$

"note 1" Spectral terms defined by L are designated by the alphabetical capital letters as shown below according to the historical usages.

L	0	1	2	3	4	5	6	7	8	9 ·····
Spectral term	S	P	D	F	G	H	I	K	L	M

For the angular momentum l of an electron, a small letter corresponding to the spectral term above is used. The states of different spectral terms are in different energy states. To tell which one corresponds to the lowest energy state, refer to **"note 3"**.

"note 2" If it is just a matter of knowing the values for the components of the synthesized angular momenta, then it is not necessary to distinguish the components as is done above by different colors. As is shown on the right, the same sets of the components of the synthesized angular momentum are found. However, it is not appropriate or incorrect in considering possible combination of the synthesized angular momenta with spin angular momentum when not only the principal quantum number but also the azimuthal quantum number are the same for the two electrons.

	2	1	0	-1	-2
2	4	3	2	1	0
1	3	2	1	0	-1
0	2	1	0	-1	-2
-1	1	0	-1	-2	-3
-2	0	-1	-2	-3	-4

Fig.87 Same sets of components of the synthesized angular momentum enclosede by dotted lines.

(a) Solution 2.

Since the two electrons cannot be distinguished from each other, the components of the synthesized angular momenta which are located below the diagonal line such as in Fig.86 are not used because they are duplicated. Possible combinations of two spins are $(\alpha,\ \alpha)$, $(\alpha,\ \beta)$, $(\beta,\ \alpha)$ and $(\beta,\ \beta)$ but the components marked by in Fig.88, only the combination of $(\alpha,\ \beta)$ is allowed because of *Pauli*'s exclusion principle. Otherwise, all of the four combinations of $(\alpha,\ \alpha)$, $(\alpha,\ \beta)$, $(\beta,\ \alpha)$ and $(\beta,\ \beta)$ are possible.

When look at the components enclosed with in Fig.88(a), only $(\alpha,\ \beta)$ is allowed for $M_L = \pm 4$ and any spin combinations are possible for $M_L = \pm 3,\ \pm 2,\ \pm 1$ and 0. These states can also be grouped in terms of the synthesized spin angular momentum. For $S = 0$ ($(\alpha,\ \beta)$), M_L can take values of $M_L = \pm 4,\ \pm 3,\ \pm 2,\ \pm 1$ and 0 and for $S = 1$ ($(\alpha,\ \alpha), (\beta,\ \alpha), (\beta,\ \beta)$) and $S = 0$ ($(\alpha,\ \beta)$), M_L can take values of $M_L =$

.

	m_{l_1}: 2	1	0	-1	-2
m_{l_2}: 2	4	3	2	1	0
1		2	1	0	-1
0			0	-1	-2
-1				-2	-3
-2					-4

Fig.88 (a) Components of synthesized angular momentum.#.

#Components below the diagonal line are not used.

combination of spin quantum numbers

	m_{l_1}: 2	1	0	-1	-2
m_{l_2}: 2	4^{1}	$3^{3,1}$	$2^{3,1}$	$1^{3,1}$	$0^{3,1}$
1		2^{1}	$1^{3,1}$	$0^{3,1}$	$-1^{3,1}$
0			0^{1}	$-1^{3,1}$	$-2^{3,1}$
-1				-2^{1}	$-3^{3,1}$
-2					-4^{1}

Fig.88(b) Components of angular momentum and possible spin quantum number (multiplicity)*.

*Upper case numbers 3 and 1 indicate $S = 1$ and $S = 0$, respectively.

$\pm 3,\ \pm 2,\ \pm 1$ and 0 and $M_L = \pm 3,\ \pm 2,\ \pm 1$ and 0, respectively. Therefore, the spectral term ^{1}G is derived from the former combination and ^{3}F and ^{1}F are from the latter combinations.

For all of the components of the synthesized angular momenta, possible combinations of the spin quantum numbers are summarized in Fig.88(b). From this table, it is derived that the possible spectral terms for the electronic configuration of nd^2 are ^{1}G, ^{3}F, ^{1}F, ^{1}D, ^{3}P, ^{1}P and ^{1}S. Furthermore, by using the values for the total angular momentum given by $J = L + S$, those spectral terms are split into the 11 terms of $^1G_4$, $^3F_4$, $^3F_3$, $^3F_2$, $^1F_3$, $^1D_2$, $^3P_2$, $^3P_1$, $^3P_0$, $^1P_1$ and $^1S_0$. Hence, similarly calculated as in Solution 1, the total number of electronic states amounts to $9+9+7+5+7+5+5+3+1+3+1=55$.

Ans. ^1G, ^3F, ^1F, ^1D, ^3P, ^1P, ^1S ($^1G_4$, $^3F_4$, $^3F_3$, $^3F_2$、$^1F_3$, $^1D_2$, $^3P_2$, $^3P_1$, $^3P_0$, $^1P_1$, $^1S_0$)

(b)The components of the synthesized angular momentum are listed in Fig.89. Since the angular momentum can have the quantum numbers of $L = 3,\ 2,\ 1$, the corresponding spectral terms are F, D and P. The two electrons in the present case originally have different angular quantum numbers and hence have already fulfilled *Pauli*'s exclusion principle. Therefore, each electron can have either spin angular momentum ($\uparrow$ *or* $\downarrow$), yielding the spectral terms of $^{3,1}F$、$^{3,1}D$、$^{3,1}P$. Furthermore, by using the values for *J*=*L*+*S*, those spectral terms are split into the 12 terms of $^3F_4$, $^3F_3$, $^3F_2$, $^1F_3$, $^3D_3$, $^3D_2$, $^3D_1$, $^1D_2$, $^3P_2$, $^3P_1$, $^3P_0$ and $^1P_1$. The total number of the electronic states are readily calculated from the combinations of the quantum numbers as equal to $(3\times 2)\times(5\times 2)=60$, which can also be found from the sum of the spectral terms as below,

$$9+7+5+7+7+5+3+5+5+5+3+1+3=60$$

and, of course, both are in agreement.

	m_{l_1} = 2	1	0	-1	-2
m_{l_2} = 1	3	2	1	0	-1
0	2	1	0	-1	-2
-1	1	0	-1	-2	-3

Fig.89 Components of the synthesized angular momentum of the two electrons specified by the different azimuthal quantum numbers of $l=1$ and $l=2$.

Ans. $^3F_4$, $^3F_3$, $^3F_2$, $^1F_3$, $^3D_3$, $^3D_2$, $^3D_1$, $^1D_2$, $^3P_2$, $^3P_1$, $^3P_0$, $^1P_1$

"note 3" In general, spectral terms have different energies because of the electronic repulsion and the spin-orbit interaction (***L*•*S*** coupling). The order of the energies can be predicted by following three *Hund*' s rules: (i) the energy is lowest for the spectral term with the largest value of *S*, (ii) when the value for *S* is the same, the spectral term with the larger value of *L* is more stable than others and (iii) when both of *L* and *S* are the same and the subshell is half filled or filled with less than a half, the energy increases with increasing *J* but for the subshell filled more than half, the energy decreases with increasing *J*.

Consequently, in the case of (b) ^3F is the lowest state. Locked at more closely, ^3F state is split into the three states of $^3F_4 > {}^3F_3 > {}^3F_2$, $^3F_2$ being most stable. In the presence of magnetic field, each of these terms is further split into a few levels with an equal interval (magnetic field effect).

"note 4" As is seen from Fig.87 and Fig.89, for a pair of electronic configurations when the sum of electrons in the subshells is equal to the maximum number of electrons to be accommodated in the subshell such as (p^1, p^5), (p^2, p^4), (d^1, d^9), (d^2, d^8), (d^3, d^7) and (d^4, d^6), the spectral terms are the same for the pair.

"note 5" Selection rules.

Interval between split spectral terms can be found by observing light absorption or emission caused by the electronic transition between these spectral terms. What conditions should be fulfilled for such transitions to occur is simply known by the selection rules. For the transition to be allowed, the following conditions should be satisfied.

$$\Delta S = 0,\ \ \Delta L = 0,\ \ \pm 1,\ \ \Delta l = \pm 1 \qquad S \text{ is synthesized spin angular momentum.}$$

$$\Delta J = 0,\ \ \pm 1,\ \ \text{however, } J = 0 \leftrightarrow J = 0 \text{ is forbidden.}$$

When only one electron is involved in the transition, the transitions of $ns \leftrightarrow n's$, $np \leftrightarrow n'p$ and $nd \leftrightarrow n'd$ are readily found to be forbidden ones from the selection rules above.

Solution 1-95b

For each atom, the spectral term for the lowest energy state is found as follows:

Electronic configuration: $_{22}\text{Ti}([\text{Ar}]4\text{s}^2 3d^2)$.

synthesized angular momentum: $L = 4,\ 3,\ 2,\ 1,\ 0$, synthesized spin angular momentum: $S = 1,\ 0$,

combination with L for which the spin multiplicity becomes largest: $L = 3,\ S = 1$,

Hund's rule, minimum J: $J = 2 \ \Rightarrow\ ^3\text{F}_2$. The found spectral term $^{2S+1}\text{L}_J$ is $^3\text{F}_2$.

Electronic configuration: $_{26}\text{Fe}([\text{Ar}]4\text{s}^2 3d^6)$.

synthesized angular momentum: out of the components of $m_l = 2,\ 1,\ 0,\ -1,\ -2$, four different values selected so as to make their sum maximum,
from the sum of $(2+1+0-1=2)$ $L = 2$ is found,

maximum synthesized spin angular momentum: $S = 2$,

combination of L and S obtained above: $L = 2,\ S = 2 \ \Rightarrow\ J = 4,\ 3,\ 2,\ 1,\ 0$,

Hund's rule, maximum J: $J = 4 \ \Rightarrow\ ^5\text{D}_4$. The found spectral term $^{2S+1}\text{L}_J$ is $^5\text{D}_4$.

Electronic configuration: $_{29}\text{Cu}([\text{Ar}]4\text{s}^2 3d^9)$. (In the case of $[\text{Ar}]4\text{s}^1 3d^{10}$, $L = 0$ and $S = 1/2$ are found.)

synthesized angular momentum: $L = 2$, synthesized spin angular momentum: $S = 1/2$,

combination of L and S obtained above: $L = 2,\ S = 1/2 \ \Rightarrow\ J = 5/2,\ 3/2,\ 1/2$,

Hund's rule, maximum J: $J = 5/2 \ \Rightarrow\ ^2\text{D}_{5/2}$. The found spectral term $^{2S+1}\text{L}_J$ is $^2\text{D}_{5/2}$.

As is seen above, even for many electron atoms, the spectral term for the lowest energy state (alternatively the state of the largest multiplicity) can easily be found. The *Tanabe-Sugano* diagrams ($\Delta_o / B = 0$) are useful to find out required spectral terms.

Ans. spectral term: $_{22}\text{Ti}:\ ^3\text{F}_2$, $_{26}\text{Fe}:\ ^5\text{D}_4$, $_{29}\text{Cu}:\ ^2\text{D}_{5/2}$

Solution 1-96

(a) When the spectral terms are searched for, the completely filled subshells need not be taken into account. In the present case, therefore, it is sufficient to find the spectral term corresponding to $3s^1$. For this state, $L(=l) = 0$ and $S = 1/2$, resulting in $^{2S+1}\text{L}_J = {}^2\text{S}_{1/2}$.

Ans. $^2\text{S}_{1/2}$

(b) In a similar manner, the spectral term for the electronic configuration of $3p^1$ is searched. Since $L(=l)=1$ and $S=1/2$, the spectral terms of $^2P_{3/2}$ and $^2P_{1/2}$ are obtained. According to *Hund*'s rules (see **"note 3"** in **solution 1-95a**) the order of the energies of the split terms is $^2P_{3/2}>^2P_{1/2}$. In the presence of magnetic field, these terms are further split into $2J+1=4$ levels and 2 levels, respectively.

Ans. $^2P_{3/2}$, $^2P_{1/2}$

(c) The selection rules (**"note 5"** in **solution 1-95a**) are applied to the transitions from the ground state to the two excited states to find that both transitions are allowed as shown below.

$$^2S_{1/2} \rightarrow {}^2P_{3/2} \;:\Delta S=0,\;\; \Delta L(=\Delta l)=1,\;\; \Delta J=1$$
$$^2S_{1/2} \rightarrow {}^2P_{1/2} \;:\Delta S=0,\;\; \Delta L(=\Delta l)=1,\;\; \Delta J=0,\;\; J\neq 0$$

The corresponding energy level diagram is shown below in Fig.90. The light emission or absorption associated with the electronic transitions is called as sodium D-lines. The energy interval is expressed by $A\cdot(J+1)$ with A constant (interval between the levels specified by J and J+1).

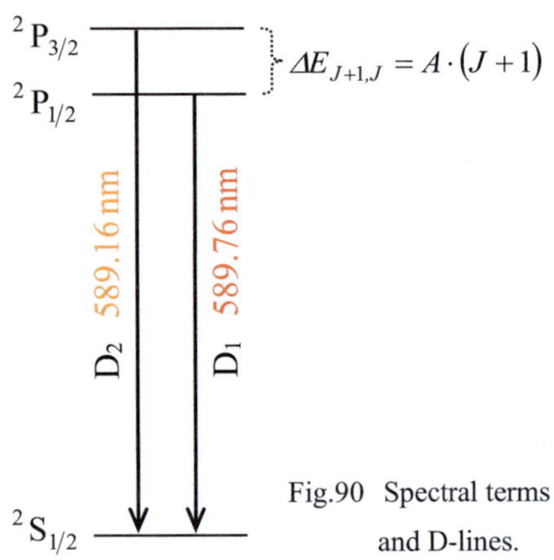

Fig.90 Spectral terms of Na and D-lines.

"note" Splitting of the spectral terms due to ***LS*** coupling. The orbital motion of an electron generates a magnetic filed. This field interacts with the spin magnet of the electron and its magnitude depends on both of the angular momentum L and spin angular momentum S. For this, the interaction is called ***LS*** coupling and its magnitude is given by $A\times(\boldsymbol{L}\cdot\boldsymbol{S})$. When the total angular momentum $\boldsymbol{J}=\boldsymbol{L}+\boldsymbol{S}$ is employed, from the relation of $(\boldsymbol{J}\cdot\boldsymbol{J})=((\boldsymbol{L}+\boldsymbol{S})\cdot(\boldsymbol{L}+\boldsymbol{S}))=(\boldsymbol{L}\cdot\boldsymbol{L})+2(\boldsymbol{L}\cdot\boldsymbol{S})+(\boldsymbol{S}\cdot\boldsymbol{S})$ $2(\boldsymbol{L}\cdot\boldsymbol{S})=|J|^2-|L|^2-|S|^2=J(J+1)-L(L+1)-S(S+1)$ is derived and, hence, the magnitude of the splitting due to ***LS*** coupling is given by $A\times(\boldsymbol{L}\cdot\boldsymbol{S})=A/2\times\{J(J+1)-L(L+1)-S(S+1)\}$. The interval between the levels specified by J and $J+1$ is given by $\Delta E_{J+1,J}=A\cdot(J+1)$, which is called *Landé*'s interval rule.

Ans. allowed transitions

(d) Whether the transition under consideration is quantum mechanically allowed or not is judged by calculating the relevant transition moment $\boldsymbol{\mu}$. In the present case, $\boldsymbol{\mu}$ is defined by eq(196). Although

$$\boldsymbol{\mu}=\int \varphi_{3s}(\boldsymbol{r})\cdot\boldsymbol{r}\cdot\varphi_{3p}(\boldsymbol{r})d\tau \qquad (196)$$

eq(196) can be evaluated by substituting the relevant eigenfunctions into eq(196), we will tell whether eq(196) is zero or not on the basis of the symmetry of each function in the integral.

When the polar coordinates are inverted against the origin, the functions in eq(196) change their signs as follows.

inversion : $x,\ y,\ z \Rightarrow -x,\ -y,\ -z$

$$\varphi_{3s}(\boldsymbol{r}) \Rightarrow \varphi_{3s}(\boldsymbol{r}) : \quad \text{totally symmetric}$$

$$\boldsymbol{r} \Rightarrow -\boldsymbol{r} : \quad \text{antisymmetric}$$

$$\varphi_{3p}(\boldsymbol{r}) :\ 3p_x,\ 3p_y,\ 3p_z \Rightarrow -3p_x,\ -3p_y,\ -3p_z : \quad \text{antisymmetric}$$

The function to be integrated changes its sign as a whole according to the result of the multiplication of each sign, that is, (totally symmetric)×(antisymmetric)×(antisymmetric)=(totally symmetric). Thus the integral of eq(195) is found totally symmetric and takes a non-zero value. Therefore, the transitions under consideration are concluded as allowed transitions. In a similar manner, the transitions of $ns \leftrightarrow n's,\ np \leftrightarrow n'p,\ ns \leftrightarrow n'd,$ etc. are found to be inhibited ones.

Ans. See (d) text.

Solution 1-97

(a) Since the *ns*-orbital is fully occupied, only *np*-orbitals need be taken into account. Furthermore, if it is noticed that the electronic configuration of np^5 is equivalent to the configuration in which one electron is removed from that of np^6, the sum of the angular momenta for the np^5 configuration is found to be the same as that for the np^1 configuration. Namely, as was mentioned before, both of the electronic configurations of np^5 and np^1 have the same spectral terms. For np^1, $L=1$ and $S=1/2$ yielding the spectral terms of $^2\mathrm{P}_{3/2}$ and $^2\mathrm{P}_{1/2}$. The *np*-orbitals are occupied more than half and *Hund*'s rule (**"note 3"** in **solution 1-95a**) indicates that the order of the energies of the split levels (terms) is $^2\mathrm{P}_{1/2} > {}^2\mathrm{P}_{3/2}$, being reverse to the case of the Na atom.

(b) The interval between the split spectral terms is given by $A\cdot(J+1)$. (*Landé*'s interval rule). For the halogen atoms a value of *J* is common but *A* is a function of the atomic number and takes larger values with increasing the atomic number. The interval in cm^{-1} is plotted against the atomic number in Fig.91.

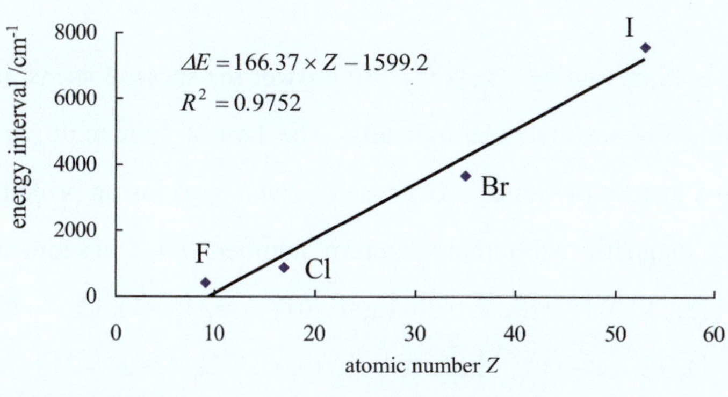

Fig.91 Correlation of the interval between the split spectral terms with the atomic number. Linearity to $Z^{1.70}$ is better than that to Z.

Ans. See text (a) and (b).

Solution 1-98a

The electronic configuration of $_{63}\mathrm{Eu}$ is [Xe]$6s^2 4f^7$. Therefore, the electronic configuration of the trivalent ion $_{63}\mathrm{Eu}^{3+}$ is given by [Xe]$4f^6$. To answer the spectral terms, it is only necessary to consider *f*-electrons alone. The largest spin multiplicity arises from $S=(1/2)\times 6=3$ leading to the septet (7) state.

Because of *Pauli*'s exclusion principle, all these six electrons should have different magnetic azimuthal quantum numbers m_l. Consequently, possible combinations of m_l are those for which any one of the seven components of $(3, 2, 1, 0, -1, -2, -3)$ is deleted, i.e., $\{(2, 1, 0, -1, -2, -3)\}$, $\{(3, 1, 0, -1, -2, -3)\}$, $\{(3, 2, 0, -1, -2, -3)\}$, $\{(3, 2, 1, -1, -2, -3)\}$, $\{(3, 2, 1, 0, -2, -3)\}$, $\{(3, 2, 1, 0, -1, -3)\}$, $\{(3, 2, 1, 0, -1, -2)\}$. From these combinations, it is found that the sum of m_l ($\sum m_{l_i}$) takes its maximum value of 3 for the last combination listed above. Therefore, $L = 3, 2, 1, 0$ are possible for the total angular momentum L, but in the present case, because of *Hund*'s rule, the state with the smallest value of $J = L + S$ as well as with the largest value of L is lowest in energy, that is, ${}^7F_0$ is the ground state. The other spectral terms which belong to ${}^7F$ are ${}^7F_6 > {}^7F_5 > {}^7F_4 > {}^7F_3 > {}^7F_2 > {}^7F_1$. See **solution 1-79** and **solution 1-95a**.

Ans. electronic configuration: $[Xe]4f^6$, ground state: ${}^7F_0$, the other spectral terms specified by ${}^7F$:

${}^7F_6 > {}^7F_5 > {}^7F_4 > {}^7F_3 > {}^7F_2 > {}^7F_1$

(b) The electronic transition is described by $\{(3, 2, 1, 0, -1, -2)\} \rightarrow \{(3, 2, 1, 0, -1, -3)\}$. Therefore, the maximum total angular momentum after the transition is $L = 2$. Since the electron responsible to the transition invert its spin upon transition, the maximum spin multiplicity is given by $S = (1/2) \times 4 = 2$, forming a quintet (5) state. According to *Hund*'s rules, the state with the largest L as well as the smallest $J = L + S$, i.e., ${}^5D_0$ is considered to be the lowest excited state. Therefore, the transition responsible for the light emission is ascribed to ${}^7F_0 \rightarrow {}^5D_0$.

Ans. electronic transition: ${}^7F_0 \rightarrow {}^5D_0$

"note" The characteristic sharp red emission observed at around 610 nm is ascribed to ${}^5D_0 \rightarrow {}^7F_1$ and ${}^5D_0 \rightarrow {}^7F_2$. These transitions are basically electronic transitions and are not accompanied by vibrational transitions. Therefore, even at room temperatures, the emission is observed as a sharp band like a spectral line.

Solution 1-98b

(a), (b) The electronic configuration of ${}_{60}Nd^{3+}$ is expressed by $[Xe]4f^3$. To answer the spectral terms, it is only necessary to consider *f*-electrons occupying the subshell incompletely. The largest spin multiplicity arises from $S = 1/2 + 1/2 + 1/2 = 3/2$, leading to a quartet (4) state. Because of *Pauli*'s exclusion principle, all these three electrons should have different magnetic azimuthal quantum numbers m_l. Consequently, possible combinations of m_l are $\{(3, 2, 1)\}$, $\{(3, 2, 0)\}$, $\{(3, 2, -1), (3, 1, 0)\}$, $(3, 1, -1), (2, 1, 0)\}$, $\{(3, 2, -2), \{(3, 2, -3), (3, 1, -2), (3, 0, -1), (2, 1, -1)\}$, $\{(3, 1, -3), (3, 0, -2), (2, 1, -2)\}$ and $\{(3, 0, -3), (3, -1, -2), (2, 1, -3), (1, 0, -1)\}$.

The combination or combinations in { } have the same value for the total angular momentum and the values are $L = 6, 5, 4, 3, 2, 1, 0$ in the order of above appearance. Because of *Hund*'s rule, the state with the smallest value of $J = L + S$ as well as with the largest value of L is lowest in energy, that is, ${}^4I_{9/2}$ is considered to be the ground state. See **solution 1-9** and **1-95a**.

Ans. electronic configuration : $[Xe]4f^3$, ground state: ${}^4I_{9/2}$

(c) If the one-electron excitation is expressed as $4f \rightarrow 6d$, then the electronic configuration of the excited state is given by $[\mathrm{Xe}]4f^2 6d^1$. Following the procedure given in (a), the synthesized angular momenta L are found as $L = 7$, 6, 5, 4, 3, 2, 1, 0. Therefore, the state with $L = 3$ and $S = 1/2 + 1/2 + 1/2 = 3/2$, i.e., ${}^4\mathrm{F}_{3/2}$ is found possible as a state resulting from the one-electron excitation.

Ans. See text (c).

Solution 1-99

The atomic number of the K atom is 19 and the shielding effect (constant) of 1*s*-electrons on a 2*s*-electon is 0.85/each. The effective nuclear charge, therefore, is calculated to be $Z_{eff} = Z - 0.85 \times 2 = 19 - 1.70 = 17.30$. From this value, the energy of the 2*s*-orbital is estimated by eq(197), in which R_H is the *Rydberg* constant (in the unit of J).

$$E_{2s} = -\frac{e^4 m_e}{8\varepsilon_0^2 h^2} \cdot \frac{Z_{eff}^2}{(n)^2} = -R_H \cdot \frac{17.3^2}{(2)^2} = -2.178 \times 10^{-18} \times \frac{17.3^2}{4} = -1.630 \times 10^{-16}\,\mathrm{J} \qquad (197)$$

The ionization energy of the 2*s*-electron is given by $E_\infty - E_{2s}$ and the wave number of the corresponding light is calculated to be $8.205 \times 10^6\,\mathrm{cm^{-1}}$. It is $1.219\,\mathrm{nm}$ in wavelength and corresponds to the *X*-ray.

$$\overline{\nu} = \frac{E_\infty - E_{2s}}{hc} = \frac{1.630 \times 10^{-16}}{6.626 \times 10^{-34} \times 2.998 \times 10^{10}\left(\mathrm{cm \cdot s^{-1}}\right)} = 8.205 \times 10^6\,\mathrm{cm^{-1}} \qquad (198)$$

The value given by eq(198) is readily obtained by means of conversion coefficient as shown below. The value in eV calculated using a relevant conversion coefficient is also given below.

$$1.630 \times 10^{-16}\,\mathrm{J} \times 5.034 \times 10^{22}(\text{conversion coefficient}: \mathrm{cm^{-1}/J}) = 8.205 \times 10^6\,\mathrm{cm^{-1}}$$

$$1.630 \times 10^{-16}\,\mathrm{J} \times 6.242 \times 10^{18}(\text{conversion coefficient}: \mathrm{eV/J}) = 1.02 \times 10^3\,\mathrm{eV}$$

Ans. ionization energy: $1.630 \times 10^{-16}\,\mathrm{J}$, light wavelength: 1.219 nm

Solution 1-100

Even when the strength (Sv or Bq) of radioactive radiation is known or measured, we cannot tell exactly what radioactive species is responsible for the radiation. However, not only the kinds of radiation such as α, β, γ, or neutron but also the energies of radiation are uniquely determined by each radioactive nuclear species. For instance, both of ${}^{60}\mathrm{Co}$ and ${}^{40}\mathrm{K}$ radiate γ-ray but since their energies are different, it is possible to identify the species by measuring the energy of radiation. The energies of γ-ray from ${}^{60}\mathrm{Co}$, ${}^{40}\mathrm{K}$ and ${}^{137}\mathrm{Cs}$ are 1.173×10^6, 1.46×10^6 and $0.662 \times 10^6\,\mathrm{eV}$, respectively. Of course, for any other radioactive radiations, if their energies are measured, it is possible to identify the radioactive nuclear species. Instead of a simple measurement of the strength (or counts) of radioactive radiation, a sophisticated apparatus is required to measure the energy of radioactive radiation. Although the half-life of a radioactive species is an important property intrinsic to a radioactive species, it is not practical to employ it to identify the species.

Ans. See text.

Solution 1-101

The coordinates of the point charge and the homogeneously charged sphere are taken as shown in Fig.92. Within the sphere an arbitrarily point at (x, y, z) is taken and the charge there is considered to be given by δQ. Then, the coulomb force between the point charge located outside of the sphere and the point thus selected is obtained. The arrangement shown in Fig.92 is redrawn more specifically in Fig.93 in order to make the calculation easier.

The distance between these two pint charges R' given by eq(199) is rewritten as eq(200) by substituting the relation of $z = r\cos\vartheta$ into eq(199).

$$R'^2 = x^2 + y^2 + (R - z)^2 \qquad (199)$$
$$= R^2 - 2Rz + x^2 + y^2 + z^2$$
$$= R^2 - 2Rr\cos\vartheta + r^2 \qquad (200)$$

The coulomb force between these two charges is expressed by a vector whose direction is indicated by the red arrow in Fig.93. The vector is decomposed into two orthogonal components indicated by the green and blue arrows. After summing the component vectors over the whole charges in the sphere, the component corresponding to the vector in blue becomes zero because of a cylindrical symmetry of the arrangement. For

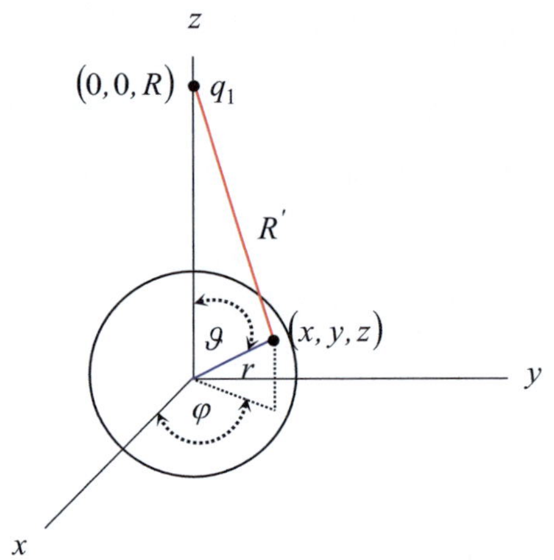

Fig.92 Arrangement of the point charge and the charged sphere in polar coordinate.

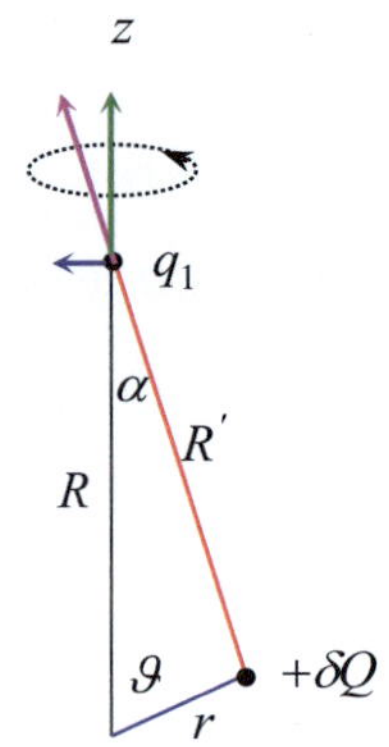

Fig.93 More specified arrangement. In the case of attraction, the directions of the vectors are exactly opposite.

this, the component corresponding to the vector in green, i.e., z-component, needs to be calculated. The force along the z-axis expressed by F_z is derived as being equal to eq(201) from eq(200) by referring to the configuration shown in Fig.92.

$$F_z = \frac{q_1 \times \delta Q}{4\pi\varepsilon_0}\frac{1}{R'^2} \times \cos\alpha$$
$$= \frac{q_1 \times \delta Q}{4\pi\varepsilon_0}\frac{1}{R'^2} \times \frac{R - r\cos\vartheta}{R'}$$
$$= \frac{q_1 \times \delta Q}{4\pi\varepsilon_0}\frac{R - r\cos\vartheta}{(R^2 - 2Rr\cos\vartheta + r^2)^{3/2}} \qquad (201)$$

The total coulomb force denoted by F_{tot} is obtained by summing F_z over the whole sphere as given by eq(202) (integration of eq(201)).

$$F_{tot} = \iiint_{\tau} F_z d\tau = \iiint_{r,\vartheta,\varphi} F_z r^2 \sin\vartheta \, dr d\vartheta d\varphi \qquad (202)$$

$$= \frac{q_1 \times \delta Q}{4\pi\varepsilon_0} \iiint \frac{R - r\cos\vartheta}{(R^2 - 2Rr\cos\vartheta + r^2)^{3/2}} r^2 \sin\vartheta \, dr d\vartheta d\varphi$$

$$= \frac{q_1 \times \delta Q}{4\pi\varepsilon_0} \times 2\pi \iint_{r,\vartheta} \frac{R - r\cos\vartheta}{(R^2 - 2Rr\cos\vartheta + r^2)^{3/2}} r^2 \sin\vartheta \, dr d\vartheta$$

Employing a new variable X defined by $\cos\vartheta = X$ and its differential $-\sin\vartheta d\vartheta = dX$, the above equation is rewritten as eq(203).

$$F_{tot} = \frac{q_1 \times \delta Q}{2\varepsilon_0} \iint_{r,X} \frac{R - rX}{(R^2 - 2RrX + r^2)^{3/2}} r^2 dr dX \qquad (203)$$

Over the range of $X[-1, \ +1]$, eq(203) is integrated by means of the partial integration as follows and the final result is given by eq(204).

$$F_{tot} = \frac{q_1 \times \delta Q}{2\varepsilon_0} \int_r \left\{ \left[(R - rX) \times \frac{-2}{-2Rr} (R^2 - 2RrX + r^2)^{-1/2} \right]_{-1}^{+1} - \frac{1}{Rr} \int_{-1}^{+1} -r(R^2 - 2RrX + r^2)^{-1/2} dX \right\} r^2 dr$$

$$= \frac{q_1 \times \delta Q}{2\varepsilon_0} \int \left\{ \frac{1}{Rr} \left(\frac{(R-r)}{(R-r)} - \frac{(R+r)}{(R+r)} \right) + \frac{1}{R} \cdot \frac{2}{-2Rr} \left[(R^2 - 2RrX + r^2)^{1/2} \right]_{-1}^{+1} \right\} \cdot r^2 dr$$

$$= \frac{q_1 \times \delta Q}{2\varepsilon_0} \int -\frac{1}{R^2 r} \{(R-r) - (R+r)\} \cdot r^2 dr$$

$$= \frac{q_1 \times \delta Q}{\varepsilon_0 R^2} \int_0^a r^2 dr$$

$$= \frac{q \times \delta Q}{3\varepsilon_0 R^2} a^3 \qquad (204)$$

On the other hand, the total charge Q is related to δQ by eq(205).

$$\delta Q \times \frac{4\pi}{3} a^3 = Q \ \Rightarrow \ \delta Q = \frac{3Q}{4\pi a^3} \qquad (205)$$

Substituting eq(205) into eq(204), eq(206) (eq(15) in the problem) is obtained (*Q.E.D.*).

$$F_{tot} = \frac{q_1 \times Q}{4\pi\varepsilon_0 R^2} \qquad (206)$$

Ans. See text.

"note" (1) When *Gauss*' theorem (eq(207) below) is employed, the electric field E at the position where a E_n is the electric field normal to the sphere surface. It is assumed that the charge distribution is spherically

$$\int_S \boldsymbol{E} \cdot d\boldsymbol{S} = \frac{Q}{\varepsilon_0} \qquad (207)$$

$$\int_S \boldsymbol{E} \cdot d\boldsymbol{S} = \int_S E_n \cdot dS = E_n \int_S dS = E_n \times 4\pi R^2$$

$$E_n = \frac{Q}{4\pi\varepsilon_0 R^2} \ \Rightarrow \ F_{tot} = q_1 \cdot E_n = \frac{q_1 \times Q}{4\pi\varepsilon_0 R^2} \qquad (206')$$

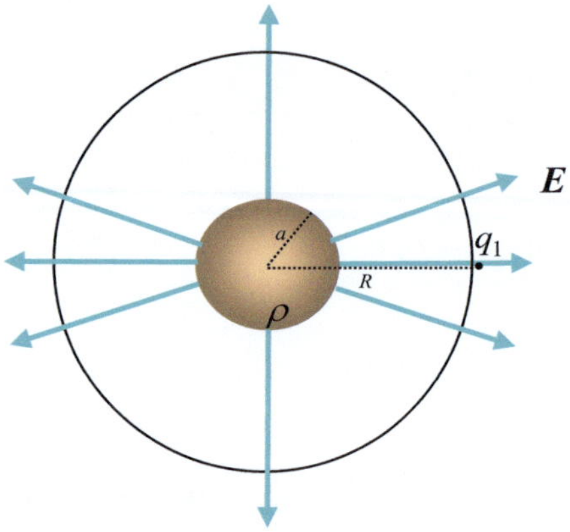

Fig.94 *Gauss*' theorem applied to derive coulomb force working between the point charge of q_1 and the homogeneously charged sphere of radius *a*.

symmetric.point charge is located can be obtained easily, eq(206) is directly derived as is shown below. The surface integral is carried out over the sphere surface (radius *R*) indicated by a circle in black in Fig.94.

(2) The relation used here (and the result) also holds for the force which is inversely proportional to the distance squared. Hence, the gravitational force acting between two mass objects can be expressed by the force between two total masses accumulated at each center of mass. In another word, we can say that *Gauss*' theorem holds also with respect to masses in the gravitational field.

Solution 1-102

Among the 3*d*-orbitals, those under discussion are listed below in eq(208). (For all of them see **solution 1-24.**)

$$\left.\begin{aligned} 3d_{xy} &= \frac{1}{81\sqrt{2\pi}}\left(\frac{Z}{a_0}\right)^{\frac{3}{2}}\left(\frac{Zr}{a_0}\right)^2 e^{-\frac{Zr}{3a_0}}\sin^2\vartheta\sin2\varphi = F\times\left(\frac{Zr}{a_0}\right)\sin^2\vartheta\sin2\varphi \;\propto\; xy \\ 3d_{x^2-y^2} &= \frac{1}{81\sqrt{2\pi}}\left(\frac{Z}{a_0}\right)^{\frac{3}{2}}\left(\frac{Zr}{a_0}\right)^2 e^{-\frac{Zr}{3a_0}}\sin^2\vartheta\cos2\varphi = F\times\left(\frac{Zr}{a_0}\right)\sin^2\vartheta\cos2\varphi \;\propto\; x^2-y^2 \\ 3d_{z^2} &= \frac{1}{81\sqrt{6\pi}}\left(\frac{Z}{a_0}\right)^{\frac{3}{2}}\left(\frac{Zr}{a_0}\right)^2 e^{-\frac{Zr}{3a_0}}(3\cos^2\vartheta-1) = G\times\left(\frac{Zr}{a_0}\right)(3\cos^2\vartheta-1) \;\propto\; (3z^2-r^2) \end{aligned}\right\} \quad (208)$$

Here, *F* and *G* are constants independent of angles ϑ and φ. When the operator $\hat{l}_z$ (see **"note"**) is applied to each of the functions in eq(208), the results of operation are as follows.

$$\begin{aligned} \hat{l}_z(3d_{xy}) &= -i\frac{\partial}{\partial\varphi}F\times\left(\frac{Zr}{a_0}\right)\sin^2\vartheta\sin2\varphi \\ &= -2iF\times\left(\frac{Zr}{a_0}\right)\sin^2\vartheta\cos2\varphi = -2i\times(3d_{x^2-y^2}) \qquad (209) \\ &\neq -2\times(3d_{xy}) \qquad \Rightarrow \quad \underline{\text{being not an eigenfunction, same for } 3d_{yz} \text{ and } 3d_{zx}.} \end{aligned}$$

$$\begin{aligned} \hat{l}_z(3d_{x^2-y^2}) &= -i\frac{\partial}{\partial\varphi}F\times\left(\frac{Zr}{a_0}\right)\sin^2\vartheta\cos2\varphi \\ &= 2iF\times\left(\frac{Zr}{a_0}\right)\sin^2\vartheta\sin2\varphi = 2i\times(3d_{xy}) \qquad (210) \\ &\neq -2\times(3d_{xy}) \qquad \Rightarrow \quad \underline{\text{not an eigenfunction.}} \end{aligned}$$

$$\hat{l}_z(3d_{z^2}) = -i\frac{\partial}{\partial\varphi} G \times \left(\frac{Zr}{a_0}\right)(3\cos^2\vartheta - 1) \qquad (211)$$
$$= 0$$
$$= 0 \times (3d_{z^2}) \qquad \Rightarrow \quad \underline{\text{an eigenfunction with eigenvalue of 0.}}$$

Namely, it is found that the $3d$-orbitals other than $3d_{z^2}$ are not the eigenfunctions of the operator of $\hat{l}_z$. As to the other eigenfunctions of $\hat{l}_z$, see **solution 1-103**.

"note" For simplicity, $\hat{l}_z$ is put equal to $-i\frac{\partial}{\partial\varphi}$. instead of $-i\hbar\frac{\partial}{\partial\varphi}$.

Ans. See text.

Solution 1-103

Two of the $3d$-orbitals and the operator of the magnetic quantum number are given in eq(212). For simplicity $\hbar$ is omitted from the operator.

$$3d_{xy} = F \times \left(\frac{Zr}{a_0}\right)\sin^2\vartheta\sin 2\varphi, \quad 3d_{x^2-y^2} = F \times \left(\frac{Zr}{a_0}\right)\sin^2\vartheta\cos 2\varphi, \quad \hat{l}_z = -i\frac{\partial}{\partial\varphi} \qquad (212)$$

By operating $\hat{l}_z$ on the function of d_{-2}-orbital, which is a linear combination of $3d_{x^2-y^2}$ and $3d_{xy}$, it is found via the following calculations that the function d_{-2} is the eigenfunction of $\hat{l}_z$ with the eigenvalue of -2 for the magnetic quantum number.

$$\hat{l}_z(d_{-2}) = \hat{l}_z(3d_{x^2-y^2}) - i\hat{l}_z(3d_{xy})$$
$$= -i\frac{\partial}{\partial\varphi}F \times \left(\frac{Zr}{a_0}\right)\sin^2\vartheta\sin 2\varphi - i \times i\frac{\partial}{\partial\varphi}F \times \left(\frac{Zr}{a_0}\right)\sin^2\vartheta\cos 2\varphi$$
$$= 2iF \times \left(\frac{Zr}{a_0}\right)\sin^2\vartheta\cos 2\varphi - 2F \times \left(\frac{Zr}{a_0}\right)\sin^2\vartheta\sin 2\varphi$$
$$= -2\left\{F \times \left(\frac{Zr}{a_0}\right)\sin^2\vartheta\sin 2\varphi - iF \times \left(\frac{Zr}{a_0}\right)\sin^2\vartheta\cos 2\varphi\right\}$$
$$= -2(3d_{x^2-y^2} - i3d_{xy})$$
$$= -2 \times d_{-2} \qquad (213)$$

How does the d_{-2}-orbital look like then? We cannot draw the function because it is not a real-valued but a complex-valued function. Therefore, the electron density distribution of d_{-2}, i.e., its absolute value squared, is calculated as below and found to be proportional to $\sin^4\vartheta$.

$$d_{-2} \times \bar{d}_{-2} = (3d_{x^2-y^2} - i3d_{xy}) \times \overline{(3d_{x^2-y^2} - i3d_{xy})} = 3d^2_{x^2-y^2} + 3d^2_{xy}$$
$$\propto \sin^4\vartheta \qquad (214)$$

Its approximate shape is drawn in Fig.95 below.

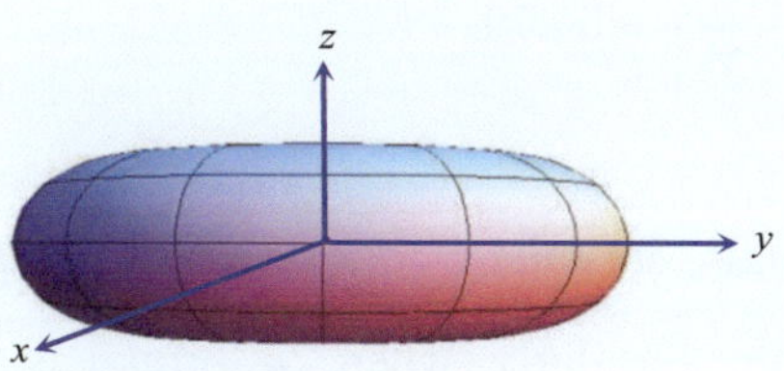

Fig.95 Approximate shape of the electron density of d_{-2}-orbital. $\sin^4\vartheta$ is plotted (see eq(214)).

As is clearly seen from Fig.94 above, it looks like a thick disc featureless in spatial orientation. In the discussions of hybrid orbitals, chemical bonds of metal complexes, ligand field effects, etc., therefore, the orbitals of d_{xy}, d_{yz}, d_{zx}, $d_{x^2-y^2}$ and d_{z^2} which are not only the real functions and eigenfunctions of energy but also characteristic in their space orientations are employed much more frequently than d_{-2}, d_{-1}, etc., though the formers are not eigenfunctions of the magnetic angular momentum. The ligand-field splitting of the *d*-orbitals are discussed in Chapter 7 in detail.

Ans. See text.

Solution 1-104

It is readily known that the radiative transitions expressed by $1s \leftarrow 2p$ and $1s \leftarrow 3p$ are allowed transitions judging from the parities of the states associated with the transitions. However, this is confirmed here by actually calculating the transition moments for the transitions mentioned above.

The wave functions which correspond to the initial and final states of the transition are expressed as $\psi_i(r,\vartheta,\varphi)$ and $\psi_f(r,\vartheta,\varphi)$, respectively. Then the transition moment $\bar{\boldsymbol{\mu}}$ is defined by eq(215).

$$\bar{\boldsymbol{\mu}} = \iiint_{r,\vartheta,\varphi} \psi_i^*(r,\vartheta,\varphi)\boldsymbol{\mu}\psi_f(r,\vartheta,\varphi)d\tau \qquad (215)$$

For the transition expressed by $1s \leftarrow 2p_x$, those wave functions are given by eqs(216) and (217).

$$\psi_i^*(r,\vartheta,\varphi) = \psi_{2p_x} = \frac{1}{4\sqrt{2\pi}}\left(\frac{1}{a_0}\right)^{3/2}\frac{r}{a_0}e^{-r/2a_0}\sin\vartheta\cos\varphi \qquad (216)$$

$$\psi_f(r,\vartheta,\varphi) = \psi_{1s} = \frac{1}{\sqrt{\pi}}\left(\frac{1}{a_0}\right)^{3/2}e^{-r/a_0} \qquad (217)$$

Consequently, when the dipole moment is expressed by $\boldsymbol{\mu} = x\boldsymbol{i} + y\boldsymbol{j} + z\boldsymbol{k}$ in the orthogonal coordinates, the *x*-component of the transition moment is expressed by eq(218) and yield a value of 0.279 via the calculation given below.

$$\bar{\mu}_x = \iiint_{r,\vartheta,\varphi} \frac{1}{\sqrt{\pi}}\left(\frac{1}{a_0}\right)^{3/2}e^{-r/a_0}\cdot x\cdot\frac{1}{4\sqrt{2\pi}}\left(\frac{1}{a_0}\right)^{3/2}\frac{r}{a_0}e^{-r/2a_0}\sin\vartheta\cos\varphi\cdot r^2\sin\vartheta\, dr d\vartheta d\varphi \qquad (218)$$

$$= \frac{1}{4\sqrt{2}\pi}\left(\frac{1}{a_0}\right)^4 \iiint_{r,\vartheta,\varphi} r^3 e^{-3r/2a_0}\sin^3\vartheta\cos^2\varphi\, dr d\vartheta d\varphi$$

$$= \frac{1}{4\sqrt{2}\pi}\left(\frac{1}{a_0}\right)^4 \times \pi\iint_{r,\vartheta} r^3 e^{-3r/2a_0}\sin^3\vartheta\, dr d\vartheta$$

$$= \frac{1}{4\sqrt{2}\pi}\left(\frac{1}{a_0}\right)^4 \times \pi\times\frac{4}{3}\int_r r^3 e^{-3r/2a_0}dr$$

$$= \frac{1}{3\sqrt{2}}\left(\frac{1}{a_0}\right)^4 \times \frac{3!}{\left(\dfrac{3}{2a_0}\right)^4} = \frac{1}{3\sqrt{2}}\times\frac{3!}{\left(\dfrac{3}{2}\right)^4} = \frac{16\sqrt{2}}{3^4} = 0.279$$

When $2p_y$-orbital is employed as one of the $2p$-orbitals, only the *y*-component of the transition moment is different from zero and is equal to $\bar{\mu}_x$ obtained above. Similarly, when $2p_z$-orbital is used, the transition moment is found equal to $\overline{\mu}_x$ as shown below in eq(219).

$$\overline{\mu}_z = \iiint_{r,\vartheta,\varphi} \frac{1}{\sqrt{\pi}}\left(\frac{1}{a_0}\right)^{3/2} e^{-r/a_0} \cdot z \cdot \frac{1}{4\sqrt{2\pi}}\left(\frac{1}{a_0}\right)^{3/2} \frac{r}{a_0} e^{-r/2a_0} \cos\vartheta \cdot r^2 \sin\vartheta \, dr d\vartheta d\varphi \qquad (219)$$

$$= \frac{1}{4\sqrt{2\pi}}\left(\frac{1}{a_0}\right)^4 \times 2\pi \iint_{r,\vartheta} r^3 e^{-3r/2a_0} \cos^2\vartheta \sin\vartheta \, dr d\vartheta$$

$$= \frac{1}{4\sqrt{2\pi}}\left(\frac{1}{a_0}\right)^4 \times \frac{4\pi}{3} \int_r r^3 e^{-3r/2a_0} dr$$

$$= \overline{\mu}_x$$

Therefore, it is found that for the transition of $1s \leftarrow 2p$, there is a relation among the components of the transition moment given by eq(220).

$$\overline{\mu}_x = \overline{\mu}_y = \overline{\mu}_z \qquad (220)$$

Next, a similar calculation is performed for the transition of $1s \leftarrow 3p_x$. The wave functions employed are those given by eqs(221) and (217).

$$\psi_i^*(r,\vartheta,\varphi) = \psi_{3p_x} = \frac{1}{81\sqrt{\pi}}\left(\frac{1}{a_0}\right)^{3/2}\left(\frac{6r}{a_0} - \frac{r^2}{a_0^2}\right) e^{-r/3a_0} \sin\vartheta \cos\varphi \qquad (221)$$

$$\psi_f(r,\vartheta,\varphi) = \psi_{1s} = \frac{1}{\sqrt{\pi}}\left(\frac{1}{a_0}\right)^{3/2} e^{-r/a_0} \qquad (217)$$

After the calculation of eq(222), a value for $\overline{\mu}_x$ is found as 0.211.

$$\overline{\mu}_x = \iiint_{r,\vartheta,\varphi} \frac{1}{\sqrt{\pi}}\left(\frac{1}{a_0}\right)^{3/2} e^{-r/a_0} \cdot x \cdot \frac{1}{81\sqrt{\pi}}\left(\frac{1}{a_0}\right)^{3/2}\left(\frac{6r}{a_0} - \frac{r^2}{a_0^2}\right) e^{-r/3a_0} \sin\vartheta \cos\varphi \cdot r^2 \sin\vartheta \, dr d\vartheta d\varphi \qquad (222)$$

$$= \frac{1}{81\pi}\left(\frac{1}{a_0}\right)^4 \iiint_{r,\vartheta,\varphi} r^3 \left(\frac{6r}{a_0} - \frac{r^2}{a_0^2}\right) e^{-4r/3a_0} \sin^3\vartheta \cos^2\varphi \, dr d\vartheta d\varphi$$

$$= \frac{1}{81\pi}\left(\frac{1}{a_0}\right)^4 \times \pi \iint_{r,\vartheta} r^3 \left(\frac{6r}{a_0} - \frac{r^2}{a_0^2}\right) e^{-4r/3a_0} \sin^3\vartheta \, dr d\vartheta$$

$$= \frac{1}{81\pi}\left(\frac{1}{a_0}\right)^4 \times \pi \times \frac{4}{3} \int_r r^3 \left(\frac{6r}{a_0} - \frac{r^2}{a_0^2}\right) e^{-4r/3a_0} dr$$

$$= \frac{4 \times 4!}{81}\left(\frac{3^6 a_0^6}{a_0{}^6 4^6}\right) = \frac{3^2 \times 6}{4^4} = \boxed{0.211}$$

In this case, it is evident for their functional forms that the relation expressed by eq(220) also holds for the 3*p*-orbitals. The components of the transition moment become zero when the direction of a *p*-orbital (*x*, *y*, *z*) is not the same as that of a component of the dipole moment.

Ans. See text.

"note" How do we know the transition moments obtained above are small or large. In comparison of those with the transition moment calculated for the simple system in which a certain transition is fully allowed, we can tell how easily (or strongly) the transition under discussion can occur. For such standard system, the harmonic oscillation of an electron is chosen and the transition moment for the transition from the ground state to a vibrationally excited state is employed. (See **solution 7-11** for the oscillator strength.)

Solution 1-105

The molar volume V_m is given in terms of density ρ and formula weight M by eq(223).

$$V_m = \frac{M}{\rho} \qquad (223)$$

On the other hand, when looked at the unit cubic lattice, each corner of the cube is shared by 8 cubes including itself and, hence, one eighth of an atom at each corner contributes to the unit cube under consideration. Therefore, it should be regarded that just one half of NaCl unit belongs to the unit lattice. The volume of the cube is d^3 and, hence, the volume of one unit of NaCl is $2d^3$. Since the molar volume is given by eq(224),

$$N_A \times 2d^3 = V_m \qquad (224)$$

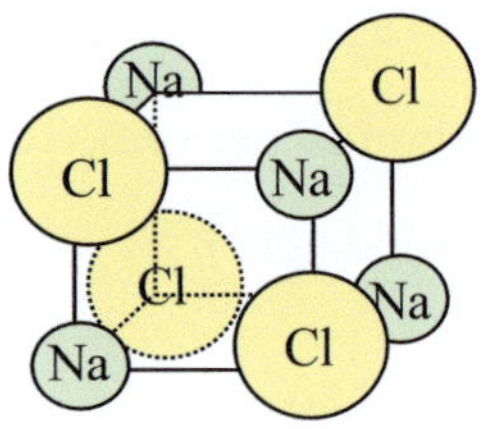

Fig.96 Unit lattice of NaCl.

Avogadro's number is calculated from eqs(223) and (224) by eq(225).

$$N_A = \frac{V_m}{2d^3} = \frac{1}{2d^3} \times \frac{M}{\rho} \qquad (225)$$

Substituting the following values into eq(225) with paying attention to the units of each value, *Avogadro*'s number is obtained as $N_A = 6.07 \times 10^{23}\,\mathrm{mol}^{-1}$.

$$d = 2.81 \times 10^{-10}\mathrm{m} = 2.81 \times 10^{-8}\mathrm{cm}, \qquad \rho = 2.17\left[\mathrm{g \cdot cm^{-3}}\right], \qquad M = 58.44\left[\mathrm{g \cdot mol^{-1}}\right]$$

$$N_A = \frac{1}{2\left(2.81 \times 10^{-8}\mathrm{cm}\right)^3} \times \frac{58.44\mathrm{g \cdot mol^{-1}}}{2.17\mathrm{g \cdot cm^{-3}}} = \frac{58.44}{2 \times \left(2.81 \times 10^{-8}\right)^3 \times 2.17} \times \frac{\cancel{\mathrm{g}} \cdot \mathrm{mol^{-1}}}{\cancel{\mathrm{cm^3}} \cdot \cancel{\mathrm{g}} \cdot \cancel{\mathrm{cm^{-3}}}} = 6.06\overset{7}{\cancel{8}} \times 10^{23}\,\mathrm{mol^{-1}}$$

Ans. $N_A = \frac{1}{2d^3} \times \frac{M}{\rho}$, $6.07 \times 10^{23}\,\mathrm{mol}^{-1}$

"note" *Perrin*'s experiments and *Avogadro*'s number. (*J.B.Perrin,* Nobel Prize in Physics in 1926.) See **problem 2-73**.

[advanced problem] How or by what method is *Avogadro*'s number determined at present as an SI unit?

Solution 1-106

The dimension of the *Planck* constant h is $[h] = \mathrm{J \cdot s}$. When some physical constants which contain either mass or electric charge and whose dimensions are J are searched for, kinetic energy and coulomb energy are found to meet the conditions. Consequently, the equations from (226) to (228) in terms of dimensions are obtained.

$$[h] = \mathrm{J \cdot s} \qquad (226)$$

$$[m_e] \times [\upsilon^2] = \mathrm{J} \ \Rightarrow\ [m_e] = \mathrm{J}/[\upsilon^2] = \mathrm{J \cdot m^{-2} \cdot s^2} \qquad (227)$$

$$[e^2] \times [r^{-1}] \times [\varepsilon_0^{-1}] = \mathrm{J} \ \Rightarrow\ [e^2] = \mathrm{J} \times [r] \times [\varepsilon_0] = \mathrm{J \cdot m} \cdot [\varepsilon_0] \qquad (228)$$

From eqs(227) and (228), eq(229) is derived.

$$[m_e] \times [e^2] = \mathrm{J^2 \cdot s^2 \cdot m^{-1}} \cdot [\varepsilon_0] = h^2 \cdot \mathrm{m^{-1}} \cdot [\varepsilon_0] \qquad (229)$$

Eq(230) is derived from eq(229) as a quantity which has the dimension of length ([m]) and is given by a simple combination of the fundamental physical quantities of m_e and e. Apart from a constant ($1/\pi$),

$$[\mathrm{m}] = \frac{h^2 \times \varepsilon_0}{m_e \times e^2} \qquad (230)$$

eq(230) corresponds to the *Bohr* radius. It is considered that a "quantum mechanically intrinsic length" is automatically determined through the introduction of h.

Ans. eq(230)

[advanced problem] How fundamental physical quantities can be combined in order to generate a quantity with the dimension of [time]? Give one example.

Solution 1-107

The velocity of an electron in the *Bohr* orbital with the quantum number n of a hydrogen-like atom is given by eq(231) (see **solution 1-16**).

$$v = \frac{nh}{2\pi m_e} \cdot \frac{\pi m_e Z e^2}{\varepsilon_0 h^2 n^2} = \frac{Ze^2}{2\varepsilon_0 nh} \qquad (231)$$

From this equation, it is found that the electron with the principal quantum number $n=1$ moves fastest. Its velocity, however, cannot exceed the velocity of light c. For simplicity, assuming that there does not exist electric shielding from the other electrons if any, from the condition set by eq(232),

$$\frac{Ze^2}{2\varepsilon_0 h} < c \;\Rightarrow\; Z < \frac{2\varepsilon_0 hc}{e^2} = \frac{2\times 8.854\times 10^{-12} \times 6.626\times 10^{-34} \times 2.998\times 10^{8}}{(1.602\times 10^{-19})^2} \approx 137 \qquad (232)$$

$Z \leq 137$ is derived for the upper limit of the atomic number. (See **solution 1-36**). It would be worth considering whether there exists any physical quantity other than velocity which sets the upper limit of the atomic number.

Ans. $Z \leq 137$

Solution 1-108

(a), (b) The simplest way of dividing the space into two parts is to divide it by one of the xy-、yz- or zx-planes which pass through the origin as are illustrated below in Fig.97. There are three way of this sort. In addition, the space can be divided by a spherical surface. The figure on the bottom and on the most right shows the manner

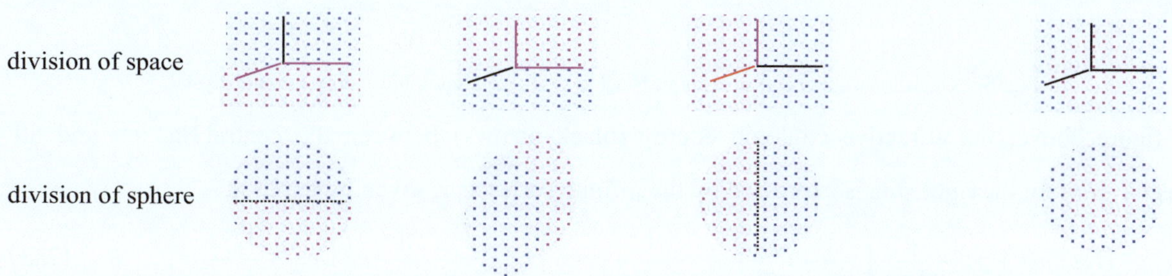

Fig.97 Illustration how to divide a homogeneous space or sphere to make the divided spaces orthogonal to each other.

of the division of a spherical space. As is readily understood, therefore, there are four ways of dividing a space (spherical space) and they correspond from left to right to $2p_z$-, $2p_x$-, $2p_y$- and $2s$-orbitals, respectively.

Ans. There are four ways of division. See text and Fig.97.

(c) It is easy to answer the problem by following (a) and (b) above. There are 9 ways of division in total. For example, by adding the division by spherical surface to the division by planes illustrated in (a), the divisions corresponding to the three $3p$-orbitals are obtained. Similarly, by adding the division by planes to the division by planes, the divisions corresponding to $3d$-orbitals are obtained. Repeating the division by a spherical surface twice, the division corresponding to $3s$-orbital results.

A sort of manipulation may be required, but when the space is divided by a double corn surface into four parts, then the division corresponding to $3d_{z^2}$ -orbital is obtained. All those divisions are illustrated below. It is easy to see from the figures in Fig.98 given below that those divisions are orthogonal to each other, in other words, independent of each other.

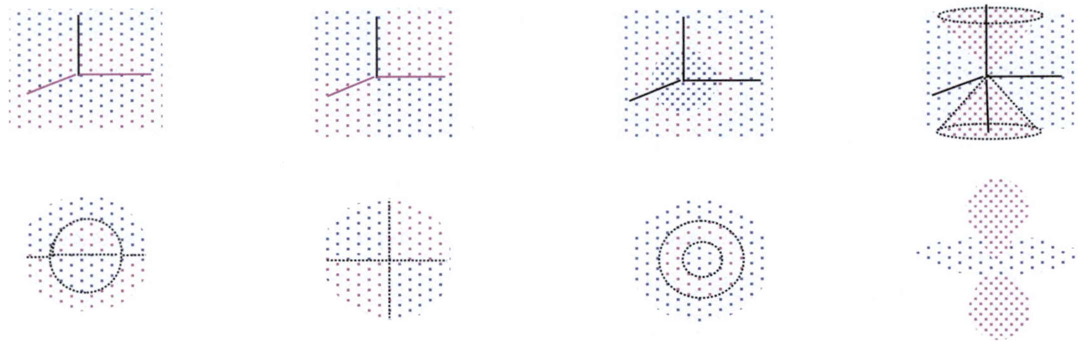

$3p_z$ $3d_{xy}$ $3s$ $3d_{z^2}$

Fig.98 Correspondece of the way of space division to the sign (positive or negative) of an atomic orbital.

Ans. See text (c) and Fig.98.

Solution 1-109

One pair of NaCl is selected from a one-dimensional crystal and then, as is indicated by two-way arrows in

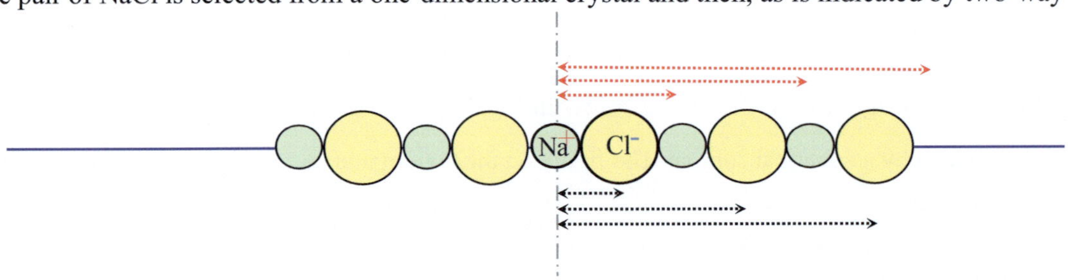

Fig.99 One-dimensional crystal array of NaCl to work out coulombic energy in ionic crystal.

the figure above, the attractive coulomb energy (black arrows) between the central Na^+ ion and all the other Cl^- ions on the right side is expressed by the infinite series as is given by eq(233).

$$(Na^+)\text{-}(Cl^-):\ -\frac{e^2}{4\pi\varepsilon_0}\left(\frac{1}{r_{NaCl}}+\frac{1}{3r_{NaCl}}+\frac{1}{5r_{NaCl}}+\cdots\right)=-\frac{e^2}{4\pi\varepsilon_0}\frac{1}{r_{NaCl}}\times\left(1+\frac{1}{3}+\frac{1}{5}+\cdots\right) \tag{233}$$

When looked at from the Cl^- ion on the right next to the central Na^+ ion, the attractive coulomb energy due to the all other Na^+ ions is given by eq(233').

$$(Cl^-)\text{-}(Na^+):\ -\frac{e^2}{4\pi\varepsilon_0}\left(\frac{1}{r_{NaCl}}+\frac{1}{3r_{NaCl}}+\frac{1}{5r_{NaCl}}+\cdots\right)=-\frac{e^2}{4\pi\varepsilon_0}\frac{1}{r_{NaCl}}\times\left(1+\frac{1}{3}+\frac{1}{5}+\cdots\right) \tag{233'}$$

In a similar manner, the repulsive coulomb energies between ions with the same charge are given by eqs(234) and (234').

$$(\mathrm{Na}^+)\text{-}(\mathrm{Na}^+):\ \frac{e^2}{4\pi\varepsilon_0}\left(\frac{1}{2r_{\mathrm{NaCl}}}+\frac{1}{4r_{\mathrm{NaCl}}}+\frac{1}{6r_{\mathrm{NaCl}}}+\cdots\right)=\frac{e^2}{4\pi\varepsilon_0}\frac{1}{r_{\mathrm{NaCl}}}\times\left(\frac{1}{2}+\frac{1}{4}+\frac{1}{6}+\cdots\right) \tag{234}$$

$$(\mathrm{Cl}^-)\text{-}(\mathrm{Cl}^-):\ \frac{e^2}{4\pi\varepsilon_0}\left(\frac{1}{2r_{\mathrm{NaCl}}}+\frac{1}{4r_{\mathrm{NaCl}}}+\frac{1}{6r_{\mathrm{NaCl}}}+\cdots\right)=\frac{e^2}{4\pi\varepsilon_0}\frac{1}{r_{\mathrm{NaCl}}}\times\left(\frac{1}{2}+\frac{1}{4}+\frac{1}{6}+\cdots\right) \tag{234'}$$

From these, it is found that the total coulomb energy for the selected pair of NaCl due to all the ions on the right-hand side is given by the sum of eqs(233), (233'), (234) and (234') as follows.

$$-\frac{e^2}{4\pi\varepsilon_0}\frac{1}{r_{\mathrm{NaCl}}}\left\{\left(1-\frac{1}{2}\right)+\left(\frac{1}{3}-\frac{1}{4}\right)+\left(\frac{1}{5}-\frac{1}{6}\right)+\cdots+\left(\frac{1}{2n-1}-\frac{1}{2n}\right)+\cdots\right\}\times 2 \tag{235}$$

Since the total sum in the curly parentheses of eq(235) can be expressed by eq(236),

$$\sum_{n=1}^{\infty}\left(\frac{1}{2n-1}-\frac{1}{2n}\right)=\frac{1}{2}\int_1^{\infty}\left(\frac{1}{2x-1}-\frac{1}{2x}\right)dx=\frac{1}{2}\left[\ln\frac{2x-1}{2x}\right]_1^{\infty}=\frac{1}{2}\ln 2 \tag{236}$$

the coulomb energy for one pair of $\mathrm{Na}^+\,\mathrm{Cl}^-$ in the center due to all the ions on the right-hand side is given by eq(236').

$$-\frac{e^2}{4\pi\varepsilon_0}\frac{1}{r_{\mathrm{NaCl}}}\times\ln 2 \tag{236'}$$

Furthermore, by taking into account of the contribution from all the ions on the left-hand side, the coulomb energy which one pair of $\mathrm{Na}^+\,\mathrm{Cl}^-$ in the one-dimensional crystal has must be two times of eq(236) and hence is found to be given by eq(237).

$$-\frac{e^2}{4\pi\varepsilon_0}\frac{M}{r_{\mathrm{NaCl}}}=-\frac{e^2}{4\pi\varepsilon_0}\frac{1}{r_{\mathrm{NaCl}}}\times 2\ln 2 \tag{237}$$

As compared with the coulomb energy of an isolated $\mathrm{Na}^+\,\mathrm{Cl}^-$, the found energy is $2\ln 2$ times more. As a result, the *Madelung* constant for the one-dimensional 1:1 ionic crystal becomes $2\ln 2=1.386\cdots$.

Ans. *Madelung* constant: $M(1\,\text{dimension})=1.386$

[advanced problem] Find the *Madelung* constants for the tow-dimensional and three dimensional NaCl crystals. The constant known for the three-dimensional lattice is 1.7476.

Solution 1-110

(a) The potential $V(r)$ is zero everywhere within the sphere. Whether the function $\psi(r)$ is an eigenfunction of eq(17) given in the problem is judged by confirming whether $\psi(r)$ satisfies eq(238) given below.

$$-\frac{\hbar^2}{2m}\frac{1}{r^2}\frac{d}{dr}\left(r^2\frac{d}{dr}\right)\psi(r)=E\psi(r)\quad E\text{ is constant.} \tag{238}$$

Actually, substituting $\psi(r)=\sin\alpha r/r$ into eq(238), the calculation is carried out to yield eq(239') via eq(239). Consequently, $\psi(r)=\sin\alpha r/r$ is considered to be an eigenfunction representing a free motion of the particle within the sphere ($V(r)=0$ and $l=0$).

$$-\frac{\hbar^2}{2m}\frac{1}{r^2}\frac{d}{dr}\left(r^2\frac{d}{dr}\right)\psi(r)=-\frac{\hbar^2}{2m}\frac{1}{r^2}\frac{d}{dr}\left(r^2\frac{d}{dr}\right)\left(\frac{1}{r}\cdot\sin\alpha r\right) \tag{239}$$

$$=-\frac{\hbar^2}{2m}\frac{1}{r^2}\frac{d}{dr}(-\sin\alpha r+\alpha r\cos\alpha r)$$

$$= -\frac{\hbar^2}{2m}\frac{1}{r^2}(-\alpha\cos\alpha r + \alpha\cos\alpha r - \alpha^2 r\sin\alpha r)$$

$$= \frac{\alpha^2\hbar^2}{2m}\frac{\sin\alpha r}{r} = \frac{\alpha^2\hbar^2}{2m}\times\psi(r) \qquad (239')$$

Ans. It is an eigenfunction.

(b) The boundary condition set by eq(240) is solved to yield the value for α as is given by eq(241).

$$\psi(a) = \frac{\sin\alpha\cdot a}{a} = 0 \qquad (240)$$

$$\alpha\cdot a = n\pi \;\Rightarrow\; \alpha = \frac{n\pi}{a} \qquad (241)$$

Substituting the result into eq(239'), eq(242) is obtained as the eigenenergies. n here is called specifically as the radial quantum number.

$$E_n = \frac{\alpha^2\hbar^2}{2m} = \frac{h^2n^2}{8ma^2} \qquad (242)$$

Eq (242) is exactly the same as the eigenenergies of a particle in a one-dimensional box with the length of a. The reason for this is very clear when the motion along the radius is regarded as the one-dimensional motion (see **"note"** below).

Ans. $E_n = h^2n^2/8ma^2 \quad n = 1,2,3,\cdots$

(c) The eigenfunction $\frac{1}{r}\sin\frac{n\pi}{a}r$ and probability density function $4\pi\sin^2\frac{n\pi}{a}r$ are plotted against radius r in Figs. 100 and 101, respectively.

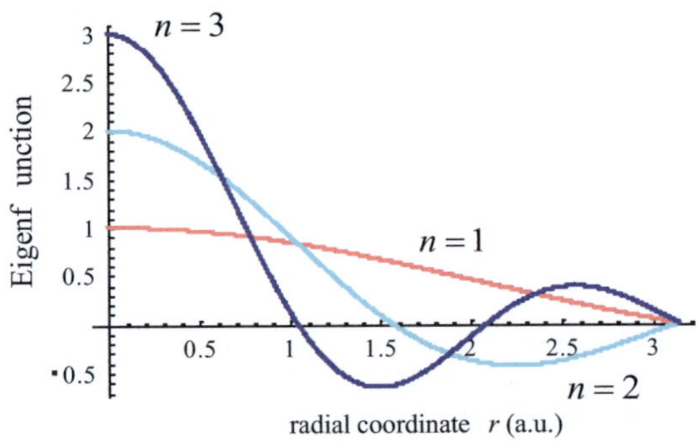

Fig.100 Eigenfunction plotted against r.

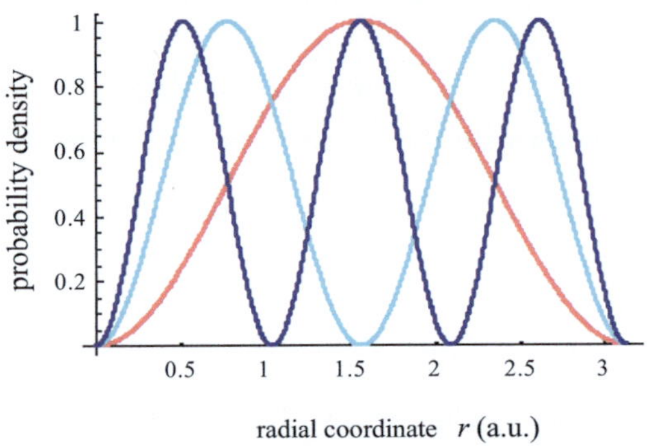

Fig. 101 Probability density plotted against r.

Ans. Figs.100 and 101

(d) The two models are analogous with respect to a feature that spherical shell structures are formed (Fig.100) along the radial coordinate, but distinctly different in the points that the eigenenergy does not converge to a constant but can take any large value, since no binding force to the particle exists in the present model. For the eigenfunctions and eigenenergies in the case that the azimuthal quantum number is not zero ($l \neq 0$), see, for instance, *L.D.Landau and E.M.Lifhitz, Quantum Mechanics-non-relativistic theory*, §§ 33. *Translated from Russian into English by J.B.Skyes and J.S.Bell.* The radial quantum number n is not related the azimuthal quantum number l in a simple manner as is known for the electronic sate of hydrogen, i.e., $l = n-1,\ n-2,\ \cdots,\ 1,\ 0$.

Ans. See text.

"note" The Hamiltonian expressed by eq(238) is a part representing the kinetic energy given by eq(243), i.e., in the case of the azimuthal quantum number $l = 0$,

$$-\frac{\hbar^2}{2m}\frac{1}{r^2}\frac{d}{dr}\left(r^2\frac{d}{dr}\right) \qquad (243)$$

which can be rewritten as eq(245), in terms of the operator of the momentum along the radial coordinate defined by eq(244).

$$-i\hbar\frac{1}{r}\frac{\partial}{\partial r}\cdot r \qquad (244)$$

As is seen from the equation in the middle of eq(245), the kinetic energy given by eq(243) is expressed by the squared momentum along the radial coordinate.

$$-\frac{\hbar^2}{2m}\frac{1}{r^2}\frac{d}{dr}\left(r^2\frac{d}{dr}\right)=\frac{1}{2m}\left(-i\hbar\frac{1}{r}\frac{\partial}{\partial r}\cdot r\right)\times\left(-i\hbar\frac{1}{r}\frac{\partial}{\partial r}\cdot r\right)=-\frac{\hbar^2}{2m}\left\{\frac{\partial^2}{\partial r^2}+\frac{2}{r}\frac{\partial}{\partial r}\right\} \qquad (245)$$

The eigenfunction and eigenvalue of eq(244) are written as $\varphi(r)$ and p, respectively so as for the operator to satisfy eq(246). Eq (247) is obtained as the solutions of eq(246).

$$\left(-i\hbar\frac{1}{r}\frac{\partial}{\partial r}\cdot r\right)\varphi(r)=p\cdot\varphi(r) \;\Rightarrow\; -i\hbar\frac{\partial}{\partial r}\cdot r\varphi(r)=p\cdot r\varphi(r) \qquad (246)$$

$$r\varphi(r)=e^{i\frac{p}{\hbar}r} \;\Rightarrow\; \varphi(r)=\frac{1}{r}e^{i\frac{p}{\hbar}r} \qquad (247)$$

From eqs(246) and(247), it is easily understood that the motion of a particle bound within a sphere resembles the motion of the particle in a one-dimensional box. See also "Degeneracy and Symmetry" given in the end of this chapter.

[advanced problem] The quantized energy levels discussed here and in **Solution 1-43** are realized in various nanoparticles (quantum dots). Work out the lowest electronic transition energy in the unit of cm^{-1} for a spherical quantum dot of the radius of 1 nm as well as for a cubic quantum dot of the edge of the length of 2nm

Solution 1-111

(a) The molecular orbitals of the CO molecule is considered as given below by referring to the figure given in **solution 1-45** and can accommodate 14 electrons in total.

$$1\sigma_{1s}^{\ 2}1\sigma_{1s}^{*\ 2}2\sigma_{2s}^{\ 2}2\sigma_{2s}^{*\ 2}1\pi_{2pa}^{2}1\pi_{2pb}^{2}3\sigma_{2p}^{\ 2}$$

From this configuration, it is seen that there are three bonding orbitals. They are $1\pi_{2pa}=1\pi_{2pb}$ and $3\sigma_{2p}$ in the order of increasing energy. Therefore, the band of the lowest ionization energy is ascribed to the ionization of an electron in the $3\sigma_{2p}$ orbital and the band of the next lowest ionization energy is ascribed to the ionization of an electron in either $1\pi_{2pa}$ or $1\pi_{2pb}$ orbital. As is estimated from the atomic orbitals which construct these molecular orbitals, the ionization energies (without vibrational excitation) of group 1 and group 2 are close to the sum of the first ionization energy of the O atom or the C atom (11-14 eV) and the bond energy (ca.3eV). Group 3 is considered to be due to the ionization of an electron in the antibonding $2\sigma_{2s}^{*}$ orbital. The first ionization energy of the C atom or the O atom is due to the ionization of an electron in 2*p*-orbitals. In order to estimate the ionization energy from the $2\sigma_{2s}^{*}$ orbital, the ionization energy from the

2*s*-orbital is required. For this, the electronic excitation energy of the ion expressed by $C^+(2s^2 2p^1) \rightarrow C^+(2s^1 2p^2)$, which is of the order of 11 - 14eV, is added to the first ionization energy of C. Thus, the ionization energy corresponding to the process of $C(2s^2 2p^2) \rightarrow C^+(2s^1 2p^2) + 1e$ is estimated to be $11.3 + 13.7 = 25$ eV. Subtracting the antibonding energy of 3 - 4eV from this, a value of 21 - 22 eV is obtained, which is close to the ionization energy associated with the orbital of $2\sigma_{2s}^*$.

Including the ionization bands originating from the shell electrons in the orbitals of $1\sigma_{1s}$, $1\sigma_{1s}^*$, $2\sigma_{2s}$ and $2\sigma_{2s}^*$, six band groups in total are observed for the photoelectron spectrum of the CO molecule.

Ans. In the order of increasing energy: $1\sigma_{1s}$, $1\sigma_{1s}^*$, $2\sigma_{2s}$, $2\sigma_{2s}^*$, $1\pi_{2pa} = 1\pi_{2pb}$, $3\sigma_{2p}$

(b) Before ionization, the molecules are vibrationally in the ground state. The ionized molecules, however, are found in various vibrational levels. Since the vibrational energies are included in the total energy (rotational and translation energy of the molecule neglected), the vibrational energy should be included in the energy of the ionized molecule as is expressed in eq(19) given in the problem. Generally speaking, a change in the vibrational state seen between the molecule and its product ion is the greater the stronger the bond responsible for the ionization. As a consequence, a difference in the vibrational energy becomes larger, causing the vibrational structure more complex. On the other hand, when the ionization from the weaker bonds such as π-bond orbitals or non-bonding orbitals, there is less difference in the vibrational states between the molecule and its ion. Therefore, the difference in the vibrational energies between the molecule and its ion is approximately equal to the difference in the vibrational enegy of the molecule in the ground state, resulting in rather a simple vibrational structure (*Franck-Condon* principle). In the case of the ionization from the bonding orbitals, several vibrational bands are observed depending on how largely the equilibrium position shifts upon ionization. The vibrational structures observed for the band group 2 reflect the vibrational features of the CO bond.

Ans. See text (c).

[advanced problem] Referring to the energy level diagram and the molecular orbitals of the NH_3 molecule given in **solution 7-81a**, predict the photoelectron spectrum of NH_3.

Solution 1-112

The problem will be discussed from the following several aspects.

(i) Almost all elements given in the problem are alkaline metals or alkaline earth metals and their electronic configurations are represented by ns^1 or ns^2. Even for Cu which does not belong to the two families above, its electronic configuration is $4s^1 3d^{10}$ and resembles well that of an alkaline metal

(ii) It is seen from the figure of the ionization energy vs.*Z* (see, for instance, **solution 1-33**) that all those elements in the problem are readily ionized. However, even for the most easily ionized $_{37}$Rb among them, the ionization energy of ca. 400 kJ · mol^{-1} is required and is far greater than the thermal energy of $RT \approx 10$ kJ · mol^{-1} available in the flame reaction.

(iii) When the emission wavelength is assumed to be 600 nm, its energy becomes $200\ \mathrm{kJ \cdot mol^{-1}}$ and is less than a half of the ionization energy.

(iv) Summarizing those facts given above, the direct ionization process has nothing to do with the flame reaction but the thermally excited *s*-electron is considered responsible for the phenomena. The figure which shows the relation between the atomic orbital and the atomic number (**problem 1-23** indicates that the atomic orbital which lies most closely to but higher than the *ns*-orbital in energy is *np*-orbital. From the figure alone, the difference in energies between those orbitals is not known, but it seems not to vary significantly with the atomic number.

(v) Actually, when the emission wave number of the lowest energy (hence the longest wavelength) is plotted against the atomic number, including the emissions from Sr and Ta, Fig.102 is obtained.

It is recognized from this figure that the emission wavelength shifts gradually to the shorter wavelength with the atomic number. Also, the emission wavelengths of Ca and Cu can be predicted. For instance, the emission from Cu is to be found at around $17800\ \mathrm{cm^{-1}}$. Under the condition of much higher energy than that available for the flame reaction, the emission from the ionized atom can naturally be observed.

(See Principles of atomic emission analysis.)

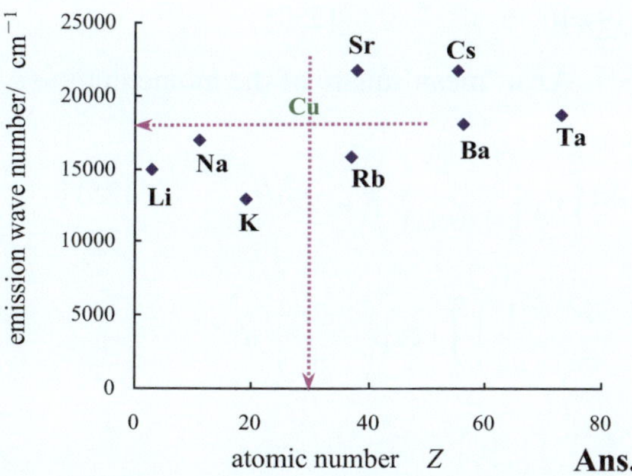

Fig.102 Emission wave number vs. atomic number.

Ans. $s \leftarrow p$ transition after the thermal excitation. See text.

Solution 1-113

The mean value of an operator $\hat{O}$ associated with a certain system is defined by eq(248) through an eigenfunction of the system ψ_a and can be calculated by this equation.

$$<O>=\int \psi_a^* \hat{O} \psi_a d\tau \qquad (248)$$

It should be noted that ψ_a is not necessary the eigenfunction of the operator $\hat{O}$.

(a) We can answer the problem in a simple way other than calculating the mean by eq(248). This is because the mean of the momentum (physical quantity) must be a real number but the operator of momentum is imaginary. For an eigenfunction expressed by a real number such as for the 1*s*-orbital, the mean value of the momentum cannot take a value other than zero (see **"note"** below). Of course, this can be directly confirmed by calculation as is performed for (b) and (c) below.

Ans. See text (a).

(b) When we put $\psi_a = \psi_{1s}$ and $\hat{O} = -i\hbar \frac{1}{r}\frac{\partial}{\partial r} \cdot r$, the mean square of the momentum $< p^2 >$ is defined by eq(249) and after performing the integration, the result is given by eq(249'). As is readily confirmed, the dimension of the obtained value is equal to that of momentum squared.

$$< p^2 >= \int_0^\infty \psi_{1s}^* \left(-i\hbar \frac{1}{r}\frac{\partial}{\partial r} \cdot r \right)^2 \psi_{1s} d\tau = 4\pi \int_0^\infty \psi_{1s} \left(-i\hbar \frac{1}{r}\frac{\partial}{\partial r} \cdot r \right)^2 \psi_{1s} \cdot r^2 dr \qquad (249)$$

$$= -4\pi \hbar^2 \int_0^\infty \psi_{1s} \left(\frac{\partial^2}{\partial r^2} + \frac{2}{r}\frac{\partial}{\partial r} \right) \psi_{1s} \cdot r^2 dr$$

$$= -4\hbar^2 \left(\frac{1}{a_0} \right)^3 \int_0^\infty r^2 \exp\left(-\frac{r}{a_0} \right) \left(\frac{\partial^2}{\partial r^2} + \frac{2}{r}\frac{\partial}{\partial r} \right) \exp\left(-\frac{r}{a_0} \right) \cdot dr$$

$$= -4\hbar^2 \left(\frac{1}{a_0} \right)^3 \int_0^\infty \left(\frac{r^2}{a_0^2} - \frac{2r}{a_0} \right) \exp\left(-\frac{2r}{a_0} \right) \cdot dr$$

$$= -4\hbar^2 \left(\frac{1}{a_0} \right)^3 \left(\frac{a_0}{4} - \frac{a_0}{2} \right)$$

$$= \left(\frac{\hbar}{a_0} \right)^2 = \left[(\mathrm{M \cdot L \cdot s^{-1}})^2 \right] \qquad (249')$$

Consequently, the velocity of the electron is given by eq(250), where the *Bohr* radius $a_0 = \dfrac{\varepsilon_0 h^2}{\pi m_e e^2}$ is used.

$$m_e \upsilon = \frac{\hbar}{a_0} \Rightarrow \upsilon = \frac{h}{2\pi m_e a_0} = \frac{e^2}{2\varepsilon_0 h} = 2.19 \times 10^6\,\mathrm{m \cdot s^{-1}} \qquad (250)$$

Ans. mean square of the momentum: $(\hbar/a_0)^2$

(c)

$$< -\frac{e^2}{4\pi\varepsilon_0 r} >= \int_0^\infty \psi_{1s}^* \left(-\frac{e^2}{4\pi\varepsilon_0 r} \right) \psi_{1s} d\tau = 4\pi \int_0^\infty \psi_{1s} \left(-\frac{e^2}{4\pi\varepsilon_0 r} \right) \psi_{1s} \cdot r^2 dr \qquad (251)$$

$$= -\frac{e^2}{\pi\varepsilon_0} \left(\frac{1}{a_0} \right)^3 \int_0^\infty r \exp\left(-\frac{2r}{a_0} \right) \cdot dr$$

$$= -\left(\frac{1}{a_0} \right)^3 \left(\frac{a_0}{2} \right)^2$$

$$= -\frac{e^2}{4\pi\varepsilon_0 a_0} = -\frac{m_e e^4}{4\varepsilon_0^2 h^2} \qquad (251')$$

Ans. mean value of coulomb energy: $-\dfrac{m_e e^4}{4\varepsilon_0^2 h^2}$

(d) From eq(250) (eq(249')) and eq(251'), the total energy of the 1*s*-electron expressed by eq(252) is found as eq(252').

$$E = T + V = \frac{< p^2 >}{2m_e} + < -\frac{e^2}{4\pi\varepsilon_0 r} >= \frac{1}{2m_e} \left(\frac{m_e e^2}{2\varepsilon_0 h} \right)^2 - \frac{m_e e^4}{4\varepsilon_0^2 h^2} \qquad (252)$$

$$= -\frac{m_e e^4}{8\varepsilon_0^2 h^2} = E_{1s} \qquad (252')$$

Eq(252') shows that the total energy obtained by adding the mean of coulomb potential and the mean kinetic energy is equal to the eigenenergy of 1*s*-orbital. Although there is no wonder about the result obtained, of course, the introduction of the concept of the mean of an operator makes it possible to match the quantum mechanical picture (eigenvalues) to the kinetical one (velocity and potential). Specifically, what the *Bohr* model and the *Bohr* radius rely on will be understood more precisely. Refer to the next problem and its solution.

Ans. See text.

"note" The operators of the momentum in the orthogonal coordinates are expressed as $-i\hbar(\partial/\partial x)\cdots$, etc. The mean of each component of the angular momentum as well as the momentum is zero for any *ns*-orbital because of the same reason. When the eigenfunction of an atomic orbital is a complex function, the means of those momenta are not necessary zero.

Solution 1-114

The momentum operator along the radial coordinate, as has been discussed above, is expressed by eq(253),

$$-i\hbar\frac{1}{r}\frac{\partial}{\partial r}\cdot r \qquad (253)$$

and its eigenfunction with the eigenvalue of *p* is given by eq(254).

$$\frac{1}{r}\exp\left(-i\frac{p}{\hbar}\cdot r\right) \qquad (254)$$

The eigenfunction of 1*s*-orbital denoted by ψ_{1s} is written as eq(255).

$$\psi_{1s}=\frac{1}{\sqrt{\pi}}\left(\frac{1}{a_0}\right)^{3/2}e^{-r/a_0} \qquad (255)$$

The distribution of momentum of the 1*s*-orbital is nothing but to know how widely the function ψ_{1s} given by eq(255) contains the component expressed by eq(254). In other words, when the function ψ_{1s} is expanded in terms of the function in eq(254) by allowing all values of *p* for it, it is to determine the coefficient of the eigenfunction of the eigenvalue *p* (see **"note"** below). When the coefficient is expressed by $\psi(p)$, it is given by eq(256),

$$\psi(p)=A\iiint_{r,\vartheta,\varphi}\frac{1}{r}\exp\left(i\frac{p}{\hbar}\cdot r\right)\cdot\psi_{1s}(r)\cdot d\tau \qquad (256)$$

where *A* is a constant for normalization. Eq(256) is readily integrated via eq(256') and eq(256'') is obtained as a result.

$$\psi(p)=4\pi A\int_0^\infty\frac{1}{r}\exp\left(i\frac{p}{\hbar}\cdot r\right)\psi_{1s}(r)\cdot r^2dr \qquad (256')$$

$$=4\pi A\times\frac{1}{\sqrt{\pi}}\left(\frac{1}{a_0}\right)^{3/2}\int\exp\left(i\frac{p}{\hbar}\cdot r\right)\exp\left(-\frac{r}{a_0}\right)\cdot rdr$$

$$=4\sqrt{\pi}A\left(\frac{1}{a_0}\right)^{3/2}\int\exp\left\{-\left(\frac{1}{a_0}-i\frac{p}{\hbar}\right)\cdot r\right\}\cdot rdr$$

$$=4\sqrt{\pi}A\left(\frac{1}{a_0}\right)^{3/2}\frac{1}{\left(\frac{1}{a_0}-i\frac{p}{\hbar}\right)^2}$$

$$=4\sqrt{\pi}A\left(\frac{1}{a_0}\right)^{3/2}\frac{1}{\left(\frac{1}{\hbar}\right)^2\left(\frac{\hbar}{a_0}-ip\right)^2} \qquad (256'')$$

In quantum mechanics, the squared value of the absolute value of the coefficient ($\psi(p)$ in the present case) is regarded as the probability of its occurrence. Consequently, the probability for the electron in $1s$-orbital to have the momentum of p is given by eq(257).

$$|\psi(p)|^2 = 16\pi A^2\left(\frac{1}{a_0}\right)^3 \hbar^4 \frac{1}{\left(p^2+\left(\frac{\hbar}{a_0}\right)^2\right)^2} \propto \frac{1}{\left(p^2+\left(\frac{\hbar}{a_0}\right)^2\right)^2} \qquad (257)$$

What is known from eq(257) is that the distribution of p is symmetric against $p=0$. For this, it is evident that the mean of p is zero.

Next, the mean of the squared values of p is calculated from eq(258) by employing the distribution function of eq(257).

$$<p^2>=\frac{\int_{-\infty}^{\infty} p^2|\psi(p)|^2 dp}{\int_{-\infty}^{\infty}|\psi(p)|^2 dp} = \int_{-\infty}^{\infty} p^2\left(p^2+\left(\frac{\hbar}{a_0}\right)^2\right)^2 dp \Big/ \int_{-\infty}^{\infty}\left(p^2+\left(\frac{\hbar}{a_0}\right)^2\right)^2 dp \qquad (258)$$

By equating $x=\frac{a_0}{\hbar}p$ eq(258) is rewritten as eq(258') and in order to perform integration, the variable x is further replaced by $\tan\vartheta$. After integration over ϑ eq(258'') is obtained.

$$<p^2>=\int_{-\infty}^{\infty}\left(\frac{a_0}{\hbar}\right)\frac{x^2}{(x^2+1)^2}dx \Big/ \int_{-\infty}^{\infty}\left(\frac{a_0}{\hbar}\right)^3\frac{1}{(x^2+1)^2}dx \qquad (258')$$

$$=\left(\frac{\hbar}{a_0}\right)^2 \qquad (258'')$$

This is exactly the same as eq(249') in the preceding solution, though it is evident from a comparison of the definitions given in each case.

What is made very clear from the discussion about the momentum distribution is that the $1s$-electron is moving along the radial coordinate to and fro with the same probability (eq(257)), but is not rotating in the *Bohr* radius and, hence, its movement does not produce any magnetic momentum, which is naturally consistent with the fact that the magnetic quantum number of the $1s$-electron is zero.

Ans. See text and eq(258'').

Solution 1-115

(a) The normalization constants N_0 and N_1 are calculated as in eqs(259) and (260), respectively.

$$\int_{-\infty}^{+\infty}\phi_0^2 dx = 1 = N_0^2\int_{-\infty}^{+\infty}e^{-x^2/\alpha^2}dx = N_0^2\cdot\alpha\cdot\sqrt{\pi} \Rightarrow N_0=\frac{1}{\alpha^{1/2}\pi^{1/4}} \qquad (259)$$

$$\int_{-\infty}^{+\infty}\phi_1^2 dx = 1 = \frac{4N_1^2}{\alpha^2}\int_{-\infty}^{+\infty}e^{-x^2/\alpha^2}x^2 dx = \frac{4N_1^2}{\alpha^2}\cdot\frac{\alpha^2}{4}\cdot\alpha\cdot\sqrt{\pi} \Rightarrow N_1=\frac{1}{\alpha^{1/2}\pi^{1/4}} \qquad (260)$$

Ans. $N_0=N_1=1/\alpha^{1/2}\pi^{1/4}$

(a) The transition moment for the transition of $v = 0 \to v = 1$ is defined by eq(261), which is calculated by employing the result obtained in (a) above to give eq(261').

$$\mu = \int_{-\infty}^{+\infty} \phi_0 \cdot ex \cdot \phi_1 dx \qquad (261)$$

$$= \frac{2N_0 N_1 e}{\alpha} \int_{-\infty}^{+\infty} e^{-x^2/\alpha^2} x^2 h dx = \frac{2e}{\alpha^2 \pi^{1/2}} \cdot \frac{\alpha^2}{4} \cdot \alpha \cdot \sqrt{\pi} = \frac{e\alpha}{2} \qquad (261')$$

Since the transition probability is given by the square of the transition moment μ^2, eq(262) is obtained from eq(261) by substituting the constants α and κ (eq(26) in the problem) with care.

$$\mu^2 = \frac{e^2 a^2}{4} = \frac{e^2}{4}\left(\frac{h^2}{m_e \kappa}\right)^{1/2} = \frac{e^2 h}{8\pi^2 m_e \nu} \qquad (262)$$

Ans. transition probability: $\mu^2 = e^2 h / 8\pi^2 m_e \nu$

"note" In order to compare the magnitudes of transition probabilities associated with the transitions between quantum states upon light absorption or emission, three times of the transition probability obtained above is taken as the standard (three-dimensional oscillator) and its ratio to the square μ^2_{sample} of the transition moment obtained for a sample (or a system) defined by eq(263) is called the <u>oscillator strength</u> f of the sample.

$$f = \frac{\mu^2_{sample}}{3 \times e^2 h / 8\pi^2 m_e \nu} = \frac{8\pi^2 m_e \nu \cdot \mu^2_{sample}}{3e^2 h} \qquad (263)$$

Solution 1-116

(a) An electron with the magnetic moment of μ_e placed in the magnetic field given by $B(x)$ has the potential energy expressed as eq(264).

$$U(x) = -\mu_e \cdot B(x) \qquad (264)$$

Then, the electron receives the force described by eq(265).

$$F(x) = -\left(\frac{\partial U(x)}{\partial x}\right) = \left(\frac{\partial \mu_e \cdot B(x)}{\partial x}\right) \qquad (265)$$

When the slope of the magnetic field is constant, by equating $F(x) = F(\text{constant})$, the electron, and hence a Ag atom starts to move with the acceleration of a (eq(266)) along the x-axis at the entrance to the magnet. If the initial velocity along the x-axis of the Ag atom is zero, then it continues to move with the velocity υ_x given by eq(267) until it reaches the exit point of the magnet.

$$F = m_{\text{Ag}} \times a \qquad (266)$$

$$\upsilon_x = a \times t \qquad (267)$$

$$x = \frac{1}{2} a t^2 \qquad (268)$$

Since the mean kinetic energy and mean squared velocity per a Ag atom are related by eq(269), as an approximation eq(270) is obtained by putting $\overline{|\upsilon_z|} \approx \sqrt{\overline{\upsilon_z^2}}$.

$$\frac{1}{2} m \overline{\upsilon_z^2} = \varepsilon \qquad (269), \qquad \overline{|\upsilon_z|} \approx \sqrt{\overline{\upsilon_z^2}} = \sqrt{\frac{2\varepsilon}{m}} \qquad (270)$$

Therefore, the time required for the atom pass through the magnet of the length L is given by eq(271) during which the distance the Ag atom travels along the x-axis is calculated from eq(268) and is given by eq(272).

$$t=\frac{L}{|v_z|}=L\sqrt{\frac{m_{Ag}}{2}} \quad (271), \qquad x=\frac{1}{2}\cdot\frac{F}{m_{Ag}}\cdot\frac{L^2 m_{Ag}}{2\varepsilon}=\frac{\mu_e L^2}{4\varepsilon}\left(\frac{\partial B(x)}{\partial x}\right) \quad (272)$$

When $(\partial B(x)/\partial x)>0$, the Ag atom with the electron spin quantum number of $1/2$ or the one with $-1/2$ aligned along the direction of the atomic beam travels oppositely along the x-axis by the same distance and, hence, the distance between the two split beams under the magnetic field will be twice of x given by eq(272). It should also be mentioned that the Ag atom in the electronic ground state is expressed by $^2S_{1/2}$ and has no orbital angular momentum. Therefore, it is not necessary to take the angular momentum as well as the nuclear spin momentum into account in the present case.

Ans. distance between two beams: $2\times\frac{\mu_e L^2}{4\varepsilon}\left(\frac{\partial B(x)}{\partial x}\right)$

(b) The distance between the two split beams after passing the inclining magnetic field is given by eq(273).

$$2\times\frac{\mu_e L^2}{4\varepsilon}\left(\frac{\partial B(x)}{\partial x}\right)_{y,z}=1.0\times10^{-3}\,\text{m} \quad (273)$$

Substituting the following values into this equation,

$$\mu_e=\mu_B(Bohr\ \text{magneton})=\frac{eh}{4\pi m_e}=\frac{1.602\times10^{-19}\times6.626\times10^{-34}}{4\pi\times9.109\times10^{-31}}=9.273\times10^{-24}\,\text{J}\cdot\text{T}^{-1}$$

$$2\varepsilon=k_BT=1.381\times10^{-23}\times1000=1.381\times10^{-20}\,\text{J}$$

the slope of the magnetic field is found to be given by eq(274).

$$\left(\frac{\partial B(x)}{\partial x}\right)_{y,z}\geq\frac{1.0\times10^{-3}\times1.381\times10^{-20}}{9.273\times10^{-24}\times0.5^2}=5.96\,\text{T}\cdot\text{m}^{-1} \quad (274)$$

Besides、from eq(273) it is known that the higher the temperature of the furnace is the larger slope of the magnetic field or the longer magnet is required to have the beam split by the same distance.

Ans. slope of the magnetic field: $5.96\ \text{T}\cdot\text{m}^{-1}$

Solution 1-117

(a) The spin angular momentum for the electronic state expressed by the spectral term of $^1\text{D}_2$ or $^1\text{P}_1$ is zero in either case. However, the total angular momentum is $L=2$ $(M=2,\ 1,\ 0,-1,-2)$ for the former and $L=1\,(M=1,\ 0,-1)$ for the latter. When $S\neq0$, the *Zeeman* effect becomes complex.

As is shown in Fig.103 given below, due to the interaction with the magneticg field dependent on M, the $^1\text{D}_2$ state is split into five levels and the $^1\text{P}_1$ state into three levels either with a constant interval. Since the direction of the angular momentum (vector) is opposite to the magnetic dipole moment, the level with the larger value of M has the lower energy. The interval of the splitting is given by eq(275) as $4.66\ \text{cm}^{-1}$.

$$-\mu_z\cdot B_z=M\times\frac{eh}{4\pi m_e}\cdot B_z \Rightarrow \Delta\varepsilon=\frac{eh}{4\pi m_e}\cdot B_z \quad (275)$$

$$\Delta\varepsilon=\frac{1.602\times10^{-19}\times6.626\times10^{-34}\times10}{4\pi\times9.109\times10^{-31}}=0.925\times10^{-22}\,\text{J}=4.66\ \text{cm}^{-1}$$

As is seen from the result obtained, the interval of the splitting is large enough to be observed under strong magnetic fields. Note that $\frac{eh}{4\pi m_e}$ is the *Bohr* magneton.

Ans. $^1D_2$ into 5 levels, $^1P_1$ into 3 levels. splitting interval: $0.925\times10^{-22}\,J = 4.66\ cm^{-1}$

(b) According to the selection rule for the magnetic quantum number, which is cited again below (eq(276)),

$$\Delta M = 0,\ \pm 1 \qquad (276)$$

the allowed transitions between the split levels are as shown in Fig.102 given below.

As is seen from the figure, there are nine possible transitions in total but the transitions marked by the arrows of the same color have the same energy. Consequently, the observed spectrum consists of three bands. As compared with the thermal energy, when energy difference between levels ($^1P_1$ in the figure) is small, the thermal populations of the lower levels can be regarded comparable and, hence, the band heights of the three bands will be about the same. Even when the $^1D_2$ state is the lower state, three split bands will be observed.

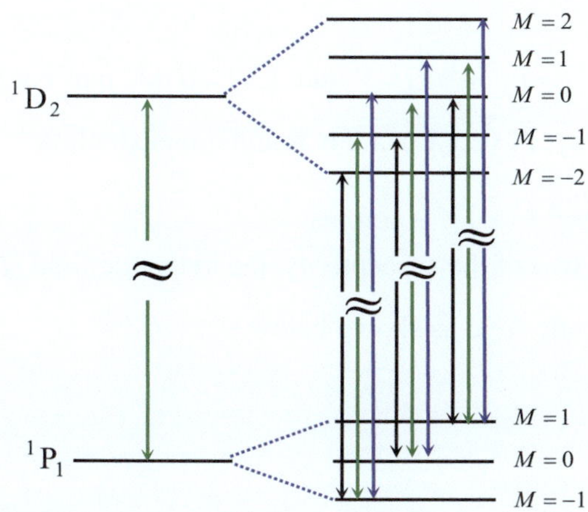

Fig.103 *Zeeman* splitting of $^1D_2$ and $^1P_1$ states and allowed electronic transitions.

Ans. See Fig.103 for the spectrum.

Solution 1-118a

Magnetic properties of atoms and molecules are principally decided by the spin and angular momentum of the electrons associated with them. A contribution from nuclear spins is small enough to be neglected.

For atoms:

(1) The one which has non-zero total spin angular momentum S is paramagnetic.

(2) The one for which $S = 0$ but L is not zero shows paramagnetic property.

(3) The one for which $S = 0$ and $L = 0$ exhibit diamagnetic property.

For molecules:

To be considered principally conforming to the guidelines given above for atoms, but in addition,

(4) The one which has no unpaired electron and is in the singlet state is diamagnetic.

Judging based on the guidelines from (1) through (4),

Diamagnetic substances (with no paramagnetic property) are as follows:

$Li^+(1s^2)$: ^1S

$Ne(1s^2 2s^2 2p^6)$: ^1S

$F^-(1s^2 2s^2 2p^6)$: ^1S

$H_2(\sigma_{1s}{}^2)$: $^1\Sigma$

$N_2(\sigma_{1s}{}^2\sigma_{1s}^{*\,2}\sigma_{2s}{}^2\sigma_{2s}^{*\,2}\sigma_{2p_z}{}^2\pi_{2p_x}^2\pi_{2p_y}^2)$: $^1\Sigma$,

HCl, CO_2, NH_3, CH_4: no unpaired electron

The number of the valence electrons of NO_2 is 17 and the molecule has one unpaired electron.

The O_2 molecule in the ground state is electronically in the triplet state and, hence, the molecule is paramagnetic, as is seen from the magnetism shown by liquid oxygen (see **solution 1-41a**).

As to the electronic states of diatomic molecules, see **solution 1-9**.

Ti $(1s^2 2s^2 2p^6 3s^2 3p^6 4s^2 3d^2)$ exhibits a large paramagnetism because of its non-zero spin as well as angular momentum.

Methyl radical has one unpaired electron.

Ans. substances which have no paramagnetism and are diamagnetic:

Li^+, Ne, F^-, H_2, N_2, HCl, CO_2, NH_3, CH_4

Solution 1-118b

(a) Substituting the dimensions of the relevant constants $[m_e c^2] = \mathrm{J}$, $[e^2/r] = \mathrm{J}$ and $[N] = 1/\mathrm{m}^3$ into eq(29) given in the problem, the dimension of the susceptivity χ is reduced to eq(1′) and is found dimensionless.

$$[\chi] = \frac{1}{\mathrm{m}^3} \cdot \frac{\mathrm{J}}{\mathrm{J}} \cdot \mathrm{m}^3 = \text{dimensionless} \qquad (29')$$

This is also directly derived simply from the relations between two of the susceptivity, the magnetic field (*H*), the energy of magnetization (χH^2) and the energy of the magnetic field as shown below.

$$\mu = \chi H, \qquad \mu H = \chi H^2$$

$$[\chi] = \frac{[\mu H]}{[H^2]} = \frac{\mathrm{J}}{\mathrm{J}} \Rightarrow \text{dimensionless}$$

(b) Refer to **"note"** in **solution 1-12** for the estimation of the effective atomic number and also to **solution 1-16** for the relation between the *Bohr* radius and the nuclear charge.

The *Bohr* radius is calculated by putting the principal quantum number $n = 1$ as in eq(277).

$$r = \frac{\varepsilon_0 h^2 n^2}{\pi m_e Z_{eff} e^2} = \frac{8.834 \times 10^{-12} \times (6.626 \times 10^{-34})^2 \times 1}{\pi \times 9.109 \times 10^{-31} \times 1.70 \times (1.602 \times 10^{-19})^2} = 3.11 \times 10^{-11} \mathrm{m} \qquad (277)$$

Since the units in cgs are employed for the *Langevin-Debye* equation, the susceptivity is calculated in those units as below and the value of -5.47×10^{-7} is found.

$$\chi = -\frac{N_A e^2}{6 m_e c^2} \times 2 \times (\overline{r_{1s}})^2 = -\frac{6.022 \times 10^{23} \times (4.803 \times 10^{-10})^2}{6 \times 9.109 \times 10^{-28} \times (2.998 \times 10^{10})^2} \times 2 \times (3.11 \times 10^{-9})^2 \qquad (277')$$

$$= -2.83 \times 10^{10} \times (3.11 \times 10^{-9})^2 = -5.47 \times 10^{-7}$$

The value observed experimentally is $\chi(\mathrm{He}) = -1.88 \times 10^{-6}$ (see **"note"** below).

In order to calculate the value in SI units, eq(29) given in the problem should be rewritten as eq(277′) beforehand with attention to $1\mathrm{C} = c \times 10^{-1}$(esu), $1\mathrm{kg} = 10^3 \mathrm{g}$, etc. The molar susceptivity is obtained by putting $N = N_A$.

$$\chi = -\frac{N e^2 \times 10^{-2}}{6 m_e \times 10^3} \sum_{i=1}^{\text{all electrons}} \overline{r_i^2} \times 10^4 \qquad (277'')$$

Ans. molar magnetic susceptivity: $\chi(\mathrm{He}) = -5.47 \times 10^{-7}$

[advanced problem] Explain why the (absolute) value of χ(He) calculated above is significantly smaller than the experimental value.

"note" *J. H. van Vleck, The Theory of Electric and Magnetic Susceptibilities,* Oxford University Press. Japanese translation by *M.Kotani and K.Kanda.*

The diamagnetic susceptivity is less than one percent of the paramagnetic susceptivity. When the value of $\overline{r_i^2}$ calculated by employing the electronic wave function of He (*Slater* orbital) is used, the theoretical value of χ_{mol}(He) shows a good agreement with the experimental value.

Degeneracy and Symmetry.

Quantization of the motion of a particle within a space, such as one-dimensional box, a square, a cube or a sphere and associated energy levels are treated extensively in **problems 1-4**, **1-42**, **1-43** and **1-110**. The Hamiltonian which describes the motion of the particle quantum mechanically should have the same symmetry as the space (plus field) in which the particle is moving. Therefore, the eigenfunctions of the Hamiltonian should have the same symmetry as the space (plus field). As a result of this, the eigenfunctions are expressed by relevant irreducible expressions of the point group to which the space (plus field) belongs and the degeneracy of the eigenstate is given by the "order" of the irreducible expression (the number equal to that given under the operation E in the character table). For instance, the space (no field) of a rectangle or a cuboid belongs to D_{2h}, whose irreducible expressions are only A ($E=1$) or B ($E=1$). Therefore, in either case, the degeneracy is always one ($E=1$). On the other hand, a cube belongs to T_d for which there are four irreducible expressions of *A*, *B*, *E* and *T* and corresponding to these expressions, the eigenstates are degenerate singly (*A* and *B*), doubly (*E*) and triply (*T*). Even when the eigenstates are specified by different irreducible expressions, they may have the same energy. In such a case, the degeneracy of the state is given by the sum of degeneracy for each irreducible expression. Except for these cases, even for the states characterized by different irreducible expressions, their energies may be the same, which is called "accidental degeneracy". As is discussed in **problem 1-43**, the energy levels of a particle moving in a cube expressed by $(5,5,n_z)$ and $(1,7,n_z)$ are degenerate and in total the levels of this energy are nine-fold degenerate. Therefore, actual degeneracy of 9 is much larger than the degeneracy of 3 expected from the symmetry of T_d. For the point group a sphere belongs, the traces of matrixes for the irreducible expressions are given by $2l+1$ in which l corresponds to either the angular momentum or the number of nodal planes. The degeneracy, therefore, is given by $2l+1$ and naturally equal to the degeneracy of the atomic orbitals of the electron moving in the spherically symmetric field. In **problem 1-108**, the shapes and degeneracy of the eigenfunction in three dimensional space are discussed from a viewpoint of how to divide the space into parts so as for the original space is orthogonal to the space divided.

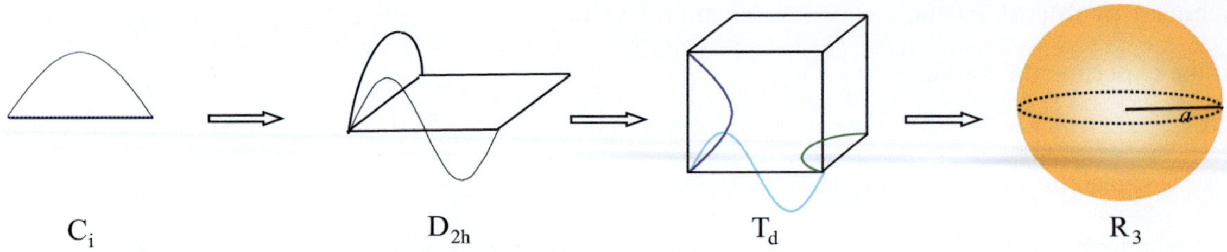

Fig.104 Space symmetry (point group) of the space in which a particle moves freely and degeneracy of quantized states.

III-2 Gas-state Molecular Kinetics and State Equations

Part of Worked Solutions

Solution 2-1a

The mole numbers of methane (CH_4) 320 mg, Ar 175 mg and Ne 225 mg are calculated as follows.

$$n(\text{Ne}) = \frac{225\times10^{-3}(\text{g})}{20(\text{g/mol})} = 1.13\times10^{-2}\,\text{mol},\quad n(\text{Ar}) = \frac{175\times10^{-3}}{40} = 4.38\times10^{-3}\,\text{mol},$$

$$n(\text{CH}_4) = \frac{320\times10^{-3}}{16} = 2.0\times10^{-2}\,\text{mol}$$

Assuming all the gases behave ideally at this temperature, the total pressure P_{total} and the partial pressure of each gas are found from eq(1) as follows.

$$P_iV = n_iRT,\qquad P_i = \frac{n_i}{\sum_j n_j}\times P_{total},\qquad P_{total} = \sum_j P_j \tag{1}$$

$$P_{\text{Ne}} = 15.2\,\text{kPa},\quad P_{\text{Ar}} = 15.2\times\frac{n(\text{Ar})}{n(\text{Ne})} = 5.9\,\text{kPa},\quad P_{\text{CH}_4} = 15.2\times\frac{n(\text{CH}_4)}{n(\text{Ne})} = 27.0\,\text{kPa}$$

$$P_{total} = P_{\text{Ne}} + P_{\text{Ar}} + P_{\text{CH}_4} = 15.2 + 5.9 + 27.0 = 48.1\,\text{kPa}$$

Also, the volume of the gas mixture is found from eqs(1) and (2) as below.

$$V = \frac{n_1}{P_1}RT = \frac{n_1}{\frac{n_1}{\sum_j n_j}\times P_{total}}RT = \frac{\sum_j n_j}{P_{total}}\times RT \tag{2}$$

$$= \frac{(1.13 + 0.438 + 2.0)\times10^{-2}}{48.1\times10^3}\times8.31\times300 = 1.85\,\text{dm}^3$$

Ans. $P_{\text{Ne}} = 15.2\,\text{kPa}$, $P_{\text{Ar}} = 5.9\,\text{kPa}$, $P_{\text{CH}_4} = 27.0\,\text{kPa}$, $P_{total} = 48.1\,\text{kPa}$, volume $1.85\,\text{dm}^3$

"remark" A value for the molecular weight is usually given without dimension. However, it should be treated as the value in the units of g/mol for calculation. When the SI units system is employed, mass is expressed in the unit of kg and, hence, the value for the molecular weight should be converted so as to be consistent with the unit.

"note" It is not easy to measure partial pressures of a gas mixture. How can you prove experimentally that the law of partial pressures holds. For real gases, as is considered in **problem 2-37**, the law of partial pressures comes to fail as the pressure is increased.

Solution 2-1b

Let the mass of 1 mole of the mixed gas be m(g) and express its apparent molecular weight by M_{app}(g/mol). Then these values are related by eq(3).

$$PV = \frac{m}{M_{app}}RT = RT \tag{3}$$

Since the oxygen and the nitrogen molecules behave ideally, their volume ratio is equal to the mole ratio.

S. Hirayama and M. Yamamoto, *Physical Chemistry Problems and Solutions*,
https://doi.org/10.1007/978-981-97-0165-0_10

For this, the mass of 1 mole of the mixed gas is given as below.

$$m = n_{O_2} \times M_{O_2} + n_{N_2} \times M_{N_2} = 0.2 \times 32.00 + 0.8 \times 28.03 = 28.82$$

By putting $m/M_{app} = 1$, the apparent molecular weight is calculated from eq(3) as $M_{app} = 28.82$. Similarly, when the CO_2 coexists 1% by volume, the apparent molecular weight of that mixed gas is calculated to be 28.98 as below.

$$m = n_{O_2} M_{O_2} + n_{N_2} M_{N_2} + n_{CO_2} M_{CO_2} = (0.2 \times 32.00 + 0.8 \times 28.03) \times 0.99 + 0.01 \times 12.01$$
$$= 28.98$$
$$\therefore M_{app} = 28.98$$

Ans. app.mol.wt.: $M_{app} = 28.82$; when CO_2 coexists, $M_{app} = 28.98$

"note" The fact that the air contains other gases than oxygen and nitrogen can be easily revealed, as is seen above, by knowing the apparent molecular weight (or the mass of the air per volume) of the air.

W.Ramsay (Nobel Prize in Chemistry in 1904) discovered a new element of Ar (later Ne、Kr and Xe as well) in principle in a similar way. (*Royal Institution Proceedings,* **14**, 524 (1895).) *Ramsay* and *Lord Rayleigh* (Nobel Prize in Physics in 1904 for his work of finding Ar) came to know that the newly found gas was far more inert than nitrogen and the emission spectrum observed from the gas in a discharge tube was entirely different from those of any gases known by that time. *Ramsay* correctly placed the new element in the periodic table, however, to know its atomic number exactly we had to wait for a young scientist *H.Moseley* (see **problem 1-1**) to propose *Moseley* 's law based on *X*-ray analysis of elements.

Solution 2-2

Pressure is the force applied per unit area. By expressing the force by F and the cross section of the mercury column by S, the pressure is given by eq(4), in which the vector expression is not employed.

$$P = \frac{F}{S} \qquad (4) \qquad F \text{ in N}, \quad S \text{ in m}^2, \quad P \text{ in Pa}$$

$$F = M \times g = \langle 0.76 \times 13.595 \times 10^3 \times \pi(0.01)^2 \rangle \times 9.807 \text{ (N)}$$

$$S = \pi(0.01)^2$$

$$P = 0.76 \times 13.595 \times 10^3 \times 9.807 = 101.3 \text{ kPa}$$

Since the density of water is $1/13.595$ of that of mercury, the mass equivalent to the height of 0.76 m of mercury is gained by increasing the height of the water column 13.595 times of that of mercury column, i.e., $13.5951 \times 0.7600 = 10.33$ m (four digits are taken as the significant figures).

Ans. 101.3 kPa, 10.33 m

Solution 2-3a

(a), (b) The *van der Waals* equation of state ($n = 1$ mole) is expressed by eq(5),

$$P = \frac{RT}{(V-b)} - \frac{a}{V^2} = \frac{RT}{(V - 4.267 \times 10^{-2})} - \frac{3.592}{V^2} \qquad (5)$$

where a is given in the units of $\text{dm}^6 \cdot \text{atm} \cdot \text{mol}^{-2}$. Therefore, the gas constant R given in the units of g $R = 0.08206 \text{ dm}^3 \cdot \text{atm} \cdot \text{K}^{-1}\text{mol}^{-1}$ should be used.

Employing eq(5), the *P*-*V* relation is drawn as presented in Fig.1. The curve exhibits a minimum *M* as well as a maximum *L* and between these points the volume of the system increases with increasing pressure, i.e., $(\partial P/\partial V)_T > 0$. Such phenomena do not occur naturally and are not seen in nature (thermodynamical instability or inequality). Therefore, the liquefaction by increasing pressure at a constant temperature is considered to start at a pressure smaller than that at *L* as well as larger than that at *M*. The liquefaction is assumed to start at the pressure P_A at *A* and end at *B* and then, since the liquefaction is a process in equilibrium, a change in the free energy associated with the process must be zero, that is $(\Delta G)_T = 0$ (*Maxwell*'s rule). On the other hand, when the liquefaction is carried out hypothetically along the *P*-*V* curve described by the *van der Waals* equation of state, starting from *A* and ending at *B* via $A \to L \to M \to B$, *Maxwell*'s rule indicates that the work done in that process must be equal to the work to be done from *A* to *B* at the constant pressure of P_A (see **solution 2-4**). The pressure P_A is determined so as for eq(6) to be satisfied.

$$\int_{A\to L\to M\to B} -PdV = \int_{A\to B} -P_A dV \;\Rightarrow\; \int_{A\to L\to M\to B} -(P-P_A)dV \;=\; \int_{B\to M\to L\to A} (P-P_A)dV = 0 \qquad (6)$$

To perform this, as is shown in Fig.2, an initial value for P_A is chosen first (for instance 56 atm in this problem) and then the values for the volume at which P_A crosses the curve, say $V_1 = V_l$, V_2, $V_3 = V_g$, are employed to check how closely eq(6') is satisfied. If the integral is not sufficiently close to zero,

$$\int_{V_1}^{V_3} (P-P_A)dV = 0 \qquad (6')$$

then chose another value for P_A and repeat the calculation until the satisfactory result is achieved. For the repetitive calculation of this sort, software such as *Mathematica* or *Theorist* is useful.

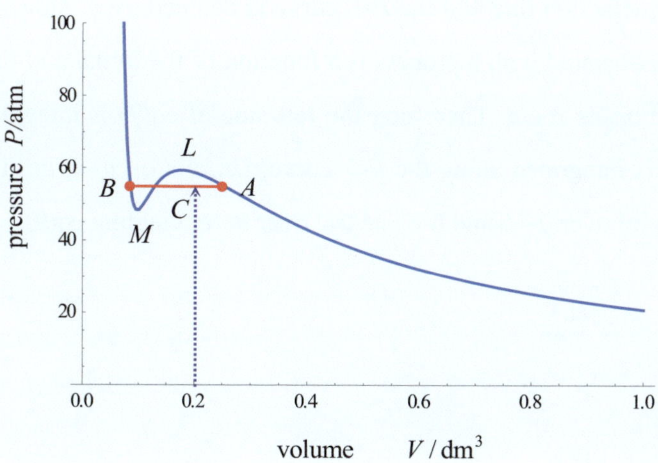

Fig.1 *P-V* curve expressed by the *van der Waals* equation of state for CO_2 at *T*=285 K.

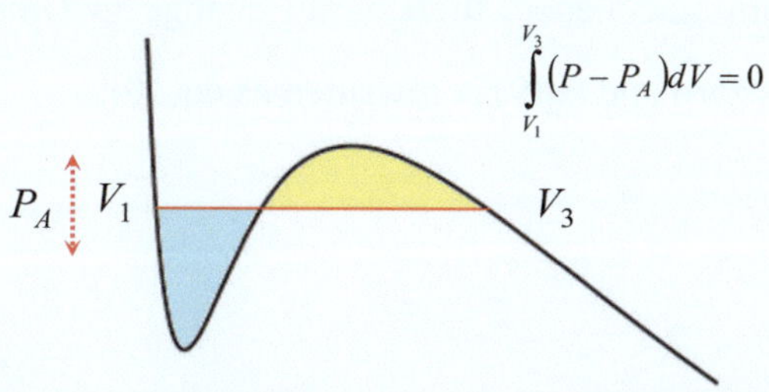

Fig.2 Expanded curve near the liquefaction line (see Fig.1).
At the liquefaction pressure P_A, the areas specified by blue and yellow colors are equal.

For this problem, the values finally obtained for the pressure P_A and the volumes V_l and V_g are $P_A = 56.02$ atm, $V_l = 84.23 \times 10^{-3}\,\mathrm{dm}^3$ and $V_g = 240.6 \times 10^{-3}\,\mathrm{dm}^3$. From these results, it can be said that at the pressure of $P_A = 56.02$ atm and temperature of $T = 285$, the volume of the gas at which the liquefaction starts is $V_g = 240.6 \times 10^{-3}\,\mathrm{dm}^3$ and the volume at which all gas is liquefied is $V_l = 84.23 \times 10^{-3}\,\mathrm{dm}^3$.

Next, we consider the region where the gas and the liquid coexist and the total volume is given by $\upsilon_g + \upsilon_l = 200 \times 10^{-3}\,\mathrm{dm}^3$, in which the volumes of the gas and the liquid are denoted by υ_g and υ_l, respectively. Relying on the lever rule (see **solution 2-27**), the mole ratio of the gas to liquid ($x_g : x_l$) is calculated as below.

$$x_g : x_l = BC : CA = (200 - 84.23) : (240.6 - 200) = 0.740 : 0.260$$

Ans. $240.6 \times 10^{-3}\,\mathrm{dm}^3$, gas : liquid = 0.740 : 0.260 (mole ratio)

Solution 2-3b

In a similar manner as employed in the preceding solution, the liquefaction line $P = P_0$ is determined by iterating calculations so as for the two areas enclosed by the P-V curve given by the *van der Waals* equation of state for CO_2 and this line to have the same area. By using *Mathematica*, the result obtained is 46.94 atm in the accuracy of 0.01 atm.

Ans. 46.94 atm

Solution 2-4

There may be various ways to prove *Maxwell*'s rule. Three of them are given here.

(1) To employ the relation $dG = VdP$.

An infinitesimal change in *Gibbs* free energy at a constant temperature is given by eq(7). It is assumed that the liquefaction starts at the point a and all is converted to the liquid at the point c. The pressures at both points are the same. The crossing point of the liquefaction line and the P-V curve is denoted by b. Since the *Gibbs* free energy is a state function, its change associated with a process is a function of the initial and final points of the process alone and does not depend on its route. Therefore, the left side of eq(7) is integrated along the liquefaction line ac and the right side is integrated along the P-V curve, i.e., via $a-b-c$ in Fig.3 and the latter is divided into two parts of the integral over $a-b$ and the integral over $b-c$, yielding eq(9).

$$dG = VdP \qquad (7)$$

$$\int_g^l dG = \int_a^b VdP \qquad (8)$$

$$G^l - G^g = 0 = \int_a^b VdP + \int_b^c VdP \qquad (9)$$

The liquefaction treated here is a process in equilibrium and, hence, there is no change in *Gibbs* free energy during the liquefaction. Then the left side of eq(8) is zero and eq(9) is rewritten as eq(10).

$$\int_a^b VdP = -\int_b^c VdP \qquad (10)$$

Usually, the *P-V* curve is drawn taking *V* as a horizontal coordinate. In this case, however, the integrals in eq(10) are a little bit hard to grasp visually. By inverting the horizontal and vertical coordinates, that is *P* as the horizontal and *V* as vertical coordinates, the *V-P* curve can be drawn as shown in Fig.3. Then it is easily understood that eq(10) means that the areas designated by ① and ② must be the same. This is just what *Maxwell*'s rule tells us.

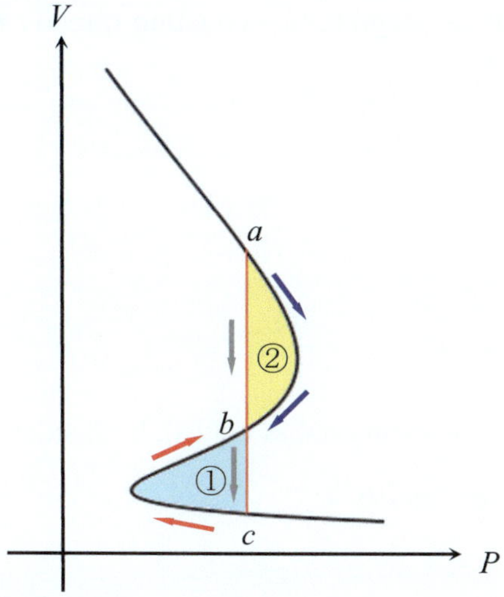

Fig.3 Expanded *V-P* curve (*van der Waals* equation of state) near the liquefaction line. Vertical and horizontal axes are exchanged.

(2) To replace the variable *P* with *V*.

In eq(7), $d(PV)$ is subtracted from both sides in order to replace the variable *P* with *V*, yielding eq(11) which is equal to the work done by the process (*Helmholtz* energy).

$$dG - d(PV) = VdP - d(PV) = -PdV \qquad (11)$$

Noticing that since *PV* as well as *G* is a state function, the left side of eq(11) is independent of an integration path, the left side is integrated along the liquefaction line *ac* and the right side along the *P-V* curve via *c–b–a* to yield eqs(12) and (13). In Fig.2, the rectangular areas enclosed by gray lines are denoted by S_1 and S_2 and the areas colored blue and yellow are denoted by ① and ②, respectively.

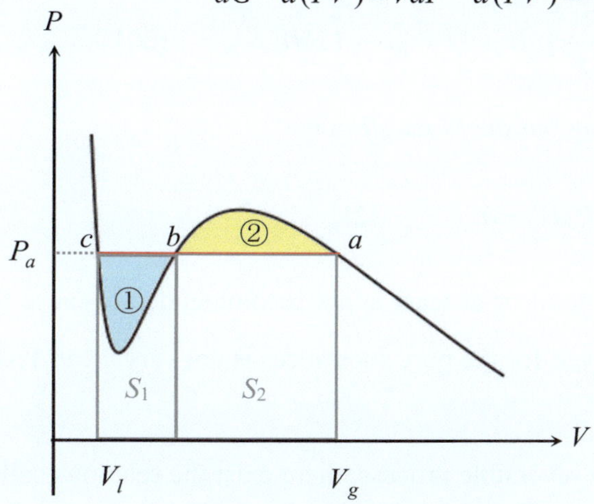

Fig.4 Expanded *V–P* curve (*van der Waals* equation of state) near the liquefaction line.

$$\int_g^l dG - \int_g^l d(PV) = G^l - G^g - [PV]_g^l = 0 - P_a(V^l - V^g) = S_1 + S_2 \qquad (12)$$

$$-\int_a^c PdV = -\int_a^b PdV - \int_b^c PdV = (S_1 - ①) + (S_2 + ②) \qquad (13)$$

By equating eq(12) to eq(13), eq(14) is obtained from which it is proved that area ① must be equal to the area ② as is given by eq(14).

$$S_1 + S_2 = (S_1 - ①) + (S_2 + ②) \qquad (14)$$

$$① = ② \qquad (15)$$

(3) To employ the partial derivative of chemical potential with respect to pressure $(\partial\mu/\partial P)_T = \overline{V}$ (from the problem in *McQuarrie-Simon*).

The relation to be employed is eq(16). However, it is a bit difficult to integrate this equation directly by using the ordinary *P*-*V* curve.

$$\left(\frac{\partial\mu}{\partial P}\right)_T = \overline{V} \qquad (16)$$

For this, the variable *P* is replaced with *V* by means of eq(17).

$$\left(\frac{\partial\mu}{\partial V}\right)_T\left(\frac{\partial V}{\partial P}\right)_T = \overline{V} \qquad (17)$$

Since the partial derivative of chemical potential with respect to volume is expressed as eq(18), both sides of eq(18) are integrated by volume over the range $[V_g,\ V_l]$ as are given by eq(19).

$$\left(\frac{\partial\mu}{\partial V}\right)_T = \overline{V}/\left(\frac{\partial V}{\partial P}\right)_T = \overline{V}\left(\frac{\partial P}{\partial V}\right)_T \qquad (18)$$

$$\int_{V_g}^{V_l}\left(\frac{\partial\mu}{\partial V}\right)_T dV = \int_{V_g}^{V_l}\overline{V}\left(\frac{\partial P}{\partial V}\right)_T dV \qquad (19)$$

The right side of eq(19) is integrated by means of partial integration as is expressed by eq(20).

$$\int_{\overline{V}_g}^{\overline{V}_l} d\mu = \int_{\overline{V}_g}^{\overline{V}_l}\overline{V}\left(\frac{\partial P}{\partial V}\right)_T dV = [P\overline{V}]_{\overline{V}_g}^{\overline{V}_l} - \int_{\overline{V}_g}^{\overline{V}_l} P\left(\frac{\partial \overline{V}}{\partial V}\right)_T dV = [P\overline{V}]_{\overline{V}_g}^{\overline{V}_l} - \int_{\overline{V}_g}^{\overline{V}_l{}'} PdV \qquad (20)$$

Gathering these together, eq(21) is obtained which is nothing but the *Maxwell* rule.

$$\mu_l - \mu_g = P_0(\overline{V}_l - \overline{V}_g) - \int_{\overline{V}_g}^{\overline{V}_l} PdV = \int_{\overline{V}_g}^{\overline{V}}(P_0 - P)dV = 0 \qquad (21)$$

However, this way of proving the *Maxwell* rule is not good or at least is not recommended because the reason why the chemical potentials are intentionally employed for the pure substances is not very clear. Proof (1) or (2) above is recommended.

"note" In thermodynamics which presumes equilibrium or reversible process, there exist the relations called thermodynamic inequality and they must normally be fulfilled. A few examples are given below.

For pure substances,

$$\left(\frac{\partial S}{\partial T}\right)_V > 0 \ \Rightarrow\ \frac{C_V}{T} > 0 \ \Rightarrow\ C_V > 0 \qquad (22)$$

$$\left(\frac{\partial P}{\partial V}\right)_T < 0 \qquad (23)$$ (important inequality for *van der Waals* equation)

$$\left(\frac{\partial V}{\partial T}\right)_P\left(\frac{\partial T}{\partial P}\right)_V > 0 \qquad (24)$$

For solutions,

$$\left(\frac{\partial\mu}{\partial c}\right)_{P,T} > 0 \qquad (25)$$ (for dilute solutions $\left(\frac{\partial\mu}{\partial c}\right)_{P,T} = \frac{Nk_BT}{c} > 0$)

By applying eq(23) to *Euler*'s chain equation $\left(\frac{\partial P}{\partial V}\right)_T\left(\frac{\partial V}{\partial T}\right)_P\left(\frac{\partial T}{\partial P}\right)_V=-1$, eq(24) is derived. To derive eqs(22) – (25) directly, more detailed explanations are required. See, for instance, *Landau- Lifshits, Statistical Physics* sections 21 and 94. See also **problems 3-46b** and **3-103**.

Solution 2-5

The collision number at which an oxygen molecule moving with the velocity of v collides with other oxygen molecules per second is given by eq(26).

$$v\times\pi d^2\times\frac{N(\mathrm{O_2})}{V} \qquad (26)$$

The number of oxygen molecules per unit volume $N(\mathrm{O_2})/V$ is calculated by the state equation of the gas, i.e., eq(27).

$$PV=N(\mathrm{O_2})k_BT \;\Rightarrow\; \frac{N(\mathrm{O_2})}{V}=\frac{P}{k_BT} \qquad (27)$$

Assuming that the velocity distribution of the oxygen molecule obeys the *Maxwell-Boltzmann* distribution, eq(26) is integrated over a velocity range [0, ∞] to find the collision number of $4.84\times10^{10}\mathrm{s}^{-1}$.

$$\int_0^\infty v\times\pi d^2\times\frac{N(\mathrm{O_2})}{V}\times\left(\frac{2M_{\mathrm{O_2}}}{\pi RT}\right)^{1/2}\exp\left(-\frac{M_{\mathrm{O_2}}v^2}{2RT}\right)dv \quad \text{(one - dimensional velocity distribution)}$$

$$=\pi d^2\times\frac{P}{k_BT}\int_0^\infty v\left(\frac{2M_{\mathrm{O_2}}}{\pi RT}\right)^{1/2}\exp\left(-\frac{M_{\mathrm{O_2}}v^2}{2RT}\right)dv \quad \text{(see II-4-6 for integration)}$$

$$=\pi d^2\times\frac{P}{k_BT}\times\left(\frac{2RT}{\pi M_{\mathrm{O_2}}}\right) \qquad (28)$$

$$=\frac{3.14\times(3.57\times10^{-10})^2\times0.2\times101\times10^3}{1.38\times10^{-23}\times298}\times\left(\frac{2\times8.31\times298}{3.14\times32.0\times10^{-3}}\right)^{1/2}$$

$$=4.84\times10^{10}\mathrm{s}^{-1}$$

Ans. collision number:$4.84\times10^{10}\ \mathrm{s}^{-1}$

Solution 2-6

(a) Based on the equation of state for an ideal gas,

$$PV=\frac{m}{M}RT \;\Rightarrow\; 101.32\times10^3\times V=\frac{(24.9100-24.8015)\times10^{-3}}{28.02\times10^{-3}}\times8.3145\times298.15.$$

From the equation above, the capacity (volume) of the glass vessel is found as below.

$$V=\frac{9.599}{101.32\times10^3}=0.09474\ \mathrm{dm}^3$$

Ans. capacity (volume): $0.09474\,\mathrm{dm}^3$

(b) The number of N_2 molecules in the vessel is calculated as below from the molecular weight M of N_2.

$$N=\frac{m}{M}\times N_A=\frac{0.1085\times10^{-3}}{28.02\times10^{-3}}\times6.022\times10^{23}=2.332\times10^{21}$$

Ans. the number of N_2 molecules: 2.332×10^{21}

(c) The equation of state for the ideal gas mixture in the problem can be written as eq(29), where the masses of the methane and the ethane are given by m_1 and m_2, and their molecular weights are $M_1=16.04$ and $M_2=30.07$, respectively.

$$PV = P_1V_1 + P_2V_2 = n_1RT + n_2RT = \left(\frac{m_1}{M_1} + \frac{m_2}{M_2}\right) \times 8.3145 \times 298.15$$
$$= 101.32 \times 10^3 \times 0.09474 \times 10^{-3} \qquad (29)$$

The measurement of the weight gives eq(29). From eq (29), eq(31) is obtained and from eq(30), eq(32).

$$m_1 + m_2 = (24.8768 - 24.8015) \times 10^{-3} = 0.0753 \times 10^{-3}\,\text{kg} \qquad (30)$$

$$\frac{m_1}{M_1} + \frac{m_2}{M_2} = 3.869 \times 10^{-3} \qquad (31)$$

$$\frac{m_1 + m_2}{M_1} = 6.147 \times 10^{-3} \qquad (32)$$

After subtracting eq(31) from eq(32) and rearranging the result, the mass m_2 of the ethane gas is found as below.

$$\frac{m_2}{M_1} - \frac{m_2}{M_2} = (6.147 - 3.869) \times 10^{-3} = m_2\left(\frac{1}{M_1} - \frac{1}{M_2}\right) = 2.278 \times 10^{-3}$$

$$m_2 = \frac{16.04 \times 30.07 \times 2.278 \times 10^{-3}}{14.03 \times 10^{-3}} = 78.31 \times 10^{-6}\,\text{kg}$$

From this value, the mole number of each gas and, hence, the molar ratio is calculated as below.

$$\frac{m_2}{M_2} = \frac{78.31 \times 10^{-6}}{30.07 \times 10^{-3}} = 2.604 \times 10^{-3}$$

$$\frac{m_1}{M_1} = 3.869 \times 10^{-3} - 2.604 \times 10^{-3} = 1.256 \times 10^{-3}$$

$$n_1 : n_2 = \frac{m_1}{M_1} : \frac{m_2}{M_2} = 1.256 \times 10^{-3} : 2.604 \times 10^{-3} = 1 : 2.073$$

Ans. methane:ethane = 1:2.073 (molar ratio)

Solution 2-7

i) those which increase with temperature

(B) mean square velocity

From *Maxwell*'s velocity distribution law, it is obvious that the mean square velocity $\overline{c^2}$ increases with temperature. Alternatively, you can consider that since the mean kinetic energy is proportional to k_BT, the mean square velocity $\overline{c^2}$ will increase with temperature.

formula: $\overline{c^2} = \dfrac{3k_BT}{m}$ (see **problem 2-33**).

(D) viscosity coefficient of a gas

The viscosity coefficient is associated with the transport of momentum arising from the velocity gradient, so that for gases the viscosity will increase with increasing molecular velocity. (On the contrary, the viscosity of liquids generally decreases with temperature.)

formula: $\eta = \dfrac{m\bar{c}}{3\sqrt{2}\pi d^2} = \dfrac{m}{3\sqrt{2}\pi d^2} \times \left(\dfrac{8k_BT}{\pi m}\right)^{1/2} \propto (k_BT)^{1/2}$, d : collision radious

For the ideal gas, the equation of state is derived by assuming that (i) the intermolecular force is zero and (ii) the size of a molecule is zero. If these assumptions are taken rigorously, there would exist neither molecular collision nor energy transfer. However, even in the limit where ideal gas behavior is recognized, there exists no gas which does not show viscosity.

(F) pressure of the gas

In a container whose volume is constant, the pressure of the gas in it increases with temperature.

formula: $P = \dfrac{nRT}{V}$ or $P = \dfrac{nRT}{(V-nb)} - \dfrac{n^2a}{V^2}$, $\left(\dfrac{\partial P}{\partial T}\right)_V > 0$ (see **solution 2-1**)

(H) sound velocity in the gas

The sound velocity in a gas is given in **problem 2-14**. As is obvious from that equation, the sound velocity in the gas increases with temperature.

ii) those which decrease with increasing temperature

Nothing applicable to this category.

iii) those which are independent of temperature

(A) mean free path λ of a gas at a constant volume

formula: $\lambda = \dfrac{\bar{c}}{z} = \dfrac{\bar{c}}{\sqrt{2}\bar{c}\pi d^2 N} = \dfrac{1}{\sqrt{2}\pi d^2 N}$ z: collision number, N : number of molecules in a unit V

(C) heat capacity

The internal energy U of the ideal gas is proportional to RT and the heat capacity is given by $(\partial U/\partial T)_V$.

(E) the number of molecules with velocities less than the mean velocity $\bar{u}$ (one-dimensional in the range $[0, \infty]$)

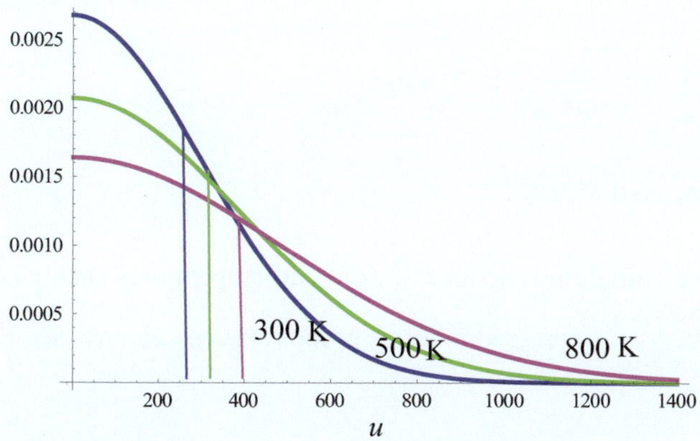

Fig.5 One-dimensional velocity distributions and mean velocities $\bar{u}$ at $T = 300$, 500 and 800 K (indicated by corresponding vertical colored bars).

$u \Rightarrow u_R = u/\bar{u}$

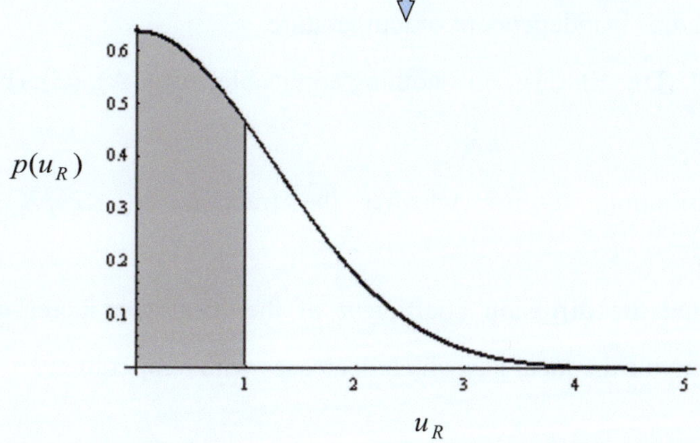

Fig.6 Reduced velocity distribution. The reduced velocity distribution function $p(u_R)$ becomes independent of temperature as well as molecular weight. The areas in the range $[0, \bar{u}]$ in Fig.5 are given by the area colored gray in Fig.6. See **[alternative solution]** below.

To answer the above question intuitively is difficult, since not only $\bar{u}$ is expected to increase with temperature but also the fraction of the molecules which have large u does. Therefore, we need to calculate the number exactly. Putting $m = 28\times10^{-3}/N_A$, the mean velocities at $T = 300$, 500 and 800 K are calculated and the required fraction (to the total number, say 1 mole) at each temperature is calculated from the formula given below to find the fraction is ca. 0.575 at each temperature and, hence, practically the number of the molecules in question is concluded to be independent of temperature.

formula: $\dfrac{N(u \le \bar{u})}{N_{total}} = 2\times\displaystyle\int_0^{\bar{u}}\left(\frac{m}{2\pi k_B T}\right)^{1/2} e^{-\frac{mu^2}{2k_BT}}du$, $\quad \bar{u} = 2\displaystyle\int_0^{\infty} u\times\left(\frac{m}{2\pi k_B T}\right)^{1/2} e^{-\frac{mu^2}{2k_BT}}du = \left(\frac{2k_BT}{\pi m}\right)^{1/2}$

The velocity distribution at each temperature is illustrated in Fig.5, where $p(u)$ represents the probability of the molecules which have velocity u.

[alternative solution]

When the reduced velocity $u_R = u/\bar{u}$ which is defined by dividing a velocity u by the mean velocity $\bar{u}$ is used, u can be expressed as $u = u_R\bar{u} = u_R\times(2k_BT/\pi m)^{1/2}$, so that we can show readily that the fraction required is independent of either temperature or molecular mass and takes a value of 0.57506, as is shown below, upon integration of the given equation by paying attention to the relations given by $du = \bar{u}du_R$ and $u_R = 1$ for $u = \bar{u}$.

$$\frac{N(u \le \bar{u})}{N_{total}} = 2\times\int_0^{\bar{u}}\left(\frac{m}{2\pi k_B T}\right)^{1/2} e^{-\frac{mu^2}{2k_BT}}du$$

$$= 2\times\int_0^{1}\left(\frac{m}{2\pi k_B T}\right)^{1/2}\times\left(\frac{2k_BT}{\pi m}\right)^{1/2} e^{-\frac{u_R{}^2}{\pi}}du_R$$

$$= 2\times\int_0^{1}\frac{1}{\pi}e^{-\frac{u_R{}^2}{\pi}}du_R = 0.57506$$

In other words, this shows that the velocity distribution functions at different temperatures such as those shown in Fig.5 assume exactly the same form when they are expressed in terms of the reduced velocity (see Fig.6).

(G) collision radius

The gaseous molecules for which their intermolecular forces can be neglected, each molecules can be treated as a hard sphere, so that the collision radius is given by the sum of the molecular radii of the colliding two molecules. Because of this, the collision radius is independent of temperature.

Ans. i: (B), (D), (F), (H); ii: nothing applicable; iii: (A), (C), (E), (G)

[advanced problem]

For the three-dimensional velocity distribution, discuss whether the fraction $N(c \le \bar{c})/N_{total}$ is temperature independent or not.

"note" Both of the thermal conductivity and the diffusion coefficient of the ideal gas increase with temperature. Then, is there any physical property of the ideal gas which decreases with temperature?

Solution 2-8

The *van der Waals* equation of state is treated (see **problem 2-8**, **2-11** and **2-20**) in many ways so often that a few other simple equations of state are considered here. Since the strategy to derive the critical values and the corresponding equation of state is the same in principle, the present problem will be a good exercise for handling the more complex equations of state.

When an inflection point is recognized in the *P*-*V* curve (of a gas), the inflection point is the critical point. Therefore, the point which satisfies the conditions set by eqs(33) and (34) is searched for in each of the given equations of state.

$$\left(\frac{\partial P}{\partial V}\right)_T = 0 \qquad (33), \qquad \left(\frac{\partial^2 P}{\partial V^2}\right)_T = 0 \qquad (34)$$

i) $PV = RT$

$$\left(\frac{\partial P}{\partial V}\right)_T = -\frac{RT}{V^2} \neq 0.$$ Therefore, the critical point does not exist.

ii) $PV = RT(1+BP) \Rightarrow P = \dfrac{RT}{V-BRT}$ $\quad B$: constant

$$\left(\frac{\partial P}{\partial V}\right)_T = -\frac{RT}{(V-BRT)^2} \neq 0.$$ Therefore, the critical point does not exist.

iii) $P = \dfrac{RT}{V} + \dfrac{B}{V^2} + \dfrac{C}{V^3}$, B and C are constants. Eqs(35) and (36) are derived from eqs(33) and (34).

$$\left(\frac{\partial P}{\partial V}\right)_T = -\frac{RT}{V^2} - \frac{2B}{V^3} - \frac{3C}{V^4} = -\frac{1}{V^4}(RTV^2 + 2BV + 3C) = 0 \Rightarrow (RTV^2 + 2BV + 3C) = 0 \qquad (35)$$

$$\left(\frac{\partial^2 P}{\partial V^2}\right)_T = \frac{2RT}{V^3} + \frac{6B}{V^4} + \frac{12C}{V^5} = \frac{2}{V^5}(RTV^2 + 3BV + 6C) = 0 \Rightarrow (RTV^2 + 3BV + 6C) = 0 \qquad (36)$$

Subtracting eq(35) from eq(36),

$$BV + 3C = 0 \Rightarrow V = V_c = -\frac{3C}{B} \qquad (37)$$

and the result is substituted into eq(35), leading to eq(38).

$$RT\left(-\frac{3C}{B}\right)^2 + 2B\left(-\frac{3C}{B}\right) + 3C = 0 \Rightarrow T_c = \frac{B^2}{3RC} \qquad (38)$$

Here, T_c and V_c are the critical temperature and critical volume, respectively. The critical pressure P_c is obtained by substituting eqs(37) and eq(38) into the equation of state as below.

$$P_c = \frac{R(B^2/3RC)}{(-3C/B)} + \frac{B}{(-3C/B)^2} + \frac{C}{(-3C/B)^3}$$

$$= \frac{1}{(-3C/B)^3}\{R(B^2/3RC)\times(-3C/B)^2 + B(-3C/B) + C\} = \frac{1}{(-3C/B)^3}\times C$$

$$= -\frac{B^3}{27C^2} = \frac{1}{3}\times\frac{T_c}{V_c} \qquad (39)$$

From eqs(37) and (38), the constants B and C are expressed in terms of the critical pressure, volume and temperature, as are given by eqs(40) and (41).

$$V_c \times T_c = -\frac{3C}{B}\times\frac{B^2}{3RC} = -\frac{B}{R} \Rightarrow B = -RT_cV_c \qquad (40)$$

$$C = -\frac{BV_c}{3} = \frac{RT_cV_c^2}{3} \qquad (41)$$

Then, eqs(40) and (41) are substituted into the equation of state to eliminate the constants B and C. The obtained result is arranged in terms of the reduced pressure $(P_R = P/P_c)$, the reduced volume $(V_R = V/V_c)$ and the reduced temperature $(T_R = T/T_c)$ to construct the reduced equation of state eq(42) via following equations.

$$P = \frac{RT}{V} + \frac{-RT_cV_c}{V^2} + \frac{RT_cV_c^2}{3V^3}$$

$$P \times \frac{V_c}{T_c} = \left(\frac{RT}{V} + \frac{-RT_cV_c}{V^2} + \frac{RT_cV_c^2}{3V^3}\right) \times \frac{V_c}{T_c} = R\left\{\frac{T/T_c}{V/V_c} - \left(\frac{1}{V/V_c}\right)^2 + \frac{1}{3}\left(\frac{1}{V/V_c}\right)^3\right\}$$

Rewriting the above equation using eq(39), eq(42) is derived.

$$\frac{P}{3P_c} = R\left\{\frac{T/T_c}{V/V_c} - \left(\frac{1}{V/V_c}\right)^2 + \frac{1}{3}\left(\frac{1}{V/V_c}\right)^3\right\} \Rightarrow P_R = R\left(\frac{3T_R}{V_R} - \frac{3}{V_R^{\,2}} + \frac{1}{V_R^{\,3}}\right) \tag{42}$$

Ans. iii) alone can describe the critical state. However, since $P_c > 0$, $V_c > 0$ and $T_c > 0$, the constants should satisfy $B < 0$ and $C > 0$. $P_c = -\frac{B^3}{27C^2}$, $V_c = -\frac{3C}{B}$, $T_c = \frac{B^2}{3RC}$, $P_R = R\left(\frac{3T_R}{V_R} - \frac{3}{V_R^{\,2}} + \frac{1}{V_R^{\,3}}\right)$

Solution 2-9

There are various ways to derive equations of state for gaseous molecules.

1) Dynamical viewpoint. Forces exerted by gaseous molecules on a vessel wall upon the collisions with the wall are summed up to find the total force $F(N)$ exerted. The force per unit area $F(N)/S(\mathrm{m}^2)$ is regarded as the pressure $P(\mathrm{Pa})$ the gas in the vessel exhibits. Kinetics of gaseous molecules (velocity distribution) allows one to calculate $F(N)$.

2) Based on the statistical mechanics, the partition function obtained for a gas yields pressure of the gas straightforward as a thermodynamical quantity.

3) On the basis of the virial theorem (see **problem 2-77**).

The reason why the equation of state for ideal gases does not reflect any physical property (for instance, the mass of a molecule) specific to an individual gas is provided in a different way depending on the theory relied on. Only the answer to 1) is given here. See **problem 5-103** about 2).

For simplicity, we consider a rectangular vessel with a volume of V containing N gaseous molecules and assume that the molecules can move along one-dimensional direction alone with distributed velocities (see **"note"**). The pressure the vessel wall receives arises from the collisions of the molecules with the wall. Therefore, we need to calculate the magnitude of the force exerted by the colliding molecules.

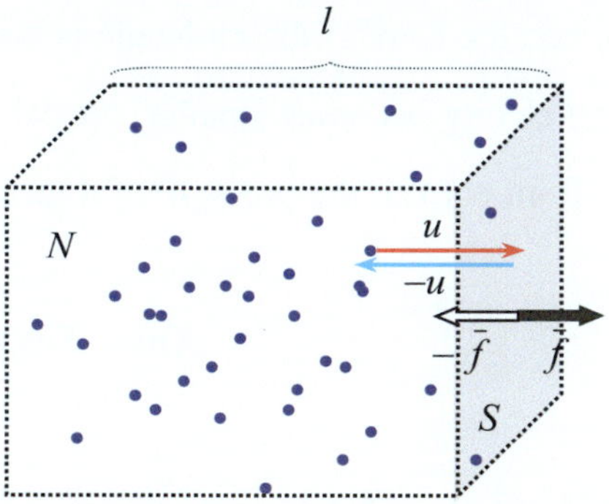

[important concept of mechanics]

variation rate of momentum $\boldsymbol{mu} = \boldsymbol{p}$ is force $\boldsymbol{f}(N)\,(= \mathrm{kg \cdot m \cdot s^{-2}})$.

impulse = force × period

variation of momentum = impulse

Fig.7 Molecular collisions with wall and the pressure of a gas.

Variation rate of the momentum of a colliding molecule upon a collision with the wall is given by eq(43).

$$\frac{d\boldsymbol{mu}}{dt} = \boldsymbol{f} \;\Rightarrow\; m\int_{t_1}^{t_2} d\boldsymbol{u} = \int_{t_1}^{t_2} \boldsymbol{f}(t)dt \qquad (43)$$

Employing a mean force $\bar{f}$, the integral on the right side of eq(43) is rewritten as eq(44) (see Fig.8),

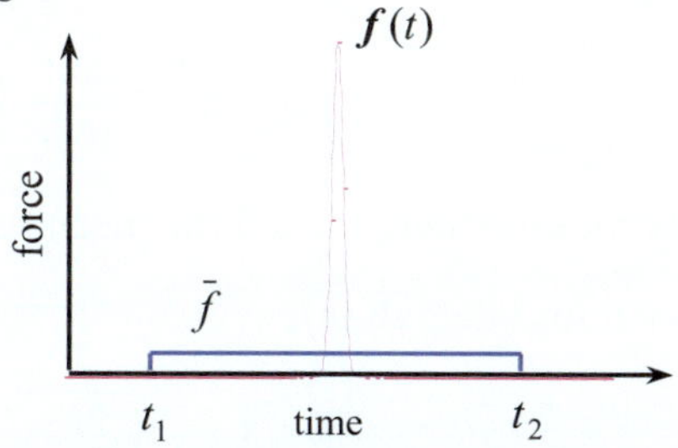

Fig.8 Instantaneous $\boldsymbol{f}(t)$ and mean $\bar{f}$ forces.

$$\int_{t_1}^{t_2} \boldsymbol{f}(t)dt = -\bar{f} \times (t_2 - t_1) \qquad (44)$$

From eqs(43) and (44), eq(45) is derived.

$$m \times (-u - u) = -\bar{f} \times (t_2 - t_1) \;\Rightarrow\; 2mu = \bar{f}(t_2 - t_1) \qquad (45)$$

Eq(45) is appropriate for one collision of a molecule for a short period, but for a longer time, the same molecule collides with the wall many times. Putting $t_2 - t_1 = t$, the collision number with the wall (colored gray in Fig.7) per molecule is given by eq(46), since the same molecule can travel round between the walls that number of times for the duration of t.

$$\frac{u \times t}{2l} \qquad (46)$$

Therefore, eq(45) applicable to the impulse of a single collision can be rewritten as eq(47).

$$2mu \times \frac{u \times t}{2l} = \bar{f} \times t \qquad (47)$$

Since the pressure is the force per unit area, the pressure P exerted by a single molecule is given by eq(48),

$$P = \frac{\bar{f}}{S} = \frac{mu^2}{l \times S} = \frac{mu^2}{V} \qquad (48)$$

and then the total pressure P_{total} produced by the collisions of N molecules with the wall is derived as eq(49) by taking the distribution of u into account.

$$\begin{aligned} P_{total} &= \int_0^{\infty} \left(\frac{mu^2}{V}\right) \times N\sqrt{\frac{m}{2\pi k_B T}} \cdot \exp\left(-\frac{mu^2}{2k_B T}\right) du \\ &= \frac{m}{V} \times \overline{u^2} = \frac{\cancel{m}}{V} \times \frac{Nk_B T}{\cancel{m}} = \frac{Nk_B T}{V} \end{aligned} \qquad (49)$$

It is clearly shown in eq(49) where a physical property inherent to a gas (mass in the present case) disappears from the equation of state (crossed out by red lines in eq(49)). What is meant by this is that the two factors of

(i) the pressure exerted by a molecule of the same velocity is the greater, the heavier the molecule is and (ii) the greater the mass is the fraction of molecules which have larger velocity becomes smaller, cancel each other resulting in the mass independent equation (49). As is obvious from eq(49), the pressure of a gas is an average value over duration and distribution of velocities.

Ans. See text.

Solution 2-10

(a) The (normalization) constant A is determined by equating the integral of eq(50) to one. Notice that the range of the one-dimensional velocity is take as $[0,\ \infty]$.

$$\int_0^\infty A\cdot\exp\left(-\frac{mu^2}{2k_BT}\right)du=1 \;\Rightarrow\; A\times\sqrt{\frac{\pi k_BT}{2m}}=1 \;\Rightarrow\; A=\sqrt{\frac{2m}{\pi k_BT}} \qquad (50)$$

The mean velocity $\bar{u}$ and the mean square velocity $\overline{u^2}$ are calculated from eqs(51) and (52), respectively (see **II-4-6** for integration).

$$\bar{u}=\int_0^\infty u\times\sqrt{\frac{2m}{\pi k_BT}}\cdot\exp\left(-\frac{mu^2}{2k_BT}\right)du=\sqrt{\frac{2m}{\pi k_BT}}\times\frac{2k_BT}{2m}=\sqrt{\frac{2k_BT}{\pi m}} \qquad (51)$$

$$\begin{aligned}\overline{u^2}&=\int_0^\infty u^2\times\sqrt{\frac{2m}{\pi k_BT}}\cdot\exp\left(-\frac{mu^2}{2k_BT}\right)du\\&=\sqrt{\frac{2m}{\pi k_BT}}\int_0^\infty -\frac{2k_BT}{m}\times u\times\exp\left(-\frac{mu^2}{2k_BT}\right)du\\&=\sqrt{\frac{2m}{\pi k_BT}}\times\left(-\frac{2k_BT}{m}\right)\left\{\left[u\times\exp\left(-\frac{mu^2}{2k_BT}\right)\right]_0^\infty-\int_0^\infty\exp\left(-\frac{mu^2}{2k_BT}\right)du\right\}\\&=\sqrt{\frac{2m}{\pi k_BT}}\times\left(-\frac{2k_BT}{m}\right)\times\left(-\frac{1}{2}\sqrt{\frac{2\pi k_BT}{m}}\right)=\frac{2k_BT}{m}\end{aligned} \qquad (52)$$

Ans. $\bar{u}=\sqrt{\dfrac{2k_BT}{\pi m}}$, $\overline{u^2}=\dfrac{2k_BT}{m}$

(b) The number of atoms per unit volume is N/V. From the mean velocity of the atoms denoted by $\bar{u}$, the number of Ge atoms effluent from the orifice of an area given by πr^2 during the time of t is given by eq(53).

$$\pi r^2\times\bar{u}\times t\times\frac{N}{V}=\frac{4.3\times10^{-5}\times6.02\times10^{23}}{72.5\times10^{-3}} \qquad (53)$$

Substituting the following values for $\bar{u}$, t and πr^2 into eq(53), the value for N/V is calculated first and then from eq(54), the vapor pressure P of Ge in the furnace is found as eq(54).

$$\bar{u}=\sqrt{\frac{2\times1.38\times10^{-23}\times1000\times6.02\times10^{23}}{3.14\times72.5\times10^{-3}}}=270.1\,\mathrm{ms^{-1}},\quad t=2\times60\times60,\quad \pi r^2=3.14\times(1.0\times10^{-3})^2$$

$$P=\frac{Nk_BT}{V}=5.84\times10^{19}\times1.38\times10^{-23}\times1000=0.807\ \mathrm{Pa} \qquad (54)$$

Ans. vapor pressure of Ge: 0.807 Pa

Solution 2-11

The *van der Waals* equation of state is given by eq(55) for $n = 1$.

$$\frac{PV}{RT} = \frac{V}{(V-b)} - \frac{a}{RTV} \qquad (55)$$

When the right side of eq(55) is expanded into the power series of $b/V(<<1)$, eq(56) is obtained.

$$\frac{V}{(V-b)} - \frac{a}{RTV} = \frac{1}{(1-b/V)} - \frac{a}{RTV}$$

$$= 1 + \frac{b}{V} + \left(\frac{b}{V}\right)^2 + \frac{1}{2}\left(\frac{b}{V}\right)^3 + \cdots - \frac{a}{RTV}$$

$$= 1 + \frac{(b - a/RT)}{V} + \left(\frac{b}{V}\right)^2 + \frac{1}{2}\left(\frac{b}{V}\right)^3 + \cdots \qquad (56)$$

Eq(56) is compared term by term with the virial equation of state to find the relations between the *van der Waals* parameters of a and b and the virial coefficients of B and C. They are given by eq(57).

$$B(T) = (b - a/RT), \qquad C(T) = b^2 \qquad (57)$$

Since the *Boyle* temperature is the temperature at which a real gas behaves like an ideal gas in the first order of approximation, by putting $B(T)$ equal to zero, the *Boyle* temperature is found as eq(58).

$$(b - a/RT) = 0 \;\Rightarrow\; T_B = a/bR \qquad (58)$$

Values for the *Boyle* temperature are calculated from eq(58) for several gases and are listed in the table below.

Table A comparison of the *Boyle* temperatures (K) calculated and measured*.

	He	Ne	Ar	O_2	CO_2	CH_4
calculated	17.5	150	509	521	1026	643
measured	22.6	122.1	411.5	405.9	714.8	510

* Gases with higher *Boyle* temperatures seem to behave more non-ideally.

Ans. $B(T) = (b - a/RT)$, $C(T) = b^2$, $T_B = a/bR$

Solution 2-12

The volume V of the air flowed through the tube and the number of moles n of naphthalene lost in the flow experiment are calculated as below.

$$V = 60 \times 1.000 \times 10^{-3} \times \frac{298.15}{273.15} = 65.6\ \mathrm{dm}^3, \quad n = \frac{m}{M_{C_{10}H_8}} = \frac{0.0475 \times 10^{-3}}{126.1 \times 10^{-3}} = 3.77 \times 10^{-4}\ \mathrm{mol}$$

From these values, the vapor pressure P of naphthalene is calculated as 14.2 Pa.

$$P = \frac{nRT}{V} = \frac{3.77 \times 10^{-4} \times 8.31 \times 298.15}{65.6 \times 10^{-3}} = 14.2\ \mathrm{Pa}$$

Ans. vapor pressure of naphthalene: 14.2 Pa

Solution 2-13

(a) The mean velocity $\bar{c}$ in the three dimensions is calculated by using the velocity distribution function given in the problem (see **solution 2-33**) as is given by eq(59) below.

$$\bar{c} = 4\pi\left(\frac{m}{2\pi k_B T}\right)^{3/2} \int_0^\infty c \times \exp\left(-\frac{mc^2}{2k_B T}\right) c^2 dc$$

$$= \left(\frac{8k_B T}{\pi m}\right)^{1/2} \propto \sqrt{\frac{T}{m}} \qquad (59)$$

From this result, it is seen that the mean velocity is inversely proportional to the mass of a gas molecule at a given temperature. Therefore, the ratios to be found are $\bar{c}_{H_2}:\bar{c}_{O_2}:\bar{c}_{CO_2}=\sqrt{1/2}:\sqrt{1/32}:\sqrt{1/44}=1:0.25:0.23$ which are naturally independent of temperature.

Ans. the ratios of the mean velocity: $\sqrt{1/2}:\sqrt{1/32}:\sqrt{1/44}=1:0.25:0.23$ at either temperature

(b) Since the collision number z between molecules of the same kind is proportional to $\bar{c}\times\pi d^2\times N/V$ (more precisely, proportional to the mean relative velocity $\bar{c}_{rel}$), the ratio of the collision numbers is obtained by multiplying the ratio found in (a) with the square of the molecular diameter for a relevant molecule as is done below.

$$z_{H_2}:z_{O_2}:z_{CO_2}=1\times(218)^2:0.25\times(296)^2:0.23\times(460)^2=1:0.33:0.68$$

Ans. the ratios of the collision numbers: $1:0.33:0.68$

(c) The mean free path λ is defined by $\bar{c}/z\propto 1/d^2$ and is inversely proportional to the square of the molecular diameter. Therefore, the ratios are found as below.

$$\lambda_{H_2}:\lambda_{O_2}:\lambda_{CO_2}=\frac{1}{(218)^2}:\frac{1}{(296)^2}:\frac{1}{(460)^2}=1:0.68:0.22$$

Ans. the ratios of the mean free paths: $1:0.68:0.22$

Solution 2-14

(a) On the right side of eq(1) given in the problem, γ is dimensionless since it is the ratio of the heat capacities with the same dimension. Therefore, the dimension of the right side is given by dividing $RT=$[J/mol] by $M=$[g/mol], which is equal to that of [kinetic energy:J]=[velocity]$^2\times$[mass] divided by [mass:g], i.e., [velocity]^2.

(b) The molecular weights M of He, N_2, 1:1 He-N_2 mixture and the air are 2.0, 28.0, $(2.0+28.0)/2=15.0$ and $(28.0\times4+32.0)/5=28.8$, respectively. For the mixed gas the molecular weight is an apparent one. The heat capacity at constant pressure C_P and the one at constant volume C_V for the atomic gas are $5R/2$ and $3R/2$, respectively and for the diatomic gas, $7R/2$ and $5R/2$, respectively. For the mixed gases the arithmetically averaged values of C_P and C_V are employed. Hence, the values of γ for the gases are He: $\gamma=5/3$, N_2: $\gamma=7/5=1.40$, 1 : 1 mixture : $\gamma=(5+7)/(3+5)=1.5$ and air : $\gamma=1.40$.

The sound velocity υ_{sound} in each gas is calculated by eq(60) which is derived for the ideal gas from the equation given in the problem.

$$\upsilon_{sound}=\sqrt{\frac{\gamma RT}{M}} \qquad (60)$$

The results obtained are given below.

He: $\upsilon_{sound}=\sqrt{\dfrac{5\times8.31\times298}{3\times2\times10^{-3}}}=1437\ \mathrm{m\cdot s^{-1}}$, N_2: $\upsilon_{sound}=\sqrt{\dfrac{1.4\times8.31\times298}{28\times10^{-3}}}=352\ \mathrm{m\cdot s^{-1}}$,

1:1 mixed gas: $\upsilon_{sound}=\sqrt{\dfrac{1.5\times8.31\times298}{15\times10^{-3}}}=498\ \mathrm{m\cdot s^{-1}}$, air : $\upsilon_{sound}=\sqrt{\dfrac{1.4\times8.31\times298}{28.8\times10^{-3}}}=347\ \mathrm{m\cdot s^{-1}}$

Ans. He: $1437\ \mathrm{m\cdot s^{-1}}$, N_2: $352\ \mathrm{m\cdot s^{-1}}$, 1:1 mixed gas : $498\ \mathrm{m\cdot s^{-1}}$, air : $347\ \mathrm{m\cdot s^{-1}}$

(c) The mean velocity $\bar{c}$ is given by eq(61) (see **solution 2-33**).

$$\bar{c} = \sqrt{\frac{8RT}{\pi M}} \qquad (61)$$

The results calculated from eq(61) are He: $\bar{c} = 1776\ \mathrm{m \cdot s^{-1}}$ and N_2: $\bar{c} = 475\ \mathrm{m \cdot s^{-1}}$. It is evident from a comparison of eqs(60) and (61) that the mean velocity takes a value close to the sound velocity. The sound velocity provides one of the examples in which a microscopic quantity such as the molecular velocity can be estimated from an easily measurable macroscopic quantity known as the sound velocity.

Ans. He: $\bar{c} = 1776\ \mathrm{m \cdot s^{-1}}$, N_2: $\bar{c} = 475\ \mathrm{m \cdot s^{-1}}$

(d) Writing the frequency of the sound wave as ν, the wavelength of the sound wave λ is related to the sound velocity υ_{sound} by eq(62).

$$\nu = \frac{\upsilon_{sound}}{\lambda} \qquad (62)$$

On the other hand, the mean free path $\bar{l}_{free}$ is given by eq(63) (see **solutions 2-52** and **2-55**), where d is the collision diameter of a gas.

$$\bar{l}_{free} = \frac{\bar{c}}{z_{coll}} = \frac{k_B T}{\sqrt{2}\pi d^2 P} \qquad (63)$$

Substituting the values of d of $2.6 \times 10^{-10}\,\mathrm{m}$ and $3.8 \times 10^{-10}\,\mathrm{m}$ for He and N_2, respectively, the values for the mean free path are calculated as $6.33 \times 10^{-8}\,\mathrm{m}$ and $1.35 \times 10^{-7}\,\mathrm{m}$, respectively. From these values and those for the sound velocity obtained in (b), the frequencies which satisfy the relation of $\lambda = \bar{l}_{free}$ are calculated as $1.06 \times 10^{10}\,\mathrm{s^{-1}}$ and $2.60 \times 10^{9}\,\mathrm{s^{-1}}$ for He and N_2, respectively. Since the audible frequency (audible limit) to human being is in the range $20\ \mathrm{s^{-1}} \sim 2 \times 10^{2}\ \mathrm{s^{-1}}$, the frequencies obtained here are too high to be detected by our ears

Ans. He: $1.06 \times 10^{10}\,\mathrm{s^{-1}}$; N_2: $2.60 \times 10^{9}\,\mathrm{s^{-1}}$. Inaudible.

(e) When the absolute temperature $T\,\mathrm{K}$ in eq(60) is rewritten by the *Celsius* temperature $t^\circ\mathrm{C}$, eq(60') is obtained. Then eq(60') is expanded at $0^\circ\mathrm{C}$ to yield eq(60'') as shown below, into which the values of $M = 28.8$ and $\gamma = 1.4$ are substituted to derive the final equation.

$$\upsilon_{sound} = \sqrt{\frac{\gamma RT}{M}} = \sqrt{\frac{\gamma R(273 + t^\circ\mathrm{C})}{M}} \qquad (60')$$

$$= \sqrt{\frac{273\gamma R}{M}} \times \sqrt{\left(1 + \frac{t^\circ\mathrm{C}}{273}\right)}$$

$$\approx \sqrt{\frac{273\gamma R}{M}} \times \left(1 + \frac{1}{2} \times \frac{t^\circ\mathrm{C}}{273}\right) \qquad (60'')$$

$$= 332.1 + 0.608 \times t^\circ\mathrm{C}$$

It is realized from this that the sound travels faster as the temperature gets higher and, hence, the sound is heard at higher tone at higher temperatures (see next problem).

Ans. See text.

"note" For the derivation of eq(60), see *Molecular Theory of Gases and Liquids, J.O.Hirschfelder, C.F.Curtiss, R.B.Bird,* P728. Also, for the derivation of the *P-V-T* equations associated with the adiabatic expansion and compression processes, see **solution 3-42a.**

Solution 2-15

Writing the velocity of the sound in helium as $\upsilon_{\rm He}$ (eq(64)), the sound in this gas is propagated as a wave described by eq(65).

$$\upsilon_{\rm He} = \sqrt{\frac{\gamma RT}{M}} \qquad (64)$$

$$A \cdot \sin\frac{2\pi}{\lambda}(z - \upsilon_{\rm He}t) \qquad (65) \qquad \text{(propagation along } z\text{-axis)}$$

The sound (wave) emitted from the source in the resonator propagates in the resonator to and fro with the velocity of $\upsilon_{\rm He}$ and eventually resonates with the resonating wavelength of λ. The wavelength λ is determined by the shape of the resonator (see **"note"**) and is basically independent of the kind of a gas in the resonator. However, since the frequency ν of the sound is given by eq(66), it varies with the kind of a gas in

$$\nu = \frac{\upsilon_{\rm He}}{\lambda} \qquad (66)$$

the resonator, making an audible tone different from gas to gas. In the case of He, as is seen in the previous problem, the velocity of $\upsilon_{\rm He}$ is approximately four times larger as compared with that of the air and, hence, the frequency of the sound propagated is also four times larger. Since the sound with this higher frequency is propagated from the resonator, the tone we hear becomes higher as compared with that resonated in the air which we usually hear. This is the reason why our voice sounds in a higher tone when we speak after breathing in a gas mixture of He and air.

"note" For a resonator with one-end open such as illustrated in the figure (flute, etc.), the resonance wavelength λ is given by $(n+1/2)\lambda = a$ $(n = 0,\ 1,\ 2,\ 3,\ \cdots)$ with a as the length of the resonator.

Ans. We hear the sound in a higher tone. The reason for this is given in text.

Solution 2-16

<u>Solution 1</u>**.**

The *van der Waals* equation of state is given by eq(67) for $n = 1$.

$$\frac{PV}{RT} = \frac{V}{(V-b)} - \frac{a}{RTV} \qquad (67)$$

The right side is expanded in the power series of b/V to obtain eq(68). So far the procedure is the same as that employed in **solution 2-11**.

$$\frac{V}{(V-b)} - \frac{a}{RTV} = \frac{1}{(1-b/V)} - \frac{a}{RTV}$$

$$= 1 + \frac{b}{V} + \left(\frac{b}{V}\right)^2 + \frac{1}{2}\left(\frac{b}{V}\right)^3 + \cdots - \frac{a}{RTV}$$

$$= 1 + \frac{(b - a/RT)}{V} + \left(\frac{b}{V}\right)^2 + \frac{1}{2}\left(\frac{b}{V}\right)^3 + \cdots \qquad (68)$$

In order to convert eq(68) into the power series of P, we will try to find a simple way of expressing $1/V$ as a function of P. For this, the left side of eq(67) is approximated by the first two terms of eq(68), i.e., by eq(69).

$$\frac{PV}{RT}=1+\frac{(b-a/RT)}{V} \qquad (69)$$

Eq(69) is rewritten as eq(70) and, when solved for V, eq(71) is obtained.

$$PV^2 = RTV + RTb - a$$
$$PV^2 - RTV - (RTb-a) = 0 \qquad (70)$$
$$V=\frac{RT\pm\sqrt{(RT)^2+4P(RTb-a)}}{2P} \qquad (71)$$

The inverse of eq(71) is rewritten as below and its approximate form is obtained as eq(72).

$$\frac{1}{V}=\frac{2P}{RT\pm RT\sqrt{1+\dfrac{4P(RTb-a)}{(RT)^2}}}$$
$$\approx\frac{2P}{RT\pm RT\left(1+\dfrac{1}{2}\dfrac{4P(RTb-a)}{(RT)^2}\right)}=\frac{P}{RT\left(1+\dfrac{P(RTb-a)}{(RT)^2}\right)}$$
$$\approx\frac{P}{RT}\left\{1-\frac{P}{RT}\left(b-\frac{a}{RT}\right)\right\} \qquad (72)$$

Substituting eq(72) into eq(68), the obtained power series function is arranged according to the power of P to yield eq(73).

$$\frac{PV}{RT}=1+(b-a/RT)\times\frac{P}{RT}\left\{1-\frac{P}{RT}\left(b-\frac{a}{RT}\right)\right\}+b^2\times\left\langle\frac{P}{RT}\left\{1-\frac{P}{RT}\left(b-\frac{a}{RT}\right)\right\}\right\rangle^2+\cdots$$
$$=1+\frac{(b-a/RT)}{RT}P+\frac{\{b^2-(b-a/RT)^2\}}{(RT)^2}P^2+\cdots \qquad (73)$$

Comparing eq(73) with the equation given in the problem, the second and third vrial coefficients are found to be expressed as below. The first coefficient is 1.

$$B'(T)=\frac{(b-a/RT)}{RT}\left(=\frac{B(T)}{RT}\right),\qquad C'(T)=\frac{\{b^2-(b-a/RT)^2\}}{(RT)^2} \qquad (74)$$

Ans. $B'(T)=\dfrac{(b-a/RT)}{RT}\left(=\dfrac{B(T)}{RT}\right),\qquad C'(T)=\dfrac{\{b^2-(b-a/RT)^2\}}{(RT)^2}$

Solution 2.

The virial equation of state is a power series expansion of a certain function. Therefore, we can answer the problem by using a general way of expanding a relevant function into the power series of P. The left side of eq(67) is the compression factor and is regarded as a function of P. This is expressed by eq(75) and when expanded into the *Tayler* series at the pressure P^0, eq(76) is obtained.

$$\frac{PV}{RT}=f(P) \qquad (75)$$
$$=f(P^0)+\left(\frac{\partial f}{\partial P}\right)_{P=P^0}(P-P^0)+\frac{1}{2}\left(\frac{\partial^2 f}{\partial P^2}\right)_{P=P^0}(P-P^0)^2+\cdots \qquad (76)$$

In the limit of $P^0 \to 0$, the first term of eq(76) approaches to 1 and can be written as eq(77).

$$f(P^0) = f(P^0 \to 0) = 1 \qquad (77)$$

The coefficient of the second term (first derivative) can be expressed as eq(78),

$$\left(\frac{\partial f}{\partial P}\right)_{P=P^0,T} = \left\{\left(\frac{\partial f}{\partial V}\right)_{P,T}\left(\frac{\partial V}{\partial P}\right)_T\right\}_{P=P^0} \qquad (78)$$

and each partial derivative on the right is found for the *van der Waals* equation of state as eq(79) and eq(80).

$$\left(\frac{\partial f}{\partial V}\right)_{P,T} = \frac{1}{V-b} - \frac{V}{(V-b)^2} + \frac{a}{RTV^2} = \frac{RT(V-b)V^2 - RTV^3 + a(V-b)^2}{RT(V-b)^2V^2}$$
$$= \frac{-bRTV^2 + a(V-b)^2}{RT(V-b)^2V^2} \qquad (79)$$

$$\left(\frac{\partial V}{\partial P}\right)_T = \frac{1}{\left(\frac{\partial P}{\partial V}\right)_T} = \frac{(V-b)^2V^3}{-RTV^3 + 2a(V-b)^2} \qquad (80)$$

To derive eq(80), eq(81) is employed.

$$\left(\frac{\partial P}{\partial V}\right)_T = -\frac{RT}{(V-b)^2} + \frac{2a}{V^3} = \frac{-RTV^3 + 2a(V-b)^2}{(V-b)^2V^3} \qquad (81)$$

From the product of eq(79) and eq(80), eq(82) is obtained as the coefficient of the second term.

$$\left(\frac{\partial f}{\partial V}\right)_{P,T}\left(\frac{\partial V}{\partial P}\right)_T = \frac{-bRTV^2 + a(V-b)^2}{RT(V-b)^2V^2} \times \frac{(V-b)^2V^3}{-RTV^3 + 2a(V-b)^2}$$
$$= \frac{V}{RT} \times \frac{-bRTV^2 + a(V-b)^2}{-RTV^3 + 2a(V-b)^2}$$
$$= \frac{1}{RT} \times \frac{-bRT + a\left(1 - \frac{b}{V}\right)^2}{-RT + 2a\left(1 - \frac{b}{V}\right)^2 \times \frac{1}{V}} \qquad (82)$$

In the limit of $P^0 \to 0$, the volume approaches to infinity ($V \to \infty$). Thus eq(83) is derived from eq(82) as the answer to be given.

$$\left\{\left(\frac{\partial f}{\partial V}\right)_{P,T}\left(\frac{\partial V}{\partial P}\right)_T\right\}_{P=P^0} = \frac{1}{RT} \times \frac{bRT - a}{RT} = \frac{1}{RT} \times (b - a/RT) \qquad (83)$$

The third virial coefficient $C'(T)$ is also obtained similarly from $\frac{1}{2}\left(\frac{\partial^2 f}{\partial P^2}\right)_{P=P^0}$ (calculation omitted).

Ans. $B'(T) = \frac{1}{RT} \times (b - a/RT)$

Solution 2-17

Eq(73) given in **solution 2-16** is reproduced below. This equation is an approximate expression when the compression factor $z = PV/RT$ of the *van der Waals* gas is expanded into the power series of *P* in terms of its parameters *a* and *b*.

$$\frac{PV}{RT} = 1 + \frac{(b - a/RT)}{RT}P + \frac{\left\{b^2 - (b - a/RT)^2\right\}}{(RT)^2}P^2 + \cdots \qquad (73) \quad \text{(reposted)}$$

Substituting the given values for *a* and *b* of each gas into eq(73) (up to the third term) above, *z* is plotted against *P* at each temperature for the two gases as shown below.

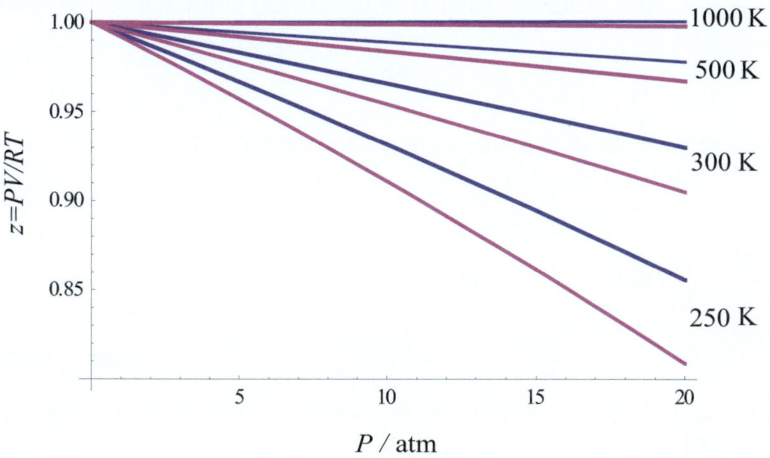

Fig.9 Pressure dependence of z.

—— : carbon dioxide,

—— : ammonia.

From this figure, it is expected that both gases can easily be liquefied at temperatures lower than room temperature but ammonia would be more readily liquefied than carbon dioxide. At higher temperatures, however, the compression factor z tends to approach to unity for either gas and, hence, it can be said that even these readily liquefied gases behave like ideal gases at high temperatures.

Ans. See text and Fig.9

Solution 2-18

Equating the first and second derivatives by volume of the *van der Waals* equation of state to zero, eqs(84) and (85) are obtained. The set of these equations is solved simultaneously to obtain the critical pressure, volume and temperature, P_c, V_c and T_c, respectively.

$$P = \frac{RT}{(V-b)} - \frac{a}{V^2} : \textit{van der Waals} \text{ equation of state}$$

$$\left(\frac{\partial P}{\partial V}\right)_T = -\frac{RT}{(V-b)^2} + \frac{2a}{V^3} = 0 \qquad (84)$$

$$\left(\frac{\partial^2 P}{\partial V^2}\right)_T = \frac{2RT}{(V-b)^3} - \frac{6a}{V^4} = 0 \qquad (85)$$

An easy way to solve it is to calculate $(84)\times 3/V + (85)$ as shown below.

$$\left\{-\frac{RT}{(V-b)^2} + \frac{2a}{V^3}\right\} \times \frac{3}{V} + \frac{2RT}{(V-b)^3} - \frac{6a}{V^4} = -\frac{3RT}{(V-b)^2 V} + \frac{2RT}{(V-b)^3}$$

$$= \frac{-3RT(V-b) + 2RTV}{(V-b)^3 V}$$

$$= \frac{-RT(V-3b)}{(V-b)^3 V} = 0$$

From this, the critical volume V_c is obtained as eq(86).

$$(V-3b) = 0 \Rightarrow \ V_c = 3b \qquad (86)$$

Substituting eq(86) into eq(84), eq(87) is obtained for T_c.

$$-\frac{RT}{(2b)^2} + \frac{2a}{27b^3} = 0 \ \Rightarrow \ T_c = \frac{8a}{27bR} \qquad (87)$$

Substituting eqs(86) and (87) into the *van der Waals* equation of state at the critical point, eq(88) is obtained.

$$P_c = \frac{RT_c}{(V_c - b)} - \frac{a}{V_c^2} = \frac{4a}{27b^2} - \frac{a}{9b^2} = \frac{a}{27b^2} \qquad (88)$$

Eqs(86) to (88) are the well known relations, from which the compression factor at the critical point z_c is calculated as $3/8$.

$$z_c = \frac{P_c V_c}{RT_c} = \frac{\dfrac{a}{27b^2} \times 3b}{R \times \dfrac{8a}{27bR}} = \frac{3}{8} = 0.375 \qquad (89)$$

Ans. See text.

"note" As is seen from the values in the table below, z_c is close to 0.3 for many real gases. In spite of only two parameters employed, the *van der Waals* equation of state can describe the behaviors of real gases quite well under various situations.

Table Critical constants and parameters of *van der Waals* equation of state for several gases

	T_c/K	P_c/MPa	$V_c \times 10^3/\text{m}^3$	z_c	$a^{\#}$	b^{*}
He	5.20	0.2275	5.75	0.303	0.03414	0.02373
H_2	32.98	1.293	6.42	0.303	0.2432	0.02667
N_2	126.2	3.400	8.92	0.289	1.3483	0.03858
O_2	154.6	5.043	7.34	0.288	1.3639	0.03186
Ar	150.7	4.865	7.45	0.289	1.3307	0.03183
CO_2	304.2	7.38	9.44	0.276	3.6073	0.04281
CH_4					2.2725	0.04307

$^{\#}(\text{dm}^6 \cdot \text{atm} \cdot \text{mol}^{-2})$, $^{*}(\text{dm}^3 \cdot \text{mol}^{-1})$

Solution 2-19

The equations (eq(3) to eq6)) given in the present problem are shown again below with different equation numbers. All are related to the *van der Waals* equation of state.

$$P = \frac{RT}{(V - b)} - \frac{a}{V^2} \qquad (90)$$

$$P_R = P/P_c,\ V_R = V/V_c,\ T_R = T/T_c \qquad (91)$$

$$P_c = \frac{a}{27b^2},\ T_c = \frac{8a}{27bR},\ V_c = 3b \qquad (92)$$

Using eq(91), P, V and T in eq(90) are replaced with $P_R \cdot P_c$, $V_R \cdot V_c$ and $T_R \cdot T_c$, respectively, giving eq(90').

$$P_R \cdot P_c = \frac{RT_R \cdot T_c}{(V_R \cdot V_c - b)} - \frac{a}{(V_R \cdot V_c)^2} \qquad (90')$$

This equation is rewritten by using the relations in eq(92) so as for the critical constants and the parameters a and b to be eliminated. Finally it leads to eq(93).

$$P_R = \frac{RT_R \cdot T_c/P_c}{(V_R \cdot V_c - b)} - \frac{a/P_c}{(V_R \cdot V_c)^2}$$
$$= \frac{RT_R \cdot T_c/P_c \cdot V_c}{(V_R - b/V_c)} - \frac{a/P_c \cdot V_c^2}{V_R^2}$$
$$= \frac{8T_R}{(3V_R - 1)} - \frac{3}{V_R^2} \qquad (93)$$

As is obvious from these procedures adopted here, any equation of state with only two parameters can always be rewritten into the reduced equation of state.

Ans. See text.

Solution 2-20

Since the reduced pressure is positive, the reduced equation of state (eq(94)) must satisfy the inequality of the right side of eq(94), namely the equivalently rewritten eq(94').

$$P_R = \frac{8T_R}{(3V_R - 1)} - \frac{3}{V_R^2} > 0 \qquad (94)$$

$$8T_R V_R^2 - 3(3V_R - 1) = 8T_R V_R^2 - 9V_R + 3 > 0 \qquad (94')$$

Consequently, when eq(94') is regarded as a quadratic equation of V_R, its discriminant D must be negative.

$$D = 81 - 96T_R < 0 \qquad (95) \Rightarrow T_R > \frac{27}{32} = 0.84375 \qquad (96)$$

Namely, T_R must satisfy eq(96). The reduced equation of state is plotted in Fig.10 against V_R at several values for T_R. As is evident from this figure, P_R comes to take a negative value in some range of V_R at $T_R = 0.80$ (curve in black), though it does not become negative at $T_R = 0.85$. Needless to say, T_R = 1.0 indicates the critical temperature.

Summarizing the results obtained above, it is concluded that in the case of carbon dioxide, the temperature range where the *van der Waals* equation of state is meaningful is $T > \frac{27}{32} \times 304.2 = 256.7$ K.

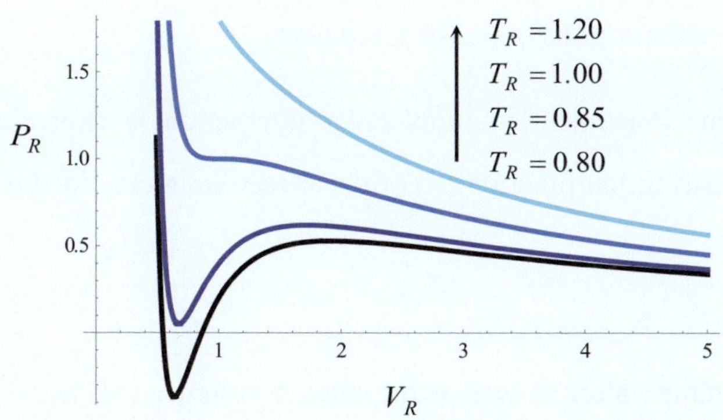

Fig.10 The reduced *van der Waals* equation of state plotted at several reduced temperatures.

Ans. $T_R > \frac{27}{32} = 0.84375$, temperatures above 256.7 K

Solution 2-21

(a), (b) As is readily seen from the values listed in the table, the molecules in the table are classified into the two groups of non-polar molecules and polar molecules in terms of the value of z_c. In the case of non-polar molecules, though the value for T_c varies widely for five molecules from He to O_2, the values for z_c of these five molecules gather around 0.3. It is rather surprising to know that the *van der Waals* equation of state with only two parameters of a and b can yield the value for z_c close to the experimental values.

In order to improve the agreement between the experimental values and the calculated one based on the *van der Waals* equation of state, the parameter a which reflects the intermolecular interaction can be treated from a modified viewpoint of intermolecular force, since the value for the other parameter b associated with the excluded volume does not vary much from molecule to molecule at least among those listed in the table. Also, b is rather insensitive to both temperature and pressure. From these reasons, it would not be a good solution to modify b somehow (see **"note" 1**).

The averaged intermolecular force, as is discussed in detail in Chapter 5, are dependent on temperature or volume (intermolecular distance) and, hence, it would be one of the solutions needed to make the term a dependent on temperature or volume. Actually, there are two such well-known equations of state.

One of them is;

Berthelot's equation: The intermolecular interactions become more significant at lower temperatures as compared with the kinetic energy. For this, *Berthelot*'s equation of state (eq(97)) is derived by replacing the term $\frac{a}{V^2}$ with the term $\frac{a}{TV^2}$, reflecting temperature dependent intermolecular interaction.

$$P = \frac{RT}{(V-b)} - \frac{a}{TV^2} \qquad (97)$$

From this equation, the critical pressure, temperature and volume are obtained as eq(98) (derivation omitted).

$$P_c = \frac{1}{12}\left(\frac{2aR}{3b^3}\right)^{1/2}, \quad T_c = \frac{2}{3}\left(\frac{2a}{3bR}\right)^{1/2}, \quad V_c = 3b \qquad (98)$$

Then, the critical compression factor z_c is calculated as eq(99) and is found the same as that of the *van der Waals* equation of state.

$$z_c = \frac{P_c V_c}{RT_c} = \frac{1}{12}\left(\frac{2aR}{3b^3}\right)^{1/2} \cdot 3b \Big/ \frac{2}{3}\left(\frac{2a}{3bR}\right) \cdot R = \frac{3}{8} \qquad (99)$$

Therefore, unfortunately eq(97) is concluded to be inappropriate to make z_c smaller.

Dieterict's equation: The temperature and volume dependent intermolecular interaction is expressed as eq(100), which is consistent with the *Boltzmann* distribution function. For the attractive interaction ($a > 0$), eq(100)

$$P = \frac{RT \cdot e^{-a/RTV}}{(V-b)} \qquad (100)$$

predicts that the pressure gets lower for the lower temperature as well as the smaller volume. This behavior is appropriate in taking the intermolecular interaction between polar molecules into account.

From that equation, the critical pressure, temperature and volume are obtained as eq(101) (see **solution 2-49**).

$$P_c = \frac{a}{4e^2b^2}, \quad T_c = \frac{a}{4Rb}, \quad V_c = 2b \qquad (101)$$

Then, the critical compression factor z_c is calculated as eq(102) and is found to be in good agreement with

$$z_c = \frac{P_c V_c}{RT_c} = \frac{a}{4e^2b^2} \cdot 2b \bigg/ \frac{a}{4Rb} \cdot R = \frac{2}{e^2} = 0.2707 \qquad (102)$$

those measured for the polar gases listed in the table.

Ans. See text (a) and (b).

"note 1" Taking the effect of the excluded volume into account, the *van der Waals* equation of state may be modified as eq(103). Unfortunately, however, since eq(103) does not have a critical point, such modification

$$P = \frac{RT}{(V-b)} - \frac{a}{(V-b)^2} \qquad (103)$$

is meaningless as far as the present issue is concerned. See also **problem and solution 2-50**.

"note 2" *Dieterict*'s equation can be expanded as eq(104), from which it can be predicted what kind of modification is effective in making z_c smaller.

$$P = \frac{RT \cdot e^{-a/RTV}}{(V-b)} \approx \frac{RT}{(V-b)}\left\{1 - \frac{a}{RTV} + \frac{1}{2}\left(\frac{a}{RTV}\right)^2 + \cdots\right\} \qquad (104)$$

Solution 2-22

(a) One-dimensional mean velocity $\bar{u}$ (see **problem 2-7**).

$$\bar{u} = \sqrt{\frac{2k_BT}{\pi m}} = \left(\frac{2 \times 1.38 \times 10^{-23} \times 300}{3.14 \times 32.0 \times 10^{-3}/6.02 \times 10^{23}}\right)^{1/2} = 223\ \mathrm{m \cdot s^{-1}}$$

Ans. $223\ \mathrm{m \cdot s^{-1}}$

(b) Three-dimensional mean velocity $\bar{c}$ (see **problem 2-10**).

$$\bar{c} = \sqrt{\frac{8k_BT}{\pi m}} = \left(\frac{8 \times 1.38 \times 10^{-23} \times 300}{3.14 \times 32.0 \times 10^{-3}/6.02 \times 10^{23}}\right)^{1/2} = 445\ \mathrm{m \cdot s^{-1}}$$

Ans. $445\ \mathrm{m \cdot s^{-1}}$

(c) The three-dimensional velocity c is given by eq(105), where u, v and w represent one-dimensional velocity along x-, y- and z-axis, respectively.

$$c = \sqrt{u^2 + v^2 + w^2} \geq u,\ (v,\ w) \qquad (105)$$

When both sides of eq(105) are integrated over the velocity distribution of u, v and w, the inequality of eq(105) leads to eq(106). All of the distribution functions (each indicated by a differently colored dotted lines) are normalized to unity.

$$\iiint_{u,v,w} c \times \sqrt{\frac{2m}{\pi k_BT}} \cdot \exp\left(-\frac{mu^2}{2k_BT}\right) \cdot \sqrt{\frac{2m}{\pi k_BT}} \cdot \exp\left(-\frac{mv^2}{2k_BT}\right) \cdot \sqrt{\frac{2m}{\pi k_BT}} \cdot \exp\left(-\frac{mw^2}{2k_BT}\right) dudvdw \geq$$

$$\int_u u \times \sqrt{\frac{2m}{\pi k_BT}} \cdot \exp\left(-\frac{mu^2}{2k_BT}\right) du \times \int_v \sqrt{\frac{2m}{\pi k_BT}} \cdot \exp\left(-\frac{mv^2}{2k_BT}\right) dv \times \int_w \sqrt{\frac{2m}{\pi k_BT}} \cdot \exp\left(-\frac{mw^2}{2k_BT}\right) dw \qquad (106)$$

Since the left side is equal to $\bar{c}$ and right side is equal to $\bar{u}$ by definition, it is obvious that inequality of eq(107) always holds.

$$\bar{c} \geq \bar{u},\ (\bar{v},\ \bar{w}) \qquad (107) \qquad \text{(trivial from eq(105))}$$

Ans. See text.

Solution 2-23

(a) The universal gravitation force works between a molecule and the earth. The masses of the molecule and the earth are denoted by m and M, respectively. The distance from the center of the earth to the molecule is r

and the universal gravitation constant is written as G.. Then this gravitational force $\boldsymbol{F}$ is given by eq(108).

$$\boldsymbol{F}=-G\frac{m\times M}{r^2}\cdot\hat{\boldsymbol{r}} \qquad (\hat{\boldsymbol{r}}: \text{ unit vector along the radial coordinate}) \qquad (108)$$

Consequently, the potential energy of the molecule is given by eq(109).

$$U(r)=-\int_{\infty}^{r}\boldsymbol{F}\cdot d\boldsymbol{r}=\int_{\infty}^{r}G\frac{m\times M}{r^2}dr=-G\frac{m\times M}{r} \qquad (109)$$

For the molecule on the surface of the earth at $r=R_E$ to be able to escape from the gravitational field, "the kinetic energy of the molecule" must be larger than "its gravitational potential energy". This condition is expressed by eq(110), where υ is the velocity of the center of gravity of the molecule.

$$\frac{1}{2}m\upsilon^2>U(\infty)-U(R_E)=\int_{R_E}^{\infty}G\frac{m\times M}{r^2}dr=\left[-G\frac{m\times M}{r}\right]_{R_E}^{\infty}$$

$$=G\frac{m\times M}{R_E}=m\times\frac{G\times M}{R_E^2}\times R_E \qquad \left(g=\frac{G\times M}{R_E^2},\ g: \text{acceralation of gravity}\right)$$

$$=m\times g\times R_E \qquad (110)$$

Therefore、by putting $R_E=6378$ km, the minimum velocity is found as below.

$$\upsilon^2>2g\,R_E \Rightarrow \upsilon>\sqrt{2gR_E}=\sqrt{2\times 9.807(\text{m}\cdot\text{s}^{-2})\times 6378\times 10^3(\text{m})}=11.18\ \text{km}\cdot\text{s}^{-1}$$

The obtained velocity of $11.18\ \text{km}\cdot\text{s}^{-1}$ is called the escape velocity from the earth and is independent of mass. The same velocity is required for rockets.

Ans. velocity: larger than $11.18\ \text{km}\cdot\text{s}^{-1}$

(b) The fraction of the molecules which have the velocities greater than $c=11.18\ \text{km}\cdot\text{s}^{-1}$ is given by eq(111) from the three-dimensional velocity distribution function $p(c)$, since c can be regarded as the velocity along the radial coordinate.

$$\int_{11180}^{\infty}p(c)dc=4\pi\left(\frac{m}{2\pi k_BT}\right)^{3/2}\int_{11180}^{\infty}c^2\exp\left(-\frac{mc^2}{2k_BT}\right)dc \qquad (111)$$

Substituting the mass of each molecule into eq(111), the fraction at 300 K is obtained as below for each molecule.

$H_2: \approx 9.6\times 10^{-22}$, $O_2: \approx 0$

(The integrated values of eq(111) may be too small for a conventional calculation software to be executed properly. In that case, it is suggested to calculate $p(c)$ at selected points first by putting $c=11.18+j\times 0.1$ $\text{km}\cdot\text{s}^{-1}$ with $j=1,2\cdots$ and then to sum the results up to approximate the integral. The values of $p(c)$ at $c=11.18\ \text{km}\cdot\text{s}^{-1}$ are ca.8.05×10^{-24} and ca.1.56×10^{-348} for H_2 and O_2, respectively.)

Ans. ratio for $H_2: \approx 9.6\times 10^{-22}$, for $O_2: \approx 0$

(c) From the results obtained above, it is understood that the H_2 molecule can escape from the gravitational field of the earth eventually, though it would take long time. In fact, the percentage of H_2 in the atmosphere is practically zero, which is a surprisingly small number to know when it is noticed that the percentage of H_2 in the universe is more than 93% in the ratio of the numbers of atoms. Incidentally, the velocity of O_2 above which the fraction of the molecules becomes the same as that of H_2 whose velocities are above the escaping velocity is only $1970\ \mathrm{m \cdot s^{-1}}$.

Ans. See text (c).

Well then, if the temperature on the surface of the earth is at $T = 1000\,\mathrm{K}$ or $10000\,\mathrm{K}$, can O_2 molecules escape from the earth?

Solution 2-24

The compression factors z of CO_2 plotted against P at various temperatures are shown in Figs.11 and 12. The former plot is for high temperatures and the latter is for the lower temperatures in the range higher than the critical temperature but up to 500 K.

The compression factors derived by using the first and second approximations of the virial expression as well as that derived directly from the *van der Waals* equation of state are compared in Fig.13. From this figure, it is seen that the second order approximation can represent the *van der Waals* equation of state quite well. (When some sort of *Parametric plot* is not available, calculated values for z and P at various values of V first by *Excel* and then plot the obtained values of z against those of P.)

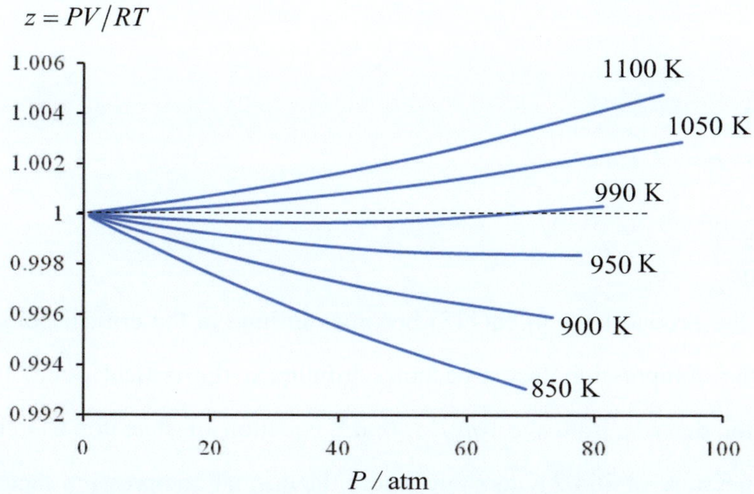

Fig.11 z *vs.* P plot in a high temperature range.

The *van der Waals* equation of state for CO_2 is employed. The *Boyle* temperature is 1026 K. At this temperature, the slope of the curve at the origin is zero.

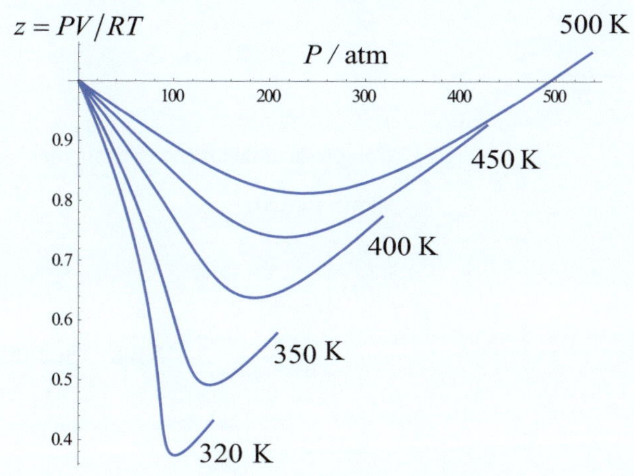

Fig.12 z *vs.* P plot in a low temperature range.

The *van der Waals* equation of state for CO_2 is employed. At the critical point, the slope of the curve becomes infinite (see **problem 2-25**).

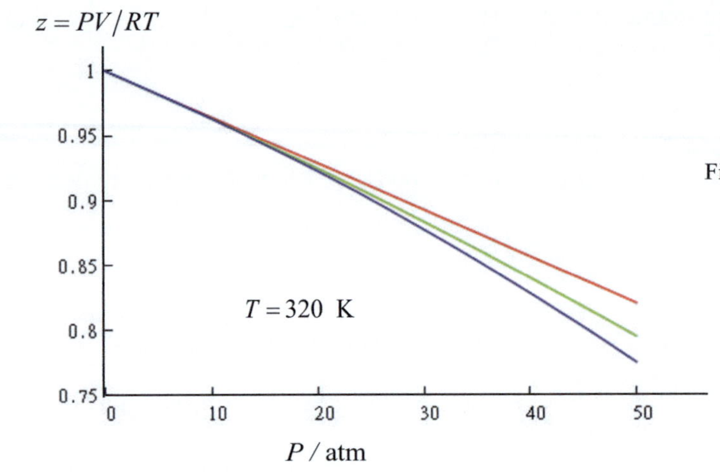

Fig.13 A comparison of three approximate equations of state.

— $1+\frac{(b-a/RT)}{RT}P$

— $1+\frac{(b-a/RT)}{RT}P+\frac{b^2-(b-a/RT)^2}{(RT)^2}P^2$

— z vs. P (parametric plot, see Figs.11 and 12.)

Ans. See Figs.11 through 13.

Solution 2-25

For any gas the compression factor z is defined by eq(112). The pressure derivative of eq(112) at constant temperature is given by eq (113).

$$z=\frac{PV}{RT} \qquad (112)$$

$$\begin{aligned}\left(\frac{\partial z}{\partial P}\right)_T &= \frac{V}{RT}+\frac{P}{RT}\left(\frac{\partial V}{\partial P}\right)_T \\ &= \frac{V}{RT}+\frac{P}{RT}\frac{1}{\left(\frac{\partial P}{\partial V}\right)_T}\end{aligned} \qquad (113)$$

Since $(\partial P/\partial V)_T = 0$ at the critical point, the second term in eq(113) becomes infinite at the critical point. Consequently, the pressure variation of the compression factor becomes infinite at the critical point. In Fig.14 given below, the compression factor derived from the *van der Waals* equation of state for CO_2 is plotted against pressure at the critical temperature of 304.2 K to show where the critical compression factor ($z_c = 0.375$) is located on that curve.

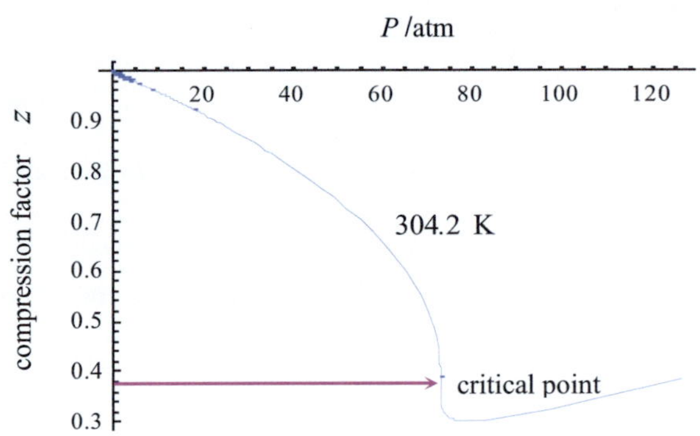

Fig.14 Compression factor and critical point shown for CO_2.

$$\left(\frac{\partial z}{\partial P}\right)_{T=304.2} \to \infty$$

Ans. See text.

Solution 2-26

For the virial equation of state expanded in terms of $1/V$ (eq(114)), the partial derivative of z with respect to pressure is given by eq(115).

$$z = \frac{PV}{RT} = 1 + \frac{B(T)}{V} + \frac{C(T)}{V^2} + \frac{D(T)}{V^3} + \cdots \qquad (114)$$

$$\left(\frac{\partial z}{\partial P}\right)_T = -\frac{B(T)}{V^2}\left(\frac{\partial V}{\partial P}\right)_T - 2\frac{C(T)}{V^3}\left(\frac{\partial V}{\partial P}\right)_T - 3\frac{D(T)}{V^4}\left(\frac{\partial V}{\partial P}\right)_T + \cdots$$

$$= -\left(\frac{B(T)}{V^2} + 2\frac{C(T)}{V^3} + 3\frac{D(T)}{V^4} + \cdots\right)\frac{1}{\left(\frac{\partial P}{\partial V}\right)_T} \qquad (115)$$

Since the partial derivative in the denominator of eq(115) becomes zero at the critical point, the left side of eq(115) approaches to infinity near the critical point (see the preceding problem).

$$\left(\frac{\partial z}{\partial P}\right)_{T=T_c} \rightarrow \infty$$

On the other hand, in eq(116) obtained by expanding with respect P, all of the coefficients should take finite values, otherwise eq(116) will not converge.

$$\frac{PV}{RT} = 1 + B'(T)P + C'(T)P^2 + D'(T)P^3 + \cdots \qquad (116)$$

$$\left(\frac{\partial z}{\partial P}\right)_T = B'(T) + 2C'(T)P + 3D'(T)P^2 + \cdots \qquad (117)$$

Therefore, the partial derivative of z with respect to P never takes an infinite value. In other words, eq(116) cannot describe the critical state. Thus the two virial equations of state, eqs(114) and (116) (eqs(9) and (10) in the problem, respectively) are decisively different in such an essential point.

Ans. See text.

Solution 2-27

As is shown in Fig.15 given below, the volumes at the points indicated by the blue, red and yellow dots are denoted by V_G, V_a and V_L, respectively. Since the figure is given for one mole of a pure substance (V_G and V_L are, therefore, molar volumes), when the mole numbers of the gas and the liquid at the red point are denoted by x and 1-x, respectively, the volume V_a is given by eq(118). Solving eq(118) for x, eqs(119) and (120) are obtained. Therefore, the ratio is found as eq(121). The relation of eq(121) indicates that the ratio of

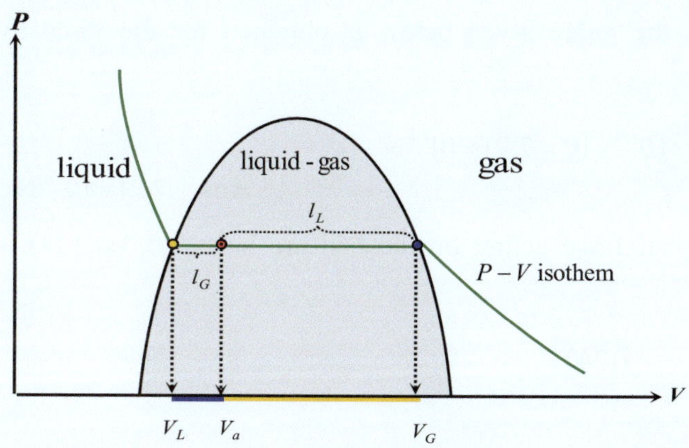

$$V_a = xV_G + (1-x)V_L \qquad (118)$$

$$x = \frac{V_a - V_L}{V_G - V_L} \quad (119), \quad 1 - x = \frac{V_G - V_a}{V_G - V_L} \qquad (120)$$

$$x : 1 - x = (V_a - V_L) : (V_G - V_a) = l_G : l_L \qquad (121)$$

Fig.15 *P-V* isotherm of gas and liquid in equilibrium illustrating the lever rule.

Ans. See text.

the amounts of the two components (in moles) is given by the ratio of the lengths of l_G and l_L, which are called lever arms. Because of this, the rule to determine the ratio of each component is called lever rule. You will not misunderstand which length represents which component if you realize to which of gas or liquid the lever length at the blue (pure gas) dot or yellow dot (pure liquid) corresponds.

Solution 2-28

The mean square velocity $\overline{c^2}$ is calculated from the three-dimensional velocity distribution function $p(c)$.

$$\overline{c^2} = \int_0^\infty c^2 p(c)dc = 4\pi\left(\frac{m}{2\pi k_B T}\right)^{3/2}\int_0^\infty c^2 \times \exp\left(-\frac{mc^2}{2k_B T}\right)c^2 dc \qquad (122)$$

The integration of eq(122) is readily carried out by following a standard procedure. For this, eq(122) is rewritten as eq(122') by replacing $\frac{m}{2k_B T}$ with α.

$$\overline{c^2} = 4\pi\left(\frac{m}{2\pi k_B T}\right)^{3/2}\int_0^\infty c^4 \times \exp(-\alpha c^2)dc \qquad (122')$$

Then using the formula $\int_0^\infty x^{2n+1}e^{-ax^2}dx = \frac{1\cdot 3\cdot 5\cdots(2n-1)}{2^{n+1}a^n}\sqrt{\frac{\pi}{\alpha}}$ (see **II-4-6**) the integral above is reduced to eq(123),

$$\overline{c^2} = 4\pi\left(\frac{m}{2\pi k_B T}\right)^{3/2}\times\frac{3}{8\alpha^2}\sqrt{\frac{\pi}{\alpha}} = 4\pi\left(\frac{m}{2\pi k_B T}\right)^{3/2}\times\frac{3}{8}\left(\frac{2k_B T}{m}\right)^{5/2}\sqrt{\pi}$$

$$= \frac{3k_B T}{m} \qquad (123)$$

from which the mean translational kinetic energy per molecule is found as below.

$$\frac{1}{2}\overline{mc^2} = \frac{1}{2}m\overline{c^2} = \frac{3}{2}k_B T$$

Ans. See text.

Solution 2-29

(a) The value of $\bar{c} = \sqrt{8k_B T/\pi m} = 469\ \mathrm{ms^{-1}}$ and the other known values are substituted into the equation given in the problem (see **"note"**) to find the value for η as below.

$$\eta = \frac{m\bar{c}}{2\sqrt{2}\pi d^2} = \frac{28.8\times10^{-3}\times469}{2\sqrt{2}\times3.14\times9\times10^{-20}\times6.02\times10^{23}} = 2.81\times10^{-5}\mathrm{kg\cdot m^{-1}\cdot s^{-1}} = 281\ \mu\mathrm{P}$$

Ans. $2.81\times10^{-5}\mathrm{kg\cdot m^{-1}\cdot s^{-1}}$

(b) From the *Stokes* equation (see **solution 2-75**), the value given below is obtained for the viscosity resistance. The value of η found in (a) is used.

$$6\pi\eta rv = 6\times3.14\times2.81\times10^{-5}\times1.5\times10^{-1}\times10 = 7.91\times10^{-4}\mathrm{N}$$

Ans. $7.91\times10^{-4}\mathrm{N}$

(c) When the viscosity resistance and the gravitational force acting on the ball are balanced, eq(124) is satisfied, from which eq(125) is derived.

$$6\pi\eta rv = M\times g = \frac{4}{3}\pi r^3\times\rho\times g \qquad (124)$$

$$v = \frac{2}{9\eta}r^2\times\rho\times g \qquad (125)$$

$$= \frac{2\times 0.15^2[\mathrm{m}^2]\times 1.5\times 10^3[\mathrm{kg\cdot m^{-3}}]\times 9.81\ [\mathrm{m\cdot s^{-2}}]}{9\times 2.81\times 10^{-5}[\mathrm{kg\cdot m^{-1}\cdot s^{-1}}]}$$

$$= 2.61\times 10^6\,\mathrm{m\cdot s^{-1}}$$

Substituting the given values into eq(125) with due attention to their units, the value above is obtained. The found velocity is enormously large but it is easily imagined that a flying object in space burns out when it plunges into the atmosphere to reach such a high speed. Falling rain drops with a constant velocity can be explained in a similar manner. Their speeds are usually in the range $2-10\ \mathrm{m\cdot s^{-1}}$.

Ans. $2.61\times 10^6\,\mathrm{m\cdot s^{-1}}$

"note" When the formula for η given in the problem is multiplied by $2/3$ to give $\eta = m\bar{c}/3\sqrt{2}\,\pi d^2$, the obtained formula yields values of η closer to the observed ones.

Solution 2-30

To describe molecular collisions with the wall, the one-dimensional velocity distribution function $p(u)$ given by eq(126) is employed, since only the velocity component perpendicular to the wall matters.

$$p(u)du = \left(\frac{m}{2\pi k_B T}\right)^{1/2} e^{-\frac{mu^2}{2k_B T}}du, \qquad [u:-\infty,\ \infty] \qquad (126)$$

We denote by dZ the collision number which the molecules with the velocity of u make with a part of the wall (area S) per unit time. Then, this is given by the number of those molecules within the volume of $u\times S$ whose velocities are in the range $[u,\ u+du]$ (positive toward the wall) as is expressed by eq(127).

$$dZ = (u\times S)\times \frac{N}{V}\times p(u)du \qquad (127)$$

Therefore, the total number of the collisions Z_{wall} with the area S of the wall is found by substituting eq(126) into eq(127) and then integrating it over the velocity toward the wall in the range $[0,\ \infty]$ as is done in eq(128), in which $\bar{c} = (8k_B T/\pi m)^{1/2}$ (see **solution 2-13**) is used.

$$Z_{wall} = \int_0^\infty (u\times S)\times \frac{N}{V}\times p(u)du$$

$$= \frac{N}{V}\times S\times \left(\frac{m}{2\pi k_B T}\right)^{1/2}\int_0^\infty ue^{-\frac{mu^2}{2k_B T}}du = \frac{N}{V}\times S\times \left(\frac{k_B T}{2\pi m}\right)^{1/2} = \frac{N}{V}\times S\times \frac{1}{4}\bar{c} \qquad (128)$$

Eq(128) can also be rewritten as a function of P as eq(129) by employ the relation of $N/V = P/k_B T$.

$$Z_{wall} = \frac{P}{k_B T}\times S\times \left(\frac{k_B T}{2\pi m}\right)^{1/2} = \frac{P}{(2\pi m k_B T)^{1/2}}\times S \qquad (129)$$

Summarizing the results、the collision number with the wall per unit area and unit time z_{wall} is given in a simple form as eq(130) (see **solution 2-54a**).

$$z_{wall} = \frac{Z_{wall}}{S} = \frac{1}{4}\bar{c}\times \frac{N}{V} = \frac{P}{(2\pi m k_B T)^{1/2}} \qquad (130)$$

Ans. See text.

Solution 2-31

Employing eq(129) obtained in the preceding solution, the equation is rewritten as eq(131) below so as to meet the requirement of the problem. S is the collision cross section between the oxygen molecule and the metal atom.

$$Z_{wall} = \frac{P}{(2\pi mk_B T)^{1/2}} \times S < 1 \;\Rightarrow\; P < \frac{(2\pi mk_B T)^{1/2}}{S} \qquad (131)$$

From eq(131), the maximum pressure of the oxygen is calculated as below.

$$P < \frac{(2\pi mk_B T)^{1/2}}{S} = \frac{(2\times 3.14\times 32.0\times 10^{-3}\times 1.38\times 10^{-23}\times 300/6.02\times 10^{23})^{1/2}}{3.14\times\left\langle(2.0+1.5)\times 10^{-10}\right\rangle^2} = 9.66\times 10^{-5}\,\text{Pa}$$

In atm or in Torr, the pressure is found as below, which is rather easily achieved in laboratories.

$$P < 9.66\times 10^{-5}/1.01\times 10^{5} = 9.6\times 10^{-10}\,\text{atm} = 7.3\times 10^{-7}\,\text{Torr}$$

Ans. pressure of oxygen: below $9.6\times 10^{-10}\,\text{atm} = 7.3\times 10^{-7}\,\text{Torr}$

Solution 2-32

When the velocities of two different molecules are denoted by $c_1(u_1, v_1, w_1)$ and $c_2(u_2, v_2, w_2)$ and their masses are by m_1 and m_2, the sum of their kinetic energies ε is given by eq(132).

$$\varepsilon = \frac{1}{2}m_1c_1^2 + \frac{1}{2}m_1c_1^2 = \frac{1}{2}m_1(u_1^2 + v_1^2 + w_1^2) + \frac{1}{2}m_2(u_2^2 + v_2^2 + w_2^2) \qquad (132)$$

The velocities of the center of the gravity of the two molecules denoted as u_c, v_c and w_c are given by eq(133).

$$u_c = \frac{m_1u_1 + m_2u_2}{m_1 + m_2},\quad v_c = \frac{m_1v_1 + m_2v_2}{m_1 + m_2},\quad w_c = \frac{m_1w_1 + m_2w_2}{m_1 + m_2} \qquad (133)$$

Also, the components of the relative velocity denoted as α, β and γ are given by eq(134).

$$\alpha = u_2 - u_1,\quad \beta = v_2 - v_1,\quad \gamma = w_2 - w_1 \qquad (134)$$

In order to rewrite eq(132) in terms of the velocities at the center u_c, v_c and w_c and the relative velocities α, β and γ, the molecular velocities (u_1, v_1, w_1) and (u_2, v_2, w_2) are expressed by using u_c, v_c and w_c and α, β and γ as eq(135).

$$\left.\begin{array}{ll} u_1 = u_c - \dfrac{m_2\alpha}{m_1 + m_2}, & u_2 = u_c - \dfrac{m_1\alpha}{m_1 + m_2} \\ v_1 = v_c - \dfrac{m_2\beta}{m_1 + m_2}, & v_2 = v_c - \dfrac{m_1\beta}{m_1 + m_2} \\ w_1 = w_c - \dfrac{m_2\gamma}{m_1 + m_2}, & w_2 = w_c - \dfrac{m_1\gamma}{m_1 + m_2} \end{array}\right\} \qquad (135)$$

Substituting eq(135) into eq(132) and after some rearrangements, eq(136) is obtained as the expression of the kinetic energy ε.

$$\begin{aligned}\varepsilon &= \frac{1}{2}(m_1 + m_2)\times(u_c^2 + v_c^2 + w_c^2) + \frac{1}{2}\left(\frac{m_1m_2}{m_1 + m_2}\right)(\alpha^2 + \beta^2 + \gamma^2) \\ &= \frac{1}{2}(m_1 + m_2)\times c^2 + \frac{1}{2}\left(\frac{m_1m_2}{m_1 + m_2}\right)\chi^2 \qquad (136)\end{aligned}$$

Here, $u_c^2 + v_c^2 + w_c$ and $\alpha^2 + \beta^2 + \gamma^2$ are replaced by c^2 and χ^2, respectively.

It should be noticed that c and χ are independent to each other. When we look at eq(136), therefore, it is known that the velocity distribution of the two different molecules is divided into two parts, one is the velocity distribution of the center of gravity and the other is the distribution of the relative velocity (see **"note"**) and each distribution is governed independently by the *Maxwell-Boltzmann* distribution law. Consequently, the relative velocity distribution is derived just by replacing the mass m which appears in the *Maxwell-Boltzmann* velocity distribution function of one kind of free molecule with the reduced mass of $\mu = m_1 m_2/(m_1 + m_2)$. Accordingly, the mean relative velocity is obtained by replacing the mass which appears in $\bar{c}$ with the reduced mass μ.

Therefore, in the case of the molecules of the same kind, the relative velocity is obtained as eq(137) by equating $m_1 = m_2 \equiv m$.

$$\bar{c}_{rel} = \sqrt{\frac{8k_B T}{\pi\mu}} = \sqrt{\frac{8k_B T}{\pi \times \frac{1}{2}m}} = \sqrt{2}\sqrt{\frac{8k_B T}{\pi m}} = \sqrt{2} \times \bar{c} \qquad (137)$$

Ans. See text and eq(137).

[advanced problem] Confirm that the mean kinetic energy given by eq(136) is $2 \times 3k_B T/2$.

Solution 2-33

(a) The three-dimensional velocity distribution function is given by eq(138).

$$p(c) = 4\pi \left(\frac{m}{2\pi k_B T}\right)^{3/2} \exp\left(-\frac{mc^2}{2k_B T}\right) c^2 \qquad (138)$$

The velocity at which eq(138) reaches a maximum is found by equating the first derivative of $p(c)$ to zero as is given by eq(139) and solving it for c.

$$\frac{dp(c)}{dc} = 4\pi \left(\frac{m}{2\pi k_B T}\right)^{3/2} \left(2c - \frac{mc}{k_B T} \times c^2\right) \exp\left(-\frac{mc^2}{2k_B T}\right) c^2 = 0 \Rightarrow c_{\max} = \sqrt{\frac{2k_B T}{m}} \qquad (139)$$

Substituting the given values into eq(139), the velocity at the maximum distribution c_{max} is found as below.

$$c_{\max} = \sqrt{\frac{2 \times 1.38 \times 10^{-23} \times 300}{32.0 \times 10^{-3}/6.02 \times 10^{23}}} = 395 \text{ m} \cdot \text{s}^{-1}$$

Ans. velocity at the maximum distribution $c_{max} : \sqrt{2k_B T/m}$, $395 \text{ m} \cdot \text{s}^{-1}$

(b) The mean velocity $\bar{c}$ is defined by eq(140).

$$\bar{c} = 4\pi \left(\frac{m}{2\pi k_B T}\right)^{3/2} \int_0^\infty c \times \exp\left(-\frac{mc^2}{2k_B T}\right) c^2 dc \qquad (140)$$

The eq(140) is easily integrated by means of the formula (see **II-4**). However, the integration is carried out here directly by using the integration by part, since it serves as a purpose of practicing integration as well as proving the formula.

Since eq(140) can be rewritten as eq(140'), if you carry out the integration by part twice carefully as is shown below, then you will finally be able to derive eq(140'').

$$\bar{c} = 4\pi \left(\frac{m}{2\pi k_B T}\right)^{3/2} \int_0^\infty -\frac{k_B T}{m} \left\{\exp\left(-\frac{mc^2}{2k_B T}\right)\right\}' c^2 dc \qquad (140')$$

$$= 4\pi\left(\frac{m}{2\pi k_BT}\right)^{3/2}\left\langle -\frac{k_BT}{m}\left\{\left[\exp\left(-\frac{mc^2}{2k_BT}\right)c^2\right]_0^\infty - 2\int_0^\infty \exp\left(-\frac{mc^2}{2k_BT}\right)cdc\right\}\right\rangle$$

$$= 4\pi\left(\frac{m}{2\pi k_BT}\right)^{3/2}\left\langle \frac{2k_BT}{m}\int_0^\infty \exp\left(-\frac{mc^2}{2k_BT}\right)cdc\right\rangle$$

$$= 4\pi\left(\frac{m}{2\pi k_BT}\right)^{3/2}\left\langle \frac{2k_BT}{m}\times\left(-\frac{k_BT}{m}\right)\int_0^\infty \left\{\exp\left(-\frac{mc^2}{2k_BT}\right)\right\}' dc\right\rangle$$

$$= 4\pi\left(\frac{m}{2\pi k_BT}\right)^{3/2}\left\langle \frac{2k_BT}{m}\times\left(-\frac{k_BT}{m}\right)\left[\exp\left(-\frac{mc^2}{2k_BT}\right)\right]_0^\infty\right\rangle$$

$$= 4\pi\left(\frac{m}{2\pi k_BT}\right)^{3/2}\frac{2k_BT}{m}\times\frac{k_BT}{m}$$

$$= \left(\frac{8k_BT}{\pi m}\right)^{1/2} \qquad (140'')$$

Substituting the given values into eq(140''), the mean velocity is found as below.

$$\bar{c} = \left(\frac{8k_BT}{\pi m}\right)^{1/2} = \left(\frac{8\times1.38\times10^{-23}\times300}{3.14\times32.0\times10^{-3}/6.02\times10^{23}}\right)^{1/2} = 445\ \text{m}\cdot\text{s}^{-1}$$

Ans. $\bar{c} = \left(\frac{8k_BT}{\pi m}\right)^{1/2}$, $445\ \text{m}\cdot\text{s}^{-1}$

(c) The mean square velocity $\overline{c^2}$ is defined by eq(141).

$$\overline{c^2} = 4\pi\left(\frac{m}{2\pi k_BT}\right)^{3/2}\int_0^\infty c^2\times\exp\left(-\frac{mc^2}{2k_BT}\right)c^2dc \qquad (141)$$

Again, eq(141) is readily integrated relying on the formula (see **II-4**). However, following the procedure described in (b) above, eq(141) is directly integrated based on the integ- ration by part.

Eq(141) can be rewritten as eq(141') and then it is integrated by part carefully to yield eq(141''). The underlined part is the integral of the distribution function (eq(138)) and is normalized to unity.

$$\overline{c^2} = 4\pi\left(\frac{m}{2\pi k_BT}\right)^{3/2}\int_0^\infty -\frac{k_BT}{m}\left\{\exp\left(-\frac{mc^2}{2k_BT}\right)\right\}' c^3dc \qquad (141')$$

$$= 4\pi\left(\frac{m}{2\pi k_BT}\right)^{3/2}\left\langle -\frac{k_BT}{m}\left\{\left[\exp\left(-\frac{mc^2}{2k_BT}\right)c^3\right]_0^\infty - 3\int_0^\infty \exp\left(-\frac{mc^2}{2k_BT}\right)c^2dc\right\}\right\rangle$$

$$= 4\pi\left(\frac{m}{2\pi k_BT}\right)^{3/2}\left\langle \frac{3k_BT}{m}\int_0^\infty \exp\left(-\frac{mc^2}{2k_BT}\right)c^2dc\right\rangle$$

$$= \frac{3k_BT}{m}\times\underline{4\pi\left(\frac{m}{2\pi k_BT}\right)^{3/2}\int_0^\infty \exp\left(-\frac{mc^2}{2k_BT}\right)c^2dc}$$

$$= \frac{3k_BT}{m}\times1 \qquad (141'')$$

From this, the value for the square root of the mean square velocity $\sqrt{\overline{c^2}}$ is found as below.

$$\sqrt{\overline{c^2}} = \sqrt{\frac{3k_BT}{m}} = \sqrt{\frac{3\times1.38\times10^{-23}\times300}{32.0\times10^{-3}/6.02\times10^{23}}} = 483\ \mathrm{m\cdot s^{-1}}$$

The increasing order of the three specific velocities obtained in (a), (b) and (c) is $c_{max} < \bar{c} < \sqrt{\overline{c^2}}$ (see Fig.15 below).

Ans. $\sqrt{\overline{c^2}} = \sqrt{3k_BT/m}$, $483\ \mathrm{m\cdot s^{-1}}$

(d) The total number of the molecules, the number of the molecules in the velocity range required and its fraction are denoted as N_{total}, ΔN and $\Delta N/N_{total}$, respectively. Then the fraction to the total number of the molecules is given by integrating the distribution function over the required velocity range as shown below.

$$\frac{\Delta N(390-400)}{N_{total}} = 4\pi\left(\frac{m}{2\pi k_BT}\right)^{3/2}\int_{390}^{400}\exp\left(-\frac{mc^2}{2k_BT}\right)c^2dc = 0.021$$

Calculation like this is easily performed by using such software as *Mathematica*. If not, the integration over a small range can be approximated by the area of a relevant trapezoid or a rectangle as is explained below.

Ans. fraction: 0.021

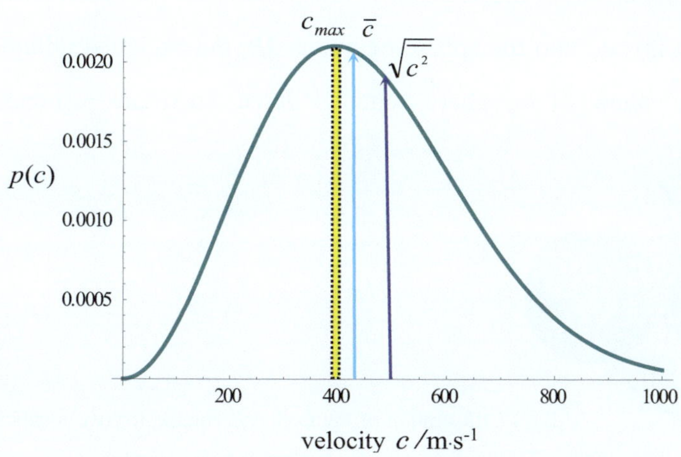

Fig.16 Three-dimensional velocity distribution function at 300 K.

For the present problem, as is seen from the distribution function shown on the left, the median of the range [390, 400] 395 $\mathrm{m\cdot s^{-1}}$ is close to the velocity at the maximum c_{max}. Therefore, it is sufficient to approximate $\Delta N/N_{total}$ by $p(395)\times10 = 0.0021\times10 = 0.021$.

Solution 2-34

Since the number of molecules leaking out per unit time is equal to the number of molecules lost per unit time from the vessel, the latter is expressed by $-dN(t)/dt$ in which the number of the molecules in the vessel at time t is denoted $N(t)$. By equating this number to the number of molecules which collide with the hole of the area S (see **problem 2-27**), eq(142) is derived.

$$-\frac{dN(t)}{dt} = \frac{1}{4}\bar{c}\times S\times\frac{N(t)}{V} \qquad (142)$$

On the other hand, eq(142) can be rewritten in terms of the time variation of pressure $P(t)$ by using the ideal gas law $PV = Nk_BT$ as eq(143).

$$-\frac{V}{k_BT}\frac{dP(t)}{dt} = \frac{1}{4}\bar{c}\times S\times\frac{P(t)}{k_BT} \Rightarrow -\frac{dP(t)}{dt} = \frac{1}{4}\bar{c}\times S\times\frac{P(t)}{V} \qquad (143)$$

Eq(143) is regarded as the first order rate equation with respect to $P(t)$ and by denoting its proportional constant as k_{eff}, its value is obtained as below (see **solution 2-30** for the value for $\bar{c}$).

$$k_{eff} = \frac{1}{4}\bar{c} \times S \times \frac{1}{V} = \frac{445 \times (10^{-3})^2}{4 \times 10^{-2}} = 1.1 \times 10^{-2}\,\mathrm{s}^{-1}$$

Then, the time required for the initial pressure to drop to one half of it is found as below.

$$\frac{\ln 2}{k_{eff}} = \frac{0.692}{1.1 \times 10^{-2}} = 63\,\mathrm{s}$$

Also, since the value for k_{eff} is proportional to S and is reduced to the same extent when the area of the hole is reduced to one tenth, the time to elapse in this case before the pressure reaches one half of the initial pressure needs to be multiplied by ten.

Ans. time required for the pressure to be halved: 63 s, 630 s

"note" It is assumed that no gas outside flows into the vessel. The leaking process is considered to be the first order process, so that the time required for any change in pressure by a constant fraction does not depend on the initial pressure. See **problems 2-64**, **2-65** and **2-66**.

Solution 2-35

The figure below illustrates the two (spherical) molecules of the same kind being in contact with each other. Since the center of one of the molecules cannot invade into the sphere of radius $2R$, the excluded volume per a pair of molecules is regarded as equal to the volume of the sphere etched by green. Therefore, just one

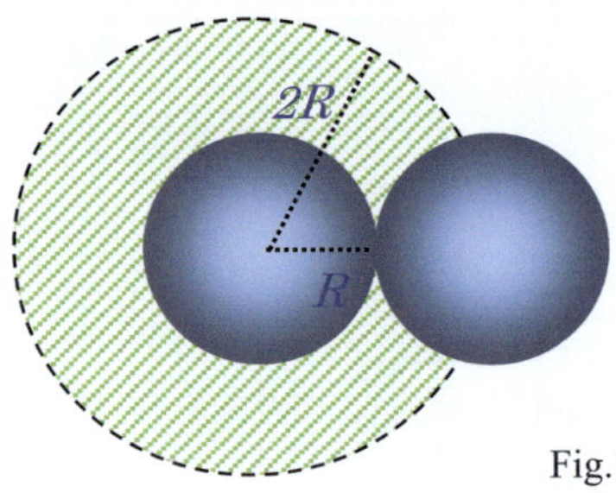

Fig.17 Illustration of the excluded volume for two identical spherical molecules (Fig.7 in the problem).

half of this volume is equal to the excluded volume per molecule. Hence, the excluded volume per molecule υ_{excl} is expressed by the following equation.

$$\upsilon_{excl} = \frac{4}{3}\pi(2R)^3 \times \frac{1}{2} = \frac{16}{3}\pi R^3 = 4 \times \left(\frac{4}{3}\pi R^3\right) = 4 \times [\text{volume of one molecule}]$$

The parameter b in the *van der Waals* equation of state corresponds to the excluded volume per mole and can be expressed by $b = 4N_A \times (4\pi R^3/3)$ when the molecular radius is R.

Ans. excluded volume: $b = 4N_A \times \left(\frac{4}{3}\pi R^3\right)$

Solution 2-36

There are three different cases in which two molecules A with radius R_1 and B with radius R_2 are in contact with each other as shown below.

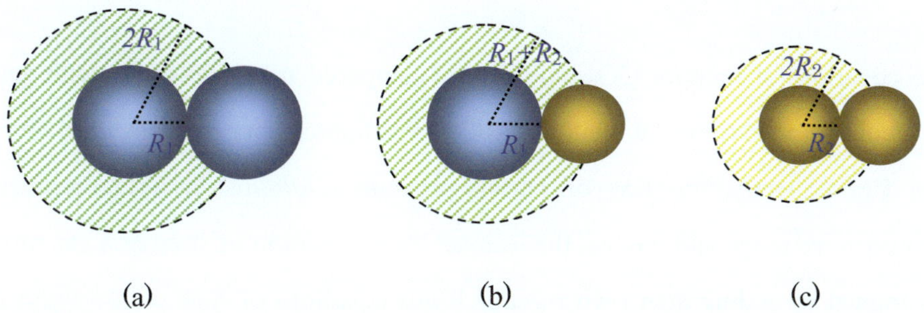

Fig.18 Three different cases of the exclusion volume in which two identical or different molecules are in contact with each other

As is discussed in the preceding problem,

(a) : the excluded volume per mole is $b_1 = 4N_A \times \left(\frac{4}{3}\pi R_1{}^3\right)$,

(c) : the excluded volume per mole is $b_2 = 4N_A \times \left(\frac{4}{3}\pi R_2{}^3\right)$.

In the case of (b), the excluded volume differs depending on to which molecule the central molecule refers to, A or B. When the numbers of A and B molecules are denoted N_1 and N_2, respectively, the number of molecular pairs is given by $N_1 \times N_2$. Taking it into account, the excluded volume with respect to A (molecule 1) is given by eq(144), which can be approximated by eq(145).

$$\upsilon_{12} = \frac{4}{3}\pi(R_1 + R_2)^3 \times \frac{N_1 \times N_2}{N_1} = \frac{4}{3}\pi(R_1 + R_2)^3 N_A \times \frac{N_2}{N_A}$$

$$= \frac{n_2}{4}\left(b_1 + b_2 + 3b_1\frac{R_2}{R_1} + 3b_2\frac{R_1}{R_2}\right) \qquad (144)$$

$$\approx \frac{n_2}{4}\{4b_1 + 4b_2 + 3(b_1 - b_2)x\} \qquad \left(\text{approximated by } \frac{R_1}{R_2} \approx 1 - x \text{ and } \frac{R_2}{R_1} \approx 1 + x\right)$$

$$\approx n_2(b_1 + b_2), \quad n_2 = N_2/N_A \qquad (145)$$

In a similar manner, the excluded volume with respect to B (molecule 2) is expressed as eq(146) and is approximated by eq(147).

$$\upsilon_{21} = \frac{4}{3}\pi(R_1 + R_2)^3 \times \frac{N_1 \times N_2}{N_2} = \frac{4}{3}\pi(R_1 + R_2)^3 N_A \times \frac{N_1}{N_A}$$

$$= \frac{n_1}{4}\left(b_1 + b_2 + 3b_1\frac{R_2}{R_1} + 3b_2\frac{R_1}{R_2}\right) \qquad (146)$$

$$\approx \frac{n_1}{4}\{4b_1 + 4b_2 + 3(b_1 - b_2)x\}$$

$$\approx n_1(b_1 + b_2), \quad n_1 = N_1/N_A \qquad (147)$$

Taking account of the three cases in which two molecules A and B are in contact, the parameter to be used for the *van der Waals* equation of state of a mixed gas is expressed by eq(148) for A and by eq(149) for B, in which n_1 and n_2 are the numbers of moles for A and B, respectively.

$$n_1 b_1 + n_2(b_1 + b_2) \qquad (148)$$

$$n_2 b_2 + n_1(b_1 + b_2) \qquad (149)$$

Ans. for A and B, $n_1 b_1 + n_2(b_1 + b_2)$ and $n_2 b_2 + n_1(b_1 + b_2)$, respectively

Solution 2-37

The *van der Waals* equations of state for CO_2 and NH_3 are compared at low pressures with that of an ideal gas in Fig.19. As is recognized from Fig.19, neither gas deviates appreciably from an ideal gas at pressures lower than 10 atm. Consequently, the law of partial pressure (*Dalton*'s law) is considered to hold approximately in that pressure range and, hence, the *van der Waals* equation of state of a gas mixture of CO_2 and NH_3 can be constructed by adding their own *van der Waals* equations of state with weights proportional to moles of each gas.

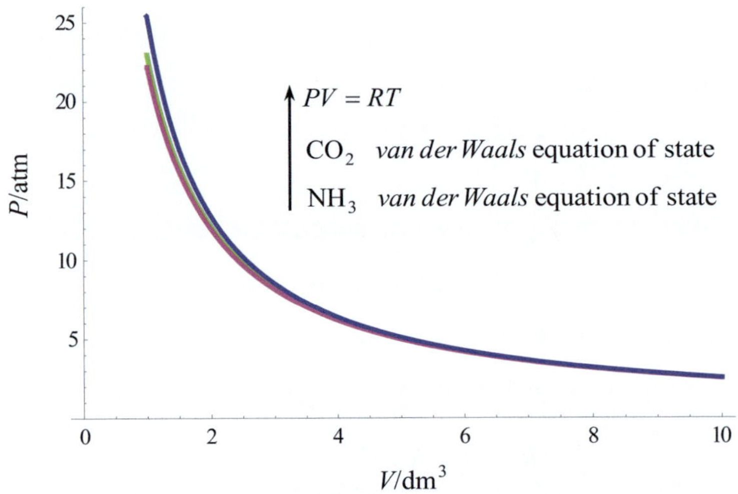

Fig.19 A comparison of the *van der Waals* equations of state with the ideal gas law at low pressures. $T = 310$ K.

When the pressure increases further, the deviation of the *van der Waals* gases from the ideal gas law becomes significant and the law of partial pressure comes to fail. In Figs.20 and 18, the *P-V* curves of a 1:1 gas mixture of CO_2 and NH_3 obtained for the following three cases are displayed: (i) treated as an ideal gas mixture, (ii) assuming the law of partial pressure, the *van der Waals* equation of state obtained just by adding the two *van der Waals* equations of state in proportion to the moles of each gas, and (iii) the *van der Waals* equation of state with the parameters a and b both of which are estimated for the gas mixture:

The parameters employed in (iii) are estimated for b as is discussed in the previous solution and for a by employing the geometrical mean since intermolecular force is proportional to the number of molecular pairs in the first approximation. Namely a's estimated for the molecules 1 and 2 are $n_1^2 a_1 + n_1 n_2 \sqrt{a_1 a_2}$ and $n_2^2 a_2 + n_1 n_2 \sqrt{a_1 a_2}$, respectively.

What is known from a comparison of the curves in Fig.20 is that the law of partial pressure does not hold for the *van der Waals* equations of state, since the molecular interaction as well as the excluded volume not only varies from molecule to molecule but also depends on the kind of molecular pairs, i.e., pairs of the same molecules or pairs of different molecules. Of course, at low pressures or at high temperatures where the molecular interactions become relatively less effective, the law of partial pressure can be considered to be effective.

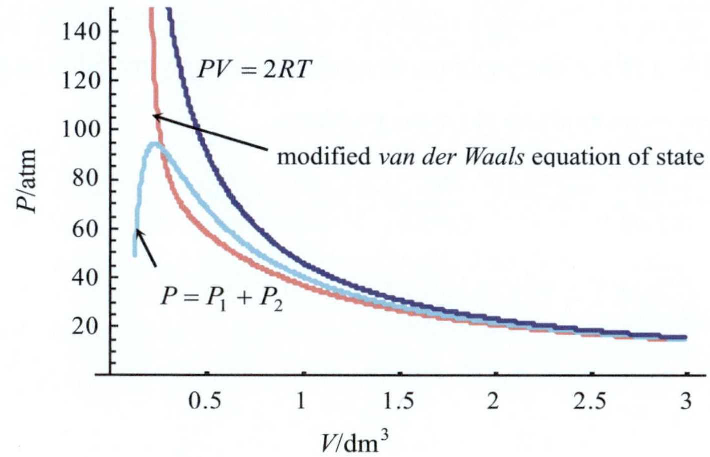

Fig.20 Equations of state for the equimolar gas mixture of CO_2 and NH_3. Comparison at 280 K.

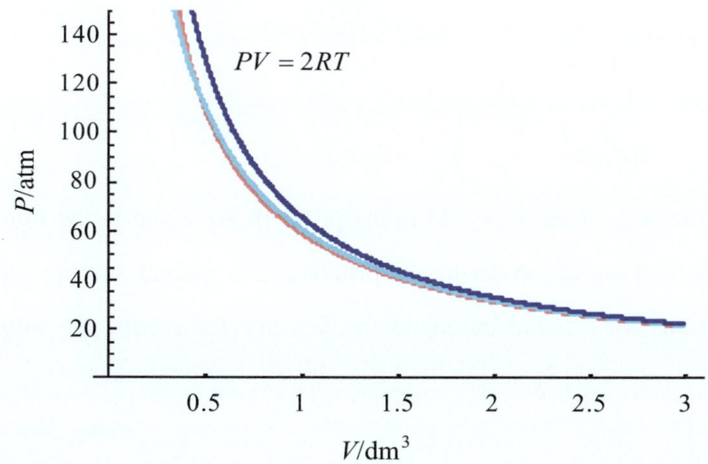

Fig.21 Equations of state for the equimolar gas mixture of CO_2 and NH_3. Comparison at 400K. See Fig.20 for the equations. At this temperature the law of partial pressure appears to be effective up to relatively high pressures.

It is of a great interest to know whether or not what is predicted in the solution is realized.

Ans. See text and Figs.19-21.

Solution 2-38

Employing the given values, the *van der Waals* equation of state and the ideal gas law both at 300 K are expressed by eq(150) and (151), respectively. Since a value of $0.0821\ \mathrm{dm^3 \cdot atm \cdot K^{-1} \cdot mol^{-1}}$ is used for the gas constant R, the volumes are obtained in the units of dm^3.

$$P = \frac{RT}{(V - 3.785 \times 10^{-2})} - \frac{4.248}{V^2} \Rightarrow 1 \text{ (or 5)} - \frac{0.082 \times 300}{(V - 3.785 \times 10^{-2})} + \frac{4.248}{V^2} = 0 \qquad (150)$$

$$V = \frac{RT}{P} = \frac{0.082 \times 300}{P} \qquad (151)$$

The volumes of the *van der Waals* gas at 1 and 5 atm are calculated to be 24.46 dm^3 and 4.78 dm^3, respectively and for the ideal gas, 24.60 dm^3 and 4.29dm^3 are obtained under the same conditions from the equations above. From these values it is found that the ammonia as the *van der Waals* gas occupies 99.4% at 1 atm and 97.2% at 5 atm of the volume the ideal gas exhibits under each condition.

Ans. 99.4% at 1 atm、97.2% at 5 atm

Solution 2-39a

(a) Whether or not eq(12) given in the problem is a correct solution is confirmed directly by substituting it into both sides of the one-dimensional diffusion equation (eq(11) in the problem).

$$[\text{left side}] = \frac{\partial}{\partial t}N(x,t) = -\frac{N_0}{4\sqrt{\pi D}}t^{-3/2}e^{-\frac{x^2}{4Dt}} + \frac{N_0}{2\sqrt{\pi D}}t^{-1/2}\times\frac{x^2}{4D}t^{-2}e^{-\frac{x^2}{4Dt}}$$
$$= \frac{N_0}{4\sqrt{\pi D}}t^{-3/2}e^{-\frac{x^2}{4Dt}}\left(-1+\frac{x^2}{2Dt}\right) \qquad (152)$$

$$D\frac{\partial}{\partial x}N(x,t) = \frac{D\times N_0}{2\sqrt{\pi D}}t^{-1/2}e^{-\frac{x^2}{4Dt}}\times\left(-\frac{x}{2Dt}\right) = -\frac{N_0}{4\sqrt{\pi D}}t^{-3/2}\cdot x\cdot e^{-\frac{x^2}{4Dt}}$$

$$[\text{right side}] = D\frac{\partial^2}{\partial x^2}N(x,t) = -\frac{N_0}{4\sqrt{\pi D}}t^{-3/2}\times e^{-\frac{x^2}{4Dt}} + \frac{N_0}{4\sqrt{\pi D}}t^{-3/2}\cdot x\times\frac{x}{2Dt}\times e^{-\frac{x^2}{4Dt}}$$
$$= \frac{N_0}{4\sqrt{\pi D}}t^{-3/2}e^{-\frac{x^2}{4Dt}}\left(-1+\frac{x^2}{2Dt}\right) \qquad (153)$$

From eqs(152) and (153), it is shown that the both sides of eq(11) in the problem are equal to the function given by eq(12). Therefore, it is confirmed that eq(12) given in the problem is a correct solution of the diffusion equation of eq(11). For the solution to be found based on the *Laplace* transform, see **solution 5-88b**.

Ans. See text.

(b) Since the distribution of distance x at time t is known, the mean square $\overline{x^2}$ of the distance x by which a molecule (or a particle) has migrated is given by eq(154).

$$\overline{x^2} = \frac{1}{N_0}\int_{-\infty}^{\infty} x^2\times\frac{N_0}{2\sqrt{\pi Dt}}e^{-\frac{x^2}{4Dt}}dx \qquad (154)$$

The integration of eq(154) is easily performed based on the formula (see **II-4-6**), yielding the result given by eq(155).

$$\overline{x^2} = \frac{1}{2\sqrt{\pi Dt}}\times 2Dt\sqrt{4\pi Dt} = 2Dt \ \Rightarrow\ \sqrt{\overline{x^2}} = \sqrt{2Dt} \qquad (155)$$

Ans. $\sqrt{\overline{x^2}}$: eq(155)

(c) Substituting the given values into eq(155), the following values for t are found for each D.

$$t = \frac{0.1^2}{2D} = \frac{0.01}{2\times10^{-5}} = 5\times10^2\ \text{s},\quad t = \frac{1.0^2}{2\times10^{-5}} = 5\times10^4\ \text{s};\quad t = \frac{0.1^2}{2\times10^{-9}} = 5\times10^6\ \text{s},\quad t = \frac{1.0^2}{2\times10^{-9}} = 5\times10^8\ \text{s}$$

Ans. $5\times10^2\,\text{s}$, $5\times10^4\,\text{s}$ for $D = 10^{-5}\,\text{m}^2\cdot\text{s}^{-1}$; $5\times10^6\,\text{s}$, $5\times10^8\,\text{s}$ for $D = 10^{-9}\,\text{m}^2\cdot\text{s}^{-1}$

A time-dependence of the one-dimensional molecular diffusion is illustrated in Fig 22 below.

"note" During the diffusional process, the diffusion coefficient D is assumed constant. This assumption indicates that the diffusion process is taking place in a homogeneous medium. Namely, except for the diffusing molecule A, the existence of B molecule as a medium is indispensable as far as the diffusion in vacuum is not in mind (see **"note"** in **solution 2-40**). For this, when the ratio of the component gases varies largely before and after the diffusion, such as eq(11) given in the problem cannot be employed.

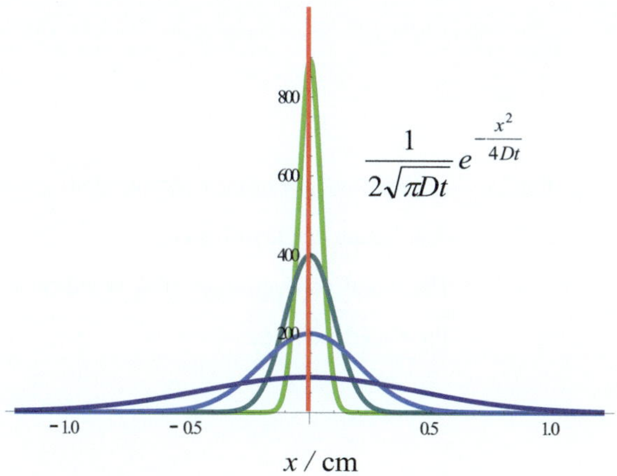

Fig.22 Time and distribution profile of one-dimensional molecular diffusion.
The plots are for $D = 10^{-5}\,\mathrm{m^2 \cdot s^{-1}}$. The vertical red line indicates the distribution at $t = 0\,\mathrm{s}$. From the next high curve to the bottom one, the distributions are at t = 0.01 s, 0.05 s, 0.2 s and 1.0 s, respectively The vertical scale is proportional to the number of molecules.

Solution 2-39b

A general solution of eq(156) under the condition of the initial concentration distribution $C_A(x,\ 0) \equiv C_A^0(x)$ at $t = 0$ is given by eq(157). Refer to *The Mathematics of Physics and Chemistry*, by *H.Margenau* and *G.M.Murphy,* D. van Nostrand Company, Inc., 1955, section 7.14 or a solution of partial differential equation by *Fourier* transformation given in **solution 5-88b** for the derivation of eq(157).

$$\frac{\partial C_A}{\partial t} = D_A\left(\frac{\partial^2 C_A}{\partial x^2}\right) \qquad (156)$$

$$C_A(x,\ t) = \frac{1}{2\sqrt{\pi D_A t}}\int_{-\infty}^{\infty} C_A^0(x')\cdot e^{-\frac{(x-x')^2}{4D_A t}}\,dx' \qquad (157)$$

It should be noted here that eq(157) is a convolution of$(1/2\sqrt{\pi D_A t})\cdot e^{-\frac{x^2}{4D_A t}}$, which is a special solution of eq(156) (the solution in the preceding problem), by the initial concentration distribution $C_A^0(x)$ (see **II-4 [Useful Mathematical Expressions]**).

The initial conditions given in the problem are expressed as eq(158).

$$\left.\begin{array}{ll} C_A^0(x) = 1 & x \le 0 \\ C_A^0(x) = 0 & x > 0 \end{array}\right\} \qquad (158)$$

Substituting these initial conditions into eq(157) and integrating it, eq(159) is obtained. Here, *erf* represents an error function.

$$\begin{aligned} C_A(x,\ t) &= \frac{1}{2\sqrt{\pi D_A t}}\int_{-\infty}^{0} e^{-\frac{(x-x')^2}{4D_A t}}\,dx' \qquad x' - x = z \ \Rightarrow\ dx' = dz \\ &= \frac{1}{2\sqrt{\pi D_A t}}\int_{-\infty}^{-x} e^{-\frac{z^2}{4D_A t}}\,dz \\ &= \frac{1}{2\sqrt{\pi D_A t}}\int_{-\infty}^{0} e^{-\frac{z^2}{4D_A t}}\,dz + \frac{1}{2\sqrt{\pi D_A t}}\int_{0}^{-x} e^{-\frac{z^2}{4D_A t}}\,dz \\ &= \frac{1}{2} - \frac{1}{2}erf\left(\frac{x}{2\sqrt{D_A t}}\right) \qquad (159) \end{aligned}$$

In Fig.23, eq(159) is plotted as a function of x at the times of 1.0, 4.0 and 20.0 after the start of diffusion.

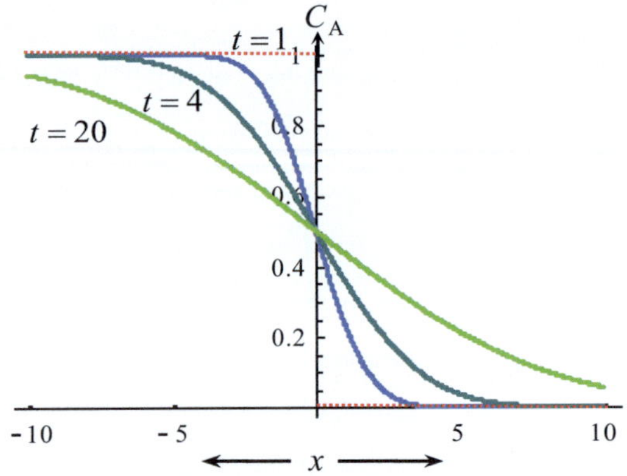

Fig.23 Time-dependence of the concentration distribution due to diffusion.
The initial concentration of A is indicated by the dotted red lines.

As to the diffusion of the other molecule B as an ideal gas, eq(160) must hold at any time during the process since the pressure and temperature of the mixed ideal gas should be kept constant. Consequently, the

$$C_A(x,\ t)+C_B(x,\ t)=\text{constant }(=1) \qquad (160)$$

concentration profile of B, $C_B(x,\ t)$, is given by inverting Fig.23 against the vertical axis. Thus its function is expressed by (159'). Naturally, the diffusion coefficient for each molecule must be equal, i.e., $D_A = D_B$ (see **"note"**).

$$C_B(x,\ t)=\frac{1}{2}+\frac{1}{2}erf\left(\frac{x}{2\sqrt{D_A t}}\right) \qquad (159')$$

Ans. eq(159), eq(159') and Fig.23

"note 1" Mutual diffusion coefficient

From a view point of molecular dynamics, it is natural to regard a diffusion coefficient as specific to a molecular species and different from molecule to molecule. Now, by assuming they are different and hence by putting $D_A \neq D_B$, the diffusion equation is solved for each of A and B independently and the results obtained at a certain time after the commencement of diffusion are plotted in Fig.24. The green line shows

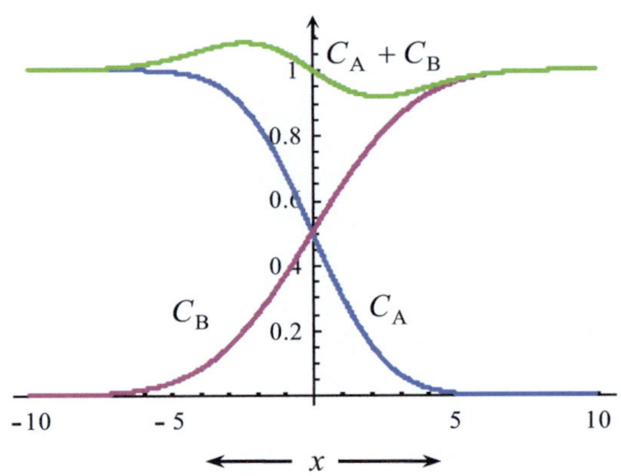

Fig.24 Concentration distributions of A and B in the case of $D_A/D_B = 3$.
Sum of A and B is shown by the green curve.

a sum of the concentrations of A and B, $C_A(x,\ t)+C_B(x,\ t)$. It is obvious from the curve that the total concentration, and hence the number of molecules, is not constant along x-axis. Namely, a difference in pressure arises everywhere.

If we consider that both of A and B molecules move in order to reduce this pressure difference, then, we are led naturally to the conclusion that apparent diffusion coefficients for A and B become identical, i.e., $D_A = D_B \equiv D_{AB}$. D_{AB} is called mutual diffusion coefficient (see **problem and solution 2-53**).

"note 2" A particular solution of eq(156) is *Dirac*'s delta function $\delta(x)$ in the limit $t \to 0$.

$$\delta(x) = \lim_{t\to 0} (1/2\sqrt{\pi D_A t}) \cdot e^{-\frac{x^2}{4D_A t}} \qquad (157')$$

Solution 2-40

(a) One step toward the right ($+x$) is denoted λ_+ and one step toward the left ($-x$) is denoted λ_-. After the time of $3\Delta t$ three steps have been made in total. Where and with what probability the molecule is to be found after the three steps can be answered by using the binomial expansion to the third power given by eq(161). For instance, the term $\lambda_+{}^2\lambda_-$ represents a walk of two steps to the right and one step to the left and

$$(\lambda_+ + \lambda_-)^3 = \lambda_+{}^3 + 3\lambda_+{}^2\lambda_- + 3\lambda_+\lambda_-{}^2 + \lambda_-{}^3 \qquad (161)$$

its coefficient 3 indicates that there are three cases of such a walk, i.e., (r., r., l.), (r., l., r.) and (l., r., r.). Since the total number of such cases is given by $2^3 = 8$, a probability for each case is found by dividing each coefficient of eq(161) by 8.

Ans. position and probability: $(3\lambda : 1/8)$, $(\lambda: 3/8)$, $(-\lambda: 3/8)$ and $(-3\lambda : 1/8)$

(b) Suppose that m steps to the right, $n-m$ steps to the left and hence n steps in total are made. As is readily deduced from the example discussed in (a), the position and its probability for the molecule to locate are found by the binomial expansion to the nth-power given by eq(162).

$$(\lambda_+ + \lambda_-)^n = \lambda_+{}^n + C_1^n\lambda_+{}^{n-1}\lambda_- + C_2^n\lambda_+{}^{n-2}\lambda_-{}^2 + \cdots + C_m^n\lambda_+{}^{n-m}\lambda_-{}^m + \cdots + \lambda_-{}^n \qquad (162)$$

The coefficients of the expanded equation are given by eq(163) and the positions for the molecule to be able to reach are by eq(164). Eq(163) represents the number of events in which the molecule reaches the relevant position.

$$C_m^n = \frac{n!}{m!(n-m)!} \qquad (163)$$

$$\lambda^{n-m}\lambda^n : \quad (n-m)\lambda - m\lambda = (n-2m)\lambda \qquad (164)$$

The total number of events is 2^n as is seen from eq(165).

$$\sum_{m=0}^{n} C_m^n = 1 + n + \cdots + \frac{n!}{m!(n-m)!} + \cdots + n + 1 = (1+1)^n = 2^n \qquad (165)$$

Consequently, the probability p of finding the molecule at the position specified by eq(164) is expressed by eq(166).

$$\frac{{}_nC_m}{2^n} = \frac{1}{2^n} \times \frac{n!}{m!(n-m)!} \equiv p \qquad (166)$$

Ans. probability and position: ${}_nC_m/2^n$ for $(n-2m)\lambda$, eq(166)

(c) We consider a case in which n is extremely large. Taking the logarithm of the probability p in order to use *Stirling*'s approximation, $\ln p$ an be approximated by eq(166').

$$\ln p = \ln n! - \ln m! - \ln(n-m)! - n\ln 2$$

$$= n\ln n - n - m\ln m + m - (n-m)\ln(n-m) + (n-m) - n\ln 2$$

$$= n\ln\frac{n}{n-m} - m\ln\frac{m}{n-m} - n\ln 2 \qquad (166')$$

A distance traveled by the molecule is denoted x (position) and then x is given by eq(167). Replacing $(n-2m)$ with s, m is expressed as eq(168).

$$x=(n-m)\lambda-m\lambda=(n-2m)\lambda \qquad (167)$$

$$n-2m=s \;\Rightarrow\; m=\frac{n-s}{2} \qquad (168)$$

Then, eq(166′) is rewritten as eq(169).

$$\begin{aligned}\ln p &= n\ln\frac{n}{n+s}-\frac{n-s}{2}\ln\frac{n-s}{n+s}\\ &= n\ln\left(\frac{1}{1+\frac{s}{n}}\right)-\frac{n-s}{2}\ln\left(\frac{1-\frac{s}{n}}{1+\frac{s}{n}}\right)\end{aligned} \qquad (169)$$

Using the approximation of $\ln(1\pm x)=\pm x-x^2/2$, eq(169) is expanded to give eq(169′).

$$\begin{aligned}\ln p &\approx -n\times\left(\frac{s}{n}-\frac{s^2}{2n^2}\right)-\frac{n-s}{2}\left(-\frac{s}{n}-\frac{s^2}{2n^2}-\frac{s}{n}+\frac{s^2}{2n^2}\right)\\ &= -\frac{s^2}{2n}\end{aligned} \qquad (169')$$

Therefore, the probability of finding the molecule at the position of x ($=s\lambda$) is proportional to eq(170).

$$p\propto e^{-\frac{s^2}{2n}}=e^{-\frac{x^2}{2n\lambda^2}} \qquad (170)$$

Using the relations of eqs(171) and (172), eq(170) is further rewritten as eq(173).

$$n=\frac{t}{\Delta t}=\frac{t\bar{u}}{\lambda} \;\Rightarrow\; n\lambda^2=t\bar{u}\lambda \qquad (171)$$

$$D=\frac{1}{2}\bar{u}\lambda \qquad (172)$$

$$p\equiv p(x)=Ae^{-\frac{x^2}{4Dt}}=\frac{1}{2\sqrt{\pi Dt}}e^{-\frac{x^2}{4Dt}} \qquad (173) \qquad \left(\because \int_{-\infty}^{\infty}p(x)dx=1\right)$$

It should be noticed that the functional form of eq(173) is the same as that of the solution of the one-dimensional diffusion equation. For this it is important to realize that microscopically speaking molecules are not driven by a concentration gradient but that as a consequence of free molecular movements (random walk), a macroscopic concentration distribution given by eq(173) or a concentration gradient as is shown in solution **2-39a,b** arises.

Ans. When the number of steps (random walk) n is extremely large.

"remark" A magnitude of one step of the random walk (λ) is assumed to be constant here all through the random process. This assumption demands the random walk to be in a homogeneous media. We should be cautious, therefore, in treating random walk or diffusion when a nature of the medium varies significantly due to mass transfer.

Solution 2-41

The diffusion coefficient of a gas is given by eq(174) using the mean velocity $\bar{c}$ and the mean free path λ. This equation is obtained by multiplying the equation derived from a simple molecular dynamic of a gas by 2/3 (see **solution 2-44**).

$$D=\frac{1}{3}\bar{c}\lambda=\frac{1}{3}\left(\frac{8RT}{\pi M}\right)^{1/2}\left(\frac{RT}{PN_A}\right)\cdot\frac{1}{\sqrt{2}\sigma} \qquad (174)$$

The mean velocity and mean free path are given by the following equations. The suffix 1 refers to oxygen.

$$\bar{c}_1=\left(\frac{8RT}{\pi M_1}\right)^{1/2}, \qquad \lambda_1=\frac{\bar{c}_1}{z_{11}}=\frac{\bar{c}_1}{(N_1/V)\sqrt{2}\sigma\cdot\bar{c}_1}=\left(\frac{RT}{P_1N_A}\right)\cdot\frac{1}{\sqrt{2}\sigma}$$

The diffusion coefficient D at 1 atm is calculated as below from these values.

$$D=\frac{1}{3}\left(\frac{8\times8.31\times298}{3.14\times32\times10^{-3}}\right)^{1/2}\left(\frac{8.31\times298}{1.01\times10^{5}\times6.02\times10^{23}}\right)\cdot\frac{1}{\sqrt{2}\times0.40\times10^{-18}}=1.07\times10^{-5}\text{m}^2\text{s}^{-1}$$

At 10 atm, the value for D is just one tenth of that at 1 atm, i.e., $1.07\times10^{-6}\text{m}^2\cdot\text{s}^{-1}$.

As is seen from eq(174), the diffusion coefficient of a gas is inversely proportional to the pressure. Representable values of gases at 1 atm are of the order of $10^{-5}\text{m}^2\cdot\text{s}^{-1}$.

Ans. at 1 atm: $1.07\times10^{-5}\text{m}^2\cdot\text{s}^{-1}$, at 10 atm: $1.07\times10^{-6}\text{m}^2\cdot\text{s}^{-1}$

Solution 2-42

If we assume that the diffusion coefficient of aroma molecules in air at 1 atm is of the same magnitude as those for pure oxygen or pure nitrogen at 1 atm and, hence, is of the order of $D=10^{-5}\text{m}^2\text{s}^{-1}$, then we are able to know from the calculation given in **solution 2-39a** that it will take about 500 s for the aroma molecules to reach a place 0.1 m away from the initial place on the average. If this is the case, the situation is not consistent with our daily experience that we can smell an odorous stuff quite far away from it. Therefore, we assume that we can smell it if even a few parts of million of aroma molecules reach us and calculate the distance that fraction of molecules reaches us after 10 s.

We substitute 10 s for t in eq(175) which gives the concentration distribution at time t and x, from which the distance at which eq(175) takes a value of 10^{-6} is found as 8.3×10^{-2} m.

$$\frac{1}{2\sqrt{\pi Dt}}e^{-\frac{x^2}{4Dt}} \qquad (175)$$

$$\frac{1}{2\sqrt{\pi\times10^{-5}\times10}}e^{-\frac{x^2}{4\times10^{-5}\times10}}=10^{-6} \Rightarrow -x^2=4\times10^{-5}\times10\times\ln\left\langle(10^{-4}\times4\pi)^{1/2}\times10^{-6}\right\rangle$$

$$x^2=68.62\times10^{-4} \Rightarrow x=8.3\times10^{-2}\text{m}$$

From this result, it is concluded that even with a very sensitive detector, to detect molecules transported by diffusion alone quickly at a distant place is difficult. When we smell an odorous stuff, we wave air to our noses or place our noses close to the stuff. A slight fragrance is brought to a nose maybe by convection but not by diffusion. If we are able to detect a single aroma molecule with a sophisticated apparatus, then it would be interesting to know how far away we can detect it after 10 s which is only diffusing to us in the air.

Ans. Brought to us not by diffusion but by convection or with the wind.

Solution 2-43

(a) From the known speed of the rotation of the discs, the time required for the discs to rotate 30 degrees is calculated to be $t_{30^\circ} = \frac{1}{300} \times \frac{30}{360} = 1/(3.6 \times 10^4)\,\text{s}$. During this period, the molecules (atoms) which pass through the selector can move $1.0\,\text{cm}$. Consequently the velocity of the molecules is calculated to be

$$\upsilon_{300} = 0.01/t_{30^\circ} = 0.01/1/3.6 \times 10^4 = 360\ \text{m} \cdot \text{s}^{-1}$$

Ans. the velocity of the molecules to be detected: $360\ \text{m} \cdot \text{s}^{-1}$

(b) Similarly calculated as in (a), the velocity required is found to be $\upsilon_{400} = 360\ \text{m} \cdot \text{s}^{-1} \times 400/300 = 480$ $\text{m} \cdot \text{s}^{-1}$. A ratio of the distributions for two fixed velocities is given by the three-dimensional velocity distribution function as is shown in the following equation. By putting the temperature of the beam source as 350 K and assuming the beam is composed of K atoms (M=39.1), the ratio is found to be 0.902 as is calculated below. It should be noted here that the distribution function to be employed is not one-dimensional but three-dimensional one (see **problem 2-33**). Why is that?

$$\upsilon_{400}^2 \exp\left(-\frac{m\upsilon_{400}^2}{2k_BT}\right) \Big/ \upsilon_{300}^2 \exp\left(-\frac{m\upsilon_{300}^2}{2k_BT}\right) = 480^2 \exp\left(-\frac{39.1 \times 10^{-3} \times 480^2}{2 \times 8.31 \times 350}\right) \Big/ 360^2 \exp\left(-\frac{39.1 \times 10^{-3} \times 360^2}{2 \times 8.31 \times 350}\right)$$
$$= 0.902$$

Ans. amount to be detected: 0.902 times

Solution 2-44

The diffusion, thermal conduction and viscosity of gases are the phenomena of transporting substances (N), kinetic energies ($1/2\,m\upsilon^2$) and momenta ($m\upsilon$). These are treated under the same framework because it is basically migration of molecules that is responsible for those transports. After introducing a general theory of transport phenomena, the problem associated with the thermal conduction will be discussed.

[Molecular dynamics and transport phenomena of gases]

A quantity to be transported is denoted Γ and it is assumed that Γ is dependent only on the x-direction and homogeneous along the y- and z-directions. For that, the transport of Γ occurs along x-direction alone. Now, let us consider the quantity Γ transported by molecular translation across a cross section O at the position of x_0 and denote its amount J_+. To make the situation simpler, we assume that (i) all molecules are moving with the mean velocity of $\bar{u}$ and (ii) within the distance shorter than the mean free path, no molecules collide each other and, hence, Γ varies only when x changes in the unit of λ. With these assumptions, the quantity to be transported per unit time is calculated as below:

Γ toward the right direction in Fig.22 is given by eq(176) (denoted J_+),

$$J_+ = \left\{\Gamma - \lambda\left(\frac{\partial \Gamma}{\partial x}\right)_{x=x_0}\right\} \times \frac{1}{2}\left(\frac{N}{V}\right) \times \bar{u} \times \text{O} \qquad (176)$$

Γ toward the left direction in Fig.25 is given by eq(177) (denoted J_-),

$$J_- = \left\{\Gamma + \lambda\left(\frac{\partial \Gamma}{\partial x}\right)_{x=x_0}\right\} \times \frac{1}{2}\left(\frac{N}{V}\right) \times \bar{u} \times \text{O} \qquad (177)$$

Consequently, the net quantity, denoted J_{net} (called flux hereafter), transported across O is given by eq(178).

$$J_{net} = J_+ - J_- = -\lambda \left(\frac{\partial \Gamma}{\partial x} \right)_{x=x_0} \times \left(\frac{N}{V} \right) \times \bar{u} \times \mathrm{O} \qquad (178)$$

Eq(178) is expressed as eq(179) with α as a proportional constant and is called *Fick*'s first law.

$$J_{net} / \mathrm{O} = -\alpha \left(\frac{\partial \Gamma}{\partial x} \right)_{x=x_0} \qquad (179)$$

When Γ is energy, the transport phenomenon is observed as a thermal conduction. Denoting the energy of one molecule by ε and putting $\Gamma = \varepsilon$ in eq(178), the flux at $x = x_0$ is given by eq(180).

$$J(x_0) = -\mathrm{O}\bar{u} \left(\frac{N}{V} \right) \lambda \left(\frac{\partial \varepsilon}{\partial x} \right)_{x=x_0} \qquad (180)$$

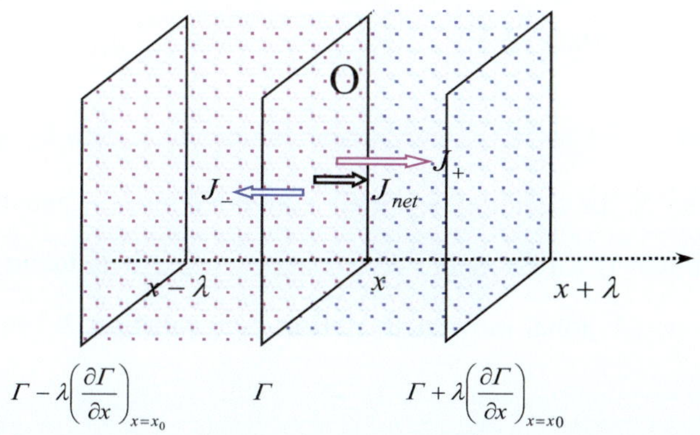

Fig.25 Illustration of transport phenomena due to substance migration.
All molecules are assumed to move with the velocity of either $\bar{u}$ or $-\bar{u}$. That is why a factor of 1/2 appears in eqs(176) and (177).

Since the energy gradient can be expressed in terms of the temperature gradient as in eq(181),

$$\left(\frac{\partial \varepsilon}{\partial x} \right) = \left(\frac{\partial \varepsilon}{\partial T} \right) \left(\frac{\partial T}{\partial x} \right) = \frac{C_V}{N_A} \left(\frac{\partial T}{\partial x} \right) \qquad (181)$$

eq(180) is rewritten as eq(180') by using eq(181).

$$J(x_0) = -\mathrm{O}\bar{u} \left(\frac{N}{V} \right) \lambda \frac{C_V}{N_A} \left(\frac{\partial T}{\partial x} \right)_{x=x_0} \qquad (180')$$

When eq(180') is compared with eq(182) which is derived directly from *Fick*'s first law for the thermal conduction, the coefficient κ is found to be expressed in terms of properties specific to the gas molecule.

$$J_{net}(\text{energy}) / \mathrm{O} = -\kappa \left(\frac{\partial T}{\partial x} \right) \qquad (182)$$

$$\kappa = \bar{u} \lambda \frac{N}{V} \frac{C_V}{N_A} \qquad (183)$$

Now, a variation of temperature brought about by the migration of energy is examined. As is seen from Fig.26, the energy stored in a cuboid of a volume of $\mathrm{O}dx$ with the surface O and a height dx is given by the difference of two fluxes derived from eq(180') as is expressed by eq(184).

$$J(x_0) - J(x_0 + dx) = \mathrm{O}\bar{u}\lambda \frac{N}{V} \frac{C_V}{N_A} \left\{ \left(\frac{\partial T}{\partial x} \right)_{x=x_0+dx} - \left(\frac{\partial T}{\partial x} \right)_{x=x_0} \right\}$$

$$= \mathrm{O}\bar{u}\lambda \frac{N}{V} \frac{C_V}{N_A} \left(\frac{\partial^2 T}{\partial x^2} \right)_{x=x_0} dx \qquad (184)$$

This energy stored is converted into a change in the energy per molecule (underlined part in blue) and then by expressing it in terms of the heat capacity C_V, a change in temperature per unit time (underlined part in gray) is expressed by eq(185).

$$\frac{J(x_0)-J(x_0+dx)}{\mathrm{O}dx}=\frac{\{J(x_0)-J(x_0+dx)\}}{\mathrm{O}dx\times(N/V)}\times(N/V)=\frac{N}{V}\frac{C_V}{N_A}\left(\frac{\partial T}{\partial t}\right)_{x=x_0} \qquad (185)$$

By equating eq(184) with eq(185), eqs(186) and (187) are finally obtained as the equation of thermal conduction of gases, in which the relation $\bar{u}=\bar{c}/2$ is used.

$$\frac{\partial T}{\partial t}=\bar{u}\lambda\left(\frac{\partial^2 T}{\partial x^2}\right)=\frac{\kappa}{(NC_V/N_AV)}\left(\frac{\partial^2 T}{\partial x^2}\right)$$
$$=\frac{1}{2}\bar{c}\lambda\left(\frac{\partial^2 T}{\partial x^2}\right)=D\left(\frac{\partial^2 T}{\partial x^2}\right) \qquad (186)$$

$$D=\frac{1}{2}\bar{c}\lambda \qquad (187)$$

In the equation thus obtained, when *T* is replaced by *N*, the equation is exactly converted into the equation of diffusion. This may be an unexpected result, but it is a reasonable consequence when it is taken into account that the energy migration is caused by the translational molecular motion. See **solution 2-39a** and **2-45** for the solution of eq(186).

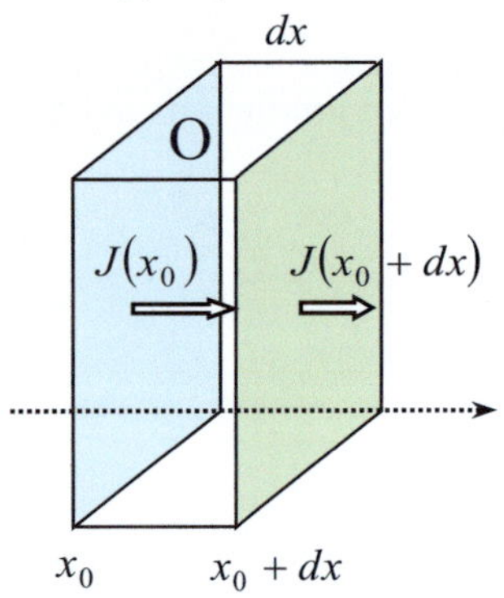

Fig.26 Energy accumulation and temperature change due to energy transport.

Ans. See text and eq(186) for proof.

"note 1" The reason why a minus sign is added to eq(179) is to make a proportional constant α positive as is recognized from the relation between the direction of the flow (positive) and the gradient (negative).

"note 2" In order to take the velocity distribution into account, $\bar{u}$ in eqs(176) and (177) must be replaced by the integral over velocity. However, when the integration yields the mean velocity, the same conclusion is obtained even if a relevant quantity is assumed to be transported with the mean velocity.

Denoting the velocity distribution function $p(u)$, the following relations are used.

$$\int_0^\infty up(u)du=\frac{1}{2}\bar{u}=\frac{1}{4}\bar{c}, \qquad \left(\because\int_{-\infty}^{\infty}p(u)du=1\right)$$

"note 3" As is clear from the procedure used to derive eqs(176) and (177), the concentration (N/V) of a gas in which thermal conduction occurs is assumed constant throughout the gas. This is also evident from the assumption that the mean free path is constant.

Solution 2-45

A general solution to the one-dimensional equation of thermal conduction eq(188) is given by eq(189) in which the temperature at the time t and at the position x is denoted $T(x,\ t)$ (see **solution 2-39b**).

$$\frac{\partial T(x,\ t)}{\partial t}=D\frac{\partial^2 T(x,\ t)}{\partial x^2} \qquad (188)$$

$$T(x,\ t)=A+\frac{B}{2\sqrt{\pi Dt}}\int_{-\infty}^{\infty}T^0(x')\cdot e^{-\frac{(x-x')^2}{4Dt}}dx' \qquad (189)$$

Here $T^0(x')$ is the temperature of the heat reservoir given by the initial conditions for the reservoir at $t=0$ and, in the present case, it is given by eq(190).

$$\text{for } x'\le 0,\quad T^0(x')=500\text{ K}$$
$$\text{for } x'>0,\quad T^0(x')=0\text{ K} \qquad (190)$$

Constants A and B are also determined by the initial conditions and they are determined so as to satisfy eq(191).

$$\text{for } x=0,\quad T(0.\ t)=500\text{ K}$$
$$\text{for } x>>1,\quad T(x,\ t)=300\text{ K} \qquad (191)$$

The solution satisfying these conditions is given by eq(192).

$$T(x,\ t)=300+\frac{225}{2\sqrt{\pi Dt}}\int_{-a}^{0}e^{-\frac{(x-x')^2}{4Dt}}dx' \qquad (192)$$

In eq(192) the range of integration is taken as $[-a,\ 0]$ and if a takes a large value, then it corresponds to the reservoir of infinite heat capacity.

Substituting $1.0\times10^{-5}\text{m}^2\text{s}^{-1}$ for D into eq(192), the temperature profiles at some times after the contact are obtained as shown in Fig.27. The vertical scale is the temperature and the horizontal scale represents the distance (m) from the contact surface of the reservoir. The variation of the temperature of the gas at the distance of 0.1 m from the surface is shown in Fig.28 as a function of time.

From Figs.27 and 28, it is understood that it takes quite a long time for the gas to reach higher temperature

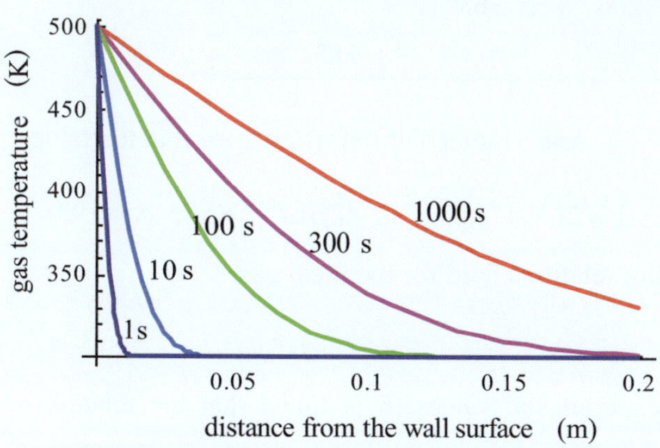

Fig.27 Temperature gradients in the gas after several seconds.

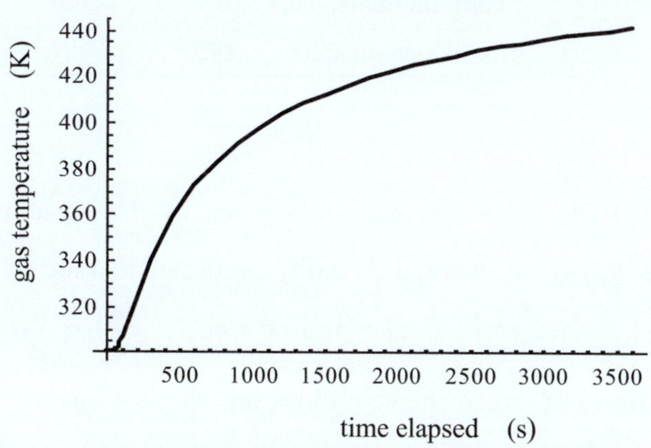

Fig.28 Temperature variation at the distance of 0.1m.

against our expectation. In practice, therefore, it is necessary to devise a way to achieve the thermal equilibrium at 500 K faster by stirring the gas or by using radiant heat. A large number of problems in thermodynamic are connected with the processes by which the temperature of a gas as a working substance is changed in many ways. We now know from these results that any experiment associated with temperature variation requires various ideas to achieve thermal equilibrium efficiently.

Ans. See text and Figs.27 and 28.

Solution 2-46

The theory of molecular dynamics of gases gives the viscosity coefficient η ($\mathrm{kg \cdot m^{-1} \cdot s^{-1} \times 10^6}$), diffusion coefficient D ($\mathrm{m^2 \cdot s^{-1}}$) and coefficient of thermal conduction κ ($\mathrm{J \cdot K^{-1} \cdot m^{-1} \cdot s^{-1} \times 10^3}$) by eqs(193) to (195), respectively. (a factor of $2/3$ is not multiplied. See **"note".**) The mean velocity, mean free path and molecular diameter are denoted by $\bar{c}$, λ and *d*, respectively. Also, the density is given by $\rho / \mathrm{kg \cdot m^{-3}}$ and C_V is the specific heat at constant volume.

$$D = \frac{1}{2}\lambda\bar{c} = \frac{m\bar{c}}{2\sqrt{2}\rho\pi d^2} \qquad (193)$$

$$\eta = \frac{1}{2}\left(\frac{mN}{V}\right)\lambda\bar{c} = \frac{1}{2}\rho\lambda\bar{c} = \frac{m\bar{c}}{2\sqrt{2}\pi d^2} \qquad (194)$$

$$\kappa = \frac{1}{2}\left(\frac{mN}{V}\right)C_V\lambda\bar{c} = \frac{1}{2}\rho C_V\lambda\bar{c} = \frac{C_V m\bar{c}}{2\sqrt{2}\pi d^2} \qquad (195)$$

$$\bar{c} = \sqrt{\frac{8k_BT}{\pi m}}$$

(a) Values for the molecular diameter calculated from eqs(193) to (195) for the given gases are summarized in the table below. The temperature is at 298 K and the results are compared with the values for the collision diameter.

Molecular diameters (pm) of gases calculated from the data of transport phenomena.

	H_2	CO_2	O_2	NH_3	CH_4
viscosity coef.	282	480	370	447	428
diffusion coef.	240	391	314		
coef. thermal cond.	200	372	266	363	
collision diameter	293	407	360		383

Ans. molecular diameter: Listed in the table.

(b) The mean free path λ is given by the equation $\lambda = 1/\sqrt{2}(N/V)\pi d^2 = m/\sqrt{2}\rho\pi d^2$. From eqs(193) to (195) given above, it is readily confirmed that the following relations hold for the ideal gas.

$$D = \frac{\eta}{\rho}, \quad \kappa = \eta C_V, \quad \frac{\kappa}{\rho C_V} = \frac{\eta}{\rho} = D.$$

However, when these relations are checked against the experimental values, it is found that the relation of $\kappa = \eta \cdot C_V$ is not filled sufficiently as is seen from the values given in the table below.

	H_2	CO_2	O_2	NH_3	CH_4
$\kappa / \mathrm{J \cdot K^{-1} \cdot m^{-1} \cdot s^{-1} \times 10^3}$	168	14.5	24.5	24.6	30.2
$\eta \cdot C_V / \mathrm{J \cdot K^{-1} \cdot m^{-1} \cdot s^{-1} \times 10^3}$	84	8.7	12.7	16.3	

Ans. See text (b) and the table.

Solution 2-47

The equation given in the problem will be derived here first. The initial state of the gas is given by (P_0, V_0, T_0) and the state after expansion is given by (P, V, T). The internal energies of the gas at the two states are denoted by U_0 and U, respectively. The difference $U - U_0$ is compensated by the work done by the gas. Therefore, eq(196) is satisfied (equivalent to the *Joule-Thomson* process, see Chapter 3). Namely, the expansion process here is the isenthalpic process. Furthermore, taking it into account that the gas molecules expanded move with a high speed u in one direction, the energy conservation law is written as eq(197) by adding the kinetic energy of the expanded gas. Using the relation $dH = C_P dT$ for the ideal gas, the enthalpy of the gas is expressed by eq(198).

$$U - U_0 = P_0V_0 - PV \;\Rightarrow\; U + PV = U_0 + P_0V_0 \;\Rightarrow\; H = H_0 \qquad (196)$$

$$H(T_0) = H(T) + \frac{1}{2}mu^2 \times N_A \qquad (197)$$

$$H(T) = C_P T + \text{const.} \qquad (198)$$

On the other hand, the sound velocity in the gas employed is given by eq(199) (see **problem 2-15**).

$$\upsilon_{sound} = \sqrt{\frac{\gamma RT}{N_A m}} \qquad (199)$$

Defining the *Mach* number by eq(200) and by noticing the relation between C_P and C_V given by eq(201),

$$M = \frac{u}{\upsilon_{sound}} \qquad (200)$$

$$C_P = \frac{R\gamma}{\gamma - 1} \quad \left(\gamma = \frac{C_P}{C_V}、\quad C_P - C_V = R\right) \qquad (201)$$

eq(203) is finally derived from eq (197) via eq(202). This is the equation given in the problem.

$$C_P T_0 = C_P T + \frac{1}{2}N_A m \times M^2 \frac{\gamma RT}{N_A m} \qquad (202)$$

$$\frac{T}{T_0} = \frac{1}{1 + \dfrac{M^2\gamma R}{2C_P}} = \frac{1}{1 + \dfrac{M^2(\gamma - 1)}{2}} = \left\{1 + \frac{M^2(\gamma-1)}{2}\right\}^{-1} \qquad (203)$$

(a) From eq(202), the following value is found for the *Mach* number.

$$\frac{400}{10} = 1 + \frac{M^2(5/3 - 1)}{2} \;\Rightarrow\; \frac{39 \times 2 \times 3}{2} = M^2 \;\Rightarrow\; M = 3\sqrt{13} = 10.82$$

Ans. *Mach* number: 10.82

(b) Using the *Mach* number obtained in (a), the position at which the required temperature is reached is found at 0.24 cm downstream.

$$10.82 = 3.26 \times (x/0.04)^{2/3} \;\Rightarrow\; x = \left(\frac{10.82}{3.26}\right)^{3/2} \times 0.04 = 0.24\ \text{cm}$$

Ans. 0.24 cm downstream

Solution 2-48

(a) Since the argument a/RTV of the exponential function must be dimensionless, the dimension of a is the same as that of RTV, which is $\mathrm{J \cdot mol^{-1} \cdot m^3}$.

Ans. $\mathrm{J \cdot mol^{-1} \cdot m^3}$

(b) Since the pressure is an intensive variable, it does not vary with the number of moles as far as the system per unit volume is identical. However, any extensive variable such as V which is increased by a factor of n when the number of moles is increased n times. The equation for P, therefore, must be modified so as for the factor of n to be canceled somehow as is given in the solution. The simplest way to do this is to multiply a, b and RT, all of which are values per mole, by n.

Ans. $P = nRTe^{-\frac{n\cdot a}{RTV}} \Big/ (V - nb)$

(c) For the *Dieterici* equation of state given by eq(204), the volume at which the pressure takes an extremum is found.

$$P = \frac{RTe^{-\frac{a}{RTV}}}{(V-b)} \qquad (204)$$

Solving eqs(205) and (206) both of which should be satisfied at the extremum,

$$\left(\frac{\partial P}{\partial V}\right)_T = -\frac{RTe^{-\frac{a}{RTV}}}{(V-b)^2} + \frac{RTe^{-\frac{a}{RTV}}}{V-b} \times \frac{a}{RTV^2} = \frac{RTe^{-\frac{a}{RTV}}}{V-b}\left(-\frac{1}{V-b} + \frac{a}{RTV^2}\right) = 0 \qquad (205)$$

$$\left(\frac{\partial^2 P}{\partial V^2}\right)_T = \frac{RTe^{-\frac{a}{RTV}}}{V-b}\left(-\frac{1}{V-b} + \frac{a}{RTV^2}\right)^2 + \frac{RTe^{-\frac{a}{RTV}}}{V-b}\left(\frac{1}{(V-b)^2} - \frac{2a}{RTV^3}\right)$$

$$= \frac{RTe^{-\frac{a}{RTV}}}{V-b}\left\{\left(-\frac{1}{V-b} + \frac{a}{RTV^2}\right)^2 + \frac{1}{(V-b)^2} - \frac{2a}{RTV^3}\right\} = 0 \qquad (206)$$

eq(207) is derived from eq(205) and also eq(208) from eq(206).

$$\frac{1}{V-b} = \frac{a}{RTV^2} \qquad (207)$$

$$\left(\frac{a}{RTV^2}\right)^2 = \frac{2a}{RTV^3} \qquad (208)$$

Both equations are satisfied at the critical point at $T = T_c$ and $V = V_c$. Substituting these into eq(207), eq(209) is obtained and from simplified eqs(208) and (209), eq(210) is derived.

$$\frac{1}{V_c - b} = \frac{2RT_cV_c}{RT_cV_c^2} \Rightarrow \frac{1}{V_c - b} = \frac{2}{V_c} \Rightarrow V_c = 2b \qquad (209)$$

$$\frac{a}{RT_cV_c} = 2 \Rightarrow T_c = \frac{a}{4bR} \qquad (210)$$

Substituting these results into eq(204), eq(211) is obtained.

$$P_c = \frac{1}{b} \times \frac{a}{4b} \times e^{-2} = \frac{a}{4b^2e^2} \qquad (211)$$

Ans. $P_c = \dfrac{a}{4b^2e^2}$, $V_c = 2b$, $T_c = \dfrac{a}{4bR}$

[alternative solution] At the critical point, eq(207) must have a double root. From the rewritten eq(207'), therefore, the answers required are readily obtained.

$$RTV^2 - aV + ab = 0 \qquad (207')$$

$$\text{(discriminant)}\ \mathrm{D} = a^2 - 4RTab = 0 \Rightarrow T_c = a/4bR \Rightarrow (V_c - 2b)^2 = 0 \Rightarrow V_c = 2b$$

(d) From eqs(209) to (211), the answer to the question is obtained as below.

$$\frac{P_c V_c}{RT_c} = \frac{\dfrac{a}{4b^2e^2}\times 2b}{R\times\dfrac{a}{4bR}} = \frac{2}{e^2} = 0.2707 \qquad (212)$$

The critical compression factor z_c for the *van der Waals* equation of state is $P_c V_c / RT_c = 3/8 = 0.375$. The experimentally observed values for $RT_c / P_c V_c$ are listed for several gases in the table below. It is seen that the *Dieterici* equation of state yields better agreements with the experimental values as compared with the *van der Waals* equation of state.

Ans. See text (d) and **solution 2-21**.

Table A comparison of values for $RT_c / P_c V_c$.

substanc	RT_c/P_cV_c	substanc	RT_c/P_cV_c	substanc	RT_c/P_cV_c
He	0.324	Ar	0.291	H_2	0.327
O_2	0.292	CO_2	0.275	H_2O	0.228
NH_3	0.243	CH_4	0.289	C_6H_{14}	0.261
C_2H_5OH	0.249	CH_3COOCH_3	0.254	C_6H_6	0.266

E.A.Moelwyn-Hughes, Physical Chemistry, p586 Table 1, extracted and revised.

Solution 2-49

In order to examine the critical phenomena from the data given in the problem alone, we can rely on the facts that at the critical point, (i) the enthalpy change upon the gas-liquid conversion becomes zero and (ii) the densities of liquid and gas are the same.

(i) The enthalpy of vaporization ΔH_{vap} is given by *Clausius-Clapeyron*'s equation (eq(213)) and at the critical temperature where the liquid and the gas cannot be distinguished, it becomes zero.

$$\frac{d\ln P}{dT} = \frac{\Delta H_{vap}}{RT^2} \qquad (213)$$

Consequently, the temperature derivative of the logarithmic vapor pressure P approaches to zero at the critical point (eq(214)).

$$\Delta H_{vap} \to 0 \;\Rightarrow\; \frac{d\ln P}{dT} \to 0 \qquad (214)$$

From the data given in the problem, the logarithmic values of the vapor pressure is plotted against temperature in Fig.29. As is seen from this figure, the vapor pressure stops increasing with temperature at

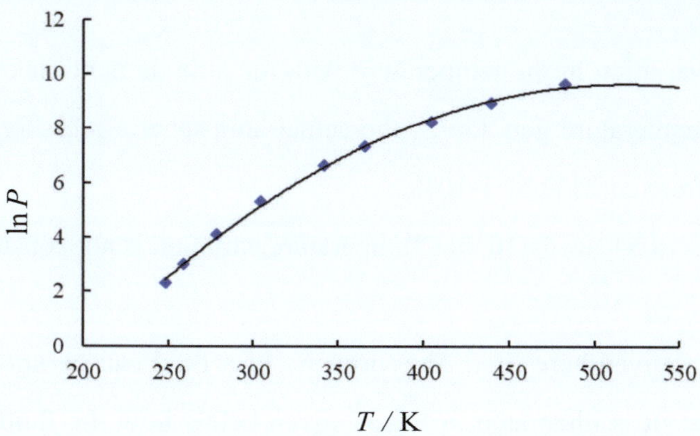

Fig.29 Vapor pressure and temperature of *n*-hexane in the gas-liquid equilibrium .

around 500 K. The plots are best-fitted to a quadratic function using the non-linear least-squares method. As a result, the temperature at which the obtained function reaches a maximum ($d \ln P/dT = 0$) is found to be $T = 501$ K. And from the vapor pressure at this temperature, the critical pressure is obtained as $P = 19.5$ atm. (ii) In Fig.30 the densities of the liquid (red dots) and the densities of the gas (blue dots) which are calculated from the vapor pressures based on the ideal gas law are plotted against temperature. When the former data are best-fitted to a straight line and the latter to a quartic function, eqs(215) and (216) are obtained, respectively.

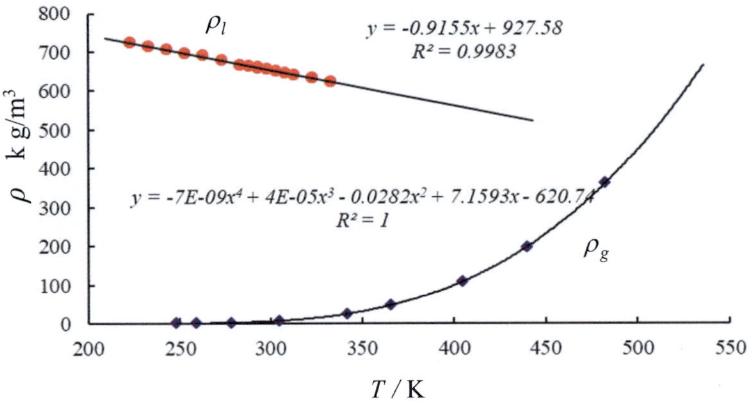

Fig.30 Temperature variations of the densities of *n*-hexane in the gas and liquid states.

$$\rho_l = -0.9155T + 927.58 \qquad (215)$$

$$\rho_g = -7\times10^{-8}T^4 + 4\times10^{-5}T^3 - 0.0282\,T^2 + 7.1593\,T - 620.74 \qquad (216)$$

From the intersect of these two lines, the critical temperature is found as $T = 499.8$ K and from the critical density of $\rho = 470\ \mathrm{kg\cdot m^{-3}}$ (eq(215)), the critical volume is obtained as $V = 1.83\times10^{-4}\,\mathrm{m^3}$.

At the obtained critical temperature and pressure, the ideal gas occupies the volume of 2.12×10^{-3} $\mathrm{m^3\cdot mol^{-1}}$. As compared with this, the critical volume of *n*-hexane obtained here is much smaller. Although the calculated critical temperature of *n*-hexane is found to be close to the literature value, the agreement is not so good for the other critical variables (pressure and volume).

Ans. critical temp.: $500 - 501$K, critical pressure: 19.5 atm, critical volume: $V = 1.83\times10^{-4}\,\mathrm{m^3\cdot mol^{-1}}$

Solution 2-50

Taking the average molecular weight of the air as 28.8 and the temperature as 298 K, the atmospheric pressure is calculated by eq(217) to be $P = 1.0\times\exp(-28.8\times10^{-3}\times10^3/8.31\times298) = 0.89$ atm at the altitude of

$$P = P_0 e^{-Mgh/RT} \qquad (217) \quad \text{(barometric equation)}$$

10^3 m and 0.54 atm at 5×10^3 m.

As is seen from the calculation above, the variation in the temperature with the altitude is not taken into account. Since the pressure decreases as the temperature gets lower, the actual atmospheric pressure at the altitude *h* would be lower than that calculated from eq(217) by 1-10%.

Ans. 0.89 atm at 10^3 m, 0.54 at 5×10^3 m. Temperature variation is not considered.

Solution 2-51a

The equation given in the problem will be derived here first. The viscosity of a fluid can be known by measuring how easily the fluid flows in a tube. It is illustrated in Fig.31 given below how the fluid flows through the tube (pipe) in a laminar form. The velocity of the laminar flow is largest at the center of the tube and the velocity gradient from the center toward the tube wall is generated. Denoting the viscosity of the fluid η, the generated velocity gradient causes a force acting on the fluid given by eq(218) in which S is the side area of the tube at the radius of r. The force F acts oppositely to the flow direction.

$$F(r) = -\eta\left(\frac{\partial v}{\partial r}\right) \times S, \qquad S = 2\pi r l \qquad (218)$$

The difference between the pressures of P_1 and P_2 at both sides of the pipe is related to the force by eq(219) and substituting it into eq(218), eq(220) is obtained.

$$F(r) = (P_1 - P_2)\pi r^2 \qquad (219)$$

$$\left(\frac{\partial v}{\partial r}\right) = -\frac{(P_1 - P_2)\pi r^2}{2\pi r l \eta} = -\frac{(P_1 - P_2) r}{2 l \eta} \qquad (220)$$

Solving eq(220) for $v(r)$, eq(220') is obtained.

$$v(r) = -\frac{(P_1 - P_2) r^2}{4 l \eta} + C \qquad (220')$$

A constant C of the integration can be determined by using that the velocity at the tube wall ($r = R_0$) is zero.

$$C = \frac{(P_1 - P_2) R_0^2}{4 l \eta}$$

With this, eq(220') is expressed as eq(220'').

$$v(r) = \frac{(P_1 - P_2)(R_0^2 - r^2)}{4 l \eta} \qquad (220'')$$

The flow at the radius r per unit time is given by $v(r) \times 2\pi r dr$ and, hence, the total flow per unit time through the whole cross section of the tube is obtained by integrating it over r in the range $[0,\ R_0]$ as is given by eq(221).

$$\frac{dV}{dt} = \int_0^{R_0} v(r) \times 2\pi r dr = \int_0^{R_0} \frac{(P_1 - P_2)(R_0^2 - r^2)}{4 l \eta} \times 2\pi r dr$$

$$= \frac{\pi (P_1 - P_2) R_0^4}{8 l \eta} \qquad (221)$$

The pressure of the gas outside of the tube is denoted as $P_\circ$ and its volume as V^g, and the volume of the gas V measured inside the tube is assumed to be related approximately to V^g by eq(222). Then the time dependence of the volume of the gas measured outside the tube can be expressed by eq(221'). This is the equation given in the problem.

$$V^g P_\circ = \frac{(P_1 + P_2) V}{2} \qquad (222)$$

$$\frac{dV^g}{dt} = \frac{\pi (P_1 - P_2) R_0^4}{8 l \eta} \times \frac{(P_1 + P_2)}{2 P_\circ} = \frac{\pi ({P^2}_1 - P_2^2) R_0^4}{16 l \eta P_\circ} \qquad (221')$$

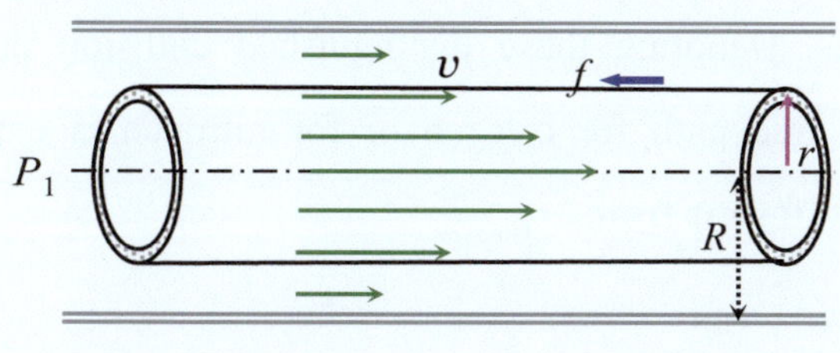

Fig.31 Illustration of fluid-flow through a tube (pipe) in a laminar form.

Substituting the given values into eq(221') and solving it for η in the SI units, a value of 1.81×10^{-5} $\mathrm{kg\cdot m^{-1}\cdot s^{-1}}$ is obtained as shown below.

$$\frac{dV^g}{dt}=\frac{9.02\times10^{-5}}{100}$$

$$\frac{\pi(P^2{}_1-P_2{}^2)R_0{}^4}{16l\eta P_\circ}=\frac{\pi\left\langle(765/760)^2-(760/760)^2\right\rangle(0.50\times10^{-3})^4\times(1.01\times10^5)^2}{16\times1.00\times\eta\times(760/760)\times1.01\times10^5}$$

$$\eta=\frac{\pi\left\langle(765/760)^2-1^2\right\rangle(0.50\times10^{-3})^4\times1.01\times10^5}{16\times1.00\times9.02\times10^{-7}}=1.81\times10^{-5}\,\mathrm{kg\cdot m^{-1}\cdot s^{-1}}$$

Since 1 cP is $10^{-3}\,\mathrm{kg\cdot m^{-1}\cdot s^{-1}}$, the obtained value is equivalent to $1.81\times10^{-2}\,\mathrm{cP}$.

Ans. viscosity of the air: $1.81\times10^{-5}\,\mathrm{kg\cdot m^{-1}\cdot s^{-1}}$

Solution 2-51b

The equation employed for the evacuation method will be derived here first. The *Poiseuille* equation derived in the preceding **solution 2-51** is applied for the ideal gas and when the pressure P_2 at the outlet is quite low (high vacuum), eq(223) can be rewritten as eq(224) by putting $V=nRT/P_2$.

$$\frac{dV}{dt}=\frac{\pi(P_1-P_2)R_0{}^4}{8l\eta} \qquad (223)$$

$$\frac{d\,1/P_2}{dt}=\frac{\pi(P_1-P_2)R_0{}^4}{8l\eta nRT}\approx\frac{\pi P_1R_0{}^4}{8l\eta nRT}\equiv\frac{\alpha}{\eta} \qquad (224)$$

Integrating eq(224) over t for the initial pressure of P_0, eq(225) is obtained. This is the equation employed for the evacuation method in the present problem.

$$\frac{1}{P_2}=\frac{\alpha}{\eta}t+C \;\Rightarrow\; \frac{1}{P_2}=\frac{\alpha}{\eta}t+\frac{1}{P_0} \qquad (225)$$

The slope is given by α/η and α is a constant dependent on the apparatus used. Using the given values measured, the relative viscosity of Ar is found as $\eta_{\mathrm{Ar}}=2.24\times10^{-2}\,\mathrm{cP}$.

$$\frac{(\alpha/\eta)_{air}}{(\alpha/\eta)_{\mathrm{Ar}}}=\frac{2.50}{2.02}=\frac{\eta_{\mathrm{Ar}}}{\eta_{\mathrm{air}}}=\frac{\eta_{\mathrm{Ar}}}{1.81\times10^{-2}\,\mathrm{cP}} \;\Rightarrow\; \eta_{\mathrm{Ar}}=2.24\times10^{-2}\,\mathrm{cP}$$

Ans. viscosity of Ar: $\eta_{\mathrm{Ar}}=2.24\times10^{-2}\,\mathrm{cP}$

Solution 2-52

The mean free path λ is defined by eq(226).

$$\lambda=\frac{\bar{c}(\text{mean velocity})}{z(\text{collision number/ s / }N)} \qquad (226)$$

For a pure gas of a single component, λ is calculable as is defined. However, for a gas mixture we need to take care in calculating the collision numbers. In the present case, the collision number for oxygen or nitrogen is given by a sum of the collision number between the molecules of the same kind and the collision number between the molecules of different kinds. Denoting these distinguished collision numbers by $z_{\mathrm{O_2-O_2}}$, $z_{\mathrm{O_2-N_2}}$, $z_{\mathrm{N_2-O_2}}$ and $z_{\mathrm{N_2-N_2}}$, the mean free path for oxygen or for nitrogen is expressed as eq(227). Here, it should be noticed that $z_{\mathrm{O_2-N_2}}$ is different from $z_{\mathrm{N_2-O_2}}$.

$$\lambda_{O_2} = \frac{\bar{c}_{O_2}}{z_{O_2-O_2} + z_{O_2-N_2}}, \qquad \lambda_{N_2} = \frac{\bar{c}_{N_2}}{z_{N_2-N_2} + z_{N_2-O_2}} \qquad (227)$$

The mean velocity is given by eq(228) and the relevant collision numbers by eqs(229) and (229'). The suffixes 1 and 2 indicate molecular species 1 and 2, respectively.

$$\bar{c} = \left(\frac{8k_BT}{\pi m}\right)^{1/2} \qquad (228)$$

$$z_{11} = N_1 \times \pi d_1^{\,2} \sqrt{2}\left(\frac{8k_BT}{\pi m_1}\right)^{1/2} \qquad (229)$$ (multiplied by $\sqrt{2}$ since the relative mean velocity is used.)

$$z_{12} = N_2 \times \pi\left(\frac{d_1+d_2}{2}\right)^2\left(\frac{8k_BT}{\pi\mu}\right)^{1/2}, \quad z_{21} = N_1 \times \pi\left(\frac{d_1+d_2}{2}\right)^2\left(\frac{8k_BT}{\pi\mu}\right)^{1/2} \qquad (229')$$

The numbers of oxygen and nitrogen molecules in the air are calculated from the ideal gas law as follows.

$$N_{O_2} = \frac{n \times N_A}{V} = \frac{P \times (1/5)}{k_BT} = \frac{1.01\times10^5}{5\times1.38\times10^{-23}\times298} = 4.91\times10^{24}\,\text{m}^{-3}, \qquad N_{N_2} = 1.96\times10^{25}\,\text{m}^{-3}$$

Employing these values, the distinguished collision numbers are calculated as follows.

$$z_{O_2-O_2} = 4.91\times10^{24}\times3.14\times(0.353\times10^{-9})^2\times\sqrt{2}\times444 = 1.21\times10^9\,\text{m}^{-3}\cdot\text{s}^{-1}$$

$$z_{O_2-N_2} = 1.96\times10^{25}\times3.14\times\left(\frac{0.353+0.373}{2}\times10^{-9}\right)^2\times651 = 5.28\times10^9\,\text{m}^{-3}\cdot\text{s}^{-1}$$

$$z_{N_2-O_2} = 4.91\times10^{24}\times3.14\times\left(\frac{0.353+0.373}{2}\times10^{-9}\right)^2\times651 = 1.32\times10^9\,\text{m}^{-3}\cdot\text{s}^{-1}$$

$$z_{N_2-N_2} = 1.96\times10^{25}\times3.14\times(0.373\times10^{-9})^2\times\sqrt{2}\times475 = 5.75\times10^9\,\text{m}^{-3}\cdot\text{s}^{-1}$$

The mean velocities of oxygen and nitrogen and the relative velocity are also calculated as follows.

$$\bar{c}_{O_2} = \left(\frac{8\times1.31\times10^{-23}\times6.02\times10^{23}\times298}{3.14\times32\times10^{-3}}\right)^{1/2} = 444\,\text{m}\cdot\text{s}^{-1}$$

$$\bar{c}_{N_2} = \left(\frac{8\times1.31\times10^{-23}\times6.02\times10^{23}\times298}{3.14\times28\times10^{-3}}\right)^{1/2} = 475\,\text{m}\cdot\text{s}^{-1}$$

$$\bar{c}_{rel} = \left(\frac{8k_BT}{\mu}\right)^{1/2} = 651\,\text{m}\cdot\text{s}^{-1}, \qquad \mu = \frac{32\times28}{32+28} = 14.9$$

From these values, the mean free paths for oxygen and nitrogen are calculated by eq(227) as follows.

$$\lambda_{O_2} = \frac{444}{1.21\times10^9 + 5.28\times10^9} = 6.84\times10^{-8}\,\text{m}$$

$$\lambda_{N_2} = \frac{475}{1.32\times10^9 + 5.75\times10^9} = 6.72\times10^{-8}\,\text{m}$$

Ans. $\lambda_{O_2} = 6.84\times10^{-8}\,\text{m}, \quad \lambda_{N_2} = 6.72\times10^{-8}\,\text{m}$

For your reference, the diffusion coefficient of each molecule calculated based on its definition given in **solution 2-44** is given below

$$D_{O_2} = \frac{1}{2}\lambda\bar{c} = \frac{6.84\times10^{-8}\times444}{2} = 1.52\times10^{-5}\,\text{m}^2\cdot\text{s}^{-1}, \qquad D_{N_2} = \frac{6.72\times10^{-8}\times475}{2} = 1.60\times10^{-5}\,\text{m}^2\cdot\text{s}^{-1}$$

Ans. $D_{O_2} = 1.52\times10^{-5}\,\text{m}^2\cdot\text{s}^{-1}$、$D_{N_2} = 1.60\times10^{-5}\,\text{m}^2\cdot\text{s}^{-1}$

"note" The relation between the diffusion coefficient D and the mean free path λ is used here without any modification, but there is a way to modify it by taking the orientational distribution of λ into account. In that case λ is replaced with $2\lambda/3$. A factor of $2/3$ arises from averaging λ over its homogeneous orientational distribution, leading to the relation of $D = 1/3 \cdot \lambda\bar{c}$.

"reference" Atkins 6th edition Sec.4 in Chapter 24.

Solution 2-53

The molecular collisions in a gas mixture of two components 1 and 2 are those between 1 and 1 and 1 and 2, and those between 2 and 2 and 2 and 1. Denoting the former number z_{11}, z_{12}, and the latter number z_{22}, z_{21}, the mean free path of the molecule 1 is given by eq(230).

$$\lambda_{12} = \frac{\bar{c}_1}{z_{11} + z_{12}} \qquad (230)$$

$$= \frac{\left(\frac{8k_BT}{\pi m_1}\right)^{1/2}}{N_1 \times \pi d_1^{\ 2} \times \sqrt{2}\left(\frac{8k_BT}{\pi m_1}\right)^{1/2} + N_2 \times \pi\left(\frac{d_1 + d_2}{2}\right)^2\left(\frac{8k_BT}{\pi\mu}\right)^{1/2}}$$

$$= \frac{\left(\frac{8k_BT}{\pi m_1}\right)^{1/2}}{N_1 \times \pi d_1^{\ 2} \times \sqrt{2}\left(\frac{8k_BT}{\pi m_1}\right)^{1/2} + N_2 \times \pi\left(\frac{d_1 + d_2}{2}\right)^2\left(\frac{8k_BT(m_1 + m_2)}{\pi m_1 m_2}\right)^{1/2}}$$

$$= \frac{1}{N_1 \times \pi d_1^{\ 2} \times \sqrt{2} + N_2 \times \pi\left(\frac{d_1 + d_2}{2}\right)^2\left(\frac{m_1 + m_2}{m_2}\right)^{1/2}} \qquad (230')$$

Consequently, the diffusion coefficient of the component 1 in the gas mixture should be expressed by eq(231).

$$D_{12} = \frac{1}{2}\bar{c}_1\lambda_{12} = \frac{\left(\frac{8k_BT}{\pi m_1}\right)^{1/2}}{N_1 \times \pi d_1^{\ 2} \times \sqrt{2} + N_2 \times \pi\left(\frac{d_1 + d_2}{2}\right)^2\left(\frac{m_1 + m_2}{m_2}\right)^{1/2}} \qquad (231)$$

In a similar manner, the mean free path of the component 2 is given by eq(232') via eq(232).

$$\lambda_{21} = \frac{\bar{c}_2}{z_{22} + z_{21}} \qquad (232)$$

$$= \frac{\left(\frac{8k_BT}{\pi m_2}\right)^{1/2}}{N_2 \times \pi d_2^{\ 2} \times \sqrt{2}\left(\frac{8k_BT}{\pi m_2}\right)^{1/2} + N_1 \times \pi\left(\frac{d_1 + d_2}{2}\right)^2\left(\frac{8k_BT}{\pi\mu}\right)^{1/2}}$$

$$= \frac{1}{N_2 \times \pi d_2^{\ 2} \times \sqrt{2} + N_1 \times \pi\left(\frac{d_1 + d_2}{2}\right)^2\left(\frac{m_1 + m_2}{m_1}\right)^{1/2}} \qquad (232')$$

Therefore, the diffusion coefficient of the molecular species 2 is given by eq(233).

$$D_{21} = \frac{1}{2}\bar{c}_2\lambda_{21} = \frac{\left(\dfrac{8k_BT}{\pi m_2}\right)^{1/2}}{N_2 \times \pi d_2^{\ 2} \times \sqrt{2} + N_1 \times \pi\left(\dfrac{d_1+d_2}{2}\right)^2\left(\dfrac{m_1+m_2}{m_1}\right)^{1/2}} \qquad (233)$$

Comparing eq(231) with eq(233), it is found that $D_{12} \neq D_{21}$. In practice, however, such a situation does not occur because the mass transport driven by a difference in pressure proceeds so as to make the relation $D_{12} = D_{21}$ be satisfied (see **problem 2-39b**).

The diffusion coefficient of a gas mixture is given in terms of molar ration by the following equation called *H.F.Meyer*'s equation (see **"note"**). The interdiffusion coefficient is common to both component molecules.

$$D_{12} = D_{21} = \frac{N_1\bar{c}_2\lambda_{21} + N_2\bar{c}_1\lambda_{12}}{2(N_1+N_2)} = X_1 \times \frac{1}{2}\bar{c}_2\lambda_{21} + X_2 \times \frac{1}{2}\bar{c}_1\lambda_{12}$$

"note" *E.A.Moelwyn-Hughes, Physical Chemistry,* 2nd revised ed., Pergamon Press, p.60,1965

The expression of $D = X_AD_A + X_BD_B$ is employed in *Physical Chemistry* by *W.J.Moore* 4th edition.

Solution 2-54a

As is shown in Fig.32 (a) below, we consider a case in which a molecule with a velocity *c* collides with a part of vessel wall with an area *S* at the angles of ϑ (from the *z*-axis vertical to the surface) and φ. Imagine a square pillar with a height of $c \cdot \cos\vartheta$ and with a bottom area of *S*, then it is seen that all the molecules which are within the square pillar and move with the velocity of *c* distributed within the solid angel of $\sin\vartheta \cdot d\vartheta \cdot d\varphi$ collide with the base area *S*.

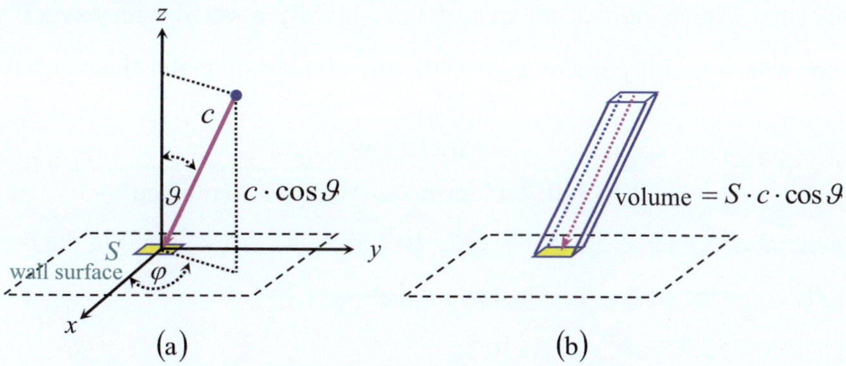

Fig.32 Molecular collisions with a unit area of a container wall of a gas.

Since the fraction of the molecules with such velocities is given by eq(234),

$$4\pi\left(\frac{m}{2\pi k_BT}\right)^{3/2} c^2 \exp\left(-\frac{mc^2}{2k_BT}\right)dc \times \frac{\sin\vartheta \cdot d\vartheta \cdot d\varphi}{4\pi} \qquad (234)$$

the number of the colliding molecules is given by multiplying eq(234) by the number of the molecules in the square pillar as eq(235).

$$4\pi\left(\frac{m}{2\pi k_BT}\right)^{3/2} c^2 \exp\left(-\frac{mc^2}{2k_BT}\right)dc \times \frac{\sin\vartheta \cdot d\vartheta \cdot d\varphi}{4\pi} \times c \cdot \cos\vartheta \times S \times \frac{N}{V} \qquad (235)$$

Consequently, the total collision number with the area *S* is found by integrating eq(235) with respect to *c* and the solid angle. As is seen in the figure (a), the ranges of integration for ϑ and φ are $[0, \pi/2]$ and $[0, 2\pi]$, respectively. Carrying out the integration, a simple equation of eq(236) is obtained as the total collision number Z_{wall} with the wall of the surface area S.

$$Z_{wall} = \int_0^{\infty}\int_0^{\pi/2}\int_0^{2\pi} 4\pi\left(\frac{m}{2\pi k_B T}\right)^{3/2} c^2 \exp\left(-\frac{mc^2}{2k_B T}\right) dc \times \frac{\sin\vartheta \cdot d\vartheta \cdot d\varphi}{4\pi} \times c\cdot\cos\vartheta \times S \times \frac{N}{V}$$

$$= 4\pi\left(\frac{m}{2\pi k_B T}\right)^{3/2}\int_0^{\infty} c^3 \exp\left(-\frac{mc^2}{2k_B T}\right) dc \times \frac{1}{4} \times \frac{N}{V} \times S$$

$$= \frac{1}{4}\bar{c} \times \frac{N}{V} \times S = \frac{1}{4}\left(\frac{8k_B T}{\pi m}\right)^{1/2} \times \frac{P}{k_B T} \times S \qquad (236)$$

The collision number per unit area of the vessel wall is given by eq(236'), from which the collision number per 1cm^2 per second is found as below.

$$Z_{wall}/S = \frac{1}{4}\left(\frac{8k_B T}{\pi m}\right)^{1/2} \times \frac{P}{k_B T} \qquad (236')$$

$$= \frac{1}{4}\left(\frac{8\times 8.31\times 298}{\pi\times 32\times 10^{-3}}\right)^{1/2} \times \frac{1.01\times 10^5 \times 6.02\times 10^{23}}{8.31\times 298}$$

$$= 2.72\times 10^{27}\,\text{s}^{-1}\text{m}^{-2} = 2.72\times 10^{23}\,\text{cm}^{-2}\cdot\text{s}^{-1}$$

Ans. collision number of oxygen with wall: $2.72\times 10^{23}\,\text{cm}^{-2}\cdot\text{s}^{-1}$

Solution 2-54b

The oxygen molecule is regarded as a hard sphere and then its cross section is one fourth of the collision cross section of 0.40 nm^2. Assuming that an area of $1\ \text{cm}^2$ is fully covered with oxygen molecules in the close-packed structure with a packing fraction of 0.907, the number of molecules needed is calculated as below.

$$1\,\text{cm}^2 = 1.0\times 10^{-4}\,\text{m}^2 \Rightarrow \frac{1.0\times 10^{-4}\,\text{m}^2 \times 0.907}{0.10\times 10^{-18}\,\text{m}^2\ \text{(cross section of one molecule)}} = 9.07\times 10^{14}\ \text{molecules}$$

Since the collision number per sec per $1\ \text{cm}^2$ is $2.72\times 10^{23}\,\text{cm}^{-2}\cdot\text{s}^{-1}$, as is found in the preceding problem, the time needed to fully cover the surface is calculated as below.

$$9.07\times 10^{14}/2.72\times 10^{23} = 3.3\times 10^{-9}\,\text{s}$$

It is seen from this that the surface is fully covered with oxygen molecules (monolayer) instantaneously. The fraction of the oxygen molecules adsorbed to those in the gas phase is approximately 20 ppm.

Ans. time needed to fully cover wall surface: $3.3\times 10^{-9}\,\text{s}$

"note" Calculation of packing fraction. In order to cover a plane with a circle of a radius 1 in close-packed manner, circles must be placed in contact with one another as is shown in Fig.33. Since any plane can be covered with a hexagon shown by orange lines without any gap, the packing fraction is calculated from the ratio of the area of the

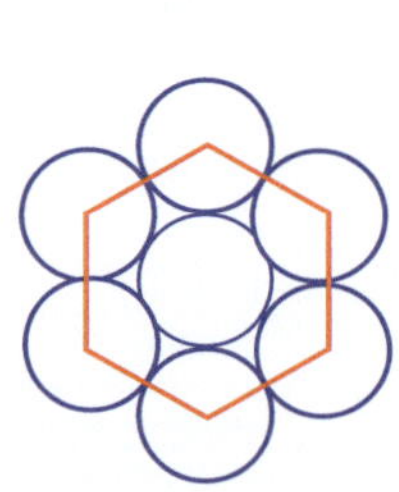

Fig.33 An image of most compact alignment of oxygen molecules on a solid surface.

hexagon to the area of circles inside the hexagon. The former is $6\times\sqrt{3}$ and the latter is given by $(1+6\times 1/3)\pi = 3\pi$. From these values the packing fraction is calculated as $3\pi/6\sqrt{3} = 0.9069$ (see **problem 7-24a**).

Solution 2-55

The distance which oxygen molecules can fly with no collision is denoted l, which varies from molecule to molecule. It is assumed that the distribution of l is governed by the normal distribution and is expressed by eq(237) (see **"note"**). The function $f(l)$ is normalized to one over the range $[0,\ \infty]$ and a is a constant determined by a width of the distribution.

$$f(l) = \frac{2}{a}\sqrt{\frac{1}{\pi}}\cdot \exp\left(-\frac{l^2}{a^2}\right) \qquad \left(\int_0^\infty f(l)dl = 1\right) \qquad (237)$$

Employing the function $f(l)$, the mean free path λ is defined by the left side of eq(238). On the other hand, λ is given by the right side of eq(238) based on the molecular dynamics of gases.

$$\int_0^\infty l\times f(l)dl = \lambda = \frac{k_B T}{\sqrt{2}\sigma P} \qquad (238)$$

From this, the constant a in eq(237) is determined as follows.

$$\int_0^\infty l\times\frac{2}{a}\sqrt{\frac{1}{\pi}}\cdot \exp\left(-\frac{l^2}{a^2}\right)dl = \frac{2}{a}\sqrt{\frac{1}{\pi}}\int_0^\infty l\times \exp\left(-\frac{l^2}{a^2}\right)dl = \frac{2}{a}\sqrt{\frac{1}{\pi}}\times\frac{a^2}{2} = a\times\sqrt{\frac{1}{\pi}}$$

$$a\times\sqrt{\frac{1}{\pi}} = \frac{k_B T}{\sqrt{2}\sigma P} \ \Rightarrow\ a = \sqrt{\pi}\lambda \qquad (237')$$

Using a thus determined, $f(l)$ is rewritten as eq(237'').

$$f(l) = \frac{2}{\pi\lambda}\cdot \exp\left\{-\frac{1}{\pi}\left(\frac{l}{\lambda}\right)^2\right\} \qquad (237'')$$

With eq(237''), it is obvious that the probability with which the molecules can fly a distance of x without colliding once (probability of free path being longer than x), is given by eq(239).

$$\int_x^\infty \frac{2}{\pi\lambda}\cdot \exp\left\{-\frac{1}{\pi}\left(\frac{l}{\lambda}\right)^2\right\}dl \qquad (239)$$

Based on this equation, substituting 2λ, 3λ and 10λ for x and with $\lambda = k_B T/\sqrt{2}\sigma P = 66$ nm, the values for the probability of free path are calculated as 0.11, 0.017 and 1.5×10^{-15}, respectively. It can be seen from this that it is very unlikely that a molecule in a gas at the standard state can fly a distance of ten times the mean free path without a single collision. The values obtained are reasonable.

$$\int_{132}^\infty \frac{2}{\pi\lambda}\cdot \exp\left\{-\frac{1}{\pi}\left(\frac{l}{\lambda}\right)^2\right\}dl = 0.11, \quad \int_{198}^\infty \frac{2}{\pi\lambda}\cdot \exp\left\{-\frac{1}{\pi}\left(\frac{l}{\lambda}\right)^2\right\}dl = 0.017, \quad 1.5\times 10^{-15}$$

Ans. probabilities for flying with no collision: 2λ : 0.11, 3λ : 0.017, 10λ : 1.5×10^{-15}

"note" The assumption adopted here is appropriate from a viewpoint of random walk.

Solution 2-56

(a) From the equation for a one-dimensional random walk, the one-dimensional velocity distribution law is derived here. After many (drunken) steps through a distance λ to the right or left for one step, the probability of finding the drunker at the distance of x is given by eq(240).

$$p(x) \propto e^{-\frac{s^2}{2n}} = e^{-\frac{x^2}{2n\lambda^2}} \qquad (240) \qquad \text{(eq(170) in **solution 2-40**)}$$

On the other hand, a molecule in the gas phase flying with the velocity of u can travel a distance of x in the time t sec and these are related by eq(241).

$$x = ut \qquad (241)$$

Substituting this relation into eq(240), the distribution of x can be rewritten as the distribution of u as is given by eq(242), in which a constant A is determined by using the condition that the mean one-dimensional kinetic

$$p(u) \propto e^{-\frac{u^2t^2}{2n\lambda^2}} = e^{-Au^2} \qquad (242)$$

energy per molecule in the gas phase is equal to $1/2 \times k_BT$ as is shown in eq(243).

$$< \frac{1}{2}mu^2 > = \frac{1}{2}m \int_{-\infty}^{+\infty} u^2 e^{-Au^2} du \bigg/ \int_{-\infty}^{+\infty} e^{-Au^2} du = \frac{1}{2}k_BT \qquad (243)$$

After integrating eq(243) (see **II-4-6**), the constant A is determined as is given in eq(244).

$$\frac{1}{2}m \times \frac{1}{2A}\sqrt{\frac{\pi}{A}} \bigg/ \sqrt{\frac{\pi}{A}} = \frac{m}{4A} = \frac{1}{2}k_BT \;\Rightarrow\; A = \frac{m}{2k_BT} \qquad (244)$$

Substituting the result for A into eq(242), eq(242') is obtained, which is exactly the equation to give the one-dimensional velocity distribution.

$$p(u) \propto e^{-\frac{m}{2k_BT}u^2} \qquad (242')$$

It is realized that the premise of eq(243) is important in deriving the velocity distribution law.

Ans. See text (a).

(b) Based on the fact that the velocity distribution is isotropic in space, the probability of finding a set of the velocities $(u,\ \upsilon,\ w)$ denoted by $F(u,\ \upsilon,\ w)$ must be a function of speed c ($c^2 = u^2 + \upsilon^2 + w^2$) alone. Furthermore, the distribution of each component of $(u,\ \upsilon,\ w)$ must be independent of one another. Taking those into account, $F(u,\ \upsilon,\ w)$ is expressed as eq(245).

$$F(u,\ \upsilon,\ w) = F(c) = f(u) \times f(\upsilon) \times f(w) \qquad (245)$$

After taking the logarithm of each side of eq(245), each resultant equation is partially differentiated by u, yielding eq(246).

$$\left(\frac{\partial \ln F(c)}{\partial u}\right)_{\upsilon,w} = \frac{d \ln F(c)}{dc}\left(\frac{\partial c}{\partial u}\right)_{\upsilon,w} = \frac{u}{c}\frac{d \ln F(c)}{dc} = \frac{d \ln f(u)}{du} \qquad (246)$$

In a similar manner, the logarithmic function of eq(245) is partially differentiated by υ or w, the function in which u is replaced with υ or w is obtained. Summarizing these three results, the equality expressed by eq(247) is obtained.

$$\frac{1}{c}\frac{d \ln F(c)}{dc} = \frac{1}{u}\frac{d \ln f(u)}{du} = \frac{1}{\upsilon}\frac{d \ln f(\upsilon)}{d\upsilon} = \frac{1}{w}\frac{d \ln f(w)}{dw} = -\alpha = \text{const.} \qquad (247)$$

Since each function in the above equality is a function of a single variable, a value of the equality must be a constant. Putting this constant as $-\alpha < 0$ and integrating one of the equations above, say with respect to u, eq(248) is derived (same as for υ or w).

$$f(u) = A\exp(-\alpha u^2) \qquad (248)$$

A is a normalization constant and takes a value of $\sqrt{\alpha/\pi}$. A value of α can be determined to be $m/2k_BT$ as shown below based again on the condition that the mean one-dimensional kinetic energy per molecule in the gas phase is equal to $1/2 \times k_BT$.

$$\frac{1}{2}m\overline{u^2} = \sqrt{\frac{\alpha}{\pi}}\int_{-\infty}^{\infty}\frac{1}{2}m\overline{u^2}\exp(-\alpha u^2)du = \sqrt{\frac{\alpha}{\pi}} \times \frac{1}{2}m \times \frac{1}{2\alpha}\sqrt{\frac{\pi}{\alpha}} = \frac{1}{2}k_BT \;\Rightarrow\; \alpha = \frac{m}{2k_BT}$$

Finally, eq(249) is obtained for the one-dimensional velocity distribution function.

$$f(u) = \sqrt{\frac{m}{2\pi k_BT}}\exp\left(-\frac{mu^2}{2k_BT}\right) \qquad (249)$$

Ans. See text (b).

Solution 2-57

(a) The parameters a and b of the *van der Waals* equation of state are to take into consideration the deviation of real gases from the ideal gas. The parameter a is associated with the intermolecular force and b with the excluded volume. For the three equations given in the problem, the excluded volume is taken into account in the same manner. However, eqs(ii) and (iii) are largely different from the *van der Waals* equation of state in the point that the temperature dependence of the intermolecular interaction is taken into account. Since the effect of temperature decreases relatively with increasing temperature, its effect is expressed in the form proportional to $1/T$ in eq(ii) and in the form of the *Boltzmann* factor of $e^{-a/RTV}$. In the sense that the temperature effect is of the *Arrhenius* type, the *Dieterici* equation of state is considered to describe real gases most closely among the three. In fact, the critical compression factor calculated from eq(iii) is, as is seen in **solution 2-18** , close to those of real gases (polar molecules in particular).

Ans. See text (a).

(b) By expanding the exponential term of the *Dieterici* equation of state (eq(ii)), eq(250) is obtained.

$$P = \frac{RTe^{-\frac{a}{RTV}}}{(V-b)} \qquad \text{(ii)} \qquad \text{(reposted)}$$

$$e^{-\frac{a}{RTV}} = 1 - \frac{a}{RTV} + \frac{1}{2}\left(\frac{a}{RTV}\right)^2 - \cdots \qquad (250)$$

Substituting eq(250) into eq(ii) and rearranging it, eq(251') is obtained via eq(251).

$$P = \frac{RT}{(V-b)}\left\{1 - \frac{a}{RTV} + \frac{1}{2}\left(\frac{a}{RTV}\right)^2 - \cdots\right\} \qquad (251)$$

$$= \frac{RT}{(V-b)} - \frac{a}{(V-b)V} + O\left(\frac{1}{V^3}\right)$$

$$= \frac{RT}{(V-b)} - \frac{a}{V^2} \qquad (251')$$

Ans. See text (b).

(c) Most practical way to draw *P-V* curves in the three cases is to express the parameters *a* and *b* in terms of the critical constants derived for each equation of state and then to use the critical values measured for a real gas (CO_2 in the present case) to determine practical values for *a* and *b* needed for drawing a *P-V* curve.

The critical constants measured for CO_2 are $T_c = 304.1\,\mathrm{K}$, $P_c = 72.88\,\mathrm{atm}$ and $V_c = 9.40\times10^{-2}\,\mathrm{L\cdot mol^{-1}}$, and they are expressed in terms of the parameters *a* and *b* for each equation as listed in the table below. The parameters expressed by using the relevant critical constants are given in the margin of the table. Since P_c and T_c are more accurately determined experimentally as compared with V_c, the values for *a* and *b* are determined by using the values for P_c and T_c.

	P_c	V_c	T_c	a	$b\,/\mathrm{L\cdot mol^{-1}}$
van der Waals	$a/27b^2$	$3b$	$8a/27bR$	3.61#	4.28×10^{-2}
Dieterici	$a/4e^2b^2$	$2b$	$a/4bR$	4.62#	4.63×10^{-2}
Berthelot	$(2aR/3b^2)^{1/2}/12$	$3b$	$2(2a/3bR)^{1/2}/3$	1096&	4.28×10^{-2}

$\mathrm{atm\cdot L^2\cdot mol^{-2}}$, & $\mathrm{atm\cdot L^2\cdot mol^{-2}\cdot K}$.

(i) $\begin{cases} b = RT_c/8P_c \\ a = 27(RT_c)^2/64P_c \end{cases}$ (ii) $\begin{cases} b = RT_c/P_c e^2 \\ a = 4R^2T_c^2/P_c e^2 \end{cases}$ (iii) $\begin{cases} b = RT_c/8P_c \\ a = 27R^2T_c^3/64P_c \end{cases}$

Employing the values for *a* and *b* given in the table above, *P-V* curves obtained for the three equations are plotted at three temperatures in Figs 34(a) to (c). From these figures, it is seen that as compared with *van der Waals,* the *Berthelot* equation of state can describe a tendency of being liquefied more readily but, on the contrary, the *Dieterici* equation of state does it less easily (see **problem 2-21**).

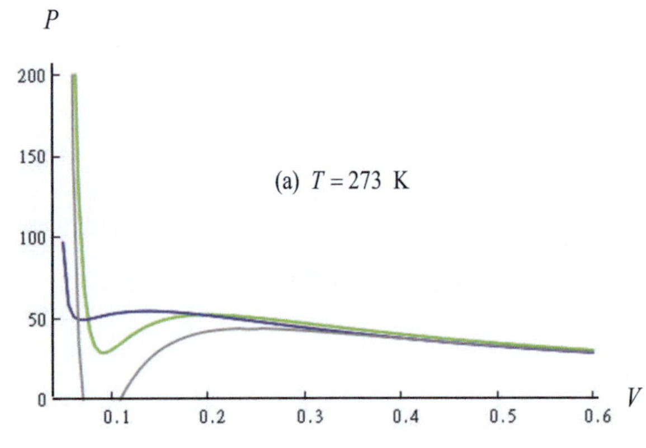

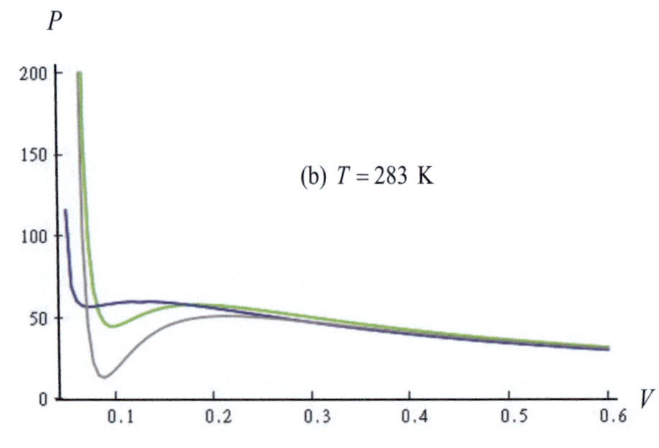

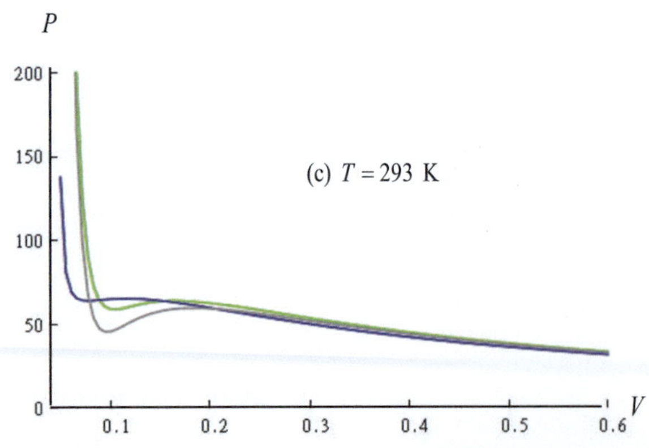

Figs.34 (a)-(c) *P-V* curves of the three equations of state plotted at three different temperatures.
(——) *van der Waals,* (——) *Dieterici,*
(——) *Berthelot.*

Employing the values for a and b given in the table above, P-V curves obtained for the three equations are plotted at three temperatures in Figs 34(a) to (c). From these figures, it is seen that as compared with *van der Waals,* the *Berthelot* equation of state can describe a tendency of being liquefied more readily but, on the contrary, the *Dieterici* equation of state does it less easily (see **problem 2-21**).

Ans. See text (c) and Figs.34(a)-(c).

Solution 2-58

(a) We should confirm first whether the thermodynamic variables, apart from the constants, in eq(16) given in the problem are extensive or intensive variables. Then, we should notice that the intensive variable of pressure remains constant when all extensive variables (volume alone in the present case) are increased (or decreased) n (number of moles) times. To keep P invariant, each term of eq(16) should be multiplied by either n or n^2 as is shown in eq(16'), resulting in the *Redlich-Kwong* equation of state valid for the gas of n moles.

$$P = \frac{nRT}{V - nb} - \frac{n^2 a}{T^{1/2} V(V + nb)} \qquad (16')$$

Ans. equation of state for n moles: eq(16')

(b) The three equations of state dealt with in the preceding problems are featured in that the relation between temperature and effective intermolecular force is taken into account from different points of view. For the *Redlich-Kwong* equation of state, the following two points are noticed as its features; (i) the effect of the excluded volume is taken into account but in a way to weaken it by adding the second term, (ii) the second term is proportional to $1/\sqrt{T}$ but not to $1/T$ so as to make the effect of temperature remain until at higher temperatures.

For the *Redlich-Kwong* equation of state, the compression factor z_c at the critical point is calculated to be $1/3$, which is smaller than that for the *van der Waal* equation of state but larger than that of the *Dieterici* equation of state. Judging from this, the *Redlich-Kwong* equation of state is expected to fit well non-polar gases. As is explained in detail in the preceding solution, the expressions for the parameters a and b in terms of the critical constants are acquired first and then the numerical values for them are calculated from the obtained expressions to draw its P-V curves at given temperatures (see Fig.35).

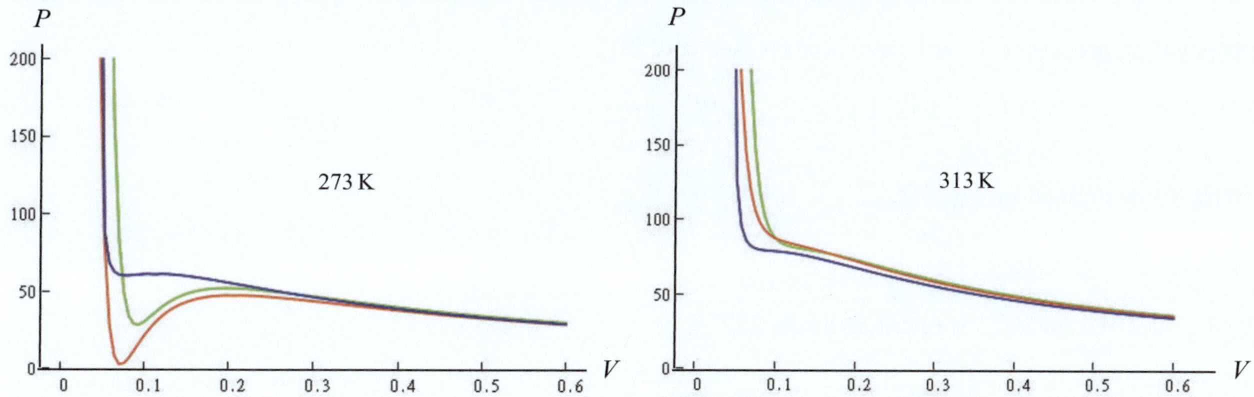

Fig.35 P-V curves drawn for the three equations of state at two temperatures.
(——) *van der Waal,* (——) *Dieterici,* (——) *Delich-Kwong.*

You can compare them with those of other equations of state as are shown in the figure above. For your reference, the parameters expressed in terms of the critical constants are given below for the *Redlich-Kwong* equation of state (derivation omitted).

$$b = 0.26 \times \frac{RT_c}{3P_c}\ , \qquad a = 0.427 \times \frac{R^2 T_c^{5/2}}{P_c}$$

Ans. See text.

Solution 2-59

(a) The equations derived by referring to Fig.7a and Fig.7b illustrated in **problem 2-59** and reposted here are given as eq(252) and eq(253) on the right side of Fig.7, respectively.

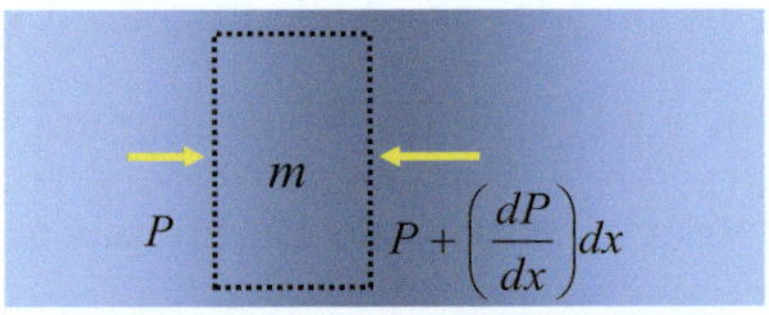

Fig.7a

$$m\left(\frac{\partial \upsilon}{\partial t}\right) = \left(\frac{\partial P}{\partial \boldsymbol{x}}\right) dx \times A \Rightarrow \frac{m}{A \times dx}\left(\frac{\partial \upsilon}{\partial t}\right) = \left(\frac{\partial P}{\partial x}\right)$$

$$\Rightarrow \rho\left(\frac{\partial \upsilon}{\partial t}\right) = \left(\frac{\partial P}{\partial x}\right) \qquad (252)$$

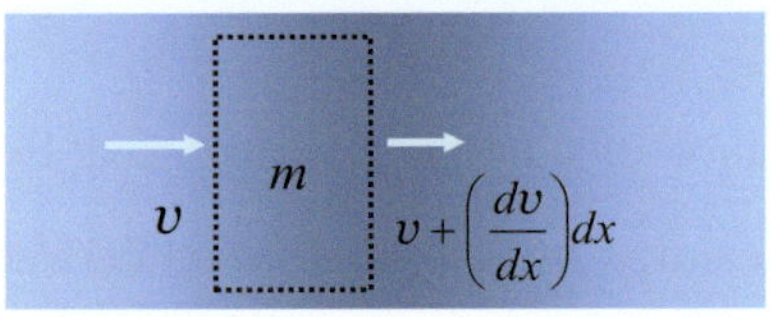

Fig.7b

$$\frac{\partial m}{\partial t} + m\left(\frac{\partial \upsilon}{\partial x}\right) = 0 \Rightarrow \frac{1}{dx \times A}\frac{\partial m}{\partial t} + \frac{m}{dx \times A}\left(\frac{\partial \upsilon}{\partial x}\right) = 0$$

$$\Rightarrow \frac{\partial \rho}{\partial t} + \rho\left(\frac{d\upsilon}{dx}\right) = 0 \qquad (253)$$

In eqs(252) and (253), the density ρ which appears as coefficients of relevant partial differentials can satisfactorily be approximated by a constant density of ρ_0 of a medium, so that eqs(252') and (253') can be employed instead of eqs(252) and (253) as good approximations.

$$\rho_0\left(\frac{\partial \upsilon}{\partial t}\right) = \left(\frac{\partial P}{\partial x}\right) \quad (252') \qquad \frac{\partial \rho}{\partial t} + \rho_0\left(\frac{d\upsilon}{\partial x}\right) = 0 \quad (253')$$

Ans. eqs(252) and (253) or eqs(252') and (253')

(b) The right side of eq(252') expresses pressure variation with distance due to the compression-expansion process associated with the sound propagation. For sound waves, this process occurs very rapidly, i.e., adiabatically with constant entropy (see **"note"**). Consequently, the right side of eq(252') can be rewritten as eq(254), in which the suffix S is added to relevant partial differentials to emphasize that the sound propagation process is the adiabatic process (see eq(258)).

$$\left(\frac{\partial P}{\partial x}\right)_S = \left(\frac{\partial P}{\partial \rho}\right)_S\left(\frac{\partial \rho}{\partial x}\right)_S = \gamma\left(\frac{\partial P}{\partial \rho}\right)_T\left(\frac{d\rho}{\partial x}\right)_S = \upsilon_0^2\left(\frac{d\rho}{\partial x}\right)_S \qquad (254)$$

Here, υ_0^2 is defined by eq(255).

$$\upsilon_0^2 = \gamma\left(\frac{\partial P}{\partial \rho}\right)_T, \qquad \gamma = \frac{C_P}{C_V} \qquad (255)$$

Using eq(254), eq(252') is rewritten as eq(252'').

$$\rho_0\left(\frac{\partial \upsilon}{\partial t}\right) = \upsilon_0^2\left(\frac{d\rho}{\partial x}\right)_S \qquad (252'')$$

Both sides of eq(253') are partially differentiated by t and substituting eq(252'') into the equation thus obtained, eq(256') is finally derived via eq(256).

$$\frac{\partial^2 \rho}{\partial t^2} + \rho_0 \frac{d}{dx}\left(\frac{\partial v}{\partial t}\right) \Rightarrow \frac{d^2 \rho}{\partial t^2} - \frac{\rho_0 v_0^2}{\rho_0}\frac{\partial}{\partial x}\left(\frac{\partial \rho}{\partial x}\right) = 0 \qquad (256)$$

$$\frac{\partial^2 \rho}{\partial t^2} = v_0^2 \frac{\partial^2 \rho}{\partial x^2} = 0 \qquad (256')$$

A solution of eq (256') is given by eq(257), which represents a sine wave of wavelength λ propagating with velocity v_0.

$$\rho = \rho_0\left\{1 + \phi_0 \sin\frac{2\pi}{\lambda}(x - v_0 t)\right\} \qquad (257)$$

Ans. See text.

(c) It could be a lengthy equation but eq(254) can be rewritten as eq(259) by using eq(258). The derivation of eq(258) is given in **solution 3-51** in detail.

$$\left(\frac{\partial P}{\partial V}\right)_S = \gamma\left(\frac{\partial P}{\partial V}\right)_T \qquad (258)$$

$$\left(\frac{\partial P}{\partial \rho}\right)_S = \left(\frac{\partial P}{\partial V}\right)_S\left(\frac{\partial V}{\partial \rho}\right)_S = \left(\frac{\partial P}{\partial V}\right)_S \Big/ \left(\frac{\partial \rho}{\partial V}\right)_S = -\gamma\left(\frac{\partial P}{\partial V}\right)_T \frac{V}{\rho} = \gamma\left(\frac{\partial P}{\partial \rho}\right)_T \qquad (259)$$

Consequently, the sound velocity is given by eq(255').

$$v_0 = \sqrt{\gamma\left(\frac{\partial P}{\partial \rho}\right)_T} = \sqrt{-\gamma\left(\frac{\partial P}{\partial V}\right)_T \frac{V}{\rho}} \qquad (255')$$

Using the relation of $\rho = nM/V$, the sound velocity in the ideal gas is found to be given by a simple eq(255'').

$$v_0 = \sqrt{\frac{\gamma RT}{M}} \qquad (255'')$$

See **solution 2-14** for the practical calculations of sound velocities in a few gases other than the ideal gas.

"note" It is reasonable to consider that the compression-expansion cycles associated with the sound propagation proceed adiabatically since the frequencies of sounds in the audible range are comparatively high. However, if the cycles are slow enough for them to proceed isothermally, then instead of $(\partial P/\partial V)_S$ $(\partial P/\partial V)_T$ appears in eq(254). For this, the sound velocity given by eqs(255') and (255'') will not contain γ.

Ans. See text.

Solution 2-60a

The parameters a and b of the *Dieterici* equation of state (eq(260) given below again) can be expressed by eq(261), as is obtained in **solution 2-57**.

$$P = \frac{RTe^{-\frac{a}{RTV}}}{(V - b)} \qquad (260)$$

$$a = 4R^2T_c^2 / P_c e^2 = 2RV_cT_c, \qquad 2b = V_c, \qquad \frac{P_cV_c}{RT_c} = \frac{2}{e^2} \qquad (261)$$

Using eq(261), the relation of eq(262) is derived and is used to replace the parameter a in eq(260) to derive eq(263).

$$\frac{a}{RTV} = \frac{2RV_cT_c}{RTV} = \frac{2}{T_RV_R} \qquad (262)$$

$$P \cdot \frac{RT_c}{P_cV_c} = \frac{RTe^{-\frac{2}{T_RV_R}}}{(V - V_c/2)} \times \frac{e^2}{2} \qquad (263)$$

Rewriting eq(263) so as to express P/P_c, eq(263') is obtained, and the result is rewritten by using the

$$\frac{P}{P_c}=\frac{V_c}{T_c}\frac{Te^{-\frac{2}{T_R V_R}}}{(V-V_c/2)}\frac{e^2}{2} \qquad (263')$$

reduced variables of $P_R=P/P_c$, $V_R=V/V_c$ and $T_R=T/T_c$, finally obtaining eq(263'').

$$P_R=\frac{T_R e^{-\frac{2}{T_R V_R}}}{(V_R-1/2)}\frac{e^2}{2} \qquad (263'')$$

At three reduced temperatures T_R of 0.6, 1.0、and 1.4, P_R is plotted against V_R as shown in the figure below. It is characteristic of this equation that the reduced pressure does not take a negative value as far as $V_R \geq 1/2$. See Fig.36.

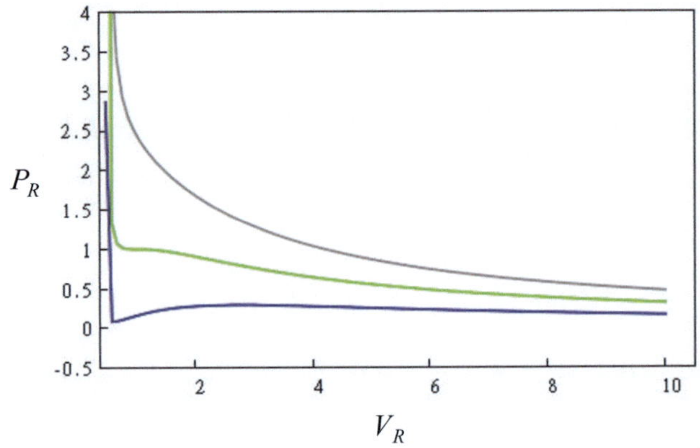

Fig.36 P_R-V_R plot for *Dieterici*'s reduced equation of state.
T_R = 0.6 , 1.0 and 1.4 (upward).
T_R = 1.0 and V_R = 1.0 represent the critical point.

Ans. *Dieterici*'s reduced equation of state eq(263'')

Solution 2-60b

The *Dieterici* equation of state given by eq(264) is expanded in powers of $1/V$ and the function obtained up to the first power is given by eq(264').

$$\frac{PV}{RT}=\frac{V}{(V-b)}\times e^{-\frac{a}{RTV}}=\frac{1}{(1-b/V)}\times e^{-\frac{a}{RTV}} \qquad (264)$$

$$=\left\{1+\frac{b}{V}+\left(\frac{b}{V}\right)^2+\cdots\right\}\times\left\{1-\frac{a}{RTV}+\frac{1}{2}\left(\frac{a}{RTV}\right)^2+\cdots\right\}$$

$$=1+\frac{b}{V}-\frac{a}{RTV}+O\left(\frac{1}{V^2}\right)$$

$$=1+\frac{1}{V}\left(b-\frac{a}{RT}\right)+O\left(\frac{1}{V^2}\right) \qquad (264')$$

From this, the second virial coefficient is found to be given by eq(265).

$$B(T)=b-\frac{a}{RT} \qquad (265)$$

The right side of eq(265) is identical in form to that found for the *van der Waals* equation of state (see **solution 2-11**). However, since the values for the constants of a and b are not the same as those for the latter equation of state, they are not the same equations of $B(T)$. Naturally, the *Boyle* temperatures T_{Boyle} acquired from the second virial coefficient are different for these equations of state (see **problem 2-61**).

Ans. second virial coefficient of *Dieterici*'s equation of state: eq(265)

Solution 2-61

(a) Eq(21) given in the problem, when rewritten using r in the unit of nm, is expressed by eq(265') via eq(265) for He, for instance.

$$B(T) = -2\pi N_A \times 10^3 \int_0^\infty \left\{ \exp\left\{ -\frac{4}{T} \times 10.22 \left[\left(\frac{0.256}{r} \right)^{12} - \left(\frac{0.256}{r} \right)^6 \right] \right\} - 1 \right\} \left(\frac{r}{10^9} \right)^2 d\left(\frac{r}{10^9} \right) \qquad (265)$$

$$= -2\pi \times 6.02 \times 10^{-1} \int_0^\infty \left\{ \exp\left\{ -\frac{4}{T} \times 10.22 \left[\left(\frac{0.256}{r} \right)^{12} - \left(\frac{0.256}{r} \right)^6 \right] \right\} - 1 \right\} r^2 dr \qquad (265')$$

The integral is multiplied by a factor of 10^3 in order to express $B(T)$ in the units of $\mathrm{L \cdot mol^{-1}}$. By using *Mathematica*, for instance, the integration of eq(265') is readily performed as a numerical integration.

The values of $B(T)$ thus obtained for the three gases are plotted against T as well as $1/T$ in Figs.37 and 38, respectively.

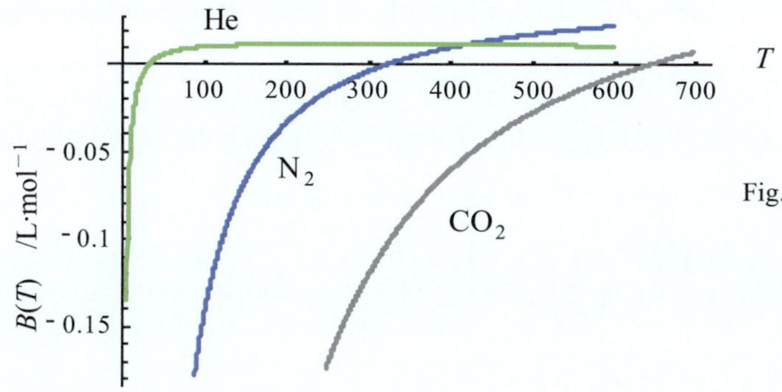

Fig.37 The second virial coefficient $B(T)$. The temperature at which the curve crosses the T-axis is the *Boyle* temperature of the gas. See the table below.

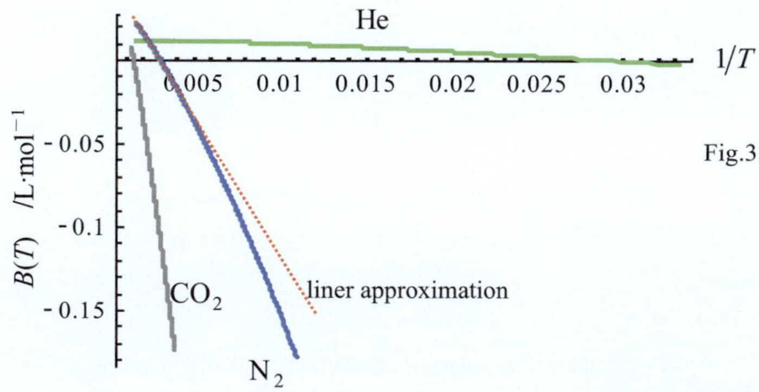

Fig.38 $B(T)$ in Fig.31 are plotted against $1/T$. The relation of $B(T) = \beta - \alpha/T$ is an approximation realized in a certain temperature range.

Ans. temperature dependence of $B(T)$: Figs. 37 and 38

(b)The values for the *Boyle* temperature experimentally measured and those calculated from the equation of $B(T) = 0$ for the *van der Waals*, *Dieterici,* and virial equations of state are summarized in the table below

Table A comparison of *Boyle* temperatures measured and calculated.

	T_{Boyle}	T_{Boyle}^{vdW}	$T_{Boyle}^{Dieterici}$	T_{Boyle}^{virial}	σ (pm)	ε/k_B (K)	$a^{\#}$	b^{*}
He	22.64	29.3	20.8	34.9	256	10.22	0.057	2.370
N_2	327.2	438	505	325	370	95.1	1.408	3.913
CO_2	714.8	1039	1218	646	449	189	3.64	4.267

van der Waals parameters、units $^{\#}\mathrm{atm \cdot L^2 \cdot mol^{-2}}$, $^{*}10^{-2}\mathrm{L \cdot mol^{-1}}$

For the parameters of the *Dieterici* equation of state, see **solution 2-57**.

together with the values for several parameters. From a comparison of the *Boyle* temperatures in the table, the virial equation of state is seen to show a good agreement with the experimental values.

Ans. *Boyle* temperature: values in the table

(c) If the function $B(T)$ calculated from eq(265') is approximated by a liner function of $1/T$ as is given by

$$B(T) = \beta - \alpha/T \qquad (266)$$

eq(266), the parameters associated with the *van der Waals* or *Dietric* equation of state are equated to those in eq(266) as are given by $\beta = b$ and $\alpha = a$ and then they may be correlated to the *Lennard-Jones* potential parameters of ε and σ. However, as is seen from Fig.38, eq(266) is an approximation valid only in a narrow range of temperature. Therefore, a discussion on the relation between the intermolecular force and the parameter values of such as those of the *van der Waals* equation of state remains only qualitative.

Ans. See text (c).

"note" In Fig.37 the second virial coefficient is plotted for each of the three gases. However, since eq(21) given in the problem can be rewritten as eq(21'') via eq(21') by putting $x = \sigma/r$ and $T^* = k_B T/4\varepsilon$, the left side of eq(21'') can be regarded as a general virial coefficient $B(T^*)$ (eq(267)) which does not contain any value specific to a given molecule. $B(T^*)$ is plotted against T^* in Fig39 given below.

$$B(T) = -2\pi\sigma^3 N_A \int_0^\infty \left\{ \exp\left\{ -\frac{1}{T^*}\left[\left(\frac{1}{x}\right)^{12} - \left(\frac{1}{x}\right)^6 \right] \right\} - 1 \right\} x^2 dx \qquad (21')$$

$$\frac{B(T)}{2\pi\sigma^3 N_A/3} = -3\int_0^\infty \left\{ \exp\left\{ -\frac{1}{T^*}\left[\left(\frac{1}{x}\right)^{12} - \left(\frac{1}{x}\right)^6 \right] \right\} - 1 \right\} x^2 dx \qquad (21'')$$

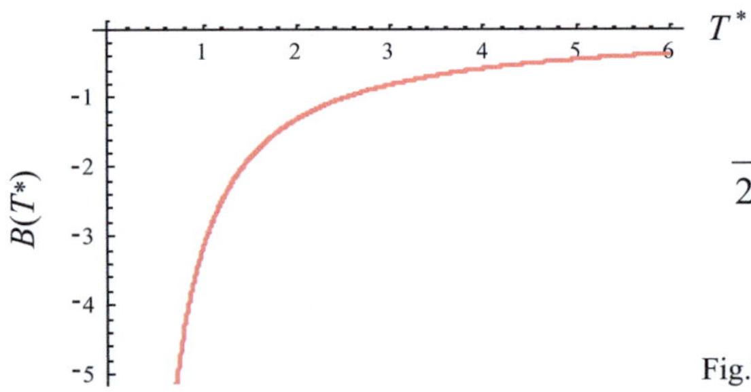

$$\frac{B(T)}{2\pi\sigma^3 N_A/3} = B(T^*) \qquad (267)$$

Fig.39 Plot of the general virial coefficient $B(T^*)$ against T^*

Solution 2-62

The expansion coefficient of volume with temperature is expressed by eq(268).

$$\alpha = \frac{1}{V}\left(\frac{\partial V}{\partial T}\right)_P \qquad (268)$$

For each equation of state, the right side of eq(268) is specifically obtained as followings.

(i) $PV = nRT$

$$\frac{1}{V}\left(\frac{\partial V}{\partial T}\right)_P = \frac{P}{nRT}\cdot\frac{nR}{P} = \frac{1}{T} \qquad (269)$$

(ii) $P(V - nb) = nRT$

Rewriting the equation as eq(270),

$$V - nb = \frac{nRT}{P} \qquad (270)$$

both sides of eq(270) are differentiated with pressure at constant temperature to lead to eq(271).

$$\frac{1}{V}\left(\frac{\partial V}{\partial T}\right)=\frac{nR}{VP}=\frac{1}{T}\left(1-\frac{b}{\overline{V}}\right) \qquad (271)$$

Here, $\overline{V}$ is a molar volume given by $\overline{V}=V/n$.

(iii) *van der Waals*

The equation is rewritten as eq(272),

$$P+\frac{n^2a}{V^2}=\frac{nRT}{V-nb} \qquad (272)$$

and both sides of eq(272) are differentiated with pressure at constant temperature and the resultant equations are rearranged to lead to eq(273).

$$-\frac{n^2a}{2V^3}\left(\frac{\partial V}{\partial T}\right)_P=\frac{nR}{V-nb}-\frac{nRT}{(V-nb)^2}\left(\frac{\partial V}{\partial T}\right)_P \Rightarrow \left\{\frac{nRT}{(V-nb)^2}-\frac{n^2a}{2V^3}\right\}\left(\frac{\partial V}{\partial T}\right)_P=\frac{nR}{V-nb} \qquad (273)$$

From eq(273), the expansion coefficient for the *van der Waals* equation of state is given by eq(274).

$$\frac{1}{V}\left(\frac{\partial V}{\partial T}\right)_P=\frac{V\cdot\dfrac{nR}{V-nb}}{V^2\cdot\left\{\dfrac{nRT}{(V-nb)^2}-\dfrac{n^2a}{2V^3}\right\}}=\frac{\dfrac{1}{1-nb/V}}{\dfrac{T}{(1-nb/V)^2}-\dfrac{na}{RV}}=\frac{\dfrac{1}{1-b/\overline{V}}}{\dfrac{T}{(1-b/\overline{V})^2}-\dfrac{a}{R\overline{V}}} \qquad (274)$$

(iv) *Dieterici*

Taking natural logarithms of both sides of eq(275), the equation of state is expressed as eq(276).

$$P=\frac{nRTe^{-\frac{na}{RTV}}}{V-nb} \qquad (275)$$

$$\ln P=\ln nRT-\frac{na}{RTV}-\ln(V-nb) \qquad (276)$$

Both sides of eq(276) are differentiated with pressure at constant temperature and the resultant equations are rearranged to lead to eq(277).

$$0=\frac{1}{T}+\frac{na}{RT^2V}+\frac{na}{RTV^2}\left(\frac{\partial V}{\partial T}\right)-\frac{1}{V-nb}\left(\frac{\partial V}{\partial T}\right) \Rightarrow \frac{1}{T}+\frac{na}{RT^2V}=\left(\frac{1}{V-nb}-\frac{na}{RTV^2}\right)\left(\frac{\partial V}{\partial T}\right)_P \qquad (277)$$

Rearranging eq(277), the expansion coefficient for the *Dieterici* equation of state is obtained as eq(278).

$$\frac{1}{V}\left(\frac{\partial V}{\partial T}\right)_P=\frac{\dfrac{1}{T}\left(1+\dfrac{na}{RTV}\right)}{V\cdot\left(\dfrac{1}{V-nb}-\dfrac{na}{RTV^2}\right)}=\frac{\left(1+\dfrac{a}{RT\overline{V}}\right)}{T\left\{\dfrac{1}{(1-b/\overline{V})}-\dfrac{a}{RT\overline{V}}\right\}} \qquad (278)$$

Summarizing above, eqs(269), (271), (274) and (278) are what are called expansion coefficients for the corresponding equations of state. In order to answer the questions, their molar volumes at 300 K are required, which are directly known (apart from practical calculations) from the corresponding equation of state. The obtained results are summarized in the table below.

Table Equations of state of a gas, molar volumes at 300 K and their temperature variations.

	$\overline{V}_{1\,\text{atm}}$ (L)	% volume change	$\overline{V}_{10\,\text{atm}}$ (L)	% volume change
$PV = nRT$	24.62	0.3333	2.462	0.3333
$P(V - nb) = nRT$	24.66	0.3328	2.504	0.3277
*van der Waals**	24.51	0.3348	2.353	0.3484
Dieterici#	24.48	0.3379	2.316	0.3837

* $a = 3.64\ \text{atm} \cdot \text{L}^2 \cdot \text{mol}^{-2}$, $b = 4.267\ 10^{-2}\,\text{L} \cdot \text{mol}^{-1}$.

\# $a = 4.63\ \text{atm} \cdot \text{L}^2 \cdot \text{mol}^{-2}$, $b = 4.63\ 10^{-2}\,\text{L} \cdot \text{mol}^{-1}$.

As is clear from eq(269), for the ideal gas the temperature variation of a volume is simply inversely proportional to temperature and independent of pressure. However, for the other equations of state, the % volume change at 300 K starts to deviate significantly at high pressures from the value of 0.3333% expected from the ideal gas law.

Ans. molar volumes of a gas and their temperature variations:. See text and table.

Solution 2-63

(a) The molar volume of a gas at the standard state is denoted $\overline{V}$. Since the space one molecule occupies is given by $\overline{V}/N_A$, the mean intermolecular distance is regarded as equal to its cubic root as is shown below.

$$\frac{\overline{V}}{N_A} = \frac{k_B T}{P} = \frac{1.38\times10^{-23}\times273}{1.01\times10^{5}} = 37.5\times10^{-27}\,\text{m}^3 \quad r = \sqrt[3]{37.5}\times10^{-9}\,\text{m} = 3.4\,\text{nm}$$

Ans. mean intermolecular distance : 3.4 nm

(b) The molar volume of the gas is compressed to the one tenth of that at 1atm 10 atm and, hence, the mean intermolecular distance obtained in (a) should be divided by $\sqrt[3]{10}$ to give the mean distance at 10 atm.

$$r = \sqrt[3]{37.5/10}\times10^{-9}\,\text{m} = 1.6\,\text{nm}$$

The given value for *b*, $3.22\times10^{-2}\text{L}\cdot\text{mol}^{-1}$, is divided by N_A to yield the excluded volume per one molecule and then its cubic root is calculated as below. Compared with the mean intermolecular distance at 10 atm, it is found that the value obtained thus is just of the order of a quarter of the mean distance.

$$\sqrt[3]{\frac{b}{N_A}} = \sqrt[3]{\frac{3.22\times10^{-2}\times10^{-3}}{6.02\times10^{23}}} = 3.8\times10^{-10}\,\text{m} = 0.38\ \text{nm}$$

Ans. mean intermolecular distance : 1.6 nm, about four times larger

(c), (d) The dimension of a/V is $[\text{L}\cdot\text{atm}\cdot\text{mol}^{-1}]$.

$$[a/V] = \frac{[\text{L}^2\cdot\text{atm}\cdot\text{mol}^{-1}]}{[\text{L}]} = [\text{L}\cdot\text{atm}\cdot\text{mol}^{-1}]$$

On the other hand, the dimension of *PV* which represents work due to volume change is $[\text{L}\cdot\text{atm}]$. Comparing these two dimensions, it is easy to see that a/V represents energy per mole. This is also seen by rewriting the *van der Waals* equation of state as the right side of eq(279), which not only indicates that a/V has the same dimension as *RT*, i.e., energy per mole but also represents the deviation of *P*-*V* work from that of the ideal gas (see **"note"**).

$$P+a/V^2=\frac{RT}{V-b} \;\Rightarrow\; PV+\frac{a}{V}=RT\left(\frac{V}{V-b}\right)\approx RT \qquad (279)$$

The molar volume of the gas at which the energy of a/V becomes approximately one tenth of the dipole-dipole interaction energy is calculated as below to yield $2.29\ \mathrm{L\cdot mol^{-1}}$.

$$\frac{a}{V}=0.6\ \mathrm{kJ\cdot mol^{-1}}\times 0.1 \;\Rightarrow\; V=\frac{a}{600\times 0.1}=\frac{1.36[\mathrm{dm^6\cdot atm\cdot mol^{-2}}]}{600\times 0.1[\mathrm{J\cdot mol^{-1}}]}$$

$$=\frac{1.36\times 10^{-6}\times 1.01\times 10^{5}[\mathrm{m^6\cdot Pa\cdot mol^{-2}}]}{600\times 0.1[\mathrm{J\cdot mol^{-1}}]}$$

$$=2.29\times 10^{-3}[\mathrm{m^3\cdot mol^{-1}}]=2.29\ \mathrm{L\cdot mol^{-1}}$$

Also, the length of an edge of a cube whose volume (volume per molecule) is equal to the volume obtained above divided by *Avogadro*'s number is calculated as $1.56\ \mathrm{nm}$.

$$\sqrt[3]{2.29\times 10^{-3}/6.02\times 10^{23}}=1.56\ \mathrm{nm}$$

To know that the thermal energy or the work of an ideal gas done by pressure is of the order of $RT=8.31\times 300=2493\ \mathrm{J\cdot mol^{-1}}$ will provide us with an important viewpoint in judging how large the intermolecular interaction must be for a gas to deviate significantly from the ideal gas law.

Ans. volume: $2.29\ \mathrm{L\cdot mol^{-1}}$, mean intermolecular distance : $1.56\ \mathrm{nm}$

"note" The internal pressure defined by $(\partial U/\partial V)_T$ is given by a/V^2 for the *van der Waals* equation of state. Therefore, a/V can be said a work done by internal pressure (see **problem 3-31**).

Solution 2-64

It is important to know the reason why the intermolecular attractive force (or repulsive force) causes the pressure of a gas to decrease (or increase) from a view point of kinetic theory of molecular collisions with the wall of a container.

You may argue "When the intermolecular attractive force works between two molecules, the velocity of the molecule attracted by the other molecule is reduced resulting in the collision with the wall with the reduced velocity, i.e., reduced momentum change, leading to a reduced pressure." Or, "Because of the symmetry of motion against the inversion of time, the molecule attracted by the other molecule will escape from the attractive force field with the same velocity as the one with which the molecule entered the attractive force field. For this, there is no change in the velocity." Or you may say that "The velocity which matters is the one along the *z*-axis (perpendicular to the wall) but not along the intermolecular axis and, hence, some molecules are accelerated but the same number of molecules are decelerated."

Making the problem simpler, let us consider the one-dimensional case. Denoting the potential energy due to the intermolecular force by $U(x)\le 0$, the total energy of a molecule is given by eq(280).

$$\varepsilon=\frac{1}{2}mu^2+U(x) \qquad (280)$$

Considering that the velocity distribution follows the *Boltzmann* distribution, the mean velocity $\bar{u}$ is given by eq(281).

$$\bar{u} = \int_{-l}^{l}\int_{0}^{\infty} u \exp\left(-\frac{1}{2}mu^2 - U(x)\Big/k_BT\right)dxdu \Bigg/ \int_{-l}^{l}\int_{0}^{\infty} \exp\left(-\frac{1}{2}mu^2 - U(x)\Big/k_BT\right)dxdu \qquad (281)$$

However, the molecules which can collide with the wall are considered those which satisfy the condition of $\varepsilon \geq 0$ in eq(280). In other words, those which have the velocity larger than $u_{\min}$ satisfying $\varepsilon \geq 0$. Then the mean velocity of the molecules which can collide with the wall, denoted $\bar{u}_{coll}$, is given by eq(282),

$$\bar{u}_{coll} = \int_{-l}^{l}\int_{u_{\min}}^{\infty} u \exp\left(-\frac{1}{2}mu^2 - U(x)\Big/k_BT\right)dxdu \Bigg/ \int_{-l}^{l}\int_{0}^{\infty} \exp\left(-\frac{1}{2}mu^2 - U(x)\Big/k_BT\right)dxdu \qquad (282)$$

and, as compared with eq(281), it is clear that eq(283) is satisfied.

$$\bar{u}_{coll} < \bar{u} \qquad (283)$$

After this, following to the procedure in which the equation of state for the ideal gas is derived based on the kinetic theory (see **solution 2-9**) $\bar{u}$ is replaced by $\bar{u}_{coll}$ to show that the pressure of a real gas is reduced as compared with that for the ideal gas. Alternatively, we can say that when the attractive intermolecular force is working, some molecules remain within the attractive intermolecular force field and, hence, the number of molecules which can collide with the wall is reduced by that number, resulting in the reduced pressure.

For a general method to derive the equation of state of a real gas on the basis of statistical mechanics, see Chapter 5.

Ans. See the text.

Solution 2-65

Employing the data given in the table in the problem, P is plotted against $\bar{V}$ in Fig.40 for the two gases. Since the temperatures for the two gases are different, it is hard to tell how closely or how differently they behave from the Fig.40 alone.

Next, a value for the compression factor $z = P\bar{V}/RT$ is calculated at each pressure given in the table and then it is plotted against the reduce pressure $P_R = P/P_C$ in Fig.41. When the compression factor is expressed in terms of the reduced pressure, even gases of different properties are seen to behave rather closely, as far as

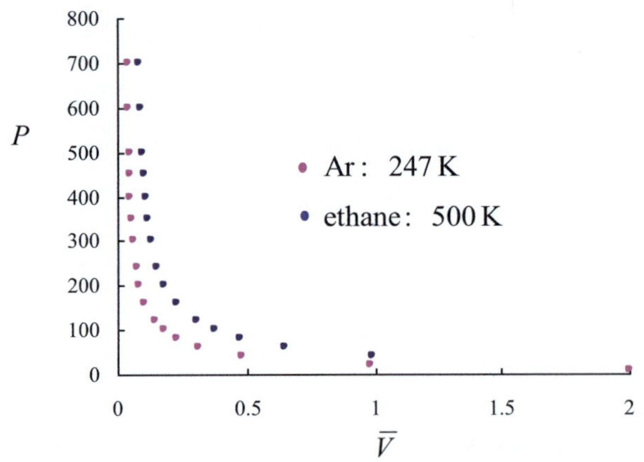

Fig.40 P vs. $\bar{V}$ plots for Ar and ethane.

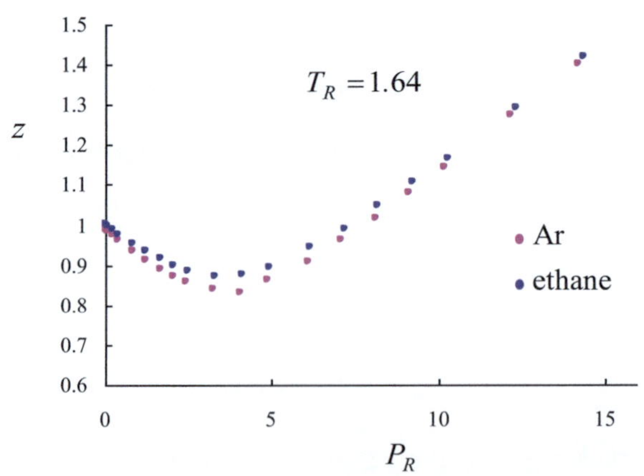

Fig.41 Compression factor z plotted against reduced P (P_R).

the reduced temperature is the same. Compared with Ar and ethane, gases of much more different natures one another, such as polar and non-polar gases, are worth examining along this line to understand a role of the reduced equation of state. Although Z can simply be plotted against P to obtain plots not much different from those in Fig.41, the meaning for using the reduced variables is lost.

Ans. See the text.

Solution 2-66

(a) Under the same pressure (the same number of molecules), the rate of effusion is proportional to the collision number per the cross section of the orifice. From eq(22) given in the problem, the collision number is seen to be inversely proportional to the square root of the molecular weight (see *Graham*'s law **solution 2-71**). Consequently, the ratio of the numbers of He and N_2 molecules flowing out per unit time is calculated as below.

$$\frac{\text{He (amount of outflow)}}{\text{N}_2\text{ (amount of outflow)}}=\frac{\sqrt{M_{\text{N}_2}}}{\sqrt{M_{\text{He}}}}=\sqrt{\frac{28.02}{4.00}}=2.65$$

Ans. ratio of the numbers of outflow molecules: $\text{He}/\text{N}_2=2.65$

(b) The mean free path of gaseous molecules is given by eq(284) below, in which σ is the collision cross section between molecules of the same species.

$$\lambda=\frac{k_BT}{2^{1/2}\sigma P} \qquad (284)$$

For the same number of molecules, the mean free path of N_2 is shorter than that of He. For this, employing the value of $\sigma=0.43\text{ nm}^2$ for N_2, the mean free path is calculated as below.

$$\lambda=\frac{1.38\times10^{-23}\times300}{2^{1/2}\times0.43\times10^{-18}\times1.01\times10^5}=67\text{ nm}$$

Ans. diameter of the orifice: less than 67 nm

[advanced problem] The law of *Graham* was established in 1831. Needless to say, nanotechnology or even micro-technology was not available at that time. How large do you guess the diameter of the orifice he employed for his experiments was?

Solution 2-67

A change in the number of molecules in the vessel with time due to the effusion can be described by eq(285) as is introduced in **problem 2-66**.

$$\frac{dN}{dt}=-Z \qquad (285)$$

Employing the equation of state for the ideal gas, $PV=Nk_BT$, the change of the number of the molecules is expressed by a change in the pressure as is given by eq(286).

$$\frac{V}{k_BT}\frac{dP}{dt}=-\left(\frac{1}{2\pi mk_BT}\right)^{1/2}\times P\times S_0 \qquad (286)$$

Putting the constants in eq(286) together and denoting it as k_{leak}, eq(286) can be expressed as eq(286'), which is the first order rate equation with respect to the pressure P and k_{leak} is an apparent rate constant of the equation (eq(287)). Therefore, the pressure change with time is given by eq(286'').

$$\frac{dP}{dt} = -k_{leak}P \qquad (286'), \qquad P = P_9 e^{-k_{leak}t} \qquad (286'')$$

$$k_{leak} = \left(\frac{1}{2\pi m k_B T}\right)^{1/2} \frac{k_B T \cdot S_0}{V} = \left(\frac{k_B T}{2\pi m}\right)^{1/2} \frac{S_0}{V} \qquad (287)$$

Substituting the given values into eq(287), the value for k_{leak} is calculated as follows, and from it the time required for the pressure to drop to one half of the initial pressure, half-life $t_{1/2}$, is calculated as 176 h.

$$k_{leak} = \left(\frac{1.381\times10^{-23}\times300\times6.02\times10^{23}}{2\pi\times4.00\times10^{-3}}\right)^{1/2} \times \frac{\pi\times(10^{-6})^2}{10^{-3}} = 9.90\times10^{-7}\,\mathrm{s}^{-1}$$

$$t_{1/2} = \frac{0.629}{k_{leak}} = 176\ \mathrm{h}$$

Ans. rate equation for pressure: eq(287''), half-life of pressure: 176 h

Solution 2-68

The rate equation of effusion expressed in terms of pressure and given in the preceding solution can also be expressed in terms of the volume of the balloon under the atmospheric pressure by eq(288)..

$$\frac{dN}{dt} = -Z$$

$$= -\frac{P}{k_B T}\frac{dV}{dt} = -\left(\frac{1}{2\pi m k_B T}\right)^{1/2} \times P \times S_0 \qquad (288)$$

Putting the constants in eq(288) together and denoting it as $k_{leak(P)}$, eq(288) can be expressed by eq(289), which is the zeroth order rate equation of the volume. Here, no gas is assumed to flow into the balloon from outside.

$$\frac{dV}{dt} = -k_{leak(P)} \qquad (289), \qquad k_{leak(P)} = \left(\frac{1}{2\pi m k_B T}\right)^{1/2} k_B T \cdot S_0 = \left(\frac{k_B T}{2\pi m}\right)^{1/2} S_0 \qquad (290)$$

The values consistent with the given conditions are substituted into eq(290) to find the value for $k_{leak(P)}$ as below.

$$k_{leak(P)} = \left(\frac{1.381\times10^{-23}\times300\times6.02\times10^{23}}{2\pi\times4.00\times10^{-3}}\right)^{1/2} \times\pi\times(10^{-6})^2 = 9.90\times10^{-10} \qquad (\mathrm{m}^3\cdot\mathrm{s}^{-1})$$

Solving eq(289) for V, the volume of the balloon after t-second is found as eq(291).

$$V(t) = V_0 - k_{leak(P)}t \qquad (291)$$

Denoting the density of the air as ρ_{air} and that of He as ρ_{He}, the buoyancy of the balloon is proportional to the left side of eq(292). Since the buoyancy must be greater than the sum of the weights of the balloon and the He gas inside the balloon, eq(292) must be satisfied.

$$V\times(\rho_{air} - \rho_{\mathrm{He}}) \geq m_b \qquad (292)$$

The volume of an ideal gas of 1 mole under 1 atm at 300 K is 24.6 dm^3 and using this value to calculate the densities in eq(292), the volume of the balloon is found to be greater than 2 L as given below.

$$V \geq \frac{m_b}{\rho_{air} - \rho_{\mathrm{He}}} = \frac{m_b\ [\mathrm{g}]}{\dfrac{(28.8-4)\ [\mathrm{g}]}{24.6\ [\mathrm{L}]}} = \frac{24.6\times2}{24.8} = 2.0\ \mathrm{L}$$

From eq(291), the time having elapsed just before the volume of the balloon becomes smaller than 2 L is calculated as 11.7 days. Namely, the balloon will fall to the ground in 11-12 days, though if the diameter of the small hole is doubled, the flow rate will become four times greater, and the balloon will fall down in 2-3 days.

$$t = \frac{1\ [\mathrm{L}]}{9.90\times10^{-7}[\mathrm{L\cdot s^{-1}}]} = 11.7\ \text{days}$$

Ans. The balloon will fall down to the ground in 11.7days.

Solution 2-69

(a) According to the *Graham* law, the number of effluent molecules through an orifice is inversely proportional to its molecular weight. Therefore, the ratio of oxygen to nitrogen molecules effluent is calculated as below to yield $O_2 : N_2 = 1:1.069$. Eventually, the molar ratio of the gases in the container approaches to this value.

$$O_2 : N_2 = \frac{1}{\sqrt{32.00}} : \frac{1}{\sqrt{28.02}} = 1:1.069$$

Ans. molar ratio: $O_2 : N_2 = 1:1.069$

(b) After one cycle of the effusion-compression, the molar ratio of the two gases becomes $O_2 : N_2 = 1:1.069$. Then by repeating the cycle *n* times, the ratio achieved will be $O_2 : N_2 = 1:(1.069)^n$. In order to make this ratio 1:100, the number of the cycles needed is calculated as below to find more than 69 cycles.

$$n\cdot\ln 1.069 = \ln 100 \ \Rightarrow\ n = \frac{\ln 100}{\ln 1.069} = 69.01$$

Ans. more than 69 cycles

(c)The molecular weight of each fluorine compound is calculated as $^{238}UF_6 = 238.141 + 6\times19.004 = 352.149$ and $^{235}UF_6 = 235.117 + 6\times19.004 = 349.141$, from which the molar ratio of these compounds after one cycle of effusion-compression is calculated as below.

$$^{238}UF_6 : {}^{235}UF_6 = \frac{1}{\sqrt{3352.149}} : \frac{1}{\sqrt{349.141}} = 1:1.0043$$

Assuming that the natural abundance of ^{235}U is 1% of ^{238}U and the initial molar ratio of the two gases is $0.01:0.99$, then to achieve the ratio of 10 to 1 after *n* cycles, *n* is found, as calculated below, to be ca.230800.

$$0.99\times1 : 0.01\times(1.0043)^n = 1:10 \ \Rightarrow\ n\cdot\ln 1.0043 = 990 \ \Rightarrow\ n \approx 230800 \quad \text{cycles}$$

Ans. more than 230800 cycles

"note" There are two major methods employed to separate gases effectively. One of them is, as are seen in **problems 2-66 and 2-67**, to use the fact that the rate of effusion or diffusion of a gas is inversely proportional to the molecular weight. The other is to use the fact that the solubility of a gas in a liquid or the adsorption of a gas on solid surface varies from gas to gas, which is essentially based on principle of chromatography.

Solution 2-70

(a) Since the effusion rate is inversely proportional to the square root of the molecular weight, when the rate constant for the gas 1 is given by $k_1 = a/\sqrt{M_1}$, the rate constant for the gas 2 can be expressed as $k_2 = a/\sqrt{M_2}$ with the same constant *a*.

Ans. $k_2 = a/\sqrt{M_2}$

(b) After a while since the commencement of effusion, the gas 1 in the container A flows out but at the same time the gas 1 in B flows in A. Just opposite to this holds for the gas 2. Denoting the number of gas 1 in A as N_1^A and that in B as N_1^B, etc., eqs(293) and (294) hold for the gas 1. The superscript ^{0} indicates values under the initial conditions.

$$N_1^A + N_1^B = N_1^0 \qquad (293), \qquad P_1^0 V = N_1^0 RT \qquad (294)$$

The rate equation for the gas 1 in A due to the flow out and in is given by eq(295') via eq(295).

$$\frac{dN_1^A}{dt} = -k_1 N_1^A + k_1 N_1^B \qquad (295)$$

$$= -2k_1 N_1^A + k_1 N_1^0$$

$$= -2k_1 (N_1^A - \frac{1}{2} N_1^0) \qquad (295')$$

Solving eq(295') for N_1^A and N_1^B, they are given by eq(296) and the partial pressures in the container A and B denoted by P_1^A and P_1^B are given by eq(296').

$$N_1^A = \frac{1}{2} N_1^0 e^{-2k_1 t} + \frac{1}{2} N_1^0 , \qquad N_1^B = -\frac{1}{2} N_1^0 e^{-2k_1 t} + \frac{1}{2} N_1^0 \qquad (296)$$

$$P_1^A = \frac{1}{2} P_1^0 (e^{-2k_1 t} + 1) , \qquad P_1^B = \frac{1}{2} P_1^0 (1 - e^{-2k_1 t}) \qquad (296')$$

Solving for the gas 2 similarly, the numbers of the molecules and the partial pressures in the container A and B are found to be given by eq(297) and eq(297'), respectively.

$$N_2^A = -\frac{1}{2} N_2^0 e^{-2k_2 t} + \frac{1}{2} N_2^0 , \qquad N_2^B = \frac{1}{2} N_2^0 e^{-2k_2 t} + \frac{1}{2} N_2^0 \qquad (297)$$

$$P_2^A = \frac{1}{2} P_2^0 (1 - e^{-2k_2 t}) , \qquad P_2^B = \frac{1}{2} P_2^0 (1 + e^{-2k_2 t}) \qquad (297')$$

Consequently, the total pressures in the container A and B are given by eq (298) and eq(298'), respectively.

$$P_1^A + P_2^A = \frac{1}{2} P_1^0 (e^{-2k_1 t} + 1) + \frac{1}{2} P_2^0 (1 - e^{-2k_2 t}) \qquad (298)$$

$$P_1^B + P_2^B = \frac{1}{2} P_1^0 (1 - e^{-2k_1 t}) + \frac{1}{2} P_2^0 (1 + e^{-2k_2 t}) \qquad (298')$$

Ans. rate equation: eq(295) or (295'); pressure change : eq(298) and eq(298')

(c) Putting the gas 1 as N_2 and gas 2 as He, the ratio of the effusion rate constants is calculated to be 2.65 as below

$$k_1/k_2 = \sqrt{M_2/M_1} = \sqrt{4.01/28.02} = 2.65$$

Employing this value and the given initial pressure of each gas, the pressures in the containers A and B are plotted against time in Fig.42. The time scale is in the unit of $1/k_1$. In either case, both pressures finally approach to the same equilibrium pressure. When the gas in A is replaced with CH_4 at 2 atm, the pressures in the containers vary with time as shown in Fig.43.

The conditions under which the pressure in either container takes a maximum or a minimum are readily found from eq(298) or (298').

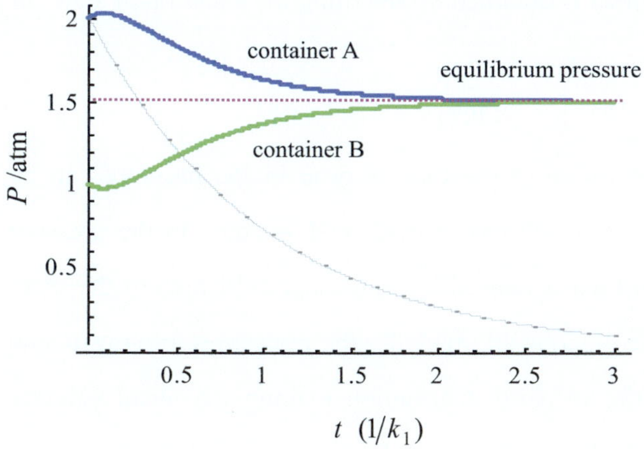

Fig.42 Pressure variation due to flow of gases through a small hole. The curve in grey shows a pressure change in A due to the flow of the gas into vacuum.

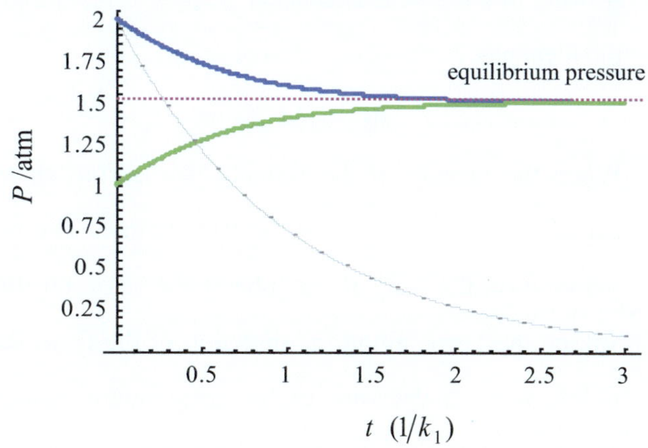

Fig.43 Same as Fig.42 except for the gas in A replaced by CH_4 at 2 atm.

Ans. pressure variations with time: Fig.42

Solution 2-71

Bernoulli's equation is derived first. When a fluid flows through a tube whose diameter is position-dependent, the cross sections of A_1 and A_2 marked by blue-dotted circles at the positions 1 and 2, respectively move per unit time to the positions indicated by red-dotted circles shown in Fig.44.

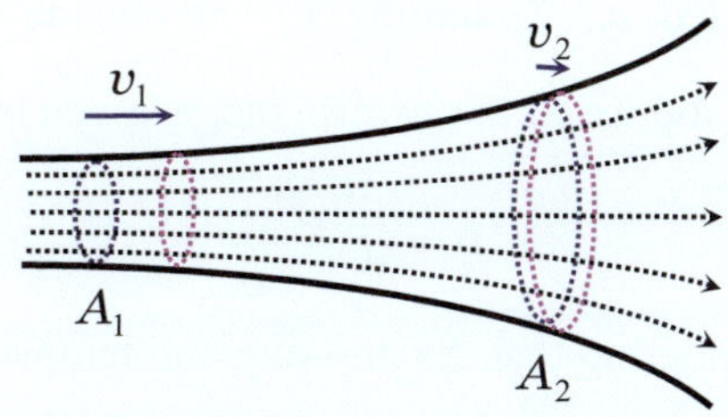

Fig,44 Illustration of fluid flow in a tube whose diameter is position dependent.

Denoting the densities of the fluid at the positions 1 and 2 by ρ_1 and ρ_2, respectively, eq(299) is derived from the conservation law of the mass and eq(300) from the conservation law of energy.

$$A_1 v_1 \times \rho_1 = A_2 v_2 \times \rho_2 = M \qquad (299)$$

$$P_1 A_1 \times v_1 + \frac{1}{2} M v_1^2 = P_2 A_2 \times v_2 + \frac{1}{2} M v_2^2 \qquad (300)$$

Substituting the relation in eq(299) into eq(300), eq(301) is obtained. If the fluid is assumed incompressible

$$P_1 A_1 \times v_1 + \frac{1}{2} A_1 v_1 \rho_1 v_1^2 = P_2 A_2 \times v_2 + \frac{1}{2} A_2 v_2 \rho_1 v_2^2 \qquad (301)$$

(see**"note"**), then eq(302) is satisfied since the density of the fluid is position independent. In other words,

$$\rho_1 = \rho_2 \;\Rightarrow\; A_1 v_1 = A_2 v_2 \qquad (302)$$

the volume of the fluid which has flowed per unit time is constant. Applying eq(302) to eq(301), eq(301') is obtained which is the same equation given in the problem, i.e., *Bernoulli*'s equation.

$$P_1 + \frac{1}{2} \rho v_1^2 = P_2 + \frac{1}{2} \rho v_2^2 \qquad (301')$$ same as eq(23) given in the problem

(a) When a gas is moving rapidly in a thin tube, the pressure on a cross section of the tube may erroneously be considered to be large. However, the cross section is a hypothetical one and no gas molecules are reflected by the surface of the cross section to generate pressure. The reason why the pressure of the gas flowing in the tube is decreased is clear from eq(303) which is obtained by rewriting the equation of state for the ideal gas.

$$P_1V_1 = N_1k_BT_1 \;\Rightarrow\; P_1 = \frac{N_1k_BT_1}{V_1} = \frac{\rho k_BT_1}{m} \qquad (303)$$

When the density of the flowing gas is constant everywhere in the tube, a drop in the pressure can be explained by a decrease in the temperature. The pressure of the gas at position 1 is equal to the pressure working on the wall of the tube at the same position which is caused by molecular collisions to the wall. Taking the z-axis along the direction of the flow, such collisions are made by the molecules moving x- and y-directions. A decrease of the temperature changes the velocity distribution making the mean velocity smaller and leading to the lower pressure. Of course, the velocity distribution along the z-axis is also changed in a similar manner though.

Ans. See text (a).

"note" A gas can be regarded as an incompressible fluid when its motion is slower than the sound velocity.

(b) The position of the *Pitot* tube is indicated by the subscript 1 and the flow speed of the gas at that position is denoted υ_1. Assuming $\upsilon_2 = 0$ outside the outlet of the tube, the velocity of the gas in the tube υ_1, namely the speed of the airplane, is given by eq(304) from eq(301').

$$\upsilon_1 = \sqrt{\frac{2(P_2 - P_1)}{\rho}} \qquad (304)$$

Considering that the pressure and temperature of the atmosphere in which the airplane is flying are 0.5 atm and 273 K and that the apparent molecular weight is 28.8 for the air, the density of the gas is calculated as below.

$$\rho = \frac{28.8\times10^{-3}}{2\times22.4\times10^{-3}} = 0.643\ \text{kg}\cdot\text{m}^{-3}$$

Employing this density and given conditions, υ_1 is calculated to be $638\ \text{km}\cdot\text{h}^{-1}$ as shown below.

$$\upsilon_1 = \sqrt{\frac{2\times0.1\times1.01\times10^5}{0.643}}\times3600 = 638\ \text{km}\cdot\text{h}^{-1} \qquad (305)$$

Ans. speed of the airplane: $638\ \text{km}\cdot\text{h}^{-1}$

Solution 2-72

(a) Substituting the solution (eq(25) in the problem) into the left side of eq(24) given in the problem, eq(24') is obtained.

$$\frac{\partial\rho(t,\ x)}{\partial t} = \exp\left\{-\frac{(x+ut)^2}{4Dt}\right\}\left\langle -\frac{N_0}{2\sqrt{4\pi D}}t^{-3/2} - \frac{N_0}{\sqrt{4\pi Dt}}\frac{u(x+ut)}{2Dt} + \frac{N_0}{\sqrt{4\pi Dt}}\frac{(x+ut)^2}{4Dt^2}\right\rangle$$

$$= \frac{N_0}{\sqrt{4\pi Dt}}\exp\left\{-\frac{(x+ut)^2}{4Dt}\right\}\left\langle -\frac{1}{2t} - \frac{u(x+ut)}{2Dt} + \frac{(x+ut)^2}{4Dt^2}\right\rangle \qquad (24')$$

Similarly, each term of the right side of eq(24) in the problem is calculated as eq(24'') and eq(24''').

$$u\frac{\partial\rho(t,\ x)}{\partial x} = \frac{N_0}{\sqrt{4\pi Dt}}\exp\left\{-\frac{(x+ut)^2}{4Dt}\right\}\times -\frac{u(x+ut)}{2Dt} \qquad (24'')$$

$$D\frac{\partial^2 \rho(t,\ x)}{\partial x^2}=\frac{N_0}{\sqrt{4\pi Dt}}\exp\left\{-\frac{(x+ut)^2}{4Dt}\right\}\left\langle-\frac{1}{2t}+\frac{(x+ut)^2}{4Dt^2}\right\rangle \qquad (24''')$$

As is obvious from a comparison of eqs(24') to (24'''), it is now confirmed by substituting the solution (eq(25)) into the both sides of the equation of diffusion (eq(24)) individually that both sides of eq(24) in the problem are equal.

Ans. See text (a).

(b) When the diffusion has reached an equilibrium, the partial differential equation of eq(26) given in the problem can be rewritten as eq(26'), which is a second-order ordinary differential equation.

$$u\frac{\partial \rho(x)}{\partial x}+D\frac{\partial^2 \rho(x)}{\partial x^2}=0 \Rightarrow \frac{d^2\rho(x)}{dx^2}+\frac{u}{D}\frac{d\rho(x)}{dx}=0 \qquad (26')$$

Then, it is obvious that a solution of eq(26') is given by eq(26''). For the significance of the result see **2-74**.

$$\rho(x)=\rho(0)\exp\left(-\frac{u}{D}x\right) \qquad (26'')$$

Ans. solution at equilibrium: eq(26'')

Solution 2-73

The particles in the water column are assumed here to be distributed as a function of the gravitational potential energy according to the *Boltzmann* distribution law (see next problem). The number of the particles at the height *h* along the column is denoted by $N(h)$ and the number at the reference height of zero is denoted by N_0. Then these numbers are related by eq(306), in which *m* is the mass of one particle and *g* is the gravitational acceleration. Notice that eq(306) takes the same form as the barometric equation discussed in **2-50.**

$$\frac{N(h)}{N_0}=\exp\left(-\frac{mgh}{k_BT}\right)=\exp\left(-\frac{N_A\times mgh}{RT}\right) \qquad (306)$$

Taking logarithms of both sides of eq(306), it is rewritten as eq(306'). Plotting the left side of eq(306') against *h*, its slope is given by $-(N_Amg/RT)$.

$$\ln\frac{N(h)}{N_0}=-\frac{N_A\times mgh}{RT}=-\left(\frac{N_A\times mg}{RT}\right)\times h \qquad (306')$$

Data in the table are plotted in Fig.45 and the slope of the best-fitted straight line is given there. The apparent mass of the particle taking buoyancy into account is calculated as below.

$$m=\frac{4\pi r^3}{3}\times d=\frac{4\pi}{3}\times(0.212\times10^{-6})^3\times(1.206-1)\times10^{-3}\times10^6$$
$$=8.22\times10^{-18}\text{kg}$$

Substituting this for *m* in the equation of the slope,

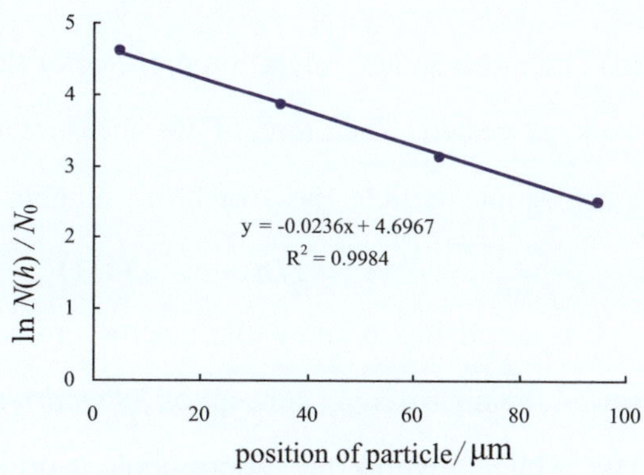

Fig.45 Logarithmic plot of particle distribution.

$$0.0236 = -\left(\frac{N_A \times mg}{RT}\right)$$
$$= -\left(\frac{N_A \times 4\pi(0.212\times10^{-6})^3/3\times1.206\times10^{-3}\times10^6\times9.8}{8.31\times288}\right) \qquad (306'')$$

Avogadro's number is calculated as 7.01×10^{23}.

$$N_A = \frac{0.0236\times10^6\times8.31\times288}{9.8\times8.22\times10^{-18}} = 7.01\times10^{23}$$

Ans. *Avogadro*'s number: 7.01×10^{23}

Solution 2-74

(a) The mean potential energy is expressed by eq(307) using the *Boltzmann* distribution law and $-fu/D$ is obtained as its value.

$$<-f\cdot z> = \frac{\int_0^\infty -f\cdot z\rho_0\exp\left(-\frac{u}{D}z\right)dz}{\int_0^\infty \rho_0\exp\left(-\frac{u}{D}z\right)dz} = -f\times\frac{D}{u} \qquad (307)$$

Ans. mean potential energy: $-fD/u$

(b) Equating $f=-m\cdot g$ since the force is opposite to the positive direction of z-axis, eq(308) is obtained.

$$m\cdot g\times\frac{D}{u} = k_BT \Rightarrow \frac{u}{D} = \frac{m\cdot g}{k_BT} = \frac{m\cdot g\cdot N_A}{RT} \qquad (308)$$

Substituting this into eq(307) given in the problem, eq(309) is obtained, which is the same in the form as eq(306) given in the preceding solution.

$$\frac{\rho(z)}{\rho_0} = \frac{N(z)/V}{N_0/V} = \frac{N(z)}{N_0} = \exp\left(-\frac{N_Am\cdot gz}{RT}\right) \qquad (309)$$

Ans. See text (b).

(c) Putting $\mu=1/6\pi\eta a$, the relation of $u=m\cdot g/6\pi\eta a$ is obtained. Substituting this relation into eq(308), the following eq(310) is derived.

$$D = \frac{RT}{6\pi\eta aN_A} \qquad (310)$$

Also, it is obvious from the relation of $P=Nk_BT/V$ that eq(309) is identical to the barometric equation.

Ans. See text (c).

(d) The mean square migration distance of the particle is given by eq(311) as is discussed when the random walk is treated. Therefore, if we measure the migration distance $\overline{x^2}$ during t sec and the radius of the macroscopic particle, the *Avogadro*'s number can be determined from eq(310) through the value of D.

$$\overline{x^2} = 2Dt \qquad (311)$$

It is surprising to know that macroscopic measurements of such as the migration distance of particles can reveal the microscopic concept of *Avogadro*'s number. It may be more surprising to find that the distribution law which governs the microscopic world of atoms and molecules is considered to control also the distribution and mean energy of macroscopic particles.

Ans. See text (d).

Solution 2-75

Since the *Stokes* equation of $6\pi\eta av$ is so frequently used to describe the motion of molecules in fluid, its derivation is described here in detail. The following paper is refereed to for the answer given here.

H.Hayakawa, *Mathematical Science* 54, 2007, ch.6. exercise 6.1 and its answer (in Japanese).

Instead of considering that a sphere is moving in a stationary fluid with a constant velocity υ, the sphere is regarded as being placed stationary in the fluid flowing with the constant velocity υ along the x-axis. And then for the latter situation, the viscosity resistant force working on the sphere is worked out.

For incompressible fluids, the *Stokes* equation is generally given by eq(312), where $\boldsymbol{u}$ and P represent the velocity and pressure at the position (x, y, z) of the fluid, respectively The origin of the coordinate is at the center of the sphere.

$$\eta\nabla^2\boldsymbol{u} = \nabla P \qquad (312a), \qquad \nabla\cdot\boldsymbol{u} = 0 \qquad (312b)$$

We now solve eq(312) under the boundary conditions of eq(313).

$$\boldsymbol{u}(r=a) = 0 \qquad (313a), \qquad \boldsymbol{u}\to U\boldsymbol{e}_x,\ r\to\infty \qquad (313b)$$

Calculating the divergence ($div \equiv \nabla$, $div\,\boldsymbol{a} = (\nabla\cdot\boldsymbol{a})$) of eq(312$b$), eq(314) is obtained.

$$\left.\begin{aligned}\nabla\cdot\eta\nabla^2\boldsymbol{u} &= \nabla\cdot\nabla P = \nabla^2 P\\ \nabla\cdot\eta\nabla^2\boldsymbol{u} &= \eta\nabla^2\nabla\cdot\boldsymbol{u} = 0\end{aligned}\right\} \quad \Rightarrow \nabla^2 P = 0 \qquad (314)$$

Therefore, eq(312a) is regarded as *Poisson*'s equation and the solution is expressed by a linear combination of a general and special solutions for the corresponding homogeneous differential equation.

$$\boldsymbol{u} = \nabla\Phi \quad \Rightarrow \quad \nabla^2\Phi = 0 \qquad (315) \qquad \text{(incompressible fluid)}$$

This is equivalent to regard the flow of the fluid as being expressed by a sum of the laminar ($U\boldsymbol{e}_x$) and turbulent flows. Among the turbulent flows, the solution of the homogeneous equation regular at $r\to\infty$ is given by eq(316), in which $P_n(\cos\vartheta)$ is *Legendre*'s polynomial.

$$\Phi(r,\vartheta) = \sum_{n=0}^{\infty} C_n r^{-n-1} P_n(\cos\vartheta) \qquad (316)$$

The laminar flow is expressed as a potential by eq(317). U is the velocity of the laminar flow.

$$\Phi_u = Ux = Ur\cos\vartheta \qquad (317)$$

Since $\boldsymbol{u}$ has to satisfy the boundary condition of $\boldsymbol{u} = 0$ at $r = a$ (eq(313a)), in eq(316) only the term which cancels $Ur\cos\vartheta$ of eq(317) is required. Therefore, only the first term of eq(316) is needed and it is expressed as eq(318).

$$\begin{aligned}\Phi(r,\vartheta) &= C_1 r^{-2} P_1(\cos\vartheta) = C_1 r^{-2}\cos\vartheta\\ &= \frac{C_1 x}{r^3} = -C_1\frac{\partial}{\partial x}\left(\frac{1}{r}\right) \qquad (318)\end{aligned}$$

The velocity of the turbulent flow is given by $\boldsymbol{u} = \nabla\Phi$ and from eq(318), each component of the flow velocity is found as expressed by eq(319).

$$\boldsymbol{u}_\Phi = C_1\left(\frac{1}{r^3} - \frac{3x^2}{r^5},\ -\frac{3xy}{r^5},\ -\frac{3xz}{r^5}\right) \qquad (319)$$

Similarly, the pressure is expressed by a harmonic function (functions which satisfy *Laplace* differential equation), so that from the condition of $P_{r=\infty} \to P_\infty$ (constant), it is given by eq(320) just like eq(318).

$$P = \frac{\partial}{\partial x}\left(\frac{1}{r}\right) \qquad (320)$$

When $\boldsymbol{u}_S$ is defined by eq(321), by using the relations of $\nabla^2(1/r) = 0$ except for the origin and $\nabla^2 r = 2/r$, $\boldsymbol{u}_S$ is found to be a special solution of the *Stokes* equation.

$$\boldsymbol{u}_S = \frac{1}{2\eta}\nabla\left(\frac{\partial r}{\partial x}\right) - \frac{1}{\eta r}\boldsymbol{e}_x \qquad (321)$$

Proof

$$\begin{aligned} \nabla\cdot\boldsymbol{u}_S &= \frac{1}{2\eta}\nabla^2\left(\frac{\partial r}{\partial x}\right) - \frac{1}{\eta}\left(\frac{\partial 1/r}{\partial x}\right) \\ &= \frac{1}{2\eta}\frac{\partial}{\partial x}\nabla^2 r - \frac{1}{\eta}\left(\frac{\partial 1/r}{\partial x}\right) \\ &= \frac{1}{2\eta}\frac{\partial}{\partial x}\frac{2}{r} - \frac{1}{\eta}\left(\frac{\partial 1/r}{\partial x}\right) \\ &= 0 \qquad \text{corresponding to } (312a) \end{aligned}$$

$$\begin{aligned} \eta\nabla^2\boldsymbol{u}_S &= \eta\nabla^2\left\{\frac{1}{2\eta}\nabla\left(\frac{\partial r}{\partial x}\right) - \frac{1}{\eta r}\boldsymbol{e}_x\right\} \\ &= \left\{\frac{1}{2}\nabla\cdot\nabla^2\left(\frac{\partial r}{\partial x}\right) - \nabla^2\frac{1}{r}\boldsymbol{e}_x\right\} \\ &= \frac{1}{2}\nabla\cdot\frac{\partial}{\partial x}\nabla^2 r \\ &= \nabla\cdot\frac{\partial 1/r}{\partial x} = \nabla\cdot P \qquad \text{corresponding to } (312b) \end{aligned}$$

From eq(321), the x, y and z components of $\boldsymbol{u}_S$ is given by eq(322).

$$\boldsymbol{u}_S = -\frac{1}{2\eta}\left(\frac{1}{r} + \frac{x^2}{r^3}, \frac{xy}{r^3}, \frac{xz}{r^3}\right) \qquad (322)$$

Therefore, the flow velocity $\boldsymbol{u}$ is given by a sum of the three equations, eqs(313*b*), (319) and (322), with constants A and C_1 as in eq(323).

$$\boldsymbol{u} = U\boldsymbol{e}_x + \boldsymbol{u}_\Phi + A\boldsymbol{u}_S \qquad (323)$$

From this, each component of $\boldsymbol{u}$ is expressed as eq(324).

$$\left.\begin{aligned} u_x &= U + C_1\left(\frac{1}{r^3} - \frac{3x^2}{r^5}\right) - A\frac{1}{2\eta}\left(\frac{1}{r} + \frac{x^2}{r^3}\right) \\ u_y &= -\frac{3xy}{r^5}C_1 - A\frac{1}{2\eta}\frac{xy}{r^3} \\ u_z &= -\frac{3xz}{r^5}C_1 - A\frac{1}{2\eta}\frac{xz}{r^3} \end{aligned}\right\} \qquad (324)$$

To meet the boundary conditions of $\boldsymbol{u} = 0$ at $r = a$, eq(324') must be satisfied upon substitution of a for r in eq(324).

$$\begin{aligned} u_x &= U + \frac{1}{a^3}C_1 - A\frac{1}{2\eta}\frac{1}{a} - \left(\frac{3}{a^2}C_1 + A\frac{1}{2\eta}\right)\frac{x^2}{a^3} = 0 \\ u_y &= -\frac{3xy}{r^5}C_1 - A\frac{1}{2\eta}\frac{xy}{r^3} \Rightarrow \frac{3}{a^2}C_1 + A\frac{1}{2\eta} = 0 \\ u_z &= -\frac{3xz}{r^5}C_1 - A\frac{1}{2\eta}\frac{xz}{r^3} \Rightarrow \frac{3}{a^2}C_1 + A\frac{1}{2\eta} = 0 \end{aligned} \qquad (324')$$

From eq(324'), the constants A and C_1 are determined as in eq(325).

$$\left.\begin{aligned} &U-\frac{A}{2\eta a}+\frac{C_1}{a^3}=0 \\ &\frac{A}{2\eta}+\frac{3C_1}{a^2}=0 \end{aligned}\right\} \Rightarrow \left.\begin{aligned} &A=\frac{3\eta aU}{2} \\ &C_1=-\frac{Ua^3}{4} \end{aligned}\right\} \quad (325)$$

Employing these constants, the three components of the flow velocity are obtained as follow.

$$\left.\begin{aligned} u_x &= U-\frac{Ua}{4}\left(\frac{a^2}{r^3}+\frac{3}{r}\right)-\frac{3aU}{4r^3}\left(1-\frac{a^2}{r^2}\right)x^2 \\ u_y &= -\frac{3aU}{4}\frac{xy}{r^3}\left(1-\frac{a^2}{r^2}\right) \\ u_z &= -\frac{3aU}{4}\frac{xz}{r^3}\left(1-\frac{a^2}{r^2}\right) \end{aligned}\right\} \quad (326)$$

The *Stokes* equation (eq(312a)) can be expressed as eq(312'a) by using u_x in eq(326).

$$\eta\nabla^2 u_x = -\frac{3a\eta U}{2}\left(\frac{1}{r^3}-\frac{3x^2}{r^5}\right)=\frac{\partial P}{\partial x} \quad (312'a)$$

Since the secondary partial differential of $1/r$ with x yields the relation given by eq(327),

$$\frac{\partial}{\partial x}\frac{\partial}{\partial x}\left(\frac{1}{r}\right)=-\frac{\partial}{\partial x}\frac{x}{r^3}=-\frac{1}{r^3}+\frac{3x^2}{r^5} \quad (327)$$

the pressure P is found to be proportional to the left side of eq(328) and, as compared with eq(312'a), is expressed as the right side of eq(328).

$$P \propto \frac{\partial}{\partial x}\left(\frac{1}{r}\right) \Rightarrow P=-\frac{3aU\eta}{2}\frac{x}{r^3}+P_\infty \quad (328)$$

The force working on the sphere is given by integrating the stress $P_{rx|r=a}$ given by eq(329) over the spherical surface

$$P_{rx|r=a}=-\left(\frac{x}{r}P_{xx}+\frac{y}{r}P_{xy}+\frac{z}{r}P_{xz}\right)_{r=a} \quad (329)$$

Here, the components of the tensor are given by eq(330).

$$\left.\begin{aligned} P_{xx} &= P-2\eta\frac{\partial u_x}{\partial x} \\ P_{xy} &= -\eta\left(\frac{\partial u_y}{\partial x}+\frac{\partial u_x}{\partial y}\right) \\ P_{xz} &= -\eta\left(\frac{\partial u_z}{\partial x}+\frac{\partial u_x}{\partial z}\right) \end{aligned}\right\} \quad (330)$$

By working out eq(329) carefully using eq(330) (see **[exercise 3]** below), eq(329') is obtained.

$$P_{rx|r=a}=\frac{1}{a}\left(P_\infty x-\frac{3}{2}\eta U\right) \quad (329')$$

The force working on the sphere, therefore, is obtained by integrating this over the sphere surface and is given by eq(331).

$$\begin{aligned} F &= -2\pi a^2\int_0^\pi P_{rx|r=a}\sin\vartheta d\vartheta \\ &= -2\pi a\int_0^\pi\left(P_\infty r\cos\vartheta-\frac{3}{2}\eta U\right)\sin\vartheta d\vartheta = 3\pi a\eta U\int_0^\pi\sin\vartheta d\vartheta = 6\pi a\eta U \end{aligned} \quad (331)$$

Noticing the relation of $F=\zeta U$, the equation to be proved is obtained as eq(332).

$$\zeta = 6\pi a \eta \qquad (332) \qquad (Q.E.D.)$$

Ans. eq(332) and the derivation leading to it

It may be meaningful to refer to the following.

Grasping eq(328) in a rough manner, the pressure difference ΔP between the front ($x \geq 0$)and the back face ($x \leq 0$) of the sphere can be considered to be given by eq(333).

$$\Delta P = -3aU\eta \times \frac{x}{r^3}, \quad x \geq 0 \qquad (333)$$

By integrating eq(333) over the front surface of the sphere, the total pressure due to the pressure difference in eq(333) is given by the following equation.

$$\int_{\varphi=0}^{\pi}\int_{\vartheta=0,r=a}^{\pi} \frac{3aU\eta}{r^3}(\boldsymbol{x}\cdot \boldsymbol{dS}) = \frac{3aU\eta}{a^3}\times\int_0^{\pi}\int_0^{\pi} a\cos^2\vartheta\cdot a^2\sin\vartheta d\vartheta d\varphi = \frac{2}{3}\times 3\pi aU\eta = 2\pi aU\eta$$

The value calculated correctly by using the tensor of pressure is three times larger than the value obtained above, though the difference is smaller as compared with the ratio of four for the surface area to the cross section of the sphere.

[exercise 1] Derive eq($312'a$).

(solution) Without giving any explanations, only the equations crucial for derivation are given below.

$$\frac{\partial r}{\partial x} = \frac{x}{r}, \quad \frac{\partial r}{\partial y} = \frac{y}{r}, \quad \frac{\partial r}{\partial z} = \frac{z}{r}$$

$$\nabla^2 = \frac{\partial^2}{\partial x^2} + \frac{\partial^2}{\partial y^2} + \frac{\partial^2}{\partial z^2}$$

$$= \frac{1}{r^2}\frac{\partial}{\partial r}\left(r^2\frac{\partial}{\partial r}\right) + \frac{1}{r^2}\left\{\frac{1}{\sin\vartheta}\frac{\partial}{\partial\vartheta}\left(\sin\vartheta\frac{\partial}{\partial\vartheta}\right) + \frac{1}{\sin^2\vartheta}\frac{\partial^2}{\partial\varphi^2}\right\}$$

$$u_x = U - \frac{aU}{4r}\left(3+\frac{a^2}{r^2}\right) - \frac{3x^2}{4r^3}Ua\left(1-\frac{a^2}{r^2}\right) \qquad x = r\cos\vartheta$$

$$= U - \frac{1}{4}aUf(r) - \frac{3}{4}aUg(r)\cos^2\vartheta, \qquad f(r) = \left(\frac{3}{r}+\frac{a^2}{r^3}\right), \quad g(r) = \left(\frac{1}{r}-\frac{a^2}{r^3}\right)$$

$$\frac{1}{r^2}\frac{\partial}{\partial r}\left(r^2\frac{\partial}{\partial r}\right)f(r) = \frac{1}{r^2}\frac{\partial}{\partial r}r^2 f'(r) = f''(r) + \frac{2}{r}f'(r) = \left(-\frac{3}{r^2}-\frac{3a^2}{r^4}\right)' + \frac{2}{r}\left(-\frac{3}{r^2}-\frac{3a^2}{r^4}\right)$$

$$= \left(\frac{6}{r^3}+\frac{12a^2}{r^5}\right) - \frac{6}{r^3} - \frac{6a^2}{r^5} = \frac{6a^2}{r^5}$$

$$\frac{1}{r^2}\frac{\partial}{\partial r}\left(r^2\frac{\partial}{\partial r}\right)g(r)\cos^2\vartheta = \cos^2\vartheta\times\left(g''(r)+\frac{2}{r}g'(r)\right) = \cos^2\vartheta\times\left\{\left(-\frac{1}{r^2}+\frac{3a^2}{r^4}\right)' + \frac{2}{r}\left(-\frac{1}{r^2}+\frac{3a^2}{r^4}\right)\right\}$$

$$= \cos^2\vartheta\times\left(\frac{2}{r^3}-\frac{12a^2}{r^5}-\frac{2}{r^3}+\frac{6a^2}{r^5}\right) = -\cos^2\vartheta\frac{6a^2}{r^5}$$

$$\frac{1}{r^2}\frac{1}{\sin\vartheta}\frac{\partial}{\partial\vartheta}\left(\sin\vartheta\frac{\partial}{\partial\vartheta}\right)g(r)\cos^2\vartheta=\frac{g(r)}{r^2}\frac{1}{\sin\vartheta}\frac{\partial}{\partial\vartheta}\left(-2\sin^2\vartheta\cos\vartheta\right)=\frac{g(r)}{r^2}\left(-4\cos^2\vartheta+2\sin^2\vartheta\right)$$

$$=\frac{g(r)}{r^2}\left(2-6\cos^2\vartheta\right)=2\left(\frac{1}{r^3}-\frac{a^2}{r^5}\right)-6\left(\frac{1}{r^3}-\frac{a^2}{r^5}\right)\cos^2\vartheta$$

$$\frac{1}{4}\times\frac{6a^2}{r^5}+\frac{3}{4}\left\langle-\frac{6a^2}{r^5}\cos^2\vartheta-6\left(\frac{1}{r^3}-\frac{a^2}{r^5}\right)\cos^2\vartheta+2\left(\frac{1}{r^3}-\frac{a^2}{r^5}\right)\right\rangle$$

$$=\frac{3a^2}{2r^5}-\frac{9}{2}\frac{1}{r^3}\cos^2\vartheta+\frac{3}{2}\frac{1}{r^3}-\frac{3a^2}{2r^5}=\frac{3}{2}\left(\frac{1}{r^3}-\frac{3}{r^3}\cos^2\vartheta\right)=\frac{3}{2}\left(\frac{1}{r^3}-\frac{3x^2}{r^5}\right)$$

$$\nabla^2 u_x=-\frac{3aU}{2}\left(\frac{1}{r^3}-\frac{3x^2}{r^5}\right)=\frac{\partial P}{\partial x}\quad\Rightarrow\quad P\propto\frac{\partial}{\partial x}\left(\frac{1}{r}\right)=-\frac{x}{r^3}\qquad(312'b)$$

$$\frac{\partial}{\partial x}=-\frac{\sin\vartheta}{r}\frac{\partial}{\partial\vartheta}+\cos\vartheta\frac{\partial}{\partial r}\qquad(x\text{ is equivalent to }z.)$$

[exercise 2] Confirm that the solution given by eq(326) satisfies $\nabla\cdot\boldsymbol{u}=0$.

(solution)

$$\frac{\partial u_x}{\partial x}=-\frac{\partial}{\partial r}\left\{\frac{aU}{4r}\left(3+\frac{a^2}{r^2}\right)\right\}\cdot\frac{\partial r}{\partial x}-\frac{3x}{2r^3}Ua\left(1-\frac{a^2}{r^2}\right)-\frac{3x^2Ua}{4}\frac{\partial}{\partial r}\left\{\frac{1}{r^3}\left(1-\frac{a^2}{r^2}\right)\right\}\cdot\frac{\partial r}{\partial x}$$

$$=-aU\left\langle\left(-\frac{3}{4r^2}-\frac{3a^2}{4r^4}\right)\cdot\frac{x}{r}+\frac{3x}{2r^3}\left(1-\frac{a^2}{r^2}\right)+\frac{3x^2}{4}\left(-\frac{3}{r^4}+\frac{5a^2}{r^6}\right)\cdot\frac{x}{r}\right\rangle$$

$$=aU\left\langle\frac{3x}{4r^3}\left(1+\frac{a^2}{r^2}\right)\cdot x-\frac{3x}{2r^3}\left(1-\frac{a^2}{r^2}\right)+\frac{3x^3}{4}\left(\frac{3}{r^5}-\frac{5a^2}{r^7}\right)\right\rangle$$

$$\frac{\partial u_y}{\partial y}=-3xy\frac{\partial}{\partial r}\left\{\frac{1}{4r^3}Ua\left(1-\frac{a^2}{r^2}\right)\right\}\cdot\frac{\partial r}{\partial y}-\frac{3x}{4r^3}Ua\left(1-\frac{a^2}{r^2}\right)$$

$$=aU\left\langle\frac{3}{4}xy^2\left(\frac{3}{r^5}-\frac{5a^2}{r^7}\right)-\frac{3x}{4r^3}\left(1-\frac{a^2}{r^2}\right)\right\rangle$$

$$\frac{\partial u_z}{\partial z}=-3xz\frac{\partial}{\partial r}\left\{\frac{1}{4r^3}Ua\left(1-\frac{a^2}{r^2}\right)\right\}\cdot\frac{\partial r}{\partial z}-\frac{3x}{4r^3}Ua\left(1-\frac{a^2}{r^2}\right)$$

$$=aU\left\langle\frac{3}{4}xz^2\left(\frac{3}{r^5}-\frac{5a^2}{r^7}\right)-\frac{3x}{4r^3}Ua\left(1-\frac{a^2}{r^2}\right)\right\rangle$$

$$\frac{\partial u_x}{\partial x}+\frac{\partial u_y}{\partial y}+\frac{\partial u_z}{\partial z}=aU\left\langle\frac{3x}{4}\left(\frac{3}{r^5}-\frac{5a^2}{r^7}\right)(x^2+y^2+z^2)-\frac{6x}{4r^3}\left(1-\frac{a^2}{r^2}\right)-\frac{6x}{4r^3}\left(1-\frac{a^2}{r^2}\right)+\left(3+\frac{3a^2}{r^3}\right)\cdot\frac{x}{r^3}\right\rangle$$

$$=aU\left\langle\frac{x}{r^3}\left(\frac{9}{4}-\frac{12}{4}+\frac{3}{4}\right)+a^2\frac{x}{r^5}\left(-\frac{15}{4}+\frac{12}{4}+\frac{3}{4}\right)\right\rangle=0$$

$$\nabla\cdot\boldsymbol{u}=0\qquad(Q.E.D.)$$

[exercise 3] Derive eq(329') from eqs(329) and (330).

(solution)

$$\frac{\partial u_x}{\partial x}=-\frac{\partial}{\partial r}\left\{\frac{aU}{4r}\left(3+\frac{a^2}{r^2}\right)\right\}\cdot\frac{\partial r}{\partial x}-\frac{3x}{2r^3}Ua\left(1-\frac{a^2}{r^2}\right)-\frac{3x^2Ua}{4}\frac{\partial}{\partial r}\left\{\frac{1}{r^3}\left(1-\frac{a^2}{r^2}\right)\right\}\cdot\frac{\partial r}{\partial x}$$

$$=-aU\left\langle\left(-\frac{3}{4r^2}-\frac{3a^2}{4r^4}\right)\cdot\frac{x}{r}+\frac{3x}{2r^3}\left(1-\frac{a^2}{r^2}\right)+\frac{3x^2}{4}\left(-\frac{3}{r^4}+\frac{5a^2}{r^6}\right)\cdot\frac{x}{r}\right\rangle$$

$$=-aU\left\langle-\frac{3x}{4r^3}\left(1+\frac{a^2}{r^2}\right)+\frac{3x}{2r^3}\left(1-\frac{a^2}{r^2}\right)-\frac{3x^3}{4r^5}\left(3-\frac{5a^2}{r^2}\right)\right\rangle$$

$$=-aU\left\langle\frac{3x}{4r^3}\left(1-\frac{3a^2}{r^2}\right)-\frac{3x^3}{4r^5}\left(3-\frac{5a^2}{r^2}\right)\right\rangle$$

$$\frac{\partial u_x}{\partial y}=-\frac{\partial}{\partial r}\left\{\frac{aU}{4r}\left(3+\frac{a^2}{r^2}\right)\right\}\cdot\frac{\partial r}{\partial y}-\frac{3x}{2r^3}Ua\left(1-\frac{a^2}{r^2}\right)-\frac{3x^2Ua}{4}\frac{\partial}{\partial r}\left\{\frac{1}{r^3}\left(1-\frac{a^2}{r^2}\right)\right\}\cdot\frac{\partial r}{\partial y}$$

$$=-aU\left\langle\left(-\frac{3}{4r^2}-\frac{3a^2}{4r^4}\right)\cdot\frac{y}{r}+\frac{3x}{2r^3}\left(1-\frac{a^2}{r^2}\right)+\frac{3x^2}{4}\left(-\frac{3}{r^4}+\frac{5a^2}{r^6}\right)\cdot\frac{y}{r}\right\rangle$$

$$=-aU\left\langle-\frac{3y}{4r^3}\left(1+\frac{a^2}{r^2}\right)+\frac{3x}{2r^3}\left(1-\frac{a^2}{r^2}\right)+\frac{3x^2y}{4r^5}\left(-3+\frac{5a^2}{r^2}\right)\right\rangle$$

In a similar manner,

$$\frac{\partial u_y}{\partial x}=-3aU\left\langle\left(-3+\frac{5a^2}{r^2}\right)\frac{x^2y}{4r^5}+\frac{3}{4}\frac{y}{r^3}\left(1-\frac{a^2}{r^2}\right)\right\rangle$$

$$\frac{\partial u_x}{\partial z}=-aU\left\langle\left(-\frac{3}{4r^2}-\frac{3a^2}{4r^4}\right)\cdot\frac{z}{r}+\frac{3x^2}{4}\left(-\frac{3}{r^4}+\frac{5a^2}{r^6}\right)\cdot\frac{z}{r}\right\rangle$$

$$\frac{\partial u_z}{\partial x}=-\frac{3aU}{4}\left(-3+\frac{5a^2}{r^2}\right)\frac{x^2z}{r^5}-\frac{3aU}{4}\frac{z}{r^3}\left(1-\frac{a^2}{r^2}\right)$$

$$P_{xx}=P-2\eta\frac{\partial u_x}{\partial x}$$

$$=-\frac{3aU\eta}{2}\frac{x}{r^3}+P_\infty+2\eta aU\left\langle\frac{3x}{4r^3}\left(1-\frac{3a^2}{r^2}\right)-\frac{3x^3}{4r^5}\left(3-\frac{5a^2}{r^2}\right)\right\rangle$$

$$=P_\infty+\eta aU\left\langle-\frac{9xa^2}{2r^5}-\frac{3x^3}{2r^5}\left(3-\frac{5a^2}{r^2}\right)\right\rangle$$

$$=P_\infty+\frac{3x}{2r^5}\eta aU\left\langle-3a^2-x^2\left(3-\frac{5a^2}{r^2}\right)\right\rangle$$

$$P_{xy}=-\eta\left(\frac{\partial u_y}{\partial x}+\frac{\partial u_x}{\partial y}\right)$$

$$=3a\eta U\left\langle\left(-3+\frac{5a^2}{r^2}\right)\frac{x^2y}{4r^5}+\frac{3}{4}\frac{y}{r^3}\left(1-\frac{a^2}{r^2}\right)\right\rangle+a\eta U\left\langle-\frac{3y}{4r^3}\left(1+\frac{a^2}{r^2}\right)+\frac{3x^2y}{4r^5}\left(-3+\frac{5a^2}{r^2}\right)\right\rangle$$

$$=a\eta U\left\{\left\langle\left(-3+\frac{5a^2}{r^2}\right)\frac{3x^2y}{4r^5}+\frac{9}{4}\frac{y}{r^3}\left(1-\frac{a^2}{r^2}\right)\right\rangle+\left\langle-\frac{3y}{4r^3}\left(1+\frac{a^2}{r^2}\right)+\frac{3x^2y}{4r^5}\left(-3+\frac{5a^2}{r^2}\right)\right\rangle\right\}$$

$$=a\eta U\left\{\left\langle\left(-3+\frac{5a^2}{r^2}\right)\frac{3x^2y}{2r^5}+\frac{3}{2}\frac{y}{r^3}\left(1-\frac{2a^2}{r^2}\right)\right\rangle+\frac{3x}{2r^3}\left(1-\frac{a^2}{r^2}\right)\right\}$$

$$P_{xz}=-\eta\left(\frac{\partial u_z}{\partial x}+\frac{\partial u_x}{\partial z}\right)$$

$$=\frac{3a\eta U}{4}\left\langle\left(-3+\frac{5a^2}{r^2}\right)\frac{x^2z}{r^5}+\frac{z}{r^3}\left(1-\frac{a^2}{r^2}\right)\right\rangle+a\eta U\left\langle\left(-\frac{3}{4r^2}-\frac{3a^2}{4r^4}\right)\cdot\frac{z}{r}+\frac{3x^2}{4}\left(-\frac{3}{r^4}+\frac{5a^2}{r^6}\right)\cdot\frac{z}{r}\right\rangle$$

$$=\frac{3a\eta U}{4}\left\langle\left(-3+\frac{5a^2}{r^2}\right)\frac{x^2z}{r^5}+\frac{z}{r^3}\left(1-\frac{a^2}{r^2}\right)\right\rangle+\frac{3a\eta U}{4}\left\langle-\left(1+\frac{a^2}{r^2}\right)\cdot\frac{z}{r^3}+\left(-3+\frac{5a^2}{r^2}\right)\cdot\frac{x^2z}{r^5}\right\rangle$$

$$=\frac{3a\eta U}{4}\left\langle 2\left(-3+\frac{5a^2}{r^2}\right)\frac{x^2z}{r^5}+\frac{z}{r^3}\left(1-\frac{a^2}{r^2}\right)-\left(1+\frac{a^2}{r^2}\right)\cdot\frac{z}{r^3}\right\rangle$$

$$=\frac{3a\eta U}{4}\left\langle 2\left(-3+\frac{5a^2}{r^2}\right)\frac{x^2z}{r^5}-\frac{2a^2z}{r^5}\right\rangle$$

$$\left(\frac{x}{r}P_{xx}+\frac{y}{r}P_{xy}+\frac{z}{r}P_{xz}\right)/=P_\infty\frac{x}{ra\eta U}+\frac{3x^2}{2r^6}\left\langle-3a^2-x^2\left(3-\frac{5a^2}{r^2}\right)\right\rangle$$

$$+\left\{\left\langle\left(-3+\frac{5a^2}{r^2}\right)\frac{3x^2y^2}{2r^6}+\frac{3}{2}\frac{y^2}{r^4}\left(1-\frac{2a^2}{r^2}\right)\right\rangle\right\}$$

$$+\frac{3}{4}\left\langle 2\left(-3+\frac{5a^2}{r^2}\right)\frac{x^2z^2}{r^6}-\frac{2a^2z^2}{r^6}\right\rangle$$

$$=-\frac{a^2}{2r^6}\left(9x^2+3z^2+6y^2\right)-\frac{3}{2r^6}\left(-3+\frac{5a^2}{r^2}\right)x^2\left\langle-x^2-y^2-z^2\right\rangle+\frac{3}{2}\frac{y^2}{r^4}$$

$$=-\frac{a^2}{2r^6}\left(3r^2-9x^2+3y^2\right)-\frac{9x^2}{2r^4}+\frac{3}{2}\frac{y^2}{r^4}$$

$$=-\frac{3a^2}{2r^2}-\frac{9x^2}{2r^4}\left(1-\frac{a^2}{r^2}\right)+\frac{3}{2}\frac{y^2}{r^4}\left(1-\frac{a^2}{r^2}\right)$$

$$r=a,\quad\left(\frac{x}{r}P_{xx}+\frac{y}{r}P_{xy}+\frac{z}{r}P_{xz}\right)=-\frac{3}{2}a\eta U+P_\infty x\qquad (Q.E.D.)$$

Solution 2-76

(a) For the square-well potential given, the integral of eq(334) is performed by separating the range of integration into two parts and is obtained as eq(334') as shown below.

$$B(T)=-2\pi N_A\int_0^\infty[e^{-u(r)/k_BT}-1]\cdot r^2dr\qquad(334)$$

$$=2\pi N_A\int_0^\sigma 1\times r^2dr-2\pi N_A\int_\sigma^{\sigma+d}[e^{\varepsilon/k_BT}-1]\cdot r^2dr$$

$$=\frac{2\pi N_A\sigma^3}{3}-\frac{2\pi N_A}{3}[e^{\varepsilon/k_BT}-1]\left\langle(\sigma+d)^3-\sigma^3\right\rangle$$

$$\approx\frac{2\pi N_A\sigma^3}{3}-\frac{2\pi N_A}{3}\cdot\frac{\varepsilon}{k_BT}\left\langle(\sigma+d)^3-\sigma^3\right\rangle\qquad(334')$$

Ans. $B(T)$ and parameters of potential: eq(334')

(b) By comparing eq(334') with the second virial coefficient given by eq(335) for the *van der Waal* equation of state, eqs(336) and (336') are obtained for b and a, respectively.

$$B(T) = b - a/RT \qquad (335)$$

$$b = \frac{2\pi N_A \sigma^3}{3} = \frac{1}{2} N_A \left(\frac{4\pi}{3} \sigma^3 \right) \qquad (336)$$

$$a = \frac{2\pi N_A^2}{3} \cdot \varepsilon \left\{ (\sigma + d)^3 - \sigma^3 \right\} = N_A \cdot \varepsilon \times \frac{1}{2} N_A \underbrace{\left\langle \left(\frac{4\pi}{3} (\sigma + d)^3 \right) - \left(\frac{4\pi}{3} \sigma^3 \right) \right\rangle}_{\text{potential acvive volume}} \qquad (336')$$

From eq(336), it is seen that the *van der Waals* parameter b is equal to the excluded volume (per mole) between two molecules. On the other hand, it is seen that eq(336') is expressed by the product of the depth (or height for repulsive potential) of the potential times the volume in which the potential is active (or non-zero effectively). Therefore, the parameter a can be said proportional to the product of (depth of potential) $\times$ (potential active volume). The sign of a is determined by whether the potential is attractive or repulsive.

Ans. b: excluded volume (per mole), a: (depth of potential) $\times$ (potential active volume)

Solution 2-77

The virial theorem and equations of state of gases

Since no intermolecular force is operative in the present case, the force which works on a molecule is only seen at $x = (0,\ a)$, $y = (0,\ b)$ and $z = (0,\ c)$ upon collisions to the wall of the rectangle. Therefore, the first term on the right side of eq(337) (reproduced below with subscript υ)

$$\overline{T}_\upsilon = \sum_i \overline{T_{\upsilon_i}} = -\frac{1}{2} \sum_i \overline{\boldsymbol{F}_i \cdot \boldsymbol{r}_i} = -\frac{1}{2} \sum_i (\overline{X_i \cdot x_i} + \overline{Y_i \cdot y_i} + \overline{Z_i \cdot z_i}) \qquad (337) \text{ (eq(36) in the problem)}$$

can be expressed as eq(338).

$$\sum_i (\overline{X_i \cdot x_i}) = a \sum_i \overline{X_i} \qquad (338)$$

Eq(338) is equivalent to the total sum of the time-averaged force which works on each molecule moving along the x-direction at the wall at $x = a$ taken over all molecules. By denoting it as $-\overline{F_x}$ (note that the force is opposite to the direction of translational motion), $\overline{F_x}$ is related to the pressure of the gas by eq(339).

$$\sum_i \overline{X_i} = -\overline{F_x}, \qquad \frac{\overline{F_x}}{S} = \frac{\overline{F_x}}{bc} = P \qquad (339)$$

Substituting this result into the right side of eq(337), eq(340) is obtained. Furthermore, relying on the fact that the mean kinetic energy of the gas molecules at T is equal to $(3/2)Nk_BT$, eq(341) to be proved is obtained. See **solution 2-9** for a related problem.

$$\sum_i (\overline{X_i \cdot x_i}) = \sum_i (\overline{Y_i \cdot y_i}) = \sum_i (\overline{Z_i \cdot z_i}) = -P \times abc = -PV \qquad (340)$$

$$\overline{T}_\upsilon = \frac{3}{2} N k_B T = \frac{3}{2} PV \;\Rightarrow\; N k_B T = PV \qquad (341)$$

Eq(341) is also expressed as eq(341') relating the internal energy of the gas to the pressure.

$$PV = \frac{2}{3}E \qquad (341')$$

Ans. See eq(337) to eq(341) for the proof.

Derivation of eq(337)

The equation of motion for the i-th particle with the momentum $\boldsymbol{p}_i$ is given by the equation below.

$$\boldsymbol{F}_i = \dot{\boldsymbol{p}}_i$$

Now, employing the momentum and the coordinates of the i-th particle, the quantity G is defined by eq(342).

$$G = \sum_i \boldsymbol{p}_i \cdot \boldsymbol{r}_i \qquad (342)$$

Differentiating G defined by eq(342) with time, eq(343) is obtained,

$$\begin{aligned}\frac{dG}{dt} &= \sum_i \dot{\boldsymbol{p}}_i \cdot \boldsymbol{r}_i + \sum_i \boldsymbol{p}_i \cdot \dot{\boldsymbol{r}}_i \\ &= \sum_i F_i \cdot \boldsymbol{r}_i + \sum_i m_i \dot{\boldsymbol{r}}_i \cdot \dot{\boldsymbol{r}}_i \\ &= \sum_i F_i \cdot \boldsymbol{r}_i + 2\sum_i T_{\upsilon_i} \qquad (343)\end{aligned}$$

and for its time-averaged values, indicated by over lines, eq(344) is obtained from eq(343).

$$\overline{\frac{dG}{dt}} = \sum_i \overline{F_i \cdot r_i} + 2\sum_i \overline{T_{\upsilon_i}} \qquad (344)$$

The left side of eq(344) can be regarded as the mean of the integrated value of dG/dt over the time range $[0,\tau]$ ($\tau >> 1$), and it is given by eq(345).

$$\overline{\frac{dG}{dt}} = \lim_{\tau\to\infty}\frac{1}{\tau}\int_0^\tau \frac{dG}{dt}dt = \lim_{\tau\to\infty}\frac{1}{\tau}\langle G(\tau) - G(0)\rangle \qquad (345)$$

For the motions in which $G(\tau)$ takes finite values, eq(345) can be put equal to zero after a long enough time so that eq(346) holds.

$$\lim_{\tau\to\infty}\frac{1}{\tau}\langle G(\tau) - G(0)\rangle = 0 \qquad (346)$$

Consequently, eq(337) known as the virial theorem is derived from eq(344).

$$\overline{T}_\upsilon = \sum_i \overline{T_{\upsilon_i}} = -\frac{1}{2}\sum_i \overline{\boldsymbol{F}_i \cdot \boldsymbol{r}_i} = -\frac{1}{2}\sum_i (\overline{X_i \cdot x_i} + \overline{Y_i \cdot y_i} + \overline{Z_i \cdot z_i}) \qquad (337)$$

Based on this equation, it is easy to expect that equations of state for non-ideal gases can be derived by separating the term of $\boldsymbol{F}_i \cdot \boldsymbol{r}_i$ into two parts, one due to the force from the wall and the other due to the intermolecular force.

"reference" *Statistical Thermodynamics, R.Fowler and E.A.Guggenheim* ,Cambridge University Press, § 706, 1960. *Classical Mechanics, Herbert Goldstein*, Addison-Wesley Publishing Company, INC, 1957.

Solution 2-78

We consider one spherical molecule whose center is moving freely within a sphere of radius R. Since the mean intermolecular distance is R, another molecule, if any, must exist on the circumference on the average (Fig.46(a)). Taking this into account, the free space in which the molecule with a radius of a and indicated by a red circle can move freely is given by a sphere marked by a black dotted line with the radius of $R-2a$ (Fig.46(b)). On the other hand, the free space looked at from the other molecule marked by a red dotted line should be superposed on (b) as shown in Fig.46(c). Therefore, the free volume within which one molecule can move freely is found as within a sphere marked by a black dotted line and is given by eq(347).

$$\frac{4\pi}{3}(R-2a)^3\,/\,\text{molecule} \qquad (347)$$

As is expected from Fig.(c), for the most simplified liquid model, the liquid can be expressed by a medium in which free volumes represented by spheres of a constant radius (dotted circles) are dispersed randomly.

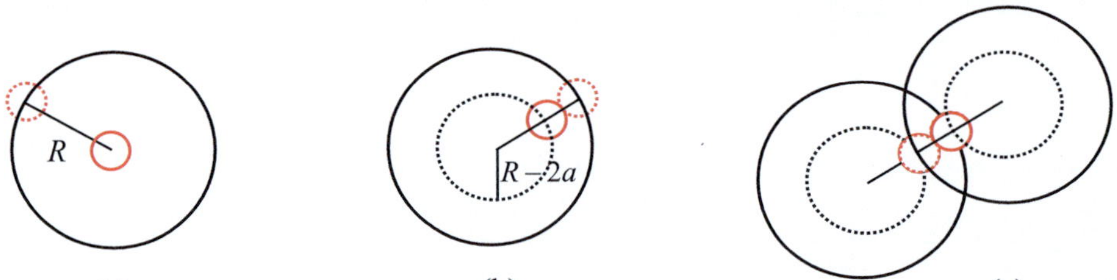

(a) (b) (c)

Fig.46 An image of free volume in which two identical molecules with the radius a and the mean intermolecular distance R can move freely.

Ans. free volume: eq(347)

“note” Free volumes of liquids can be acquired experimentally from the measurements of vapor pressures or of compression factor/expansion coefficient of a liquid.

III-3 Thermodynamics

Part of Worked Solutions

Solution 3-1

(a) The work done by a volume change of a gas is given by $-PdV$. Therefore, under the condition of a constant pressure, the work done by the gas whose volume changes from $5\ \mathrm{dm}^3$ to $50\ \mathrm{dm}^3$ is calculated as follows. A negative sign here indicates that the gas has lost the corresponding energy, namely the gas has

$$-\int_{5\times10^{-3}}^{50\times10^{-3}} 100\times10^5\,dV = -100\times10^5(50\times10^{-3}-5\times10^{-3}) = -450\times10^3 = -450\ \mathrm{kJ}$$

done the work to the surroundings (external). As is seen here, <u>in thermodynamics thermodynamic values are always measured from those of the system at an initial state which are taken as reference values</u>.

Ans. work done to the surroundings: 450 kJ

(b) The molar volume of an ideal gas at 273 K under 100 kPa is calculated as $22.7\times10^{-3}\,\mathrm{m}^3$.

$$V = \frac{RT}{P} = \frac{8.314\times273}{100\times10^3} = 22.7\times10^{-3}\,\mathrm{m}^3$$

The work done upon expanding this volume to $50\ \mathrm{dm}^3$ is found as -1792 J.

$$w(g) = -\int PdV = -\int \frac{RT}{V}dV = -RT\int_{22.7\times10^{-3}}^{50\times10^{-3}} \frac{1}{V}dV = -8.314\times273\times\ln\frac{50}{22.7} = -1792\ \mathrm{J}$$

Since the temperature of the ideal gas has not changed, there is no change in the internal energy of the gas. Therefore, the required values are calculated as follows.

$$\Delta U(g) = 0\ \mathrm{J},\ \text{also from } q(g)+w(g)=0,\quad q(g)=1792\ \mathrm{J}$$

$$\Delta H(g) = \Delta U(g)+\Delta(PV) = \Delta U(g) = 0\ \mathrm{J}$$

$$\Delta S(g) = \frac{q(g)}{T} = \frac{1792}{273} = 6.56\ \mathrm{J\cdot K^{-1}}$$

Ans. $q(g)=1792$ J, $w(g)=-1792$ J, $\Delta U(g)=0$, $\Delta H(g)=0$, $\Delta S(g)=6.56\ \mathrm{J\cdot K^{-1}}$

(c) The surroundings, which work as a heat bath kept at a constant temperature of 273 K, have lost the heat energy of $-q(g)$. Then, the entropy change in the heat bath is calculated as $-1792/273=-6.56\ \mathrm{J\cdot K^{-1}}$. Namely, a change in the total entropy for the whole system (gas + surroundings) is zero, since the expansion process here is a <u>reversible</u> one carried out at a constant temperature.

Ans. entropy change for the whole system: zero

Solution 3-2

(a) The maximum efficiency η_{max} is given by eq(1), where T_2 and T_1 are the temperatures (K) of the hot and cold heat reservoirs, respectively.

$$\eta_{max} = \frac{T_2-T_1}{T_2} \qquad (1)$$

S. Hirayama and M. Yamamoto, *Physical Chemistry Problems and Solutions*,
https://doi.org/10.1007/978-981-97-0165-0_11

From this equation, $\eta_{max} = \langle(273+150)-(273+30)\rangle/(273+150) = 0.284$ is obtained.

Ans. maximum efficiency: 0.284

(b)The maximum energy to be converted into work by using the heat energy absorbed from the hot heat reservoir is governed by the maximum efficiency given in (a). Since the ratio of the maximum work to the heat energy absorbed is equal to η_{max}, $0.284 \times 100\ \text{kJ} = 28.4\ \text{kJ}$ is found as a value of the maximum work.

Ans. maximum work : 28.4 kJ

(c) The entropy change in the hot heat reservoir is given by $-100\times10^3/423 = -236\ \text{J}\cdot\text{K}^{-1}$. Since the heat absorbed by the cold heat reservoir is $(100-28.4)\times10^3\ \text{J}$, the entropy change in the cold heat reservoir is given by $(100-28.4)\times10^3/303 = 236\ \text{J}\cdot\text{K}^{-1}$. It should be noted that during the energy conversion processes, the temperatures of the heat reservoirs are kept constant. The maximum work efficiency is achieved by reversible processes and, hence, the total entropy change in the whole system plus surroundings is zero for the present case.

Ans. entropy change in hot heat source: $-236\ \text{J}\cdot\text{K}^{-1}$, in cold heat source: $236\ \text{J}\cdot\text{K}^{-1}$

Solution 3-3

(a) Expressing the standard free energy of formation of each gas such as $\Delta G_f^\circ(\text{NH}_3)$, the standard free energy change of the reaction (1) at 298 K denoted ΔG_r° is given by the following equation. The subscript *r* signifies *reaction*.

$$\Delta G_r^\circ = \Delta G_f^\circ(\text{HCN}(g)) + \Delta G_f^\circ(\text{H}_2\text{O}(g)) - \Delta G_f^\circ(\text{NH}_3(g)) - \Delta G_f^\circ(\text{CO}(g))$$

From the values given in the table in the problem, therefore, ΔG_r° is calculated as below.

$$\Delta G_r^\circ = 124.7 + (-228.57) - (-16.45) - (-137.17) = 49.75\ \text{kJ}\cdot\text{mol}^{-1}$$

Ans. ΔG_r°: $49.75\ \text{kJ}\cdot\text{mol}^{-1}$

(b) Denoting the partial pressure of each gas when the reaction has reached an equilibrium as $P^{eq}_{\text{NH}_3}$, and so on, the equilibrium constant in terms of the partial pressures is given by eq(2).

$$K_P = \frac{P^{eq}_{\text{HCN}} \cdot P^{eq}_{\text{H}_2\text{O}}}{P^{eq}_{\text{NH}_3} \cdot P^{eq}_{\text{CO}}} \qquad (2) \qquad (eq: \textit{equilibrium})$$

Also, the equilibrium constant is related to the standard free energy change of the reaction through eq(3).

$$\Delta G_r^\circ = -RT\ln K_P \qquad (3)$$

From the values found for ΔG_r° in (a) and the relation above, a value for K_P is obtained as below. It should be noted that K_P is dimensionless.

$$K_P = \exp\left(-\frac{\Delta G_r^\circ}{RT}\right) = \exp\left(-\frac{49.75\times10^3}{8.31\times298}\right) = 1.88\times10^{-9}$$

Ans. $\Delta G_r^\circ = -RT\ln K_P$, $K_P = 1.88\times10^{-9}$

(c) Before the reaction starts, partial pressures of the gases are $P_{\text{NH}_3} = 100$ kPa, $P_{\text{CO}} = 100$ kPa, $P_{\text{HCN}} = 1$ kPa and $P_{\text{H}_2\text{O}} = 1$ kPa. Therefore, assuming the reaction proceeds to the right side, the free energy change of the reaction ΔG at this point is given by eq(4) and the value for ΔG is found as

$26.9 \times 10^3 \text{kJ} \cdot \text{mol}^{-1}$. Eq(4) is valid even if the reaction is not in equilibrium.

$$\Delta G = \Delta G_r^\circ + RT \ln \frac{P_{\text{HCN}} \cdot P_{\text{H}_2\text{O}}}{P_{\text{NH}_3} \cdot P_{\text{CO}}} \qquad (4)$$

$$= 49.75 \times 10^3 + 8.31 \times 298 \ln \frac{1 \times 1}{100 \times 100} = 26.9 \times 10^3 \text{kJ} \cdot \text{mol}^{-1}$$

Solution 1. Judging from that value, the reaction (eq(5))) (reaction (1) in the problem) proceeds in the direction of decreasing free energy, namely the second term of eq(4), decreases. Consequently, the reaction proceeds toward the left side.

$$\text{NH}_3(g) + \text{CO}(g) \Leftarrow \text{HCN}(g) + \text{H}_2\text{O}(g) \qquad (5)$$

Solution 2. The partial pressure of each gas in equilibrium is calculated from the value found for K_P. The total pressure is 202 kPa. In the equilibrium, the following equation is satisfied (see eq(2)) with x either po-

$$1.88 \times 10^{-9} = \frac{(1-x) \times (1-x)}{(100+x) \times (100+x)}$$

sitive or negative. Solving this equation, a value of 0.996 is found for x. Namely, when the the reaction system has reached equilibrium, the partial pressures of HCN and H_2O must be far smaller than 1 kP. Consequently, we can say that the reaction proceeds toward the left to reduce the partial pressures of HCN and H_2O.

Ans. $\Delta G = 26.9 \text{ kJ} \cdot \text{mol}^{-1}$, reaction proceeds toward the left side.

Solution 3-4

Values for the standard enthalpy of formation ΔH_f° of various types of compounds are usually listed in a text book of physical chemistry as the most fundamental thermodynamic data. For instance, the following values are given in the table in the data section of *Atkins' Physical Chemistry*.

$$\Delta H_f^\circ(\text{H}_2\text{O}) = -241.82 \ , \ \Delta H_f^\circ(\text{NO}) = 90.25 \ , \ \Delta H_f^\circ(\text{C}_2\text{H}_2) = 226.73 \ ,$$
$$\Delta H_f^\circ(\text{CO}_2) = -393.51 \ , \ \Delta H_f^\circ(\text{NH}_3) = -46.11 : \quad \text{unit in kJ} \cdot \text{mol}^{-1}$$

Using these values, the enthalpy of the reaction given in teh poblem ΔH_r (298K) is calculated as follows.

$$\begin{aligned}\Delta H_r(298\text{K}) &= 2\Delta H_f^\circ(\text{H}_2\text{O}(g)) + 2\Delta H_f^\circ(\text{NO}) + \Delta H_f^\circ(\text{C}_2\text{H}_2) - 2\Delta H_f^\circ(\text{CO}_2) - 2\Delta H_f^\circ(\text{NH}_3) \\ &= 2 \times (-241.82) + 2 \times 90.25 + 226.73 - 2 \times (-393.51) - 2 \times (-46.11) \\ &= 802.83 \text{ kJ} \cdot \text{mol}^{-1}\end{aligned}$$

On the other hand, in order to obtain the enthalpy of reaction at 400 K, we need to know how the enthalpy varies with temperature for every substance involved in the reaction. This is readily understood by referring to Fig.1 in which the enthalpies of reaction at two temperatures are illustrated.

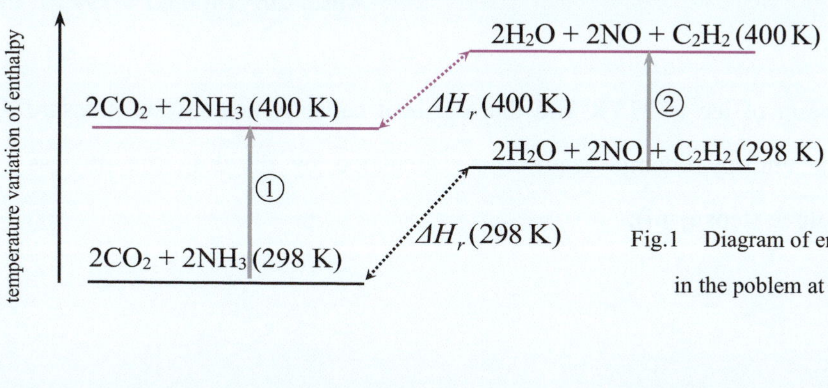

Fig.1 Diagram of enthalpy level of the reaction given in the poblem at the two temperatures.

The value for ΔH_r(298 K) is already obtained. In order to acquire a value for ΔH_r(400 K), we only need to know the enthalpy changes associated with the processes ① and ② shown in Fig.1. For this, it is necessary to find how the heat capacity varies with temperature for each one of the reactants and products. Here, we employ for the heat capacity either the following equations or constant values depending on compounds

$$C_P(CO_2) = 26.0 + 43.5\times10^{-3}T - 148.3\times10^{-7}T^2,\quad C_P(H_2O) = 30.36 + 9.61\times10^{-3}T + 11.8\times10^{-7}T^2$$

$$C_P(NH_3) = 35.06,\quad C_P(C_2H_2) = 43.93,\quad C_P(NO) = 29.844.\ \text{in unit of J}\cdot\text{K}^{-1}\text{mol}^{-1}.$$

The enthalpy change ΔH_1 associated with the process ① is given by eq(6)..

$$\Delta H_1 = \int_{298}^{400}\langle 2\times C_P(CO_2) + 2\times C_P(NH_3)\rangle dT \qquad (6)$$

$$= \int_{298}^{400}\langle 2\times(26.0 + 43.5\times10^{-3}T - 148.3\times10^{-7}T^2) + 2\times35.06\rangle dT$$

$$= 15.1_8\ \text{kJ}\cdot\text{mol}^{-1}$$

The enthalpy change ΔH_2 associated with the process ② is given by eq(7).

$$\Delta H_2 = \int_{298}^{400}\langle 2\times C_P(H_2O) + 2\times C_P(NO) + C_P(C_2H_2)\rangle dT \qquad (7)$$

$$= \int_{298}^{400}\langle 2\times(30.36 + 9.61\times10^{-3}T + 11.8\times10^{-7}T^2) + 2\times29.844 + 43.93\rangle dT$$

$$= 18.7_0\ \text{kJ}\cdot\text{mol}^{-1}$$

From these results, the enthalpy of reaction at 400 K ΔH_r(400 K) is found as follows.

$$\Delta H_r(400\ \text{K}) = \Delta H_r(298\ \text{K}) + \Delta H_2 - \Delta H_1 = 802.83 + 15.12 - 18.70 = 799\ \text{kJ}\cdot\text{mol}^{-1}$$

In conclusion, it is found that a temperature effect on the enthalpy of reaction is rather small for a change in temperature of that magnitude. Why is that for? The molecules participating in the present reaction are all small molecules consisted of less than four atoms. Molecular motions which mainly contribute to the heat capacity for such small molecules are translational and rotational motions. Since their contributions to the heat capacity are already almost 100% at room temperature, a change in the temperature, as far as it is not too large, causes only a slight change in the heat capacity and, hence, in the enthalpy. At much higher temperatures, however, vibrational motions come to contribute to the heat capacity and as a consequence of this, the enthalpy of reaction starts to depend strongly on the kind of molecules as well as temperature.

Ans. $\Delta H_r(400\ \text{K}) = 799\ \text{kJ}\cdot\text{mol}^{-1}$

Solution 3-5

(a) Assuming that the melting point of ice is 273 K and that the heat capacity at constant pressure C_P is independent of temperature, the entropy change required is calculated by dividing the process under consideration into the following three steps i)-iii);

i) entropy change upon bringing 1g of ice at 270 K into 1g of ice at 273 K: ΔS_1

$$\Delta S_1 = \frac{1}{18} \times \int_{270}^{273} \frac{C_P}{T} dT = \frac{C_P}{18} \times [\ln T]_{270}^{273} = \frac{37.5}{18} \times \ln \frac{273}{270} = 2.30 \times 10^{-2}\,\text{J} \cdot \text{K}^{-1} \tag{8}$$

ii) entropy change upon bringing 1g of ice at 273 K into 1g of water at 273 K: ΔS_2

$$\Delta S_2 = \frac{1}{18} \times \frac{\Delta H_f}{T_m} = \frac{1}{18} \times \frac{6.01 \times 10^3}{273} = 1.22\,\text{J} \cdot \text{K}^{-1} \tag{9}$$

iii) entropy change upon warming 1g of water at 273 K to 300 K: ΔS_3

$$\Delta S_3 = \frac{1}{18} \times \int_{273}^{300} \frac{C_P}{T} dT = \frac{C_P}{18} \times [\ln T]_{273}^{300} = \frac{75.3}{18} \times \ln \frac{300}{273} = 3.95 \times 10^{-1}\,\text{J} \cdot \text{K}^{-1} \tag{10}$$

The entropy change required is given by the sum of these three entropy changes and is found as $1.64\,\text{J} \cdot \text{K}^{-1}$.

$$\Delta S = \Delta S_1 + \Delta S_2 + \Delta S_3 = 2.30 \times 10^{-2} + 1.22 + 3.95 \times 10^{-1} = 1.64\,\text{J} \cdot \text{K}^{-1} \tag{11}$$

Ans. $1.64\,\text{J} \cdot \text{K}^{-1}$

(b) Denoting the constant temperature reached as T_{eq}, the heat dissipated upon cooling 100g of water at 370 K to the temperature T_{eq} is equal to the heat required upon melting 10g of ice at 270 K and warming the water obtained to the temperature T_{eq}. This balance is expressed by eq(12).

$$-\frac{100}{18} \int_{370}^{T_{eq}} C_P(\text{water}) dT = \frac{10}{18} \int_{270}^{273} C_P(ice) dT + \frac{10}{18} \times \Delta H_f + \frac{10}{18} \int_{273}^{T_{eq}} C_P(\text{water}) dT \tag{12}$$

Substituting the relevant values into eq(12) and integrating it, eq(13) is obtained.

$$-75.3 \times (T_{eq} - 370) \times 10 = 37.5 \times (273 - 270) + 6.01 \times 10^3 + 75.3 \times (T_{eq} - 273) \tag{13}$$

From eq(13), T_{eq} is found as below.

$$T_{eq} = \frac{-37.5 \times 3 - 6.01 \times 10^3 + 75.3 \times 3973}{75.3 \times 11} = 353.8\,\text{K}$$

Since the equilibrium temperature is obtained, the entropy change under consideration can be found by putting the temperature equal to 353.8 K in a similar manner as in (a). Noting the differences in mass and temperature as compared with the case in (a), the entropy changes associated with the steps i)-iii) are denoted ΔS_1, ΔS_2 and ΔS_3, respectively and the entropy change upon cooling 100g of water at 370 K to T_{eq} is denoted ΔS_4. Then, the total entropy change is given by a sum of these four entropy changes (from eq(14) to eq(17)) as shown below in eq(18).

$$\Delta S_1 = \frac{10}{18} \times \int_{270}^{273} \frac{C_P}{T} dT = \frac{C_P}{18} \times [\ln T]_{270}^{273} = \frac{37.5}{18} \times \ln \frac{273}{270} = 2.30 \times 10^{-1}\,\text{J} \cdot \text{K}^{-1} \tag{14}$$

$$\Delta S_2 = \frac{10}{18} \times \frac{\Delta H_f}{T_m} = \frac{1}{18} \times \frac{6.01 \times 10^3}{273} = 12.23\,\text{J} \cdot \text{K}^{-1} \tag{15}$$

$$\Delta S_3 = \frac{10}{18} \times \int_{273}^{353.8} \frac{C_P}{T} dT = \frac{C_P}{18} \times [\ln T]_{2730}^{353.8} = \frac{75.3 \times 10}{18} \times \ln \frac{353.8}{273} = 10.85\,\text{J} \cdot \text{K}^{-1} \tag{16}$$

$$\Delta S_4 = \frac{100}{18} \times \int_{370}^{353.8} \frac{C_P}{T} dT = \frac{100 \times C_P}{18} \times [\ln T]_{370}^{353.8} = \frac{75.3 \times 100}{18} \times \ln \frac{353.8}{370} = -18.73\,\text{J} \cdot \text{K}^{-1} \tag{17}$$

$$\Delta S = \Delta S_1 + \Delta S_2 + \Delta S_3 + \Delta S_4 = 0.23 + 12.23 + 10.85 - 18.73 = 4.58\,\text{J} \cdot \text{K}^{-1} \tag{18}$$

Ans. equilibrium temperature: 353.8 K, $\Delta S = 4.58\,\text{J} \cdot \text{K}^{-1}$

"note" 110g of water at 353.8 K is finally obtained. According to the four steps considered above, therefore, the fifth step of mixing 100 g and 10 g waters at 353.8 K together to obtain 110 g of water at the same temperature still remains. However, since this fifth mixing step does not generate any entropy (see **problems 5-4** and **5-19** for entropy of mixing), the answer given above is correct. Ice naturally melts and certainly the whole process considered above is accompanied by an increase of entropy.

Solution 3-6

The formation reaction of ethylamine $C_2H_5NH_2(g)$ (one mole) is given by eq(19).

$$2C(s)+\frac{7}{2}H_2+\frac{1}{2}N_2 \rightarrow C_2H_5NH_2(g) \qquad (19)$$

In order to find the enthalpy of formation ΔH_f of the this reaction from bond energies, the easiest way for the calculation may be to break down all chemical bonds of the reactants and products into a common state of the component atoms. For this, the reactants are decomposed into the atoms as given by eq(20) and the products are by eq(21).

$$2C(s)+\frac{7}{2}H_2+\frac{1}{2}N_2 \rightarrow 2C(g)+7H+N \qquad (20)$$

$$C_2H_5NH_2(g) \rightarrow 2C(g)+7H+N \qquad (21)$$

The conceptual energy level diagram for the reactants, products and the common state of the decomposed atoms is shown below. Then, it is straightforward to calculate a value for the enthalpy of formation of the present reaction based on this diagram.

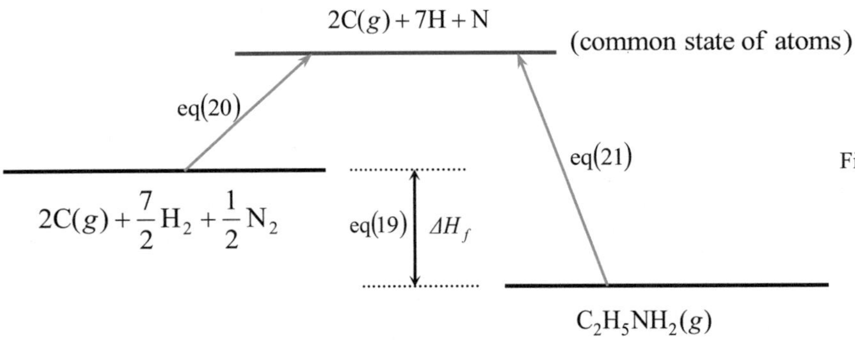

Fig.2 Bond energies measured from decomposed atoms and enthalpy of formation.

The energies for the decomposition reactions (eqs(20) and (21)) are calculated from relevant bond energies as shown below.

$2C(s) \rightarrow 2C(g)$	$2\times 717\ \mathrm{kJ\cdot mol^{-1}}$
$7/2H_2 \rightarrow 7H$	$7\times 218\ \mathrm{kJ\cdot mol^{-1}}$
$1/2N_2 \rightarrow N$	$437\ \mathrm{kJ\cdot mol^{-1}}$
total sum (eq(20))	$3433\ \mathrm{kJ\cdot mol^{-1}}$

$1\times C-C$	$1\times 348\ \mathrm{kJ\cdot mol^{-1}}$
$5\times C-H$	$5\times 413\ \mathrm{kJ\cdot mol^{-1}}$
$1\times C-N$	$1\times 292\ \mathrm{kJ\cdot mol^{-1}}$
$2\times N-H$	$2\times 391\ \mathrm{kJ\cdot mol^{-1}}$
total sum (eq(21))	$3487\ \mathrm{kJ\cdot mol^{-1}}$

As is seen from the diagram, therefore, the enthalpy of formation ΔH_f is obtained as

$3433-3487=-54\ \mathrm{kJ\cdot mol^{-1}}$.

Ans. $\Delta H_f = -54\ \mathrm{kJ\cdot mol^{-1}}$

"note" The reported experimental value for the enthalpy of formation ΔH_f° is $-47.5\ \mathrm{kJ\cdot mol^{-1}}$. The value calculated here from the bond energies can be regarded as equal to ΔH_f°.

Solution 3-7

(a) The total differential of the *Gibbs* free energy, denoted $G(P,T)$, is written as eq(22).

$$dG = VdP - SdT \qquad (22)$$

Under a constant temperature ($dT = 0$), both sides of eq(22) is divided by dP to yield eq(23).

$$\frac{(dG)_T}{(dP)_T} \equiv \left(\frac{\partial G}{\partial P}\right)_T = V \qquad (23)$$

By expressing V as a function of P, the above equation is readily integrated. For an ideal gas, V is given by $V = nRT/P$ and after substituting this relation into eq(23), the integration of eq(23) over the range [P°, P] leads to eq(24),

$$\int_{P^\circ}^{P} dG = G(P,T) - G(P^\circ,T) = \int_{P^\circ}^{P} \frac{nRT}{P} dP = nRT \ln\frac{P}{P^\circ} \qquad (24)$$

from which the *Gibbs* free energy change accompanied by varying the pressure of the gas from P° to P at the constant temperature is found to be given by $nRT \ln P / P^\circ$.

Ans. free energy change: $nRT \ln P / P^\circ$

(b) The partial pressures and mole numbers of the gas molecules at the chemical equilibrium are denoted by P(A), P(B), P(C) and P(D), and n_{A}, n_{B}, n_{C} and n_{D}, respectively. The free energy of the molecular species A is denoted G_{A} and that of B G_{B} and so on. Then from eq(24), their free energies at the equilibrium are given by the following equations and so on. Notice that these equations are for the mole numbers of n_{A}, $n_{\mathrm{B}}\cdots$.

$$\left.\begin{aligned} G_{\mathrm{A}}(P_{\mathrm{A}},T) &= G_{\mathrm{A}}(P^\circ,T) + n_{\mathrm{A}}RT \ln\frac{P_{\mathrm{A}}}{P^\circ} \\ G_{\mathrm{B}}(P_{\mathrm{B}},T) &= G_{\mathrm{B}}(P^\circ,T) + n_{\mathrm{B}}RT \ln\frac{P_{\mathrm{B}}}{P^\circ} \\ &\vdots \end{aligned}\right\} \qquad (25)$$

Writing the free energy per mole as $\overline{G}_{\mathrm{A}}$ for each species, the free energy per mole is expressed by eq(26).

$$\left.\begin{aligned} \overline{G}_{\mathrm{A}}(P_{\mathrm{A}},T) &= \overline{G}_{\mathrm{A}}(P^\circ,T) + RT \ln\frac{P_{\mathrm{A}}}{P^\circ} \\ \overline{G}_{\mathrm{B}}(P_{\mathrm{B}},T) &= \overline{G}_{\mathrm{B}}(P^\circ,T) + RT \ln\frac{P_{\mathrm{B}}}{P^\circ} \\ \overline{G}_{\mathrm{C}}(P_{\mathrm{C}},T) &= \overline{G}_{\mathrm{C}}(P^\circ,T) + RT \ln\frac{P_{\mathrm{C}}}{P^\circ} \\ \overline{G}_{\mathrm{D}}(P_{\mathrm{D}},T) &= \overline{G}_{\mathrm{D}}(P^\circ,T) + RT \ln\frac{P_{\mathrm{D}}}{P^\circ} \end{aligned}\right\} \qquad (26)$$

Let us assume that the system at equilibrium is composed of large quantities of A, B, C and D. Namely, an inequality of n_A, n_B, n_C, $n_D >> a$, b, c, d holds. Since n_A, n_B, $\cdots$etc. are not concentrations but mole numbers, they can take any large numbers. Under these conditions, when A reacts by the amount of a moles and B reacts by b moles, c moles of C and d moles of D are produced. The free energy change accompanied by this reaction is zero, because the system is in equilibrium and such a small amount of change in mole numbers does not affect the equilibrium at all. The system under consideration is at the equilibrium before and after the reaction. Namely, the relation $\Delta G = (c\overline{G}_C + d\overline{G}_D) - (a\overline{G}_A + b\overline{G}_B) = 0$ holds.

Ans. $\Delta G = (c\overline{G}_C + d\overline{G}_D) - (a\overline{G}_A + b\overline{G}_B) = 0$

(c) Using eq(26), the relation obtained above ($\Delta G = 0$) can be rewritten as eq(26').

$$\begin{aligned}\Delta G &= c\cdot\overline{G}_C(P^\circ,T) + d\cdot\overline{G}_D(P^\circ,T) + c\cdot RT\ \ln\frac{P_C}{P^\circ} + d\cdot RT\ \ln\frac{P_D}{P^\circ} \\ &\quad -\left(a\cdot\overline{G}_A(P^\circ,T) + b\cdot\overline{G}_B(P^\circ,T) + a\cdot RT\ \ln\frac{P_A}{P^\circ} + b\cdot RT\ \ln\frac{P_B}{P^\circ}\right) \\ &= \left\{ c\cdot\overline{G}_C(P^\circ,T) + d\cdot\overline{G}_D(P^\circ,T) - \left(a\cdot\overline{G}_A(P^\circ,T) + b\cdot\overline{G}_B(P^\circ,T)\right) \right\} + RT\ \ln\frac{P_C^c\cdot P_D^d}{P_A^a\cdot P_B^b}\times P^{\circ\,-(c+d-a-b)} \\ &= 0 \qquad (26')\end{aligned}$$

By denoting the free energy of reaction at the standard state as ΔG°, it is given by eq(27).

$$\Delta G^\circ = c\cdot\overline{G}_C(P^\circ,T) + d\cdot\overline{G}_D(P^\circ,T) - \left\{a\cdot\overline{G}_A(P^\circ,T) + b\cdot\overline{G}_B(P^\circ,T)\right\} \qquad (27)$$

Therefore, from eq(26') the equilibrium constant $K_P(T)$ is given by eq(28) or by eq(29) as a more common expression.

$$\Delta G^\circ = -RT\ \ln\frac{P_C^c\cdot P_D^d}{P_A^a\cdot P_B^b}\times P^{\circ\,-(c+d-a-b)} \equiv -RT\ \ln K_P(T) \qquad (28)$$

$$K_P(T) = \frac{P_C^c\cdot P_D^d}{P_A^a\cdot P_B^b}\times P^{\circ\,-(c+d-a-b)} \qquad (29)$$

When 1 atm is taken as the standard state for the pressure ($P^\circ = 1$ atm) and pressures are measured in the unit of atm but not Pa, the factor $P^{\circ\,-(c+d-a-b)}$ in eqs(28) and (29) can be put equal to one for any stoichiometric coefficients of a, b, etc.

Ans. $K_P(T) = \dfrac{P_C^c\cdot P_D^d}{P_A^a\cdot P_B^b}\times P^{\circ\,-(c+d-a-b)}$

"note 1" On deriving the relation between the equilibrium constant and the free energy, the concept of chemical potential is not utilized intentionally in the solution here. The underlined part is essentially important in discussing the equilibrium and leads to the concept of chemical potential. The answer is given for the ideal gases but when the molar free energy given by G/n_i is replaced by $\overline{G}_i = (\partial G/\partial n_i)_{P,T}$, the equation given in the answer holds generally.

(d) Writing a molar concentration as [m], the equation of state of an ideal gas is expressed as $P = \frac{n}{V}RT = [m]RT$. Substituting this relation into $K_P(T)$ of eq(29), eq(30) is obtained.

$$K_P(T) = \frac{P_C^c\cdot P_D^d}{P_A^a\cdot P_B^b}\times P^{\circ\,-(c+d-a-b)} = \frac{[m_C]^c\cdot[m_D]^d}{[m_A]^a\cdot[m_B]^b}\times\left(\frac{RT}{P^\circ}\right)^{(c+d-a-b)} \qquad (30)$$

Noticing that the equilibrium constant $K_c(T)$ defined by molar concentrations instead of partial pressures is given by $K_c(T)=\dfrac{[m_{\rm C}]^c}{[m_0]^c}\cdot\dfrac{[m_{\rm D}]^d}{[m_0]^d}\bigg/\dfrac{[m_{\rm A}]^a}{[m_0]^a}\cdot\dfrac{[m_{\rm B}]^b}{[m_0]^b}$ and comparing this with eq(30), eq(31) is derived as the relation between $K_c(T)$ and $K_P(T)$. Here $[m_0]$ is a standard molar concentration.

$$K_c(T)=\frac{[m_{\rm C}]^c}{[m_{\rm A}]^a}\cdot\frac{[m_{\rm D}]^d}{[m_{\rm B}]^b}\cdot[m_0]^{-(c+d-a-b)}=K_P(T)\left(\frac{RT\cdot[m_0]}{P^\circ}\right)^{-(c+d-a-b)} \tag{31}$$

Ans. $K_c(T)=K_P(T)\left(\dfrac{RT\cdot[m_0]}{P^\circ}\right)^{-(c+d-a-b)}$

"note 2" Both of $K_c(T)$ and $K_P(T)$ are defined so as to be dimensionless by employing the values for the standard concentration and standard pressure.

(e) Employing the result obtained in (d), eq(31') is obtained.

$$\ln K_c(T)=\ln K_P(T)-(c+d-a-b)\ln\left(\frac{RT\cdot[m_0]}{P^\circ}\right) \tag{31'}$$

Differentiating both side of this equation with T and after some rearrangements, the following relation is obtained.

$$\left(\frac{\partial\ \ln K_c(T)}{\partial T}\right)_P=\left(\frac{\partial\ \ln K_P(T)}{\partial T}\right)_P-(c+d-a-b)\left(\frac{\partial\ \ln RT\cdot[m_0]/P^\circ}{\partial T}\right)_P$$

$$=-\left(\frac{\partial\Delta G^\circ/RT}{\partial T}\right)_P-\frac{\Delta n}{T}\qquad(\Delta n\equiv(c+d-a-b))$$

$$=\frac{\Delta G^\circ+T\Delta S^\circ-\Delta nRT}{RT^2}=\frac{\Delta H^\circ-\Delta nRT}{RT^2} \tag{32}$$

Ans. $\dfrac{\Delta H^\circ-\Delta nRT}{RT^2}$ $\quad\Delta n=(c+d-a-b)$

(f) ΔG° here is defined at the standard pressure of P° and is not a function of P anymore. Therefore, the relation $\left(\dfrac{\partial\ln K_P}{\partial P}\right)_T=-\dfrac{1}{RT}\left(\dfrac{\partial\Delta G^\circ}{\partial P}\right)_T=0$ is derived. Hence, it is seen that the equilibrium constant does not vary with the total pressure. The same is true for K_c.

Ans. $(\partial\ln K/\partial P)_T=0$

"note 3" In general, the relation of $\left(\dfrac{\partial\ln K}{\partial P}\right)_T=-\dfrac{\Delta V}{RT}$ holds (see (a)). $\Delta V\neq 0$ in many cases, so that this relation may appear to be contradictory to the finding obtained in (f). This superficial disagreement arises from a difference in how the standard states are defined, in other words, how to define the equilibrium constant K.

Solution 3-8a

The partial pressures of acetone (A) and chloroform (C) are calculated from eqs(33) and (34), respectively, in which P is the total vapor pressure.

$$P_{\rm A}=P\times X_{\rm A}^{g} \quad (33),\qquad P_{\rm C}=P-P_{\rm A} \quad (34)$$

The activity and activity coefficient of acetone (or chloroform) in solution can be calculated based on the fact that in the equilibrium, the chemical potentials of acetone (or chloroform) in the gas and solution phases are equal. We take the reference states as those of pure substances and assume that the gases behave ideally. Taking the vapor pressure of pure acetone at 323 K as $P_A^\circ = 81.6$ kPa, the chemical potential of gaseous acetone is given by eq(35) (see **problem 3-41a** for derivation). It should be noted here that $\mu_A^\circ(g)$ is the molar free energy of pure gaseous acetone at the pressure of $P = P_A^\bullet$.

$$\mu_A(g) = \mu_A^\circ(g) + RT \ln \frac{P_A}{P_A^\bullet} \qquad (35)$$

On the other hand, the chemical potential of acetone in solution is given by eq(36), where $\mu_A^\circ(l)$ is the

$$\mu_A(l) = \mu_A^\circ(l) + RT \ln a_A , \qquad a_A = \gamma_A X_A^l \qquad (36)$$

chemical potential when $a_A = 1$, i.e., for pure liquid acetone. Putting eq(35) equal to eq(36), the left side of eq(37) is obtained for the activity, from which the activity coefficient is found as given on the right of eq(37) as a function of the mole fraction in solution..

$$\frac{P_A(X_A^g)}{P_A^\bullet} = a_A(X_A^l) = \gamma_A X_A^l \quad \Rightarrow \quad \gamma_A = \frac{P_A(X_A^g)}{P_A^\bullet \cdot X_A^l} \qquad (37)$$

The obtained results are summarized in the table below.

Table Activity and activity coefficients* of acetone in mixed solutions of acetone and chloroform (323K).

X_A^l	0	0.100	0.200	0.300	0.400	0.500	0.600	0.700	0.800	0.900	1.000
P_A	0	4.69	10.4	17.2	25.0	34.4	44.6	53.7	64.1	73.3	81.6
P_{total}	69.5	66.0	63.2	61.7	61.3	62.5	65.2	68.1	72.0	76.8	81.6
a_A		0.057	0.128	0.211	0.307	0.421	0.547	0.658	0.785	0.899	1.0
γ_A		0.574	0.639	0.703	0.766	0.843	0.911	0.940	0.982	0.999	1.0

* $P_A^\circ = 81.6$ kPa

Ans. values given in the table

A *P-X* diagram for the mixed solution of acetone and chloroform is shown below in Fig.3. Several important things can be learned from this diagram. See **problems 4-89**, **4-90** and their solutions for the regular solution and its vapor pressure.

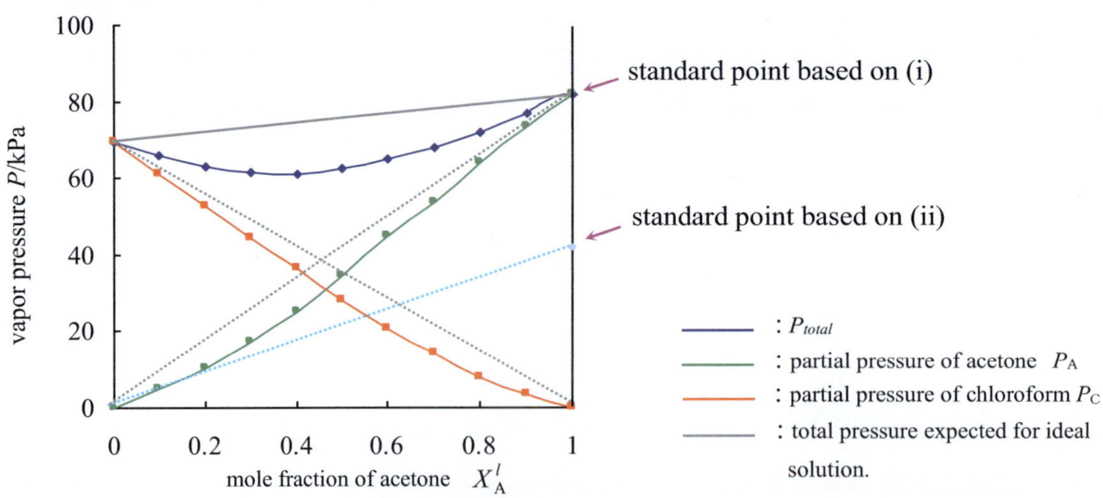

Fig.3 *P-X* diagram for acetone-chloroform mixed solution.

(I) About standard states (reference states)

For a mixture of two pure substances of A and B, there are two ways of specifying the standard state as explained below.

(i) To designate the state of a pure substance for the standard state

This is the approach employed in the **solution 3-8**. The value for $\mu_A^\circ(l)$ represents a molar free energy of the pure substance. As is seen from the values given in the table or from the deviation from the *Raoult*'s law shown in the figure, where X_A^l takes a large value the values for γ_A do not deviate largely from unity. However, where X_A^l takes a small value, the values for γ_A deviate significantly from unity. As far as the chemical potential is expressed by eq(36), this large deviation must totally be ascribed to a large variation in γ_A, which is not, of course, desirable. In order to avoid this we employ the second method of designating the standard state.

(ii) Assuming that *Henry*'s law is satisfied for dilute solutions of A, the law is extrapolated to $X_A^l = 1$ and the chemical potential thus extrapolated is taken as the standard value of A.

The value for $\mu_A^\circ(l)$ is a hypothetical molar free energy for the pure substance A acquired by assuming that *Henry*'s law is satisfied up to $X_A^l = 1$ and is shown as the extrapolated point in the figure above. Since the deviation from *Henry*'s law is not significant where X_A^l is small, the chemical potential expressed by eq(36) does not require the activity coefficient deviating from one largely.

"note" Naturally, the value of the chemical potential itself does not change depending on the manner of specifying the standard state. Only accompanied by a change of the value of $\mu_A^\circ(l)$, the value for $a_A = \gamma_A X_A^l$ changes so as to keep the value for $\mu_A(l)$ invariant.

(**II**) Three cases in which the activity coefficient is one (ideal solution), greater than one (positive deviation) and less than one (negative deviation).

The system treated in the present problem, the vapor pressures are lower than those expected from *Raoult*'s law. This fact indicates that the molecules remaining in the solution are more stable than those in the gas phase. Namely, the intermolecular attractive interaction of A-B is stronger than either of A-A or B-B. Conversely, when it is weaker than either of A-A or B-B, the vapor pressures exhibit higher values than those expected from *Raoutl*'s law. In general, there exist three cases as indicated above.

Solution 3-8b

The vapor pressures of the mixed solutions deviate largely in a convex form as compared with the vapor pressures expected when the solutions are regarded as ideal. This indicates that the substance Z in acetone or acetone in Z tends rather strongly to escape into the gas phase. Namely, this is telling that the intermolecular interaction between acetone and Z is more repulsive than those between acetones or between Z's. Compared with the system of acetone and chloroform, the difference is so remarkable that it provides a good example to understand that properties of solution are entirely dependent on the combination of component substances.

Specifically, the substance Z is carbon disulfide CS_2. (*Zawidzki, Z.Physikal.Chem.*, **35**,129 (1900).) For your reference, some physical properties of acetone and CS_2 are listed in the table below.

As is seen in the *P-X* diagram given in **problem 3-8b**, the cases in which vapor pressures deviate in a convex manner are known more frequently as compared with the cases for concave deviations. It should be noticed that the partial pressure of each component is always smaller than that of the corresponding pure component but total pressures can be larger than any of the vapor pressures of pure components. See **problems 4-89, 4-90** and their solutions for regular solutions.

	$T_b\,/\,^\circ\mathrm{C}$	$d(\mathrm{g/cm^3})$	η (cP)	ε	n_D	$\Delta H_v(T_b)(\mathrm{cal\cdot mol^{-1}})$
acetone (CH_3COCH_3)	56.29	0.7844*	0.3040*	20.70*	1.336*	6.95
carbon disulfide (CS_2)	46.23	1.270#	0.380#	2.641$	1.632#	6.39

*at 25 °C, #at 15 °C, $at 20 °C. *Techniques of Chemistry, vol.2, Organic Solvents 3rd ed.*

Ans. See text.

Solution 3-9

1.0 kg of CO_2 is $1000/44.01 = 22.7$ mols and the pressure of this gas, regarded as an ideal gas, is calculated to be 562 kPa as below.

$$P = nRT/V = (22.7\times 8.31\times 298)/0.1 = 562\ \mathrm{kPa}$$

Ans. 562 kPa

The work w to be done to the air against P_{ext} (surroundings) is given by eq(38), putting the initial pressure as P_i and the final pressure as $P_f = P_{ext}$.

$$w = -\int_{P_i}^{P_f} P_{ext}\,dV = -P_{ext}[V]_{P_i}^{P_f} = -P_{ext}\left[\frac{nRT}{P}\right]_{P_i}^{P_f} = -P_{ext}\times nRT\left(\frac{1}{P_f}-\frac{1}{P_i}\right) \qquad (38)$$

Substituting the given values into eq(38), the following value is obtained for w.

$$w = -101\times 10^3\times 22.7\times 8.31\times 298\left(\frac{1}{101}-\frac{1}{562}\right)\times 10^{-3} = -46.1\ \mathrm{kJ}$$

Ans. -46.1 kJ

For the reversible expansion at the constant temperature, the work to be done by the system (gas) to the surroundings is given by eq(39).

$$w = -\int_{P_i}^{P_f} P\,dV = -\int_{P_i}^{P_f}\frac{nRT}{V}\,dV = -nRT[\ln V]_{P_i}^{P_f} = -nRT\left(\ln\frac{nRT}{P_f}-\ln\frac{nRT}{P_i}\right) = -nRT\times\ln\frac{P_f}{P_i} \qquad (39)$$

Substituting the given values into eq(38), the following value is obtained for w. Comparing this with the value

$$w = -22.7\times 8.31\times 298\times\ln\frac{562\times 10^3}{101\times 10^3} = -96.5\ \mathrm{kJ}$$

obtained above, it is seen that the larger (maximum) work is gained by the reversible process.

Ans. -96.5 kJ

Solution 3-10

(a) For the solutions composed of two pure substances A and B, three distinct states of solution in equilibrium with the gas phase are schematically depicted in Fig.4 below. They are a dilute solution of A, a pure solvent of B and a dilute solution of B illustrated in Figs.4(a), (b) and (c), respectively. For the dilute solution of A shown in Fig.4(a), a B molecule is almost always surrounded by B molecules themselves. For this, a tendency for the B molecule to escape to the vapor phase is diminished only by a factor proportional to the number of B molecules in solution as compared with the pure solvent B shown in Fig.4(b), that is, the vaporization tendency is proportional to the mole fraction of B. Therefore, we are led to the conclusion that *Raoult*'s law holds for B in the dilute solution of A.

On the other hand, for a concentrated solution of A, as is shown in Fig.4(c), a B molecule is surrounded almost always by A molecules alone. Consequently, a tendency for the B molecule to be transferred to the vapor phase is not simply proportional to the number of B molecules (mole fraction) but highly dependent on the nature of the interaction between A and B molecules.

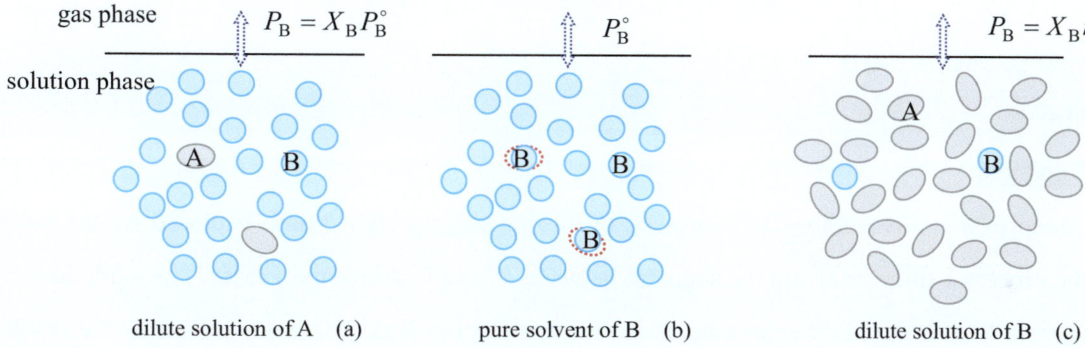

Fig.4 Three distinct states of solution composed of two components and in equilibrium with the gas phase.

Now, let us consider a hypothetical pure solvent B for which the interaction between A and B is retained and, hence, the vapor pressure of this hypothetical solvent is changed from P_B° to κ_B^H. Then, the vapor pressure of a dilute solution of B can be considered to be given by $X_B \kappa_B^H$. Namely, *Henry*'s law $P_B = X_B \kappa_B^H$ holds for B with κ_B^H as *Henry*'s constant. In conclusion, we can say that in a dilute solution of A, *Henry*'s law holds for A and *Raoult*'s law for B and vice versa for a dilute solution of B.

Ans. See text (a).

(b) We consider a solution composed of A and B, whose mole numbers are n_A and n_B and chemical potentials are μ_A and μ_B, respectively. Denoting the chemical potentials of the pure solvents A and B as μ_A° and μ_B°, respectively, μ_A and μ_B are given by eq(40).

$$\mu_A = \mu_A^\circ + RT \ln P_A / P_A^\circ, \qquad \mu_B = \mu_B^\circ + RT \ln P_B / P_B^\circ \qquad (40)$$

The *Gibbs-Duhem* relation for the two-component system can be written as eq(41) (see **problem 3-38a** for proof).

$$n_A d\mu_A + n_B d\mu_B = 0 \qquad (41)$$

Employing the relation given by eq(40), eq(41) can be rewritten as eqs(42)-(42'').

$$\frac{n_A}{P_A}dP_A + \frac{n_B}{P_B}dP_B = 0 \qquad (42)$$

$$\frac{X_A}{P_A}dP_A + \frac{X_B}{P_B}dP_B = 0 \qquad (42')$$

$$\frac{X_A}{P_A}\left(\frac{dP_A}{dX_A}\right)dX_A + \frac{X_B}{P_B}\left(\frac{dP_B}{dX_B}\right)dX_B = 0 \qquad (42'')$$

Noticing that $dX_A = -dX_B$, eq(42'') is further rewritten as eq(43).

$$\frac{X_A}{P_A}\left(\frac{dP_A}{dX_A}\right) = \frac{X_B}{P_B}\left(\frac{dP_B}{dX_B}\right) \qquad (43)$$

If *Raoul*'s law holds for A, then from the relation of $P_A = P_A^\circ X_A$, eq(44) is derived.

$$\frac{X_A}{P_A}\left(\frac{dP_A}{dX_A}\right) = 1 \qquad (44)$$

Substituting this into eq(43) to solve it for P_B, the result that P_B should be proportional to X_B is found as is shown in eq(45). Namely, it is proved that in the range of mole fraction where *Raoul*'s law holds for A,

$$\left(\frac{\partial \ln P_B}{\partial \ln X_B}\right) = 1 \Rightarrow \ln P_B = \ln X_B + \text{const.} \Rightarrow P_B = \alpha X_B \qquad (45)$$

Henry's law must be satisfied by B.

Solution 3-11a

About the *P*-*V* diagram.

(a) The work done to the surroundings is expressed by the negatively signed area enclosed by the curve representing the process 1 (blue line) and the segment between V_1 and V_2 above the *V*-axis. The work done by the process 2 (green line) is similarly expressed as is illustrated in Fig.5(a). The work gained by the system (the working gas in this case) is given by the positively signed areas between each of the two curves representing the processes 3 and 4 and the *V*-axis corresponding to the segment between V_3 and V_4 and the segment between V_4 and V_1, respectively. See Fig.5(b).

Ans. Fig.5(a).

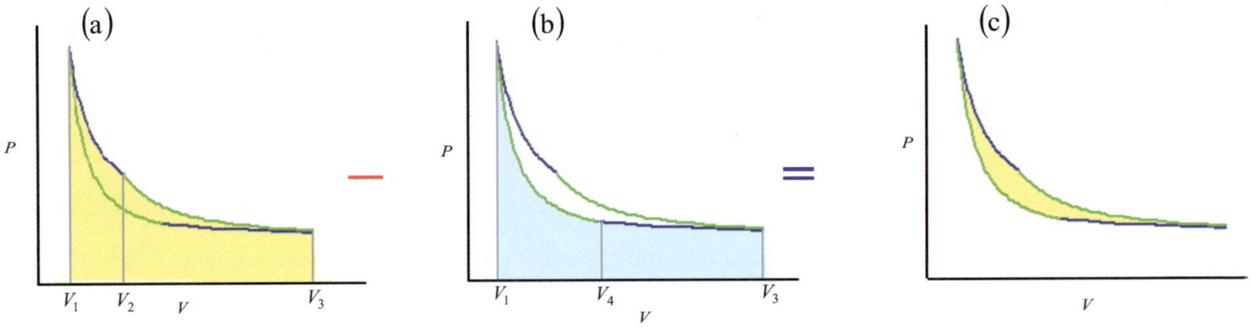

Fig.5 Work done or gained upon isothermal and adiabatic expansions or compressions of a working gas.

(a) Work done to the surroundings by the processes 1 and 2. (b) Work gained by the system during the processes 3 and 4. (c) Work done by the system to the surroundings during one cycle. Isothermal processes are shown by blue curves and adiabatic processes by green curves.

(b) The work done by one cycle is given by the negatively singed area enclosed by the four *P*-*V* curves. See Fig.5(c) and **"note"**.

Ans. Fig.5(c).

About the *T*-*S* diagram.

Answers to (c)-(g) are as follows.

(c) Noticing that the processes in question are either at constant temperature or at constant entropy, the *T*-*S* diagram is added as given in Fig.6(a) below.

(d) The area enclosed by the line expressing the process 1, the *S*-axis and vertical lines at $S = S_1,\ S_2$, i.e., the area etched by yellow lines in Fig.6(a). is the heat taken from the hot reservoir.

(e) The area enclosed by the line expressing the process 3, the *S*-axis and vertical lines at $S = S_1,\ S_2$, i.e., the area etched by green lines in Fig.6(b) is the heat discarded to the low temperature reservoir.

(f) The definition of the work efficiency is given by $\eta =$ (work done by the system)∕(thermal energy gained by the system) = (the area enclosed by the curves representing the four processes 1-4 shown in Fig.5(c)∕(the area enclosed by the lines for the four processes 1-4 shown in Fig.6(c).

(g) $PV^{\gamma} =$ constant.

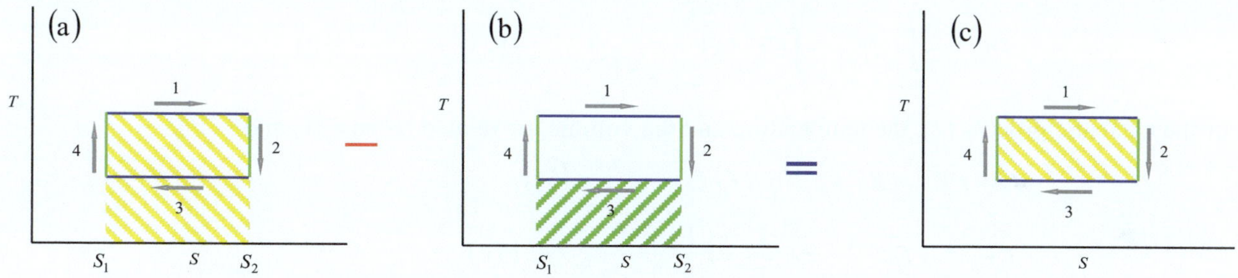

Fig.6 Thermal energy gain or loss upon isothermal and adiabatic expansions or compressions of a working gas. (a) Thermal energy gained from the high temperature heat reservoir. (b) Thermal energy emitted to the low temperature heat reservoir. (c) Thermal energy gained by the system. These quantities are indicated by the etched areas.

"note 1" With respect to the signs of areas in the above figures, they are the same as defined by the integration and for both of *P*-*V* and *T*-*S* diagrams, they are positive when the integration is carried out in the increasing direction of *V* or *S* and negative when in the decreasing direction. However, in the case of the work, it is given by $-PdV$. For this, it is necessary to put a negative sign again to the integral expressed by the area in the *P*-*V* diagram. On defining the work efficiency in terms of heat gained and work done, the absolute values of the thermal energy gained (+ sign) and the work done (energy lost : – sign) are used to avoid a negative value for the efficiency.

"note 2" Most of *P*-*V* diagrams employed in many text books in order to explain the basic idea of the *Carnot* cycle as an ideal heat engine are only schematically drawn but not precisely. The diagrams employed in **problem 3-11a** is, however, faithfully drawn according to the isothermal and adiabatic *P*-*V* curves (see **problem 3-26**) of an ideal gas.

Solution 3-11b

The work energy ΔW and the thermal energy associated with each process is calculated as follows. Each process is indicated by a subscript (number) attached to a relevant thermodynamic variable.

process (1) (isothermal expansion)

Since the temperature does not change, there is no change in the internal energy U, leading to eq(46). The work done by this process is given by eq(47).

$$\Delta U_1 = \Delta W_1 + \Delta Q_1 = 0 \;\Rightarrow\; \Delta W_1 = -\Delta Q_1 \qquad (46)$$

$$\Delta W_1 = -\int_{V_1}^{V_2} PdV = -\int_{V_1}^{V_2} \frac{RT}{V} dV = -RT_h \ln\frac{V_2}{V_1} \qquad (47)$$

process (2) (adiabatic expansion)

The relation $\Delta Q_2 = 0$ holds because of the adiabatic process. To calculate ΔW_2, the P-V relation which holds for the adiabatic process is required and is given by eq(49). Employing this relation for the integrand of the work, the right side of eq(50) is obtained as ΔW_2.

$$\Delta Q_2 = 0 \qquad (48)$$

$$P = CV^{-\gamma} \qquad (49)$$ surroundings$C$ is a constant to be determined for each

$$\Delta W_2 = -\int_{V_2}^{V_3} PdV = -\int_{V_2}^{V_3} CV^{-\gamma} dV = \frac{C}{\gamma-1}(V_3^{-\gamma+1} - V_2^{-\gamma+1}) \qquad (50)$$

For the adiabatic process (2), the temperature and the volume are related by eqs(51) and (51').

$$C = PV^{\gamma} = RT_h V_2^{\gamma-1} = RT_l V_3^{\gamma-1} \qquad (51)$$

$$T_h V_2^{\gamma-1} = T_l V_3^{\gamma-1} \;\Rightarrow\; \frac{T_l}{T_h} = \left(\frac{V_2}{V_3}\right)^{\gamma-1} \qquad (51')$$

Employing these relations, eq(50) is rewritten as eq(50').

$$\Delta W_2 = \frac{C}{\gamma-1} V_3^{-\gamma+1}\left(1 - \frac{V_2^{-\gamma+1}}{V_3^{-\gamma+1}}\right) = \frac{C}{\gamma-1} V_2^{-\gamma+1}\left(\frac{T_l}{T_h} - 1\right)$$
$$= \frac{R}{\gamma-1} T_h \left(\frac{T_l}{T_h} - 1\right) = C_V (T_l - T_h) \qquad (50')$$

process (3) (isothermal compression)

Considering similarly as for process (1), eqs(52) and (53) are obtained.

$$\Delta U_3 = 0 \;\Rightarrow\; \Delta Q_3 = -\Delta W_3 \qquad (52)$$

$$\Delta W_3 = -\int_{V_3}^{V_4} PdV = -RT_l \ln\frac{V_4}{V_3} \qquad (53)$$

process (4) (adiabatic compression)

Since the process is adiabatic, $\Delta Q_4 = 0$. The value for ΔW_4 is obtained in a similar manner as for process (2), being led to eq(55').

$$\varDelta Q_4 = 0 \qquad (54)$$

$$\varDelta W_4 = -\int_{V_4}^{V_1} PdV = -\int_{V_4}^{V_1} C'V^{-\gamma}dV = \frac{C'}{\gamma-1}(V_1^{-\gamma+1} - V_4^{-\gamma+1}) \qquad (55)$$

$$= \frac{C'}{\gamma-1}V_1^{-\gamma+1}\left(1-\frac{V_4^{-\gamma+1}}{V_1^{-\gamma+1}}\right) = \frac{C'}{\gamma-1}V_1^{-\gamma+1}\left(1-\frac{T_l}{T_h}\right)$$

$$= \frac{R}{\gamma-1}T_h\left(1-\frac{T_l}{T_h}\right) = C_V(T_h - T_l) \qquad (55')$$

Here, the relations given by eqs(56) and (56') are employed.

$$C' = RT_hV_1^{\gamma-1} = RT_lV_4^{\gamma-1} \qquad (56)$$

$$T_lV_4^{\gamma-1} = T_hV_1^{\gamma-1} \;\Rightarrow\; \frac{T_l}{T_h} = \left(\frac{V_1}{V_4}\right)^{\gamma-1} \qquad (56')$$

The efficiency of work η is defined by eq(57).

$$\eta = \frac{\text{work done by the system to the}}{\text{thermal energy gained from high temperature source}} = \frac{\varDelta W_1 + \varDelta W_2 + \varDelta W_3 + \varDelta W_4}{\varDelta Q_1} \qquad (57)$$

Noticing that the relation $\varDelta W_2 + \varDelta W_4 = 0$ holds as is seen from eqs(50') and (55'), the substitution of the relevant values obtained above into eq(57) yields eq(58) for the efficiency of work of *Carnot* cycle.

$$\eta = \frac{|\varDelta W_1 + \varDelta W_2 + \varDelta W_3 + \varDelta W_4|}{\varDelta Q_1} = \frac{|\varDelta W_1 + \varDelta W_3|}{\varDelta Q_1}$$

$$= \frac{\left|-RT_h\ln\frac{V_2}{V_1} - RT_l\ln\frac{V_4}{V_3}\right|}{RT_h\ln\frac{V_2}{V_1}}$$

$$= \frac{T_h\ln\frac{V_2}{V_1} + T_l\ln\frac{V_4}{V_3}}{T_h\ln\frac{V_2}{V_1}} = \frac{T_h - T_l}{T_h} \qquad (58)$$

Here, the relation of eq(59) which can be derived from eqs(51) and (56) (or eqs(51') and (56')) is employed.

$$\frac{T_l}{T_h} = \left(\frac{V_1}{V_4}\right)^{\gamma-1} = \left(\frac{V_2}{V_3}\right)^{\gamma-1} \;\Rightarrow\; \frac{V_1}{V_4} = \frac{V_2}{V_3} \;\Rightarrow\; \frac{V_1}{V_2} = \frac{V_4}{V_3} \;\Rightarrow\; \ln\frac{V_2}{V_1} = -\ln\frac{V_4}{V_3} \qquad (59)$$

Ans. $\eta = (T_h - T_l)/T_h$, eq(58)

"note" The relation of eqs(50') and (55') can directly be obtained from eq(60) which describes a variation

$$dU = C_VdT \;\Rightarrow\; \varDelta U = C_V\varDelta T \qquad (60)$$

of the internal energy of an ideal gas with temperature. Although it is also readily found that $\varDelta W_2 + \varDelta W_4 = 0$ from eq(60), it is recommended to calculate the relevant work energy based on its definition.

Solution 3-12

(a) The total differential of $G(P,T)$ is given by eq(61). See *Legendre* transformation in **II-2-1**(5).

$$dG(P,T) = VdP - SdT \qquad (61)$$

Ans. $dG(P,T) = VdP - SdT$

(b) Not only at T and P but also at $T+dT$ and $P+dP$ the system is in vapor-liquid phase equilibrium. Therefore, eq(62) is satisfied and from eq(61) the free energy change in each phase is given by eq(63).

$$G^g(P,T)=G^l(P,T)\ ,\quad G^g(P+dP,T+dT)=G^l(P+dP,T+dT) \qquad (62)$$

$$dG^g=V^g dP-S^g dT\ ,\quad dG^l=V^l dP-S^l dT \qquad (63)$$

From eq(62), the free energy change in each phase is found equal to each other, leading to eq(64).

$$V^g dP-S^g dT=V^l dP-S^l dT \;\Rightarrow\; (V^g-V^l)dP=(S^g-S^l)dT \qquad (64)$$

Rewriting eq(64), eq(65) is obtained for the answer.

$$\frac{dP}{dT}=\frac{(S^g-S^l)}{(V^g-V^l)}=\frac{\Delta S_{vap}}{\Delta V_{vap}} \qquad (65) \qquad \textit{vap: vaporization}$$

Ans. See text.

(c) The latent heat of vaporization ΔH_{vap} is related to the entropy of vaporization ΔS_{vap} by the following equation. Also, since the molar volume of the liquid is much smaller than that of the gas, the former volume

$$\frac{\Delta H_{vap}}{T}=\Delta S_{vap}$$

is neglected and the volume of vaporization is expressed as $\Delta V_{vap}\approx V^g=\dfrac{nRT}{P}$. Substituting these relations into eq(65) and after some rearrangements, eq(66) is derived.

$$\frac{dP}{dT}=\frac{\Delta H_{vap}P}{nRT^2} \;\Rightarrow\; \frac{dP}{PdT}=\frac{d\ln P}{dT}=\frac{\Delta H^m_{vap}}{RT^2} \qquad (66)$$

In order to indicate explicitly that the latent heat of vaporization given is a molar value, the upper case suffix m is attached to it.

Ans. See text.

(d) By integrating eq(66) with ΔH^m_{vap} as a constant, the right side of eq(67) is obtained.

$$\int_{P^0}^{P} d\ln P=\int_{T^0}^{T}\frac{\Delta H^m_{vap}}{RT^2}dT \;\Rightarrow\; \ln P/P^0=-\frac{\Delta H^m_{vap}}{R}\left(\frac{1}{T}-\frac{1}{T^0}\right) \;\Rightarrow\; \ln P=-\frac{\Delta H^m_{vap}}{R}\left(\frac{1}{T}-\frac{1}{373}\right) \qquad (67)$$

Ans. vapor pressure: $\ln P=-\dfrac{\Delta H^m_{vap}}{R}\left(\dfrac{1}{T}-\dfrac{1}{373}\right)$

(e) Employing the equation obtained above, the given values are substituted into it to find the temperature as below (eq(68)). The unit of pressure can either be in atm or in Pa, since the variable P appears in its ration, making its unit irrelevant.

$$\ln 0.34=-\frac{44.1\times10^3}{8.31}\left(\frac{1}{T}-\frac{1}{373}\right) \;\Rightarrow\; T=1/\left(-\frac{8.31\times\ln 0.34}{44.1\times10^3}+\frac{1}{373}\right)=347\ \mathrm{K} \qquad (68)$$

Ans. boiling point: 347 K

Solution 3-13

(a) The boiling temperatures given in the table are plotted against the mole fractions of oxygen as is shown in Fig.7. The vertical axis is T_b and the horizontal axis is the mole fraction. The blues dots are for (T_b, $X_{O_2}(l)$) and the green dots are for (T_b, $X_{O_2}(g)$).

Ans. 77.3 – 90.1 K, See Fig.7.

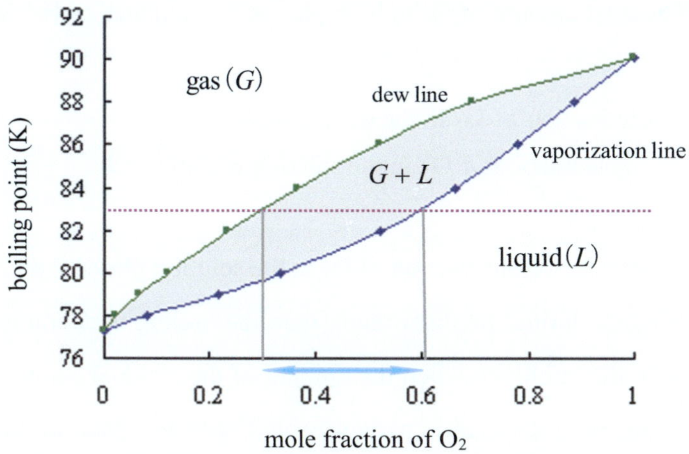

Fig.7 Phase diagram of N_2-O_2 mixed solutions. Vaporization and dew lines are equilibrium lines. Grey area indicates the region in which the gas and liquid coexist. The coexisting region at 83 K (red dotted line) is indicated by the two-way arrow (sky blue).

(b) From Fig.7, the intersections of a horizontal linear line drawn at 83 K with the gas-liquid equilibrium lines (dew and vaporization lines) are found to be located at the mole fraction of 0.3 where liquefaction starts and that of 0.6 where the vaporization starts.

Ans. range of mole fractions of O_2 which exists in solution: 0.30-0.60

(c) It is important to notice that during the whole process of distillation, the total mole fraction of O_2 in the vapor and solution phases remains constant at 0.2. Just when the distillation is started, the intersection of the vertical line at the mole fraction of 0.2 and the vaporization line is the boiling point and it is 78.9 K. To continue the distillation further, the temperature should be raised. When 80% (in mole numbers) of the initial mixed solution is removed by distillation (or brought into the gas phase), the ratio of the mole numbers in gas phase to those in solution becomes 4:1. To find the temperature reached, the lever rule is used as is shown in Fig.8. Along the vertical line at the mole fraction of 0.2, the temperature of the solution is raised until the lever arms ratio (right to left) reaches to 4:1. Thus, the boiling point is found to be 80.5 K (red dotted arrow) and the mole fraction of O_2 is found as 0.38 (blue dotted arrow).

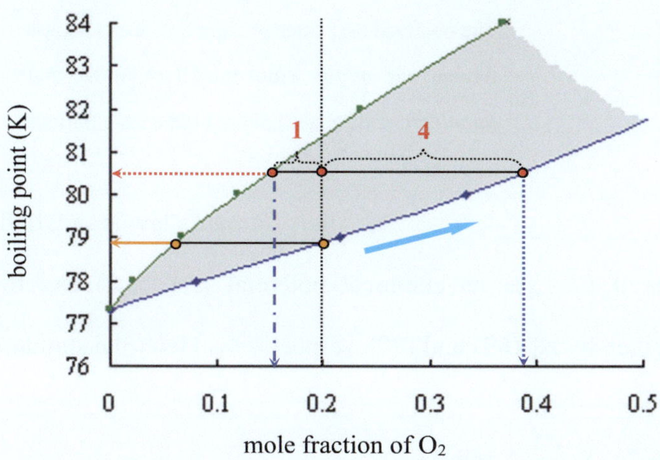

Fig.8 Expanded part of Fig.7

Boiling points under considerations are indicated by the red dotted and orange solid arrows. The sky blue thick arrow indicates a rise in the temperature accompanied by distillation.

Just to make sure, it is confirmed that the above calculation is correct. The initial solution is assumed to be composed of 0.2 mole of O_2 and 0.8 mole of N_2. By distillation, its 80% is vaporized and 20% remains in solution. Namely, 0.8 mole is in the gas phase and 0.2 mole is in the solution phase. At 80.5 K, the mole fractions of O_2 in the gas phase and in the solution phase are read from Fig.8 as 0.155 and 0.38, respectively.

The ratio of the lever arms is 1:4 (left to right). The total amount of O_2 in both phases is calculated as below by using the lever rule.

$$0.8 \text{ (moles in the gas phase)} \times 0.155 \text{ (mole fraction of } O_2 \text{ in the gas phase)} +$$
$$0.2 \text{ (moles in the solution phase)} \times 0.38 \text{ (mole fraction of } O_2 \text{ in the solution phase)} = 0.2$$

Surely, the total amount of O_2 is kept constant at 0.2 moles.

Ans. boiling point: 80.5 K, mole fraction of O_2 in the solution phase: 0.38

"note" If the amount of a solution does not vary much during the distillation, then the boiling temperature remains nearly constant. However, as is discussed in this problem, when the amount of the residual solution changes largely, the boiling temperature usually rises with distillation as is seen in Fig.8. We have to pay attention to the conditions how the gas-liquid equilibrium is achieved. Instinctively, the rise in the boiling temperature may be understood as follows. By continuing distillation, the component which has the lower boiling point (N_2 in the present case) is removed faster than the other, the mole fraction of the other component which has the higher boiling point (O_2 in the present case) increases with distillation, causing a further temperature rise upon continuing distillation.

Solution 3-14

(a) The saturated vapor pressures of water are plotted in Fig.9 below against the mole fraction X_{H_2O}. As is seen from the figure, a good linear correlation is found and, hence, we can say that *Raoult*'s law is satisfied in this concentration range of water.

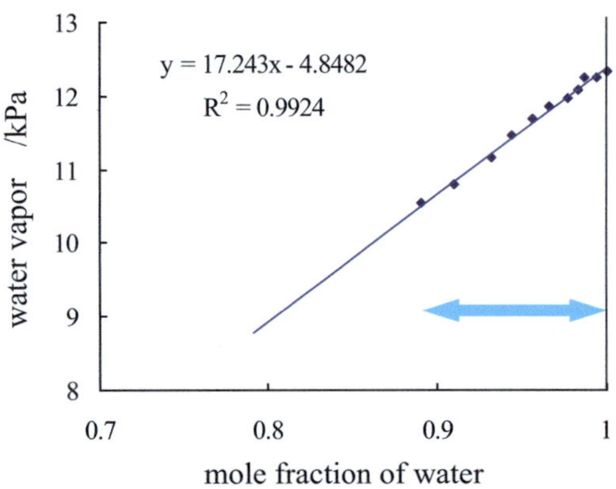

Fig.9 Correlation between water vapor pressures and mole fractions of water in water-sucrose solutions.
The obtained best fitted straight line does not pass through the origin. From this, it is known that *Raoult*'s law does not hold at high concentrations of sucrose.

Ans. *Raoult*'s law is satisfied.

(b) Denoting the activity of liquid water (A) as $a_A(X_{H_2O})$, its chemical potential $\mu_A^l(X_{H_2O})$ and the chemical potential of water vapor $\mu_A^g(X_{H_2O})$ are given by eqs(69) and (70), respectively. Here, the standard state is that of the pure water (see **solution 3-8**).

$$\mu_A^l(X_{H_2O}) = \mu_A^l(1.0) + RT \ln a_A(X_{H_2O}) \qquad (69)$$

$$\mu_A^g(X_{H_2O}) = \mu_A^g(P = P^\circ) + RT \ln \frac{P(X_{H_2O})}{P^\circ} \qquad (70)$$

The condition for the gas-liquid equilibrium is given by eq(71) in which eq(72) is satisfied since the standard

$$\mu_A^l(X_A) = \mu_A^g(X_A) \qquad (71), \qquad \mu_A^l(1.0) = \mu_A^g(P = P^\circ) \qquad (72)$$

state is the pure water. From these equations, eq(73) is derived, from which the values for the activity a_A

$$\ln a_A(X_{H_2O}) = \ln\frac{P(X_{H_2O})}{P^\circ} \Rightarrow a_A(X_{H_2O}) = \frac{P(X_{H_2O})}{P^\circ} = \gamma_A \cdot X_{H_2O} \qquad (73)$$

and the activity coefficient γ_A are found as given in the solution by substituting the relevant values given in the table and $P^\circ = 12.333$ Pa into eq(73) .

Ans. the values to be filled in: $a_A = 0.9934,\ 0.9477,\ 0.8758;\ \gamma_A = 1.007,\ 0.9914,\ 0.9626$

"note" Strictly speaking, *Raoult*'s law does not hold since the values found for the activity coefficient deviate from unity.

(c) By putting $\mu_A^l(1.0) = \Delta G_f^0 = -231.7\ \text{kJ}\cdot\text{mol}^{-1}$ in eq(69), the values for the activity found in (b) are substituted into eq(69) to find a value for the chemical potential, for instance,

as below.

$$\mu_A^l(0.9864) = -231.7\times10^3 + 8.31\times298\ln0.9934 = -231.7\ \text{kJ}\cdot\text{mol}^{-1}.$$

Ans. chemical potentials of water: $-231.7,\ -231.8$ and $-232.0\ \text{kJ}\cdot\text{mol}^{-1}$

(d) The *Gibbs* free energy of this solution is generally expressed by eq(74) with $n=1$ for one mole of solution.

$$G^l = n_A\mu_A^l + n_B\mu_B^l = n(X_A\mu_A^l + X_B\mu_B^l) \qquad (74)$$

The chemical potentials for water (A) employed in eqs(69) to (74) and for sucrose (B) are expressed by eq(75).

$$\mu_A^l(X_A) = \mu_A^l(1.0) + RT\ln a_A(X_A)\ ,\quad \mu_B^l(X_B) = \mu_B^l(1.0) + RT\ln a_B(X_B) \qquad (75)$$

In order to find $\mu_B^l(1.0)$ for sucrose, its solubility in water (solution equilibrium) is employed. The condition for the solution equilibrium is expressed as eq(76) since the chemical potential of pure solid sucrose is equal to the chemical potential of sucrose in its saturated aqueous solution.

$$\Delta G_f^\circ = \mu_B^s(1.0) = \mu_B^l(1.0) + RT\ln a_B(X_B) \qquad (76)$$

Employing the approximation of $a_B \approx X_B$ and substituting the given values for the solubility and the free energy of formation into eq(76), a value for $\mu_B^s(1.0)$ is found as below(eq(77)) .

$$-1549\times10^3 = \mu_B^l(1.0) + 8.31\times323\times\ln\frac{(72/342)}{(100/18)+(72/342)} \Rightarrow \mu_B^l(1.0) = -1540\ \text{kJ}\cdot\text{mol}^{-1} \qquad (77)$$

Substituting the values thus obtained into eq(75), the chemical potentials for water and sucrose at the concentrations required are found as below.

$$\left.\begin{aligned}\mu_A^l(0.9098) &= -231.7\times10^3 + RT\ln0.8758 = -237.5\ \text{kJ}\cdot\text{mol}^{-1}\\ \mu_B^l(0.0902) &= -1540\times10^3 + RT\ln0.2190 = -1544\ \text{kJ}\cdot\text{mol}^{-1}\end{aligned}\right\} \qquad (75')$$

Therefore, the free energy G^l required is calculated from eq(74) as below.

$$G^l = 0.9098\times(-237.5) + 0.0902\times(-1544) = -355\ \text{kJ}$$

Ans. free energy of the solution: -355 kJ

Solution 3-15

The mole fractions of B at the points of $\alpha,\ \beta$ and γ in Fig.10 are denoted X_B^l, X_B and X_B^g, respectively.

Anywhere in the region indicated by the orange straight line, the mole fractions of B in the coexisting gas and liquid are X_B^g and X_B^l, respectively, though the amounts of the gas and liquid are different. When the mole numbers of the gas and the liquid at the point β are denoted x and 1-x, respectively, the mole fraction X_B at this point is expressed by eq(78).

$$xX_B^g + (1-x)X_B^l = X_B \qquad (78)$$

By putting $X_B = xX_B + (1-x)X_B$, eq(78) is rewritten as eq(79).

$$xX_B^g - xX_B = (1-x)X_B - (1-x)X_B^l$$

$$x(X_B^g - X_B) = (1-x)(X_B - X_B^l) \qquad (79)$$

By replacing $(X_B^g - X_B)$ and $(X_B - X_B^l)$ with a and b as below, eq(79) can be expressed as eq(80).

$$(X_B^g - X_B) = a \ , \ (X_B - X_B^l) = b$$

$$x \times a = (1-x) \times b = \ \Rightarrow x : (1-x) = b : a \qquad (80)$$

$$\text{mole numbers of gas} : \text{mole numbers of liquid} = b : a$$

Eq(80) is nothing but the lever rule. As is seen from eq(78), this rule is another expression of the conservation law of mass. The reason why the mole numbers of the gas and liquid are denoted x and 1-x is because the total mole numbers are invariant during the whole vaporization or liquefaction process.

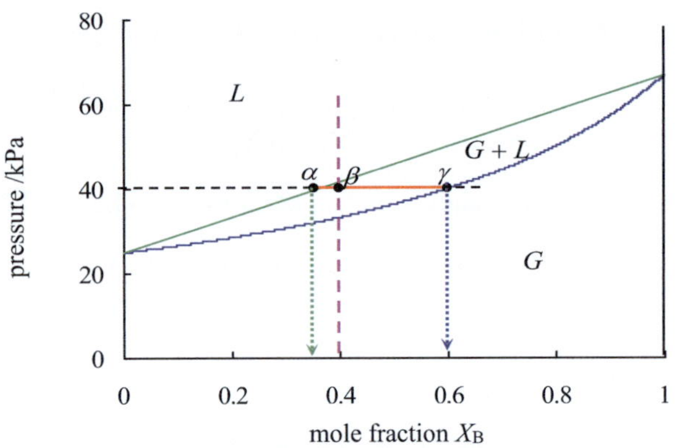

Fig.10 (shown again). See the problem.

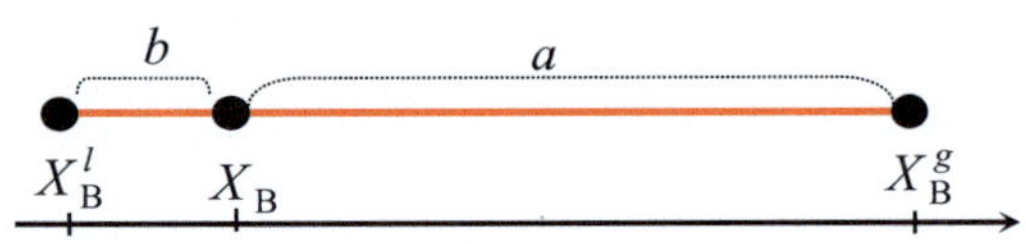

Fig.11 (shown again). See the problem.

Ans. See eq(80)and text.

Solution 3-16a

The boiling point of a mixed solution is the temperature at which the total of the partial pressures of all components in the gas phase reaches 1 atm (1.01×10^5 Pa). For each component, the dependence of the vapor pressure on temperature can be estimated from *Clausius-Clapeyron*'s equation (eq(81)) where ΔH_{vap} is the latent heat of vaporization.

$$\frac{1}{P}\frac{dP}{dT} = \frac{\Delta H_{vap}}{RT^2} \ \Rightarrow \ \ln P = -\frac{\Delta H_{vap}}{RT} + \text{const.} \qquad (81)$$

Eq(81) can be rewritten as eq(81'), where a constant α introduced upon integration of eq(81) is determined

$$P = \alpha \times \exp\left(-\frac{\Delta H_{vap}}{RT}\right) \qquad (81')$$

for each compound individually from the value of its boiling point. In the present case, A and B happen to take the same value of $\alpha = 1.06 \times 10^7$ Pa.

The boiling point of a two-component solution specified by its mole fractions of X_{A}^{l} and X_{B}^{l} is the temperature T at which the partial pressures of A and B, denoted $P_{\mathrm{A}}(T)$ and $P_{\mathrm{B}}(T)$, satisfy the relation given by eq(82).

$$P_{\mathrm{A}}(T)+P_{\mathrm{B}}(T)=1.01\times10^{5}\,\mathrm{Pa} \qquad (82)$$

Employing eq(81') and *Raoult*'s law, eq(82) can be expressed as eq(83).

$$P_{\mathrm{A}}(T)+P_{\mathrm{B}}(T)=\alpha_{\mathrm{A}}\times\exp\left(-\frac{\Delta H_{vap}^{\mathrm{A}}}{RT}\right)\times X_{\mathrm{A}}^{l}+\alpha_{\mathrm{B}}\times\exp\left(-\frac{\Delta H_{vap}^{\mathrm{B}}}{RT}\right)\times X_{\mathrm{B}}^{l}=1.01\times10^{5}\,\mathrm{Pa} \qquad (83)$$

Solving eq(83) for $X_{\mathrm{A}}^{l}\,(=1-X_{\mathrm{B}}^{l})$, eq(84) is obtained.

$$X_{\mathrm{A}}^{l}=\left\{\frac{1.01\times10^{5}}{1.06\times10^{7}}-\exp\left(-\frac{\Delta H_{vap}^{\mathrm{B}}}{RT}\right)\right\}\Big/\left\{\exp\left(-\frac{\Delta H_{vap}^{\mathrm{A}}}{RT}\right)-\exp\left(-\frac{\Delta H_{vap}^{\mathrm{B}}}{RT}\right)\right\} \qquad (84)$$

On the other hand, the mole fraction of A in the gas phase is given in terms of the partial pressures $P_{\mathrm{A}}(T)$ and $P_{\mathrm{B}}(T)$ for A and B as is given by eq(85).

$$X_{\mathrm{A}}^{g}=\frac{P_{\mathrm{A}}(T)}{P_{\mathrm{A}}(T)+P_{\mathrm{B}}(T)}=\frac{1.06\times10^{7}}{1.01\times10^{5}}\times\exp\left(-\frac{\Delta H_{vap}^{\mathrm{A}}}{RT}\right)\times X_{\mathrm{A}}^{l} \qquad (85)$$

Consequently, the *T-X* diagram for the present system can be obtained by plotting the values calculated from eqs(84) and (85) against temperature. Since the mole fractions are given as a function of temperature in these equations, a simple way of plotting the result is to take the mole fraction along the vertical axis and the temperature along the horizontal axis. The diagram thus obtained, however, is not easy to use for practical purposes. For this, the diagram with *x*- and *y*-axis exchanged by using *Parametric Plot* in *Mathematica* is shown in Fig.12

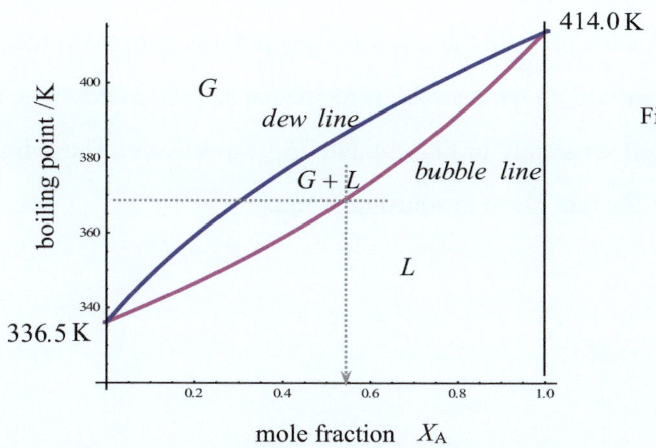

Fig.12 *T-X* diagram of the ideal solutions whose boiling points and vapor pressure curve are given.
The dotted arrow indicates the composition of the solution whose boiling point is 370.0 K.
bubble line: vaporization line,
dew line: liquefaction line.

Ans. See Fig.12.

"note 1" It is known that 2-methyl-1-propanol (bp 355.5 K) and 2-propanol (bp 381K) form nearly ideal solutions. The *T-X* diagram experimentally obtained by measuring vapor pressures of the mixed solutions is shown in Fig.13 which resembles Fig.12 very closely.

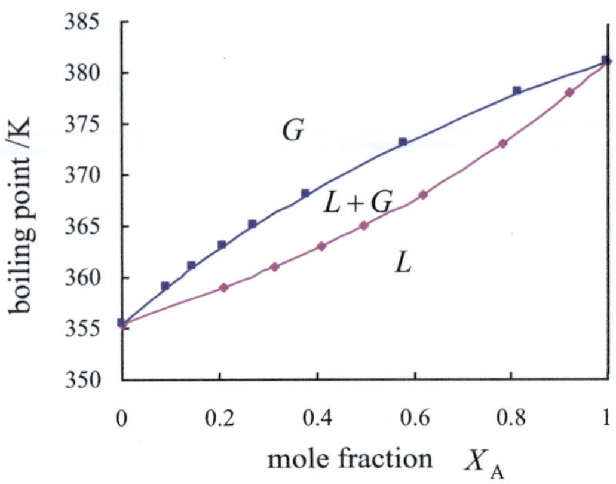

"**note 2**" For the sake of simplicity, the latent heat of vaporization is assumed constant here. A better approximation is to put $\Delta H_{vap} = a + bT$ (see **problem 3-37**).

"**note 3**" It is easy to draw the *P-X* diagram for the ideal solution.

Fig.13 *T-X* diagram for the solutions of 2-methyl-1-propanol and 2-propanol.

Solution 3-16b

The relation between the boiling point and the mole fraction of the solution is given by eq(84), which is derived in the preceding solution (given below again).

$$X_A^l = \left\{\frac{1.01\times10^5}{1.06\times10^7} - \exp\left(-\frac{\Delta H_{vap}^B}{RT}\right)\right\} \Big/ \left\{\exp\left(-\frac{\Delta H_{vap}^A}{RT}\right) - \exp\left(-\frac{\Delta H_{vap}^B}{RT}\right)\right\} \tag{84}$$

By substituting the given values into this equation, the mole fraction is found as $X_A^l = 0.556$. This corresponds to the intersection of the dotted line with the bubble line in Fig.12 given in the preceding solution. Thus, when the bubble line is known for the solution of A and B, it is possible to prepare a solution which boils at any arbitrary temperature located between the boiling points of the solvent A and solvent B (see **problem 3-37**).

Ans. $X_A^l = 0.556$

Solution 3-16c

Schematic diagrams for the two cases, one in which B with the lower boiling point is purified as much as possible and the other in which A with the higher boiling temperature is recovered as pure as possible, both from mixed solutions of A and B, are illustrated separately in Figs.14 and 15, respectively. These figures are the enlarged parts of the corresponding *T-X* diagram given in **solution 3-16a**.

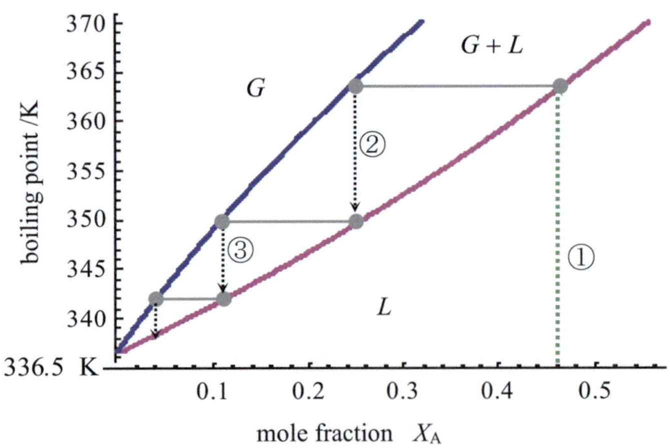

Fig.14 Schematic procedure for recovering pure B of the lower boiling point from a A-B mixed solution.

As is seen in Fig.14, when the solution initially with the mole fraction 0.46 of A is heated at the boiling point of 364 K (operation ①) and the collected vapor is condensed (operation ②), a solution with the mol fraction 0.25 of A is gained. The solution collected thus is heated again but this time at 350 K and the collected vapor is condensed (operation ③), a solution with the mole fraction 0.11 of A is gained. By repeating the operation of this sort many times, a liquid of high purity which hardly contains A is obtained.

As is illustrated in Fig.15, the solution initially of the mole fraction 0.80 of A is heated at 396 K (operation ①) and stuff in the vapor phase are removed by condensation until the mole fraction of A in the solution phase increases to 0.85. Upon continuation of heating, the boiling point rises to 402K (operation ②) and again stuff in the vapor phase is removed by condensation. Then, the solution with the mole fraction 0.90 of A is obtained. By repeating the operation of this sort (operation ③ and thereafter) many times, a liquid of high purity which hardly contains B is obtained.

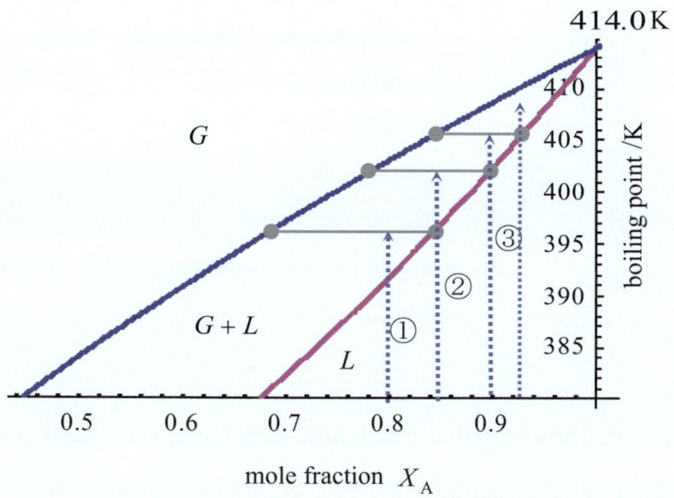

Fig.15 Schematic procedure for recovering pure A of the higher boiling point from a A-B mixed solution.

In contrast to the case in Fig.14, upon purification of the component with the higher boiling point, it is only necessary to continue heating in order to remove stuff in the vapor phase containing more B than A by condensation. Then, a liquid of almost pure A is obtained as a residual.

Ans. See text.

Solution 3-17a

The *P-X* diagram which shows a gas-liquid equilibrium for ideal solutions composed of A and B can easily be drawn in the following manner. The total pressure is denoted P and the mole fractions of a solution are given by X_{A} and X_{B}. *Raoult*'s law is expressed by the left side of eq(86), from which P is expressed as a function of X_{B} as is given by the right side of eq(86).

$$P = X_{\mathrm{A}}P_{\mathrm{A}}^{\bullet} + X_{\mathrm{B}}P_{\mathrm{B}}^{\bullet} = (1 - X_{\mathrm{B}})P_{\mathrm{A}}^{\bullet} + X_{\mathrm{B}}P_{\mathrm{B}}^{\bullet} = P_{\mathrm{A}}^{\bullet} + (P_{\mathrm{B}}^{\bullet} - P_{\mathrm{A}}^{\bullet})X_{\mathrm{B}} \tag{86}$$

On the other hand, the mole fractions of A and B in the gas phase, labeled by a superscript *g*, which are in equilibrium with the solution are given by eq(87).

$$X_{\mathrm{A}}^{g} = \frac{P_{\mathrm{A}}^{\bullet}X_{\mathrm{A}}}{P} \Rightarrow X_{\mathrm{A}} = \frac{PX_{\mathrm{A}}^{g}}{P_{\mathrm{A}}^{\bullet}}, \quad X_{\mathrm{B}}^{g} = \frac{P_{\mathrm{B}}^{\bullet}X_{\mathrm{B}}}{P} \Rightarrow X_{\mathrm{B}} = \frac{PX_{\mathrm{B}}^{g}}{P_{\mathrm{B}}^{\bullet}} \tag{87}$$

From the definition of the mole fraction and eq(87), P and X_{B}^{g} is related by eq(88), which is then solved for P as in eq(89).

$$1 = X_{\mathrm{A}} + X_{\mathrm{B}} = \frac{PX_{\mathrm{A}}^{g}}{P_{\mathrm{A}}^{\bullet}} + \frac{P(1-X_{\mathrm{A}}^{g})}{P_{\mathrm{B}}^{\bullet}} \Rightarrow P\left(\frac{X_{\mathrm{A}}^{g}}{P_{\mathrm{A}}^{\bullet}} + \frac{(1-X_{\mathrm{A}}^{g})}{P_{\mathrm{B}}^{\bullet}}\right) \tag{88}$$

$$P = \frac{1}{\left(\frac{X_{\mathrm{A}}^{g}}{P_{\mathrm{A}}^{\bullet}} + \frac{1-X_{\mathrm{A}}^{g}}{P_{\mathrm{B}}^{\bullet}}\right)} = \frac{P_{\mathrm{A}}^{\bullet}P_{\mathrm{B}}^{\bullet}}{P_{\mathrm{A}}^{\bullet}(1-X_{\mathrm{A}}^{g}) + P_{\mathrm{B}}^{\bullet}X_{\mathrm{A}}^{g}} = \frac{P_{\mathrm{A}}^{\bullet}P_{\mathrm{B}}^{\bullet}}{P_{\mathrm{A}}^{\bullet} - (P_{\mathrm{A}}^{\bullet} - P_{\mathrm{B}}^{\bullet})X_{\mathrm{A}}^{g}} = \frac{P_{\mathrm{A}}^{\bullet}P_{\mathrm{B}}^{\bullet}}{P_{\mathrm{B}}^{\bullet} + (P_{\mathrm{A}}^{\bullet} - P_{\mathrm{B}}^{\bullet})X_{\mathrm{B}}^{g}} \tag{89}$$

Eq(86) expresses a liquid-(gas-liquid) equilibrium straight line and eq(89) expresses a gas-(gas-liquid) equilibrium line. Here, (gas-liquid) indicates a gas and liquid coexisting phase. Taking $P_{\mathrm{A}}^{\bullet} = 25$ kPa and $P_{\mathrm{B}}^{\bullet} = 67$ kPa, eq(86) and eq(89) are plotted against the mole fraction of B in Fig.16 below in which G and L indicate gas and liquid, respectively.

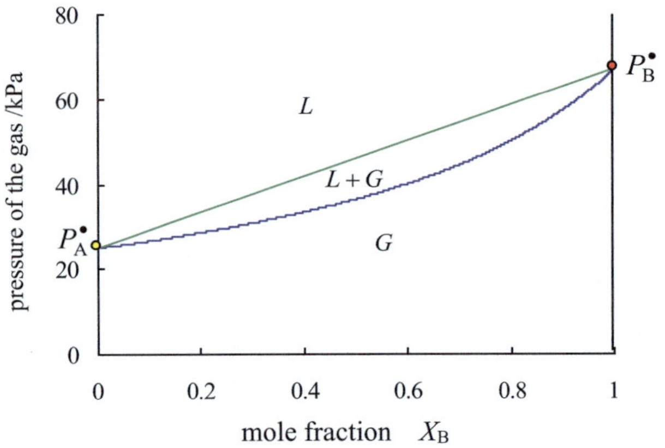

Fig.16 *P-X* diagram of gas (*G*)-liquid (*L*) equilibrium drawn for benzene (A)-toluene(B) solutions (see **problem 3-20).**
(——) eq(86)
(——) eq(89).
Gas and liquid coexist in the region (*L*+*G*) enclosed by the straight line and the concave curve.

"note" The mole fraction along the horizontal axis signifies the mole fraction in each phase. Its meaning is trivial in either gas or liquid phase. In the gas-liquid coexisting region (*L*+*G*), the mole fraction is for the total amount of the substances existing in both phases. See the lever rule given in **problem 3-15**.

Solution 3-17b

Under a constant pressure (usually 1 atm), a typical liquid-liquid *T-X* (*X*: mole fraction) diagram for a two-component system is shown in Fig.17. When a small amount of A is added to a pure liquid of B (e.g., along the black dotted horizontal arrow in Fig.17) at the temperature T_1, it solves well in B to form a homogeneous solution. Upon adding A further, however, at the point L the solution starts to separate into two phases. With adding A still more, the ratio of the amounts of two phases varies continuously until to the point M, at which the two phases suddenly mix each other into one phase. At the point P located halfway in the segment LM (tie line), where two separated solutions coexist, and the mole fraction of A in each separated solution is fixed and is given by the mole fraction of A given at the point L or M. It is easy to tell which solution is given by which mole fraction. In Fig.17, they are 0.2 and 0.7 and the ratio of the amounts of the two solutions is determined by the ratio of the lever arms, i.e., $l:m$ (lever rule) (see **solution 3-15**).

Now, in the present problem, the phase diagram is *T-wt* % but not *T-X* diagram since the horizontal axis is

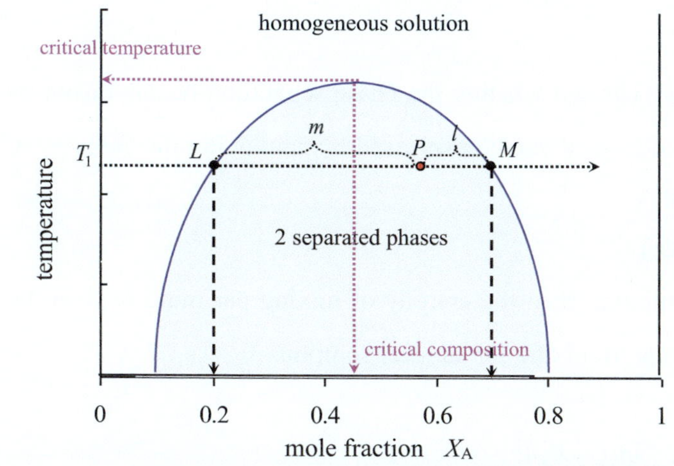

Fig.17 Phase diagram for mixed solutions of pure liquids A and B.

In the area enclosed by the blue line and the *x*-axis the mixtures exist in the two separated phases. Above the critical temperature indicated by the red dotted horizontal arrow, the mixture can exist only as a homogeneous solution in one phase.

given in the unit of *wt* %. However, *wt* % is an expression consistent with the conservation of mass, so that the lever rule can be applied to know the relative amount of each phase.

Specifically, when two-phase separated solutions with the wt % of w_l and w_m exist in the ratio of $a:b$, the mass of A is given by $(1-w_l)a+(1-w_m)b$ and that of B is by w_la+w_mb. From these, a and b are expressed as follows in eq(90) in terms of w_P and are found to correspond to the lever arms.

$$\left.\begin{array}{l} w_p=(w_l\times a+w_m\times b)=(w_l-w_m)\times a+w_m \quad\Rightarrow \\ \left.\begin{array}{l} a=(w_m-w_p)/(w_m-w_l) \\ b=(w_p-w_l)/(w_m-w_l) \end{array}\right\}\text{ lever arms} \end{array}\right\} \tag{90}$$

If it is assumed that under higher pressures the two pure liquids mix more readily, then a phase diagram such as shown in Fig.18 is expected for the mixture. With increasing pressure the region where the two phases coexist becomes narrower and narrower and above the critical pressure, the mixture can exist only as a uniform solution in one phase. Fig.18 is nothing more than a schematic representation but various types of *P-T-X* (three-dimensional) diagrams can be drawn in a similar manner.

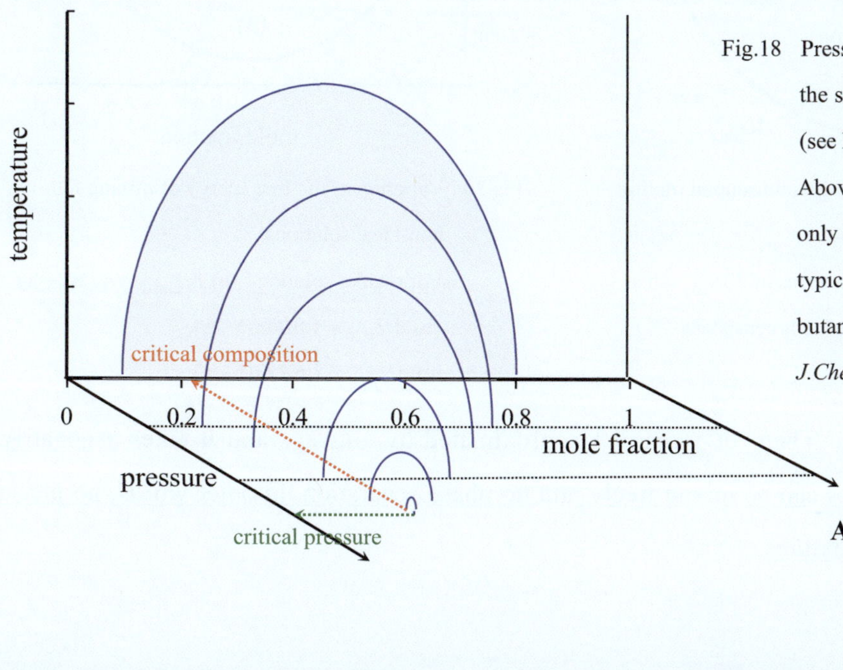

Fig.18 Pressure dependence of the phase diagram for the solutions made of two pure liquids A and B (see Fig.17). A schematic diagram.

Above the critical pressure, the mixture can only exist as a uniform solution in one phase. A typical example is given by a mixture of butanol-2 + water (see *Moriyoshi* et al. *J.Chem.Thermo.*, **7**, 537 (1973)).

Ans. See text and Figs.17 and 18.

Solution 3-17c

The answer is omitted. See Figs.17 and 18 in the preceding solution.

Solution 3-17d

When two kinds of gases A and B are mixed, we can tell whether the phase separation occurs or not by examining the free energy of mixing. Denoting it ΔG_{mix}, it can be expressed by employing the enthalpy of mixing ΔH_{mix} and entropy of mixing ΔS_{mix} as eq(91).

$$\Delta G_{mix} = \Delta H_{mix} - T\Delta S_{mix} \qquad (91)$$

If we assume that a uniform phase is formed by mixing, then the entropy of mixing per mole is given by eq(92), which is a well known equation as the entropy of mixing for an ideal solution.

$$\begin{aligned}\Delta S_{mix} &= -R(X_A \ln X_A + X_B \ln X_B) \\ &= -R\{X_A \ln X_A + (1-X_A)\ln(1-X_A)\} \qquad (92)\end{aligned}$$

The intermolecular interaction between A and B is also assumed to be totally the same as that between A and A or between B and B. Then there is no enthalpy change upon mixing, so that we can put $\Delta H_{mix} = 0$. In such a case, ΔG_{mix} is given by eq(93).

$$\Delta G_{mix} = -T\Delta S_{mix} \qquad (93)$$

Employing eq(92), ΔG_{mix} is plotted against X_A in Fig.19 below. As is clear from Fig.19, the free energy of mixing ΔG_{mix} is always negative and does not take a maximum peak anywhere. Consequently, the phase separation cannot occur.

In conclusion, in the case of gases, molecules are separated one another far apart on the average and the intermolecular interaction does not change significantly upon mixing, leading to a negligibly small value of

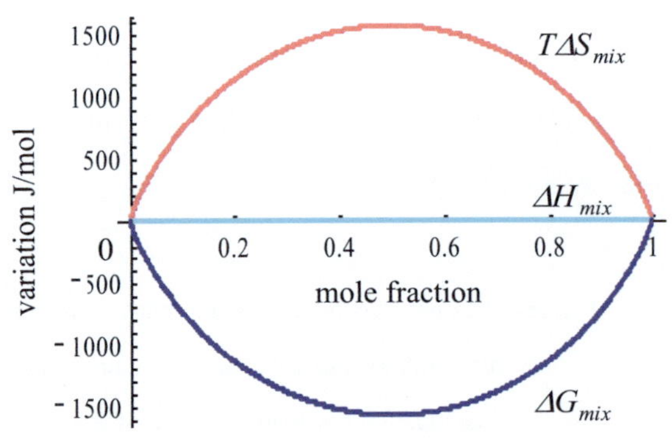

Fig.19 Variations of thermodynamic quantities upon mixing for ideal solutions. Entropy of mixing is given in terms of $T\Delta S_{mix}$ so as to make the unit of the vertical axis consistent (T = 298 K).

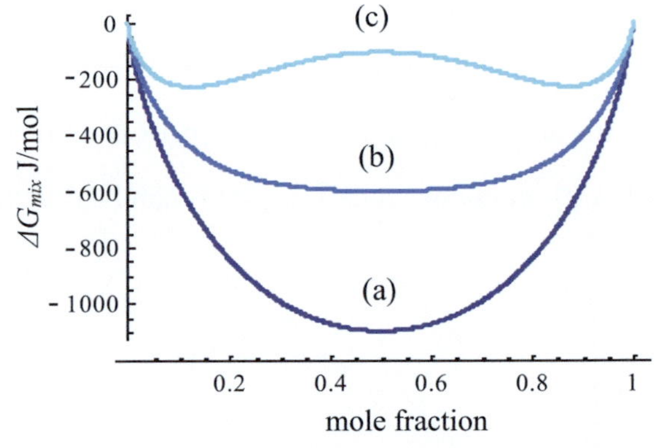

Fig.20 Variation of the free energy of mixing for non-ideal solutions. For regular solutions: (a) $\Delta H_{mix} = RTx(1-x)$, (b) $\Delta H_{mix} = 1.8RTx(1-x)$, (c) $\Delta H_{mix} = 2.6RTx(1-x)$

ΔH_{mix} as compared with $-T\Delta S_{mix}$. Then ΔG_{mix} can be approximated by $-T\Delta S_{mix}$ and it takes a negative value always. Because of this, gases can be mixed freely and no phase separation, in other words, no phase equilibrium is recognized for gas mixtures.

When ΔH_{mix} comes to have a comparatively large value, the free energy of mixing starts to take a maximum as is shown in Fig.20 and as a consequence, the phase separation comes to be recognized in some range of the mole fraction. Obviously, this kind of situation does not arises for gases.

As is evident from the discussions above, it can be said that for ideal or nearly ideal solutions, any liquid-liquid phase equilibrium is not recognized. For further discussions on phase separation and regular solutions, see **problems 4-89 - 4-91**.

Ans. See text.

"note" For real mixed solutions, the entropy of mixing ΔS_{mix}^{real} is different from that for ideal solutions ΔS_{mix}^{ideal}. The difference given by $\Delta S_{mix}^{ex} = \Delta S_{mix}^{real} - \Delta S_{mix}^{ideal}$ is called the excess entropy of mixing and its value varies depending on the conditions of mixing, that is, mixing at constant pressure or at constant volume. If there does not exist any specific intermolecular interactions such as hydrogen bonding, ΔS_{mix}^{ex} is of the magnitude of a few % of ΔS_{mix}^{ideal}.

[advanced problem] When the entropy of mixing ΔS_{mix} is given by eq(92), show that the chemical potential of A is given by the following equation and vice versa.

$$\mu_A(P,T) = \mu_A^\circ(P,T) + RT\ln X_A$$

Solution 3-18

Based on the first and second laws of thermodynamics, the total differential of the internal energy of a system is expressed by eq(94).

$$dU = TdS - PdV \qquad (94)$$

From both sides of eq(94), the differential of the product of conjugate variables *T* and *S*, i.e., *d*(*TS*), is subtracted as in eq(95). Rewriting $U - TS$ as *F* (or *A*) in eq(95) and then rearranging eq(95) from left to right as shown below,

$$dU - \underline{d(TS)} = TdS - PdV - \underline{d(TS)} = d(U - TS) = TdS - PdV - TdS - SdT = -SdT - PdV \qquad (95)$$

the relation of $dF = d(U - TS) = -SdT - PdV$ is derived. From this differential form, *F* is regarded as a newly introduced thermodynamic variable specified by *T* and *V* and is called *Helmholtz* free energy.

Ans. $F \equiv U - TS$, $dF = -SdT - PdV$

"note" V and P, T and S, and μ and *n*, each pair of which is called conjugate thermodynamic variables. The product of the conjugate variables for any pair has a dimension of energy.

Solution 3-19a

(a) Since the solutions behave ideally, the pressure *P* of the gas in equilibrium with the solution is given by eq(96) based on *Raoult*'s law.

$$P = P_A^\bullet X_A + P_B^\bullet X_B = 67 \times X_A + 25 \times (1 - X_A) = 42X_A + 25 \text{ (Torr)} \qquad (96)$$

Ans. $(42X_A + 25)$ Torr

(b) The pressure is calculated from eq(96) as $P = 42 \times 0.2 + 25 = 33.4$ (Torr). Also, denoting the mole fractions of benzene (A) and toluene (B) in the gas phase as X_A^g and X_B^g, respectively, they are readily calculated from the total pressure *P* and the partial pressure given by $X_A P_A^\bullet$ for A as follows.

$$X_A^g = \frac{X_A P_A^\bullet}{P} = \frac{0.2 \times 67}{33.4} = 0.401, \qquad X_B^g = 1 - X_A^g$$

Ans. $P = 33.4$ Torr, $X_A^g = 0.401$ and $X_B^g = 0.599$

(c) The vapor pressure at $X_A = 0.3$ is calculated as $P = 42 \times 0.3 + 25 = 37.6$ Torr. The given pressure of 33.4 Torr is lower than that but higher than the pressure of 30.8 Torr at which all are transformed into the gas. Consequently, under the given conditions, the mixture is considered to exist in the gas-liquid coexisting region. The pressure at which all are transformed into the gas is found by eq(97) upon substitution of 0.3 for

$$P = \frac{P_A^\bullet \cdot P_B^\bullet}{P_A^\bullet - (P_A^\bullet - P_B^\bullet) X_A^g} = \frac{67 \times 25}{67 - (67 - 25) X_A^g} \qquad (97)$$

X_A^g into it (see **problem 3-17**). In this equation, substituting 33.4 Torr for P and solving for X_A^g, 0.401 is found for X_A^g (or see the answer in (b)). Therefore, the mole fractions at which $P = 33.4$ intersects with eq(96) and eq(97) are found as 0.200 and 0.401. Since the mole fraction of the solution is 0.3, the left and right lever arms to those mole fractions are approximately equal, leading to the conclusion that the ratio of the amounts in the gas and liquid phases will be 1:1 in moles.

Ans. In the gas-liquid coexisting region with 1:1 gas to liquid ratio in moles.

"note" The mole fractions found here are denoted either X_A or X_A^g. It is important to understand correctly their meanings. These values are at the points where the vaporization starts or the liquefaction commences and almost all amount exists as liquid or gas alone. Therefore, it is obvious that either mole fraction is definitely for the total amount.

Solution 3-19b

(a) The chemical potentials of pure benzene (A) and pure toluene (B) at 300 K are denoted μ_A° and μ_B°. Then, assuming the solutions behave ideally, the free energy change upon mixing is given by eq(98).

$$\begin{aligned}
\Delta G_{mix}(300\text{ K}) &= (\text{free energy of solution}) - (\text{sum of the free energies of pure liquids}) \\
&= n_A\mu_A + n_B\mu_B - (n_A\mu_A^\circ + n_B\mu_B^\circ) \\
&= n_A(\mu_A^\circ + RT\ln X_A) + n_B(\mu_B^\circ + RT\ln X_B) - (n_A\mu_A^\circ + n_B\mu_B^\circ) \\
&= nRT(X_A \ln X_A + X_B \ln X_B) \qquad (n = n_A + n_B) \qquad (98) \\
&= -T\Delta S_{mix} \qquad (\because \Delta H_{mix} = 0)
\end{aligned}$$

From this, putting $X_A = 0.2$, $X_B = 0.8$ and $n = 1$, a value for $\Delta G_{mix}(300\text{ K})$ is found as -1.25×10^3 J.

Ans. $RT(X_A \ln X_A + X_B \ln X_B) = -1.25 \times 10^3$ J

(b) Denoting the vapor pressures of the pure substances A and B at 300 K as $P_A^\bullet$ and $P_B^\bullet$, the pressure P of the gas in equilibrium with the solution is given by eq(99) based on *Raoult*'s law.

$$P = X_A P_A^\bullet + X_B P_B^\bullet = X_A P_A^\bullet + (1 - X_A) P_B^\bullet = P_B^\bullet + (P_A^\bullet - P_B^\bullet) X_A \qquad (99)$$

Substituting the relevant values into eq(99), a value for P is obtained as below.

$$P = 25 + (67 - 25) \times 0.2 = 33.4 \text{ Torr}$$

The mole fractions of benzene and toluene in the gas phase written as X_A^g and X_B^g, respectively, are readily found from the total pressure P and the partial pressures for A and B as follows (see preceding solution).

$$X_{\mathrm{A}}^{g}=\frac{X_{\mathrm{A}}P_{\mathrm{A}}^{\bullet}}{P}=\frac{0.2\times 67}{33.4}=0.401\,,\qquad X_{\mathrm{B}}^{g}=\frac{X_{\mathrm{B}}P_{\mathrm{B}}^{\bullet}}{P}=\frac{0.8\times 25}{33.4}=0.599$$

Ans. $P=33.4$ Torr, $X_{\mathrm{A}}^{g}=0.401$ and $X_{\mathrm{B}}^{g}=0.599$

(c), (d) The free energy of a mixed solution is given by eq(100). To distinguish between gas and liquid, superscripts g or l are added to relevant variables.

$$G_{\mathrm{AB}}^{l}=n_{\mathrm{A}}\mu_{\mathrm{A}}^{l}+n_{\mathrm{B}}\mu_{\mathrm{B}}^{l} \qquad (100)$$

At the point where the vaporization just commences (gas-liquid equilibrium), the chemical potentials of each substance in the gas and solution phases must be equal. Thus eqs(101) and (102) are satisfied.

$$\mu_{\mathrm{A}}^{l}=\mu_{\mathrm{A}}^{g}=\mu_{\mathrm{A}}^{\circ}(g)+RT\ln\frac{P_{\mathrm{A}}}{P^{\circ}}=\mu_{\mathrm{A}}^{\circ}(g)+RT\ln\frac{X_{\mathrm{A}}P_{\mathrm{A}}^{\bullet}}{P^{\circ}} \qquad (101)$$

$$\mu_{\mathrm{B}}^{l}=\mu_{\mathrm{B}}^{g}=\mu_{\mathrm{B}}^{\circ}(g)+RT\ln\frac{P_{\mathrm{B}}}{P^{\circ}}=\mu_{\mathrm{B}}^{\circ}(g)+RT\ln\frac{X_{\mathrm{B}}P_{\mathrm{B}}^{\bullet}}{P^{\circ}} \qquad (102)$$

Consequently, for the free energy of the solution G_{AB}^{l}, eq(103) is obtained from eq(100).

$$G_{\mathrm{AB}}^{l}=n_{\mathrm{A}}\left(\mu_{\mathrm{A}}^{\circ}(g)+RT\ln\frac{X_{\mathrm{A}}P_{\mathrm{A}}^{\bullet}}{P^{\circ}}\right)+n_{\mathrm{B}}\left(\mu_{\mathrm{B}}^{\circ}(g)+RT\ln\frac{X_{\mathrm{B}}P_{\mathrm{B}}^{\bullet}}{P^{\circ}}\right) \qquad (103)$$

On the other hand, the free energy of the gas mixture obtained by transforming all substances in the solution into the gas phase is expressed by eq(104) with P_{total} as the total pressure of the gas.

$$G_{\mathrm{AB}}^{g}=n_{\mathrm{A}}\left(\mu_{\mathrm{A}}^{\circ}(g)+RT\ln\frac{X_{\mathrm{A}}P_{total}}{P^{\circ}}\right)+n_{\mathrm{B}}\left(\mu_{\mathrm{B}}^{\circ}(g)+RT\ln\frac{X_{\mathrm{B}}P_{total}}{P^{\circ}}\right) \qquad (104)$$

From eqs(103) and (104), the free energy change associated with transforming solution into gas $\Delta G_{\mathrm{AB}}(l\rightarrow g)$ is given by eq(105).

$$\begin{aligned}\Delta G_{\mathrm{AB}}(l\rightarrow g)=&\left\{n_{\mathrm{A}}\left(\mu_{\mathrm{A}}^{\circ}(g)+RT\ln\frac{X_{\mathrm{A}}P_{total}}{P^{\circ}}\right)+n_{\mathrm{B}}\left(\mu_{\mathrm{B}}^{\circ}(g)+RT\ln\frac{X_{\mathrm{B}}P_{total}}{P^{\circ}}\right)\right\}\\&-\left\{n_{\mathrm{A}}\left(\mu_{\mathrm{A}}^{\circ}(g)+RT\ln\frac{X_{\mathrm{A}}P_{\mathrm{A}}^{\bullet}}{P^{\circ}}\right)+n_{\mathrm{B}}\left(\mu_{\mathrm{B}}^{\circ}(g)+RT\ln\frac{X_{\mathrm{B}}P_{\mathrm{B}}^{\bullet}}{P^{\circ}}\right)\right\}\\=&\,n_{\mathrm{A}}RT\ln\frac{P_{total}}{P_{\mathrm{A}}^{\bullet}}+n_{\mathrm{B}}RT\ln\frac{P_{total}}{P_{\mathrm{B}}^{\bullet}} \qquad (105)\end{aligned}$$

Since P_{total} is the pressure of the gas established as a result of this transformation, it is calculated from the dew line as below.

$$P_{total}=\frac{P_{\mathrm{A}}^{\bullet}P_{\mathrm{B}}^{\bullet}}{P_{\mathrm{A}}^{\bullet}-(P_{\mathrm{A}}^{\bullet}-P_{\mathrm{B}}^{\bullet})X_{\mathrm{A}}}=\frac{67\times 25}{67-42\times 0.2}=\frac{1675}{58.6}=28.6\ \text{Torr}$$

Ans. 28.6 Torr

Substituting this and the other relevant values into eq(105), the free energy change associated with the vaporization is found as below.

$$\Delta G_{\mathrm{AB}}(l\rightarrow g)=0.2RT\ln\frac{28.6}{67}+0.8RT\ln\frac{28.6}{25}$$

The result obtained is evidently equivalent to the pressure change of each substance in the gas phase.

Ans. $\Delta G_{\mathrm{AB}}(l\rightarrow g)=0.2RT\ln\frac{28.6}{67}+0.8RT\ln\frac{28.6}{25}=-156\ \mathrm{J}$

Solution 3-20a

The densities of pure substances A and B are denoted ρ_{A} and ρ_{B}, respectively. When they are mixed together in the volume ratio of V_A to V_B, the solution of the volume of V is obtained. The following three cases are possible for the volume of the solution obtained.

(i) $V = V_A + V_B$, (ii) $V < V_A + V_B$, (iii) $V > V_A + V_B$

The density of the solution ρ_{mix} is given by eq(106).

$$\rho_{mix} = \frac{\rho_A V_A + \rho_B V_B}{V} \qquad (106)$$

For simplicity, it is assumed that one of the relations (i)-(iii) is always satisfied for any mixing ratio.

Case (i) The density of the solution is given by the right side of eq(107) from which it is seen that the density is proportional to the mixing volume ratio given by $X_A = V_A/(V_A + V_B)$ or X_B.

$$\rho_{mix} = \frac{\rho_A V_A + \rho_B V_B}{V_A + V_B} = \rho_A X_A + \rho_B X_B = (\rho_A - \rho_B)X_A + \rho_B \qquad (107)$$

A probable cause for this is that both molecules A and B are alike in many respects such as in the molecular size, the manner of occupying the space (number of neighboring molecules or coordination number), the intermolecular force, etc.

Case (ii) As is evident from eq(106), eq(108) is satisfied.

$$\rho_{mix} > (\rho_A - \rho_B)X_A + \rho_B \qquad (108)$$

There may be several possible reasons for that. For example, the size of the molecule B is much smaller than that of A and, hence, B can penetrate into the cavities (or voids) existing in the solvent A. Or the attractive intermolecular interaction between A and B molecules are significantly stronger than those between the same kind of molecules. As a result, the solution comes to assume a more compact liquid structure. Or the molecule B can play a role of solvent-structure breaking toward the solvent A, so that the solution takes a more compact structure as compared with that before mixing B.

Case (iii) It is evident that eq(108') is satisfied.

$$\rho_{mix} < (\rho_A - \rho_B)X_A + \rho_B \qquad (108')$$

This case could be rare because the solution is homogeneous but its volume is larger than the simple sum of the volumes of the two component solvents. Nevertheless, contrary to the case (ii), if a presence of B enhances the formation of a liquid structure, if any, then the solution forms a more sparse structure as compared with the solvent A , making the case (iii) possible.

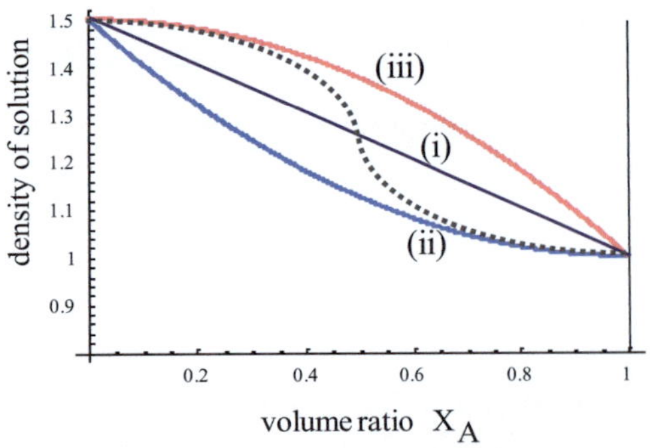

Fig.21 Variation of the density of a solution expessed in trems of the mixing ratio by volume X_A in the three distinct cases.

Based on from eq(106) to eq(108'), how the density of a solution varies approximately with the mixing ratio by volume is illustrated in Fig.21 above for the three cases discussed above

Case (iv) As a fourth case, the density may change initially following eq(108) but after a certain volume ratio, it turns to follow eq(108') or vice versa. One such case is shown by grey dotted line in Fig.21.

Of course in any case discussed above, the solvents are not necessarily expected to mix homogeneously at any mixing volume ratio. In practice, there exist many cases in which the phase separation occurs in a wide range or a certain range of the mixing ratio. It is worth of searching for what kinds of solvent combinations the cases from (i) to (iv) are realized (solubility parameters). See **4-89** - **4-91** for regular solutions.

Ans. See text.

Solution 3-20b

(a) The molecular weights of the compounds A and B are denoted M_{A} and M_{B}. A solution prepared by mixing the mass of m_{A} of A with that of m_{B} of B has a volume of V. Then the density of the solution ρ_{mix} is given by eq(109), where n_{A} and n_{B} are the mole numbers of A and B, respectively.

$$\rho_{mix}=\frac{m_{\mathrm{A}}+m_{\mathrm{B}}}{V}=\frac{n_{\mathrm{A}}M_{\mathrm{A}}+n_{\mathrm{B}}M_{\mathrm{B}}}{V} \tag{109}$$

In order to find partial molar volumes, it is simpler to employ a concept of the specific volume, i.e., the volume per unit mass. Therefore, the reciprocal of eq(109), $1/\rho_{mix}$ (eq(110)), is used. Since differentials or partial differentials are frequently used in deriving the result, it is essential to clearly specify what are and what are not variables with respect to those differential operations.

$$\frac{1}{\rho_{mix}}=\frac{V(n_{\mathrm{A}},n_{\mathrm{B}})}{m_{\mathrm{A}}+m_{\mathrm{B}}}=\frac{V(n_{\mathrm{A}},n_{\mathrm{B}})}{n_{\mathrm{A}}M_{\mathrm{A}}+n_{\mathrm{B}}M_{\mathrm{B}}} \tag{110}$$

Differentiating eq(110) with X_{B}, eq(111) is obtained. The last term of eq(111) is the partial molar volume of A.

$$\begin{aligned}\frac{d}{dX_{\mathrm{B}}}\left(\frac{1}{\rho_{mix}}\right)&=\frac{\partial}{\partial n_{\mathrm{A}}}\left(\frac{1}{\rho_{mix}}\right)\cdot\frac{\partial n_{\mathrm{A}}}{\partial X_{\mathrm{B}}}=\left\langle-\frac{M_{\mathrm{A}}V}{(n_{\mathrm{A}}M_{\mathrm{A}}+n_{\mathrm{B}}M_{\mathrm{B}})^2}+\frac{1}{n_{\mathrm{A}}M_{\mathrm{A}}+n_{\mathrm{B}}M_{\mathrm{B}}}\left(\frac{\partial V}{\partial n_{\mathrm{A}}}\right)_{n_{\mathrm{B}}}\right\rangle\cdot\frac{\partial n_{\mathrm{A}}}{\partial X_{\mathrm{B}}}\\&=-\frac{1}{n_{\mathrm{A}}M_{\mathrm{A}}+n_{\mathrm{B}}M_{\mathrm{B}}}\left\langle-\frac{M_{\mathrm{A}}}{\rho_{mix}}+\left(\frac{\partial V}{\partial n_{\mathrm{A}}}\right)_{n_{\mathrm{B}}}\right\rangle\frac{n_{\mathrm{A}}}{X_{\mathrm{A}}X_{\mathrm{B}}}\\&=-\frac{1}{X_{\mathrm{B}}}\frac{1}{X_{\mathrm{A}}M_{\mathrm{A}}+X_{\mathrm{B}}M_{\mathrm{B}}}\left\langle-\frac{M_{\mathrm{A}}}{\rho_{mix}}+\left(\frac{\partial V}{\partial n_{\mathrm{A}}}\right)_{n_{\mathrm{B}}}\right\rangle\end{aligned} \tag{111}$$

Here, the relation of eq(112) is employed.

$$\left(\frac{\partial X_{\mathrm{B}}}{\partial n_{\mathrm{A}}}\right)=-\frac{n_{\mathrm{B}}}{(n_{\mathrm{A}}+n_{\mathrm{B}})^2}=-\frac{X_{\mathrm{A}}X_{\mathrm{B}}}{n_{\mathrm{A}}},\quad\left(\frac{\partial X_{\mathrm{B}}}{\partial n_{\mathrm{B}}}\right)=\frac{n_{\mathrm{A}}}{(n_{\mathrm{A}}+n_{\mathrm{B}})^2}=\frac{X_{\mathrm{A}}X_{\mathrm{B}}}{n_{\mathrm{B}}} \tag{112}$$

Rewriting eq(111) with respect to the partial molar volume $\bar{v}_{\mathrm{A}}$, eq(111') is derived.

$$\bar{v}_{\mathrm{A}}\equiv\left(\frac{\partial V}{\partial n_{\mathrm{A}}}\right)_{n_{\mathrm{B}}}=-X_{\mathrm{B}}(X_{\mathrm{A}}M_{\mathrm{A}}+X_{\mathrm{B}}M_{\mathrm{B}})\frac{d}{dX_{\mathrm{B}}}\left(\frac{1}{\rho_{mix}}\right)+\frac{M_{\mathrm{A}}}{\rho_{mix}} \tag{111'}$$

Following totally the same procedure, the partial molar volume of B $\bar{v}_{\mathrm{B}}$ is found to be given by eq(113).

$$\bar{v}_{\mathrm{B}}\equiv\left(\frac{\partial V}{\partial n_{\mathrm{B}}}\right)_{n_{\mathrm{A}}}=X_{\mathrm{A}}(X_{\mathrm{A}}M_{\mathrm{A}}+X_{\mathrm{B}}M_{\mathrm{B}})\frac{d}{dX_{\mathrm{B}}}\left(\frac{1}{\rho_{mix}}\right)+\frac{M_{\mathrm{B}}}{\rho_{mix}} \tag{113}$$

When the specific volume is given by eq(114) as a function of the mole fraction, the partial molar volume

$$\frac{1}{\rho_{mix}} = aX_B^2 + bX_B + \frac{1}{\rho_A}, \qquad \frac{1}{\rho_B} - \frac{1}{\rho_A} = a + b \qquad (114) \quad \text{(given again)}$$

$\overline{v}_A$ is obtained by substituting this equation into eq(111') as is given by eq(115) below.

$$\overline{v}_A = -X_B(X_A M_A + X_B M_B)(2aX_B + b) + M_A(aX_B^2 + bX_B + \frac{1}{\rho_A})$$

$$= -2a(M_B - M_A)X_B{}^3 - (aM_A + b(M_B - M_A))X_B^2 + \frac{M_A}{\rho_A} \qquad (115)$$

When the following combination of the values is substituted into eq(115), $\overline{v}_A$ is expressed as eq(115').

$$M_A = 50,\ M_B = 100,\ a = 1.2\ \text{cm}^3 \cdot \text{g}^{-1},\ b = -0.4\ \text{cm}^3 \cdot \text{g}^{-1},\ \frac{1}{\rho_A} = 0.8\ \text{cm}^3 \cdot \text{g}^{-1},$$

$$\overline{v}_A = 120X_B{}^3 - 140X_B^2 + 80 \qquad (115')$$

The value for $\overline{v}_B$ can be found from eq(113) though, it is reserved for the **advanced problem**. Here, instead, it is calculated from eq(116) (or eq(116')) which holds for the partial molar volumes.

$$\frac{1}{\rho_{mix}} = \frac{V}{m_A + m_B} = \frac{n_A\overline{v}_A + n_B\overline{v}_B}{n_A M_A + n_B M_B} = \frac{(1 - X_B)\overline{v}_A + X_B\overline{v}_B}{M_A + X_B(M_B - M_A)} \qquad (116)$$

$$\frac{M_A + X_B(M_B - M_A)}{\rho_{mix}} = (1 - X_B)\overline{v}_A + X_B\overline{v}_B \qquad (116')$$

From eq(116'), $\overline{v}_B$ is expressed in terms of $\overline{v}_A$ by eq(116'').

$$\overline{v}_B = \left\{\frac{M_A}{X_B} + (M_B - M_A)\right\}\frac{1}{\rho_{mix}} - \left(\frac{1}{X_B} - 1\right)\overline{v}_A$$

$$= \left\{\frac{M_A}{X_B} + (M_B - M_A)\right\}\left(aX_B^2 + bX_B + \frac{1}{\rho_A}\right) - \left(\frac{1}{X_B} - 1\right)\overline{v}_A \qquad (116'')$$

Substituting the given values into eq(116''), eq(117) is obtained for $\overline{v}_B$.

$$\overline{v}_B = \left(\frac{100}{X_B} - 50\right)(1.2X_B^2 - 0.4X_B + 0.8) - \left(\frac{1}{X_B} - 1\right)(120X_B^3 - 140X_B^2 + 80)$$

$$= 120X_B^3 - 320X_B^2 + 280X_B \qquad (117)$$

Ans. $\overline{v}_A$: eq(115), $\overline{v}_B$: eq(116''), specific example for $\overline{v}_A$: eq(115'), for $\overline{v}_B$: eq(117)

(b) The *Gibbs-Duhem* relation for the partial molar volumes of A and B is given by eq(118).

$$n_A d\overline{v}_A + n_B d\overline{v}_B = 0 \Rightarrow X_A d\overline{v}_A + X_B d\overline{v}_B = 0 \qquad (118)$$

Therefore, in order to prove the relation in the present case, it seems enough to confirm it by acquiring the differentials of $\overline{v}_A$ and $\overline{v}_B$ given by eq(115) and eq(116'), respectively and then by substituting the results into eq(118). However, since the procedure itself is too complicated to be performed, the relation is confirmed here for a specific example chosen for which $\overline{v}_A$ and $\overline{v}_B$ are expressed by eqs(115') and (117), respectively

Noticing that both partial molar volumes are expressed by X_B alone, the differentials of $\overline{v}_A$ and $\overline{v}_B$ are obtained from eqs(115') and (117) as are given by eq(119).

$$d\overline{v}_A = (360X_B{}^2 - 280X_B)dX_B, \qquad d\overline{v}_B = (360X_B^2 - 640X_B + 280)dX_B \qquad (119)$$

Substituting the result into eq(118), it is confirmed that the *Gibbs-Duhem* relation is satisfied as is seen from eq(120).

$$(1 - X_B)(360X_B{}^2 - 280X_B) + X_B(360X_B^2 - 640X_B + 280) = 0 \qquad (120)$$

Ans. See text (b).

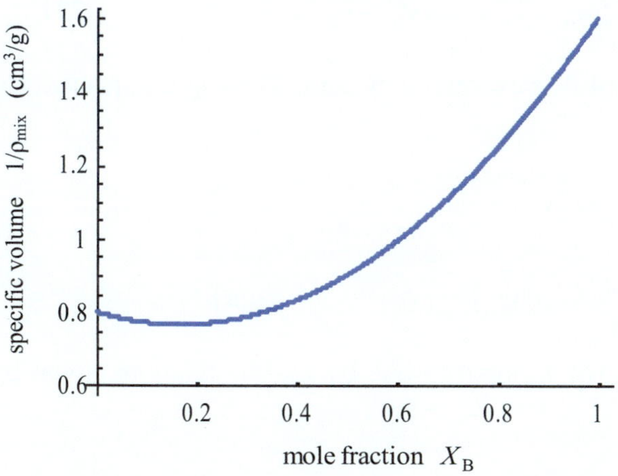

Fig.22 Specific volume and mole fraction of a mixed solvent.
The curve is given by
$1/\rho_{mix} = 1.2X_B^2 - 0.4X_B + 0.8$.

For more information, the specific volumes and the values of $\bar{v}_A$ and $\bar{v}_B$ worked out from the given specific volumes and the molecular weights are plotted against the mole fraction of X_B in Figs. 22 and 23. See also **problem 3-119** for the general proof of the *Gibbs-Duhem* relation.

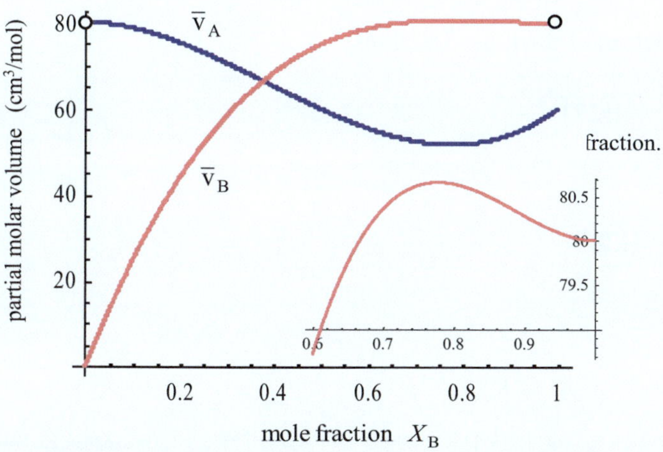

Fig.23 Partial molar volumes plotted against mole fraction.

A trend of increase or decrease in $\bar{v}_A$ is just opposite to that in $\bar{v}_B$. The open circles indicate the molar volumes of the pure substances. The inserted figure is an enlargement of a part of $\bar{v}_B$ around its maximum.

Solution 3-20c

(a) The values listed in the table for the specific volume $\bar{V}_{spc}$ are the volumes per 1 g. For pure substances, therefore, when they are multiplied by the corresponding molecular weights, the molar volumes are obtained. Consequently, the molar volume of ethyl iodide is found as $\bar{V} = 0.53622 \times 156 = 83.65\,\text{cm}^3$ and that of ethyl acetate is found as $\bar{V} = 1.15866 \times 88.10 = 102.1\,\text{cm}^3$. In contrast to a large difference in their molecular weights, their molar volumes are of a similar magnitude.

Ans. ethyl iodide: $83.65\ \text{cm}^3$, ethyl acetate: $102.1\ \text{cm}^3$

(b)The specific volume is given by eq(121) and the mass fraction of the mixed solvent is given by eq(122).

$$\bar{V}_{spc} = \frac{V}{m_A + m_B} = \frac{V}{n_A M_A + n_B M_B} \qquad (121)$$

$$x_A = \frac{m_A}{m_A + m_B} = \frac{n_A M_A}{n_A M_A + n_B M_B} \qquad (122)$$

Since the volume V of the mixed solvent is expressed in terms of the partial molar volumes as $V = \bar{v}_A n_A + \bar{v}_B n_B$, the specific volume $\bar{V}_{spc}$ can also be expressed by eq(123), in which we put

$\bar{v}_A^{sp} \equiv \bar{v}_A / M_A$ and $\bar{v}_B^{sp} \equiv \bar{v}_B / M_B$.

$$\bar{V}_{spc}(x_A) = \frac{\bar{v}_A n_A + \bar{v}_B n_B}{n_A M_A + n_B M_B} = \frac{\bar{v}_A x_A}{M_A} + \frac{\bar{v}_B x_B}{M_B} = \bar{v}_A^{sp} x_A + \bar{v}_B^{sp} x_B = (\bar{v}_A^{sp} - \bar{v}_B^{sp}) x_A + \bar{v}_B^{sp} \quad (123)$$

Form eq(123), it is seen that the partial molar volume of A is expressed by eq(124) using the specific volume $\bar{V}_{spc}(x_A)$ given as a function of the mass fraction x_A.

$$\bar{v}_A \equiv \frac{\partial V}{\partial n_A} = \frac{\partial}{\partial n_A} \left\langle \bar{V}_{spc} \times (n_A M_A + n_B M_B) \right\rangle$$

$$= (n_A M_A + n_B M_B) \frac{\partial \bar{V}_{spc}(x_A)}{\partial n_A} + M_A \bar{V}_{spc}(x_A) \quad (124)$$

In order to convert the partial differential of $\bar{V}_{spc}(x_A)$ by n_A into the one by x_A, the relations given by eqs (125) and (126) can be used.

$$\frac{\partial \bar{V}_{spc}(x_A)}{\partial n_A} = \frac{\partial \bar{V}_{spc}}{\partial x_A} \cdot \frac{\partial x_A}{\partial n_A} = \frac{\partial \bar{V}_{spc}}{\partial x_A} \cdot \frac{M_A x_B}{(n_A M_A + n_B M_B)} \quad (125)$$

$$\frac{\partial x_A}{\partial n_A} = \frac{n_B M_A M_B}{(n_A M_A + n_B M_B)^2} = \frac{M_A x_B}{(n_A M_A + n_B M_B)} \quad (126)$$

Finally, $\bar{v}_A$ (eq(124)) is given by eq(127) as a function of the mass fractions,

$$\bar{v}_A = M_A \left(x_B \frac{d\bar{V}_{spc}}{dx_A} + \bar{V}_{spc} \right) \quad (127)$$

and from eq(123) $\bar{v}_B$ is given by eq(128).

$$\bar{v}_B = \frac{M_B}{x_B} \left(\bar{V}_{spc} - \frac{x_A \bar{v}_A}{M_A} \right) \quad (128)$$

$\bar{V}_{spc}$ is plotted against x_A in Figs.24 and 25.

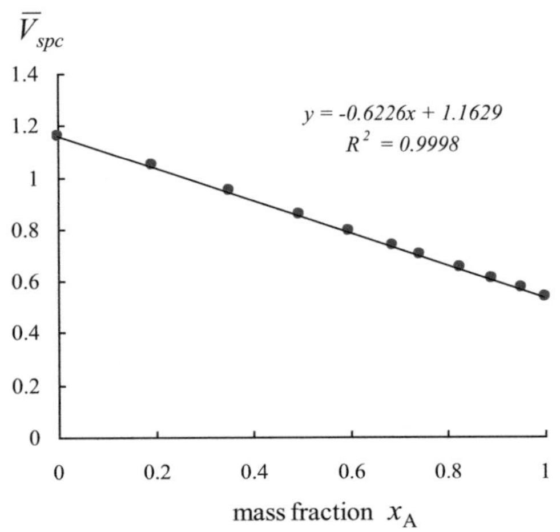

Fig.24 Specific volumes plotted against mass fraction for the mixed solvents of ethyl iodide and ethyl acetate. Linear plot over the whole range of the mass fraction.

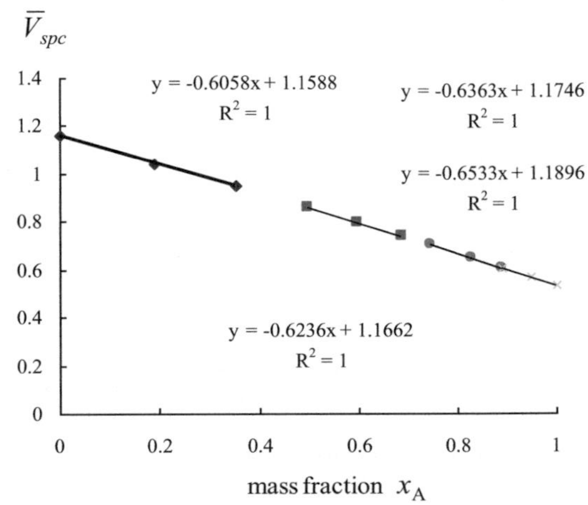

Fig.25 Specific volumes plotted against mass fraction for the mixed solvents of ethyl iodide and ethyl acetate. Linear plots for three sets of selected three points.

As is seen from Fig.24, over the whole range of the mixing ratio, a good linear correlation is recognized for the present system. Consequently, using the slope at the mass fraction of 0.59741, the partial molar volumes of the solvents A and B are found from eq(127) and eq(128), respectively, as below.

$$\bar{v}_A = 156 \times \langle (1-0.59741)(-0.6236) + 0.79377 \rangle = 84.67\ \text{cm}^3 \cdot \text{mol}^{-1}$$

$$\bar{v}_B = \frac{88.10}{(1-0.59741)}\left(0.79377 - \frac{0.59741 \times 84.67}{156}\right) = 102.7\ \text{cm}^3 \cdot \text{mol}^{-1}$$

These values are very close to the molar volumes of pure A and B found in (a). This system can be said to present a specific example of case (i) treated in **problem 3-20a.**

Ans. ethyl iodide: $84.67\ \text{cm}^3$, ethyl acetate: $102.7\ \text{cm}^3$

"note" A slight deviation from the linearity can be treated by selecting a few sets of three points and by determining the slope for each set as is shown in Fig.25. Employing the slopes thus determined, the partial molar volume dependent on the mass fraction can be determined.

Solution 3-21

(a) The volume *V* of a solution is expressed by a function of the temperature *T*, the pressure *P* and the mole numbers of the components $n_1, n_2, n_3, \cdots$, and is denoted as $V(T, P, n_1, n_2, n_3, \cdots)$. The partial molar volume of the component *i* is defined by eq(129). In the case of a pure substance, its molar volume and partial molar volume is identical since there exists only one component.

In the case of a solution, imagine the following situation to understand what is meant by eq(129). To the solution made of large number of moles of $n_1,\ n_2,\ n_3,\ \cdots >> 1$, just 1 mole of the *i*-th component is added and the mole number of the *i*-th component is changed from n_i to $n_i + 1$. A volume change (either negative or positive) observed for the solution upon this operation is equivalent to the partial molar volume of the *i*-th component at those mole concentrations.

$$\overline{V_i} = \left(\frac{\partial V}{\partial n_i}\right)_{T,P,n_j,j\neq i} \qquad (129)$$

Ans. definition of the partial molar volume: eq(129). See text (a) for the explanation.

(b) i) From the figure given in the problem, the values for the molar volumes of the pure substances are read as $58.6\ \text{cm}^3\text{mol}^{-1}$ for ethanol and $18.1\ \text{cm}^3\text{mol}^{-1}$ for water.

Ans. ethanol: $58.6\ \text{cm}^3\text{mol}^{-1}$, water: $18.1\ \text{cm}^3\text{mol}^{-1}$

ii) The mole numbers of water and ethanol are found from their volumes by dividing them with the corresponding molar volume. Thus they are given by $90/18.1 = 4.97$ and $110/58.6 = 1.88$ and the mole fractions at this mixing ratio are given by $4.97/(4.97+1.88) = 0.726$ and $1-0.726 = 0.274$. The partial molar volumes of water and ethanol at this mixing ratio are read as $17.4\ \text{cm}^3$ and $56.2\ \text{cm}^3$, respectively from the figure. Therefore, the volume of the solution is calculated to be 192 cm^3 as below.

$$V = n_w\overline{V}_w + n_e\overline{V} = 4.97 \times 17.4 + 1.88 \times 56.2 = 192\ \text{cm}^3$$

Ans. volume of the mixture: $192\ \text{cm}^3$

(c) The *Gibbs-Duhem* relation for the two components can be expressed as $n_w d\overline{V}_w + n_e d\overline{V}_e = 0$. It is seen from the figure qualitatively that the partial molar volume of one component shows an increase then the other exhibits a decrease. According to the *Gibbs-Duhem* equation given above, with respect to a given increment for the one component, the corresponding decrement for the other can be found quantitatively. For this purpose, eq(130) is rewritten as eq(131) via eq(130).

$$n_w d\overline{V}_w + n_e d\overline{V}_e = 0 \;\Rightarrow\; n_w d\overline{V}_w/dX_e + n_e d\overline{V}_e/dX_e = 0 \;\Rightarrow\; X_w d\overline{V}_w/dX_e + X_e d\overline{V}_e/dX_e = 0 \qquad (130)$$

$$\frac{d\overline{V}_e/dX_e}{d\overline{V}_w/dX_e} = -\frac{1-X_e}{X_e} \qquad (131)$$

We can check whether eq(131) is satisfied at the given five points.

i) At the both sides of the figure, i.e., at $X_e = 0$ or $X_e = 1$, eq(131) implies that the ratio of the slopes of the partial molar volumes for the two components is infinity or zero. Therefore, in the limit of $X_e \to 0$, the slope of $d\overline{V}_w/dX_e$ should approach to zero, i.e., $d\overline{V}_w/dX_e \to 0$ and in the limit of $X_e \to 1$, the slope of $d\overline{V}_e/dX_e$ should approach to zero ,i.e., $d\overline{V}_e/dX_e \to 0$. The figure given in the problem is drawn so as to satisfy these limiting values.

ii) As is indicated by eq(130), where the partial molar volume of one component takes a maximum, that of the other component assumes a minimum and vice versa. This correlation is recognized at $X_e = 0.08$ in the figure (see **"note"**).

iii) To examine the relation of eq(131), the slopes (tangents) at the two positions of $X_e = 0.4$ and $X_e = 0.6$ are read from the figure as given below.

$$\left.\begin{aligned} \left(\frac{d\overline{V}_e}{dX_e}\right)_{X_e=0.4} &= 5.63\ \text{cm}^3 \\ \left(\frac{d\overline{V}_w}{dX_e}\right)_{X_e=0.4} &= -2.78\ \text{cm}^3 \end{aligned}\right\} \quad \frac{d\overline{V}_e/dX_e}{d\overline{V}_w/dX_e} = \frac{5.67}{-2.78} = -2.04 \;\Leftrightarrow\; -\frac{1-X_e}{X_e} = -1.5 \qquad (131')$$

$$\left.\begin{aligned} \left(\frac{d\overline{V}_e}{dX_e}\right)_{X_e=0.6} &= 2.63\ \text{cm}^3 \\ \left(\frac{d\overline{V}_w}{dX_e}\right)_{X_e=0.6} &= -3.96\ \text{cm}^3 \end{aligned}\right\} \quad \frac{d\overline{V}_e/dX_e}{d\overline{V}_w/dX_e} = \frac{2.63}{-3.96} = -0.664 \;\Leftrightarrow\; -\frac{1-X_e}{X_e} = -0.667 \qquad (131'')$$

At the mole fraction of $X_e = 0.6$, it is confirmed that eq(131) is fully satisfied (eq(131'')), but at the mole fraction of $X_e = 0.4$ both sides of eq(131) do not show a good agreement (eq(131')) and furthermore, at $X_e = 0.2$ a disagreement is significant.

Summarizing i)-iii), from the **"note"** below as well as a comparison with some experimentally determined partial molar volumes obtained directly from the density measurement of the water-ethanol mixed solutions, it can be said that the reliability of the figure cited in the problem is rather low.

"note" It was in *Physical Chemistry* 4th edition by *Walter J. Moore* that one of the authors came across for the first time to a diagram of the partial molar volumes for the water-ethanol mixtures (Fig.7.2), though one of the earliest could be that (Fig.1.2) in *Chemical Thermodynamic* I by *I. Prigogenue* and *R. Defay* (Japanese translation) apart from scientific papers published earlier. Later, in the 4th edition of *Atkins Physical Chemistry*, a similar diagram for the partial molar volumes was cited. Unfortunately, however, neither diagram is drawn carefully enough to show that the maximum for one component corresponds to the minimum for the other. In this sense, they are not reliable diagrams. In the 6th edition of *Atkins Physical Chemistry*, as is seen in the figure cited, the figure is subtly adjusted so as to show the maximum and minimum at the same mole fraction, though the curves drawn do not satisfy the *Gibbs-Duhem* relation. The figure appears to have changed slightly with each subsequent edition, but there are many questions as to whether the figure is based on actual measurements or not. For this it is worth calculating the partial molar volumes for the water-ethanol system accurately based on the accurate measurements of the solution density.

Ans. *Gibbs-Duhem* relation is satisfied at the points *a, b, d* and *e* but is not at the point *c*. See text(c).

Solution 3-22

(a) Since the relation between the mole fractions and the equilibrium vapor pressure P of the solution is given by the left side of eq(132), the given values are substituted into eq(132) to find the relation required as below.

$$P = X_{\mathrm{A}}P_{\mathrm{A}}^{\bullet} + X_{\mathrm{B}}P_{\mathrm{B}}^{\bullet} = X_{\mathrm{A}}P_{\mathrm{A}}^{\bullet} + (1 - X_{\mathrm{A}})P_{\mathrm{B}}^{\bullet} = P_{\mathrm{B}}^{\bullet} + (P_{\mathrm{A}}^{\bullet} - P_{\mathrm{B}}^{\bullet})X_{\mathrm{A}} = 25 + 75X_{\mathrm{A}} \tag{132}$$

Ans. $25 + 75X_{\mathrm{A}}$

(b) Rewriting eq(132) by using the relation of $X_{\mathrm{A}}^{g} = P \cdot X_{\mathrm{A}} / P_{\mathrm{A}}^{\circ}$, where X_{A}^{g} is the mole fraction of A in the gas phase, eq(133) is derived.

$$P = \frac{P_{\mathrm{A}}^{\bullet} \times P_{\mathrm{B}}^{\bullet}}{P_{\mathrm{A}}^{\bullet} - (P_{\mathrm{A}}^{\bullet} - P_{\mathrm{B}}^{\bullet})X_{\mathrm{A}}^{g}} = \frac{100 \times 25}{100 - 75X_{\mathrm{A}}^{g}} \tag{133}$$

Ans. eq(133)

(c) The range of mole fractions is found from the intersections of the line at $P = 76$ Torr with eq(132) and eq(133). From eq(132), $X_{\mathrm{A}} = 0.68$ is found and from eq(133) $X_{\mathrm{A}}^{g} = 0.895$. Therefore, the range of the mole fractions to be obtained is $0.68 - 0.895$.

Ans. range of the mole fractions: $0.68 - 0.895$

Solution 3-23

(a) There would be several ways to answer to this kind of question. The most orthodox one is given first and then a different approach is given as well.

Solution 1. The total differential of S which is described as a function of V and T is given by eq(134).

$$dS = \left(\frac{\partial S}{\partial V}\right)_T dV + \left(\frac{\partial S}{\partial T}\right)_V dT \tag{134}$$

Both sides of eq(134) is divided by dP under the condition of a constant temperature ($dT = 0$), and then

$$\left(\frac{\partial S}{\partial P}\right)_T = \left(\frac{\partial S}{\partial V}\right)_T \left(\frac{\partial V}{\partial P}\right)_T = -\frac{nRT}{P^2}\left(\frac{\partial S}{\partial V}\right)_T = -\frac{V}{P}\left(\frac{\partial S}{\partial V}\right)_T \tag{135}$$

eq(135) is obtained. The required answer is given by the last form of eq(135). The first equality of eq(135) holds for any system which need not be ideal.

Ans. $\left(\frac{\partial S}{\partial P}\right)_T = -\frac{V}{P}\left(\frac{\partial S}{\partial V}\right)_T$

Solution 2. From the two *Maxwell*'s relations given below, nR is eliminated to give the required answer.

$$\left(\frac{\partial S}{\partial V}\right)_T = \left(\frac{\partial P}{\partial T}\right)_V = \frac{nR}{V}, \quad \left(\frac{\partial S}{\partial P}\right)_T = -\left(\frac{\partial V}{\partial T}\right)_P = -\frac{nR}{P} \tag{136}$$

Simple though, this is a kind of proof for an ideal gas.

Solution 3. Based on the equation of state $PV = nRT$ for an ideal gas, the rewritten relation $V = \frac{nRT}{P}$ is used to derive the differential $dV = -\frac{nRT}{P^2}dP$ at a constant temperature. Employing this relation, dV is replaced by dP in eq(134) under the condition of a constant temperature to yield eq(137).

$$\left(\frac{\partial S}{\partial V}\right)_T = \left(\frac{\partial S}{\partial \frac{nRT}{P}}\right)_T = \left(\frac{\partial S}{-\frac{nRT}{P^2}\partial P}\right)_T = -\frac{P^2}{nRT}\left(\frac{\partial S}{\partial P}\right)_T = -\frac{P}{V}\left(\frac{\partial S}{\partial P}\right)_T \qquad (137)$$

(b) We now prove that eq(138) is satisfied. But it is a proof only for an ideal gas.

$$\int_V^{3V}\left(\frac{\partial S}{\partial V}\right)_T dV = \int_P^{1/3P}\left(\frac{\partial S}{\partial P}\right)_T dP \qquad (138)$$

For that purpose, the integral on the left side is rewritten by using the relation obtained in (a) above and the differential dV is replaced by dP. Then, finally the last equality is derived as shown below in eq(139).

$$\int_V^{3V}\left(\frac{\partial S}{\partial V}\right)_T dV = \int_V^{3V} -\frac{P}{V}\left(\frac{\partial S}{\partial P}\right)_T dV = \int_V^{3V} -\left(\frac{\partial S}{\partial P}\right)_T PV\frac{dV}{V^2} = \int_P^{1/3P}\left(\frac{\partial S}{\partial P}\right)_T \frac{PV}{nRT}dP = \int_P^{1/3P}\left(\frac{\partial S}{\partial P}\right)_T dP \qquad (139)$$

Ans. See text (b).

Solution 3-24

(a) When the temperature is raised at a constant pressure, the vapor phase comes to dominate, that is, the gas is the more stable at the higher temperatures as compared with the liquid. Therefore, the broken line which shows larger values for $-G$ (more negative) than the solid line at higher temperatures should correspond to the gas. The intersection of these two lines is the boiling point.

Ans. Gas corresponds to the broken line.

(b)From a viewpoint of the free energy, the liquid-solid equilibrium can be considered similarly as for the case of gas and liquid. Since the more stable phase at the higher temperatures is a liquid phase, a conceptual

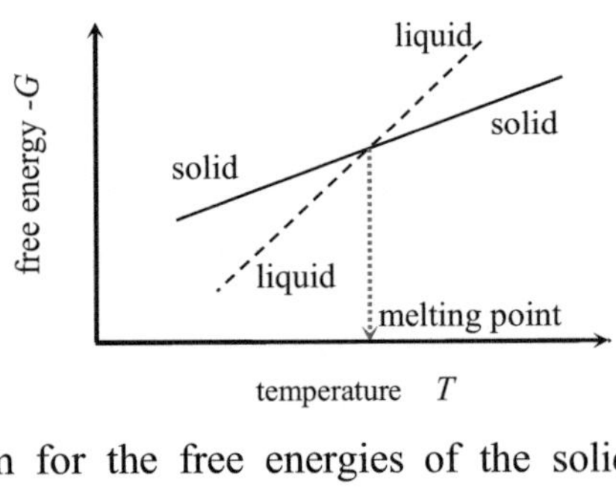

Fig.26 Free energy change of a pure solid or its liquid with temperature near the melting point.

diagram for the free energies of the solid and liquid is just like the one given above in Fig.26. The intersection of the two lines is the melting point.

Ans. See Fig.26.

Solution 3-25

(a) The mole fractions at points A, B and C are $X_A = 1$, $X_B = 1$ and $X_C = 1$ and, hence, each point of these represents a relevant pure liquid. The symbols A, B and C represent water, acetic acid and butanol, respectively.

Ans. Each point represents a pure liquid. A:water, B: acetic acid and C: butanol.

(b) The length of perpendicular line drawn from the point P to each side of the triangular diagram is denoted as l_A, l_B and l_C. Then it is seen that eq(140) is satisfied. Notice here that the symbol of the vertex located opposite to the perpendicular line from P is added as a subscript. When we consider the case in which the point P is located on one of the vertexes, the meaning of the length of the perpendicular line is trivial.

$$l_A : l_B : l_C = X_A : X_B : X_C \qquad (140)$$

Denoting the height of the triangle h, eq(141) is derived from the area of the triangle.

$$l_A + l_B + l_C = h \qquad (141)$$

As a result, eq(142) is obtained for the mole fractions from eq(140) (or each length of the segment drawn parallel to each side from the point P).

$$X_A = \frac{l_A}{h},\ X_B = \frac{l_B}{h},\ X_C = \frac{l_C}{h} \qquad (142)$$

Applying eq(142) to the given point P, the mole fractions there are found as $X_A = 0.4$, $X_B = 0.4$ and $X_C = 0.2$.

Ans. $X_A = 0.4$ (water), $X_B = 0.4$ (acetic acid) and $X_C = 0.2$ (butanol)

(c) i) As is seen readily from Fig.27, for the two combinations of (water-acetic acid) and (acetic acid-butanol) there exists no two-phase separated region at any mixing ratio. ii) On the other hand, for the combination of (water-butanol) two-phase separation occurs when the mole fraction of either component exceeds 0.1.

Ans. i) homogeneous solution: (water-acetic acid), (acetic acid-butanol)

ii) not always mixable at an arbitrary mixing ratio: (water-butanol)

(d) **Ans.** Separated into two solution phases.

(e) The point Q is located in the two-phase separated region and the compositions of the two solutions are given by the mole fractions indicated at the end points of the tie linc L and M (see **"note 1"**), being consistent with the lever rule. The compositions at the points L and M are read from Fig.27 as below.

$$L : (X_A,\ X_B,\ X_C) = (0.53,\ 0.35,\ 0.12), \qquad M : (X_A,\ X_B,\ X_C) = (0.10,\ 0.17,\ 0.73)$$

Also, the ratio of the lengths of the levers is found as

$$l_{QL} : l_{QM} = (0.72 - 0.33) : (0.33 - 0.19) = 39 : 14.$$

Consequently, the compositions of the two solutions existing separated at the point Q are found as

water: acetic acid:butanol $= 0.53 : 0.35 : 0.12$ (solution L)

water: acetic acid:butanol $= 0.10 : 0.17 : 0.73$ (solution M).

They exist in the ratio of $14:39$ and the total mole fractions of these solutions are calculated as below and,

$$X_A : X_B : X_C = \frac{14}{53}(0.53 : 0.35 : 0.12) + \frac{39}{53}(0.10 : 0.17 : 0.73)$$
$$= 0.21 : 0.22 : 0.57$$

naturally, these fractions agree with the mole fractions at the point Q.

Ans. Two solutions of water:acetic acid:butanol$=0.53:0.35:0.12$ (solution L) and water:acetic acid:butanol$=0.10:0.17:0.73$ (solution M) exist separated in the ratio of $14:39$.

(f) The phase rule is expressed by eq(143).

$$f = c - p + 2 \qquad (143)$$

Putting the number of the components $c = 3$ and the number of the phases $p = 3$ in eq(143), the number of the freedom is found as $f = 2$. Since two freedoms of temperature T and pressure P are already fixed at constant values, there is no other freedom available. Therefore, if the solution can exist in three separated phases, then the compositions of each phase are determined unambiguously.

Ans. Possible to exist but the compositions of each phase are determined unambiguously.

An example of a phase diagram in which three phases can coexist is given in Fig.28. The three solution phases, each of which is a homogeneous solution in a narrow range of concentrations, are denoted by α, β and γ, respectively. A two-phase separated region consists of one of the phases of $(\alpha,\ \beta)$, $(\beta,\ \gamma)$ and

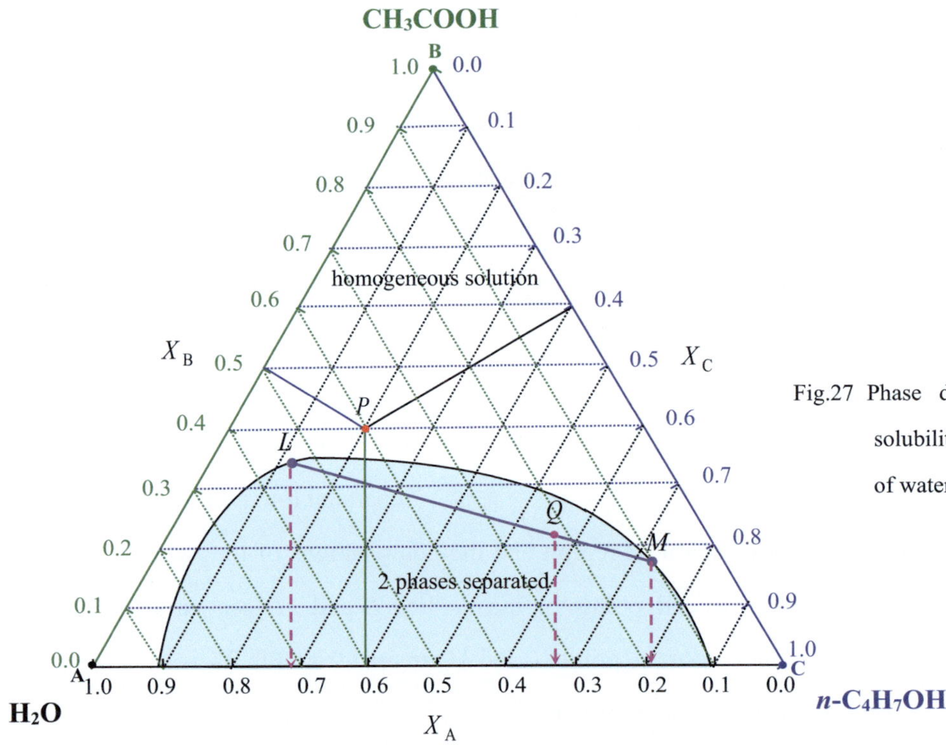

Fig.27 Phase diagram showing mutual solubility of three solvent system of water-acetic acid-butanol.

$(\alpha,\ \gamma)$ enclosed by these three tie lines in which the three separated phases of α, β and γ coexist. When the tie line in each phase comes to share either end of the line, they form a triangle (green area in Fig.28) At a first look, it seems that three phases of various compositions can exist because of rather a wide green area. However, as is discussed in (f) based on the phase rule, the composition of each phase is uniquely determined.

That is confirmed as follows. The vertexes of the triangle in which the three phases coexist are denoted L, M and N (see Fig.28), and the mole fractions of the components A, B and C are written as $(X_{\mathrm{A}}^{L}, X_{\mathrm{B}}^{L}, X_{\mathrm{C}}^{L})$,

(X_A^M, X_B^M, X_C^M) and (X_A^N, X_B^N, X_C^N). The mole fractions at the point P where the three separated phases coexist are written as (X_A^P, X_B^P, X_C^P) and the amounts of α, β and γ phases are given by l, m and n moles. Then, among these values the relations given by q(144) are found (conservation of substances).

$$\begin{aligned} l\times X_A^L + m\times X_A^M + n\times X_A^N &= (l+m+n)X_A^P \quad\Rightarrow\quad l(X_A^L - X_A^P) + m(X_A^M - X_A^P) + n(X_A^N - X_A^P) = 0 \\ l\times X_B^L + m\times X_B^M + n\times X_B^N &= (l+m+n)X_B^P \quad\Rightarrow\quad l(X_B^L - X_B^P) + m(X_B^M - X_B^P) + n(X_B^N - X_B^P) = 0 \\ l\times X_C^L + m\times X_C^M + n\times X_C^N &= (l+m+n)X_C^P \quad\Rightarrow\quad l(X_C^L - X_C^P) + m(X_C^M - X_C^P) + n(X_C^N - X_C^P) = 0 \end{aligned} \qquad (144)$$

This is equivalent to say that the point P is the center of the mole fraction (see the proof given below). In other words, when the amounts of the α, β, γ phases at a point arbitrary chosen within the triangle are to be found, their amounts l, m and n are determined so as for the point chosen to be the center of the mole fraction.

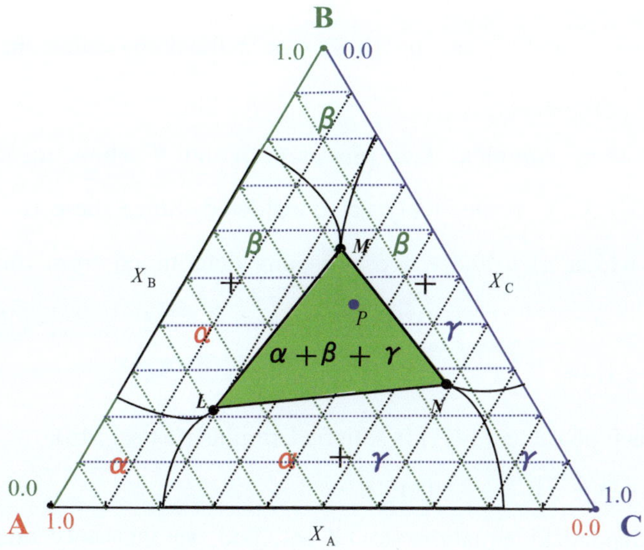

Fig.28. A phase diagram showing a mutually unmixable three phase separated region. There are three two-phase-separated regions such as $\alpha+\beta$. The three-phase-separated region is indicated by the green triangle area enclosed by the three tie lines each of which is drawn in the corresponding two-phase-sepaated region.

Ans. phase diagram for three phase separation: See Fig.28.

<u>Proof that P is the center of gravity (mole fractions)</u>

Supposing that the amounts of α, β and γ phases are given by l, m and n moles, we consider a triangle whose vertexes L, M and N have l, m and n moles as shown in Fig.29 below. In the first place, the mole fractions (a,b,c) at the center of gravity of the vertexes of L and M at the point P' are obtained from the balance of the momenta as below.

$$l(X_A^L - a) = m(a - X_A^M) \quad\Rightarrow\quad a = \frac{lX_A^L + mX_A^M}{l+m}$$

$$l(X_B^L - b) = m(b - X_B^M) \quad\Rightarrow\quad b = \frac{lX_B^L + mX_B^M}{l+m}$$

$$l(X_C^L - c) = m(c - X_C^M) \quad\Rightarrow\quad c = \frac{lX_C^L + mX_C^M}{l+m}$$

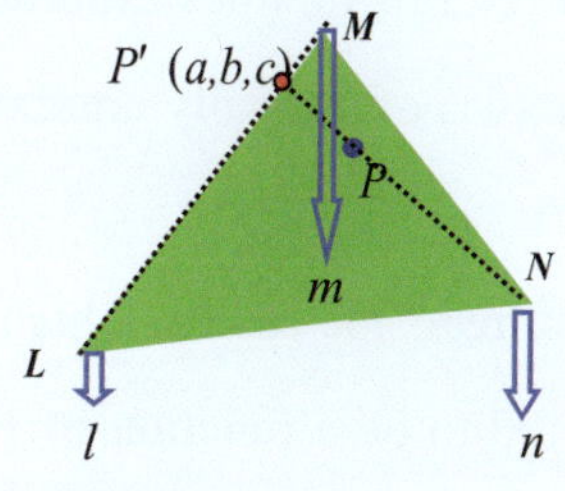

Fig.29 How to work out the center of gravity of a triangle whose vertexes are weighted differently.

In the second place, the mole fractions (X_A^P, X_B^P, X_C^P) at the center of gravity of the vertex N and the point P', denoted as P in the above figure, are obtained as follows. This point P is the center of gravity of the triangle shown in Fig.29 above. As is confirmed from the calculations below, the mole fractions at the center of gravity of the triangle satisfy eq(144).

$$(l+m)(a-X_A^P)=n(X_A^P-X_A^N)\ \Rightarrow\ (l+m)\cdot\left(\frac{lX_A^L+mX_A^M}{l+m}-X_A^P\right)=n(X_A^P-X_A^N)\ \Rightarrow\ l(X_A^L-X_A^P)+m(X_A^M-X_A^P)+n(X_A^N-X_A^P)=0$$

$$(l+m)(b-X_B^P)=n(X_B^P-X_B^N)\ \Rightarrow\ (l+m)\cdot\left(\frac{lX_B^L+mX_B^M}{l+m}-X_B^P\right)=n(X_B^P-X_B^N)\ \Rightarrow\ l(X_B^L-X_B^P)+m(X_B^M-X_B^P)+n(X_B^N-X_B^P)=0$$

$$(l+m)(c-X_C^P)=n(X_C^P-X_C^N)\ \Rightarrow\ (l+m)\cdot\left(\frac{lX_A^L+mX_A^M}{l+m}-X_A^P\right)=n(X_C^P-X_C^N)\ \Rightarrow\ l(X_C^L-X_C^P)+m(X_C^M-X_C^P)+n(X_C^N-X_C^P)=0$$

"note 1" In the present case, the freedom is 2 for a one-phase region and 1 for a two-phase region. When the mole fractions are taken as the freedom, eq(145) must be satisfied. Hence, at the point P in the one-

$$X_A+X_B+X_C=1 \qquad (145)$$

phase region in Fig.27, if we fix at $X_C=0.2$, then X_A and X_B can be chosen as 2 freedoms under the condition of $X_A+X_B=0.8$.

For the two phase region, the situation is a little more complex. Two phases of α and β whose mole fractions are given by $(X_A^\alpha,\ X_B^\alpha,\ X_C^\alpha)$ and $(X_A^\beta,\ X_B^\beta,\ X_C^\beta)$, respectively are considered. Since these two phases are in equilibrium, eq(146) is satisfied in which a, b and c are constants determined from the conditions

$$\frac{X_A^\alpha}{X_A^\beta}=a,\ \frac{X_B^\alpha}{X_B^\beta}=b,\ \frac{X_C^\alpha}{X_C^\beta}=c \qquad (146)$$

of the equilibrium. Also, for the mole fractions in each phase, eq(145') is satisfied similarly as eq(145).

$$X_A^\alpha+X_B^\alpha+X_C^\alpha=1,\quad X_A^\beta+X_B^\beta+X_C^\beta=1 \qquad (145')$$

There are six unknown variables and the five independent equations to be satisfied, so that only one undetermined variable remains. Namely, when one of the mole fractions of $(X_A^\alpha,\ X_B^\alpha,\ X_C^\alpha)$ and $(X_A^\beta,\ X_B^\beta,\ X_C^\beta)$ is determined arbitrarily, the rest of five variables are unambiguously determined from eqs(145') and (146). The relevant mole fractions in the two phase region are connected to form a tie line. In other words, the two phase region shown in Fig.27 is composed of infinite number of tie lines which certainly do not cross each other.

"note 2" How to draw tie lines experimentally.

"note 3" Examples of mutually unmixable solvents (*J.Hildebrand, The Solubility of Nonelectrolite*).

Solution 3-26

The problem treats the case in which the volume change of the gas is known. Therefore, the entropy of the gas is considered to be a function of V and T and is expressed as $S(V,T)$. Then, the partial differential at a constant entropy $(\partial V/\partial T)_S$ needs to be expressed as a unction of V and T. To achieve this as an always effective method, the total differential of $S(V,T)$ is obtained first as is given by eq(147).

$$dS=\left(\frac{\partial S}{\partial T}\right)_V dT+\left(\frac{\partial S}{\partial V}\right)_T dV \qquad (147)$$

Next, putting $dS = 0$ (reversible adiabatic process), both sides of eq(147) are divided by dV and, after some rewritings, eq(148) is derived.

$$0 = \left(\frac{\partial S}{\partial T}\right)_V \left(\frac{\partial T}{\partial V}\right)_S + \left(\frac{\partial S}{\partial V}\right)_T \Rightarrow 0 = \frac{C_V}{T}\left(\frac{\partial T}{\partial V}\right)_S + \left(\frac{\partial P}{\partial T}\right)_V \quad (148)$$

To rewrite the left side of eq(148), the relation $TdS = (dU)_V \Rightarrow \left(\frac{\partial S}{\partial T}\right)_V = \left(\frac{\partial U}{\partial T}\right)_V \frac{1}{T} = \frac{C_V}{T}$ and the *Maxwell* relation $\left(\frac{\partial S}{\partial V}\right)_T = \left(\frac{\partial P}{\partial T}\right)_V$ are used.

The obtained eq(148) is further rewritten for the ideal gas and after some rearrangements eq(149), equivalent to eq(147) in the problem, is obtained, in which C_V^m is the molar heat capacity at a constant volume.

$$\left(\frac{\partial T}{\partial V}\right)_S = -\frac{T}{C_V} \times \frac{nR}{V} = -\frac{R}{C_V^m} \times \frac{T}{V} \Rightarrow \ln T / T_i = -\frac{R}{C_V^m} \ln V / V_i \quad (149)$$

The temperature to be reached upon the required volume change in the gas is found from eq(149) by putting $C_V^m = 3R/2$ (dependent on the kind of the gas) as below,

$$T/T_i = (V/V_i)^{-\frac{2}{3}} \Rightarrow T = T_i \times (V/V_i)^{-\frac{2}{3}} = 300 \times 2^{\frac{2}{3}} = 476\ \text{K}$$

and the pressure is found from the equation of state as follows.

$$PV / P_iV_i = T / T_i \Rightarrow P / P_i = (T / T_i) \times (V_i / V) = (V / V_i)^{-\frac{5}{3}} \quad (150)$$

$$= 2^{\frac{5}{3}} = 3.17$$

It is easy to explain intuitively why the final pressure is much higher as compared with the compression at a constant temperature.

Ans. See text for proof. temperature reached: 476 K, pressure reached: 3.17 times larger

[analogous problem] When a change in the pressure is known for the adiabatic compression process, find the relation between P and T from $\left(\frac{\partial T}{\partial P}\right)_S$.

Solution 3-27

Ans. intensive variable: T, P and μ ; extensive variable: V, $U(E)$, S, H, $F(A)$ and G; neither of these variables: $\overline{V}$

"note" The chemical potential is an intensive variable. All of the extensive thermodynamic variables (denoted f here) are *Euler*'s 1st order (n=1) homogeneous functions and the following relation is satisfied.

$$x_1 \frac{\partial f}{\partial x_1} + x_2 \frac{\partial f}{\partial x_2} + x_3 \frac{\partial f}{\partial x_3} + \cdots = \text{n} \cdot f(x_1, x_2, x_3, \cdots) \quad (151)$$

(n=1 for quantitative thermodynamic variables)

Replacing f with G and putting $n_i = n_t x_i$ ($\sum n_i = n_t$), $G = \sum n_i \mu_i$ is derived from eq(151) above.

Solution 3-28

(a) The work the gas has done is given by the following equation. Since the value found is positive, the gas has gained the corresponding energy.

$$\int_V^{V/2} -PdV = -nRT \int_V^{V/2} \frac{1}{V} dV = -nRT\left[\ln V\right]_V^{V/2} = nRT \ln 2 > 0$$

Ans. $nRT \ln 2$

(c) The internal energy of the ideal gas is expressed by a function of T alone. Consequently, there is no change of the internal energy for the isothermal process and from the first law of thermodynamics, the following relations are derived. The work energy gained is lost as heat energy. Here, δ indicates an infinitesimally small change and Δ is for a finite change.

$$dU = 0 \ \Rightarrow \ \delta w = -\delta q \ \Rightarrow \ \Delta w = -\Delta q$$

Ans. change in the heat content of the gas: $\Delta q = -nRT \ln 2$

(d) From (b) above $\Delta U = 0$.

Ans. $\Delta U = 0$

(e) An entropy change of the gas is given by eq(152), since the process is the reversible isothermal one.

$$dS = \frac{\delta q}{T} = \frac{PdV}{T} = \frac{nRdV}{V} \qquad (152)$$

Integrating both sides of eq(152),

$$\int_V^{V/2} dS = \int_V^{V/2} \frac{nRdV}{V} = nR[\ln V]_V^{V/2} = -nR \ln 2$$

the entropy change ΔS accompanied by the volume change $V \to 1/2V$ is found as $-nR \ln 2$..

Ans. $-nR \ln 2$

"note" As is clearly seen from the formula given for the solution, the entropy (and its change) is an extensive variable (see **problem 3-27**).

Solution 3-29

(a) $dG = -SdT + VdP = VdP$ (isothremal change)

Ans. $dG = -SdT + VdP = VdP$ (isothremal change)

(b) The free energy change ΔG associated with the volume change of $V \to 1/2V$ is found by integrating the (total) differential dG obtained in (a). The result obtained thus (eq(153)) agree with the value calculated from $\Delta G = \Delta(U + PV - TS)$ for the isothermal process, since the free energy G is a state quantity.

$$\Delta G = \int_V^{V/2} dG = \int_V^{V/2} VdP = \int_P^{2P} \frac{nRT}{P} dP = nRT[\ln P]_P^{2P} = nRT \ln 2 \qquad (153)$$

Ans. $\Delta G = nRT \ln 2 > 0$

(c) For the processes which occur naturally, the free energy of the system should decrease always, that is $\Delta G < 0$. The value of ΔG obtained in the present case is positive, so that it is concluded that the isothermal compression does not proceed naturally. On the other hand, the isothermal expansion occurs naturally because of $\Delta G < 0$.

Ans. The process does not occur naturally.

Solution 3-30

(a) $6C(s) + 3H_2(g) \ \rightarrow \ C_6H_6(l)$

Ans. $6C(s) + 3H_2(g) \ \rightarrow \ C_6H_6(l)$

(b) By adding or subtracting eqs(154), (155) and (156) appropriately, the formation reaction of benzene, which is the same one as obtained in (a), is found as follows.

$$6\times\left(\mathrm{C}(s)+\mathrm{O}_2=\mathrm{CO}_2(g)\_\_\_\ (\mathrm{i})\right),\quad 6\times\Delta H_c(\mathrm{i}) \qquad (154)$$

$$3\times\left(\mathrm{H}_2+\frac{1}{2}\mathrm{O}_2=\mathrm{H}_2\mathrm{O}(l)\_\_\_\ (\mathrm{ii})\right),\quad 3\times\Delta H_c(\mathrm{ii}) \qquad (155)$$

$$-1\times\left(\mathrm{C}_6\mathrm{H}_6(l)+\frac{15}{2}\mathrm{O}_2=6\mathrm{CO}_2(g)+3\mathrm{H}_2\mathrm{O}\_\_\_\ (\mathrm{iii})\right),\quad -1\times\Delta H_c(\mathrm{iii}) \qquad (156)$$

+)

$$6\mathrm{C}(s)+3\mathrm{H}_2=\mathrm{C}_6\mathrm{H}_6(l),\qquad \Delta H_f=6\times\Delta H_c(\mathrm{i})+3\times\Delta H_c(\mathrm{ii})-1\times\Delta H_c(\mathrm{iii})$$
$$=-(6\times94.0+3\times68.3-783.4)\times4.18$$
$$=60.6\ \mathrm{kJ}$$

From the energetic point of view, a chemical reaction plus its heat of the reaction can be considered as an equality and, hence, the heat of combustion ΔH_c of each reaction, if necessary after being multiplied appropriately, can be added or subtracted just in the same manner as done in deriving the formation reaction of benzene, giving the heat of formation ΔH_f.

Ans. enthalpy of formation of benzene: $60.6\ \mathrm{kJ\cdot mol^{-1}}$

Solution 3-31

(a) When the internal energy U is expressed by a function of S and V, its total differential is given by eq(157).

$$dU=TdS-PdV \qquad (157)$$

Both sides of eq(157) are divided by dV at a constant temperature ($dT=0$) to yield eq(158).

$$\left(\frac{\partial U}{\partial V}\right)_T=T\left(\frac{\partial S}{\partial V}\right)_T-P \qquad (158)$$

The *Maxwell* relation given by eq(159) is used to replace the first term of the right side of eq(158) and for the

$$\left(\frac{\partial S}{\partial V}\right)_T=\left(\frac{\partial P}{\partial T}\right)_V \qquad (159)$$

ideal gas eq(160) is derived.

$$\left(\frac{\partial U}{\partial V}\right)_T=T\left(\frac{\partial S}{\partial V}\right)_T-P=T\times\frac{nR}{V}-P=0 \qquad (160)$$

Similarly, the relation $(\partial U/\partial P)_T=0$ is readily derived, though its derivation is omitted here.

Ans. See text (a).

(b) Alternatively, U can be regarded as a function of T and V (see **II-2**). Then , eq(161) is derived.

$$dU=\left(\frac{\partial U}{\partial T}\right)_V dT+\left(\frac{\partial U}{\partial V}\right)_T dV \qquad (161)$$

From eq(160), the second term of eq(161) is found to be zero for the ideal gas. For a minute change in U at a constant volume, eq(162) is satisfied since there is no change in the energy of work. Combining eq(162) with the definition of the heat capacity at a constant volume given by eq(163), eq(164) is obtained.

$$(dU)_V=(\delta q)_V \qquad (162),\qquad C_V=(\delta q)_V/dT \qquad (163)$$

$$(dU)_V=\left(\frac{\partial U}{\partial T}\right)_V dT=C_V dT\ \Rightarrow\ C_V=\left(\frac{\partial U}{\partial T}\right)_V \qquad (164)$$

Based on the discussions so far, it is seen that the internal energy of the ideal gas is a function of temperature T alone.

Ans. See text (b).

"note" For a *van der Waals* gas, the equation of state is given by eq(165).

$$P=\frac{RT}{V-b}-\frac{a}{V^2} \qquad (165) \quad (n=1)$$

Substituting this relation into eq(160) using eq(159), eq(166) is derived, from which it is known that the internal energy of the *van der Waals* gas depends on the volume, in other words, it depends on the intermolecular distance. Or conversely saying, the meaning of the parameter a becomes clear. See **solution 3-55** for the internal pressure.

$$\left(\frac{\partial U}{\partial V}\right)_T=T\left(\frac{\partial P}{\partial T}\right)_V-P=T\times\frac{R}{V-b}-P=\frac{a}{V^2} \qquad (166)$$

Solution 3-32

The entropy of the gas is expressed by $S(T,V)$ as a function of T and V. Therefore, the entropy change associated with the process specified in the problem is given by the following equation.

$$\Delta S=S(T_2,V_2)-S(T_1,V_1)=\langle S(T_2,V_1)-S(T_1,V_1)\rangle+\langle S(T_2,V_2)-S(T_2,V_1)\rangle \qquad (167)$$

By taking S along the z-axis, and T and V along the x- and y-axes, respectively, the entropy change above can graphically be expressed as is shown in Fig.30 below.

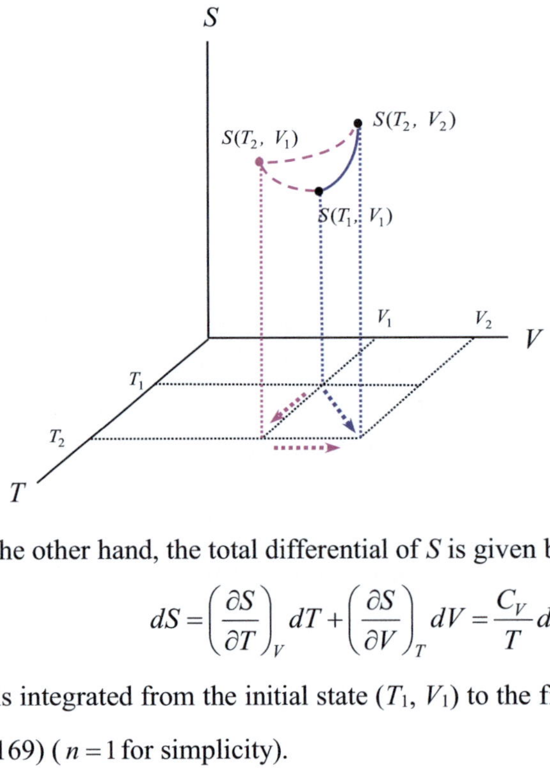

Fig.30 Illustration of route-independent entropy change and associated routes.
Entropy change does not depend on routes (e.g. a route along a blue line or one along a red broken line) connecting the initial and final states.
Notice that $\left(\frac{\partial S}{\partial T}\right)_V>0$, $\left(\frac{\partial S}{\partial V}\right)_T>0$.

On the other hand, the total differential of S is given by eq(168),

$$dS=\left(\frac{\partial S}{\partial T}\right)_V dT+\left(\frac{\partial S}{\partial V}\right)_T dV=\frac{C_V}{T}dT+\left(\frac{\partial P}{\partial T}\right)_V dV \qquad (168)$$

and it is integrated from the initial state (T_1, V_1) to the final state (T_2, V_2) along the relevant routes as is given by eq(169) ($n=1$ for simplicity).

$$\int_{T_1,V_1}^{T_2,V_2} dS=\int_{V_1}^{V_2}\frac{R}{V}dV+\int_{T_1}^{T_2}\frac{C_V}{T}dT \qquad (169)$$

As a result, eq(170) is obtained. See the figure given above. It is assumed here that C_V is independent of temperature.

$$\Delta S = \Delta S(T_1)_{V_1 \to V_2} + \Delta S(V_2)_{T_1 \to T_2} = R\ln\frac{V_2}{V_1} + C_V \ln\frac{T_2}{T_1} \qquad (170)$$

Ans. $\Delta S = R\ln V_2/V_1 + C_V \ln T_2/T_1$

"note" The analytical function of $S(T,V)$ can be obtained based on the standard procedure of the statistical mechanics. Therefore, the difference in the entropy $\Delta S = S(T_2,V_2) - S(T_1,V_1)$ (associated with any other process as well) is directly found from its function (see **problem 5-6**).

Solution 3-33

The work w (energy) done by the gas against the external pressure of P_{ex} is given by $-P_{ex}\Delta V$ and found as

$$w = -P_{ex}\Delta V = -101\times10^3 \times (V_f - V_i) = 101\times10^3 \times V_i, \quad V_i : \text{initial volume}$$
$$= -101\times10^3 \times (8.31\times298/101\times10^4) = -248 \text{ J.}$$

An accompanying change in the internal energy ΔU is calculated as below.

$$\Delta U = q + w = 0 - P_{ex}\Delta V = -248 \text{ J}$$

A change in the temperature of the gas is calculated from eq(164) given in **solution 3-31**, putting $C_V = 3R/2$ since the gas is mono-atomic.

$$\Delta T = \frac{\Delta U}{C_V} = \frac{-248}{3R/2} = -\frac{248\times2}{3\times8.31} = -19.9\ K$$

Employing these results, a change in the enthalpy ΔH and a change in the entropy ΔS are found as below.

$$\Delta H = \Delta U + \Delta(PV) = \Delta U + R\Delta T = -248 - 8.31\times19.9 = -412\ J$$

$$\Delta S = R\ \ln\frac{V_2}{V_1} + C_V \ln\frac{T_2}{T_1} = 8.31\times\ln2 + 8.31\times\frac{3}{2}\times\ln\frac{278.1}{298} = 4.90\ \text{J}\cdot\text{K}^{-1}$$

The change in the entropy is not zero $\Delta S = 0$. Why is this? Because the process is an adiabatic but irreversible one. There is no change in the entropy of the surroundings since the process is adiabatic. Therefore, the total entropy change is ascribed solely to the one in the gas arising from its adiabatic expansion. As the calculation shows, it takes a positive value indicating that the process is an irreversible one. Then, what value for ΔG will be found?

Ans. $q = 0,\ w = -248\text{ J},\ \Delta U = -248\text{ J},\ \Delta H = -412\text{ J},\ \Delta S = 4.90\text{ J}\cdot\text{K}^{-1},\ \Delta T = -19.9\text{ K}.$

See text for the comment.

Solution 3-34

Employing the *Maxwell* relation (see **II-2**), $\left(\frac{\partial S}{\partial V}\right)_T$ is expressed in terms of *P*, *T* and *V* for the *van der Waals* gas as is given by eq(171).

$$\left(\frac{\partial S}{\partial V}\right)_T = \left(\frac{\partial P}{\partial T}\right)_V = \frac{\partial}{\partial T}\left(\frac{RT}{V-b} - \frac{a}{V^2}\right) = \frac{R}{V-b} \qquad (171) \qquad (n=1)$$

Integrating eq(171) over the range [*V*, 2*V*] corresponding to the volume change of $V \to 2V$, the entropy change is given by eq(172).

$$\Delta S = \int_V^{2V} \frac{R}{V-b} dV = R\left[\ln(V-b)\right]_V^{2V} = R\ln\frac{2V-b}{V-b} > R\ln2 \qquad (172)$$

From the fact that the *van der Waals* parameter $b\ (\geq 0)$ is contained in the result of integration, it can be said that the entropy change is larger for *van der Waals* gases compared with that for the ideal gas. This is a natural consequence from a view point of molecular theory since it is known that the entropy is determined by a magnitude of the space occupied by molecules. In the limit of $b << V$, the entropy change becomes the same as that for the ideal gas. See **"note"** in **solution 3-31** for the volume dependence of the internal energy.

Ans. $\Delta S = R\ln\dfrac{2V-b}{V-b} > R\ln 2$

[advanced problem] For a gas which obeys the *Dieterici* equation of state (see **problem 2-49**), find the entropy change accompanied by a volume change at a constant temperature.

Solution 3-35

(a)The pressure is expressed by a function of $P = P(T,V)$. The total differential of P is given by eq(173). (For a general proof, the variable z is expressed as $z = f(x,y)$.)

$$dP = \left(\frac{\partial P}{\partial T}\right)_V dT + \left(\frac{\partial P}{\partial V}\right)_T dV \qquad (173)$$

Both sides of eq(173) are divided by dT under the condition of a constant pressure ($dP = 0$), leading to eq(174)

$$\left(\frac{\partial P}{\partial T}\right)_V + \left(\frac{\partial P}{\partial V}\right)_T \left(\frac{\partial V}{\partial T}\right)_P = 0 \qquad (174)$$

and eq(175) after a rearrangement.

$$\left(\frac{\partial P}{\partial V}\right)_T \left(\frac{\partial V}{\partial T}\right)_P = -\left(\frac{\partial P}{\partial T}\right)_V \qquad (175)$$

Both sides of eq(175) are divided by $\left(\dfrac{\partial P}{\partial T}\right)_V$. Noticing the identity of $\left(\dfrac{\partial P}{\partial T}\right)_V = \dfrac{(\partial P)_V}{(\partial T)_V}$, eq(176), which is nothing but an *Euler* chain relation, is obtained.

$$\left(\frac{\partial P}{\partial V}\right)_T \left(\frac{\partial V}{\partial T}\right)_P \frac{1}{\left(\dfrac{\partial P}{\partial T}\right)_V} = \left(\frac{\partial P}{\partial V}\right)_T \left(\frac{\partial V}{\partial T}\right)_P \frac{(\partial T)_V}{(\partial P)_V} = \left(\frac{\partial P}{\partial V}\right)_T \left(\frac{\partial V}{\partial T}\right)_P \left(\frac{\partial T}{\partial P}\right)_V = -1 \qquad (176)$$

These relations employed here can systematically be derived by means of *Jacobian*. Some examples are given in **"note"** below.

Ans. See text.

"note" Definition of *Jacobian* and some useful relations

The *Jacobian* of the functions x and y which are expressed in terms of two independent variable u and v is defined by eq(a.1) using the determinant.

$$J(x,y/u,v) \equiv \frac{\partial(x,y)}{\partial(u,v)} \equiv \begin{vmatrix} \left(\dfrac{\partial x}{\partial u}\right)_v & \left(\dfrac{\partial x}{\partial v}\right)_u \\ \left(\dfrac{\partial y}{\partial u}\right)_v & \left(\dfrac{\partial y}{\partial v}\right)_u \end{vmatrix} = \left(\frac{\partial x}{\partial u}\right)_v \left(\frac{\partial y}{\partial v}\right)_u - \left(\frac{\partial x}{\partial v}\right)_u \left(\frac{\partial y}{\partial u}\right)_v \qquad (a.1)$$

From this definition, the following relations are readily derived.

$$J(x,y/u,v) = -J(y,x/u,v) = -J(x,y/v,u) = J(y,x/v,u) \qquad (a.2)$$

$$J(x,y/x,y) = \frac{\partial(x,y)}{\partial(x,y)} = 1, \quad J(x,y/y,x) = \frac{\partial(x,y)}{\partial(y,x)} = -1, \quad \partial(x,y) = -\partial(y,x) \qquad (a.3)$$

Some of the *Jacobians* frequently used in thermodynamics are given below.

$$\frac{\partial(x,z)}{\partial(y,z)}=\begin{vmatrix}\left(\frac{\partial x}{\partial y}\right)_z & \left(\frac{\partial z}{\partial y}\right)_z\\ \left(\frac{\partial x}{\partial z}\right)_y & \left(\frac{\partial z}{\partial z}\right)_y\end{vmatrix}=\begin{vmatrix}\left(\frac{\partial x}{\partial y}\right)_z & 0\\ \left(\frac{\partial x}{\partial z}\right)_y & 1\end{vmatrix}=\left(\frac{\partial x}{\partial y}\right)_z \qquad (a.4)$$

$$\frac{\partial(y,z)}{\partial(x,z)}=\begin{vmatrix}\left(\frac{\partial y}{\partial x}\right)_z & \left(\frac{\partial z}{\partial x}\right)_z\\ \left(\frac{\partial y}{\partial z}\right)_x & \left(\frac{\partial z}{\partial z}\right)_x\end{vmatrix}=\begin{vmatrix}\left(\frac{\partial y}{\partial x}\right)_z & 0\\ \left(\frac{\partial y}{\partial z}\right)_x & 1\end{vmatrix}=\left(\frac{\partial y}{\partial x}\right)_z \qquad (a.5)$$

$$\frac{\partial(z,y)}{\partial(x,z)}=\begin{vmatrix}\left(\frac{\partial z}{\partial x}\right)_z & \left(\frac{\partial y}{\partial x}\right)_z\\ \left(\frac{\partial z}{\partial z}\right)_x & \left(\frac{\partial y}{\partial z}\right)_x\end{vmatrix}=\begin{vmatrix}0 & \left(\frac{\partial y}{\partial x}\right)_z\\ 1 & \left(\frac{\partial y}{\partial z}\right)_x\end{vmatrix}=-\left(\frac{\partial y}{\partial x}\right)_z \qquad (a.6)$$

Euler's chain relation expressed by means of the *Jacobian* is given below as one example.

$$\left(\frac{\partial P}{\partial V}\right)_T\left(\frac{\partial T}{\partial P}\right)_V\left(\frac{\partial V}{\partial T}\right)_P=\frac{\partial(P,T)}{\partial(V,T)}\times\frac{\partial(T,V)}{\partial(P,V)}\times\frac{\partial(V,P)}{\partial(T,P)}=-1 \qquad (a.7)$$

Noticing the relation of $\partial(P,T)=-\partial(T,P)$ given in (a.3), it is easy to show that the numerator and denominator in eq(a.7) cancel each other yielding a value of -1. Employing *Jacobian*s, not only the *Maxwell* relations shown below but also some complicated thermodynamic relations can automatically be derived, but only a reference book is cited here. (*H.Margenau and G.M.Murphy, The Mathematics of Physics and Chemistry,* ch.1, D.Van Nostrand Company, Inc. 1943.)

$\left(\frac{\partial S}{\partial V}\right)_T$ is expressed as eq(a.8) by employing *Jacobian*s.

$$\left(\frac{\partial S}{\partial V}\right)_T=\frac{\partial(S,T)}{\partial(V,T)}=\frac{\partial(S,T)}{\partial(P,T)}\times\frac{\partial(P,T)}{\partial(V,P)}\times\frac{\partial(V,P)}{\partial(V,T)}=\frac{\partial(S,T)}{\partial(V,P)}\times\left(\frac{\partial P}{\partial T}\right)_V \qquad (a.8)$$

Since S and P are first partial derivatives of the free energy F, they can be written as those given by eq(a.9).

$$S=\left(\frac{\partial F}{\partial T}\right)_V=\frac{\partial(F,V)}{\partial(T,V)},\qquad P=\left(\frac{\partial F}{\partial V}\right)_T=\frac{\partial(F,T)}{\partial(V,T)} \qquad (a.9)$$

Using this relation, $\partial(S,T)/\partial(V,P)$ in eq(a.8) can be rewritten as eq(a.10).

$$\frac{\partial(S,T)}{\partial(V,P)}=-\frac{\partial(S,T)}{\partial(P,V)}=\frac{\partial\left(\frac{\partial(F,V)}{\partial(T,V)},T\right)}{\partial\left(\frac{\partial(F,T)}{\partial(V,T)},V\right)}=-\frac{\partial\left(\frac{\partial(F,V)}{\partial(T,V)},T\right)\Big/\partial(V,T)}{\partial\left(\frac{\partial(F,T)}{\partial(V,T)},V\right)\Big/\partial(V,T)}=\frac{\left(\frac{\partial}{\partial V}\frac{\partial(F,V)}{\partial(T,V)}\right)_T}{\left(\frac{\partial}{\partial T}\frac{\partial(F,T)}{\partial(V,T)}\right)_V}=\frac{\left\langle\frac{\partial}{\partial V}\left(\frac{\partial F}{\partial T}\right)_V\right\rangle_T}{\left\langle\frac{\partial}{\partial T}\left(\frac{\partial F}{\partial V}\right)_T\right\rangle_V}=1 \qquad (a.10)$$

Namely, an interesting relation of $\partial(S,T)=\partial(V,P)$ is obtained and then from eq(a.8), one of the *Maxwell* relations is obtained as below.

$$\left(\frac{\partial S}{\partial V}\right)_T=\left(\frac{\partial P}{\partial T}\right)_V$$

(b) Since it is easy to confirm the relation in question for the ideal gas, the proof for the *van der Walls* gas is given below. The total differential of the state equation of $P=\dfrac{RT}{V-b}-\dfrac{a}{V^2}$ is written as eq(177).

$$dP=\frac{RdT}{V-b}-\frac{RT}{(V-b)^2}dV+\frac{2a}{V^3}dV \qquad (177)$$

Under the condition of a constant temperature *T*, both sides of eq(177) are divided by *dV* to yield eq(178).

$$\left(\frac{\partial P}{\partial V}\right)_T=-\frac{RT}{(V-b)^2}+\frac{2a}{V^3} \qquad (178)$$

On the other hand, when divided by *dP*, eq(179) is obtained.

$$1=\frac{R}{V-b}\left(\frac{\partial T}{\partial P}\right)_V \Rightarrow \left(\frac{\partial T}{\partial P}\right)_V=\frac{V-b}{R} \qquad (179)$$

Also, when eq(177) is divided by *dT* under a constant pressure, eq(180) is derived.

$$\left(\frac{\partial V}{\partial T}\right)_P=\frac{\dfrac{R}{V-b}}{\left(\dfrac{RT}{(V-b)^2}-\dfrac{2a}{V^3}\right)} \qquad (180)$$

The product of the three equations given by eq(178),eq(179) and eq(180) yields -1, i.e., *Euler*'s chain relation.

Ans. See text (b).

Solution 3-36

Using the relation of $\rho=\dfrac{nM}{V} \Rightarrow \left(\dfrac{\partial V}{\partial \rho}\right)_S=-\dfrac{nM}{\rho^2}=-\dfrac{V^2}{nM}$, eq(181) given in the problem is rewritten as below.

$$\upsilon=\sqrt{\left(\frac{\partial P}{\partial \rho}\right)_S}=\sqrt{\left(\frac{\partial P}{\partial V}\right)_S\left(\frac{\partial V}{\partial \rho}\right)_S}=\sqrt{\left(\frac{\partial P}{\partial V}\right)_S\left(-\frac{V^2}{nM}\right)} \qquad (181)$$

Then, the derivative $(\partial P/\partial V)_S$ corresponding to the adiabatic volume change in eq(181) is expressed as eq(182).

$$\left(\frac{\partial P}{\partial V}\right)_S=\frac{\left(\dfrac{\partial P}{\partial T}\right)_S}{\left(\dfrac{\partial V}{\partial T}\right)_S}=\frac{-\dfrac{T}{C_V}\left(\dfrac{\partial P}{\partial T}\right)_V}{\dfrac{T}{C_P}\left(\dfrac{\partial V}{\partial T}\right)_P}=-\frac{C_P}{C_V}\cdot\frac{\dfrac{nR}{V}}{\dfrac{nR}{P}}=-\frac{C_P}{C_V}\cdot\frac{P}{V} \qquad (182)$$

Substituting this into eq(181), the last term of eq(183) is obtained as the sound velocity. The heat capacities of the gas (air) are taken to be $C_P=\dfrac{5}{2}nR$ and $C_V=\dfrac{3}{2}nR.$

$$\upsilon=\sqrt{\frac{C_P}{C_V}\cdot\frac{P}{V}\cdot\frac{V^2}{nM}}=\sqrt{\frac{C_P}{C_V}\cdot\frac{PV}{nM}}=\sqrt{\frac{C_P}{C_V}\cdot\frac{RT}{M}}=\sqrt{\frac{5}{3}\cdot\frac{RT}{M}} \qquad (183)$$

In order to derive eq(182), an alternative orthodox method can be used. Starting from the premise that *P* is expressed as a function of *T* and *V*, the total differential of *P* is given by eq(184).

$$dP=\left(\frac{\partial P}{\partial T}\right)_V dT+\left(\frac{\partial P}{\partial V}\right)_T dV \qquad (184)$$

Both sides of eq(184) is divided by *dV* under the condition of the adiabatic process ($dS=0$) to obtain eq(185).

$$\left(\frac{\partial P}{\partial V}\right)_S = \left(\frac{\partial P}{\partial T}\right)_V \left(\frac{\partial T}{\partial V}\right)_S + \left(\frac{\partial P}{\partial V}\right)_T \qquad (185)$$

Employing the relation derived so far (see **problem 3-26**),

$$\left(\frac{\partial T}{\partial V}\right)_S = -\frac{\left(\frac{\partial S}{\partial V}\right)_T}{\left(\frac{\partial S}{\partial T}\right)_V} = -\frac{\left(\frac{\partial P}{\partial T}\right)_V}{\frac{C_V}{T}} = -\frac{T}{C_V}\cdot\left(\frac{\partial P}{\partial T}\right)_V$$

eq(185) is rewritten as eq(186) and for the ideal gas, eq(186') identical to eq(182) is derived.

$$\left(\frac{\partial P}{\partial V}\right)_S = \left(\frac{\partial P}{\partial T}\right)_V \left\{-\frac{T}{C_V}\left(\frac{\partial P}{\partial T}\right)_V\right\} + \left(\frac{\partial P}{\partial V}\right)_T \qquad (186)$$

$$= -\left(\frac{nR}{V}\right)^2 \frac{T}{C_V} - \frac{nRT}{V^2} = -\frac{nRT(nR + C_V)}{C_V V^2} = -\frac{C_P}{C_V}\cdot\frac{P}{V} \qquad (186')$$

By substituting this result into eq(181), the velocity of sound is obtained.

To find the sound velocity required, the apparent molecular weight of the air calculated by $\frac{4\times 28+32}{5} = 28.8\,(\mathrm{g})$ is used. Also, the heat capacities of the air are taken to be $C_P = C_V + R$ and $C_V = 3R/2 + R$, regarding that the air is composed of diatomic molecules and that the vibrational motion does not contribute to the heat capacities at room temperature. Then, the sound velocity in the air is found as below.

$$v = \sqrt{\frac{7\times 8.31\times 298}{5\times 28.8\times 10^{-3}}} = 347\ \mathrm{m\cdot s^{-1}}$$

Ans. sound velocity in the air: $347\ \mathrm{m\cdot s^{-1}}$

For the derivation of eq(181), see **solution 2-59**.

Solution 3-37

(a) Solution 1. *Clausious-Clapeyron*'s equation is employed.

A liquid and its gas are in equilibrium at T and P and their *Gibbs* free energies are denoted as $G^l(T,P)$ and $G^g(T,P)$, respectively. Here, the superscripts l and g stand for liquid and gas, respectively. Eq(187) holds in equilibrium under consideration.

$$G^l(T,P) = G^g(T,P) \qquad (187)$$

When the temperature T and the pressure P vary to $T+dT$ and $P+dP$ keeping the gas-liquid equilibrium, eq(188) is satisfied since the free energies of both phases must be equal in equilibrium.

$$G^l(T+dT, P+dP) = G^g(T+dT, P+dP) \qquad (188)$$

Writing the associated changes in the free energy as dG^l and dG^g, eq(188) is rewritten as eq(189).

$$G^l(T,P) + dG^l = G^g(T,P) + dG^g \qquad (189)$$

From eqs(188) and (189), eq(190) is obtained.

$$dG^l = dG^g \;\Rightarrow\; V^l dP - S^l dT = V^g dP - S^g dT \qquad (190)$$

Rearranging the terms in eq(190), eq(191), which is exactly *Clausious-Clapeyron*'s equation, is obtained.

$$(S^g - S^l)dT = (V^g - V^l)dP \;\Rightarrow\; \frac{dP}{dT} = \frac{(S^g - S^l)}{(V^g - V^l)} = \frac{\Delta S_{vap}}{\Delta V_{vap}} = \frac{\Delta H_{vap}}{T\Delta V_{vap}} \qquad (191)$$

Notice here that the relation of $\Delta G_{vap} = \Delta H_{vap} - T\Delta S_{vap} = 0$ is used.

The vapor is assumed to behave ideally and the molar volume of the liquid is so small as compared with that of the gas that the volume change upon vaporization can be approximated as $\Delta V_{vap} \approx \dfrac{nRT}{P}$. Then, eq(192) is obtained from eq(191).

$$\frac{1}{P}\frac{dP}{dT} = \frac{\Delta H_{vap}}{nRT^2} \Rightarrow \ln P/P_0 = \int_{T_0}^{T} \frac{\Delta H_{vap}}{nRT^2} dT \qquad (192)$$

Putting $n=1$ and substituting the given equation of $\Delta H_{vap} = H_0 + a(T - T_0)$ into eq(192), it is integrated to find the answer required as follows (eq(193)).

$$\ln P/P_0 = \int_{T_0}^{T} \frac{H_0 + a(T-T_0)}{RT^2} dT = \left[-\frac{H_0 - aT_0}{RT} + \frac{a}{R}\ln T \right]_{T_0}^{T} = \left(\frac{H_0 - aT_0}{R} \right)\left(\frac{1}{T_0} - \frac{1}{T} \right) + \frac{a}{R}\ln\frac{T}{T_0} \qquad (193)$$

Ans. $\ln P/P_0 = \left(\dfrac{H_0 - aT_0}{R} \right)\left(\dfrac{1}{T_0} - \dfrac{1}{T} \right) + \dfrac{a}{R}\ln\dfrac{T}{T_0}$

<u>Solution 2</u>. Method of using chemical potentials

The chemical potentials of a substance in the two phases which are in equilibrium are equal. The chemical potential in the liquid phase is denoted $\mu^l(T,P)$ and that in the gas phase $\mu^g(T,P)$. If the gas is assumed to behave ideally, then eq(194) is satisfied.

$$\mu^l(T,P) = \mu^g(T,P) = \mu^g(T,P^\circ) + RT\ln\frac{P}{P^\circ} \qquad (194)$$

After rewriting eq(194) as eq(195), both sides of eq(195) are differentiated with T, yielding eq(196) which

$$\frac{\mu^l(T,P) - \mu^g(T,P^\circ)}{RT} = \ln\frac{P}{P^\circ} \qquad (195)$$

assumes the same form as eq(192) (see **"note 2"**).

$$\begin{aligned} \frac{d\{\mu^l(T,P) - \mu^g(T,P^\circ)\}/RT}{dT} &= \frac{1}{n}\frac{d\{G^l(T,P) - G^g(T,P^\circ)\}/RT}{dT} \\ &= -\frac{H^l(T,P) - H^g(T,P^\circ)}{nRT^2} + \frac{V^l}{RT}\frac{dP}{dT} \\ &\approx -\frac{H^l(T,P) - H^g(T,P)}{nRT^2} \\ &\approx \frac{\Delta H_{vap}}{nRT^2} = \frac{1}{P}\frac{dP}{dT} \qquad (196) \end{aligned}$$

Since the relation identical to eq(192) is obtained, the remaining part of the calculation is the same as that in Solution 1. Notice, however, chemical potentials of pure substances are nothing but molar free energies.

(b) After integrating eq(192), the result is rewritten as eq(197), from which it is seen that the function obtained is a sort of the reduced equation and that <u>the plot of P/P_b *vs.* T/T_b assumes the same form</u> as far as the value for $\Delta H^m_{vap}/T_b$ is the same (see Fig.31).

$$\ln P/P_b = \int_{T_b}^{T} \frac{\Delta H_{vap}}{nRT^2} dT = \left[-\frac{\Delta H_{vap}}{nRT} \right]_{T_b}^{T} = \frac{\Delta H^m_{vap}}{R}\left(\frac{1}{T_b} - \frac{1}{T} \right) = \frac{\Delta H^m_{vap}}{RT_b}\left(1 - \frac{1}{T/T_b} \right) \qquad (197)$$

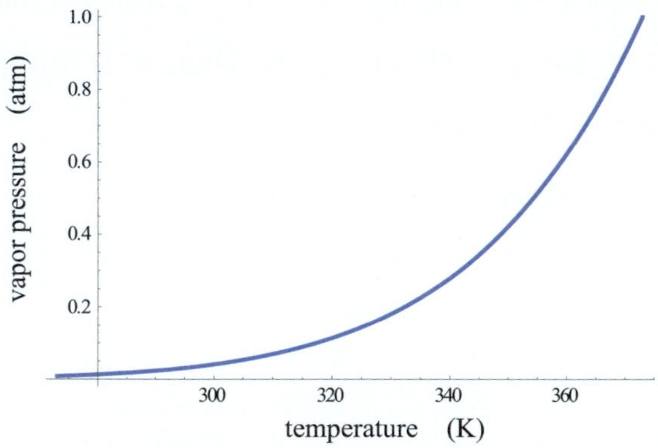

Fig.31 Vapor pressure curve of water drawn based on the heat of vaporization.

Ans. $\ln P/P_b = \dfrac{\Delta H^{m}_{vap}}{RT_b}\left(1-\dfrac{1}{T/T_b}\right)$

(c) Putting $T_b = 373\,\mathrm{K}$, the values given for water are substituted into eq(197) and P/P_b is plotted against T/T_b as is shown in the figure above. The curve is found to reproduce the actual vapor pressures of water quite well (see **"note 1"**).

"note1" Vapor pressure curves of liquids are usually approximated by an experimental equation of $\ln P = A - \dfrac{B}{T} + C\cdot \ln T$. The vapor pressure curve drawn by using the experimental values of $A = 60.32$, $B = 6851$ and $C = 5.138$ for water almost coincides with the curve obtained in (c).

"note 2" Equations necessary for deriving eq(196).

$$dG = -SdT + VdP$$

$$\frac{dG}{dT} = \left(\frac{\partial G}{\partial T}\right)_P + \left(\frac{\partial G}{\partial P}\right)_T \frac{dP}{dT}$$

$$= -S + V\frac{dP}{dT} \quad \text{(notice that the differentials are not partial differentials.)}$$

$$\frac{d\,G^l(T,P)/RT}{dT} = -\frac{G^l}{RT^2} - \frac{1}{RT}S^l + \frac{1}{RT}V^l\frac{dP}{dT} = \frac{-H^l(T,P)}{RT^2} + \frac{V^l}{RT}\frac{dP}{dT}$$

$$\frac{d\,G^g(T,P^\circ)/RT}{dT} = -\frac{G^g}{RT^2} - \frac{1}{RT}S^g = \frac{-H^g(T,P^\circ)}{RT^2} = \frac{-H^g(T,P)}{RT^2}$$

$$\left(\frac{\partial H^g}{\partial P}\right)_T = T\left(\frac{\partial S}{\partial P}\right)_T + V = -T\left(\frac{\partial V}{\partial T}\right)_P + V = 0 \quad \text{(independent of pressure for the ideal gas)}$$

$$\mu = \left(\frac{\partial G}{\partial n}\right)_{T,P} = \frac{G}{n} \quad \text{(for pure substance)}$$

Solution 3-38a

Since *Gibbs* free energy G (or all of the extensive thermodynamic variables in general) belongs to *Euler*'s homogeneous function of 1st-order, its more general expression, replacing G with f, is given by eq(198) putting $\mathrm{n} = 1$ (n to be distinguished from n).

$$x_1\frac{\partial f}{\partial x_1} + x_2\frac{\partial f}{\partial x_2} + x_3\frac{\partial f}{\partial x_3} + \cdots = \mathrm{n}\cdot f(x_1, x_2, x_3, \cdots) \qquad (198)$$

On the other hand, when the system is multiplied by λ times, the resulting free energy is related to the free energy of the system before multiplication through eq(199) because of the extensive nature of *G*. Also, from eq(198), eq(200) is obtained, in which G^{λ} represents the left side of eq(199).

$$G(T,P,\lambda n_{\mathrm{A}},\lambda n_{\mathrm{B}},\lambda n_{\mathrm{C}},\cdots)=\lambda\cdot G(T,P,n_{\mathrm{A}},n_{\mathrm{B}},n_{\mathrm{C}},\cdots) \qquad (199)$$

$$\frac{\partial G^{\lambda}}{\partial\lambda}=\sum_{i=1}\frac{\partial G^{\lambda}}{\partial\lambda n_i}\cdot\frac{\partial\lambda n_i}{\partial\lambda}=\sum_{i=1}\frac{\partial G}{\partial n_i}\times n_i=\frac{\partial\lambda G}{\partial\lambda}=G \qquad (200)$$

The latter half of eq(200) can be rewritten as eq(201).

$$n_A\frac{\partial G}{\partial n_A}+n_B\frac{\partial G}{\partial n_B}+n_C\frac{\partial G}{\partial n_C}+\cdots \quad =G(T,P,n_A,n_B,n_C,\cdots \quad) \qquad (201)$$

Since the chemical potential of the *i*-th component is defined by eq(202), eq(203), which is the equation to be proved, is obtained by replacing the partial differentials in eq(201) by the relevant chemical potentials.

$$\mu_i=\left(\frac{\partial G}{\partial n_i}\right)_{T,P,n_j,j\neq i} \qquad (202)$$

$$G(T,P,n_A,n_B,n_C,\cdots \quad)=\sum n_i\mu_i=n_A\mu_A+n_B\mu_B+n_C\mu_C+\cdots \qquad (203)$$

Ans. See text.

"note" The total differential of eq(203) is given by eq(204).

$$\begin{aligned}dG(T,P,n_A,n_B,n_C,\cdots \quad)&=dn_A\cdot\mu_A+dn_B\cdot\mu_B+dn_C\cdot\mu_C+\cdots\\&+n_A\cdot d\mu_A+n_B\cdot d\mu_B+n_C\cdot d\mu_C+\cdots\end{aligned} \qquad (204)$$

On the other hand, the right side of eq(204) can be written as eq(205).

$$\begin{aligned}dG(T,P,n_A,n_B,n_C,\cdots \quad)&=\left(\frac{\partial G}{\partial n_A}\right)_{T,P,n_j,j\neq A}dn_A+\left(\frac{\partial G}{\partial n_B}\right)_{T,P,n_j,j\neq B}dn_B+\left(\frac{\partial G}{\partial n_C}\right)_{T,P,n_j,j\neq C}dn_C+\cdots\\&=\mu_A\cdot dn_A+\mu_B\cdot dn_B+\mu_C\cdot dn_C+\cdots\end{aligned} \qquad (205)$$

Subtracting eq(204) from eq(205), eq(206) is derived.

$$n_A\cdot d\mu_A+n_B\cdot d\mu_B+n_C\cdot d\mu_C+\cdots=0 \qquad (206)$$

This is the relation called *Gibbs-Duhem*'s equation and plays an important role in knowing to which directions chemical potentials (or more generally partial molar quantities) change for multi-component systems (see **problems 3-10**, **3-20** and **3-21**).

Solution 3-38b

Since *Hermholtz* free energy *F* (or all of the extensive thermodynamic variables in general) belongs to *Euler*'s homogeneous function of 1st-order, its more general expression, replacing *G* with f, is given by eq(207) putting $\mathrm{n}=1$ (n to be distinguished from *n*).

$$x_1\frac{\partial f}{\partial x_1}+x_2\frac{\partial f}{\partial x_2}+x_3\frac{\partial f}{\partial x_3}+\cdots \quad =\mathrm{n}\cdot f(x_1,x_2,x_3,\cdots \quad) \qquad (207)$$

On the other hand, when the system (or extensive variable of *F*) is multiplied by λ, the resulting free energy denoted F^{λ} is related to the free energy of the system before multiplication through eq(208) because of the extensive nature of *F*. Also, from eq(207), eq(209) is obtained. However, being different from the previous problem, the variable V must be treated as an extensive variable. For the partial differentials in eq(209), it is clearly specified which variables are to be kept constant upon differentiation.

$$F^{\lambda}(T,\lambda V,\lambda n_{\mathrm{A}},\lambda n_{\mathrm{B}},\lambda n_{\mathrm{C}},\cdots)=\lambda F(T,V,n_{\mathrm{A}},n_{\mathrm{B}},n_{\mathrm{C}},\cdots) \qquad (208)$$

$$\left(\frac{\partial F^{\lambda}}{\partial \lambda}\right)_{T,V,n_i}=\frac{\partial F^{\lambda}}{\partial \lambda V}\cdot\frac{\partial \lambda V}{\partial \lambda}+\sum_{i=1}\frac{\partial F^{\lambda}}{\partial \lambda n_i}\cdot\frac{\partial \lambda n_i}{\partial \lambda}=V\times\left(\frac{\partial F}{\partial V}\right)_{T,n_j}+\sum_{i=1}n_i\times\left(\frac{\partial F}{\partial n_i}\right)_{T,V,n_j,j\neq i} \qquad (209)$$

$$=-PV+\sum n_i\phi_i \qquad (209')$$

On the other hand, it is evident that eq(210) is satisfied from eq(209).

$$\left(\frac{\partial F^{\lambda}}{\partial \lambda}\right)_{T,V,n_i}=\left(\frac{\partial \lambda F}{\partial \lambda}\right)_{T,V,n_i}=F \qquad (210)$$

By putting eq(210) equal to eq(209), eq(211) is derived. It should be noticed that $F\neq\sum_i n_i\phi_i$.

$$F=-PV+\sum_i n_i\phi_i \;\Rightarrow\; F+PV=G=\sum_i n_i\phi_i \qquad (211)$$

In conclusion, eq(212) is derived.

$$\phi_i=\mu_i \qquad (212)$$

Or explicitly, eq(212) is expressed as eq(213) in terms of the chemical potential.

$$\left(\frac{\partial F}{\partial n_i}\right)_{T,V,n_j,j\neq i}=\left(\frac{\partial G}{\partial n_i}\right)_{T,P,n_j,j\neq i}=\mu_i \qquad (213)$$

Ans. eq(211) and eq(213)

[advanced problem] Define the chemical potentials using the internal energy U.

Solution 3-39

The total differential of the enthalpy $H(T,P)$ is given by eq(214) and from this eq(215) is derived.

$$dH=\left(\frac{\partial H}{\partial T}\right)_P dT+\left(\frac{\partial H}{\partial P}\right)_T dP \qquad (214) \Rightarrow \left(\frac{d\,\Delta H_l}{dT}\right)_e=\left(\frac{\partial\,\Delta H_l}{\partial T}\right)_P+\left(\frac{\partial\,\Delta H_l}{\partial P}\right)_T\left(\frac{dP}{dT}\right)_e \qquad (215)$$

<u>The subscript e indicates that the differentiation is carried out by keeping the equilibrium</u> (differentiation along the specified route).

On the other hand, from the definition of the enthalpy (eq(216)), eq(217) is derived by dividing eq(216) by dP at a constant temperature. Here, the *Maxwell* relation is used.

$$dH=-TdS+VdP \qquad (216)$$

$$\left(\frac{\partial H}{\partial P}\right)_T=-T\left(\frac{\partial S}{\partial P}\right)_T+V=-T\left(\frac{\partial V}{\partial T}\right)_P+V \qquad (217)$$

Applying this relation to the latent heat of a phase transition ΔH_l, eq(218) is obtained, in which ΔV_e is a volume change upon the phase transition.

$$\left(\frac{d\,\Delta H_l}{dP}\right)_T=-T\left(\frac{\partial\,\Delta V_e}{\partial T}\right)_P+\Delta V_e \qquad (218)$$

Replacing $(dP/dT)_e$ with the *Clausius-Clapeyron* equation of $(dP/dT)_e=\Delta H_l/T\,\Delta V_t$ (**solution 3-12**) and noticing the relation of $(\partial\Delta H_l/\partial T)_P=\Delta C_P$, eq(215) is rewritten as eq(219).

$$\left(\frac{d\,\Delta H_l}{dT}\right)_e = \Delta C_P + \left\{-T\left(\frac{\partial\,\Delta V_e}{\partial T}\right)_P + \Delta V_e\right\}\frac{\Delta H_l}{T\,\Delta V_l} \tag{219}$$

Rearranging eq(219) ($\Delta V_e \equiv \Delta V_t$), eq(220) is derived and thus, *Planck*'s equation is proved.

$$\left(\frac{d\,\Delta H_l}{dT}\right)_e = \Delta C_P + \frac{\Delta H_l}{T} - \frac{\Delta H_l}{\Delta V_e}\left(\frac{\partial\,\Delta V_e}{\partial T}\right)_P = \Delta C_P + \frac{\Delta H_l}{T} - \Delta H_l\left(\frac{\partial \ln \Delta V_e}{\partial T}\right)_P \tag{220}$$

Ans. See text.

Solution 3-40

An example in which the curves representing the two adiabatic processes cross each other is illustrated in the figure below. One cycle of a heat engine in this case is expressed by a closed curve consisting of three reversible processes. Gaining or losing of work and heat energy for each process are given on the right of the figure. Repeating the cycle of process ① ⇒ process ② ⇒ process ③, a heat engine by which all of the heat energy q_1 gained from the heat source at a high temperature is converted to the work ($w_1 + w_2 + w_3$: corresponding to the area enclosed by the three curves in Fig.32) can be constructed. However, this is against the thermodynamic law (perpetual motion engine of the second kind) which says that no heat engine can

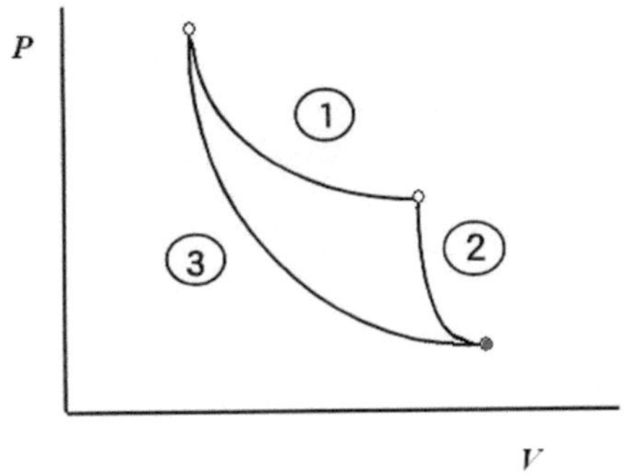

process ①: isothermal expansion q_1, w_1

process ②: adiabatic expansion w_2

process ③: adiabatic compression w_3

Fig.32 One work-cycle consisting of two crossing adiabatic processes. Two adiabatic processes are drawn so as to cross at the red point on the right end. However, this kind of cycle never happens according to the thermodynamic law.

convert all of the heat gained from a heat source into work without any loss. Consequently, it is concluded that no two adiabatic processes cross each other (see **"note"**). The perpetual motion engine of the first kind is an engine that does only work, in one cycle, and does not cause any change to the outside world or to the heat engine. Of course these two kinds of perpetual motion engine do not exist.

Ans. See text.

"note" Any function of state describes the state unambiguously and, hence, any two functions of state of the same sort never cross each other. For instance, the equations of state for an ideal gas at different temperatures, the *P-V* equations for the adiabatic processes (curves ② and ③ in Fig.32), the *van der Waals* equations of state, tie lines in the same phase, etc., never cross each other. Therefore, the answer to the present problem is trivial without relying specifically on the thermodynamic law. Regarding the present problem, however, it should be remarked that the principle of *Carathéodory* states "In every arbitrarily close neighborhood of a given initial state, there exist states that cannot be approached arbitrarily closely by adiabatic processes."

Solution 3-41a

For pure substances, the chemical potential μ is identical to the molar free energy and, hence, if the pressure dependence of the *Gibbs* free energy is known, then the chemical potential is derived readily form it.

Based on the pressure dependence of G, the pressure dependence of μ is given by eq(221).

$$\left(\frac{\partial G(P,T)}{\partial P}\right)_T = n\left(\frac{\partial \mu(P,T)}{\partial P}\right)_T = V \Rightarrow \left(\frac{\partial \mu(P,T)}{\partial P}\right)_T = \frac{V}{n} = \frac{RT}{P} \qquad (221)$$

Integrating eq(221) over the pressure range $[P^\circ, P]$, the expression required is obtained as eq(222).

$$\mu(P,T) - \mu(P^\circ,T) = \int_{P^\circ}^{P} \frac{RT}{P} dP = RT[\ln P]_{P^\circ}^{P} = RT \ln \frac{P}{P^\circ}$$

$$\mu(P,T) = \mu(P^\circ,T) + RT \ln \frac{P}{P^\circ} \qquad (222)$$

P° is the pressure at the standard state and the chemical potential there is given by $\mu(P^\circ,T)$.

Ans. See text.

Solution 3-41b

The *van der Waals* equation of state ($n = 1$ mole) is given by eq(223) and the differential dP is given by eq(224).

$$\left(P + \frac{a}{V^2}\right)(V-b) = RT \Rightarrow P = \frac{RT}{(V-b)} - \frac{a}{V^2} \qquad (223)$$

$$dP = -\frac{RT}{(V-b)^2} dV + \frac{2a}{V^3} dV \qquad (224)$$

From this, the chemical potential of the *van der Waals* gas is found by integrating the equation $\left(\frac{\partial \mu(P,T)}{\partial P}\right)_T = \frac{V}{n}$ derived in the previous problem as follows (eq(225)).

$$\mu(P,T) = \mu^\circ(P^\circ,T) + \int_{P^\circ}^{P} VdP \qquad (n = 1 \text{ mole})$$

$$= \mu^\circ(P^\circ,T) + \int_{V^\circ}^{V} -\left\{\frac{RTV}{(V-b)^2} - \frac{2a}{V^2}\right\} dV \qquad (P \to V \text{ variable interchange})$$

$$= \mu^\circ(P^\circ,T) + \int_{V^\circ}^{V} -\left\{\frac{RTb}{(V-b)^2} + \frac{RT}{(V-b)} - \frac{2a}{V^2}\right\} dV$$

$$= \mu^\circ(P^\circ,T) + \left[\frac{RTb}{(V-b)} - RT\ln(V-b) - \frac{2a}{V}\right]_{V^\circ}^{V}$$

$$= \mu^\circ(P^\circ,T) + RTb\left\{\frac{1}{(V-b)} - \frac{1}{(V^\circ-b)}\right\} - RT\ln\frac{(V-b)}{(V^\circ-b)} - 2a\left(\frac{1}{V} - \frac{1}{V^\circ}\right) \qquad (225)$$

Putting $a = 0$ and $b = 0$, eq(225) gives eq(222) found in the preceding solution.

Ans. eq(225)

Solution 3-41c

Eq(14) given in **problem 3-14c** is rewritten as eq(226) below to match the definition of the fugacity coefficient γ,

$$\frac{\mu(P,T)-\mu^\circ(P^\circ,T)}{RT}=-\ln\frac{(V-b)}{(V^\circ-b)}+b\left\{\frac{1}{(V-b)}-\frac{1}{(V^\circ-b)}\right\}-\frac{2a}{RT}\left(\frac{1}{V}-\frac{1}{V^\circ}\right) \tag{226}$$

from which eq(227) is obtained for γ.

$$\ln\gamma=\frac{\mu(P,T)-\mu^\circ(P^\circ,T)}{RT}-\ln\frac{P}{P^\circ}$$

$$=-\ln\frac{(V-b)}{(V^\circ-b)}+b\left\{\frac{1}{(V-b)}-\frac{1}{(V^\circ-b)}\right\}-\frac{2a}{RT}\left(\frac{1}{V}-\frac{1}{V^\circ}\right)-\ln\frac{P}{P^\circ} \tag{227}$$

Ans. fugacity coefficient: eq(227)

Solution 3-41d

Solution 1. Employing the reduced temperature T_R, reduced pressure P_R and reduced volume V_R defined by eq(228) as well as the relations between the *van der Waals* parameters and the critical constants (T_c, P_c

$$T_R=T/T_c,\ P_R=P/P_c,\ V_R=V/V_c\ \Rightarrow\ T=T_RT_c,\ P=P_RP_c,\ V=V_RV_c \tag{228}$$

$$a=3P_cV_c^2,\ b=\frac{1}{3}V_c,\ R=\frac{8P_cV_c}{3T_c} \tag{229}$$

and V_c), eq(229) given in the preceding solution is rewritten as eq(230) below, which does not contain any parameter specific to an individual gas (i.e., a and b in the present case).

$$\ln\gamma=-\ln\frac{(V-b)}{(V^\circ-b)}+b\left\{\frac{1}{(V-b)}-\frac{1}{(V^\circ-b)}\right\}-\frac{2a}{RT}\left(\frac{1}{V}-\frac{1}{V^\circ}\right)-\ln\frac{P}{P^\circ}$$

$$=-\ln\frac{(3V_R-1)}{(3V_R^\circ-1)}+\frac{1}{(3V_R-1)}-\frac{1}{(3V_R^\circ-1)}-\frac{6P_cV_c^2}{RT_RT_c}\left(\frac{1}{V_RV_c}-\frac{1}{V_R^\circ V_c}\right)-\ln\frac{P_RP_c}{P_R^\circ P_c}$$

$$=-\ln\frac{(3V_R-1)}{(3V_R^\circ-1)}+\frac{1}{(3V_R-1)}-\frac{1}{(3V_R^\circ-1)}-\frac{9}{4T_R}\left(\frac{1}{V_R}-\frac{1}{V_R^\circ}\right)-\ln\frac{P_R}{P_R^\circ} \tag{230}$$

Thus, eq(230) is obtained as a reduced function to express the fugacity coefficient γ for a *van der Waals* gas.

Ans. reduced fugacity coefficient γ : eq(230)

Solution 2. The reduced fugacity coefficient is derived directly form the reduced equation of state. For this purpose, the reduced *van der Waals* equation of state is obtained first (see **problem** and **solution 2-19**). Substituting the relations given by eqs(228) and (229) into eq(231),

$$\left(P+\frac{a}{V^2}\right)(V-b)=RT\ \Rightarrow\ P=\frac{RT}{(V-b)}-\frac{a}{V^2} \tag{231}$$

$$P_cP_R=\frac{RT_cT_R}{(V_cV_R-b)}-\frac{a}{(V_cV_R)^2}=\frac{RT_cT_R}{\left(V_cV_R-\frac{1}{3}V_c\right)}-\frac{3P_cV_c^2}{(V_cV_R)^2}$$

and after some rearrangements, eq(232) is obtained as the reduced *van der Waals* equation of state.

$$P_R=\frac{RT_cT_R/P_cV_c}{\left(V_R-\frac{1}{3}\right)}-\frac{3}{V_R^2}=\frac{RT_cT_R/P_cV_c}{\left(V_R-\frac{1}{3}\right)}-\frac{3}{V_R^2}=\frac{8T_R}{3\left(V_R-\frac{1}{3}\right)}-\frac{3}{V_R^2} \tag{232}$$

At a constant temperature (T_R = constant), the total differential of eq(232) is given by eq(233).

$$dP_R = -\frac{8T_R}{3}\frac{1}{(V_R - 1/3)^2}dV_R + \frac{6}{V_R^{\ 3}}dV_R \qquad (233)$$

On the other hand, the chemical potential divided by RT can be expressed as eq(234) by using the reduced variables (see **solution 3-41a**). Consequently, integrating the right side of eq(234), eq(235) is finally obtained,

$$\frac{1}{RT}\int_{P^\circ}^{P} VdP = \frac{3}{8T_R}\int_{P_R^\circ}^{P_R} V_R dP_R \qquad (234)$$

$$= \frac{3}{8T_R}\int_{V_R^\circ}^{V_R}\left\{-\frac{8T_R}{3}\frac{V_R}{(V_R-1/3)^2} + \frac{6}{V_R^{\ 2}}\right\}dV_R$$

$$= \frac{1}{3}\left[\frac{1}{(V_R-1/3)} - 3\ln(V_R-1/3)\right]_{V_R^\circ}^{V_R} - \frac{9}{4T_R}\left[\frac{1}{V_R}\right]_{V_R^\circ}^{V_R}$$

$$= -\ln\frac{(3V_R-1)}{(3V_R^\circ-1)} + \frac{1}{(3V_R-1)} - \frac{1}{(3V_R^\circ-1)} - \frac{9}{4T_R}\left(\frac{1}{V_R} - \frac{1}{V_R^\circ}\right) \qquad (235)$$

which is identical to eq(230) obtained in Solution 1 when $\ln P_R/P_R^\circ$ is subtracted from both sides of eq(235).

Ans. reduced fugacity coefficient γ : eq(235)$-\ln P_R/P_R^\circ$

"note 1" Since the reduced variables are dimensionless, any equation which can be expressed in terms of reduced variables should be dimensionless, such as the compression factor $z = PV/RT$ or the fugacity coefficient. Therefore, the chemical potential μ which has a dimension of energy cannot be expressed by reduced variables alone. However, when the chemical potential is divided by RT so as to make the quotient dimensionless, the resulting μ/RT can be expressed by a reduced equation as is shown in Solution 2.

"note 2" What is fugacity? See **solution 3-41f**.

Solution 3-41e

Employing eq(230) given in Solution 1 in the preceding solution which is shown again below, the values appropriate for CO_2 are substituted to calculate its fugacity coefficients.

$$\ln\gamma = -\ln\frac{(3V_R-1)}{(3V_R^\circ-1)} + \frac{1}{(3V_R-1)} - \frac{1}{(3V_R^\circ-1)} - \frac{9}{4T_R}\left(\frac{1}{V_R} - \frac{1}{V_R^\circ}\right) - \ln\frac{P_R}{P_R^\circ} \qquad (230)$$

The values needed are as follows.

$V_c = 92.4\times10^{-6}\,\mathrm{m}^3$, $T_c = 304.2\,\mathrm{K}$, $P_c = 7.38\times10^6\,\mathrm{Pa}$

$V_R^\circ = 22.4\times10^{-3}/92.4\times10^{-6} = 242.4$, $T_R = 373/304.2 = 1.23$, $P_R^\circ = 1.01\times10^5/7.38\times10^6 = 1.37\times10^{-2}$

The value for the reduced variable P_R is calculated from the reduced equation given below.

$$P_R = \frac{8}{3}\frac{T_R}{\left(V_R - \frac{1}{3}\right)} - \frac{3}{V_R^2} \qquad (236)$$

The pressure at the standard state is taken as $1.01\times10^5\,\mathrm{Pa}$ and CO_2 is assumed to behave ideally at this state. Substituting $V_R = 20.0$ and $T_R = 1.23$ into eq(230), $P_R = 0.166$ is obtained. The fugacity coefficient γ at the state specified by the values of these reduced variables is found by substituting them into eq(230) as below.

$$\ln\gamma = -0.05265 \;\Rightarrow\; \gamma = 0.949$$

The volume of CO_2 at the supercritical state specified by those values of the reduced variables is found as

$$V = 20.0\times 92.4\times 10^{-6} = 1.85\times 10^{-3}\,\mathrm{m}^3 .$$

In order to find the value for γ based on eq(235) given in Solution 2 in the preceding solution, the integration range $[V_R^\circ,\ V_R]$ need to be taken as $\left[242.4,\ 20\right]$. When the integration is performed by *Mathematica,* the same value of $\gamma = 0.949$ as found above is obtained (see **"note"** below).

Ans. fugacity coefficient: $\gamma = 0.949$

"note" The gas in question needs to behave ideally at the standard state, so that it might be better to take the standard pressure as 0.1 atm in the present calculation. However, even when a valued for P_R° is reduce to one tenth by taking 0.1 atm as the standard pressure instead of 1.0 atm, the value for γ calculated thus is found as 0.945. This is a reasonable consequence, when we notice the fact that the volume of CO_2 calculated from the *van der Waals* equation of state at the standard state (273 K and 1 atm) is $22.34\ \mathrm{dm}^3$, which is hardly different from that of the ideal gas.

Solution 3-41f

Taking CO_2 gas as an example, we consider the reason why the fugacity of the gas can be a measure of escaping (fugitive) tendency of molecules from one another. The state of CO_2 whose fugacity coefficient is found in **problem 3-41e** is at $T = 373\,\mathrm{K}$, $P = 0.166\times 7.38\times 10^6\,\mathrm{Pa}$ and $V = 1.85\times 10^{-3}\,\mathrm{m}^3$. As compared with an ideal gas at the same state at T and V, it is seen that the fugacity ($f = \gamma\ P$) of CO_2 is smaller than that of the ideal gas. Also, the volume which the gas of the same number of moles occupies under the same conditions is smaller for CO_2 than for the ideal gas ($V_{ideal} = 2.52\times 10^{-3}\,\mathrm{m}^3$ at the same T and P). Namely, the gas with the larger fugacity exhibits the greater tendency of expanding in space. In other words, we can say the gas with the larger fugacity has the higher tendency of escaping from the enclosed space (container). Under the present conditions, the ideal gas shows higher tendency to escape as compared with CO_2. The magnitude of the fugacity can also be related with the intermolecular force or the excluded volume.

Ans. See text.

Solution 3-42a

The total differential of the internal energy ($U(T,V)$) is generally given by eq(237).

$$dU = \left(\frac{\partial U}{\partial T}\right)_V dT + \left(\frac{\partial U}{\partial V}\right)_T dV \qquad (237)$$

Since the first term of eq(237) is equal to $C_V dT$ by the definition of the heat capacity at a constant temperature, the total differential can also be expressed by eq(238).

$$dU = C_V dT + \left(\frac{\partial U}{\partial V}\right)_T dV \qquad (238)$$

Considering that the internal energy of the ideal gas is a function of temperature alone as is expressed by $U(T)$, the second term of eq(238) becomes zero for the ideal gas. Consequently, eq(239) is always satisfied for the ideal gas (see **solution 3-31**).

$$dU = C_V dT \qquad (239)$$

For the ideal gas, eq(240) is always satisfied as well.

$$dH = C_P dT \qquad (240)$$

$$\left[\begin{aligned} dH &= \left(\frac{\partial H}{\partial T}\right)_P dT + \left(\frac{\partial H}{\partial P}\right)_T dP \\ &= C_P dT \qquad \because \text{ for ideal gas } \left(\frac{\partial H}{\partial P}\right)_T = \left(\frac{\partial U}{\partial P}\right)_T = 0 \end{aligned}\right]$$

For the real gases, however, eq(239) or eq(240) is not satisfied because of $\left(\frac{\partial U}{\partial V}\right)_T \neq 0$ or $\left(\frac{\partial H}{\partial P}\right)_T \neq 0.$ It should be noticed that these equations cannot simply be derived from the relevant heat capacity defined by eq(241) or eq(242).

$$\left(\frac{\partial U}{\partial T}\right)_V = \frac{(dU)_V}{(dT)_V} = C_V \qquad (241), \qquad \left(\frac{\partial H}{\partial T}\right)_P = \frac{(dH)_P}{(dT)_P} = C_P \qquad (242)$$

If we may say so, the relations which can be derived from the definitions of the heat capacities are eqs(241') and (242') and, hence, they are obviously different from eqs(239) and (240).

$$(dU)_V = C_V (dT)_V \qquad (241'), \qquad (dH)_P = C_P (dT)_P \qquad (242')$$

For the ideal gas, eqs(239) and (240) hold for the adiabatic process as well. Therefore, by equating the change in the internal energy (or enthalpy) equal to the change accompanied by the adiabatic process, eq(243) (or eq(244)) is obtained (for molar heat capacity, $C_P - C_V = R$). See also **"note"**.

$$dU = C_V dT = -PdV \qquad (243)$$

$$dH = C_P dT = VdP \qquad (244)$$

Integrating eq(243) or eq(244) by using the relation of $PV = RT$ $(n = 1\,\text{mole})$, eq(243'') for the relation between *T* and *V*,

$$\frac{C_V}{R}\frac{dT}{T} = -\frac{dV}{V} \Rightarrow \ln T = \ln V^{-\frac{R}{C_V}} + \text{Const.} = \ln V^{-\frac{C_P - C_V}{C_V}} + \text{Const.}$$

$$= \ln V^{-\gamma+1} + \text{Const.} \qquad (243') \Rightarrow TV^{\gamma-1} = \text{Const.}' \qquad (243'')$$

or eq(244'') for the relation between *T* and *P* is obtained via eq(243') or eq(244'), respectively..

$$\frac{C_P}{R}\frac{dT}{T} = \frac{dP}{P} \Rightarrow \ln T = \ln P^{\frac{R}{C_P}} + \text{Const.} = \ln P^{\frac{C_P - C_V}{C_P}} + \text{Const.}$$

$$= \ln P^{1-\frac{1}{\gamma}} + \text{Const.} \qquad (244') \Rightarrow TP^{\frac{1}{\gamma}-1} = \text{Const.}'' \qquad (244'')$$

Here, $\gamma = C_P / C_V$ and, of course, is not the fugacity coefficient. Although eq(243'') and eq(244'') look different superficially, they describe entirely the same functional relation among (P, V, T) for the adiabatic process and the relations are called *Poisson*'s law. They are also equivalent to the relation of $PV^\gamma = \text{Const.}'''$ What is mentioned here is trivial from the fact that two adiabatic processes do not cross each other (see **problem 3-40**).

Ans. eq(243'') and eq(244'')

"note" Eqs(243) and (244) indicate that $C_V(dT)_S = -P(dV)_S \Rightarrow (dT/dV)_S = -P/C_V$ and $C_P(dT)_S = V(dP)_S \Rightarrow (dT/dP)_S = V/C_P$, respectively.

Solution 3-42b

Integrating eq(243) given in the preceding solution (shown below again) over the range $[T_1, T_2]$, eq(245) is obtained.

$$dU = C_V dT = -PdV \quad \text{(eq(243) shown here again)}$$

$$U(T_2) - U(T_1) = C_V(T_2 - T_1) \qquad (245)$$

On the other hand, a change in the internal energy associated with this adiabatic process is ascribed to the one due to the work alone and, hence, the right side of eq(245) is given by eq(246).

$$U(T_2) - U(T_1) = -\int_{V_1}^{V_2} PdV \qquad (246)$$

Since the equation of state for the adiabatic process is given by eq(247) (see the preceding solution) and the

$$PV^{\gamma} = const. \qquad \gamma = C_P/C_V \qquad (247)$$

constant here can be written as $const. = P_1V_1^{\gamma}$ from the initial state specified by (P_1, V_1, T_1), the right side of eq(246) can be calculated as below.

$$U(T_2) - U(T_1) = -\int_{V_1}^{V_2} \frac{const.}{V^{\gamma}} dV = \left[\frac{const.}{\gamma - 1} V^{-\gamma+1}\right]_{V_1}^{V_2} = \frac{const.}{\gamma - 1}\left(\frac{1}{V_2^{\gamma-1}} - \frac{1}{V_1^{\gamma-1}}\right)$$

Consequently, from eq(245) the equation to give the temperature reached by this adiabatic process is obtained as the last term of eq(248).

$$C_V(T_2 - T_1) = \frac{P_1V_1^{\gamma}}{\gamma - 1}\left(\frac{1}{V_2^{\gamma-1}} - \frac{1}{V_1^{\gamma-1}}\right) \Rightarrow T_2 = \frac{1}{C_V} \times \frac{P_1V_1^{\gamma}}{\gamma - 1}\left(\frac{1}{V_2^{\gamma-1}} - \frac{1}{V_1^{\gamma-1}}\right) + T_1 = T_1(V_1/V_2)^{\gamma-1} \qquad (248)$$

Of course, without relying on the procedure given above, eq(247) can be rewritten in terms of T and V as below by using the equation of state $P = RT/V$ for the ideal gas to yield eq(248) directly.

$$TV^{\gamma-1} = const. = T_1V_1^{\gamma-1} \Rightarrow T_2V_2^{\gamma-1} = T_1V_1^{\gamma-1} \Rightarrow T_2 = T_1(V_1/V_2)^{\gamma-1}$$

Ans. $T_1(V_1/V_2)^{\gamma-1}$

"note" Orthodoxly, if the derivative of $(dT/dV)_S$ or $(dT/dP)_S$ is known as a function of *P*, *V* and *T* for a real gas, then the equation of state for the adiabatic process of the real gas can be obtained in general.

Solution 3-43a

From the virial equation of state expressed as eq(249), eq(19) given in the present problem is derived. The mole number of B is denoted n_B (see the problem for eqs(17) through eq(19)).

$$\frac{PV}{n_B RT} = 1 + \frac{n_B B'(T)}{V} + \frac{n_B^2 C'(T)}{V^2} + \cdots \qquad (249)$$

First, eq(249) is rewritten using the molecular weight M_B and the mass of the solute m_B as eq(250).

$$P \approx \frac{n_B RT}{V}\left\{1 + \frac{n_B B'(T)}{V} + \frac{n_B^2 C'(T)}{V^2} + \cdots\right\} = \frac{m_B}{M_B}\frac{RT}{V}\left\{1 + \frac{m_B}{M_B}\frac{B'(T)}{V} + \left(\frac{m_B}{M_B}\right)^2 \frac{C'(T)}{V^2}\right\} \qquad (250)$$

The approximation to the second term of eq(250) is given by eq(251).

$$P\Big/\left(\frac{m_{\rm B}}{V}\right)\approx\frac{RT}{M_{\rm B}}\left\{1+\frac{m_{\rm B}}{V}\frac{B'(T)}{M_{\rm B}}\right\} \qquad (251)$$

$$\Pi\Big/\left(\frac{m_{\rm B}}{V}\right)\approx\frac{RT}{M_{\rm B}}\left\{1+B(T)\frac{m_{\rm B}}{V}\right\} \qquad (19)\ \text{(reposted)}$$

By replacing the coefficient of the second term $B(T)$ of eq(19) with $B'(T)/M_{\rm B}$, it is found that eq(251) and eq(19) are identical in the functional form. It is needless to say that solute B is non-volatile, i.e., no B in the gas phase.

What is the reason why the idea of non-ideal gas is applicable to explain the osmotic pressure that a non-volatile solute exhibits? It is because the chemical potential assumes the same functional form for both of the gas and the solute. Consequently, a deviation of the chemical potential from the ideal behavior can be treated for a non-ideal solution in the same manner, at least in a functional form, as done for a non-ideal gas which deviates from the ideal gas.

Ans. See text and eq(251).

Solution 3-43b

Based on eq(19) given in the preceding **problem 3-43a**, the values for the osmotic pressure divided by the concentration $\Pi\Big/\left(\frac{m_{\rm B}}{V}\right)$ are plotted against the concentration of BSA in Fig.33. Here, the unit of concentration should be converted to $\text{kg}\cdot\text{m}^{-3}=10^3\text{g}\times(10^6\text{cm}^3)^{-1}=10^{-3}\text{g}\cdot\text{cm}^{-3}$ (see **"note"**). From the value for the intercept of the straight line acquired by the least-squares fitting $a=(35.2\pm0.9)$, the molecular weight $M_{\rm B}$ is calculated as below.

$$a=\frac{RT}{M_{\rm B}}\ \Rightarrow\ M_{\rm B}=\frac{RT}{a}=\frac{8.31\times273}{(35.2\pm0.9)}=(66.1-62.8)\ \text{kg}\cdot\text{mol}^{-1}$$

Usually, the molecular weight is expressed in the units of $\text{g}\cdot\text{mol}^{-1}$ and we can say $M_{\rm B}$ is 66100 – 62800 (or alternatively, 66.1 — 62.8 kDa in the unit of Da).

Incidentally, when it is assumed that *Raoult*'s law is effective and the values for the osmotic pressure are plotted against the concentration of BSA according to this law, Fig.34 is obtained. The linearity becomes poorer as compared with Fig.33 and the calculated molecular weight from the slope is only 45.1 kDa, much smaller than that obtained above.

"note" After converting the units of pressure to atm and and employing the value of $R=8.21\times10^{-2}\,\text{L}\cdot\text{atm}\cdot\text{K}^{-1}\text{mol}^{-1}$, the correct value for the molecular weight is not found unless the concentrations given in the table are multiplied by a factor of 1000.

[advanced problem] Real gases exhibit negative or positive values for $B(T)$ depending on temperature. The temperature at which $B(T)$ satisfies the relation $B(T)=0$ is called the *Boyle* temperature. At this temperature gases behave apparently like an ideal gas. On the other hand, the second term of the corrected formula (eq(19), **3-43a**) to give the osmotic pressure of BSA is positive. Explain why it is positive.

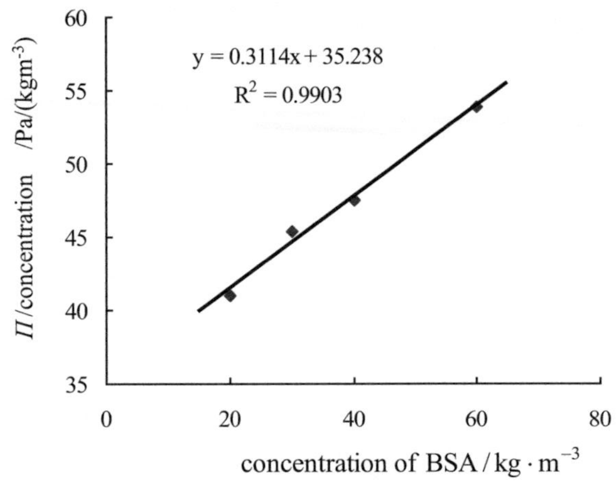

Fig.33 Osmotic pressure and concentration of BSA. Linear plot based on eq(19).

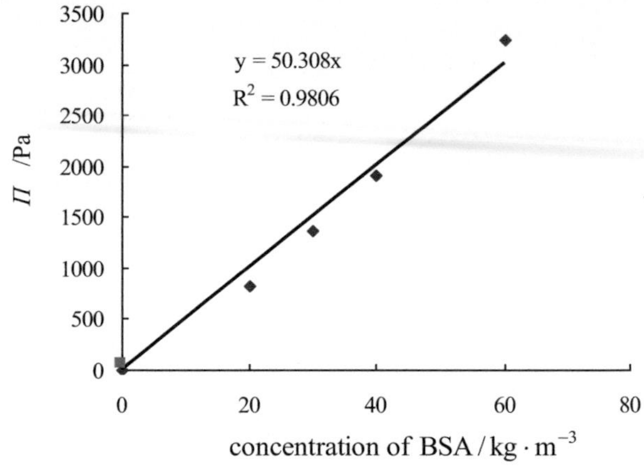

Fig.34 Osmotic pressure and concentration of BSA. Uncorrected linear plot. The red point indicates that the best-fitted straight line passes through the origin.

Ans. molecular weight: $66100 - 62800$

Solution 3-44

The *Helmholtz* free energy is denoted $F(T,V)$ and the two phases of *a* (gas) and *b* (liquid) are in equilibrium at the temperature *T*. Then eq(252) holds from the conditions of equilibrium and small variations in *T* and *V* there are related by eq(253) (see **"note"**).

$$F^a(T,V_a) = F^b(T,V_b) \qquad (252)$$

$$dF^a = dF^b \;\Rightarrow\; -S_a dT - PdV_a = -S_b dT - PdV_b \qquad (253)$$

From the definition of the compressibility β, eq(254) is obtained.

$$-\frac{1}{V_a}\left(\frac{\partial V_a}{\partial P}\right)_T = \beta_a,\; -\frac{1}{V_b}\left(\frac{\partial V_b}{\partial P}\right)_T = \beta_b \;\Rightarrow\; PdV_a = PV_a\beta_a dP,\; PdV_b = PV_b\beta_b dP \qquad (254)$$

Substituting eq(254) into eq(253), eq(255) is obtained. V_b (liquid) can be neglected as compared with V_a (gas).

$$(S_a - S_b)dT = -P(dV_a - dV_b) = P(V_a\beta_a - V_b\beta_b)dP \approx PV_a\beta_a dP \qquad (255)$$

In the cases of gas-liquid and gas-solid equilibriums, when the gas can be approximated by an ideal gas, the compressibility is given by $\beta_a = 1/P$. Then eq(255) is rewritten so as to yield a well known *Clausius-Clapeyron*'s equation given below.

$$\frac{dP}{dT} = \frac{(S_a - S_b)}{PV_a\beta_a} = \frac{\Delta S_{vap}}{PV_a\beta_a} \approx \frac{\Delta S_{vap}}{V_a} = \frac{\Delta H_{vap}}{TV_a}$$

It should be realized, however, that the volume is not an appropriate variable in discussing the equilibrium since it is not an intensive variable such as *T* or *P* and there does not exist any condition it should fulfill at the eqilibrium. Also, eq(254) is correct as an approximation (see the diffferential along the route, **solution 3-39**).

Ans. See text.

"note" Since the pressures in the two phases are in balance, the same value of pressure is used in eq(253), though the *Helmholtz* free energy is used to consider the equilibrium. As compared with the case in which the balance of the *Gibbs* free energy is considered, the point of the problem is to understand what and how is different when the equilibrium is considered by the balance of the *Helmholtz* free energy. The point is eq(255).

Solution 3-45a

The work done to (or by) the gas is given by $-PdV$, which is integrated according to the conditions given in the problem as shown below.

$$\Delta w = -\int_{100\text{kPa}}^{400\text{kPa}} PdV = -\int \frac{RT}{V-b} dV = -RT\left[\ln(V-b)\right]_{V_1}^{V_2}$$

$$= -RT\ln\frac{V_2-b}{V_1-b} = -RT\ln\frac{P_1}{P_2} = 8.31\times 300\times \ln\frac{400}{100} = 3460\ \text{J}$$

The obtained value is perfectly the same as the one expected for the ideal gas. Why is this? Because the ratio of the volume change ($V-b$) is independent of b and is the same as that for the ideal gas.

Ans. 3460 J. Same as for the ideal gas.

If, however, the conditions of compression are not to increase the pressure four times but <u>to reduce the volume to the one fourth of the initial volume</u> at the same temperature, then the work done to the ideal gas is also 3460 J but to the non-ideal gas it becomes 3457 J as is calculated below and they do not agree each other.

$$V_1 = \frac{RT}{P} - b = \frac{8.31\times 300}{10^5} - 4.27\times 10^{-6} = 2.49257\times 10^{-2}\text{m}^3,\quad V_2 = 1/4V_1 = 6.23143\times 10^{-3}\text{m}^3$$

$$-RT\ln\frac{V_2-b}{V_1-b} = -8.31\times 300\times \ln\frac{6.23143\times 10^{-3} - 4.27\times 10^{-6}}{2.49257\times 10^{-2} - 4.27\times 10^{-6}} = 3457\ \text{J}$$

Well then, how about the *van der Waals* gas (see next problem)?

Solution 3-45b

Since the work done to (or by) the gas is given by $-PdV$, it is integrated over the pressure range given in the problem by substituting the *van der Waals* equation of state for P as below, yielding eq(256).

$$\Delta w = -\int_{100\ \text{kPa}}^{400\ \text{kPa}} PdV = -\int \frac{RT}{V-b} dV + \int \frac{a}{V^2} dV = -RT\left[\ln(V-b)\right]_{V_1}^{V_2} - \left[\frac{a}{V}\right]_{V_1}^{V_2}$$

$$= -RT\ln\frac{V_2-b}{V_1-b} - a\left(\frac{1}{V_2} - \frac{1}{V_1}\right) = -RT\ln\frac{P_1}{P_2} - a\left(\frac{1}{V_2} - \frac{1}{V_1}\right) \qquad (256)$$

The volumes at the two pressures are found by solving eq(257) as $V_1 = 24.82$ L and $V_2 = 6.12$ L.

$$P = \frac{RT}{V-b} - \frac{a}{V^2} \qquad (257)$$

Substituting these values into the second term of eq(256), a value of 0.448 J is found as below.

$$a\left(\frac{1}{V_2} - \frac{1}{V_1}\right) = 3.64\times\left(\frac{1}{6.12} - \frac{1}{24.82}\right) = 0.448\ \text{atm}\cdot\text{L} = 0.448\ \text{J}$$

Putting this together with the value for the first term of eq(256) already obtained in the preceding solution, the work done to the gas is found as below. Just make sure here that [atm · L] = [J].

$$\Delta w = 8.31 \times 300 \times \ln \frac{400}{100} - 3.64 \times \left(\frac{1}{6.12} - \frac{1}{24.82} \right) = 3460 - 0.448 = 3459.5 \text{ J}$$

As a result, it has turned out that neither of the excluded volume or the intermolecular force affect the work to be done or to be gained by the gas as far as the conditions are not too extreme.

Ans. 3459.5 J. Hardly different from the work done by the ideal gas.

Solution 3-46a

Since the ideal gas always satisfies eq(258) with a negative value, it is evident that the ideal gas fulfills the inequality given in the problem.

$$\left(\frac{\partial P}{\partial V} \right)_T = -\frac{nRT}{V^2} < 0 \qquad (258)$$

However, for the gas which obeys the *van der Waals* equation of state, eq(259) is derived and its right side does not always take a negative value.

$$P = \frac{nRT}{V - nb} - \frac{n^2 a}{V^2} \Rightarrow \left(\frac{\partial P}{\partial V} \right)_T = -\frac{nRT}{(V - nb)^2} + \frac{2n^2 a}{V^3} \qquad (259)$$

The plot of $(\partial P/\partial V)_T$ given in eq(259) against V is shown in the figure below by the orange curve.

In general, a *van der Waals* gas exhibits maximum or minimum in the *P*-*V* curve in a certain range of temperature. In such a case, at any volume located between the two volumes, one where the pressure of the gas assumes a maximum and the other where the pressure assumes a minimum, the inequality $\left(\frac{\partial P}{\partial V} \right)_T > 0$ holds (see Fig.35). Such states are considered unstable thermodynamically and are not realized usually, namely a single phase of the gas cannot exist stably there. Therefore, the region where $\left(\frac{\partial P}{\partial V} \right)_T > 0$is excluded from the equation of state for the gas (see **problem 2-4** and **its solution**).

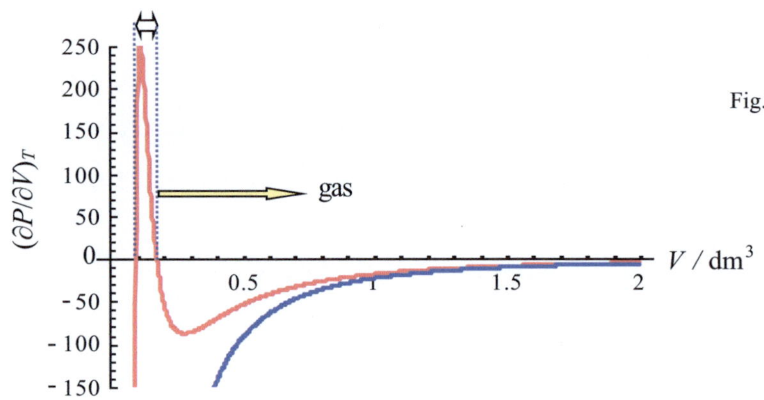

Fig.35 *van der Waals* gas and thermodynamic inequality $(\partial P/\partial V)_T < 0$. In the volume range indicated by ⇔ the themodynamic inequality does not hold. The blue line (negative always) is for the ideal gas.

Ans. eq(258) for the ideal gas. See Fig.35 for the volume range of the *van der Waals* gas.

"note" In the range where the gas-liquid equilibrium is established, the pressure does not change upon the volume change accompanied by the vaporization or liquefaction, that is, $(\partial P/\partial V)_T = 0$. The thermodynamic inequality is applicable to the phenomena in a single phase.

Solution 3-46b

The *Gibbs* free energy of the system in equilibrium with the surroundings at the temperature T_0 and at the pressure P_0 is given by eq(260).

$$G(S,V)=U(S,V)-T_0S+P_0V \qquad (260)$$

Generally, the *Gibbs* free energy of the system is expressed as $G(T,P)$. However, for the present case, it is expressed as $G(S,V)$ employing the variables *S* and *V* of the system, since the system under consideration is in equilibrium with the surroundings at the fixed temperature T_0 and pressure P_0. The number of particles (molecules) *N* in the system is assumed constant.

For infinitesimal changes in *S* and *V*, the accompanied change *dG* is acquired up to the second order of the infinitesimals as in eq(261).

$$dG=\left(\frac{\partial G}{\partial S}\right)_V dS+\left(\frac{\partial G}{\partial V}\right)_S dV$$
$$+\frac{1}{2}\left\langle\left(\frac{\partial^2 G}{\partial S^2}\right)_V (dS)^2+\left\{\frac{\partial}{\partial V}\left(\frac{\partial G}{\partial S}\right)_V\right\}_S dSdV+\left\{\frac{\partial}{\partial S}\left(\frac{\partial G}{\partial V}\right)_S\right\}_V dSdV+\left(\frac{\partial^2 G}{\partial V^2}\right)_S (dV)^2\right\rangle \qquad (261)$$

For the system in equilibrium, *G* assumes a minimum value and, hence, *G* must satisfy eq(262) against an infinitesimally small variation in either *S* or *V*.

$$\left(\frac{\partial G}{\partial S}\right)_V=0,\quad \left(\frac{\partial G}{\partial V}\right)_S=0 \qquad (262)$$

Furthermore, since *G* assumes a minimum but not maximum value in equilibrium, a small variation of *dG* must always be positive, i.e., $dG>0$. Therefore, *dG* must satisfy eq(263) in equilibrium.

$$dG=\left(\frac{\partial^2 G}{\partial S^2}\right)_V (dS)^2+2\left\{\frac{\partial}{\partial V}\left(\frac{\partial G}{\partial S}\right)_V\right\}_S dSdV+\left(\frac{\partial^2 G}{\partial V^2}\right)_S (dV)^2>0 \qquad (263)$$

Here the relation $\left\{\frac{\partial}{\partial V}\left(\frac{\partial G}{\partial S}\right)_V\right\}_S=\left\{\frac{\partial}{\partial S}\left(\frac{\partial G}{\partial V}\right)_S\right\}_V$ is used.

Regarding eq(261) as a quadratic function of *dS*, since the coefficient of the quadratic term is positive as is

$$\left(\frac{\partial^2 G}{\partial S^2}\right)_V=\left(\frac{\partial T}{\partial S}\right)_V=\frac{T}{C_V}>0 \quad \because C_V>0 \qquad (264)$$

seen from eq(264), the discriminant *D* of eq(263) should be negative when eq(263) is satisfied against any arbitrary infinitesimal variations of *dS* and *dV*, as is given by eq(265).

$$D=\left\langle\left\{\frac{\partial}{\partial V}\left(\frac{\partial G}{\partial S}\right)_V\right\}_S\right\rangle^2 (dV)^2-\left(\frac{\partial^2 G}{\partial S^2}\right)_V\left(\frac{\partial^2 G}{\partial V^2}\right)_S (dV)^2$$
$$\Rightarrow \left\langle\left\{\frac{\partial}{\partial V}\left(\frac{\partial G}{\partial S}\right)_V\right\}_S\right\rangle^2-\left(\frac{\partial^2 G}{\partial S^2}\right)_V\left(\frac{\partial^2 G}{\partial V^2}\right)_S<0 \qquad (265)$$

Employing *Jacobians*, eq(265) can be rewritten as eq(266) as a final form.

$$\left\langle\left\{\frac{\partial}{\partial V}\left(\frac{\partial G}{\partial S}\right)_V\right\}_S\right\rangle^2-\left(\frac{\partial^2 G}{\partial S^2}\right)_V\left(\frac{\partial^2 G}{\partial V^2}\right)_S=-\frac{\partial\left(\left(\frac{\partial G}{\partial S}\right)_V,\left(\frac{\partial G}{\partial V}\right)_S\right)}{\partial(S,V)}=-\begin{vmatrix}\left(\frac{\partial^2 G}{\partial S^2}\right)_V & \left\{\frac{\partial}{\partial S}\left(\frac{\partial G}{\partial V}\right)_S\right\}_V\\ \left\{\frac{\partial}{\partial V}\left(\frac{\partial G}{\partial S}\right)_V\right\}_S & \left(\frac{\partial^2 G}{\partial V^2}\right)_S\end{vmatrix}$$

$$=\frac{\partial(-T,P)}{\partial(S,V)}=-\frac{\dfrac{\partial(-T,P)}{\partial(T,V)}}{\dfrac{\partial(S,V)}{\partial(T,V)}}=\frac{\left(\dfrac{\partial P}{\partial V}\right)_T}{\left(\dfrac{\partial S}{\partial T}\right)_V}=\frac{T}{C_V}\left(\frac{\partial P}{\partial V}\right)_T<0 \qquad (266)$$

Since T/C_V is positive, eq(267), one of the thermodynamic inequalities, is obtained from eq(266).

$$\left(\frac{\partial P}{\partial V}\right)_T<0 \qquad (267)$$

Ans. See text.

"note 1" Of course, the transformation of eq(265) can be performed as below without employing *Jacobians*.

Using the relation of eq(268), eq(265) can be rewritten as eq(269).

$$\left(\frac{\partial^2 G}{\partial S^2}\right)_V=\left(\frac{\partial T}{\partial S}\right)_V,\quad \left\{\frac{\partial}{\partial S}\left(\frac{\partial G}{\partial V}\right)_S\right\}_V=-\left(\frac{\partial P}{\partial S}\right)_V=\left(\frac{\partial T}{\partial V}\right)_S,\quad \left(\frac{\partial^2 G}{\partial V^2}\right)_S=-\left(\frac{\partial P}{\partial V}\right)_S \qquad (268)$$

$$\left\langle\left\{\frac{\partial}{\partial V}\left(\frac{\partial G}{\partial S}\right)_V\right\}_S\right\rangle^2-\left(\frac{\partial^2 G}{\partial S^2}\right)_V\left(\frac{\partial^2 G}{\partial V^2}\right)_S=-\left(\frac{\partial P}{\partial S}\right)_V\left(\frac{\partial T}{\partial V}\right)_S+\left(\frac{\partial T}{\partial S}\right)_V\left(\frac{\partial P}{\partial V}\right)_S \qquad (269)$$

$$=\left(\frac{\partial T}{\partial S}\right)_V\left\{\left(\frac{\partial P}{\partial S}\right)_V\left(\frac{\partial S}{\partial V}\right)_T+\left(\frac{\partial P}{\partial V}\right)_S\right\} \qquad (269')$$

$$=\left(\frac{\partial T}{\partial S}\right)_V\left(\frac{\partial P}{\partial V}\right)_T=\frac{T}{C_V}\left(\frac{\partial P}{\partial V}\right)_T<0 \qquad (269'')$$

Substituting here *Euler*'s chain relation given by eq(270) is substituted into eq(269) , eq(269') is obtained.

$$\left(\frac{\partial V}{\partial S}\right)_T\left(\frac{\partial S}{\partial T}\right)_V\left(\frac{\partial T}{\partial V}\right)_S=-1 \Rightarrow \left(\frac{\partial T}{\partial V}\right)_S=-\frac{1}{\left(\dfrac{\partial V}{\partial S}\right)_T\left(\dfrac{\partial S}{\partial T}\right)_V}=-\frac{\left(\dfrac{\partial T}{\partial S}\right)_V}{\left(\dfrac{\partial V}{\partial S}\right)_T}=-\left(\frac{\partial T}{\partial S}\right)_V\left(\frac{\partial S}{\partial V}\right)_T \qquad (270)$$

Furthermore, by substituting the relation given by eq(271), which is to be derived from the total differential of $P(S,V)$, into eq(269), eq(269'') to be proved is obtained.

$$dP=\left(\frac{\partial P}{\partial S}\right)_V dS+\left(\frac{\partial P}{\partial V}\right)_S dV \overset{T=\text{const.}}{\Rightarrow} \left(\frac{\partial P}{\partial V}\right)_T=\left(\frac{\partial P}{\partial S}\right)_V\left(\frac{\partial S}{\partial V}\right)_T+\left(\frac{\partial P}{\partial V}\right)_S \qquad (271)$$

"note 2" Eq(263) can also be regarded as a quadratic function of dV. Then the coefficient of the term of the second order is $-\left(\dfrac{\partial P}{\partial V}\right)_S$ which is positive as is seen in the **problem 3-51**. Therefore, the discussions made above for dS can correctly be followed for dV.

"note 3" Some other thermodynamic inequalities are as follows.

$$\left(\frac{\partial \mu}{\partial c}\right)_T<0\,,\quad \left(\frac{\partial S}{\partial T}\right)_V>0\,, C_P-C_V>0,\ \ C_V>0,\ \text{etc.}$$

Solution 3-47

The entropies of the gases in the two containers before mixing are denoted by S_1 and S_2, respectively. The entropy after mixing is denoted as S_{1+2}. Then the entropy change given by $S_{1+2}-(S_1+S_2)$ can easily be calculated by considering as follows.

Writing the pressure reached after mixing the gases as P_f, the pressure of the gas 1 is varied from P_1 to P_f and similarly the pressure of the gas 2 from P_2 to P_f. Since the numbers of molecules in the containers are the same, the initial volume of each gas is changed first to $(V_1+V_2)/2$ and the pressure reached is given by eq(272).

$$P_f = \frac{2Nk_BT}{(V_1+V_2)} = \frac{2P_1P_2}{P_1+P_2} \qquad (272)$$

Then the two gases at the same pressure are mixed isothermally. In this last process there is no change in the entropy since the gases are of the same kind (see **problem 3-115**).

On the other hand, the entropy change with pressure at the constant temperature is given by eq(273).

$$\left(\frac{\partial S}{\partial P}\right)_T = -\left(\frac{\partial V}{\partial T}\right)_P = -\frac{Nk_B}{P} \qquad (273)$$

Consequently, the total entropy change ΔS is calculated as below in eq(274). Following the equations one by one, it is easy to understand how to calculate the entropy change in the present case.

$$\begin{aligned}\Delta S &= S_{1+2} - (S_1+S_2)\\ &= \int_{P_1}^{P_f} -\frac{Nk_B}{P}dP + \int_{P_2}^{P_f} -\frac{Nk_B}{P}dP = -Nk_B[\ln P]_{P_1}^{P_f} - Nk_B[\ln P]_{P_2}^{P_f}\\ &= -Nk_B\left(\ln\frac{2P_2}{P_1+P_2} + \ln\frac{2P_1}{P_1+P_2}\right) = -Nk_B\ln\frac{4P_1P_2}{(P_1+P_2)^2} = -Nk_B\ln\frac{4V_1V_2}{(V_1+V_2)^2} \geq 0 \qquad (274)\end{aligned}$$

Ans. $-Nk_B\ln\frac{4P_1P_2}{(P_1+P_2)^2} = -Nk_B\ln\frac{4V_1V_2}{(V_1+V_2)^2}$

Solution 3-48

Maximum work in the isothermal reversible process.

The gas 1 changes from (P_1, V_1) to $(P_f, V_f = Nk_BT/P_f)$ and the gas 2 from (P_2, V_2) to $(P_f, V_f = Nk_BT/P_f)$, both reversibly at the constant temperature. For each gas the work done (or gained) by the gas is given by eq(275) as below. The work done is maximum when the process is reversible.

$$\begin{aligned}w_{max} &= -\int_{V_1}^{V_f} PdV - \int_{V_2}^{V_f} PdV = -\int_{V_1}^{V_f}\frac{Nk_BT}{V}dV - \int_{V_2}^{V_f}\frac{Nk_BT}{V}dV \qquad (275)\\ &= -Nk_BT\ln\frac{V_f}{V_1} - Nk_BT\ln\frac{V_f}{V_2} = -Nk_BT\ln\frac{V_f^2}{V_1V_2} = -Nk_BT\ln\frac{(V_1+V_2)^2}{4V_1V_2} \leq 0\end{aligned}$$

Notice here that the final volume is given by $V_f = Nk_BT/P_f = (V_1+V_2)/2$. When $V_1 = V_2$, no work is done, naturally. Since the energy of work is negative, the gas has done work to the outside.

"note 1" Combined with the results of the preceding question, we can confirm that the internal energy of a gas does not change in such a mixing of gases. In other words, we can confirm that $\Delta U = \Delta w + \Delta q = \Delta w + T\Delta S = 0$. In the case of an ideal gas, the internal energy is a function of temperature alone, so this is a natural result. It is a good idea to compare the values of ΔG.

(b) Maximum work in adiabatic reversible process.

Since the maximum work is obtained in the reversible process, it is calculated in the same manner as in (a) by using $-PdV$ from eq(276).

$$w_{max} = -\int_{V_1}^{V_f} PdV - \int_{V_2}^{V_f} PdV \qquad (276)$$

Here, the gas 1 changes from (P_1, V_1, T) to (P_f, V_{f_1}, T_{f_1}) and the gas 2 from (P_2, V_2, T) to (P_f, V_{f_2}, T_{f_2}), being either compressed or expanded adiabatically (or isentropically). The P-V relation for the isentropic process is given by $\left(\frac{\partial P}{\partial V}\right)_S$ and, hence, if it can be expressed in terms of P, V and T, then the integral in eq(276) can be calculated. The procedure to acquire $\left(\frac{\partial P}{\partial V}\right)_S$ is as follows (see **problem 3-51** for a more orthodox procedure).

Expressing both of P and S as functions of V and T, the total differentials of P and S are rewritten by using the *Maxwell* relations as eq(277) and eq(278).

$$dP = \left(\frac{\partial P}{\partial V}\right)_T dV + \left(\frac{\partial P}{\partial T}\right)_V dT \overset{S=const.}{\Rightarrow} \left(\frac{\partial P}{\partial V}\right)_S = \left(\frac{\partial P}{\partial V}\right)_T + \left(\frac{\partial P}{\partial T}\right)_V \left(\frac{\partial T}{\partial V}\right)_S \qquad (277)$$

$$dS = \left(\frac{\partial S}{\partial V}\right)_T dV + \left(\frac{\partial S}{\partial T}\right)_V dT \overset{dS=0}{\Rightarrow} \left(\frac{\partial S}{\partial V}\right)_T + \left(\frac{\partial S}{\partial T}\right)_V \left(\frac{\partial T}{\partial V}\right)_S = \left(\frac{\partial P}{\partial T}\right)_V + \frac{C_V}{T}\left(\frac{\partial T}{\partial V}\right)_S = 0 \qquad (278)$$

Substituting eq(278) into eq(277), the expression needed is obtained as eq(279). Here, the relation $C_P - C_V = Nk_B$ for the ideal gas is used.

$$\left(\frac{\partial P}{\partial V}\right)_S = \left(\frac{\partial P}{\partial V}\right)_T + \left(\frac{\partial P}{\partial T}\right)_V \times \left\{-\left(\frac{\partial P}{\partial T}\right)_V \frac{T}{C_V}\right\}$$

$$= -\frac{Nk_B T}{V^2} - \frac{T}{C_V}\left(\frac{Nk_B}{V}\right)^2 = -\frac{Nk_B T}{V^2}\left(1 + \frac{Nk_B}{C_V}\right) = -\frac{Nk_B T}{V^2}\frac{C_P}{C_V} \qquad (279)$$

Via eq(280) which is derived from eq(279) by separating the V-dependent part from the P-dependent part,

$$\left(\frac{\partial P}{\partial V}\right)_S = -\frac{Nk_B T}{V^2}\frac{C_P}{C_V} = -\frac{P}{V}\frac{C_P}{C_V} \Rightarrow -\left(\frac{1}{P}dP\right)_S = \frac{C_P}{C_V}\left(\frac{1}{V}dV\right)_S \qquad (280)$$

eq(281) to describe the relation between P and V is found (*Poisson*'s equation). Here, the ratio of the specific heats is put $C_P/C_V = \gamma$.

$$\ln P/P_1 = -\frac{C_P}{C_V}\ln V/V_1 \Rightarrow P = const.\times V^{-\frac{C_P}{C_V}} = const.\times V^{-\gamma} \qquad (281)$$

Substituting this relation into eq(276), the maximum work for the adiabatic reversible process is calculated as below in eq(282).

$$w_{max} = -\int_{V_1}^{V_{f_1}} const.\times V^{-\gamma} dV - \int_{V_2}^{V_{f_2}} const.\times V^{-\gamma} dV = const.\times\left[\frac{V^{-\gamma+1}}{\gamma-1}\right]_{V_1}^{V_{f_1}} + const.\times\left[\frac{V^{-\gamma+1}}{\gamma-1}\right]_{V_2}^{V_{f_2}}$$

$$= \frac{const.}{\gamma-1}(V_{f_1}^{-\gamma+1} - V_1^{-\gamma+1}) + \frac{const.}{\gamma-1}(V_{f_2}^{-\gamma+1} - V_2^{-\gamma+1}) \qquad (282)$$

Noticing here the relations below,

$$P = const.\times V^{-\gamma} \Rightarrow P_1 = const.\times V_1^{-\gamma} = Nk_B T V_1^{-1} \Rightarrow const. = Nk_B T V_1^{\gamma-1} (or\ Nk_B T V_2^{\gamma-1})$$

the value for *const.* is substituted into the above equation and after some rearrangements the maximum work is found to be given by the following equation (eq(283)) (see **"note 2"**).

$$w_{max} = \frac{const.}{\gamma - 1}\left(V_{f_1}{}^{-\gamma+1} + V_{f_2}{}^{-\gamma+1} - V_1^{-\gamma+1} - V_2^{-\gamma+1}\right)$$

$$= \frac{Nk_BT}{\gamma - 1}\left\{\left(\frac{V_{f_1}}{V_1}\right)^{-\gamma+1} + \left(\frac{V_{f_2}}{V_2}\right)^{-\gamma+1} - 2\right\} \qquad (283)$$

Ans. $\frac{Nk_BT}{\gamma - 1}\left\{\left(\frac{V_{f_1}}{V_1}\right)^{-\gamma+1} + \left(\frac{V_{f_2}}{V_2}\right)^{-\gamma+1} - 2\right\}$

"note 2" For the adiabatic reversible processes treated in this problem, the final temperature and pressure cannot take values common to gases 1 and 2. If we start from the same temperature for both gases and try to make the final pressure the same (P_f), then, as is shown below, not only the final temperatures but also the final volumes become different for both gases. Therefore, the two gases cannot be mixed together after the adiabatic processes without changing variables. This is a natural consequence because the two adiabatic processes cannot cross each other.

$$\ln P_f / P_1 = -\frac{C_P}{C_V}\ln V_{f_1}/V_1 \Rightarrow P_f / P_1 = (V_{f_1}/V_1)^{-\gamma} \Rightarrow P_f = P_1(V_{f_1}/V_1)^{-\gamma}$$

$$\ln P_f / P_2 = -\frac{C_P}{C_V}\ln V_{f_2}/V_2 \Rightarrow P_f / P_2 = (V_{f_2}/V_2)^{-\gamma} \Rightarrow P_f = P_2(V_{f_2}/V_2)^{-\gamma}$$

$$P_f = P_1(V_{f_1}/V_1)^{-\gamma} = \frac{Nk_BT}{V_1}(V_{f_1}/V_1)^{-\gamma} = P_2(V_{f_2}/V_2)^{-\gamma} = \frac{Nk_BT}{V_2}(V_{f_2}/V_2)^{-\gamma}$$

final volume V_{f_1}, V_{f_2}: $\frac{1}{V_1}(V_{f_1}/V_1)^{-\gamma} = \frac{1}{V_2}(V_{f_2}/V_2)^{-\gamma} \Rightarrow (V_{f_1}/V_{f_2}) = (V_1/V_2)^{-\frac{\gamma+1}{\gamma}} \neq 1$

final temp. T_{f_1}, T_{f_2}:

$$P_fV_{f_1} = Nk_BT_{f_1} \Rightarrow T_{f_1} = \frac{P_fV_{f_1}}{Nk_B} = T \times \frac{V_{f_1}}{V_1} \times (V_{f_1}/V_1)^{-\gamma} = T \times (V_{f_1}/V_1)^{-\gamma+1}$$

$$P_fV_{f_2} = Nk_BT_{f_2} \Rightarrow T_{f_2} = \frac{P_fV_{f_2}}{Nk_B} = T \times \frac{V_{f_2}}{V_2} \times (V_{f_2}/V_2)^{-\gamma} = T \times (V_{f_2}/V_2)^{-\gamma+1}$$

Solution 3-49a

(a) The answers with which the equations in question are completed are given below by marking them with green squares. Isenthalpic process is specified by $dH = [0]$.

$$dH = \left(\frac{\partial H}{\partial P}\right)_T dP + \left(\frac{\partial H}{\partial T}\right)_P dT \qquad (21)$$

$$\left(\frac{\partial T}{\partial P}\right)_H = -\frac{\left(\frac{\partial H}{\partial P}\right)_T}{\left(\frac{\partial H}{\partial T}\right)_P} = -\frac{\left(\frac{\partial H}{\partial P}\right)_T}{C_P} \qquad (22)$$

$$\left(\frac{\partial H}{\partial P}\right)_T = T\left(\frac{\partial S}{\partial P}\right)_T + V = T \times \left\{-\left(\frac{\partial V}{\partial T}\right)_P\right\} + V \qquad (24)$$

Ans. isenthalpic process: $dH = [0]$. Parts marked in green in eqs(21), (22) and (24).

(b) The right side of eq(24) is calculated to be zero for the ideal gas. Hence, the *Joule-Thomson* coefficient μ_{JT} is found zero for the ideal gas.

Ans. *Joule-Thomson*'s coefficient μ_{JT} : zero

Solution 3-49b

The internal energy *U* is considered as a function of *T* and *V*. Then, its total differential is expressed as eq(284).

$$dU = \left(\frac{\partial U}{\partial V}\right)_T dV + \left(\frac{\partial U}{\partial T}\right)_V dT \qquad (284)$$

Both sides of eq(284) are divided by dV under the condition of $U = \text{constant}$ to give eq(285).

$$\left(\frac{\partial T}{\partial V}\right)_U = -\frac{\left(\frac{\partial U}{\partial V}\right)_T}{\left(\frac{\partial U}{\partial T}\right)_V} = -\frac{\left(\frac{\partial U}{\partial V}\right)_T}{C_V} \qquad (285)$$

On the other hand, the total differential of *U* can be expressed by eq(286) as well. By dividing both sides of this equation by dV at a constant temperature, eq(287) is obtained.

$$dU = TdS - PdV \qquad (286)$$

$$\left(\frac{\partial U}{\partial V}\right)_T = T\left(\frac{\partial S}{\partial V}\right)_T - P = T \times \left(\frac{\partial P}{\partial T}\right)_V - P \qquad (287)$$

It is, therefore, shown from eqs(285) and (287) that *Thomson*'s coefficient μ_T is given by eq(288).

$$\mu_T \equiv \left(\frac{\partial T}{\partial V}\right)_U = -\left\{T \times \left(\frac{\partial P}{\partial T}\right)_V - P\right\} \Big/ C_V \qquad (288)$$

The value for the ideal gas is found zero from eq(288). Since the left side of eq(287) corresponds to the internal pressure, *Thomson*'s coefficient μ_T can be said as the internal pressure divided by the specific heat of C_V.

Ans. *Thomson*'s coefficient: eq(288)

Solution 3-49c

(a) *Joule-Thomson*'s coefficient is given by eq(289) (see **problem 3-49a**).

$$\mu_{JT} = -\left\{-T\left(\frac{\partial V}{\partial T}\right)_P + V\right\} \Big/ C_P \qquad (289)$$

The *van der Waals* equation of state given by eq(290) ($n = 1$ mole) is differentiated by *T* at a constant pressure to

$$\left(P + \frac{a}{V^2}\right)(V - b) = RT \qquad (290)$$

yield eq(291) and eq(291') after rearrangement.

$$-\frac{2a}{V^3}(V-b)\left(\frac{\partial V}{\partial T}\right)_P + \left(P + \frac{a}{V^2}\right)\left(\frac{\partial V}{\partial T}\right)_P = R \qquad (291)$$

$$\left(\frac{\partial V}{\partial T}\right)_P \left\{\frac{RT}{(V-b)} - \frac{2a}{V^3}(V-b)\right\} = \left(\frac{\partial V}{\partial T}\right)_P \left\{\frac{RTV^3 - 2a(V-b)^2}{(V-b)V^3}\right\} = R \qquad (291')$$

Eq(291') is solved for $(\partial V/\partial T)_P$ to obtain eq(292).

$$\left(\frac{\partial V}{\partial T}\right)_P = R(V-b)V^3 / \left\{RTV^3 - 2a(V-b)^2\right\} \qquad (292)$$

Employing eq(292), the numerator of eq(289) is found as eq(293) bel

$$-T\left(\frac{\partial V}{\partial T}\right)_P + V = -T\times\frac{R(V-b)V^3}{RTV^3-2a(V-b)^2}+V$$

$$=\frac{-RT(V-b)V^3+RTV^4-2a(V-b)^2V}{RTV^3-2a(V-b)^2}$$

$$=\frac{-RT(V-b)+RTV-\dfrac{2a}{V^2}(V-b)^2}{RT-\dfrac{2a}{V^3}(V-b)^2} \qquad (293)$$

In order to find an approximate form of eq(293), it is expanded in powers of *1/V* and $P = RT/V$ is used as an approximation, then eq(294) is obtained up to the approximation of *1/V*.

$$\text{eq}(293) \approx \frac{\left(RTb-2a+\dfrac{4ab}{V}\right)}{RT\left(1-\dfrac{2a}{RTV}\right)} = b-\frac{2a}{RT}+\frac{2ab}{RTV}+\frac{4ab}{RTV}-\frac{4a^2}{(RT)^2V}+\cdots$$

$$\approx b-\frac{2a}{RT}+\frac{6ab}{RTV} \qquad (294)$$

Consequently, *Joule-Thomson*'s coefficient for the *van der Waals* gas is given by eq(293) divided by C_P or by its approximate function of eq(295), which is plotted in Fig.36 against *V* at various temperatures.

$$\mu_{JT}\times C_P = -\left\langle V - RT\Big/\left\{\frac{RT}{(V-b)}+\frac{2a}{V^3}(V-b)\right\}\right\rangle \approx -\left(b-\frac{2a}{RT}+\frac{6ab}{RTV}\right) \qquad (295)$$

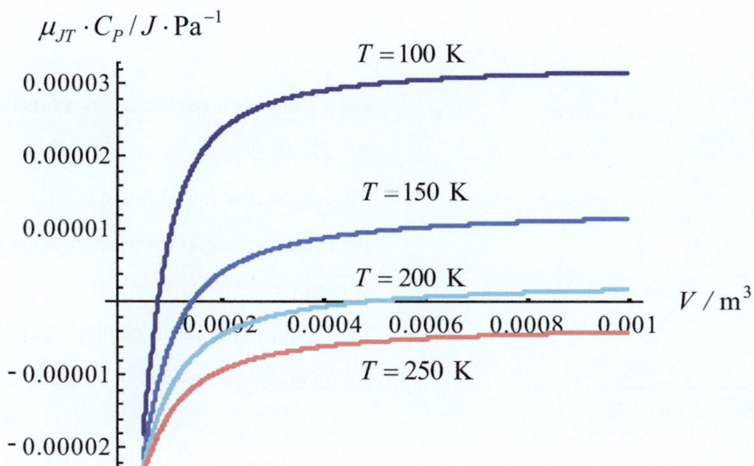

Fig.36 Volume dependence of μ_{JT} for the *van der Waals* gas at several temperatures. Following values are employed for the parameters *a* and *b* and the approximate function (eq(295)) is plotted.
$a = 24.7\times10^{-3}\,\text{m}^6\cdot\text{Pa}\cdot\text{mol}^{-2}$,
$b = 26.6\times10^{-6}\,\text{m}^3\cdot\text{mol}^{-1}$.

Ans. $\mu_{JT}\times C_P = -\left\langle V - RT\Big/\left\{\frac{RT}{(V-b)}+\frac{2a}{V^3}(V-b)\right\}\right\rangle \approx -\left(b-\frac{2a}{RT}+\frac{6ab}{RTV}\right)$

(b) To lower the temperature of a gas by *Joule-Thomson*'s process, *Joule-Thomson*'s coefficient μ_{JT} must be positive. The temperature at which the coefficient becomes zero is obtained from eq(292) via eq(296), which is solved for *T* giving eq(297) as a function of *V*.

$$RT\left(\frac{V}{(V-b)}-1\right)-\frac{2a}{V^2}(V-b)=0 \qquad (296)$$

$$T=\frac{1}{R}\cdot\frac{\frac{2a}{V^2}(V-b)}{\frac{V}{(V-b)}-1}=\frac{2a}{bR}\frac{(V-b)^2}{V^2} \qquad (297)$$

From eqs(290) and (297), the pressure of the gas can be expressed by eq(298) as a function of V.

$$P=\frac{RT}{(V-b)}-\frac{a}{V^2}$$
$$=\frac{1}{(V-b)}\times\frac{2a}{b}\frac{(V-b)^2}{V^2}-\frac{a}{V^2}=\frac{2a}{b}\frac{(V-b)}{V^2}-\frac{a}{V^2} \qquad (298)$$

Employing the given values for the *van der Waals* parameters a and b of hydrogen gas, (T, P) are calculated from eq(297) and eq(298), respectively, as a function of V. (T, P) thus expressed in terms of V as a parameter are plotted in Fig.36 giving the curve of $\mu_{JT}=0$. From this curve, it is found that when the initial pressure of the hydrogen gas is 2 Mpa, the temperature of the gas must be lower than $T=210.3\,\mathrm{K}$. A value for the coefficient depends significantly on T and P. The region where $\mu_{JT}>0$ or $\mu_{JT}<0$ can readily be known by calculating a value for μ_{JT} at an arbitrary chosen point of (T, P) (e.g., red or black point in Fig.36). At $P=1$ atm and $T=298$ K, the coefficient is calculated as $\mu_{JT}=-0.023\ \mathrm{K\cdot atm^{-1}}$ and the reference value is $\mu_{JT}=-0.03\ \mathrm{K\cdot atm^{-1}}$.

Under a constant pressure, there exists a temperature at which a value of μ_{JT} varies from positive to negative (vice versa) as is indicated by a red dotted horizontal line in Fig.37. That temperature is called inversion temperature.

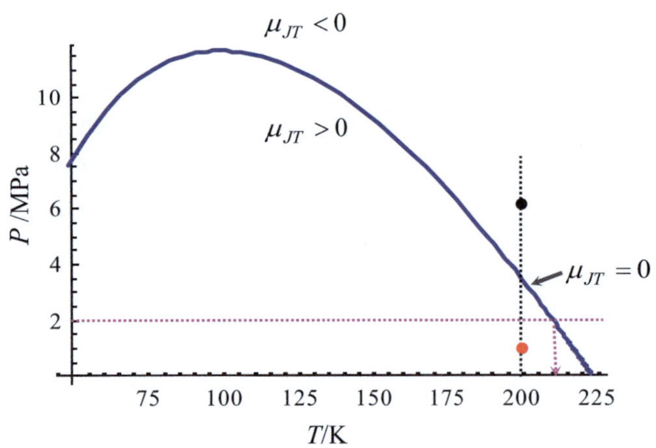

Fig.37 *Joule-Thomson*'s coefficient as a function of (T, P).
The higher the initial pressure of a gas, the lower the initial temperature must be chosen.
black point: $\mu_{JT}=-0.0079\ \mathrm{K\cdot atm^{-1}}$,
red point: $\mu_{JT}=0.0077\ \mathrm{K\cdot atm^{-1}}$.

Ans. below $T=210.3\,\mathrm{K}$

"note 1" Liquefiability of gases.

When the temperature of a gas is lower than its critical temperature, the gas can be liquefied by compressing the gas. Good examples for this are carbon dioxide and ammonium gases. If the critical temperature is quite low, then it is impossible to liquefy the gas simply by compressing it. In old days when this fact was not known, the gases which could not be liquefied were called permanent gases. Hydrogen or helium gas was among others. Nowadays, however, any kind of gases can be liquefied.

Usually, there are three ways of liquefying gases: (i) to employ *Joule-Thomson*'s process (e.g. *Linde* refrigerator), (ii) to expand gases adiabatically, and (iii) to combine the two methods of (i) and (ii). It is described below briefly how liquid hydrogen which is regarded as an important energy source for the future is produced.

"note 2" Production of liquid hydrogen.

Hydrogen gas which does not produce CO_2 upon combustion is attracting a great concern as an important combustible gas. In particular, a significant advance in the technology to obtain hydrogen gas from water by using solar energy is now strongly hoped to develop further in terms of reducing the environmental burden.

At present the photoelectrolysis of water and thermolysis of water heated by focused solar light by means of optical tools are considered promising. The resulting hydrogen gas must be liquefied for storage and transportation. A technological advance for this, together with the development of fuel cells, is an important issue in the quest for a decarbonized society based on renewable energy.

Solution 3-50

Since the derivation of each equation is given in the problem, only the answers required are given here.

Ans. $(a)\ \left(\frac{dP}{dT}\right)_e,\ (b)\ \left(\frac{\partial V}{\partial T}\right)_P,\ (c)\ \Delta V_e,\ (d)\ \Delta C_P$

Solution 3-51

(a) In a similar manner as employed in the case of *Joule-Thomson*'s coefficient, we start with obtaining the total differential of the thermodynamic variable which is to be kept constant. In the present case, it is the entropy S and is given by eq(299).

$$dS = \left(\frac{\partial S}{\partial P}\right)_T dP + \left(\frac{\partial S}{\partial T}\right)_P dT \Rightarrow \left(\frac{\partial S}{\partial P}\right)_T \left(\frac{\partial P}{\partial T}\right)_S + \left(\frac{\partial S}{\partial T}\right)_P = 0 \qquad (299)$$

Rewriting eq(299) with respect to $(\partial P/\partial T)_S$ and employing *Maxwell*'s relation, the second equation of eq(300) is obtained.

$$\left(\frac{\partial P}{\partial T}\right)_S = -\frac{\left(\frac{\partial S}{\partial T}\right)_P}{\left(\frac{\partial S}{\partial P}\right)_T} = -\frac{\frac{1}{T}\left(\frac{\partial H}{\partial T}\right)_P}{-\left(\frac{\partial V}{\partial T}\right)_P} = \frac{\frac{C_P}{T}}{\left(\frac{\partial V}{\partial T}\right)_P} = \frac{C_P}{T\left(\frac{\partial V}{\partial T}\right)_P} = \left\langle\frac{C_P}{V}\right\rangle \qquad (300)$$

Furthermore, using the relation given by eq(301), the third equation of eq(300) is derived. And <u>the last equality is satisfied for the ideal gas</u>.

$$\begin{aligned} H &= U + PV \\ dH &= TdS + VdP \end{aligned} \Rightarrow \left(\frac{\partial S}{\partial T}\right)_P = \frac{1}{T}\left(\frac{\partial H}{\partial T}\right)_P = \frac{C_P}{T} \qquad (301)$$

Ans. $\left(\frac{\partial P}{\partial T}\right)_S = \left\langle\frac{C_P}{V}\right\rangle$

(b) Starting from eq(299) , it is divided by dV under the condition of $dS = 0$ to derive eq(302).

$$dS = \left(\frac{\partial S}{\partial P}\right)_T dP + \left(\frac{\partial S}{\partial T}\right)_P dT \Rightarrow \left(\frac{\partial S}{\partial P}\right)_T \left(\frac{\partial P}{\partial V}\right)_S + \left(\frac{\partial S}{\partial T}\right)_P \left(\frac{\partial T}{\partial V}\right)_S = 0 \qquad (302)$$

Solving eq(302) for $(\partial P/\partial V)_S$ with some rearrangements, eq(303) is obtained.

$$\left(\frac{\partial P}{\partial V}\right)_S = -\frac{\left(\frac{\partial S}{\partial T}\right)_P\left(\frac{\partial T}{\partial V}\right)_S}{\left(\frac{\partial S}{\partial P}\right)_T} = \frac{\frac{1}{T}\left(\frac{\partial H}{\partial T}\right)_P\left(\frac{\partial T}{\partial V}\right)_S}{\left(\frac{\partial V}{\partial T}\right)_P} = -\frac{\frac{C_P}{T}\left(\frac{\partial P}{\partial T}\right)_V}{\left(\frac{\partial V}{\partial T}\right)_P\frac{C_V}{T}} = -\frac{C_P\left(\frac{\partial P}{\partial T}\right)_V}{C_V\left(\frac{\partial V}{\partial T}\right)_P} = \frac{C_P}{C_V}\left(\frac{\partial P}{\partial V}\right)_T \quad (303)$$

Here, the relation of eq(304) is employed.

$$dS = \left(\frac{\partial S}{\partial V}\right)_T dV + \left(\frac{\partial S}{\partial T}\right)_V dT \;\Rightarrow\; \left(\frac{\partial S}{\partial V}\right)_T + \left(\frac{\partial S}{\partial T}\right)_V\left(\frac{\partial T}{\partial V}\right)_S = 0$$

$$\left(\frac{\partial T}{\partial V}\right)_S = -\frac{\left(\frac{\partial S}{\partial V}\right)_T}{\left(\frac{\partial S}{\partial T}\right)_V} = -\frac{\left(\frac{\partial P}{\partial T}\right)_V}{\frac{C_V}{T}} \quad (304)$$

Ans. $\left(\frac{\partial P}{\partial V}\right)_S = \frac{C_P}{C_V}\left(\frac{\partial P}{\partial V}\right)_T, \quad \left(\frac{\partial T}{\partial V}\right)_S = -T\left(\frac{\partial P}{\partial T}\right)_V \Big/ C_V$

"note 1" If you use *Jacobians* (see **solution 3-35**), eq(304) can easily be derived as is shown by eq(304').

$$\left(\frac{\partial T}{\partial V}\right)_S = \frac{\partial(T,S)}{\partial(V,S)} = \frac{\frac{\partial(T,S)}{\partial(T,V)}}{\frac{\partial(V,S)}{\partial(T,V)}} = \frac{\left(\frac{\partial S}{\partial V}\right)_T}{-\left(\frac{\partial S}{\partial T}\right)_V} = -\frac{\left(\frac{\partial P}{\partial T}\right)_V}{\frac{C_V}{T}} \quad (304')$$

In a similar manner, eq(303) can also be derived, but some manipulation is required as is seen below.

$$\left(\frac{\partial P}{\partial V}\right)_S = \frac{\partial(P,S)}{\partial(V,S)} = \frac{\frac{\partial(P,S)}{\partial(T,V)}}{\frac{\partial(V,S)}{\partial(T,V)}} = \frac{\begin{vmatrix} \left(\frac{\partial P}{\partial T}\right)_V & \left(\frac{\partial S}{\partial T}\right)_V \\ \left(\frac{\partial P}{\partial V}\right)_T & \left(\frac{\partial S}{\partial V}\right)_T \end{vmatrix}}{-\left(\frac{\partial S}{\partial T}\right)_V} = -\frac{\begin{vmatrix} \left(\frac{\partial P}{\partial T}\right)_V & \frac{C_V}{T} \\ \left(\frac{\partial P}{\partial V}\right)_T & \left(\frac{\partial P}{\partial T}\right)_V \end{vmatrix}}{\frac{C_V}{T}}$$

$$= -\frac{\left(\frac{\partial P}{\partial T}\right)_V^2 - \left(\frac{\partial P}{\partial V}\right)_T\frac{C_V}{T}}{\frac{C_V}{T}} = -\frac{\left(\frac{\partial P}{\partial T}\right)_V^2 - \left(\frac{\partial P}{\partial V}\right)_T\left\{\frac{C_P}{T} - \left(\frac{\partial P}{\partial T}\right)_V\left(\frac{\partial V}{\partial T}\right)_P\right\}}{\frac{C_V}{T}}$$

$$= -\frac{\left(\frac{\partial P}{\partial T}\right)_V^2 - \left(\frac{\partial P}{\partial V}\right)_T\frac{C_P}{T} + \left(\frac{\partial P}{\partial T}\right)_V\left(\frac{\partial V}{\partial T}\right)_P\left(\frac{\partial P}{\partial V}\right)_T}{\frac{C_V}{T}} = -\frac{\left(\frac{\partial P}{\partial T}\right)_V^2 - \left(\frac{\partial P}{\partial V}\right)_T\frac{C_P}{T} - \left(\frac{\partial P}{\partial T}\right)_V^2}{\frac{C_V}{T}}$$

$$= \frac{C_P}{C_V}\left(\frac{\partial P}{\partial V}\right)_T \quad (303')$$

It is also confirmed, as is shown below, that the first equality of eq(303) is correct.

$$\left(\frac{\partial P}{\partial V}\right)_S = -\frac{\left(\frac{\partial S}{\partial T}\right)_P\left(\frac{\partial T}{\partial V}\right)_S}{\left(\frac{\partial S}{\partial P}\right)_T} = -\frac{\frac{\partial(S,P)}{\partial(T,P)}\frac{\partial(T,S)}{\partial(V,S)}}{\frac{\partial(S,T)}{\partial(P,T)}} = \frac{\partial(P,S)}{\partial(V,S)} \quad (303'')$$

"note 2" As to the *P-V* or *T-V* relation of the ideal gas in adiabatic reversible processes and problems related to it, see **problems 3-42a, 3-42b and 3-48**.

Solution 3-52a

The chemical potential of a pure solvent A at a constant temperature under the pressure P is denoted $\mu_A^l(P)$. When the solution is regarded as ideal, the chemical potential of A in the solution under the increased pressure of P' is given by $\mu_A^l(P') + RT\ln X_A$ where X_A is the mole fraction of A. These chemical potentials of A become equal when the equilibrium (stationary state) is established, leading to eq(305). It should be noted here that the chemical potential is a function of pressure.

$$\mu_A^l(P) = \mu_A^l(P') + RT\ln X_A \Rightarrow -\langle \mu_A^l(P') - \mu_A^l(P) \rangle = RT\ln X_A \qquad (305)$$

What eq(305) conveys to us is that a decrease in the chemical potential of A brought about by dissolving B in A is compensated by a corresponding increase in the chemical potential caused by rasing the pressure.

The term on the left side of the rearranged eq(305) is rewritten as eq(306) based on the pressure dependence of the chemical potential (or free energy). $\overline{V}_A$ is the partial molar volume of A at that concentration (X_A).

$$-\langle \mu_A^l(P') - \mu_A^l(P) \rangle = -\left(\frac{\partial \mu_A^l(P)}{\partial P}\right)_T dP = -\overline{V}_A dP, \qquad dP = P' - P \qquad (306)$$

The pressure difference dP generated thus is the osmotic pressure Π and it is given by eq(308) via eq(307).

$$P' - P = \Pi$$

$$\Pi \overline{V}_A = -RT\ln X_A = -RT\ln(1 - X_B) \approx RTX_B \qquad (307)$$

$$\Pi = RT\cdot\frac{X_B}{\overline{V}_A} = RT\cdot\frac{n_B}{n_A + n_B}\cdot\frac{1}{\overline{V}_A} \approx RT\cdot\frac{n_B}{n_A\overline{V}_A} \approx RT\cdot\frac{n_B}{V} = RTc_B \qquad (308)$$

The last approximation of eq(308) is derived from the volume of the solution V expressed in terms of the partial molar volumes of A and B as below. The molar concentration of B is expressed as c_B.

$$V = n_A\overline{V}_A + n_B\overline{V}_B \approx n_A\overline{V}_A \quad (n_A >> n_B)$$

Eq(308) is called *van't Hoff*'s equation (*J.H. van't Hoff*, the first *Nobel* Prize in Chemistry in 1901) and the equation is seen to assume the same functional form as the equation of state of the ideal gas. See **problem 3-68** for another *van't Hoff*'s equation.

Ans. $\Pi = RTc_B$

Solution 3-52b

The solution is considered as an ideal one. Then, both of eqs(305) and (306) given in the preceding soslution are satisfied as they are with X_A as the mole fraction of A even when more than two species coexist in the solution. Consequently, the osmotic pressure Π is given in terms of X_A and the partial molar volume $\overline{V}_A$ by eq(309). Under the condition of $X_A \approx 1$, eq(309) can be approximated by eq(309').

$$P' - P = \Pi$$

$$\Pi \overline{V}_A = -RT\ln X_A \qquad (309)$$

$$= -RT\ln(1 - X_B - X_C) \approx RT(X_B + X_C) \qquad (310)$$

By expressing eq(310) in terms of the molar concentrations of the solutes, eq(310') is obtained. Therefore, it is found that the osmotic pressure is additive with respect to the molar concentrations of solutes unless they are too high.

$$\Pi = RT\cdot\frac{X_{\mathrm{B}}+X_{\mathrm{C}}}{\overline{V}_{\mathrm{A}}} = RT\cdot\frac{n_{\mathrm{B}}+n_{\mathrm{C}}}{n_{\mathrm{A}}+n_{\mathrm{B}}+n_{\mathrm{C}}}\cdot\frac{1}{\overline{V}_{\mathrm{A}}}$$

$$\approx RT\cdot\frac{n_{\mathrm{B}}+n_{\mathrm{C}}}{n_{\mathrm{A}}\overline{V}_{A}} \approx RT\cdot\left(\frac{n_{\mathrm{B}}}{V}+\frac{n_{\mathrm{C}}}{V}\right) = RT(c_{\mathrm{B}}+c_{\mathrm{C}}) \qquad (310')$$

It is easy to generalize eq(310) to the case where n kinds of solute coexist.

Ans. See text.

Solution 3-53a

The pressure is kept constant at 1 atm, so that the chemical potentials of A in the gas phase (g) and liquid phase (l) satisfy eq(311) at the boiling temperature T_b.

$$\mu_{\mathrm{A}}^{\bullet g}(T_b) = \mu_{\mathrm{A}}^{\bullet l}(T_b) \qquad (311)$$

Here, the superscript $^{\bullet}$ is put in order to emphasize that the chemical potentials are those for the pure substances. By dissolving B in A, the chemical potential of A in the solution changes accompanied by a change in the boiling temperature. This relation is expressed by eq(312).

$$\mu_{\mathrm{A}}^{\bullet g}(T_b+\Delta T) = \mu_{\mathrm{A}}^{\bullet l}(T_b+\Delta T) + R(T_b+\Delta T)\ln X_{\mathrm{A}}^{l} \qquad (312)$$

Assuming that only a small amount of substance B is dissolved in A ($X_{\mathrm{B}}^{l} << 1$), the chemical potentials on both sides of eq(312) are expanded in terms of ΔT to the first power to obtain eqs(313) and (314).

$$\mu_{\mathrm{A}}^{\bullet g}(T_b+\Delta T) = \mu_{\mathrm{A}}^{\bullet g}(T_b) + \left(\frac{\partial \mu_{\mathrm{A}}^{\bullet g}}{\partial T}\right)_{P,T=T_b}\times \Delta T \qquad (313)$$

$$\mu_{\mathrm{A}}^{\bullet l}(T_b+\Delta T) = \mu_{\mathrm{A}}^{\bullet l}(T_b) + \left(\frac{\partial \mu_{\mathrm{A}}^{\bullet l}}{\partial T}\right)_{P,T=T_b}\times \Delta T \qquad (314)$$

Also, the second term of eq(312) can be approximated as in eq(315).

$$R(T_b+\Delta T)\ln X_{\mathrm{A}}^{l} = R(T_b+\Delta T)\ln(1-X_{\mathrm{B}}^{l}) \approx -RT_bX_{\mathrm{B}}^{l} \qquad (315)$$

Denoting the latent heat of vaporization of A at T_b by ΔH_{vap}, therefore, eq(312) can be approximated by eq(316) and eq(317).

$$\left(\frac{\partial \mu_{\mathrm{A}}^{\bullet g}}{\partial T}\right)_{P,T=T_b}\times \Delta T - \left(\frac{\partial \mu_{\mathrm{A}}^{\bullet l}}{\partial T}\right)_{P,T=T_b}\times \Delta T = \left(\frac{\partial(\mu_{\mathrm{A}}^{\bullet g}-\mu_{\mathrm{A}}^{\bullet l})}{\partial T}\right)_{P,T=T_b}\times \Delta T = -RT_bX_{\mathrm{B}}^{l} \qquad (316)$$

$$\left(\frac{\partial(\mu_{\mathrm{A}}^{\bullet g}-\mu_{\mathrm{A}}^{\bullet l})}{\partial T}\right)_{P,T=T_b}\times \Delta T = \left(\frac{-S_{\mathrm{A}}^{\bullet g}+S_{\mathrm{A}}^{\bullet l}}{T_b}\right)\times \Delta T = -\frac{\Delta H_{vap}(T_b)}{T_b}\times \Delta T \qquad (317)$$

Putting eq(316) and eq(317) together, eq(318) is obtained.

$$\frac{\Delta H_{vap}}{T_b}\Delta T = RT_bX_{\mathrm{B}}^{l} \Rightarrow \Delta T = \frac{RT_b^{\,2}}{\Delta H_{vap}}\cdot X_{\mathrm{B}}^{l} \qquad (318)$$

Since ΔH_{vap} is positive, ΔT is also positive, resulting in a rise in the boiling temperature. It is seen from eq(312) that a change in the chemical potential of A caused by dissolving substance B in A is compensated by varying the boiling temperature.

Ans. $\Delta T = \dfrac{RT_b^{\,2}}{\Delta H_{vap}}\cdot X_{\mathrm{B}}^{l}$

Solution 3-53b

Since the conclusion to be drawn will vary, when both substances are volatile, depending on what conditions are given, we will consider the limiting case in which the condition of $X_{\mathrm{A}} >> X_{\mathrm{B}}$ is satisfied.

Although the chemical potential of A in solution is not affected in principle by whether B is volatile or not, the mole fraction of B in the solution decreases by the amount of B brought into the gas phase by vaporization. When the chemical potential of A in the gas phase at the total pressure of 1 atm is approximated as equal to that of the pure gas A, the boiling point will elevate but its extent is smaller as compared with the case in which B is non-volatile as far as the same amount of B is used.

When the boiling point is lower than that of A, B hardly exists in the solution near the boiling point of A resulting in almost no change in the boiling point of A, as is seen for the boiling point of water with gases dissolved.

Of course the discussions above are effective as crude approximations. More accurate relations will be derived by examining exactly the balance of the chemical potentials of A in the solution and gas phases.

Ans. Boiling point will rise but its extent is smaller as compared with non-volatile B.

Solution 3-54

Just in the same manner as is given in the solution to **problem 3-53a**, the equation to give the depression of the freezing point can be derived as follows starting from eq(319) which holds for pure A at the freezing point T_m. The superscripts s and l stand for solid and liquid, respectively.

$$\mu_{\mathrm{A}}^{\bullet s}(T_m) = \mu_{\mathrm{A}}^{\bullet l}(T_m) \qquad (319)$$

When a small amount of substance B is dissolved in solvent A, the freezing point of the solution changes and the chemical potential of A varies to satisfy eq(320).

$$\mu_{\mathrm{A}}^{\bullet s}(T_m + \Delta T) = \mu_{\mathrm{A}}^{\bullet l}(T_m + \Delta T) + R(T_m + \Delta T)\ln X_{\mathrm{A}}^{l} \qquad (320)$$

Assuming the amount of B is small enough ($X_{\mathrm{B}}^{l} << 1$), both sides of eq(320) are expanded to the first-order term of ΔT, obtaining eqs (321) to (323).

$$\mu_{\mathrm{A}}^{\bullet s}(T_m + \Delta T) = \mu_{\mathrm{A}}^{\bullet s}(T_m) + \left(\frac{\partial \mu_{\mathrm{A}}^{\bullet s}}{\partial T}\right)_{P,T=T_m} \times \Delta T \qquad (321)$$

$$\mu_{\mathrm{A}}^{\bullet l}(T_m + \Delta T) = \mu_{\mathrm{A}}^{\bullet l}(T_m) + \left(\frac{\partial \mu_{\mathrm{A}}^{\bullet l}}{\partial T}\right)_{P,T=T_m} \times \Delta T \qquad (322)$$

$$R(T_m + \Delta T)\ln X_{\mathrm{A}}^{l} = R(T_m + \Delta T)\ln(1 - X_{\mathrm{B}}^{l}) \approx -RT_m X_{\mathrm{B}}^{l} \qquad (323)$$

Employing these relations, eq(320) can be rewritten as eqs(324) and (325) in which $\Delta H_m = T(S_{\mathrm{A}}^{\bullet l} - S_{\mathrm{A}}^{\bullet s})$ is the latent heat of melting.

$$\left(\frac{\partial \mu_{\mathrm{A}}^{\bullet s}}{\partial T}\right)_{P,T=T_m} \times \Delta T - \left(\frac{\partial \mu_{\mathrm{A}}^{\bullet l}}{\partial T}\right)_{P,T=T_m} \times \Delta T = \left(\frac{\partial (\mu_{\mathrm{A}}^{\bullet s} - \mu_{\mathrm{A}}^{\bullet l})}{\partial T}\right)_{P,T=T_m} \times \Delta T = -RT_m X_{\mathrm{B}}^{l} \qquad (324)$$

$$\left(\frac{\partial (\mu_{\mathrm{A}}^{\bullet s} - \mu_{\mathrm{A}}^{\bullet l})}{\partial T}\right)_{P,T=T_m} \times \Delta T = \left(\frac{-S_{\mathrm{A}}^{\bullet s} + S_{\mathrm{A}}^{\bullet l}}{T_m}\right) \times \Delta T = \frac{\Delta H_m(T_m)}{T_m} \times \Delta T \qquad (325) \quad \text{("note")}$$

Putting eqs(324) and (325) together, eq(326) is obtained.

$$-\frac{\Delta H_m}{T_m}\Delta T = RT_m X_B^l \;\Rightarrow\; \Delta T = -\frac{RT_m^{\ 2}}{\Delta H_m}\cdot X_B^l \qquad (326)$$

The obtained eq(326) is the equation to give the depression of the freezing point (the case of $\Delta T < 0$). The constant for the freezing point depression (cryoscopic constant) is given by $RT_m^{\ 2}/\Delta H_m$.

If you notice that the chemical potential is a function of T and P, then it is readily understood that all of the **problems 3-52 to 3-54** can be answered from the entirely same point of view. Therefore, those properties treated in the problems are called *colligative properties,* which mean that those properties solely depend on the number of mole of a substance but not on its any specific property. For their additivity see **problem 3-52b**.

"note" A positive or negative sign of the latent heat depends on which direction of the equilibrium of gas ⇔ liquid, liquid ⇔ solid, etc., is considered.

Ans. $\Delta T = -\frac{RT_m^{\ 2}}{\Delta H_m}\cdot X_B^l$

Solution 3-55

The first law of thermodynamics is expressed in terms of thermodynamic variables as eq(327). Both sides

$$dU = TdS - PdV \qquad (327)$$

of eq(327) are divided by dV at a constant temperature ($dT = 0$) to obtain eq(328) and eq(329) is derived from the *van der Waals* equation of state.

$$\left(\frac{\partial U}{\partial V}\right)_T = T\left(\frac{\partial S}{\partial V}\right)_T - P = T\left(\frac{\partial P}{\partial T}\right)_V - P \qquad (328)$$

$$\left(\frac{\partial P}{\partial T}\right)_V = \left(\frac{\partial}{\partial T}\frac{RT}{(V-b)} - \frac{a}{V^2}\right)_V = \frac{R}{(V-b)} \qquad (329)$$

Substituting eq(329) into eq(328), eq(330) is derived as the internal pressure.

$$\left(\frac{\partial U}{\partial V}\right)_T = \frac{RT}{(V-b)} - \frac{RT}{(V-b)} + \frac{a}{V^2} = \frac{a}{V^2} \qquad (330)$$

From this result, it is readily seen that the internal pressure $(\partial U/\partial T)_V$ arises from the intermolecular interaction whose magnitude is represented by a. When $a < 0$, that is in the case of repulsive interaction, the gas tself has pressure to expand (internal pressure). On the other hand, when $a > 0$, that is in the case of attractive interaction, the gas tends to execute a force on its own to condense. It should be noted that the volume excluded, i.e., the intrinsic size of a molecule, does not cause internal pressure. It is readily confirmed from eq(328) that for the ideal gas where no intermolecular force is operative, the internal pressure is zero. In contrast to the internal pressure the term $T\left(\frac{\partial P}{\partial T}\right)_V$ in eq(328) is called thermal pressure.

Ans. $\left(\frac{\partial U}{\partial V}\right)_T = \frac{a}{V^2}$, see also text.

Also, the internal pressure of the gas which is described by the *Dieterici* or virial equation of state is given by a relevant function shown in the answer below. In either case, unlike the *van der Waals* gas, the internal pressure depends on temperature.

Ans. internal pressure for *Dieterici* equation of state: $\dfrac{a\times e^{-\frac{a}{RTV}}}{V(V-b)} \approx \dfrac{a\times e^{-\frac{a}{RTV}}}{V^2}$,

for virial equation of state: $\approx \dfrac{RT^2B'(T)}{V^2}$

Solution 3-56a

The relation among three partial differentials in which the three variables change their parts cyclically is easily verified by means of *Euler*'s chain relation as is shown below in eq(331).

$$\mu_{JT}\times C_P=\left(\frac{\partial T}{\partial P}\right)_H\left(\frac{\partial H}{\partial T}\right)_P=\frac{\left(\frac{\partial T}{\partial P}\right)_H\left(\frac{\partial H}{\partial T}\right)_P\left(\frac{\partial P}{\partial H}\right)_T}{\left(\frac{\partial P}{\partial H}\right)_T}=-\frac{1}{\left(\frac{\partial P}{\partial H}\right)_T}=-\left(\frac{\partial H}{\partial P}\right)_T=-\mu_T^T \qquad (331)$$

Ans. See text.

Solution 3-56b

The *Joule-Thomson* coefficient μ_{JT} is defined by the first part of eq(332). Assuming that the value given for N_2 is constant in the range of T and P in question, eq(332) is integrated over the pressure range [1.0, 0.2] to find a change in the temperature as calculated below. (Notice the unit of the coefficient.)

$$\mu_{JT}=\left(\frac{\partial T}{\partial P}\right)_H=0.26 \qquad (332)$$

$$\Delta T=\mu_{JT}\int_{1.0}^{0.2}dP=0.26\times(0.2-1.0)=-0.208\ \text{K}$$

Ans. a decrease in temperature by the *Joule-Thomson* process: 0.21 K

On the other hand, P, V and T in the adiabatic process are related each other by the following equations. Taking a value of $\gamma=1.4$ for N_2, the volume of the expanded gas is calculated first based on the first relation below as 3.16L when the initial volume is 1L. Then the temperature reached is calculated using the volume thus obtained and the second relation as 172 K.

$$PV^\gamma=\text{Const.} \qquad (1(\text{L}))^{1.4}\times 1(\text{atm})=\text{Const.}=0.2\times V^{1.4} \Rightarrow V=3.16(\text{L})$$

$$TV^{\gamma-1}=\text{Const.'} \qquad 273\times(1(\text{L}))^{0.4}=\text{Const.'}=T\times V^{0.4} \Rightarrow T=172\ \text{K}$$

Ans. temperature reached by adiabatic process: 172 K

Solution 3-57

The chemical potential of a gas is given in a simple form by using a partition function *p.f.* as q(333). The translational partition function of the gas is expressed by eq(334) (see **II-2-2**).

$$G^{\text{m}}\equiv G/n=\mu=-N_AkT\left(\ln\frac{p.f.}{N_A}+1\right)+RT \qquad (333)$$

$$p.f.=\left(\frac{2\pi mkT}{h^3}\right)^{\frac{3}{2}}V \qquad (334)$$

Here, Ar is taken as an example for the calculation to show the relation in question.

The values necessary for the calculation are as follows:

$$h=6.63\times10^{-34}\,\text{J}\cdot\text{s},\quad k_B=1.38\times10^{-23}\,\text{J}\cdot\text{K}^{-1},\quad m=20.2\times10^{-3}/N_A\text{kg},$$

$$N_A=6.02\times10^{23}\,\text{mol}^{-1},\quad R=8.31\ \text{J}\cdot\text{K}^{-1}\cdot\text{mol}^{-1},\quad V=22.4\times10^{-3}\text{m}^3$$

Employing these values, μ evaluated by eq(333) is plotted against T and shown by a green line in Fig.38 below. As is seen from the figure, a good linear relation between μ and T is recognized over rather a wide range of temperatures. Although the chemical potentials of a solid or liquid can theoretically be calculated based on the relevant partition functions, their dependences on T can be represented schematically as the black and blue lines in the figure below, respectively,

On the other hand, The chemical potential of a solution of substance A becomes smaller by the amount of $RT\ \ln X_A$ as compared with that of a pure solvent if the solution is ideal and it is shown by a black broken line in the figure. Then when we consider the equilibrium between the solid and the solution or the solution and the gas, the intersection point between the black line and the black broken line or the one between the green line and the black broken line, both of which are indicated by red dots, gives the equilibrium temperature. By comparing the positions of these dots with the equilibrium temperatures between two pure substances indicated by open circles, it is easy to see how and why the freezing point depression (see **problem 3-54**) or the boiling point elevation (see **problem 3-53**) occurs. Except for Ar gas, the chemical potentials of the solid and liquid are shown only schematically in order to understand the concept of equilibrium from the viewpoint of a crossing of the chemical potentials.

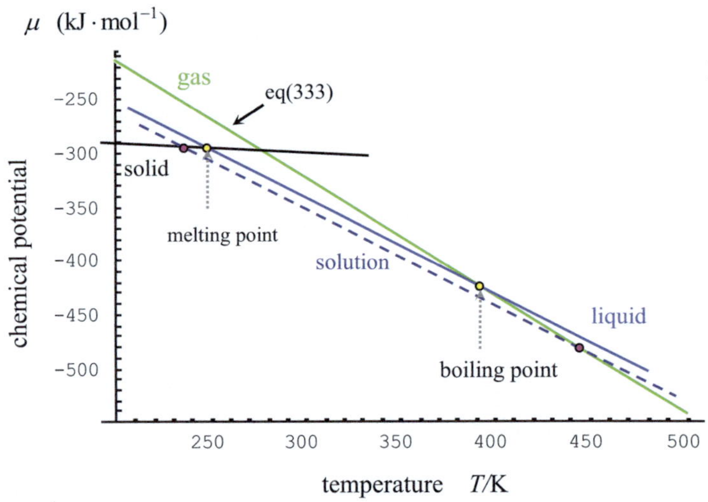

Fig.38 Temperature dependence of chemical potential. The chemical potential of the gas (Ar) shown by a green line is calculated theoretically. Those for the three condensed phases are schematically shown by three separate lines. Open circles indicate the boiling point or freezing point of a pure substance. Two red dots indicate either the freezing point depression or the boiling point elevation.

Ans. See text and Fig.38.

Solution 3-58

(a) The *Gibbs* free energy of the ideal solution composed of n_A moles of the pure substance A and n_B moles of the pure substance B is denoted by G_{AB}, which is given by eq(335).

$$G_{AB} = n_A \mu_A + n_B \mu_B \qquad (335)$$

Therefore, the free energy of mixing ΔG_{mix} (free energy change upon mixing) can be expressed by eq(336).

$$\Delta G_{mix} = G_{AB} - n_A \mu_A^\circ - n_B \mu_B^\circ \qquad (336)$$

Employing the expressions for the chemical potentials of A and B in the solution, which are given by $\mu_A = \mu_A^\circ + RT\ \ln X_A$ and $\mu_B = \mu_B^\circ + RT\ \ln X_B$, respectively, ΔG_{mix} is readily found from eq(336) as eq(337).

$$\Delta G_{mix} = n_A RT\ \ln X_A + n_B RT\ \ln X_B \qquad (337)$$

Ans. $\Delta G_{mix} = n_A RT\ \ln X_A + n_B RT\ \ln X_B$

(b) The relation $d\Delta G = -\Delta S dT + \Delta V dP$, which is directly derived from the difference of the differentials of $dG = -SdT + VdP$ at two states by noticing $\Delta G \equiv G_2 - G_1$, is used to derive eq(338) from eq(337).

$$\Delta S_{mix} = -\left(\frac{\partial \Delta G_{mix}}{\partial T}\right)_P = -(n_A \ln X_A + n_B \ln X_B) \qquad (338)$$

Writing the total mole number as n, eq(338) can be rewritten as eq(339), which is the expression to give the entropy of mixing we are seeking for .

$$\Delta S_{mix} = -nR\left(\frac{n_A}{n}\ln X_A + \frac{n_B}{n}\ln X_B\right) = -nR(X_A \ln X_A + X_B \ln X_B) \qquad (339)$$

Ans. $\Delta S_{mix} = -nR(X_A \ln X_A + X_B \ln X_B)$

(c) Since eq(337) is independent of pressure P, $\Delta V_{mix} = 0$ is derived from $\Delta V_{mix} = (\partial \Delta G_{mix}/\partial P)_T$. This is obvious because there is no volume change by mixing for the ideal solution.

Ans. $\Delta V_{mix} = 0$

Solution 3-59

(a) Writing the *Gibbs* free energy of the non-ideal solution as G_{AB}, it is given by eq(340) just the same as in the preceding problem. Therefore, the free energy of mixing is also given by eq(341).

$$G_{AB} = n_A \mu_A + n_B \mu_B \qquad (340) \qquad \Delta G_{mix} = G_{AB} - n_A \mu_A^\circ - n_B \mu_B^\circ \qquad (341) \quad \text{(reposted)}$$

From the expressions for chemical potentials of $\mu_A = \mu_A^\circ + RT \ln a_A$ and $\mu_B = \mu_B^\circ + RT \ln a_B$ together with eqs(340) and (341), ΔG_{mix} is readily found to be written as eq(342).

$$\Delta G_{mix} = n_A RT \ln a_A + n_B RT \ln a_B \qquad (342)$$

Ans. eq(342): $\Delta G_{mix} = n_A RT \ln a_A + n_B RT \ln a_B$

(b)When the relation $d\Delta G = -\Delta S dT + \Delta V dP$ derived from $dG = -SdT + VdP$ is employed, eq(343) is readily derived from eq(342). Generally, the activity coefficient is a function of T and P. The last term in the braces of eq(342) arises from taking that into account. Also, eq(342) is rewritten in terms of mole fractions using the total number of moles n.

$$\Delta S_{mix} = -\left(\frac{\partial \Delta G_{mix}}{\partial T}\right)_P = -n_A R \ln X_A - n_B R \ln X_B - n_A R\left(\frac{\partial \ln \gamma_A}{\partial T}\right)_P - n_B R\left(\frac{\partial \ln \gamma_B}{\partial T}\right)_P$$
$$= -nR(X_A \ln X_A + X_B \ln X_B) - nR\left\{X_A\left(\frac{\partial \ln \gamma_A}{\partial T}\right)_P + X_B\left(\frac{\partial \ln \gamma_B}{\partial T}\right)_P\right\} \qquad (343)$$

Ans. $\Delta S_{mix} = -nR(X_A \ln X_A + X_B \ln X_B) - nR\left\{X_A\left(\frac{\partial \ln \gamma_A}{\partial T}\right)_P + X_B\left(\frac{\partial \ln \gamma_B}{\partial T}\right)_P\right\}$

(c) The entropy of mixing for the non-ideal solution is obviously different from that for the ideal solution. When the substances A and B are mixed isothermally to make a non-ideal solution, the entropy of the solution is considered to differ by the amount of $\Delta Q_{mix}/T = \Delta H_{mix}/T$ as compared with when A and B mix ideally. As a consequence eq(343) can be rewritten as eq(344),

$$\Delta S_{mix} = \Delta S_{mix}(ideal) + \Delta H_{mix}/T \qquad (344)$$

where $\Delta S_{mix}(ideal)$ and ΔH_{mix} are given by the following equations.

$$\Delta S_{mix}(ideal) = -nR(X_{\rm A}\ln X_{\rm A} + X_{\rm B}\ln X_{\rm B}) \qquad (345)$$

$$\Delta H_{mix}/T = -nR\left\{X_{\rm A}\left(\frac{\partial \ln\gamma_{\rm A}}{\partial T}\right)_P + X_{\rm B}\left(\frac{\partial \ln\gamma_{\rm B}}{\partial T}\right)_P\right\} \qquad (346)$$

Based on eq(344), we can tell under what circumstances the entropy of mixing becomes larger or smaller as compared with that for the ideal solution.

Ans. See text (c).

(d) In terms of the partial molar volumes of A and B, the volume of mixing (volume change) is expressed as follows: $\Delta V_{mix} = (n_{\rm A}\overline{V}_{\rm A} + n_{\rm B}\overline{V}_{\rm B}) - (n_{\rm A}V_{\rm A}^{\circ} + n_{\rm B}V_{\rm B}^{\circ})$. On the other hand, ΔV_{mix} can be derived from the partial differential of ΔG_{mix} in eq(342) with pressure as is given by eq(347).

$$\Delta V_{mix} = n_{\rm A}RT\left(\frac{\partial \ln\gamma_{\rm A}}{\partial P}\right)_{T,n_{\rm A},n_{\rm B}} + n_{\rm B}RT\left(\frac{\partial \ln\gamma_{\rm B}}{\partial P}\right)_{T,n_{\rm A},n_{\rm B}} \qquad (347)$$

Ans. volume change: eq(347)

Solution 3-60

(a) Expressing the quantities for the pure substance by putting the superscript$^{\circ}$, the chemical potential is given by the relation $\mu_{\rm A}^{\circ} = G_{\rm A}^{\circ}/n$ and its temperature dependence at a constant pressure is expressed as is given in **Ans.**

Ans. $\left(\frac{\partial \mu_{\rm A}^{\circ}}{\partial T}\right)_P = \frac{1}{n}\left(\frac{\partial G_{\rm A}^{\circ}}{\partial T}\right)_P = -\frac{S_{\rm A}^{\circ}}{n}$

(b) The possible combinations of the straight lines (i) and (ii) expressing a behavior near the equilibrium are (1) liquid-gas, (2) solid-liquid and (3) solid-gas. The reason for this is that the molar entropy (see (a)) proportional to the number of ways of distribution increases in the order of solid < liquid < gas. Alternatively, since the intersection is the equilibrium point, you can answer the question on the basis of on which side of T_e one phase of the two is more stable than the other. Consequently, the intersection represents one of the boiling point, freezing (melting) point and sublimation point.

Ans. boiling, freezing or sublimation point

(c) In the case of liquid-gas equilibrium.

The condition for the equilibrium is expressed by eq(348), where the superscript$^{\circ}$ is deleted for a sake of simplicity.

$$\mu_{\rm A}^{g}(T_e, P) = \mu_{\rm A}^{l}(T_e, P) \qquad (348)$$

When the temperature and pressure vary a little as expressed as $T_e \to T_e + dT_e$ and $P \to P + dP$, the condition for the equilibrium at the newly reached temperature and pressure is given by eq(349).

$$\mu_{\rm A}^{g}(T_e + dT_e, P + dP) = \mu_{\rm A}^{l}(T_e + dT_e, P + dP) \qquad (349)$$

In the differential form, therefore, it is expressed as eq(350),

$$\left(\frac{\partial \mu^{g}}{\partial P}\right)_T dP + \left(\frac{\partial \mu^{g}}{\partial T}\right)_P dT_e = \left(\frac{\partial \mu^{l}}{\partial P}\right)_T dP + \left(\frac{\partial \mu^{l}}{\partial T}\right)_P dT_e \qquad (350)$$

which is further rewritten as eq(351).

$$-S^g dT_e + V^g dP = -S^l dT_e + V^l dP \;\Rightarrow\; (V^g - V^l)dP = (S^g - S^l)dT_e \qquad (351)$$

As a result, the *Clausius-Clapeyron* equation is derived directly from eq(351). T_e is the temperature at the equilibrium.

Ans. $\dfrac{dT_e}{dP} = \dfrac{T_e \Delta V}{\Delta H_b}$

"note" Since the pressure is also a variable determined at the equilibrium, P may be written as P_e for clarity. Or the differential here can be expressed as $(\partial T/\partial P)_e$ in order to clarify that it is along the specified route (equilibrium) just as for the *Clausius-Clapeyron* equation.

Solution 3-61

The formation reaction of ethanol is expressed as reaction(i). Notice that the reaction equation is written so as to produce one mole of ethanol.

$$2\text{C}(graphite) + 3\text{H}_2 + \frac{1}{2}\text{O}_2 = \text{C}_2\text{H}_5\text{OH} \qquad \text{(i)}, \qquad \Delta H\text{(i)}$$

The combustion reactions and the values for the heat of combustion of carbon, hydrogen gas and ethanol are given by reactions from (ii) to (iv).

$$\text{C}(graphite) + \text{O}_2 = \text{CO}_2 \qquad \text{(ii)}, \qquad \Delta H\text{(ii)} = -393.51\,\text{kJ}\cdot\text{mol}^{-1}$$

$$\text{H}_2 + \frac{1}{2}\text{O}_2 = \text{H}_2\text{O}(gas) \qquad \text{(iii)}, \qquad \Delta H\text{(iii)} = -241.83\,\text{kJ}\cdot\text{mol}^{-1}$$

$$\text{C}_2\text{H}_5\text{OH} + 3\text{O}_2 = 2\text{CO}_2 + 3\text{H}_2\text{O}(gas), \qquad \text{(iv)} \qquad \Delta H\text{(iv)} = -1366.4\,\text{kJ}\cdot\text{mol}^{-1}$$

Reactions from (ii) to (iv) are combined linearly so as to produce reaction (i). As shown below, the equation of $reaction\ \text{(ii)}\times 2 + \text{reaction (iii)}\times 3 - \text{reaction (iv)}$ is confirmed to yield reaction (i) and, hence, the value for the molar heat of the formation reaction of ethanol is calculated from those reactions as $-171.11\,\text{kJ}\cdot\text{mol}^{-1}$.

$$2\text{C}(graphite) + 2\text{O}_2 = 2\text{CO}_2, \qquad 2\times\Delta H\text{(ii)} = -787.02\,\text{kJ}\cdot\text{mol}^{-1}$$

$$3\text{H}_2 + \frac{3}{2}\text{O}_2 = 3\text{H}_2\text{O}(gas), \qquad 3\times\Delta H\text{(iii)} = -725.49\,\text{kJ}\cdot\text{mol}^{-1}$$

$$-\text{C}_2\text{H}_5\text{OH} - 3\text{O}_2 = -2\text{CO}_2 - 3\text{H}_2\text{O}(gas), \qquad -\Delta H\text{(iv)} = 1366.4\,\text{kJ}\cdot\text{mol}^{-1}$$

$$2\text{C}(graphite) + 3\text{H}_2 + \frac{1}{2}\text{O}_2 = \text{C}_2\text{H}_5\text{OH}, \qquad \Delta H\text{(i)} = 2\times\Delta H\text{(ii)} + 3\times\Delta H\text{(iii)} - \Delta H\text{(iv)} = -171.11\,\text{kJ}\cdot\text{mol}^{-1}$$

Ans. molar heat of formation of ethanol: $-171.11\,\text{kJ}\cdot\text{mol}^{-1}$

Solution 3-62

The solution to the problem is given in the following two ways.

Solution 1.

For instance, referring to the table A3 in *Basic Physical Chemistry* by *Walter J. Moore*, the values of $-(G_T^\circ - H_0^\circ)/T$ (**"note"**) for the substances $C_2H_6(g)$, $C_2H_4(g)$ and $H_2(g)$ are found as 189.4, 184 and

$102.2\ \mathrm{J\cdot K^{-1}mol^{-1}}$ at 298 K and 212.4, 203.7 and $116.9\ \mathrm{J\cdot K^{-1}mol^{-1}}$ at 500 K. Also, the values of molar standard enthalpy of formation of those substances at 0 K are -69.12×10^3, 60.73×10^3 and $0\ \mathrm{J\cdot mol^{-1}}$. From these values the relevant quantities are calculated as below from eq(352) to eq(354).

$$G^\circ_{298}(\mathrm{C_2H_6}) - H^\circ_0(\mathrm{C_2H_6}) = -298\times189.4\ \mathrm{J\cdot mol^{-1}} \qquad (352)$$

$$G^\circ_{298}(\mathrm{C_2H_4}) - H^\circ_0(\mathrm{C_2H_4}) = -298\times184\ \mathrm{J\cdot mol^{-1}} \qquad (353)$$

$$G^\circ_{298}(\mathrm{H_2}) - H^\circ_0(\mathrm{H_2}) = -298\times102.2\ \mathrm{J\cdot mol^{-1}} \qquad (354)$$

The value for the equation given by { eq(353)+eq(354)-eq(352)} is calculated obtaining eq(355).

$$\Delta G^\circ_{298} - \Delta H^\circ_0 = -298\times(184+102.2-189.4) = -28.85\ \mathrm{J\cdot mol^{-1}} \qquad (355)$$

On the other hand, using the relation given by eq(356),

$$\Delta H^\circ_0 = \left\langle \Delta H^\circ_0(\mathrm{C_2H_4}) + \Delta H^\circ_0(\mathrm{H_2}) - \Delta H^\circ_0(\mathrm{C_2H_6})\right\rangle = 129.85\times10^3\ \mathrm{J\cdot mol^{-1}} \qquad (356)$$

the value of ΔG°_{298} is calculated from eq(355) as below.

$$\Delta G^\circ_{298} = -28.85\times10^3 + \Delta H^\circ_0 = -28.85\times10^3 + 129.85\times10^3 = 101\ \mathrm{kJ\cdot mol^{-1}}$$

In a similar manner, eq(355') and (356') given below are used at 500 K to find the values of ΔG°_{500} as $75.75\ \mathrm{kJ\cdot mol^{-1}}$ (eq(357)).

$$\Delta G^\circ_{500} - \Delta H^\circ_0 = -54.1\times10^3\ \mathrm{J\cdot mol^{-1}} \qquad (355')$$

$$\Delta H^\circ_0 = \left\langle \Delta H^\circ_0(\mathrm{C_2H_4}) + \Delta H^\circ_0(\mathrm{H_2}) - \Delta H^\circ_0(\mathrm{C_2H_6})\right\rangle = 129.85\times10^3\ \mathrm{J\cdot mol^{-1}} \qquad (356')$$

$$\Delta G^\circ_{500} = -54.1\times10^3 + \Delta H^\circ_0 = -54.1\times10^3 + 129.85\times10^3 = 75.75\ \mathrm{kJ\cdot mol^{-1}} \qquad (357)$$

As a result, the equilibrium constant given by the equation $\Delta G^\circ = -RT\ln K_P$ is calculated at the two temperatures as shown below.

$$\ln K_P(298\ \mathrm{K}) = -\frac{\Delta G^\circ_{298}}{RT} = -\frac{101\times10^3}{8.31\times298} = -40.79 \Rightarrow K_P(298\ \mathrm{K}) = 1.94\times10^{-18}$$

$$\ln K_P(500\ \mathrm{K}) = -\frac{\Delta G^\circ_{500}}{RT} = -\frac{75.75\times10^3}{8.31\times500} = -18.23 \Rightarrow K_P(500\ \mathrm{K}) = 1.21\times10^{-8}$$

Ans. $\Delta G^\circ_{298} = 101\ \mathrm{kJ\cdot mol^{-1}}$, $K_P(298\ \mathrm{K}) = 1.94\times10^{-18}$; $\Delta G^\circ_{500} = 75.75\ \mathrm{kJ\cdot mol^{-1}}$, $K_P(500\ \mathrm{K}) = 1.21\times10^{-8}$

Solution 2.

The standard free energy change for the dehydrogenation of ethane at a certain temperature is given by eq(358).,

$$\Delta G^\circ = G^\circ(\mathrm{C_2H_4}) + G^\circ(\mathrm{H_2}) - G^\circ(\mathrm{C_2H_6}) \qquad (358)$$

In order to obtain the values of G°_{298} and G°_{500} from a data table for each substance, the values of $-(G^\circ - H^\circ_0)/T$ and H°_0 are required. Usually, however, the values for H°_0 are not given and instead of these, the values for $H^\circ_{298} - H^\circ_0$ are provided. The value for H°_{298} can be calculated from the standard enthalpy of formation as is shown below by eq(359).

$$\Delta H^\circ_{298}(\mathrm{C_2H_4}) = H^\circ_{298}(\mathrm{C_2H_4}) - 2H^\circ_{298}(\mathrm{C}) - 2H^\circ_{298}(\mathrm{H_2}) \qquad (359)$$

Since the value of H°_{298} for the simple substance (element) is defined as below (eq(360)), eq(361) holds for the substances in the present problem.

$$H^\circ_{298}(Element) = 0 \qquad (360)$$

$$H^\circ_{298}(\mathrm{C}) = H^\circ_{298}(\mathrm{H_2}) = 0 \qquad (361)$$

Consequently, from eq (362) to eq(364) are obtained.

$$\Delta H^\circ_{298}(\mathrm{C_2H_6}) = H^\circ_{298}(\mathrm{C_2H_6}) - 2H^\circ_{298}(\mathrm{C}) - 6H^\circ_{298}(\mathrm{H_2}) = -84.60\times 10^3\,\mathrm{J\cdot mol^{-1}} \qquad (362)$$

$$\Delta H^\circ_{298}(\mathrm{C_2H_4}) = H^\circ_{298}(\mathrm{C_2H_4}) - 2H^\circ_{298}(\mathrm{C}) - 2H^\circ_{298}(\mathrm{H_2}) = 52.25\times 10^3\,\mathrm{J\cdot mol^{-1}} \qquad (363)$$

$$\Delta H^\circ_{298}(\mathrm{H_2}) = H^\circ_{298}(\mathrm{H_2}) = 0 \qquad (364)$$

On the other hand, the following values of $H^\circ_{298} - H^\circ_0$ are given in the table mentioned above.

$$H^\circ_{298}(\mathrm{C_2H_6}) - H^\circ_0(\mathrm{C_2H_6}) = 11.95\times 10^3\,\mathrm{J\cdot mol^{-1}} \qquad (365)$$

$$H^\circ_{298}(\mathrm{C_2H_4}) - H^\circ_0(\mathrm{C_2H_4}) = 10.52\times 10^3\,\mathrm{J\cdot mol^{-1}} \qquad (366)$$

$$H^\circ_{298}(\mathrm{H_2}) - H^\circ_0(\mathrm{H_2}) = 8.47\times 10^3\,\mathrm{J\cdot mol^{-1}} \qquad (367)$$

Employing eqs(362) - (367), the values of H°_0 are obtained for the relevant substances as given below.

$$H^\circ_0(\mathrm{C_2H_6}) = -84.60\times 10^3 - 11.95\times 10^3 = -96.55\,\mathrm{kJ\cdot mol^{-1}}$$

$$H^\circ_0(\mathrm{C_2H_4}) = 52.25\times 10^3 - 10.52\times 10^3 = 41.73\,\mathrm{kJ\cdot mol^{-1}}$$

$$H^\circ_0(\mathrm{H_2}) = -8.47\,\mathrm{kJ\cdot mol^{-1}}$$

Also, as are already calculated in Solution 1 and are given again below, eqs(352) through (354) are satisfied.

$$G^\circ_{298}(\mathrm{C_2H_6}) - H^\circ_0(\mathrm{C_2H_6}) = -298\times 189.4\,\mathrm{J\cdot mol^{-1}} \qquad (352)$$

$$G^\circ_{298}(\mathrm{C_2H_4}) - H^\circ_0(\mathrm{C_2H_4}) = -298\times 184\,\mathrm{J\cdot mol^{-1}} \qquad (353)$$

$$G^\circ_{298}(\mathrm{H_2}) - H^\circ_0(\mathrm{H_2}) = -298\times 102.2\,\mathrm{J\cdot mol^{-1}} \qquad (354)$$

Substituting the vales of H°_0 obtained above into eqs(352) to (354) , eqs(366) through (368) are obtained.

$$G^\circ_{298}(\mathrm{C_2H_6}) = -256.44 - 96.55 = -152.99\,\mathrm{kJ\cdot mol^{-1}} \qquad (366)$$

$$G^\circ_{298}(\mathrm{C_2H_4}) = -54.83 + 41.73 = -13.1\,\mathrm{kJ\cdot mol^{-1}} \qquad (367)$$

$$G^\circ_{298}(\mathrm{H_2}) = -230.36 - 8.47 = -38.93\,\mathrm{kJ\cdot mol^{-1}} \qquad (368)$$

From eqs(366) to (368), the value of ΔG°_{298} is found as below.

$$\Delta G^\circ_{298} = -13.1 - 38.93 + 152.99 = 101\,\mathrm{kJ\cdot mol^{-1}}$$

In summary, the same value as that in Solution 1 is obtained. The calculation of the equilibrium constant is the same as is given in Solution 1.

Ans. $\Delta G^\circ_{298} = 101\ \mathrm{kJ\cdot mol^{-1}}$

"note" The function $-(G^\circ_T - H^\circ_0)/T$ is called *free energy function* and plays an important role in quantifying chemical reactions or calculating equilibrium constants at various temperatures. If we know the specific heat under constant pressure as a function of temperature, the function can be evaluated from the integral given below.

$$\frac{-(G^\circ_T - H^\circ_0)}{T} = -\int_0^T \frac{1}{(T')^2}\int_0^{T'} C^\circ_P(T'')dT''dT' \qquad (369)$$

Although the entropy S_T° of a pure substance at T can be calculated from the temperature dependence of the specific heat $C_P^\circ(T)$ (see **problem 3-97**), it can also be evaluated by using free energy function as is shown below.

$$S_T^\circ = \frac{(H_T^\circ - H_0^\circ)}{T} - \frac{(G_T^\circ - H_0^\circ)}{T} \qquad (370)$$

Solution 3-63

The chemical potential defined by eq(371) is partially differentiated with T or P and then the order of differentiation is interchanged with the mole number n_A. The interchanged first differential with T or P yields entropy or volume and the second differential with n_A yields partial molar entropy or partial molar volume by definition as is shown by eq(372) or eq(373). Upon interchanging the variables, you need to be careful about which variables are to be kept constant on differentiation. The derivation is given here for the two-component case, but self-evidently it can be extended to the n-component case.

$$\mu_A = \left(\frac{\partial G}{\partial n_A}\right)_{T,P,n_B} \qquad (371)$$

$$\left(\frac{\partial \mu_A}{\partial T}\right)_{P,n_A,n_B} = \frac{\partial}{\partial T}\left\{\left(\frac{\partial G}{\partial n_A}\right)_{T,P,n_B}\right\}_{P,n_A,n_B} = \frac{\partial}{\partial n_A}\left\{\left(\frac{\partial G}{\partial T}\right)_{P,n_A,n_B}\right\}_{T,P,n_B} = -\left(\frac{\partial S}{\partial n_A}\right)_{T,P,n_B} = -\bar{S}_A \qquad (372)$$

$$\left(\frac{\partial \mu_A}{\partial P}\right)_{T,n_A,n_B} = \frac{\partial}{\partial P}\left\{\left(\frac{\partial G}{\partial n_A}\right)_{T,P,n_B}\right\}_{T,n_A,n_B} = \frac{\partial}{\partial n_A}\left\{\left(\frac{\partial G}{\partial P}\right)_{T,n_A,n_B}\right\}_{T,P,n_B} = \left(\frac{\partial V}{\partial n_A}\right)_{T,P,n_B} = \bar{V}_A \qquad (373)$$

From these results, eq(374) is obtained.

$$d\mu_A = \left(\frac{\partial \mu_A}{\partial T}\right)_{P,n_A,n_B} dT + \left(\frac{\partial \mu_A}{\partial P}\right)_{T,n_A,n_B} dP = -\bar{S}_A dT + \bar{V}_A dP \qquad (374)$$

Ans. See text.

Solution 3-64

Solution 1. From the definition of the chemical potential $\mu_i = (\partial G/\partial n_i)_{T,P,n_j \neq n_i}$, dG is written as eq(375).

$$dG = \sum \mu_i dn_i \qquad (375)$$

Replacing differentials dn_i's with $n_i\, d\xi$'s as defined by eq(376), eq(375) is integrated over the range [0, 1] of ξ.

$$\frac{dn_i}{n_i} = \frac{dn_j}{n_j} = d\xi \qquad (376)$$ (ξ is the variation ratio of a mole number and takes a value in the range [0, 1].)

Then the integrations of the left and right sides of eq(375) are expressed by eq(377) and eq(378), respectively.

$$\text{left side} = \int_0^{n_1, n_2, \cdots} dG = G \qquad (377)$$

$$\text{right side} = \int_0^{n_1, n_2, \cdots} \sum \mu_i dn_i = \int_0^1 \left(\sum \mu_i n_i\right) d\xi = \sum \mu_i n_i \qquad (378)$$

The relation $G = \sum_{i=1} n_i \mu_i$ is derived by equating eq(377) with eq(378). It should be noticed here that G is a function expressed as $G = G(T, P, n_1, n_2, \cdots)$. Also, all of μ_i 's are kept constant on integration by ξ, which is readily understood if it is noticed that the chemical potentials do not depend on the amount of the solution and are invariant as far as the molar ratios of the components are the same (see Solution 2. Below). Namely, the chemical potential is an intensive thermodynamic variable just like P and T and does not depend on the total amount of the system.

An image of the integration in eqs(377) and (378) is illustrated in Fig.39 on the right. The integration route here is along which the ratio of the components of the system is kept constant but the total amount is increased with increasing ξ.

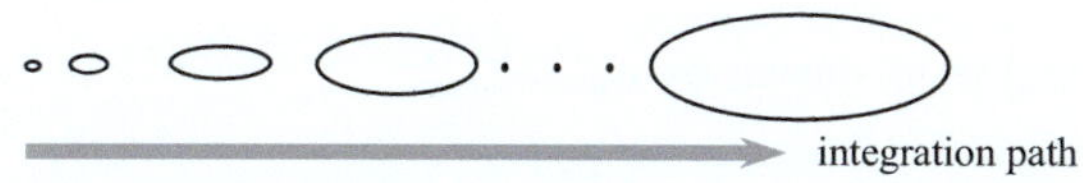

Fig.39 An image of the integration in eqs(377) and (378).

Generally, an extensive thermodynamic variable Λ is expressed by eq(379) and its functional form is totally

$$d\Lambda = \sum_{i=1}\left(\frac{\partial \Lambda}{\partial n_i}\right)_{n_j, j\neq i} dn_i = \sum_{i=1}\lambda_i dn_i \qquad (379)$$

identical to eq(375). Therefore, by following the same procedure as employed in the case of G, the relation $\Lambda = \sum_{i=1} n_i\lambda_i,\ \lambda_i = \left(\frac{\partial \Lambda}{\partial n_i}\right)_{n_j, j\neq i}$ can be confirmed.

Ans. See text.

Solution 2.

Since the *Gibbs* free energy expressed by $G = G(T, P, n_1, n_2, \cdots)$ is *Euler*'s homogeneous function of the first order (in general, all of the extensive thermodynamic variables), eq(380) is satisfied.

$$n_1\frac{\partial G}{\partial n_1} + n_2\frac{\partial G}{\partial n_2} + \cdots + n_l\frac{\partial G}{\partial n_l} = G(T, P, n_1, n_2, \cdots) \qquad (380)$$

This is proved as follows. Since G is an extensive variable, when the system is enlarged by a factor of λ without changing the component ratio, the free energy of the enlarged system is given by λG, that is, eq(381) is satisfied.

$$G(T, P, \lambda n_1, \lambda n_2, \cdots \lambda n_l) = \lambda G(T, P, n_1, n_2, \cdots n_l) \qquad (381)$$

By differentiating both sides of eq(381) by λ, eq(382) is obtained.

$$\frac{\partial G(T, P, \lambda n_1, \lambda n_2, \cdots \lambda n_l)}{\partial \lambda} = \sum_{i=1}^{l}\frac{\partial G(T, P, \lambda n_1, \lambda n_2, \cdots \lambda n_l)}{\partial \lambda n_i} \times \frac{\partial \lambda n_i}{\partial \lambda} = \sum_{i=1}^{l}\left(\frac{\partial G}{\partial n_i}\right) \times n_i$$

$$= \sum_{i=1}^{l}\mu_i \times n_i = \left(\frac{\partial \lambda G}{\partial \lambda}\right) = G \qquad (382)$$

Therefore, it is proved that eq(380) holds for G.

Ans. See text.

Solution 3-65

(a) The data necessary for the calculation are given below for each substance participating in the reaction.

Fe_2O_3 : $\Delta H_f^\circ(Fe_2O_3,\ 298K) = -824\ kJ \cdot mol^{-1}$, $C_P = 104\ J \cdot K^{-1}mol^{-1}$

H_2 : $\Delta H_f^\circ(H_2,\ 298K) = -0\ kJ \cdot mol^{-1}$, $C_P = 28.8\ J \cdot K^{-1}mol^{-1}$

Fe : $\Delta H_f^\circ(Fe,\ 298K) = -0\ kJ \cdot mol^{-1}$, $C_P = 25.2\ J \cdot K^{-1}mol^{-1}$

$H_2O(g)$: $\Delta H_f^\circ(H_2O,\ 298K) = -242\ kJ \cdot mol^{-1}$, $C_P = 33.6\ J \cdot K^{-1}mol^{-1}$

From these values, $\Delta H_r^\circ(298\text{K})$ is calculated as below.

$$\Delta H_r^\circ(298\text{K}) = 3\Delta H_f^\circ(\text{H}_2,\ 298\text{K}) + \Delta H_f^\circ(\text{Fe}_2\text{O}_3,\ 298\text{K}) - 3\Delta H_f^\circ(\text{H}_2\text{O},\ 298\text{K}) - 2\Delta H_f^\circ(\text{Fe},\ 298\text{K})$$

$$= 0\times 3 - 824 - (-242\times 3 + 0\times 2) = -98\ \text{kJ}\cdot\text{mol}^{-1}$$

Ans. $\Delta H_r^\circ(298\text{K}) = -98\ \text{kJ}\cdot\text{mol}^{-1}$

(b) The variation of ΔH_r with temperature can be calculated from the temperature dependence of the specific heat as is shown by eq(383).

$$\left(\frac{\partial H}{\partial T}\right)_P = C_P \Rightarrow \left(\frac{\partial \Delta H_r}{\partial T}\right)_P = \Delta C_P \Rightarrow d\Delta H_r = \Delta C_P dT \qquad (383)$$

Integrating eq(383) over the temperature range [298, 400] under the constant pressure,

$$\int_{298}^{400} d\Delta H_r^\circ = \int_{298}^{400} \Delta C_P dT \Rightarrow \Delta H_r^\circ(400\text{ K}) - \Delta H_r^\circ(298\text{ K}) = (28.8\times 3 + 104 - 25.2\times 2 - 33.6\times 3)\times(400-298)$$

$$= 4.00\ \text{kJ}\cdot\text{mol}^{-1}$$

a value of $4.00\ \text{kJ}\cdot\text{mol}^{-1}$ is obtained. From this, $\Delta H_r^\circ(400\text{ K}) = 4.00 - 98 = -94\ \text{kJ}\cdot\text{mol}^{-1}$ is found.

Ans. $\Delta H_r^\circ(400\text{ K}) = -94\ \text{kJ}\cdot\text{mol}^{-1}$

Solution 3-66a

Solution 1. Based on *Clausius-Clapeyron*'s equation.

Substituting the following known values for ΔV and $\Delta H_t (= \Delta H_f)$ into *Clausius-Clapeyron*'s equation

$$\Delta V = M\left(\frac{1}{\rho_d} - \frac{1}{\rho_g}\right) = 12\times 10^{-3}\left(\frac{1}{3.513\times 10^3} - \frac{1}{2.260\times 10^3}\right) = -1.89\times 10^{-6}\ \text{m}^3\text{mol}^{-1}$$

$$\Delta H_t = 1900\ \text{J}\cdot\text{mol}^{-1}$$

given by eq(384) and integrating it, eq(385) is obtained.

$$\frac{dP}{dT} = \frac{\Delta H_t}{T\Delta V} \qquad (384)$$

$$P - P_0 = -\frac{1.9\times 10^3}{1.89\times 10^{-6}}\ln\frac{T}{T_0} \qquad (385)$$

Here initial values are given by $T_0 = 298$ K and $P_0 = 1.51\times 10^9$ Pa (= the equilibrium pressure at 298 K) (see Solution 2 and **"note"**). Substituting these values into eq(385), the equilibrium pressure at 1300 K is found as below.

$$P = 1.51\times 10^9 - 1.48\times 10^9 = 3\times 10^7\ \text{Pa}$$

Ans. equilibrium pressure at 1300 K: 3×10^7 Pa

Solution 2. According to eq(386) given below, ΔG is calculated by integration along two routes, one at constant temperature and the other at constant pressure.

$$\left(\frac{\partial \Delta G}{\partial P}\right)_T = \Delta V, \qquad \left(\frac{\partial \Delta G}{\partial T}\right)_P = -\Delta S \qquad (386)$$

Assuming ΔV is constant, the first equation of eq(386) is integrated over the pressure range $[P^\circ, P]$ at 298 K, yielding eq(388) via eq(387).

$$d\Delta G = \Delta V \times dP \Rightarrow \int_{P^\circ}^{P} d\Delta G = \Delta V \int_{P^\circ}^{P} dP \qquad (387)$$

$$\Delta G(298\text{ K}, P) - \Delta G(298\text{ K}, P^\circ) = -1.89\times 10^{-6}(P - P^\circ) \qquad (388)$$

Since the system is in equilibrium under pressure P, the value for this P is found by equating $\Delta G(298\text{ K}, P) = 0$ in eq(388), from which P is found as eq(389) below.

$$P = \frac{\Delta G(298\text{ K}, P^\circ)}{1.89 \times 10^{-6}} + P^\circ = \frac{2.87 \times 10^3}{1.89 \times 10^{-6}} + 1.01 \times 10^5 = 1.51 \times 10^9 \text{Pa} \tag{389}$$

Here, the standard free energy of formation $\Delta G(298\text{ K}, P^\circ)$ is calculated as below.

$$\Delta G(298\text{ K}, P^\circ) = \Delta H_f^\circ - T\Delta S_f^\circ = 1900 - 298 \times (2.439 - 5.694) = 2.87 \times 10^3 \text{J} \cdot \text{mol}^{-1}$$

At 1300 K, on the other hand, employing the free energy change calculated at 1300 K as below, the pressure

$$\Delta G(1300\text{K}, P^\circ) = \Delta H_f^\circ - T\Delta S_f^\circ = 1900 - 1300 \times (2.439 - 5.694) = 6.13 \times 10^3 \text{J} \cdot \text{mol}^{-1}$$

P is calculated in a similar manner as for eq(389) and is given by eq(390).

$$P = \frac{\Delta G(1300\text{K}, P^\circ)}{1.89 \times 10^{-6}} + P^\circ = \frac{6.13 \times 10^3}{1.89 \times 10^{-6}} + 1.01 \times 10^5 = 3.24 \times 10^9 \text{Pa} \tag{390}$$

Ans. 1.51×10^9 Pa(298K), 3.24×10^9 Pa(1300K)

"note" Why the answers for Solution 1 and 2 are different? One of the reasons for that is, in Solution 1 it is assumed that $\Delta H_t = const.$ However, to employ the assumptions of $\Delta H_t = const.$ and $\Delta S_t = const.$ at the same time is self-contradicting since ΔS_t and ΔH_t are related each other by the equation $\Delta S_t = \frac{\Delta H_t}{T_t}$.

Eq(384) is expressed as eq(384') in terms of the entropy change. After separating variables P and T to each side of eq(384') and then both sides are integrated as shown in eq(384''). At this point it is important to notice that the integration must be carried out along the route in equilibrium (see **problem 3-39**). Pay attention to the values of the upper and lower limits of the integration.

$$\frac{dP}{dT} = \frac{\Delta S_t}{\Delta V} \quad (384') \Rightarrow \int_{P_{298}}^{P_{1300}} \Delta V dP = \int_{298}^{1300} \Delta S_t dT \quad (384'')$$

Carrying out the integration of eq(384'') as below, the same result as Solution 2 is obtained for the equilibrium pressure at 1300 K.

$$-1.894 \times 10^{-6}(P_{1300} - P_{298}) = (2.439 - 5.694)(1300 - 298)$$

$$(P_{1300} - P_{298}) = \frac{3.255 \times 1002}{1.894 \times 10^{-6}} = 1.72 \times 10^9 \text{Pa}$$

$$P_{1300} = 1.72 \times 10^9 + 1.52 \times 10^9 = 3.24 \times 10^9 \text{Pa}$$

As is seen from the argument above, due care must be taken in applying *Clausius-Clapeyron*'s equation over a wide range of temperature and pressure. Judging from the result obtained here alone, to assume $\Delta S_t = \text{const.}$ seems more practical than to employ $\Delta H_t = \text{const.}$ for the integration along the *P-T* equilibrium curve as is done in Solution 1.

Solution 3-66b

The *Gibbs* free energy changes obtained at the two temperatures under the standard pressure P° (see preceding problem) are given below, from which the values for the equilibrium constant at the two temperatures are calculated as below.

$$\Delta G(298\text{K}, P^\circ) = -RT\ln K = 2.87 \times 10^3 \text{J} \cdot \text{mol}^{-1} \Rightarrow K(298\text{ K}, P^\circ) = \exp\left(-\frac{2.87 \times 10^3}{8.31 \times 298}\right) = 0.314$$

$$\Delta G(1300\text{K}, P^\circ) = -RT\ln K = 6.13\times10^3\,\text{J}\cdot\text{mol}^{-1} \Rightarrow K(1300\text{ K}, P^\circ) = \exp\left(-\frac{6.13\times10^3}{8.31\times1300}\right) = 0.558$$

Ans. equilibrium constant: 0.314 at 298 K and 0.558 at 1300 K

"note" Diamond can be synthesized at high temperatures under high pressures ($\Delta V < 0$) (artificial diamond).

Solution 3-67

A general method of expressing partial differentials such as $\left(\frac{\partial T}{\partial P}\right)_S$, $\left(\frac{\partial V}{\partial P}\right)_S$ or $\left(\frac{\partial T}{\partial V}\right)_S$ as a function of *P*, *V* and *T* including known constants has been described already in detail (see **solution 3-51**). In the case of the ideal gas, the required functions are readily derived starting from the relation between the internal energy and the entropy or the relation between the enthalpy and the entropy. The point is its internal energy or enthalpy is a function of *T* alone and is expressed by eq(391) or eq(391').

$$dU = C_V dT \qquad (391)$$

$$dH = C_P dT \qquad (391')$$

The total differential of the internal energy or the enthalpy given by the left side of eq(392) or eq(392') can be rewritten as the total differential of the entropy as is expressed by the right side of eq(392) or eq(392'), in which eq(391) or eq(391') is used..

$$dU = TdS - PdV \Rightarrow dS = \frac{1}{T}dU + \frac{P}{T}dV = \frac{C_V}{T}dT + \frac{P}{T}dV \qquad (392)$$

$$dH = TdS + VdP \Rightarrow dS = \frac{1}{T}dH - \frac{V}{T}dP = \frac{C_P}{T}dT - \frac{V}{T}dP \qquad (392')$$

Then, eqs(393) and (393') are directly derived from eqs(392) and (392'), respectively.

$$\frac{C_V}{T}\left(\frac{\partial T}{\partial V}\right)_S + \frac{P}{T} = 0 \Rightarrow \left(\frac{\partial T}{\partial V}\right)_S = -\frac{P}{C_V} \qquad (393)$$

$$\frac{C_P}{T}\left(\frac{\partial T}{\partial P}\right)_S - \frac{V}{T} = 0 \Rightarrow \left(\frac{\partial T}{\partial P}\right)_S = \frac{V}{C_P} \qquad (393')$$

Also, both sides of eq(392) are divided by dP under the condition of $S = \text{constant}$ to obtain eq(394), from which eq(395) is derived by the substitution of eq(393').

$$\frac{C_V}{T}\left(\frac{\partial T}{\partial P}\right)_S + \frac{P}{T}\left(\frac{\partial V}{\partial P}\right)_S = 0 \qquad (394)$$

$$\left(\frac{\partial V}{\partial P}\right)_S = -\frac{C_V}{C_P}\frac{V}{P} \qquad (395)$$

The required functions are given by eqs(393), (393') and (395) and they are the same as are derived based on the general method given in **solution 3-51**. Since any function obtained here can easily be integrated, the *P-V-T* relations of the ideal gas for the adiabatic process is readily known. For instance, by integrating eq(393'), the relation of $T \propto P^{R/C_P}$ (eq(396)) is finally found.

$$\frac{(dT)_S}{T} = \frac{R}{C_P}\frac{(dP)_S}{P} \Rightarrow \int\frac{(dT)_S}{T} = \frac{R}{C_P}\int\frac{(dP)_S}{P} \Rightarrow \ln T = \frac{R}{C_P}\ln P + \text{const.} \Rightarrow T = \text{const.}\times P^{\frac{R}{C_P}} \qquad (396)$$

Ans. *P-V-T* relation for the adiabatic process: eqs(393), (393') and (395)

"note" General expressions and the relations of eqs(393), (393') and (395) derived for the ideal gas (given again).

$$\left(\frac{\partial T}{\partial P}\right)_S = \frac{T\left(\frac{\partial V}{\partial T}\right)_P}{C_P} \overset{\text{ideal gas}}{\Rightarrow} \frac{V}{C_P}$$

$$\left(\frac{\partial V}{\partial P}\right)_S = \frac{C_V}{C_P}\left(\frac{\partial V}{\partial P}\right)_T \overset{\text{ideal gas}}{\Rightarrow} -\frac{C_V}{C_P}\frac{V}{P}$$

$$\left(\frac{\partial T}{\partial V}\right)_S = -\frac{T\left(\frac{\partial P}{\partial T}\right)_V}{C_V} = \overset{\text{ideal gas}}{\Rightarrow} -\frac{P}{C_V}$$

Solution 3-68

(i) In the case of positive ΔH_r° (equilibrium reaction given in the problem).

The equilibrium constant K increases with increasing temperature, that is, the equilibrium under discussion shifts toward the right upon raising the temperature. Because of the endothermicity ($\Delta H_r^\circ > 0$) of this reaction, the equilibrium shifts to the direction so as to lower the temperature of the reaction system. On the other hand, when the temperature is lowered, K decreases and the equilibrium shifts from the right to the left so as to raise the temperature.

(ii) In the case of negative ΔH_r°.

The equilibrium constant K increases with decreasing temperature, that is, the equilibrium shifts from left to right upon lowering the temperature. Because of the exothermicity ($\Delta H_r^\circ < 0$) of this reaction, the equilibrium shifts to the direction so as to raise the temperature of the reaction system.

As is evident from these discussions, the equilibrium shifts in accordance with the principle of *Le Chatelier* which tells us that any phenomenon proceeds toward the direction opposing to a change imposed externally.

Ans. See text.

Solution 3-69

The relation between G and $F(\equiv A)$ is given by eq(397).

$$G = F + PV \qquad (397)$$

Differentiating both sides of eq(397) by n_i keeping T, P and $n_j(j \neq i)$ constant, eq(398) is obtained as the definition of the chemical potential of μ_i.

$$\mu_i = \left(\frac{\partial G}{\partial n_i}\right)_{T,P,n_j,j\neq i} = \left(\frac{\partial F}{\partial n_i}\right)_{T,P,n_j,j\neq i} + P\left(\frac{\partial V}{\partial n_i}\right)_{T,P,n_j,j\neq i} \qquad (398)$$

On the other hand, the total differential of F can be written as eq(399).

$$dF = \left(\frac{\partial F}{\partial T}\right)_{V,n_i} dT + \left(\frac{\partial F}{\partial V}\right)_{T,n_i} dV + \sum_{i=1}^{l}\left(\frac{\partial F}{\partial n_i}\right)_{T,V,n_j,j\neq i} dn_i \qquad (399)$$

Under the conditions of $T =$ constant, $dn_j = 0\ (j \neq i)$ and $P =$ constant, both sides of eq(399) are divided by dn_i, leading to eq(400).

$$\left(\frac{\partial F}{\partial n_i}\right)_{T,P,n_j,j\neq i} = \left(\frac{\partial F}{\partial V}\right)_{T,n_i}\left(\frac{\partial V}{\partial n_i}\right)_{T,P,n_j,j\neq i} + \left(\frac{\partial F}{\partial n_i}\right)_{T,V,n_j,j\neq i} = -P\left(\frac{\partial V}{\partial n_i}\right)_{T,P,n_j,j\neq i} + \left(\frac{\partial F}{\partial n_i}\right)_{T,V,n_j,j\neq i} \quad (400)$$

Substituting this result into eq(398), eq(401) is obtained as the definition of the chemical potential.

$$\mu_i = \left(\frac{\partial G}{\partial n_i}\right)_{T,P,n_j,j\neq i} = -P\left(\frac{\partial V}{\partial n_i}\right)_{T,P,n_j,j\neq i} + \left(\frac{\partial F}{\partial n_i}\right)_{T,V,n_j,j\neq i} + P\left(\frac{\partial V}{\partial n_i}\right)_{T,P,n_j,j\neq i} = \left(\frac{\partial F}{\partial n_i}\right)_{T,V,n_j,j\neq i_i} \quad (401)$$

Ans. See text and eq(401).

"note" See **problem 3-38b** for a proof based on that the extensive thermodynamic variables are *Euler*'s homogeneous function of the first order.

Solution 3-70

A diagram showing the entropy of each state at the relevant temperature is given in Fig.40.

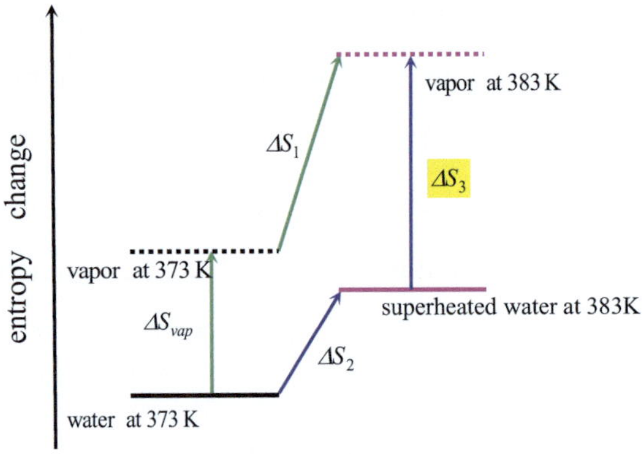

Fig.40. Entropy change associated with each relevant process.

$$\Delta S_1 = \int_{373}^{383} \frac{(30.36 + 9.61\times10^{-3}T + 11.8\times10^{-7}T^2)}{T}dT$$

$$= 0.90\ \mathrm{J\cdot K^{-1}\cdot mol^{-1}}$$

$$\Delta S_2 = \int_{373}^{383} \frac{C_P(l)}{T}dT = \int_{373}^{383} \frac{75.4}{T}dT = 75.4[\ln T]_{373}^{383}$$

$$= 1.97\ \mathrm{J\cdot K^{-1}\cdot mol^{-1}}$$

$$\Delta S_{vap} = \frac{\Delta H_{vap}}{T} = \frac{47.3\times10^3}{373} = 126.8\ \mathrm{J\cdot K^{-1}\cdot mol^{-1}}$$

Although the superheated water at 1 atm is in an unstable state、it is assumed that this state can be reached by a suitable reversible process. Then, to answer the question, it is only necessary to find ΔS_3 marked in yellow in Fig.38. The entropy change associated with each process is calculated as below.

Consequently、$\Delta S_3 = \Delta S_1 + \Delta S_{vap} - \Delta S_2 = 0.90 + 126.8 - 1.97 = 125.7\ \mathrm{J\cdot K^{-1}\cdot mol^{-1}}$ is found.

Ans. entropy change of superheated water: $\Delta S_3 = 125.7\ \mathrm{J\cdot K^{-1}\cdot mol^{-1}}$, entropy change of the surroundings: $-\Delta S_3 = -125.7\ \mathrm{J\cdot K^{-1}\cdot mol^{-1}}$, the total entropy change : zero (reversible vaporization)

Solution 3-71

Clausius-Clapeyron's equation (eq(402)) (see **solution 3-12**) is integrated assuming that ΔH_{vap} is independent of temperature. The obtained eq(403) is used to plot $\ln P$ against $1/T$.

$$\frac{d\ln P}{dT} = \frac{\Delta H_{vap}}{RT^2} \quad (402)$$

$$\ln P = -\frac{\Delta H_{vap}}{RT} + const. \quad (403)$$

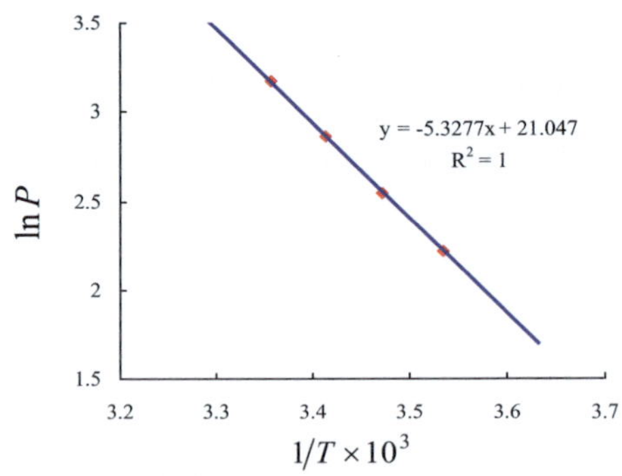

Fig.41 A plot of ln*P* against 1/*T* to acquire heat of vaporization.

As is shown in Fig.41, a good straight line is obtained and from its slope, the heat of vaporization of water ΔH_{vap} is calculated as $\Delta H_{vap} = 5328\times8.31 = 44.3\ \mathrm{kJ\cdot mol^{-1}}$.

Ans. heat of vaporization of water: $44.3\ \mathrm{kJ\cdot mol^{-1}}$

Solution 3-72a

(a) Since the amount of the oxygen dissolved in water is very minute, 1 kg of the solution is regarded equal the mass of the water and, hence, the mole number of the water is 55.5. Therefore, the mole fraction of the dissolved oxygen is given by $X_{O_2} = n_{O_2}/(n_{O_2} + 55.5)$ where n_{O_2} is the mole number of the dissolved oxygen. According to *Henry*'s law, X_{O_2} is related to the pressure of oxygen above the water by eq(404).

$$1.01\ \text{bar} = 4.4\times10^4\ \text{bar}\times\frac{n_{O_2}}{n_{O_2}+55.5} \qquad (404)$$

Solving eq(404), $n_{O_2} = 1.27\times10^{-3}\,\text{mol}$ is obtained, from which the solubility of oxygen is calculated to be $1.27\times10^{-3}\,\text{mol}\cdot\text{kg}^{-1}$ at ambient temperature.

Ans. solubility: $1.27\times10^{-3}\,\text{mol}\cdot\text{kg}^{-1}$

(b) In a similar manner as above, the solubility (mole numbers/kg) of carbon dioxide is calculated from eq(405) to be $n_{CO_2} = 6.76\times10^{-2}\,\text{mol}$.

$$2\times760\ \text{Torr} = 1.25\times10^6\,\text{Torr}\times\frac{n_{CO_2}}{n_{CO_2}+55.5} \qquad (405)$$

From this it is found that $3.38\times10^{-2}\,\text{mol}$ of carbon dioxide is dissolved in 500 ml of water. Upon opening the cap suddenly, therefore, the amount of $24.5\,\text{L}\cdot\text{mol}^{-1}\times3.38\times10^{-2}\,\text{mol}\approx0.83\ \text{L}$ of carbon dioxide at ambient temperature is released. It is assumed here that the partial pressure of carbon dioxide in the air is zero.

Ans. volume of the gas released: ca. 0.83 L

Solution 3-72b

The relative density of blood is assumed to be 1.0. Denoting the number of moles of a dissolve nitrogen per 1 L of blood n_{N_2}, eq(406) is satisfied from *Henry*'s law. Here, the partial pressure of the nitrogen gas in a lung is taken as $4\ \text{atm}\times4/5$ at a depth of 30m.

$$3.2\ \text{atm}\times760 = 6.51\times10^7\,\text{Torr}\times\frac{n_{N_2}}{n_{N_2}+55.5} \qquad (406)$$

Solving eq(406), $n_{N_2} = 2.07\times10^{-3}\,\text{mol}$ is found. Similarly, the mole number of nitrogen saturated in the blood under the air is calculated from eq(406') to be $n'_{N_2} = 5.18\times10^{-4}\,\text{mol}$.

$$0.8\ \text{atm}\times760 = 6.51\times10^7\,\text{Torr}\times\frac{n'_{N_2}}{n'_{N_2}+55.5} \qquad (406')$$

The difference given by $n_{N_2} - n'_{N_2} = 1.56\times10^{-3}\,\text{mol}$ is to be released per 1 L of the blood. Its volume approximately amounts to 38 mL/L.

Ans. nitrogen gas to be released: ca. 38 mL/L

Solution 3-73

(a) The *Henry*'s law constant κ^H is defined by eq(407), where P_A is the partial pressure of gas A above water and X_A is the mole fraction of A in the solution.

$$P_A = \kappa_A^H\cdot X_A \qquad (407)$$

In Fig.42 are plotted the partial pressures against the mole fraction at each temperature based on the data in the table in the problem. At any temperatures, good straight lines satisfying eq(407) are found. From the slopes of the straight lines, values for the *Henry*'s law constant κ^H (see the equations in Fig.40) are obtained as follows. The constants are in the units of atm for all.

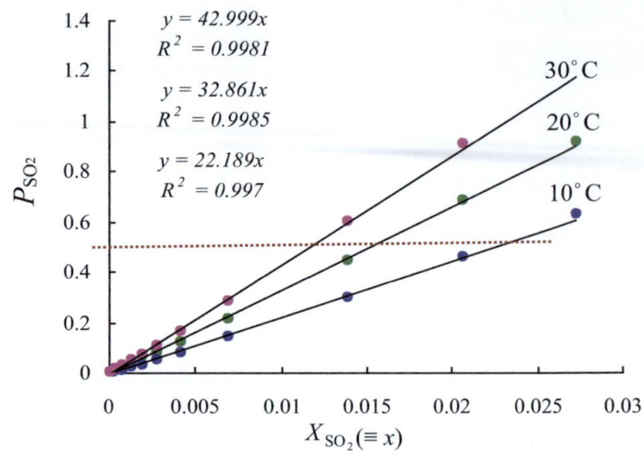

Fig.42 Dissolution of gaseous SO_2 in water.

$$10^\circ C : 22.2,\ 20^\circ C : 32.9,\ 30^\circ C : 43.0 .$$

As is readily seen from Fig.40, the constant increases with temperature, which means that the gas dissolvesthe more the lower the temperature is under the same partial pressure of the gas as is indicated by an orange dotted line.

Ans. κ^H : $10^\circ C : 22.2,\ 20^\circ C : 32.9,\ 30^\circ C : 43.0$ in the units of atm

(b) The logarithmic values of the *Henry*'s law constants obtained above, $\ln \kappa^H$, are plotted against the reciprocal of temperature in Fig.43. A reasonably good straight line is obtained. Since the *Henry*'s law constant can be regarded as a sort of the equilibrium constant, the slope of a straight line is proportional to the enthalpy of disso- lution ΔH_{solv} as is given by eq(408).

$$\frac{\partial \ln \kappa^H}{\partial (1/T)} = -\frac{\Delta H_{solv}}{R} \qquad (408)$$

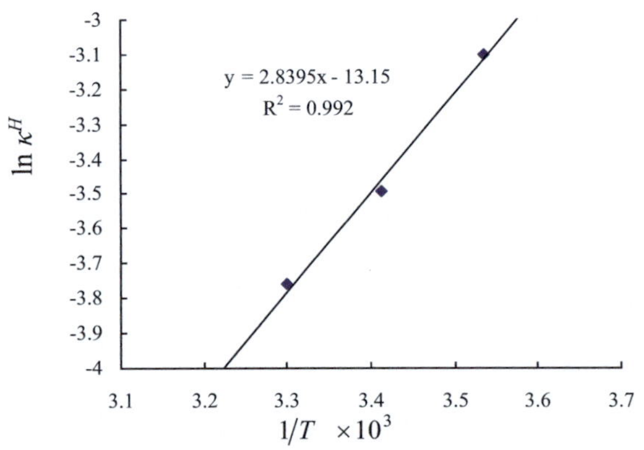

Fig.43 Temperature dependence of the constant of *Henry*'s law.

From the slope of the best-fitted straightline, the value of $\Delta H_{solv} = -2.84 \times 8.31 \times 10^3\,\mathrm{J \cdot mol^{-1}}$ is obtained. Namely, the dissolution process here is an exothermic one and when the temperature is raised, it proceeds to the direction so as to suppress the dissolution.

From a molecular theoretical point of view, the interaction between SO_2 and H_2O is attractive enough to overcome a decrease in entropy caused by the dissolution and drive the gas-solution equilibrium to the direction in favor of dissolution. In fact, from eq(409),

$$\Delta G^\circ = -RT \ln \kappa^H, \qquad \Delta G^\circ = \Delta H^\circ - T\Delta S^\circ \qquad (409)$$

the entropy of dissolution at $20^\circ C$ is negative as is given below.

$$\Delta S^\circ = \frac{-(23.6 - 8.51) \times 10^3}{293} = -51.5\,\mathrm{J \cdot K^{-1} \cdot mol^{-1}}$$

Upon the dissolution of SO_2 in water, it is assumed here that SO_2 does not react with H_2O and, in fact, the good straight line shown in Fig.43 indicates such a reaction, if any, can be neglected. Of course, in the case of NH_3 or HCl, ionic reactions in aqueous solutions come to play an important role. See **problem 4-66** for such examples.

Ans. temperature dependence of the *Henry*'s law constant: See text and Fig.43.

Solution 3-74

(a) See **solution 3-12**.

(b) From *Clausis-Clapeyron*'s equation which gives the relation between the vapor pressure and the temperature, the vapor pressures (kPa) of octanol (A) and propanol (B) at 400 K are calculated as below.

$$\text{A}: \quad \ln\frac{P_2(\text{A})}{101} = -\frac{52.3\times10^3}{8.31}\left(\frac{1}{400}-\frac{1}{468}\right) \Rightarrow \frac{P_2(\text{A})}{101} = 0.102 \Rightarrow P_2(\text{A}) = 10.3\ \text{kPa}$$

$$\text{B}: \quad \ln\frac{P_2(\text{B})}{101} = -\frac{40.1\times10^3}{8.31}\left(\frac{1}{400}-\frac{1}{371}\right) \Rightarrow \frac{P_2(\text{B})}{101} = 2.57 \Rightarrow P_2(\text{B}) = 260\ \text{kPa}$$

When a mixture of the two solvents boils at 400 K, the pressure of the mixed vapor reaches 1 atm (the mixed vapor pressure in the bubble is also 1 atm). From this, the mole fraction of each solvent is found according to *Rault*'s law by the following relation, where X_A^l denotes the mole fraction of octanol (A).

$$0.102\times101\times X_\text{A}^l + 2.57\times101\times(1-X_\text{A}^l) = 101\ (\text{kPa}) \Rightarrow X_\text{A}^l = 0.636$$

Ans. mixing ratio: $X_\text{A}^l = 0.636$, $X_\text{B}^l = 0.364$

(c) On the other hand, the mole fractions in the gas phase in equilibrium with the solution at the pressure of 1 atm are calculated from the partial pressure of each gas according to the law of partial pressures (*Dalton*'s law) as below.

$$X_\text{A}^g = \frac{0.102\times101}{(2.57+0.102)\times101} = 0.038$$

$$X_\text{B}^g = \frac{2.57\times101}{(2.57+0.102)\times101} = 0.962$$

Since these mole fractions also correspond to the mole fractions in the gas phase when all solution is brought into the gas phase by keeping the equilibrium at 400 K, the range of the mole fractions where both phases coexist is concluded to be in the range [0.038, 0.636] of X_A by combining the result obtained in (b).

Ans. range of X_A: 0.038-0.636

Solution 3-75a

(a) The definition of fugacity is given first before answering the question. In general, the chemical potential of a pure gas varies with pressure as is given by eq(410). When the molar volume $\overline{V}$ is known as a function of pressure, then by integrating eq(410) with pressure, the chemical potential of the gas can be expressed as a function of pressure.

$$\left(\frac{\partial\mu}{\partial P}\right)_T = \left(\frac{\partial G/n}{\partial P}\right)_T = \frac{V}{n} = \overline{V} \qquad (410)$$

Thus the chemical potential of the ideal gas μ^{ideal} is found as eq(411) by using $\overline{V} = RT/P$ and integrating eq(410). P° is the standard pressure at which the chemical potential of the gas is given by $\mu^\circ(T)$.

$$\mu^{ideal}(P,T) = \mu^\circ(T) + RT\ln\frac{P}{P^\circ} \qquad (411)$$

On the other hand, the chemical potential of a non-ideal gas μ^{real} depends on the function of $\overline{V}$ differently from the ideal gas and, hence, comes to be expressed by a complex function. On this account, therefore, in order to express μ^{real} in the same functional form as eq(411), a quantity called fugacity $(= f)$ is introduced and μ^{real} is expressed by eq(412). This is the same methodology as is employed to cope with the deviation of a real solution from the ideal solution by introducing the concept of the activity and activity coefficient.

$$\mu^{real}(P,T) = \mu^\circ(T) + RT\ln\frac{f}{f^\circ} \qquad (412)$$

The fugacity has the same units as the pressure and is considered as an effective pressure. f° is the standard fugacity and usually is put equal to the standard pressure of the ideal gas, i.e., $f^\circ = P^\circ$, which is usually taken as 1 atm. In other words, when $f = P^\circ$, the chemical potential of the gas is equal to μ° of the ideal gas at the standard state.

As a measure to know how largely the actual pressure differs from the effective pressure of fugacity, the fugacity coefficient γ is defined by eq(413).

$$f = \gamma \times P \qquad (413)$$

When $\gamma > 1$、the effective pressure is larger than the actual pressure and, hence, the gas tends to expand more readily (or more fugitive). When $\gamma < 1$, vice versa, tending to contract more easily.

Now, the fugacity and fugacity coefficient of the gas given in the problem are to be found. The molar volume of this gas is given by eq(414).

$$\left(\frac{\partial \mu}{\partial P}\right)_T = \overline{V} = \frac{RT}{P} + b \qquad (414)$$

Integrating eq(414) to find the pressure dependence of the chemical potential, eq(415) is obtained.

$$\mu^{real} = \mu^\circ + RT\ln\frac{P}{P^\circ} + b(P - P^\circ) \qquad (415)$$

From eqs(411) and (412), the difference of the chemical potentials between the ideal and real gases is given by eq(416).

$$\mu^{real} - \mu^{ideal} = RT\ln\frac{f/P}{f^\circ/P^\circ} = RT\ln\frac{f}{P} \qquad (416)$$

On the other hand, from eqs(411) and (415) the difference is given by eq(417), which is equated to eq(416) giving eq(418) as the required answer.

$$\mu^{real} - \mu^{ideal} = b(P - P^\circ) \qquad (417)$$

$$RT\ln\frac{f}{P} = b(P - P^\circ) \;\Rightarrow\; f = Pe^{\frac{b(P-P^\circ)}{RT}}, \qquad \gamma = e^{\frac{b(P-P^\circ)}{RT}} \qquad (418)$$

The obtained fugacity coefficient is always larger than unity and because of this, it is known that the larger the excluded volume is the more fugitive (tending to expand) the gas is.

Ans. $f = Pe^{\frac{b(P-P^\circ)}{RT}}, \quad \gamma = e^{\frac{b(P-P^\circ)}{RT}}$

(b) Substituting $b = 4.3\times10^{-2}\,\mathrm{dm^3\cdot mol^{-1}}$ and $P^\circ = 1.01\times10^5\,\mathrm{Pa}$ into eq(418), the fugacity and fugacity coefficient at each pressure are found as below.

$$f = 5\times1.01\times10^5 e^{\frac{4.50\times10^{-5}\times4\times1.01\times10^5}{8.31\times300}} = 5.087\times10^5\,\mathrm{Pa},\qquad \gamma = e^{\frac{4.50\times10^{-5}\times4\times1.01\times10^5}{8.31\times300}} = 1.007$$

$$f = 10\times1.01\times10^5 e^{\frac{4.50\times10^{-5}\times9\times1.01\times10^5}{8.31\times300}} = 10.27\times10^5\,\mathrm{Pa},\qquad \gamma = e^{\frac{4.50\times10^{-5}\times9\times1.01\times10^5}{8.31\times300}} = 1.017$$

$$f = 20\times1.01\times10^5 e^{\frac{4.50\times10^{-5}\times19\times1.01\times10^5}{8.31\times300}} = 20.91\times10^5\,\mathrm{Pa},\qquad \gamma = e^{\frac{4.50\times10^{-5}\times19\times1.01\times10^5}{8.31\times300}} = 1.035$$

Ans. fugacity: 5.037, 10.17, 20.70 atm; fugacity coefficient γ: 1.007, 1.017, 1.035

Solution 3-75b

(a) The equation of state of the gas under the discussion is expressed by eq(419) using its molar volume $\overline{V}$.

$$P = \frac{RT}{\overline{V}} - \frac{a}{\overline{V}^2} \qquad (419)$$

The integration to obtain the chemical potential is readily carried out by replacing the variable P with $\overline{V}$ as is shown below in eq(420).

$$\mu^{real} - \mu^\circ = \int_{P^\circ}^{P} \overline{V} dP = \int_{\overline{V}^\circ}^{\overline{V}} \overline{V}\left(-\frac{RT}{\overline{V}^2} + \frac{2a}{\overline{V}^3}\right) d\overline{V} = \int_{\overline{V}^\circ}^{\overline{V}} \left(-\frac{RT}{\overline{V}} + \frac{2a}{\overline{V}^2}\right) d\overline{V},\quad \left(dP = -\frac{RT}{\overline{V}^2} d\overline{V} + \frac{2a}{\overline{V}^3} d\overline{V}\right)$$

$$= \left[-RT\ln\overline{V} - \frac{2a}{\overline{V}}\right]_{\overline{V}^\circ}^{\overline{V}} = -RT\ln\frac{\overline{V}}{\overline{V}^\circ} - 2a\left(\frac{1}{\overline{V}} - \frac{1}{\overline{V}^\circ}\right) \qquad (420)$$

The fugacity f is defined by eq(421),

$$\mu^{real} - \mu^{ideal} = RT\ln\frac{f/P}{f^\circ/P^\circ} = RT\ln\frac{f}{P} \qquad (\because f^\circ = P^\circ) \qquad (421)$$

and the chemical potential of the ideal gas is given by eq(422) in terms of the molar volume.

$$\mu^{ideal} = \mu^\circ + RT\ln\frac{P}{P^\circ} = \mu^\circ + RT\ln\frac{\overline{V}^\circ}{\overline{V}} \qquad (422)$$

From eqs(421) and (422), the functions of the fugacity and the fugacity coefficient are found as given by eq(423).

$$RT\ln f/P = -2a\left(\frac{1}{\overline{V}} - \frac{1}{\overline{V}^\circ}\right) \Rightarrow f = P\times e^{-2a\left(\frac{1}{\overline{V}} - \frac{1}{\overline{V}^\circ}\right)/RT},\quad \gamma = e^{-2a\left(\frac{1}{\overline{V}} - \frac{1}{\overline{V}^\circ}\right)/RT} \qquad (423)$$

Ans. $f = P\times e^{-2a\left(\frac{1}{\overline{V}} - \frac{1}{\overline{V}^\circ}\right)/RT},\quad \gamma = e^{-2a\left(\frac{1}{\overline{V}} - \frac{1}{\overline{V}^\circ}\right)/RT}$

(b) The molar volumes of the gas at 300 K under the pressures of 1, 5, 10 and 20 atm are calculated first from eq(424) and substituting the obtained values into eq(423) above, the values listed in the table below are found.

$$a\left(\frac{1}{\overline{V}}\right)^2 - RT\left(\frac{1}{\overline{V}}\right) + P = 0,\ \Rightarrow\ ax^2 - RTx + P = 0 \qquad (424)$$

P / atm	1	5	10	20
$\frac{1}{\overline{V}}$ / $\mathrm{m^{-3}}$	40.76	208.7	431.2	932.3
f / 10^5 Pa	1.01	4.82	9.05	15.7
γ	1.0	0.954	0.896	0.779

Since the value of a given in the problem is positive, the intermolecular force is attractive. In such cases, it is intuitively understandable that the gas is less fugitive as compared with the ideal gas under the same conditions. The values obtained above for the fugacity and fugacity coefficient tell us this tendency quantitatively.

Ans. See text and the table.

Solution 3-76

It is important to know the critical temperature first before drawing *P*-*V* curves. Using the given values, the critical temperature T_c of the *van der Waals* equation of state is found as below.

$$T_c = \frac{8a}{27bR} = \frac{8\times 3.50\times 10^{-1}}{27\times 4.50\times 10^{-5}\times 8.31} = 277.3\ \mathrm{K}$$

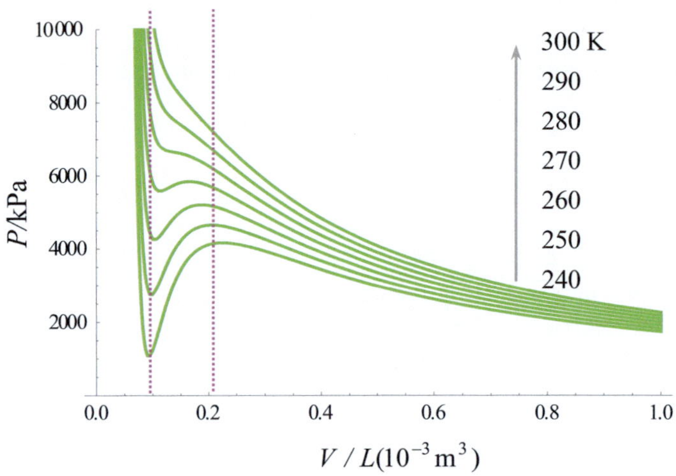

Fig.44 *P*-*V* curves of the *van der Waals* equation of state at several temperatures. Notice the scale of the vertical axis.

The temperature range below T_c, the gas is liquefied by compression. Therefore, when the *P*-*V* curve is drawn it is essential to select a volume range small enough as compared with the volume of the gas so that the liquefaction can be recognized. Otherwise, it is not possible to see detailed behaviors specific in the volume range located between the two vertical red dotted lines shown in Fig.44. As is known from the scale along the horizontal axis, the volume range is of the magnitude of one-hundredth of the volume of the gas at the standard state. Alternatively, working out the critical volume first and then the vertical scale can be selected so as to include the critical volume.

Ans. See Fig.44.

Solution 3-77

The premise here is that the working substance is a monoatomic gas and behaves ideally. Particularly, it is required to know that the internal energy of the ideal gas is a function of temperature alone. Specifically, in order to calculate work or heat associated with the three steps, the following three things are needed.

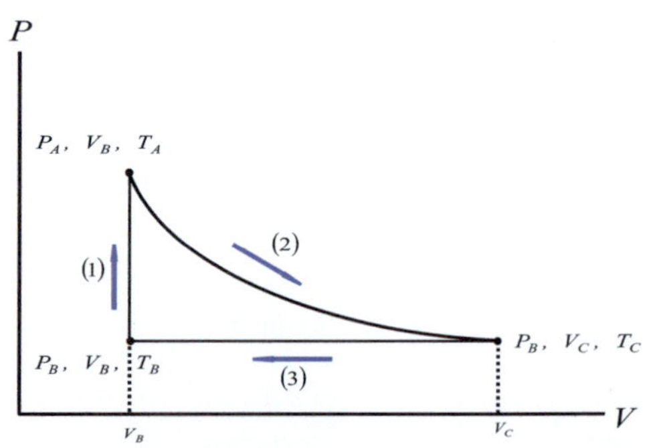

Fig.45 Cyclic process composed of three steps of isovolumetric heating–adiabatic expansion–isobaric compression.

(i) equation of state of ideal gas : $PV = RT$

(ii) $P-V$ or $P-T$ relation for the adiabatic prosess :

$$PV^{\gamma} = \text{const.}\ , \quad TV^{\gamma-1} = \text{const.}', \quad \gamma = C_P / C_V$$

(iii) temperature dependence of internal energy of ideal gas :

$$dU = C_V dT$$

Also, for the mono-atomic gas, $C_V = 3R/2$ and $C_P = 5R/2$ are used for calculation.

With these preparations, changes in the heat energy : ΔQ and work : $\Delta W = -\int PdV$ are obtained for each step, noticing the relation of $T_A > T_C > T_B$ derived from $PV = RT$.

(a) step(1) (isovolumetric heating, see Fig.45)

Based on the relation given in (iii), the change in the internal energy is found as eq(425). For this step, since there is no volume change, eq(426) holds and then from eqs(425) and (426), eq(427) is derived.

$$\Delta U_1 = \Delta W_1 + \Delta Q_1 = C_V (T_A - T_B) \qquad (425)$$

$$\Delta W_1 = -\int_{V_A}^{V_B} PdV = 0 \qquad (426)$$

$$\Delta Q_1 = C_V (T_A - T_B) \qquad (427)$$

step(2) (adiabatic expansion, see Fig.45)

Because of an adiabatic nature of the process, there is no heat exchange in step(426), i.e., $\Delta Q_2 = 0$. In order to work out ΔW_2, the *P-V* relation given by eq(428) is needed (see (ii) above) and after the relevant integration performed, eq(429) is obtained.

$$P = CV^{-\gamma} \qquad (428) \qquad C: \text{ constant}$$

$$\Delta W_2 = -\int_{V_B}^{V_C} PdV = -\int_{V_B}^{V_C} CV^{-\gamma} dV = \frac{C}{\gamma - 1}(V_C^{-\gamma+1} - V_B^{-\gamma+1})$$

$$= \frac{P_A V_B^{\gamma}}{\gamma - 1}\left\{\left(\frac{P_A}{P_B}\right)^{\frac{-\gamma+1}{\gamma}} V_B^{-\gamma+1} - V_B^{-\gamma+1}\right\}$$

$$= \frac{P_A V_B}{\gamma - 1}\left\{\left(\frac{P_A}{P_B}\right)^{\frac{-\gamma+1}{\gamma}} - 1\right\} \qquad (429)$$

Here, it should be noticed that V_C *and* T_C satisfy the relations given by eqs(430) and (431), respectively.

$$P_A V_B^{\gamma} = P_B V_C^{\gamma} = C \ \Rightarrow\ V_C = \left(\frac{P_A}{P_B}\right)^{\frac{1}{\gamma}} V_B \qquad (430)$$

$$T_A V_B^{\gamma-1} = T_C V_C^{\gamma-1} \ \Rightarrow\ T_C = T_A \left(\frac{V_B}{V_C}\right)^{\gamma-1} = T_A \left(\frac{P_B}{P_A}\right)^{\frac{\gamma-1}{\gamma}} \qquad (431)$$

step(3) (isobaric compression, see Fig.45)

The change in the internal energy is found first as eq(432) and then the work gained as eq(433).

$$\Delta U_3 = C_V (T_B - T_C) = C_V \left(T_B - T_A \frac{V_B^{\gamma-1}}{V_C^{\gamma-1}}\right) = C_V \left\{T_B - T_A \left(\frac{P_B}{P_A}\right)^{\frac{\gamma-1}{\gamma}}\right\}$$

$$= C_V T_B \left\{ 1 - \left(\frac{T_A}{T_B} \right) \cdot \left(\frac{T_B}{T_A} \right)^{\frac{\gamma-1}{\gamma}} \right\} = C_V T_B \left\{ 1 - \left(\frac{T_A}{T_B} \right)^{\frac{1}{\gamma}} \right\} \quad (432)$$

$$\Delta W_3 = -P_B \int_{V_C}^{V_B} dV = P_B (V_C - V_B) = P_B V_B \left\{ \left(\frac{P_A}{P_B} \right)^{\frac{1}{\gamma}} - 1 \right\} = R T_B \left\{ \left(\frac{T_A}{T_B} \right)^{\frac{1}{\gamma}} - 1 \right\} \quad (433)$$

Consequently, the change in the heat energy is given by eq(434).

$$\Delta Q_3 = \Delta U_3 - \Delta W_3 = C_V (T_B - T_C) - R T_B \left\{ \left(\frac{T_A}{T_B} \right)^{\frac{1}{\gamma}} - 1 \right\}$$

$$= \frac{3}{2} R T_B \left\{ 1 - \left(\frac{T_A}{T_B} \right)^{\frac{1}{\gamma}} \right\} - R T_B \left\{ \left(\frac{T_A}{T_B} \right)^{\frac{1}{\gamma}} - 1 \right\}$$

$$= \frac{5}{2} R T_B \left\{ 1 - \left(\frac{T_A}{T_B} \right)^{\frac{1}{\gamma}} \right\} \quad (434)$$

Ans. $\Delta W_1 = 0,\ \Delta Q_1 = \frac{3}{2} R (T_A - T_B);\quad \Delta W_2 = \frac{P_A V_B}{\gamma - 1} \left\{ \left(\frac{P_A}{P_B} \right)^{\frac{-\gamma+1}{\gamma}} - 1 \right\},\ \Delta Q_2 = 0;$

$\Delta W_3 = P_B V_B \left\{ \left(\frac{P_A}{P_B} \right)^{\frac{1}{\gamma}} - 1 \right\} = R T_B \left\{ \left(\frac{T_A}{T_B} \right)^{\frac{1}{\gamma}} - 1 \right\},\quad \Delta Q_3 = \frac{5}{2} R T_B \left\{ 1 - \left(\frac{T_A}{T_B} \right)^{\frac{1}{\gamma}} \right\}$

(b)The efficiency of work is defined by $\eta = \frac{\text{work done by the system to the surroundin gs}}{\text{heat gained from hot reservior}}$.

Then, it is calculated as below.

$$\eta = \frac{\Delta W_2 + \Delta W_3}{\Delta Q_1} = \frac{\frac{R T_B}{\gamma - 1} \left(\left(\frac{T_A}{T_B} \right)^{\frac{-\gamma+1}{\gamma}} - 1 \right) + R T_B \left(\left(\frac{T_A}{T_B} \right)^{\frac{1}{\gamma}} - 1 \right)}{C_V (T_A - T_B)}$$

$$= \frac{R T_B \left\{ \frac{1}{\gamma - 1} \left(\left(\frac{T_A}{T_B} \right)^{\frac{-\gamma+1}{\gamma}} - 1 \right) + \left(\left(\frac{T_A}{T_B} \right)^{\frac{1}{\gamma}} - 1 \right) \right\}}{\frac{3R}{2} (T_A - T_B)}$$

$$= \frac{\frac{2}{3} T_B \left\{ \frac{3}{2} \left(\left(\frac{T_A}{T_B} \right)^{-2/5} - 1 \right) + \left(\left(\frac{T_A}{T_B} \right)^{3/5} - 1 \right) \right\}}{(T_A - T_B)}$$

$$= \frac{\left\{ \left(\left(\frac{T_A}{T_B} \right)^{-2/5} - 1 \right) + \frac{2}{3} \left(\left(\frac{T_A}{T_B} \right)^{3/5} - 1 \right) \right\}}{\left(\frac{T_A}{T_B} - 1 \right)} \quad (435)$$

Here the value of $\gamma = C_P/C_V = 5/3$ is used. As a final form, eq(435) is obtained. It is seen that the efficiency of work is expressed by a function of the ratio of the temperatures of the heat reservoirs, i.e., T_A/T_B. A value for η for $T_A = 900$ K and $T_B = 300$ K is calculated to be 0.133, which is much smaller than 0.667 to be found for the *Carnot* cycle working at the same two temperatures

Ans. efficiency of work: eq(435)

Just to make sure, it is confirmed below that the sum of the change in the internal energy of the working substance calculated for each step is zero for one cycle.

$$\begin{aligned}\varDelta U_1 + \varDelta U_2 + \varDelta U_3 &= \varDelta Q_1 + \varDelta W_2 + \varDelta W_3 + \varDelta Q_3 \\ &= C_V(T_A - T_B) + \varDelta W_2 + C_V(T_B - T_C) \\ &= C_V(T_A - T_C) + \varDelta W_2 \\ &= C_V\left(T_A - T_A\left(\frac{P_B}{P_A}\right)^{\frac{\gamma-1}{\gamma}}\right) + \frac{P_A V_B}{\gamma - 1}\left(\left(\frac{P_A}{P_B}\right)^{\frac{-\gamma+1}{\gamma}} - 1\right) \\ &= C_V T_A\left(1 - \left(\frac{T_B}{T_A}\right)^{\frac{\gamma-1}{\gamma}}\right) + \frac{1}{\gamma-1} R T_A\left(\left(\frac{T_A}{T_B}\right)^{\frac{-\gamma+1}{\gamma}} - 1\right) \\ &= \frac{3}{2} R T_A\left(1 - \left(\frac{T_B}{T_A}\right)^{2/5}\right) + \frac{3}{2} R T_A\left(\left(\frac{T_A}{T_B}\right)^{-2/5} - 1\right) \\ &= \frac{3}{2} R T_A - \frac{3}{2} R T_A^{3/5} T_B^{2/5} + \frac{3}{2} R T_A^{3/5} T_B^{2/5} - \frac{3}{2} R T_A = 0 \qquad (436)\end{aligned}$$

Solution 3-78

Writing the ratio of the temperatures of a hot heat reservior at T_A and a cold resevior at T_B as x, the efficiency of work for the cyclic process in the present case is given by eq(437), which is obtained as eq(435) in the preceding solution. Also, the efficiency for the *Carnot* cycle is expressed by eq(438).

$$\eta = \frac{\left\{\frac{2}{3} x^{\frac{3}{5}} + x^{-\frac{2}{5}} - \frac{5}{3}\right\}}{(x-1)}, \qquad x = T_A/T_B \qquad (437), \qquad \eta_{Carnot} = \frac{T_A - T_B}{T_A} = \frac{x-1}{x} \qquad (438)$$

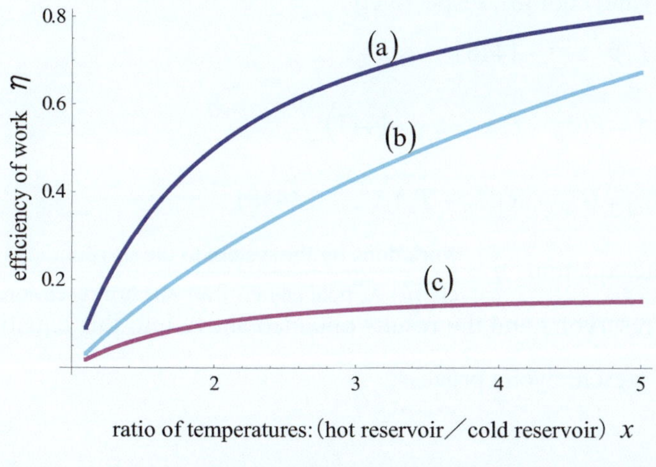

Fig.46 A comparison of the efficiencies of work.
(a) *Carnot* cycle,
(b) three steps of isovolumetric compression – isothermal expansion – isobaric compression (see **problem 3-80**),
(c) three steps of isometric compression – adiabatic expansion – isobaric compression (see **problem 3-77**).

In Fig.44 above are plotted the efficiencies given by eqs(437) and (438) against x. It is seen from Fig.46 that the cyclic process consisting of the three steps treated in the **problem 3-77** is significantly less efficient as compared with the *Carnot* cycle composed of the four steps of isothermal expansion – adiabatic expansion – isothermal compression – adiabatice compression.

Ans. comparison of the efficiency of work: See Fig.46.

Solution 3-79

step (1) (isovolumetric heating)

A change in the internal energy of the ideal gas is always given by eq(439). Therefore, the change in the internal energy for the step (1) is given by eq(440). Since there is no volume change during this step the change in the energy of work (eq(441)) is zero. From eqs(440) and (441), eq(442) is obtained for the change of heat.

$$dU = C_V dT \qquad (439)$$

$$\Delta U_1 = \Delta W_1 + \Delta Q_1 = C_V (T_A - T_B) \qquad (440)$$

$$\Delta W_1 = -\int_{V_B}^{V_B} PdV = 0 \qquad (441)$$

$$\Delta Q_1 = C_V (T_A - T_B) \qquad (442)$$

step (2) (isothermal expansion)

The internal energy of the ideal gas does not change for the isothermal process as is given by eq(443). Work done to the surroundings is calculated by eq(444). Putting the two results together, q(445) is obtained for the change in the heat energy.

$$\Delta U_2 = \Delta W_2 + \Delta Q_2 = 0 \qquad (443)$$

$$\Delta W_2 = -\int_{V_A}^{V_B} PdV = -\int_{V_A}^{V_B} \frac{RT_A}{V} dV = -RT_A \ln\frac{V_B}{V_A} = -RT_A \ln\frac{T_A}{T_B}, \quad (n = 1\,\text{mole}) \qquad (444)$$

$$\Delta Q_2 = RT_A \ln\frac{V_B}{V_A} = RT_A \ln\frac{T_A}{T_B} \qquad (445)$$

step (3) (isobaric compression)

The change in the internal energy is found first by eq(446) as is clear from eq(439). Also, the changes in the energy of work and heat are found by eq(447) and eq(448), respectively.

$$\Delta U_3 = \Delta W_3 + \Delta Q_3 = C_V (T_B - T_A) \qquad (446)$$

$$\Delta W_3 = -\int_{V_B}^{V_A} PdV = -P_B (V_A - V_B) = R(T_A - T_B) \qquad (447)$$

$$\Delta Q_3 = C_V (T_B - T_A) - R(T_A - T_B) = (C_V + R)(T_B - T_A) \qquad (448)$$

Since the efficiency of work η is defined by the equation $\eta = \dfrac{\text{work done by the system to the surroundings}}{\text{energy of heat gained from the hot reservior}}$, substituting the temperatures of the hot and cold reservoirs and the results obtained above into this equation, eq(449) is found for the efficiency of work of the present cyclic process.

$$\eta = \frac{|\Delta W_1 + \Delta W_2 + \Delta W_3|}{\Delta Q_2} = \frac{RT_A \ln\frac{T_A}{T_B} - R(T_A - T_B)}{RT_A \ln\frac{T_A}{T_B}} = 1 - \frac{1}{\ln x} + \frac{1}{x \ln x}, \quad x = \frac{T_A}{T_B} \qquad (449)$$

Ans. See eq(449) and Fig.44 in **solution 3-78**

Solution 3-80

The work W, the internal energy U and the heat Q change for each step of the cycle are given below.

① $\Delta W_1 = -\int_{V_1}^{V_2} P dV = -\int_{V_1}^{V_2} \frac{RT_2}{V} dV = -RT_2 \ln\frac{V_2}{V_1}, \quad \Delta U_1 = 0, \quad \Delta Q_1 = RT_2 \ln\frac{V_2}{V_1}$

② $\Delta W_2 = 0, \quad \Delta Q_2 = \int_{T_2}^{T_1} \frac{C_V}{T} dT = \frac{3R}{2} \ln\frac{T_1}{T_2}$

③ $\Delta W_3 = -\int_{V_2}^{V_1} P dV = -\int_{V_2}^{V_1} \frac{RT_1}{V} dV = -RT_1 \ln\frac{V_1}{V_2}, \quad \Delta U_3 = 0, \quad \Delta Q_3 = RT_1 \ln\frac{V_1}{V_2}$

④ $\Delta W_4 = 0, \quad \Delta Q_4 = \int_{T_1}^{T_2} \frac{C_V}{T} dT = \frac{3R}{2} \ln\frac{T_2}{T_1}$

After completing one cycle, the changes of work and heat are given by eqs(450) and (451), respectively,

$$\Delta W = \Delta W_1 + \Delta W_2 + \Delta W_3 + \Delta W_4 = -RT_2 \ln\frac{V_2}{V_1} - RT_1 \ln\frac{V_1}{V_2} = -R \ln\frac{V_2}{V_1}(T_2 - T_1) \qquad (450)$$

$$\Delta Q = \Delta Q_1 + \Delta Q_2 + \Delta Q_3 + \Delta Q_4 = RT_2 \ln\frac{V_2}{V_1} + \frac{3R}{2}\ln\frac{T_1}{T_2} + RT_1 \ln\frac{V_1}{V_2} + \frac{3R}{2}\ln\frac{T_2}{T_1} = R \ln\frac{V_2}{V_1}(T_2 - T_1) \qquad (451)$$

and, naturally, their sum must be zero (eq(452)).

$$\Delta W + \Delta Q = 0 \qquad (452)$$

The efficiency of work is expressed by eq(453) from which it is found that the efficiency of work of this cycle is <u>the same as that for the *Carnot* cycle</u>.

$$\eta = \frac{|\Delta W|}{Q_1} = \frac{-(\Delta W_1 + \Delta W_3)}{Q_1}\left(RT_2 \ln\frac{V_2}{V_1} + RT_1 \ln\frac{V_1}{V_2}\right) \Big/ \left(RT_2 \ln\frac{V_2}{V_1}\right) = \frac{T_2 - T_1}{T_2} = 1 - \frac{T_1}{T_2} \qquad (453)$$

Ans. eq(453), same as the *Carnot* cycle

"note" For a cycle where temperatures of heat reservoirs are not constant, see **problem 3-106**. In such a case, the efficiency of work cannot be defined in terms of the temperatures of the heat reservoirs.

Solution 3-81

(a), (b) Assuming that the refrigerator is an ideal heat engine, the temperatures of a hot and cold reservoirs can be regards as 298 K and 273 K, respectively. The efficiency η (maximum performance) of the refrigerator, therefore, is related to the amount of heat obtained from the hot reservoir Q_h and the one discarded to the cold reservoir Q_c through eq(454). See **problem 3-11** for an explanation for the *Carnot* cycle.

$$\eta = \frac{298 - 273}{298} = 0.0839 = 1 - \frac{Q_c}{Q_h} \qquad (454)$$

The refrigerator is a machine which operates by reversing entirely the cycle of this ideal heat engine and, hence, gains heat Q_c from the cold reservoir i.e., freezer and emits heat Q_h to the hot reservoir i.e., the room. In order to convert 1 kg of water into 1 kg of ice both at $0\,^\circ\mathrm{C}$, it is necessary to remove heat of $79.8 \times 4.18 \times 10^3\,\mathrm{J} = Q_c$ from the water and to discard heat Q_h to the room. Consequently, Q_h necessary to be discarded is calculated using the efficiency η and is found from eq(454) to be 364 kJ as is shown below.

$$Q_h = \frac{Q_c}{1-\eta} = \frac{298}{273} \times 79.8 \times 4.18 \times 10^3 = 364\ \mathrm{kJ} \qquad (454')$$

For the refrigerator to function properly in cycle, the energy deficient should be constantly supplied externally as energy of work. The minimum work W_{min} required in the present case is calculated to be 30.5 kJ from the equation given below.

$$W_{min} = \eta \times Q_h = Q_h - Q_c = 0.0839 \times 364 = 30.5\ \mathrm{kJ}$$

Ans. (a) $W_{min} = 30.5\,\mathrm{kJ}$, (b) heat to be discarded: 364 kJ

Solution 3-82

(a) By employing the non-linear least squares fitting provided by the free software gnuplot, not only the best-fitted values for the parameters but also the errors associated with the parameters can be obtained. The best-fitted function for the equation given in the problem is obtained as eq(455).

$$P = (-251968 \pm 2.789 \times 10^4) + \frac{1.0452 \times 10^7 \mp 1.256 \times 10^6}{T} + (38103 \mp 4154) \times \ln T \qquad (455)$$

Values written after double sings ($\pm$, $\mp$) denote the errors associated with the corresponding parameters. Since the errors are correlated one another and the double sings must be used consistently in the same order, i.e., double-sign correspondence. Employing the obtained values, eq(455) is plotted in the temperature range $[\,270, 400\,]\,\mathrm{K}$ in Fig.48 below. The brown curve located in the middle is drawn neglecting the errors of all parameters. The blue curve is drawn employing the errors given by the upper-side sings and the green curve employing the errors given by the lower-side signs. In the temperature range experimentally employed, the three fitted curves almost overlap one another. However, when the temperature range is extended either way, the three curves diverge from one another significantly reflecting the importance of the errors associated with the parameters. See **II-3** for a primitive usage of the free software gnuplot.

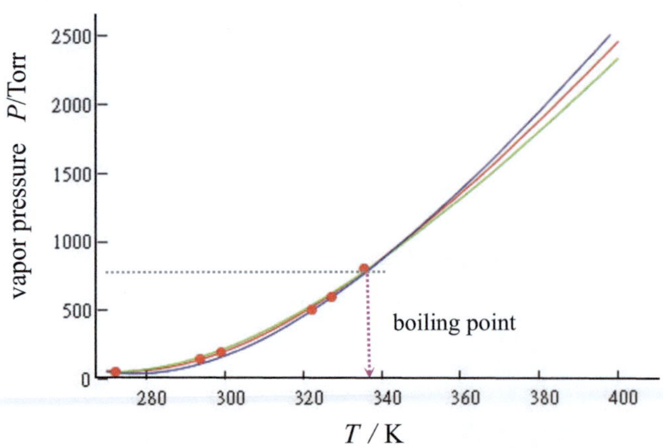

Fig.48 Vapor pressure curve — best-fitted curves by gnuplot (eq(455)).

● : experimental values in the table,

—— : best-fitted curve without errors,

—— / —— : with errors indicated by double signs.

(b)The pressure is given in the unit of Torr and, hence, the boiling point T_b is the temperature at which the vapor pressure reaches 760 Torr. Thus from eq(456), T_b is found as 334.6 K when the errors are not taken into account.

$$760 = (-251968 \pm 2.789 \times 10^4) + \frac{1.0452 \times 10^7 \mp 1.256 \times 10^6}{T} + (38103 \mp 4154) \times \ln T \qquad (456)$$

Instead, employing the functions for which the errors are taken into account, the boiling point is found as $T_b = 335.2$ K or 334.1K, revealing in this case that the effect of the errors associated with the best-fitted parameters on T_b is very limited.

The heat of vaporization is found by applying eq(455) to *Clausius-Clapeyron*'s equation given by eq(457).

$$\frac{1}{P}\frac{dP}{dT} = \frac{\Delta H_{vap}}{RT^2} \qquad (457)$$

Differentiating eq(455) with T, eq(458) is obtained and combining it with eq(457), eq(459) is derived to give ΔH_{vap}. By equating $P = 760$ Torr in eq(459), the heat of vaporization at 1atm ΔH°_{vap} is found as below.

$$\frac{dP}{dT} = -\frac{b}{T^2} + \frac{c}{T} \qquad (458)$$

$$\frac{\Delta H_{vap}}{RT^2} = \frac{1}{P}\left(-\frac{b}{T^2} + \frac{c}{T}\right) \Rightarrow \Delta H^\circ_{vap} = \frac{R}{760}(-b + cT) \qquad (459)$$

$$\Delta H^\circ_{vap} = \frac{8.31}{760}(-1.0452 \times 10^7 \mp 1.256 \times 10^6 + (38103 \mp 4154) \times 334.6) = 25.1 \mp 1.5\,\text{kJ} \cdot \text{mol}^{-1}$$

Ans. $T_b = 334.6 \pm 0.5\,\text{K}$, $\Delta H^\circ_{vap} = 25.1 \mp 1.5\ \text{kJ} \cdot \text{mol}^{-1}$

Solution 3-83

Substituting the values given into the rewritten *Clausius-Clapeyron*'s equation (right side of eq(460))), the required pressure change dP is found. Also, it is seen from eq(460) that a reduction of pressure from 1 atm to high vacuum, practically zero, is not enough to cause any significant change in the transition temperature.

$$\frac{dP}{dT} = \frac{\Delta H_l}{T\Delta V_{r \to m}} \Rightarrow dP = \frac{\Delta H_l}{T\,\Delta V_{r \to m}} dT \qquad (460)$$

$$dP = \frac{2.52 \times 4.18 \times 10^3}{368.6 \times 1.26 \times 10^{-5}} \times 3.0 = 6.80 \times 10^6\,\text{J} \cdot \text{m}^{-3} = 6.80\,\text{MPa}$$

Ans. pressure to be applied: 6.80 MPa ; impossible to lower the transition temperature by reducing pressure.

Solution 3-84

The difference of the molar heat capacities $C_P - C_V$ is expressed as a function of P, V and T by eq(461).

$$C_P - C_V = T\left(\frac{\partial P}{\partial T}\right)_V \left(\frac{\partial V}{\partial T}\right)_P = -T\left(\frac{\partial P}{\partial V}\right)_T \left(\frac{\partial V}{\partial T}\right)_P^2 \qquad (461)$$

The equation is applicable not only to gas but also to liquid or solid. In the case of solid, it is convenient to express it by employing the cubic expansion coefficient α and the compressibility coefficient β as eq(462), in which M is the molecular (atomic) weight and ρ is the density.

$$\alpha = \frac{1}{V}\left(\frac{\partial V}{\partial T}\right)_P, \quad \beta = -\frac{1}{V}\left(\frac{\partial V}{\partial P}\right)_T, \quad V = \frac{M}{\rho}$$

$$C_P - C_V = \frac{\alpha^2 VT}{\beta} \qquad (462)$$

$$= \frac{(2.13 \times 10^{-5})^2 \times 298 \times 26.98 \times 10^{-3}}{1.32 \times 10^{-11} \times 2.70 \times 10^3} = 0.102\,\text{J} \cdot \text{mol}^{-1} \cdot \text{K}^{-1}$$

Substituting the given values into eq(462), $0.102\ \mathrm{J\cdot mol^{-1}\cdot K^{-1}}$ is found for the answer required as shown above. Just to make sure, the unit of the heat capacity given by the right side of eq(462) is confirmed below. Certainly, it turns out to be $\mathrm{J\cdot mol^{-1}\cdot K^{-1}}$.

$$\frac{\mathrm{K^{-2}\cdot K\cdot kg\cdot mol^{-1}}}{\mathrm{Pa^{-1}\cdot kg\cdot m^{-3}}}=\frac{\mathrm{Pa\cdot m^{3}\cdot mol^{-1}}}{\mathrm{K}}=\frac{\mathrm{N\cdot m\cdot mol^{-1}}}{\mathrm{K}}=\mathrm{J\cdot mol^{-1}\cdot K^{-1}}$$

Judging from the result obtained, for solids in general, the approximation of $C_P \approx C_V$ is valid. In theoretical treatment, it is easier to treat the volume as constant, but when dealing with experimental values, the condition of constant pressure is achieved more readily and convenient. For that, occasionally it becomes necessary to convert the quantities to each other using eq (461) or eq(462)

Ans. difference in heat capacities: $0.102\ \mathrm{J\cdot mol^{-1}\cdot K^{-1}}$, for solids: $C_P \approx C_V$

Solution 3-85

The *T-V* relation of the adiabatic process is given by eq(463) for the ideal gas. (See **solution 3-42a** for derivation.)

$$TV^{\gamma-1}=\text{const.} \qquad \gamma=C_P/C_V \qquad (463)$$

A value of const. can be determined from the temperature and pressure before compression. Since both gases are diatomic molecules, their molar heat capacities can be taken as $C_V=5R/2$ and $C_P=7R/2$ if the contribution of the rotational motion is included and, hence, their ratio becomes $\gamma=7/5$.

Consequently, denoting the temperature and volume after compression as T_f and V_f, respectively and equating $V_f=1/10\ \mathrm{dm^3}$, T_f is found from eq(464) as $T_f=749\ \mathrm{K}$.

$$T_f V_f^{\,2/5}=298\times(1.0)^{2/5} \qquad (464)$$

Also, the pressure after compression is calculated from $P=nRT/V$ (for $n=0.0408$ moles) equal to $2.54\times10^{6}\ \mathrm{Pa}$, i.e., ca.25 atm. Because the found temperature and pressure well exceed the explosion limit, it is quite natural that the mixture exploded by the sudden compression. Then, what will happen if the initial pressure of the mixture is 0.1 atm?

Ans. temperature and pressure after compression: $749\ \mathrm{K}$ and $2.54\times10^{6}\ \mathrm{Pa}$

"note" For gas mixtures of hydrogen, carbon monoxide or carbon disulfide with oxygen, when their temperature and pressure exceed certain limits (explosion limits), chemical reactions proceed explosively. There exist more than one explosion regions such as the lowest or the highest pressure. For further details, refer to, for instance, *S.W.Benson, The Foundations of The Chemical Kinetics,* p.452.

Solution 3-86

Two types of heat of formation are considered. One is calculated from the thermodynamic values such as the heat of combustion and the other is obtained from bond energies. The difference between these two is ascribed to the resonance energy as is illustrated in Fig.49. In order to acquire the heat of formation based on the latter type, it is necessary to assume a type of bonding to each bond of the molecule under discussion. In the case of toluene, its *Kekulé* structure shown in Fig.50 is employed. Also, since a theoretical value is compared with an experimental one, in contrast to the case in which thermochemically the standard enthalpy of formation is obtained, it is assumed that all substances are in their atomic states in the initial state of the reaction and that in the final state all products are in the gas state. Namely, the heat of formation of the reaction given below is obtained in the two different manners.

$$7C(g) + 8H(g) \rightarrow C_7H_8(g)$$

i) <u>heat of formation of toluene acquired from thermodynamical values such as heat of combustion</u>

The reactions given in the problem are shown below again.

$$C_7H_8(l) + 9O_2(g) \rightarrow 7CO_2(g) + 4H_2O(l) \qquad \Delta H^\circ_{conv.} = -3910 \text{ kJ} \qquad (51)$$

$$C_7H_8(l) \rightarrow C_7H_8(g) \qquad \Delta H^\circ_{vap.} = 28.1 \text{ kJ} \qquad (52)$$

$$H_2(g) \rightarrow 2H(g) \qquad \Delta H^\circ_{diss.} = 436 \text{ kJ} \qquad (53)$$

$$C(s) \rightarrow C(g) \qquad \Delta H^\circ_{sub.} = 715 \text{ kJ} \qquad (54)$$

$$H_2(g) + \frac{1}{2}O_2(g) \rightarrow H_2O(l) \qquad \Delta H^\circ_{conv.} = -286 \text{ kJ} \qquad (55)$$

$$C(s) + O_2(g) \rightarrow CO_2(g) \qquad \Delta H^\circ_{conv.} = -394 \text{ kJ} \qquad (56)$$

In order to eliminate $CO_2(g)$ and $4H_2O(l)$ from reaction (51), the operation (a) below is performed using reactions (51), (56) and (55), giving reaction (a').

$$(1) - 7\times(6) - 4\times(5) \quad : \quad (a)$$

$$C_7H_8(l) - 7C(s) - 4H_2(g) \rightarrow 0 \qquad (a')$$

Employing this equation together with reactions (53) and (54), the operation (b) below is performed, giving reaction (58).

$$(7) + 7\times(4) + 4\times(3) \quad : \quad (b)$$

$$C_7H_8(l) \rightarrow 7C(g) + 8H(g) \qquad (b')$$

From reactions (58) and (52), the operation (c) yields finally the required reaction (c').

$$(2) - (8) \quad : \quad (c)$$

$$0 \rightarrow C_7H_8(g) - 7C(g) - 8H(g) \Rightarrow 7C(g) + 8H(g) \rightarrow C_7H_8(g) \qquad (c')$$

The enthalpy of reaction (c') is obtained by performing the same operations (a)-(c) used above upon the enthalpy of reaction for each reaction involved. Denoting the value to be obtained as $\Delta H^\circ_f(c')$, it is calculated as shown below.

$$\begin{aligned}\Delta H^\circ_f(c') &= \Delta H^\circ_{vap.}(52) - \{\Delta H^\circ_{conv.}(51) - 7\times\Delta H^\circ_{conv.}(56) - 4\Delta H^\circ_{conv.}(55) + 7\times\Delta H^\circ_{sub.}(54) + 4\times\Delta H^\circ_{diss.}(53)\}\\ &= 28.1 - \{-3910 - 7\times(-394) - 4\times(-286) + 7\times 715 + 4\times 436\}\\ &= -6725 \text{ kJ}\cdot\text{mol}^{-1}\end{aligned}$$

ii) <u>heat of formation of toluene acquired from bond energies</u>

The *Kekulé* structure of toluene is composed of eight C–H bonds, three C=C bonds and four C–C bonds. Therefore, the reaction energy calculated from bond energies is equal to the sum of the bond energies of these bonds. Since the reaction (c') is a bond formation reaction (exothermic), a negative sign is put to the calculated value as shown below.

$$-(413\times 8 + 610\times 3 + 346\times 3) = -6518 \text{ kJ}\cdot\text{mol}^{-1}$$

The resonance energy is the difference between the values obtained in i) and ii), so that the value of $-6725-(-6518)=-207\ \mathrm{kJ\cdot mol^{-1}}$ is found. Since the product is stabilized because of resonance, thermodynamically speaking, the resonance energy should be negative but conventionally, the stabilized energy is employed as it is for the resonance energy.

Ans. resonance energy: $207\,\mathrm{kJ\cdot mol^{-1}}$

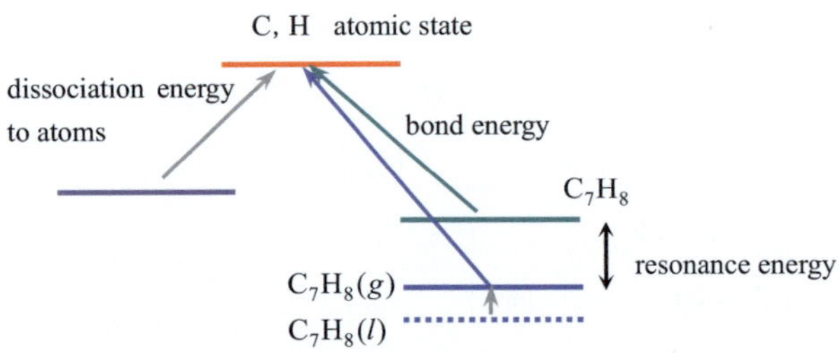

Fig.49 Energy-level diagram showing the relation among the dissociation energy, bond energy and resonance energy.

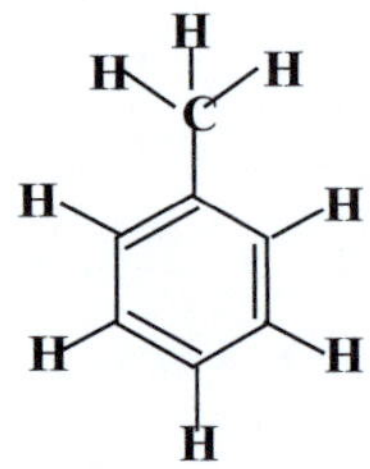

Fig.50 *Kekulé* structure.

Solution 3-87

The heat of dissolution produced when sulfuric acid is dissolved in water is plotted against the mole fraction of sulfuric acid X in Fig.51 below. The amount of sulfuric acid to be added is always 1.0 mole. The curve in Fig.51 is given by an approximate function obtained by gnuplot so as for the four data points to fit the curve most closely.

$$\text{heat of dissolution (kcal)} = 19.27(1-\sqrt{X})\exp(1.16X) \qquad (465)$$

<u>Eq(465) is only meaningful in reproducing data well but lacks any physico-chemical basis.</u>

The mole number n of water in the aqueous sulfuric acid before the second dilution is 55.6 and, hence, the mole fraction of sulfuric acid is given by $X=1/(55.6+1)=0.0177$. Substituting this value into eq(465), 17.04 kcal is found as the heat of dissolution. Namely, one litter of the aqueous sulfuric acid of the concentration of 1.0 mole/L has already produced this amount of heat.

When that solution is further diluted tenfold with water, the mole fraction of sulfuric acid of the diluted solution becomes $X=1/(556+1)=0.00180$. The amount of heat to be produced is calculated as 18.49 kcal

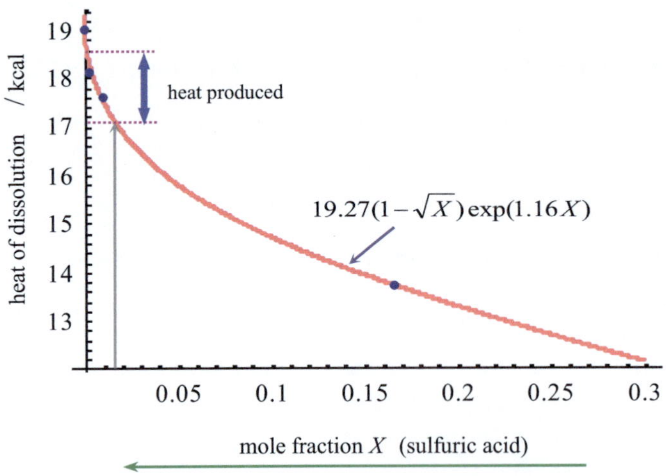

Fig.51 Heat of dissolution (dilution) of sulfuric acid and its mole fraction.

The grey vertical arrow indicates the position of $1.0\,\mathrm{dm^3}$ of $1.0\,\mathrm{mol\cdot dm^{-3}}$ aqueous sulfuric acid. The amount of heat produced upon tenfold dilution of that solution is indicated by the blue double arrows.

from eq(465). Consequently, the heat of dilution when one litter of the solution at the concentration of 1.0 mole/L is diluted tenfold is given by the difference between the heats of dissolution calculated above, i.e., $18.49 - 17.04 = 1.45$ kcal. Assuming the specific heat of the solution is unity, a change in the temperature of the solution is calculated from the equation below as $0.15\,\mathrm{K}$.

$$1(\mathrm{cal/g \cdot K}) \times 10^4(\mathrm{g}) \Delta T = 1.45 \times 10^3(\mathrm{cal}) \;\Rightarrow\; \Delta T = 0.15\,\mathrm{K}$$

Ans. heat produced: 1.45 kcal, temperature rise: 0.15 K

Solution 3-88

The reduction reaction of CdS with H_2 is expressed as eq(466).

$$\mathrm{CdS}(g) + \mathrm{H_2}(g) \rightarrow \mathrm{Cd}(g) + \mathrm{H_2S}(g) \qquad (466)$$

This equation is derived by subtracting eq(58) from eq(57) given in the problem as shown below.

$$(57) - (58)$$
$$\mathrm{H_2}(g) - \mathrm{Cd}(g) \rightarrow \mathrm{H_2S}(g) - \mathrm{CdS}(g) \;\Rightarrow\; \mathrm{CdS}(g) + \mathrm{H_2}(g) \rightarrow \mathrm{Cd}(g) + \mathrm{H_2S}(g)$$
$$\Delta G_3 = \Delta G_1 - \Delta G_2 = 78.6\,\mathrm{kJ \cdot mol^{-1}}$$

Therefore, the free energy change of the reaction expressed as eq(466) is also given by the difference between those for eqs(57) and (58). The obtained free energy change is a large positive value, indicating the equilibrium of the reduction reaction of eq(466) is largely shifted to the left side. Because of this, it is concluded that the reduction reaction of CdS with H_2 at $1100\,^\circ\mathrm{C}$ is not possible.

Ans. not possible by a simple reduction reaction with H_2

Solution 3-89

The total differential of the internal energy U is expressed by eq(467) or eq(468). Eq(467) is relevant when U is considered as a function of S and V and eq(468) is for a function of T and V.

$$dU = TdS - PdV \qquad (467)$$

$$dU = \left(\frac{\partial U}{\partial T}\right)_V dT + \left(\frac{\partial U}{\partial V}\right)_T dV \qquad (468)$$

From eq(467), the partial differentials of eq(468) can be expressed as eq(469) and eq(470).

$$\left(\frac{\partial U}{\partial T}\right)_V = T\left(\frac{\partial S}{\partial T}\right)_V = C_V \qquad (469)$$

$$\left(\frac{\partial U}{\partial V}\right)_T = T\left(\frac{\partial S}{\partial V}\right)_T - P = T\left(\frac{\partial P}{\partial T}\right)_V - P \qquad (470)$$

For the *van der Waals* equation of state expressed as eq(471) ($n = 1$ mole), the left side of eq(470) is found as eq(472).

$$P = \frac{RT}{V-b} - \frac{a}{V^2} \qquad (471)$$

$$\left(\frac{\partial U}{\partial V}\right)_T = T \times \frac{R}{(V-b)} - P = \frac{a}{V^2} \qquad (472)$$

Employing eqs(469) and (472), therefore, the total differential of U expressed as eq(472) can be written as eq(473).

$$dU = C_V dT + \frac{a}{V^2} dV \qquad (473)$$

Since the first partial differential coefficient of eq(473) is a function of T alone and the second one is a function of V alone, the integration of eq(473) can be performed along the two separate routes, first along T and then followed by along V as is expressed by

$(T_0, V_0) \xrightarrow{\text{integration along } T} (T, V_0) \xrightarrow{\text{integration along } V} (T, V)$ and is given by eq(473') as a result.

$$U(T,V) - U(T_0,V_0) = \int_{T_0}^{T} C_V dT + \int_{V_0}^{V} \frac{a}{V^2} dV \qquad (473')$$

When C_V is independent of temperature, the internal energy of the *van der Waals* gas is expressed by eq(473'').

$$U(T,V) - U(T_0,V_0) = C_V (T - T_0) - a\left(\frac{1}{V} - \frac{1}{V_0}\right) \qquad (473'')$$

With respect to the entropy, writing the total differential of S as eq(474), similar calculations as performed for U are repeated and then eqs(478) and (478'') are obtained as functional forms of S required.

$$dS = \left(\frac{\partial S}{\partial T}\right)_V dT + \left(\frac{\partial S}{\partial V}\right)_T dV \qquad (474)$$

$$\left(\frac{\partial S}{\partial T}\right)_V = \frac{C_V}{T} \qquad (475), \qquad \left(\frac{\partial S}{\partial V}\right)_T = \left(\frac{\partial P}{\partial T}\right)_V \qquad (476)$$

$$\left(\frac{\partial S}{\partial V}\right)_T = \frac{R}{(V-b)} \qquad (477)$$

$$dS = \frac{C_V}{T} dT + \frac{R}{(V-b)} dV \qquad (478)$$

$$S(T,V) - S(T_0,V_0) = \int_{T_0}^{T} \frac{C_V}{T} dT + \int_{V_0}^{V} \frac{R}{(V-b)} dV \qquad (478')$$

$$S(T,V) - S(T_0,V_0) = C_V \ln\frac{T}{T_0} + R\ln\frac{(V-b)}{(V_0-b)} \qquad (478'')$$

It should be noted here that out of the two *van der Waals* parameters of a and b, only a appears for U and on the other hand, only b for the expression of S. The reason for this is obvious. When $b = 0$, eq(478'') is reduced to the equation to give the entropy of the ideal gas. It should also be remarked that without employing partition functions described in detail in Chapter 5, the internal energy or entropy can be found from the thermodynamical approach, though it is not the absolute value but a difference from the standard value that can be found. See also **problem 3-10**.

Ans. internal energy: eqs(473) and (473''), entropy: eqs(478) and (478'')

Solution 3-90

It is shown first that the temperature variation of the solubility of a solute is related to the heat of fusion of the solute.

The chemical potential of pure solid picric acid (s) is denoted as $\mu^\circ(s)$ and the chemical potential of picric acid in benzene solution is denoted as $\mu(l, X)$, in which X is the mole fraction of picric acid. The solution is regarded as an ideal solution composed of <u>liquid picric acid</u> and liquid benzene. At the equilibrium of dissolution the chemical potentials of solid picric acid and dissolved picric acid must be equal and, hence, eq(479) is satisfied,

$$\mu^\circ(s,T) = \mu(l,T,X) = \mu(l,T,1.0) + RT\ln X = \mu^\circ(l) + RT\ln X \qquad (479)$$

where $\mu^\circ(l)$ is the chemical potential of pure liquid picric acid.

Rewriting eq(479) as eq(479'), both sides are differentiated by T as is given by eq(480).

$$\frac{\mu^\circ(s,T)-\mu^\circ(l,T)}{RT} = \ln X \qquad (479')$$

$$\frac{d}{dT}\left(\frac{\mu^\circ(s,T)-\mu^\circ(l,T)}{RT}\right) = \frac{d\ln X}{dT} \qquad (480)$$

The left side of eq(480) is calculated as eq(481), in which $\bar{S}$ and $\bar{H}$ represent molar values.

$$\text{differntiation of left sid by } T: \ -\frac{\mu^\circ(s)-\mu^\circ(l)}{RT^2} + \frac{\bar{S}(s)-\bar{S}(l)}{RT} = -\frac{\bar{H}(s,T)-\bar{H}(l,T)}{RT^2} = \frac{\Delta H_{fus.}(T)}{RT^2} \qquad (481)$$

Consequently, eq(480) can be written as eq(482) and $\Delta H_{fus.}(T)$ is the heat of fusion at the temperature T. It is seen from this equation that the temperature variation of the solubility of a solute is related to its heat of fusion.

$$\frac{\Delta H_{fus.}(T)}{RT^2} = \frac{d\ln X}{dT} \qquad (482)$$

When $\Delta H_{fus.}(T)$ can be regarded constant, the integration of eq(482) leads to eq(483),

$$\ln X = -\frac{\Delta H_{fus.}}{RT} + \text{const.} \qquad (483)$$

and, hence, by plotting $\ln X$ against $1/T$, the heat of fusion is found from the slope of the plot.

In the present problem, the molar concentration of picric acid is not so high that its mole fraction can be regarded proportional to the solubility. Therefore, to plot the logarithm of the solubility against $1/T$ is

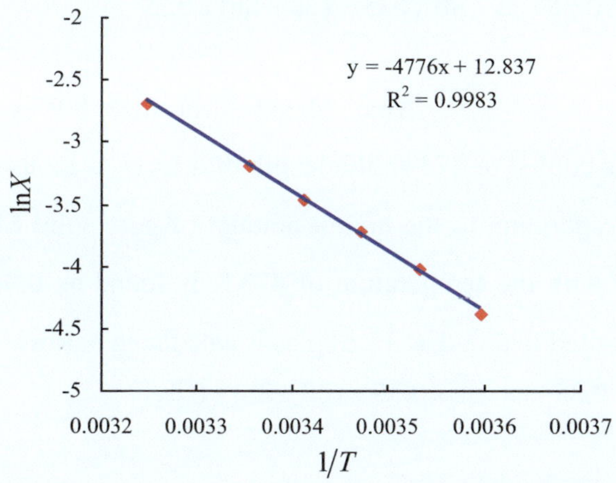

Fig.52 Temperature dependence of the solubility of picric acid in benzene.
The logarithm of the solubility is plotted against the inverse of the absolute temperature (see text). The heat of fusion of picric acid is found from the slope.

sufficient to calculate the heat of fusion (see Fig.52). The value found thus is

$$\Delta H_{fus.} = R \times (4776 \pm 99) = 39.7 \pm 0.8 \ \text{kJ}\cdot\text{mol}^{-1}.$$

Ans. heat of fusion: 39.7 ± 0.8 kJ·mol^{-1}

Solution 3-91

(a) The melting point is plotted against w% in Fig.53 below in which each phase is explained. At Pb 94 w%, an inflection in the curve is recognized, which is considered due to the existence of a eutectic compound (alloy). When the alloy of this composition is heated from room temperature, the alloy behaves as if it were a pure substance and melts at 473 K. The liquid region is simply a mixture of melted Pb and Sb. For the solid phase, however, there are three kinds of solid and they are pure Sb, pure Pb and the eutectic alloy. Their relative amounts are determined by the temperature and the composition ratio.

Ans. See text and Fig.53.

(b) The solid-liquid equilibrium line (brown lines in Fig.53) is obtained first. For this purpose, such lines couldbe drawn with visual measurement. If possible, however, to avoid arbitrariness in data analysis, it is recommended to find the best-fitted line(s). Assuming that the data can be fitted satisfactorily by a quadratic function, it is found by using gnuplot that the melting point T is best-fitted by eq(484),

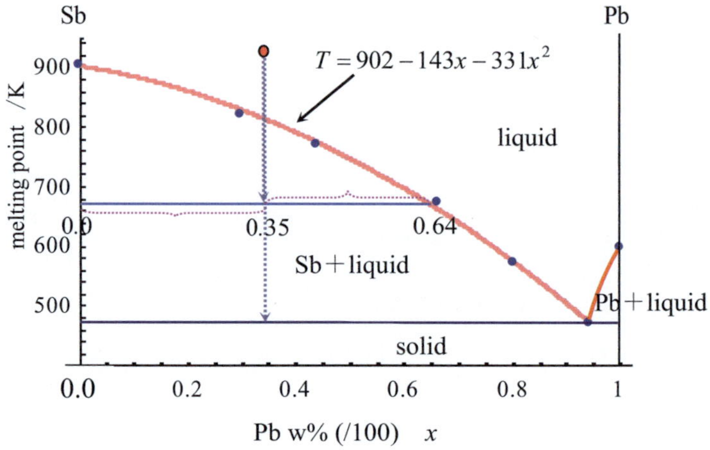

Fig.53 *T-x* diagram of Sb-Pb mixtures.

$$T = 902 - 143x - 331x^2 \qquad (484)$$

where x is given by w%/100. At the composition of 35% at 673 K, solid Sb and liquid co-exist (see Fig.53). The composition x where the equilibrium line intersects with the temperature horizontal line (blue line) is found as $x = 0.64\,(64\%)$ from eq(484) by solving $673 = 902 - 143x - 331x^2$. From this value, using the lever rule (applicable to w% as well), the amount of Sb to be recovered is found as 8.9 kg as shown below.

$$\text{Sb} : \text{liquid} = (0.64 - 0.35) : 0.35 = 0.29 : 0.35 \;\Rightarrow\; \text{Sb} \times 035 = (20 - \text{Sb}) \times 0.29$$

$$\text{Sb} = \frac{20 \times 0.29}{0.64} = 8.9\ \text{kg}$$

Ans. Sb to be deposited: 8.9 kg

(c) From the obtained phase diagram, when the mixture is cooled to 473 K, the length of a lever to be used for the lever rule becomes longest on the right side, corresponding to the largest amount of pure solid Sb. The composition at which the equilibrium line intersects with the temperature of 473 K is found as 0.94. From the lever rule, therefore, the amount of Sb to be deposited is found as 12.6 kg as is calculated below.

$$\text{Sb} : \text{liquid} = (0.94 - 0.35) : 0.35 = 0.59 : 0.35 \;\Rightarrow\; \text{Sb} \times 0.35 = (20 - \text{Sb}) \times 0.59$$

$$\text{Sb} = \frac{20 \times 0.59}{0.94} = 12.6\ \text{kg}$$

The mass of Sb contained in the starting alloy is 13.0 kg. Therefore, through the operation of melting followed by cooling the fraction of $12.6/13.0 = 0.969$, i.e., 96.9% of Sb is recovered as pure Sb.

Ans. pure Sb to be obtained: 12.6 kg

Solution 3-92

Employing *Jacobians*, the variables (T, S) can easily be replaced with the variables (T, V) as is shown in eq(485).

$$\left(\frac{\partial U}{\partial T}\right)_S = \frac{\partial(U,S)}{\partial(T,S)} = \frac{\dfrac{\partial(U,S)}{\partial(T,V)}}{\dfrac{\partial(T,S)}{\partial(T,V)}} = \frac{\begin{vmatrix} \left(\dfrac{\partial U}{\partial T}\right)_V & \left(\dfrac{\partial U}{\partial V}\right)_T \\ \left(\dfrac{\partial S}{\partial T}\right)_V & \left(\dfrac{\partial S}{\partial V}\right)_T \end{vmatrix}}{\left(\dfrac{\partial S}{\partial V}\right)_T} \tag{485}$$

$$= \frac{\left(\dfrac{\partial U}{\partial T}\right)_V\left(\dfrac{\partial S}{\partial V}\right)_T - \left(\dfrac{\partial U}{\partial V}\right)_T\left(\dfrac{\partial S}{\partial T}\right)_V}{\left(\dfrac{\partial S}{\partial V}\right)_T} = \left(\frac{\partial U}{\partial T}\right)_V \tag{485'}$$

The determinant of the numerator of eq(485) is calculated and using $(\partial U/\partial V)_T = 0$ for the ideal gas, the left side of eq(485) is found equal to $(\partial U/\partial T)_V$ as is given in eq(485').

Next, replacing the variables (T,P), with (T,V), in a similar manner as above, eq(486) is derived. As a result, it is found that $(\partial U/\partial T)_P = C_V$ for the ideal gas.

$$\left(\frac{\partial U}{\partial T}\right)_P = \frac{\partial(U,P)}{\partial(T,P)} = \frac{\dfrac{\partial(U,P)}{\partial(T,V)}}{\dfrac{\partial(T,P)}{\partial(T,V)}} = \frac{\left(\dfrac{\partial U}{\partial T}\right)_V\left(\dfrac{\partial P}{\partial V}\right)_T - \left(\dfrac{\partial U}{\partial V}\right)_T\left(\dfrac{\partial P}{\partial T}\right)_V}{\left(\dfrac{\partial P}{\partial V}\right)_T} = \left(\frac{\partial U}{\partial T}\right)_V = C_V \tag{486}$$

For the heat capacity at the constant pressure C_P, the equation given in the problem can be proved similarly, though instead of $(\partial U/\partial V)_T = 0$, formally the relation $(\partial H/\partial P)_T = 0$ is used as is shown in eq(487).

$$\left(\frac{\partial H}{\partial P}\right)_T = \left(\frac{\partial(U+PV)}{\partial P}\right)_T = \left(\frac{\partial U}{\partial P}\right)_T = \left(\frac{\partial U}{\partial V}\right)_T\left(\frac{\partial V}{\partial P}\right)_T = 0 \tag{487}$$

Practical calculations are given by eqs(488) and (489).

$$\left(\frac{\partial H}{\partial T}\right)_S = \frac{\partial(H,S)}{\partial(T,S)} = \frac{\dfrac{\partial(H,S)}{\partial(T,P)}}{\dfrac{\partial(T,S)}{\partial(T,P)}} = \frac{\left(\dfrac{\partial H}{\partial T}\right)_P\left(\dfrac{\partial S}{\partial P}\right)_T - \left(\dfrac{\partial H}{\partial P}\right)_T\left(\dfrac{\partial S}{\partial T}\right)_P}{\left(\dfrac{\partial S}{\partial P}\right)_T} = \left(\frac{\partial H}{\partial T}\right)_P = C_P \tag{488}$$

$$\left(\frac{\partial H}{\partial T}\right)_V = \frac{\partial(H,V)}{\partial(T,V)} = \frac{\dfrac{\partial(H,V)}{\partial(T,P)}}{\dfrac{\partial(T,V)}{\partial(T,P)}} = \frac{\left(\dfrac{\partial H}{\partial T}\right)_P\left(\dfrac{\partial V}{\partial P}\right)_T - \left(\dfrac{\partial H}{\partial P}\right)_T\left(\dfrac{\partial V}{\partial T}\right)_P}{\left(\dfrac{\partial V}{\partial P}\right)_T} = \left(\frac{\partial H}{\partial T}\right)_P = C_P \tag{489}$$

Ans. See text and eqs(485'), (486), (488) and (489).

Solution 3-93

The equation (490) to be proved is the *Maxwell* relation derived from eq(491) which expresses the conservation of the internal energy.

$$\left(\frac{\partial V}{\partial T}\right)_S = -\left(\frac{\partial S}{\partial P}\right)_V \tag{490}$$

$$dU = TdS - PdV \tag{491}$$

Dividing both sides of eq(491) by dP under the condition of $V = \text{constant}$, eq(492) is obtained. As a result, the right side of eq(490) is expressed in terms of P, V and T and experimentally measurable C_V alone.

$$\left(\frac{\partial S}{\partial P}\right)_V = \frac{1}{T}\left(\frac{\partial U}{\partial P}\right)_V = \frac{1}{T}\left(\frac{\partial U}{\partial T}\right)_V\left(\frac{\partial T}{\partial P}\right)_V = \frac{C_V}{T}\left(\frac{\partial T}{\partial P}\right)_V \qquad (492)$$

On the other hand, the left side of eq(490) can be expressed by eq(493) using *Euler*'s chain relation.

$$\left(\frac{\partial V}{\partial T}\right)_S\left(\frac{\partial T}{\partial S}\right)_V\left(\frac{\partial S}{\partial V}\right)_T = -1 \Rightarrow \left(\frac{\partial V}{\partial T}\right)_S = -\left(\frac{\partial S}{\partial T}\right)_V\left(\frac{\partial V}{\partial S}\right)_T \qquad (493)$$

Noticing eq(494) (another *Maxwell* relation for the second equation) is valid,

$$\left(\frac{\partial S}{\partial T}\right)_V = \frac{C_V}{T}, \qquad \left(\frac{\partial V}{\partial S}\right)_T = \left(\frac{\partial T}{\partial P}\right)_V \qquad (494)$$

eq(493) can be rewritten as eq(495) and then from eqs(492) and (495), it is proved that eq(490) holds generally.

$$\left(\frac{\partial V}{\partial T}\right)_S = -\frac{C_V}{T}\left(\frac{\partial T}{\partial P}\right)_V \qquad (495)$$

"note" It may not be convincing to employ another *Maxwell*'s relation in order to prove a specific *Maxwell*'s relation. It may be more straightforward to rephrase the problem as a problem "Express both sides of eq(490) only in terms of *P, V, T* and the specific heat." In the case of the ideal gas, however, it is not necessary to use the *Maxwell* relation. This is because eqs (497) and (498) are satisfied as is derived from eq(496) (see **problem 3-92** for the relation $(\partial U/\partial T)_S = C_V$).

$$dU = C_V dT = TdS - PdV \qquad (496)$$

$$\left(\frac{\partial V}{\partial T}\right)_S = -\frac{1}{P}\left(\frac{\partial U}{\partial T}\right)_S = -\frac{C_V}{P} \qquad (497)$$

$$-\left(\frac{\partial S}{\partial P}\right)_V = -\frac{C_V}{T}\frac{V}{R} = -\frac{C_V}{P} \qquad (498)$$

Also, for mono-atomic gases, eqs(497) and (498) are found equal to $-3V/2T$.

Solution 3-94

Writing the temperature at which the thermal equilibrium is established as T_x, the heat transferred from the hot substance (I) to the cold substance (II) is given by eq(499) (expressed as negative) and heat obtained by the substance (II) is given by eq(500) (expressed as positive). The amount of substance is taken as one mole for each block.

$$\Delta Q(\mathrm{I}) = \int_{400}^{T_x}(A + 2BT)dT = A(T_x - 400) + B(T_x^2 - 400^2) \qquad (499)$$

$$\Delta Q(\mathrm{II}) = \int_{200}^{T_x}(A + 2BT)dT = A(T_x - 200) + B(T_x^2 - 200^2) \qquad (500)$$

Putting substance (I) and (II) together to regard them as one isolated system, there is no heat loss or gain for the system as a whole. Consequently, eq(501) is satisfied.

$$\Delta Q(\mathrm{I}) + \Delta Q(\mathrm{II}) = 2BT_x^2 + 2AT_x - 600A - 20\times 10^4 B = 0 \qquad (501)$$

Solving eq(501) for T_x, after substituting the given values into A and B, $T_x = 315\,\mathrm{K}$ is found.

As is seen from this example, when the specific heat depends on temperature, the equilibrium temperature reached is not equal to a median between the hot and cool temperatures. The heat transfer treated in this problem is an automatic irreversible phenomenon. If the heat transfer is carried out reversibly, then the equilibrium temperature achieved will become lower than 315 K (see next problem).

Ans. equilibrium temperature : 315 K

(b) The entropy changes in the substances (I) and (II) are given by eq(502), where the equilibrium temperature is put $T_x = 315\,\mathrm{K}$.

$$\Delta S(\mathrm{I}) = \int_{400}^{T_x} \frac{A+2BT}{T} dT, \qquad \Delta S(\mathrm{II}) = +\int_{200}^{T_x} \frac{A+2BT}{T} dT \tag{502}$$

Therefore, the total change in entropy ΔS_{tot} is given by eq(503). Substituting the relevant values into eq(503),

$$\begin{aligned}\Delta S_{tot} = \Delta S(\mathrm{I}) + \Delta S(\mathrm{II}) &= \int_{400}^{315} \frac{A+2BT}{T} dT + \int_{200}^{315} \frac{A+2BT}{T} dT \qquad (503)\\ &= 9.6 \times \left(\ln \frac{315}{400} + \ln \frac{315}{200} \right) + 2 \times 0.24 \times (315 \times 2 - 400 - 200)\\ &= 16.5\ \mathrm{J \cdot mol^{-1} \cdot K^{-1}}\end{aligned}$$

$\Delta S_{tot} = 16.5\ \mathrm{J \cdot mol^{-1} \cdot K^{-1}}$ is obtained. Needless to say, this heat transfer process is an irreversible one accompanied by an increase in the entropy.

Ans. change in entropy: $16.5\ \mathrm{J \cdot mol^{-1} \cdot K^{-1}}$

Solution 3-95

The heat transfer treated in the preceding problem is an irreversible one and since nothing happens other than the heat transfer, there is no work available. Maximum work is gained by a reversible process. Its amount depends on what type of reversible process is employ. The following two cases are examined here. In order to achieve the heat transfer reversibly, a working substance (ideal gas) which works reversibly between hot (T_h) and cold (T_l) blocks is introduced.

Approach 1.

Accompanied by heat transfer, not only the temperature of the hot block drops but also the temperature of the cold block rises. When the temperatures of heat sources vary, it is difficult to consider of a working substance to operate between such heat sources. For this, a hypothetical heat reservoir of infinite capacity at temperature T_x is introduced and two (quasi) *Carnot* cycles, one operating between T_h and T_x and the other between T_x and T_l, are considered under the condition that when the heat transfer in question has been completed, there will be neither gain nor loss of heat in the hypothetically introduced heat reservoir.

heat engine (I) working between T_h and T_x

When the temperature of the hot heat source is T and that of the cold reservoir is T_x, the efficiency of the *Carnot* cycle η working between these temperatures is given by eq(504). Also, the temperature of the hot heat source is lowered to $T-dT$ due to the heat transfer to the reservoir, the amount of heat lost is $-C_VdT$ and the work done is given by eq(505).

$$\eta=\frac{T-T_x}{T} \quad (504), \qquad -C_VdT\times\eta=-C_V\left(\frac{T-T_x}{T}\right)dT \quad (505)$$

Consequently, the total amount of work $W(\mathrm{I})$ done while the temperature of the hot heat source varies from T_h to finally T_x is obtained by integrating eq(505) over the temperature range $[T_h, T_x]$ and is given by eq(506), in which the relation $C_V=A+2BT$ is used.

$$|W(\mathrm{I})|=-\int_{T_h}^{T_x}C_V\left(\frac{T-T_x}{T}\right)\cdot dT=-\int_{T_h}^{T_x}(A+2BT)\left(\frac{T-T_x}{T}\right)\cdot dT \quad (506)$$

$$=-\int_{T_h}^{T_x}\left\langle(A+2BT)-\frac{AT_x}{T}-2BT_x\right\rangle\cdot dT$$

$$=-(A-2BT_x)(T_x-T_h)-B(T_x^2-T_h^2)+AT_x\ln\frac{T_x}{T_h}$$

$$=-(9.6-0.48\times T_x)(T_x-400)-0.24(T_x^2-400^2)+9.6T_x\ln\frac{T_x}{400} \quad (507)$$

The heat energy discarded to the hypothetical heat reservoir at T_x is equal to $-C_VdT\times(1-\eta)$ and, hence, the total amount is given by eq(508).

$$Q(\mathrm{I})=-\int_{T_h}^{T_x}\frac{C_VT_x}{T}\cdot dT=-\int_{T_h}^{T_x}\frac{(A+2BT)T_x}{T}\cdot dT=-AT_x\ln\frac{T_x}{T_h}-2BT_x(T_x-T_h) \quad (508)$$

$$=-9.6T_x\ln\frac{T_x}{400}-0.48T_x(T_x-400)$$

heat engine (II) working between T_x and T_l

With this heat engine the temperature of the cold heat source rises to T_x as the quasi *Carnot* cycle is repeated. When the temperature of the hot heat source is kept constant at T_x and the temperature of the cold heat source is T', the efficiency of this heat engine is given by eq(504') and the work done is given by eq(505'),

$$\eta'=\frac{T_x-T'}{T_x} \quad (504'), \qquad C_VdT'\times\eta'=C_V\left(\frac{T_x-T'}{T_x}\right)dT' \quad (505')$$

from which the total amount of work done is given by eq(506') (= eq(507')).

$$|W(\mathrm{II})|=\int_{T_l}^{T_x}C_V\left(\frac{T_x-T'}{T_x}\right)\cdot dT'=\int_{T_l}^{T_x}(A+2BT')\left(\frac{T_x-T'}{T_x}\right)\cdot dT' \quad (506')$$

$$=A(T_x-T_l)+B(T_x^2-T_l^2)-\frac{A}{2T_x}(T_x^2-T_l^2)-\frac{2B}{3T_x}(T_x^3-T_l^3)$$

$$=9.6(T_x-200)+B(T_x^2-200^2)-\frac{9.6}{2T_x}(T_x^2-200^2)-\frac{0.48}{3T_x}(T_x^3-200^3) \quad (507')$$

On the other hand, in order to warm the true cold heat source to the temperature T_x by repeating the quasi *Carnot* cycle, the heat energy given by eq(509) is needed.

$$Q(\mathrm{II})=\int_{T_l}^{T_x}C_V\cdot dT'=\int_{T_l}^{T_x}(A+2BT')\cdot dT'=A(T_x-T_l)+B(T_x^2-T_l^2) \quad (509)$$

$$=9.6(T_x-200)+0.24(T_x^2-200^2)$$

Since the heat energy disposed to the hypothetical heat reservior (eq(508)) is used to warm the cold heat source as well as for the heat engine (II) to do work, eq(510) must be satisfied.

$$Q(\mathrm{I}) = Q(\mathrm{II}) + W(\mathrm{II}) \qquad (510)$$

Substituting the given values into eq(510), it is rewritten as eq(510').

$$-9.6T_x \ln\frac{T_x}{400} - 0.48T_x(T_x - 400) = 9.6(T_x - 200) + B(T_x^2 - 200^2) - \frac{9.6}{2T_x}(T_x^2 - 200^2) - \frac{0.48}{3T_x}(T_x^3 - 200^3)$$
$$+ 9.6(T_x - 200) + 0.24(T_x^2 - 200^2) \qquad (510')$$

A value for T_x which satisfies eq (510') is found by searching for a point at which the curves on both sides of eq(510') cross each other as is illustrated in Figs.54 and 55. The found temperature is 300.8 K.

Work done is found from eqs(507) and (507') and is given by eq(511). Substituting $T_x = 300.8\,\mathrm{K}$ into it, a value of 4.55 kJ is found as the work done by heat transfer.

$$|W(\mathrm{I})| + |W(\mathrm{II})| = -(9.6 - 0.48\times T_x)(T_x - 400) - 0.24(T_x^2 - 400^2) + 9.6T_x \ln\frac{T_x}{400}$$
$$+ 9.6(T_x - 200) + 0.24(T_x^2 - 200^2) - \frac{9.6}{2T_x}(T_x^2 - 200^2) - \frac{0.48}{3T_x}(T_x^3 - 200^3) \qquad (511)$$

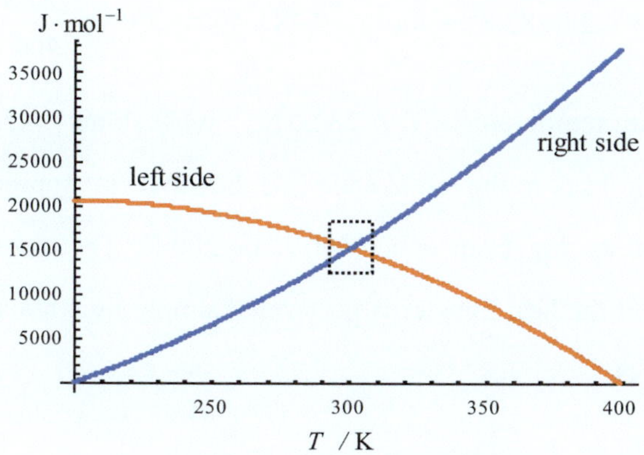

Fig.54 Intersection point of the two curves given either on the left or right side of eq(510').

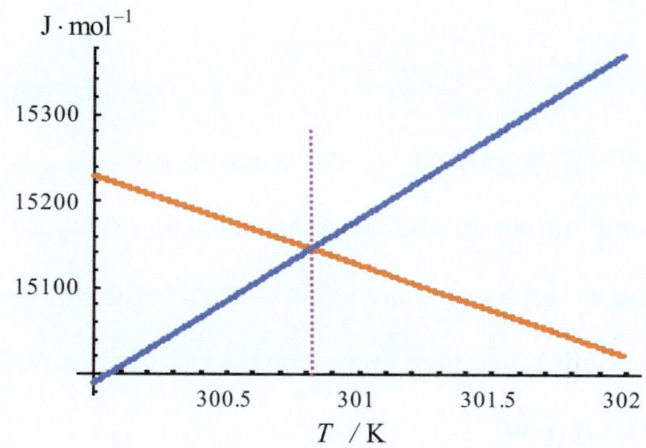

Fig.55 Enlarged dotted square section shown in Fig.54.

Ans. equilibrium temperature: $T_x = 300.8\,\mathrm{K}$, work done: 4.55 kJ

Approach 2.

Assuming the temperature of the hot heat source is T and the one of the cold heat source is T_x, when the temperatrure of the former varies by $-dT$ through the *Carnot* cycle, the heat energy disposed to the cold heat source, as has been derived so far in Approach 1, yields eq(512) relating the two temperatures.

$$-C_V\frac{T_x}{T}dT = C_V dT_x \;\Rightarrow\; -\frac{dT_x}{T_x} = \frac{dT}{T} \qquad (512)$$

Integrating eq(512), eq(513) is obtained.

$$-\ln T_x = \ln T + \text{const.} \;\Rightarrow\; T_x T = \text{const.} \qquad (513)$$

Since eq(513) holds for the heat sources at the temperatures of T_h and T_l at the start, substituting these two temperatures into eq(513), its const. is found as $T_h \cdot T_l$ leading to the relation given by eq(514).

$$T_x = \frac{\text{const.}}{T} \;\Rightarrow\; T_x = \frac{T_h \cdot T_l}{T} \qquad (514)$$

Consequently, the equilibrium temperature T_e at which the quasi *Carnot* cycle terminates is found by equating $T_e = T_x = T$ in eq(514) as 282.8K.

$$T_e = \frac{T_h \cdot T_l}{T_e} \Rightarrow T_e = \sqrt{T_h \cdot T_l} = 282.8\,\text{K}$$

The maximum work (absolute value) to be achieved is obtained by operating the quasi *Carnot* cycle until the equilibrium temperature is reached and is found by eq(515).

$$W_{max} = -\int_{T_h}^{T_e} C_V \frac{T - T_x}{T} dT = -\int_{T_h}^{T_e} (A + 2BT) \frac{T - \frac{T_h \cdot T_l}{T}}{T} dT \qquad (515)$$

Integrating it, eq(515') is derived.

$$W_{max} = -A\int_{T_h}^{T_e}\left(1 - \frac{T_h \cdot T_l \cdot}{T^2}\right)dT - 2B\int_{T_h}^{T_{ex}}\left(T - \frac{T_h \cdot T_l \cdot}{T}\right)dT$$

$$= -A\left\{(T_e - T_h) + T_h \cdot T_l\left(\frac{1}{T_e} - \frac{1}{T_h}\right)\right\} - B(T_e^2 - T_h^2) + 2BT_hT_l \ln\frac{T_e}{T_h} \qquad (515')$$

Inserting the given values and temperatures into eq(515'), a value for W_{max} is found as 6.22 kJ (**"note"**).

$$W_{max} = -9.6\left\{(282.8 - 400) + 400 \cdot 200\left(\frac{1}{282.8} - \frac{1}{400}\right)\right\} - 0.24(282.8^2 - 400^2) + 0.48 \times 400 \times 200 \cdot \ln\frac{282.8}{400}$$

Ans. equilibrium temperature: $T_e = 282.6\,\text{K}$, work done: $6.22\,\text{kJ}$

"note" In Approach 1, the working substance is brought back to the initial state (T_x), however, in Approach 2 the temperature of the working substance changes to Te starting from either 400 K or 200 K. Therefore, the process is not a complete cycle. Instead of cyclic process, if the heat transfer is performed simply by a reversible (quasi static) one, then there would exist various possibilities.

Solution 3-96

(a) The difference between the two heat capacities is given by eq(516) and is rewritten as eq(516') by using the relation of $H = U + PV$.

$$C_P - C_V = \left(\frac{\partial H}{\partial T}\right)_P - \left(\frac{\partial U}{\partial T}\right)_V \qquad (516)$$

$$= \left(\frac{\partial U}{\partial T}\right)_P + P\left(\frac{\partial V}{\partial T}\right)_P - \left(\frac{\partial U}{\partial T}\right)_V \qquad (516')$$

Expressing the internal energy as a function of T and P, its total differential is given by eq(517).

$$dU = \left(\frac{\partial U}{\partial T}\right)_P dT + \left(\frac{\partial U}{\partial P}\right)_T dP \qquad (517)$$

Dividing both sides of eq(517) by dT at constant volume ($dT \equiv (dT)_V$), eq(518) is obtained.

$$\left(\frac{dU}{dT}\right)_V = \left(\frac{\partial U}{\partial T}\right)_P + \left(\frac{\partial U}{\partial P}\right)_T \left(\frac{dP}{dT}\right)_V \qquad (518)$$

On the other hand, dividing eq(519) derived from the first and second thermodynamic laws by $dP \equiv (dP)_T$ at constant temperature and then employing one of the *Maxwell* relations, eq(520) is derived.

$$dU = TdS - PdV \qquad (519)$$

$$\left(\frac{\partial U}{\partial P}\right)_T = T\left(\frac{\partial S}{\partial P}\right)_T - P\left(\frac{\partial V}{\partial P}\right)_T = -T\left(\frac{\partial V}{\partial T}\right)_P - P\left(\frac{\partial V}{\partial P}\right)_T \qquad (520)$$

Substituting eqs(518) and (520) into eq(516') and after some rearrangements, eq(521) is obtained.

$$C_P - C_V = \left(\frac{\partial P}{\partial T}\right)_V \left\{T\left(\frac{\partial V}{\partial T}\right)_P + P\left(\frac{\partial V}{\partial P}\right)_T\right\} + P\left(\frac{\partial V}{\partial T}\right)_P$$

$$= T\left(\frac{\partial V}{\partial T}\right)_P\left(\frac{\partial P}{\partial T}\right)_V + P\left(\frac{\partial V}{\partial P}\right)_T\left(\frac{\partial P}{\partial T}\right)_V + P\left(\frac{\partial V}{\partial T}\right)_P \qquad (521)$$

Partial differentials in eq(521) are with respect to the variables T, P and V and these variables appear in a cyclic order in each of them. Therefore, it is seen that *Euler*'s chain relation can be used in order to simplify eq(521) further. In fact, employing eq(522), eqs(523) and (524) are obtained.

$$\left(\frac{\partial P}{\partial T}\right)_V\left(\frac{\partial T}{\partial V}\right)_P\left(\frac{\partial V}{\partial P}\right)_T = -1 \qquad (522) \quad (\textit{Euler}\text{'s chain relation})$$

$$\left(\frac{\partial V}{\partial T}\right)_P\left(\frac{\partial P}{\partial T}\right)_V = -\frac{\left(\frac{\partial V}{\partial T}\right)_P}{\left(\frac{\partial T}{\partial V}\right)_P\left(\frac{\partial V}{\partial P}\right)_T} = -\frac{\left(\frac{\partial V}{\partial T}\right)_P^2}{\left(\frac{\partial V}{\partial P}\right)_T} = \frac{V\alpha^2}{\kappa_T} \qquad (523)$$

$$\left(\frac{\partial V}{\partial P}\right)_T\left(\frac{\partial P}{\partial T}\right)_V = -\left(\frac{\partial V}{\partial T}\right)_P \qquad (524)$$

Substituting these equations into eq(521), the relation to be proved $C_P - C_V = TV\alpha^2/\kappa_T$ is obtained. It is trivial that the relation of $C_P - C_V = R$ holds for the ideal gas though, it should be noticed here that each specific heat can be dependent on temperature (see **solution 3-105**).

Ans. proof: See text.

(b) It is confirmed specifically for a liquid or a solid that $C_P - C_V \approx 0$ by substituting its values for α, κ_T and molar volume ($V = \overline{V}$) into the equation derived above. For instance, for metal copper whose specific heat is given by $C_P = 24.4\ \mathrm{J\cdot K^{-1}mol^{-1}}$, the following values of $\alpha = 5.01\times10^{-5}\,\mathrm{K^{-1}}$, $\kappa_T = 7.35\times10^{-7}\,\mathrm{atm^{-1}}$ and $V = 7.12\times10^{-3}\,\mathrm{L\cdot mol^{-1}}$ are substituted finding a value of $0.0072\ \mathrm{J\cdot K^{-1}\cdot mol^{-1}}$ as the difference between the two heat capacities. And for water, substituting $\alpha = 2.1\times10^{-4}\mathrm{K^{-1}}$, $\kappa_T = 4.96\times10^{-5}\mathrm{atm^{-1}}$, $V = 1.8\times10^{-2}\,\mathrm{L\cdot mol^{-1}}$, the difference is found as $0.0048\ \mathrm{J\cdot K^{-1}\cdot mol^{-1}}$. In either case the difference found is negligibly small as compared with the difference $R = 8.31\ \mathrm{J\cdot K^{-1}\cdot mol^{-1}}$ known for the ideal gas.

Ans. See text.

Solution 3-97

The thermodynamic entropy (absolute entropy, third-law entropy) $S(T)$ is obtained by integrating eq(525) over the temperature from the absolute zero to T.

$$S(T) - S(0) = \int_0^T \frac{C_P}{T}dT \qquad (525)$$

The values calculated for C_P/T from the data given in the problem are plotted against T in Fig.56. The integral of eq(525) is equal to an area under the curve in Fig.56 and it is worked out either directly by measuring the area in Fig.54 or by integrating a function obtained by best-fitting the data.

Here, the latter method is employed. The best-fitted function is searched for in the form of a quartic equation of T by employing free software gnuplot. In the temperature range of $13.74\ \text{K} \le T \le 100\ \text{K}$,

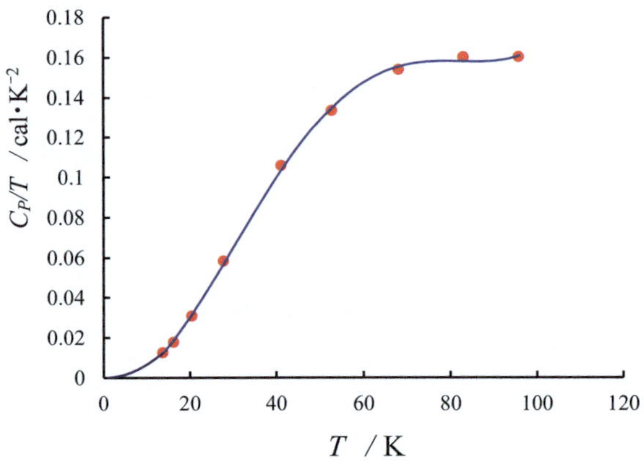

Fig.56 Temperature dependence of the heat capacity of solid Na_2SO_4 expressed in the form C_P/T.

eq(526) is obtained as the best-fitted polynomial.

$$\frac{C_P}{T} = 1.244\times10^{-8}T^4 - 2.788\times10^{-6}T^3 + 1.820\times10^{-4}T^2 - 1.115\times10^{-3}T \qquad (526)$$

Also, in the range of $0\ \text{K} \le T \le 13.74\ \text{K}$, eq(527) is used as an approximation (**"note"**).

$$\frac{C_P}{T} = 6.592\times10^{-5}T^2 \qquad (527)$$

Consequently, the entropy at 100K $S(100\ \text{K})$ is given by an integral of eq(528) and the difference $S(100) - S(0)$ is found as $8.72\times10^3\ \text{cal}\cdot\text{K}^{-1}$ as shown below.

$$S(100) - S(0) = \int_{13.74}^{T}(1.244\times10^{-8}T^4 - 2.788\times10^{-6}T^3 + 1.820\times10^{-4}T^2 - 1.115\times10^{-3}T)dT$$

$$+ \int_{0}^{13.74} 6.592\times10^{-5}T^2 dT \qquad (528)$$

$$= 8.72\times10^3\,\text{cal}\cdot\text{K}^{-1}$$

According to the third law of thermodynamics, we can put $S(0\ \text{K}) = 0$ and, hence, as the absolute entropy of Na_2SO_4 at 100 K, $S(100\ \text{K}) = 8.72\times10^3\,\text{cal}\cdot\text{K}^{-1}$ is found.

Ans. $S(100\ \text{K}) = 8.72\times10^3\,\text{cal}\cdot\text{K}^{-1}$

"note" For a large number of substances, the heat capacity at constant pressure can be expressed by a cubic formula of T given by eq(529) in a low temperature region.

$$C_P = \alpha\times T^3 + \beta\times T \qquad (529)$$

However, in order to reproduce the data in the table, the polynomial such as eq(526) containing a term of the higher order is better. Furthermore, in order to improve the approximation of integration below 13.74 K, eq(527) which not only approaches monotonously to zero as T decreases to zero but also reflects the functional form of eq(529) is employed. The reason why the units of entropy (or heat capacity) given for the answer are not per mass or per mole is because the entropy obtained here represents the one for the whole substance employed for the measurements.

Solution 3-98

Solution 1. Denoting the mole number of He in container 1 as n_1 and that in container 2 as n_2, they are given by eq(530).

$$n_1 = \frac{P_1 V}{RT_1} = \frac{0.5\times 10}{R\times 293}\ , \qquad n_2 = \frac{P_2 V}{RT_2} = \frac{1.0\times 10}{R\times 353} \qquad (530)$$

After opening the valve fully, we assume the temperature of the gas has reached a homogeneous temperature T_e(K). Since there is no heat flow in or out from the outside, the following equations (eq(531) – eq(533)) are satisfied, where C_V^m is the molar specific heat at constant volume of the gas and T_e and P_e are the reached temperature and pressure, respectively.

$$n_1 \times C_V^m (T_e - 293) = n_2 \times C_V^m (353 - T_e) \qquad (531)$$

$$T_e(n_1 + n_2) = n_2 \times 353 + n_1 \times 293 \ \Rightarrow\ T_e = \frac{n_2 \times 353 + n_1 \times 293}{(n_1 + n_2)} = 330.4\ \text{K} \qquad (532)$$

$$T_e = \frac{n_2 \times 353 + n_1 \times 293}{(n_1 + n_2)} \ \Rightarrow\ \frac{P_2 V + P_1 V}{(n_1 + n_2)R} = \frac{(P_2 + P_1)V}{(n_1 + n_2)R} \qquad (532')$$

$$P_e = \frac{(n_1 + n_2)RT_e}{2V} = \frac{(P_2 + P_1)}{2} = \frac{1.0 + 0.5}{2} = 0.75\ \text{atm} \qquad (533)$$

Solution 2. Since neither work nor heat energy is accompanied with the process, the internal energy of the gas does not change before and after opening the valve. Noticing this fact, the internal energy of the gas in each container before opening the valve is given by eq(534) ,

$$\frac{3}{2}RT_1 \times n_1 = \frac{3}{2}P_1 V\ , \quad \frac{3}{2}RT_2 \times n_2 = \frac{3}{2}P_2 V \qquad (534)$$

and after the equilibrium has been reached, the sum of the internal energies is expressed by eq(535).

$$\frac{3}{2}RT_e \times (n_1 + n_2) = \frac{3}{2}P_1 V + \frac{3}{2}P_2 V \qquad (535)$$

Also, denoting the reached pressure as P_e, eq(536) is derived from eq(535), from which P_e is found as eq(537).

$$(n_1 + n_2)RT_e = (P_1 + P_2)V = P_e \times 2V \qquad (536)$$

$$P_e = \frac{P_1 + P_2}{2} = 0.75\ \text{atm} \qquad (537)$$

Ans. equilibrium pressure: 0.75 atm, equilibrium temperature: 330.4 K

Solution 3-99

The equilibrium pressures of the gas-liquid and gas-solid equilibrium of water near the triple point are given by eqs(538) and (539) shown below again.

$$\ln P = 2.303 \times 20.9586 - 4.0843 \times \ln T - 2.303 \times \frac{2825.4}{T} \qquad (538)$$

$$\ln P = 2.303 \times 11.5041 - 0.4001 \times \ln T - 2.303 \times \frac{2694.2}{T} \qquad (539)$$

The temperature derivative of the equilibrium pressure at the triple point is given by eq(540) for the gas-liquid equilibrium and by eq(541) for the gas-solid equilibrium.

$$\left(\frac{dP}{dT}\right)_{T=T_{triple}} = P_{triple} \times \left(-\frac{4.0843}{T} + \frac{2.303 \times 2825.4}{T^2}\right)_{T=T_{triple}} \qquad (540)$$

$$\left(\frac{dP}{dT}\right)_{T=T_{triple}} = P_{triple} \times \left(-\frac{0.4001}{T} + \frac{2.303\times 2694.2}{T^2}\right)_{T=T_{triple}} \quad (541)$$

Based on *Clausius-Clapeyron*'s approach, the latent heat for the gas (g)-liquid (l) or gas (g)-solid (s) equilibrium at the triple point is given by eq(542).

$$\left(\frac{dP}{dT}\right)_{T=T_{triple}} = \frac{P_{triple}\times \Delta H_{latent}}{RT_{triple}^2} \quad (542)$$

Putting $T_{triple} = 273.16\,\text{K}$, the relevant latent heats are obtained from eqs(540) and (541) as given below. Notice here signs for the latent heats.

$$\Delta H_{g\to l} = -RT_{triple}^2 \times \left(-\frac{4.0843}{T} + \frac{2.303\times 2825.4}{T^2}\right)_{T=T_{triple}} = -44.8\,\text{kJ}\cdot\text{mol}^{-1}$$

$$\Delta H_{s\to g} = RT_{triple}^2 \times \left(-\frac{0.4001}{T} + \frac{2.303\times 2694.2}{T^2}\right)_{T=T_{triple}} = 50.7\,\text{kJ}\cdot\text{mol}^{-1}$$

On the other hand, in the case of the liquid (l)-solid (s) equilibrium the volume change accompanied by the phase transition is so small that original *Clausius-Clapeyron*'s eq(542') for which the exact volume change is kept as it is must be employed for the calculation.

$$\left(\frac{dP}{dT}\right)_{T=T_{triple}} = \frac{\Delta H_{l\to s}}{T_{triple}\Delta V_{l\to s}} \quad (542')$$

The value for $(dP/dT)_{T=T_{triple}}$ is obtained from the phase diagram shown in Fig.57 and the volume change is calculated as $\Delta V_{l\to s} = 1.59\times 10^{-6}\,\text{m}^3$. Substituting these values into eq(542'), the latent heat $\Delta H_{l\to s}$ is found as below.

$$\Delta H_{l\to s} = -\frac{(760-4.64)\times 1.01\times 10^5}{0.01\times 760}\times 273.16\times 1.59\times 10^{-6} = -4.36\,\text{kJ}\cdot\text{mol}^{-1}$$

Meanwhile, the three latent heats obtained at the triple point should naturally satisfy the relation given by eq(543).

$$\Delta H_{g\to l} + \Delta H_{l\to s} + \Delta H_{s\to g} = 0 \quad (543)$$

Out of these three, $\Delta H_{g\to l}$ and $\Delta H_{s\to g}$ are considered more accurately determinable than does $\Delta H_{l\to s}$ since the experimental equations of the equilibrium pressures for these two phase transitions are accurately given here. Because of this, the value obtained from eq(543) for $\Delta H_{l\to s}$ is chosen as the answer here as is given below.

$$\Delta H_{l\to s} = -50.7 + 44.8 = -5.9\,\text{kJ}\cdot\text{mol}^{-1}$$

Ans. condensation: $\Delta H_{g\to l} = -44.8\,\text{kJ}\cdot\text{mol}^{-1}$, sublimation: $\Delta H_{s\to g} = 50.7\,\text{kJ}\cdot\text{mol}^{-1}$, solidification: $\Delta H_{l\to s} = -5.9\,\text{kJ}\cdot\text{mol}^{-1}$

"note 1" The literature value of the heat of solidification of water is $-6.0\,\text{kJ}\cdot\text{mol}^{-1}$ at $T = 273.15\,\text{K}$ and is very close to the value found here.

"note 2" The equilibrium lines near the triple point of water is often drawn schematically as shown in Fig.57 below. However, when they are drawn actually by using the functions representing the equilibriums, it is found that the lines expressing *S*- *G* and *L*- *G* equilibriums cross each other rather smoothly as shown in Fig.17 in the problem.

"note 3" Water behaves anomalously among other popular pure substances in the sense that the melting point of normal ice decreases with increasing pressure, in other words, $\Delta V_{l\to s}$ takes a positive value. However, under high pressures, the normal ice phase-transfers into one of the polymorphic crystals of water and $\Delta V_{l\to s}$ turns to take a positive value. Accordingly the melting point of ice increases with increasing pressure, just like other popular pure substances.

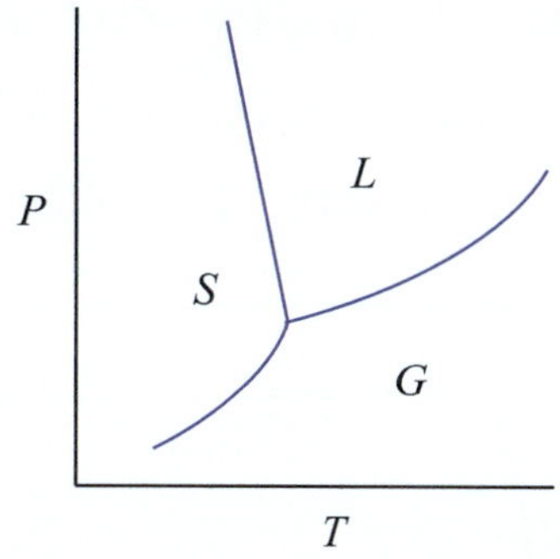

Fig.57 *P-T* phase diagram near triple point.

In fact, the melting point of ice at 22,000 atm is 81.6 °C ! See, for instance, *S.D.Hammann, Physico-Chemical Effects of Pressure,* p.68, Butterworth Scientific Publications, 1957. *P.W.Bridgman*, Nobel Prize in Physics in 1946 for his work under high pressures.

Solution 3-100

(a) The boiling point of O_2 at 1 atm is obtained by finding the temperature at which the vapor pressure calculated by the *van der Waals* equation of state of O_2 reaches 1 atm. When an appropriate temperature is given, the pressure of the gas at which it starts to liquefy is, as is explained in detail in **problem 2-4**, found by drawing *Maxwell*'s *tie line.* If a trial temperature is too high, the *van der Waals* equation of state of O_2 takes neither minimum nor maximum. Therefore, the temperature of liquefaction should be searched for below the critical temperature of 154.8 K so as for the vapor pressure to reach surely 1 atm.

Here, the boiling temperature of O_2 is found in the following procedures.

(i) At several temperatures, *P-V* diagrams based on the *van der Waals* equation of state are drawn to find an approximate temperature at which *Maxwell's tie line* can be drawn near at 1 atm. See Fig.58.

(ii) At that temperature, the intersections of $P = 1$ atm and the *P-V* curve are found as V_1, V_2, V_3 ($V_1 < V_2 < V_3$) and then the integral given by eq(544) is calculated in the range [V_1, V_3].

$$\int_{V_1}^{V_3}\left\{\left(\frac{RT}{V-b}-\frac{a}{V^2}\right)-P_0\right\}dV=\int_{V_1}^{V_3}\left\{\left(\frac{0.082\times T}{V-3.183\times10^{-2}}-\frac{1.378}{V^2}\right)-1.0\right\}dV \qquad (544)$$

(iii) By varying temperature by a small increment (or decrement) step by step, the integration of eq(544) is repeated to find the temperature at which the integral approaches to zero. For this type of repetitive calculations computer software such as *Mathematica* is useful. As a result of the calculation in three significant figures, it is found that the tie line of the gas-liquid equilibrium can be drawn at $T = 74.5$ K under $P = 1$ atm. Therefore, the boiling temperature of O_2 is said to be at 74.5 K.

Ans. boiling temperature: 74.5 K

(b) The latent heat of liquefaction ΔH_l at the boiling point can be found by *Clausius-Clapeyron*'s equation (eq(545)).

$$\frac{dP}{dT}=\frac{P\times\Delta H_l}{RT^2} \qquad (545)$$

In order to find a value for dP/dT on the left side of eq(545) near at the boiling temperature of 74.5 K, the equilibrium vapor pressure at the temperature higher than the boiling point by 1.0K is obtained first. Putting $T = 75.5\,\mathrm{K}$ in eq(544), the equilibrium vapor pressure at this temperature is found this time by varying

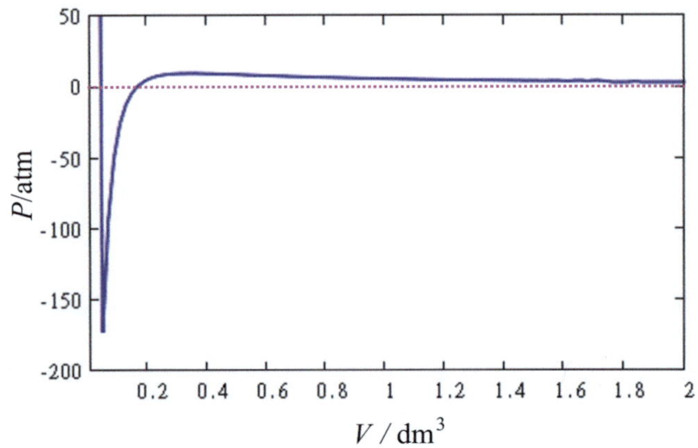

Fig.58 *P-V* curve of O_2 at 80K calculated from its *van der Waals* equation of state.
The red dotted line indicates the position of 1.0 atm.

pressure starting from $P_0 = 1.0$ gradually until the integral of eq(544) approaches to zero as small as possible. As a result of these repetitive calculations, $P_0 = 1.098$ atm is found as the equilibrium vapor pressure at . $T = 75.5\,\mathrm{K}$. Employing these values to calculate the slope of dP/dT of eq(545), the latent heat is calculated from eq(545) as shown below finding $\varDelta H_l = 4.5\ \mathrm{kJ \cdot mol^{-1}}$. (Note that the value is independent of the unit of pressure.)

$$\left(\frac{dP}{dT}\right)_{T=74.5} = \frac{0.098}{1.0} \Rightarrow (\varDelta H_l)_{P=1.0} = \frac{0.098 \times 8.31 \times 74.5^2}{1.0 \times 1.0} = 4.5\ \mathrm{kJ \cdot mol^{-1}}$$

Ans. latent heat of liquefaction: $\varDelta H_l = 4.5\ \mathrm{kJ \cdot mol^{-1}}$

"note" The literature values of the boiling point and the latent heat of O_2 at that temperature are $T_b = 90.1$ K and $\varDelta H_l = 6.5\ \mathrm{kJ \cdot mol^{-1}}$, respectively. These values are a little different from those calculated here. However, it is surprising to know that such physical properties can be obtained from the *van der Waals* equation of state alone. The actual calculation is difficult to do without using some calculation software, so it would have been unthinkable as a physical chemistry problem before the widespread use of personal computers. Nevertheless, if you can draw a *P-V* diagram at various temperatures, you can analyze it graphically to obtain a rough estimate of the liquefaction pressure, so it is a good idea to give it a try.

Solution 3-101

(a) The hot heat reservoir of a constant capacity initially at the temperature T_h lowers its temperature by changing a part of heat energy into work continuously until it reaches to the temperature equal to T_0 outside. If we consider a *Carnot* cycle working between the temperatures T and T_0, then the heat disposed to the outside while the temperature of the hot reservoir varies form T to $T - dT$ is given by $-C_V dT$ and the efficiency of work η is given by eq(546). Consequently, the work to be obtained from this process is given by eq(547).

$$\eta = \frac{T - T_0}{T} \quad (546), \quad -C_V dT \times \eta = -C_V\left(\frac{T - T_0}{T}\right) \cdot dT \quad (547)$$

The total amount of work accumulated when the temperature of the hot reservoir has changed from T_h to T_0 is obtained simply by integrating eq(547) over the temperature range $[T_h,\ T_0]$ as is given by eq(548), which is the same equation as given in the problem for proof.

$$W_{max} = -\int_{T_h}^{T_0} C_V\left(\frac{T-T_0}{T}\right)\cdot dT \qquad (548)$$

Ans. See text.

(b) Substituting the given values into eq(548) in (a) and executing the integration, the maximum work per one mole of water is found as $962\ \mathrm{J\cdot mol^{-1}}$ as is shown below.

$$W_{max} = -75.3\int_{371}^{278}\frac{T-278}{T}dT = -75.3\left\{(278-371)+278\times\ln\frac{278}{371}\right\} = 962\ \mathrm{J\cdot mol^{-1}}$$

For 10kg of water, therefore, $962\times(10^4/18) = 534\ \mathrm{kJ}$ is obtainable.

Ans. maximum work: 534 kJ

(c) Writing the mass of the object as m and the acceleration of gravity as g, the potential energy at the height h is given by mgh. Putting this potential energy equal to the maximum work obtained in (b), h is found as below.

$$mgh = 534\times 10^3 \;\Rightarrow\; h = \frac{534}{1\times 9.81} = 54.4\ \mathrm{km}$$

Ans. height: 54.4 km

"note" See **problem 3-95** for a related one.

Solution 3-102

A cooling machine is a heat engine which reversely performs a cycle of gaining heat from the hot reservoir and discarding a part of the heat to the cold reservoir. Therefore, the efficiency of work of the cooling machine is found from that of the normal heat engine which works in the manner exactly reverse to the cycle of the cooling machine.

The efficiency of work of the ideal heat engine which works between 298 K (T_h) and 0.001 K (T_l) is given by eq(549).

$$\eta_{ideal} = \frac{T_h - T_l}{T_h} = \frac{298-0.001}{298} = 0.999997 \qquad (549)$$

This is also expressed in terms of heats flowing in the cold reservoir (Q_l) and out of the hot reservoir (Q_h) as eq(550).

$$\frac{Q_h - Q_l}{Q_h} = 1 - \frac{Q_l}{Q_h} \qquad (550)$$

The heat Q_h gained from the hot reservoir is converted to the work with the efficiency given by eq(549). When the heat to be discarded to the cold reservoir as unusable energy is 1 J, the heat to be gained from the hot reservoir is found as $2.98\times 10^5\ \mathrm{J}$ and almost all of it can be converted to work.

$$\frac{Q_l}{Q_h} = 1-\eta \;\Rightarrow\; Q_h = \frac{Q_l}{1-\eta} = 2.98\times 10^5\ \mathrm{J\cdot s^{-1}}$$

On the other hand, when the cooling machine is operated between the same temperatures as above but in the reverse manner, the heat of $2.98\times 10^5\ \mathrm{J}$ must be discarded to the hot reservoir in order to take the heat of 1 J away from the cold reservoir. Therefore, if there is no energy loss at all in the operation of the device, the work equivalent to the heat of 2.98×10^{5} J (in this sense minimum work) is required. In other words, to take heat away from a cold body to discard it to a hot body is equivalent to making heat flow oppositely against a temperature gradient, which is only possible when more than or equal to the minimum work required is supplied to the heat flowing system. And it works most ideally when the normal ideal heat engine is operated reversely. Since $1\,\mathrm{W}=1\,\mathrm{J}\cdot\mathrm{s}^{-1}$, the electric power of 298 kW is required at least.

As is evident from the above explanation, the lower is the temperature of the cold reservoir, the efficiency of the normal heat engine becomes higher and, hence, the larger is the maximum work obtainable. In the case of the cooling machine, the lower is the temperature to be cooled, the larger is the minimum energy required to achieve it. Also the lower is the temperature of the hot reservoir, the smaller work is required to take the same amount of heat away from the cold reservoir. Thus, for example, if the temperature environment of liquid nitrogen (77 K) or liquid helium (4 K) is used as the high heat source, the power consumption to reach the further lower cryogenic temperature will be smaller, though total cost is another thing.

Ans. electric power required: 298 kW

Solution 3-103

(a) When the entropy *S* is regarded as a function of *T* and *P*, its total differential is expressed by eq(551).

$$dS=\left(\frac{\partial S}{\partial T}\right)_P dT+\left(\frac{\partial S}{\partial P}\right)_T dP=\frac{C_P}{T}dT-\left(\frac{\partial V}{\partial T}\right)_P dP \qquad (551)$$

From this, the temperature variation with pressure for the adiabatic process is given by eq(552).

$$\left(\frac{\partial T}{\partial P}\right)_S=\frac{T}{C_P}\times\left(\frac{\partial V}{\partial T}\right)_P=\frac{T\cdot V}{C_P}\times\frac{1}{V}\left(\frac{\partial V}{\partial T}\right)_P$$

$$=\frac{T\cdot V\cdot\alpha}{C_P} \qquad (552)$$

Consequently, the equation given in the problem is expressed by eq(553). *T* and *V* are always positive, so that by using $C_P>0$, which is one of the thermodynamic inequalities, it is seen that eq(553) is always positive.

$$\frac{1}{\alpha}\cdot\left(\frac{\partial T}{\partial P}\right)_S=\frac{T\cdot V}{C_P}>0 \qquad (553)$$

Ans. inequality: eq(553)

(b) When it is tried to compress the substances which satisfy the relation of $\alpha\propto(\partial V/\partial T)_P>0$ by applying pressure adiabatically, a change against this process takes place so as to make their volumes increase by raising their temperatures, i.e., $(\partial T/\partial P)_S>0$. Conversely, for the substances which satisfy the relation of $\alpha\propto(\partial V/\partial T)_P<0$, their volumes decrease with increasing temperature at constant pressure. Therefore, when it is tried to compress their volumes with pressure adiabatically, a change against this action takes place so as to make the volumes increase by lowering their temperatures, i.e., $(\partial T/\partial P)_S<0$. In either case, the inequality given by eq(553) $(\partial T/\partial P)_S/\alpha>0$ is consistent with the principle of *Le Chatelier* which states "the direction of change caused by an added influence is the direction in which the added influence is held back."

Ans. See text.

Solution 3-104

Free expansion of the gas under the given conditions is accompanied by any change neither in heat nor in work. Therefore, the present question is equivalent to requiring us to find the enthalpy change of the gas brought about along the route on which the internal energy is kept constant (see *Thomson* process). The total differential of the enthalpy expressed by a function of T and V is given by eq(554).

$$dH = \left(\frac{\partial H}{\partial V}\right)_T dV + \left(\frac{\partial H}{\partial T}\right)_V dT \qquad (554)$$

Using eq(555) below,

$$dH = TdS + VdP \qquad (555)$$

the partial differential coefficients in eq(554) can be expressed by eqs(556) and (557).

$$\left(\frac{\partial H}{\partial V}\right)_T = T\left(\frac{\partial S}{\partial V}\right)_T + V\left(\frac{\partial P}{\partial V}\right)_T = T\left(\frac{\partial P}{\partial T}\right)_V + \frac{1}{\frac{1}{V}\left(\frac{\partial V}{\partial P}\right)_T} = \frac{\alpha T}{\kappa_T} - \frac{1}{\kappa_T} \qquad (556)$$

$$\left(\frac{\partial H}{\partial T}\right)_V = T\left(\frac{\partial S}{\partial T}\right)_V + V\left(\frac{\partial P}{\partial T}\right)_V = C_V + \frac{\alpha V}{\kappa_T} \qquad (557)$$

Here, the expression for $(\partial P/\partial T)_V$ given by eq(558) is derived from *Euler*'s chain relation rewritten as eq(559).

$$\left(\frac{\partial P}{\partial T}\right)_V = \frac{\alpha}{\kappa_T} \qquad (558)$$

$$-1 = \left(\frac{\partial P}{\partial T}\right)_V \left(\frac{\partial T}{\partial V}\right)_P \left(\frac{\partial V}{\partial P}\right)_T = \left(\frac{\partial P}{\partial T}\right)_V \cdot V\left(\frac{\partial T}{\partial V}\right)_P \cdot \frac{1}{V}\left(\frac{\partial V}{\partial P}\right)_T = \left(\frac{\partial P}{\partial T}\right)_V \cdot \frac{1}{\alpha}\cdot(-\kappa_T) \qquad (559)$$

Using eqs(556) and (557), eq(554) is rewritten as eq(560), which is expressed as eq(560') when the variables of T and V change along the route on which the internal energy is kept constant.

$$dH = \left(\frac{\alpha T - 1}{\kappa_T}\right)\cdot dV + \left(C_V + \frac{\alpha V}{\kappa_T}\right)\cdot dT \qquad (560)$$

$$(dH)_U = \left(\frac{\alpha T - 1}{\kappa_T}\right)\cdot (dV)_U + \left(C_V + \frac{\alpha V}{\kappa_T}\right)\cdot (dT)_U \qquad (560')$$

The relation between $(dV)_U$ and $(dT)_U$ is found form the total differential of U to be given by eq(561).

$$dU = \left(\frac{\partial U}{\partial V}\right)_T dV + \left(\frac{\partial U}{\partial T}\right)_V dT \Rightarrow C_V\cdot(dT)_U = \left(P - \frac{\alpha T}{\kappa_T}\right)\cdot(dV)_U \qquad (561)$$

Here eq(562) is employed.

$$\left(\frac{\partial U}{\partial V}\right)_T = T\left(\frac{\partial S}{\partial V}\right)_T - P = T\left(\frac{\partial P}{\partial T}\right)_V - P = \frac{\alpha T}{\kappa_T} - P \qquad (562)$$

Expressing $(dT)_U$ by $(dV)_U$ using eq(561) and rewriting eq(560'), eq(563) is obtained.

$$(dH)_U = \left(\frac{\alpha T - 1}{\kappa_T}\right)\cdot(dV)_U + \left(C_V + \frac{\alpha V}{\kappa_T}\right)\cdot\frac{1}{C_V}\cdot\left(P - \frac{\alpha T}{\kappa_T}\right)\cdot(dV)_U$$

$$= \left\{\frac{\alpha T - 1}{\kappa_T} + \left(1 + \frac{\alpha V}{\kappa_T C_V}\right)\left(P - \frac{\alpha T}{\kappa_T}\right)\right\}\cdot(dV)_U \qquad (563)$$

Under the given conditions, changes in T and P due to the free expansion are considered small. Consequently, putting $T = 100$ K and $P = 0.2$ atm in eq(563) and substituting the other given values into it, eq(563) is integrated over the volume range [1.0, 1.01] (L) to calculate the enthalpy change, obtaining eq(564), though the integration is not necessary because its range is small enough. Here the units of 1 atm · L = 101 J is used.

$$\Delta H = \int_{1.0}^{1.01} (dH)_U = -3.17 \text{ J} \qquad (564)$$

In the case of the free expansion of the ideal gas for which there is no change in the internal energy, the change in the enthalpy becomes zero also as is seen from the relation $\Delta H = \Delta U + \Delta(PV) = 0$. This fact is also confirmed on the basis of eq(563) by substituting the relations of $\alpha = 1/T$ and $\kappa_T = 1/P$ found for the ideal gas into eq(563), which yields $(dH)_U = 0$.

Ans. enthalpy change: -3.17 J

Solution 3-105

When the internal energy U is considered as a function of S and V, the relation of $(\partial U/\partial V)_S = -P$ is derived as below from the total differential of U given by the left side of eq(565).

$$dU = TdS - PdV \Rightarrow \left(\frac{\partial U}{\partial V}\right)_S = -P \qquad (565)$$

Replacing the variables of U with V and T by using *Jacobians*, $(\partial U/\partial V)_S$ can also be expressed by eq(566).

$$\left(\frac{\partial U}{\partial V}\right)_S = \frac{\dfrac{\partial(U,S)}{\partial(V,T)}}{\dfrac{\partial(V,S)}{\partial(V,T)}} = \frac{\begin{vmatrix} \left(\frac{\partial U}{\partial V}\right)_T & \left(\frac{\partial S}{\partial V}\right)_T \\ \left(\frac{\partial U}{\partial T}\right)_V & \left(\frac{\partial S}{\partial T}\right)_V \end{vmatrix}}{\left(\frac{\partial S}{\partial T}\right)_V} = -\frac{\left(\frac{\partial U}{\partial T}\right)_V \cdot \left(\frac{\partial S}{\partial V}\right)_T}{C_V/T} = -T\left(\frac{\partial S}{\partial V}\right)_T = -P \qquad (566)$$

As a result, the first part of eq(567) is derived generally but its second part must be satisfied for the equation of state of the ideal gas.

$$\left(\frac{\partial S}{\partial V}\right)_T = \frac{P}{T} = \frac{nR}{V} \qquad (567)$$

Using the second equality, eq(567) is integrated to yield eq(568), in which $f(T)$ is an integration constant given by an arbitrary function of T.

$$S = nR \ln V + f(T) \qquad (568)$$

In conclusion, when the entropy of a system is given by a function in the form of q(568), P, V and T of the system are related one another by the equation of state of the ideal gas and the system expressed by eq(568) behaves as an ideal gas. The specific heat at constant volume is readily derived from eq(568) and is found generally to be given by a function of temperature, not necessarily a constant.

Ans. condition of thermodynamic function: eq(568)

Solution 3-106

Otto cycle is a heat engine consisting of two adiabatic and two isovolumetric processes with an ideal gas as a working substance which works between hot and cold heat reservoirs. As is shown in Fig.59 on the right, starting from (P_1,V_1,T_1), heat Q_{12} is gained via isovolumetric heating to (P_2,V_1,T_2). Then, the working substance does work W_{23} by the adiabatic expansion to (P_3,V_3,T_3) followed by the isovolumetric cooling to (P_4,V_3,T_4) giving up heat Q_{34}, which is not used for doing work but is discarded to the cold reservoir. After that, work W_{41} is gained via adiabatic compression to the initial state (P_1,V_1,T_1). In order to calculate the efficiency of work for the cycle, it is sufficient to know the amounts of the heat gained from the hot reservoir and the work done to the surroundings.

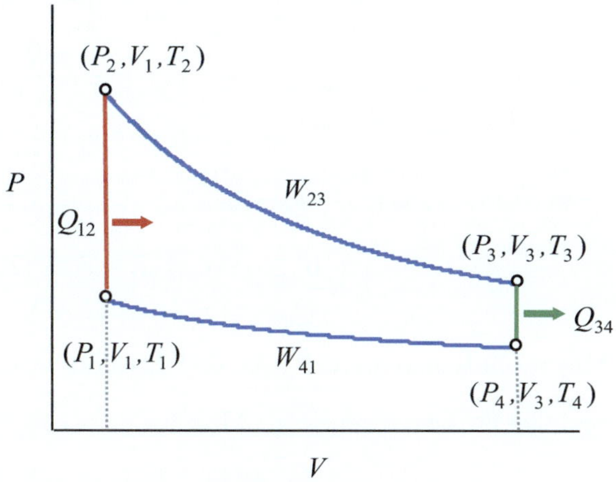

Fig.59 Illustration of *Otto* cycle.

Before the calculation, it is checked first how P, V and T at each starting point for the four processes are related one another. They are related by eqs(569) and (570) which are satisfied for the adiabatic processes of the ideal gas as well as eq(571) derived from the equation of state of the ideal gas.

$$\left.\begin{aligned} &TV^{\gamma-1} = \text{const.} \;\Rightarrow\; T_2V_1^{\gamma-1} = T_3V_3^{\gamma-1} \;\Rightarrow\; T_3 = T_2\left(\frac{V_1}{V_3}\right)^{\gamma-1} \\ &T_1V_1^{\gamma-1} = T_4V_3^{\gamma-1} \;\Rightarrow\; T_4 = T_1\left(\frac{V_1}{V_3}\right)^{\gamma-1} \end{aligned}\right\} \tag{569}$$

$$PV^{\gamma} = \text{const.} \;\Rightarrow\; P_1V_1^{\gamma} = P_4V_3^{\gamma} \;\Rightarrow\; P_4 = P_1\left(\frac{V_1}{V_3}\right)^{\gamma} \tag{570}$$

$$P_3V_3 = nRT_3\,, \quad P_4V_3 = nRT_4 \;\Rightarrow\; \frac{P_4}{T_4} = \frac{P_3V_3}{V_3T_3} = \frac{P_3}{T_3} \tag{571}$$

The heat gained or lost upon isovolumetric processes are given by eq(572).

$$Q_{12} = C_V(T_2 - T_1)\,, \quad Q_{34} = C_V(T_4 - T_3) \tag{572}$$

On the other hand, the work done by each adiabatic process is expressed by using the relation of $PV^{\gamma} = \text{const.}$ as is given by eq(573) or eq(574).

$$\begin{aligned} W_{23} &= -\int_{V_1}^{V_3} PdV = -P_2V_1^{\gamma}\int_{V_1}^{V_3} V^{-\gamma}dV = \frac{P_2V_1^{\gamma}}{\gamma-1}(V_3^{-\gamma+1} - V_1^{-\gamma+1}) \\ &= \frac{P_2V_1}{\gamma-1}\left(\frac{V_3^{-\gamma+1}}{V_1^{-\gamma+1}} - 1\right) = \frac{P_2V_1}{\gamma-1}\left(\frac{T_3}{T_2} - 1\right) = \frac{nRT_2}{\gamma-1}\left(\frac{T_3}{T_2} - 1\right) \end{aligned} \tag{573}$$

$$\begin{aligned} W_{41} &= -\int_{V_3}^{V_1} PdV = -P_1V_1^{\gamma}\int_{V_3}^{V_1} V^{-\gamma}dV = \frac{P_1V_1^{\gamma}}{\gamma-1}(V_1^{-\gamma+1} - V_3^{-\gamma+1}) \\ &= \frac{P_1V_1}{\gamma-1}\left(1 - \frac{V_3^{-\gamma+1}}{V_1^{-\gamma+1}}\right) = \frac{P_1V_1}{\gamma-1}\left(1 - \frac{T_4}{T_1}\right) = \frac{nRT_1}{\gamma-1}\left(1 - \frac{T_4}{T_1}\right) \end{aligned} \tag{574}$$

Net work W_{out} usable for work, therefore, is given by eq(575).

$$\begin{aligned} W_{out} &= |W_{23} + W_{41}| = -W_{23} - W_{41} \\ &= -\frac{nR}{\gamma-1}(T_3 - T_2) - \frac{nR}{\gamma-1}(T_1 - T_4) \qquad C_V = \frac{nR}{\gamma-1} \\ &= -C_V\{(T_1 - T_2) + (T_3 - T_4)\} \end{aligned} \tag{575}$$

From eqs(572) and (575), the efficiency of work of the *Otto* cycle is found as eq(576).

$$\eta = \frac{W_{out}}{Q_{12}} = \frac{-C_V\{(T_1 - T_2) + (T_3 - T_4)\}}{C_V(T_2 - T_1)} = 1 - \frac{T_3 - T_4}{T_2 - T_1} \tag{576}$$

This result is also derived from eq(572) alone since the efficiency of work is equivalent to the fraction of the heat that the heat engine can utilize for work (eq(577)).

$$\eta = \frac{Q_{12} + Q_{34}}{Q_{12}} = \frac{C_V(T_2 - T_1) + C_V(T_4 - T_3)}{C_V(T_2 - T_1)} = 1 - \frac{T_3 - T_4}{T_2 - T_1} \tag{577}$$

Eq(577) is also expressed as eq(577') by using the relation given in eq(569). See **solution 3-80** for the *Stirling* cycle.

$$\eta = 1 - \left(\frac{V_1}{V_3}\right)^{\gamma-1} \tag{577'}$$

As is seen from eq(577'), the efficiency of work of the *Otto* cycle depends on the kind (γ) of the ideal gas.

Ans. efficiency of *Otto* cycle: See text and eq(576).

Solution 3-107

The function given in the problem to describe the entropy S in terms of U, V and N is differentiated to yield the total differential as given by eq(578).

$$dS = \left(\alpha \ln\frac{UV}{\beta N^2} - 2\alpha\right)dN + \alpha N\frac{dU}{U} + \alpha N\frac{dV}{V} \tag{578}$$

On the other hand, the total differential of U derived from the first and second laws of thermodynamics is expressed by eq(579).

$$dU = \left(\frac{\partial U}{\partial S}\right)_{V,N} dS + \left(\frac{\partial U}{\partial V}\right)_{S,N} dV + \left(\frac{\partial U}{\partial N}\right)_{S,V} dN = TdS - PdV + \mu dN \tag{579}$$

Obtaining $\left(\frac{\partial U}{\partial V}\right)_{S,N}$ and $\left(\frac{\partial U}{\partial S}\right)_{V,N}$ from eq(578) and putting them equal to the corresponding partial differential coefficients, eqs(580) through (582) are derived.

$$\left(\frac{\partial U}{\partial V}\right)_{S,N} = -\frac{U}{V} = -P \tag{580}$$

$$\left(\frac{\partial U}{\partial S}\right)_{V,N} = \frac{U}{\alpha N} = T \tag{581}$$

$$\left(\frac{\partial U}{\partial N}\right)_{S,V} = -\frac{U}{N}\left(\ln\frac{UV}{\beta N^2} - 2\right) = \mu \tag{582}$$

Eliminating U from eqs(580) and (581), eq(583), which is the equation of state of the ideal gas, is obtained. Here, the constant α is put equal to R/N.

$$PV = \alpha NT \overset{\alpha \equiv R/N}{\Rightarrow} PV = RT \qquad (583)$$

Also, eq(584) is derived from eq(582) to express the chemical potential.

$$\mu = -\frac{U}{N}\left(\ln\frac{UV}{\beta N^2} - 2\right) = -\frac{U}{N}\left(\ln\frac{P^2V^2}{P\beta N^2} - 2\right) = \alpha T\left(\ln P/P^\circ - \ln\frac{(\alpha T)^2}{\beta P^\circ} + 2\right) \qquad (584)$$

However, the heat capacity at constant volume C_V derived from eq(581) is just R as is seen from eq(585) below.

$$C_V = T\left(\frac{\partial S}{\partial T}\right)_{V,N} = \left(\frac{\partial U}{\partial T}\right)_{V,N} = \alpha N = R \qquad (585)$$

In order to achieve $C_V = 3R/2$ without changing the relation given by eq(583), just the multiplication of eq(585) by $3/2$ seems to be sufficient judging from the form of eq(585). For this, α is multiplied by $3/2$ to put $\alpha = (R/N) \times 3/2$. However, with this modification alone, the relation of $PV = U$ still remains unchanged. Therefore, by replacing V in the fundamental equation with $V^{2/3}$, the third term of eq(578) is replaced with $\frac{2}{3} \times \alpha N \frac{dV}{V}$ and, accompanied with this, eq(580) is replaced by eq(580').

$$\left(\frac{dU}{\partial V}\right)_{S,N} = -\frac{2}{3} \times \frac{U}{V} \qquad (580')$$

Consequently, eq(586) is obtained as the one of the simplest modification, in which β is replaced with β' so as to keep the dimension of the equation unchanged. See also **solution 3-105** for the fundamental function for the entropy.

$$S(U,V,N) = \alpha N \ln\frac{UV^{2/3}}{\beta' N^2}, \qquad \alpha = \frac{3R}{2N} \qquad (586)$$

Ans. eq(583), expression for S: eq(586)

Solution 3-108

(a) Solution 1. The equation of state of the ideal gas is used to find $\alpha_{(P)}$ as given by eq(587) and the V-T relation for the adiabatic process given by eq(589) below is used to find $\alpha_{(S)}$ as is given by eq(588). (See also **solutions 3-26 and 3-36** for the direct derivation.)

$$\alpha_{(P)} = \frac{1}{V}\left(\frac{\partial V}{\partial T}\right)_P = \frac{1}{T} \qquad (587)$$

$$\alpha_{(S)} = \frac{1}{V}\left(\frac{\partial V}{\partial T}\right)_S = \frac{1}{V}\cdot\frac{dV}{dT} = \frac{1}{A^{\frac{1}{\gamma-1}}T^{-\frac{1}{\gamma-1}}} \times \left(-\frac{1}{\gamma-1}\right)A^{\frac{1}{\gamma-1}}T^{-\frac{1}{\gamma-1}-1} = \left(-\frac{1}{\gamma-1}\right)\cdot\frac{1}{T} = -\frac{C_V}{R}\cdot\frac{1}{T} \qquad (588)$$

$$TV^{\gamma-1} = \text{const.} \equiv A \Rightarrow V = (A\cdot T^{-1})^{\frac{1}{\gamma-1}} = A^{\frac{1}{\gamma-1}}T^{-\frac{1}{\gamma-1}} \qquad (589)$$

For the ideal gas, the T-V relation is given by eq(590) for the isobaric process and by eq(591) for the adiabatic process and they are plotted in Fig.60 below. The superscript $^\circ$ signifies the value at the standard state.

$$T = \frac{P^\circ V}{R} = \frac{V}{0.0821} \quad (590), \qquad T = 694/V^{\gamma-1} \quad (591)$$

The constant in eq(591) is found to be 694 as calculated below.

$$T^\circ = \frac{P^\circ V^\circ}{R} \Rightarrow T^\circ (V^\circ)^{\gamma-1} = \text{const.} = 273 \times (22.4)^{0.3} = 694 \text{ (L, atm, K)}$$

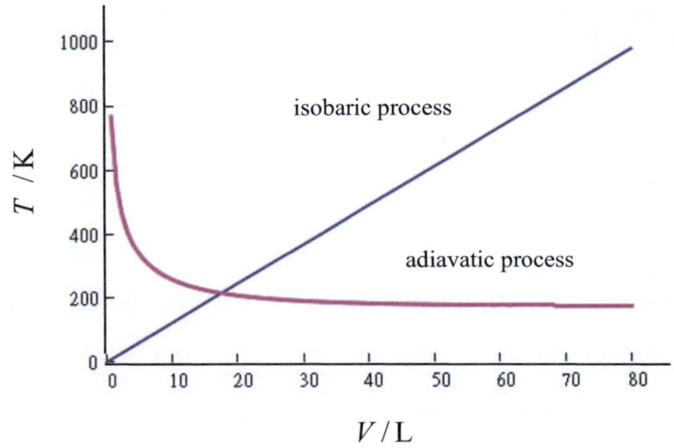

Fig.60 *V-T* relations of the ideal gas upon the isobaric and adiabatic processes. From this figure, the thermal expansion coefficient is seen to be negative for the adiabatic process.

Solution 2. In the above derivation, the premise employed is that the specific heats C_V and C_P are independent of temperature. Employing *Euler*'s chain relation, however, $\alpha_{(S)}$ can readily be expressed by the known values (see **solution 3-36**) as is given by eq(592).

$$\alpha_{(S)} = \frac{1}{V}\left(\frac{\partial V}{\partial T}\right)_S = -\frac{C_V}{TV} \cdot \left(\frac{\partial T}{\partial P}\right)_V = -\frac{C_V}{T \cdot R} \qquad (592)$$

Although eq(592) obtained assumes the same form as eq(588) found in Solution 1, the specific heat in eq(592) need not be a constant but can be a function of temperature.

Ans. isobaric process: eq(587), adiabatic process: eq(588) or eq(592)

(b) For the ideal gas, the relation of $C_P - C_V = R$ is always satisfied, even though each specific is dependent on temperature. Consequently, substituting the relation of $C_V = C_P - R$ and eq(589) given in the problem into eq(592), eq(593) is obtained as the answer.

$$\alpha_{(S)} = -\frac{C_V}{T \cdot R} = -\frac{(6.26 + 2.746 \times T - 0.770 \times T^2) \times 4.184 - R}{T \cdot R} \qquad (593)$$

Ans. adiabatic expansion coefficient : eq(593)

Solution 3-109a

The isothermal compressibility of the ideal gas is inversely proportional to the pressure as is given by eq(594).

$$\kappa_T = -\frac{1}{V}\left(\frac{\partial V}{\partial P}\right)_T = \frac{1}{P} \qquad (594)$$

On the other hand, the adiabatic compressibility is found as eq(596) by substituting the *P-V* relation given by eq(595) for the adiabatic process into the definition.

$$PV^\gamma = \text{const.} \equiv B \Rightarrow V = (B \cdot P^{-1})^{\frac{1}{\gamma}} = B^{\frac{1}{\gamma}} P^{-\frac{1}{\gamma}} \qquad (595)$$

$$\kappa_S = -\frac{1}{V}\left(\frac{\partial V}{\partial P}\right)_S = -\frac{1}{V} \cdot \frac{dV}{dP} = -\frac{1}{B^{\frac{1}{\gamma}} P^{-\frac{1}{\gamma}}} \times \left(-\frac{1}{\gamma}\right) B^{\frac{1}{\gamma}} P^{-\frac{1}{\gamma}-1} = \frac{1}{\gamma} \cdot \frac{1}{P} = \frac{C_V}{C_P} \cdot \frac{1}{P} \qquad (596)$$

Since the inequality $C_V < C_P$ is always satisfied, it is seen from eq(596) that the adiabatic compressibility is smaller than the isothermal one.

More generally, the result of eq(596) is derived by expressing the definition of the adiabatic compressibility as a function of *P*, *V* and *T* including heat capacities. This task is easily achieved by means of *Jacobian*s as is performed below.

Employing *Jacobian*s, the variation of volume with pressure for the adiabatic process is expressed by eq(597).

$$\left(\frac{\partial V}{\partial P}\right)_S = \frac{\partial(V,S)}{\partial(P,S)} = \frac{\dfrac{\partial(V,S)}{\partial(V,T)}}{\dfrac{\partial(P,S)}{\partial(V,T)}} = \frac{\dfrac{1}{T}C_V}{\begin{vmatrix} \left(\dfrac{\partial P}{\partial V}\right)_T & \left(\dfrac{\partial S}{\partial V}\right)_T \\ \left(\dfrac{\partial P}{\partial T}\right)_V & \left(\dfrac{\partial S}{\partial T}\right)_V \end{vmatrix}} = \frac{\dfrac{1}{T}C_V}{\begin{vmatrix} -\dfrac{RT}{V^2} & \dfrac{R}{V} \\ \dfrac{R}{V} & \dfrac{C_V}{T} \end{vmatrix}} = -\frac{V^2 C_V}{RT(C_V + R)} = -\frac{VC_V}{PC_P} \tag{597}$$

To derive the last equality of eq(597), the relations of $PV = RT$ and $C_P - C_V = R$ for the ideal gas is used. From eq(597), the adiabatic compressibility is given by eq(598).

$$\kappa_S = -\frac{1}{V}\left(\frac{\partial V}{\partial P}\right)_S = \frac{C_V}{C_P}\cdot\frac{1}{P} \tag{598}$$

As a result, the same equation as q(596) is obtained. However, it should be noted that in deriving eq(598), it is not an essential prerequisite that the heat capacities are temperature-independent constants.

Ans. adiabatic compressibility: eq(596) or eq(598)

Solution 3-109b

Here, the answer is given simply by employing *Jacobian*s. The ratio of the adiabatic compressibility κ_S to the isothermal compressibility κ_T is given by the first equality of eq(599) by definition, so that transforming it using *Jacobian*s easily leads to the last equality of eq(599).

$$\frac{\kappa_S}{\kappa_T} = \frac{-\dfrac{1}{V}\left(\dfrac{\partial V}{\partial P}\right)_S}{-\dfrac{1}{V}\left(\dfrac{\partial V}{\partial P}\right)_T} = \frac{\left(\dfrac{\partial V}{\partial P}\right)_S}{\left(\dfrac{\partial V}{\partial P}\right)_T} = \frac{\dfrac{\partial(V,S)}{\partial(P.S)}}{\dfrac{\partial(V,T)}{\partial(P.T)}} = \frac{\dfrac{\partial(V,S)}{\partial(V,T)}}{\dfrac{\partial(P.S)}{\partial(P.T)}} = \frac{\left(\dfrac{\partial S}{\partial T}\right)_V}{\left(\dfrac{\partial S}{\partial T}\right)_P} = \frac{C_V}{C_P} \tag{599}$$

Ans. ratio of the compressibility: eq(599)

Solution 3-110

Writing the entropy at $T = 0$ as S_0, the entropy $S(T_a)$ at the temperature T_a can be expressed by eq(600).

$$S(T_a) = S_0 + \int_0^{T_a}\left(\frac{\partial S}{\partial T}\right)_V dT \;\Rightarrow\; S(T_a) = S_0 + \int_0^{T_a}\frac{C_V}{T}dT \tag{600}$$

For $S(T_a)$ to be finite, the integrand of eq(560) should be finite at the lower limit of integration. Consequently, C_V should be proportional to *T* at least at low temperatures, i.e., $C_V \to \alpha \times T$ and, hence, approaches to zero ($C_V \to 0$) when the temperature gets really closer to zero K.

Exactly the same argument can be made about C_P and, hence, it should approach to zero ($C_P \to 0$) as well.

Now then, does this mean that the result is inconsistent with the relation given by eq. (601) which is established for the ideal gas?

$$C_P - C_V = R \qquad (601)$$

Generally speaking, the left side of eq(601) can be written as eq(601') (see **solution 3-97)** and for substances

$$C_P - C_V = \frac{\alpha^2 VT}{\beta} \qquad (601') \quad \text{, where} \quad \alpha = \frac{1}{V}\left(\frac{\partial V}{\partial T}\right)_P \text{,} \quad \beta(\equiv \kappa_T) = -\frac{1}{V}\left(\frac{\partial V}{\partial P}\right)_T$$

other than gases, the thermal expansion coefficient α, the compressibility coefficient β and V remain constant even when $T \to 0$. Consequently, for these cases the results that $C_V \to 0$ and $C_P \to 0$ when $T \to 0$ are consistent with the general relation given by eq(601'). For the ideal gas, however, when $T \to 0$, either $P \to 0$ or $V \to 0$ must be realized, indicating that the ideal gas itself ceases to exist or at least becomes meaningless.

Ans. See text.

Solution 3-111

In the expanded form of $d(\varDelta G_{A\to B})_{eq}$, the first-order term is zero as is obvious from eq(603) given in the problem and the second-order terms are obtained as eq(602') via eq(602) below,

$$d(\varDelta G_{A\to B})_{eq} = \left(\frac{\partial^2 \varDelta G_{A\to B}}{\partial T^2}\right)_{eq}(dT_{eq})^2 + 2\left(\frac{\partial^2 \varDelta G_{A\to B}}{\partial P \partial T}\right)_{eq}(dT_{eq})(dP_{eq}) + \left(\frac{\partial^2 \varDelta G_{A\to B}}{\partial P^2}\right)_{eq}(dP_{eq})^2 \qquad (602)$$

$$= -\left(\frac{\partial \varDelta S_{A\to B}}{\partial T}\right)_{eq}(dT_{eq})^2 - 2\left\{-\left(\frac{\partial V_B}{\partial T}\right)_P + \left(\frac{\partial V_A}{\partial T}\right)_P\right\}(dT_{eq})(dP_{eq}) + \left(\frac{\partial \varDelta V_{A\to B}}{\partial P}\right)_{eq}(dP_{eq})^2$$

$$= -\frac{1}{T}\varDelta C_P (dT_{eq})^2 + 2V\varDelta\alpha(dT_{eq})(dP_{eq}) - V\varDelta\kappa_T(dP_{eq})^2$$

$$= -\left\{\frac{1}{T}\varDelta C_P(dT_{eq}) - V\varDelta\alpha(dP_{eq})\right\}(dT_{eq}) + V\left\{\varDelta\alpha(dT_{eq}) - \varDelta\kappa_T(dP_{eq})\right\}(dP_{eq}) \qquad (602')$$

in which the following relations are used.

$$\left(\frac{\partial S_B}{\partial P}\right)_T - \left(\frac{\partial S_A}{\partial P}\right)_T = -\left(\frac{\partial V_B}{\partial T}\right)_P + \left(\frac{\partial V_A}{\partial T}\right)_P = -V\Delta\alpha \qquad (603)$$

As far as the equilibrium is maintained during the change, the equality of $d(\varDelta G_{A\to B})_{eq} = 0$ must always be satisfied. Therefore, the coefficients of dT_{eq} and dP_{eq} in eq(602') must be zero as is given by eq(564).

$$\left.\begin{aligned} &\left\{\frac{1}{T}\varDelta C_P(dT_{eq}) - V\varDelta\alpha(dP_{eq})\right\} = 0 \\ &\left\{\varDelta\alpha(dT_{eq}) - \varDelta\kappa_T(dP_{eq})\right\} = 0 \end{aligned}\right\} \qquad (604)$$

From eq(604), eqs(605) and (606) are derived for the functions describing dP_{eq}/dT_{eq}.

$$\frac{dP_{eq}}{dT_{eq}} = \frac{\varDelta C_P}{TV\varDelta\alpha} \qquad (605) \text{,} \qquad \frac{dP_{eq}}{dT_{eq}} = \frac{\varDelta\alpha}{\varDelta\kappa_T} \qquad (606)$$

These relations can also be derived by expanding $\varDelta V_{A\to B}$ and $\varDelta S_{A\to B}$ to the first order of dP_{eq} and dT_{eq} as are given by eq(607) and eq(608), respectively,

$$\left(\frac{\partial \Delta V_{A\to B}}{\partial P}\right)_{eq} dP_{eq} + \left(\frac{\partial \Delta V_{A\to B}}{\partial T}\right)_{eq} dT_{eq} = -(V_B\kappa_T^B - V_A\kappa_T^A)dP_{eq} + (V_B\alpha^B - V_A\alpha^A)dT_{eq}$$
$$= V\,(-\Delta\kappa_T dP_{eq} + \Delta\alpha dT_{eq}) = 0 \qquad (607)$$
$$\left(\frac{\partial \Delta S_{A\to B}}{\partial P}\right)_{eq} dP_{eq} + \left(\frac{\partial \Delta S_{A\to B}}{\partial T}\right)_{eq} dT_{eq} = -(V_B\alpha^B - V_A\alpha^A)dP_{eq} + \frac{1}{T}(C_P^B - C_P^A)dT_{eq}$$
$$= -V\Delta\alpha dP_{eq} + \frac{1}{T}\Delta C_P dT_{eq} = 0 \qquad (608)$$

and by equating them equal to zero, resulting in eq(607') and eq(608'). These functions derived thus are called *Ehrenfest*'s equation. (Note that $\kappa_T \equiv \beta$ here.)

$$\frac{dP_{eq}}{dT_{eq}} = \frac{\Delta\alpha}{\Delta\kappa_T} \qquad (607'), \qquad \frac{dP_{eq}}{dT_{eq}} = \frac{\Delta C_P}{TV\Delta\alpha} \qquad (608')$$

Ans. *Ehrenfest*'s equation: eqs (605) and (606) or eqs(607') and (608')

Solution 3-112a

In Fig.61 the *P-T* diagram illustrating the phase equilibrium among the three phases is shown again. Where two phases coexist, the freedom allowed by the phase rule is just one and its region is given by a line in the *P-T* diagram. At the temperature T_1 in Fig.61, when the pressure of the gas is raised along the orange arrow, the volume of the gas decreases until the liquefaction comes in at the pressure of the intersection of the orange arrow and the *G-L* equilibrium line. The volume at that point is denoted V_a. Keeping the pressure constant, the liquefaction can be progressed further making the volume smaller until the whole gas is liquefied to the volume of V_b.

As is seen from this, once *P* or *T* is determined in the *P-T* diagram, the other is unambiguously determined. In the *T-V* diagram, however, the volume which satisfies the *G-L* equilibrium is not unique but can take values in a certain range such as $[V_a, V_b]$. This process, which is indicated by the vertical orange dotted arrow in Fig.61, is expressed by the horizontal orange dotted arrow in the *T-V* diagram shown in Fig.62.

As is indicated by *G* or *L* in Fig.62, a region on the left side of the convexed curve is liquid and one on the right is gas. A region in the middle under the convexed curve is where gas and liquid coexist. With increasing the temperature, the coexisting region gets smaller and, finally, at the critical point, the two phases emerges into one phase. As is clear from this explanation, the coexisting region is composed of infinite number of

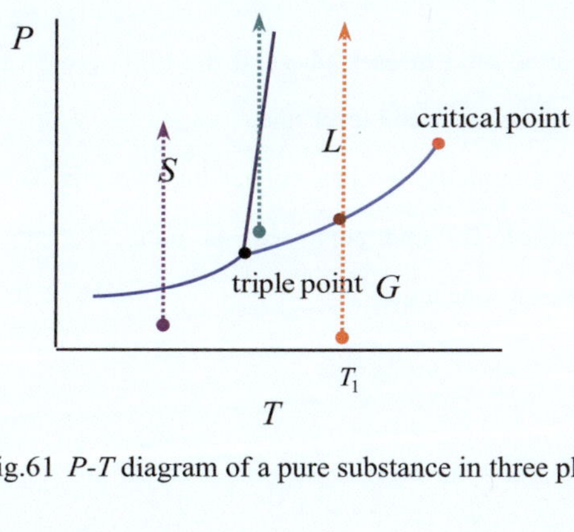

Fig.61 *P-T* diagram of a pure substance in three phases.

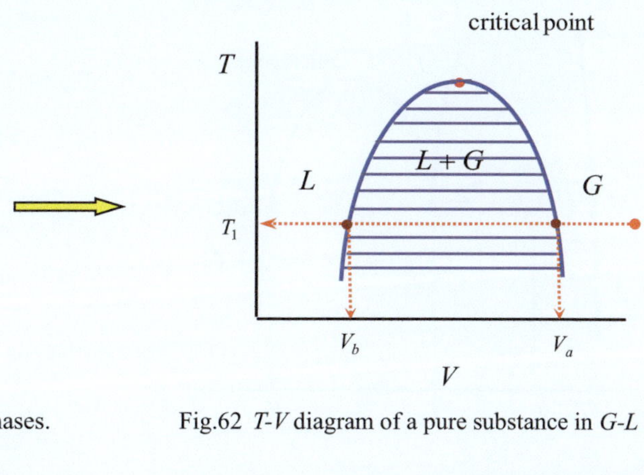

Fig.62 *T-V* diagram of a pure substance in *G-L* equilibrium.

tie lines (horizontal lines in Fig.62). In this sense, it is appropriate to express the two phases coexisting region as an area etched with horizontal lines.

In order to present all of the phase equilibriums consisting of three phases in one *T-V* diagram, the same procedure as employed for giving Fig.62 is applied along not only the green dotted arrow but also the violet dotted arrow given in Fig.61. The diagram obtained thus is shown in Fig.63. It should be noted here that the *P-T* diagram is a plot of the intensive vs. intensive variables but that the *T-V* diagram is a plot of intensive vs. extensive variables.

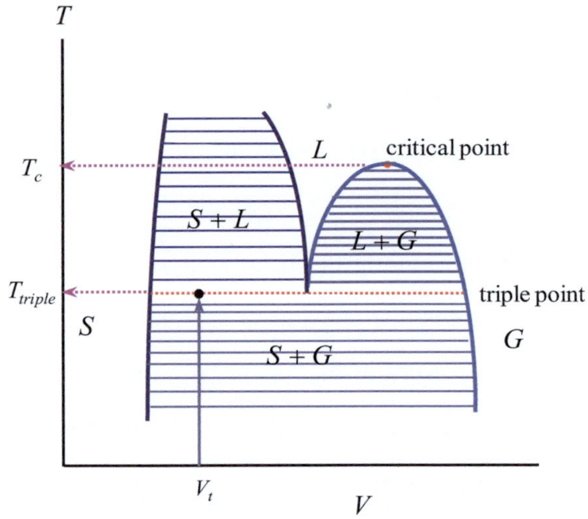

Fig.63 *T-V* phase diagram of a pure substance.
S: solid phase, *L*: liquid phase, *G*: gas phase.
In the region etched with horizontal lines, two phases can coexist. V_t is the volume at the triple point. As is clear from this diagram, V_t can take arbitrary values in a certain range (see the following problem).

Ans. See text and Fig.63.

Solution 3-112b

The volumes of the gas, liquid and solid coexisting at the triple point 3 are denoted V_G, V_L and V_S, respectively. Then the total volume V_t is given by eq(609).

$$V_t = V_G + V_L + V_S \qquad (609)$$

Since each phase is composed of the pure substance, if the phase diagram is drawn for one mole of the substance in total, the volumes such as V_1, V_2 and V_3 given in Fig.64 below represent the relevant molar volumes of *S, L* and *G*. Consequently, taking the lengths of the lever arm *a*, *b* and *c* as shown in the figure below, the relations expressed in eqs(610) and (611) are obtained (see **"note"**).

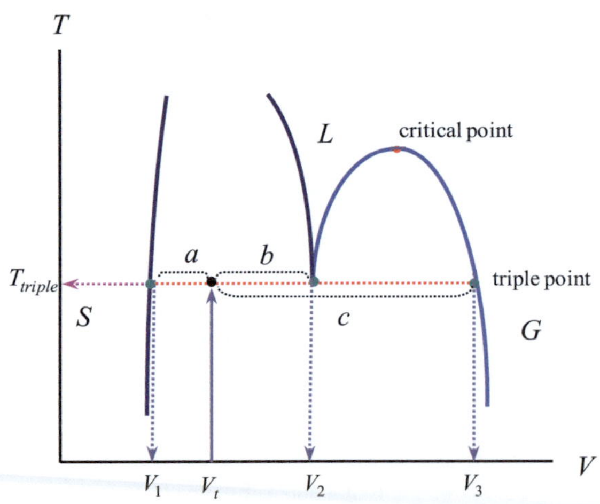

Fig.64 Volume ratio of each phase at the triple point of a pure substance and lever rule.
For simplicity, tie-lines connecting two phases are omitted. The total amount of the pure substance is taken as one mole.

$$(V_t - V_1):(V_3 - V_t) = a:c = V_G : V_S \qquad (610)$$

$$(V_2 - V_t):(V_t - V_1) = b:a = V_S : V_L \qquad (611)$$

From these, V_G, V_L and V_S can be expressed as eq(612).

$$\frac{V_G}{V_S} = \frac{a}{c}, \quad \frac{V_L}{V_S} = \frac{a}{b} \qquad (612)$$

Substituting eq(612) into eq(609) to obtaine V_G, V_L and V_S, they are found as eq(614) via eq(613) below.

$$V_t = V_G + V_L + V_S = \frac{ab+bc+ca}{bc} \times V_S \qquad (613)$$

$$V_G = \frac{ab}{ab+bc+ca} V_t, \quad V_L = \frac{ca}{ab+bc+ca} V_t, \quad V_S = \frac{bc}{ab+bc+ca} V_t \qquad (614)$$

Ans. volume of each phase at the triple point: eq(614) and Fig.64

"note" Since the volume is not governed by the conservation law, it is neither inevitable nor rational to apply the lever rule (see **solutions 2-27** and **3-15**).

Solution 3-113

The specific heat at constant pressure C_P is defined by eq(615).

$$C_P = \left(\frac{\partial H}{\partial T}\right)_P = T\left(\frac{\partial S}{\partial T}\right)_P \qquad (615)$$

Therefore, its partial derivative with pressure is expressed as eq(616) by using entropy.

$$\left(\frac{\partial C_P}{\partial P}\right)_T = \left\{\frac{\partial}{\partial P} T\left(\frac{\partial S}{\partial T}\right)_P\right\}_T = T\left\{\frac{\partial}{\partial P}\left(\frac{\partial S}{\partial T}\right)_P\right\}_T$$

$$= T\left\{\frac{\partial}{\partial T}\left(\frac{\partial S}{\partial P}\right)_T\right\}_P \qquad (616)$$

$$= -T\left\{\frac{\partial}{\partial T}\left(\frac{\partial V}{\partial T}\right)_P\right\}_P = -T\left\{\frac{\partial}{\partial T}\left(\frac{V}{T}\right)\right\}_P \qquad (616')$$

$$= -T\left\{-\frac{V}{T^2} + \frac{1}{T}\left(\frac{\partial V}{\partial T}\right)_P\right\} = -T\left(-\frac{V}{T^2} + \frac{V}{T^2}\right) = 0$$

Employing *Maxwell's* relation of $(\partial S/\partial P)_T = -(\partial V/\partial T)_P$, eq(616) is rewritten as eq(616'). Carrying out the differentiation of eq(576') with temperature, the equation $(\partial C_P/\partial P)_T = 0$ to be proved is derived.

Ans. See text.

Solution 3-114

For the ideal gas, eq(80) given in the problem can be written as eq(617) and, hence, the relation between entropy and volume is given by eq(617') after integration of eq(617).

$$(\partial S/\partial V)_T = \frac{nR}{V} \qquad (617)$$

$$S = nR\ln V + \text{const.} \qquad (617')$$

The mixing process is, as illustrated below, regarded as proceeding in the manner that each gas independently expands to the volume of $V_A + V_B$ (for instance, through a hole in the separating wall). The entropy change of each gas upon expansion is given by eq(618).

$$\left.\begin{aligned} \Delta S_{\mathrm{A}}(V_{\mathrm{A}} \to V_{\mathrm{A}} + V_{\mathrm{B}}) &= n_{\mathrm{A}} R \ln \frac{V_{\mathrm{A}} + V_{\mathrm{B}}}{V_{\mathrm{A}}} \\ \Delta S_{\mathrm{B}}(V_{\mathrm{B}} \to V_{\mathrm{A}} + V_{\mathrm{B}}) &= n_{\mathrm{B}} R \ln \frac{V_{\mathrm{A}} + V_{\mathrm{B}}}{V_{\mathrm{B}}} \end{aligned}\right\} \quad (618)$$

Therefore, the entropy of mixing ΔS_{mix} is given by eq(619).

$$\Delta S_{mix} = \Delta S_{\mathrm{A}} + \Delta S_{\mathrm{B}} \qquad (619)$$

Noting that the mole number of a gas is proportional to its volume under the same temperature and pressure, eq(579) can be expressed as eq(579') and the obtained equation is the same as eq(81) given in the problem.

$$\begin{aligned} \Delta S_{mix} &= \left(n_{\mathrm{A}} R \ln \frac{V_{\mathrm{A}} + V_{\mathrm{B}}}{V_{\mathrm{A}}} + n_{\mathrm{B}} R \ln \frac{V_{\mathrm{A}} + V_{\mathrm{B}}}{V_{\mathrm{B}}} \right) \\ &= -(n_{\mathrm{A}} + n_{\mathrm{B}}) R \left\{ \frac{n_{\mathrm{A}}}{(n_{\mathrm{A}} + n_{\mathrm{B}})} \ln \frac{n_{\mathrm{A}}}{(n_{\mathrm{A}} + n_{\mathrm{B}})} + \frac{n_{\mathrm{B}}}{(n_{\mathrm{A}} + n_{\mathrm{B}})} \ln \frac{n_{\mathrm{B}}}{(n_{\mathrm{A}} + n_{\mathrm{B}})} \right\} \\ &= -nR(X_{\mathrm{A}} \ln X_{\mathrm{A}} + X_{\mathrm{B}} \ln X_{\mathrm{B}}) \qquad (619') \qquad \text{(eq(81) in the problem)} \end{aligned}$$

It is clearly seen from the derivation above that under the condition of constant temperature, the entropy of mixing arises simply from a change in the volume of each gas.

Ans. See text and eq(619').

Solution 3-115

(a) As is given in the preceding solution, in the case of two different ideal gases, by putting $X_A = X_B = 1/2$ in eq(620), $nR \ln 2$ is found for the entropy of mixing.

$$\Delta S_{mix} = -nR(X_{\mathrm{A}} \ln X_{\mathrm{A}} + X_{\mathrm{B}} \ln X_{\mathrm{B}}) = nR \ln 2 \qquad (620)$$

However, in the case of two (ideal) gases of the same kind, it is totally impossible in a usual sense to tell whether each gas has expanded through the hole in the separating wall to double its volume or nothing has happened at all. In another word, since the two states before and after mixing cannot be distinguished, there exists no difference in the thermodynamic quantities. See Fig.65.

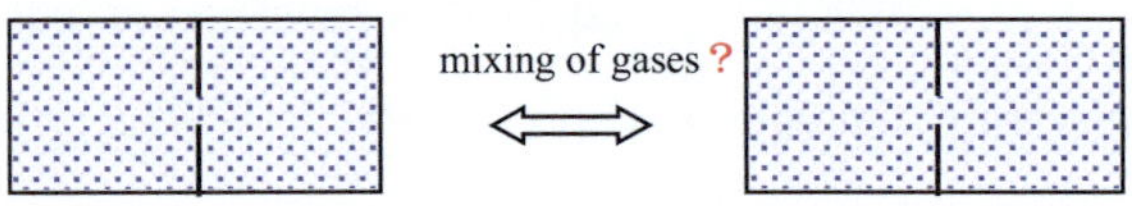

Fig.65 An image of mixing of two gases of the same kind.

When a separation process opposite to the mixing is considered, the difference between the mixture of two gases of different kinds and that of the same kind is self-evident. The former mixture cannot be separated into each component just by inserting a separating wall to divide it into two equal portion. The latter gas, however, can be separated simply by the same operation into the two portions each of which is identical to the one before mixing. Since the insertion of the separating wall does not cause any change in the thermodynamic quantities of the gas, it is concluded that the entropy of mixing of the same gases is zero. In other words, in the case of a single pure substance, the extensive thermodynamic quantities are proportional to the amount of the substance. For this, mixing and multiplying are the same thing naturally. However, from a molecular theoretical point of view, more strict argument is required (see **problems 5-4 and 5-19)**.

Ans. See text.

(b) For the two gases of different kinds, the entropy of mixing per mole, when mixed in the mole ration of 1 to n, is given by eq(621).

$$\Delta S_{mix} = -R\left\{\frac{1}{1+\mathrm{n}}\times\ln\frac{1}{1+\mathrm{n}}+\frac{\mathrm{n}}{1+\mathrm{n}}\times\ln\frac{\mathrm{n}}{1+\mathrm{n}}\right\} \qquad (621)$$

In the case of the gases of the same kind, since it cannot be distinguished between the addition of two parts and the addition of n+1 parts, as is discussed above, the entropy of mixing becomes zero($\Delta S_{mix}=0$). Again, from a molecular theoretical point of view, whether or not the entropy change upon mixing two gases of different kinds in the ratio of 1 to n is given by eq(621) must be examined more carefully.

Ans. See text.

Solution 3-116

Since the gases behave ideally, the gas in each vessel obeys eq(622) and, also, the final pressure P_f after mixing the gases is given by eq(623).

$$P_1V_1 = n_1RT\,, \qquad P_2V_2 = n_2RT \qquad (622)$$

$$P_f(V_1+V_2) = (n_1+n_2)RT \;\Rightarrow\; P_f = \frac{(n_1+n_2)RT}{(V_1+V_2)} \qquad (623)$$

For the mixing of the gases of the same kind, the mixing process is considered in the following two steps.

(i) One or the other gas is <u>expanded</u> or <u>compressed</u> <u>reversibly at constant temperature</u> to the final pressure P_f. Denoting the volumes of the gases after the reversible processes by V_x and V_y, they are given by eq(624) and satisfy eq(624').

$$P_fV_x = n_1RT, \qquad P_fV_y = n_2RT \;\Rightarrow\; V_x = \frac{n_1RT}{P_f}, \qquad V_y = \frac{n_2RT}{P_f} \qquad (624)$$

$$V_x + V_y = V_1 + V_2 \qquad (624')$$

Since these processes are carried out reversibly, the work done or gained by the gas upon expansion or compression is expressed by eq(625) or eq(625').

$$W_1(T,P_1 \to T,P_f) = -\int_{P_1}^{P_f} PdV = -n_1RT\int_{V_1}^{V_x}\frac{1}{V}dV = -n_1RT\,\ln\frac{V_x}{V_1} \qquad (625)$$

$$W_2(T,P_2 \to T,P_f) = -\int_{P_2}^{P_f} PdV = -n_2RT\int_{V_2}^{V_y}\frac{1}{V}dV = -n_2RT\,\ln\frac{V_y}{V_1} \qquad (625')$$

Consequently, the total amount of work done by the gases is given by eq(626).

$$\Delta W = W_1 + W_2 = -n_1RT\ln\frac{V_x}{V_1} - n_2RT\ln\frac{V_y}{V_1} \qquad (626)$$

Also, the entropy changes associated with this concomitant isothermal compression and expansion are given by eqs(627) and (627') ($(\partial S/\partial V)_T = nR/V$).

$$S_1(T,P_1 \to T,P_f) = S_1(T,V_1 \to T,V_x) = n_1 R \ln \frac{V_x}{V_1} \qquad (627)$$

$$S_2(T,P_2 \to T,P_f) = S_2(T,V_2 \to T,V_y) = n_2 R \ln \frac{V_y}{V_2} \qquad (627')$$

From these equations, the sum of the entropy changes of the two gases is expressed by eq(628).

$$\Delta S = S_1 + S_2 = n_1 R \ln \frac{V_x}{V_1} + n_2 R \ln \frac{V_y}{V_2} \qquad (628)$$

(ii) Two gases of the same kind at the same T and pressure (P_f) are mixed in the mole numbers of n_1 and n_2. As is already discussed in the case of mixing substances of the same kind, the thermodynamic extensive quantities of the resulting mixture is given simply by a sum of the corresponding quantity for each mixing component. Therefore, the work done for the process of **A**→**B**, which is separated into processes (i) and (ii), is given by eq(626) and the entropy change for the same process is given by eq(628).

The total change in the internal energies of the gases ΔU is given by eq(629) and the change in heat ΔQ is expressed by the second equality of eq(629) for the isothermal reversible process. Thus, using eqs(586) and (628), it is derived from eq(629) that ΔU is zero, which is of course consistent with the well known fact that the internal energy of the ideal gas is a function of T alone.

$$\Delta U = \Delta Q + \Delta W = T\Delta S + \Delta W \qquad (629)$$

Ans. work: eq(626), entropy change: eq(628)

(b) Using V_x and V_y in eq(624), eq(628) can be rewritten as eq(628').

$$\Delta S = n_1 R \ln \frac{n_1(V_1+V_2)}{V_1(n_1+n_2)} + n_2 R \ln \frac{n_2(V_1+V_2)}{V_2(n_1+n_2)} \qquad (628')$$

This equation is again rewritten by using the mole fractions and the volume fractions defined by eq(630),

$$\left.\begin{aligned} x &= \frac{n_1}{n_1+n_2}, & 1-x &= \frac{n_2}{n_1+n_2} \\ a &= \frac{V_1}{V_1+V_2}, & 1-a &= \frac{V_2}{V_1+V_2} \end{aligned}\right\} \qquad (630)$$

and, as a result, eq(628') is simply expressed by eq(628'').

$$\frac{\Delta S}{n_1+n_2} = x \ln \frac{x}{a} + (1-x) \ln \frac{1-x}{1-a} > 0 \qquad (628'')$$

It is confirmed that the right side of eq(588'') takes a minimum at $x = a$ for an arbitrary value of a in the range $0 < a < 1$ (volume fraction) and otherwise is always positive by examining a trend of increment or decrement of the first and second derivatives of eq(588'') based on the conventional method. A couple of functions with different values of a are exemplified in Fig.66 above.

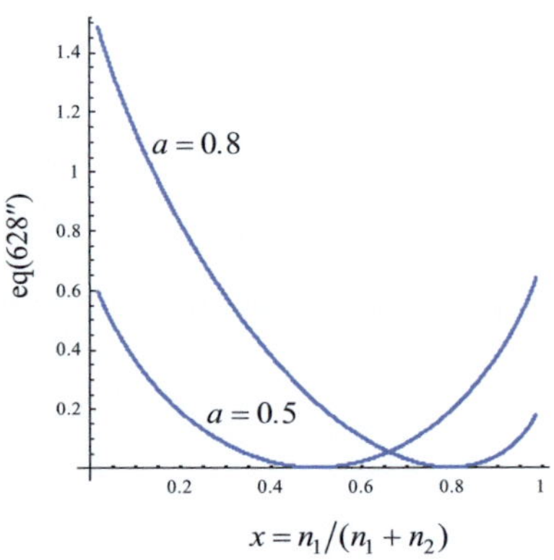

Fig.66 Entropy of mixing. See eq(628'').

From the result obtained thus, it is shown that the entropy change of the gases accompanied by the process **A**→**B** is always positive.

Ans. See text (b).

(c) When the process **A→B** occurs by free expansion with no work and reaches the same state as discussed in (a), the internal energy change is zero because not only the initial state but also the final state is the same as that in (a). From this the change in heat energy ΔQ is also found as zero. Consequently, since there is neither heat flow in to nor out from the thermostat, the total entropy change of the system including the thermostat is ascribed only to the entropy change in the gases due to the free expansion. This value is always positive apart from minimum zero as is seen in (b), leading to the conclusion that the process **A→B** is irreversible. It should be noticed that starting from the common initial state, as far as the final states reached are the same, the entropy changes (of the gases) are also the same irrespective of whether the process is reversible or irreversible. For the irreversible process, the inequality of $\Delta Q \neq T\Delta S$ holds.

Ans. irreversible process **A→B**: See text (c).

Solution 3-117

When the first-order phase transition was considered, the *Gibbs* energy change upon transition $\mathrm{I} \leftrightarrow \mathrm{II}$ was obtained based on the first-order differential of G given by eq(631) as is expressed by eq(632).

$$dG(P,T) = -SdT + VdP \qquad (631)$$

$$dG^{\mathrm{II}} - dG^{\mathrm{I}} = -(S^{\mathrm{II}} - S^{\mathrm{I}})\cdot dT + (V^{\mathrm{II}} - V^{\mathrm{I}})dP = -\Delta S \cdot dT + \Delta V dP \qquad (632)$$

However, in the case of the second-order phase transition, nothing about the behavior of the transition point at ($P_\circ$, $T_\circ$) can be known from eq(632), since both of ΔS and ΔV are zero. Therefore, it is necessary to examine the second-order differential of G given by eq(633) as has been done so far for the study of the critical phenomena or the fluctuation near the equilibrium.

$$\begin{aligned}
G(P,T) - G(P_\circ,T_\circ) = dG(P,T) &= \left(\frac{\partial G}{\partial T}\right)_{P_\circ} dT + \left(\frac{\partial G}{\partial P}\right)_{T_\circ} dP + \frac{1}{2}\left\{\left(\frac{\partial^2 G}{\partial T^2}\right)_{P_\circ} (dT)^2 + 2\left(\frac{\partial^2 G}{\partial T \partial P}\right)(dT)(dP) + \left(\frac{\partial^2 G}{\partial P^2}\right)_{T_\circ} (dP)^2\right\} \\
&= -SdT + VdP + \frac{1}{2}\left\{-\left(\frac{\partial S}{\partial T}\right)_{P_\circ} (dT)^2 - 2\left(\frac{\partial S}{\partial P}\right)_{T_\circ} (dT)(dP) + \left(\frac{\partial V}{\partial P}\right)_{T_\circ} (dP)^2\right\} \\
&= -SdT + VdP + \frac{1}{2}\left\{-\frac{C_P}{T_\circ}(dT)^2 + 2\left(\frac{\partial V}{\partial T}\right)_{T_\circ} (dT)(dP) + \left(\frac{\partial V}{\partial P}\right)_{T_\circ} (dP)^2\right\} \\
&= (-SdT + VdP) + \frac{1}{2}\left\{-\frac{C_P}{T_\circ} + 2V\beta\left(\frac{dP}{dT}\right) - V\alpha\left(\frac{dP}{dT}\right)^2\right\}(dT)^2 \qquad (633)
\end{aligned}$$

Here, the relation of eq(634) is used.

$$\frac{C_P}{T} = \left(\frac{\partial S}{\partial T}\right)_P, \qquad \beta - \frac{1}{V}\left(\frac{\partial V}{\partial T}\right)_P, \qquad \alpha = -\frac{1}{V}\left(\frac{\partial V}{\partial P}\right)_T \qquad (634)$$

Employing the fact that the changes of G, S and V upon the second-order phase transition $\mathrm{I} \leftrightarrow \mathrm{II}$ are zero, eq(635) is derived from eq(633).

$$\left\{\frac{C_P^{\mathrm{II}} - C_P^{\mathrm{I}}}{T_\circ} - 2V(\beta^{\mathrm{II}} - \beta^{\mathrm{I}})\left(\frac{dP}{dT}\right) + V(\alpha^{\mathrm{II}} - \alpha^{\mathrm{I}})\left(\frac{dP}{dT}\right)^2\right\} = 0 \qquad (635)$$

It is seen that eq(635) is a quadratic equation with respect to dP/dT. Therefore, dP/dT is generally expressed as the solutions of eq(635). However, if the change in G assumes a minimum at the solution, then the equilibrium will be realized more stably than otherwise against a differential change in T or P. Therefore, using the condition for a double root given by eq(636), the double root of eq(635) is obtained by eq(637).

$$V^2(\beta^{\text{II}} - \beta^{\text{I}})^2 - \frac{(C_P^{II} - C_P^{\text{I}})(\alpha^{\text{II}} - \alpha^{\text{I}})V}{T_\circ} = 0 \qquad (636)$$

$$\left(\frac{dP}{dT}\right)_e = \frac{(\beta^{\text{II}} - \beta^{\text{I}})}{(\alpha^{\text{II}} - \alpha^{\text{I}})} \qquad (637)$$

The subscript e added to the differential of eq(637) indicate that the differential with T is along the equilibrium.

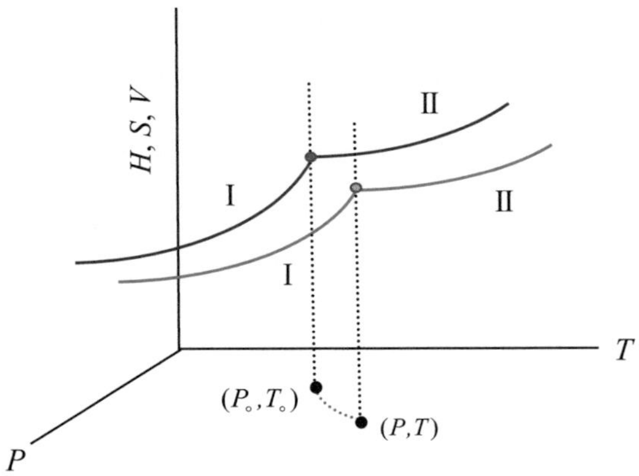

Fig.67 Schematic illustration to show how the thermodynamic variables change near the transition point against P and T.

Fig.67 shown above schematically illustrates how the thermodynamic variables change near the transition point against P and T. The red dotted line can be drawn based on the obtained $(dP/dT)_e$.

Ans. $(dP/dT)_e$: eq(637)

Solution 3-118

(a) When the heat flow is stationary, eq(638) is solved by putting Q as constant and eq(598') is obtained..

$$Q = -\lambda \frac{dT}{dx} \qquad (638) \qquad \text{(reposted)}$$

$$T = -\frac{Qx}{\lambda} + \text{const.} \qquad (638')$$

For the temperature of 278 K outside, the constant of eq(638') is found as $\text{const.} = 278\ \text{K}$. Denoting the thickness of the window glass as Δx and the temperature difference as ΔT, the amount of heat flowing through the window glass can be expressed as eq(639) from eq(638').

$$\Delta T = -\frac{Q \cdot \Delta x}{\lambda} \qquad (639)$$

Putting the thickness of the window glass equal to 3 mm and the difference in the temperatures equal to 15 K, the amount of heat flowing out to the outside per second is found as $4.75\ \text{kJ} \cdot \text{s}^{-1}$.

$$-15 = \frac{Q \times 3 \times 10^{-3}}{\lambda} \Rightarrow Q = -\frac{15 \times 0.95}{3 \times 10^{-3}} = 4.75\ \text{kJ} \cdot \text{s}^{-1}$$

Ans. amount of heat flowing out: $4.75\ \text{kJ} \cdot \text{s}^{-1}$

(b) The temperatures at the two interfaces, one between the glass and the air layer and the other between the air layer and glass, are denoted T_a and T_b as are shown in Fig.68 below. Noticing that the amount of heat flowing through the double glazed window is constant Q at the stationary state, the three relations given in eq(600) are derived.

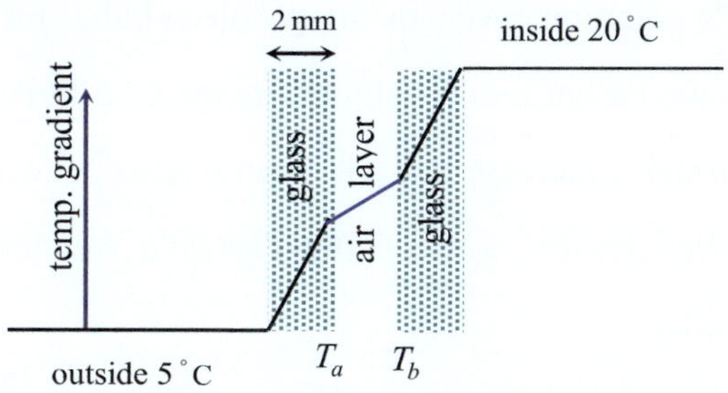

Fig.68 Temperature and its gradient at the interface of double glazed window.

$$\left.\begin{aligned} T_a - 278 &= -\frac{Q\times 2\times 10^{-3}}{0.95} \\ T_b - T_a &= -\frac{Q\times 2\times 10^{-3}}{0.025} \\ 293 - T_b &= -\frac{Q\times 2\times 10^{-3}}{0.95} \end{aligned}\right\} \quad (640)$$

Eliminating T_a and T_b from above relations, the following equation is obtained.

$$293-278 = -Q\times 2\times 10^{-3}\left(\frac{1}{0.95}+\frac{1}{0.025}+\frac{1}{0.95}\right)$$

From this equation, the values for Q, T_a and T_b are found as below.

$$Q = -\frac{15}{2\times 10^{-3}\times 42.1} = 178\ \mathrm{J\cdot s^{-1}}\ ,\quad T_a = 278+\frac{178\times 2\times 10^{-3}}{0.95} = 278.3_8\ ,\quad T_b = 293-\frac{178\times 2\times 10^{-3}}{0.95} = 292.6_3$$

Compared with the result obtained in (a), it is found that the heat flow per unit time is reduced approximately to $178/4750 \approx 4/100$.

Ans. temperatures at the interfaces: $T_a = 15.4^\circ\mathrm{C}$, $T_b = 19.6^\circ\mathrm{C}$; heat flow: reduced to ca. 4/100

(c) Replacing the thickness of the air layer of 2×10^{-3} m with 6×10^{-3} m in the middle relation of eq(640) and solving for Q in the same manner as in (b), $Q = 61.4$ J is found from the following equation. The obtained value is seen to be reduced to one third of the amount obtained for the air layer of 2 mm in thickness.

$$293-278 = -Q\times 2\times 10^{-3}\left(\frac{1}{0.95}+\frac{3}{0.025}+\frac{1}{0.95}\right) \Rightarrow Q = 61.4\ J$$

Ans. Reduced further to one third.

Solution 3-119

The partial molar volumes $\bar{v}_A$ and $\bar{v}_B$ are defined by eq(641). The following procedure employed in the case of the partial molar volume is applicable as it is for any other partial molar variables in general when any extensive thermodynamic quantities are employed as a substitute for V. V is a function of the mole numbers but the partial molar volume is expressed as a function of the mole fractions.

$$\bar{v}_A(X_A, X_B) = \left(\frac{\partial V}{\partial n_A}\right)\ ,\qquad \bar{v}_B(X_A, X_B) = \left(\frac{\partial V}{\partial n_B}\right) \qquad (641)$$

The differentials of $\bar{v}_A$ and $\bar{v}_B$, therefore, can be expressed as eqs(642) and (643).

$$d\bar{v}_A = \left(\frac{\partial \bar{v}_A}{\partial X_A}\right)dX_A = \left(\frac{\partial \bar{v}_A}{\partial n_B}\right)\left(\frac{\partial n_B}{\partial X_A}\right)dX_A = -\frac{n_B}{X_A X_B}\left(\frac{\partial^2 V}{\partial n_B \partial n_A}\right)dX_A \qquad (642)$$

$$d\bar{v}_B = \left(\frac{\partial \bar{v}_B}{\partial X_A}\right)dX_A = \left(\frac{\partial \bar{v}_B}{\partial n_A}\right)\left(\frac{\partial n_A}{\partial X_A}\right)dX_A = \frac{n_A}{X_A X_B}\left(\frac{\partial^2 V}{\partial n_A \partial n_B}\right)dX_A \qquad (643)$$

Here, in order to derive the last equalities, either one of eq(644) is used.

$$\left(\frac{\partial X_A}{\partial n_B}\right)_{n_A} = -\frac{n_A}{(n_A+n_B)^2} = -\frac{X_A X_B}{n_B},\qquad \left(\frac{\partial X_A}{\partial n_A}\right)_{n_A} = \frac{n_B}{(n_A+n_B)^2} = \frac{X_A X_B}{n_A} \qquad (644)$$

From eqs(642) and (643), it is evident that *Gibbs-Duhem*'s equation given in the problem holds for the partial molar volumes. Based on the equation, when one of the partial molar volumes increases a little, the other partial molar volume decreases to compensate the former's increase and vice versa. Except for this correlation, there are several other important features to be satisfied when *Gibbs-Duhem*'s equation is applied to partial molar variables practically (see **solution 3-21**).

Ans. See text.

Solution 3-120

(a)The distribution of radiation energy from the sun (eq(84') in the problem) is plotted against wave number in the figure below. If it is assumed that the whole radiation energy in the wave number range $15000-30000\ \text{cm}^{-1}$ (the area between two green vertical dotted lines in the figure) is converted into the electricity, then the efficiency of conversion is given by eq(645).

$$\int_{15000}^{30000}\varepsilon(\bar{\nu})\mathrm{d}\bar{\nu}\Big/\int_{0}^{\infty}\varepsilon(\bar{\nu})\mathrm{d}\bar{\nu}=\int_{15000}^{30000}\frac{\bar{\nu}^3}{e^{hc\bar{\nu}/k_BT}-1}\mathrm{d}\bar{\nu}\Big/\int_{0}^{\infty}\bar{\nu}^3\cdot\frac{\bar{\nu}^3}{e^{hc\bar{\nu}/k_BT}-1}\mathrm{d}\bar{\nu}=0.4157 \tag{645}$$

Substituting the given values into eq(605) and being integrated by using *Mathematica,* the efficiency is found as 41.6%.

Ans. achievable efficiency: 41.6%

(b) When the upper limit of the integration of eq(605) is extended to $40000\ \text{cm}^{-1}=250\ \text{nm}$, the efficiency of conversion is found as 47.0%, just a little improvement. On the other hand, when the lower limit is extended to $10000\ \text{cm}^{-1}$ making the light in the range $10000-30000\ \text{cm}^{-1}$ usable, the efficiency is enormously improved to 67.3%. Consequently, although the energy of light itself is small, the efficiency of conversion significantly depends on how efficiently the sun light in the infrared range can be converted into electricity by the solar cell, as is readily seen from Fig.69 shown above.

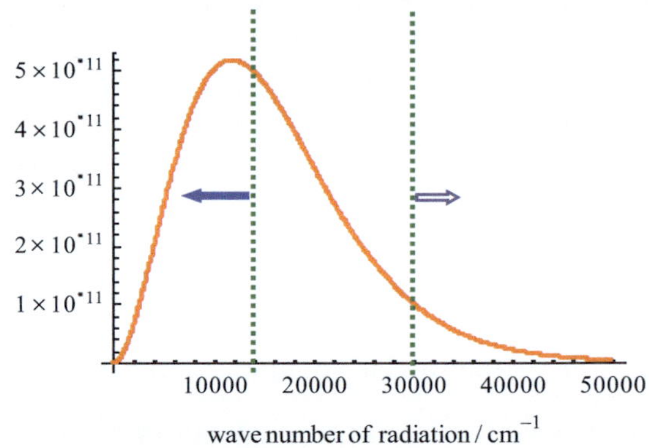

Fig.69 Radiationenergy distribution from the sun (6000 K).

Ans. To make sun light in the infrared range usable.

(c) If there exists an ideal heat engine (*Carnot* cycle) which can work between the hot reservoir (the sun at 6000 K) and the cold reservoir (the earth at 295 K), its efficiency of work is calculated as 95.1% as shown below.

$$\eta=\frac{T_h-T_l}{T_h}=\frac{6000-295}{6000}=95.1\%$$

Ans. efficiency of work η : 95.1 %

Solution 3-121

Expressing the internal energy U as a function of T and V, the total differential of $U(T,V)$ is given by eq(646).

$$dU = \left(\frac{\partial U}{\partial T}\right)_V dT + \left(\frac{\partial U}{\partial V}\right)_T dV \qquad (646)$$

Therefore, eq(646) can be rewritten by using $\delta q = dU + pdV$ as eq(647).

$$\delta q = \left(\frac{\partial U}{\partial T}\right)_V dT + pdV + \left(\frac{\partial U}{\partial V}\right)_T dV \qquad (647)$$

Dividing both sides of eq(647) by T, eq(648') is obtained via eq(648).

$$\frac{\delta q}{T} = \frac{1}{T}\left(\frac{\partial U}{\partial T}\right)_V dT + \frac{p}{T} dV + \frac{1}{T}\left(\frac{\partial U}{\partial V}\right)_T dV \qquad (648)$$

$$\frac{\delta q}{T} = \frac{1}{T}\left(\frac{\partial U}{\partial T}\right)_V dT + \frac{1}{T}\left\{P + \left(\frac{\partial U}{\partial V}\right)_T\right\} dV \qquad (648')$$

In eq(648'), $\delta q/T$ is expressed by a sum of two terms, one is proportional to dT and the other is proportional to dV. For this type of differential, it is easy to tell whether the differential is a total differential or not by comparing the results obtained by partially differentiating the coefficient of dT of eq(648') by V and the coefficient of dV by T. When the results are the same, then $\delta q/T$ is said to be a total differential. Differentiating the coefficient of dT of eq(648') by V at constant temperature, eq(649) is obtained,

$$\left\{\frac{\partial}{\partial V}\frac{1}{T}\left(\frac{\partial U}{\partial T}\right)_V\right\}_T = \frac{1}{T}\left\{\frac{\partial}{\partial V}\left(\frac{\partial U}{\partial T}\right)_V\right\}_T = \frac{1}{T}\left\{\frac{\partial}{\partial T}\left(\frac{\partial U}{\partial V}\right)_T\right\}_V \qquad (649)$$

and differentiating the coefficient of dV by T at constant volume、eq(650) is derived.

$$\left(\frac{\partial}{\partial T}\frac{1}{T}\left\{P + \left(\frac{\partial U}{\partial V}\right)_T\right\}\right)_V = -\frac{1}{T^2}\left\{P + \left(\frac{\partial U}{\partial V}\right)_T\right\} + \frac{1}{T}\frac{\partial}{\partial T}\left\{P + \left(\frac{\partial U}{\partial V}\right)_T\right\}_V$$

$$= -\frac{1}{T^2}\left\{P + \left(\frac{\partial U}{\partial V}\right)_T\right\} + \frac{1}{T}\left(\frac{\partial P}{\partial T}\right)_V + \frac{1}{T}\left\{\frac{\partial}{\partial T}\left(\frac{\partial U}{\partial V}\right)_T\right\}_V \qquad (650)$$

On the other hand, when the total differential of the internal energy is expressed by $dU = TdS - PdV$, $(\partial U/\partial V)_T$ is given by eq(651), which is rearranged by transferring the right side to the left side to yield eq(651).

$$dU = TdS - PdV \;\Rightarrow\; \left(\frac{\partial U}{\partial V}\right)_T = T\left(\frac{\partial S}{\partial V}\right)_T - P$$

$$= T\left(\frac{\partial P}{\partial T}\right)_V - P \qquad (651)$$

$$-\frac{1}{T^2}\left\{P + \left(\frac{\partial U}{\partial V}\right)_T\right\} + \frac{1}{T}\left(\frac{\partial P}{\partial T}\right)_V = 0 \qquad (651')$$

Substituting this into eq(650), it is found that eq(650) is given by eq(650'), which is identical to eq(649).

$$\left(\frac{\partial}{\partial T}\frac{1}{T}\left\{P + \left(\frac{\partial U}{\partial V}\right)_T\right\}\right)_V = \frac{1}{T}\left\{\frac{\partial}{\partial T}\left(\frac{\partial U}{\partial V}\right)_T\right\}_V \qquad (650')$$

Namely, $\delta q/T$ is concluded to be a total differential (quantity of state) based on the equality proved and given by eq(652). This state quantity expressed by $\delta q/T$ is entropy.。

$$\left(\frac{\partial}{\partial T}\frac{1}{T}\left\{P+\left(\frac{\partial U}{\partial V}\right)_T\right\}\right)_V=\left\{\frac{\partial}{\partial V}\frac{1}{T}\left(\frac{\partial U}{\partial T}\right)_V\right\}_T \qquad (652)$$

Ans. proof of total differential: eq(652)

[advanced problem] Show that $\delta q/T^2$ is not a total differential.

Solution 3-122

A change in the *Gibbs* free energy, when T or P varies, is given by eq(653).

$$dG(P,T)=-SdT+VdP \qquad (653)$$

Based on this equation, the change of the *Gibbs* free energy for each step of the *Carnot* cycle is obtained. Each step is indicated by numbers added to G as subscripts. The temperatures of the hot and cold reservoirs are denoted as T_h and T_l, respectively and the working substance is 1 mole of an ideal gas.

(i) step($1\rightarrow 2$) (isothermal expansion ($dT=0$)): $\Delta G_{1\rightarrow 2}$

$$\Delta G_{1\rightarrow 2}=\int_{P_1}^{P_2}VdP=-\int_{V_1}^{V_2}\frac{RT_h}{V}dV=-RT_h\ln\frac{V_2}{V_1} \qquad (654)$$

(ii) step($2\rightarrow 3$) (adiabatic expansion ($S=$ constant)): $\Delta G_{2\rightarrow 3}$

The integral of eq(653) along the process expressed by eq(655) and the result of integration is given by eq(655').

$$\Delta G_{2\rightarrow 3}=\int_{P_2}^{P_3}VdP-\int_{T_h}^{T_l}SdT=-\int_{V_2}^{V_3}\gamma CV^{-\gamma}dV-S_2(T_l-T_h) \qquad (655)$$

$$=\frac{\gamma C}{\gamma-1}(V_3^{-\gamma+1}-V_2^{-\gamma+1})-S_2(T_l-T_h),\qquad \gamma=\frac{C_P}{C_V}$$

$$=\frac{\gamma R}{\gamma-1}(T_l-T_h)-S_2(T_l-T_h),\qquad \frac{\gamma R}{\gamma-1}=C_P \qquad (655')$$

Here, , it is used that the temperature and pressure of the working ideal gas are related by eqs(656) and (656') for the adiabatic process (see **solution 3-11b**). C is a constant.

$$C=PV^{\gamma}=RT_hV_2^{\gamma-1}=RT_lV_3^{\gamma-1} \qquad (656)$$

$$dP=-\gamma CV^{-\gamma-1}dV \qquad (656')$$

Also, the value of the entropy is denoted as $S_2(=S_3)$ since the process is an isentropic one.

(iii) step($3\rightarrow 4$) (isothermal expansion ($dT=0$)): $\Delta G_{3\rightarrow 4}$

In a similar manner to step (i), $\Delta G_{3\rightarrow 4}$ is obtained by eq(657).

$$\Delta G_{3\rightarrow 4}=\int_{P_3}^{P_4}VdP=-\int_{V_3}^{V_4}\frac{RT_l}{V}dV=-RT_l\ln\frac{V_4}{V_3} \qquad (657)$$

(iv) step($4\rightarrow 1$) (adiabatic expansion ($S=$ constant)): $\Delta G_{4\rightarrow 1}$

Similarly as for step($2\rightarrow 3$), $\Delta G_{4\rightarrow 1}$ is calculated by eq(658) and the result of the integration is given by eq(658'). Here the relation given by eq(656'') is used. C' is a constant. Also, the entropy is put $S=S_1(=S_4)$ because of the isentropic process.

$$\Delta G_{4\rightarrow 1}=\int_{P_4}^{P_1}VdP-\int_{T_l}^{T_h}SdT=-\int_{V_4}^{V_1}\gamma C'V^{-\gamma}dV-S_1(T_h-T_l) \qquad (658)$$

$$= \frac{\gamma C'}{\gamma - 1}(V_1^{-\gamma+1} - V_4^{-\gamma+1}) - S_1(T_h - T_l)$$

$$= \frac{\gamma R}{\gamma - 1}(T_h - T_l) - S_1(T_h - T_l) \qquad (658')$$

$$C' = PV^{\gamma} = RT_l V_4^{\gamma-1} = RT_h V_1^{\gamma-1} \qquad (656'')$$

Summarizing the results from (i) to (iv), the change in the *Gibbs* free energy for the one cycle, $\Delta G_{1\to1} = \Delta G_{1\to2} + \Delta G_{2\to3} + \Delta G_{3\to4} + \Delta G_{4\to1}$, is found to be given by eq(659) or eq(659').

$$\Delta G_{1\to1} = -RT_h \ln\frac{V_2}{V_1} + \frac{\gamma R}{\gamma - 1}(T_l - T_h) - S_2(T_l - T_h) - RT_l \ln\frac{V_4}{V_3} + \frac{\gamma R}{\gamma - 1}(T_h - T_l) - S_1(T_h - T_l) \qquad (659)$$

$$= -RT_h \ln\frac{V_2}{V_1} - S_2(T_l - T_h) - RT_l \ln\frac{V_4}{V_3} - S_1(T_h - T_l) \qquad (659')$$

The entropy change associated with the isothermal step($1 \to 2$) is brought about by the volume change and is given by eq(660).

$$\Delta S(V_1 \to V_2) = S_2 - S_1 = RT \ln\frac{V_2}{V_1} \qquad (660)$$

From eq(620), the left side of eq(621) can be expressed as its right side.

$$T_h(S_2 - S_1) + T_l(S_1 - S_2) = T_h(S_2 - S_1) + T_l(S_4 - S_3) = RT_h \ln\frac{V_2}{V_1} + RT_l \ln\frac{V_4}{V_3} \qquad (661)$$

Substituting this relation into eq(619'), the equality of $\Delta G_{1\to1} = 0$ is derived, though this is a trivial consequence, since the *Gibbs* free energy a quantity of state.

Work done is equal to the *Gibbs* free energy change associated with the two isothermal processes and thus it is understood that free energy has a meaning of maximum usable energy (see **solution 3-11b**).

$$\Delta G_{1\to2} + \Delta G_{3\to4} = -RT_h \ln\frac{V_2}{V_1} - RT_l \ln\frac{V_4}{V_3}$$

What does $\Delta G_{2\to3} + \Delta G_{4\to1}$ correspond to then? For a deeper understanding of the meaning of "free enerrgy", it is recommended to perform the same as above for the *Helmholtz* free energy.

Ans. *Gibbs* free energy change for each step: See text and eqs(654), (655'), (657) and (658').

"note" The calculation of the *Gibbs* free energy change is carried out above by separating the integration route into two routes of *dT* and *dV*. In order to execute the integration directly along the adiabatic route, it is necessary to express a partial differential such as $(\partial G/\partial V)_S$ or $(\partial G/\partial T)_S$ as a function of *V* (or *P*) and *T*. This might be a bit tedious, though good exercise of thermodynamics (see **solution 3-48**).

Solution 3-123

(a) In order to prove that the chemical potential is *Euler*'s homogeneous function of the zeroth-order, we will show that eq(662) is satisfied, namely the definition of the zeroth-order function.

$$\mu_i(T, P, \lambda n_1, \lambda n_2, \cdots, \lambda n_i) = \lambda^0 \mu_i(T, P, n_1, n_2, \cdots, n_i) = \mu_i(T, P, n_1, n_2, \cdots, n_i) \qquad (662)$$

$\mu_i(T, P, \lambda n_1, \lambda n_2, \cdots, \lambda n_i)$ is expressed as eq(663) using the partial differential of the *Gibbs* free energy and since the *Gibbs* free energy is *Euler*'s homogeneous function of the first-order, the first equality of eq(663) can be rewritten as the last equality and finally eq(663') is obtained.

$$\mu_i(T,P,\lambda n_1,\lambda n_2,\cdots,\lambda n_l)=\left(\frac{\partial G(T,P,\lambda n_1,\lambda n_2,\cdots,\lambda n_l)}{\partial \lambda n_i}\right)_{T,P,\lambda n_j,j\neq i}=\left(\frac{\partial \lambda G(T,P,n_1,n_2,\cdots,n_l)}{\partial \lambda n_i}\right)_{T,P,n_j,j\neq i} \quad (663)$$

$$=\mu_i(T,P,n_1,n_2,\cdots,n_l) \quad (663')$$

The last relation is nothing but eq(662) and, hence, it is shown that the chemical potential μ_i is *Euler*'s homogeneous function of the zeroth-order.

(b) The *Gibbs* free energy can also be expressed by eq(664).

$$G(T,P,n_1,n_2,\cdots,n_l)=\sum_{i=1}^{l} n_i\mu_i \quad (664)$$

Both sides of eq(664) are partially differentiated with n_k to obtain eq(665).

$$\left(\frac{\partial G}{\partial n_k}\right)_{T,P,n_j,j\neq k}=\mu_k+\sum_{i=1}^{l} n_i\left(\frac{\partial \mu_i}{\partial n_k}\right)_{T,P,n_j,j\neq k}=\mu_k \quad (665)$$

For the second equality of eq(665) to hold, eq(666) must be satisfied. In other words, the chemical potential varies with mole numbers under the restriction of eq(666), which is equivalent to the *Gibbs-Duhem* equation.

$$\sum_{i=1}^{l} n_i\left(\frac{\partial \mu_i}{\partial n_k}\right)_{T,P,n_j,j\neq k}=0 \quad (666)$$

(c) Applying eq(666) for the two component regular solution, eq(666') must be satisfied.

$$n_{\mathrm{A}}\left(\frac{\partial \mu_{\mathrm{A}}^{non}}{\partial n_{\mathrm{A}}}\right)_{T,P,n_{\mathrm{B}}}+n_{\mathrm{B}}\left(\frac{\partial \mu_{\mathrm{B}}^{non}}{\partial n_{\mathrm{A}}}\right)_{T,P,n_{\mathrm{B}}}=0 \quad (666')$$

Noticing that the partial differentials in eq(666') are with mole numbers (**"note"**), the chemical potentials of the regular solution given in terms of the mole fractions are rewritten as eq(667) using the mole numbers.

$$\mu_{\mathrm{A}}^{non}=\mu_{\mathrm{A}}^{\circ}+RT\ \ln\frac{n_{\mathrm{A}}}{(n_{\mathrm{A}}+n_{\mathrm{B}})}+\alpha\cdot\frac{n_{\mathrm{B}}^{2}}{(n_{\mathrm{A}}+n_{\mathrm{B}})^2},\qquad \mu_{\mathrm{B}}^{non}=\mu_{\mathrm{B}}^{\circ}+RT\ \ln\frac{n_{\mathrm{B}}}{(n_{\mathrm{A}}+n_{\mathrm{B}})}+\alpha\cdot\frac{n_{\mathrm{A}}^{2}}{(n_{\mathrm{A}}+n_{\mathrm{B}})^2} \quad (667)$$

The chemical potentials thus rewritten are partially differentiated by the relevant mole numbers and then the calculated results are substituted into eq(666') to find that eq(666') is actually satisfied.

Ans. See text (a), (b) and (c).

"note" Superficially, it may seem correct that when eq(626) is divided by the total mole number n ($n=\sum n_i$), the following relation can be derived.

$$\sum_{i=1}^{l} X_i\left(\frac{\partial \mu_i}{\partial X_k}\right)_{T,P,n_j,j\neq k}=0 \Rightarrow X_1\left(\frac{\partial \mu_1}{\partial X_1}\right)_{T,P}+X_2\left(\frac{\partial \mu_2}{\partial X_1}\right)_{T,P}=0\ ?$$

However, you must be careful about this rearrangement because mole numbers are independent variables but the all of the mole fractions are not independent one another. Consequently, you will not be able to obtain the correct answer by replacing carelessly the partial differentials by mole numbers with the partial differentials by the mole fraction simply relying on the fact that the chemical potentials of the regular solution are expressed in terms of the mole fraction, though in the case of the ideal solution, such an erroneous calculation happens to lead to a correct answer zero. This is because it is not fully understood what variables are kept constant upon partial differentiation. The present problem will be a good exercise for partial differentiation.

Solution 3-124

As we have already seen in Chapter 2, the second derivative of P of the *van der Waals* equation of state with volume V can be expressed as in eq (668).

$$P=\frac{RT}{(V-b)}-\frac{a}{V^2} \Rightarrow \left(\frac{\partial^2 P}{\partial V^2}\right)_T=\frac{2RT}{(V-b)^3}-\frac{6a}{V^4} \qquad (668) \qquad (n=1\,\text{mole})$$

The third derivative can be expressed as the first equation of eq (669).

$$\left(\frac{\partial^3 P}{\partial V^3}\right)_T=-\frac{6RT}{(V-b)^4}+\frac{24a}{V^5}=-\frac{6RT}{(V-b)^4}+\frac{8RT}{V(V-b)^3}=\frac{2RT}{V(V-b)^4}(V-4b) \qquad (669)$$

Since eq(668) given in the problem holds at the critical point, eq(669) can be rewritten as in the last equation. Furthermore, at the critical point, since the critical volume is $V_c=3b$ and $(V-4b)_{V=V_c}=-b<0$, it can be shown that eq(670) is satisfied.

$$\left(\frac{\partial^3 P}{\partial V^3}\right)_T<0 \qquad (670) \qquad \text{(reposted)} \quad \text{inequality at the critical point}$$

The fact that the inequality in eq. (670) holds at the critical point is obvious if we consider the common behavior of a gas in general as follows. Expressing the pressure P near the critical point as a function of volume V and then expanding P with volume near the critical point $V=V_c$, it can be expressed as eq(671) (*Taylor* expansion) .

$$P(V)=P(V_c)+P'(V_c)(V-V_c)+\frac{1}{2!}P''(V_c)(V-V_c)^2+\frac{1}{3!}P'''(V_c)(V-V_c)^3+\cdots \qquad (671)$$

At the critical point, its first and second order derivatives should be zero, i.e., $P'(V_c)=0$ and $P''(V_c)=0$, so that it is found that $P(V)$ can be approximated by a cubic function of $(V-V_c)$ near the critical point. Since the pressure must be a decreasing function of the volume (thermodynamic inequality), it is obvious that the coefficient $P'''(V_c)$ must be negative. In other words, if we recall the form of P expanded as a function of V (the negative value of the third derivative at the critical point), the proof of eq (671) will be straightforward.

Ans. See eq(670) and text.

Solution 3-125

(a) Since the chemical potentials of an ideal solution consisting of the two components 1 and 2 are expressed by eq(672),

$$\mu_1=\mu_1^\circ+RT\ln X_1\,,\quad \mu_2=\mu_2^\circ+RT\ln X_2 \qquad (672)$$

the relation expressed by eq(673) is obtained via the following equation,

$$\mu_{12}=\left(\frac{\partial\mu_1}{\partial n_2}\right)_{T,P,n_1}=\left(\frac{\partial\mu_1}{\partial X_2}\right)_{T,P}\left(\frac{\partial X_2}{\partial n_2}\right)_{n_1}=\left(\frac{\partial\mu_1}{\partial X_2}\right)_{T,P}\left(\frac{n_1}{(n_1+n_2)^2}\right)=\frac{X_1}{n}\left(\frac{\partial\mu_1}{\partial X_2}\right)_{T,P}$$

$$\mu_{12}=\frac{X_1}{n}\left(\frac{\partial\mu_1}{\partial X_2}\right)_{T,P}=-\frac{RTX_1}{n(1-X_2)}=-\frac{RT}{n}<0 \qquad (673)$$

Similarly, eq(673') can be obtained for μ_{21}.

$$\mu_{21}=\left(\frac{\partial\mu_2}{\partial n_1}\right)_{T,P,n_2}=\frac{X_2}{n}\left(\frac{\partial\mu_2}{\partial X_1}\right)_{T,P}=-\frac{RTX_2}{n(1-X_1)}=-\frac{RT}{n}<0 \qquad (673')$$

From the above, it can be seen that eq(674), which is given in the problem as eq(91), always holds for an ideal solution.

$$\mu_{12} = \mu_{21} < 0 \qquad (674)$$

In other words, this shows that in an ideal solution, the two components (more than two as well) can always mix in any arbitrary proportion to form a homogeneous solution.

Ans. See text (a).

(b) The chemical potentials of a two-component regular solution are generally expressed as eq(675),

$$\left.\begin{aligned} \mu_1 &= \mu_1^\circ + RT\ln X_1 + \alpha X_2^2 \\ \mu_2 &= \mu_2^\circ + RT\ln X_2 + \alpha X_1^2 \end{aligned}\right\} \qquad (675) \quad \text{(eq(88) in the problem)}$$

so that their first derivatives derived by differentiating the above chemical potential by the counterpart mole fraction are given by eqs(676) and (676').

$$\mu_{12} = X_1\left(\frac{\partial \mu_1}{\partial X_2}\right) = X_1\left\langle -\frac{RT}{X_1} + 2\alpha(1 - X_1)\right\rangle = R\left\langle -T + \frac{2\alpha}{R}(1 - X_1)X_1\right\rangle \qquad (676)$$

$$\mu_{21} = X_2\left(\frac{\partial \mu_2}{\partial X_1}\right) = X_2\left\langle -\frac{RT}{X_2} + 2\alpha(1 - X_2)\right\rangle = R\left\langle -T + \frac{2\alpha}{R}(1 - X_1)X_1\right\rangle \qquad (676')$$

Since the condition for the mixed solution to form one phase stably can be expressed in the first half of eq (677),

$$\mu_{12} = \mu_{21} = R\left\langle -T + \frac{2\alpha}{R}(1 - X_1)X_1\right\rangle < 0 \;\Rightarrow\; T > \frac{2\alpha}{R}(1 - X_1)X_1 \qquad (637)$$

it is seen that the temperature must be in the range satisfying the second half of eq(677). The range satisfying eq(677) is illustrated in Fig.70 given below in which the upper side of the red curve (or the blue curve as well) is drawn so as to reach a maximum value of 300 K by putting $\alpha/R = 600\,\text{K}$.

On the other hand, in a regular solution described by eq(675), the components in the two phases are symmetrical with respect to the mole fraction when separated into phases I and II, so that when the mole fractions of component 1 in phases I and II are written as X_1^{I} and X_1^{II}, respectively, then, X_1^{II} and X_2^{II} are given by eq(678).

$$X_1^{\text{II}} = 1 - X_1^{\text{I}}, \qquad X_2^{\text{II}} = X_1^{\text{I}} \qquad (678)$$

In equilibrium, the chemical potentials of each component in phases I and II are equal, so there exists a relationship expressed by eq(679) or (679').

$$\mu_1^\circ + RT\ln X_1^{\text{I}} + \alpha(X_2^{\text{I}})^2 = \mu_1^\circ + RT\ln X_1^{\text{II}} + \alpha(X_2^{\text{II}})^2 \qquad (679)$$

$$RT\ln X_1^{\text{I}} - RT\ln X_1^{\text{II}} = \alpha(X_2^{\text{II}})^2 - \alpha(X_2^{\text{I}})^2 \qquad (679')$$

Rearranging eq(679') using the relationship in eq(678), eq(680) can be obtained via the following equation.

$$RT\ln\frac{X_1^{\text{I}}}{1\text{-}X_1^{\text{I}}} = \alpha(X_2^{\text{II}} + X_2^{\text{I}})(X_2^{\text{II}} - X_2^{\text{I}}) = \alpha(X_1^{\text{I}} + X_2^{\text{I}})(X_1^{\text{I}} - X_2^{\text{I}}) = \alpha(2X_1^{\text{I}} - 1)$$

$$T = \frac{\alpha}{R}\frac{(2X_1 - 1)}{\ln(X_1/1\text{-}X_1)} \qquad (680)$$

In eq(680), the symbols indicating phases are omitted for simplicity. The blue curve in the figure depicts eq(680) for $\alpha/R = 600\,\text{K}$.

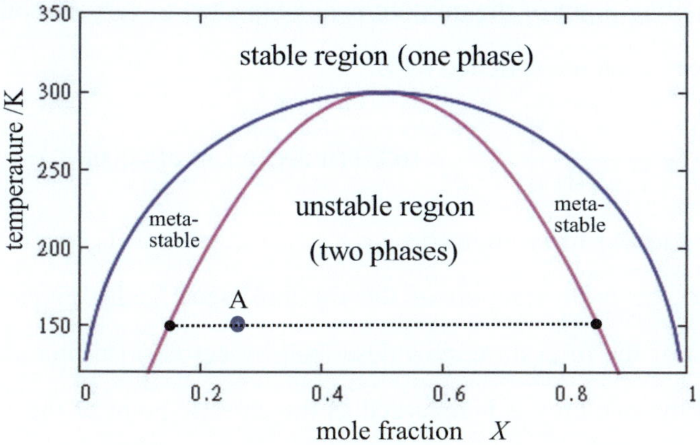

Fig.70 Phase stability of two-component regular solution against T and X.

curve in red: $T = \frac{2\alpha}{R}(1-X)X$, see eq(677)

curve in blue: $T = \frac{\alpha}{R}\frac{(2X-1)}{\ln(X/1-X)}$, see eq(680)

Relative ratio of the two phases at point A indicated by a dot can be calculated based on the lever rule.

From the value of chemical potential, it is seen to be more stable for a regular solution to separate into two phases below the blue curve than to remain in one phase, so even above the red curve, the region bounded by the red and blue curves can be considered a metastable solution state.

Ans. See Fig.70 and text (b).

Solution 3-126

(a) The *Clausius-Clapeyron* equation derived for the liquid-vapor equilibrium of a pure substance (and mixture as well) is written as eq(681).

$$\frac{\partial \ln P}{\partial T} = \frac{\Delta H^m_{vap}}{RT^2} \quad (681)$$

We can formally mimic the case of the solid-liquid equilibrium of A achieved by dissolving some amount of non-reactive B into A (melting point depression) to the vapor-liquid equilibrium described by eq(681). Provided that $\ln P$ in eq(681) is replaced by $\ln X_A$ and the heat of vaporization per mole, ΔH^m_{vap}, is replaced by the heat of fusion of A per mole, ΔH^m_{fusion}, as is given by eq(682),

$$\frac{\partial \ln X_A}{\partial T} = \frac{\Delta H^m_{fusion}(A)}{RT^2} \quad (682)$$

then the mole fraction of A in the solid-liquid equilibrium at temperature T is given by eq(683). Of course, eq(683) can be derived by following the standard procedure of treating the solid-liquid equilibrium of a binary mixture.

$$\ln X_A = -\Delta H^m_{fusion}(A) \times \left(\frac{1}{T} - \frac{1}{T_{m.p.}(A)} \right) \quad (683)$$

Eq(643) can be solved for X_A as is given by eq(684) or eq(684').

$$X_A = \exp\left\{ -\Delta H^m_{fusion}(A) \times \left(\frac{1}{T} - \frac{1}{T_{m.p.}(A)} \right) \right\} \quad (684)$$

$$1 - X_A = 1 - \exp\left\{ -\Delta H^m_{fusion}(A) \times \left(\frac{1}{T} - \frac{1}{T_{m.p.}(A)} \right) \right\} \quad (684')$$

Looking at the solid-liquid equilibrium of the same binary mixture of A and B from the standpoint of B, the relation between the mole fraction X_B and the temperature T of the mixture is described by eq(685), which is in principle independent of eq(684), since the melting point of the mixture is not expected to vary smoothly with increasing the mole fraction of B to those near mixtures enriched by B.

$$X_B = \exp\left\{-\Delta H^m_{fusion}(B) \times \left(\frac{1}{T} - \frac{1}{T_{m.p.}(B)}\right)\right\} \qquad (685) \text{ (independent of eq(684'))}$$

Applying eq(684) to *p*-dinitrobenzene (A) and eq(685) to *m*-dinitrobenzene (B), the relation between the temperature (melting point of the mixture) and the mole fraction of the mixture, specifically given by eqs(646) and (687), is shown in Fig.71 in terms of the relevant curves described by eq(686) (in blue) and (687) (in pink), respectively. The intersection of the two curves is regarded as the eutectic point of the

$$\left\{1 - \exp\left[-\frac{14.0\times10^3}{8.31}\times\left(\frac{1}{T} - \frac{1}{446.7}\right)\right],\ T\right\} \qquad (686)$$

$$\left\{\exp\left[-\frac{17.60\times10^3}{8.31}\times\left(\frac{1}{T} - \frac{1}{363.0}\right)\right],\ T\right\} \qquad (687)$$

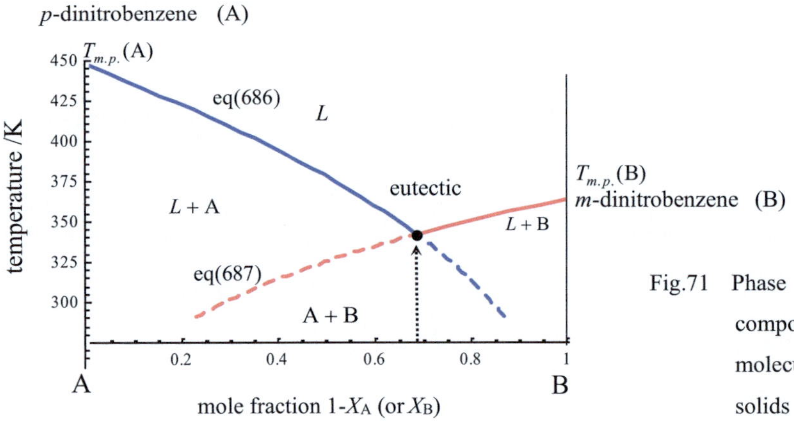

Fig.71 Phase diagram (*T-X* diagram) of two component system consisting of alike molecules which form homogeneous mixed solids (A+B) for any mole fraction.

mixture along which the A+B mixture behaves as if it were a pure compound when the temperature is varied. The eutectic point of the mixture is calculated as 341.5 K and the mole fraction of 1-X_B as 0.715 from the intersection of these two curves.

(b) In a totally similar manner employed in (a), two-dimensional phase diagrams can be drawn for the C-A and B-C systems using the relations given by a pair of eqs(688) and (689) and a pair of eqs(690) and (691), respectively, in which relevant values for the melting point and molar heat of fusion are employed.

$$\left\{\exp\left[-\frac{14.0\times10^3}{8.31}\times\left(\frac{1}{T} - \frac{1}{446.7}\right)\right],\ T\right\} \qquad (688)$$

$$\left\{1 - \exp\left[-\frac{16.33\times10^3}{8.31}\times\left(\frac{1}{T} - \frac{1}{390.1}\right)\right],\ T\right\} \qquad (689)$$

$$\left\{\exp\left[-\frac{16.33\times10^3}{8.31}\times\left(\frac{1}{T} - \frac{1}{390.1}\right)\right],\ T\right\} \qquad (690)$$

$$\left\{1 - \exp\left[-\frac{17.60\times10^3}{8.31}\times\left(\frac{1}{T} - \frac{1}{363.0}\right)\right],\ T\right\} \qquad (691)$$

The phase diagrams thus obtained for the systems of A-B, B-C and C-A are displayed in Fig.72 in an array so as for the axes for a common component, A or B in the present case, to overlap each other. Three-dimensional display of the phase diagram of *o*-, *m*-, and *p*-dinitrobenzene mixtures in solid-liquid equilibrium can be constructed from Fig.72 by folding it so as for the C axes in Fig.72 superpose each other as is shown in Fig.73.

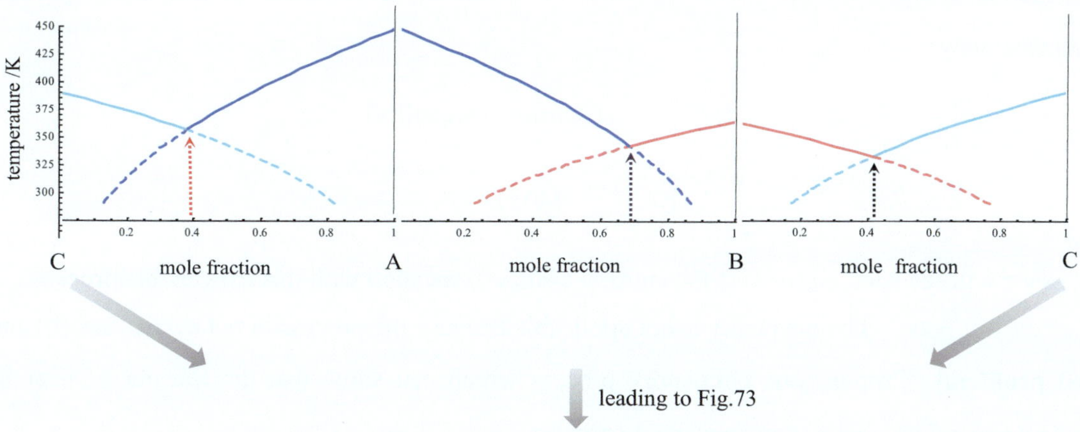

Fig.72 Phase diagrams (*T-X* diagrams) of two-component systems conceivable for the three combinations chosen from A, B and C.

If we assume that the mixed A-B solid of the mole fraction at the eutectic point will behave as if it were a pure compound, then it is expected that the eutectic point for the A-B system will vary when a small amount of C is added, decreasing with increasing the amount of C. The same would be true the other two-component

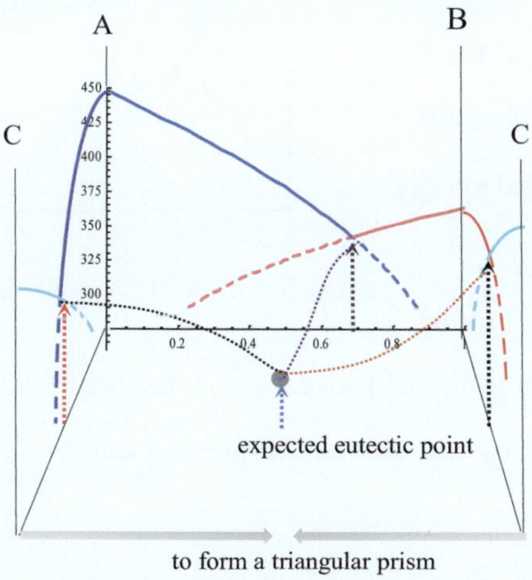

Fig.73 Three-dimensional phase diagram constructed from a set of three two-dimensional phase diagrams illustrated in Fig.72 by superposing the vertical axes of C on each other so as to form a triangular prism. The dot in the figure indicates an eutectic point expected for the three- component system.

systems. Just for illustration, how it varies is depicted in Fig.73 by each of the three dotted curves. Then, the eutectic point for the A-B-C three component system would be given by the intersection of the three dotted lines as is shown in Fig.73. Can you propose how to draw one of these dotted curves by assuming only if the information about the compounds provided so far is available and the *Clausius-Clapeyron* equation is applicable to the solid-liquid equilibrium for the three-component system? See **problem** and **solution 3-25** for another three-component phase diagram.

Ans. See Fig.73.

Solution 3-127

(a) We assume that the heat energy ($q_1 = -q$) flowing out of the object at the higher temperature is equal to the heat energy ($q_2 = q$) that is received by the object at the lower temperature and nothing else happens. Then, by simply substituting these into the left side of *Clausius*' inequality eq(692), eq(693) is obtained, which shows that this phenomenon does not satisfy *Clausius*' inequality (eq(692)). How can this be, since the outflow of thermal energy from the warmer object to the colder object should be an irreversible process that occurs spontaneously?

$$\frac{q_1}{T_1} + \frac{q_2}{T_2} \leq 0 \qquad (692) \qquad (\textit{Clausius}\text{' inequality})$$

$$\frac{q_1}{T_1} + \frac{q_2}{T_2} = q\left(\frac{1}{T_2} - \frac{1}{T_1}\right) > 0 \qquad (693)$$

In fact, as we will see later, eq (653) is the entropy change associated with this irreversible process.

Ans. The inequality is not applicable because the process in not cyclic. See (b) and (e).

[advanced problem] Propose one circulatory process which can show that the transfer of heat from a warmer object to a colder object is an irreversible process.

(b) As we have already seen, the *Carnot* cycle consists of two reversible isothermal processes and two reversible adiabatic processes (see **solution 3-11a**). Since heat is only transferred to and from via the isothermal processes, let us look at the temperature and the heat transferred in and out upon these processes.

In the *Carnot* cycle shown in Fig.74 on the right, if the thermal energy obtained from a high-temperature heat source at temperature T_1 is q_1 (>0) in the isothermal expansion of process (①), the relationship in eq(694) holds for an ideal gas, since the change in the internal energy of the gas due to the isothermal process is zero.

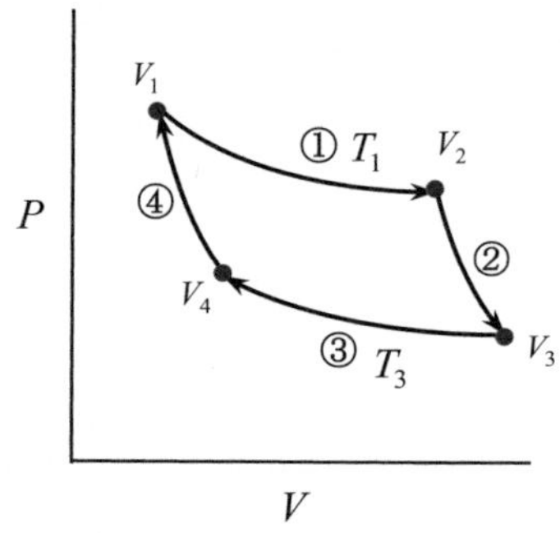

Fig.74 *Carnot* cycle working at T_1 and T_2.

$$\Delta U_1 = \Delta W_1 + q_1 = 0 \Rightarrow \Delta W_1 = -q_1 \qquad (694)$$

In the adiabatic expansion of process (②) and adiabatic compression of process (④), the heat energy in and out is zero ($q_2 = q_4 = 0$). On the other hand, the work done by the gas to the outside due to the volume expansion can be expressed by eq(695) as

$$\Delta W_1 = -\int_{V_1}^{V_2} PdV = -\int_{V_1}^{V_2} \frac{RT}{V} dV = -RT_1 \ln\frac{V_2}{V_1} \qquad (695).$$

In the same way, if q_3 (<0) is the thermal energy released to the low-temperature heat source at T_3 during the isothermal compression process (③), it can be related to the work energy given by eq (696).

$$-q_3 = \Delta W_3 = -\int_{V_3}^{V_4} PdV = -RT_3 \ln\frac{V_4}{V_3} \qquad (696)$$

Applying the *Clausius* equation to the cyclic process consisting of these four processes with care, we can express it as in eq(697),

$$\sum_{i=1}^{4}\frac{q_i}{T_i}=\frac{q_1}{T_1}+\frac{q_3}{T_3}=R\ln\frac{V_2}{V_1}+R\ln\frac{V_4}{V_3}=R\ln\frac{V_2V_4}{V_1V_3} \qquad (697)$$

In the reversible *Carnot* cycle we are considering now, the starting and ending volumes of the adiabatic processes are related by $V_4/V_1=V_3/V_2$ (see **solution 3-11a**), so we can use this to obtain the *Clausius* equation as an equality (eq(697')), which suggests that $\frac{q}{T}$ can be a state quantity for a reversible process.

$$\sum_{i=1}^{4}\frac{q_i}{T_i}=0 \quad (q_2=q_4=0) \qquad (697')$$

This means that the *Clausius* equality holds for a circulatory process consisting of reversible processes. However, it should be noted that the simple transfer of heat between heat sources at different temperatures is not a circulatory process, so the *Clausius* equation is not applicable to such a process (see (a)).

Ans. Heat engine: $\frac{q_1}{T_1}+\frac{q_3}{T_3}=0$, heat source: $\frac{-q_1}{T_1}+\frac{-q_3}{T_3}=0$

(c) The exchange of thermal energy with an external heat source is an irreversible process because it is a transfer of thermal energy between two objects at different temperatures. Therefore, as can be seen from the diagram shown in Fig.23 in the problem, the whole process can be divided into an irreversible inflow of heat from the external high-temperature heat source at temperature T_1 to the heat engine at temperature T_1', a reversible *Carnot* cycle working between temperatures T_1' and T_3', and an irreversible outflow of heat from the heat engine at temperature T_3' to the external heat source at temperature T_3. However, *Clausius*' inequality cannot be applied to the total process because it is not a cyclic process. Since the cyclic process of the heat engine working between temperatures T_1' and T_3' is a reversible process, *Clausius*' inequality, eq(698), holds for the heat engine itself, as is seen in (b).

$$\frac{q_1'}{T_1'}+\frac{q_3'}{T_3'}=0 \qquad (698)$$

The work efficiency of this heat engine is expressed by eq(699),

$$\eta'=\frac{T_1'-T_3'}{T_1'}=1-\frac{T_3'}{T_1'} \qquad (699)$$

$$1-\frac{T_3'}{T_1'}<1-\frac{T_3}{T_1} \qquad (699')$$

which is lower as is seen from eq(699') than the efficiency of a heat engine working directly between T_1 and T_3 as heat-source temperatures.

Generally speaking, the work efficiency of a *Carnot* cycle involving irreversible process(es) is lower than that of a reversible *Carnot* cycle.

Ans. *Clausius*' inequality is not applicable because the total process is not cyclic process. Eq(698) is valid for the heat engine.

(d) If work energy other than the work energy associated with the volume change exists and the amount of its change is expressed as δW^{other}, the internal energy change in the working substance can generally be expressed as in eq(700).

$$dU = \delta q - PdV + \delta W^{other} \quad \Rightarrow \quad \Delta U = q + \Delta W + \Delta W^{other} \qquad (700)$$

Specifically for the process ① in Fig.25 given in the present problem, it can be written as in eq(701) by putting $\Delta W^{other} = \Delta W_1^{friction}$. Note that q_1 and $\Delta W_1^{friction}$ take positive values.

$$\Delta U_1 = q_1 + \Delta W_1 + \Delta W_1^{friction} = 0 \quad \Rightarrow \quad q_1 = -\Delta W_1 - \Delta W_1^{friction} \qquad (701)$$

$$q_1 > 0, \quad \Delta W_1^{friction} > 0$$

$$\frac{q_1}{T_1} = \frac{-\Delta W_1 - \Delta W_1^{friction}}{T_1}$$

The work energy associated with the volume change in this process is expressed by eq(702).

$$\Delta W_1 = -\int_{V_1}^{V_2} PdV = -\int_{V_1}^{V_2} \frac{RT}{V} dV = -RT_1 \ln\frac{V_2}{V_1} \qquad (702)$$

Similarly, eqs(703) and (704) hold for the process ③. Note that q_3 is negative in this process but $\Delta W_3^{friction}$ takes a positive value.

$$\Delta U_3 = \Delta W_3 + q_3 + \Delta W_3^{friction} = 0 \quad \Rightarrow \quad q_3 = -\Delta W_3 - \Delta W_3^{friction} \qquad (703)$$

$$q_3 < 0, \quad \Delta W_3^{friction} > 0$$

$$\frac{q_3}{T_3} = \frac{-\Delta W_3 - \Delta W_3^{friction}}{T_3}$$

$$\Delta W_3 = -\int_{V_3}^{V_4} PdV = -\int_{V_1}^{V_2} \frac{RT}{V} dV = -RT_3 \ln\frac{V_4}{V_3} \qquad (704)$$

Substituting these relations into the *Clausius* equation, we obtain the relation in eq (705), which shows that *Clausius*' inequality holds, suggesting that the heat engine shown in Fig.24 in the problem works irreversibly.

$$\frac{q_1}{T_1} + \frac{q_3}{T_3} = -\left(\frac{\Delta W_1}{T_1} + \frac{\Delta W_3}{T_3}\right) - \left(\frac{\Delta W_1^{other}}{T_1} + \frac{\Delta W_3^{other}}{T_3}\right) = 0 - \left(\frac{\Delta W_1^{other}}{T_1} + \frac{\Delta W_3^{other}}{T_3}\right) < 0 \qquad (705)$$

Ans. eq (705): *Clausius*' inequality holds.

(e) Calculation o f entropy change for each case of (b) to (e).

Case (b): Since entropy is a state quantity, the entropy change for the heat engine that has cycled back to the initial state must be zero. In fact, for a reversible process, the entropy change is found to be zero as calculated in eq(706).

$$dS = \frac{\delta q}{T} \quad \Rightarrow \Delta S = \oint \frac{\delta q}{T} = \frac{q_1}{T_1} + \frac{q_3}{T_3} = 0 \qquad (706)$$

The entropy change of the heat sources can also be found to be zero from eq(706').

$$\frac{-q_1}{T_1} + \frac{-q_3}{T_3} = 0 \qquad (706')$$

We can say that the work efficiency is maximum in a reversible circulating process that does not produce entropy.

Ans. Entropy change of heat engine: zero, entropy change of external heat sources: zero.

Case (c): The entropy change of the heat engine, which has been restored through the cyclic process, is obviously zero. On the other hand, the heat sources irreversibly transfer heat from the high-temperature heat source to the low-temperature heat source, and the entropy change associated with this process can be calculated by considering the reversible (quasi-static) process of transferring heat from the high-temperature heat source to the low-temperature heat source, and calculating the entropy change along that path. Specifically, we transfer heat energy q isothermally from the high-temperature heat source to a working substance (e.g., an ideal gas), then adiabatically expand the gas to lower its temperature to that of the low-temperature heat source, and then transfer heat q to the low-temperature heat source, reversibly. In this way, the entropy change can be calculated based on the left side of eq(666) since the entire process is reversible. Consequently, the entropy change of the external high-temperature heat source, when heat $q_1(>0)$ is transferred from the external high-temperature heat source to the high-temperature heat source of the heat engine, is given by $-q_1/T_1$ and the entropy change of the external low-temperature heat source, when heat $q_3(<0)$ is transferred from the low-temperature heat source of the heat engine to the external low-temperature heat source, is given by $-q_3/T_3$. For the external heat sources as a whole, therefore, the entropy change is given by eq(707).

$$\Delta S = \frac{-q_1}{T_1} + \frac{-q_3}{T_3} \qquad (707)$$

Here, since the heat engine itself returns to its original state in a reversible cyclic process, $q_1(>0)$ and $q_3(<0)$ are related by the *Clausius* equality (eq(708)).

$$\frac{q_1}{T_1'} + \frac{q_3}{T_3'} = 0 \qquad (708)$$

Using this relationship and rewriting equation (707) in terms of q_1, we obtain the result in eq(709), which shows that the entropy of the external heat sources of the heat engine increases as a whole.

$$\begin{aligned}\Delta S &= \frac{-q_1}{T_1} + \frac{-q_3}{T_3} \\ &= \frac{-q_1}{T_1} + q_1 \times \frac{T_3'}{T_1'} \times \frac{1}{T_3} = \frac{q_1}{T_1}\left(\frac{T_3'}{T_1'} \times \frac{T_1}{T_3} - 1\right) = \frac{q_1}{T_1}\left(\frac{T_3'}{T_3} \times \frac{T_1}{T_1'} - 1\right) > 0 \end{aligned} \qquad (709)$$

Since this heat engine involves the generation of entropy throughout the cycle, the work efficiency will be lower as compared with that in the ideal case.

Ans. entropy change of heat engine: zero, entropy change of external heat sources: eq(669)

Case (d): Since the heat engine cycles and returns to its original state, its entropy change is zero. On the other hand, since the transfer of heat from the heat source to the heat engine or from the heat engine to the heat source is reversible, the entropy change of the heat sources is given by eq(710) and is found to increase.

Therefore, the work efficiency of the heat engine of this type will naturally be lower than that in a reversible *Carnot* cycle.

$$\Delta S = \frac{-q_1}{T_1} + \frac{-q_3}{T_2} = -\left(\frac{\Delta W_1}{T_1} + \frac{\Delta W_3}{T_2}\right) + \left(\frac{\Delta W_1^{other}}{T_1} + \frac{\Delta W_3^{other}}{T_2}\right) = 0 + \left(\frac{\Delta W_1^{other}}{T_1} + \frac{\Delta W_3^{other}}{T_2}\right) > 0 \qquad (710)$$

Ans. entropy change of heat engine: zero, entropy change of external heat source: eq (710)

"note 1" *Clausius*' Statement: Heat can never pass from colder to warmer body without some other change.

Kelvin-Planck's Statement: It is impossible to transform heat into work with no other effect.

R.Neito et al., *J.Chem.Educ.* 2011, 88, 597.

The *Clausius* inequality is an expression of these statements and holds for cyclic processes which can include irreversible processes. For a general proof of *Clausius*' inequality based on *Clausius*' Statement, please refer to the above literature or reference books dealing with the laws of thermodynamics (e.g., *H.Yukawa* and *M.Tamura*'s *General Physics*, vol.1, in Japanese).

"note 2" Reasons why *Clausius*' inequality is applicable only to cyclic processes

1. Considering that at the dawn of thermodynamics, the main focus of research was how heat and work could be efficiently converted by a heat engine, so that there is no need to say that the heat engine must be able to repeatedly return to its original state and operate as a cycle as many times as desired. The *Clausius* equation clarifies the relationship between the heat energy that flows in and out in a cycle and the temperature of the heat source involved. Then, we can understand why *Clausius*' inequality should be applied only to the cyclic process. Even when only a small number of terms such as $\frac{q_1}{T_1} + \frac{q_2}{T_2} \le 0$ as in the *Carnot* cycle are involved, but it is still a cyclic process that the inequality refers to and it is just due to a smaller number of processes that make up the cycle. It is also appropriate to say that the reason why each term in the equation is in the form of q/T is because a set of *Carnot* cycles is used in general proof of *Clausius*' inequality. The equality $\sum_{i=1}^{n} \frac{q_i}{T_i} = 0$ or $\oint \frac{\delta q}{T} = 0$ which holds for a reversible cyclic process suggests that the quantity q/T is conserved in a cycle and, hence, can be a state quantity, leading to the concept of entropy (see **solution 3-121**).
2. For an open path connecting two states, there are innumerable ways of doing so, and therefore the heat energy entering and leaving along the path varies from path to path. In such cases, it is not possible to establish any law about δq or $\delta q/T$ associated with an irreversible path. Instead, some thermodynamic quantities which are state quantities and, hence, are path-independent are employed. These quantities do not depend on whether the path connecting the initial and final states is reversible or irreversible, but if the path is reversible, a change in the state quantity can be determined along the path. In a cyclic process, the change in state quantity in one cycle is naturally zero.

Solution 3-128

A reversible cyclic process given by the closed curve is divided into a number of reversible *Carnot* cycles as illustrated in Fig.75 below. The paths representing isothermal processes of each *Carnot* cycle should be located near the closed curve as closely as possible. The adiabatic processes of adjacent *Carnot* cycles are opposite to each other (compression (red arrow) vs. expansion (blue arrow) and vice versa) and, hence, cancel out each other. Therefore, the reversible path of the closed curve can be approximated by the sum of many thin ribbon-like *Carnot* cycles, and ultimately by the integral of countless thin *Carnot* cycles.

Assuming that the thin ribbon-like *Carnot* cycle indicated by a in Fig.75 works between the high-temperature heat source at T_1 and the low-temperature heat source at T_3 so as to contain the tangent points of the *Carnot* cycle shown by the dotted lines, its work efficiency can be expressed by eq(711).

$$\eta_a = \frac{T_1 - T_3}{T_1} = 1 - \frac{T_3}{T_1} \qquad (711)$$

Since a thin ribbon-like *Carnot* cycle adjacent to the cycle denoted a will operate between T_1', lower than T_1, and T_3', higher than T_3, the work efficiency of this cycle given by eq(712) will be smaller than that of a.

$$\eta' = 1 - \frac{T_3'}{T_1'} < 1 - \frac{T_3}{T_1} \qquad (712)$$

The total work efficiency of the closed curve cycle can be obtained by multiplying the work efficiency of each ribbon-like component cycle by its contribution ratio and adding them together over the entire cycle. The contribution ratio can be expressed as follows: if the area of the ribbon-like i-th component cycle is dS_t and the area of the closed curve is S_0, then the contribution ratio of the i-th cycle will be dS_t/S_0. Therefore, the total work efficiency of the closed curve ($c.c.$) cycle can be expressed as in eq(713).

$$\eta_{c.c.} = \sum \eta_t \times \frac{dS_t}{S_0} = \frac{1}{S_0} \int_{c.c.} \eta(S) dS < \frac{1}{S_0} \int_{c.c.} \eta_a dS = \eta_a \qquad (713)$$

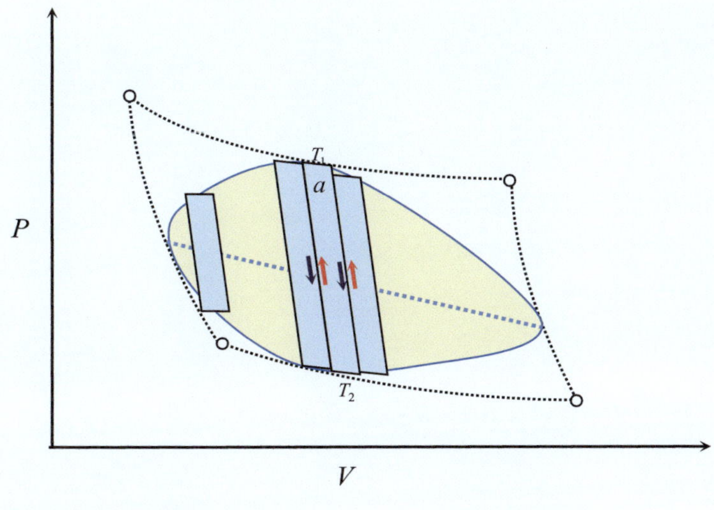

Fig.75. Decomposition of the circular cycle into a number of *Carnot* cycles. Thin ribbon-like component *Carnot* cycles are depicted only schematically. The blue dotted line indicates a continuous sequence of them.

"note" The temperature of the heat source for each ribbon-like component *Carnot* cycle can be worked out from the known values of P and V of an ideal gas, since the temperature of the working substance (ideal gas) and the heat source are the same for reversible processes. Therefore, by numerically integrating eq(673), the value of work efficiency of any reversible cyclic process can be obtained (see **3-77~3-81** and **3-106** for work efficiency of various heat engines). For example, it can be easily shown that the work efficiency of the *Stirling* cycle (see **3-106**) is equal to that of the *Carnot* cycle based on the method used here.

Ans. The work efficiency of the closed cycle shown in Fig.75 is less than that of the *Carnot* cycle shown by the dotted lines.

[advanced problem] By applying the method employed here to a closed cycle whose part or parts are composed of irreversible processes (see the preceding problem and its solution), show that *Clausius*' inequality $\sum_{i=1}^{n}\frac{q_i}{T_i}<0$ or $\oint\frac{\delta q}{T}<0$ can readily be derived.

III-4 Equilibriums in Solution and Interface

Part of Worked Solutions

Solution 4-1

The dissociation equilibrium reactions and dissociation constants K_1 and K_2 are expressed as eqs(1) and (2).

$$\begin{array}{llll} H_2SO_4 & \rightleftarrows & H^+ & + \; HSO_4^- \\ C_0(1-x) & & z & \;\; C_0(x-y) \end{array} \qquad K_1 = \frac{[H^+][HSO_4^-]}{[H_2SO_4]} = \frac{z \times (x-y)}{1-x} = 10^2 \qquad (1)$$

$$\begin{array}{llll} HSO_4^- & \rightleftarrows & H^+ & + \; SO_4^{2-} \\ C_0(x-y) & & z & \;\; C_0 y \end{array} \qquad K_2 = \frac{[H^+][SO_4^{2-}]}{[HSO_4^-]} = \frac{z \times y}{x-y} = 2 \times 10^{-2} \qquad (2)$$

There are two important points to be noticed when multi-step dissociation constants are expressed in terms of concentrations. (i) The equilibrium constants generally given are dimensionless. Any relevant concentration, therefore, must be normalized to the standard concentration (for instance $1\ \mathrm{mol \cdot L^{-1}}$). Namely, in the above case, it is expressed in such a way as "x fraction out of one part of sulfuric acid dissociates." (ii) The concentration of every chemical species involved in the equilibrium reactions is determined so as to fulfill the law of conservation of mass. In the above case, since the solution is electrically neutral, eq(3) is satisfied and for the chemical species containing SO_4, eq(4) is satisfied.

$$[H^+] = [HSO_4^-] + 2[SO_4^{2-}] \iff z = C_0(x-y) + 2C_0 y = C_0(x+y), \ (\mathrm{OH^-\ is\ neglected.}) \qquad (3)$$

$$[H_2SO_4] + [HSO_4^-] + [SO_4^{2-}] = C_0 \iff C_0(1-x) + C_0(x-y) + C_0 y = C_0 \qquad (4)$$

A value for x can be found by solving the set of eqs(1), (2) and (3) simultaneously (see **"note"**). However, this is a bit tedious and an approximate solution for x is found here by solving approximate eqs($1'$) and ($2'$) obtained by putting $[H^+] = z \approx C_0 x$ in eqs(1) and (2).

$$\frac{C_0 \times x \times (x-y)}{1-x} = 10^2 \qquad (1'), \qquad \frac{C_0 \times x \times y}{x-y} = 2 \times 10^{-2} \qquad (2')$$

From eq($2'$), y is expressed as a function of x and then it is substituted into eq($1'$). After some rearrangements, eq(5) is obtained for x.

$$5x^3 + 10^4 x^2 - 6 \times 10^3 x - 4 \times 10^3 = 0 \qquad (5)$$

Solving eq(5) for x, $x = 0.99964$ is obtained. Thus the mole number of sulfuric acid in 100 mL of the 0.1N aqueous solution is found as below.

$$(1-x) \times 0.05(\mathrm{mol \cdot L^{-1}}) \times 0.1(\mathrm{L}) = 1.8 \times 10^{-5}\ \mathrm{mol}$$

When the second dissociation expressed by eq(2) is not taken into account, eq(1) is solved by putting $y = 0$ yielding $x = 0.99950$. Then the mole number of the undissociated sulfuric acid is obtained as 2.5×10^{-5} mol. Naturally, which approximation is to be adopted depends on to what accuracy the answer is required. For instance, eq(5) can also be approximated as $10^4 x^2 - 6 \times 10^3 x - 4 \times 10^3 + 5.0 = 0$ by putting $x \approx 1.0$.

Ans. undissociated sulfuric acid: 1.8×10^{-5} mol

S. Hirayama and M. Yamamoto, *Physical Chemistry Problems and Solutions*,
https://doi.org/10.1007/978-981-97-0165-0_12

"note" When the set of simultaneous equations of eqs(1), (2) and (3) are rigorously solved for x, $x = 0.99953$ is found as a solution of the following equation.

$$-(2\times10^3-1.6)x^3+(4\times10^3+1.2\times10^3-4\times10^6)x^2+(8\times10^6-2.88\times10^3)x-(0.32+4\times10^3)10^3=0$$

Solution 4-2

(a) The equilibrium reactions of the dissociation of acetic acid is expressed by eq(6). Sodium acetate is fully ionized in aqueous solution as is expressed by eq(7). The concentration of each species is written under it in eqs(6) and (7). C_s is the concentration of sodium acetate used to prepare the solution.

$$\underset{C_0(1-x)}{CH_3COOH} \rightleftarrows \underset{C_0x}{H^+} + \underset{C_0x\ (+C_s)}{CH_3COO^-} \quad (6)$$

$$\underset{0}{CH_3COONa} \rightarrow \underset{C_s}{Na^+} + \underset{C_s}{CH_3COO^-} \quad (7)$$

Therefore, the equilibrium constant K_a is given using these concentrations by eq(8),

$$K_a=\frac{[H^+][CH_3COO^-]}{[CH_3COOH]}=\frac{C_0x\times(C_0x+C_s)}{C_0(1-x)} \quad (8)$$

which is solved for the fraction of the concentration of hydrogen ion, denoted x, and eq(9) is obtained.

$$x=\frac{-(C_s+K_a)+\sqrt{(C_s+K_a)^2+4C_0K_a}}{2C_0} \quad (9)$$

Substituting given values of $K_a=1.8\times10^{-5}$, $C_0=1.0\times10^{-1}$ and $C_s=1.0\times10^{-1}$ into eq(9), $x=1.7994\times10^{-4}$ is obtained, giving $pH=-\log_{10}C_0x=-\log_{10}(0.1\times1.7994\times10^{-4})=4.74_4$. The pH value of the solution containing acetic acid alone is found as $pH=2.88$ by putting $C_s=0$ in eq(9).

Ans. $pH = 4.74$

(b) Using eq(9), $pH=-\log_{10}C_0x$ is plotted in the figure below against C_s/C_0 at those concentrations of $C_0=1.0$, 0.5, 0.25 and 0.1. For any value of C_0, the same curve is obtained and, hence, it is concluded that the value of pH does not depend on the concentration C_0 itself.

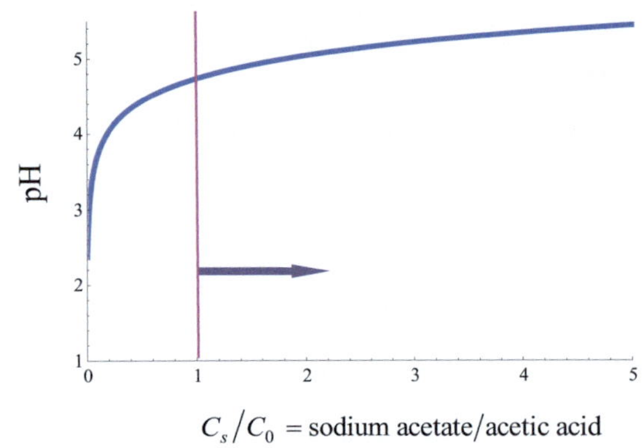

Fig.1 Variation of pH of versatile buffer solutions made up of acetic acid and sodium acetate with the concentration ratio of the components.
In the region where the concentration ratio is larger than unity, the value of pH varies only gradually with the ratio. The concentration of acetic acid is in the range 0.1–1.0 mol/L.

Ans. mixing ratio and pH: See Fig.1

(c) The result obtained above (Fig.1) can be proved by finding an approximate expression of eq(9). Employing the concentration ratio of sodium acetate to acetic acid $C_s/C_0=y$, eq(9) can be written as eq(9').

$$x = \frac{-(C_s/C_0 + K_a/C_0) + \sqrt{(C_s/C_0 + K_a/C_0)^2 + 4K_a/C_0}}{2}$$

$$= \frac{-(y + K_a/C_0) + \sqrt{(y + K_a/C_0)^2 + 4K_a/C_0}}{2} \qquad (9')$$

Neglecting the term of K_a^2 in the square root in eq(9'), the resulting function is expanded by using an approximation of $\sqrt{1+x} \approx 1 + x/2$, to find eq(9'') as shown below.

$$(4') \approx \frac{-(y + K_a/C_0) + \sqrt{y^2 + 2yK_a/C_0 + 4K_a/C_0}}{2} = \frac{-(y + K_a/C_0) + y\sqrt{1 + 2K_a/yC_0 + 4K_a/y^2C_0}}{2}$$

$$x \approx \frac{-(y + K_a/C_0) + y\left\{1 + \frac{1}{2}(2K_a/yC_0 + 4K_a/y^2C_0)\right\}}{2} = \frac{K_a}{yC_0} \qquad (9'')$$

From this result, a well-known eq(10), which is called *Henderson-Hasselbalch*'s equation is obtained as the equation to give pH values of buffer solutions.

$$\text{pH} = -\log_{10}[\text{H}^+] = -\log_{10}C_0x = -\log_{10}K_a + \log_{10}\frac{C_s}{C_0} \qquad (10)$$

It is seen from eq10) that the pH value of a buffer solution is determined by the dissociation constant K_a of a weak acid and the ratio of the concentrations of the acid to the salt employed C_s/C_0.

Ans. pH of buffer solution: eq(10)

Solution 4-3

(a) Solution 1.

i) in the case of $x < 10$ mL: When x mL of 0.1 N solution of a strong base is added to 10 mL of 0.1 N solution of a strong acid, the mole number of the hydrogen ion is given by eq(11) and, hence, its concentration is given by eq(12).

$$0.1 \times \frac{10}{1000} - 0.1 \times \frac{x}{1000} = (10 - x) \times 10^{-4} \text{mol} \qquad (11)$$

$$[\text{H}^+] = \frac{(10 - x) \times 10^{-4}(\text{mol})}{(10 + x) \times 10^{-3}(\text{L})} = \frac{(10 - x) \times 10^{-1}}{(10 + x)} \text{mol} \cdot \text{L}^{-1} \qquad (12)$$

Then, the value of pH of the solution is given by eq(13) from eq(12).

$$\text{pH} = -\log_{10}\frac{(10 - x) \times 10^{-1}}{(10 + x)} = 1 - \log_{10}\frac{(10 - x)}{(10 + x)} \qquad (13)$$

The blue curve given in Fig.2 shows a plot of eq(13) against the titrant amount x mL As is evident from eq(13), however, this equation cannot give a pH value at the equivalence point of $x = 10$ mL.

ii) in the case of $x > 10$ mL: When the solution of NaOH is added more over the equivalence point, the mole number of hydroxide ion is given by eq(14) and, hence, its concentration is given by eq(15).

$$0.1 \times \frac{x}{1000} - 0.1 \times \frac{10}{1000} = (x - 10) \times 10^{-4} \text{mol} \qquad (14)$$

$$[\text{OH}^-] = \frac{(x - 10) \times 10^{-4}(\text{mol})}{(10 + x) \times 10^{-3}(\text{L})} = \frac{(x - 10) \times 10^{-1}}{(10 + x)} \text{mol} \cdot \text{L}^{-1} \qquad (15)$$

The pH value of the solution is found as eq(16) from eq(15) by using the ion product of $[H^+][OH^-]=10^{-14}$.

$$pH=-\log_{10}\frac{10^{-14}}{[OH^-]}=13+\log_{10}\frac{(x-10)}{(10+x)} \tag{16}$$

A plot of eq(16) is shown in Fig.2 by the curve in red.

In order to give pH values correctly by a single function spanning a whole range of x over the equivalence point, the treatment given below in Solution 2 is required.

Solution 2.

During the titration of a strong acid with a strong base and vice versa, the concentration of $[H^+]$ in the solution being titrated is easily found by using the ion product (eq(17)) and the requirement of electroneutrality (charge balance, eq(18)). These two equations are satisfied always during the whole titration process from beginning to end.

$$[H^+][OH^-]=10^{-14} \tag{17}$$

$$[H^+]+[Na^+]=[OH^-]+[Cl^-] \tag{18}$$

When x mL of the NaOH solution is added, the concentrations $[Na^+]$ and $[Cl^-]$ are given by eq(19). Notice here that the total volume of the solution being titrated has increased to $(10+x)$ mL.

$$[Na^+]=\frac{x\times C_0\times 10^{-3}}{(10+x)\times 10^{-3}}=\frac{x\times C_0}{(10+x)}, \qquad [Cl^-]=\frac{(10-x)\times C_0}{(10+x)} \tag{19}$$

Next, the concentrations of $[H^+]$ and $[OH^-]$ are treated as unknown variables by putting them as eq(20),

$$[H^+]=y, \quad [OH^-]=z \tag{20}$$

and then the concentration of each ion in eqs(19) and (20) are substituted into eq(18) to find eq(21).

$$z=y+\frac{(x-10)\times C_0}{10+x} \tag{21}$$

From eqs(17) and (21), eq(22) is obtained as a quadratic equation and its positive solution is given by eq(23).

$$zy=y\left\{y+\frac{(x-10)C_0}{10+x}\right\}=10^{-14} \Rightarrow y^2+\left(\frac{x-10}{10+x}\right)\times C_0 y-10^{-14}=0 \tag{22}$$

$$y=\frac{-\left(\frac{x-10}{10+x}\right)\times C_0+\sqrt{\left(\frac{x-10}{10+x}\right)^2\times C_0^2+4\times 10^{-14}}}{2} \tag{23}$$

Since the concentration of $[H^+]=y$ is now expressed as a function of x, a titration curve can be drawn by plotting $-\log_{10}y$ against x. One such plot is shown in Fig.3 for the case of $C_0=0.1$.

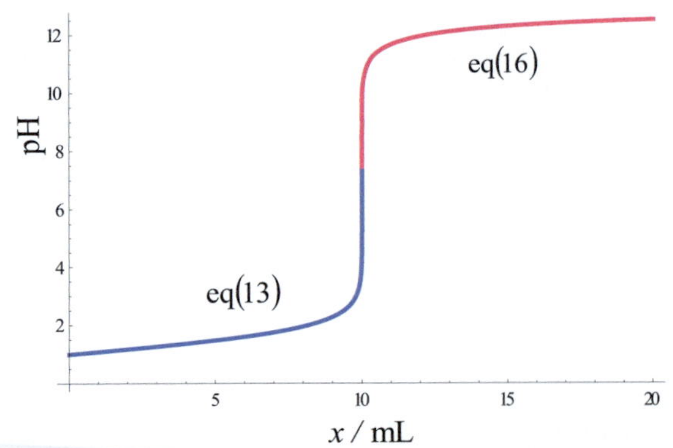

Fig.2 Titration curve based on Solution 1.

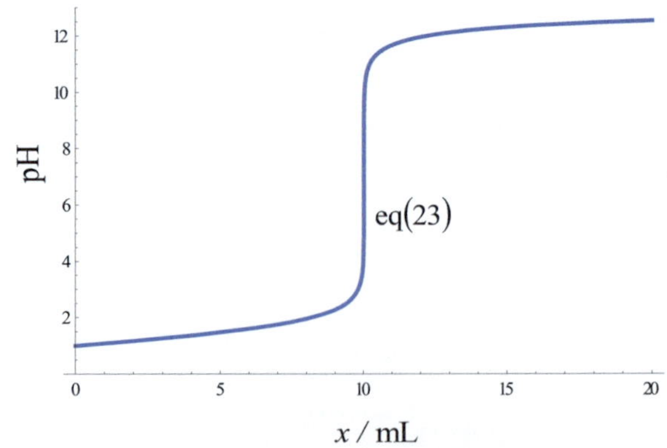

Fig.3 Titration curve based on Solution 2.

Ans. See text and Figs.2 and 3.

(b) The chemical equilibrium and the species existing in solution are shown in Scheme I below.

$$\left.\begin{array}{l} CH_3COOH \rightleftarrows H^+ + CH_3COO^- \\ CH_3COONa \rightarrow Na^+ + CH_3COO^- \\ H_2O \rightleftarrows H^+ + OH^- \end{array}\right\} \quad \text{Scheme I}$$

After adding x mL of the titrant NaOH solution to the solution being titrated, the concentrations of all ions and chemical species in the resulting solution are governed by eq(24) (electroneutrality) and eq(25) (conservation of substance).

$$[H^+]+[Na^+]=[CH_3COO^-]+[OH^-] \qquad (24)$$

$$[CH_3COOH]+[CH_3COO^-]=\frac{10\times C_0}{10+x}=\frac{1}{10+x} \qquad (25)$$

The concentration of Na^+ is given by eq(26) using the titrant amount x since NaOH is considered fully ionized in aqueous solution.

$$[Na^+]=\frac{x}{10+x}\times C_0=\frac{x}{10+x}\times 0.1 \qquad (26)$$

The concentrations of the other ions are unknown at this stage and they are put y, z and u as in eq(27).

$$[H^+]=y, \qquad [OH^-]=z, \qquad [CH_3COO^-]=u \qquad (27)$$

Substituting these unknowns and eq(26) into eq(24), eq(24) is rewritten as eq(24').

$$y+\frac{x}{10+x}\times 0.1=u+z \qquad (24')$$

Next, the dissociation constant K_a of acetic acid is expressed as eq(28) by using the concentrations of ions and neutral species defined above.

$$K_a=\frac{[H^+][CH_3COO^-]}{[CH_3COOH]}=\frac{y\times u}{\dfrac{1}{10+x}-u} \qquad (28)$$

Solving eq(28) for u, eq(28') is obtained.

$$u(y+K_a)=\frac{K_a}{10+x} \Rightarrow u=\frac{K_a}{(10+x)(y+K_a)} \qquad (28')$$

Substituting eq(28') into eq(24') and solving the resultant equation for z, eq(24'') is obtained.

$$z=y+\frac{x}{10+x}\times 0.1-\frac{K_a}{(10+x)(y+K_a)}=y+\frac{0.1x(y+K_a)-K_a}{(10+x)(y+K_a)} \qquad (24'')$$

This is substituted into eq(29) for the ion product of water to find eq(29'), which contains only x and y as variables and can provide a relation between the concentration of hydrogen ion and the amount of titration.

$$yz=K_w=10^{-14} \qquad (29)$$

$$\left\{y+\frac{0.1x(y+K_a)-K_a}{(10+x)(y+K_a)}\right\}\times y=K_w \Rightarrow y^2(10+x)(y+K_a)+0.1xy(y+K_a)-K_a y$$
$$=K_w(10+x)(y+K_a) \qquad (29')$$

Here, the above equation is solved not for y but for x, giving eq(30) as the solution.

$$x=\frac{-10y^2(y+K_a)+K_a y+10K_w(y+K_a)}{y^2(y+K_a)-K_w(y+K_a)+0.1y(y+K_a)} \qquad (30)$$

Expressing the pH of the solution by χ, the variable y can be written as $y=10^{-\chi}$, with which y is replaced to derive eq(30').

$$x=\frac{-10\times10^{-2\chi}(10^{-\chi}+K_a)+K_a10^{-\chi}+10K_w(10^{-\chi}+K_a)}{10^{-2\chi}(10^{-\chi}+K_a)-K_w(10^{-\chi}+K_a)+0.1\times10^{-\chi}(10^{-\chi}+K_a)},\qquad \chi=\mathrm{pH} \tag{30'}$$

Taking pH along the vertical axis and the titrant amount x along the horizontal axis, eq(30) (or eq(30')) is plotted in Fig.4 for $K_a=1.8\times10^{-5}$. The equivalence point is located at $x=10$ mL and marked by the full green circle. The pH value at that point can be read as 8.6-8.7 from the (titration) curve in Fig.4, however, the exact value can be found from eq(30) by putting $x=10$ mL to calculate the value of y and then corresponding pH. Thus, the pH value at the equivalence point is found as $\mathrm{pH}=8.72$. Although eq(30) is a cubic equation with respect to y, an approximate value of pH can be found without solving it directly. At the equivalence point, eq(30) is rearranged into eq(31). And near the equivalence point, since the magnitude of y

$$y^3+\left(K_a+\frac{C_0}{2}\right)y^2-K_wy-K_wK_a=0 \tag{31}$$

is of the order of 10^{-7}, the cubic term y^3 can be neglected as compared with the other terms. Therefore, eq(31) can be approximated by eq(31') and it is sufficient to solve eq(31') to find the pH value at the equivalence point,

$$\left(K_a+\frac{C_0}{2}\right)y^2-K_wy-K_wK_a=0 \tag{31'}$$

and its solution is given by the first equality of eq(32), which is further approximated as the last equality.

$$y=\frac{K_w+\sqrt{K_w^2+4K_wK_a\left(K_a+\frac{C_0}{2}\right)}}{2\left(K_a+\frac{C_0}{2}\right)}\approx\frac{\sqrt{2K_wK_aC_0}}{C_0}=\left(\frac{2K_wK_a}{C_0}\right)^{1/2} \tag{32}$$

From this, the pH value at the equivalence point is given by eq(33).

$$\mathrm{pH}=-\log_{10}y=-\frac{1}{2}\left(\log_{10}K_a+\log_{10}K_w-\log_{10}\frac{C_0}{2}\right) \tag{33}$$

Substituting the given values into eq(33), the value for pH is found as $\mathrm{pH}=8.7$, which agrees reasonably well with the value found above by solving the cubic equation precisely as far as triple significant figures are concerned.

Apart from the equivalence point, the approximate equation to find the pH values at the points indicated by the solid yellow and red circles in Fig.4 are given in the figure caption. In the region marked by ⟺, the pH varies only gradually and, hence, buffer action is recognized.

Another simple approach to derive eq(33) is as follows. The concentration of CH_3COO^- at the equivalence point is equated as $[CH_3COO^-]=C_0/2-\Delta$ where Δ is a small quantity and then from eqs(25) and (27) the dissociation constant can be written as eq(28'').

$$K_a=\frac{[H^+][CH_3COO^-]}{[CH_3COOH]}=\frac{y\times(C_0/2-\Delta)}{\Delta} \tag{28''}$$

Also, since eq(24) can be expressed as eq(24'''), the concentration of hydrogen ion squared is approximated by eq(29'').

$$y+\frac{C_0}{2}=\frac{C_0}{2}-\Delta+z \;\Rightarrow\; z=y+\Delta=y+\frac{y}{K_a}\frac{C_0}{2} \qquad (24''')$$

$$yz=y^2\left(1+\frac{C_0}{2K_a}\right)=K_w \;\Rightarrow\; y^2=\frac{K_w}{\left(1+\frac{C_0}{2K_a}\right)}\approx\frac{K_aK_w}{C_0/2} \qquad (29'')$$

Consequently, the pH value at the equivalence point is given by eq(33) which is identical to that obtained above as an approximation.

$$y=\left(\frac{K_aK_w}{C_0/2}\right)^{1/2} \;\Rightarrow\; \mathrm{pH}=-\log_{10}y=-\frac{1}{2}\left(\log_{10}K_a+\log_{10}K_w-\log_{10}\frac{C_0}{2}\right) \qquad (33)$$

In Fig.5 is shown an enlarged titration curve near the equivalence point given in Fig.4. As is seen from this figure, the pH value at $x=9.99$ mL, just before the equivalence point, is pH = 7.73 and at $x=10.01$ mL, just over the equivalence point, pH = 9.70. Near the equivalence point, therefore, the pH value jumps enormously when the titrant amount is varied only slightly such as by 0.01mL. This is the reason why the equivalence point is found by an appropriate pH indicator.

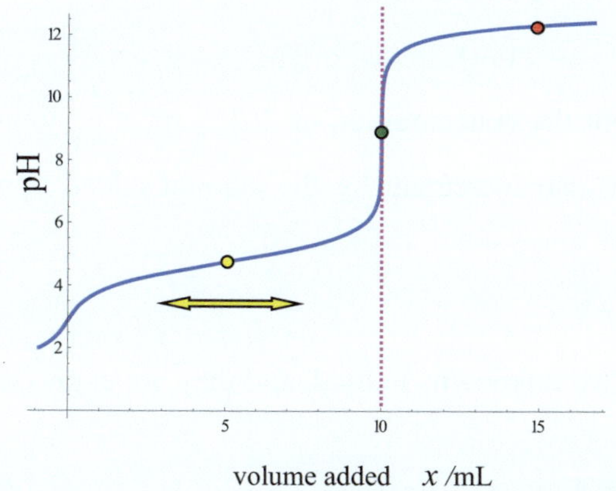

Fig.4 Titration curve of weak acid –strong base.
The pH values at the colored full circles are given by the equation below.

- ● (green) $\mathrm{pH}=-\frac{1}{2}\left(\log_{10}K_a+\log_{10}K_w-\log_{10}\frac{C_0}{2}\right)$
- ● (yellow) $\mathrm{pH}=-\log_{10}K_a$
- ● (red) $\mathrm{pH}=-\log_{10}K_w+\log_{10}\frac{C_0}{3}$

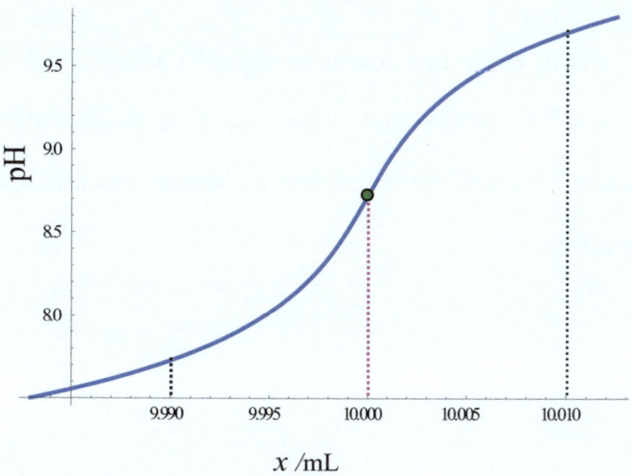

Fig.5 Enlarged curve near the equivalence point shown in Fig.4.
Shifting the amount of titration from the equivalence point only by ± 0.01 mL, pH value varies approximately by ± 1.

Ans. titration curve for weak acid-strong base: See text and Fig.4.

Solution 4-4

The acid-base reaction in the aqueous solution of Na_2CO_3 is expressed as below.

$$\frac{1}{2}\mathrm{Na_2CO_3} + \mathrm{HCl} \rightarrow \mathrm{NaCl} + \frac{1}{2}\mathrm{H_2CO_3} \qquad (34)$$

Unreacted Na_2CO_3 is considered to be fully ionized as expressed by eq(35).

$$\frac{1}{2}\mathrm{Na_2CO_3} \rightarrow \mathrm{Na^+} + \frac{1}{2}\mathrm{CO_3}^{2-} \qquad (35)$$

However, not only acetic acid but also hydro-carbonate and carbonate ions produced are only partially ionized according to eqs(36) (first dissociation) and (37) (second dissociation).

$$\mathrm{H_2CO_3} \rightleftarrows \mathrm{H^+} + \mathrm{HCO_3}^- \qquad K_1 = 4.3\times10^{-7} \qquad (36)$$

$$\mathrm{HCO_3}^- \rightleftarrows \mathrm{H^+} + \mathrm{CO_3}^{2-} \qquad K_2 = 5.6\times10^{-10} \qquad (37)$$

When x mL of the 0.1 N HCl aqueous solution is added and the volume of the solution being titrated becomes $10+x$ mL, among the concentrations of the ions existing in the solution, the equation for the electroneutrality expressed by eq(38) is satisfied. Also, from the conservation of substances, eq(39) is satisfied for the substances containing CO_3 group and its ions.

$$[\mathrm{H^+}]+[\mathrm{Na^+}]=[\mathrm{HCO_3}^-]+2[\mathrm{CO_3}^{2-}]+[\mathrm{OH^-}]+[\mathrm{Cl^-}] \qquad (38)$$

$$[\mathrm{H_2CO_3}]+[\mathrm{HCO_3}^-]+[\mathrm{CO_3}^{2-}]=C_0\times\frac{10}{10+x} \qquad (39)$$

The concentration of $[\mathrm{Na^+}]$ is simply diluted with titration, but the concentration of $[\mathrm{Cl^-}]$ must be found by taking into account of not only dilution with titration but also increment by the amount added upon titration as eq(40).

$$[\mathrm{Na^+}]=\frac{C_0\times20}{10+x}, \qquad [\mathrm{Cl^-}]=\frac{C_0 x}{10+x} \qquad (40)$$

The concentrations of the other ions are considered unknown, five unknowns in total, and they are expressed by using alphabets as follows.

$$[\mathrm{H^+}]=y, \qquad [\mathrm{OH^-}]=z$$

$$[\mathrm{H_2CO_3}]=u, \qquad [\mathrm{HCO_3}^-]=v, \qquad [\mathrm{CO_3}^{2-}]=w$$

The dissociation constants of acetic acid are expressed by using these unknowns as eqs(41) and (42). And also, the ion product is given by eq(43). Although there are five unknowns, there are five independent equations (eqs(38), (39) and (41)-(43)) relating those five unknowns and, hence, when the titrant amount x is given, these equations are simultaneously solved for the five unknowns.

$$\frac{[\mathrm{H^+}][\mathrm{HCO_3}^-]}{[\mathrm{H_2CO_3}]}=\frac{y\times v}{u}=K_1 \qquad (41)$$

$$\frac{[\mathrm{H^+}][\mathrm{CO_3}^{2-}]}{[\mathrm{HCO_3}^-]}=\frac{y\times w}{v}=K_2 \qquad (42)$$

$$[\mathrm{H^+}][\mathrm{OH^-}]=K_w=10^{-14} \qquad (43)$$

From eqs(39) and (40), eqs(39') and (40') are obtained.

$$u+v+w=C_0\times\frac{10}{10+x} \qquad (39')$$

$$y+\frac{C_0\times20}{10+x}=v+2w+z+\frac{C_0 x}{10+x} \qquad (40')$$

Rewriting eqs(41) and (42) as eqs(41') and (42'), respectively, the resulting equations are substituted into eq(39'),

$$u=\frac{y\times v}{K_1}=\frac{y^2\times w}{K_1K_2} \quad (41'), \qquad v=\frac{y\times w}{K_2} \quad (42')$$

which is then solved for w as is given by eq(44).

$$\left(\frac{y^2}{K_1K_2}+\frac{y}{K_2}+1\right)w=\frac{C_0\times 10}{10+x} \Rightarrow w=\left(\frac{C_0\times 10}{10+x}\right)\Big/\left(\frac{y^2}{K_1K_2}+\frac{y}{K_2}+1\right) \quad (44)$$

Rewriting eq(40') for z with some rearrangements, the last equality of eq(45) is obtained.

$$\begin{aligned} z &= y+\frac{C_0\times 20}{10+x}-v-2w-\frac{C_0x}{10+x} \\ &= y-\frac{y\times w}{K_2}-2w+\frac{C_0(20-x)}{10+x}=y-\left(\frac{y}{K_2}+2\right)w+\frac{C_0(20-x)}{10+x} \end{aligned} \quad (45)$$

After multiplying eq(45) by y and substituting w given by eq(44) into the resultant equation, eq(46) is obtained.

$$yz=y^2-\left(\frac{y^2}{K_2}+2y\right)\left(\frac{C_0\times 10}{10+x}\right)\Big/\left(\frac{y^2}{K_1K_2}+\frac{y}{K_2}+1\right)+\frac{C_0(20-x)y}{10+x}=10^{-14} \quad (46)$$

Eliminating two denominators in the second equality of eq(46), eq(46) can be expressed as eq(46'), which contains the titrant amount x and the unknown concentration of hydrogen ion y alone.

$$\begin{aligned} & y^2\left(\frac{y^2}{K_1K_2}+\frac{y}{K_2}+1\right)(10+x)-\left(\frac{y^2}{K_2}+2y\right)C_0\times 10 \\ & \quad +C_0(20-x)y\left(\frac{y^2}{K_1K_2}+\frac{y}{K_2}+1\right)-10^{-14}\left(\frac{y^2}{K_1K_2}+\frac{y}{K_2}+1\right)(10+x)=0 \end{aligned} \quad (46')$$

This equation is further rearranged into the sum of two parts, one containing x and the other without x, resulting in eq(46'').

$$\begin{aligned} & y^2\left(\frac{y^2}{K_1K_2}+\frac{y}{K_2}+1\right)\times 10-\left(\frac{y^2}{K_2}+2y\right)C_0\times 10-10^{-13}\left(\frac{y^2}{K_1K_2}+\frac{y}{K_2}+1\right)+C_0\times 20y\left(\frac{y^2}{K_1K_2}+\frac{y}{K_2}+1\right) \\ & +x\left\{\left(\frac{y^2}{K_1K_2}+\frac{y}{K_2}+1\right)\left(y^2-C_0y-10^{-14}\right)\right\}=0 \end{aligned} \quad (46'')$$

Solving eq(46'') for x, eq(47) is obtained via the following equation.

$$x=\frac{y^2\left(\frac{y^2}{K_1K_2}+\frac{y}{K_2}+1\right)\times 10-\left(\frac{y^2}{K_2}+2y\right)C_0\times 10-10^{-13}\left(\frac{y^2}{K_1K_2}+\frac{y}{K_2}+1\right)+C_0\times 20y\left(\frac{y^2}{K_1K_2}+\frac{y}{K_2}+1\right)}{\left(\frac{y^2}{K_1K_2}+\frac{y}{K_2}+1\right)\left\langle y(C_0-y)+10^{-14}\right\rangle}$$

$$x=\frac{y^2\times 10+C_0\times 20y-10^{-13}}{y(C_0-y)+10^{-14}}-\frac{\left(\frac{y^2}{K_2}+2y\right)C_0\times 10}{\left(\frac{y^2}{K_1K_2}+\frac{y}{K_2}+1\right)\left\langle y(C_0-y)+10^{-14}\right\rangle} \quad (47)$$

The final result is eq(47), however, it is difficult to plot pH against the titrant amount x by using eq(47) as it is. In such a case, a method named *Parametric Plot* is versatile. For this purpose, the unknown y is rewritten as χ, which is equivalent to pH, as is given by eq(48).

$$[H^+] = y = 10^{-\chi} \;\Rightarrow\; pH = -\log_{10} y = \chi \qquad (48)$$

Then, eq(47) can be expressed as eq(47'),

$$x = \frac{10^{-2\chi}\times 10 + C_0 \times 20 \times 10^{-\chi} - 10^{-13}}{10^{-\chi}(C_0 - 10^{-\chi}) + 10^{-14}} - \frac{\left(\dfrac{10^{-2\chi}}{K_2} + 2\times 10^{-\chi}\right) C_0 \times 10}{\left(\dfrac{10^{-2\chi}}{K_1K_2} + \dfrac{10^{-\chi}}{K_2} + 1\right)\left\langle 10^{-\chi}(C_0 - 10^{-\chi}) + 10^{-14}\right\rangle} \qquad (47')$$

which shows that x is given as a function of the pH value (χ). Although x can be plotted against χ in a usual way, the figure to be obtained becomes an unfamiliar one since the vertical axis is x and horizontal axis is pH. It is better, therefore, to interchange the vertical and horizontal axes in order to make the curve a conventional titration curve given by pH vs. x plot. This is easily performed by employing, for instance, *Parametric Plot* of *Mathematica*.

The curve plotted in this way is shown in Fig.6 given below. The following values are used for plotting the curve: $K_1 = 4.3\times 10^{-7}$, $K_2 = 5.6\times 10^{-10}$, $C_0 = 0.1$ (blue line) and $C_0 = 0.01$ (green line). Two inflection points

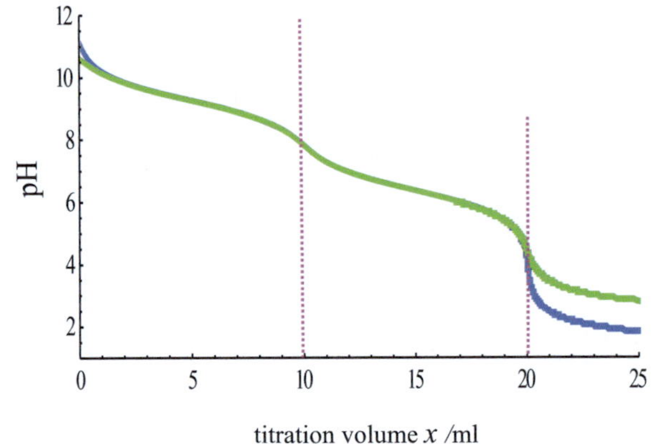

Fig.6 Titration curve of aqueous solution of Na_2CO_3 Titrated with aqueous HCl solution. $K_1 = 4.3\times 10^{-7}$, $K_2 = 5.6\times 10^{-10}$, blue line: $C_0 = 0.1$, green line: $C_0 = 0.01$.

are clearly recognized at $x = 10$ mL and $x = 20$ mL in these curves, the former corresponding to the equivalence point of the first dissociation and the latter to the second dissociation of carbonic acid. See the next problem and its answer for the functions which give pH values at those equivalence points.

Ans. See text and Fig.6

Solution 4-5

The relation between the concentration of hydrogen ion y and the titrant amount x is given by eq(46'') in **solution 4-4.** It may be orthodox to solve eq(46'') given below under the given conditions. However, it is convenient if some approximate functions are available to know the pH values needed. A shortcut to meet this is to find approximate functions of eq(46''), which is a quartic function of y, just by keeping terms up to the first or second order of y.

$$y^2\left(\frac{y^2}{K_1K_2} + \frac{y}{K_2} + 1\right)\times 10 - \left(\frac{y^2}{K_2} + 2y\right)C_0\times 10 - 10^{-13}\left(\frac{y^2}{K_1K_2} + \frac{y}{K_2} + 1\right) + C_0\times 20y\left(\frac{y^2}{K_1K_2} + \frac{y}{K_2} + 1\right)$$
$$+ x\left\{\left(\frac{y^2}{K_1K_2} + \frac{y}{K_2} + 1\right)(y^2 - C_0 y - 10^{-14})\right\} = 0 \qquad (46'') \qquad \text{(reposted)}$$

(a) As is seen from Fig.6 in the preceding solution, at the first equivalence point, y is of the order of 10^{-8}. Employing the values of $K_1 = 4.3\times10^{-7}$ and $K_2 = 5.6\times10^{-10}$, the order of the magnitude is calculated for each term of eq(46'') by noticing eq(49).

$$\frac{y^2}{K_1K_2}\approx 1\ ,\qquad \frac{y}{K_2}\approx 10 \qquad (49)$$

The terms smaller than 10^{-8} being neglected, eq(50) is obtained, which is then rearranged as eq(50'). From eq(50'), the approximate function required is found as eq(51).

$$-\left(\frac{y^2}{K_2}+2y\right)C_0\times10 + C_0\times20y\left(\frac{y^2}{K_1K_2}+\frac{y}{K_2}+1\right)-\left(\frac{y^2}{K_1K_2}+\frac{y}{K_2}+1\right)10C_0y\approx 0 \qquad (50)$$

$$-\left(\frac{y^2}{K_2}+2y\right)+\left(\frac{y^2}{K_1K_2}+\frac{y}{K_2}+1\right)y\approx 0 \;\Rightarrow\; \frac{y^2}{K_1K_2}-1\approx 0 \qquad (50')$$

$$y\approx\sqrt{K_1K_2}\;\Rightarrow\;\mathrm{pH}=-\frac{1}{2}(\log_{10}K_1+\log_{10}K_2)=\frac{1}{2}(\mathrm{pK}_1+\mathrm{pK}_2) \qquad (51)$$

(b) From Fig.6, the second equivalence point is seen to be located at around $y\approx10^{-4}$. In a similar manner as above, the order of magnitude of each term of eq(46'') is evaluated and only the terms of the order of 1 are kept to obtain eq(52). After rearranging eq(52), it is simplified to eq(52'), from which eq(53) is obtained as an approximate function to give the pH value at the second equivalence point.

$$\frac{y^4\times10}{K_1K_2}-\frac{y^2C_0\times10}{K_2}+\frac{20C_0y^3}{K_1K_2}+\frac{20C_0y^2}{K_2}+\frac{20y^4}{K_1K_2}-\frac{20C_0y^3}{K_1K_2}-\frac{20C_0y^2}{K_2}\approx 0 \qquad (52)$$

$$\frac{y^4\times3}{K_1K_2}-\frac{y^2C_0}{K_2}\;\Rightarrow\;\frac{y^2\times3}{K_1}-C_0\approx 0 \qquad (52')$$

$$y=\sqrt{\frac{K_1C_0}{3}}\;\Rightarrow\;\mathrm{pH}=-\frac{1}{2}\left(\log_{10}K_1+\log_{10}\frac{C_0}{3}\right)=\frac{1}{2}\left(\mathrm{pK}_1-\log_{10}\frac{C_0}{3}\right) \qquad (53)$$

Based on eqs(51) and (53), the pH values at the first and second equivalence points are found as below and they are practically equal to the values calculated precisely.

$$\text{first equivalence point: }\ \mathrm{pH}=\frac{1}{2}(\mathrm{pK}_1+\mathrm{pK}_2)=7.81$$

$$\text{second equivalence point: }\ \mathrm{pH}=\frac{1}{2}\left(\mathrm{pK}_1-\log_{10}\frac{0.1}{3}\right)=3.92$$

The pH value at the first equivalence point is independent of the concentration of a solution, however, the one at the second equivalence point does depend on the concentration, as is seen from eq(53) and the titration curves shown in Fig.6 as well.

Ans. pH values at the first and second equivalence points: eq(51) and eq(53)

Solution 4-6

The ionic species present in the solution being titrated are the same as those given in **problem 4-4**. The concentration of each ionic species is different from the one given there because of the presence of NaOH, though each concentration is found in the totally same manner as before.

Eqs(39) and (40) given in **solution 4-4** must be replaced with the following ones, in which the replaced parts are marked in color.

$$[H_2CO_3]+[HCO_3^-]+[CO_3^{2-}]=C_0\times\frac{10}{20+x} \qquad (54)$$

$$[Na^+]=\frac{C_0\times 30}{20+x}, \quad [Cl^-]=\frac{C_0 x}{20+x} \qquad (55)$$

Using these equations, the calculation similar to that given in the previous solution is carried out to find x (amount of titrant) as a function of y ($pH=-\log_{10}y$). As a result, eq(56) is obtained.

$$x=\frac{y^2\times 20+C_0\times 30y-10^{-13}}{y(C_0-y)}-\frac{\left(\frac{y^2}{K_2}+2y\right)C_0\times 10}{\left(\frac{y^2}{K_1K_2}+\frac{y}{K_2}+1\right)\left\langle y(C_0-y)+10^{-14}\right\rangle} \qquad (56)$$

This equation differs from eq(47) given in **solution 4-4** only in the parts marked in color. Employing *Mathematica*, eq(14′) is parametrically plotted as is done in **solution 4-4** and the obtained titration curve is shown in Fig.7. From Fig.7, it is seen that the pH value jumps at three places, i.e., at the equivalence point of NaOH and those of the first and the second dissociations of Na_2CO_3 (from the left to the right). When phenolphthalein is used as an indicator in practice, the equivalence point of NaOH cannot be recognized. For this, in order to determine the concentration of NaOH and Na_2CO_3 two indicators of phenolphthalein and BPB are used to find the second and third equivalence points (marked as ①, ② and ③ in Fig.7). Then the volume of the HCl solution to titrate NaOH is given by $(A-B)$ mL as is seen from Fig.7. This is the method used in the differential titration.

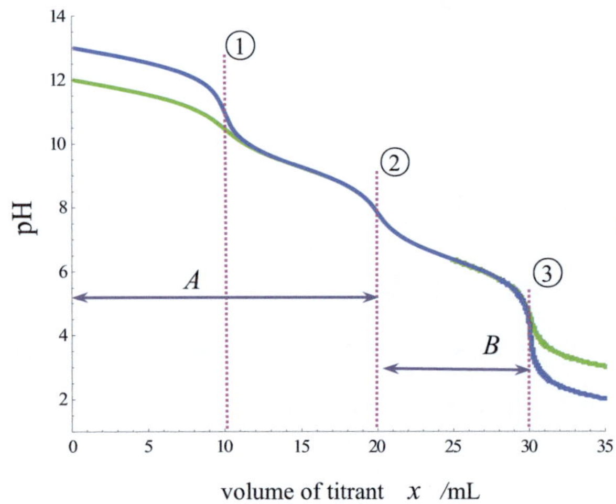

Fig.7 Titration curve (analytical function) of an equimolar mixture of aqueous solutions of Na_2CO_3 and NaOH titrated with aqueous solution of HCl.
$K_1=4.3\times10^{-7}$, $K_2=5.6\times10^{-10}$ and $C_0=0.1$ are employed. For comparison, a titration curve for $C_0=0.01$ is shown by the green line.

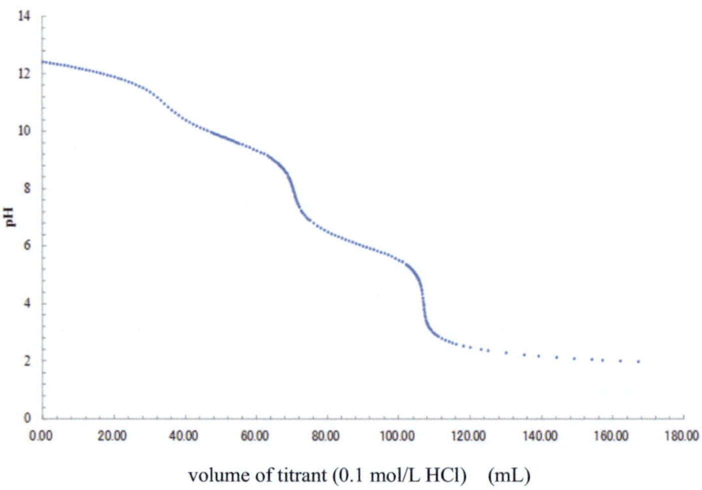

Fig.8 Experimentally obtained titration curve (courtesy of Prof.Karatani).

Ans. (differential) titration curve: See text and Figs.7 and 8.

"note" Real titration curve obtained experimentally

Under the same conditions as employed for Fig.7, the experimentally acquired pH values are plotted against the volume of titrant in Fig.8. Compared with Fig.7, almost identical curve is obtained experimentally. In particular, it should be noted that the three equivalence points are clearly recognized for the latter as well.

Solution 4-7

The dissociation equilibrium of acetic acid is expressed by eq(57) and sodium acetate is fully ionized as is given by eq(58). The concentration of each species is denoted under the relevant species in eq(57) or eq(58). C_0 and C_s are the molar concentrations of acetic acid and sodium acetate, respectively used for the preparation of solutions. Both are taken as 0.1 (mol/L) here.

$$\underset{u}{CH_3COOH} \rightleftarrows \underset{y}{H^+} + \underset{v}{CH_3COO^-} \qquad (57)$$

$$\underset{0}{CH_3COONa} \rightarrow \underset{C_0'}{Na^+} + \underset{v}{CH_3COO^-} \qquad (58)$$ (C_0' after consideration of volume change upon titration)

Consequently, the equilibrium constant K_a, the charge balance (electroneutrality) and the conservation of substances are expressed by from eq(59) to eq(61), respectively.

$$K_a = \frac{[H^+][CH_3COO^-]}{[CH_3COOH]} = \frac{y \times v}{u} \qquad (59)$$

$$[H^+]+[Na^+]=[CH_3COO^-]+[OH^-]+[Cl^-] \qquad (60)$$

$$[CH_3COOH]+[CH_3COO^-]=(C_0+C_s)\times 100/101 \qquad (61)$$

Noticing that the concentration of chloride ion $[Cl^-]$ originating from HCl added mistakenly is given by $0.1\times 1/101$, eqs(60) and (61) are expressed by relevant concentrations as eq(60') and (61'), respectively.

$$y+\frac{10}{101}=v+z+\frac{1}{101}\times 0.1 \qquad (60')$$

$$u+v=\frac{20}{101} \qquad (61')$$

Solving for u and v, $u=\frac{20}{101}-v=z-y+\frac{10.1}{101}$ and $v=y-z+\frac{9.9}{101}$ are obtained, and the equilibrium constant is given by eq(59').

$$K_a=\frac{y\left(y-z+\frac{9.9}{101}\right)}{z-y+\frac{10.1}{101}} \Rightarrow K_a\left(z-y+\frac{10.1}{101}\right)-y\left(y-z+\frac{9.9}{101}\right)=0 \qquad (59')$$

Rewriting eq(59') as the equation for z, eq(62) is obtained.

$$z=\frac{K_a\left(y-\frac{10.1}{101}\right)+y\left(y+\frac{9.9}{101}\right)}{(K_a+y)} \qquad (62)$$

Consequently, using the ion product of water given by $yz=10^{-14}$ and eq(62), the cubic equation of y is derived below as given by eq(63).

$$K_a y\left(y-\frac{10.1}{101}\right)+y^2\left(y+\frac{9.9}{101}\right)-(K_a+y)\times 10^{-14}=0 \Rightarrow$$

$$y^3+\left(K_a+\frac{9.9}{101}\right)y^2-\left(10^{-14}+\frac{10.1}{101}\times K_a\right)y-K_a\times 10^{-14}=0 \qquad (63)$$

Solving eq(63), one of the solutions is found as $y = 1.854 \times 10^{-5}$, from which the pH value is obtained as 4.732, a value very close to the value of $\mathrm{pH} = 4.745$ found for the buffer solution itself.

Although the buffer action of the solution is confirmed from the result obtained above, it is much easier to understand it if considered as follows. Eq(63) can be approximated by eq(63').

$$y^3 + \frac{9.9}{101}y^2 - \frac{10.1}{101} \times K_a y = 0 \;\Rightarrow\; y^2 + \frac{9.9}{101}y - \frac{10.1}{101} \times K_a = 0 \qquad (63')$$

Solving eq(63') for *y*, eq(64) is obtained from which an approximate pH value is calculated by eq(65) and is found as 4.736, which is practically identical to the exact solution of eq(63). When eq(65) is compared with eq(66) that gives the pH value of the buffer solution, it is found that the effect of adding 0.1mL of the HCl solution to the buffer solution can be regarded as equivalent to change the concentration of the acid from 10.0 to 10.1 and the concentration of the salt from 10.0 to 9.9 in the same concentration unit. On the other hand, the pH value of the buffer solution given by eq(66) varies only slightly with a change in the concentration ratio of acid to salt made of its conjugate base. Putting these two facts together, the buffer action will be understood more easily.

$$y = \frac{-\frac{9.9}{101} + \sqrt{\left(\frac{9.9}{101}\right)^2 + \frac{10.1}{101} \times 4K_a}}{2} \approx \left(\frac{101}{9.9}\right)\frac{10.1}{101} \times K_a = \frac{10.1}{9.9} \times K_a \qquad (64)$$

$$\mathrm{pH} = -\log_{10} y = \mathrm{pK_a} + \log_{10}\left(\frac{9.9}{10.1}\right) = 4.736 \qquad (65)$$

$$\mathrm{pH} = \mathrm{pK_a} + \log_{10}\frac{C_s}{C_0} \qquad (66)$$ (*Henderson-Hasselbach*'s equation, cited again)

From the reasoning above, it is also understood that the higher the concentrations of acid and salt are, the greater is the buffer action.

Ans. exact solution: $\mathrm{pH} = 4.732$, approximate solution: $\mathrm{pH} = 4.736$

[advanced problem]

Find the pH value of 0.01M aqueous solution of ammonium acetate. Here, the dissociation constant of acetic acid is given by $K_a = 10^{-4.65}$ and the dissociation constant of ammonium hydroxide is $K_b = 10^{-9.24}$.

[clue] Denoting the concentration of hydrogen ion *x*, the following quartic equations must be solved to find an exact solution for *x*.

$$K_b x^4 + \langle (C_b + K_a)K_b + K_w \rangle x^3 - \langle (C_a - C_b)K_a K_b - K_w(K_a - K_b) \rangle x^2 - K_w \langle (C_a + K_b)K_a + K_w \rangle x - K_w^2 K_a = 0 \qquad (67)$$

As is shown below, its approximate equation is given by the quadratic equation (zero for the linear term), but

$$\begin{aligned} &\langle (C_b + K_a)K_b + K_w \rangle x^3 - \langle (C_a - C_b)K_a K_b - K_w(K_a - K_b) \rangle x^2 - K_w \langle (C_a + K_b)K_a + K_w \rangle x = 0 \\ &\langle (C_b + K_a)K_b + K_w \rangle x^2 - \langle (C_a - C_b)K_a K_b - K_w(K_a - K_b) \rangle x - K_w \langle (C_a + K_b)K_a + K_w \rangle = 0 \\ &\langle (C_b + K_a)K_b + K_w \rangle x^2 - K_w \langle (C_a + K_b)K_a + K_w \rangle = 0 \end{aligned} \qquad (68)$$

how accurate is it?

$$x = \sqrt{\frac{K_w \langle (C_a + K_b)K_a + K_w \rangle}{\langle (C_b + K_a)K_b + K_w \rangle}} \approx \sqrt{\frac{K_w (C_a + K_b)K_a}{(C_b + K_a)K_b}} \qquad (69)$$

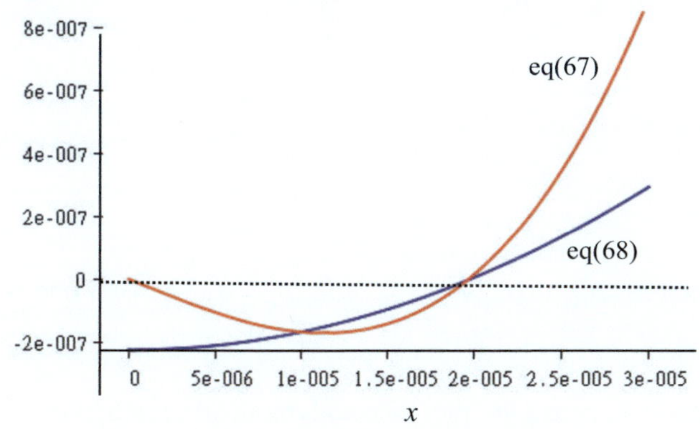

Fig.9. Plots of eq(67) and eq(68) against x.

[summary solution]

Employing the following values, both solutions of the quadratic and quartic equations are compared in Fig.9 on the left. It is seen from the figure, each equation yields one real root almost identical. ($e^{-006} = 10^{-6}$, e.g., for gnuplot expression.)

$$K_a = 10^{-4.65},\quad K_b = 10^{-9.24},$$
$$K_w = 10^{-14},\quad C_a = C_b = 10^{-2}\,\text{M}$$

Solution 4-8

(a) On the basis of *Henderson-Hasselbalch*'s' equation (see **solution 4-2**) which gives pH values of buffer solutions, the pH value of the blood is given by eq(70).

$$\text{pH} = -\log_{10} K_a + \log_{10}\frac{C_s}{C_0} = -\log_{10} K_a + \log_{10}\frac{[\text{HCO}_3^-]}{[\text{H}_2\text{CO}_3]} \qquad (70)$$

Substituting $K_a = 4.3\times10^{-7}$ into the above equation, the range of the concentration ratio of $[\text{HCO}_3^-]/[\text{H}_2\text{CO}_3]$ which yields pH values in the range $7.35-7.45$ is found by eq(71), from which the range of the concentration range is calculated as $9.55 < [\text{HCO}_3^-]/[\text{H}_2\text{CO}_3] < 12.0$.

$$7.35-6.37 < \log_{10}\frac{[\text{HCO}_3^-]}{[\text{H}_2\text{CO}_3]} < 7.45-6.37 \;\Rightarrow\; 10^{0.98} < \frac{[\text{HCO}_3^-]}{[\text{H}_2\text{CO}_3]} < 10^{1.08} \qquad (71)$$

Ans. $9.55 < [\text{HCO}_3^-]/[\text{H}_2\text{CO}_3] < 12.0$

(b) For simplicity, let us assume that the initial value of the concentration ratio (hydrogen carbonate/carbonic acid) is 10. Denoting the concentration of carbonic acid x $\text{mol}\cdot\text{L}^{-1}$, a change in pH ($\Delta\text{pH}$) when 0.1 mL of 0.1 N HCl is added is given by eq(72) from eq(70). Note that the volume of the total solution is now 1.1 mL.

$$\Delta\text{pH} = \Delta\log_{10}\frac{C_s}{C_0} = \log_{10}\frac{10x}{x} - \log_{10}\frac{10x/1.1}{(x+0.1\times0.1)/1.1} \qquad (72)$$

$$= \log_{10}\frac{(x+0.1\times0.1)}{x} \;\Rightarrow\; 10^{0.1} = \frac{(x+0.1\times0.1)}{x}$$

Solving this for x, $x = 3.86\times10^{-2}\,\text{mol}\cdot\text{L}^{-1}$ is found.

Ans. concentration of carbonic acid: $3.86\times10^{-2}\,\text{mol}\cdot\text{L}^{-1}$

"note" This problem is provided only for the purpose of understanding the buffer action of blood and, hence, the concentrations both used and found are different from actual ones.

Solution 4-9

The electrode reaction of hydrogen ion is expressed as eq(73).

$$\text{H}^+ + e \;\rightarrow\; \frac{1}{2}\text{H}_2 \qquad (73)$$

The free energy change associated with the above electrode reaction is given by eq(74) in terms of chemical potentials and eq(74') is derived after its rearrangement.

$$\Delta G=\frac{1}{2}(\mu^{\circ}_{\mathrm{H_2}}+RT\ln f_{\mathrm{H_2}})-\left\langle(\mu^{\circ}_{\mathrm{H^+}}+RT\ln a_{\mathrm{H^+}})+\mu_e\right\rangle \tag{74}$$

$$=\left(\frac{1}{2}\mu^{\circ}_{\mathrm{H_2}}-\mu^{\circ}_{\mathrm{H^+}}-\mu_e\right)+RT\ln\frac{f^2_{\mathrm{H_2}}}{a_{\mathrm{H^+}}}$$

$$=\Delta G^{\circ}+RT\ln\frac{f^2_{\mathrm{H_2}}}{a_{\mathrm{H^+}}} \tag{74'}$$

Here, a few of the relevant chemical potentials are defined as $\mu^{\circ}_{\mathrm{H_2}}=0$, $\mu^{\circ}_{\mathrm{H^+}}=0$ and $\mu_e=0$. The first one is based on the thermodynamic definition but the other two are newly defined specifically in relation to the electrode reaction. These definitions are not problematic because in the thermodynamical treatments of chemical reactions or equilibriums, only differences in thermodynamic quantities between the two specified states are significant and are dealt with.

When a change in the free energy is due to the change in the electrochemical energy, the electric potential E and the change in the free energy are related by eq(75) where n is the number of electrons involved in the electrode reaction (see **"note"**). Putting $n=1$ in the present case, eq(74') is rewritten as eq(76) by using E. Here, the fugacity of H_2 $f_{\mathrm{H_2}}$ is equated as $f_{\mathrm{H_2}}=1$.

$$\Delta G=-n\cdot F\cdot E \tag{75}$$

$$E=-\frac{RT}{F}\ln\frac{f^2_{\mathrm{H_2}}}{a_{\mathrm{H^+}}}=-\frac{RT}{F}\ln\frac{1}{a_{\mathrm{H^+}}}=-\frac{RT}{F}\ln\frac{1}{\gamma_{\mathrm{H^+}}[\mathrm{H^+}]} \tag{76}$$

$$\frac{E\times F}{RT}=2.303\log_{10}a_{\mathrm{H^+}}\Rightarrow \mathrm{pH}=-\log_{10}a_{\mathrm{H^+}}=-\frac{E\times F}{2.303RT}=-16.9_1E \tag{76'}$$

Substituting $\gamma_{\mathrm{H^+}}=1$ and $[\mathrm{H^+}]=5.0\times10^{-5}\,\mathrm{mol\cdot L^{-1}}$ into eq(76), the electric potential E at 25°C is found as 0.2543 V. Also, from eq(76'), a value for pH is found as 4.301. This value accords with the value (apart from an error associated with the calculation) obtained by substituting the concentration of hydrogen ion into the equation $\mathrm{pH}=-\log_{10}[\mathrm{H^+}]$ which is commonly used to define pH.

Correctly, the electric potential measured against a reference electrode is related, as is seen from eq(76), to the activity of hydrogen ion in solution. Therefore, the electric potential corresponding to the activity of $a_{\mathrm{H^+}}=0.9854\times5\times10^{-5}$ becomes 0.2547 V. The difference arises from $\ln\gamma_{\mathrm{H^+}}$ and the correct value for pH is obtained as 4.308. The higher the concentration of hydrogen ion is, the greater the activity coefficient deviates from unity. For this, enough care should be taken for relating the concentration of hydrogen ion with the pH value when its concentration is high.

Ans. for $\gamma_{\mathrm{H^+}}=1$, $E=0.2543$ V and pH = 4.301; for $\gamma_{\mathrm{H^+}}=0.985$, $E=0.2547$ V and pH = 4.308

"note" Upon writing an electrode reaction, if the stoichiometrical coefficients are normalized to the number of electrons n involved, that is n is always put equal to one, then you would not be lost in the relation between the electric potential and the number of electrons n.

Solution 4-10

(a) Since HCl is fully ionized in solution, the concentration of hydrogen ion in the HCl solution of 10^{-8} M must be $[\mathrm{H^+}]=10^{-8}\,\mathrm{mol\cdot L^{-1}}$ and, hence, $\mathrm{pH}=8$???? Why is the HCl solution *basic* in spite of the presence of acid HCl?? In the cases like this where the concentration of hydrogen ion is extremely low ($C_0 < 10^{-7}$), an approximate approach does not yield a correct answer.

In order to get the correct answer, the concentrations of the all ions present in the solution are expressed as eq(77) and, as has been practiced many times so far, the relations of eqs(78) and (79) relevant to those ions are combined with eq(77) to solve for y simultaneously. As a result, a quadratic equation of y (eq(80)) is obtained and its meaningful solution is given under eq(80).

$$\left.\begin{array}{lll} HCl & \rightarrow & \underset{y}{H^+} + \underset{C_0}{Cl^-} \\ H_2O & \rightleftarrows & \underset{y}{H^+} + \underset{z}{OH^-} \end{array}\right\} \quad (77)$$

$$[H^+] = [Cl^-] + [OH^-] \;\Rightarrow\; y = C_0 + z \qquad (78)$$

$$[H^+]\cdot[OH^-] = yz = y(y - C_0) = 10^{-14} \qquad (79)$$

$$y^2 - C_0 y - 10^{-14} = 0 \qquad (80)$$

$$y = \frac{C_0 + \sqrt{C_0^2 + 4\times10^{-14}}}{2} = \frac{10^{-8} + \sqrt{10^{-16} + 4\times10^{-14}}}{2} = 1.051\times10^{-7} \;\Rightarrow\; \text{pH} = 6.978$$

Putting $C_0 = 10^{-8}$, pH = 6.978 is obtained and it is seen from this that the solution is certainly *acidic*.

Ans. pH of 10^{-8} M HCl solution: $\text{pH} = 6.978$

(b)The dissociation constant of acetic acid is expressed by eq(81) with α as the degree of dissociation. Here, the concentration of acetic acid is so low ($C_0 < 10^{-7}$) that $[H^+]$ is treated as an unknown variable y instead of equating it to $C_0\alpha$.

$$K_a = \frac{[CH_3COO^-]\cdot[H^+]}{[CH_3COOH]} = \frac{C_0\alpha \times y}{C_0(1-\alpha)} = \frac{\alpha y}{(1-\alpha)} = 1.8\times10^{-5} \qquad (81)$$

From the balance of electric charges, eq(82) holds and eq(81) is rewritten as eq(81').

$$[H^+] = [CH_3COO^-] + [OH^-] \;\Rightarrow\; y = C_0\alpha + z \qquad (82)$$

$$K_a(1-\alpha) = \alpha y \;\Rightarrow\; \alpha = \frac{K_a}{y + K_a} \qquad (81')$$

Using these equations, the ion product of water, eq(83), can be rewritten as eq(83') or eq(83'').

$$[H^+]\cdot[OH^-] = yz = y(y - C_0\alpha) = 10^{-14} \qquad (83)$$

$$y\left(y - C_0\frac{K_a}{y + K_a}\right) = 10^{-14} \qquad (83')$$

$$y^3 + K_a y^2 - (C_0K_a + 10^{-14})y - 10^{-14}K_a = 0 \qquad (83'')$$

Substituting $C_0 = 10^{-8}$ and $K_a = 1.8\times10^{-5}$ into eq(83''), it is solved for y finding $y = 1.0509\times10^{-7}$ and $\text{pH} = 6.978$. In addition, the degree of dissociation is found from eq(81') as 0.9942, which is close to unity and, hence, the pH value obtained is the same as found for the HCl aqueous solution considered in (a).

Ans. pH of 10^{-8} M acetic acid solution: $\text{pH} = 6.978$

The degree of dissociation for acetic acid is a function of the concentration of C_0. Since eq(82) can be rewritten as eq(82'), y in this equation is eliminated by using eq(81) or (83), yielding eq(82''). Solving it for C_0, q(82'') is obtained, which relates the concentration C_0 with the degree of dissociation α.

$$y = C_0\alpha + \frac{10^{-14}}{y} \qquad (82')$$

$$\frac{K_a(1-\alpha)}{\alpha} = C_0\alpha + \frac{\alpha\times 10^{-14}}{K_a(1-\alpha)} \qquad (82'')$$

$$C_0 = \frac{K_a^2(1-\alpha)^2 - \alpha^2\times 10^{-14}}{K_a\alpha^2(1-\alpha)} \qquad (82''')$$

Employing *Parametric Plot* in *Mathematica*, α in eq(82''') is plotted against C_0 instead of the other way round. The result obtained is shown in Fig.10.

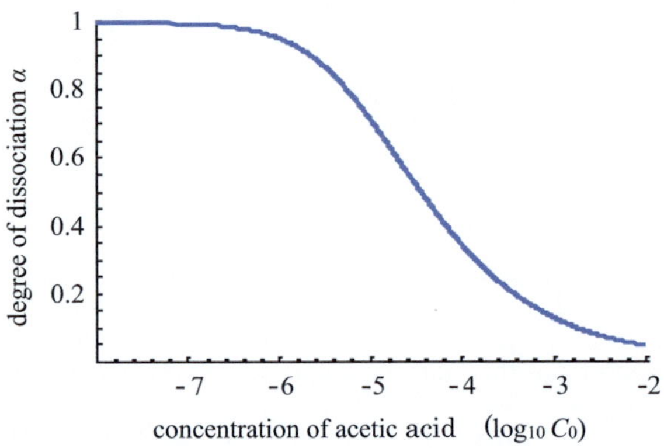

Fig.10 Variation of the degree of dissociation with the concentration of aqueous acetic acid.

When the degree of dissociation is related with the concentration of the acid, eq(84) is often used by neglecting the ionic dissociation of water and taking the dissociation equilibrium of acetic acid alone into account.

$$K_\alpha = \frac{[CH_3COO^-]\cdot[H^+]}{[CH_3COOH]} = \frac{C_0\alpha\times C_0\alpha}{C_0(1-\alpha)} = \frac{C_0\alpha^2}{(1-\alpha)} = 1.8\times10^{-3} \Rightarrow C_0\alpha^2 + \alpha\times1.8\times10^{-5} - 1.8\times10^{-5} = 0 \qquad (84)$$

Solving eq(84) for α, eq(85) is obtained to describe its relation with the concentration of acetic acid.

$$\alpha = \frac{-(1.8\times10^{-5}) + \sqrt{(1.8\times10^{-5})^2 + 4C_0(1.8\times10^{-5})}}{2C_0} \qquad (85)$$

In the low concentration range, the degree of dissociation given by eq(82''') (blue line) is compared with that of eq(85) (red line) in Fig.11. From this figure, it is seen that eq(85) gives too large a degree of dissociation

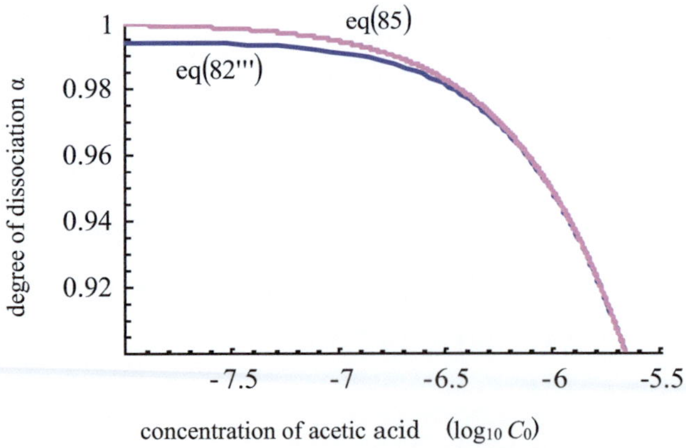

Fig.11 Comparison of eq(82''') with eq(85) when the degree of dissociation is varied with the concentration of aqueous acetic acid in the low concentration range.

in the low concentration range. The reason for this is evident because the hydrogen ion produced by the ionic dissociation of water participates in the dissociation equilibrium of acetic acid.

"note" *Parametric Plot* used for plotting curves shown in Figs.10 and 11 is explained here specifically. When α is to be plotted it is not necessary to solve eq(82''') for α. Generally speaking, when we have two functions $f(t)$ and $g(t)$, each of which is a function of a common variable *t*, *Mathematica* enables us to make a two-dimensional plot by equating $x = f(t)$ and $y = g(t)$, that is, to draw a figure expressed by $\{f(t),\ g(t)\}$. In the present case, by equating $x = \log_{10} C_0 = \log_{10}((82''')) \equiv f(\alpha)$ and $y = \alpha \equiv g(\alpha)$ and executing the command of *ParametricPlot* for an appropriate plot range, curves such as those given Figs.10 and 11 are obtained.

Solution 4-11

(a) Writing the concentrations of ions in the saturated AgCl solution as $[Ag^+] = [Cl^-] = x\ (\text{mol} \cdot \text{kg}^{-1})$, the solubility product is expressed by eq(86), from which $x = 1.26 \times 10^{-5}\,\text{mol} \cdot \text{kg}^{-1} (\approx \text{mol} \cdot \text{L}^{-1})$ is found as the saturated concentration.

$$K_{sp} = 1.6 \times 10^{-10} = x^2 \qquad (86)$$

The activity of each ion is assumed to be unity and the density of the solution is approximated by1.0.

Ans. $1.26 \times 10^{-5}\,\text{mol} \cdot \text{kg}^{-1} (\approx \text{mol} \cdot \text{L}^{-1})$

(b) The solubility product of Hg_2Cl_2 is expressed by $K_{sp} = [Hg^+]^2 \times [Cl^-]^2$. In a similar manner to (a), writing the concentration of each ion as $[Hg^+] = [Cl^-] = x\ (\text{mol} \cdot \text{kg}^{-1})$, from the solubility product expressed by eq(87), $x = 3.78 \times 10^{-5}\,\text{mol} \cdot \text{kg}^{-1} (\approx \text{mol} \cdot \text{L}^{-1})$ is found as the saturated concentration.

$$K_{sp} = 2.04 \times 10^{-18} = x^4 \qquad (87)$$

Ans. $3.78 \times 10^{-5}\,\text{mol} \cdot \text{kg}^{-1} (\approx \text{mol} \cdot \text{L}^{-1})$

(c) (i) approximate calculation taking the effect of common ion(s) alone into account

There exists Cl^- in the solution as a common ion and, hence, the solubility products of AgCl and Hg_2Cl_2 must be expressed by eq(86') and eq(87'), respectively.

$$\text{AgCl}:\ K_{sp} = 1.6 \times 10^{-10} = (x + 0.01) \times x \qquad (86')$$

$$\text{Hg}_2\text{Cl}_2:\ K_{sp} = 2.04 \times 10^{-18} = x^2 (x + 0.01)^2 \qquad (87')$$

Solving these equations, the saturated concentrations of AgCl and Hg_2Cl_2 are found as 1.60×10^{-8} $\text{mol} \cdot \text{L}^{-1}$ and $1.43 \times 10^{-7}\,\text{mol} \cdot \text{L}^{-1}$, respectively.

Ans. AgCl: $1.60 \times 10^{-8}\,\text{mol} \cdot \text{L}^{-1}$, Hg_2Cl_2: $1.43 \times 10^{-7}\,\text{mol} \cdot \text{L}^{-1}$

(ii) calculation taking the activity coefficients (ionic strength) into account (see **problem 4-18**)

Neglecting the concentrations of almost insoluble salts, the ionic strength *I* of the sodium chloride solution at the concentration of $0.1\ \text{mol} \cdot \text{L}^{-1}$ is calculated as eq(88).

$$I = \frac{1}{2} \sum_i^{\text{all ions}} m_i Z_i^2 = \frac{1}{2} (0.01 \times 1^2 + 0.01 \times 1^2) = 0.01 \qquad (88)$$

The activity coefficient of the ions is calculated from the *Debye-Hückel* limiting equation (eq(89)) (see

solution 4-18) as below, where it is assumed the all ions have the same mean activity coefficient.

$$\ln\gamma_{\pm} = -A|Z_{+}Z_{-}|I^{1/2} = -1.172\sqrt{0.01} \Rightarrow \gamma_{\pm} = 0.889 \qquad (89)$$

Now, expressing each solubility product by using the activity coefficient thus calculated, the saturated concentrations of the salts are calculated from eqs(86'') and (87'') as $2.02\times10^{-8}\,\text{mol}\cdot\text{L}^{-1}$ and 1.81×10^{-7} $\text{mol}\cdot\text{L}^{-1}$, respectively.

$$\text{AgCl}: \ K_{sp} = 1.6\times10^{-10} = \gamma_{\pm}^{2}\times(x+0.01)\times x \qquad (86'')$$

$$\text{Hg}_2\text{Cl}_2: \ K_{sp} = 2.04\times10^{-18} = \gamma_{\pm}^{4}x^{2}(x+0.01)^{2} \qquad (87'')$$

Comparing these values with those obtained as an approximation in (i), it is seen that the correction for the ionic strength is large enough to be recognized.

Ans. AgCl: $2.02\times10^{-8}\,\text{mol}\cdot\text{L}^{-1}$, Hg_2Cl_2: $1.81\times10^{-7}\,\text{mol}\cdot\text{L}^{-1}$

"note" Definition of the solubility product.

The solubility is usually defined by the mole number of a solute dissolved in 1000 g of a solvent (mole/weight concentration). Because of this, to express the solubility product, the concentrations used are in the unit of mole/weight, i.e., mola<u>l</u>ity (<u>l</u>-mola<u>l</u>ity but not r-molarity). However, sometime the unit of mole/volume, i.e., mola<u>r</u>ity (r-molarity) is used instead. As is generally the equilibrium constant expressed as dimensionless, so is the solubility product because the concentrations are normalized to a standard concentration. (For daily use, the solubility is often given by the gram number of a solute dissolved in 100g of a solvent.)

Solution 4-12a

The equilibrium constant for the dissolution of a salt $\text{MX}(s)$ expressed as eq(90) is given by eq(91).

$$\text{MX}(s) \rightleftarrows \text{M}^{+}(aq) + \text{X}^{-}(aq) \qquad (90)$$

$$K_{sp} = \frac{a_{\text{M}^{+}(aq)}\times a_{\text{X}^{-}(aq)}}{a_{\text{MX(s)}}} \approx [\text{M}^{+}(aq)]\cdot[\text{X}^{-}(aq)] \qquad (91)$$

The activity of a pure salt can always be put equal to unity and, hence, the equilibrium constant is equivalent to the solubility product. Therefore, it is related to the standard free energy change ΔG° of the equilibrium reaction in eq(90) and is given by eq(92).

$$\Delta G^{\circ} = -RT\ln K_{sp} \qquad (92)$$

On the other hand, the equilibrium reaction in eq(90) can be expressed by separating it into two electrode reactions given by eqs(93) and (94). As a result, the free energy of the dissolution reaction ΔG_1 is given by eq(95).

$$\text{M}^{+}(aq) + e^{-} \rightleftarrows \text{M}(s) \qquad (93) \qquad \Delta G_4$$

$$\text{MX(s)} + e^{-} \rightleftarrows \text{M}(s) + \text{X}^{-}(aq) \qquad (94) \qquad \Delta G_5$$

$$\Delta G_1 = \Delta G_5 - \Delta G_4 \qquad (95)$$

The molar free energy changes for the reactions in eq(93) and eq(94) can be expressed as eq(96) and eq(97), respectively by using the sandard potential E° for each reaction. Here, F is *Faraday*'s constant ($F = N_{A}e$).

(See *Nernst*'s equation and **"note"**).

$$\Delta G_4 = -N_A eE_4^\circ + RT\ln\frac{a_{M(s)}}{[M^+]} = -FE_4^\circ + RT\ln\frac{a_{M(s)}}{[M^+]} \quad (96)$$

$$\Delta G_5 = -N_A eE_5^\circ + RT\ln\frac{a_{M(s)}\cdot[X^-]}{a_{MX(s)}} = -FE_5^\circ + RT\ln\frac{a_{M(s)}\cdot[X^-]}{a_{MX(s)}} \quad (97)$$

Substituting eqs(96) and (97) into eq(95) and using that $\Delta G_1 = 0$ in equilibrium , eq(98) , which relates the solu- bility product with the standard potentials, is derived as shown below.

$$\Delta G_1 = 0 = -FE_5^\circ + RT\ln\frac{a_{M(s)}\cdot[X^-]}{a_{MX(s)}} + FE_4^\circ + RT\ln\frac{[M^+]}{a_{M(s)}}$$

$$F\cdot(E_5^\circ - E_4^\circ) = RT\ln\frac{[M^+]\cdot[X^-]}{a_{MX(s)}} \approx RT\ln K_{sp}$$

$$K_{sp} = \exp\left\{\frac{F\cdot(E_5^\circ - E_4^\circ)}{RT}\right\} \quad (98)$$

For the silver halides in the problem, their solubility products are calculated from eq(98) by using the standard potential $E_4^\circ = 0.800$ V for metal silver Ag and E_5°'s for the salts AgF: 0.779 , AgCl: 0.222 , AgBr: 0.071 and AgI: -0.152 (unit in V). The obtained values are given in **Ans.**

Ans. AgF: 0.44, AgCl: 1.7×10^{-10}, AgBr: 4.7×10^{-13} and AgI: 7.9×10^{-17}

"note 1" *Nernst*'s equation (*Walter Nernst* Nobel Prize in Chemistry in 1920), which is so often employed in electrochemistry, expresses the equilibrium in eq(93) as eq(99) using the electric potential *E*.

$$E = E^\circ - \frac{RT}{nF}\ln\frac{a_{M(s)}}{a_{M^+}} \quad (99)$$

Although there are a few ways to derive this equation, it is important to understand it on the basis of the definition of the free energy. For this, if it is looked at from the following viewpoint, nothing special is associated with the electric cells or electrode reactions. The point is to understand that the (free) energy of an electron at the electric potential *E* is given by $-eE$ (electric potential energy). Consequently, the equations from (93) to (95) used above are expressed by using the chemical potentials (free energy) as follows.

$$\Delta G_4 = \mu(M(s)) - (-N_A eE) - \mu(M^+(aq)) \quad (96')$$
$$= \mu^\circ(M(s)) + RT\ln a_{M(s)} - \mu^\circ(M^+(aq)) - RT\ln a_{M^+} - (-N_A eE)$$
$$= \mu^\circ(M(s)) - \mu^\circ(M^+(aq)) - RT\ln a_{M^+} + N_A eE$$

$$\Delta G_5 = \mu(M(s)) + \mu(X^-(aq)) - (-N_A eE) - \mu(MX(s)) \quad (97')$$
$$= \mu^\circ(M(s)) + RT\ln a_{M(s)} + \mu^\circ(X^-(aq)) + RT\ln a_{X^-}$$
$$- \mu^\circ(MX(s)) - RT\ln a_{MX(s)} - (-N_A eE)$$
$$= \mu^\circ(M(s)) + \mu^\circ(X^-(aq)) + RT\ln a_{X^-} - \mu^\circ(MX(s)) + N_A eE$$

Then, the free energy change of the reaction in eq(90) can be expressed as eq(90') from eqs(96') and (97') and, of course, it is identical to eq(90) which has nothing to do with the electrode reactions, though.

$$\Delta G_1 = \Delta G_5 - \Delta G_4 = \mu(X^-(aq)) + \mu(M^+(aq)) - \mu(MX(s)) \quad (90')$$
$$= \mu^\circ(X^-(aq)) + \mu^\circ(M^+(aq)) - \mu^\circ(MX(s)) + RT\ln a_{X^-} + RT\ln a_{M^+}$$

Here, the standard electric reduction potential is defined by the electrode potential at which all of the reactants and products of an electrode reaction are in their standard states and the electrode reaction is in equilibrium. Then, taking the activities of the substances involved as unity, the standard electric reduction potentials become equal to the electric potentials at which the electron is found when the free energy changes in the relevant reduction reactions (eqs(96') and (97')) are zero. More specifically, they are given by eqs(96'') and (97'').

$$\mu^{\circ}(\mathrm{M}(s))-(-N_A eE_4^{\circ})-\mu^{\circ}(\mathrm{M}^+(aq))=0 \;\Rightarrow\; -N_A eE_4^{\circ}=\mu^{\circ}(\mathrm{M}(s))-\mu^{\circ}(\mathrm{M}^+(aq)) \qquad (96'')$$

$$\mu^{\circ}(\mathrm{M}(s))+\mu^{\circ}(\mathrm{X}^-(aq))-(-N_A eE_5^{\circ})-\mu^{\circ}(\mathrm{MX}(s))=0$$
$$\Rightarrow\; -N_A eE_5^{\circ}=\mu^{\circ}(\mathrm{M}(s))+\mu^{\circ}(\mathrm{X}^-(aq))-\mu^{\circ}(\mathrm{MX}(s)) \qquad (97'')$$

From these, eq(100) is derived.

$$\mu^{\circ}(\mathrm{X}^-(aq))+\mu^{\circ}(\mathrm{M}^+(aq))-\mu^{\circ}(\mathrm{MX}(s))=-N_A eE_5^{\circ}+N_A eE_4^{\circ}=-F\cdot(E_5^{\circ}-E_4^{\circ}) \qquad (100)$$

Substituting this into eq(95), eq(98) derived above is obtained.

$$\begin{aligned}\Delta G_1&=\Delta G_5-\Delta G_4\\&=\mu^{\circ}(\mathrm{X}^-(aq))+\mu^{\circ}(\mathrm{M}^+(aq))-\mu^{\circ}(\mathrm{MX}(s))+RT\ln a_{X^-}+RT\ln a_{M^+}\\&=-F\cdot(E_5^{\circ}-E_4^{\circ})+RT\ln a_{X^-}a_{M^+}\\&=-F\cdot(E_5^{\circ}-E_4^{\circ})+RT\ln K_{sp}=0 \;\Rightarrow\; K_{sp}=\exp\left(\frac{F\cdot(E_5^{\circ}-E_4^{\circ})}{RT}\right)\end{aligned} \qquad (98)$$

Rather a lengthy explanation is given above, but it is important to put the concepts of electrode reduction and oxidation potentials or electrode reactions in order in terms of thermodynamics (see **solution 4-49** for the definition of electrochemical potential).

"note 2" The ratio of the solubilities of the two silver halides each of which is composed of a different halide ion (X^- or Y^-) is given by eq(101) as is readily derived from eq(98).

$$\frac{K_{sp}(\mathrm{AgX})}{K_{sp}(\mathrm{AgY})}=\exp\left(\frac{F(E_5^{\circ}(\mathrm{AgX})-E_5^{\circ}(\mathrm{AgY}))}{RT}\right) \qquad (101)$$

Substituting the standard reduction potential of 0.564 V for $AgNO_3$ into the above equation, the ratio of the solubility against AgF is calculated as below and, hence, it is seen that $AgNO_3$ dissolves in water to some extent.

$$\frac{K_{sp}(\mathrm{AgNO_3})}{K_{sp}(\mathrm{AgF})}=\exp\left(\frac{F\cdot(0.564-0.779)}{RT}\right)=2.3\times10^{-4}$$

In this manner, it is possible to estimate solubilities of silver salts from their standard reduction potentials.

Solution 4-12b

The saturated concentrations calculated are as follows: AgF : 0.68, AgCl : 1.3×10^{-5}, AgBr : 6.9×10^{-7} and AgI : 9.3×10^{-8} in the units of $\mathrm{mol\cdot L^{-1}}$. Out of these four silver halides, only AgF is practically soluble in water. It is seen from the values given in the table below that the ionic radius of fluoride ion F^- is smallest among the halide ions and its solvation energy is remarkably large, reflecting the largest number of solvated water molecules to the anion. In another word, the fluoride ion of AgF tends most readily to be a solvated ion in water, resulting in the largest solubility in water among the silver halides.

These features are confirmed based on the relation between the standard free energy change ΔG°_{sol}(AgX) for the dissolution equilibrium of a silver halide (eq(102)) and its solubility product K_{sp}(AgX) (eq(103)).

$$\mathrm{AgX}(s) \rightleftarrows \mathrm{Ag}^+(aq)+\mathrm{X}^-(aq) \qquad (102) \qquad \Delta G^\circ_{sol}(\mathrm{AgX})$$

Table Properties of silver halides and halide ions.

	standard free energy of formation ΔG°_f(AgX)		solvation energy $\Delta G^\circ_{sol}(\mathrm{X}^-)$	ionic radius /pm	number of solvated water molecules
AgF	-257.5[a)]	F^-	-278.79	133	4 ± 1
AgCl	-109.79	Cl^-	-131.23	181	2 ± 1
AgBr	-96.90	Br^-	-103.96	196	2 ± 1
AgI	-69.86[b)]	I^-	-51.57	220	1 ± 1
$AgNO_3$	-33.41	Ag^+(aq)	77.11		
		NO_3^-	-108.74		

a), b) Calculated from the standard reduction potentials.

$$\Delta G^\circ_{sol} = -RT\ \ln\frac{a_{\mathrm{Ag}^+}\cdot a_{\mathrm{X}^-}}{a_{\mathrm{AgX}}}$$
$$= -RT\ \ln[\mathrm{Ag}^+]\cdot[\mathrm{X}^-] = -RT\ \ln K_{sp}(\mathrm{AgX}) \qquad (103)$$

The standard free energy of formation of AgX, ΔG°_f(AgX), is given by eq(104) and the standard free energies of formation of silver ion $\Delta G^\circ_{sol}(\mathrm{Ag}^+)$ and halide ions $\Delta G^\circ_{sol}(\mathrm{X}^-)$ are given by eq(105) and eq(106), respectively.

$$\mathrm{Ag}(s)+\frac{1}{2}\mathrm{X}_2(g) \rightarrow \mathrm{AgX}(s) \qquad (104) \qquad \Delta G^\circ_f(\mathrm{AgX})$$

$$\mathrm{Ag}(s)\ -e \rightarrow \rightleftarrows \mathrm{Ag}^+(aq) \qquad (105) \qquad \Delta G^\circ_{sol}(\mathrm{Ag}^+)$$

$$\frac{1}{2}\mathrm{X}_2(g)\ +e \rightarrow \mathrm{X}^-(aq) \qquad (106) \qquad \Delta G^\circ_{sol}(\mathrm{X}^-)$$

The chemical reaction in eq(102) is expressed by a combination of the three reactions above as eq(105) + eq(106) − eq(104). Correspondingly, the free energy change for eq(102) is found from eq(107).

$$\Delta G^\circ_{sol}(\mathrm{AgX}) = \Delta G^\circ_{sol}(\mathrm{Ag}^+) + \Delta G^\circ_{sol}(\mathrm{X}^-) - \Delta G^\circ_f(\mathrm{AgX}) \qquad (107)$$

Substituting the relevant values given in Table into eq(107), the values for ΔG°_{sol}(AgX) are calculated as below. The found values are in close agreement with those given in the problem.

$$K_{sp} = \exp\left\{-\frac{\Delta G^\circ_{sol}(\mathrm{AgX})}{RT}\right\}$$

$$\Delta G^\circ_{sol}(\mathrm{AgCl}) = 55.67\ \mathrm{kJ\cdot mol^{-1}} \Rightarrow K_{sp}(\mathrm{AgCl}) = 1.73\times10^{-10}$$

$$\Delta G^\circ_{sol}(\mathrm{AgBr}) = 70.05\ \mathrm{kJ\cdot mol^{-1}} \Rightarrow K_{sp}(\mathrm{AgBr}) = 5.19\times10^{-13}$$

From these results, it is found that when ΔG°_{sol}(AgX) is positive, K_{sp} assumes a small value and when it is negative, then K_{sp} assumes a large value. The fact that silver nitrate is relatively soluble in water can be explained likewise (see **"note 2"** in the preceding solution).

Ans. solubility of AgX : See text and Table.

Solution 4-13

(a) The reduction reactions and standard reduction potentials of silver ion Ag^+ and silver halide AgCl are given by eq(108) and eq(109), respectively.

$$Ag^+(aq)+e^- \rightarrow Ag(s) \qquad E_1 = 0.800\ \mathrm{V} \qquad (108)$$

$$AgCl + e^- \rightarrow Ag(s)+Cl^-(aq) \qquad E_2 = 0.222\ \mathrm{V} \qquad (109)$$

Subtracting eq(108) from eq(109), the dissolution reaction of silver chloride is obtained as below.

$$AgCl \rightarrow Ag^+(aq)+Cl^-(aq) \qquad \Delta G^\circ_{sol} \qquad \text{(dissolution reaction)}$$

Consequently, the standard free energy change of this reaction ΔG°_{sol} is given by $-F\times(E_2-E_1)$ and the equilibrium constant, which is equivalent to the solubility product K_{sp} in the present case, is given by eq(110),

$$\ln K_{sp} = -\frac{\Delta G^\circ_{sol}}{RT}$$

$$= \frac{F\times(E_2-E_1)}{1\times RT} = -\frac{96490\times 0.572}{8.31\times 298} = \ \Rightarrow\ K_{sp} = 1.66\times 10^{-10} \qquad (110)$$

from which $K_{sp} = 1.66\times 10^{-10}$ is obtained.

Ans. solubility product: $K_{sp} = 1.66\times 10^{-10}$

(b) The standard free energies of formation of Ag^+ and Cl^- in water are given by eq(111) and eq(112), respectively, and that of silver halide is given by eq(113).

$$Ag(s) \rightarrow Ag^+(aq)+e^- \qquad \Delta G^\circ_f(Ag^+) = 77.11\ \mathrm{kJ\cdot mol^{-1}} \qquad (111)$$

$$\frac{1}{2}Cl_2(g)+e^- \rightarrow Cl^-(aq) \qquad \Delta G^\circ_f(Cl^-) = -131.23\ \mathrm{kJ\cdot mol^{-1}} \qquad (112)$$

$$Ag(s)+\frac{1}{2}Cl_2(g) \rightarrow AgCl(s) \qquad \Delta G^\circ_f(AgCl) = -109.79\ \mathrm{kJ\cdot mol^{-1}} \qquad (113)$$

Then, the dissolution equilibrium reaction shown below is given by a combination of eq(111)+eq(112)−eq(113).

$$AgCl(s) \rightarrow Ag^+(aq)+Cl^-(aq) \qquad \Delta G^\circ_{sol} \qquad \text{(dissolution reaction)}$$

Consequently, the standard free energy change of this reaction is expressed as eq(114) and the solubility product is given by eq(115).

$$\Delta G^\circ_{sol} = \Delta G^\circ_f(Ag^+)+\Delta G^\circ_f(Cl^-)-\Delta G^\circ_f(AgCl) = 55.67\ \mathrm{kJ\cdot mol^{-1}} \qquad (114)$$

$$\ln K_{sp} = -\frac{\Delta G^\circ_{sol}}{RT} = -\frac{55.67\times 10^3}{8.31\times 298} \ \Rightarrow\ K_{sp} = 1.73\times 10^{-10} \qquad (115)$$

Ans. solubility product: $K_{sp} = 1.73\times 10^{-10}$

Solution 4-14

Problems of this kind can be solved in a similar manner as employed for obtaining acid-base titration curves. The dissolution equilibrium of AgCl is expressed as eq(116). The concentrations of ions in the solution are related by equations from (117*a*) to (117*c*).

$$\mathrm{AgCl} \rightleftarrows \mathrm{Ag^+} + \mathrm{Cl^-}, \qquad K_{sp} = 1.6\times10^{-10} \qquad (116)$$

$$[\mathrm{Na^+}]+[\mathrm{Ag^+}]=[\mathrm{Cl^-}]+[\mathrm{NO_3^-}] \qquad (117a)$$

$$[\mathrm{Ag^+}]\cdot[\mathrm{Cl^-}]=1.6\times10^{-10} \qquad (117b)$$

$$\mathrm{NO_3^-}]=\frac{0.10\times x}{10+x}, \quad [\mathrm{Na^+}]=\frac{0.10\times10}{10+x} \qquad (117c)$$

Denoting $[\mathrm{Ag^+}]$ and $[\mathrm{Cl^-}]$ y and z, respectively as in eq(118),

$$[\mathrm{Ag^+}]=y、\quad [\mathrm{Cl^-}]=z \qquad (118)$$

the concentration of each ion is substituted into eq(117*a*) to find eq(119).

$$\frac{0.1\times10}{10+x}+y=z+\frac{0.1\times x}{10+x} \Rightarrow y=z+\frac{0.1\times(x-10)}{10+x} \qquad (119)$$

Substituting this into eq(117*b*), eq(120) is obtained and is solved for z. Thus the concentration $[\mathrm{Cl^-}]=z$ is expressed as a function of x. The concentration $[\mathrm{Ag^+}]=y$ is found by substituting eq(121) into eq(119).

$$z\left\{z+\frac{0.1\times(x-10)}{10+x}\right\}=1.6\times10^{-10} \qquad (120)$$

$$z^2+\frac{0.1\times(x-10)z}{10+x}-1.6\times10^{-10}=0$$

$$z=\frac{-0.1\times(x-10)+\sqrt{\{0.1\times(x-10)\}^2+4(10+x)^2\times1.6\times10^{-10}}}{2(10+x)} \qquad (121)$$

Common logarithmic values of these concentrations are plotted in Fig.12 against x.

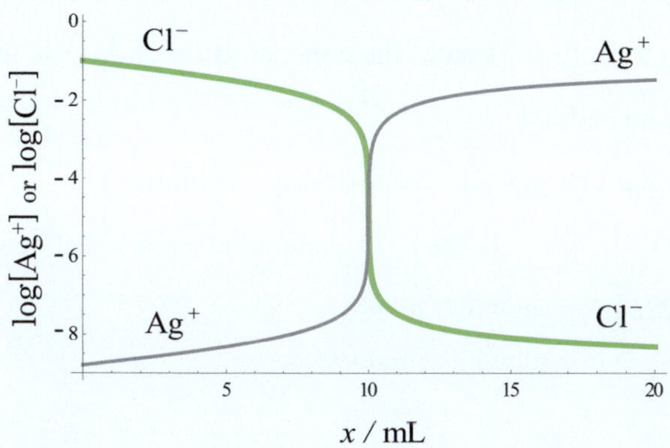

Fig.12 Titration curve of NaCl solution with AgNO3 solution.
The curves resemble those obtained for the acid-base titration.

The concentrations of the ions at the equivalence point are found as $z=y=1.26\times10^{-5}(\mathrm{ML^{-1}})=\sqrt{K_{sp}}$. As is seen from Fig.12, the concentration $[\mathrm{Ag^+}]$ varies significantly ($dC/C=d\log C$) with x. One of the methods to monitor a change in the concentration of ion is to measure the electric potential difference by

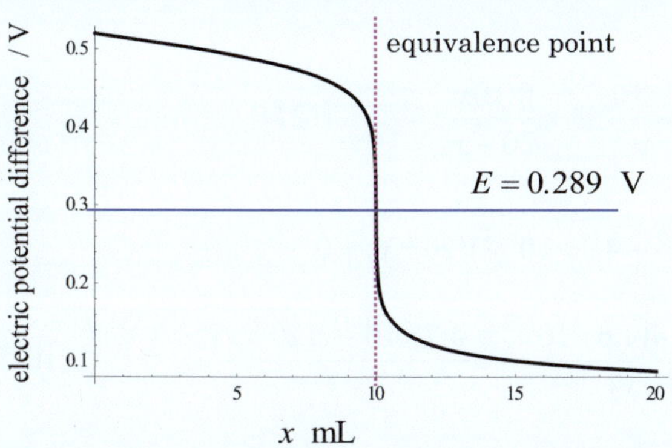

Fig.13 Electric potential difference and amount of AgNO3 solution added.

regarding the solution being titrated as a part of concentration cell (potentiometric titration). The electric potential difference is given by $-59.1\times10^{-3}(\text{V})\log([\text{Ag}^+(x)]/[\text{Ag}^+]_0)$ and, putting $[\text{Ag}^+]_0 = 1.0\,\text{mol}\cdot\text{L}^{-1}$, the electric potential difference to be measured is plotted in Fig.13 against x.

Alternatively, an appropriate indicator can be used with which an end point of the reaction is detected when the concentration of the silver ion exceeds a certain value, e.g., $[\text{Ag}^+] > 10^{-5}\text{mol}\cdot\text{L}^{-1}$. One such method is *Fajans*'s titration and the indicator employed is called adsorption indicator. Passing over the equivalence point, the surface of solid AgCl precipitated as colloidal particles starts to be covered with the excess Ag^+ ion, with which the indicator (In) fluoresceine reacts turning its color from fluorescent yellow to pale pink. The equivalence point is at the point where the pink color stays without disappearing for the first time.

$$\text{Ag}^+ + \text{In} \rightleftarrows \text{AgIn}^+$$

$$\frac{[\text{AgIn}^+]}{[\text{Ag}^+]\cdot[\text{In}]} = K \Rightarrow [\text{AgIn}^+] = K\times[\text{Ag}^+]\cdot[\text{In}] \approx K\times10^{-5}\times10^{-4} > 10^{-4}$$

$$\approx \text{(detection limit of color))}$$

$$\therefore K > 10^5 \quad \text{(magnitude of equilibrium constant)}$$

Ans. eqs(119) and (121). Equivalence point is detected by potentiometric or precipitation titration.

Solution 4-15

When the concentration of Ag^+ in the solution is known, the concentrations of Cl^- and Br^- are found readily from their solubility products with Ag^+. Therefore, to start with, the concentration of Ag^+ is found as a function of the amount x of the $AgNO_3$ solution (titrant) added.

(i) conventional approach (to use the balance of electrical charges (electroneutrality of solution))

$$[\text{Ag}^+]+[\text{Na}^+]=[\text{Cl}^-]+[\text{Br}^-]+[\text{NO}_3^{\ -}] \qquad (122a) \quad \text{balance of electrical charges}$$

$$[\text{Ag}^+][\text{Cl}^-]=1.6\times10^{-10} \qquad (122b) \quad \text{solubility product}$$

$$[\text{Ag}^+][\text{Br}^-]=4.7\times10^{-13} \qquad (122c) \quad \text{solubility product}$$

The concentration of each ion is given by eq(123).

$$\left.\begin{aligned} &[\text{Na}^+]=\frac{0.10\times20}{20+x}, \quad [\text{NO}_3^{\ -}]=\frac{0.10x}{20+x} \\ &[\text{Ag}^+]=y, \quad [\text{Cl}^-]=u, \quad [\text{Br}^-]=v \end{aligned}\right\} \qquad (123)$$

Substituting these into eq(122), eq(124) is derived,

$$\begin{aligned} y+\frac{0.10\times20}{20+x} &= u+v+\frac{0.10x}{20+x} \\ &= \frac{1.6\times10^{-10}}{y}+\frac{4.7\times10^{-13}}{y}+\frac{0.10x}{20+x} \end{aligned} \qquad (124)$$

which is solved for y to obtain eq(125).

$$y^2(20+x)+(2.0-0.10x)y-(1.6\times10^{-10}+4.7\times10^{-13})(20+x)=0$$

$$y=\frac{-(2.0-0.10x)+\sqrt{(2.0-0.10x)^2+4(1.6\times10^{-10}+4.7\times10^{-13})(20+x)^2}}{2(20+x)} \qquad (125)$$

Eq(125) thus obtained expresses the concentration of $[Ag^+] = y$ as a function of x mL of the $AgNO_3$ titrant added to the solution being titrated. In Fig.14, not only eq(125) but also the concentrations of Cl^- and Br^- are plotted against x. As is clear from eqs(122*b*) and (122*c*), the ratio $[Cl^-]/[Br^-]$ is equal to the ratio of the solubility products and, hence, it remains constant ($1.6\times10^{-10}/4.7\times10^{-13}$) all through the titration.

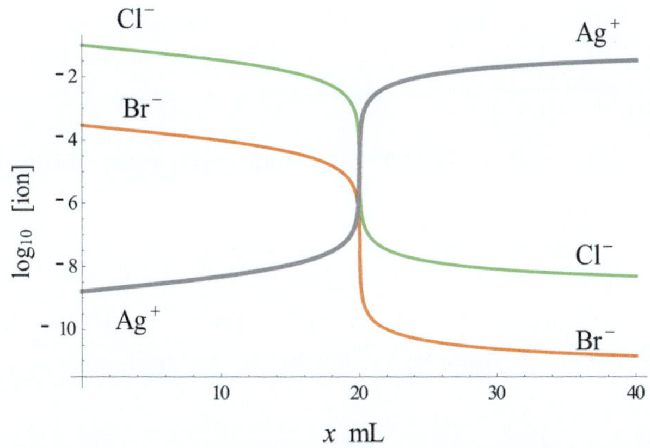

Fig.14 Titration of a mixed solution of NaCl-NaBr with the $AgNO_3$ solution. Relation between the concentrations of the ions and the amount of x mL of the $AgNO_3$ solution added. Note that $[Cl^-]/[Br^-] = \text{constant} = 1.6\times10^{-10}/4.7\times10^{-13}$.

For a comparison, when the three solutions, containing either NaCl or NaBr alone and both, are titrated with the $AgNO_3$ solution, how the concentration of Ag^+ varies with x is shown in Fig.15. Even though the solubility product of AgCl differs from that of AgBr nearly 3000 times, only the single equivalence point can be detected for any solution of the three. This shows that it is not possible to measure the concentrations of Cl^- and Br^- ions separately in the solution containing both ions, making a remarkable contrast with the acid-base titration,

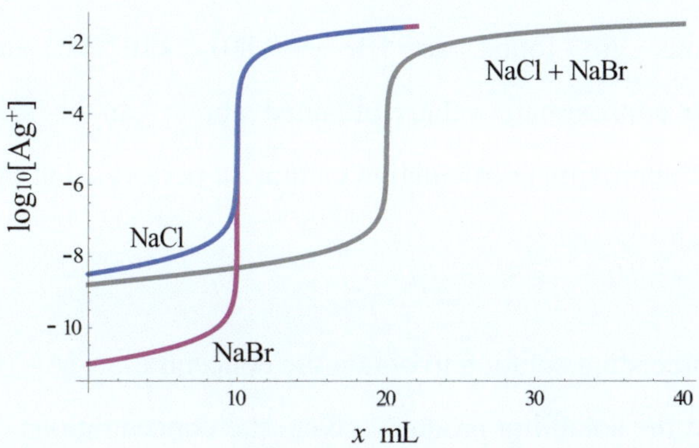

Fig.15 Comparison of the precipitation titration curves for three different solutions titrated with the $AgNO_3$ solution.
—— : NaCl,
—— : NaBr,
—— : NaCl+NaBr.

for which two or three equivalence points are usually detected when the dissociation constants differ from each other significantly (see, e.g., **solution 4-6**). The reason for this difference is that the former is based on the solid-solution equilibrium but the latter is the equilibrium in a homogeneous solution.

Ans. eq(125) and Figs.14 and 15

(ii) simple approximate calculation

As an approximation, it is assumed that NaBr (smaller solubility product) alone reacts with $AgNO_3$. Denoting the amount of the $AgNO_3$ solution added x mL, and putting the initial concentrations of the sodium halides $[NaCl]_0 = 0.05\times0.02 = 10^{-3}\,\text{mol}$, $[NaBr]_0 = 10^{-3}\,\text{mol}$, the concentrations of Br^- and Cl^- ions are expressed approximately by eqs(126) and (127), respectively.

$$\frac{[\text{NaBr}]_0 - 0.10\times x\times 10^{-3}}{(20+x)\times 10^{-3}} = [\text{Br}^-] \qquad (126)$$

$$\frac{[\text{NaCl}]_0}{(20+x)\times 10^{-3}} = [\text{Cl}^-] \qquad (127)$$

The concentrations of these ions, however, cannot take arbitrary values and they are restricted by the ratio of the solubility products as is given by eq(128).

$$\frac{[\text{Ag}^+][\text{Br}^-]}{[\text{Ag}^+][\text{Cl}^-]} = \frac{K_{sp}(\text{AgBr})}{K_{sp}(\text{AgCl})} = \frac{4.7\times 10^{-13}}{1.6\times 10^{-10}} = \frac{[\text{Br}^-]}{[\text{Cl}^-]} \qquad (128)$$

Substituting eqs(126) and (127) into eq(128) to solve for x, eq(129) is obtained going through the calculations below.

$$\frac{[\text{Br}^-]}{[\text{Cl}^-]} = \frac{\dfrac{[\text{NaBr}]_0 - 0.10\times x\times 10^{-3}}{(20+x)\times 10^{-3}}}{\dfrac{[\text{NaCl}]_0}{(20+x)\times 10^{-3}}} = \frac{[\text{NaBr}]_0 - 0.10\times x\times 10^{-3}}{[\text{NaCl}]_0} = \frac{10^{-3} - x\times 10^{-4}}{10^{-3}} = \frac{4.7}{1.6}\times 10^{-3}$$

$$1 - 0.1x = 10^{-3}\times\frac{4.7}{1.6} \Rightarrow x = 10 - 10^{-2}\times\frac{4.7}{1.6} = 9.970_{63}\text{ mL} \qquad (129)$$

From this, approximate concentrations of Br^- and Cl^- are found as eq(130) and eq(131), respectively.

$$[\text{Br}^-] = \frac{10^{-3} - 0.10\times x\times 10^{-3}}{(20+x)\times 10^{-3}} = 9.80\times 10^{-5}\text{M} \qquad (130)$$

$$[\text{Cl}^-] = \frac{[\text{Br}^-]\times 1.6\times 10^3}{4.7} = 0.0336\text{ M} \qquad (131)$$

On the other hand, $x = 9.970_{63}$ mL (eq(129)) is substituted into eq(125) to find the concentration of Ag^+, from which the concentrations of the halide ions are found as $[\text{Br}^-] = 9.801_{26}\times 10^{-5}\text{M}$ and $[\text{Cl}^-] = 0.0333_{66}$ M. Compared with these values, the approximate values obtained above can be said reasonably accurate, though it is evident that the present approximate calculation cannot be performed at the exact equivalence point at $x = 10.00$ mL.

Solution 4-16

Substituting $x = 10.00$ mL into eq(125) given in the preceding solution to obtain the concentration $[\text{Ag}^+]$, $[\text{Ag}^+] = 4.81\times 10^{-9}\text{M}$ is found. Based on this value and the solubility products given, the concentrations of the halide ions are calculated as $[\text{Cl}^-] = 3.324\times 10^{-2}\text{M}$ and $[\text{Br}^-] = 9.763\times 10^{-5}\text{M}$. Since the total amount of each of chlorine and bromine atoms is conserved, eqs(132) and (133) are satisfied, which are rewritten as eqs(132') and (133') to find the amounts of the silver halides in 30.00 mL of the solution. They are found as below.

$$[\text{Cl}^-] + [\text{AgCl}] = \frac{20\times 0.05}{30} \quad (132) \Rightarrow [\text{AgCl}] = \frac{20\times 0.05}{30} - [\text{Cl}^-] \qquad (132')$$

$$[\text{Br}^-] + [\text{AgBr}] = \frac{20\times 0.05}{30} \quad (133) \Rightarrow [\text{AgBr}] = \frac{20\times 0.05}{30} - [\text{Br}^-] \qquad (133')$$

$$[\text{AgCl}] = 9.33\times 10^{-5}\text{M} \Rightarrow 9.33\times 10^{-5}\times 58.45\times 0.030 = 1.6\times 10^{-4}\text{g}$$

$$[\text{AgBr}] = 3.32\times 10^{-2}\text{M} \Rightarrow 3.32\times 10^{-2}\times 102.9\times 0.030 = 1.02\times 10^{-1}\text{g}$$

As is seen in Fig.15 given in the preceding solution, the titration curve showing $\log[Ag^+]$ vs. x plot reveals neither inflection nor steep change at the first equivalence point. However, from the above calculation, it is understood that the precipitates formed before this hidden first equivalence point are mostly ascribed to AgBr.

Ans. AgCl : 1.6×10^{-4}g, AgBr : 1.02×10^{-1}g

Solution 4-17

(a) When the sea water is assumed to behave ideally in the given concentration range, the magnitude of the freezing point depression is proportional to the twice of the molality m (same value for Na^+ and Cl^-) with the proportional constant of K_f. Therefore, it is given by $\Delta T_m / K = 1.853 \times m \times 2$ (depressed temperature). The calculated values are listed in the table below and plotted against the concentration in Fig.16 below.

m / mol · kg^{-1}	0.01	0.02	0.05	0.1	0.20	0.5
ΔT_m / K (observed)	0.0361	0.0714	0.1758	0.3470	0.6850	1.677
ΔT_m / K (calculated)	0.0371	0.0741	0.1853	0.3706	0.7412	1.853

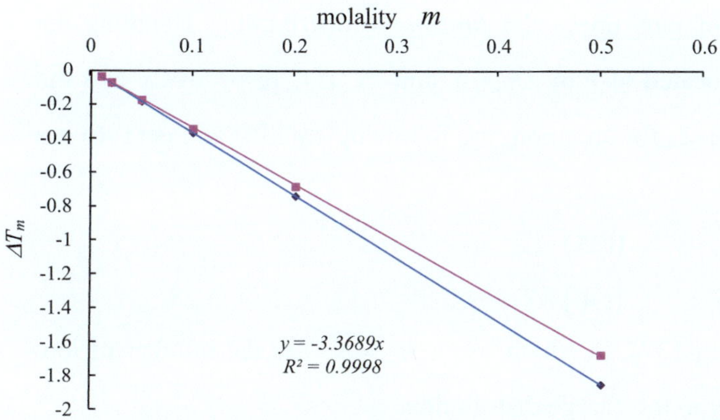

Fig.16 Relation between freezing point depression and concentration.
blue line: calculated values
red line: observed values
The measured values are proportional to the concentration in the given range.

Ans. freezing point depression: See table and Fig.16

(b) Employing the given cryoscopic constant, the concentration (in molality) at which the sea water freezes is obtained first and then the grams of NaCl contained in 1L of the sea water whose relative density is 1.0 is found as 5.9 g as is calculated below.

$$\Delta T_m / K = 0.5 = 3.369 \times m \;\Rightarrow\; m = \frac{0.5}{3.369} = 0.148$$

$$0.148 \times 40 = 5.9\ \text{g}$$

Ans. number of grams converted to NaCl: 5.9 g

Solution 4-18 Before answering the question.

The *Debye-Hückel* theory.

As is revealed in **problem 4-17**, ionic solutions, even when they are very dilute, exhibit significant deviations from ideal behaviors such as colligative properties expected for the ideal solution. In early days when the ionic dissociation of salts (electrolytes) was discovered for the first time, the deviations from ideality were ascribed to their insufficient, less than 100%, dissociations because of ion-pair formation.

However, since after the finding that strong electrolytes dissociate almost 100% over a wide range of concentration, the electrostatic interactions between ions which distribute inhomogeneously, that is, more positive ions around a negative ion tending to gather and vice versa, have been treated properly in order to explain quantitatively the non-ideal behaviors of electrolyte solutions. The most fundamental theory employed for this is the *Debye-Hückel* theory. Here, the equations which consist the backbone of the theory will be explained to the extent enough to solve relevant problems.

The electrostatic potential $\phi(r)$ due to the charges, distributed angle-independently in space and given by $\rho(r)$, is described by *Poisson*'s equation (eq(134)). Here, ε_0 and ε_r are the dielectric constants in vacuum and in a medium, respectively. For this homogeneous charge distribution, eq(134) can be rewritten as eq(134') (see **II-4-9**).

$$\nabla^2\phi(r) = -\frac{\rho(r)}{\varepsilon_0\varepsilon_r} \qquad (134)$$

$$\nabla^2\phi(r) = \frac{1}{r^2}\frac{d}{dr}\left(r^2\frac{d\phi}{dr}\right) = -\frac{\rho(r)}{\varepsilon_0\varepsilon_r} \qquad (134')$$

Here, the important thing is to know how the charge distribution in the electrolyte solution is described. *Debye* and *Hückel* assumed that the distribution of ions obeys the *Boltzmann* distribution. Denoting the numbers of cations and anions in a unit volume located at r by $N_+(r)$ and $N_-(r)$, respectively and the positive and negative charges, Z_+ for a cation and Z_- for an anion, the following eq(135) and eq(136) are satisfied for $N_+(r)$ and $N_-(r)$, respectively.

$$N_+(r) = \overline{N}_+\exp[-eZ_+\phi(r)/k_BT] \qquad (135)$$

$$N_-(r) = \overline{N}_-\exp[-eZ_-\phi(r)/k_BT] \qquad (136)$$

Consequently, the charge distribution is given by eq(137), in which $\overline{N}_+ = \overline{N}_- = \overline{N}$ are the number of ions per unit volume when there exists no inhomogeneity in the charge distribution.

$$\begin{aligned}\rho(r) &= N_+(r)eZ_+ + N_-(r)eZ_- \\ &= \overline{N}e\langle Z_+\exp[-eZ_+\phi(r)/k_BT] + Z_-\exp[-Z_-e\phi(r)/k_BT]\rangle \qquad (137)\\ &= \overline{N}e\langle Z_+ - eZ_+^2\phi(r)/k_BT + Z_- - Z_-^2\phi(r)/k_BT + \cdots\rangle\end{aligned}$$

For simplicity, a solution of a 1:1 strong electrolyte is considered. Then, using the following relations,

$$Z_+ = -Z_- = Z, \qquad m_+ = m_- = m \qquad \overline{N} = \overline{N}_+ = \overline{N}_- = nN_A = m\cdot d\cdot N_A$$

n: molar concnetration、 m: molal concnetration ($\mathrm{mol\cdot kg^{-1}}$)、 d: density of the solution

the approximation of eq(137) to the first order is given by eq(137'),

$$\begin{aligned}\rho(r) \approx -\overline{N}e^2(Z_+^2 + Z_-^2)\phi(r)/k_BT &= -d\cdot N_A^2e^2\times(m_+Z_+^2 + m_-Z_-^2)\phi(r)/RT \\ &= -2d\cdot N_A^2e^2\times I\phi(r)/RT \qquad (137')\end{aligned}$$

in which I is the ionic strength defined by eq(138). Notice that the concentrations of ions m_i's are expressed in the units of molality ($\mathrm{mol\cdot kg^{-1}}$).

$$I = \frac{1}{2}\sum_i^{\text{all ions}} m_iZ_i^2 \qquad (138)$$

Employing eq(137'), eq(134') can be expressed as eq(139), which is called *Poisson-Boltzmann*'s equation.

$$\frac{1}{r^2}\frac{d}{dr}\left(r^2\frac{d\phi}{dr}\right)=\frac{2dN_A^2e^2I\phi(r)}{\varepsilon_0\varepsilon_r RT}\equiv\Lambda\phi(r)\ ,\quad \Lambda=\frac{2dN_A^2e^2I}{\varepsilon_0\varepsilon_r RT} \qquad (139)$$

There exist two special solutions as given below which satisfy eq(139), however, since the electrostatic potential

$$\frac{C_1}{r}e^{-r/r_D},\ \frac{C_2}{r}e^{r/r_D} \qquad (C_1, C_2 : \text{constants})$$

should approach to zero when $r\to\infty$, eq(140) is employed as a solution of eq(139).

$$\phi(r)=\frac{C_1}{r}e^{-r/r_D} \qquad (140)$$

Substituting eq(140) into eq(139), the constant r_D is determined as eq(142) via eq(141). r_D is called *Debye length* and, as is evident from eq(140), the smaller the value of r_D is, the smaller the shielded potential becomes as compared with that without shielding.

$$\frac{1}{r^2}\frac{d}{dr}\left(r^2\frac{d}{dr}\frac{C_1}{r}e^{-r/r_D}\right)=\frac{1}{r_D^2}\frac{C_1e^{-r/r_D}}{r}=\Lambda\phi(r) \qquad (141)$$

$$r_D^2=\frac{1}{\Lambda}=\frac{\varepsilon_0\varepsilon_r RT}{2dN_A^2e^2I} \qquad (142)$$

The proportional constant C_1 is determined by integrating the charge density distribution over the whole space excluding a positive ion itself with a radius a located at the center of the charge density distribution. From the electroneutrality, the integrated value should counterbalance the charge of the positive ion at the center, from which eq(143) (or eq(144)) is obtained.

$$\int_a^\infty 4\pi r^2\rho(r)dr=Z\_e \qquad (143)$$

$$=-\frac{4\pi\times 2dN_A^2e^2I}{RT}\int_a^\infty r^2\times\frac{C_1}{r}e^{-r/r_D}dr$$

$$=-\frac{4\pi\times 2dN_A^2e^2I\times C_1}{RT}\int_a^\infty re^{-r/r_D}dr \qquad \left(\int_a^\infty re^{-r/r_D}dr=(r_D+a)r_De^{-a/r_D}\right)$$

$$=-\frac{4\pi\times 2dN_A^2e^2I\times C_1}{RT}\times(r_D+a)r_De^{-a/r_D}=Z\_e \qquad (144)$$

Furthermore, in dilute solutions, the central ion is approximated as a point charge and, hence, the radius a is put equal to zero. Then, the constant $C_1$is determined as eq(145).

$$C_1=-\frac{Z\_e}{4\pi(r_D+a)r_D}e^{a/r_D}\times\frac{RT}{2dN_A^2e^2I}\times\frac{\varepsilon_0\varepsilon_r}{\varepsilon_0\varepsilon_r}=-\frac{Z\_er_D}{4\pi\varepsilon_0\varepsilon_r(r_D+a)}e^{a/r_D} \qquad (145)$$

Consequently, as the electrostatic potential around a positive ion in solution, eq(140') or eq(140'') is derived.

$$\phi(r)=-\frac{Z\_er_D}{4\pi\varepsilon_0\varepsilon_r(r_D+a)}\frac{e^{-(r-a)/r_D}}{r},\qquad r\ge a \qquad (140') \qquad (a\text{: ion radius})$$

$$\phi(r)\approx-\frac{Z\_e}{4\pi\varepsilon_0\varepsilon_r}\frac{e^{-r/r_D}}{r},\qquad a<<r_D \qquad (140'') \qquad (\text{point charge approximation})$$

Debye-Hückel's limiting law: dilute solutions (point charge approximation for the central ion: eq(140'')).

As is evident from the equations derived so far, eqs(140') and (140'') describe the electrostatic potential which is generated at the distance r from a positive ion located at the center around which the other positive and negative ions in solution distribute inhomogeneously obeying the *Boltzmann* distribution. If the distribution of the positive as well as negative ions is homogeneous in total, then the charges around the central ion cancel each other and, hence, the electrostatic potential at the distance r from the central positive ion is given by eq(146).

$$\frac{Z_+e}{4\pi\varepsilon_0\varepsilon_r}\frac{1}{r} \qquad (146)$$

Consequently, the subtraction of eq(146) from eq(140") yields eq(147) which is considered to express the electrostatic potential arising solely from the inhomogeneity of the ions' distribution called *ionic atmosphere*.

$$\phi'(r) = -\frac{Z_-e}{4\pi\varepsilon_0\varepsilon_r}\cdot\frac{e^{-r/r_D}}{r} - \frac{Z_+e}{4\pi\varepsilon_0\varepsilon_r}\cdot\frac{1}{r} \qquad (147)$$

The potential at the central ion is found as eq(148) by expanding the exponential term of eq(147) in r and by approximating the expansion to the first order since $Z_+ + Z_- = 0$ (1:1 electrolyte).

$$\phi'(0) = \frac{Z_-e}{4\pi\varepsilon_0\varepsilon_r}\frac{1}{r_D} \qquad (148)$$

Therefore, the energy required to create a positive ion with a charge Z_+e at the center ($r = 0$) is considered to be equal to the electrostatic energy needed to charge a point at the center from zero to Z_+e ($0 \rightarrow Z_+e$) under the electrostatic potential given by eq(148), and it is given by eq(149). Naturally, any contribution other than the ionic atmosphere is not included in this energy.

$$\int_0^e \phi'(0)de = \frac{1}{4\pi\varepsilon_0\varepsilon_r r_D}\int_0^{Z_+e} Z_-ed(Z_+e) = -\frac{1}{4\pi\varepsilon_0\varepsilon_r r_D}\int_0^{Z_+e} Z_+ed(Z_+e) = -\frac{(Z_+e)^2}{8\pi\varepsilon_0\varepsilon_r r_D} \qquad (149)$$

Although eq(149) is derived for creating a positive ion, the energy required is proportional to the charge squared and, hence, it is seen that the energy needed to create a negative ion with the Z_-e at the center is also given by eq(149).

Now, we consider a solution at the concentration of $m\,\text{mol}\cdot\text{kg}^{-1}$ of 1:1 electrolyte AB which undergoes complete dissociation in solution. It is assumed that the dissociated ions A^{Z_+} and B^{Z_-} distribute homogeneously in solution and that their chemical potentials are expressed by eq(150).

$$\left.\begin{aligned} \mu_{\text{A}^{Z_+}} &= \mu^0_{\text{A}^{Z_+}} + RT\ln m_{\text{A}^{Z_+}} \\ \mu_{\text{B}^{Z_-}} &= \mu^0_{\text{B}^{Z_-}} + RT\ln m_{\text{B}^{Z_-}} \end{aligned}\right\} \qquad (150)$$

$\mu^0_{\text{A}^{Z_+}}$ and $\mu^0_{\text{B}^{Z_-}}$ are the chemical potentials (free energies) at the concentration at which $m_{\text{A}^{Z_+}} = m_{\text{B}^{Z_-}} = m$ is $1.0\,\text{mol}\cdot\text{kg}^{-1}$. (In order to make the arguments of logarithm dimensionless, the concentrations such as $m_{\text{A}^{Z+}}$ should be replaced by the normalized concentrations by dividing the concentrations by the standard concentration m_0, i.e., $m_{\text{A}^{Z+}}/m_0$, etc., however, for simplicity, m_0 is omitted in eq(150).) Formally, eq(150) expresses the chemical potentials of the ideal solution. However, as is seen in the *Debye-Hückel* theory, ions in solution are under the influence of an ionic atmosphere and have free energies (eq(149)) different from those in the case of a totally homogeneous ion distribution.

When the correction for this is expressed by the terms of $RT\ln\gamma_{\text{A}^{Z_+}}$ and $RT\ln\gamma_{\text{B}^{Z_-}}$ using the activity coefficients, the chemical potential of each ion in the real solution can be expressed by eq(151).

$$\left.\begin{aligned} \mu_{A^{Z_+}} &= \mu^0_{A^{Z_+}} + RT\ln(\gamma_{A^{Z_+}} \times m_{A^{Z_+}}) = \mu^0_{A^{Z_+}} + RT\ln(\gamma_{A^{Z_+}} \times m) \\ \mu_{B^{Z_-}} &= \mu^0_{B^{Z_-}} + RT\ln(\gamma_{B^{Z_-}} \times m_{B^{Z_-}}) = \mu^0_{B^{Z_-}} + RT\ln(\gamma_{B^{Z_-}} \times m) \end{aligned}\right\} \qquad (151)$$

Employing the result of eq(149), these correction terms are found as eq(152).

$$RT\ln\gamma_{A^{Z_+}} = -\frac{N_A Z_+^2 e^2}{8\pi\varepsilon_0\varepsilon_r r_D}, \qquad RT\ln\gamma_{B^{Z_-}} = -\frac{N_A Z_-^2 e^2}{8\pi\varepsilon_0\varepsilon_r r_D} \qquad (152)$$

Substituting r_D defined by eq(141) into eq(152), the activity coefficients of the ions are obtained after some rearrangements as eq(153) for 1:1 electrolytes.

$$\ln\gamma_{A^{Z_+}} = \ln\gamma_{B^{Z_-}} = -\frac{N_A Z_+^2 e^2}{8\pi\varepsilon_0\varepsilon_r RT}\frac{\sqrt{2dN_A I}\,e}{(\varepsilon_0\varepsilon_r k_B T)^{1/2}} = -\frac{N_A{}^2 e^3\sqrt{2dI}}{8\pi(\varepsilon_0\varepsilon_r RT)^{3/2}} \qquad (153)$$

In general, for $p:q$ electrolyte A_pB_q, the ion dissociation is written as eq(154) and the ionic charges are related by eq(155).

$$A_pB_q \rightarrow pA^{Z_+} + qB^{Z_-}, \quad (Z_+ \neq -Z_-) \qquad (154)$$

$$pZ_+ + qZ_- = 0 \qquad (155)$$

For the solutions of the strong electrolytes of this type, since the deviation from the ideal solution cannot be ascribed to each of the positive and negative ions, a mean value of the contributions of both ions to the deviation is employed to define the mean activity coefficient as is expressed in eq(156).

$$(p+q)\ln\gamma_\pm = p\ln\gamma_+ + q\ln\gamma_- \Leftrightarrow \gamma_\pm^{p+q} = \gamma_+^p \times \gamma_-^q \qquad (156)$$

Employing the expressions given by eq(152) for γ_+ and γ_-, the right side of eq(156) can be rearranged as eq(157).

$$p\ln\gamma_+ + q\ln\gamma_- = A\sqrt{I}\left\{p(Z_+)^2 + q(Z_-)^2\right\}, \quad A \equiv \frac{N_A{}^2 e^3\sqrt{2d}}{8\pi(\varepsilon_0\varepsilon_r RT)^{3/2}} \qquad (157)$$

From the balance of the charges given by eq(155), the ionic charges and the compositions are related by eq(158).

$$p(Z_+)^2 + q(Z_-)^2 = -(p+q)Z_+Z_- \qquad (158)$$

Employing this relation, the mean activity coefficient is found to be expressed by eq(159).

$$\ln\gamma_\pm = \frac{1}{p+q}(p\ln\gamma_+ + q\ln\gamma_-) = -|Z_+Z_-|\frac{N_A{}^2 e^3\sqrt{2dI}}{8\pi(\varepsilon_0\varepsilon_r RT)^{3/2}} = -A|Z_+Z_-|\sqrt{I} \qquad (159)$$

It is eq(159) that is employed in general to express the (mean) activity coefficient of ions in solution.

A value for A of an aqueous electrolyte solution at $25\,^\circ\mathrm{C}$ is calculated to be $A = 1.172,$ by using the values of $T = 298\,\mathrm{K},\ d = 10^3\,\mathrm{kg\cdot m^{-3}}$ and $\varepsilon_r = 78.5.$ See **problem 4-18**. (In old papers and books common logarithm is used instead of natural logarithm more often. In this case, the constant A in eq(159) is given by 0.509 as are compared specifically as follows.

$$\ln\gamma_\pm = -1.172|Z_+Z_-|\sqrt{I} \ \ vs.\ \ \log_{10}\gamma_\pm = -0.509|Z_+Z_-|\sqrt{I}.$$

The extended Debye-Hückel theory: finite ion radius a of the central ion (eq(140')).

As the concentration of ions is increased, the central ion cannot be approximated as a point charge since it is too large upon treating the electrostatic interaction between ions. This is visualized when the *Debye-length*

r_D (eq(142)) is compared with a physical size of the ion (ionic radius). When r_D is larger than the ionic radius, the central ion can be approximated as a point charge but, if not, taking a finite magnitude of the ionic radius into account, eq(140') is used to express the electrostatic potential around the central ion.

Employing eq(140'), the mean activity coefficient is obtained in a similar manner as is used for the dilute solutions and is given by eq(160).

$$\ln\gamma_{\pm} = -\frac{A|Z_+Z_-|I^{1/2}}{1+BaI^{1/2}} \qquad (160)$$

$$A \equiv \frac{N_A^2 e^3\sqrt{2d}}{8\pi(\varepsilon_0\varepsilon_r RT)^{3/2}} \ (=1.172 \text{ at } 298\text{ K}), \qquad B = \frac{\sqrt{2d}eN_A}{(\varepsilon_0\varepsilon_r RT)^{1/2}} \ (=3.29\times10^9 \text{ at } 298\text{ K})$$

The dimensions of A and B are $[\text{mol/kg}]^{-1/2}$ and $[\text{mol/kg}]^{-1/2}\cdot\text{m}^{-1}$, respectively. Since eq(160) can be approximated by eq(160'), this equation is regarded as an approximation for the higher-order terms of the ionic strength.

$$\ln\gamma_{\pm} \approx -A|Z_+Z_-|I^{1/2}(1-BaI^{1/2}) = -A|Z_+Z_-|I^{1/2} + ABa|Z_+Z_-|I \qquad (160')$$

"note" See **problem 4-83** for the solution for the electric potential of the one-dimensional *Poisson* equation and its relation to the electric double layer and the electrode potential. The *Debye-Hückel* treatment of the electrostatic ionic interactions in solution in which the magnitude of the central ion is taken into consideration properly is called the *extended Debye-Hückel theory* and is employed when the concentrations of ions are too high to use eq(159).

Solution 4-18

The ionic strength I of each solution is calculated at each relevant concentration first and then the mean

$$\ln\gamma_{\pm} = -1.172|Z_+Z_-|\sqrt{I} \qquad ((159) \text{ reposted})$$

activity coefficient $\gamma_{\pm}$ is calculated by using eq(159) with $A=1.172$ (see the above section of ***Debye-Hückel*****'s limiting law**).

The results of calculation are summarized in Tables 1 and 2 below.

The obtained results in the low concentration range ($\leq 0.1 \text{ mol}\cdot\text{kg}^{-1}$) are plotted in Figs.17 and 18. As is seen from Figs.18(a) and (b), *Debye-Hückel*'s limiting law (eq(159)) is not good enough as an approximation unless the concentration is lower than 10^{-2} M. Also, the extended *Debye-Hückel* theory is considered to be a

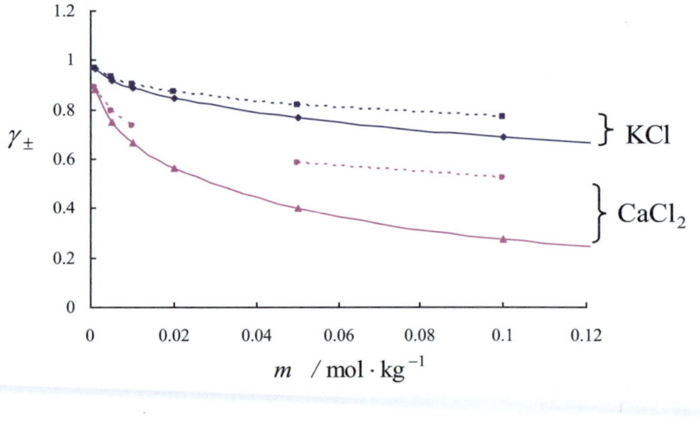

Fig.17 Concentration dependence of the activity coefficients of aqueous solutions of 1:1 and 1:2 strong electrolytes. broken lines: observed values, solid lines: calculated values.

ood approximation at the concentrations lower than $\leq 0.1\ \mathrm{mol \cdot kg^{-1}}$. See the section of **Before answering the question given** in **solution 4-18** for the *Debye-Hückel* theory.

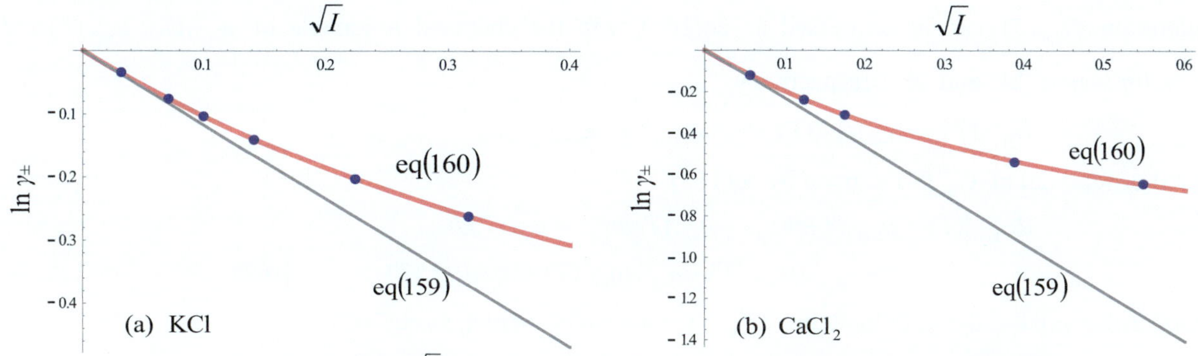

(a) $\ln\gamma_{\pm} = -\dfrac{1.17\times\sqrt{I}}{1+1.30\sqrt{I}}$ (b) $\ln\gamma_{\pm} = -\dfrac{2.38\times\sqrt{I}}{1+1.86\sqrt{I}}$

Figs.18 (a) and (b) Dependence of the activity coefficient on the ionic strength.
Linear (eq(159)) and non-linear (eq(160)) least-squares-fittings.

Table 1 Concentration dependence of the activity coefficient of aqueous solutions of KCl.

m (mol/kg)	$\sqrt{I}$	$-1.172\sqrt{I}$	$\gamma_{\pm}$ (calculaed)	$\gamma_{\pm}$ (observed)
0.001	0.03162	−0.03706	0.9636	0.966
0.005	0.07071	−0.08287	0.9205	0.927
0.01	0.1000	−0.1172	0.8894	0.902
0.02	0.1414	−0.1657	0.8473	0.869
0.05	0.2236	−0.2621	0.7695	0.816
0.1	0.3162	−0.3706	0.6903	0.770
0.2	0.4472	−-0.5241	0.5921	—
0.3	0.5477	−0.6419	0.5263	—
0.4	0.6324	−0.7412	0.4765	—
0.5	0.7071	−0.8287	0.4366	0.652

Table 2 Concentration dependence of the activity coefficient of aqueous solutions of $CaCl_2$

m (mol/kg)	$\sqrt{I}$	$-1.172\times 2\sqrt{I}$	$\gamma_{\pm}$(calculated)	$\gamma_{\pm}$(observed)
0.001	0.0548	−0.07420	0.8795	0.888
0.005	0.1225	−0.1666	0.7505	0.789
0.01	0.1732	−0.2367	0.6663	0.732
0.02	0.2449	−0.3381	0.5632	—
0.05	0.3873	−0.5497	0.4034	0.584
0.1	0.5477	−0.8120	0.2770	0.524
0.2	0.7746	−1.240	0.1627	—
0.3	0.9487	−1.623	0.1082	—
0.4	1.095	−1.989	0.0767	—
0.5	1.225	−2.344	0.0567	0.5

Ans. calculated and observed values of $\gamma_{\pm}$: Tables 1 (KCl) and 2 ($CaCl_2$)

Solution 4-19

(a) Under atmospheric pressure (constant pressure) at temperature *T*, the free energy of the solution under consideration $G_{\text{sol}}(T)$ can be expressed as eq(161) with the chemical potentials of $\mu_{\text{H}_2\text{O}}(T)$, $\mu_{\text{M}^+}(T)$ and $\mu_{\text{X}^-}(T)$ for water, M^+ and X^-, respectively.

$$G_{\text{H}_2\text{O}}(T) = n_{\text{H}_2\text{O}}\mu_{\text{H}_2\text{O}}(T) + n_{\text{M}^+}\mu_{\text{M}^+}(T) + n_{\text{X}^-}\mu_{\text{X}^-}(T) \qquad (161)$$

The total differential of $G_{\text{sol}}(T)$ is given by eq(162).

$$dG_{\text{H}_2\text{O}}(T) = \mu_{\text{H}_2\text{O}}(T)dn_{\text{H}_2\text{O}} + \mu_{\text{M}^+}(T)dn_{\text{M}^+} + \mu_{\text{X}^-}(T)dn_{\text{X}^-}$$
$$+ n_{\text{H}_2\text{O}}d\mu_{\text{H}_2\text{O}}(T) + n_{\text{M}^+}d\mu_{\text{M}^+}(T) + n_{\text{X}^-}d\mu_{\text{X}^-}(T) \qquad (162)$$

The total differential of the state function $G_{\text{sol}}(T)$ can also be written as eq(163).

$$dG_{\text{sol}}(T) = \sum_i \left(\frac{\partial G_{\text{sol}}}{\partial n_i}\right)_{T,P,n_{j\neq i}} dn_i$$
$$= \mu_{\text{H}_2\text{O}}(T)dn_{\text{H}_2\text{O}} + \mu_{\text{M}^+}(T)dn_{\text{M}^+} + \mu_{\text{X}^-}(T)dn_{\text{X}^-} \qquad (163)$$

Since eq(162) and eq(163) are equivalent, eq(164) should always be satisfied, which is derived conventionally by applying the so-called *Gibbs-Duhem* equation ($\sum_i n_i d\mu_i = 0$) to the aqueous solution of MX.

$$n_{\text{H}_2\text{O}}d\mu_{\text{H}_2\text{O}}(T) + n_{\text{M}^+}d\mu_{\text{M}^+}(T) + n_{\text{X}^-}d\mu_{\text{X}^-}(T) = 0 \qquad (164)$$

Expressing the chemical potential by using the activity as $\mu(T) = \mu^\circ(T) + RT\ln a$, the differential of μ is given by eq(165).

$$d\mu = RTd\ln a + R\ln a \cdot dT \approx RTd\ln a \qquad (165)$$

Here, noticing the relations of $n_{\text{M}^+} = n_{\text{X}^-} = n_{\text{MX}}$ and $\ln a_{\text{M}^+} = \ln a_{\text{X}^-} \equiv \ln a_\pm$, eq(165) is substituted into eq(164) to find eq(167) as an expression of $d\ln a_{\text{H}_2\text{O}}(T)$ via eq(166).

$$(164) \Rightarrow n_{\text{H}_2\text{O}}d\ln a_{\text{H}_2\text{O}}(T) + 2n_{MX}d\ln a_\pm(T) = 0 \qquad (166)$$

$$d\ln a_{\text{H}_2\text{O}}(T) = -\frac{2n_{MX}}{n_{\text{H}_2\text{O}}}d\ln a_\pm(T) = -\frac{2n_{MX}}{n_{\text{H}_2\text{O}}}d\ln\{n_{MX}\cdot\gamma_\pm(T)\} \qquad (167)$$

Since the right side of eq(167) is given as a function of the concentration of MX, both sides are integrated against $n_{\text{MX}} \equiv m$ over the range [0, m] to find eqs(168) and (168'), in which the relations of $\ln a_{\text{H}_2\text{O}}(0) = 0$ and $n_{\text{H}_2\text{O}} = 1000/18$ are used.

$$\int_{m=0}^{m} d\ln a_{\text{H}_2\text{O}} = -\int_{m=0}^{m}\frac{2m}{n_{\text{H}_2\text{O}}}d\ln(m\cdot\gamma_\pm) \qquad (168)$$

$$\ln a_{\text{H}_2\text{O}}(m) = -\frac{36}{1000}\int_{m=0}^{m} m\cdot d\ln(m\cdot\gamma_\pm) \qquad (168')$$

Employing *Debye-Hückel*'s limiting law, the activity coefficient is expressed by $\gamma_\pm = e^{-1.172\sqrt{m}}$ and, hence, $d\ln(m\cdot\gamma_\pm)$ can be written as eq(169).

$$d\ln(m\cdot\gamma_\pm) = d\ln m + d\ln\gamma_\pm = \frac{1}{m}dm - \frac{1.172}{2}m^{-1/2}dm \qquad (169)$$

Substituting this into eq(168') to integrate it against *m*, eq(170) is obtained as the expression to give the activity of water in which the electrolyte MX is dissolved at the concentration of $m\ \text{mol}\cdot\text{kg}^{-1}$.

$$\ln a_{H_2O}(m) = -\frac{36}{1000}\int_0^m \left(1-\frac{1.172}{2}m^{1/2}\right)dm = -\frac{36}{1000}\left(m-\frac{1.172}{3}m^{3/2}\right) \qquad (170)$$

Ans. activity of water: eq(170)

(b) Denoting the chemical potential of a pure substance i as $\mu_i^\circ(T)$, eq(171) holds for water at its freezing point T_f under a constant pressure

$$\mu_{ice}^\circ(T_f) = \mu_{H_2O}^\circ(T_f) \qquad (171)$$

On the other hand, the chemical potential of water $\mu_{H_2O}(T)$ in solution is given by eq(172) using its activity a_{H_2O}.

$$\mu_{H_2O}(T) = \mu_{H_2O}^\circ(T) + RT\ln a_{H_2O} \qquad (172)$$

Therefore, when the freezing point of a solution is considered to change from T_f for pure water by x, then the chemical potential of the solution at the freezing point is given by eq(173) using eq(172).

$$\mu_{ice}^\circ(T_f+x) = \mu_{H_2O}(T_f+x) = \mu_{H_2O}^\circ(T_f+x) + R(T_f+x)\ln a_{H_2O} \qquad (173)$$

For pure substances, e.g., ice, a change in the chemical potential associated with a small change x in temperature can be approximated by eq(174).

$$\mu_{ice}^\circ(T_f+x) = \mu_{ice}^\circ(T_f) + \left(\frac{\partial \mu_{ice}^\circ}{\partial T}\right)_{T=T_f} \times x$$

$$= \mu_{ice}^\circ(T_f) - S_{ice}^\circ(T_f)\cdot x \qquad (174)$$

Similarly eq(175) is derived for water.

$$\mu_{H_2O}^\circ(T_f+x) = \mu_{H_2O}^\circ(T_f) - S_{H_2O}(T_f)\cdot x \qquad (175)$$

Substituting eqs(174) and (175) into eq(173) with some rearrangement using eq(171), eq(176) is obtained.

$$\left\langle S_{H_2O}^\circ(T_f) - S_{ice}^\circ(T_f)\right\rangle \cdot x = R(T_f+x)\ln a_{H_2O} \approx RT_f \ln a_{H_2O} \qquad (176)$$

From these results, the freezing point depression x of the solution is found to be given by eq(177) using the

$$x = \frac{RT_f \ln a_{H_2O}}{\left\langle S_{H_2O}^\circ(T_f) - S_{ice}^\circ(T_f)\right\rangle} = \frac{RT_f^2 \ln a_{H_2O}}{\Delta_f H} = \mathrm{K}_f \times \ln a_{H_2O}, \quad \mathrm{K}_f \equiv \frac{RT_f^2}{\Delta_f H} \qquad (177)$$

activity of water as solvent. Here, the entropy change associated with the phase transition is expressed by using the heat of fusion given by $\Delta_f H/T_f = \left\langle S_{H_2O}^\circ(T_f) - S_{ice}^\circ(T_f)\right\rangle$. Substituting the values known for water, the constant K_f is calculated to be 103.26 K.

$$\frac{RT_f^2}{\Delta_f H} = \frac{8.315\times 273.15^2}{6.008\times 10^3} = 103.26\ \mathrm{K} = K_f$$

Ans. depression of freezing point of water: $\mathrm{K}_f \times \ln a_{H_2O}$, $\mathrm{K}_f = 103.26$ K

Putting the results obtained in (a) and (b) together, the depression of the freezing point of the aqueous solution of the 1:1 electrolyte of MX at the concentration of $m\ \mathrm{mol\cdot kg^{-1}}$ is given by eq(178) as far as *Debye-Hückel*'s limiting law is satisfied.

$$-103.12\times\frac{36}{1000}\left(m-\frac{1.172}{3}m^{3/2}\right) \qquad (178)$$

Based on this equation, the values for the freezing point depression at the concentrations of 0.01, 0.02, 0.05, 0.1, 0.2 and 0.5 $\mathrm{mol\cdot kg^{-1}}$ are found as -0.036, -0.07, -0.17, -0.33, -0.61 and -1.34 K in this order.

These results are plotted against the concentration of the salt in Fig.19.

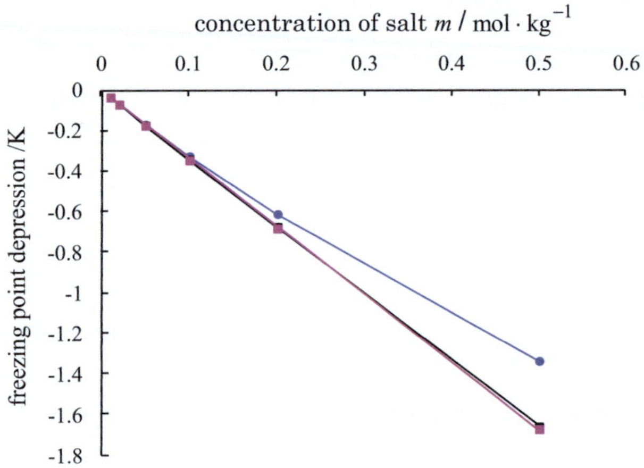

Fig.19 Freezing point depression and the concentration of 1:1 electrolyte.
blue line: *Debye-Hückel*'s limiting law,
black line: extended *Debye-Hückel* theory,
red line: observed values.

Ans. Freezing point depresses by -0.036, -0.07, -0.17, -0.33, -0.61 and -1.34 K in the relevant order. See Fig.19 for a comparison with the observed values.

Solution 4-20

To know how to calculate values for the freezing point depression based on *Debye-Hückel*'s limiting law, see **solution 4-19** .

The extended *Debye-Hückel* theory yields the mean activity coefficient given by eq(179), in which A and B

$$\ln\gamma_\pm = -\frac{A|Z_+Z_-|I^{1/2}}{1+BaI^{1/2}} \qquad (179)$$

are constants given by the theoretical values but a is to be experimentally determined. Here, therefore, a is assumed to be an adjustable parameter so as to reproduce the experimental values of $\ln\gamma_\pm$ most satisfactorily. It is eq(179') that is obtained so for the aqueous solutions of KCl at concentrations of m mol · kg^{-1}.

$$\ln\gamma_\pm = -\frac{1.16I^{1/2}}{1+1.23I^{1/2}} = -\frac{1.16m^{1/2}}{1+1.23m^{1/2}} \qquad (179')$$

The activity of water in this solution at the concentration m is given by eq(180) as is shown in the previous solution.

$$\ln a_{H_2O} = -\int_{m=0}^{m} \frac{2m}{n_{H_2O}} d\ln a_\pm \qquad (180)$$

From eq(179') for $\ln\gamma_\pm$, its differential is given by eq(181).

$$d\ln\gamma_\pm = -\frac{1.16m^{-1/2}}{2(1+1.23m^{1/2})^2} dm \qquad (181)$$

Since $d\ln a_\pm = d(\ln m\cdot\gamma_\pm)$, eq(181) is rewritten as eq(182) by using eq(181) and the value of $n_{H_2O} = 1000/18$.

$$\ln a_{H_2O} = -\frac{36}{1000}\int_0^m m\left(\frac{1}{m} - \frac{1.16m^{-1/2}}{2(1+1.23m^{1/2})^2}\right)dm = -\frac{36}{1000}\int_0^m \left(1 - \frac{1.16m^{1/2}}{2(1+1.23m^{1/2})^2}\right)dm \qquad (182)$$

The integration of eq(182) is readily carried out by using software such as *Mathematica* and thus the activity of water can be known at any concentration of the salt.

The freezing point depression ΔT is given by $\Delta T = K_f \ln a_{H_2O}$ as has been shown before and, hence, the required answer is given by eq(183).

$$\Delta T = 103.12\times\ln a_{H_2O} = -\frac{103.12\times 36}{1000}\int_0^m \left(1 - \frac{1.16m^{1/2}}{2(1+1.23m^{1/2})^2}\right)dm \qquad (183)$$

The employed eq(179') is the relation to reproduce the experimental data for KCl though, it is considered that the same equation can be used for NaCl. Thus, the activity of water at the concentration of $m = 0.5\ \text{mol}\cdot\text{kg}^{-1}$ is found as $\ln a_{H_2O} = -0.0161$. Therefore, the freezing point depression at this concentration is calculated to be $103.2 \times (-0.0161) = -1.65$ K. The observed value is -1.677 K and is in a remarkable agreement with the calculated one. The freezing point depression expected for the ideal aqueous solution and the one predicted by eq(183) are compared in Fig.20. See Fig.19 given in **solution 4-19** for a comparison of eq(183) with the experimental values.

Ans. freezing point depression against concentration: eq(183)

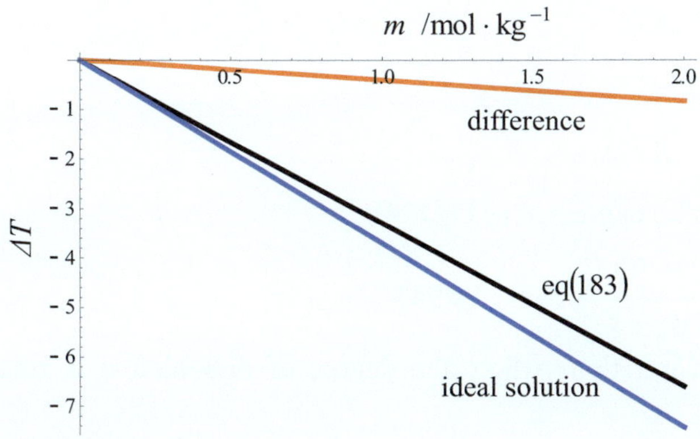

Fig.20 Freezing point depression ΔT of aqueous NaCl solutions against molality. Prediction based on the extended *Debye-Hückel* theory.
The blue line is for the ideal solution and the orange line shows the difference from eq(183).

Solution 4-21

The activities of the ions in the equilibrium MX $\rightleftarrows$ M^+ + X^- in solution are written as a_{M^+} and a_{X^-}. Reminding the activity of the solid MX is unity, the solubility product K_{sp} is expressed by eq(184) and the activity of each ion by eq(185), where $\gamma_\pm$ is the mean activity coefficient.

$$K_{sp} = a_{M^+} \times a_{X^-} \qquad (184)$$

$$a_{M^+} = [M^+] \times \gamma_\pm, \qquad a_{X^-} = [X^-] \times \gamma_\pm \qquad (185)$$

The mean activity coefficient for the ion with a unit electric charge ($Z_+ = -Z_- = 1$) is given by eq(186) (see **solution 4-18**).

$$\ln\gamma_\pm = -1.172 \times \sqrt{I} \ \Rightarrow\ \gamma_\pm = e^{-1.172\sqrt{s}} \qquad (186)$$

Here, it is employed that the ionic strength I is related to the solubility S by eq(187).

$$I = \frac{1}{2}\left(1^2[M^+] + 1^2[X^-]\right) = S \qquad (187)$$

Substituting eqs(185) and (186) into eq(184), it is shown that the relation between the solubility product K_{sp} and the solubility S is given by eq(188) in the present case.

$$K_{sp} = ([M^+] \times \gamma_\pm) \times ([X^-] \times \gamma_\pm) = S^2 \gamma_\pm{}^2 \ \Rightarrow\ S = K_{sp}{}^{1/2} / \gamma_\pm = K_{sp}{}^{1/2} e^{1.172\sqrt{s}} \qquad (188)$$

Ans. See text.

Solution 4-22

The dissociation equilibrium of acetic acid AH is given by eq(189) and its equilibrium constant is expressed by eq(190).

$$\text{AH} \rightleftarrows \text{A}^- + \text{H}^+ \qquad (189)$$

$$K_a = \frac{a_{\text{A}^-} \times a_{\text{H}^+}}{a_{\text{A}}} = \frac{[\text{A}^-]\gamma_\pm \times [\text{H}^+]\gamma_\pm}{[\text{A}]} \qquad (190)$$

Denoting the initial acid concentration $[\text{AH}]_0$ and the degree of dissociation α, the concentration of each chemical species in eq(189) is given by eq(191).

$$\left.\begin{array}{l} [\text{AH}]_0 = [\text{AH}] + [\text{A}^-] \\ [\text{AH}] = [\text{AH}]_0(1-\alpha) \\ [\text{A}^-] = \alpha[\text{AH}]_0 \\ [\text{H}^+] = \alpha[\text{AH}]_0 \quad (\text{autodissoc iation of water is neglected.}) \end{array}\right\} \qquad (191)$$

According to *Debye-Hückel*'s liming law, the activity coefficient of the dissociated ions is given by eq(192) and the ionic strength is given by eq(193).

$$\ln\gamma_\pm = -1.172\sqrt{I} \qquad (192)$$

$$I = \frac{1}{2}([\text{A}^-] + [\text{H}^+]) = \alpha[\text{A}]_0 \qquad (193)$$

Consequently, the equilibrium constant of eq(190) can be expressed as eq(194).

$$K_a = \frac{(\alpha[\text{AH}]_0 \times e^{-1.172\sqrt{I}})^2}{[\text{AH}]_0(1-\alpha)} = \frac{(x \times e^{-1.172\sqrt{x}})^2}{[\text{AH}]_0 - x} \qquad (194)$$

Since $x = \alpha[\text{AH}]_0$, it can be put $x << 1$ at the concentrations where the degree of dissociation is small, eq(194) is approximated as eq(194').

$$K_a([\text{AH}]_0 - x) = \left\langle x(1 - 1.172\sqrt{x})\right\rangle^2 \qquad (194')$$

Putting $K_a = 1.75 \times 10^{-5}$ and $[\text{AH}]_0 = 1.0 \times 10^{-3}$, eq(194') is solved for x to find $\alpha = 1.25 \times 10^{-1}$ as below.

$$x = 1.25 \times 10^{-4} \Rightarrow \alpha = 1.25 \times 10^{-1}$$

On the other hand, for the ideal solution where ion-ion interactions are neglected the equilibrium constant is expressed by eq(195) from which the degree of dissociation is calculated to be $\alpha = 1.24 \times 10^{-1}$ as is shown below. It is found that the difference between the two values, one based on *Debye-Hückel*'s liming law and the other by neglecting ionic interactions, is less than 1%.

$$K_a = \frac{(\alpha[\text{AH}]_0)^2}{[\text{AH}]_0(1-\alpha)} = \frac{x^2}{[\text{AH}]_0 - x} \qquad (195)$$

$$K_a([\text{AH}]_0 - x) = x^2 \Rightarrow x = 1.24 \times 10^{-4} \Rightarrow \alpha = 1.24 \times 10^{-1}$$

Ans. degree of dissociation: $\alpha = 1.25 \times 10^{-1}$, difference: less than 1%

Solution 4-23

It is assumed that the concentration of acetic acid is known accurately to the sufficient extent.

(1) Based on the knowledge of the dissociation constant of acetic acid, the degrees of dissociation α need to be known accurately at the temperatures near 0°C.

(2) Since the freezing point depression is given by $K_f \times \ln a_{\text{H}_2\text{O}}$ with $K_f = 103.12$ K (see **solution 4-19**), the activities of water near 0°C need to be known accurately.

(3) *Debye-Hückel*'s liming law is accurate enough to evaluate the activities of hydrogen and acetate ions.

(4) It is adequate to regard the activity coefficient of undissociated acetic acid as unity.

Ans. See text (1)-(4).

Taking the items from (1) to (4) into account, the freezing point of the dilute vinegar is calculated as (calculation omitted).

Solution 4-24

(a) The standard reduction potentials of copper ions (I) and (II) are as given by eq(196) and eq(197), respectively.

$$\mathrm{Cu}^+(aq)+e^- \rightarrow \mathrm{Cu}(s) \qquad E_1=+0.52\ \mathrm{V} \qquad (196)$$

$$\mathrm{Cu}^{2+}(aq)+e^- \rightarrow \mathrm{Cu}^+(aq) \qquad E_2=+0.16\ \mathrm{V} \qquad (197)$$

Since eqs(196) and (197) correspond to the elementary processes of the disproportionation reaction given in eq(196) in the present problem, being based on the relation between the potential difference and the equilibrium constant (*Nernst*'s equation), the equilibrium constant K is calculated by eq(198).

$$\ln K=\frac{nFE}{RT}=\frac{1\times 96500\times 0.36}{8.31\times 298}=14.0 \qquad (198)$$

$$K=e^{14.0}=1.20\times 10^6$$

Ans. equilibrium constant: $K=1.20\times 10^6$

(b) The standard free energies of formation of copper ions (I) and (II) are as given by eq(199) and eq(200), respectively.

$$\mathrm{Cu}(s)\rightarrow \mathrm{Cu}^+(aq) \qquad \Delta G_f^\circ=49.98\ \mathrm{kJ\cdot mol^{-1}} \qquad (199)$$

$$\mathrm{Cu}(s)\rightarrow \mathrm{Cu}^{2+}(aq) \qquad \Delta G_f^\circ=65.49\ \mathrm{kJ\cdot mol^{-1}} \qquad (200)$$

Since the reaction expressed by $\mathrm{eq}(200)-2\times\mathrm{eq}(199)$ corresponds to the disproportionation reaction (eq(196) in the problem), the standard free energy change $\Delta G_f^\circ(1)$ of this reaction is given by eq(201).

$$\Delta G_f^\circ(1)=65.49-2\times 49.98=-34.47\ \mathrm{kJ\cdot mol^{-1}} \qquad (201)$$

From this, the equilibrium constant K is given by eq(202) and is found to be $K=1.11\times 10^6$ at 298 K.

$$\ln K=-\frac{\Delta G_f^\circ}{RT}=\frac{34.47\times 10^3}{8.31\times 298}=13.92 \qquad (202)$$

$$K=e^{13.92}=1.11\times 10^6$$

Ans. equilibrium constant: $K=1.11\times 10^6$

The standard free energies of formation of ions in water (hydration energies) are generally acquired by means of electrochemical methods and, hence, the value used in eq(199) is obtained by $-F\times E_1$ ($-F\times E_1=-96460\times 0.52=50.1\ \mathrm{kJ\cdot mol^{-1}}$). The reason why the values found in (a) and (b) did not agree is because the number of significant digits of the electrode potential used was too small. It needs to be measured more accurately.

Solution 4-25

In order to estimate a temperature variation of the solubility product (equilibrium constant) of AgCl, minimum requirement is to know a value of the heat of dissolution ΔH_{sol}° for the dissolution reaction given by eq(203).

$$\mathrm{AgCl} \rightarrow \mathrm{Ag}^+(aq)+\mathrm{Cl}^-(aq) \qquad \Delta H_{sol}^\circ \qquad (203)$$

In a similar manner as is used for estimating the free energy change ΔG°_{sol} associated with eq(203) in the calculation of the solubility product of AgCl in **problem 4-13**, ΔH°_{sol} is found from the values for the standard enthalpy of formation for the constituent reactions given by eqs(204)-(206).

$$\mathrm{Ag(s)} \rightarrow \mathrm{Ag}^+(aq)+e^- \qquad \Delta H^\circ_f(\mathrm{Ag}^+)=105.58\ \mathrm{kJ\cdot mol^{-1}} \qquad (204)$$

$$\frac{1}{2}\mathrm{Cl_2}(g)+e^- \rightarrow \mathrm{Cl}^-(aq) \qquad \Delta H^\circ_f(\mathrm{Cl}^-)=-167.16\ \mathrm{kJ\cdot mol^{-1}} \qquad (205)$$

$$\mathrm{Ag}(s)+\frac{1}{2}\mathrm{Cl_2}(g) \rightarrow \mathrm{AgCl}(s) \qquad \Delta H^\circ_f(\mathrm{AgCl})=-127.07\ \mathrm{kJ\cdot mol^{-1}} \qquad (206)$$

Combining these reactions (and hence ΔH°_f's) in a way of $(204)+(205)-(206)$ to match the dissolution reaction (eq(203)), ΔH°_{sol} is found to be $65.49\ \mathrm{kJ\cdot mol^{-1}}$ as shown below.

$$105.58-167.16+127.07=65.49\ \mathrm{kJ\cdot mol^{-1}}$$

Assuming that ΔH°_{sol} itself is independent of temperature, the solubility product of AgCl at 80℃ is calculated from eq(207) as shown below. Since $[\mathrm{Ag}^+]=[\mathrm{Cl}^-]$, the solubility ratio at the two temperatures is found as 7.8 from the ratio of the solubility products thus calculated.

$$\frac{\partial \ln K_{sp}}{\partial T}=\frac{\Delta H^\circ_{sol}}{RT^2} \Rightarrow \ln\frac{K_{sp}(T_2)}{K_{sp}(T_2)}=-\frac{\Delta H^\circ_{sol}}{R}\left(\frac{1}{T_2}-\frac{1}{T_1}\right) \qquad (207)$$

$$-\frac{65.49\times10^3}{8.31}\left(\frac{1}{353}-\frac{1}{298}\right)=4.12$$

$$\frac{K_{sp}(80^\circ\mathrm{C})}{K_{sp}(25^\circ\mathrm{C})}=e^{4.12}=61.6 \Rightarrow \frac{[\mathrm{Ag}^+](80^\circ\mathrm{C})}{[\mathrm{Ag}^+](25^\circ\mathrm{C})}=\sqrt{61.6}\approx 7.8$$

Ans. Concentration of Ag^+ increases only 7.8 times.

In order to perform a similar calculation for $AgNO_3$ as is done above for AgCl, a value for the enthalpy of dissolution ΔH°_{sol} of $AgNO_3$ (eq(203')) is required.

$$\mathrm{AgNO_3}(s) \rightarrow \mathrm{Ag}^+(aq)+\mathrm{NO_3}^-(aq) \qquad \Delta H^\circ_{sol} \qquad (203')$$

The enthalpy change for the reaction in eq(203') can be worked out by breaking down eq(203') into the following reactions from (*a*) to (*d*).

$$\mathrm{Ag(s)} \rightarrow \mathrm{Ag}^+(aq)+e^- \qquad \Delta H^\circ_f(\mathrm{Ag}^+)=105.58\ \mathrm{kJ\cdot mol^{-1}} \qquad (a)$$

$$\mathrm{Ag(s)}+\mathrm{HNO_3}(aq) \rightarrow \mathrm{AgNO_3(s)}+\frac{1}{2}\mathrm{H_2} \qquad \Delta H^\circ_r(b)=? \qquad (b)$$

$$\mathrm{HNO_3}(aq) \rightarrow \mathrm{H}^+ + \mathrm{NO_3^-} \qquad \Delta H^\circ_r(c)=? \qquad (c)$$

$$\frac{1}{2}\mathrm{H_2} \rightarrow \mathrm{H}^+ + e^- \qquad \Delta H^\circ_r(d)=0\ \mathrm{kJ\cdot mol^{-1}} \qquad (d)$$

Since the reaction (203') can be expressed by the combination of $(203')=(a)+(c)-(b)-(d)$, ΔH°_{sol} can be calculated if the enthalpy change is known for each of the reactions (*a*) to (*d*). The enthalpy changes for the reactions (*b*) and (*c*) are calculated as is shown by eq(204') from the enthalpy of formation for each compound given below.

HNO_3: $\Delta H^\circ_f(\mathrm{HNO_3})=-207.36$, NO_3^-: $\Delta H^\circ_f(\mathrm{NO_3^-})=-205.0$, H^+: $\Delta H^\circ_f(\mathrm{H^+})=0$

$AgNO_3(s)$: $\Delta H^\circ_f(\mathrm{AgNO_3})=-129.39$ (in units of $\mathrm{kJ\cdot mol^{-1}}$)

$$\left.\begin{aligned}\Delta H_r^\circ(b) &= \Delta H_f^\circ(\mathrm{AgNO_3}) - \Delta H_f^\circ(\mathrm{HNO_3}) = -129.39 + 207.36 = 77.97\\ \Delta H_r^\circ(c) &= \Delta H_f^\circ(\mathrm{NO_3^-}) - \Delta H_f^\circ(\mathrm{HNO_3}) = -205.0 + 207.36 = 2.36\end{aligned}\right\} \qquad (204')$$

Employing the values found by eq(204'), the enthalpy change $\Delta H_{sol}^\circ(203')$ of eq(203') is found from eq(205') as $29.97\ \mathrm{kJ\cdot mol^{-1}}$.

$$\Delta H_{sol}^\circ(1') = \Delta H_f^\circ(\mathrm{Ag^+}) + \Delta H_r^\circ(c) - \Delta H_r^\circ(b) - \Delta H_r^\circ(d) \qquad (205')$$

$$= 105.58 + 2.36 - 77.97 - 0 = 29.97\ \mathrm{kJ\cdot mol^{-1}}$$

Substituting this value into eq(207), the ratio of the ion concentrations at the two temperatures is found as $[\mathrm{Ag^+}](80^\circ\mathrm{C})/[\mathrm{Ag^+}](25^\circ\mathrm{C}) = \sqrt{6.59} \approx 2.6$.

Ans. for $\mathrm{AgNO_3}$: ca. 2.6 times

Solution 4-26

(a) Substituting $c = 2$ for the number of components and $p = 2$ for the number of phases in the equation of *Gibbs*' phase rule (eq(208)), $f = 2$ is found as the degrees of freedom. Consequently, under the fixed *T* and *P*

$$f = c + 2 - p \qquad (208)$$

conditions, there is no remaining degrees of freedom and the state of the system is unambiguously determined. Namely, the ratio of the concentrations of water/benzene in each phase takes a fixed value.

(b) *Gibbs*' phase rule (eq(208)) is applied to the system in which benzoic acid distributes between the two phases of water and benzene, the number of degrees of freedom is found as $f = 3$ by substituting $c = 3$ for the number of components and $p = 2$ for the number of phases in eq(208). Consequently as an independent variable other than temperature and pressure, one more degree of freedom is available. However, when the amount of benzoic acid is fixed as the last degree of freedom, the state of the system is unambiguously determined. In another word, including the associated dimmer, the concentration of every chemical species distributed in the two phases is determined. It should be noticed here that the number of components is taken as three, i.e. benzene, water and benzoic acid and that the associated dimmer is not counted as an independent component since its concentration is determined through the association constant.

The association constant can be obtained as follows.

(i) From the total amount of benzoic acid in the water phase, the amount of undissociated benzoic acid is calculated using the dissociation constant of benzoic acid as shown below.

$$0.0214\,(1-\alpha) = 0.0124 \times (1 - 0.170) = 0.0103\ \mathrm{g/100\ mL}$$

(ii) The amount of monomeric benzoic acid in the benzene phase is calculated using the distribution coefficient as shown below.

$$0.0103 \times 0.705 = 0.00726\ \mathrm{g/100\ mL}$$

(iii) The amount of the associated dimmer is calculated by subtracting the amount of the monomeric benzoic acid from the total amount of benzoic acid in the benzene phase as shown below.

$$0.0724 - 0.00726 = 0.0651\ \mathrm{g/100\ mL}$$

The value for the association constant K_{assoc} can be obtained from the concentrations determined in (ii) and

(iii) but the unit of the concentrations must be converted into $\mathrm{mol \cdot L^{-1}}$ not only by multiplying them by ten but also by dividing them by the relevant molecular weights, i.e., $M = 122.1$ for benzoic acid and $2M$ for the associated dimmer as shown below.

$$K_{assoc} = \frac{[(C_6H_5COOH)_2]}{[C_6H_5COOH]^2} = \frac{(0.651/2M)}{(0.0726/M)^2} = \frac{M}{4} \times \frac{(0.651)}{(0.0726)^2} = 3.77 \times 10^3$$

Ans. association constant : $K_{assoc} = 3.77 \times 10^3$

Solution 4-27

Gibbs' phase rule is expressed as eq(208).

$$f = c + 2 - p \qquad (208)$$

(a) Substituting $c = 1$ for the number of components and $p = 2$ for the number of phases in the above equation, $f = 1$ is obtained as the number of degrees of freedom. However, since the vapor pressure is already fixed at 1 atm, there is no other degree of freedom and, hence, the temperature is fixed at the boiling point.

(b) For $c = 1$ and $p = 2$, $f = 1$ is obtained. If the temperature is taken as a variable and fixed constant, then the vapor pressure of water is fixed. On the other hand, if the vapor pressure of water is taken as a variable and fixed constant, then the temperature of the system is unambiguously determined. It should be noted the atmospheric pressure of 1 atm here has nothing to do with the degrees of freedom. To fix the total pressure including the vapor pressure of water at 1 atm, however, is equivalent to specify the vapor pressure.

(c) For $c = 2$ and $p = 2$, $f = 2$ is obtained. Three possible intensive variables are <u>temperature</u>, <u>vapor pressure of water</u> and <u>vapor pressure of the volatile substance</u>. When any two of these three variables are fixed, then the other is automatically determined.

(d) For $c = 2$ and $p = 2$, $f = 2$ is obtained. Three possible intensive variables are <u>temperature</u>, <u>vapor pressure of water</u> and <u>concentration of the non-volatile substance</u>. When any two of these three variables are fixed, then the other is automatically determined. The present case is distinguished from (c).

(e) For $c = 4$ and $p = 1$, $f = 5$ is obtained. Fixing temperature and pressure, three degrees of freedom remain. Furthermore, when the volume of the system, which is nearly equal to the volume of water, is fixed, the degrees of freedom become two. Any two of the concentrations of the substance A, B and C can be varied freely as independent variables but the other remaining is determined unambiguously by the equilibrium constant.

(f) For $c = 4$ and $p = 2$ (gas and liquid phases), $f = 4$ is obtained. If temperature and pressure (gas phase) are fixed, the remaining degrees of freedom become two. The amounts of four chemicals of H_2, N_2, NH_3 and H_2O can be varied but considering that they are not only in the liquid-gas equilibrium but also the formation reaction of ammonium is in equilibrium, only two out of the four material amounts can be varied freely. The total pressure of the gas phase is already fixed.

Ans. See text (a)-(f).

Solution 4-28

(a) Denoting the mole fraction of nitrogen dissolved in water (blood) χ_{N_2}, *Henry*'s law is expressed as eq(209).

$$\kappa^H \cdot \chi_{N_2} = P_{N_2} \qquad (209) \qquad \chi_{N_2} = \frac{n_{N_2}}{n_{N_2} + n_{H_2O}}$$

The partial pressure of nitrogen in the atmosphere at 1.0 atm is taken as 0.8 atm. The mole fraction of nitrogen in solution which is in equilibrium with that partial pressure is given by eq(210) using eq(209).

$$\chi_{N_2} = \frac{n_{N_2}}{n_{N_2} + n_{H_2O}} \approx \frac{n_{N_2}}{\frac{1000}{18.0}} = \frac{P_{N_2}}{\kappa^H} = \frac{0.80 \times 760}{6.80 \times 10^7} \qquad (210)$$

From eq(210), moles of nitrogen in 1L of the blood are found as $n_{N_2} = 4.97 \times 10^{-4}$ mol.

Ans. moles of nitrogen in 1L of blood: ca. $5 \times 10^{-4}\,\text{mol} \cdot \text{L}^{-1}$

(b) After multiplying the moles of nitrogen obtained above by five (5L), the calculated result is substituted into the ideal gas equation of state to find the volume of the gas at room temperature under 1 atmospheric pressure as calculated below.

$$V = \frac{n_{N_2} RT}{P} = \frac{5 \times 4.97 \times 10^{-4} \times 8.31 \times 298}{101 \times 10^3} = 6.09 \times 10^{-5}\,\text{m}^3 = 6.09 \times 10^{-2}\,\text{L}$$

Ans. volume of nitrogen at room temp. and 1 atm: 6.09×10^{-2} L

(c) Under the pressure of 10 atm, the amount of ten times more nitrogen is dissolved in the blood. As a consequence, when the surrounding pressure rapidly drops to 1 atm, the nitrogen gas corresponding to the pressure difference of 9 atm is released as a gas into blood vessels and thus it amounts to $6.09 \times 10^{-2} \times 9 = 0.55$ L.

Ans. amount released as a gas into blood vessels: 0.55 L

(d) In principle, a gas mixture of a nontoxic gas with a large *Henry*'s constant and oxygen ($\kappa^H = 3.27 \times 10^7$ /Torr) can be employed to control gas evolution as little as possible while keeping healthy breathing. Practically, a gas mixture of helium He ($\kappa^H = 1.12 \times 10^8$ / Torr) and oxygen is used.

Ans. See text (d).

Solution 4-29

Compared with oxygen or nitrogen, gaseous ammonium is known more readily soluble in water at low pressures as is judged from the magnitude of X_{NH_3}. Furthermore, the amount dissolved in water is not proportional to the pressure of the gas but is greater than that. One of the reasons for this could be as follows. Since gaseous ammonia not only dissolves in water as neutral molecules but also can react with water to form compounds other than ammonia, there exist in the solution the substances produced from ammonia which is directly in equilibrium with ammonia in the gas phase. Because of this reaction, the apparent solubility of ammonia is expected to increase by the amount having reacted. In the following, therefore, the solubility is calculated by assuming that ammonium reacts with water to produce ammonium ion to some extent in water.

Denoting the gaseous ammonia in the dissolution equilibrium A(g), the neutral ammonia in solution A(aq) and the ionic molecule produced by the reaction with proton in solution AH^+, the concentrations of these species are related by the equilibriums expressed as eqs(211) and (212).

$$A(g) \rightleftarrows A(aq) \qquad \frac{[A(aq)]}{[A(g)]} = K_1 \qquad (211)$$

$$A(aq) + H^+ \rightleftarrows AH^+ \qquad \frac{[AH^+]}{[A(aq)][H^+]} = K_2 \qquad (212)$$

The concentrations of the ions in solution must fulfill eq(213) because of the electroneutrality as well as eq(214) which is the ion product of water.

$$[AH^+] + [H^+] = [OH^-] \qquad (213)$$

$$[H^+] \cdot [OH^-] = K_w \qquad (214)$$

Substituting eq(214) into eq(213), eq(215) is obtained, and applying the relation of eq(212) to this, eq(216) is derived.

$$[AH^+] + [H^+] = \frac{K_w}{[H^+]} \qquad (215)$$

$$K_2[A(aq)][H^+] + [H^+] = \frac{K_w}{[H^+]} \qquad (216)$$

From eq(216), the concentration of hydrogen ion is given by eq(217).

$$[H^+] = \sqrt{\frac{K_w}{(K_2[A(aq)]+1)}} \qquad (217)$$

Also, the concentration of ammonium ion is given by eq(218).

$$[AH^+] = K_2[A(aq)]\sqrt{\frac{K_w}{(K_2[A(aq)]+1)}} \qquad (218)$$

As a result, the ratio of the concentration of ammonia in the gas phase to the sum of the concentrations of neutral ammonia and ammonium ion in solution is expressed by eq(219).

$$\frac{[A(aq)]+[AH^+]}{[A(g)]} = K_1 + K_2K_1\sqrt{\frac{K_w}{(K_2K_1[A(g)]+1)}} \qquad (219)$$

Consequently, the mole fraction X_{NH_3}, whose values are listed in the table in the problem, is considered to be given by eq(220).

$$X_{NH_3} \approx \frac{[A(aq)]+[AH^+]}{[H_2O]} = \frac{[A(aq)]+[AH^+]}{55.6} \qquad (220)$$

Employing the values of $K_1 = 1.0,\ K_2 = 10^{13}$ and $K_w = 10^{-14}$ in eq(219) and using the relation of $P_{NH_3} = [A(g)] \cdot 0.082 \cdot T$, P_{NH_3} is expressed as a function of $[A(aq)]+[AH^+]$ and plotted in Fig.21 against $[A(aq)]+[AH^+]$. Apart from a quantitative agreement with the experimental values, Fig.21 is considered to reproduce well a feature of the experimental data. If there exists no reaction of ammonia in solution, then dissolution equilibrium of ammonia is simply expressed by eq(211) alone, namely by *Henry*'s law and, as is shown in Fig.21, the amount of ammonia to be dissolved in water would be reduced significantly.

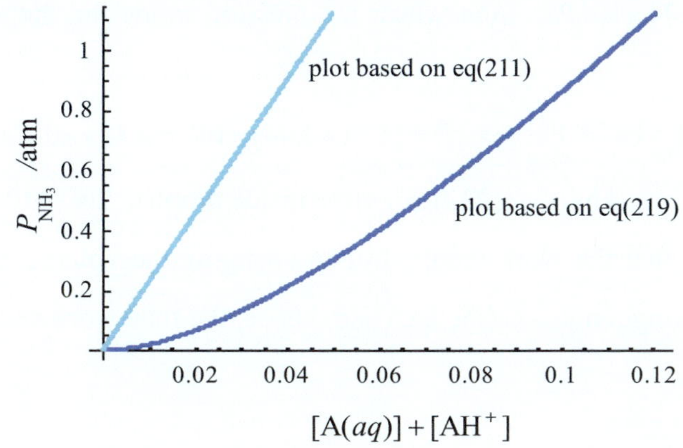

Fig.21 Dissolution of ammonia into water – effect of protonation.

On the other hand, when a much larger value is taken for the equilibrium constant in eq(211) (a far more soluble gas with smaller *Henry*'s constant), a difference between eq(211) and eq(219) becomes less remarkable.

Ans. See text.

Solution 4-30

To consider the balance of pressure in a droplet, a plane which passes through the center of a spherical droplet is chosen as is shown in yellow in Fig.22. On this plane, the pressure of the same magnitude as the external pressure P_0 is exerted as is explained in Fig.23 and indicated by a grey arrow in Fig.22. The downward force due to the surface tension of the droplet surface is indicated by brown arrows and is given by $\gamma \times 2\pi r$. Hence, the force per unit area of the plane is given by $(\gamma \times 2\pi r)/\pi r^2 = 2\gamma/r$.

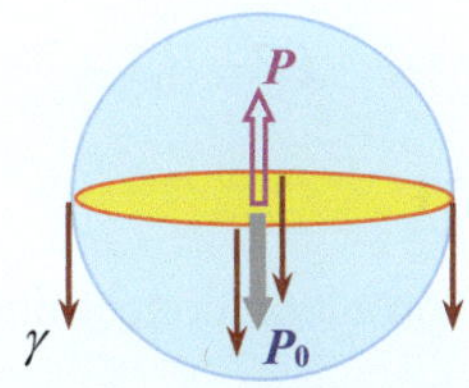

Fig.22 Balance of pressure within a droplet.

Consequently, the pressure balance at the plane under consideration is expressed as eq(221) where *P* is the pressure inside the droplet. (The gravitational force is neglected.) From this, the pressure in the droplet is given by eq(222) and is found always higher than the external pressure P_0.

$$P - P_0 - \frac{2\pi a \times \gamma}{\pi a^2} = 0 \qquad (221)$$

$$P \ \text{(inside)} = P_0\text{(outside)} + \frac{2\gamma}{a} \qquad (222)$$

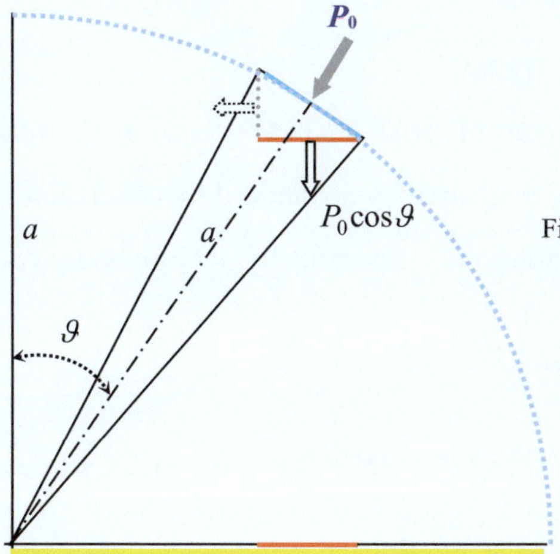

Fig.23 Isotropy of static pressure illustrated for a quarter cross section of a droplet.
Although the vertical component of the pressure is multiplied by $\cos\vartheta$, the area is also multiplied by $\cos\vartheta$ and the pressure exerted within a droplet is P_0 anywhere (principle of *Pascal*).

Substituting the given values for the radius of the water droplet and the surface tension into eq(222), the pressure due to the surface tension is found as 1.44×10^5 Pa, from which the pressure inside the droplet is calculated to be 2.45×10^5 Pa.

$$\frac{2\gamma}{a}=\frac{2\times72.0\times10^{-3}(\mathrm{N\cdot m^{-1}})}{1.0\times10^{-6}(\mathrm{m})}=1.44\times10^5\,\mathrm{Pa}\;\Rightarrow\; P=(1.01+1.44)\times10^5=2.45\times10^5\,\mathrm{Pa}$$

Ans. eq(222), pressure inside droplet: 2.45×10^5 Pa

Eq(222) can be derived readily by considering that the work required upon changing the volume of the droplet is balanced with the work required for a change in its surface area and, hence, the total work does not change, leading to eq(223).

$$-(P-P_0)dV+\gamma dA=0 \qquad (223)$$

Here, the volume change dV and the surface area change dA are related by the following equation,

$$dV=d\left(\frac{4}{3}\pi a^3\right)=4\pi a^2 da,\quad dA=d(4\pi a^2)=8\pi a da$$

and substituting this into eq(223), eq(222'), identical to eq(222), is derived.

$$(P-P_0)=\frac{2\gamma}{a} \qquad (222')$$

Solution 4-31

This is a problem concerning the two-phase equilibrium of a pure liquid. From the phase rule of $f=C+2-P$, the degree of the freedom of the system is one. For this, when either of temperature or pressure is specified, the system reaches an equilibrium in which the chemical potentials of the gas (g) and the liquid (l) are equal. Denoting their chemical potentials at the equilibrium pressure P_0 by $\mu_g(T,P_0)$ and $\mu_l(T,P_0)$, respectively, eq(224) is satisfied,

$$\mu_l(T,P_0)=\mu_g(T,P_0)=\mu_g^\circ(T)+RT\ln\frac{P_0}{P^\circ} \qquad (224)$$

in which the vapor gas is assumed to be an ideal gas whose chemical potential is given by eq(225).

$$\mu_g(T,P_0)=\mu_g^\circ(T)+RT\ln\frac{P_0}{P^\circ} \qquad (225)$$

Similarly, when the external pressure ΔP is applied and the vapor pressure changes to P' at a new equilibrium reached, eqs(224') and (225') are satisfied.

$$\mu_l(T,P'+\Delta P)=\mu_g(T,P') \qquad (224')$$

$$\mu_g(T,P')=\mu_g^\circ(T)+RT\ln\frac{P'}{P^\circ} \qquad (225')$$

The chemical potential of the liquid varies by the amount of $\mu_l(T,P'+\Delta P)_T-\mu_l(T,P_0)$ upon applying the external pressure ΔP. The amount of this kind can be evaluated by a standard thermodynamic procedure of "the integration of a state quantity along the route of variation". Specifically, it is given by eq(226).

$$\mu_l(T,P'+\Delta P)_T-\mu_l(T,P_0)=\int_{P_0}^{P'+\Delta P}\left(\frac{\partial\mu_l}{\partial P}\right)_T dP \qquad (226)$$

$$=\int_{P_0}^{P'+\Delta P}\overline{V}_l dP\approx\overline{V}_l\times(P'+\Delta P-P_0) \qquad (226')$$

When it is assumed that $\overline{V}_l=(\partial\mu_l/\partial P)_T$ does not vary with pressure, eq(226) can be approximated as eq(226'). Employing this, the condition to be met at the newly reached equilibrium is found as eq(227) via the following equation.

$$\overline{V}_l \cdot (P' + \Delta P - P_0) + \mu_g^\circ(T) + RT\ln\frac{P_0}{P^\circ} = \mu_g^\circ(T) + RT\ln\frac{P'}{P^\circ}$$

$$\overline{V}_l \cdot (P' + \Delta P - P_0) = RT\ln\frac{P'}{P_0} \qquad (227)$$

Since the applied external pressure ΔP is much larger than the vapor pressure (otherwise negligible vapor pressure change), eq(227) is further approximated as eq(227').

$$\frac{\overline{V}_l \Delta P}{RT} = \ln\frac{P'}{P_0} \quad \Rightarrow \quad \frac{P'}{P_0} = \exp\left(\frac{\overline{V}_l \Delta P}{RT}\right) \qquad (227')$$

Eq(227) indicates that the vapor pressure increases by applying the external pressure. This fact is explained on the basis of the principle of *Le Chatelier* by saying that "the compressed liquid tries to expand its volume resisting against the compression, namely tends to vaporize more," resulting in the increased vapor pressure.

Ans. See text, in particular the sentence in the double quotation marks and eq (227').

Solution 4-32

When the equilibrium vapor pressure of a bulk liquid at temperature T is P_0, the saturated vapor pressure of the liquid under the applied external pressure, ΔP is expressed by eq(228) as is derived in **solution 4-31**.

$$\frac{P'}{P_0} = \exp\left(\frac{\overline{V}_l \Delta P}{RT}\right) \qquad (228) \qquad (\text{eq}\,(227')\ \text{in}\ \textbf{solution 4-31})$$

On the other hand, the pressure inside a droplet of the radius a is increased by the amount given by eq(229) as is found in **solution 4-30**. Therefore, by substituting eq(229) into eq(228), the ratio of the saturated vapor pressure over the droplet P' to that over the bulk liquid P_0 is found to be given by eq(230).

$$\Delta P = \frac{2\gamma}{a} \qquad (229)$$

$$\frac{P'}{P_0} = \exp\left(\frac{2\gamma \overline{V}_l}{aRT}\right) \qquad (230)$$

For mercury, the molar volume $\overline{V}_l$ is calculated from its density and atomic weight and the value of ΔP for a mercury droplet of the radius of $1.0\ \mu\text{m}$ is calculated from eq(229) as are given below.

$$\overline{V}_l = \frac{M}{\rho} = \frac{200.6}{13.6} = 14.75\ \text{mL}\cdot\text{mol}^{-1}, \qquad \frac{2\gamma}{a} = \frac{2\times 472\times 10^{-3}(\text{N}\cdot\text{m}^{-1})}{1.0\times 10^{-6}(\text{m})} = 9.44\times 10^5\,\text{Pa}$$

For several values of the radius, the values for P'/P_0 are found as follows.

$1.0\ \mu\text{m}: 1.0056,\ 0.1\ \mu\text{m}: 1.058,\ 0.01\ \mu\text{m}: 1.75$

As is seen from the result obtained above, the vapor pressure over the droplet increases rapidly with decreasing its radius and as a consequence microdroplets usually cannot exist stably. This is one of the reasons why the finer mercury droplets are more dangerous. This is not limited to mercury but is true for other poisonous substances when they exist in fine particles. For this, their harmful nature should be recognized correctly.

Ans. values of P'/P_0 for relevant radii: $1.0\ \mu\text{m}: 1.0056,\ 0.1\ \mu\text{m}: 1.058,\ 0.01\ \mu\text{m}: 1.75$

"note 1" In some text books of physical chemistry, a similar calculation as above is extended to a droplet of the radius of $0.001\ \mu m = 1\ nm$. A due care must be taken because the number of molecules inside of the droplet of the radius of $1\ nm$ is of the order of a few hundreds, For instance, in the case of water, the average number of water molecules in a droplet of the radius of $0.01\ \mu m = 10\ nm$ is 1.4×10^5 as is calculated below and, hence, not more than a few hundred molecules are arrayed along the radius.

$$\left(\frac{4}{3}\pi(10^{-8})^3 \Big/ 18.0\times10^{-6}\right)\times 6.02\times10^{23} = 1.4\times10^5\ \text{(molecules)}$$

Needless to say, for a droplet of $1\ nm$ in diameter, the total number of molecules is about 140 to which the concept of pressure (thermodynamical concept) cannot obviously be applied as it is.

"note 2" In the case of gas-solid equilibrium, a relation similar to eq(230) is derived. Consequently, the sublimation pressure over fine particles is considered a few times higher than that over the bulk solid.

[advanced problem] Consider a closed container whose volume is V and is kept at a constant temperature T. Initially it contains only a single water droplet of the radius of a_0 but the droplet starts to vaporize partially becoming a smaller droplet until the vaporization stops at the radius of a. Afterwards the droplet remains stable in the container. Putting $V = 10^{-3}\ m^3$, find the radius a of the stable droplet thus generated.

Ans. $\ln\left\langle \frac{4\pi(a_0^3 - a^3)}{3\overline{V}_l}\frac{RT}{V}\cdot\frac{1}{P_0}\right\rangle = \frac{2\gamma\overline{V}_l}{a}$ (Assuming $a_0 = 2\times10^{-3}\ m$ at 298K, $a \approx 65\ \mu m$ is obtained.)

[reference] *E.A.Moelwyn-Hughes, Physical Chemistry,* 2nd revised ed., p.982, 1965.

Solution 4-33a

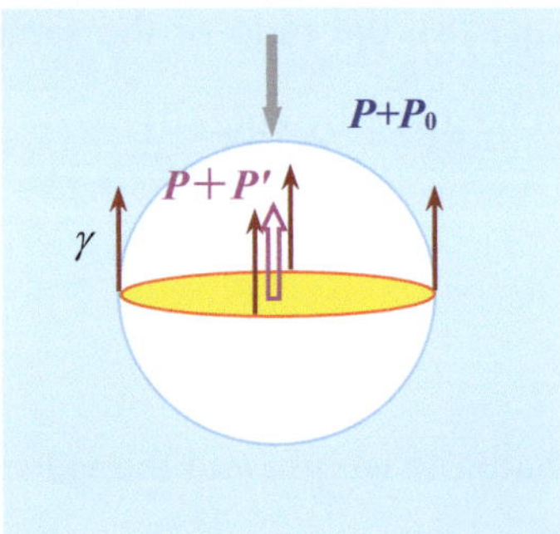

Fig.24 Forces acting on and in a bubble – dynamic balance.

In Fig.24 on the left, the forces working on or in the bubble are shown. From the balance of these forces, eq(231) must be satisfied. As a result, the vapor pressure inside the bubble P' is related to the vapor pressure of the bulk P_0 by eq(232).

$$P + P_0 = P + P' + \frac{2\gamma}{a} \qquad (231)$$

$$P_0 - P' = \frac{2\gamma}{a} \qquad (232)$$

The relation of eq(232) is also readily derived from the balance of the work required for changing the size of the bubble, which is given by eq(233) and eq(233'). In contrast to the case of droplets, since $P_0 - P' > 0$, a positive or negative sign of each term is determined in the case of decreasing volume (surface area as well).

$$-(P_0 - P')dV + \gamma dA = 0 \qquad (233)$$

$$-(P_0 - P')\cdot(-4\pi a^2 da) + \gamma\cdot(-8\pi a da) = 0 \qquad (233')$$

$$P_0 - P' = \frac{2\gamma}{a} \qquad (232) \qquad \text{(reposted)}$$

From eq(232), it is found that the vapor pressure inside the bubble is always smaller than that over the bulk liquid and also that the radius of the bubble should satisfy eq(234). Consequently, it is seen that there

$$P_0 > \frac{2\gamma}{a} \Rightarrow a > \frac{2\gamma}{P_0} \qquad (234)$$

exists the lower limit for the radius of the stable bubble at a given temperature at which P_0 and γ are determined. However, eq(232) derived from the dynamic balance (gravitational force neglected) alone does not necessary express the thermodynamic balance. See the solutin of **problem 4-35** for the method of acquiring the vapor pressure inside the bubble in a thermodynamically correct way.

If we try to explain the result based on the principle of *Le Chatelier*, it can be said as follows: "The formation of bubbles increases the volume of the system and, hence, a force works to the direction so as for the volume to decrease by decreasing the vapor pressure inside the bubbles to promote liquefaction".

Ans. See text. $P_0(\text{bulk}) - P'(\text{bubble}) = 2\gamma/a$ (eq(232))

Solution 4-33b

(a) The chemical potential of a gas (water vapor) at temperature T and pressure P is given by eq(235) in which $\mu_g^\circ(T)$ is the chemical potential of the gas at the standard pressure of P_0.

$$\mu_g(T,P) = \mu_g^\circ(T) + RT\ln\frac{P}{P^\circ} \qquad (235)$$

On the other hand, the chemical potential of the aqueous solution in equilibrium with the gas at the pressure P is given by eq(236), in which $\mu_l^\circ(T,P_0)$ is the chemical potential of the pure liquid (water) under the saturated vapor pressure P_0 and a_{H_2O} is the activity of water in the solution (salt solution).

$$\mu_l(T,P) = \mu_l^\circ(T,P_0) + RT\ln a_{H_2O} \qquad (236)$$

For the pure liquid under its saturated vapor pressure P_0, eq(237) is satisfied.

$$\mu_l^\circ(T,P_0) = \mu_g(T,P_0) = \mu_g^\circ(T) + RT\ln\frac{P_0}{P^\circ} \qquad (237)$$

Using this relation, eq(238), which holds for the gas-solution equilibrium at the saturated vapor pressure

$$\mu_l(T,P_{sol}) = \mu_g(T,P_{sol}) \qquad (238)$$

P_{sol}, is rewritten as eq(239) and it is further rearranged as the first part of eq(240).

$$\mu_g^\circ(T) + RT\ln\frac{P_{sol}}{P^\circ} = \mu_g^\circ(T) + RT\ln\frac{P_0}{P^\circ} + RT\ln a_{H_2O} \qquad (239)$$

$$\ln\frac{P_{sol}}{P_0} = \ln a_{H_2O} \;\Rightarrow\; P_{sol} = P_0 a_{H_2O} \qquad (240)$$

As a result, the saturated vapor pressure P_{sol} of the solution is given by eq(240).

Assuming that the salt solution under consideration is an ideal solution, the activity of water in the solution is equal to its mole fraction and, hence, it is given by $a_{H_2O} \approx X_{H_2O} = 0.99928$ as is calculated below.

$$a_{H_2O} \approx X_{H_2O} = \frac{1000/18}{1000/18 + 0.02\times 2} = 0.99928$$

Substituting the activity calculated thus into eq(240), the vapor pressure of the salt solution at 25°C is found as $0.99928\times 60 = 59.96$ Torr.

Ans. saturated vapor pressure: 59.96 Torr

(b) For pure liquids, the radius a of a droplet and the vapor pressure P' over the droplet is related by eq(241) as has already been discussed in the solution of **problem 4-32**.

$$\ln\frac{P'}{P_0} = \frac{2\gamma\overline{V}_l}{a} \;\Rightarrow\; \frac{P'}{P_0} = \exp\left(\frac{2\gamma\overline{V}_l}{a}\right) \qquad (230) \quad \text{(reposted)}$$

It should be remarked here that for the case under discussion, the saturated vapor pressure of the solution is P_{sol} and the vapor pressure over the droplet is the saturated vapor pressure of pure water. Consequently, upon employing eq(230), the relevant pressures are to be replaced as follows: $P_0 \Rightarrow P_{sol}$ and $P' \Rightarrow P_0$. And then substituting the relation of eq(240) into eq(230) thus replaced, the last relation of eq(241) is obtained.

$$RT\ln\frac{P'}{P_0} \Rightarrow RT\ln\frac{P_0}{P_{sol}} = RT\ln\frac{P_0}{P_0 a_{H_2O}} \approx -RT\ln X_{H_2O} \approx \frac{2\gamma\overline{V}_{H_2O}}{a} \qquad (241)$$

From this, the radius of the droplet of the salt solution is found as 1.45×10^{-5} m as is shown below. Here, the values of the surface tension γ and the molar volume $\overline{V}_{H_2O}$ used for the calculation are those of pure water.

$$a \approx -\frac{2\gamma\overline{V}_{H_2O}}{RT\ln X_{H_2O}} = -\frac{2\times72\times10^{-3}\times1.8\times10^{-5}}{8.31\times298\times\ln0.999928} = 1.45\times10^{-5}\,\text{m}$$

Ans. radius of the droplet: ca. 15 μm

(c) There may be two reasons for the stable existence of the droplets made of solutions dissolving non-volatile substances such as salts. Firstly, as is evident from what is meant by eq(240), the saturated vapor pressure of the solution is lower than that of the pure solvent. For this, the solution droplets of certain sizes which satisfy eq(241) (or equivalent one) can always exist stationary. Secondly, when the vapor pressure is lower than the saturated vapor pressure of the pure liquid, naturally the droplet of the solution partially vaporizes resulting in the smaller droplet but correspondingly the concentration of the solute in the droplet gets higher accompanied by the lower activity of water (further decrease in $a_{H_2O} < 1$). Correspondingly, the saturated vapor pressure of the droplet decreases with shrinking until for it to reach the vapor pressure outside and finally the droplet stops shrinking to exist stably. On the other hand, when the vapor pressure outside is higher than the equilibrium vapor pressure $P_0 a_{H_2O}$, the droplet gets larger (increase in $a_{H_2O} \to 1$) so as to increase the saturated vapor pressure until it reaches the vapor pressure outside.

On the contrary, for the droplet of a pure liquid, its saturated vapor pressure is always higher than that of the pure liquid, the droplet tries to reach this higher saturated vapor pressure by shrinking until it finally disappears.

Ans. See text (c).

Solution 4-34

(a) The saturated vapor pressure of water P_0 at 300 K is found as 3543 Pa by substituting the given temperature of 300 K into eq(242).

$$\log_{10}P_0 = 20.9586 - 4.0843\times\log_{10}T - 2825/T \qquad (242)$$

$$= 20.9586 - 4.0843\times\log_{10}300 - 2825/300 = 1.4246$$

$$P_0 = 10^{1.4246} = 26.64\ \text{mmHg}\langle\times(101.3\times10^3/760)\rangle = 3543\ \text{Pa}$$

Consequently, assuming that the surface tension of water at this temperature is $\gamma = 72\times10^{-3}\,\text{N}\cdot\text{m}^{-1}$, the smallest radius of the bubble formed at this temperature is calculated to be 41 μm as is shown below from the relation (eq(234)) given in **solution 4-33a**.

$$a > \frac{2\gamma}{P_0} = \frac{72\times10^{-3}(\text{N}\cdot\text{m}^{-1})}{3533(\text{N}\cdot\text{m}^{-2})} = 4.1\times10^{-5}\,\text{m} \qquad \text{(eq(234) reposted)}$$

Ans. smallest radius of the bubble: ≥ 41 μm

However, this value is obtained based solely on the mechanical balance and, hence, does not answer whether any bubble of that size can exist stably in water. For a correct treatment, see **problem 4-35** and its **solution**.

(a) As is obvious from eq(234), the radius of the smallest bubble decreases with increasing the vapor pressure of water. The surface tension of water, as is readily expected, decreases with increasing temperature and is sufficiently described by eq(243). It is plotted against T in Fig.25, from which a variation of the surface

$$\log_{10}\gamma = 2.062 + 0.766 \times \log_{10}(1 - T/647) \qquad (243)$$

tension over the temperature range 273 – 373 K is found as to an extent of ca.20 %.

Using eqs(242) and (243), the radius of the smallest bubble is obtained as $a = 2\gamma/P_0$, and it is plotted in Fig.26 against T. From Fig.26, it is seen that the radius rapidly decreases with increasing temperature, which is Well recognized through experience in boiling water. Observe this carefully when you boil water next time. The radius of the smallest bubble at 368 K near the boiling point is 1.4 μm.

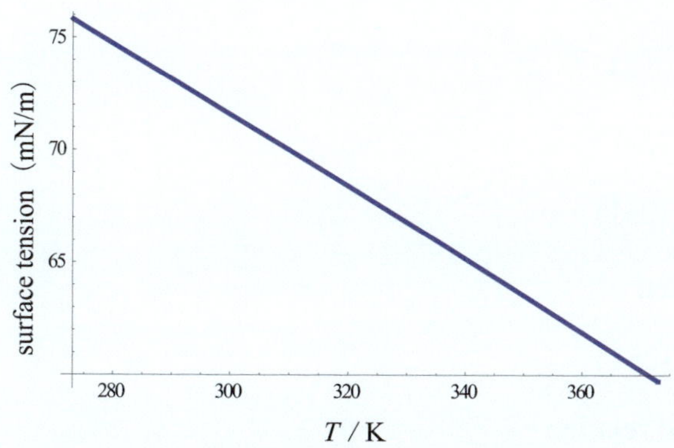

Fig.25 Variation of surface tension of water with temperature.

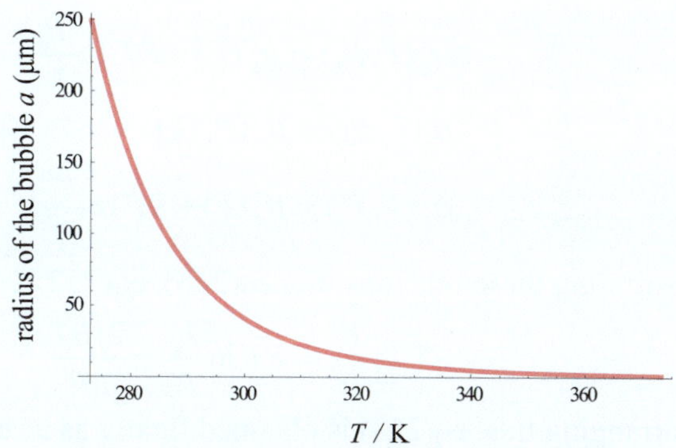

Fig.26 The radius of the smallest bubble formed in water as a function of temperature.

Ans. The radius of the smallest bubble rapidly decreases with temperature. See Fig.26.

[advanced problem] So far the gravitational force has not been taken into consideration in solving the problems related to bubbles. Describe how the boiling phenomena of water are observed in the space with no gravitational force.

Solution 4-35

Dividing eq(244), which expresses the condition that the bubble in equilibrium with the liquid can exist stably, by dn_g, eq(244') is obtained.

$$dG_{T,P} = \mu_g dn_g + \mu_l dn_l + \gamma d\mathrm{A} = 0 \qquad (244)$$

$$\mu_g + \mu_l \frac{dn_l}{dn_g} + \gamma \times \frac{d\mathrm{A}}{dn_g} = 0 \qquad (244')$$

Noticing here that the sum of the moles of the molecules in the gas and liquid phases is constant (eq(245))

$$n_g + n_l = n_0 \;\Rightarrow\; dn_g = -dn_l \qquad (245)$$

and putting the radius of the bubble equal to a, eqs(246) and (247) are derived using the relations of

$V_g = 4\pi a^3/3$, $A = 4\pi a^2$ and $P'V_g = n_g RT$. (Any change in the number of molecules in the interface is not taken into account. See **"remark"**.)

$$dn_g = \frac{d(P'V_g)}{RT} = \frac{P'dV_g}{RT} = \frac{P'4\pi a^2 da}{RT} \qquad (246)$$

$$\frac{dA}{dn_g} = \frac{8\pi a da}{\dfrac{4\pi a^2 da P'}{RT}} = \frac{2RT}{aP'} \qquad (247)$$

Substituting these relations into eq(244'), eq(248) is obtained.

$$\mu_g - \mu_l + \gamma \times \frac{2RT}{aP'} = 0 \qquad (248)$$

The chemical potential of the gas in equilibrium with the liquid at the saturated vapor pressure P_0 is given by eq(249) and the chemical potential of the liquid is given by eq(250), where $\mu_g^\circ(T)$ is the chemical potential of the gas at the standard pressure of P°. Also, the chemical potential of the gas in the bubble is given by eq(251).

$$\mu_g(T, P_0) = \mu_g^\circ(T) + RT \ln\frac{P_0}{P^\circ} \qquad (249)$$

$$\mu_l(T, P_0) = \mu_g(T, P_0) \qquad (250)$$

$$\mu_g(T, P') = \mu_g^\circ(T) + RT \ln\frac{P'}{P^\circ} \qquad (251)$$

Substituting these relations into eq(248), eq(252) is derived.

$$RT \ln\frac{P'}{P^\circ} = RT \ln\frac{P_0}{P^\circ} - \frac{2\gamma RT}{aP'} \qquad (252)$$

Rearranging this, eq(253) is obtained finally as a required relation

$$\ln\frac{P'}{P_0} = -\frac{2\gamma}{aP'} \qquad (253) \qquad P_0 > P'$$

This relation is also written as eq(253') and when $P' \approx P_0$, it can be expressed as eq(254),

$$\ln\left(1 + \frac{P' - P_0}{P_0}\right) = -\frac{2\gamma}{aP'} \qquad (253')$$

$$\frac{P' - P_0}{P_0} \approx -\frac{2\gamma}{aP'} \quad \Rightarrow \quad P_0 - P' \approx \frac{2\gamma P_0}{aP'} \approx \frac{2\gamma}{a} \qquad (254)$$

which is identical to the equation derived in **problem 4-32** based on the mechanical balance of the bubble. When the radius of the bubble gets larger, the equation shows the vapor pressure inside the bubble becomes not much different from that of the bulk liquid.

It has been shown now that for a given P_0, there exists a lower limit of a which satisfies eq(254). When a is relatively large, P' is determined unambiguously for that a as is seen from eq(254). In Figs.27 and 28 are plotted the functions of $\ln P'/P_0$ (brown) and $-2\gamma/aP'$ (light blue) against P' for $a = 1.0$ μm and $a = 0.1$ μm using $P_0 = 3.139\times10^3$ Pa for the saturated vapor pressure of water and $\gamma = 72.0$ mN · m^{-1} for the surface tension both at 298K. The intersection of the two curves in each figure corresponds to the solution of eq(253).

From Fig.27, the pressure at the intersection is obtained as $P' = 2.999\times10^3$ Pa, but in Fig.28, no intersection is found because the curve of $-2\gamma/aP'$ is shifted too low. In order to make the two curves cross, the curve for $-2\gamma/aP'$ must be lifted up by employing a larger value for a. From these results, it can be concluded that there exists the lower limit of a for the bubble to exist stably and also that the lower limit of a decreases with increasing temperature since the saturated vapor pressure increases with increasing temperature.

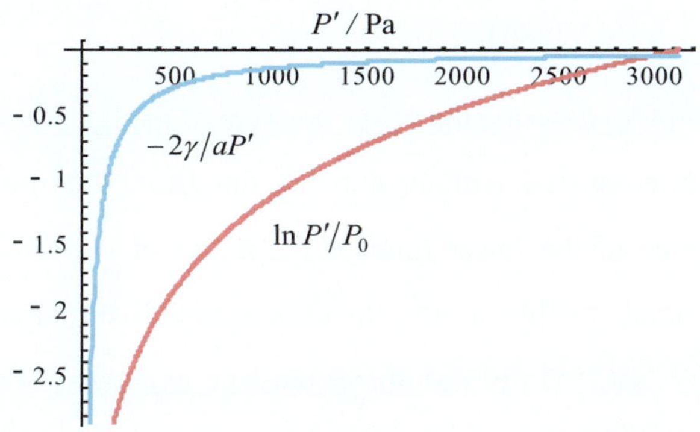

Fig.27 The intersection indicates the stable existence of a bubble of the radius $a = 1.0\,\mu\text{m}$.

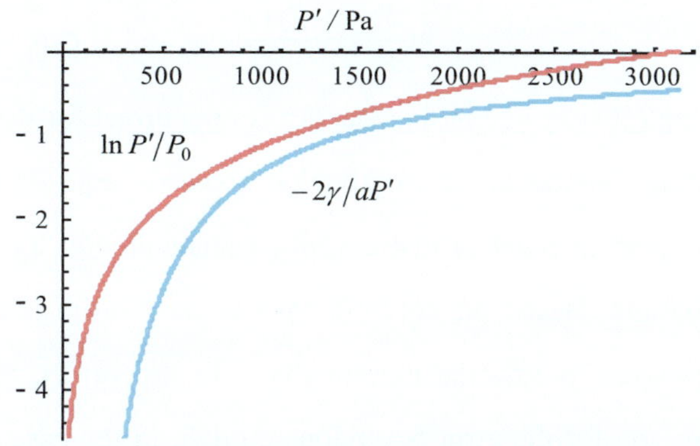

Fig.28 The absence of the intersection shows that the bubble of the radius $a = 0.1\,\mu\text{m}$ cannot exist stably.

Ans. See eq(253) and text associated with Figs.27 and 28.

"remark" The *Gibbs* free energy of the system consisting of bubbles (for simplicity a single bubble) and its surrounding liquid is given by eq(255), in which $\gamma(T,P)$ is the surface tension, μ_l and μ_g are the chemical potentials of the liquid and the bubble, respectively, n_l and n_g are the moles of the molecules in the liquid and the bubble, respectively, and A is the surface area of the bubble.

$$G(T,P,n_g,n_l) = n_g\mu_g + n_l\mu_l + \gamma \mathrm{A} \qquad (255)$$

When the system is in equilibrium at constant T and P, the change in the *Gibbs* free energy is zero and, hence, eq(244) is satisfied.

$$dG_{T,P} = \mu_g dn_g + \mu_l dn_l + \gamma d\mathrm{A} = 0 \qquad (244)$$

Dividing eq(244) by dn_l, eq(244) can be expressed as eq(244').

$$\mu_g \frac{dn_g}{dn_l} + \mu_l + \gamma \times \frac{d\mathrm{A}}{dn_l} = 0 \qquad (244')$$

Noting here that the sum of the moles of the molecules in the gas and liquid phase is constant (eq(245)) and

$$n_g + n_l = n_0 \;\Rightarrow\; dn_g = -dn_l \qquad (245)$$

putting the radius of the bubble equal to a and the molar volume of the liquid equal to $\overline{V}_l$, a change in the moles is derived as eq(256) from eqs(244') and (246) using the relations of $V_g = 4\pi a^3/3$ and $A = 4\pi a^2$. From eq(256), eq(257) is derived.

$$-dn_g = dn_l = -\frac{dV_g}{\overline{V}_l} = -\frac{4\pi a^2 da}{\overline{V}_l} \qquad (256)$$

$$\frac{d\mathrm{A}}{dn_l} = -\frac{8\pi a da}{\dfrac{4\pi a^2 da}{\overline{V}_l}} = -\frac{2\overline{V}_l}{a} \qquad (257)$$

Substituting these relations into eq(244') with some arrangements, eq(258) is obtained.

$$\mu_g = \mu_l - \frac{2\gamma\overline{V}_l}{a} \qquad \mu_g(T,P') = \mu_l(T,P_0) - \frac{2\gamma\overline{V}_l}{a} \tag{258}$$

Substituting eqs(249) and (251) given here before, eq(260) is obtained via eq(259).

$$RT\ln\frac{P'}{P^\circ} = RT\ln\frac{P_0}{P^\circ} - \frac{2\gamma\overline{V}_l}{a} \tag{259}$$

$$\ln\frac{P'}{P_0} = -\frac{2\gamma\overline{V}_l}{a} \Rightarrow \frac{P'}{P_0} = \exp\left(-\frac{2\gamma\overline{V}_l}{a}\right) \tag{260}$$

Eq(260) derived in this manner is not correct for the equation to describe the vapor pressure of the bubble. In many textbooks or workbooks, however, eq(260) as it is employed without noticing the error. The most important point is that eq(260) cannot predict the existence of the lower limit of the radius of the stable bubble. Based on eq(260), instead, any infinitesimally small bubble is able to exist stably if the vapor pressure is lowered unlimitedly. To be correct, therefore, eq(253) derived above must be employed. The reason for the error is obvious though, in the case of the bubble, a change in the moles cannot be obtained from eq(256) by assuming that a volume change of the bubble is equal to the volume change of the liquid (opposite sign). Eq(246) must be used instead.

Solution 4-36

The *Gibbs* free energy of the system consisting of micro particles of a pure substance S and a bulk solvent around them is given by eq(261), in which $\gamma(T,P)$ is the surface tension, μ_c and μ_{sol} are the chemical potentials of the pure solid and the solution, respectively, n_c and n_{sol} are the moles of S in the pure solid (micro particles) and in the solution, respectively, and *A* is the surface area of the micro particles. The free energy of the solvent which is assumed to remain constant during dissolution is not included in eq(261).

$$G(T,P,n_c,n_{sol}) = n_c\mu_c + n_{sol}\mu_{sol} + \gamma\mathrm{A} \tag{261}$$

In the equilibrium under constant *T* and *P*, a change in the *Gibbs* free energy is zero and, hence, eq(262) is satisfied.

$$dG_{T,P} = \mu_c dn_c + \mu_{sol}dn_{sol} + \gamma d\mathrm{A} = 0 \tag{262}$$

Dividing eq(262) by dn_c, eq(262') is obtained.

$$\mu_c + \mu_{sol}\frac{dn_{sol}}{dn_c} + \gamma\times\frac{d\mathrm{A}}{dn_c} = 0 \tag{262'}$$

Noticing here that the sum of the moles of S in the micro particles and in the solution is constant (eq(263)), eqs(264) and (265) are derived, where $\overline{V}_\mathrm{S}$ is the molar volume of S and *a* is the radius of a micro particle, by using the relations of $V_c = 4\pi a^3/3$ and $A = 4\pi a^2$ (see **"note"**).

$$n_c + n_{sol} = n_0 \tag{263}$$

$$dn_c = -dn_{sol} = \frac{dV_c}{\overline{V}_\mathrm{S}} = \frac{4\pi a^2 da}{\overline{V}_\mathrm{S}} \tag{264}$$

$$\frac{d\mathrm{A}}{dn_c} = \frac{8\pi a da}{\dfrac{4\pi a^2 da}{\overline{V}_\mathrm{S}}} = \frac{2\overline{V}_\mathrm{S}}{a} \tag{265}$$

Substituting these relations into eq(262'), eq(266) is obtained.

$$\mu_c = \mu_{sol} - \frac{2\gamma\overline{V}_S}{a} \quad \Rightarrow \quad \mu_c(T,P) = \mu_{sol}(T,P) - \frac{2\gamma\overline{V}_S}{a} \qquad (266)$$

When the saturation is achieved by dissolving the solid of ordinary size, eq(267) is satisfied since the chemical potentials of the solid and the solution are equal. (It should be noted here that μ_c has nothing to do with the size of the crystal.) Also, the chemical potential of the solution saturated by dissolving the microparticles is given by eq(268).

$$\mu_c(T,P) = \mu_{sol}^{\circ}(T,P) + RT\ln\frac{S_{bulk}}{S^{\circ}} \qquad (267)$$

$$\mu_{sol}(T,P) = \mu_{sol}^{\circ}(T,P) + RT\ln\frac{S_{micro}}{S^{\circ}} \qquad (268)$$

Substituting these equations into eq(266), eq(269) is obtained and rewriting it, eq(270) is derived as an equation to express S_{bulk}/S_{micro}.

$$RT\ln\frac{S_{bulk}}{S^{\circ}} = RT\ln\frac{S_{micro}}{S^{\circ}} - \frac{2\gamma\overline{V}_S}{a} \qquad (269)$$

$$\ln\frac{S_{micro}}{S_{bulk}} = \frac{2\gamma\overline{V}_S}{a} \quad \Rightarrow \quad \frac{S_{micro}}{S_{bulk}} = \exp\left(\frac{2\gamma\overline{V}_S}{a}\right) \qquad (270)$$

When the surface energy of the solid is small, namely when γ is small, it is a natural consequence that the relation $S_{bulk}/S_{micro} \approx 1$ is derived.

Ans. solubility and particle size: eq(270)

"note" The molar volume of the solid is assumed to be nearly equal to that in the liquid state. See **"remark"** in **solution 4-35**.

Solution 4-37

In the case of the substance which does not dissociate in solution, the solubility of its micro particles (sphere) S_{micro} is given by eq(270) as is shown in the preceding solution.

$$\frac{S_{micro}}{S_{bulk}} = \exp\left(\frac{2\gamma\overline{V}_S}{a}\right) \qquad (270) \qquad \text{(reposted) (undissociated and spherical)}$$

Also, the chemical potential of the crystal or solution is given by eq(266) (cited again).

$$\mu_c = \mu_{sol} - \frac{2\gamma\overline{V}_S}{a} \quad \Rightarrow \quad \mu_c(T,P) = \mu_{sol}(T,P) - \frac{2\gamma\overline{V}_S}{a} \qquad (266) \quad \text{(reposted)}$$

However, in the case of the crystal which dissociates into ions $AX \rightarrow A^+ + X^=$ like NaCl, the chemical potentials given by eqs(267) and (268) in the previous solution should be replaced with eqs(267') and (268'), respectively, in which the activity coefficient is put equal to one for simplicity.

$$\mu_c(T,P) = \mu_{sol}^{+\circ}(T,P) + \mu_{sol}^{-\circ}(T,P) + RT\ln\frac{[A^+]_{bulk}}{S^{\circ}} + RT\ln\frac{[X^-]_{bulk}}{S^{\circ}} \qquad (267')$$

$$\mu_{sol}(T,P) = \mu_{sol}^{-\circ}(T,P) + \mu_{sol}^{+\circ}(T,P) + RT\ln\frac{[A^+]_{micro}}{S^{\circ}} + RT\ln\frac{[X^-]_{micro}}{S^{\circ}} \qquad (268')$$

Here, expressing the concentrations of the ions by using the solubility products as eq(271) and substituting

$$\left.\begin{aligned} [A^+]_{bulk} &= [X^-]_{bulk} \equiv S_{bulk} \\ [A^+]_{micro} &= [X^-]_{micro} \equiv S_{micro} \end{aligned}\right\} \qquad (271)$$

eqs(267') and (268') into eq(266), eq(272) is obtained via eq(10′) after some arrangements.

$$2RT\ln\frac{S_{bulk}}{S^{\circ}} = 2RT\ln\frac{S_{micro}}{S^{\circ}} - \frac{2\gamma\overline{V}_S}{a} \qquad (270')$$

$$\ln\frac{S_{micro}}{S_{bulk}} = \frac{\gamma\overline{V}_S}{a} \Rightarrow \frac{S_{micro}}{S_{bulk}} = \exp\left(\frac{\gamma\overline{V}_S}{a}\right) \qquad (272)$$ (for spherical particle of ionic crystal)

On the other hand, for the micro crystal not spherical but cubic of an edge with the length l, eqs (264') and (265') are derived by noticing the relation between the crystal volume and surface area.

$$dn_c = -dn_{sol} = \frac{dV_c}{\overline{V}_S} = \frac{3l^2dl}{\overline{V}_S} \qquad (264')$$ (corresponding to eq(264) in the previous solution)

$$\frac{dA}{dn_c} = \frac{6\times 2ldl}{\frac{3l^2dl}{\overline{V}_S}} = \frac{4\overline{V}_S}{l} \qquad (265')$$

Following the same procedure given in the preceding solution with these relations, eq(270) in the preceding solution is found to be replaced by eq(270'').

$$\frac{S_{micro}}{S_{bulk}} = \exp\left(\frac{4\gamma\overline{V}_S}{l}\right) \qquad (270'')$$ (undissociated cubic crystal)

Summarizing eqs(270), (270'') and (272), for the 1:1 ionic cubic crystals AX in which the dissociation is given by $AX \rightarrow A^{z+} + X^{z-}$, eq(273) similar eq(270) is derived.

$$\frac{S_{micro}}{S_{bulk}} = \exp\left(\frac{2\gamma\overline{V}_S}{l}\right) \qquad (273)$$ (1:1 ionic cubic crystal)

Depending on the type of crystal, one of eqs(270), (270''), (272) and (273) is employed and in the present case, eq(273) is applicable. Substituting the given values into it, the solubility is found to increase 6.95 times, though, instead of the surface tension for the solid-liquid interface, the value for the gas-liquid interface is used.

$$\frac{S_{micro}}{S_{bulk}} = \exp\left(\frac{2\times 72.0\times 10^{-3}\times\frac{58.44}{2.17}\times 10^{-6}}{1.0\times 10^{-6}}\right) = 6.95 \quad \text{times}$$

Ans. eq(273) applicable, 6.95 times increase; See **"note"** for the reason.

"note" When we are told to dissolve a solid in water, which is not so readily soluble, then we, even small children, try to dissolve it after breaking it into fine pieces as much as possible. It is interesting to know whether or not such intuition is consistent with what the thermodynamics tells us. Over-saturated state of an ordinary solid may be regarded as analogous to the saturated state of its fine powders. Then it is easy to understand that the over-saturated state is not exceptional. However, if it happens for any large crystal of the same kind to coexist in the over-saturated solution, then the solubility around it must decrease greatly and, hence, the over-saturated state would become unstable.

Solution 4-38

(a) Where the three interfaces contact, the forces (per unit length) due to the surface tensions work in the three directions as is illustrated in Fig.29 given below. Each force works in the direction so as to reduce the area of the relevant interface. Since these three forces must be in balance mechanically, eq(274) is satisfied.

$$\gamma_{VS} + \gamma_{LV} \cdot \cos\vartheta - \gamma_{SL} = 0 \;\Rightarrow\; \gamma_{VS} - \gamma_{SL} = -\gamma_{LV} \cdot \cos\vartheta \qquad (274)$$

On the other hand, when the liquid level in the capillary is raised by *dh*, a change in the surface energy is given by eq(275), in which the liquid surface area (air-liquid) is assumed invariant.

$$\sum^{\text{all interfaces}} \gamma_{ij} dA_{ij} = \gamma_{VS} dA_{VS} + \gamma_{SL} dA_{SL} + \gamma_{LV} dA_{LV}$$

$$= -\gamma_{VS}(2\pi r dh) + \gamma_{SL}(2\pi r dh) \qquad (275)$$

Also, a change in the potential energy associated with the rise in the liquid level in the capillary is given by eq(276), in which the densities of the liquid and the air are denoted ρ_L and ρ_V, respectively.

$$\sum ghdm_i = gh \times (\rho_L \times \pi r^2 - \rho_V \times \pi r^2) \cdot dh \qquad (276)$$

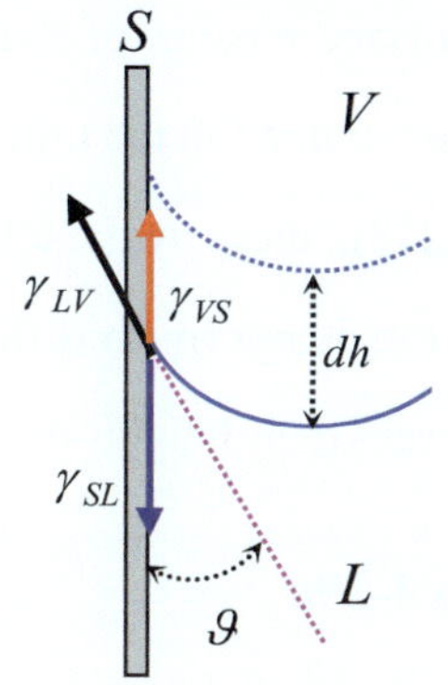

Fig.29 Illustration of surface tension and contact angle where three interphases meet.

From the conservation of the energy, these energies are equal and putting eq(275) equal to eq(276), eq(277') is obtained via eq(277).

$$(\rho_L - \rho_V) \times gh \times \pi r^2 dh = -\gamma_{VS}(2\pi r dh) + \gamma_{SL}(2\pi r dh) \qquad (277)$$

$$(\rho_L - \rho_V) \times gh \times r = -2(\gamma_{VS} - \gamma_{SL}) \qquad (277')$$

Substituting eq(274) into eq(277'), it is found the surface tension of the liquid or the height of the liquid level in the capillary is given by eq(278).

$$\gamma_{LV} = \frac{(\rho_L - \rho_V) \times rgh}{2\cos\vartheta}, \qquad h = \frac{2\gamma_{LV}\cos\vartheta}{(\rho_L - \rho_V) \times rg} \approx \frac{2\gamma_{LV}\cos\vartheta}{\rho_L \times rg} \qquad (278)$$

Looking at the resulting eq(278) alone, it may appear that the height in the capillary has nothing to do with the material of which the capillary is made, but as is evident from eq(274), the height *h* is related with the material through the contact angle.

Ans. rise in capillary: eq(278)

"note" Definition of contact angle.

At the point where the three interfaces meet a tangential line is drawn along the liquid surface and out of the two angles spanned by this tangential line and the interface, <u>the one spanning the liquid area</u> is defined as the contact angle. Therefore, when the liquid surface in the capillary assumes a concave shape, the contact angle α takes a value smaller than 90° and when it is convex, the contact angle becomes larger than 90°.

(b) Substituting the given values into eq(278), the height *h* is found as below.

$$h = \frac{2 \times 72.0 \times 10^{-3} \times \cos(\pi \times 10/180)}{10^6 \times 0.1 \times 10^{-3} \times 9.8} = 14.5\ \text{cm}$$

Even when an error of $\pm 2^\circ$ is accompanied by the value of the contact angle, the resulting error in the height is only $\pm 0.6\%$ of its value as is calculated below.

$$\pm \frac{\sin(\pi \times 10/180) \times (\pi \times 2/180)}{\cos(\pi \times 10/180)} \times 100 = \pm 0.6\%$$

Ans. height: 14.5 cm, error: $\pm 0.6\%$

(c) The surface tension of sea water which dissolves significant amounts of ionic substances is larger than that of pure water. On the other hand, the density of the sea water is greater than that of pure water. As is seen from eq(278), these two factors cancel each other and, as a result, it is expected that the height of the capillary rise is not much different between sea water and pure water. It would be interesting to test actually how different their heights are.

Ans. See text (c).

Solution 4-39a

The magnitude of the surface tension decides which form the substance B assumes when it is placed between the surfaces of A and O. As is illustrated in Fig.30, when the surface tension γ_{AO} is larger than the sum of γ_{OB} and γ_{BA}, the substance B spreads along the interface between A and O (right side of Fig.30). On the other hand, when it is smaller than the sum, the substance B is seen to contract to form a convex lens as shown on the left side of Fig.30.

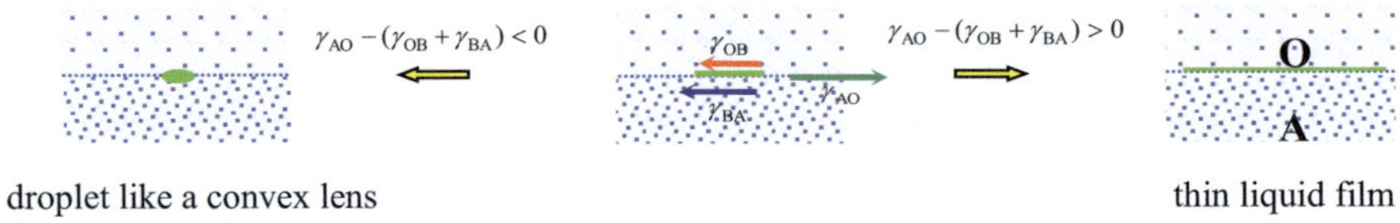

Fig.30 Stable three-phased states determined by the magnitudes of the three surface tensions. See Fig.31 for further details.

Using the given values for the surface tensions, the relation below is checked and is found negative.

$$\gamma_{AO} - (\gamma_{OB} + \gamma_{BA}) = 72.5 - (30.0 + 52.1) = -9.6 < 0$$

Consequently, it is concluded that a droplet of B like a small convex lens is formed. In fact, more exactly, how the lens looks like is known by calculation as is shown below. The contact angles where the three phases meet are denoted ϑ_1 and ϑ_2 as is shown in Fig.31. From the mechanical balances of the forces working there, $\gamma_{OB}\cos\vartheta_1 + \gamma_{BA}\cos\vartheta_2 = \gamma_{AO}$ and $\gamma_{OB}\sin\vartheta_1 = \gamma_{BA}\sin\vartheta_2$, eqs(279) and (280) are derived.

$$\cos\vartheta_1 = \frac{\gamma_{AO}^2 - \gamma_{BA}^2 + \gamma_{OB}^2}{2\gamma_{AO}\gamma_{OB}} \qquad (279)$$

$$\cos\vartheta_2 = \frac{\gamma_{AO} - \gamma_{OB}\cos\vartheta_1}{\gamma_{BA}} = \frac{\gamma_{BA}^2 - \gamma_{OB}^2 + \gamma_{AO}^2}{2\gamma_{AO}\gamma_{BA}} \qquad (280)$$

From these equations, the contact angles are determined as follows:

$$\cos\vartheta_1 = 0.791 \Rightarrow \vartheta_1 = 37.7^\circ \text{ and } \cos\vartheta_2 = 0.936 \Rightarrow \vartheta_2 = 20.6^\circ.$$

These values correspond to the system of water (A), air (O) and *n*-decane (B). When octylalcohol is employed as B, a thin liquid film will be formed.

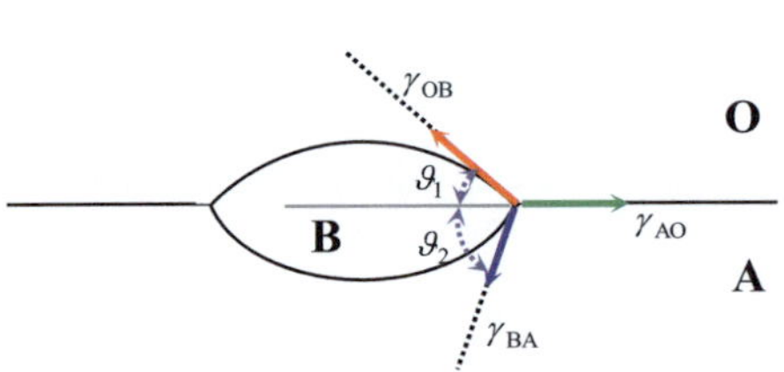

Fig.31 Balance of tensional forces and contact angles where three phases meet.

For simplicity, the gravitational force on B is neglected.

Ans. (ii), droplet like a convex lens

"note" When the phase of A is a solid phase like metal or glass, the contact angle is put $\vartheta_2 = 0$. Then, from eq(280), ϑ_1 is given by eq(279').

$$\cos\vartheta_1 = \frac{\gamma_{AO} - \gamma_{BA}}{\gamma_{OB}} \qquad (279')$$

Solution 4-39b

Which form of a lens or a film the substance B dropped on the water surface (A) develops is determined by the magnitude (positive or negative) of the spreading factor S_{BA} defined below (see preceding problem). When it takes a positive value, the substance B assumes a liquid film and when negative, it forms a droplet like a lens.

$$S_{BA} = \gamma_{AO} - (\gamma_{OB} + \gamma_{BA})$$

Based on this criterion, *n*-octane ($S_{BA} = 0.2$), *n*-octanol (36.8), benzene (8.9) and oleic acid (24.6) are found to form a liquid film but *n*-hexadecane (-9.3) forms a droplet like a lens. However, since the value S_{BA} for *n*-octane is close to zero, it must be quite sensitive to temperature which form *n*-octane actually takes on the water surface. Also, in the case of benzene, soon after it is dropped on the water surface it spreads over the surface rapidly but after a while, it comes to form a droplet because of the surface tension reduced by its non-negligible solubility in water. When a detergent which has a far greater spreading factor coexists, even the substance which intrinsically spreads over the water surface comes to coagulate to form lenses

Ans. See text for shapes.

"note" There exist solid materials which tend to form a thin film on water surface. Well-known example is camphor. One of the authors still remembers, at a night stall on a summer holiday as a child he was curiously looking at a number of tiny boats of a few cm in length made of celluloid moving around rapidly in a small water tank in which a lot of colorful golden fish for sale were swimming. What is the driving force given to those boats? Is there any other substance which can be used in the replacement of camphor?

Solution 4-40

The surface (excess) concentration is found based on the *Gibbs* isothermal adsorption equation (eq(281)) (see "note" in **solution 4-45** for the definition of thermodynamic quantities at the interface). In the case of two component systems, the main component is indicated by the suffix 1 and the minor component by the suffix 2. Denoting the surface tension γ, the surface (excess) concentration $\Gamma_{2,1}$ of the component 2 is given by eq(281), where X_2 is the mole fraction and c_2 is the concentration in the units of weight %.

$$\Gamma_{2,1} = -\frac{1}{RT}\left(\frac{\partial\gamma}{\partial\mu_2}\right)_T = -\frac{1}{RT}\left(\frac{\partial\gamma}{\partial\ln X_2}\right)_T = -\frac{1}{RT}\left(\frac{\partial\gamma}{\partial\ln c_2}\right)_T \times \left(\frac{\partial\ln c_2}{\partial\ln X_2}\right) \qquad (281)$$

The concentration c_2 is given by eq(282) using the molecular weights of M_1 and M_2 and the moles of n_1 and n_2 for the two components.

$$c_2 = \frac{n_2M_2}{n_1M_1 + n_2M_2} \times 100 = \frac{X_2M_2}{X_1M_1 + X_2M_2} \times 100 \qquad (282)$$

$$\ln c_2 \propto \ln X_2M_2 - \ln(X_1M_1 + X_2M_2) + C = \ln X_2M_2 - \ln(M_1 + X_2(M_2 - M_1)) + C \qquad (282')$$

Therefore, the term $(\partial \ln c_2/\partial \ln X_2)$ in eq(281) can be written as eq(283).

$$\frac{\partial \ln c_2}{\partial \ln X_2} = \left(\frac{\partial \ln c_2}{\partial X_2}\right)\left(\frac{\partial X_2}{\partial \ln X_2}\right) = \left\{\frac{1}{X_2} - \frac{(M_2 - M_1)}{M_1 + X_2(M_2 - M_1)}\right\} \times X_2 \qquad (283)$$

The values for the surface tension given in the problem is plotted against $\ln c_2$ in Fig.32 and the best-fitted quadratic function is given by eq(284).

$$\gamma = -0.8296 \times x^2 - 5.7208 \times x + 62.698\ , \quad x \equiv \ln c_2 \qquad (284)$$

From this, we only need to find a value for the first derivative of eq(284) as well as a value for eq(283) at $c_2 = 0.1$. Employing the values of $M_2 = 94.12,\ \ M_2 = 94.12$ and $x = \ln(0.1) = -2.303$, the values necessary for the calculation of the surface concentration are found as follows.

$$\left(\frac{\partial \gamma}{\partial \ln c_2}\right)_T = -1.6592 \times x - 5.7208 = \boxed{-1.90\ \mathrm{mN \cdot m^{-1}}}$$

$$0.1 = \frac{X_2 \times 94.12}{(1 - X_2) \times 18.02 + X_2 \times 94.12} \times 100 \ \Rightarrow$$

$$X_2 = \frac{18.01}{(94120 - 94.12 + 18.01)} = \boxed{1.92 \times 10^{-4}}$$

$$\left(\frac{\partial \ln c_2}{\partial \ln X_2}\right) = 0.9992 = \boxed{1.0}$$

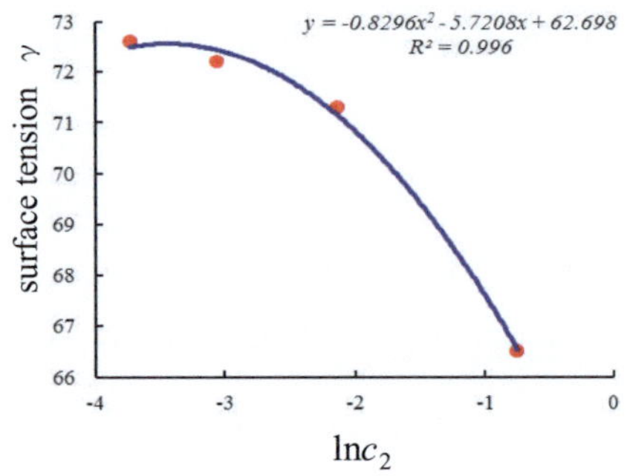

Fig.32 Concentration dependence of the surface tension for the two-component system.

The obtained values are substituted into eq(281) to find the value for $\Gamma_{2,1}$ as given below.

$$\Gamma_{2,1} = -\frac{1}{8.31 \times 303} \times (-1.90 \times 10^{-3}) \times 1.0 = 7.55 \times 10^{-7}\ \mathrm{mol \cdot m^{-2}}$$

The dimension of $\Gamma_{2,1}$ is, as is clear from eq(281), $(\mathrm{N \cdot m^{-1}})/(\mathrm{J \cdot K \cdot mol^{-1} \cdot K}) = \mathrm{m^{-2}}/\mathrm{mol^{-1}} = \mathrm{mol \cdot m^{-2}}$.

Ans. surface concentration: $7.55 \times 10^{-7}\ \mathrm{mol \cdot m^{-2}}$

"note" It is trivial that $(\partial \ln c_2/\partial \ln X_2) = 1$ for the ideal solution, since the chemical potential of the second component is given by $\mu_2 = \mu_2^{\circ} + RT \ln X_2 = \mu_2^{\bullet} + RT \ln c_2$. See **problems** and **solutions** from **4-41** to **4-43** for the meaning of the surface excess concentration.

Solution 4-41

(a) The surface tensions γ of the NaCl aqueous solutions are plotted in Fig.33 against the logarithmic value of the concentration $\ln c$. Since the solutions under consideration are not ideal, the activity coefficient at each concentration is calculated by using the extended *Debye-Hückel* theory (eq(285), see **solution 4-18**) and in Fig.34, the same surface tensions employed in Fig.33 are plotted against the logarithmic value of the activity thus calculated.

$$\ln \gamma_{\pm} = \frac{-1.17\sqrt{I}}{1 + 1.30\sqrt{I}}, \qquad a_{\pm} = \gamma_{\pm} c \qquad (285)$$

In the case of a 1:1 salt like NaCl, the ionic strength (eq(286)) becomes equal to the molal concentration (moles per weight).

$$I = \frac{1}{2} \sum_{i}^{all\ ions} m_i Z_i^2 = m_{\mathrm{NaCl}} \approx c \qquad (286)$$

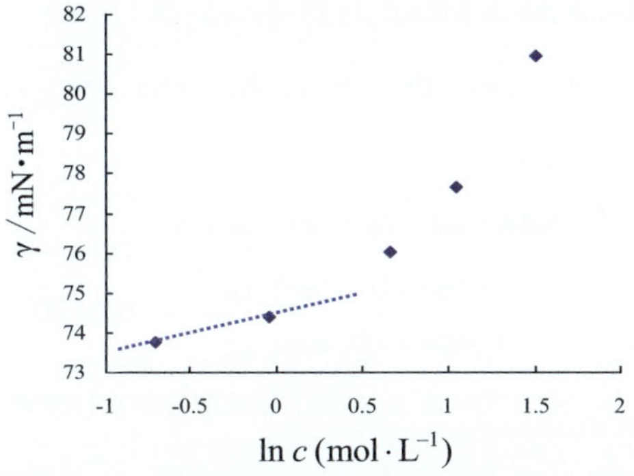

Fig.33 Concentration dependence of surface tension – NaCl aqueous solutions. ln c plot.

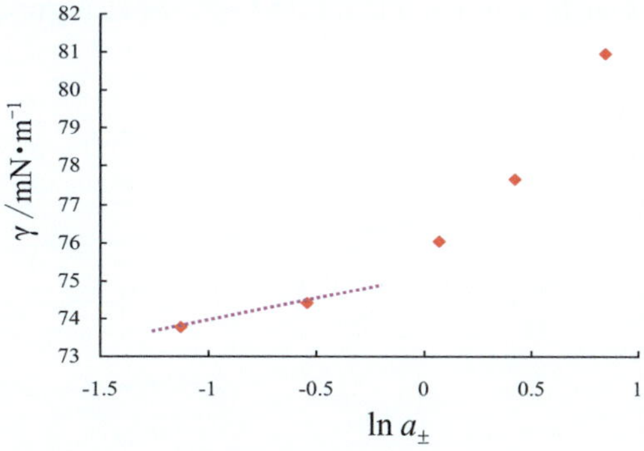

Fig.34 Activity dependence of surface tension – NaCl aqueous solutions. ln $a_{\pm}$ plot.

When the solutions behave ideally, the surface (excess) concentrations Γ_{NaCl} are found based on the definition from eq(287).

$$\Gamma_{\mathrm{NaCl}} \equiv \Gamma_2 = -\frac{1}{RT}\left(\frac{d\gamma}{d\ \ln X_2}\right) = -\frac{1}{RT}\left(\frac{d\gamma}{d\ \ln c}\right) \qquad (287)$$

Therefore, by assuming that the solutions are ideal, the surface concentration is calculated at the lowest concentration and is found as below.

$$\Gamma_{\mathrm{NaCl}} = -\frac{1}{8.31\times 293}\frac{\Delta\gamma}{\Delta \ln c} = -\frac{0.965\times 10^{-3}}{8.31\times 293} = -3.96\times 10^{-7}\ \mathrm{mol\cdot m^{-2}}$$

For the non-ideal solutions, the surface excess concentrations are given by eq(287') and at the lowest bulk concentration it is calculated as given below.

$$\Gamma_{\mathrm{NaCl}} \equiv \Gamma_2 = -\frac{1}{RT}\left(\frac{d\gamma}{d\ \ln a_{\pm}}\right) \qquad (287')$$

$$\Gamma_{\mathrm{NaCl}} = -\frac{1}{8.31\times 293}\frac{\Delta\gamma}{\Delta \ln a_{\pm}} = -\frac{1.09\times 10^{-3}}{8.31\times 293} = -4.48\times 10^{-7}\ \mathrm{mol\cdot m^{-2}}$$

What is known from Figs.33 and 34 is that the surface excess concentrations take negative values since the surface tension increases with increasing concentration or activity. Their magnitudes can be acquired more accurately by employing the activity rather than the concentration, but it is seen that they take the larger negative values at the higher concentrations from Fig.33.

From the definition of the surface excess concentration, it is evident that it can take either a positive or a negative value. The meaning of the negative concentration obtained here is that the ions tend to avoid the interface between water and air where they may be partially surrounded by gaseous (water) molecules but try to stay in the bulk where they are stabilized more under the influence of the stronger electrostatic attractive force due to water molecules around them in the bulk.

Ans. surface (excess) concentration: $-4.5\times 10^{-7}\ \mathrm{mol\cdot m^{-2}}$ at 0.496 $\mathrm{mol\cdot L^{-1}}$

(b) The thickness x of the interface is defined by eq(288) using the absolute value of the surface excess concentration.

$$x = \frac{|\Gamma|}{c} = \frac{1}{RT}\left|\frac{d\gamma}{dc}\right| \qquad (288)$$

When the surface tension is plotted against c, a good linear correlation is found as is shown in Fig.35.

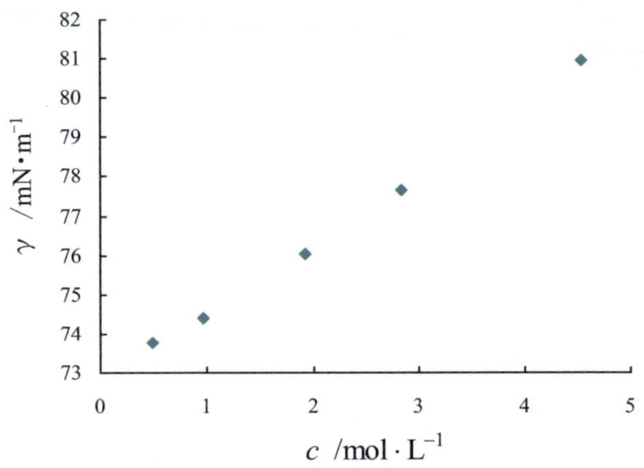

Fig.35 Concentration dependence of surface tension – linear plot (cf. Fig.33).

From its slope, therefore, the value for x is found as given below.

$$4.48\times10^{-7}\,\mathrm{mol\cdot m^{-2}} = c\times x$$

$$x=\frac{4.48\times10^{-7}\,\mathrm{mol\cdot m^{-2}}}{0.496\times10^{3}\,\mathrm{mol\cdot m^{-3}}}=9.3\times10^{-10}\,\mathrm{m}$$

It is seen from a good linear correlation of the plot in Fig.35 that the effective thickness thus defined takes a constant value over a wide range of the concentration of NaCl. However, the thickness has no meaning more than a measure to tell that in that thickness range the concentration of NaCl is significantly different from that in the bulk.

Ans. thickness: 9.3×10^{-10} m

Solution 4-42

In Fig.36 are plotted the ratios of the surface tensions (eq(7) in the problem) against the mole fraction and in Fig.37 the same ratios but against the logarithmic value of the mole fraction.

$$\frac{\gamma}{\gamma_W}=1-0.178\times\ln\left(1+\frac{X_2}{\alpha}\right) \qquad (289) \qquad \text{(reposted)}$$

Assuming that the solutions are ideal ones, the surface concentration Γ_2 of n-valeric acid (component 2) at the mole fraction X_2 can be obtained from eq(290) (see **solution 4-40**).

$$\Gamma_2=-\frac{1}{RT}\left(\frac{d\gamma}{d\ln X_2}\right) \qquad (290)$$

Substituting γ in eq(289) into eq(290), the surface concentration is calculated as below and the value is found as in eq(291).

$$\Gamma_2=-\frac{X_2}{RT}\left(\frac{d\gamma}{dX_2}\right)\times\gamma_W=\frac{X_2}{RT}\times0.178\times\frac{\gamma_W}{a+X_2}\approx\frac{0.178\times\gamma_W}{RT}$$

$$=\frac{0.178\times72.0\times10^{-3}[\mathrm{N\cdot m^{-1}}]}{8.31\times298\ [\mathrm{J\cdot mol^{-1}}]}=5.17\times10^{-6}[\mathrm{mol\cdot m^{-2}}] \qquad (291)$$

Since the obtained surface excess concentration is positive, it can be said that n-valeric acid tends to assemble in the interface as compared with the mean distribution in the bulk solution. However, it should be remarked here that the surface concentration calculated from eq(291) is no more than the excess concentration and, hence, can tell quantitatively neither the number of n-valeric acid molecules in the interface nor an area occupied by a single molecule of n-valeric acid in the interface. The obtained values may easily lead to such erroneous quantitative interpretation, which may be because the surface excess concentration is given in the units of mole/m^2.

As is seen from Fig.37, the plot to acquire the surface concentration is satisfactorily approximated by a straight line. Therefore, it is concluded that the surface (excess) concentration of *n*-valeric acid takes a constant value independent of its mole fraction in solution.

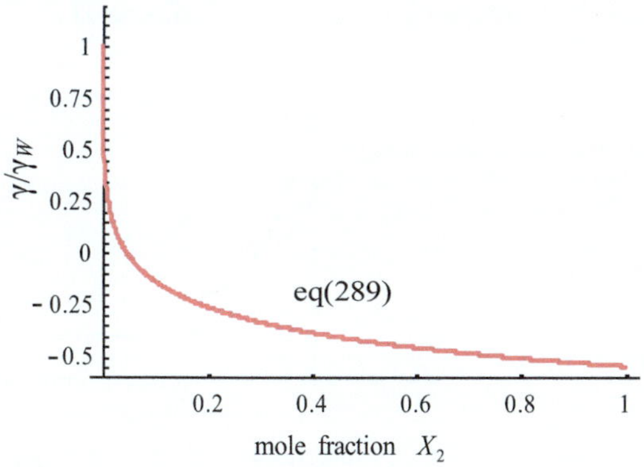

Fig.36 Surface tension and mole fraction of *n*-valeric acid.

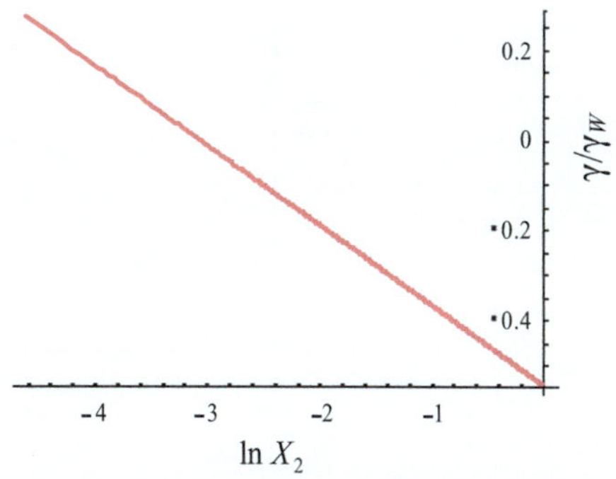

Fig.37 Logarithmic plot to acquire surface (excess) concentration.

Ans. surface (excess) concentration: $5.17\times10^{-6}[\mathrm{mol}\cdot\mathrm{m}^{-2}]$; *n*-valeric acid tends to assemble in the interface as compared in the bulk.

Solution 4-43

The surface (excess) concentration Γ_i of the component *i* which distributes over the phases I and II in equilibrium is defined by eq(292), in which A_s is the interface area, n_i is the total moles of the component *i*, and n_i^I, n_i^{II} are the moles of the same component distributed in the phases I and II, respectively.

$$\Gamma_i \cdot A_s = n_i - n_i^I - n_i^{II} \qquad (292)$$

By the definition the units of the surface (excess) concentration are $\mathrm{mol}\cdot\mathrm{m}^{-2}$. The fundamental definition of eq(292) itself is straightforward, but the interpretation of Γ_i is quite complicated. Taking the two-component system as an example, the true meaning of the surface (excess) concentration is discussed here from a viewpoint of the concentration distribution of each component in the vicinity of the interface.

How the concentrations of the component 1 (solvent) and the component 2 (solute) vary near the interface (e.g., liquid-gas interface) are schematically illustrated for the two cases in Figs.38 and 39. In each figure, the phase I is a liquid and the phase II is a gas. In the vicinity of the interface, the concentrations of both components take values different from those in the bulks (left or right side in each of Figs.) and their distributions are considered to look like those curves shown in Fig.38a or Fig.39a. Near the interface, the concentration of the main component (solvent) on the side of the phase I gradually decreases from the bulk value toward the interface. After crossing the interface, the concentration of the component 1 on the side of the phase II continues to decrease until it reaches the bulk value of zero. In either Figs., the position of the interface is determined with respect to the concentration of the component 1 so as for the area corresponding to the negative deviation from the bulk value on the side of the phase I to cancel the area Ⓑ showing the

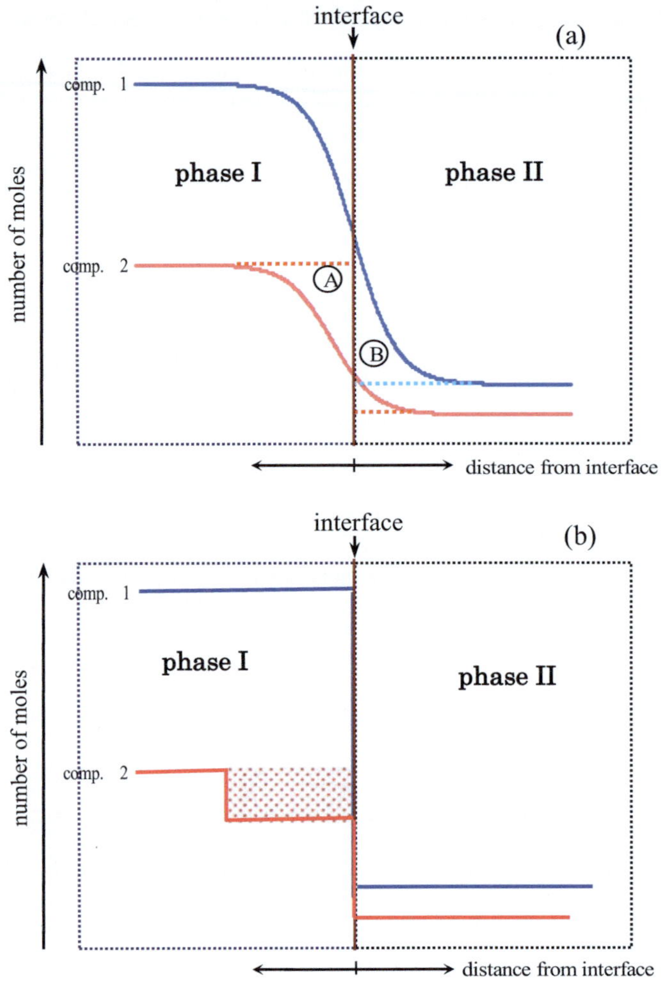

Figs.38a,b Two-phase equilibrium and concentration profile near the interface. In the case of a negative surface excess concentration.

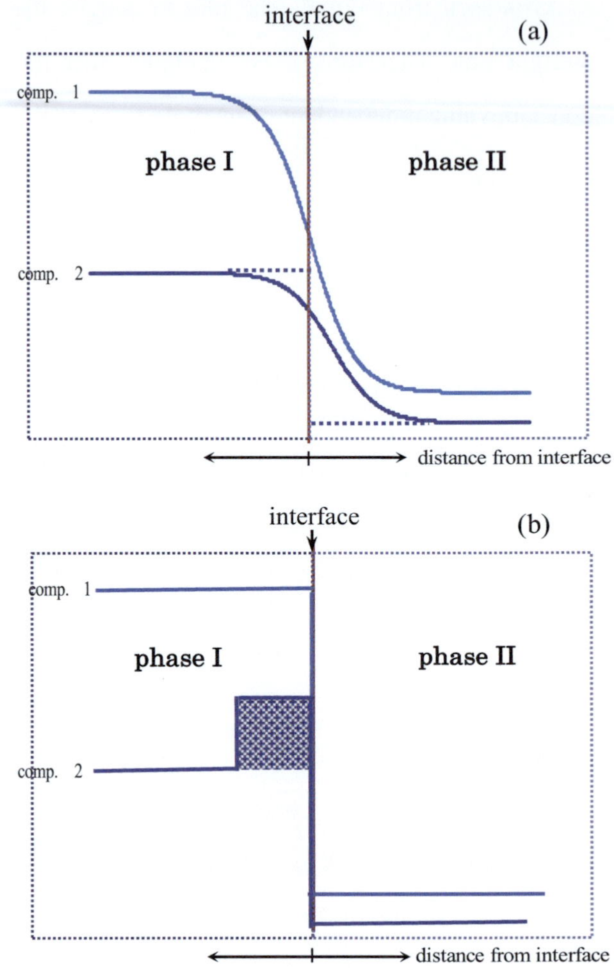

Figs.39a,b Two-phase equilibrium and concentration profile. In the case of a positive surface excess concentration.

concentration of the main component (solvent) on the side of the phase I gradually decreases from the bulk value toward the interface. After crossing the interface, the concentration of the component 1 on the side of the phase II continues to decrease until it reaches the bulk value of zero. In either Figs., the position of the interface is determined with respect to the concentration of the component 1 so as for the area corresponding to the negative deviation from the bulk value on the side of the phase I to cancel the area Ⓑ positive deviation on the phase II side as is indicated by the black vertical lines in Figs.38a and 39a. For the interface thus determined, the surface (excess) concentration of the component 1 becomes zero as is readily seen from the definition by eq(292), since there exists neither excess nor deficiency in its concentration.

However, when the concentration (moles) of the component 2 at the interface thus determined is assigned to the phase I, then there arises the region in which the concentration of the component 2 is lower than that in the bulk as is illustrated in Fig.38b. (There exists neither its excess nor deficiency in the phase II.) On the other hand, when the concentration profile of the component 2 is shifted toward to the phase II as is illustrated in Fig.39a, the concentration of the component 2 in the phase I near the interface becomes larger than that in the bulk if the concentration on the right side of the interface (phase II in Fig.39b) is totally assigned to the phase I. This is illustrated in Fig.39b. By definition, the surface (excess) concentration is either the excess or deficient moles of the component 2 (any component other than main in general) divided by the area of the interface and as a result it can take either a positive or negative value. We now see why the value thus defined is called "excess" concentration even when it takes a negative value.

It was *Gibbs* who treated fundamental interface phenomena correctly in terms of thermodynamics for the first time. A number of concepts such as the position of the interface, the surface excess concentration or thermodynamics of the interface explained here were introduced by *Gibbs* (see **"note"** in the next problem).

The definition and the concept of the surface excess concentration is explained here by using illustrative figures. A few important points are summarized below.

(i) The magnitude of the surface (excess) concentration depends on the position of the interface.

(ii) The magnitude represents nothing more than a qualitative feature of the concentration distribution near the interface.

(iii) The magnitude is simply given by dividing the excess concentration illustrated in Figs.38 and 39 by the surface area and, hence, does not describe truly how the molecules (of minor components) located near the interface are distributed.

(iv) The surface (excess) concentration should clearly be distinguished from other physicochemical quantities such as the adsorption amount on the solid surface or the thickness of a thin film formed at the interface both of which are quantitatively measurable (see next problem).

Ans. See text and Figs. 38a,b and 39a,b.

Solution 4-44

(a) Since water and ethanol are mixable each other to any ratio, the molecular interaction between a water molecule (A, component 1) and an ethanol molecule (E, component 2) is not much different from thosc between the molecules of the same kind. The strength of the intermolecular interaction between these molecules is considered to be in the order of A-A>A-E>E-E.s bulk as is seen from the illustration given in Fig.40. Because of this difference, the number of the unstable molecules in the surface is forced to decrease by shrinking the surface area. This tendency is considered to become the more prominent, the stronger the intermolecular interaction is. Consequently, the magnitude of the surface tension becomes in the same order as that of the intermolecular force, i.e., in the order of A-A>A-E>E-E. Namely, the surface tension of the mixed solution is considered to decrease with increasing the fraction of ethanol, as is practically observed (see table).

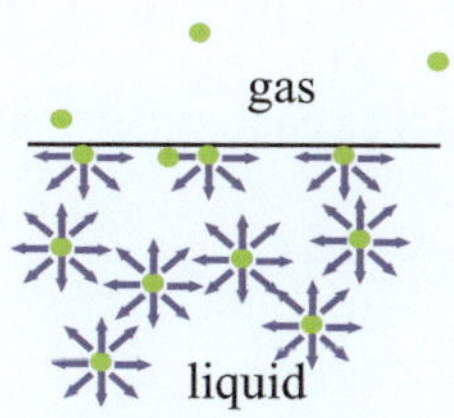

Fig.40 Illustration to show molecules located in the gas-liquid surface are less stable than those in the bulk.

Ans. cause of the decrease in the surface tension: See text.

(b) The surface concentration of ethanol Γ_2 is calculated by using eq(293) from the values of γ for the mixed solutions and their mole fractions. Since the solutions are not ideal ones, the surface tension must be differentiated by the logarithmic value of the partial vapor pressure of ethanol, i.e., the chemical potential of component 2, instead of the logarithmic value of the corresponding mole fraction.

$$\Gamma_2 = -\frac{1}{RT}\left(\frac{\partial \gamma}{\partial \ln P_2}\right)_{T,P,X_1} \qquad (293)$$

Eq(293) can be replaced by the difference ratio as an approximation and, hence, the surface (excess) concentration Γ_2 at the mole fraction of 0.10 is found as $6.63\times10^{-6}\,\mathrm{mol\cdot m^{-2}}$ as shown below.

$$-\frac{1}{8.31\times298}\times\frac{(36.72-34.42)\times10^{-3}}{2.89-3.03} = 6.63\times10^{-6}\,\mathrm{N\cdot m^{-1}\cdot mol\cdot J^{-1}} = 6.63\times10^{-6}\,\mathrm{mol\cdot m^{-2}}$$

From this value, the area occupied by a single ethanol molecule in the interface is found to correspond to a square of $a = 5.0\times10^{-10}\,\mathrm{m}$ side.

$$a = \sqrt{\frac{1.0}{6.63\times10^{-6}\times6.02\times10^{23}}} = 5.0\times10^{-10}\,\mathrm{m}$$

On the other hand, the moles of water are given by $1000/18$ and the moles x of ethanol at the mole fraction of 0.10 are calculated as below.

$$\frac{x}{1000/18+x} = 0.1 \;\Rightarrow\; x = 6.17\ \mathrm{mol\cdot L^{-1}} = 6.17\times10^{3}\,\mathrm{mol\cdot m^{-3}}$$

From the value obtained for x, the volume of a single ethanol molecule occupies in the bulk solution is found to correspond to a cube of an edge a' of $6.5\times10^{-10}\,\mathrm{m}$ as calculated below.

$$a' = \sqrt[3]{\frac{1.0}{6.17\times10^{3}\times6.02\times10^{23}}} = 6.5\times10^{-10}\,\mathrm{m}$$

Consequently, if it is assumed that a single ethanol molecule in the interface occupies on the average a square of $5.0\times10^{-10}\,\mathrm{m}$ in side, then it is speculated that the ethanol molecules in the interface are located slightly closer each other than those in the bulk (see **"note"**).

Ans. surface concentration Γ_2: $6.63\times10^{-6}\,\mathrm{mol\cdot m^{-2}}$, ethanol molecules are expected to be arranged more compactly in the interface.

"note" We often come across the calculations in which the interface average area occupied by a single molecule (minor component) is evaluated by using its surface (excess) concentration. However, a due care should be taken in extracting any quantitative value based on the surface (excess) concentration and then drawing a conclusion from the result thus obtained, as is seen in **problem 4-43**.

(c) As an example, the surface concentrations of ethanol at the mole fractions of 0.30 and 0.60 are calculated and found as $6.0\times10^{-6}\,\mathrm{mol\cdot m^{-2}}$ and $3.84\times10^{-6}\,\mathrm{mol\cdot m^{-2}}$, respectively as are shown below.

$$-\frac{1}{8.31\times298}\times\frac{(27.6-28.49)\times10^{-3}}{3.44-3.38} = 6.00\times10^{-6}\,\mathrm{mol\cdot m^{-2}}$$

$$-\frac{1}{8.31\times298}\times\frac{(24.67-25.43)\times10^{-3}}{3.68-3.60} = 3.84\times10^{-6}\,\mathrm{mol\cdot m^{-2}}$$

In Fig.41, the values for the surface tension obtained for the mixed solutions of ethanol are plotted against the logarithmic values of the partial pressures of ethanol in the gas phase. From Fig.41, it is seen that the surface (excess) concentration calculated form eq(293) tends to decrease at larger mole fractions of ethanol. However, this fact does not mean that the number of ethanol molecules in the interface decreases with increasing the mole fraction of ethanol. What it truly means is that the number of ethanol molecules in the unit area of the interface is greater as compared with that in the bulk and that such a feature becomes less prominent with increasing the mole fraction of ethanol. If the surface area occupied by a single ethanol molecule is worked out from the surface (excess) concentration of ethanol without taking due care, it will lead to an unaccountable result that the average surface area per ethanol molecule increases with increasing the mole fraction of ethanol.

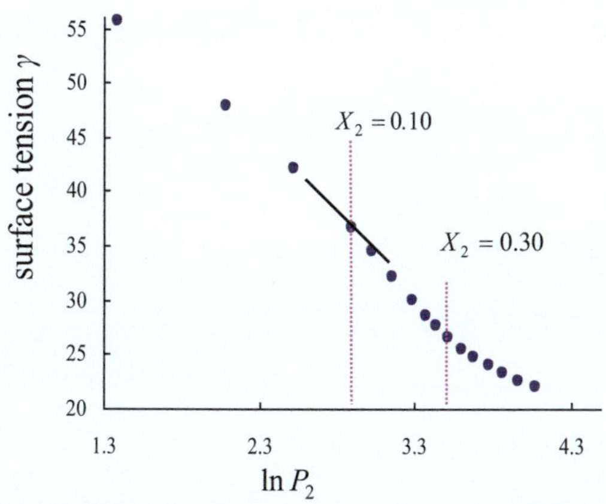

Fig.41 Partial pressure of ethanol and surface tension of the aqueous solution of ethanol.
The surface concentration is calculated from slopes of the plot by using eq(293). **"note"** The curve representing the surface tension as a function of the mole fraction X_2 exhibits neither minimum nor maximum at around $X_2 = 0.08$ in contrast to the partial molar volumes of the ethanol-water mixed solution for which a prominent maximum (or minimum) .is recognized at $X_2 = 0.08$. See **problem** and **solution 3-21**.

Ans. See text (c).

Solution 4-45

(a) Denoting the adsorbed concentration (per unit area) of the gaseous molecules $\Gamma_{2,1}$ on a solid surface, the *Gibbs* adsorption isotherm is given by eq(294), in which γ is the surface tension (see **"note"**).

$$\Gamma_{2,1} = -\left(\frac{\partial \gamma}{\partial \mu_2}\right)_T = -\frac{1}{RT}\left(\frac{\partial \gamma}{\partial \ln P}\right)_T \qquad (294)$$

The subscript 1 stands for the solid and 2 for the gas whose chemical potential is given by $\mu_2 = \mu^\circ + RT \ln P$ as the ideal gas. On the other hand, putting the number of the adsorbed molecules equal to υ/V_m, the adsorbed concentration is expressed by eq(295).

$$\Gamma_{2,1} = \frac{\upsilon}{V_m} \cdot \frac{1}{S} \quad ([\mathrm{mol \cdot m^{-2}}]) \qquad (295)$$

Denoting the surface tension of the clean solid surface γ_0, the surface pressure is expressed as eq(296).

$$f = \gamma_0 - \gamma \Rightarrow df = -d\gamma \qquad (296)$$

Consequently, from eq(294) to eq(296), the relation of eq(297) to be proved is derived.

$$df = \Gamma_{2,1} \times RT \times d\ln P = \frac{RT\upsilon}{SV_m} d\ln P \qquad (297)$$

Ans. See text and eq(297).

(b) Eq(297) obtained above is expressed in either manner that υ is a function of P or that f is a function of υ. For this, it is not possible to integrate eq(297) unless either function is exactly known. The solution to the problem, therefore, is given here in the opposite way, that is, by finding how the surface pressure is related with the pressure of the gas when *Langmuir*'s adsorption isotherm (eq(295) in the problem) is satisfied.

The relation of eq(297) is rewritten as eq(297') by using the relation $\vartheta = \upsilon/\upsilon_{max}$ and eq(298) given below.

$$\frac{1}{\vartheta} = 1 + \frac{1}{bP} \quad \text{(reposted)} \Rightarrow -\frac{1}{\vartheta^2}d\vartheta = -\frac{1}{bP^2}dP \qquad (298)$$

$$df = \frac{RTv_{max}\vartheta}{SV_m}\frac{dP}{P} = \frac{RTv_{max}\vartheta}{SV_m}\frac{P}{\vartheta^2}d\vartheta = \frac{RTv_{max}\vartheta}{SV_m}\frac{P}{\vartheta^2}d\vartheta = \frac{RTv_{max}}{SV_m b}\frac{1}{1-\vartheta}d\vartheta \qquad (297')$$

Since eq(297') can be integrated, its integration leads to eq(299), where C is an integration constant.

$$f = \frac{RTv_{max}}{SV_m b}\ln(1-\vartheta) + C \qquad (299)$$

Putting the surface pressure equal to zero at $\vartheta = 0$ in eq(299), $C = 0$ is found for C and, hence, eq(300) is obtained finally.

$$f = \frac{RTv_{max}}{SV_m b}\ln(1-\vartheta) = \frac{RTv_{max}}{SV_m b}\ln\left(\frac{1}{1+bP}\right) \qquad (300)$$

Unless b is negative, the obtained surface pressure is always negative which indicates that the adsorbed molecules tend to gather together.

Ans. function to give surface pressure: eq(300)

"note" Definition of the thermodynamic quantities in the interface. *Gibbs*' treatment.

We consider the two phases I and II composed of multi-components and they are in equilibrium under constant pressure and temperature. The mole number of the component i is written as n_i and an arbitrary thermodynamic quantity (extensive quantity) is denoted Y. Any value in each phase is indicated by writing I or II on the right shoulder of the corresponding symbol. Denoting the surface area of the interface A_s and the quantity Y per unit surface Y_s, then Y_s is defined by eq(301). Y with no suffix is the thermodynamic quantity of the whole system (phases I, II and the interface).

$$Y_s A_s = Y - Y^I - Y^{II} = Y - \sum n_i^I \left(\frac{\partial Y^I}{\partial n_i^I}\right)_{T,P,\cdots} - \sum n_i^{II} \left(\frac{\partial Y^{II}}{\partial n_i^{II}}\right)_{T,P,\cdots} \qquad (301)$$

For instance, when Y is regarded as the *Gibbs* free energy, the free energy of the interface G_s is given by eq(301') and when Y is regarded as the moles of the i-th component n_i, the surface concentration Γ_i is given by eq(301''). Since the two phases are in equilibrium, the chemical potential of each phase must be equal, i.e., $\mu_i^I = \mu_i^{II} \equiv \mu_i$.

$$G_s A_s = G - \sum n_i^I \mu_i^I - \sum n_i^{II} \mu_i^{II} = G - G^I - G^{II} \qquad (301')$$

$$\Gamma_i A_s = n_i - n_i^I - n_i^{II} \qquad (301'')$$

Putting $P =$ constant and $A_s =$ constant $= 1$, eq(301') is differentiated to yield eq(302),

$$dG_s = dG - \sum \mu_i (dn_i^I + dn_i^{II}) - \sum d\mu_i (n_i^I + n_i^{II})$$
$$= -SdT + \sum \mu_i dn_i - \sum \mu_i (dn_i^I + dn_i^{II}) - \sum d\mu_i (n_i^I + n_i^{II}) \qquad (302)$$

in which the relation $dG = -SdT + VdP + \gamma dA_s + \sum \mu_i dn_i = -SdT + \sum \mu_i dn_i$ is employed.

Here, by denoting the surface tension γ, the free energy G of the whole system is given by eq(303) with T, P, n_i and A_s as independent variables.

$$G(T, P, n_i, A_s) = \sum \mu_i n_i + \gamma A_s \qquad (303)$$

Also, from the differential of eq(301''), eq(301''') is obtained.

$$d\Gamma_i = dn_i - dn_i^I - dn_i^{II} \Rightarrow dn_i = d\Gamma_i + dn_i^I + dn_i^{II} \qquad (301''')$$

Using this, eq(302) is rewritten as eq(304) and further substituting the relation $d\mu_i = -\bar{S}_i dT$ ($P =$ constant, $\bar{S}_i$: partial molar value), eq(304') is derived.

$$dG_s = -SdT + \sum \mu_i (d\Gamma_i + dn_i^I + dn_i^{II}) - \sum \mu_i (dn_i^I + dn_i^{II}) - \sum d\mu_i (n_i^I + n_i^{II})$$
$$= -SdT + \sum \mu_i d\Gamma_i - \sum d\mu_i (n_i^I + n_i^{II}) \qquad (304)$$
$$= -SdT + \sum \mu_i d\Gamma_i + \sum \bar{S}_i (n_i^I + n_i^{II}) dT$$
$$= -\left\{ S - \sum \bar{S}_i (n_i^I + n_i^{II}) \right\} dT + \sum \mu_i d\Gamma_i \qquad (304')$$

The function in the curly parentheses of eq(304') is equivalent to the definition of the interface entropy $S - \sum \bar{S}_i (n_i^I + n_i^{II}) = S_s \ (A_s = 1)$ (eq(301)) and, hence, eq(304') is further rewritten as eq(304'').

$$dG_s = -S_s dT + \sum \mu_i d\Gamma_i \qquad (304'')$$

On the other hand, under the conditions of constant Γ_i and n_i, the differential of eq(301'') is given by eq(305).

$$d(\Gamma_i A_s) = A_s d\Gamma_i + \Gamma_i dA_s$$
$$= \Gamma_i dA_s = -dn_i^I - dn_i^{II} \qquad (305)$$

And under the conditions of $T, P, \gamma =$ constant, the differential of $G_s A_s$ is given by eq(306), for whose derivation eqs(301''), (305) and the equation of *Gibbs-Duhem* $\sum n_i d\mu_i = 0$ are employed, though that is a lengthy one.

$$d(G_s A_s) = dG - \sum \mu_i d(n_i^I + n_i^{II}) - \sum d\mu_i (n_i^I + n_i^{II})$$
$$= \gamma dA_s - \sum \mu_i d(n_i^I + n_i^{II}) - \sum d\mu_i (n_i^I + n_i^{II})$$
$$= (\gamma + \sum \mu_i \Gamma_i) dA_s + A_s \sum \mu_i d\Gamma_i - \sum d\mu_i (n_i^I + n_i^{II})$$
$$= (\gamma + \sum \mu_i \Gamma_i) dA_s + A_s \sum \mu_i d\Gamma_i - \sum (n_i - \Gamma_i A_s) d\mu_i$$
$$= (\gamma + \sum \mu_i \Gamma_i) dA_s + A_s (\sum \mu_i d\Gamma_i + \Gamma_i d\mu_i)$$
$$= (\gamma + \sum \mu_i \Gamma_i) dA_s + A_s d(\sum \mu_i \Gamma_i)$$
$$= d\langle A_s (\gamma + \sum \mu_i \Gamma_i) \rangle \qquad (\gamma = \text{constant}) \qquad (306)$$

As a result, eq(307) is derived from eq(306).

$$G_s A_s = A_s(\gamma + \sum \mu_i \Gamma_i) \Rightarrow G_s = (\gamma + \sum \mu_i \Gamma_i) \qquad (307)$$

The differential of eq(307) is given by eq(307'), which is put equal to eq(304'') derived above and then eq(308) is obtained by eliminating dG_s.

$$dG_s = d\gamma + \sum \mu_i d\Gamma_i + \sum \Gamma_i d\mu_i \qquad (307')$$

$$d\gamma = -S_s dT - \sum \Gamma_i d\mu_i \qquad (308)$$

From eq(308), eqs(309) and (310) are derived straightforward.

$$\Gamma_i = -\left(\frac{\partial \gamma}{\partial \mu_i}\right)_{T,P,\mu_{j\neq i}} = -\frac{1}{RT}\left(\frac{\partial \gamma}{\partial \ln a_i}\right)_{T,P,X_{j\neq i}} \left\langle = -\frac{1}{RT}\left(\frac{\partial \gamma}{\partial \ln X_i}\right)_{T,P,X_{j\neq i}} \right\rangle \qquad (309)$$

$$S_s = -\left(\frac{\partial \gamma}{\partial T}\right)_{P,\mu_i} \qquad (310)$$

It is seen from eq(310) that the surface entropy is found from the temperature dependence of the surface tension, which is experimentally measurable. On the other hand, since the surface entropy can be calculated theoretically on the basis of the statistical thermodynamics, the surface tension is theoretically found also from eq(310). Other thermodynamic quantities can be found as well by using the following relations.

$$G_s = (\gamma + \sum \mu_i \Gamma_i) = H_s - TS_s \qquad (311)$$

$$\gamma + \sum \mu_i \Gamma_i = \gamma + \sum \mu_i \left(\frac{\partial \gamma}{\partial \mu_i}\right) = H_s + T\left(\frac{\partial \gamma}{\partial T}\right)_{P,\mu_i} \qquad (312)$$

Solution 4-46

From the equation for the *Gibbs* free energy which includes the surface energy (see **"note"** in **solution 4-41**), the surface pressure π is given by the difference between the two surface tensions at the contacting two interfaces as is shown in eq(313) (see **"note 1"**).

$$G_1 = \sum \mu_i n_i + \gamma_1 A_{1s}, \qquad G_2 = \sum \mu_i n_i + \gamma_2 A_{2s} \Rightarrow -\left(\frac{\partial (G_1 + G_2)}{\partial A_s}\right)_{P,T,\gamma} = \pi = -\Delta\gamma \qquad (313)$$

Eq(313) assumes the same form as $(\partial F/\partial V)_{V,T} = -P$ and, just as is the equation of state for the ideal gas $PV = nRT$ derived, if it is assumed that the monomolecular membrane behaves ideally against the external force, eq(314) is deduced.

$$\pi A_s = nRT \qquad (314)$$

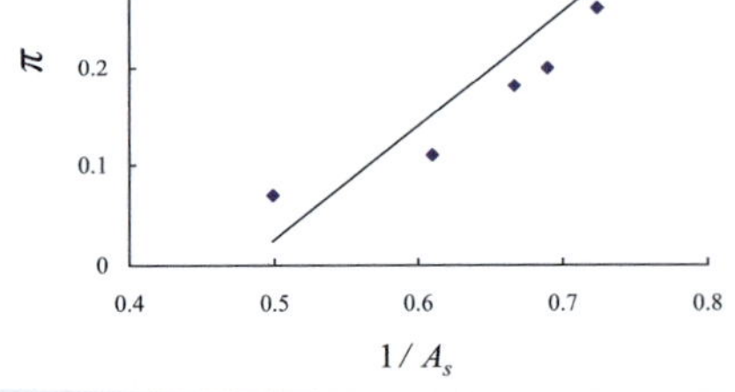

Fig.42 Ideal membrane without excluded area.

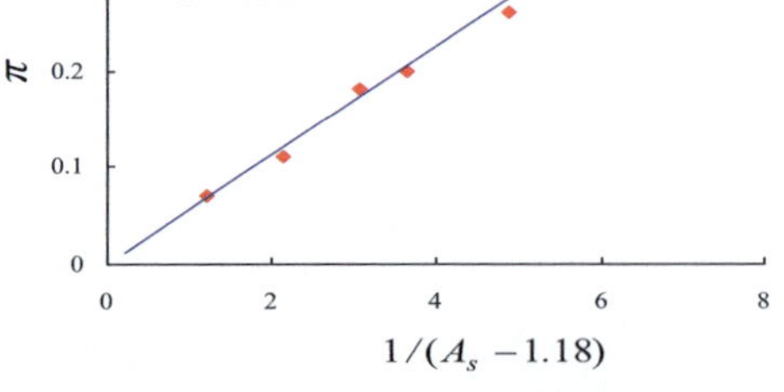

Fig.43 Membrane with excluded area.

The data in the table are plotted in Fig.42 according to eq(313). The fitting to a straight line is very poor, though it is obvious without plotting the data since the values calculated for πA_s are not constant.

When considered by comparing with the ideal gas, the constituent molecules in the monomolecular membrane are rather tightly assembled and, for this, it is hard to reduce its area beyond a certain minimum area, which corresponds to the existence of an intrinsic volume of a real gas. Taking this fact into consideration, eq(314) can be modified as eq(315). The proportional constant in eq(315) is not n but is denoted α because the unit of the constant depends on the unit of A_s.

$$\pi(A_s - b) = \alpha RT \qquad (315)$$

Employing $b = 1.18$ obtained by the liner least squares method, eq(315) is plotted in Fig.43. The best fitted straight line which passes through the origin is given by eq(316),

$$\pi = \frac{\alpha RT}{(A_s - b)} = \frac{0.0562}{(A_s - 1.18)} \qquad (316)$$

from which α is found as $2.27\times10^{-5}\,\text{mol}\cdot\text{g}^{-1}$ as is shown below.

$$\alpha = \frac{0.0562\times10^{-3}}{RT} = \frac{0.0562\times10^{-3}\times10^{3}\ [\text{N}\cdot\text{m}\cdot\text{g}^{-1}]}{8.31\times298\ [\text{J}\cdot\text{mol}^{-1}]} = 2.27\times10^{-5}\,\text{mol}\cdot\text{g}^{-1}$$

Consequently, the molecular weight of albumin M_{alb} is found as below.

$$M_{alb} = \frac{1}{\alpha} = 44100\ \text{g}\cdot\text{mol}^{-1}(=\text{Da})$$

Also, the area occupied by a single molecule of albumin is obtained by multiplying b with the mass of the single molecule expressed in the unit of mg as shown below.

$$b\times\frac{M_{alb}\times10^{3}}{N_A} = \frac{1.18\times44100\times10^{3}}{6.02\times10^{23}} = 8.6\times10^{-17}\,\text{m}^2/\text{molecule} = 86\ \text{nm}^2/\text{molecule}$$

Ans. molecular weight: 44.1 kDa , molecular area: 86 $\text{nm}^2/\text{molecule}$

"note 1" Surface pressure.

As is illustrated in Fig.44 below, when the two interfaces with different surface tensions as is indicated by the arrows contact each other (for instance, clean water surface and the water surface covered with insoluble membrane), the interface on the left in the figure tends to expand. The force arising in such situation is divided by the side length of the interface, i.e., the force per unit length, is the surface pressure and is obviously expressed as $\pi = \gamma_1 - \gamma_2 = -(\gamma_2 - \gamma_1) = -\Delta\gamma$.

Denoting the surface tension of clean water γ_1, since $\gamma_1 > \gamma_2$ in ordinary cases, the surface pressure is defined by putting a negative sign to the difference of $\Delta\gamma = \gamma_2 - \gamma_1$. Therefore, the surface pressure is totally different in its origin from the pressure of a gas which is the consequence of molecular collisions with the wall of a container. However, the equation analogous in its form either to the state of equation of the two-dimensional (eq(314)) or three-dimensional gas for which the excluded volume is taken into account is applicable successfully in certain cases. Some more explanations for eq(313) are added below.

$$G_1 = \sum \mu_i n_i + \gamma_1 A_{1s}, \qquad G_2 = \sum \mu_i n_i + \gamma_2 A_{2s}$$

$$d(G_1 + G_2)_{P,T,\gamma} = \gamma_1 dA_{1s} + \gamma_2 dA_{2s} = (-\gamma_1 + \gamma_2) dA_s$$

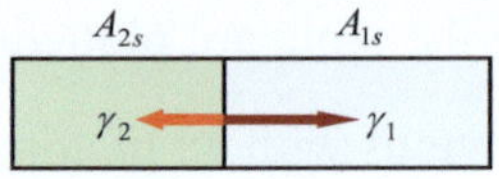

Fig.44 A plane view of two surfaces in contact with each other.

Here, the relation $-dA_{1s} = dA_{2s} = dA_s$ is used.

$$\pi = -\left(\frac{\partial(G_1 + G_2)}{\partial A_s}\right)_{P,T,\gamma} = \gamma_1 - \gamma_2 = -\Delta\gamma \qquad (313)$$

"note 2" Preparation of a black membrane.

After making a hole of diameter of ca. 1.5 mm in a sheet of polyethylene or Teflon, a liquid film is prepared by dropping a tiny amount of a synthetic detergent solution (surfactant, phosphorous lipid, or long chain carboxylic acid) on the hole and the sheet is immersed into water and is left alone for a while. Gradually the liquid film formed at the hole becomes thinner and starts to show rainbow color produced by its interference with light. Being left further, the interference pattern disappears and finally the film becomes black and for this it is called black membrane (see Fig.45 given below). This is because the finally formed membrane is the bimolecular membrane which is much thinner than the light wavelengths. The formation of bimolecular membranes is a phenomena frequently recognized for biomolecular membranes or synthetic long carbon chain molecules of amphiphilic nature. Most general method for the preparation of bimolecular membranes is the method of *Langmuir-Blodget* (*I.Langmuir*, Nobel Prize in Chemistry in 1932).

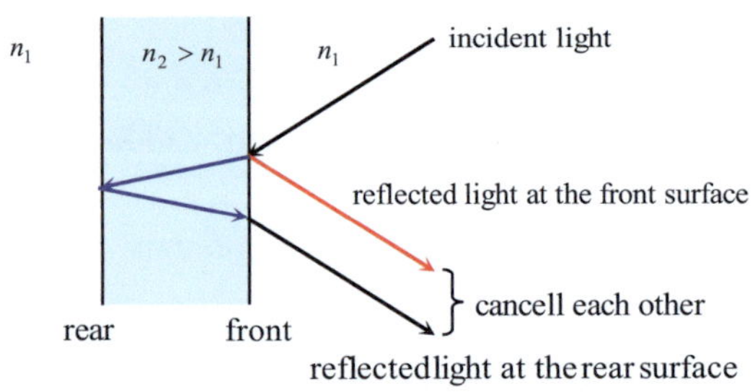

Fig.45 Black membrane to be formed by the interaction of the reflected incident lights at the front and the rear surfaces of a membrane.

For thin membranes such as a bimolecular membrane whose thickness is much smaller than the wavelength of visible light, the lights reflected at the <u>front</u> and <u>rear</u> surfaces of the membrane differ in phase just by a half wave-length of the light on the front side. Consequently, both reflected lights cancel each other on the front side (observer's side) and we cannot see the membrane and recognize it just as a dark spot called black membrane.

Solution 4-47

The *Schrödinger* equation for a particle moving in the one-dimensional potential field is given by eq(317).

$$\frac{d^2\varphi(x)}{dx^2} = -\frac{2m}{\hbar^2}(E - V_0)\varphi(x) \qquad (317)$$

Since the special solutions of eq(317) are given by eq(318), a general solution is expressed by a linear combination of these two functions.

$$\varphi(x) = \exp\left(i\frac{\sqrt{2m(E - V_0)}}{\hbar}x\right), \qquad \exp\left(-i\frac{\sqrt{2m(E - V_0)}}{\hbar}x\right) \qquad (318)$$

The given potential filed is divided into the three regions denoted *I*, *II* and *III* as is shown in the figure given in the problem. Writing the wave functions of the general solution in each region thus classified as $\varphi_I(x)$, $\varphi_{II}(x)$ and $\varphi_{III}(x)$, then they can be expressed by eqs(319), (320) and (321).

$$\varphi_I(x) = \exp\left(i\frac{\sqrt{2mE}}{\hbar}x\right) + A\exp\left(-i\frac{\sqrt{2mE}}{\hbar}x\right) = e^{ikx} + Ae^{-ikx} \qquad (319)$$

$$\varphi_{II}(x) = B\exp\left(i\frac{\sqrt{2m(E-V_0)}}{\hbar}x\right) + C\exp\left(-i\frac{\sqrt{2m(E-V_0)}}{\hbar}x\right) = Be^{ik'x} + Ce^{-ik'x} \qquad (320)$$

$$\varphi_{III}(x) = D\exp\left(i\frac{\sqrt{2mE}}{\hbar}x\right) = De^{ikx} \qquad (321)$$

Notice here that the function e^{ikx} describes the motion to the positive direction and the function e^{-ikx} to the negative direction, as is readily seen from the eigenvalue of the momentum given by $-i\hbar\,\partial/\partial x(e^{ikx}) = ke^{ikx}$. From this, what is meant by the linear combinations in eqs(319) and (320) is obvious.

At a boundary where the potential energy changes discontinuously, the following boundary conditions must be satisfied: (i) the value of the wave function and (ii) the value of its first derivative are continuous at the boundary. Consequently, the following four equations are satisfied.

$$\left.\begin{aligned} &\varphi_I(0) = \varphi_{II}(0)\ , \qquad \varphi_{II}(a) = \varphi_{III}(a) \\ &\left(\frac{d\varphi_I(x)}{dx}\right)_{x=0} = \left(\frac{d\varphi_{II}(x)}{dx}\right)_{x=0}, \qquad \left(\frac{d\varphi_{II}(x)}{dx}\right)_{x=a} = \left(\frac{d\varphi_{III}(x)}{dx}\right)_{x=a} \end{aligned}\right\} \qquad (322)$$

The boundary conditions of eq(322) are applied to eqs(319), (320) and (321) and the coefficients of *A, B, C* and *D* are found to satisfy the following four equations.

$$1 + A = B + C \quad (323), \qquad k - kA = k'B - k'C \quad (324)$$

$$Be^{ik'a} + Ce^{-ik'a} = De^{ika} \quad (323'), \qquad k'Be^{ik'a} - k'Ce^{-ik'a} = kDe^{ika} \quad (324')$$

Since the number of the unknown constants is four and there exist four independent equations to be satisfied, the coefficients of *A, B, C* and *D* are unambiguously determined. Then, from eqs(323) and (324), eq(325) is derived and from eqs(323') and (324'), eqs(326) and (327) are obtained.

$$A = \frac{(k-k')B}{2k} + \frac{(k+k')C}{2k} \qquad (325)$$

$$B = \frac{(k+k')D}{2k'}e^{-i(k'-k)a} \qquad (326)$$

$$C = \frac{(k'-k)D}{2k'}e^{i(k'+k)a} \qquad (327)$$

The coefficient *A* is eliminated first by using eqs(323) and (324). The coefficients *B* and *C* in the equation thus obtained are replaced by the right sides of eqs (326) and (327), respectively to yield *D* of eq(328') via eq(328).

$$2k = \frac{(k'+k)^2 D}{2k'}e^{-i(k'-k)a} - \frac{(k'-k)^2 D}{2k'}e^{i(k'+k)a} \qquad (328)$$

$$D = \frac{4kk'}{(k'+k)^2 e^{-ik'-k)a} - (k'-k)^2 e^{ik'+k)a}} \qquad (328')$$

In the case of the incidence of an electron whose energy E is smaller than the potential height of V_0, k' is replaced by κ defined by eq(329).

$$k' = i\kappa, \qquad \kappa = \sqrt{2m(V_0 - E)} / \hbar \qquad (329)$$

Then, D is rewritten as eq(330).

$$D = \frac{4ik\kappa e^{-ika}}{(i\kappa + k)^2 e^{\kappa a} - (i\kappa - k)^2 e^{-\kappa a}} = \frac{4ik\kappa e^{-ika}}{(k^2 - \kappa^2)(e^{\kappa a} - e^{-\kappa a}) + 2ik\kappa(e^{\kappa a} + e^{-\kappa a})}$$

$$= \frac{2ik\kappa e^{-ika}}{\sinh \kappa a \times (k^2 - \kappa^2) + 2ik\kappa \cdot \cosh ka} \qquad (330)$$

Since the probability for the electron to pass through the potential barrier is given by DD^*, it is given from eq (330) by eq(331).

$$DD^* = \frac{4k^2\kappa^2}{(k^2 - \kappa^2)^2 \sinh^2 \kappa a + 4k^2\kappa^2 \cosh^2 ka} = \frac{4k^2\kappa^2}{(k^2 + \kappa^2)^2 \sinh^2 \kappa a + 4k^2\kappa^2} \qquad (331)$$

Here, the relation $\cosh^2 x - \sinh^2 x = 1$ is used.

When $\kappa a >> 1$ is satisfied, $\sinh \kappa a \approx e^{\kappa a}/2$ is a sufficient approximation and, hence, the transmission probability (coefficient) is given by eq(332).

$$DD^* \approx \frac{16k\kappa^2 e^{-2\kappa a}}{(k^2 + \kappa^2)^2} = \frac{16(2mE)2m(V_0 - E)\exp[-2a\sqrt{2m(V_0 - E)}/\hbar\ ]}{\langle 2mE + 2m(V_0 - E)\rangle^2}$$

$$= 16\left(\frac{E}{V_0}\right)\left(1 - \frac{E}{V_0}\right)\exp\left(-\frac{2a\sqrt{2m(V_0 - E)}}{\hbar}\right) \qquad (332)$$

Employing eq(331), the transmission probability is calculated as 0.219for the following given values.

$$E = 0.1 \text{ eV}, \ V_0 = 1.0 \text{ eV}, \ m = m_e = 9.109 \times 10^{-31} \text{kg}, \ a = 0.20 \text{ nm}$$

When eq(332) is employed as an approximation, though it is a very poor approximation for $\kappa a = 0.97$ in the present case, a value of 0.206 is found for the transmission probability.

$$\begin{array}{l}\text{transmission}\\ \text{probability}\end{array} = 16\left(\frac{0.1}{1.0}\right)\left(1 - \frac{0.1}{1.0}\right)\exp\left(-\frac{2 \times \sqrt{2 \times 9.109 \times 10^{-31} \times (1.0 - 0.1) \times 1.602 \times 10^{-19}} \times 2.0 \times 10^{-10} \times 2\pi}{6.626 \times 10^{-34}}\right)$$

$$= 0.206$$

When the thickness of the barrier is increased to $a = 1.0$ nm, the transmission probability decreases to 8.6×10^{-5}. On the other hand, if the incident particle is replaced with a positron, then the probability is calculated for $m = m_p = 1.673 \times 10^{-27}$ kg, yielding a value of 1.89×10^{-36} for the transmission probability. It is seen from this result that positron whose mass is 1800 times larger that that of the electron cannot practically tunnel through the barrier.

The coefficient A corresponding to the reflection probability is given by eq(325) and substituting B and C into it, eq(333) is obtained for the case of $E - V_0 > 0$.

$$A = \frac{(k - k')B}{2k} + \frac{(k + k')C}{2k}$$

$$= \frac{(k - k')}{2k} \times \frac{(k' + k)D}{2k'} e^{-i(k' - k)a} + \frac{(k + k')}{2k} \times \frac{(k' - k)D}{2k'} e^{i(k' + k)a}$$

$$= -\frac{(k^2 - k'^2)D}{4kk'}\left(e^{i(k'+k)a} - e^{-i(k'-k)a}\right) = -\frac{(k^2 - k'^2)De^{ika}}{4kk'}\left(e^{ik'a} - e^{-ik'a}\right)$$

$$= -i\frac{(k^2 - k'^2)De^{ika}}{2kk'}\sin kl \qquad (333)$$

The reflection probability given by AA^* is calculated as shown below and is found to be given by eq(334) or eq(334').

$$AA^* = \frac{(k^2 - k'^2)^2 DD^*}{4(kk')^2}\sin^2 ka$$

$$= \frac{(k^2 - k'^2)^2 \sin^2 ka}{(4kk')^2} \times \frac{(4kk')^2}{(4kk')^2 + 4(k'^2 - k^2)^2 \sin^2 ka}$$

$$= \frac{(k^2 - k'^2)^2 \sin^2 ka}{4(kk')^2 + (k'^2 - k^2)^2 \sin^2 ka} \qquad (334)$$

$$= \frac{m^2 V_0^2 \sin^2 kl}{4m^2 E(E - V_0) + m^2 V_0^2 \sin^2 kl} = \frac{V_0^2 \sin^2 ka}{4E(E - V_0) + V_0^2 \sin^2 ka} \qquad (334')$$

In the case of the electron with $E = 1.1$ eV, employing the same values for the other constants as the previous ones, the reflection probability is found as 0.187 from eq(334'). It is seen that even though the energy of the electron is larger than the potential height, the electron is reflected by the barrier to some extent.

Ans. transmission probability: 0.219, eq(15) or eq(332); reflection probability: 0.187, eq(334')

"note" The reflection probability for the case $E - V_0 < 0$ is obtained as follows. By putting $k' = i\kappa$ and $\kappa = \sqrt{2m(V_0 - E)}/\hbar$ and rewriting eq(333), A is found to be expressed by eq(333').

$$A = \frac{(k - i\kappa)}{2k} \times \frac{(i\kappa + k)D}{2i\kappa} e^{-i(i\kappa - k)a} + \frac{(k + i\kappa)}{2k} \times \frac{(i\kappa - k)D}{2i\kappa} e^{i(i\kappa + k)a}$$

$$= \frac{(k^2 + \kappa^2)D}{4ik\kappa} e^{\kappa a} e^{ika} - \frac{(k^2 + \kappa^2)D}{4ik\kappa} e^{-\kappa a} e^{ika}$$

$$= \frac{(k^2 + \kappa^2)D}{4ik\kappa} e^{ika}(e^{\kappa a} - e^{-\kappa a})$$

$$= \frac{(k^2 + \kappa^2)D}{2ik\kappa} e^{ika} \sinh \kappa a \qquad (333')$$

As a result, the reflection probability (coefficient) in the case of $E - V_0 < 0$ is given by eq(334'').

$$AA^* = \frac{(k^2 + \kappa^2)^2 DD^*}{4(k\kappa)^2} \sinh^2 \kappa a \qquad (334'')$$

The reflection probability of the electron whose transmission probability (coefficient) has been obtained previously is calculated to be 0.780 from eq(18''). The sum of the calculated probabilities of the transmission and reflection is $0.219 + 0.780 = 0.999$ and, of course, should become strictly unity as is evident from eq(335) given below,

$$AA^* + DD^* = \frac{(k^2 + \kappa^2)^2 DD^*}{4(k\kappa)^2} \sinh^2 \kappa a + DD^*$$

$$= \left\{ \frac{(k^2 + \kappa^2)^2 \sinh^2 \kappa a + 4(k\kappa)^2}{4(k\kappa)^2} \right\} DD^* = \frac{DD^*}{DD^*} = 1 \qquad (335)$$

[advanced problem 1] Even when the energy of the incident electron exceeds the potential height of a barrier, the electron is reflected with a certain probability. In another word, the transmission probability is not unity. In Fig.46 on the right the transmission probability is plotted against (incident energy)/(potential height) $= E/V_0$ for $a = 0.20$ nm. Draw yourself a curve similar to that shown in Fig.46.

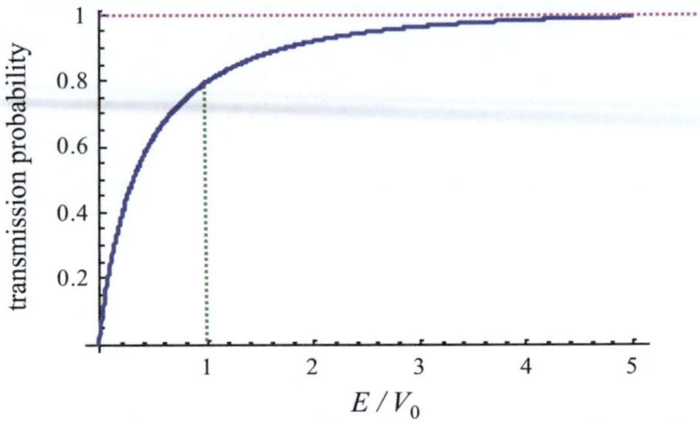

Fig.46 Transition probability of an electron incident with energy E on a potential barrier of height V_0.

[advanced problem 2] When the incident particle is not an electron but a photon, how are its reflection and transmission through a barrier treated?

Solution 4-48

(a) As is postulated in the problem, if it is assumed that the probability for a particle to transmit the potential barrier $V(x)$ is given by the product of the relevant probabilities of a number of rectangular potentials into which $V(x)$ is decomposed, the net probability is given by the left side of eq(336). Since the product of several exponential functions can be replaced by a single exponential function whose argument is given by the sum over the arguments of component exponentials, the left side of eq(334)) can be expressed as the right side of eq(336). Putting $a = dx$ and $V_i = V(x_i)$ here, eq(337) is derived.

$$\prod_{i=1}^{n} \exp\left(-\frac{2a\sqrt{2m(V_i - E)}}{\hbar}\right) = \exp\left(-\sum_{i=1}^{n}\frac{2a\sqrt{2m(V_i - E)}}{\hbar}\right) \qquad (336)$$

$$= \exp\left\{-\sum_{i=1}^{n}\frac{2dx\sqrt{2m(V(x_i) - E)}}{\hbar}\right\} \qquad (337)$$

By replacing the sum of eq(337) with the corresponding integral, it is found that the probability for the particle to pass through any given potential barrier $V(x)$ is proportional to eq(338), which is the same as the one given in the problem.

$$\text{transmission probability} \propto \exp\left(-\int_{x_1}^{x_2}\frac{2\sqrt{2m(V(x) - E)}}{\hbar}dx\right) \qquad (338) \qquad \text{(eq(16) in the problem)}$$

Here, x_1 and x_2 are the values for the x coordinate which satisfy $V(x) - E = 0$.

Ans. See text (a).

(b) Substituting the given values into eq(338) and gathering the constants together, eq(338') is found. Here, the employed units are nm, $\mathrm{V = J \cdot C^{-1}}$ and $1\mathrm{eV} = 1.602 \times 10^{-19}$ J.

$$\exp\left\{-\int_{x_1}^{x_2}\frac{2\sqrt{2m(V(x) - E)}}{\hbar}dx\right\} = \exp\left\{-\int_{x_1}^{x_2}\frac{4\pi \times 10^{-9}\sqrt{2 \times 9.109 \times 10^{-31}(V(x) - E) \times 1.602 \times 10^{-19}}}{6.626 \times 10^{-36}}dx\right\}$$

$$= \exp\left\{-4.357\int_{x_1}^{x_2}\sqrt{(V(x) - E)}dx\right\} \qquad (338')$$

Putting $E = 6\text{V}$ or 2V and substituting eq(339) for $V(x)$ (eq(337) in the problem) into eq(338') to calculate the

$$V(x) = 10.0 - 5.0 \times x - \frac{1.440}{4x} \qquad (339)$$

transmission probability, the following values are found.

$$\text{transmission probability } (E = 6\text{V}) \propto \exp\left\{-4.357 \int_{0.104}^{0.696} \sqrt{\left(10.0 - 5.0 \times x - \frac{1.440}{4x} - 6.0\right)} dx\right\} = 0.103$$

$$\text{transmission probability } (E = 2\text{V}) \propto \exp\left\{-4.357 \int_{0.0464}^{1.553} \sqrt{\left(10.0 - 5.0 \times x - \frac{1.440}{4x} - 2.0\right)} dx\right\} = 1.52 \times 10^{-5}$$

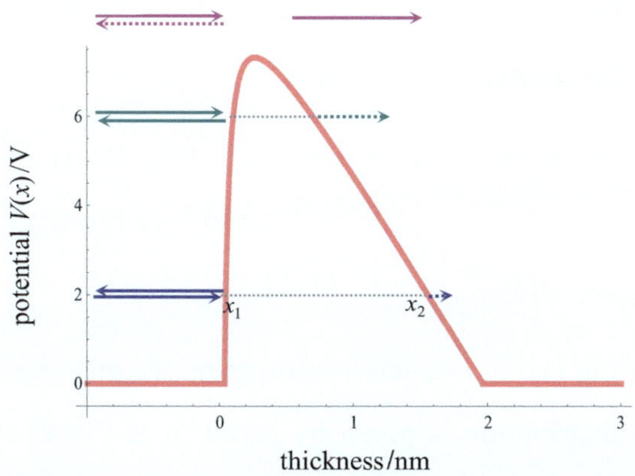

Fig.47 Potential barrier and transmission probability.

When the energy of the incoming electron is lower than the potential height, the transmission through the barrier is known as the tunneling effect. However, even when the electron energy is higher than the potential height, the incident electron is reflected with a nonzero probability. The potential curve in Fig.47 mimics a potential in the vicinity of an electrode surface.

Ans. transmission probability ($E = 6\text{V}$): 0.103

"note" Method of mirror image potential.

Against an earthed metal plate of an infinite area, when a unit charge of $-e$ is placed at the distance x from the plate, the opposite charge will be induced on the plate. The strength of the electric field due to the induced charge is found by solving the *Poisson* equation in general. More simply, however, based on the method known as mirror image method not only the strength of the electric field but also the charge distribution on the plate can easily be figured out.

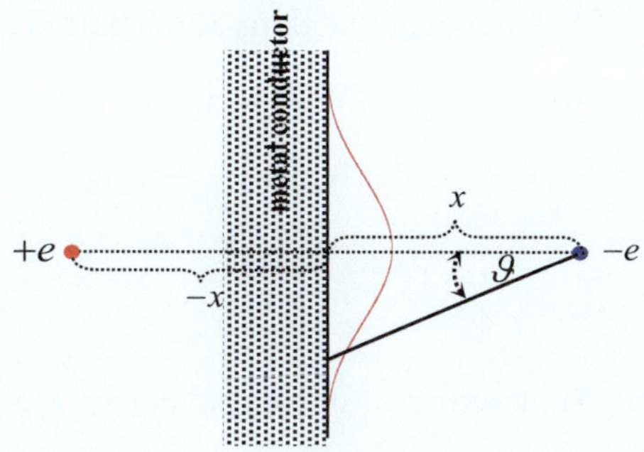

Fig.48 Induced charge and mirror image potential.

In this method, the metal plate is regarded as a symmetry plane and a charge opposite to the unit charge is placed at the position symmetrical on reflection against the plane as is shown by the red dot against the blue dot in Fig.48. The electric field generated by this mirror charge (red dot) is considered the same as that to be produced by the charge distribution induced on the plate.

For this, the net electric field is given by the superposition of each electric field, one by the initially placed charge and the other by the opposite mirror image charge. Therefore, an electron ejected from the plate and located at the position x is under the electric field equal to the one produced by the positive charge of $+e$ placed at the position of $-x$, i.e., $-\frac{1}{4\pi\varepsilon_0} \cdot \frac{e^2}{4x^2}$. Consequently, the potential of that electron $V(x)$ is given by eq(340) with V_0 as a constant.

$$V(x) = -\frac{1}{4\pi\varepsilon_0} \cdot \frac{e^2}{4x} + V_0 \qquad (340)$$

The charge density induced on the plate is considered to be proportional to the horizontal component of the electric field produced by the point charge of $-e$ at the position $-x$. The electric field at the position on the plate which makes an angle of ϑ to the perpendicular line to the plate (see Fig.48) is proportional to eq(341) and, hence, its component perpendicular to the plate is proportional to eq(341').

$$\text{electric filed}: \frac{\cos^2\vartheta}{x^2} \qquad (341), \qquad \frac{\cos^2\vartheta}{x^2}\times\cos\vartheta \qquad (341')$$

Since the area of the ring on the plate specified by the angles between ϑ and $\vartheta+d\vartheta$ is given by eq(342), the

$$2\pi r dr = 2\pi x \tan\vartheta\times\frac{x}{\cos^2\vartheta}d\vartheta \qquad (342)$$

amount of charges in this ring is given by eq(343) with A as a constant.

$$A\times\frac{\cos^2\vartheta}{x^2}\times\cos\vartheta\times 2\pi x\tan\vartheta\times\frac{x}{\cos^2\vartheta}d\vartheta = 2\pi A\sin\vartheta d\vartheta \qquad (343)$$

Noting that the total amount of the charges induced on the metal plate is $+e$, the constant A is calculated as below.

$$A\int_0^{\pi/2} 2\pi\sin\vartheta d\vartheta = 2\pi A[-\cos\vartheta]_0^{\pi/2} = 2\pi A = e \;\Rightarrow\; A = \frac{e}{2\pi} \qquad (344)$$

Taking the foot of the perpendicular line as the coordinate origin $(0,\ 0,\ 0)$, and a position on the plate being expressed by $(0,\ y,\ z)$, it is found that the charge density distribution is given by eq(345). In Fig.48 the charge density distribution based on eq(341') is shown by the red curve.

$$\cos\vartheta = \frac{x}{\sqrt{x^2+y^2+z^2}} \;\Rightarrow\; A\times\frac{\cos^3\vartheta}{x^2} = \frac{e}{2\pi}\times\frac{x}{(x^2+y^2+z^2)^{3/2}} \qquad (345)$$

Solution 4-49

(a) The chemical reactions at the half electric cells of $Ag|Ag^+(aq, a_{Ag^+})$ and $AgCl(s)|Ag$ can be described by eq(346) and eq(347), respectively, both of which are expressed as the (electron-accepting) reduction reactions

$$Ag^+ + e \rightarrow Ag, \qquad E_1^\circ = 0.7991\ \text{V} \qquad (346)$$

$$AgCl + e \rightarrow Ag + Cl^-, \qquad E_2^\circ = 0.2192\ \text{V} \qquad (347)$$

Ans. electrode reduction reactions: eq(346) and eq(347)

(b) Subtracting eq(346) from eq(347), eq(348) is obtained as the net chemical reaction of the electric cell.

$$AgCl \rightarrow Ag^+ + Cl^-, \qquad \Delta E^\circ = E_2^\circ - E_1^\circ = -0.580\text{V} \qquad (348)$$

Ans. net chemical reaction of the electric cell: eq(348)

(c) The free energy change ΔG of the reaction expressed by eq(348) is found from eq(349) in terms of the solubility product K_{sp}. Here ΔG° is given by $-F\times\Delta E^\circ(n=1)$ using the difference between the standard electrode potentials ΔE° (see **"note 1"**).

$$\Delta G = \Delta G^\circ + RT\ln K_{sp} \qquad (349)$$

When the dissolving reaction $AgCl \rightleftarrows Ag^+ + Cl^-$ has reached an equilibrium, the free energy change is zero, i.e., $\Delta G = 0$ and, hence, from eq(349) the solubility product is obtained as below via left side of eq(350).

$$\ln K_{sp} = \frac{F\times\Delta E^\circ}{RT} = -\frac{0.580\times 96500}{8.31\times 298} \;\Rightarrow\; K_{sp} = 1.53\times 10^{-10} \qquad (350)$$

Ans. solubility product: $K_{sp} = 1.53\times 10^{-10}$

More generally, starting with the expressions for the electromotive force for both half-cells, which are given by eqs(346') and (347') where a_{Ag}, a_{Ag^+}, .. are activities of the relevant chemical species,

$$E_1 = E_1^\circ - \frac{RT}{F}\ln\frac{a_{\mathrm{Ag}}}{a_{\mathrm{Ag}^+}} \qquad (346')$$

$$E_2 = E_2^\circ - \frac{RT}{F}\ln\frac{a_{\mathrm{Cl}^-}\cdot a_{\mathrm{Ag}}}{a_{\mathrm{AgCl}}} \qquad (347')$$

the subtraction of eq(346') from eq(347') yields eq(350'). Consequently, the solubility product is readily obtained from eq(350') just in the same manner as given above.

$$\Delta E = E_2 - E_1 = \Delta E^\circ - \frac{RT}{F}\ln\frac{a_{\mathrm{Ag}^+}\times a_{\mathrm{Cl}^-}}{a_{\mathrm{AgCl}}} = \Delta E^\circ - \frac{RT}{F}\ln K_{sp} \qquad (350')$$

As is shown above, the solubility products of salts hard to be dissolved can indirectly be obtained from the values of the standard reduction potentials of half-cells and vice versa (at least for either of the two half-cells).

"note 1" The *Nernst* equation (*Hermann Walter Nernst*, Nobel Prize in Chemistry in 1920) and the electrochemical potential μ_{ele}.

When the electrode reactions (eq(351), always written as the reduction reactions with the electron on the left side) are in equilibrium,

$$\mathrm{M}^{z+}(\text{solution}) + ze \rightleftarrows \mathrm{M}\,(\text{electrode}) \qquad (351)$$

the electrode potential E_{M} is given by eq(352) and this equation is usually called *Nernst*' s equation. Here, E_{M}° is the electrode potential at $a_{\mathrm{M}^{z+}} = 1$.

$$E_{\mathrm{M}} = E_{\mathrm{M}}^\circ + \frac{RT}{zF}\ln a_{\mathrm{M}^{z+}} \qquad (352)$$

Nernst' s equation is not a specifically derived equation based on a certain theory but can be derived from the general thermodynamic method applied to the chemical equilibrium. If the reactions of eq(351) are homogeneous reactions in solution, then it is enough to put their free energy change equal to zero in the equilibrium, i.e., $\Delta G_r = 0$, which can be expressed as $\mu_{\mathrm{M}} = \mu_{\mathrm{M}^{z+}} + z\mu_{\mathrm{e}}$ by employing the chemical potentials μ_{M}, $\mu_{\mathrm{M}^{z+}}$ and μ_{e} of the relevant substances involved in the reactions. When electrode reactions are involved, however, the electrode and the solution must be considered at different constant electric potentials (see **"note 2"**). Consequently, the potential energy due to this difference ((potential energy/mole) = (electric potential) × (charge) × N_A) should be added to the relevant chemical potentials before considering the balance at the equilibrium. *Faraday* constant is given by (unit charge)×$N_A = F$.

$$\mu_{\mathrm{M}} \Rightarrow \mu_{\mathrm{M}} \equiv \mu_{ele}(\mathrm{M})$$

$$\mu_{\mathrm{M}^{z+}} \Rightarrow \mu_{\mathrm{M}^{z+}} + ze\times N_A \times E_{sol} \equiv \mu_{ele}(\mathrm{M}^{z+}) \qquad (353) \qquad (e\times N_A = F)$$

$$\mu_{\mathrm{e}} \Rightarrow \mu_{\mathrm{e}} - e\times N_A \times E_{met} \equiv \mu_{ele}(e) \qquad (met\text{: }metal)$$

To the chemical potentials defined in homogeneous solution, the corresponding electric energies per unit charge are added and the sum thus defined for each substance is called electrochemical potential.

$$\mu_{ele}(\mathrm{M}) = \mu_{ele}(\mathrm{M}^{z+}) + z\mu_{ele}(e) \qquad (354)$$
$$= \mu_{\mathrm{M}^{z+}} + z \times F \times E_{sol} + z \times (\mu_{\mathrm{e}} - \mathrm{e} \times F \times E_{met})$$
$$= \mu_{\mathrm{M}^{z+}} + z\mu_{\mathrm{e}} + zF(E_{sol} - E_{met}) \qquad (354')$$

When the electrode reactions are in equilibrium at a given electrode potential, the electrochemical potentials of the substances participating the reactions must be equal before and after the reactions. Namely, eqs(354) and (354') are satisfied. Each of these chemical potentials is expressed as eq(355) using the activity corresponding to each relevant substance.

$$\left.\begin{array}{ll} \mu_{\mathrm{M}} = \mu_{\mathrm{M}}^{\circ} + RT\ln a_{\mathrm{M}} = \mu_{\mathrm{M}}^{\circ} & (\because a_{\mathrm{M}} = 1) \\ \mu_{\mathrm{M}^{z+}} = \mu_{\mathrm{M}^{z+}}^{\circ} + RT\ln a_{\mathrm{M}^{z+}} & \\ \mu_{\mathrm{e}} = \mu_{\mathrm{e}}^{\circ} + RT\ln a_{\mathrm{e}} = \mu_{\mathrm{e}}^{\circ} & (\because a_{\mathrm{e}} = 1) \end{array}\right\} \qquad (355)$$

Substituting eq(355) into eq(354') and after some rearrangements, eqs(356) and (357) are obtained,

$$\mu_{\mathrm{M}}^{\circ} - \mu_{\mathrm{M}^{z+}}^{\circ} - z\mu_{\mathrm{e}}^{\circ} = RT\ln a_{\mathrm{M}^{z+}} + zF(E_{sol} - E_{met}) \qquad (356)$$

$$(E_{met} - E_{sol}) = \frac{z\mu_e^{\circ} + \mu_{\mathrm{M}^{z+}}^{\circ} - \mu_{\mathrm{M}}^{\circ}}{zF} + \frac{RT}{zF}\ln a_{\mathrm{M}^{z+}} \qquad (357)$$

and from which by putting $(E_{met} - E_{sol}) \equiv E_{\mathrm{M}}$ and $(z\mu_{\mathrm{e}}^{\circ} + \mu_{\mathrm{M}^{z+}}^{\circ} - \mu_{\mathrm{M}}^{\circ})/zF \equiv E_{\mathrm{M}}^{\circ}$, eq(357') is derived.

$$E_{\mathrm{M}} = E_{\mathrm{M}}^{\circ} + \frac{RT}{zF}\ln a_{\mathrm{M}^{z+}} \qquad (357')$$

This is the equation called *Nernst* equation and is the same as eq(352). See next problem.

"note 2" The electric potential is defined as the potential energy due to the electric force per a unit electrostatic charge. For this it is an intensive quantity.

Solution 4-50

(a) The free energy changes of the electrode reactions given in the problem are expressed in terms of the chemical potentials of the relevant substances by eqs(358) and (359), respectively.

$$\Delta G_1 = \frac{1}{2}\mu_{\mathrm{H}_2} - \mu_{\mathrm{H}^+} - \mu_{\mathrm{e}}\left(\mathrm{H}^+, \mathrm{H}_2 \middle| \mathrm{Pt}\right) \qquad (358)$$
$$\Delta G_2 = \mu_{\mathrm{Ag}} + \mu_{\mathrm{Cl}^-} - \mu_{\mathrm{AgCl}} - \mu_{\mathrm{e}}\left(\mathrm{AgCl} \middle| \mathrm{Ag}\right) \qquad (359)$$

Notice here that the chemical potential of electron is given individually for each electrode, since its values are different under the influence of each electrode interface (electrical double layer).

When the electrode reactions are in equilibrium, the free energy change associated with each reaction is zero and this is expressed in terms of the chemical potentials by each of the following equations.

$$\mu_{\mathrm{e}}\left(\mathrm{H}^+, \mathrm{H}_2 \middle| \mathrm{Pt}\right) = \frac{1}{2}\mu_{\mathrm{H}_2} - \mu_{\mathrm{H}^+} \qquad (358')$$
$$\mu_{\mathrm{e}}\left(\mathrm{AgCl} \middle| \mathrm{Ag}\right) = \mu_{\mathrm{Ag}} + \mu_{\mathrm{Cl}^-} - \mu_{\mathrm{AgCl}} \qquad (359')$$

The left side of each equation can be expressed by using the corresponding electric potential of electron. Noting that the electric potential is an intensive quantity and that the energy of an electron at the electric potential E is given by $-e \times E$, it is found that the conversion of this energy to the one per mole gives the electrode potential for each electrode as shown below by eqs(358'') and (359'').

$$E_1 = -\frac{\mu_e(H^+, H_2|Pt)}{F} = -\frac{1}{F}\left(\frac{1}{2}\mu_{H_2} - \mu_{H^+}\right) \qquad (358'')$$

$$E_2 = -\frac{\mu_e(AgCl|Ag)}{F} = -\frac{1}{F}(\mu_{Ag} + \mu_{Cl^-} - \mu_{AgCl}) \qquad (359'')$$

Since the chemical potentials on the right side of each equation above are given by eq(360) using the relevant activities, substituting these relations into eqs(358'') and (359''),

$$\left.\begin{array}{l} \mu_{H_2} = \mu^\circ_{H_2} + RT \ln\dfrac{f_{H_2}}{f^\circ} = \mu^\circ_{H_2} + RT \ln P_{H_2} \\ \mu_{H^+} = \mu^\circ_{H^+} + RT \ln\dfrac{a_{H^+}}{a^\circ_{H^+}} = \mu^\circ_{H^+} + RT \ln a_{H^+} \\ \mu_{Ag} = \mu^\circ_{Ag}, \quad \mu_{AgCl} = \mu^\circ_{AgCl}, \quad \mu_{Cl^-} = \mu^\circ_{Cl^-} + RT \ln a_{Cl^-} \end{array}\right\} \qquad (360)$$

the electrode potentials are found as eq(361) and (362), respectively.

$$E_1 = E^\circ_1 - \frac{RT}{F}\ln\frac{(P_{H_2})^{1/2}}{a_{H^+}}, \qquad E^\circ_1 \equiv -\frac{1}{F}\left(\frac{1}{2}\mu^\circ_{H_2} - \mu^\circ_{H^+}\right) \qquad (361)$$

$$E_2 = E^\circ_2 - \frac{RT}{F}\ln a_{Cl^-}, \qquad E^\circ_2 = -\frac{1}{F}(\mu^\circ_{Ag} + \mu^\circ_{Cl^-} - \mu^\circ_{AgCl}) \qquad (362)$$

Ans. electrode potentials: eq(361) and eq(362)

Here, it is seen that from eqs(361) and (362), E°_1 and E°_2 are the electric potentials when the activities or fugacity is unity (i.e., $a_{H^+} = 1$, $P_{H_2} = 1$ and $a_{Cl^-} = 1$). Normally, these values, however, are not accessible experimentally. For this, the chemical potentials of hydrogen and hydrogen ion are put equal to zero as in eq(363) and resulting eq(364) is employed as the definition of E°_1.

$$\mu^\circ_{H_2} = 0, \ \mu^\circ_{H^+} = 0, \ \text{resulting in } \mu_e(H^+, H_2|Pt) = 0 \qquad (363)$$

$$E^\circ_1 = 0 \qquad (364)$$

After determining a value of E°_1 in this manner, the electric potential difference ΔE°_{21} between the two electrodes at the standard state (in the present case, $a_{H^+} = 1$ $P_{H_2} = 1$ and $a_{Cl^-} = 1$) is found equal to E°_2 from eq(365).

$$\Delta E_{21} = E_2 - E_1 = E^\circ_2 - E^\circ_1 = \Delta E^\circ_{21} \ \Rightarrow \ \Delta E^\circ_{21} = E^\circ_2 \qquad (365)$$

The electrode shown in Fig.11 in the present problem is called standard hydrogen electrode (SHE) and its electric potential is set at zero. The electric potential difference between SHE and any other half-cell electrode at the standard state (see Fig.49) is defined as the standard reduction potential of that electrochemical cell. In this manner, the standard reduction potential is determined for every half-cell. This approach is equivalent to the procedure taken in order to acquire the thermodynamic extensive quantities associated with the chemical reactions by setting the relevant thermodynamic quantities of simple substances at the standard state equal to zero. For the electrode reactions in which ions play major roles, this is to say that once

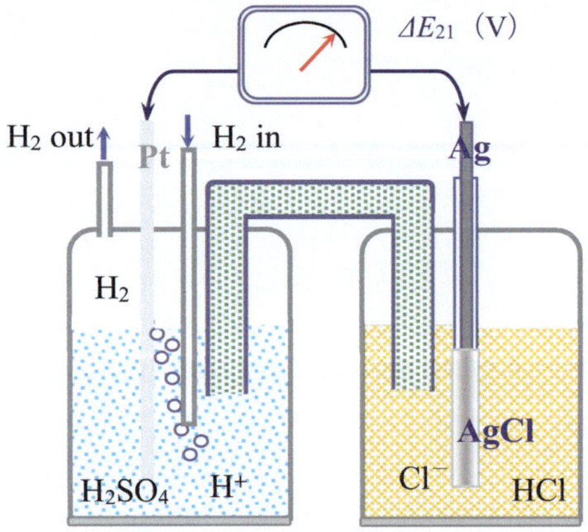

Fig.49 Measurable electrical potential between two electrodes.

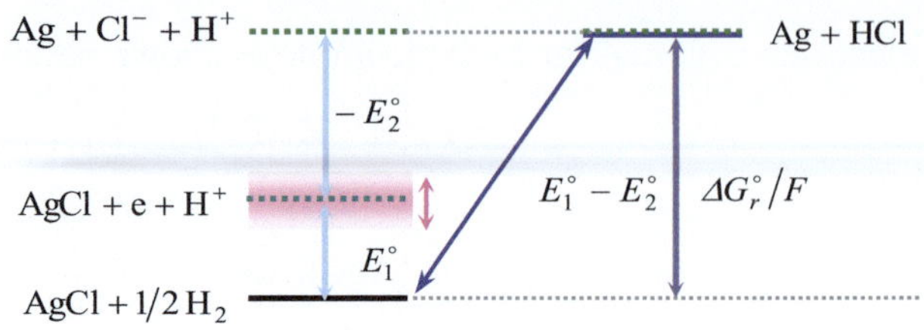

Fig.50 Indefinite electrical potential of half-cell. The difference between the standard state potentials are equal to the standard free energy change and is determined unambiguously.

the chemical potential of hydrogen ion is set to zero, i.e., $\mu^{\circ}_{\mathrm{H^+}}=0$, the thermodynamic quantities of the other ions are unambiguously determined.

(b) Subtracting eq(358) from eq(359), the reaction for the electrochemical cell is obtained as eq(366).

$$\mathrm{AgCl}+\frac{1}{2}\mathrm{H_2}\rightarrow \mathrm{Ag}+\mathrm{HCl}\quad (=\ \mathrm{Ag}+\mathrm{Cl^-}+\mathrm{H^+}) \qquad (366)$$

Consequently, the free energy change ΔG_r for this reaction can be expressed as eq(367).

$$\begin{aligned}\Delta G_r &= \mu_{\mathrm{Ag}}+\mu_{\mathrm{HCl}}-\mu_{\mathrm{AgCl}}-\frac{1}{2}\mu_{\mathrm{H_2}}\\ &=\mu^{\circ}_{\mathrm{Ag}}+\mu_{\mathrm{H^+}}+\mu_{\mathrm{Cl}}-\mu^{\circ}_{\mathrm{AgCl}}-\frac{1}{2}\mu^{\circ}_{\mathrm{H_2}}-\frac{1}{2}RT\ \ln P_{\mathrm{H_2}} \qquad (367)\end{aligned}$$

Employing here the relations of $\mu_{\mathrm{H^+}}=\mu^{\circ}_{\mathrm{H^+}}+RT\ln a_{\mathrm{H^+}}$ and $\mu_{\mathrm{Cl^-}}=\mu^{\circ}_{\mathrm{Cl^-}}+RT\ln a_{\mathrm{Cl^-}}$, eq(367) is rewritten as eq(368).

$$\begin{aligned}\Delta G_r &= \mu^{\circ}_{\mathrm{Ag}}+\mu^{\circ}_{\mathrm{H^+}}+RT\ln a_{\mathrm{H^+}}+\mu^{\circ}_{\mathrm{Cl^-}}+RT\ln a_{\mathrm{Cl^-}}-\mu^{\circ}_{\mathrm{AgCl}}-\frac{1}{2}\mu^{\circ}_{\mathrm{H_2}}-\frac{1}{2}RT\ \ln P_{\mathrm{H_2}}\\ &=\mu^{\circ}_{\mathrm{H^+}}+\mu^{\circ}_{\mathrm{Cl^-}}-\mu^{\circ}_{\mathrm{AgCl}}-\frac{1}{2}\mu^{\circ}_{\mathrm{H_2}}+\frac{1}{2}RT\ \ln\frac{(a_{\mathrm{H^+}}\cdot a_{\mathrm{Cl^-}})^2}{P_{\mathrm{H_2}}} \qquad (368)\end{aligned}$$

Ans. chemical reaction: eq(366), ΔG_r: eq(368)

(c) Rearranging eq(368) so as to divide it into two parts, eq(368') is obtained, from which the relation between the electric potential difference ΔE and the free energy change ΔG_r associated with the chemical reaction is derived. Here, F is the *Faraday* constant.

$$\begin{aligned}\Delta G_r &=\left\{(\mu^{\circ}_{\mathrm{Ag}}+\mu^{\circ}_{\mathrm{Cl^-}}-\mu^{\circ}_{\mathrm{AgCl}})+RT\ln a_{\mathrm{Cl^-}}\right\}-\left\{\left(\frac{1}{2}\mu^{\circ}_{\mathrm{H_2}}-\mu^{\circ}_{\mathrm{H^+}}\right)+RT\ \ln\frac{P_{\mathrm{H_2}}}{a_{\mathrm{H^+}}}\right\}\\ &=-F\times\langle \mathrm{eq}(5)-\mathrm{eq}(4)\rangle=-F\times(E_2-E_1) \qquad (368')\end{aligned}$$

Furthermore, once the standard free energy of formation of hydrogen ion is set equal to zero, by equating $\mu^{\circ}_{\mathrm{Ag}}=\mu^{\circ}_{\mathrm{H_2}}=\mu^{\circ}_{\mathrm{H^+}}=0$, it is seen that the standard reduction potentials of the hydrogen ion and the silver-silver chloride electrode are given unambiguously by eq(369).

$$E^{\circ}_1=0\ ,\qquad E^{\circ}_2=-\frac{1}{F}(\mu^{\circ}_{\mathrm{Cl^-}}-\mu^{\circ}_{\mathrm{AgCl}}) \qquad (369)$$

From a different viewpoint, it is said that the chemical potential of chloride ion can be determined from the value of E°_2.

Ans. proof: eq(368')

(d) The chemical potential of an electron in the electrode can theoretically be defined. However, it is not only difficult but also of no practical use to measure it actually. In principle, since the potential difference is defined by the difference between the two electric potentials, unless the potential of a half-cell is defined against some standard, the potential of the half-cell is not practically measurable. Therefore, as has been discussed above, the electric potential of the standard hydrogen electrode is set at zero first and then the electric potential difference between the half-cell and the standard hydrogen electrode is defined as the standard electric (reduction) potential of that half-cell. When the value defined thus is positive, the half-cell works as a positive electrode and when negative, it works as a negative electrode. Against an electrode other than the hydrogen electrode, the electrode which has the higher electric potential works as a positive electrode and the electrode with the lower electric potential as a negative electrode.

Ans. See text (d).

(e) See text (b) and eq(363).

The chemical potential of hydrogen, being a simple substance, is defined as zero. Setting the chemical potential of hydrogen ion at zero as well, the chemical potential of an electron in the standard hydrogen electrode (SHE) also becomes zero as is seen from eq(367). Explicitly, those values are fixed at zero as follows: $\mu^{\circ}_{H_2} = 0$, $\mu^{\circ}_{H^+} = 0$ and $\mu_e\left(H^+, H_2|Pt\right) = 0$. Naturally, the values for $[H^+]$ and $[H_2]$ are those at the standard sate. Because of this, for any half-cell, its electromotive force is unambiguously determined as the electric potential difference against SHE.

Ans. $\mu^{\circ}_{H_2} = 0$, $\mu^{\circ}_{H^+} = 0$ and $\mu_e\left(H^+, H_2|Pt\right) = 0$

Solution 4-51

(a) The reaction at the positive electrode $Pt(s)|H_2(g, f = 1\,atm)$ is expressed by eq(370) and its electrode reduction potential E_1 is given by eq(370').

$$H^+ + e \rightarrow \frac{1}{2}H_2 \qquad (370)$$

$$E_1 = E_1^{\circ} - \frac{RT}{F}\ln\frac{\left\langle f_{H_2}(=1)\right\rangle^{1/2}}{a_{H^+}} = E_1^{\circ} + \frac{RT}{F}\ln\gamma_{\pm}m \qquad (n=1) \qquad (370')$$

On the other hand, the reaction at the negative electrode $HCl(aq, m)|AgCl(s)|Ag$ is expressed by eq(371) and its electrode reduction potential E_2 is given by eq(371').

$$AgCl + e \rightarrow Ag + Cl^- \qquad (371)$$

$$E_2 = E_2^{\circ} - \frac{RT}{F}\ln\frac{a_{Ag}a_{Cl^-}}{a_{AgCl}} = E_2^{\circ} - \frac{RT}{F}\ln\gamma_{\pm}m \qquad (n=1) \qquad (371')$$

Since the chemical reaction in the electrochemical cell is found by eliminating electrons from eqs(370) and (371), eq(372) is obtained from these half-cell reactions.

$$AgCl + \frac{1}{2}H_2 \rightarrow H^+ + Cl^- + Ag \qquad (372)$$

Consequently, the electromotive force ΔE of this cell is given by eq(373) by putting $E_2^{\circ} - E_1^{\circ} \equiv \Delta E_{21}^{\circ}$.

$$\Delta E = E_2 - E_1 = \Delta E^\circ_{21} - \frac{2RT}{F}\ln \gamma_{\pm} m$$
$$= \Delta E^\circ_{21} - \frac{2RT}{F}\ln m - \frac{2RT}{F}\ln \gamma_{\pm} \qquad (373)$$

Ans. electromotive force of the cell: eq(373); See also eqs(373') and (373'') in (b).

(b) When the mean activity coefficient of the hydrochloric acid is approximated by $\gamma_{\pm} = 1$, ΔE is expressed by eq(373').

$$\Delta E = \Delta E^\circ_{21} - \frac{2RT}{F}\ln m = \Delta E^\circ_{21} - 0.05138 \times \ln m \qquad (373')$$

Plotting ΔE against $-0.05138 \ln m$ (liner least squares fitting with a fixed slope), the best-fitted intercept is found as $\Delta E^\circ_{21} = 0.2295$ V, as is shown in Fig.51.

In the higher-order approximation, the *Debye*-*Hückel* limiting equation (374) (see **problem 4-18**) is employed at the hydrochloric acid concentrations lower than $0.01\ \mathrm{mol \cdot kg^{-1}}$ to acquire the activity coefficients.

$$\ln \gamma_{\pm} = -|Z_+ Z_-| \frac{N_A{}^2 e^3 \sqrt{2dI}}{8\pi(\varepsilon_0 \varepsilon_r RT)^{3/2}} = -1.172\sqrt{m} \qquad (374)$$

Then, eq(373) can be replaced by eq(373''),

$$\Delta E = \Delta E^\circ_{21} - 0.05138 \times \ln m + 0.0602 \times \sqrt{m} \qquad (373'')$$

and its non-linear least squares fitting by using gnuplot yields the best-fitted value for a constant in eq(373'') as $\Delta E^\circ_{21} = 0.2220 \pm 0.0002$ V, as is shown in Fig.52.

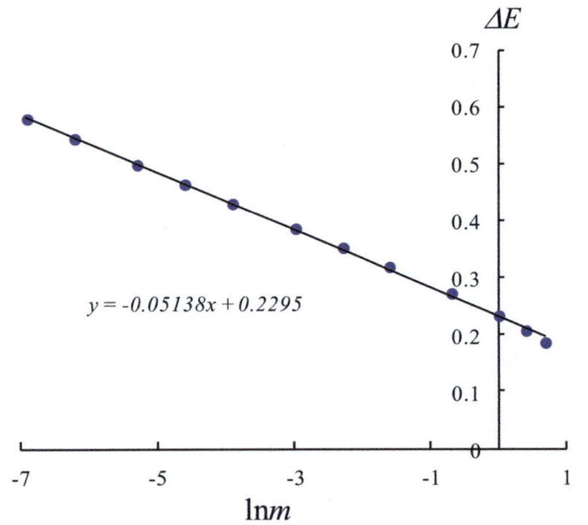

Fig.51 Liner least squares fitting with a fixed slope. The linear line is given by eq(373').

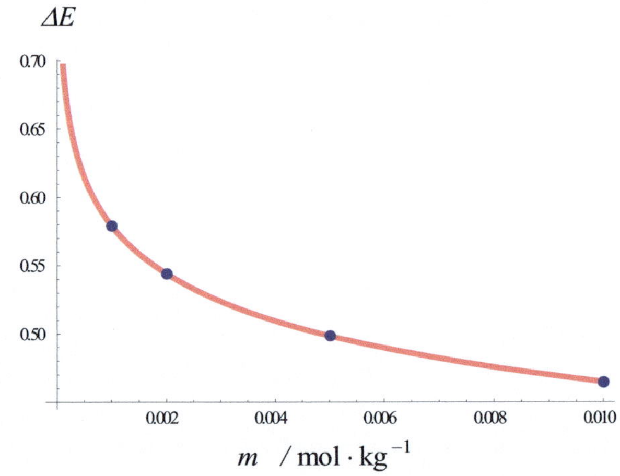

Fig.52 Non-linear least squares fitting. The fitted curve is given by eq(373'') with $\Delta E^\circ_{21} = 0.2220$.

Ans. approximate value: $\Delta E^\circ = 0.2295 \pm 0.0014$ V, more accurate value: $\Delta E^\circ_{21} = 0.2220 \pm 0.0002$ V

"note" Since the fitting parameter in either of eq(373') or (373'') is only a constant term, it is not necessary to employ software to acquire a reliable value for ΔE°_{21}. A value for ΔE°_{21} is calculated at each measured point and then the average value of ΔE°_{21} over all these points is accurate enough. In fact, employing eq(373'), the average value of ΔE°_{21} is found as 0.22952 V.

(c) The mean activity coefficient of the hydrochloric acid at 298 K is found from eq(375), a rewritten form of eq(373) for the mean activity coefficient, by employing the value of $\Delta E^\circ_{21} = 0.2220$ V found above for the standard reduction potential of the cell. The calculated values are summarized in the table below.

$$\ln \gamma_{\pm} = \frac{\Delta E^{\circ}_{21} - \Delta E}{0.05138} - \ln m \qquad (375)$$

m / mol · kg^{-1}	0.001	0.01	0.1	1.0	1.22
$\gamma_{\pm}$	0.9575	0.8975	0.7901	0.8029	0.8197

Ans. mean activity coefficient: values in the table

(d) The concentration of the hydrochloric acid at which the activity $a = \gamma_{\pm} m$ takes a value of 1.0 is determined by finding the concentration at which ΔE becomes equal to the standard electrical potential ΔE°_{21}.

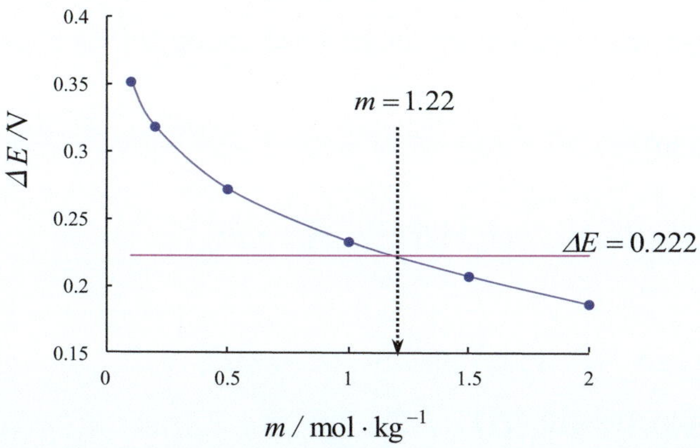

Fig.53 Calculation of the concentration of hydrochloric acid at which the activity becomes equal to 1.0.
The curve is drawn through the actually measured points. Since the concentration is close to 1.0, eq(373'') cannot be used.

This is performed by plotting first the values of ΔE against m as is shown in Fig.53 and then by finding a value of m at which the curve crosses the horizontal line drawn at $\Delta E = \Delta E^{\circ}_{21} = 0.222$, yielding a value of 1.22 for m.

Ans. concentration of hydrochloric acid at $a = 1.0$: $m = 1.22$ / mol · kg^{-1}

Solution 4-52

(a) Usually the electrode at which an oxidation reaction occurs is defined as the positive (anode) electrode and the electrode at which a reduction reaction occurs as the negative (cathode) one. Looking at a half-cell reaction, the electrode reaction expressed by a reduction reaction by placing an electron on the left side of the reaction equation, the electrode with the higher reduction potential becomes a negative electrode (cathode). Consequently, the electrode of the half-cell of $|H_2SO_4(aq,m)|PbO_2(s)|Pb(s)$ of the Pb storage battery becomes the negative electrode (cathode) and $Pb(s)|PbSO_4(s)|H_2SO_4(aq,m)|$ becomes the positive electrode (anode). The net chemical reaction to occur in the battery is found by eliminating electrons from the half-cell reactions given by eqs(376) and (377).

$$\frac{1}{2}PbO_2(s) + 2H^+ + \frac{1}{2}SO_4^{2-} + e \rightarrow \frac{1}{2}PbSO_4(s) + H_2O, \quad E_1^{\circ} \qquad (376)$$

$$\frac{1}{2}PbSO_4(s) + e \rightarrow \frac{1}{2}Pb(s) + \frac{1}{2}SO_4^{2-}, \quad E_2^{\circ} \qquad (377)$$

Thus, subtracting eq(377) from eq(376), the net reaction is found as eq(378),

$$\frac{1}{2}PbO_2(s) + 2H^+ + SO_4^{2-} + \frac{1}{2}Pb(s) \rightarrow PbSO_4(s) + H_2O \qquad (378)$$

and from which the standard electrode potential ΔE° of this battery is calculated as $\Delta E^\circ = E_1^\circ - E_2^\circ = 2.04$ V.

Ans. negative electrode: $|H_2SO_4(aq,m)|PbO_2(s)|Pb(s)$, reaction: eq(378), standard electrode potential: $\Delta E^\circ = 2.04$ V

(b) The reaction in the battery can be regarded as a formation reaction of $PbSO_4(s)$ $(n=1)$. Then, the free energy change and the electromotive force of the reaction are related by eq(379) from which $\Delta G = -188\,\mathrm{kJ\cdot mol^{-1}}$ is found at $t = 25$ °C. From this, the reaction in the battery is found to occur spontaneously (spontaneous reaction).

$$\Delta G = -F\Delta E \qquad (379)$$
$$= -96500 \times (1.917 + 56.1\times10^{-6}t + 108\times10^{-8}t^2) = -188.4\,\mathrm{kJ\cdot mol^{-1}}$$

Employing eqs(380) and (381), the following values are found for the enthalpy and entropy of the reaction, respectively.

$$\Delta H = -T^2\left(\frac{\partial \Delta G/T}{\partial T}\right)_P \qquad (380)$$
$$= 96500\times T^2 \frac{\partial}{\partial T}\left\{\frac{1.917 + 56.1\times10^{-6}(T-273) + 108\times10^{-8}(T-273)^2}{T}\right\}$$
$$= 96500\times\left\langle \begin{array}{l} -\left\{1.917 + 56.1\times10^{-6}(T-273) + 108\times10^{-8}(T-273)^2\right\} \\ +56.1\times10^{-6}T + 2\times108\times10^{-8}T(T-273) \end{array}\right\rangle$$
$$= -182.0\,\mathrm{kJ\cdot mol^{-1}}$$

$$\Delta S = -\frac{\Delta G - \Delta H}{T} \qquad (381)$$
$$= -\frac{-188.4\times10^3 + 182.0\times10^3}{298} = 21.5\ \mathrm{J\cdot mol^{-1}K^{-1}}$$

Ans. $\Delta G = -188.4\ \mathrm{kJ\cdot mol^{-1}}$, $\Delta H = -182.0\ \mathrm{kJ\cdot mol^{-1}}$, $\Delta S = 21.5\ \mathrm{J\cdot mol^{-1}K^{-1}}$

(c) A current of 2.5 A runs through the bulb at 2V with 5W and, for an hour, the total electricity amounts to (2.5×3600) coulombs. As a result the moles of $PbSO_4(s)$ equivalent to $(2.5\times3600)/96500$ are formed and correspondingly a half of the same moles of $Pb(s)$ is consumed as is seen from eq(378), yielding $(2.5\times3600)/96500\times\frac{1}{2}\times207 = 9.7$ g as the amount of Pb consumed.

Ans. $Pb(s)$ consumed: 9.7 g

(d) Based on the equation of reaction (378), the electromotive force of this battery is given by eq(382) in which the activity of water or solid is put equal to one. This is further rewritten as eq(382') in terms of the mean activity coefficient $\gamma_\pm$ and the molal concentration m.

$$\Delta E = \Delta E^\circ - \frac{RT}{F}\ln\frac{a_{PbSO_4}a_{H_2O}}{(a_{PbO_2})^{1/2}(a_{H^+})^2 a_{SO_4^{2-}} a_{Pb}} = \Delta E^\circ - \frac{RT}{F}\ln\frac{1}{(a_{H^+})^2 a_{SO_4^{2-}}} \qquad (382)$$
$$= \Delta E^\circ + \frac{3RT}{F}\ln(\gamma_\pm m) \qquad (382')$$

Here, the mean activity coefficient $\gamma_\pm$ of an electrolyte of the type of A_pX_q is given by the left side of eq(383) and, hence, for the ions of H^+ and SO_4^{2-} arising from sulfuric acid H_2SO_4 ($p=2,\ q=1$) it is given by the right side of eq(383).

$$\gamma_{\pm} = (\gamma_{+}^{p}\gamma_{-}^{q})^{1/(p+q)} \Rightarrow \gamma_{\pm} = (\gamma_{+}^{2}\gamma_{-})^{1/3} \qquad (383)$$

Using eq(383), the activities of H^+ and SO_4^{2-} are given by the left side of eq(384) and their relevant product is given by its right side. Substituting this relation into eq(382), eq(382') is derived.

$$a_{H^+} = \gamma_{+}m, \qquad a_{SO_4^{2-}} = \gamma_{-}m \Rightarrow (a_{H^+})^2 a_{SO_4^{2-}} = \gamma_{+}^{2}\gamma_{-}m^3 = (\gamma_{\pm}m)^3 \qquad (384)$$

After discharging the battery for one hour, the concentration of H^+ is found to decrease from $2.0\ \mathrm{mol \cdot kg^{-1}}$ to $1.91\ \mathrm{mol \cdot kg^{-1}}$ based on the result obtained in (c), a change in the electromotive force $\Delta(\Delta E)$ is calculated from eq(382') as $-3.5 \times 10^{-3}\ \mathrm{V}$.

$$\Delta(\Delta E) = \frac{3 \times 8.31 \times 298}{96500} \ln \frac{1.91}{2.0} = -3.5 \times 10^{-3}\ \mathrm{V}$$

Ans. drop of electromotive force: $3.5 \times 10^{-3}\ \mathrm{V}$

Solution 4-53

(a) Eliminating electrons from eqs(17) and (18) given in the problem by subtracting eq(18) from eq(17),

$$Ni_2O_3(s) + H_2O(l) + 2e \rightleftarrows 2NiO(s) + 2OH^-(aq), \qquad E_1^\circ = 0.40\ \mathrm{V} \qquad (17)$$

(reposted)

$$FeO(s) + H_2O(l) + 2e \rightleftarrows Fe(s) + 2OH^-(aq), \qquad E_2^\circ = -0.87\ \mathrm{V} \qquad (18)$$

the net chemical reaction seen in the battery is obtained as eq(385). The electrode with the higher standard electrical potential is a positive electrode.

$$Ni_2O_3(s) + Fe(s) \rightleftarrows 2NiO(s) + FeO(s), \qquad \Delta E_{12}^\circ = 1.27\ \mathrm{V} \qquad (385)$$

Ans. chemical reaction in the battery: eq(385)

(b) In the net battery reaction (eq(385)), KOH is not involved explicitly. Therefore, the activity of KOH has nothing to do with the electromotive force of the battery, though it affects the half-cell electrode reactions and the electric conductivity of the battery solution.

Ans. electromotive force of the battery: independent of the activity of KOH

(c) Noticing that the two electrons are involved in the net reaction, the electrical work to be gained per one mole of Fe is obtained by multiplying two as below,

$$2 \times F \times \Delta E_{12}^\circ = 2 \times 96500 \times 1.27 = 2.45 \times 10^5\ \mathrm{J \cdot mol^{-1}}$$

and, hence, for one kg of Fe, the electrical work amounts to $2.45 \times 10^5 \times 1000/55.85 = 4.39\ \mathrm{MJ}$.

Ans. electrical work: 4.39 MJ/kg

"note" It is worth comparing the result with that calculated from the standard free energy of formation ΔG_f°.

Solution 4-54

(a) The electrode reaction for each ion and its standard reduction potential is given by eq(386) or eq(387).

$$Zn^{2+} + 2e \rightarrow Zn, \qquad E_1^\circ = -0.76\ \mathrm{V} \qquad (386)$$

$$Cu^{2+} + 2e \rightarrow Cu, \qquad E_2^\circ = 0.34\ \mathrm{V} \qquad (387)$$

$$Zn + Cu^{2+} \rightarrow Zn^{2+} + Cu, \qquad \Delta E_{21}^\circ = 1.10\ \mathrm{V} \qquad (388)$$

Ans. $Zn + Cu^{2+} \rightarrow Zn^{2+} + Cu$, $\Delta E_{21}^\circ = 1.10\ \mathrm{V}$

(b) Substituting $t = 25$ °C into eq(386) given in the problem, $\Delta G^\circ = -210.6\ \mathrm{kJ\cdot mol^{-1}}$ is found as shown below.

$$\Delta G^\circ = -2 \times F\Delta E^\circ \qquad (388')$$
$$= -2 \times 96500 \times (11.1028 - 0.641 \times 10^{-3} t + 0.72 \times 10^{-5} t^2) = -210.6\ \mathrm{kJ\cdot mol^{-1}}$$

Judging from this value, the battery reaction given by eq(388) is said to a spontaneous one.

The standard enthalpy ΔH° and entropy ΔS° of the reaction are calculated by employing eq(389) and eq(390), respectively and the following values are found.

$$\Delta H^\circ = -T^2 \left(\frac{\partial \Delta G^\circ / T}{\partial T} \right)_P \qquad (389)$$
$$= 2 \times 96500 \times T^2 \frac{\partial}{\partial T} \left\{ \frac{1.1028 - 0.641 \times 10^{-3}(T-273) + 0.72 \times 10^{-5}(T-273)^2}{T} \right\}$$
$$= 2 \times 96500 \times \left\langle \begin{array}{l} -\{1.1028 - 0.641 \times 10^{-3}(T-273) + 0.72 \times 10^{-5}(T-273)^2\} \\ -0.641 \times 10^{-3} T + 2 \times 0.72 \times 10^{-5} T(T-273) \end{array} \right\rangle$$
$$= -226.8\ \mathrm{kJ\cdot mol^{-1}}$$

$$\Delta S^\circ = -\frac{\Delta G^\circ - \Delta H^\circ}{T} \qquad (390)$$
$$= -\frac{-210.6 \times 10^3 + 226.8 \times 10^3}{298} = -54.4\ \mathrm{J\cdot mol^{-1}K^{-1}}$$

Ans. $\Delta G^\circ = -210.6\ \mathrm{kJ\cdot mol^{-1}}$, $\Delta H^\circ = -226.8\ \mathrm{kJ\cdot mol^{-1}}$, $\Delta S^\circ = -54.4\ \mathrm{J\cdot mol^{-1}K^{-1}}$

(c) The electromotive force of this battery is given by eq(391).

$$\Delta E = \Delta E^\circ_{21} - \frac{RT}{2F} \ln \frac{a_{\mathrm{Cu}} \cdot a_{\mathrm{Zn^{2+}}}}{a_{\mathrm{Zn}} \cdot a_{\mathrm{Cu^{2+}}}} = 1.10 - \frac{RT}{2F} \ln \frac{a_{\mathrm{Zn^{2+}}}}{a_{\mathrm{Cu^{2+}}}} \qquad (391)$$

Therefore, the electromotive force is raised by increasing the concentration of Cu^{2+} and decreasing the concentration of Zn^{2+}.

Ans. See text(c).

Solution 4-55

(a), (b) Being different from the *Daniel* cell treated in **problem 4-54**, the *Volta* cell does not contain metal ions in the solution initially and one of the half-cell reactions is the dissolution of zinc metal expressed by eq(392) (opposite direction) and the other is the evolution of hydrogen given by eq(393). The net reaction of the cell and its standard electromotive force are given by eq(394).

$$Zn^{2+} + 2e \rightarrow Zn \qquad E^\circ_1 = -0.76\ \mathrm{V} \qquad (392)$$
$$2H^+ + 2e \rightarrow H_2 \qquad E^\circ_2 = 0.0\ \mathrm{V} \qquad (393)$$
$$Zn + 2H^+ \rightarrow Zn^{2+} + H_2 \qquad \Delta E^\circ_{21} = 0.76\ \mathrm{V} \qquad (394)$$

By replacing the *Volta* cell with the *Daniell* cell, the standard electromotive force of the cell can be raised to 1.10 V. The electromotive force of *Daniel* cell ΔE_{21} varies with the concentration of each cell solution as is given by eq(395),

$$\Delta E_{21} = \Delta E^\circ_{21} - \frac{RT}{2F} \ln \frac{a_{\mathrm{Zn^{2+}}} \cdot a_{\mathrm{Cu}}}{a_{\mathrm{Zn}} \cdot a_{\mathrm{Cu^{2+}}}} = 1.10 - 0.0128 \times \ln \frac{a_{\mathrm{Zn^{2+}}}}{a_{\mathrm{Cu^{2+}}}} \qquad (395)$$

and, for instance, taking the ratio of the ion concentrations as low as the one given by eq(396), the electromotive force can be increased by 0.12 V but not by a significant amount.

$$\frac{a_{Zn^{2+}}}{a_{Cu^{2+}}} \approx \frac{[Zn^{2+}]}{[Cu^{2+}]} \approx 10^{-4} \qquad (396)$$

Ans. electrode reaction and potential: eqs(392)-(394), maximum electromotive force: 0.76 V

(c) In order to turn on the LED in red, if its emission wavelength is located at 700 nm, the minimum energy required amounts to $\frac{10^7}{700\ \text{nm}} \times 1.986 \times 10^{-23} = 2.84 \times 10^{-19}$ J. On the other hand, the electrical energy the *Volta* cell can supply is found as $2 \times 0.76\ \text{V} \times 1.602 \times 10^{-19} = 2.44 \times 10^{-19}$ J, lower than the photon energy. Consequently, it is concluded that the LED cannot be turned on with the *Volta* cell. The minimum electromotive force required for this purpose is calculated as

$$\frac{2.84 \times 10^{-19}\ \text{J}}{2 \times 1.602 \times 10^{-19}\ \text{C}} = 0.89\ \text{V},$$

and, therefore, with the *Daniell* cell whose standard electromotive force is known as 1.1V the LED can be turned on stably anyway.

Ans. *Volta* cell is not enough to turn on red LED.

Solution 4-56

(a) The formation reaction of water and its standard free energy of formation is given by eq(397).

$$H_2 + \frac{1}{2}O_2 \rightarrow H_2O \qquad \Delta G_f^\circ = -237\ \text{kJ} \cdot \text{mol}^{-1} \qquad (397)$$

This reaction is also expressed as eq(399) by using the reduction reaction of hydrogen ion in solution given by eq(398) and, from this, it is seen that the production of one mole of water requires two moles of electron.

$$2H^+ + 2e \rightarrow H_2 \qquad (398)$$

$$\frac{1}{2}O_2 + 2e + 2H^+ \rightarrow H_2O \qquad (399)$$

Consequently, the standard electromotive force of the fuel cell based on the reaction in eq(397) is found as $\Delta E^\circ = 1.23$ V from eq(400).

$$\Delta E^\circ = -\frac{\Delta G_f^\circ}{2F} = \frac{237 \times 10^3}{2 \times 96500} = 1.23\ \text{V} \qquad (n = 2) \qquad (400)$$

Ans. standard electromotive force: $\Delta E^\circ = 1.23$ V

(b) The electrical reduction reactions and their standard reduction potentials of the substances involved in the combustion reaction of hydrogen (eq(397)) are given by eqs(398'), (401) and (402). The values for ΔE° are taken from the data given in *Physical Chemistry* by *P. Atkins*.

$$2H^+ + 2e \rightarrow H_2 \qquad \Delta E^\circ = 0\ \text{V} \qquad (398')$$

$$2H_2O + 2e \rightarrow H_2 + 2OH^- \qquad \Delta E^\circ = -0.83\ \text{V} \qquad (401)$$

$$O_2 + 2H_2O + 4e \rightarrow 4OH^- \qquad \Delta E^\circ = 0.40\ \text{V} \qquad (402)$$

These reactions are normalized by dividing each of them by the number of electrons involved as follows (see **"note"**).

$$H^+ + e \rightarrow \frac{1}{2}H_2 \qquad \Delta E^\circ = 0\ \text{V} \qquad (398'')$$

$$H_2O + e \rightarrow \frac{1}{2}H_2 + OH^- \qquad \Delta E^\circ = -0.83\ \text{V} \qquad (401')$$

$$\frac{1}{4}O_2+\frac{1}{2}H_2O+e\rightarrow OH^- \qquad \Delta E^\circ=0.40\ \text{V} \qquad (402')$$

Calculating $(402')-(401')+(398'')(\text{or}(402')-(401'))$, the standard electrode potential of the formation of water is found as $\Delta E^\circ=1.23$ V as shown below,

$$H^+ + e + \frac{1}{4}O_2 \rightarrow \frac{1}{2}H_2O \left(\frac{1}{2}H_2+\frac{1}{4}O_2 \rightarrow \frac{1}{2}H_2O\right) \qquad \Delta E^\circ=1.23\ \text{V}$$

which agrees with the value obtained in (a).

Ans. standard electrode potential : $\Delta E^\circ=1.23$ V which agrees with the value obtained in (a)

(c) The theoretical efficiency of the fuel cell is calculated as

$$\left|\Delta G_f^\circ/\Delta H_f^\circ\right|=273/286=0.829.$$

On the other hand, the efficiency of the ideal heat engine is found as 0.702 by putting $T_l=298$ K, which is

$$(T_h-T_l)/T_h=(1000-298)/1000=0.702$$

lower than the theoretical efficiency of the fuel cell calculated above.

Ans. higher efficiency for fuel cell

(d) The combustion reaction of methane and its standard enthalpy of combustion are given by eq(403).

$$CH_4+2O_2 \rightarrow CO_2+2H_2O \qquad \Delta H_c^\circ=-890\ \text{kJ}\cdot\text{mol}^{-1} \qquad (403)$$

On the other hand, the free energy change ΔG_r° of this reaction is found as given below from the standard free energy of formation of each component of the reaction.

$$\begin{aligned}\Delta G_r^\circ &= 2\Delta G_f^\circ(H_2O)+\Delta G_f^\circ(CO_2)-\Delta G_f^\circ(CH_4)-\Delta G_f^\circ(O_2)\\ &=2\times(-237.13)-394.36+50.72\\ &=-817.9\ \text{kJ}\cdot\text{mol}^{-1}\end{aligned}$$

From these two values, the theoretical efficiency of the fuel cell with methane as fuel is calculated as below.

$$\left|\Delta G_r^\circ/\Delta H_c^\circ\right|=817.9/890=0.919$$

If the product water is assumed in the gas state instead of the liquid state, then the efficiency becomes 0.900.

Ans. theoretical efficiency of methane fuel cell: 0.919

"note" Since the oxidation-reduction potential is an intensive variable, it has nothing to do with the stoichiometric coefficients of the reaction. Therefore, if the electrochemical reactions are always written by putting the number of electrons involved equal to one to normalize the coefficients, then it will help to avoid puzzling at the number of electrons involved in the reaction upon calculating the free energy change of the reaction.

Solution 4-57

(a) By adding the two electrode reactions so as for the electrons involved to be eliminated, the following eq(404) is found as the chemical reaction of the cell. Note that the half cell reactions are not given as the reduction reactions. Mostly they are given as reduction reactions (see **4-49**, **4-50**, **4-52** and **4-53**)

$$\begin{aligned}&Zn(s)+2OH^-(aq) \rightarrow ZnO(s)+H_2O+2e \qquad \text{A (oxidation: anode)}\\ +)\ &2MnO_2(s)+H_2O(l)+2e \rightarrow Mn_2O_3(s)+2OH^-(aq) \qquad \text{B (reduction: cathode)}\\ \hline &Zn(s)+2MnO_2(s) \rightarrow ZnO(s)+Mn_2O_3(s) \qquad (404)\end{aligned}$$

Consequently, the standard electromotive force (emf) is found as $\Delta E_2^\circ + \Delta E_1^\circ = 1.1 + 0.76 = 1.86$ V. For practical use, batteries of this type are commercially available as a battery with the standard emf of 1.5 V. Usually the electrode at which the reduction reaction occurs is defined as the positive electrode and, hence, in the above case B is the positive electrode.

Ans. overall cell reaction: eq(404), anode: B electrode, standard emf: 1.86 V

(b) Similarly as in (a), the elimination of the electrons from the half-cell reactions yields the following overall cell reaction (eq(405)).

$$\mathrm{NiO(OH)(s) + H_2O(\mathit{l}) + \mathit{e} \rightarrow Ni(OH)_2(s) + OH^-(\mathit{aq})} \quad \text{A (reduction: cathode)}$$

$$\mathrm{+)\ MH(s) + OH^-(\mathit{aq}) \rightarrow M + H_2O(\mathit{l}) + \mathit{e}} \quad \text{B (oxidation: anode)}$$

$$\mathrm{NiO(OH)(s) + MH(s) \rightleftarrows M + Ni(OH)_2(s)} \qquad (405)$$

Consequently, the standard electromotive force of the cell is found as $\Delta E_3^\circ + \Delta E_4^\circ = 0.49 + 0.52 = 1.01$ V. Based on the same reasoning as given in (a), the electrode A of this cell becomes an anode (positive electrode). Practically, batteries of this type are on the market as a battery with the standard emf of 1.2 V.

The products of the electrode reactions are solids for either electrode and they stay on each solid electrode even after being produced by the reaction. Because of this, the electrode materials are not lost by being dissolved in the electrolyte medium of the cell or by some deteriorating reactions, if any, hard to occur in solid states. These greatly help in making the cell reaction almost reversible.

Ans. overall cell reaction: eq(405), anode: A electrode, standard emf: 1.01 V , reason: See text (b).

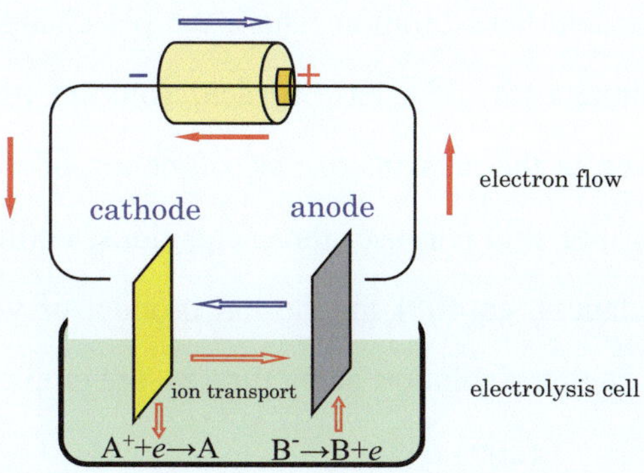

Fig.54 Electron flow in a battery and an electrolysis cell connected to the battery.

"note" In the case of electrolysis, the sign of the electrode (positive or negative) is determined, as was defined by *Faraday*, in the manner that the electrode toward which anions move is the anode and the electrode toward which cations do is the cathode. In the case of electrical cells, the electricity is considered to run from a positive electrode to a negative electrode, but the direction of the electricity is just the opposite to the flow of electrons. For this the electrons run from the negative to the positive electrode. As a result, the electrode at which the reduction reaction is seen becomes the positive (electron deficient) electrode. To clarify the naming of an electrode, see the figure given above. For (rechargeable secondary) batteries, it is recommended to use a positive (+) and a negative electrode (–), so if the battery reaction is a discharge reaction, the anode should be written as a positive electrode and the cathode as a negative electrode.

Michael Faraday, Experimental Researches in Electricity, 2004, Dover.

[advanced problem] What are the overall cell reactions and electromotive forces of button type batteries or lithium ion batteries ? Are they primary or secondary batteries?

Solution 4-58

(a) The molar concentration of an aqueous solution of a 1:1 strong electrolyte of the type of AX is denoted C_{AX}, the equivalent conductivities of A^+ and X^- are λ_{A^+} and λ_{X^-}, respectively. Then, the conductivity Λ of the solution is given by eq(406). To derived eq(406), each ion is assumed to move freely, being independent of one another. Also, contributions from hydrogen ions as well as hydroxide ions are included in eq(406) since the aqueous solution is now considered.

$$\Lambda = C_{AX}(\lambda_{A^+} + \lambda_{X^-}) + [H^+]\lambda_{H^+} + [OH^-]\lambda_{OH^-} \qquad (406)$$

Eq(406) can be expressed as eq(406') when the molar concentrations of ions are used, though the first term is unnecessary for hydrochloric acid.

$$\Lambda = [A^+]\lambda_{A^+} + [X^-]\lambda_{X^-} + [H^+]\lambda_{H^+} + \frac{K_W}{[H^+]}\lambda_{OH^-} \qquad (406')$$

Substituting the given values for the molar concentration and conductivity into eq(406'), the following values given in **Ans.** are found for the conductivities of the three solutions (see **solution 4-69** for *Kohllaush*'s law).

Ans. 0.100NHCl: $0.426\,mS\cdot m^2\cdot mol^{-1}$, 0.100 N NaOH: $0.249\,mS\cdot m^2\cdot mol^{-1}$, 0.100 MCaCl₂: $0.272\,mS\cdot m^2\cdot mol^{-1}$

(b) Exactly in the same manner as is used for the acid-base titration where the pH change of the solution being titrated is obtained as a function of a volume x mL of a base or acid solution (titrant) added, the concentration of each ion in the solution is acquired in the present case as a function of x mL of the NaOH aqueous solution added (conductometric titration). For this purpose, three equations, eq(407) derived from the electrical neutrality of the solution (charge balance), eq(408) for the ion product of water and eq(409) which gives the concentrations of $[Na^+]$ and $[Cl^-]$ in the solution being titrated, are employed.

$$[H^+] + [Na^+] = [Cl^-] + [OH^-] \qquad (407)$$

$$[H^+]\cdot[OH^-] = K_W \qquad (408)$$

$$[Na^+] = \frac{x}{25+x}\times 0.1, \quad [Cl^-] = \frac{25}{25+x}\times 0.1 \qquad (409)$$

Since the electrical conductivity is given by eq(406'), it is necessary to express the concentration $[H^+]$ as a function of x (mL).

$$\Lambda = [Na^+]\lambda_{Na^+} + [Cl^-]\lambda_{Cl^-} + [H^+]\lambda_{H^+} + \frac{K_W}{[H^+]}\lambda_{OH^-} \qquad (406')$$

For this, solving eqs(407) and (408) for $[H^+]$, a quadratic equation (410) is obtained.

$$[H^+] - [OH^-] = [Cl^-] - [Na^+] \Rightarrow [H^+] - \frac{K_W}{[H^+]} = \frac{(25-x)}{25+x}\times 0.1 \qquad (410)$$

As a solution of eq(410), eq(411) is found.

$$[H^+] = \frac{\dfrac{(25-x)}{25+x}\times 0.1 + \sqrt{\dfrac{(25-x)^2}{(25+x)^2}\times 0.01 + 4K_W}}{2} \qquad (411)$$

Substituting the concentration of each ion given by eqs(409) and (411) into eq(406'), the conductivity of the solution being titrated is found by eq(412) as a function of x (mL) (see **"note"**). How it varies with x is shown in Figs.55 and 56. Also, Fig.57 shows for all of the ionic species in the solution how their concentrations vary with x.

$$\Lambda = \frac{x}{25+x}\times 0.1\times 5.01 + \frac{25}{25+x}\times 0.1\times 7.64 + \frac{\dfrac{(25-x)}{25+x}\times 0.1 + \sqrt{\dfrac{(25-x)^2}{(25+x)^2}\times 0.01 + 4\times 10^{-14}}}{2}\times 34.96$$

$$+ \frac{10^{-14}\times 2}{\dfrac{(25-x)}{25+x}\times 0.1 + \sqrt{\dfrac{(25-x)^2}{(25+x)^2}\times 0.01 + 4\times 10^{-14}}}\times 19.91 \qquad (412)$$

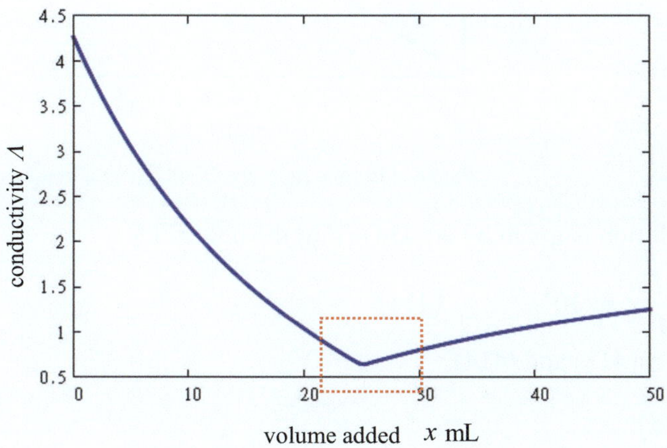

Fig.55 Conductometric titration curve.
1:1 strong electrolyte.

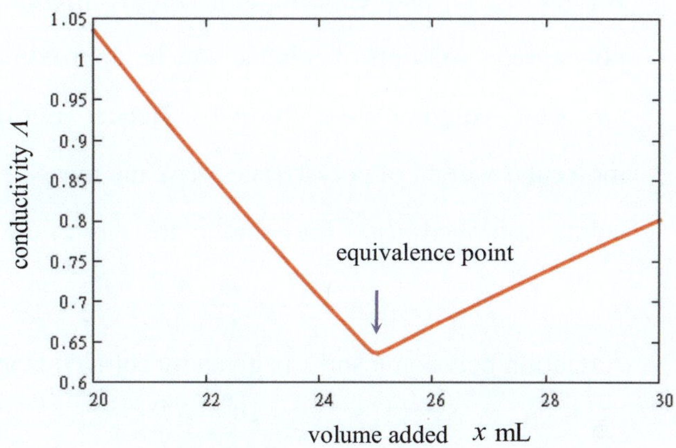

Fig.56 Expansion of the area squared by a dotted orange line in Fig.55.

Ans. conductivity: eq(412) and Figs.55and 56

"note 1" Strictly speaking, eq(412) is not a correct equation since the concentration dependence of ion-ion interactions is not properly taken into account (see **solution 4-69** for *Kohllaush*'s law). However, as is seen from Fig.57 where the concentration of every ionic species is plotted against x, a change in the conductivity of the solution near the equivalence pointis mainly determined by the extremely low concentrations of $[H^+]$ and $[OH^-]$, and, hence, it is considered that the behavior of the conductivitynear the equivalence point can reasonably be described by eq(412).

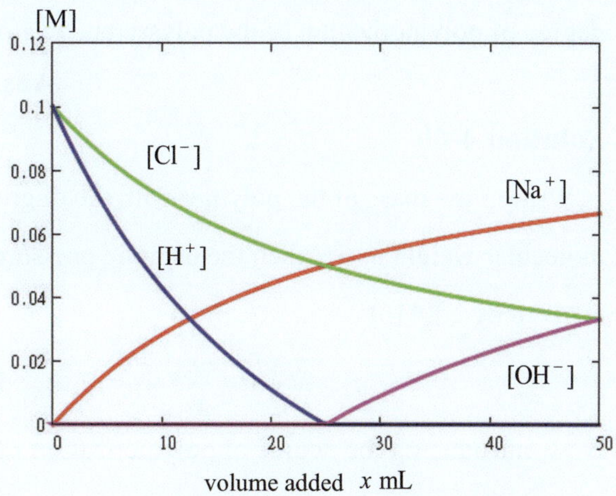

Fig.57 Variation of concentration of relevant ions upon conductomeric titiration.

"note 2" In this problem, the titration of a strong acid with a strong base is treated, but, of course, the equivalence point upon the titration of a weak acid such as acetic acid with a strong base and vice versa can also be studied by measuring the conductivity (conductometric titration).

Solution 4-59

The osmotic pressure π of this solution can be expressed by eq(413) using the density of the solution ρ and the height of the liquid column h.

$$\pi(\text{Pa} = \text{kg}\cdot\text{m}^{-1}\cdot\text{s}^{-2}) = 9.80(\text{m/s}^2)\times\rho(\text{kg/m}^3)\times h(\text{m}) = 9.80\times\rho(\text{g/L})\times h(\text{cm})\times 10^{-2} \quad (413)$$

Assuming that the density of the solution is constant, at least in a narrow range of concentration, the osmotic pressure is proportional to the height of the column.

As is seen in Fig.58, when the given column heights are plotted against the concentration x, a satisfactorily straight line which passes through the origin is obtained. As far as in that concentration range, therefore, the polystyrene solution of toluene can be approximated as an ideal solution (see **"note"**). Hence, putting the molecular weight of polystyrene M(g), the mass of the solute employed m(g), the osmotic pressure of the solution is given by eq(414) (V in the unit of L).

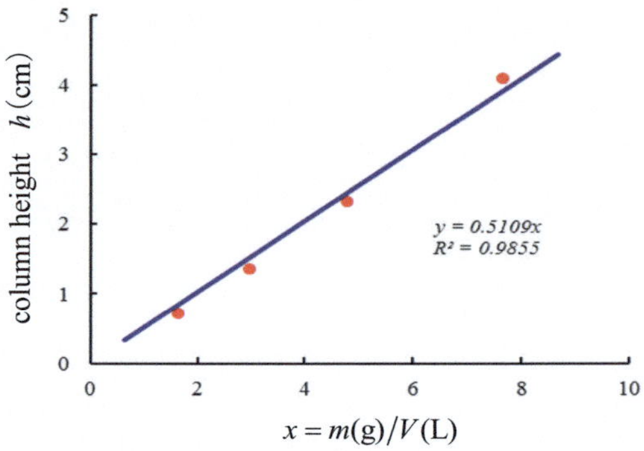

Fig.58 Osmotic pressure of polystyrene in toluene.

$$\pi = \frac{nRT}{V} = \frac{m}{M}\cdot\frac{RT}{V} = \frac{RT}{M}\cdot\frac{m}{V} = \frac{RT}{M}\times x\times 10^3 \quad (414)$$

A relation between h and x is given by eq(415) from eqs(413) and (414).

$$h(\text{cm}) = \frac{10^5}{9.80\rho}\frac{RT}{M}\times x \quad (415)$$

Putting the slope of 0.511 in Fig.58 equal to the coefficient of eq(415),

$$0.511 = \frac{10^5}{9.80\rho}\frac{RT}{M} \Rightarrow M = \frac{10^5\times 8.31\times 298}{0.511\times 9.80\times 860} = 57500\ (\text{g}\cdot\text{mol}^{-1})$$

the molecular weight of polystyrene is found as 57500 (see next problem). Since that of styrene is 104.1, the degree of polymerization of the polystyrene examined here is estimated to be ca.550 (see **problem 4-17**).

Ans. molecular weight: 57500, degree of polymerization: 550

Solution 4-60

Writing the mass of the polymer with the degree of polymerization i as m_i, its number of moles as n_i and the molecular weight as M_i, then the osmotic pressure of the solution, treated as an ideal solution (dilute solution), is given by eq(416).

$$\pi_i = \frac{n_i RT}{V} = \frac{m_i}{M_i}\cdot\frac{RT}{V} \quad (416)$$

Now, introducing the degree of polymerization i and its distribution p_i, they are related by eq(417) through the relevant molecular weights. C is a constant specific to a sample.

$$M_i = i\times M_1\,, \qquad m_i = p_i\times M_i\times C \quad (417)$$

In the case of the ideal solution, the osmotic pressure of any one of the solutes is independent of one another and additive. For this, the total osmotic pressure π is given by summing up the osmotic pressure π_i ascribed to the polymer of the degree of polymerization *i* over all *i*'s as is expressed by eq(418).

$$\pi = \sum_i \pi_i = \sum_i \frac{n_i RT}{V} = \sum_i \frac{m_i}{M_i}\frac{RT}{V} = \left(\sum_i p_i\right) \times \frac{CRT}{V} = \frac{CRT}{V} \qquad (418)$$

When the distribution of the degree of polymerization p_i is given by eq(419) as is assumed in this problem,

$$p_i = A \cdot \exp\left\{-\frac{(i-\bar{i})^2}{\sigma^2}\right\} \qquad (419) \qquad \text{(reposted)}$$

then, the mass of the sample examined is seen to be given by the rightest side of eq(420),

$$m = \sum_i m_i = \sum_i p_i \times M_i \times C = C \times M_1 \sum_i i \times A \cdot \exp\left\{-\frac{(i-\bar{i})^2}{\sigma^2}\right\} = C \times \bar{i} M_1 \qquad (420)$$

in which the definition of the mean degree of polymerization given by eq(421) is employed.

$$\bar{i} = \sum_{i=1} i \times A \cdot \exp\left\{-\frac{(i-\bar{i})^2}{\sigma^2}\right\} \qquad (421)$$

Substituting eq(420) into eq(418) and after rearrangement, eq(422) is derived. It is understood from this

$$\pi = \frac{m}{\bar{i}M_1}\frac{RT}{V} = \frac{RT}{\bar{i}M_1}\frac{m}{V} = \frac{RT}{\overline{M}} \times x \times 10^3 \qquad (422)$$

equation that the molecular weight to be found by a linear plot of the observed osmotic pressure against the concentration *x* is the mean molecular weight.

As is evident from the fact that the specific distribution function of the degree of polymerization (eq(419)) has not been used so far except for defining $\bar{i}$, any distribution function can be adapted to extract the same conclusion as above as far as the mean degree of polymerization and the mean molecular weight are appropriately defined. $\overline{M}$ is named as number average molecular weight and is simply defined by eq(423)

$$\overline{M} = \frac{\sum_i n_i \times M_i}{\sum_i n_i} = \frac{\sum_i n_i \times iM_1}{\sum_i n_i} = M_1 \times \frac{\sum_i i \times n_i}{\sum_i n_i} = \bar{i}M_1 \qquad (423)$$

Ans. proof: see text and eq(422).

"note" To find the molecular weights of a polymer sample experimentally, the method of osmotic pressure is superior to other methods which similarly employ colligative properties such as the depression of freezing point or the elevation of boiling point with respect to the accuracy and range of molecular weight measurable. *P. J. Flory, "Principles of Polymer Chemistry"*, p.272. Cornell University Press, 1953. (*P. J. Flory,* Nobel Prize in Chemistry, in 1974).

Solution 4-61

(a) A *t-X* diagram drawn based on the data given in the table in the problem is shown in Fig.59 (*t* in°C). The number of the data points is not large enough to produce a detailed diagram and to draw borderlines separating two phases precisely, but some of solidification or melting curves which can be drawn are shown by dotted lines. Existing states of mixtures are indicated in Fig.59 by notations $\boldsymbol{L}$ (liquid), $\boldsymbol{L}+\boldsymbol{S}$ (liquid + solid), $\boldsymbol{S}$ (solid), etc. The states indicated by $\boldsymbol{S}_1 \sim \boldsymbol{S}_3$ do not necessary correspond to a pure substance. $\boldsymbol{S}_3$

represents a solid solution with a constant mole fraction of 0.57.

Ans. possible phase diagram: Fig.59.

(b) In Fig.60 a schematic cooling curve starting from the liquid state at the mole fraction of 0.55 is shown. When a liquid sample is cooled with a constant cooling rate, the temperature of the liquid lowers with an approximately constant rate but when a solidification comes in, the temperature lowers more slowly due to generation of the heat of solidification. Because of this an inflection point is seen in the cooling curve. Until the solidification ends the temperature lowers with the same slow rate, but as soon as the solidification ends the temperature starts to lower at a different rate since no heat due to the solidification is supplied any more. As a result of this another inflection point appears in the cooling curve as is illustrated in Fig.60. When the heat capacity of the resultant solid is smaller than that of the initial liquid, the temperature lowers more rapidly than before.

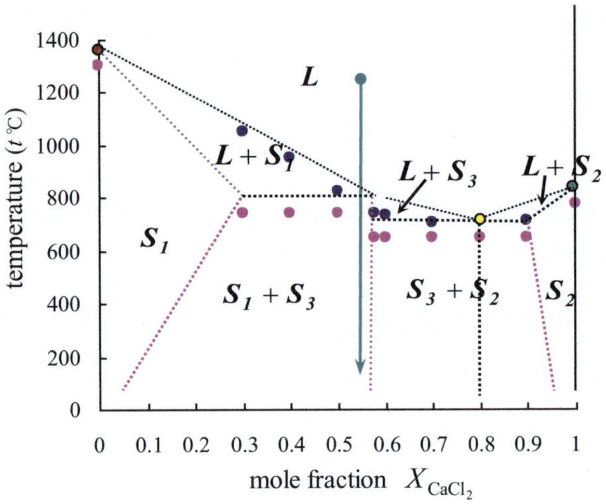

Fig.59 Phase diagram (($t-X$) of CaF_2-$CaCl_2$ mixtures.

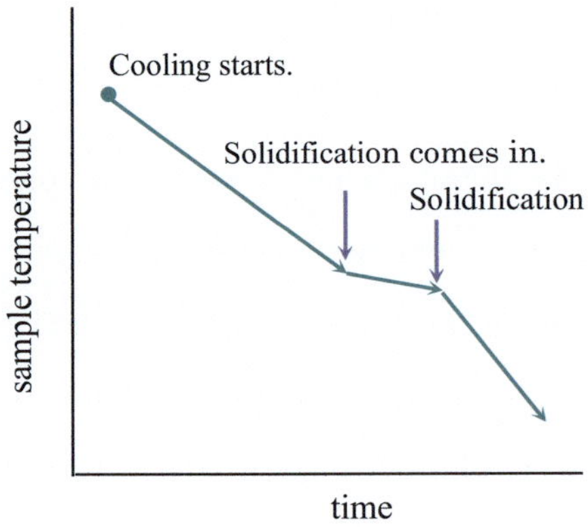

Fig.60 Illustrative cooling curve indicating the start and completion points of solidification.

Ans. illustrative cooling curve: Fig.60

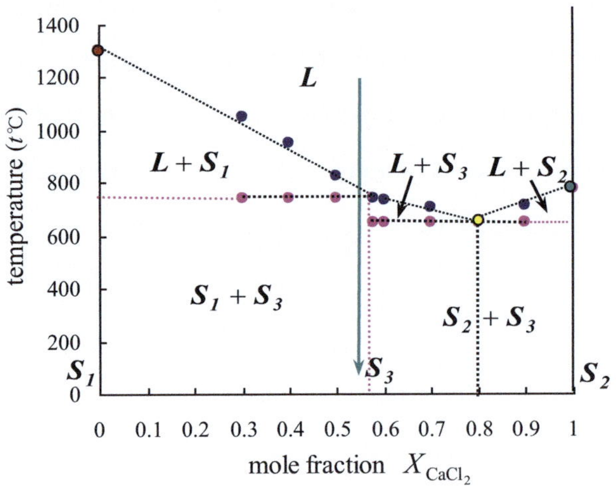

Fig.61 A possible phase diagram alternative to Fig.59.

As another possibility alternative to the one in Fig.59, pure solids can be deposited when the solid solution is cooled down. One such example is shown in Fig.61 where S_1 and S_2 represent a pure solid CaF_2 and a pure solid $CaCl_2$, respectively.

Solution 4-62

(a) An electric current is transported by ions in solution. How easily a positive or a negative ion can move in solution under a constant electric field is measured by mobility and is denoted u_{z+} for the positive ion and u_{z-} for the negative ion. The mobility thus defined is naturally different from ion to ion. Although positive and negative ions move in opposite directions under the electric field, the mobility itself is defined by using an absolute value, since the electric current is given by the sum of each contribution expressed using an absolute value.

For simplicity, both ions are assumed singly charged and then the current density $i = I/A$, which is proportional to the sum of the mobility for each ion, is given by eq(424).

$$i \propto u_+ + u_- \qquad (424)$$

Therefore, the ratios of the electric currents transported by the positive and negative ions, respectively are expressed by eq(425) in terms of the mobility. The ratios t_+ and t_- thus defined are called transport number.

$$t_+ = \frac{u_+}{u_+ + u_-}, \qquad t_- = \frac{u_-}{u_+ + u_-} \qquad (425)$$

In the present problem, the total electric charges (coulombs) passed through the system are calculated from the amount of Ag deposited in the coulometer and the molecular weight of 107.9 for Ag as given below.

$$0.2508/107.9 = 2.324 \times 10^{-3} F$$

On the other hand, the amount of charges carried by the Cl^- ion is found as

$$(0.177 - 0.163)/35.45 = 3.949 \times 10^{-4} F,$$

and, hence, the transport numbers of each ion are obtained as given below.

$$t_- = \frac{u_-}{u_+ + u_-} = \frac{3.949 \times 10^{-4}}{2.324 \times 10^{-3}} = 0.1699, \qquad t_+ = 1 - t_- = 0.8301$$

Ans. transport number: $t_- = 0.1699$ for Cl^-, $t_+ = 0.8301$ for H^+

(b) The molar conductivity λ of an ion is proportional to its mobility. The transport number, therefore, can be expressed in terms of the molar conductivity for the positive and negative ions as is given by eq(426).

$$t_+ = \frac{\lambda_+}{\lambda_+ + \lambda_-}, \qquad t_- = \frac{\lambda_-}{\lambda_+ + \lambda_-} \qquad (426)$$

Consequently, values for the transport number predicted from the relevant molar conductivity are found as below for the Cl^- and H^+ ions and show a moderate agreement with the experimental values obtained in (a).

$$t_- = \frac{\lambda_-}{\lambda_+ + \lambda_-} = \frac{7.64}{34.96 + 7.64} = 0.1793, \qquad t_+ = 1 - t_- = 0.8207$$

Ans. Supporting the result in (a).

"note 1" The precision of transport number. Although a value of the transport number of an ion varies with the concentration of the ion, the value is usually given in four significant digits. On the other hand, a value for the mobility is expressed in most cases in three significant digits. Why is this difference?

"note 2" Technical terms and definitions.

Electric current ($I = i \cdot A$, i: density of electricity), voltage (V), resistance (R) are related by a basic equation of $V = I \cdot R$.

Conductance ($1/R$ (Ω^{-1})), $1/R = \sigma \cdot A/l$.

Conductivity ($l/A \cdot R$ ($\mathrm{m}^{-1}\Omega^{-1}$)), specific conductivity ($\sigma = l/A \cdot R$).

The specific conductivity of a solution is equal to the conductance $1/R$ of a cube with an edge of $l = 1$ cm and an area of $A = 1 \times 1$ cm^2.

Conductivity or electric conductivity:

molar conductivity ($\Lambda_m = \sigma/c$, c in mol/cm^3),

equivalent conductivity ($\Lambda_{eq} = \Lambda_m/z = \sigma/cz$).

Mobility (u_{z+}, u_{z-}, $\mathrm{cm \cdot s^{-1}/V \cdot cm^{-1} = cm^2 s^{-1} \cdot V^{-1}}$):

The mobility of an ion is defined by its absolute drift velocity per unit electric field.

$\Lambda_m = \sigma/c = zF(u_{z+} + u_{z-})$.

Solution 4-63

The molar conductivity Λ_m of the solution is given by adding the molar conductivities of λ_+ for Na^+ and λ_- for F^- as in eq(427).

$$\Lambda_m = \lambda_+ + \lambda_- \qquad (427)$$

On the other hand, the molar conductivity of each ion is given by eq(428) employing the mobility of each ion, u_+ for Na^+ and u_- for F^-. Here, F is the *Faraday* constant.

$$\lambda_+ = u_+ \times F, \qquad \lambda_- = u_- \times F \qquad (428)$$

The transport numbers, t_+ for the positive ion and t_- for the negative ion, are defined by eq(429) and they take values 0.45 and 0.55, respectively in the present case.

$$t_+ = \frac{u_+}{u_+ + u_-}, \qquad t_- = \frac{u_-}{u_+ + u_-} \qquad (429)$$

Since the degree of electrolytic dissociation of NaF is known as 92%, the molar conductivity expected for the solution in which the degree of electrolytic dissociation is 100% would become $1/0.92$ times larger than the practically measured value. Taking this into consideration, eqs(427) and (429) are rewritten as eqs(427') and (429'),

$$83.5/0.92 \times (\text{observed value}) = (u_+ + u_-) \times F \qquad (427'), \qquad 0.45 \times (u_+ + u_-) = u_+ \qquad (429')$$

which are solved for u_+ and u_-, yielding the values of $u_+ = 4.23 \times 10^{-4}$ $\mathrm{cm^2 \cdot s^{-1} \cdot V^{-1}}$ and $u_- = 5.16 \times 10^{-4}$ $\mathrm{cm^2 \cdot s^{-1} \cdot V^{-1}}$. The unit of the mobility is already given before (see **"note 2"** in the preceding solution) but it can also be confirmed as shown below.

$$[\mathrm{S \cdot cm^2 \cdot F^{-1}}] = [\frac{\mathrm{cm^2}}{\mathrm{R \cdot F}}] = [\frac{\mathrm{cm^2}}{\mathrm{R \cdot i \cdot s}}] = [\mathrm{cm^2 \cdot s^{-1} \cdot V^{-1}}]$$

Ans. Na^+: $u_+ = 4.23 \times 10^{-4}$ $\mathrm{cm^2 \cdot s^{-1} \cdot V^{-1}}$, F^-: $u_- = 5.16 \times 10^{-4}$ $\mathrm{cm^2 \cdot s^{-1} \cdot V^{-1}}$

Solution 4-64

The mobility of hydrogen ion u_+ is related to its conductivity λ_+ by $\lambda_+ = u_+ F,$ from which a value for the mobility is found as below.

$$u_+ = 349.5/96450 = 3.624\times10^{-3}\text{cm}^2\cdot\text{s}^{-1}\cdot\text{V}^{-1} = 3.624\times10^{-7}\text{m}^2\cdot\text{s}^{-1}\cdot\text{V}^{-1}$$

Assuming that the mobility is given by *Stokes*' mobility expressed in eq(430), in which R is *Stokes*' radius,

$$u_+ = \frac{e}{6\pi\cdot R\cdot\eta} \qquad (430)$$

the ion radius (*Stokes*' radius) of a hydrogen ion is calculated from eq(430) as follows.

$$R = \frac{e}{6\pi\cdot\eta\cdot u_+} = \frac{1.602\times10^{-19}}{6\pi\times8.99\times10^{-4}\times3.624\times10^{-7}} = 2.61\times10^{-11}\text{m}$$

Ans. mobility: $3.624\times10^{-3}\text{cm}^2\cdot\text{s}^{-1}\cdot\text{V}^{-1}$, *Stokes*' radius: $2.61\times10^{-11}\text{m}$

"note" See **solution 2-75** for the derivation of the *Stokes'* equation $F(\text{force}) = 6\pi\eta R\cdot v$.

Solution 4-65

It is reasonable to consider that sodium acetate dissociates into ions completely in an aqueous solution. The pH value of the solution under consideration is determined by four equations derived from the dissociation equilibrium of acetic acid (eq(431)), the conservation of substances (eq(432)), the balance of charges (electroneutrality, eq(433)) and the ion product of water (eq(434)) , as has been discussed so far in many cases.

$$K_a = \frac{[\text{CH}_3\text{COO}^-][\text{H}^+]}{[\text{CH}_3\text{COOH}]} = 1.8\times10^{-5} \qquad (431)$$

$$[\text{CH}_3\text{COOH}] + [\text{CH}_3\text{COO}^-] = 0.1 \qquad (432)$$

$$[\text{Na}^+] + [\text{H}^+] = 0.1 + [\text{H}^+] = [\text{CH}_3\text{COO}^-] + [\text{OH}^-] \qquad (433)$$

$$[\text{H}^+]\cdot[\text{OH}^-] = K_w \qquad (434)$$

Eliminating $[\text{CH}_3\text{COOH}]$ from eq(431) using eq(432), eq(431) can be rewritten as eq(431').

$$K_a = \frac{[\text{CH}_3\text{COO}^-][\text{H}^+]}{0.1 - [\text{CH}_3\text{COO}^-]} \qquad (431')$$

Solving eqs(433) and (434) for $[\text{CH}_3\text{COO}^-]$ and substituting the result into eq(431') with some arrangement, eq(435) is derived.

$$[\text{H}^+]^3 + (0.1 + K_a)[\text{H}^+]^2 - K_w[\text{H}^+] - K_wK_a = 0 \qquad (435)$$

Eq(435) can be solved directly, but the term of $[\text{H}^+]^3$ which is much smaller than the other terms in the alkaline solution can be neglected and, hence, it is sufficient to solve an approximate equation given by eq(435') to find a value of $[\text{H}^+]$.

$$(0.1 + K_a)[\text{H}^+]^2 - K_w[\text{H}^+] - K_wK_a = 0 \qquad (435')$$

One of the solutions of eq(435') and its approximate value are given by eq(436).

$$[\text{H}^+] = \frac{K_w + \sqrt{K_w^{\ 2} + 4(0.1 + K_a)K_wK_a}}{2\times(0.1 + K_a)} \approx \sqrt{\frac{K_wK_a}{0.1}} \qquad (436)$$

Substituting the given values into the last equation in eq(436), the value for $[\text{H}^+]$ is found as below.

$$[\text{H}^+] = 1.34\times10^{-9}\ \text{M} \ \ (\text{cf. the exact solution of eq(5) is} 1.330\times10^{-9}\text{M.})$$

Denoting the dissociation constant of phenolphthalein (InH) α, its equilibrium constant K_{In} for the dissociation $\text{InH} \rightleftarrows \text{In}^- + \text{H}^+$ is given by eq(437).

$$K_{In} = \frac{[\text{In}^-][\text{H}^+]}{[\text{InH}]} = \frac{[\text{InH}]_0 \alpha \times [\text{H}^+]}{[\text{InH}]_0(1-\alpha)} = \frac{\alpha}{1-\alpha} \times [\text{H}^+] \qquad (437)$$

Then, substituting the given value for K_{In} and the concentration of H^+ into eq(437), a value for α is found as below. From this, it is seen that ca. 19.1% of phenolphthalein assume a dissociated form (colored form) in that solution.

$$\alpha = \frac{K_{In}}{K_{In} + [\text{H}^+]} = \frac{3.16 \times 10^{-10}}{3.16 \times 10^{-10} + 1.34 \times 10^{-9}} = 0.191$$

Ans. dissociated form: 19.1%

"note" Dissociation equilibrium of phenolphthalein (InH). A solution of phenolphthalein changes its color distinctly in the pH range $\text{pH} = 8.2 - 9.8$. Its proton dissociation is illustrated using chemical structures of the two forms, i.e., associated and dissociated ones. A part, colored pink in the dissociated form shown in Fig.62, is responsible for visible light absorption, presenting from a pinkish to reddish violet color in basic solution.

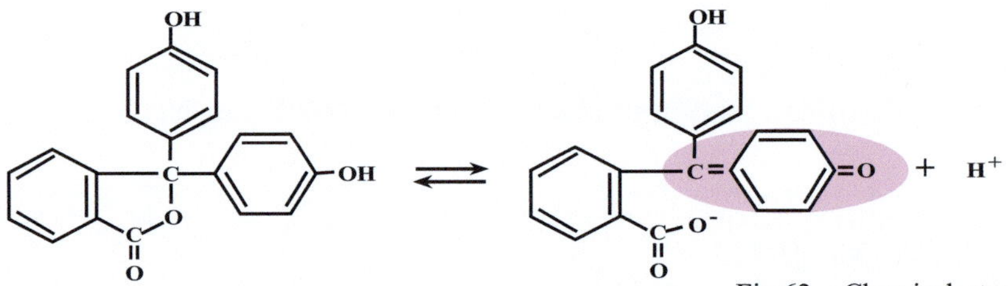

Fig.62 Chemical structure of phenolphthalein (InH) and its acid-base equilibrium.

Solution 4-66

An amphoteric electrolyte has at least two functional groups such as carboxyl or amino groups which can work as an acid or a base. In an aqueous solution at a constant concentration of hydrogen ion, these functional groups of the amphoteric electrolyte are in the acid-base equilibriums as shown below,

$$\underset{x}{\text{NH}_2-\text{C}\cdots\text{C}-\text{COOH}} + \text{H}^+ \rightleftarrows \underset{y}{\text{NH}_3^+-\text{C}\cdots\text{C}-\text{COOH}}, \quad K_1 \qquad (438)$$

$$\underset{y}{\text{NH}_3^+-\text{C}\cdots\text{C}-\text{COOH}} \rightleftarrows \underset{z}{\text{NH}_3^+-\text{C}\cdots\text{C}-\text{COO}^-} + \text{H}^+, \quad K_2 \qquad (439)$$

$$\underset{x}{\text{NH}_2-\text{C}\cdots\text{C}-\text{COOH}} \rightleftarrows \underset{w}{\text{NH}_2-\text{C}\cdots\text{C}-\text{COO}^-} + \text{H}^+, \quad K_3 \qquad (440)$$

in which K_1, K_2 and K_3 are the relevant equilibrium constants and the concentrations of the chemical species participating in the equilibriums are denoted x, y, z and w, and are labeled under the corresponding species. The following reaction is, however, neglected as unlikely to occur.

$$\text{NH}_2-\text{C}\cdots\text{C}-\text{COO}^- + \text{H}^+ \rightleftarrows \text{NH}_3^+-\text{C}\cdots\text{C}-\text{COO}^-$$

Writing the initial concentration of the amphoteric electrolyte as C_0, the other concentrations are related by eq(441) in which z corresponds to the concentration of the zwitter-ion.

$$x + y + z + w = C_0 \qquad (441)$$

By expressing each equilibrium constant in terms of the relevant concentrations including that of hydrogen ion, the concentrations of x, y and w are found by eq(442), eq(443) and eq(444), respectively.

$$K_1 = \frac{y}{x \cdot [\mathrm{H}^+]} \Rightarrow x = \frac{y}{K_1 \cdot [\mathrm{H}^+]} = \frac{z}{K_1 K_2} \qquad (442)$$

$$K_2 = \frac{z \cdot [\mathrm{H}^+]}{y} \Rightarrow y = \frac{z \cdot [\mathrm{H}^+]}{K_2} \qquad (443)$$

$$K_3 = \frac{w \cdot [\mathrm{H}^+]}{x} \Rightarrow w = \frac{x \cdot K_3}{[\mathrm{H}^+]} = \frac{z \cdot K_3}{K_1 K_2 [\mathrm{H}^+]} \qquad (444)$$

Substituting these concentrations into eq(441), eq(445) is derived after arrangement.

$$\left(1 + \frac{1}{K_1 K_2} + \frac{[\mathrm{H}^+]}{K_2} + \frac{K_3}{K_1 K_2 [\mathrm{H}^+]}\right) \cdot z = C_0 \qquad (445)$$

In order to maximize z, the value in the parentheses needs to be minimized. Namely, we need to find a value of $[\mathrm{H}^+]$ which minimizes $([\mathrm{H}^+]/K_2 + K_3/K_1 K_2 [\mathrm{H}^+])$, from which eq(446) is derived as an equation to give the $[\mathrm{H}^+]$ value at the isoelectric point.

$$\frac{1}{K_2} - \frac{K_3}{K_1 K_2 [\mathrm{H}^+]^2} = 0 \Rightarrow [\mathrm{H}^+] = \sqrt{\frac{K_3}{K_1}} \qquad (446)$$

Ans. pH at isoelectric point: eq (446)

Solution 4-67

When a solution can be regarded ideal, the osmotic pressure of the solution is proportional to the concentration $c = m/V$ of a solute as is given by eq(447).

$$\Pi = \frac{n}{V} RT = \frac{m}{M} \cdot \frac{1}{V} RT = \frac{RT}{M} \cdot c \qquad (447)$$

The data given in the problem are plotted on the basis of eq(447) in Fig.63. The best-fitted straight line is, however, not good enough to reproduce the data. In addition, it does not pass through the origin as is predicted by eq(447). Consequently, to regard the solutions as ideal is not a good approximation. In another word, the chemical potential of the solution cannot be expressed as $\mu = \mu^\circ + RT \ln c$. For this, it is assumed that that the chemical potential can be expressed as eq(448) by introducing the activity coefficient which is given by a power series of c.

$$\mu = \mu^\circ + RT \ln c \Rightarrow \mu = \mu^\circ + RT \ln \gamma \cdot c \qquad (448)$$

$$\gamma = (1 + a \cdot c + b \cdot c^2 + \cdots) \qquad (449)$$

Referring to the discussion given so far in deriving the osmotic pressure (eq(447)) as a colligative property, it is seen that the osmotic pressure is expressed as eq(450) by a virial expansion in c when the chemical potential of a solute is given by eqs(448) and (449).

$$\Pi = \frac{RT}{M} \cdot c \Rightarrow \frac{RT}{M} \cdot c \left\langle 1 + B'(T) \cdot c + C(T) \cdot c^2 + \cdots \right\rangle \qquad (450)$$

Eq(450) is rewritten as eq(450'). To approximate eq(450'), only the first and second terms are kept and a

$$\frac{\Pi}{c} = \frac{RT}{M} \left\langle 1 + B'(T) \cdot c + C(T) \cdot c^2 + \cdots \right\rangle \qquad (450')$$

linear function thus obtained is used to plot the data against c in Fig.63. (Since the unit of concentration employed is converted as follows $\mathrm{mg/cm^3} = 10^{-6}\mathrm{kg}/10^{-6}\mathrm{m^3} = \mathrm{kg/m^3}$, eqs(4) and (450') yield a plot in the SI unit system whichever concentration unit is employed.)

A satisfactorily linear correlation ($R^2 = 0.9997$) is found and from a value of 1.975 ± 0.02 for the intercept of the best-fitted straight line in Fig.64, a value of $1274 \pm 12\ \mathrm{kg \cdot mol^{-1}}$ is obtained for the molecular weight of the polymer.

$$\frac{RT}{M} = 1.975 \pm 0.02 \ \Rightarrow\ M = \frac{8.31 \times 303}{1.975 \pm 0.02} = 1274 \pm 12\ \mathrm{kg \cdot mol^{-1}}$$

The second virial coefficient is found as $20.09\ \mathrm{cm^3 \cdot mg^{-1}}$, from which it is understood that even the most dilute solution employed in the measurement cannot be regarded as an ideal solution. In fact, the molecular weight calculated from eq(447) using a value obtained in Fig.63 for the slope of the straight line which passes through the two points, the origin and the osmotic pressure at the most dilute concentration, is only $111\ \mathrm{kg \cdot mol^{-1}}$, far from a correct value.

$$\frac{RT}{M} = \frac{30}{1.33 \times 10^{-6} \times 10^{6}} \ \Rightarrow\ M = \frac{8.31 \times 303 \times 1.33}{30} = 111\ \mathrm{kg \cdot mol^{-1}}$$

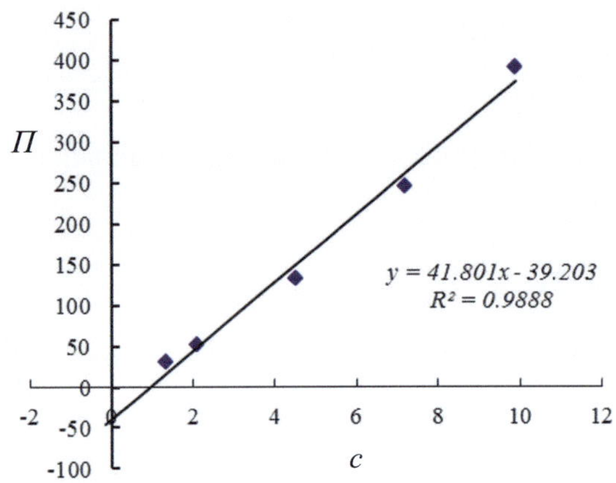

Fig.63 Concentration dependence of osmotic pressure of polymer solutions. Linear plot based on eq(447).

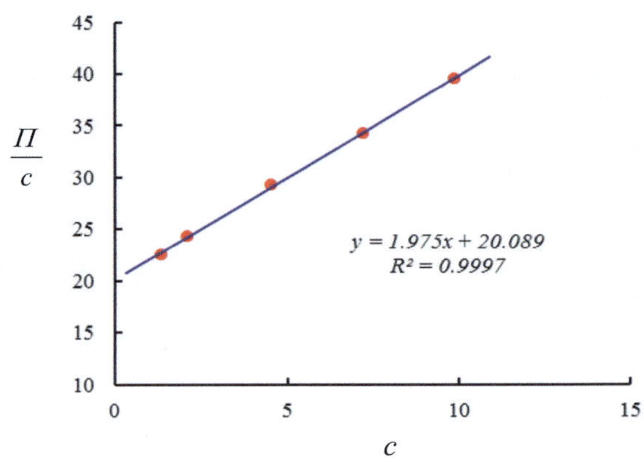

Fig.64 Concentration dependence of osmotic pressure of polymer solutions. Plot based on the introduction of the second virial coefficient (eq(450')).

Ans. molecular weight: $(1274 \pm 12) \times 10^3$

Solution 4-68

(a) Denoting monochloroacetic acid AH, its dissociation equilibrium is written as $\mathrm{AH} \rightleftarrows \mathrm{A^-} + \mathrm{H^+}$. Then, employing the degree of dissociation α, the equilibrium constant is expressed as eq(451), where $[\mathrm{AH}]_0$ is the initial concentration.

$$K_a = \frac{[\mathrm{A^-}][\mathrm{H^+}]}{[\mathrm{AH}]} = \frac{[\mathrm{AH}]_0^{\,2}\alpha^2}{[\mathrm{AH}]_0(1-\alpha)} = \frac{\alpha^2}{1-\alpha}[\mathrm{AH}]_0 \qquad (451)$$

Based on the assumption of the law of the independent migration of ions, the conductivity Λ of solutions, which should be distinguished from the molar conductivity or equivalent conductivity, is given by eq(452), where $\lambda_{\mathrm{A^-}}$ and $\lambda_{\mathrm{H^+}}$ are the molar conductivities for the corresponding ions and $\Lambda_0 = \lambda_{\mathrm{A^-}} + \lambda_{\mathrm{H^+}}$ is the molar conductivity of the solution at infinite dilution (limiting molar conductivity).

$$\Lambda = \lambda_{A^-}[A^-] + \lambda_{H^+}[H^+] = \alpha[AH]_0(\lambda_{A^-} + \lambda_{H^+}) = \alpha[AH]_0\Lambda_0 \qquad (452)$$

Since the molar conductivity Λ_m is the conductivity per mole, it is expressed by eq(453).

$$\Lambda_m = \Lambda/[AH]_0 \qquad (453)$$

Consequently, from eqs(452) and (453), it is readily seen that the degree of dissociation α is given by eq(454).

$$\alpha = \frac{\Lambda_m}{\Lambda_0} \qquad (454)$$

The derived equation is the definition of α of electrolytes by *Arrhenius*. Substituting eq(454) into eq(451) and being arranged, the dissociation constant can be expressed as eq(455) by employing the molar conductivity of solutions.

$$K_a = \frac{\left(\frac{\Lambda_m}{\Lambda_0}\right)^2}{1-\frac{\Lambda_m}{\Lambda_0}}[AH]_0 \Rightarrow K_a = \frac{\Lambda_m^2[AH]_0}{\Lambda_0(\Lambda_0 - \Lambda_m)} \qquad (455)$$

Ans. equilibrium constant expressed by molar conductivity: $K_a = \frac{\Lambda_m^2[AH]_0}{\Lambda_0(\Lambda_0 - \Lambda_m)}$

(b) Eq(455) can be rewritten as $\Lambda_m[AH]_0/K_a\Lambda_0 = (\Lambda_0 - \Lambda_m)/\Lambda_m$ and the latter is further rewritten as eq(456) so that eq(456) can be used for linear plotting. Eq(456) represents *Ostwald*'s dilution law.

$$\frac{1}{\Lambda_m} = \frac{1}{\Lambda_0} + \frac{\Lambda_m[AH]_0}{K_a\Lambda_0^2} \qquad (456)$$

In Fig.65 the obtained values of $1/\Lambda_m$ are plotted against $\Lambda_m[AH]_0$. A satisfactorily straight line is obtained and from its intercept of $2.97\times10^{-3} \mp 5.5\times10^{-5}$ and slope of $2.97\times10^{-3} \mp 5.5\times10^{-5}$, values for Λ_0 and K_a are found as below. Compared with acetic acid, the equilibrium constant of monochloroacetic acid is ca. 100 times greater

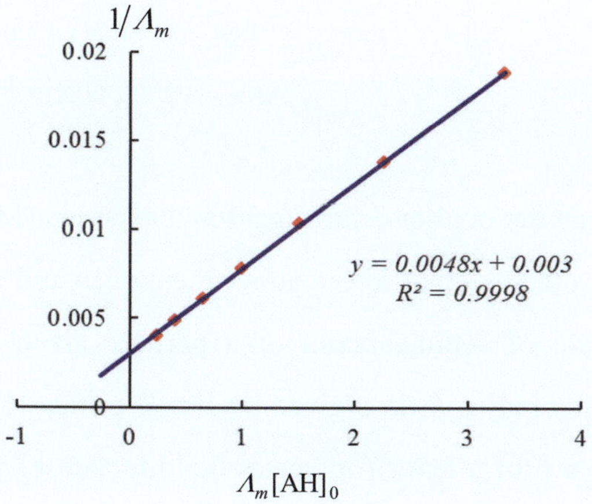

Fig.65 A plot based on eq(456) (*Ostwald* dilution law).

$$\frac{1}{\Lambda_0} = 2.97\times10^{-3} \mp 5.5\times10^{-5}$$

$$\Rightarrow \Lambda_0 = 336 \pm 6.5\ \mathrm{S\cdot cm^2\cdot mol^{-1}}$$

$$\frac{1}{K_a\Lambda_0^2} = 4.80\times10^{-3} \pm 3.3\times10^{-5}$$

$$\Rightarrow K_a = \frac{1}{\Lambda_0^2} \times \frac{1}{4.80\times10^{-3} \pm 3.3\times10^{-5}} = 1.86\times10^{-3}$$

Ans. $\Lambda_0 = 336 \pm 6.5\ \mathrm{S\cdot cm^2\cdot mol^{-1}}$, $K_a = 1.86\times10^{-3}$

(c)Taking the molar conductivity of H^+ as $349.6\ \mathrm{S\cdot cm^2\cdot mol^{-1}}$, and approximating that of monochloroacetic acid ion CH_2ClCOO^- by $40.9\ \mathrm{S\cdot cm^2\cdot mol^{-1}}$ known for acetic ion, the molar conductivity of monochloroacetic acid at the infinite dilution Λ_0 (limiting molar conductivity) is estimated as $390.5\ \mathrm{S\cdot cm^2\cdot mol^{-1}}$, which is larger than the value obtained above.

When a value for the degree of dissociation is estimated at the highest given concentration, 0.158 is found for α. The *Ostwald* dilution law is valid when the concentration is lower than 0.1.

Ans. See text (c).

Solution 4-69

The molar conductivity Λ_m of a strong electrolyte is, as is predicted by *Kohlrausch*'s law, given by eq(457). On the other hand, for a weak electrolyte, it varies as is given by eq(458) in accordance with *Ostwald*'s dilution law (see preceding solution).

$$\Lambda_m = \Lambda_0 - K \cdot \sqrt{c} \qquad (457)$$

$$\frac{1}{\Lambda_m} = \frac{1}{\Lambda_0} + \frac{\Lambda_m c}{K_a \Lambda_0^2} \qquad (458)$$

The data given in the problem are plotted in Figs.66 and 67 according to eqs(457) and (458), respectively. As is obvious from these figures, the electrolyte in question follows *Kohlrausch*'s law satisfactorily but it does not yield any good linear correlation matching *Ostwald*'s dilution law. Consequently, the electrolyte in question is concluded to be a strong electrolyte.

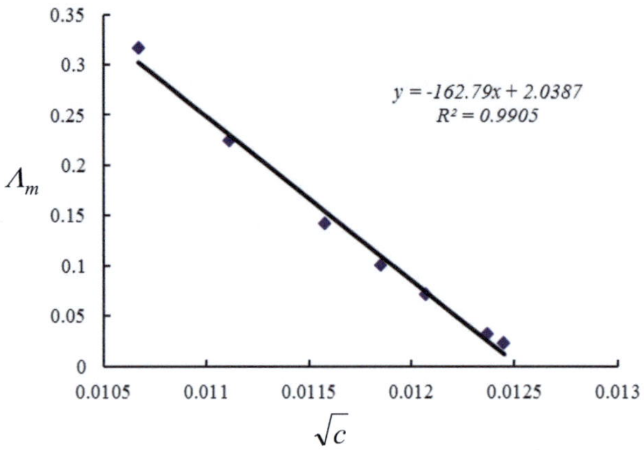

Fig.66 Linear plot based on *Kohlraush*'s law (eq(457)).

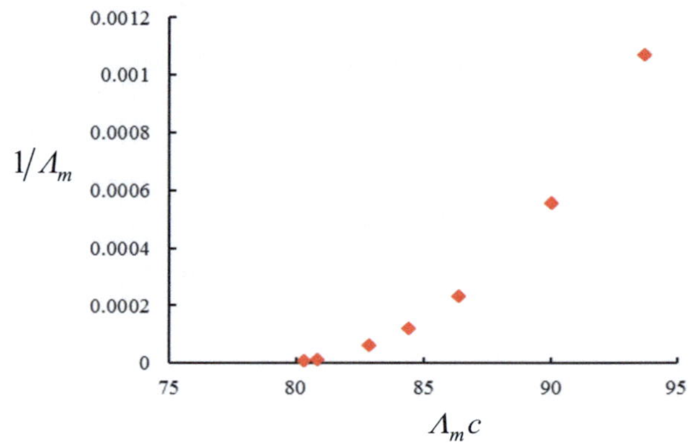

Fig.67 Linear plot based on *Ostwald's* dilution law.

Ans. strong electrolyte

Solution 4-70

Particles migrating or sedimenting in a solution under the gravitational or centrifugal force experience the viscous resistance arising from solution viscosity in addition to the external forces above mentioned and the buoyancy. This viscous resistance is proportional to the rate of sedimentation of a particle given by $\upsilon = (dx/dt)$ and when the particle is regarded as a sphere of the radius *R*, it is given by $6\pi\eta R \cdot \upsilon$ (eq(459), see **solution 2-75**) using *Stokes*' equation. An arrow on the top of a letter *f* indicates the direction of the force *f*.

$$\overleftarrow{f} = 6\pi\eta R \cdot \upsilon \qquad (459)$$

On the other hand, denoting the mass of a single particle *m*, the external force $\overrightarrow{f}$ working on the particle is given by eq(460), in which the buoyancy to the particle is taken into account. Here, *M* is the molecular weight (mass in $\text{kg} \cdot \text{mol}^{-1}$) of the particle, *a* is the gravitational acceleration *g* or the centrifugal acceleration $r\omega^2$.

$$\vec{f} = (m - \rho \times \frac{\overline{V} \times M}{N_A}) \cdot a = \frac{M}{N_A}(1 - \rho \cdot \overline{V}) \cdot a \qquad (460)$$

The force given by eq(460) is balanced by the viscous resistance given by eq(459), leading to eq(461) and as a result, the particle migrates in solution (downward in the case of sedimentation) with a constant velocity.

$$\frac{M}{N_A}(1 - \rho \cdot \overline{V}) \cdot a = 6\pi\eta R \cdot \upsilon = \frac{k_B T}{D} \cdot \upsilon \qquad (461)$$

Here, a constant S called sedimentation constant is introduced and defined by eq(462).

$$S \equiv \frac{\upsilon}{a} \quad \text{(unit is [s].} \quad a \text{ is either } g \text{ or } r\omega^2\text{, etc.)} \qquad (462)$$

Then, the molecular weight is given by eq(463) using S,

$$M = \frac{RT}{D(1 - \rho \cdot \overline{V})} \times S \qquad (463)$$

and substituting the given values into eq(463) with a due care to the units of each value, the following value is found for the molecular weight of the ribosome.

$$M = \frac{8.31 \times 293}{1.52 \times 10^{-11}(1 - 1.0 \times 0.61)} \times 82.6 \times 10^{-13} = 3.45 \times 10^3\,\text{kg} \cdot \text{mol}^{-1} = 3.45 \times 10^3\,\text{kDa}$$

Ans. molecular weight of ribosome: $M = 3.45 \times 10^3\,\text{kg} \cdot \text{mol}^{-1} = 3.45 \times 10^3\,\text{kDa}$

Solution 4-71

Denoting asparagine ($C_4H_8N_2O_4$) as AH, the ionic equilibria of asparagine in HCl acidic aqueous solution are given by eqs(464) and (465).

$$\text{H}^+\text{AH} + \text{Cl}^- \rightleftarrows \text{H}^+\text{A}^- + \text{H}^+ + \text{Cl}^- \qquad (464)$$

$$\text{H}^+\text{A}^- \rightleftarrows \text{A}^- + \text{H}^+ \qquad (465)$$

When x mL of the 0.1N NaOH aqueous solution is added and the volume of the solution has become $10 + x$ mL, the concentrations of the ions are related by eq(466) from the electroneutrality of the solution as well as by eq(467) from the conservation of substances.

$$[\text{H}^+] + [\text{Na}^+] + [\text{H}^+\text{AH}] = [\text{A}^-] + [\text{OH}^-] + [\text{Cl}^-] \qquad (466)$$

$$[\text{H}^+\text{AH}] + [\text{H}^+\text{A}^-] + [\text{A}^-] = C_0 \times \frac{10}{10 + x} \qquad (467)$$

Also, the concentrations of Na^+ and Cl^- are given by eq(468).

$$[\text{Na}^+] = \frac{0.1 \times x}{10 + x}, \qquad [\text{Cl}^-] = \frac{0.1 \times 10}{10 + x} \qquad (468)$$

The concentrations of all the ions other than Na^+ and Cl^- are treated as unknown variables and the concentration of each ion is denoted *y, z, u,* etc., as given by eq(469). Notice that there are five unknowns.

$$[\text{H}^+] = y, \qquad [\text{OH}^-] = z$$

$$[\text{H}^+\text{AH}] = u, \qquad [\text{H}^+\text{A}^-] = v, \qquad [\text{A}^-] = w \qquad (469)$$

From eqs(464) and (465), eqs(470) and (471) are derived and eq(472) is the ion product of water,

$$\frac{[\text{H}^+][\text{H}^+\text{Asp}^-]}{[\text{H}^+\text{Asn}]} = \frac{y \times v}{u} = K_{a_1} \qquad (470)$$

$$\frac{[\text{H}^+][\text{Asp}^-]}{[\text{H}^+\text{Asn}]} = \frac{y \times w}{v} = K_{a_2} \qquad (471)$$

$$[\mathrm{H}^+][\mathrm{OH}^-]=K_w=10^{-14} \qquad (472)$$

Furthermore, eqs(473)and(474) are obtained from eqs(466) and (467) and in total, the five independent equations,

$$u+v+w=C_0\times\frac{10}{10+x} \qquad (473)$$

$$y+\frac{0.1\times x}{10+x}+u=w+z+\frac{0.1\times 10}{10+x} \qquad (474)$$

the same number as that of the independent variables, are derived and, hence, can be solved for the variables as is shown in the following.

Rewriting eqs(471) and (472) as eqs(471') and (472'), respectively, and substituting the rewritten equations into eq(473), the obtained equation is solved for w as is given by eq(475).

$$u=\frac{y\times v}{K_{a_1}} \quad (471'), \qquad v=\frac{y\times w}{K_{a_2}} \quad (472')$$

$$\left(\frac{y^2}{K_{a_1}K_{a_2}}+\frac{y}{K_{a_2}}+1\right)w=\frac{C_0\times 10}{10+x} \Rightarrow w=\left(\frac{C_0\times 10}{10+x}\right)\Big/\left(\frac{y^2}{K_{a_1}K_{a_2}}+\frac{y}{K_{a_2}}+1\right) \qquad (475)$$

Eq(474) is rewritten for z as is shown below and, after its arrangement, eq(476) is obtained.

$$z=y+\frac{0.1\times x-0.1}{10+x}+u-w$$

$$=y+\frac{0.1\times x-0.1}{10+x}+\left(\frac{y^2}{K_{a_1}K_{a_2}}-1\right)w \qquad (476)$$

Multiplying both sides of eq(476) by y and then substituting w derived in eq(475) into the multiplied equation, eq(476') is obtained after some arrangements.

$$y^2\left(\frac{y^2}{K_{a_1}K_{a_2}}+\frac{y}{K_{a_2}}+1\right)(10+x)-y\left(\frac{y^2}{K_{a_1}K_{a_2}}-1\right)C_0\times 10$$
$$+(0.1\times x\times y-0.1y)\left(\frac{y^2}{K_{a_1}K_{a_2}}+\frac{y}{K_{a_2}}+1\right)-10^{-14}\left(\frac{y^2}{K_{a_1}K_{a_2}}+\frac{y}{K_{a_2}}+1\right)(10+x)=0 \qquad (476')$$

Eq(476') contains only the two variables, the volume added x and the concentration of hydrogen ion y. This equation is grouped into two types of terms, one containing x and the other containing none of x and then the rearranged equation is solved for x, yielding eq(477).

$$x=\frac{-y^2\times 10+0.1y+10^{-13}}{y^2-0.1\times y-10^{-14}}+\frac{y\left(\frac{y^2}{K_{a_1}K_{a_2}}-1\right)C_0\times 10}{\left(\frac{y^2}{K_{a_1}K_{a_2}}+\frac{y}{K_{a_2}}+1\right)(y^2-0.1\times y-10^{-14})} \qquad (477)$$

The final result for the problem is eq(477), nevertheless, it is difficult to plot y (pH values) against x as it is. Therefore, a method of *Parametric Plot*, which has been used so far from time to time, is employed here. To start with, the variable y is replaced by another variable χ which is equivalent to pH as is given by eq(478).

$$[\mathrm{H}^+]=y=10^{-\chi} \Rightarrow \mathrm{pH}=-\log_{10}y=\chi \qquad (478)$$

Then, eq(477) can be expressed as eq(477'). Further details of the procedure employed here is given in

solution 4-5 in which the titration curve for the weak acid vs. strong base titration is treated.

$$x = \frac{-10^{-2\chi}\times 10 + 10^{-13}}{10^{-2\chi} - 0.1\times 10^{-\chi} - 10^{-14}} - \frac{10^{-\chi}\left(\dfrac{10^{-2\chi}}{K_{a_1}K_{a_2}} - 10^{-\chi}\right)C_0\times 10}{\left(\dfrac{10^{-2\chi}}{K_{a_1}K_{a_2}} + \dfrac{10^{-\chi}}{K_{a_2}} + 1\right)\left(10^{-2\chi} - 0.1\times 10^{-\chi} - 10^{-14}\right)} \tag{477'}$$

In Fig.68 pH (χ) is plotted against x based on eq(477'). For comparison, the titration curve to be obtained is shown in Fig.69 when a neutral instead of acidic solution of asparagine is titrated with the same NaOH solution. In Fig.69, however, neither the first equivalent point nor the isoelectric point is recognized clearly. It is now obvious from a comparison of Fig.68 with Fig.69, why the acidic but not neutral aqueous solution is employed for the titration.

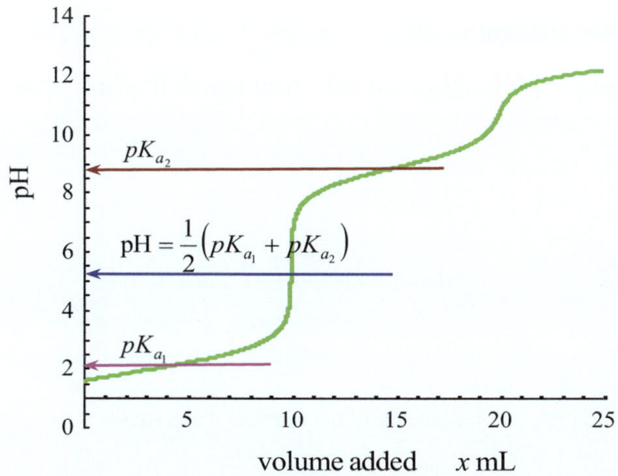

Fig.68 Titration curve of the acidic solution of asparagine.

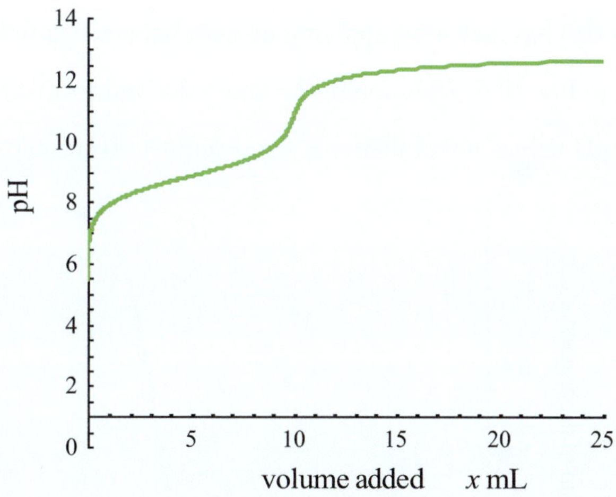

Fig.69 Titration curve of a neutral solution of asparagine.

Ans. titration curve: eq(477') and Fig.68

(b) The isoelectric point is given by a pH value at which the concentration of a cation H^+AH becomes equal to the concentration of an anion A^-. Putting $u = w$ in eqs(471') and (472'), eq(479) is obtained from which the pH value is calculated as 5.43.

$$y^2 = K_{a_1}\cdot K_{a_2} \quad \Rightarrow \quad \mathrm{pH} = \frac{1}{2}(pK_{a_1} + pK_{a_2}) \tag{479}$$

$$\mathrm{pH} = \frac{1}{2}(2.02 + 8.84) = 5.43$$

Also, noticing that at the isoelectric point, the total volume of the NaOH solution added till then is approximately 10 mL (see Fig.68), the concentration of the zwitter ion is calculated from eqs(472') and (473) and is found by eq(480).

$$v \approx \frac{1}{2}C_0\times y\Big/K_{a_2}\left(\frac{y^2}{K_{a_1}K_{a_2}} + \frac{y}{K_{a_2}} + 1\right)$$

$$= \frac{1}{2}C_0\times y\Big/\left(\frac{y^2}{K_{a_1}} + y + K_{a_2}\right)$$

$$= \frac{1}{2} C_0 \times y \Big/ (y + 2K_{a_2})$$

$$= \frac{1}{2} C_0 \times (K_{a_1} K_{a_2})^{1/2} \Big/ \{(K_{a_1} K_{a_2})^{1/2} + 2K_{a_2}\} \qquad (480)$$

$$\approx \frac{1}{2} C_0$$

In the present case, the volume of the solution at the isoelectric point has become approximately twice the initial volume and, hence, the fact that the concentration of the zwitter ion is approximately a half of the initial concentration (eq(480)) indicates that almost all of asparagine molecules exist as zwitter ions. From the equation to predict the isoelectric point (eq(479)), it is seen that this point varies from amino acid to amino acid. Generally speaking, amino acids exist almost as zwitter ions at their isoelectric points. Electrophoresis is the method based on this fact and most amino acids are separated, identified and measured by driving their ions applying an external electric field under gradient of pH.

In Fig.70 the fraction of the anion derived from asparagine is plotted against pH, from which it is seen that at pH values lower than $\text{pH} = 6.0$ almost no anion exists.

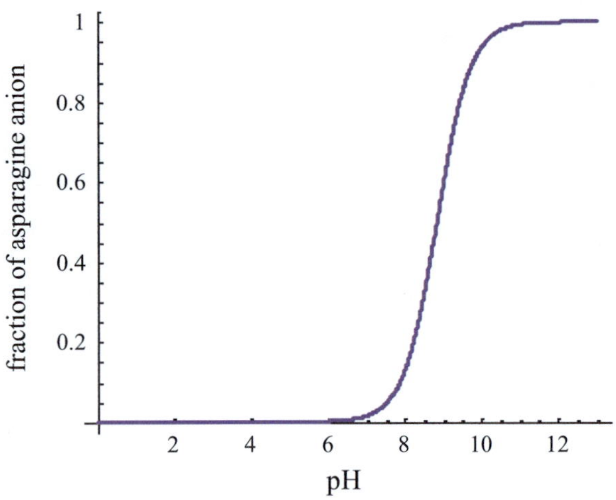

Fig.70 pH dependence of the fraction of asparagine anion existing in solution.

Ans. Almost all exist in the form of the zwitter ion at the isoelectric point.

"note" Isoelectric points for some amino acids: glycine 5.97, tryptophan 5.89, serine 5.68 and phenylalanine 5.48.

[advanced problem] Aspartic acid (Asp: 2-aminobutanedioic acid) is an amino acid with two carboxyl groups. Its acid-base dissociation consists of three steps and the dissociation are given by $pK_1 = 1.88$, $pK_2 = 3.65$ and $pK_3 = 9.60$. Find a pH value at the isoelectric point of aspartic acid.

Solution 4-72

(a) Taking logarithms of both sides of $[\eta_{in}] = K \cdot M^{\alpha}$, eq(481) is obtained.

$$\ln[\eta_{in}] = \alpha \ln M + \ln K \qquad (481)$$

Employing the values given in the table, $\ln[\eta_{in}]$ is plotted against $\ln M$ and a good linear correlation is found as is shown in Fig.71 below. From the best fitted parameters, $\alpha = 0.52 \pm 0.03$ and $\ln K = 1.22 \pm 0.21$ are found, yielding $K = 3.4 \pm 0.7$. The dimension of K ($\text{m}^3 \cdot \text{kg}^{-1}$) is the same as that of the intrinsic viscosity. By measuring the intrinsic viscosity of an amylose whose molecular weight (mass) is not known, its molecular weight is acquired on the basis of this linear relationship.

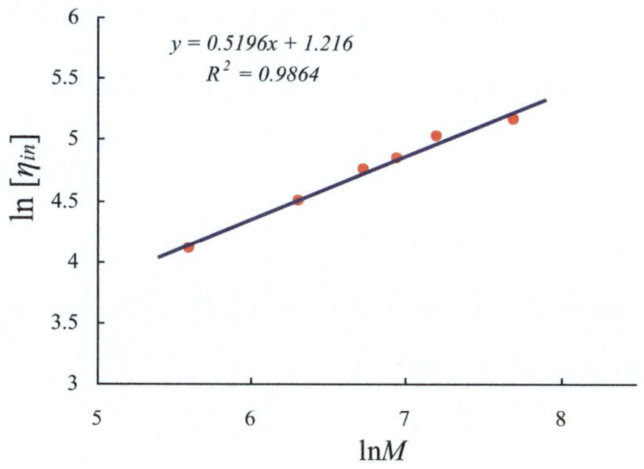

Fig.71 Plot of *Mark-Houwink-Sakurada*'s equation (eq(481)).

Ans. $\alpha = 0.52 \pm 0.03$, $K = 3.4 \pm 0.7$

(b) The chemical formula of amylose is expressed by $(C_6H_{10}O_5)_n$. Since the molecular weight found for the amylose is 3913 ± 290, the chain length n is estimated as 24 ± 1. Notice that the chain length of most amyloses is in the range 10^2-10^3.

Ans. See text.

"note" Definition of intrinsic viscosity η_{in}.

$$\eta_{in} = \lim_{c \to 0} \frac{\eta - \eta_0}{\eta_0} \cdot \frac{1}{c} = \lim_{c \to 0} \frac{\eta_{sp}}{c}, \qquad \eta_{sp} = \frac{\eta - \eta_0}{\eta_0} : \text{specific viscosity}, \qquad \eta_{rel} = \frac{\eta}{\eta_0} : \text{relative viscisity}$$

Solution 4-73

The dispersion and equilibrium of acetic acid in the two-phase separated liquids composed of benzene and water is schematically illustrated in Fig.72 on the right. The dissociation of acetic acid in water is small enough to be neglected and the dispersion into the two phase is described by eq(482) with K_1 as the dispersion coefficient.

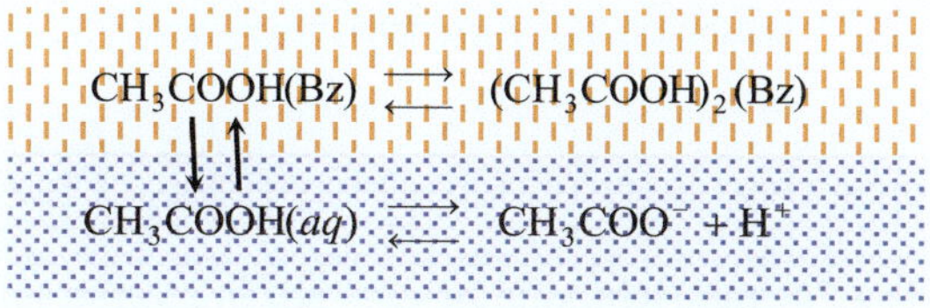

Fig.72 Dispersion of acetic acid in the two-phase separated liquids.

$$\underset{x}{CH_3COOH(aq)} \rightleftarrows \underset{y}{CH_3COOH(Bz)} \qquad K_1 \qquad (482)$$

Also, the association constant $K_{assoc.}$ of acetic acid in benzene is expressed by eq(483).

$$\underset{y}{2CH_3COOH(Bz)} \rightleftarrows \underset{z}{(CH_3COOH)_2(Bz)} \qquad K_{assoc.} \qquad (483)$$

Denoting the molarities of substances by x, y and z as are indicated in eqs(482) and (483), they are related to K_1 and $K_{assoc.}$ through eqs (484), (485) and (486).

$$x + y + 2z = C_0 \qquad (484) \qquad (C_0 : \text{total molarity})$$

$$\frac{y}{x} = K_1 \Rightarrow y = K_1 x \qquad (485)$$

$$\frac{z}{y^2} = K_{assoc.} \Rightarrow z = K_{assoc.} y^2 = K_{assoc.} K_1^2 x^2 \qquad (486)$$

From these relations, the ratio of the molarities of acetic acid in the two phases given in the table, which is expressed as R, is given by eq(487).

$$R \equiv \frac{y+2z}{x} = \frac{K_1 x + 2K_{assoc.}K_1^2 x^2}{x} = K_1 + 2K_{assoc.}K_1^2 x \qquad (487)$$

Consequently, the values of R calculated from the data in the table are plotted against x (molarity of acetic acid in water phase) to determine K_1 and $K_{assoc.}$. From the intercept and slope of the plot shown in Fig.73, the values for K_1 and $K_{assoc.}$ are found as given below.

$$K_1 = 8.06\times10^{-3}\ , \qquad K_{assoc.} = \frac{slope}{2K_1^2} = \frac{0.0227}{2\times(8.06)^2\times10^{-6}} = 1.75\times10^2$$

Also, the degree of association of acetic acid in the benzene phase is defined by the left side of eq(488) and then it can be expressed as a function of x by the right side of eq(488).

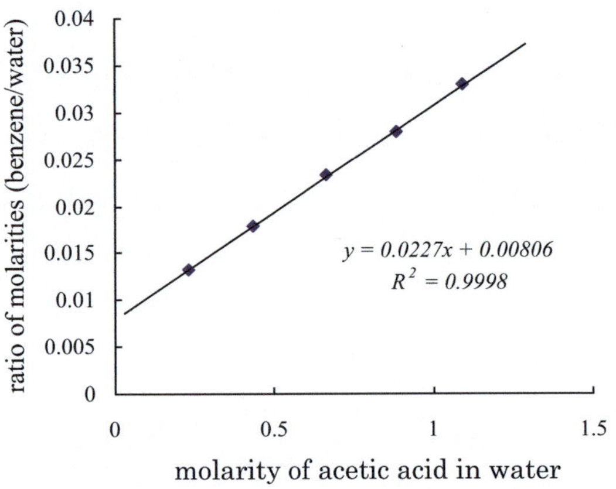

Fig.73 Ratio of the concentrations of acetic acid in two phases and its concentration dependence.

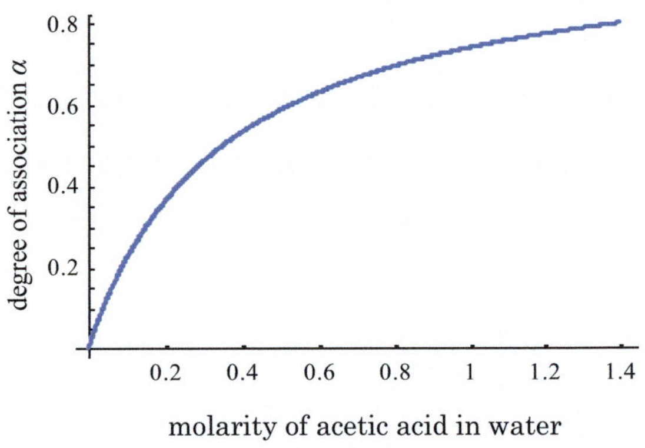

Fig.74 Concentration dependence of the degree of association of acetic acid in benzene.

$$\alpha = \frac{2z}{y+2z} = \frac{2K_{assoc.}K_1^2x^2}{K_1x + 2K_{assoc.}K_1^2x^2} = \frac{2\times1.75\times10^2\times8.06\times10^{-3}x}{1+2\times1.75\times10^2\times8.06\times10^{-3}x} \qquad (488)$$

In Fig.74, the right side of eq(488) is plotted against x. As is clear from this figure, most of the acetic acid molecules in benzene associate to from dimmers and this tendency becomes stronger with increasing the concentration of acetic acid.

Ans. $K_1 = 8.06\times10^{-3}$, $K_{assoc.} = 1.75\times10^2$: degree of association: Fig,74

"note" Association constant can alternatively be defined by $K_{assoc.} = \dfrac{C\alpha/2}{C^2(1-\alpha)^2}$ in which C is the total concentration of acetic acid, $C(1-\alpha)$ is the concentration of acetic acid and $C\alpha/2$ is that of the dimmer, all values in the benzene phase.

Solution 4-74

(a) Relation to the temperature.

The chemical potentials of a salt at any two different depths specified by temperatures T and T_x must be equal as are expressed by eq(489). Writing the concentration of the salt at the temperature T as $c(T)$, the chemical potential at each depth is given by eq(490).

$$\mu(T) = \mu(T_x) \qquad (489)$$

$$\left.\begin{aligned} \mu(T) &= \mu^\circ(T) + RT\ln c(T) \\ \mu(T_x) &= \mu^\circ(T_x) + RT_x \ln c(T_x) \end{aligned}\right\} \qquad (490)$$

Assuming that $\mu^\circ(T) \approx \mu^\circ(T_x)$, eq(491) is derived from eqs(489) and (490).

$$\frac{T}{T_x} = \frac{\ln c(T_x)}{\ln c(T)} \qquad (491)$$

Solving eq(491) for $c(T_x)$, eq(491') is obtained, in which the concentration at 298 K is put $c(298\text{K}) = 36.5 \times 10^{-3}$ (salinity divided by the standard concentration).

$$c(T_x) = \exp\left(\frac{T}{T_x} \times \ln c(T)\right) = \exp\left(\frac{298}{T_x} \times \ln(36.5 \times 10^{-3})\right) \qquad (491')$$

In Fig.75, $c(T_x)$ of eq(491') is plotted against T_x. It is seen from this figure that the salt concentration $c(T_x)$ decreases as the temperature becomes lower.

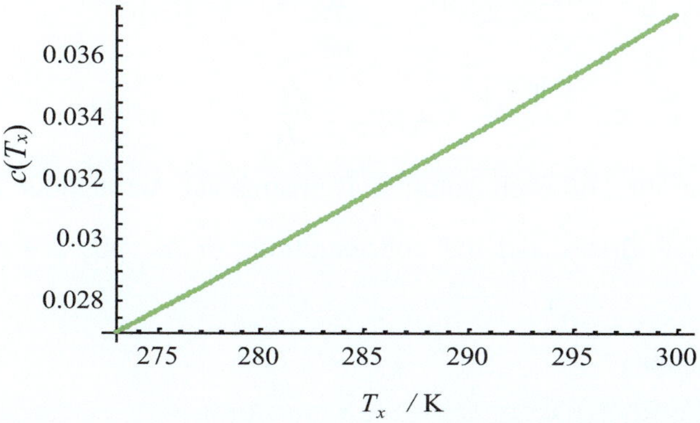

Fig.75 Relation between sea temperature and equilibrium salt concentration. Plot based on eq(491').

Ans. saturated salt concentration and temperature: eq(491') and Fig.75

(b) Relation to the depth (at constant temperature).

For simplicity, assuming that the temperature of sea water does not vary with the depth, the relation between the salt concentration and the depth is found by taking the potential energy change with the depth into account. Taking the sea level as the reference point of the potential energy, the potential energy of an ion with a mass m at the depth d is given by $-m \cdot g \cdot d$. Consequently, the balance of the chemical potentials is given by eq(492), in which the concentration of the salt at the depth d is denoted $c(d)$.

$$k_B T \ln c(0) = k_B T \ln c(d) - m \cdot g \cdot d \qquad (492)$$

Solving eq(492) for $c(d)/c(0)$, eq(493) is found and it is plotted against the depth in Fig.76.

$$\frac{c(d)}{c(0)} = \exp\left(\frac{m \cdot g \cdot d}{k_B T}\right) = \exp\left(\frac{35.5 \times 10^{-3} \times 9.8 \times d}{1.38 \times 10^{-23} \times 6.02 \times 10^{23} \times 298}\right) \qquad (493)$$

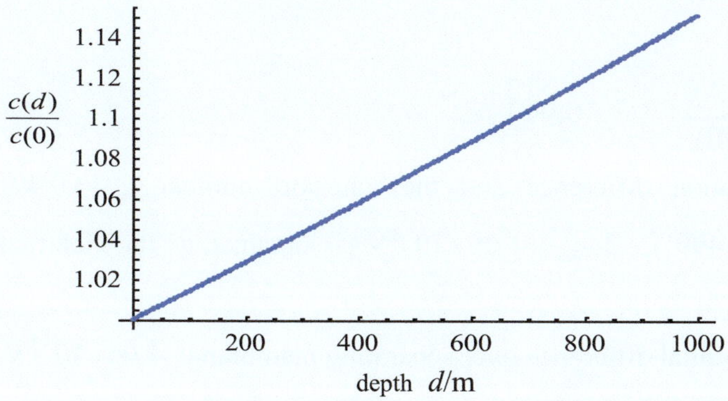

Fig.76 Relation between sea depth and equilibrium salt concentration. The sea temperature is fixed at 298 K. The slope gets larger as the temperature becomes lower.

Ans. saturated salt concentration and depth: eq(493) and Fig.76

Solution 4-75

Ions in the ion-permeable membrane-separated solutions at different concentrations migrate through the membrane because of the two causes, one is the diffusion due to the concentration gradient and the other is the electric field arising from the separation of ions by membrane. Consequently, when the migrating ions are K^+ and Cl^- and their fluxes through the unit area of the membrane are written as J_{K^+} and J_{Cl^-}, they are given by eq (494) and eq(495), respectively. Here, the first term is arising from the diffusion and the second term is due to the driving force of electric field $E = F \cdot d\varphi/dx$, and u_{K^+}, u_{Cl^-} are the mobilities of the relevant

$$J_{K^+} = -D_{K^+}\frac{\partial c_{K^+}}{\partial x} + u_{K^+}c_{K^+}E = -D_{K^+}c_{K^+}\frac{\partial \ln c_{K^+}}{\partial x} - u_{K^+}c_{K^+}F\frac{d\varphi}{dx} \qquad (494)$$

$$J_{Cl^-} = -D_{Cl^-}\frac{\partial c_{Cl^-}}{\partial x} - u_{Cl^-}c_{Cl^-}E = -D_{Cl^-}c_{Cl^-}\frac{\partial \ln c_{Cl^-}}{\partial x} + u_{Cl^-}c_{Cl^-}F\frac{d\varphi}{dx} \qquad (495)$$

ions. In spite of ion migrations, the electric neutrality of each solution is maintained throughout the migration processes, eq(496) holds between the ion fluxes and the concentrations of the positive and negative ions themselves are related by eq(497).

$$(+1)\times J_{K^+} + (-1)J_{Cl^-} = 0 \qquad (496)$$

$$c_{K^+} = c_{Cl^-} \qquad (497)$$

Also, the mobility and the diffusion coefficient of each ion are related by eq(498).

$$D_{K^+} = u_{K^+}\times RT, \qquad D_{Cl^-} = u_{Cl^-}\times RT \qquad (498)$$

Using all these equations, eq(496) can be rewritten as eq(496'), and it is further rearranged as eq(496'').

$$-u_{K^+}\cdot RT\frac{\partial \ln c_{K^+}}{\partial x} - u_{K^+}c_{K^+}F\frac{d\varphi}{dx} + u_{Cl^-}\cdot RT\frac{\partial \ln c_{Cl^-}}{\partial x} - u_{Cl^-}c_{Cl^-}F\frac{d\varphi}{dx} = 0 \qquad (496')$$

$$-(u_{K^+} + u_{Cl^-})c_{K^+}F\frac{d\varphi}{dx} = RT(u_{K^+} - u_{Cl^-})c_{K^+}\frac{\partial \ln c_{K^+}}{\partial x} \qquad (496'')$$

As a result, the potential gradient $d\varphi/dx$ at the separating membrane is found to be given by the transport numbers of the ions migrating and the gradient of the chemical potential of the ion ($RT\ln c_{K^+}$) as is presented by eq(499).

$$-\frac{d\varphi}{dx} = \frac{(u_{K^+} - u_{Cl^-})}{(u_{K^+} + u_{Cl^-})}\frac{RT}{F}\frac{\partial \ln c_{K^+}}{\partial x} = (2t_{K^+} - 1)\frac{RT}{F}\frac{\partial \ln c_{K^+}}{\partial x} \qquad (499)$$

(a) Employing eq(499), the potential difference found for the KCl solutions between the two sides of the separating membrane is given by eq(499'),

$$-\Delta\varphi_{AB} = (2t_{K^+} - 1)\frac{RT}{F}\ln\frac{c_{K^+}(A)}{c_{K^+}(B)} \qquad (499')$$

Substituting the value given for the concentration difference and the transport number $t_{K^+} = 0.491$ calculated from the mobility of each ion into eq(499'), $\Delta\varphi_{AB} = 1.06\times10^{-3}$ V is obtained as the electrical potential difference between the two solutions.

Ans. potential difference over separating membrane: 1.06×10^{-3} V

(b) When the electrolyte is replaced by NaCl while keeping the concentration the same, it is only necessary to replace the corresponding transport number in eq(499') with that of Na^+. From the mobilities of Na^+ and Cl^-, the transport numbers of these ions are found as below.

$$t_{Na^+} = \frac{u_{Na^+}}{u_{Na^+} + u_{Cl^-}} = \frac{5.19}{5.19 + 7.91} = 0.396, \qquad t_{Cl^-} = \frac{u_{Cl^-}}{u_{Na^+} + u_{Cl^-}} = 0.604$$

From $t_{Na^+} = 0.396$, the separating potential is found as $\Delta\varphi = 24.6 \times 10^{-3}\,\mathrm{V}$.

In the absence of the ion migration through the membrane, the equilibrium potential difference is solely determined by the difference of the chemical potentials. Putting the ratio of the concentrations 10:1, the potential difference is calculated from the *Nernst* equation as $-5.9 \times 10^{-2}\,\mathrm{V}$ for either electrolyte as shown below.

$$-\Delta\varphi_{Nernst} = \frac{RT}{F}\ln\frac{1.0}{0.1} = 5.9 \times 10^{-2}\,\mathrm{V} \;\Rightarrow\; \Delta\varphi_{Nernst} = -5.9 \times 10^{-2}\,\mathrm{V}$$

Ans. potential difference over separating membrane: $24.6 \times 10^{-3}\,\mathrm{V}$

Solution 4-76

The reaction rate of an electron transfer reaction $A + e \rightleftarrows A^-$ at the electrode whose electrode potential is set at zero is expressed by eq(500) or eq(501), where $\Delta\vec{G}^{\neq}$ and $\Delta\overleftarrow{G}^{\neq}$ represent activation free energies for the forward and backward reactions, respectively, whose direction is indicated by an arrow on top of the corresponding symbol.

$$\vec{k}_0 = \frac{k_B T}{h}\exp\left(-\frac{\Delta\vec{G}^{\neq}}{RT}\right) \qquad (500)$$

$$\overleftarrow{k}_0 = \frac{k_B T}{h}\exp\left(-\frac{\Delta\overleftarrow{G}^{\neq}}{RT}\right) \qquad (501)$$

These activation free energies are related to the standard free energy change ΔG° of the electron transfer reaction by eq(502).

$$\Delta\overleftarrow{G}^{\neq} - \Delta\vec{G}^{\neq} = \Delta G^\circ \qquad (502)$$

When electric potential is applied to the electrode, the potential difference $\Delta\phi$ (electrostatic potential difference) arises and it is expressed by the equilibrium potential $\Delta\phi_e$ and the overvoltage η as in eq(503).

$$\Delta\phi = \Delta\phi_e + \eta \qquad (503)$$

For such cases, it is illustrated in Fig.77 how the activation free energy varies along the reaction coordinate. The curve in blue represents the free energy change along the reaction coordinate when the electrode potential is zero. Considering that the electrode is located on the left side of the reaction coordinate, the electrostatic potential decreases linearly toward the solution, i.e., to the right, as is indicated by the thick black line. It is now assumed that the electrostatic potential at the coordinate of the activated complex is given by $\beta\Delta\phi$ using symmetry factor β ($0 \le \beta \le 1$) (see **"note"**).

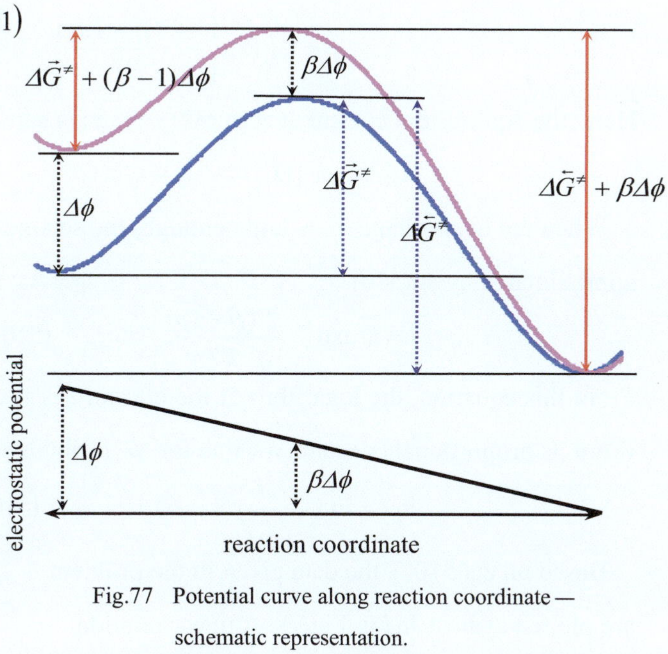

Fig.77 Potential curve along reaction coordinate — schematic representation.

Under those assumptions, the activation free energy when the electrode potential is applied is found to be given by the curve in pinkish red in Fig.77. Consequently, the rate constants for the forward and backward electron transfer reactions via electrode are given by eqs(504) and (505), respectively

$$\vec{k} = \frac{k_B T}{h}\exp\left\{-\frac{\Delta\vec{G}^{\neq} + F(\beta-1)\Delta\phi}{RT}\right\}$$
$$= \frac{k_B T}{h}\exp\left\{-\frac{\Delta\vec{G}^{\neq} + F(\beta-1)\Delta\phi_e + F(\beta-1)\eta}{RT}\right\} = \vec{k}_0 \exp\left\{-\frac{F(\beta-1)\eta}{RT}\right\} \qquad (504)$$

$$\overleftarrow{k} = \frac{k_B T}{h}\exp\left(-\frac{\Delta\overleftarrow{G}^{\neq} + F\beta\Delta\phi}{RT}\right)$$
$$= \frac{k_B T}{h}\exp\left(-\frac{\Delta\overleftarrow{G}^{\neq} + F\beta\Delta\phi_e + F\beta\eta}{RT}\right) = \overleftarrow{k}_0 \exp\left(-\frac{F\beta\eta}{RT}\right) \qquad (505)$$

When the reaction is one-electron transfer, the electric currents associated with the forward and backward reactions are given by eq(506) and eq(507), respectively.

$$\vec{i} = \vec{k}[\mathrm{A}] = \vec{k}_0[\mathrm{A}]\exp\left\{-\frac{F(\beta-1)\eta}{RT}\right\} = \vec{i}_0 \exp\left\{-\frac{F(\beta-1)\eta}{RT}\right\} = i_0 \exp\left\{-\frac{F(\beta-1)\eta}{RT}\right\} \qquad (506)$$

$$\overleftarrow{i} = \overleftarrow{k}[\mathrm{D}] = \overleftarrow{k}_0[\mathrm{D}]\exp\left\{-\frac{F\beta\eta}{RT}\right\} = \overleftarrow{i}_0 \exp\left\{-\frac{F\beta\eta}{RT}\right\} = i_0 \exp\left\{-\frac{F\beta\eta}{RT}\right\} \qquad (507)$$

As a result, the electric current i running through a circuit is given by the difference between these forward and backward currents as is expressed by eq(508).

$$i = \vec{i} - \overleftarrow{i} = i_0\left\langle \exp\left\{-\frac{F(\beta-1)\eta}{RT}\right\} - \exp\left\{-\frac{F\beta\eta}{RT}\right\}\right\rangle \qquad (508)$$

Here, the equilibrium current at zero overvoltage is named exchange current and is denoted i_0 as in eq(509).

$$\vec{k}_0[\mathrm{A}] = \overleftarrow{k}_0[\mathrm{D}] \;\Rightarrow\; \vec{i}_0 = \overleftarrow{i}_0 \equiv i_0 \qquad (509)$$

When the overvoltage η is large enough, the second term of eq(508) can be neglected and the current i is approximated by eq(510).

$$i = i_0 \exp\left\{-\frac{F(\beta-1)\eta}{RT}\right\} \qquad (510)$$

From this equation, the logarithm of the current i is found to be proportional to η as is given by eq(510').

From its proportional constant, a value for β is obtained.

$$\ln i = \ln i_0 + \frac{F}{RT}(1-\beta)\eta \qquad (510')$$

Based on eq(510'), the data given in the problem are plotted against $\ln i$ in Fig.78. A linear relation as is predicted by eq(510') is seen where the overvoltage is large and as the best-fitted *Tafel*'s plot, the following straight line is obtained.

$$\eta = 0.0489 \ln i + 0.0082$$

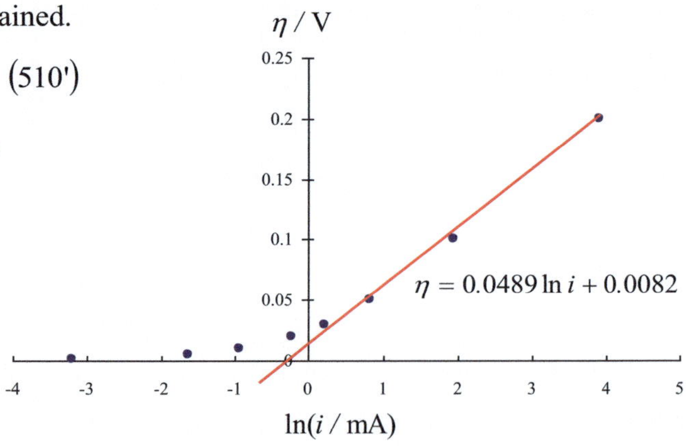

Fig.78 *Tafel*'s plot.

From this fitted line, the exchange current and the symmetry factor are found as below.

$$(1-\beta)=\frac{RT}{F}\times\frac{1}{slope}\quad\Rightarrow\quad \beta=1-\frac{8.31\times298}{96500}\times\frac{1}{0.0489}=0.475$$

Ans. exchange current: $i_0=0.846$ mA, symmetry factor: $\beta=0.475$

"note" The reason why β ($0\le\beta\le1$) is called symmetry factor is clear from Fig.77. It reflects a position at which the activated complex is located between the reactant and the product along the reaction coordinate.

Solution 4-77

Since the rate of deposition of Fe metal is given in the unit of $\mathrm{mol\cdot s^{-1}}$, the electric current used for the electrolysis is found by multiplying the rate by the *Faraday* constant *F.* Negative values of electrode potential *E* are plotted in Fig.79 against the logarithmic values of the electrolysis current thus obtained.

This corresponds to *Tafel*'s plot though, the vertical axis is not the overvoltage η. However, the true *Tafel*'s plot would be obtained just by shifting the one given in Fig.79 parallel along the vertical axis by a constant value. Therefore, from the slope of the best-fitted line shown in Fig.79 the symmetry factor β is obtained just as is done for true *Tafel*'s plot. Where $-E$ takes large values, a liner correlation is satisfactory as is seen in Fig.79 and a best fitted one is given by eq(511).

$$-E=\frac{RT}{F(1-\beta)}\times\ln i+Const. \qquad (511)$$

From this, the symmetry factor β is found as 0.619 as is calculated below.

$$(1-\beta)=\frac{RT}{F}\times\frac{1}{slope}=\frac{8.31\times298}{96500}\times0.0692=0.381\quad\Rightarrow\quad\beta=0.619$$

The exchange current i_0 (see the preceding solution) can be found, if the electric potential at which the current through the circuit becomes zero is known somehow, by substituting that electrical potential and the value for β found above into eq(511)

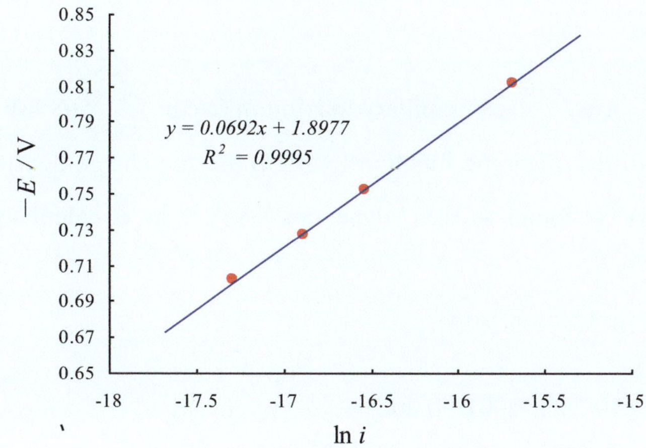

Fig.79 *Tafel*'s plot derived from the rate of deposition.

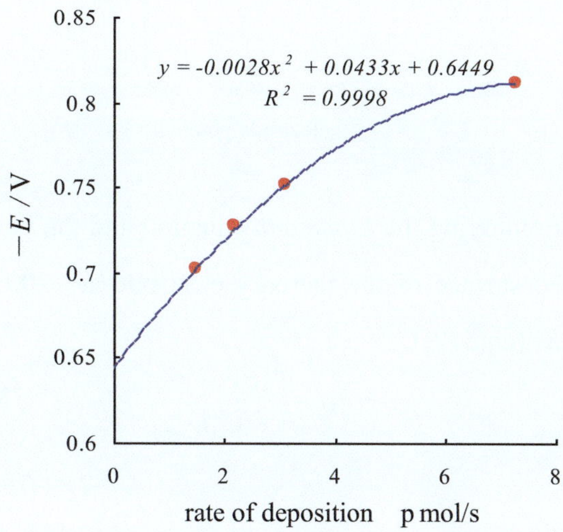

Fig.80 Rate of deposition of Fe and electric potential.

In Fig.80, $-E$ is plotted against the rate of deposition. The electric potential at which the deposition rate becomes zero is found from the best-fitted quadratic function inserted in Fig.80 as $-E=0.645$. Substituting this value into the best-fitted function (eq(511)) given in Fig.79, a value for the exchange current is found as

$i_0 = 1.33 \times 10^{-8}$ A from the calculation given below.

$$0.645 = 0.0692 \times \ln i_0 + 1.90 \Rightarrow \ln i_0 = -\frac{1.90 - 0.645}{0.0692} = -18.14$$

Ans. symmetry factor β: 0.619, exchange current i_0: 1.33×10^{-8} A

Solution 4-78

The data given in Table 1 in the problem are plotted in Fig.81. The surface tension is seen to vary with the applied voltage. Ideally, the electrocapillary curve is considered to be given by a parabola (see **[advanced problem]**) and, hence, only the values for the surface tension obtained in the voltage range $-(300-700)$ mV are employed to find a best-fitted quadratic function against the applied voltage. The obtained one is shown in Fig.81 by the green curve and is expressed by the following formula.

$$\gamma = -(1.78 \pm 0.073) \times 10^{-4} E_V^2 - (0.2023 \pm 0.0081) E_V + 365.85 \pm 2.1$$

From this function, it is found that the surface tension takes the maximum of $\gamma_{max} = 423.3 / \text{dyne} \cdot \text{cm}^{-1}$ at the voltage of $E_V = -568$ mV, which is call the electrocapillary maximum voltage. According to *Lippmann*'s equation (eq(30) in the problem), the surface charge density becomes zero at that voltage. A curve representing the relation between γ and E_V is called electrocapillary curve.

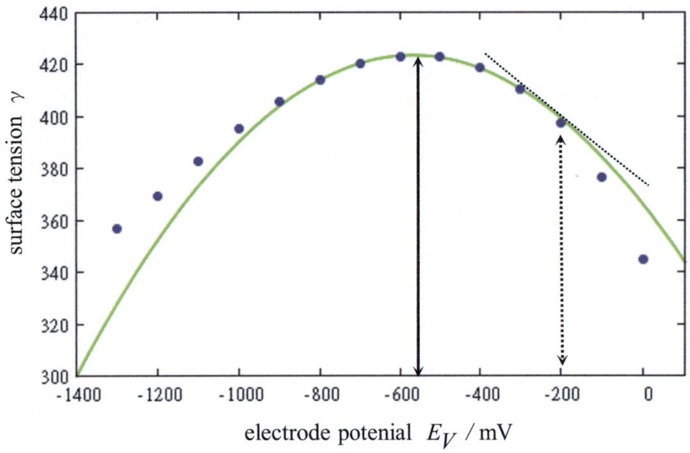

Fig.81 Electrocapillary curve for 1.0 N CsCl aqueous solution ((γ vs. E_V curve). dots: values in Table 1 in the problem, green line: best-fitted quadratic function.

Ans. electrocapillary maximum (e.c.m.) : -568 mV

(b) Employing the *Lippmann* equation and the best-fitted quadratic function found in (a), the charge density in the surface of the mercury electrode at -200 mV is found as $0.313 \text{ dyne} \cdot \text{cm}^{-1} \cdot \text{mV}^{-1}$ as is calculated below (eq(512)).

$$-q_M = \left(\frac{\partial \gamma}{\partial E_V}\right)_{\mu_i, T, P} = (-2 \times 1.78 \times 10^{-4} E_V - 0.2023)_{-200}$$

$$= -2 \times 1.78 \pm \times 10^{-4} \times (-200) - 0.2023$$

$$= -0.131 \text{ dyne} \cdot \text{cm}^{-1} \cdot \text{mV}^{-1} \qquad (512)$$

Since the units employed above are converted exactly to $\text{C} \cdot \text{m}^{-2}$ as is shown below,

$\text{dyne} \cdot \text{cm}^{-1} \cdot \text{mV}^{-1} = 10^{-5} \text{N} \cdot 100 \times \text{m}^{-1} \cdot 1000 \times \text{V}^{-1} = \text{N} \cdot \text{m}^{-1} \text{V}^{-1} = \text{N} \cdot \text{m} \cdot \text{m}^{-2} (\text{J/C})^{-1} = \text{J} \cdot (\text{J/C})^{-1} \text{m}^{-2} = \text{C} \cdot \text{m}^{-2}$,

the charge density of the mercury electrode is found as $q_M = 0131 \text{ C} \cdot \text{m}^{-2}$.

Regarding an electric double layer at the mercury electrode surface as a condenser, its electric capacitance

C per unit area is given by eq(513) in terms of the voltage E_V and the charge density q_M and is found as $0.655\ \mathrm{F}$ (see **"note"**).

$$C \cdot E_V = q_M \qquad (513) \Rightarrow \quad C = 0.131/0.200 = 0.655\ \mathrm{C \cdot V^{-1}} = 0.655\ \mathrm{F}$$

Similarly, at the voltage of $-800\ \mathrm{mV}$, the charge density is found as $q_M = -0.083\ \mathrm{C \cdot m^{-2}}$, from which the electric capacitance is calculated as $C = 0.103\ \mathrm{F}$. However, it should be noticed that the polarity of the capacitor is inverted since the electrode is now negatively charged.

Ans. at $-200\ \mathrm{mV}$: $+0.131\ \mathrm{C \cdot m^{-2}}$, $0.655\ \mathrm{F}$; at $-800\ \mathrm{mV}$: $-0.083\ \mathrm{C \cdot m^{-2}}$, $C = 0.103\ \mathrm{F}$

"note" The differential capacitance of a capacitor is defined by $(\partial Q/\partial V)$. For the mercury electrode under consideration, it is given by $(\partial q_M/\partial E_V) = 3.56 \times 10^{-4}\ \mathrm{F}$.

(c) A number of equations required for finding the excess surface concentration at the electrode are summarized below first. Denoting the electrode area *A*, the surface charge Q_M and the surface potential $\Delta\phi_{MS}$, the *Gibbs* free energy at the electrode surface is given by eq(514) (first order homogeneous function), where *M* (labeled M_A later) indicates mercury and *S* does solution, being added as subscripts.

$$G = \sum n_i \mu_i + \gamma A + Q_M \Delta\phi_{MS} \qquad (514)$$

On the other hand, the total differential of the free energy *G* expressed by a set of the independent variable of T, P, n_i, A and Q_M is given by eq(515).

$$dG = -SdT + VdP + \sum \mu_i dn_i + \gamma dA + \Delta\phi_{MS} dQ_M \qquad (515)$$

Also, the total differential of eq(514) can be expressed as eq(516),

$$dG = \sum \mu_i dn_i + \sum n_i d\mu_i + \gamma dA + A d\gamma + Q_M d\Delta\phi_{MS} + \Delta\phi_{MS} dQ_M \qquad (516)$$

Comparing eq(515) with eq(516) under the conditions of constant *T* and *P*, the differentials of the variables other than *T* and *P* are found to be related by eq(517).

$$A d\gamma + Q_M d\Delta\phi_{MS} + \sum n_i d\mu_i = 0 \qquad (517)$$

Eq(517) is rewritten with respect to $d\gamma$ as eq(517'), where q_M is the charge density defined by eq(518).

$$d\gamma = -\frac{Q_M}{A} d\Delta\phi_{MS} - \sum \frac{n_i}{A} d\mu_i = -q_M d\Delta\phi_{MS} - \sum \frac{n_i}{A} d\mu_i \qquad (517')$$

$$\frac{Q_M}{A} \equiv q_M \qquad (518)$$

Denoting the concentration of the *i*-th component in a bulk solution n_i°, and the corresponding surface concentration n_i, the excess surface concentration Γ_i is defined by eq(519) (see **solution 4-43**).

$$\Gamma_i = \frac{n_i}{A} - \frac{n_i^\circ}{A} \qquad (519)$$

After multiplying both sides of eq(519) by $d\mu_i$, each of the resulting terms is summed over *i* and then the resultant equation is rearranged so as to yield eq(519').

$$\sum_i \frac{n_i}{A} d\mu_i = \sum_i \Gamma_i d\mu_i + \sum_i \frac{n_i^\circ}{A} d\mu_i \qquad (519')$$

Since eq(520) is satisfied for the bulk solution in equilibrium,

$$\sum_i n_i^\circ d\mu_i = 0 \qquad (520)$$

the substitution of this relation into eq(519') yields eq(519'').

$$\sum_i \frac{n_i}{A} d\mu_i = \sum_i \Gamma_i d\mu_i \qquad (519'')$$

Then, substituting this into eq(517'), eq(517'') is derived.

$$d\gamma = -q_M d\Delta\phi_{MS} - \sum \Gamma_i d\mu_i \Rightarrow d\gamma_{M_A} = -q_{M_A} d\Delta\phi_{M_A S} - \sum \Gamma_i d\mu_i \qquad (517'')$$

We now consider that the system is composed of the mercury electrode (M_A) and another different electrode (M_B). Then, the phase boundary potential of the mercury electrode $\Delta\phi_{M_A S}$ (mercury (M_A)–solution(S)) is expressed by eq (521) using the other two phase boundary potentials and the applied voltage E_V, and the differential of E_V can be approximated as in eq(521') by putting $d\Delta\phi_{M_B M_A} \approx 0$.

$$E_V = \Delta\phi_{M_A S} + \Delta\phi_{SM_B} + \Delta\phi_{M_B M_A} \qquad (521)$$

$$dE_V = d\Delta\phi_{M_A S} + d\Delta\phi_{SM_B} + d\Delta\phi_{M_B M_A}$$
$$= d\Delta\phi_{M_A S} + d\Delta\phi_{SM_B} \qquad (521')$$

Employing a calomel electrode (M_B) as the other electrode against the mercury electrode, $d\Delta\phi_{SM_B}$ can be expressed by eq(522),

$$d\Delta\phi_{SM_B} = -\frac{1}{z_j F} d\mu_j = \frac{1}{F} d\mu_{Cl^-} \left(= \frac{1}{F} d\mu_- \right) \qquad (522) \qquad (cf.\ \mu_{ele} = \mu + zF\phi)$$

and then the substitution of eqs(521') and (522) into eq(517'') leads to eq(517''').

$$d\gamma_{M_A} = -q_{M_A} dE_V + \frac{q_{M_A}}{F} d\mu_- - \sum \Gamma_i d\mu_i \qquad (517''')$$

By keeping the compositions of the solution constant under a constant T and P, the division of eq(517''') by dE_V will lead to the *Lippmann* equation as is given below again.

$$\left(\frac{\partial \gamma}{\partial E_V} \right)_{\mu_i, T, P} = -q_{M_A} \qquad (\textit{Lippmann}\text{'s equation}) \qquad \text{(reposted)}$$

For a 1 : 1electrolyte CsCl aqueous solution, the chemical potential of CsCl can be expressed by eq(524) in terms of the chemical potentials of the component ions and the mean activity of the ions given by eq(523).

$$\mu_+ = RT \ln a_+, \quad \mu_- = RT \ln a_- \ \text{(constant term neglected)}, \ \sqrt{a_+ a_-} = a_\pm \qquad (523)$$

$$\mu_{CsCl} = \mu_+ + \mu_- = RT \ln a_+ a_- = 2RT \ln a_\pm \approx 2RT \ln [CsCl] \qquad (524)$$

Under the condition of a constant E_V, eq(517''') can be rewritten as eq(525) using the relation of $d\mu_{CsCl} = d\mu_+ + d\mu_-$. The coefficient of $d\mu_-$, $(q_{M_A}/F + \Gamma_+ - \Gamma_-)$, becomes zero because the interface as a whole is electrically neutral.

$$d\gamma_{M_A} = \frac{q_{M_A}}{F} d\mu_- - \Gamma_+ d\mu_+ - \Gamma_- d\mu_- = \left(\frac{q_{M_A}}{F} + \Gamma_+ - \Gamma_-\right) d\mu_- - \Gamma_+ d\mu_{CsCl} \qquad (525)$$

Consequently, eq(526) is derived as an equation to give the surface excess concentration.

$$\Gamma_+ = -\left(\frac{\partial \gamma_{M_A}}{\partial \mu} \right)_{E_V, T, P} = -\left(\frac{\partial \gamma_{M_A}}{2RT \partial \ln a_\pm} \right)_{E_V, T, P} \approx -\left(\frac{\partial \gamma_{M_A}}{2RT \partial \ln [CsCl]} \right)_{E_V, T, P} \qquad (526)$$

The results (Table 2) obtained by measuring the surface tension as a function of the electrolyte concentration are plotted in Fig.82 against the logarithm of the concentration. A satisfactorily good linear correlation is produced and from the best-fitted slope the surface excess concentration of Cs^+ is found from the calculation below as $7.1\times10^{-7}\,\mathrm{mol\cdot m^{-2}}$.

$$\Gamma_+ = \frac{3.533}{2\times8.31\times298}\times10^{-3}\,\mathrm{N\cdot m^{-1}\cdot J^{-1}\cdot mol}$$
$$= 7.1\times10^{-7}\,\mathrm{mol\cdot m^{-2}}$$

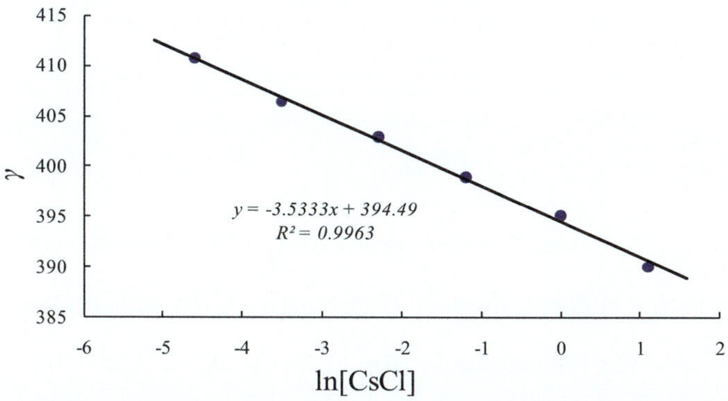

Fig.82 Electrolyte concentration dependence of surface tension.
electrode potential: -1.0V,
reference electrode: calomel electrode.

At this electrode potential (−1.0V), the mercury electrode is negatively charged (see (b)) and, hence, the present result supports that an electrified double layer is formed at the electrode interface. When a hydrogen ion electrode is employed as the reference electrode, it is possible to show that the surface excess concentration of Cl^- can be found by following procedures similar to those employed above.

Ans. surface excess concentration of Cs^+: $7.1\times10^{-7}\,\mathrm{mol\cdot m^{-2}}$

[advanced problem] Assuming that the electrode interface can be modeled as a parallel plate capacitor with an area *A* and an interval *d*, show that the electrocapillary curve is given by a parabola according to the *Lippmann* equation. Experimentally observed electrocapillary curves are, however, left-right unsymmetrical (see Fig.81), reflecting the fact that the electric double layers have more complicated structures than parallel plate capacitors.

Solution 4-79

(a) It is assumed that *Kohlrauch*'s law is valid to describe the molar conductivity of individual ions. Then it is proportional to the square root of the concentration of each ion and is given by eq(527), in which K^+ and K^- are relevant proportional constants, and α is the degree of dissociation.

$$\left.\begin{aligned}\lambda^+ &= \lambda_0^+ - K^+\sqrt{C_0\alpha}\\ \lambda^- &= \lambda_0^- - K^-\sqrt{C_0\alpha}\end{aligned}\right\}\qquad(527)$$

Employing the law of independent migration of ions, from eq(527) the molar conductivity of acetic acid is given by eq(528), in which the constants are replaced by $\Lambda_0 = \lambda_0^+ + \lambda_0^-$ and $K^\pm = K^+ + K^-$.

$$\begin{aligned}\Lambda &= \lambda^+ + \lambda^-\\ &= \lambda_0^+ - K^+\sqrt{C_0\alpha} + \lambda_0^- - K^-\sqrt{C_0\alpha}\\ &= \lambda_0^+ + \lambda_0^- - (K^+ + K^-)\sqrt{C_0\alpha}\\ &= \Lambda_0 - K^\pm\sqrt{C_0\alpha}\qquad(528)\end{aligned}$$

Since the equilibrium constant K_a is given by eq(529) in terms of α, eq(529) is solved for α, yielding eq(530) as a meaningful solution.

$$K_a = \frac{(C_0\alpha)^2}{C_0(1-\alpha)} = \frac{C_0\alpha^2}{(1-\alpha)} \qquad (529)$$

$$\alpha = \frac{-K_a + \sqrt{K_a^2 + 4K_aC_0}}{2C_0} \qquad (530)$$

Consequently, the molar conductivity of acetic acid in water is found by substituting eq(530) into eq(528) to be given by eq(531).

$$\Lambda = \Lambda_0 - K^{\pm} \times \left(\frac{-K_a + \sqrt{K_a^2 + 4K_aC_0}}{2} \right)^{1/2} \qquad (531)$$

Ans. molar conductivity of acetic acid: eq(531)

(b) Λ given by eq(531) is plotted against the molar concentration of acetic acid C_0 by substituting the value of the dissociation constant $K_a = 1.8 \times 10^{-5}$ and the limiting molar conductivity $\Lambda_0 = 39.05 \times 10^{-3}$ into eq(531) for a few values for $K^{\pm}$ and the obtained results are shown in Fig.83.

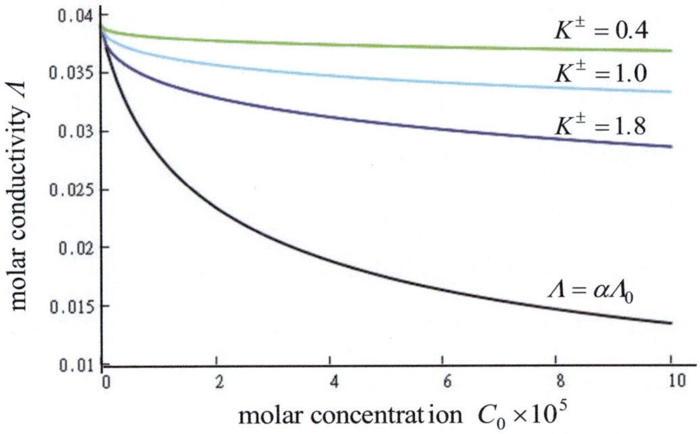

Fig.83 Λ vs. C_0 plot.

Ans. concentration dependence of molar conductivity of acetic acid: Fig.83

(c) According to eq(531), the inverse of the molar conductivity can be approximated by eq(532) when $C_0 << 1$ and, therefore, it is found approximately proportional to the concentration.

$$\frac{1}{\Lambda} = \frac{1}{\Lambda_0 - K^{\pm} \times \left(\frac{-K_a + \sqrt{K_a^2 + 4K_aC_0}}{2} \right)^{1/2}} \approx \frac{1}{\Lambda_0} \left\{ 1 + \frac{K^{\pm}}{\Lambda_0} \times \left(\frac{-K_a + \sqrt{K_a^2 + 4K_aC_0}}{2} \right)^{1/2} \right\}$$

$$\approx \frac{1}{\Lambda_0} \left\{ 1 + \frac{C_0K^{\pm}}{\Lambda_0} \right\} \qquad (532)$$

Ans. eq(532)

Based on eq(531), the inverse of the molar conductivity is plotted in Figs. 84 and 85 against the concentration in its extremely low range.

Ostwald's dilution law (*W.Ostwald,* Nobel Prize in Chemistry in 1909) is addressed in **problem 4-68** and whether the law is applicable or not is judged by examining a proportionality between $1/\Lambda$ and ΛC_0. Assuming that the conductivity is given by eq(531), the plot of $1/\Lambda$ against ΛC_0 not only gives a very poor linearity but also yields an intercept far from $1/\Lambda_0$. It is concluded, therefore, that there does not exist consistency between *Kohlrausch*'s law and *Ostowald*'s dilution law. See **[additional problem]** given at the end of this chapter.

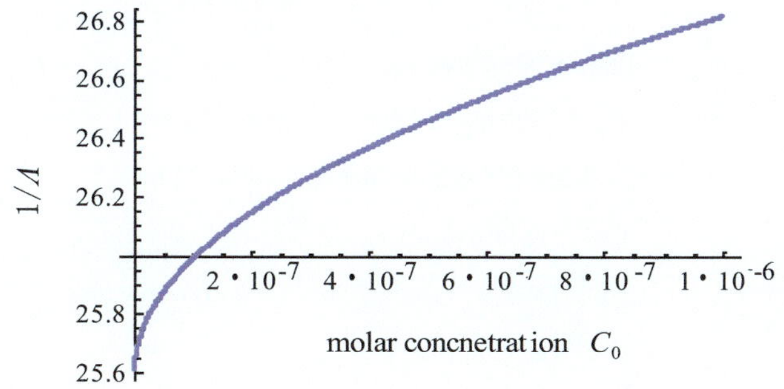

Fig.84 $1/\Lambda$ vs. C_0 plot of extremely dilute acetic acid solutions. The dissociation of water is not considered.

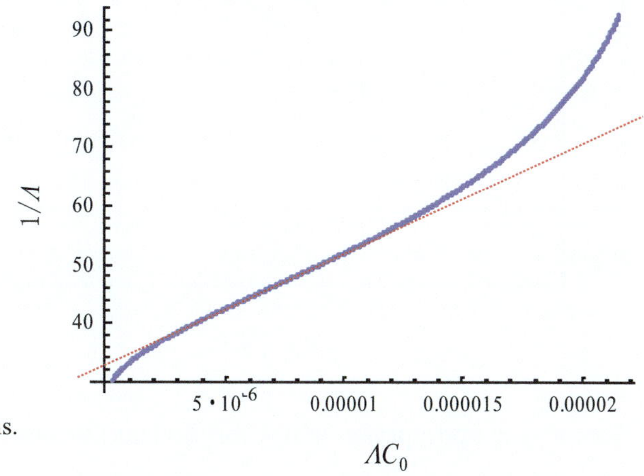

Fig.85 $1/\Lambda$ vs. ΛC_0 plot of dilute acetic acid solutions. In the concentration range higher than that in Fig.84.

Solution 4-80

The concentrations of Cu^{2+}, $Cu^{2+}(NH_3)$, $Cu^{2+}(NH_3)_2$, $Cu^{2+}(NH_3)_3$, $Cu^{2+}(NH_3)_4$ and NH_3 are expressed as y_0, y_1, y_2, y_3, y_4 and x, respectively. Based on the four-step equilibrium given in the problem, these concentrations are related through the four equilibrium constants as are given in eq(533).

$$\left.\begin{aligned} K_1 &= \frac{y_1}{y_0 \cdot x} \\ K_2 &= \frac{y_2}{y_1 \cdot x} = \frac{y_2}{y_0 \cdot K_1 \cdot x^2} \\ K_3 &= \frac{y_3}{y_2 \cdot x} = \frac{y_3}{y_0 \cdot K_1 \cdot K_2 \cdot x^3} \\ K_4 &= \frac{y_4}{y_3 \cdot x} = \frac{y_4}{y_0 \cdot K_1 \cdot K_2 \cdot K_3 x^4} \end{aligned}\right\} \Longrightarrow \left.\begin{aligned} y_1 &= y_0 \cdot K_1 \cdot x \\ y_2 &= y_0 \cdot K_1 \cdot K_2 \cdot x^2 \\ y_3 &= y_0 \cdot K_1 \cdot K_2 \cdot K_3 \cdot x^3 \\ y_4 &= y_0 \cdot K_1 \cdot K_2 \cdot K_3 \cdot K_4 \cdot x^4 \end{aligned}\right\} \quad (533)$$

Since the total concentration of the copper ion is constant as is given by eq(534), the substitution of eq(534) into eq(533) leads to eq(535).

$$y_0 + y_1 + y_2 + y_3 + y_4 = C_0 \quad (534)$$

$$y_0(1 + K_1 \cdot x + K_1 \cdot K_2 \cdot x^2 + K_1 \cdot K_2 \cdot K_3 \cdot x^3 + y_0 \cdot K_1 \cdot K_2 \cdot K_3 \cdot K_4 \cdot x^4) = C_0 \quad (535)$$

Then, from eqs(533) and (535), the concentration of each coordination complex is found as given by eq(536).

$$\left.\begin{aligned} y_0 &= \frac{C_0}{(1 + K_1 \cdot x + K_1 \cdot K_2 \cdot x^2 + K_1 \cdot K_2 \cdot K_3 \cdot x^3 + y_0 \cdot K_1 \cdot K_2 \cdot K_3 \cdot K_4 \cdot x^4)} \\ y_1 &= \frac{C_0 \cdot K_1 \cdot x}{(1 + K_1 \cdot x + K_1 \cdot K_2 \cdot x^2 + K_1 \cdot K_2 \cdot K_3 \cdot x^3 + y_0 \cdot K_1 \cdot K_2 \cdot K_3 \cdot K_4 \cdot x^4)} \\ &\ \ \vdots \\ y_4 &= \frac{C_0 \cdot K_1 \cdot K_2 \cdot K_3 \cdot K_4 \cdot x^4}{(1 + K_1 \cdot x + K_1 \cdot K_2 \cdot x^2 + K_1 \cdot K_2 \cdot K_3 \cdot x^3 + y_0 \cdot K_1 \cdot K_2 \cdot K_3 \cdot K_4 \cdot x^4)} \end{aligned}\right\} \quad (536)$$

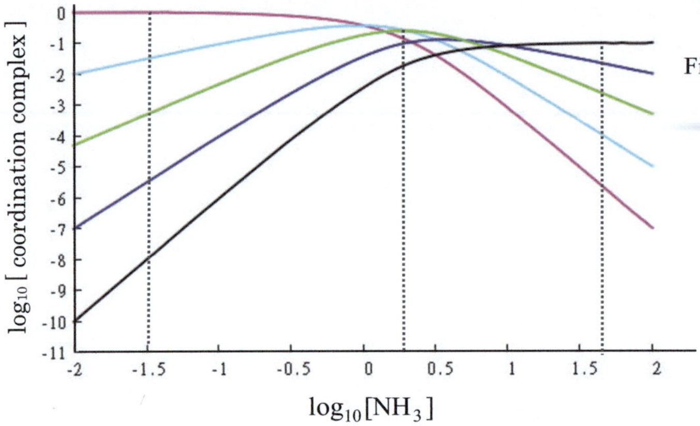

Fig.86 Concentrations of Cu-ammonium complexes plotted against the ammonium concentration. Employed values: $C_0 = 1.0$, $K_1 = 1.0,\ K_2 = 0.5,\ K_3 = 0.2,\ K_4 = 0.1.$ Each species is indicated by a colored line. Cu^{2+} (reddish purple), $Cu^{2+}(NH_3)$ (sky blue), $Cu^{2+}(NH_3)_2$ (green), $Cu^{2+}(NH_3)_3$ (navy blue), $Cu^{2+}(NH_3)_4$ (black).

Employing appropriate values for the equilibrium constants, the concentration of each complex is plotted against the logarithm of the concentration of ammonium in Fig.86 (log-log plot). From this figure it is easy to tell with which number of ammoniums coordinated complex dominates in solution in a certain concentration range. At the three concentrations of ammonium indicated by dotted vertical lines in Fig.86, the dominant species is Cu^{2+}, $Cu^{2+}(NH_3) + Cu^{2+}(NH_3)_2$ and $Cu^{2+}(NH_3)_4$ from left to right.

Ans. concentration variation of Cu-NH$_3$ complex with NH$_3$: eq(536) and Fig.86

Solution 4-81

Under a constant *T* and *P*, the energy of work put in and out of a system is equal to the *Gibbs* free energy change. Therefore, the difference of the free energies between the two systems of (the saturated solution) and (pure solvent + pure solute) becomes equal to the minimum energy of work required or the maximum energy gained upon the preparation of the saturated solution. The free energy of each system is readily found from the chemical potentials of the substances involved in the preparation process.

The chemical potentials of the pure solvent and the pure solute are nothing more than the corresponding molar free energies and are denoted μ_l^0 and μ_s^0, with the subscript *l* standing for liquid and the subscript *s* for solute. The mole numbers of the solvent and solute of the solution are n_l and n_s. The chemical potential μ_s of the solute in the saturated solution is expressed by eq(537) with the activity denoted a_s. On the other hand, the chemical potential μ_l of the solvent of the saturated solution can be approximated by eq(538).

$$\mu_s = \mu_s^0 = \mu^\bullet + RT\ln a_s \qquad (537)$$

$$\mu_l = \mu_l^0 + RT\ln\frac{n_l}{n_l + n_s} \approx \mu_l^0 - RT\frac{n_s}{n_l} \qquad (538)$$

Employing these equations, the free energy difference $\varDelta G$ between the two systems of (the saturated solution) and (pure solvent + pure solute) is calculated from eq(539) and is found equal to $-n_sRT$.

$$\begin{aligned}\varDelta G &= n_l\mu_l + n_s(\mu^\bullet + RT\ln a_s) - n_l\mu_l^0 - n_s\mu_s^0 \qquad (539)\\ &= n_s(\mu^\bullet + RT\ln a_s) + n_l\left(\mu_l^0 - RT\frac{n_s}{n_l}\right) - (n_s\mu_s^0 + n_l\mu_l^0)\\ &= n_s(\mu^\bullet + RT\ln a_s - RT - \mu_s^0) = -n_sRT\end{aligned}$$

Since the found free energy change is negative, the system can do work via dissolution and the maximum work obtainable is given by eq(540), where V is the volume of the solution and c_s is the concentration saturated.

$$w_{\max} = |\Delta G| = n_s RT = c_s VRT \qquad (540)$$

Ans. maximum work : $w_{max} = n_s RT = c_s VRT$

Solution 4-82

For a solution of concentration C, the chemical potentials μ_s for the solute and μ_l for the solvent are given by eq(541) and eq(542), respectively, in which n_s and n_l are the mole numbers for the solute and the solvent, respectively.

$$\mu_s = \mu^\circ + RT\ln C \qquad (541)$$

$$\mu_l = \mu_l^0 + RT\ln\frac{n_l}{n_l + n_s} \approx \mu_l^0 - RT\frac{n_s}{n_l} \qquad (542)$$

Then, the free energy G_1 of the solution is given by eq(543).

$$G_1 = n_l\mu_l + n_s(\mu^\circ + RT\ln C) = n_l\mu_l^0 - n_s RT + n_s(\mu^\circ + RT\ln C) \qquad (543)$$

After removing a part of the solvent by the amount Δn_l, the concentration of the remained solution has become C' and the chemical potential of the solution has become μ_l'. Then the free energy G_2 of a system composed of (the remained solution of the concentration C') plus (the removed solvent) is given by eq(544') via eq(544).

$$G_2 = (n_l - \Delta n_l)\mu_l' + n_s(\mu^\circ + RT\ln C') + \Delta n_l\mu_l^0 \qquad (544)$$

$$= (n_l - \Delta n_l)\left(\mu_l^0 - RT\frac{n_s}{n_l - \Delta n_l}\right) + n_s(\mu^\circ + RT\ln C') + \Delta n_l\mu_l^0$$

$$= n_l\mu_l^0 - n_s RT + n_s(\mu^\circ + RT\ln C') \qquad (544')$$

As a consequence, the free energy change upon removing a part of the solvent is given by $G_2 - G_1$ and is expressed by eq(545).

$$\Delta G = G_2 - G_1 = n_s RT\ln\frac{C'}{C} > 0 \qquad (545)$$

A positive value of ΔG signifies that some work is required to remove a part of solvent and the minimum work necessary is given by $w_{min} = n_s RT\ln C'/C$.

Ans. minimum work: $w_{min} = n_s RT\ln C'/C$

Solution 4-83

(a) *Gauss*' law is given by eq(546) in which $\boldsymbol{S}$ is a closed surface around a charged body, $\boldsymbol{E}$ is an electric field at a surface segment $\boldsymbol{dS}$ and is perpendicular (nominal) to it, Q is the total charge of the charged body and ε_r is a specific dielectric constant of a medium.

$$\oint_S (\boldsymbol{E}\cdot\boldsymbol{dS}) = \frac{Q}{\varepsilon_0\varepsilon_r} \qquad (546)$$

Now the total charge in the electrode is Q_1 and a cuboid $(a, b >> c)$ which is just large enough to enclose the electrode plate entirely is considered and its surface is denoted $\boldsymbol{S}_1$. Then, eq(546*a*) is obtained from eq(546).

$$\oint_{S_1} E_1 \cdot dS_1 = \frac{Q_1}{\varepsilon_0 \varepsilon_r} \quad (546a), \qquad \oint_{S_1} E_2 \cdot dS_2 = \frac{Q_2}{\varepsilon_0 \varepsilon_r} \quad (546b)$$

Integrating eq(546*a*) over areas of two cuboid planes, denoted $A (= a \times b)$ each and parallel to the electrode plate, eq($547a$) is obtained by neglecting the integral over the planes perpendicular to the electrode since their areas are small enough to be neglected $(a \times b >> a \times c,\ b \times c)$.

$$E_1 = \frac{1}{2A} \frac{Q_1}{\varepsilon_0 \varepsilon_r} = \frac{\sigma_1}{2\varepsilon_0 \varepsilon_r} \quad (547a), \qquad E_2 = \frac{1}{2A} \frac{Q_2}{\varepsilon_0 \varepsilon_r} = \frac{\sigma_2}{2\varepsilon_0 \varepsilon_r} \quad (547b)$$

Similarly, another cuboid $(a,b >> c)$ is considered which encloses the negatively charged layer alone thoroughly. Then, it is obvious that eq($547b$) is derived from eq($546b$). Since the electric double layer is electrically neutral as a whole, the relation $Q_1 = -Q_2$ is satisfied and, hence, the surface density is given by $\sigma = \sigma_1 = -\sigma_2$. Consequently, the electric field inside the double layer is given by eq(548).

$$E = E_1 - E_2 = \frac{\sigma}{\varepsilon_0 \varepsilon_r} \quad (548)$$

The reason why the electric field is given by $E_1 - E_2$ is because the direction of the field is taken as from the electrode toward the ion layer. (Consider which way the field given by eq($546a$) or eq($546b$) is directed, inward or outward of the cuboid?)

From eq(548), it is seen that the electric field within the electric double layer is constant and the associated potential is obtained by integrating eq(549) at constant *E* as is given by eq(550).

$$\nabla \phi(x) = -E \quad (549)$$

$$\phi(x) = -Ex + Const. \quad (550) \quad (\text{a solution for } \rho(x) = 0)$$

Eqs(549) and (550) are solutions of the one dimensional *Poisson*'s equation given by eq(551) when the charge density within the electric double layer is put $\rho(x) = 0$.

$$\nabla^2 \phi(x) = -\frac{\rho(x)}{\varepsilon_0 \varepsilon_r} \quad (551)$$

Ans. electric field within double layer: eq(548), potential: eq(550)

(b) Three dimensional *Poisson*'s equation is given by eq(552). The coordinate *x* is taken along a direction vertical to the electrode

$$\nabla^2 \phi(x,y,z) = -\frac{\rho(x,y,z)}{\varepsilon_0 \varepsilon_r} \quad (552)$$

The charge density within the electrified double layer depends on neither *y* nor *z* coordinate. Therefore, the charge density and the potential there can be expressed as eq(553).

$$\phi(x,y,z) = \phi(x,\alpha,\beta) \equiv \phi(x), \qquad \rho(x,y,z) = \rho(x,\alpha,\beta) \equiv \rho(x) \quad (553)$$

Then, *Poisson*'s equation can be written as an equation containing the variable *x* alone as is given by eq(552').

$$\frac{d^2}{dx^2} \phi(x) = -\frac{\rho(x)}{\varepsilon_0 \varepsilon_r} \quad (552') \quad (\text{see "note"})$$

On the other hand, both positive and negative ions within the electrified double layer distribute according to the *Boltzmann* distribution under the potential $\phi(x)$. The number of positive or negative ions per unit volume at the position *x,* labeled $N_+(x)$ or $N_-(x)$, is given by eq(554) or eq(554'), where $\overline{N}_+$ or $\overline{N}_-$ represents the number of relevant ions in the bulk.

$$N_+(x) = \overline{N}_+ \exp[-eZ_+\phi(x)/k_BT] \qquad (554)$$

$$N_-(x) = \overline{N}_- \exp[-eZ_-\phi(x)/k_BT] \qquad (554')$$

In a case of a 1:1 electrolyte, putting $\overline{N}_+ = \overline{N}_- \equiv \overline{N}$, the charge density $\rho(x)$ at the position *x* is given by eq(555), which can be approximated by eq(555') because $\phi(x)$ is small as compared with the thermal energy k_BT.

$$\begin{aligned}\rho(x) &= N_+(x)eZ_+ + N_-(x)eZ_- \\ &= \overline{N}e\langle Z_+ \exp[-eZ_+\phi(x)/k_BT] + Z_- \exp[-Z_-e\phi(x)/k_BT]\rangle \qquad (555) \\ &= \overline{N}e\langle Z_+ - eZ_+^2\phi(x)/k_BT + Z_- - eZ_-^2\phi(x)/k_BT + O((\phi(x)/k_BT)^2)\rangle \\ &\approx -\overline{N}e^2(Z_+^2 + Z_-^2)\phi(x)/k_BT \qquad (555')\end{aligned}$$

Denoting the molar concentration *n* and the molal concentration *m*, eq(555') is rearranged using the relation $\overline{N} = nN_A = mdN_A$ to express the charge density as is given by eq(555'').

$$\rho(x) \approx -2N_A{}^2e^2d \cdot I \cdot \phi(x)/RT \qquad (555'')$$

Here, *I* is the ionic strength defined by the left sided of eq(556) using the molal concentration m_i of the *i*-th ion in general and in the present case given by the right side.

$$I = \frac{1}{2}\sum_i^{\text{all ionic species}} m_i Z_i^2 = \frac{1}{2} m \cdot (Z_+^2 + Z_-^2) \qquad (556)$$

Employing eq(555''), eq(552') can be rewritten as eq(557).

$$\frac{d^2\phi(x)}{dx^2} = \frac{2dN_A^2e^2I}{\varepsilon_0\varepsilon_rRT}\phi(x) = \Lambda\phi(x), \qquad \Lambda \equiv \frac{2dN_A^2e^2I}{\varepsilon_0\varepsilon_rRT} \equiv \frac{1}{r_D^2} \qquad (557)$$

Solving the above differential equation, two functions of $\exp(-\sqrt{\Lambda}x)$ and $\exp(\sqrt{\Lambda}x)$ are obtained as its special solutions. However, since the potential function as a solution should not diverge in the limit of $x \to \infty$, the former function is adapted as the solution and $\phi(x)$ is given by eq(558) and the electric field *E* is given by eq(559). A proportional constant α is determined by the boundary conditions as is described later.

$$\phi(x) = \alpha\exp(-\sqrt{\Lambda}x) = \alpha\exp\left(-\frac{x}{r_D}\right) \qquad (558)$$

$$E(x) = -\frac{d\phi(x)}{dx} = \frac{\alpha}{r_D}\exp\left(-\frac{x}{r_D}\right) \qquad (559)$$

The parameter r_D defined in eq(557) in terms of various constants such as the ionic strength, the dielectric constant, etc., is named *Debye* length and is an important measure to know how the potential and the electric field increase or decrease with the distance *x* from the positively or negatively charged electrode (see Fig.87).

Looking at eqs(555''), (558) and (559), it is seen that all of $\rho(x)$, $\phi(x)$ and $E(x)$ vary exponentially with $1/r_D$ as their changing rate parameter. This is confirmed below more specifically by obtaining approximate values for the constant.

First of all, the ion concentration distribution is normalized by eq(560) using eq (555''). This is because the total excess charge distributed in solution near the electrode, when converted per unit area ($\int dydz = 1$), should balance in the absolute value, but opposite in sign, to the electrode charge density.

$$\int_0^\infty \rho(x)dx = -\int_0^\infty 2N_A^{\;2}e^2 d\cdot I\cdot\phi(x)/RTdx = -\sigma \qquad (560)$$

Substituting $\phi(x) = \alpha e^{-x/r_D}$ into eq(560) to find α, eq(561) is obtained.

$$\int_0^\infty \rho(x)dx = -\frac{2\alpha N_A^{\;2}e^2 d\cdot I}{RT}\int_0^\infty e^{-x/r_D}dx$$

$$= -\frac{2\alpha N_A^{\;2}e^2 dIr_D}{RT} = -\sigma \;\Rightarrow\; \alpha = \frac{\sigma RT}{2N_A^{\;2}e^2 dIr_D} \qquad (561)$$ (see supplementary problem)

Putting $d = 1$ and $I = 1$ in eq(561), eq(561') is obtained as an approximate value for α.

$$\alpha \approx \frac{\sigma RT}{2N_A^{\;2}e^2 r_D} = \frac{\sigma\sqrt{2RT}}{2N_A e\sqrt{\varepsilon_0\varepsilon_r}} \qquad (561')$$

Also、an approximate value for $1/r_D$ is given by eq(562).

$$\frac{1}{r_D} = \frac{N_A e\sqrt{2d}}{\sqrt{\varepsilon_0\varepsilon_r RT}} \approx \frac{N_A e\sqrt{2}}{\sqrt{\varepsilon_0\varepsilon_r RT}} \qquad (562)$$

From these two equations, α/r_D can be expressed as eq(563).

$$\frac{\alpha}{r_D} \approx \frac{\sigma\sqrt{2RT}}{2N_A e\sqrt{\varepsilon_0\varepsilon_r}} \times \frac{N_A e\sqrt{2}}{\sqrt{\varepsilon_0\varepsilon_r RT}} = \frac{\sigma}{\varepsilon_0\varepsilon} \qquad (563)$$

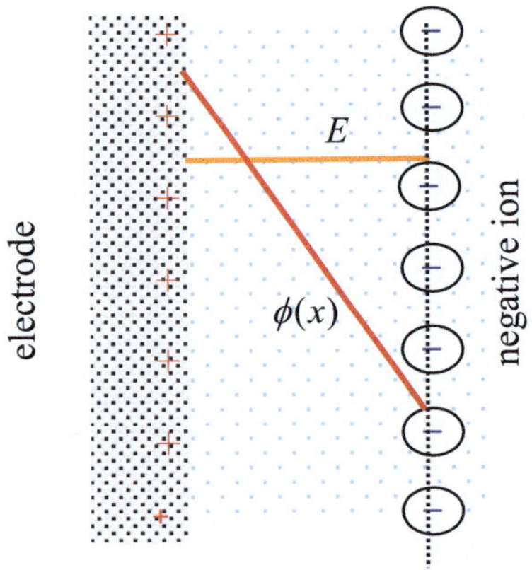

Fig.87(a) Electrical potential and field of electrical double layer.

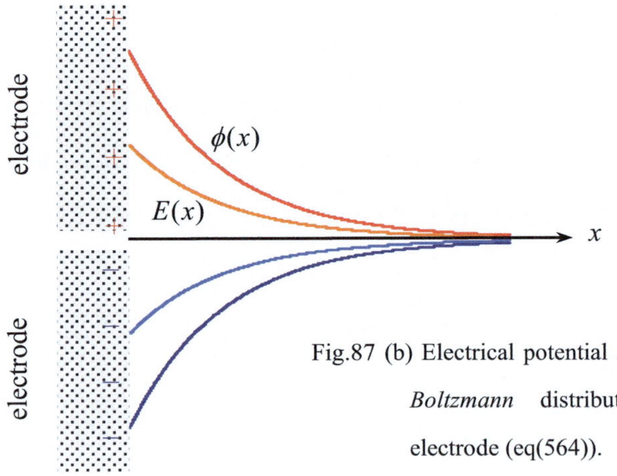

Fig.87 (b) Electrical potential and field arising from *Boltzmann* distributed ions near the electrode (eq(564)).
curves in warm color: + charged electrode,
curves in cold color: – charged electrode.

Based on these results, it is now seen that all of $\rho(x)$, $\phi(x)$ and $E(x)$ are simply expressed in terms of σ and r_D as is given by eq(564).

$$\rho(x) = -\frac{\sigma}{r_D}e^{-x/r_D}, \qquad \phi(x) = \frac{\sigma\cdot r_D}{\varepsilon_0\varepsilon}e^{-x/r_D}, \qquad E(x) = \frac{\sigma}{\varepsilon_0\varepsilon}e^{-x/r_D} \qquad (564)$$

For the two types of double layers addressed in (a) and (b) above, variations of the potential and the electric field with the distance from electrodes are shown schematically in Figs.87 (a) and (b), respectively. The (double) layer treated in (b) is called diffusion (electrical double) layer.

Ans. electrical field in the diffusion (double) layer : eq(559), potential: eq(558) or eq(564)

"note" Superficially, $\phi(x)$ appears to be a one-dimensional potential but it is not. This is readily seen when the potential arising from a one-dimensional charge distribution along the x-axis is derived and compared with $\phi(x)$.

[supplementary problem] Confirm that the dimension of α in eq(561) is $\mathrm{J \cdot C^{-1}}$.

[answer]

$$[\alpha]=\left[\frac{\sigma RT}{2N_A^{\,2}e^2 dIr_D}\right]=\left[\frac{\mathrm{C \cdot m^{-2} \cdot J \cdot mol}}{\mathrm{mol^{-2} \cdot C^2 \cdot kg \cdot m^{-3} \cdot mol \cdot kg^{-1} \cdot m}}\right]=\left[\frac{\mathrm{J}}{\mathrm{C}}\right]=[\mathrm{V}]$$

Solution 4-84

(a) Regarding the cell membrane as a parallel-plate capacitor (condenser), its capacitance C is given by eq(565). Then, substituting the given values into it, the following value is found for C.

$$C=\frac{\varepsilon_0\varepsilon_r A}{d} \quad (565) \;\Rightarrow\; \frac{C}{A}=\frac{\varepsilon_0\varepsilon_r}{d}=\frac{8\times 8.85\times 10^{-12}}{10^{-8}}=7.08\times 10^{-3}\mathrm{C^2 \cdot J^{-1} \cdot m^{-2}}$$

$$=7.08\times 10^{-3}\mathrm{F \cdot m^{-2}}$$

Thus, the required value per $1.0\,\mathrm{cm^2}$ becomes $7.08\times 10^{-7}\,\mathrm{F \cdot cm^{-2}} \approx 0.71\,\mu\mathrm{F \cdot cm^{-2}}$.

Ans. capacitance of the cell membrane: $0.71\,\mu\mathrm{F \cdot cm^{-2}}$

(b) The electrical energy which a capacitor with a capacitance C is able to store at a voltage V is given by eq(566). The answer is readily found from this equation as is given below.

$$[\text{electrical energy}]=\frac{1}{2}CV^2 \quad (566) \;\Rightarrow\; \frac{1}{2}\times 7.08\times 10^{-3}\times 0.1^2=3.54\times 10^{-5}\mathrm{J \cdot m^{-2}}$$

Ans. electrical energy to be stored in the membrane: $3.54\,\mathrm{nJ \cdot cm^{-2}}$

(c) Under a constant electric field E, the potential difference between two points separated by a distance d is given by Ed and, in the present problem, this is equal to 0.1V. Then, the electric field is found as $E=10^7\,\mathrm{V \cdot m^{-1}}$ as is calculated below.

$$E\times d=E\times 10^{-8}=0.1 \Rightarrow \quad E=0.1/10^{-8}=10^7\,\mathrm{V \cdot m^{-1}}$$

Also, an amount of charges on the surface of the cell membrane is obtained by substituting the value of the capacitance of the membrane found above into the following equation or directly from the equation given under it. The found value is $0.71\times 10^{-7}\,\mathrm{C \cdot cm^{-2}}$.

$$C\cdot V=Q \;\Rightarrow\; \frac{Q}{A}=\frac{C}{A}\times V=0.71\times 10^{-6}\times 0.1=0.71\times 10^{-7}\,\mathrm{C \cdot cm^{-2}}$$

$$\sigma=\varepsilon_0\varepsilon_r E=8.85\times 10^{-12}\times 8\times 10^7=7.1\times 10^{-4}\,\mathrm{C \cdot m^{-2}}=0.71\times 10^{-7}\,\mathrm{C \cdot cm^{-2}}$$

Ans. electric field strength within the cell: $10^7\,\mathrm{V \cdot m^{-1}}$, surface charge: $0.71\times 10^{-7}\,\mathrm{C \cdot cm^{-2}}$

(d) The potential difference given is 0.1 V and noting that its units are given by J/C, the energy (J) is converted to the one per a hydrogen ion as is given below.

$$\frac{0.1\times 96500}{6.02\times 10^{23}} = 1.6\times 10^{-20}\,\text{J}$$

Ans. energy required or dissipated per hydrogen ion: 1.6×10^{-20} J

(e) Since the electrical potential difference is generated because of a difference of the chemical potentials of the solutions outside and inside of the membrane, its absolute value is given by eq(567).

$$|\Delta E| = \left|\frac{RT}{F}\right| \left|\ln\frac{[\text{H}^+]_{out}}{[\text{H}^+]_{in}}\right| \qquad (567)$$

Adapting $|\Delta E| = 0.1\text{V}$, $\left|\ln[\text{H}^+]_{out}/[\text{H}^+]_{in}\right| = 3.90$ is obtained from eq(567) at room temperature and, from this, the ratio of the concentrations is found as 49. Notice that the units of RT/F are J/C, namely equivalent to the potential (difference) V.

Ans. ratio of concentrations (inside vs. outside): 49

Solution 4-85

(a) The problem is concerned with *Einstein*'s A and B coefficients about the radiation. The coefficient B is associated with the induced radiation transition probability. It is obvious that the transition probabilities between the states n (lower in energy) and m (higher in energy), denoted $B_{n\to m}$ for the absorption and $B_{n\leftarrow m}$ for the emission, are the same. A transition probability from the excited state is given by the sum of the induced radiation $B_{n\leftarrow m}$ and the spontaneous radiation given by $A_{n\leftarrow m}$. Since the latter is proportional to the transition frequency cubed ($\nu^3_{n\leftarrow m}$), the frequency of the transition energy labeled $\nu_{n\leftarrow m}$, $A_{n\leftarrow m}$ can be neglected as compared with $B_{n\leftarrow m}$ in the infrared and far infrared range where the transition frequencies are small. There is no need, therefore, to take the A coefficient into account for the matters in which phenomenon associated with the absorption and emission of the radiation in such low frequencies as in the infrared and far infrared range. As a result of this, we can safely put $\beta_e = \beta_a$. See **solution 7-12** for more details. *Arrhenius*' premise on the absorption coefficient had been presented quite long before the quantum theory was established for the first time.

Ans. See text (a).

(b) In order to find the rate of the temperature change against β, eq(39') given in the problem and cited below again is differentiated with β and then divided by T^4,

$$T^4 = \frac{C}{\langle 1+(1-\beta)\nu\rangle} \qquad (39') \qquad \text{(reposted)}$$

obtaining eq(568). From its rightest equation, the rate is seen to be expressed by an increasing function of ν (Notice here ν is not frequency.)

$$\frac{1}{T^4}\left(\frac{\partial T^4}{\partial \beta}\right)_\nu = \frac{\langle 1+(1-\beta)\nu\rangle}{C}\times\frac{C\nu}{\langle 1+(1-\beta)\nu\rangle^2} = \frac{\nu}{\langle 1+(1-\beta)\nu\rangle} \qquad (568)$$

Namely, eq(568) indicates that the larger the value of ν is, the greater becomes the rate of temperature rise against β.

In Fig.88, eq(568) is plotted against β for a few values of ν. From this figure too, it is seen that the temperature rise at the earth surface becomes greater for the larger values of ν. In another word, the global warming rate becomes faster, the more the warming progresses.

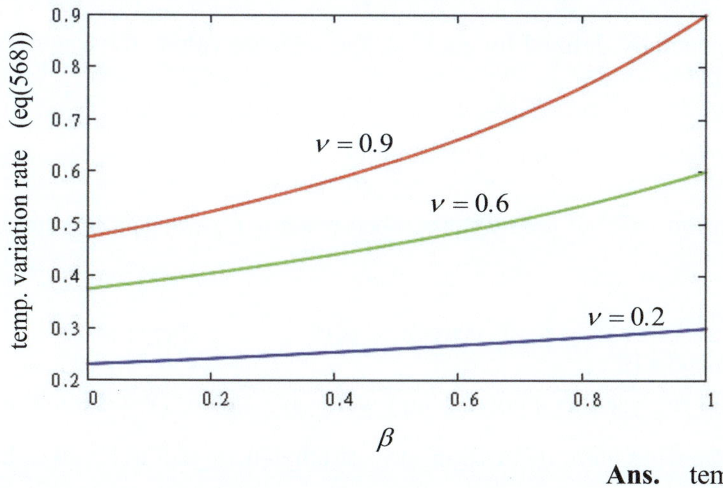

Fig.88 Variation rate of the temperature at the earth surface against absorption coeffi- cient β of radiation.

Ans. temperature variation rate: eq(568) and Fig.88

(c) When the temperature at the earth surface initially found at $T = 288$ varies to T' accompanied by a change of the absorption coefficient from β to β', eq(569) is derived from eq(39').

$$\frac{T'^4}{T^4} = \frac{(1+(1-\beta)\nu)}{(1+(1-\beta')\nu)} \qquad (569) \Rightarrow \frac{T'}{288} = \left(\frac{(1+(1-0.6)\times 0.9)}{(1+(1-0.7)\times 0.9)}\right)^{1/4} = 1.017$$

Substituting the given conditions of $\beta = 0.6$, $\beta' = 0.7$ and $\nu = 0.9$ into the above equation, $T = 293$ K is found. Consequently, it is seen that doubling the concentration of carbon dioxide in the air will result in a temperature rise by ca. 5° C.

Ans. magnitude of temperature rise: ca. 5° C

(d) About 70% of the earth surface is occupied by waters and their water vapor in the air is in equilibrium. with the waters. For this reason, the concentration of the water vapor in the air does not constantly decrease or increase greatly. On the other hand, the concentration of carbon dioxide is easy to increase constantly because of expanding human activities (see **"note 2"** in the problem). The presence of water vapor in the air, therefore, affects the magnitude of β but does not contribute to its variation. In other words, the water plays an important role in determining the temperature of the earth surface but it does not become a direct cause of the global warming because of its huge amount already existing on the earth. The paper published by *Arrhenius* more than 100 years ago is surprisingly suggestive indeed in considering the nature of "*the greenhouse gas effect*" (see **"note 4"** given in the problem).

Ans. See text (d) for the reason why water vapor is not called *greenhouse gas.*

Solution 4-86

(a) The mixed solution is assumed to behave ideally.

(i) If water (W) is regarded as a solvent and ethylene glycol (E) as a solute, then the freezing point depression ΔT of water is given by eq(570). Since the mole fraction of either component is not necessarily small, an approximate equation of logarithm $\ln X_W^l = \ln(1-X_E^l) \approx X_E^l$ is not applied (see **"note"**).

$$\frac{\Delta H_m}{T_m}\Delta T = RT_m \ln X_{\mathrm{W}}^l \;\Rightarrow\; \Delta T = \frac{RT_m^2}{\Delta H_m}\times \ln X_{\mathrm{W}}^l \qquad (570)$$

For ΔH_m and T_m here, the values for the latent heat and the freezing point of water, respectively are employed. Since the cryoscopic constant is generally defined by eq(571), the relevant values for water are substituted into eq(571) to find a value for K_f^{W}.

$$\frac{RT_m^2}{\Delta H_m} \equiv K_f \qquad (571)$$

Then, from eq(570), the freezing point depression ΔT^{W} of the mixture, when water is regarded as solvent, is given by eq(572).

$$\Delta T^{\mathrm{W}} \equiv K_f^{\mathrm{W}} \times \ln X_{\mathrm{W}}^l \left(= \frac{8.315\times 273.15^2}{6.008\times 10^3}\times \ln X_{\mathrm{W}}^l = 103.26\times \ln X_{\mathrm{W}}^l \right) \qquad (572)$$

(ii) Alternatively, ethylene glycol (E) can be regarded as a solvent and water as a solute (W). Following entirely the same discussion given in (i), the freezing point depression ΔT^{E} of ethylene glycol is found to be given by eq(572').

$$\Delta T^{\mathrm{E}} \equiv K_f^{\mathrm{E}} \times \ln X_{\mathrm{E}}^l \left(= \frac{8.315\times (273.15-13.0)^2}{2.778\times 10^3\times 4.184}\times \ln X_{\mathrm{E}}^l = 48.42\times \ln X_{\mathrm{E}}^l \right) \qquad (572')$$

The mole fractions, either of which is used in eq(572) or eq(572'), are given below.

$$X_{\mathrm{W}}^l = \frac{40/18.016}{40/18.016+100/62.07} = 0.579\ , \qquad X_{\mathrm{E}}^l = 1-X_{\mathrm{W}}^l = 0.421$$

Ans. freezing point depression: eq(572) for water, eq(572') for ethylene glycol

(b) The substitution of the mole fraction of water into eq(572) yields a value of −56.4 K as the freezing point depression of water.

$$\Delta T^{\mathrm{W}} = 103.26\times \ln X_{\mathrm{W}}^l = 103.26\times \ln 0.579 = -56.4\ \mathrm{K}$$

On the other hand, supposing that the same temperature depression is achieved by regarding ethylene glycol as the solvent, the cryoscopic constant K_f^{E} of ethylene glycol is calculated as given below.

$$\Delta T^{\mathrm{E}} = -56.4\ \mathrm{K} = K_f^{\mathrm{E}} \times \ln X_{\mathrm{E}}^l \;\Rightarrow\; K_f^{\mathrm{E}} = -\frac{56.4}{\ln 0.421} = 65.19\ \mathrm{K}$$

In contrast, its theoretical value calculated by eq(571) is 48.42 K as shown below (see **"note"**).

$$K_f^{\mathrm{E}} = \frac{8.315\times (273.15-13.0)^2}{2.778\times 10^3\times 4.184} = 48.42\ \mathrm{K}$$

Ans. freezing point depression: −56.4 K, cryoscopic constant K_f^{E}: 65.19 K, theoretical value: 48.42 K

"note 1" Dimension of the cryoscopic constant and its values. For dilute solutions, the cryoscopic constant is given as a proportional constant to the molal concentration (l molality) of a solute. Taking water as an example, when *m* mole of a solute is dissolved in 1 kg of water as a solvent, the freezing point depression is expressed by the equation below and its proportional constant is defined as the cryoscopic constant. Consequently, the constant employed in the present problem (eq(572)) is related to a well-known value of 1.860 given below, whose dimension is $\mathrm{K\cdot kg\cdot mol^{-1}}$. See **solutions 4-17**, **4-19** and **4-20** for more details.

$$\Delta T^{\mathrm{W}} \equiv K_f \times X_{\mathrm{B}}^l = K_f \times \frac{m_{\mathrm{B}}}{1000/18.016+m_{\mathrm{B}}} \approx K_f \times \frac{18.016}{1000}\times m_{\mathrm{B}} \;\Rightarrow\; K_f \times \frac{18.016}{1000} = 1.860\ \mathrm{K\cdot kg\cdot mol^{-1}}$$

"note 2" Employing the theoretically calculated cryoscopic constant for each solvent, the calculated freezing point curves (eqs(572) and (572')) are plotted against the mole fraction in Fig.89 below. At the mole fraction of $X_W^l = 0.585$, the two curves cross each other, giving the same freezing point.

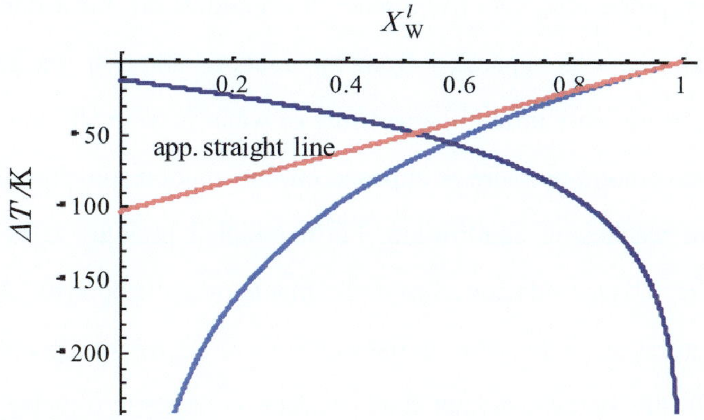

Fig.89 $\Delta T = K_f \ln X$ vs. X plot.

The orange straight line is an approximate straight line of $\ln X$.

The approximate linear relation is sufficient enough where the mole fraction of water is $\geq$ ca.0.8.

Solution 4-87

(a) In 1 L $(=1\,\text{kg})$ of the 3.3wt% seawater, all salts being regarded as NaCl, 33 g of NaCl are dissolved. From its number of moles, the molar concentration of the seawater is found as $0.565\times10^3\,\text{mol}\cdot\text{m}^{-3}$. Considering the osmotic pressure of the seawater, both ions of Na^+ and Cl^- contribute to it to the same extent and, hence, the concentration obtained above has to be doubled in calculating the osmotic pressure Π. Thus, its value is found as 27.7 atm from the following calculation.

$$\frac{33}{22.99+35.45} = 0.565\text{ mol} \Rightarrow c = 0.565\times10^3\,\text{mol}\cdot\text{m}^{-3}$$

$$\Pi = 2\times cRT = \frac{2\times0.565\times10^3\times8.31\times298}{101\times10^3} = 27.7\text{ atm}$$

Ans. osmotic pressure Π : 27.7 atm

(b) The concentration c' of the seawater at which the osmotic pressure reaches 50.0 atm is found as $1.02\times10^3\,\text{mol}\cdot\text{m}^{-3}$ from the calculation given below. The seawater is assumed incompressible.

$$c' = \frac{50(\text{atm})\times101\times10^3}{2\times RT} = \frac{50\times101\times10^3}{2\times8.31\times298} = 1.02\times10^3\,\text{mol}\cdot\text{m}^{-3}$$

Ans. concentration of the seawater under the osmotic pressure 50.0 atm: $1.02\times10^3\,\text{mol}\cdot\text{m}^{-3}$

(c) In order to change the concentration of the seawater initially provided to the concentration calculated above in (b), some amount of fresh water has to be removed from it. When a semi permeable osmosis membrane through which water alone passes freely is employed, it is possible to make the concentration of the seawater higher by transferring the more water from the seawater side to the fresh water side. This transfer continues until the osmotic pressure of the seawater becomes equilibrated with the applied hydrostatic pressure. Putting the volume of water of the seawater equilibrated at 50 atm equal to x, the volume x is found as $0.554\,\text{m}^3$ from the following calculation.

$$\frac{0.565\times10^3(\text{mol})}{x(\text{m}^3)} = 1.02\times10^3(\text{mol}\cdot\text{m}^{-3}) \Rightarrow x = 0.554\,\text{m}^3$$

Consequently the volume of water to be removed from 1 L of the seawater becomes 0.446 L and this is nothing more than the amount of fresh water to be obtained per 1L of seawater.

Ans. amount of pure water to be obtained per 1L of seawater: 0.446 L

(d) The osmotic pressure is one of the colligative properties. As has already been addressed, for instance, in **solution 3-52a**, in the system where a fresh water is separated from an aqueous solution via semi permeable osmosis membrane which only allows water pass through, a net flow of water is generally toward to the solution and when the pressure equal to the osmotic pressure is imposed on the solution somehow, the net flow of water becomes zero and the system reaches an equilibrium. Further higher pressure is being applied on the solution, the flow of water to the fresh water side comes to predominate over the opposite flow toward to the solution side. As a result, the concentration of the solution rises until its rising osmotic pressure balances with the applied hydrostatic pressure. In this process the net flow of water is opposite (reverse) to the direction of the net transport of water seen in ordinary osmosis phenomenon. This is the reason why the method discussed above is called "reverse" osmosis method. However, it is seen the principle itself is the same as the ordinary osmosis. Nevertheless, since for technological applications the semi-permeable membranes need to endure a few tens of atmospheric pressure, various types of membrane properties and shapes such as composed of micro hollow fibers have been invented so far. It is worth to mention that semi-permeable membranes are now in use in various technological applications other than purification of water.

Ans. reason for the naming: See text (d).

Solution 4-88

(a) The equilibrium constants of the multi-step equilibriums of citric acid given in the problem by from eq(42*a*) to eq(42*c*) are expressed as below.

$$\frac{[\mathrm{H_2Cit^-}][\mathrm{H^+}]}{[\mathrm{H_3Cit}]} = K_{a_1}, \quad \frac{[\mathrm{HCit^{2-}}][\mathrm{H^+}]}{[\mathrm{H_2Cit^-}]} = K_{a_2}, \quad \frac{[\mathrm{Cit^{3-}}][\mathrm{H^+}]}{[\mathrm{HCit^{2-}}]} = K_{a_3}$$

From these equations, the equilibrium constant for the reaction $\mathrm{H_3Cit} \rightleftarrows \mathrm{Cit^{3-}} + 3\mathrm{H^+}$, which is expressed as $K_a = [\mathrm{Cit^{3-}}]\cdot[\mathrm{H^+}]^3/[\mathrm{H_3Cit}]$, is found to be given by $K_a = K_{a_1} \times K_{a_2} \times K_{a_3} = 4.95 \times 10^{-15}$.

Ans. equilibrium constant for overall dissociation: $K_a = K_{a_1} \times K_{a_2} \times K_{a_3} = 4.95 \times 10^{-15}$

(b) From eq (40) to eq(42) given in the problem, the relevant equilibrium constants are expressed in terms of the concentrations of the substances involved as from eq(573) to eq(575), respectively.

$$K_{sp} = \frac{[\mathrm{Fe^{3+}}]\cdot[\mathrm{OH^-}]^3}{[\mathrm{Fe(OH)_3(s)}]} \quad (573), \quad K_f = \frac{[\mathrm{FeCit}]}{[\mathrm{Fe^{3+}}]\cdot[\mathrm{Cit^{3-}}]} \quad (574), \quad K_a = \frac{[\mathrm{Cit^{3-}}]\cdot[\mathrm{H^+}]^3}{[\mathrm{H_3Cit}]} \quad (575)$$

Since the dissolution equilibrium reaction of iron rust $\mathrm{Fe(OH)_3}$ is expressed as eq(43) given in the problem, its equilibrium constant is found by eq(576).

$$K_4 = \frac{[\mathrm{FeCit}]\cdot[\mathrm{H_2O}]^3}{[\mathrm{Fe(OH)_3(s)}]\cdot[\mathrm{H_3Cit}]} \quad (576)$$

Multiplying eqs(573), (574) and (575) all together, eq(577) is obtained and from this the equilibrium constant K_4 is found as eq(578).

$$K_f \times K_a \times K_{sp} = \frac{[\mathrm{FeCit}]\cdot([\mathrm{H^+}]\cdot[\mathrm{OH^-}])^3}{[\mathrm{Fe(OH)_3(s)}]\cdot[\mathrm{H_3Cit}]} = \frac{[\mathrm{FeCit}]\cdot[\mathrm{H_2O}]^3}{[\mathrm{Fe(OH)_3(s)}]\cdot[\mathrm{H_3Cit}]}\left(\frac{K_w}{[\mathrm{H_2O}]}\right)^3 \quad (577)$$

$$K_4 = \frac{K_f \cdot K_a \cdot K_{sp} \cdot [H_2O]^3}{K_w^3} \qquad (578)$$

Substituting the known values $K_w = 10^{-14}$ and $[H_2O] = 55.5\ mol \cdot L^{-1}$ as well as the given values for the relevant equilibrium constants and $K_a = 4.95 \times 10^{-15}$ obtained in (a) into eq(578), $K_4 = 1.37 \times 10^6$ is found.

Ans. equilibrium constant for iron rust dissolution: $K_4 = 1.37 \times 10^6$

(c) Putting the molar concentration of citric acid in the solution *a*, eq(579) is satisfied approximately from the law of conservation of substance since almost all of Cit^{-3} is considered to form the complex with Fe^{3+}.

$$[H_3Cit] + [H_2Cit^{-1}] + [HCit^{-2}] + [Cit^{-3}] + [FeCit] \approx [H_3Cit] + [FeCit] = a \qquad (579)$$

The concentration of ferric ion in the solution, denoted *x*, is related to the equilibrium constant K_4 by eq(580),

in which the activity of $Fe(OH)_3(s)$ is put equal to 1.

$$K_4 = \frac{[x] \cdot [H_2O]^3}{1 \times [0.1 - x]} \qquad (580)$$

Substituting the value for K_4 found in (b) into eq(580), $x = 0.089\ mol \cdot L^{-1}$ is found when $a = 0.1$. Consequently, the mass of iron rust that the citric acid solution can dissolve is found as $0.089 \times 106.9 = 9.5$ g for the molecular weight $Fe(OH)_3 = 106.9$.

Ans. mass of iron rust dissolvable in 1L of the citric acid solution: 9.5 g

[related problem] Silver chloride AgCl, hard to dissolve in solution, forms a complex ion with ammonium and a part of AgCl becomes readily dissolvable in solution. The solubility product of AgCl is given by eq(581) and its complex formation constant is given by eq(582).

$$AgCl(s) \rightleftarrows Ag^+ + Cl^- \qquad (581) \qquad K_{sp} = 1.8 \times 10^{-10}$$

$$Ag^+ + 2NH_3 \rightleftarrows Ag(NH_3)_2^+ \qquad (582) \qquad K_f = 1.7 \times 10^7$$

Find the molar concentration of AgCl dissolvable in solution as a function of the molar concentration of ammonium.

[brief solution]

The amount of AgCl (expressed in molar concentration) dissolvable is plotted against the molar concentration of ammonium in Fig.90 on the right.

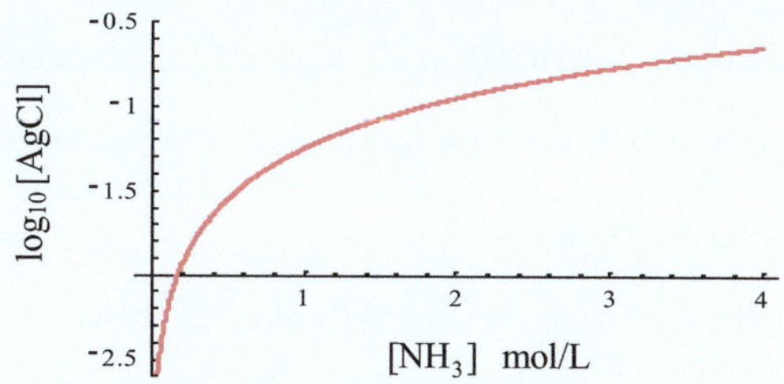

Fig.90 Dissolvable amount of AgCl plotted against $[NH_3]$.

Solution 4-89

For the (two-component) ideal solutions, the chemical potentials of the components A and B, denoted μ_A^{ide} and μ_B^{ide}, respectively, are given by eq(583).

$$\mu_A^{ide} = \mu_A^\circ + RT \ln X_A, \qquad \mu_B^{ide} = \mu_B^\circ + RT \ln X_B \qquad (583)$$

Therefore, the free energy G^{ide} of the ideal solution composed of n_A and n_B moles can be expressed as eq(584') via eq(584).

$$G^{ide} = n_A(\mu_A^\circ + RT\ln X_A) + n_B(\mu_B^\circ + RT\ln X_B) \qquad (584)$$
$$= n_A\mu_A^\circ + n_B\mu_B^\circ + (n_A + n_B)RT\langle X_A\ln X_A + X_B\ln X_B\rangle$$
$$= n_A\mu_A^\circ + n_B\mu_B^\circ + (n_A + n_B)\varDelta G_{ex}^{ide} \qquad (584')$$

From this, for the non-ideal solution whose excess free energy per mole is given by eq(585),

$$\varDelta G_{ex}^{non} = RT\langle X_A\ln X_A + X_B\ln X_B\rangle + \alpha\cdot X_A X_B \qquad (585) \qquad \text{(eq(45) in the problem)}$$

its free energy G^{non} is found to be given by eq(586) (in the case of regular solutions).

$$G^{non} - G^{ide} = (n_A + n_B)\times\alpha\cdot X_A X_B \;\Rightarrow\; G^{non} = G^{ide} + (n_A + n_B)\times\alpha\cdot X_A X_B \qquad (586)$$

Eq(586) can be rewritten as eq(586') by using eq(584').

$$G^{non} = n_A\mu_A^\circ + n_B\mu_B^\circ + (n_A + n_B)\langle RT(X_A\ln X_A + X_B\ln X_B) + \alpha\cdot X_A X_B\rangle \qquad (586')$$

Ans. free energy of non-ideal solution: eq(586')

(a) Solution 1. The chemical potential is given by $(\partial G/\partial n_i)_{P,T,n_j}$ and, hence, in order to make differentiation easier, eq(586') is rewritten as eq(586'') so as to be expressed in terms of mole numbers instead of mole fractions. Then, performing the partial differentiation as given below,

$$G^{non} = n_A\mu_A^\circ + n_B\mu_B^\circ + RT\left(n_A\ln\frac{n_A}{(n_A + n_B)} + n_B\ln\frac{n_B}{(n_A + n_B)}\right) + \alpha\cdot\frac{n_A n_B}{(n_A + n_B)} \qquad (586'')$$

$$\left(\begin{aligned}\mu_A^{non} = \left(\frac{\partial G^{non}}{\partial n_A}\right)_{P,T,n_B} &= \mu_A^\circ + RT\left\langle\ln\frac{n_A}{(n_A + n_B)} + 1 - \frac{n_A}{(n_A + n_B)} - \frac{n_B}{(n_A + n_B)}\right\rangle + \alpha\cdot\frac{n_B}{(n_A + n_B)} - \frac{n_A n_B}{(n_A + n_B)^2}\\ &= \mu_A^\circ + RT\ln\frac{n_A}{(n_A + n_B)} + \alpha\cdot\frac{n_B^{\,2}}{(n_A + n_B)^2}\end{aligned}\right)$$

the chemical potentials of the non-ideal solution are found as eqs(587) and (588) (or as eqs(587') and (588')).

$$\mu_A^{non} = \mu_A^\circ + RT\ln X_A + \alpha\cdot X_B^2 \qquad (587) \qquad \mu_B^{non} = \mu_B^\circ + RT\ln X_B + \alpha\cdot X_A^2 \qquad (588)$$
$$= \mu_A^\circ + RT\ln X_A e^{\alpha\cdot X_B^2/RT} \qquad (587') \qquad = \mu_B^\circ + RT\ln X_B e^{\alpha\cdot X_A^2/RT} \qquad (588')$$

Ans. chemical potentials of the non-ideal solution: eqs(587) and (588) (or eqs(587') and (588'))

Solution 2. The activity coefficients of A and B in the non-ideal solution are denoted γ_A and γ_B, and the chemical potentials μ_A^{non} and μ_B^{non}, respectively. Then the free energy of the solution is given by the equation below. This is put equal to eq(586) and after performing the following calculations, eq(586''') is finally derived.

$$G^{non} = n_A\mu_A^{non} + n_B\mu_B^{non} = n_A(\mu_A^\circ + RT\ln\gamma_A X_A) + n_B(\mu_B^\circ + RT\ln\gamma_B X_B)$$
$$= n_A\mu_A^\circ + n_B\mu_B^\circ + (n_A + n_B)\langle RT(X_A\ln X_A + X_B\ln X_B) + \alpha\cdot X_A X_B\rangle$$
$$= n_A\mu_A^\circ + n_B\mu_B^\circ + (n_A + n_B)\langle RT(X_A\ln X_A + X_B\ln X_B) + \alpha\cdot X_A X_B(X_A + X_B)\rangle$$
$$= n_A\mu_A^\circ + n_B\mu_B^\circ + (n_A + n_B)\langle RT(X_A\ln X_A + X_B\ln X_B) + \alpha\cdot X_A X_B^2 + \alpha\cdot X_A^2 X_B\rangle$$
$$= n_A(\mu_A^\circ + RT\ln X_A + \alpha\cdot X_B^2) + n_B(\mu_B^\circ + RT\ln X_B + \alpha\cdot X_A^2) \qquad (586''')$$

Although a slightly tricky calculation is involved in the half-way through of the calculations, it is approved if we consider how the excess energy of $\alpha\cdot X_A X_B$ is partitioned to A and B. Comparing eq(586'') with eq(586'''), the relation given below is obtained from which eq(587'), the same one as given in Solution 1 is derived.

$$RT\ln\gamma_A X_A = RT\ln X_A + \alpha\cdot X_B^2$$

$$\ln\gamma_A = \frac{\alpha\cdot X_B^2}{RT} \Rightarrow \gamma = e^{\alpha\cdot X_B^2/RT} \Rightarrow \mu_A^{non} = \mu_A^{\circ} + \ln X_A e^{\alpha\cdot X_B^2/RT} \qquad (587')$$

(b) The partial pressures of A and B in the gas phase are denoted P_A and P_B, and the vapor pressures of pure liquids P_A° and P_B°, respectively. The gases are assumed to behave ideally and then the chemical potentials of the gases are given by eqs(589) and (590).

$$\mu_A^{gas} = \mu_A^{gas}(P_A^{\circ}) + RT\ln\frac{P_A}{P_A^{\circ}} \quad (589), \qquad \mu_B^{gas} = \mu_B^{gas}(P_B^{\circ}) + RT\ln\frac{P_B}{P_B^{\circ}} \quad (590)$$

From the relation of eq(591) satisfied under the gas–liquid equilibrium, eq(592) is obtained by equating eq(587') to eq(589) and eq(588') to eq(590).

$$\left.\begin{array}{l}\mu_A^{gas} = \mu_A^{non}\\ \mu_B^{gas} = \mu_B^{non}\end{array}\right\} \quad (591), \qquad \left.\begin{array}{l}\mu_A^{gas}(P_A^{\circ}) + RT\ln\dfrac{P_A}{P_A^{\circ}} = \mu_A^{\circ} + RT\ln X_A e^{\alpha\cdot X_B^2/RT}\\ \mu_A^{gas}(P_A^{\circ}) + RT\ln\dfrac{P_A}{P_A^{\circ}} = \mu_A^{\circ} + RT\ln X_A e^{\alpha\cdot X_B^2/RT}\end{array}\right\} \quad (592)$$

Also, since the chemical potential of a liquid and the chemical potential of its gas in equilibrium with the liquid are equal、eqs(593) and (594) are obtained for the vapor pressures of A and B, respectively.

$$\mu_A^{gas}(P_A^{\circ}) = \mu_A^{\circ} \Rightarrow RT\ln\frac{P_A}{P_A^{\circ}} = RT\ln X_A e^{\alpha\cdot X_B^2/RT} \Rightarrow P_A = X_A P_A^{\circ} e^{\alpha\cdot X_B^2/RT} \qquad (593)$$

$$\Rightarrow P_B = X_B P_B^{\circ} e^{\alpha\cdot X_A^2/RT} \qquad (594)$$

Ans. partial vapor pressures of non-ideal solution: eqs(593) and (594)

(d) Since the excess energy is given by eq(585), the other excess thermodynamic quantities can be calculated from the relevant equations given below. After each equation, the corresponding excess thermodynamic quantity for an ideal solution is given.

$$\Delta H_{ex}^{non} = \Delta G_{ex}^{non} + T\Delta S_{ex}^{non} = \alpha\cdot X_A X_B \qquad (595) \qquad \Delta H_{ex}^{ide} = 0$$

$$\Delta S_{ex}^{non} = -\left(\frac{\partial \Delta G_{ex}^{non}}{\partial T}\right)_{P,X_i} = -(X_A\ln X_A + X_B\ln X_B) \qquad (596) \qquad \Delta S_{ex}^{ide} = \Delta S_{ex}^{non}$$

$$\Delta V_{ex}^{non} = \left(\frac{\partial \Delta G_{ex}^{non}}{\partial P}\right)_{T,X_i} = 0 \qquad (597) \qquad \Delta V_{ex}^{ide} = \Delta V_{ex}^{non} = 0$$

For the non-ideal solution considered in this problem, both of the excess entropy and volume are the same as the ones for the ideal solution, though the excess enthalpy is not zero as is for the ideal solution. Hence, upon mixing the two solvents, an emission or absorption of heat energy is to be observed.

Ans. excess thermodynamic quantities for non-ideal solution: $\Delta H_{ex}^{non} = \alpha\cdot X_A X_B$, $-\Delta S_{ex}^{non} = X_A\ln X_A + X_B\ln X_B$ and $\Delta V_{ex}^{non} = 0$

(e) As is obvious from the answers given in (d), if the regular solution is defined by eq(584) given in the problem, the consequence is that there is no volume change upon mixing and that a change in entropy is the same as for the ideal solution of the same concentration. This reflects that the mixing is nothing more than to distribute solvent molecules randomly without any specific orientational preference between A and B.

However, since the enthalpy of mixing is not zero for the non-ideal solution, the molecular interaction energy between A and B is different from either of between A and A or B and B. Summarizing these, the definition of the regular solution will emerge. See also **solution 3-17b**.

Ans. definition of regular solution: eq(585) or eqs(595) – (597). See text (e).

Solution 4-90

(a) As is derived in the preceding problem, the vapor pressure P of a regular solution consisting of two components A and B at temperature T is given by eq(598), where X_A and X_B are mole fractions in solution

$$P = P_A + P_B = X_A P_A^\circ e^{\alpha \cdot X_B^2/RT} + X_B P_B^\circ e^{\alpha \cdot X_A^2/RT}$$

$$= X_A P_A^\circ e^{a'(1-X_A)^2} + (1-X_A) P_B^\circ e^{a' \cdot X_A^2}, \quad a' = a/RT \qquad (598)$$

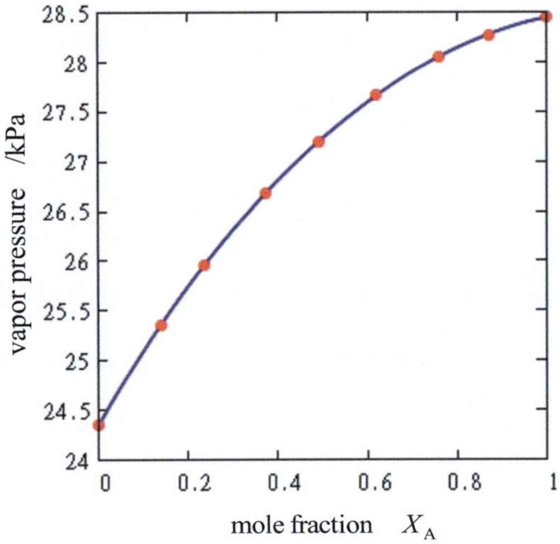

Fig.91 Vapor pressures of benzene-carbon tetrachloride mixed solutions. The curve in blue is the best-fitted one obtained by assuming regular solutions.

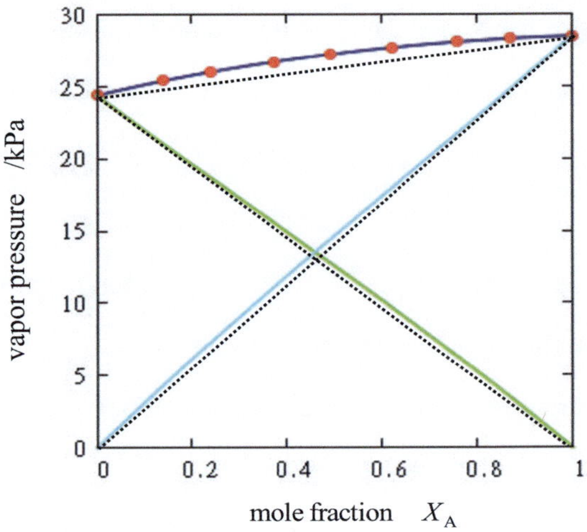

Fig92. The same mixtures as in Fig.91. Best-fitted curves for the partial vapor pressures of the components. —— : benzene, —— : carbon tetra chloride, ······ : ideal solution.

vapor pressures of the pure solvents A and B at the same temperature. Eq(598) is a function of X_A alone and, hence, is best-fitted using gnuplot obtaining the best-fitted parameter of $a' = 0.1233 \pm 0.0005$. The best-fitted curve is shown in Fig.91 together with the data points. In Fig.92, the partial pressure of each component calculated from the best-fitted eq(598) is plotted against the mole fraction. The dotted lines indicate the total and partial vapor pressures expected for the ideal solutions of the same mole fractions. As compared with these vapor pressures, the calculated partial pressures for the regular solutions, as are seen from Fig.92, display convex deviations for either component, even though slightly.

Ans. best-fitted curve: eq(598) with $a' = 0.1233 \pm 0.0005$. See Fig.91.

As is evident from Figs.91 and 92, the total and partial vapor pressures of the mixed solutions of benzene and carbon tetrachloride can sufficiently be expressed by eq(598) which is derived for the purpose of describing the regular solution. Also, the coefficient $a' = a/RT$ of the correction term, which takes a deviation from the ideal solution into account, is small and actually the deviation is small (Fig.92). Putting these two facts together, the mixed solution of benzene and carbon tetrachloride can safely be regarded as the regular solution. Judging from a positive value of the coefficient a', the intermolecular interaction between benzene and carbon tetrachloride molecules is regarded as repulsive.

To examine how the vapor pressure of the mixture varies against the mole fraction when the coefficient for the correction term is made larger positively or negatively, the vapor pressures are calculated in the two cases of $a' = 1.0$ and $a' = -1.0$, and the results obtained are shown in Fig.93 and Fig.94, respectively. For regular solutions in general, the premise is that there does not exist very large differences in the intermolecular interaction among component molecules and, hence, not much different vapor pressures of pure solvents, 30 kPa and 25 kPa, are employed in Figs.93 and 94. When the absolute value of the coefficient $|a'|$ is increased, it is seen that the deviation from the ideal behavior becomes more conspicuous. For many two-component mixtures, the deviation from the ideal behavior is either as shown in Fig.93 (convex deviation) or as shown in Fig.94 (concave deviation). However, whether or not a given mixture which exhibits either feature can be regarded as a regular solution is another thing. See next problem with respect to this.

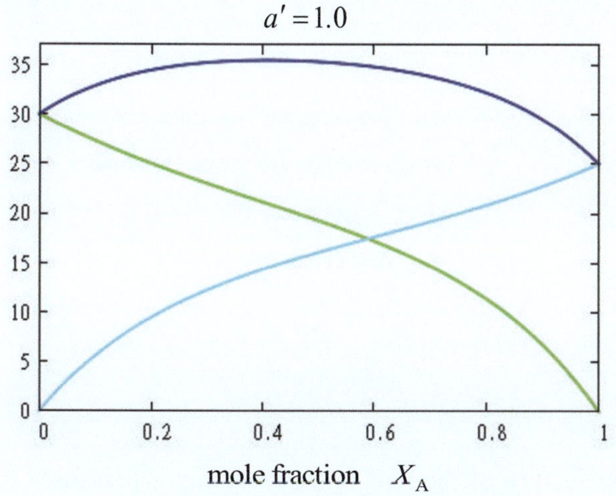

Fig.93 Positive correction coefficient.

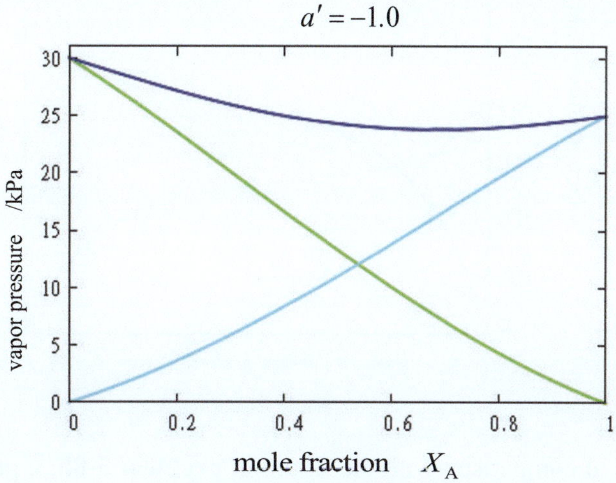

Fig.94 Negative correction coefficient.

Ans. Mixed solution of benzene and carbon tetrachloride can be regarded as regular solution.

Solution 4-91

(i) Mixed solutions of chloroform-acetone (A)

In a similar manner as employed in the preceding problem, the vapor pressure of the mixed solution is described by eq(599) and the best-fitted coefficient a' is found as $a' = -0.755 \pm 0.013$. The data used and the best-fitted curve are shown in Fig.95.

$$P = P_A + P_B = X_A P_A^\circ e^{\alpha \cdot X_B^2/RT} + X_B P_B^\circ e^{\alpha \cdot X_A^2/RT}$$

$$= X_A P_A^\circ e^{a'(1-X_A)^2} + (1 - X_A) P_B^\circ e^{a' \cdot X_A^2}, \quad a' = a/RT \qquad (599)$$

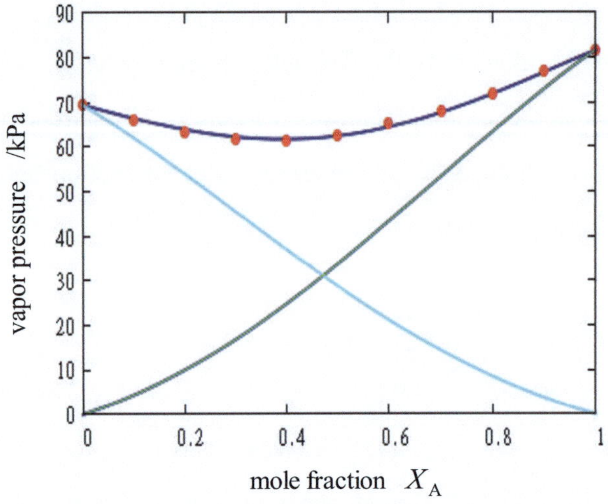

Fig.95 Vapor pressures and best-fitted curve for chloroform and acetone mixtures.
——: chloroform, ——: acetone.

(ii) Mixed solutions of <u>carbon disulfide – acetone (A)</u>

Similarly as above, the best-fitted coefficient for the correction term is found as $a' = 1.619 \pm 0.018$. The data used and the best-fitted curve are shown in Fig.96.

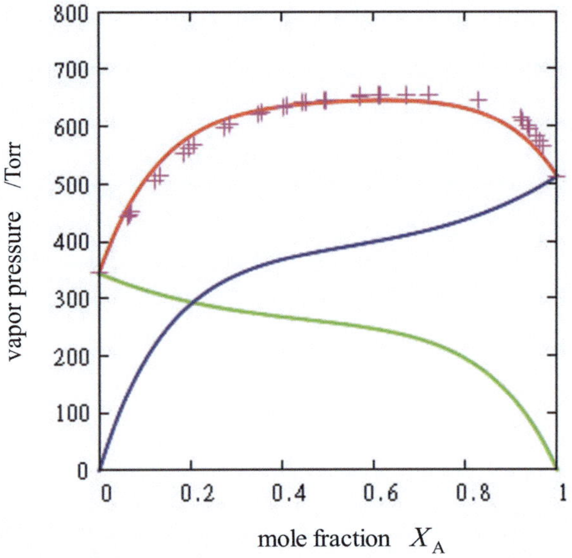

Fig.96 Vapor pressures and best-fitted curve for carbon disulfide and acetone mixtures.
——: carbon disulfide, ——: acetone,
+: data points.

In comparison with the figure in **problem 3-8b**, a gross feature of the total and partial vapor pressures is well reproduced, however, since the agreement with the data is not so satisfactory and, furthermore, the coefficient for the correction term is too large, it is not appropriate to treat the mixture as a regular solution.

Basically, the regular solution premises that there exists no specific intermolecular interactions between A and B in solution. For this, mixed solutions prepared by mixing two polar solvents are not the object of the regular solution from the start. On the other hand, for the system treated in (i), the vapor pressure is extremely well reproduced by eq(599) and, hence, it is appropriate to express the excess free energy by the term of $a \cdot X_A \cdot X_B$. Strictly speaking, however, since the excess entropy of mixing must be zero based on the rigorous definition of the regular solution as is described in **solution 4-89**, before judging, it is necessary to check whether or not the excess entropy of mixing for given solutions is negligibly small by acquiring actually the excess entropy through the measurements of the temperature dependence of the equilibrium vapor pressure. It should be noted that the correction coefficient is a function of temperature.

Ans. regular solution: chloroform-acetone. Carbon disulfide-acetone does not fit regular solution. For a more strict judgment, a value for the excess entropy of mixing is required.

[additional problem]

The molar conductivity of acetic acid in aqueous solution Λ was measured as a function of the concentration of acetic acid C and the values summarized in the table below were obtained. Denoting the degree of dissociation of acetic acid in water by α, and the molar conductivity at the infinite dilution $(C \to 0)$ by Λ_0, the *Arrhenius* equation (eq(600)) is satisfied.

$$\Lambda = \alpha\Lambda_0 \qquad (600) \Rightarrow \alpha = \frac{\Lambda}{\Lambda_0} \qquad (600')$$

We write the dissociation constant of acetic acid as K_a and assume that *Ostwald*'s dilution law (eq(601)) holds. Then, find the values for the dissociation constant of acetic acid K_a and the electrical conductivity at the infinite dilution Λ_0 (limiting molar conductivity) of acetic acid based on the data given in the table.

$$K_a = \frac{(C\alpha)^2}{C(1-\alpha)} \qquad (601)$$

Table Electrical conductivity Λ as a function of molar concentration C of acetic acid in aqueous solution.

$C \times 10^5/\text{mol}\cdot\text{L}^{-1}$	2.8	11.135	15.32	21.84	102.83	136.34	241.4	334.065	591.15
$\Lambda/\text{S}\cdot\text{cm}^2$	210.32	127.71	112.02	96.47	48.13	42.215	32.21	27.19	20.96

(*E.A.Moelwyn-Hughes, Physical Chemistry,* 2nd revised ed., p.854,1965).

[brief solution]

Substituting eq(600) into eq(601), eq(602') is obtained via eq(602).

$$K_a(1-\Lambda/\Lambda_0) = C(\Lambda/\Lambda_0)^2 \Rightarrow \left(\frac{\Lambda_0}{\Lambda}\right)K_a(1-\Lambda/\Lambda_0) = C\left(\frac{\Lambda}{\Lambda_0}\right) \qquad (602)$$

$$\left(\frac{\Lambda_0}{\Lambda}\right)(1-\Lambda/\Lambda_0) = \frac{C}{K_a}\left(\frac{\Lambda}{\Lambda_0}\right) \Rightarrow \frac{1}{\Lambda} = \frac{1}{\Lambda_0} + \frac{1}{K_a\Lambda_0^2} \times C\Lambda \qquad (602')$$

Eq(602') indicates that when the *Arrhenius* equation (eq(600)) and *Ostwald*'s dilution law are satisfied, there exists a linear relationship between $1/\Lambda$ and $C\Lambda$. The values for these variables were obtained from the data given in the table and they are plotted in Fig.97 below.

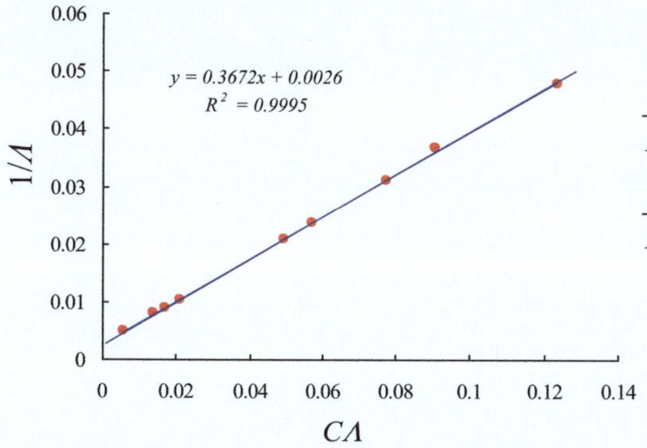

$$\frac{1}{K_a\Lambda_0^2} = 0.3672 \pm 2.6\times10^{-3} \Rightarrow K_a = \frac{(2.64\times10^{-3})^2}{0.3672} = 1.89\times10^{-5}$$

$$\frac{1}{\Lambda_0} - 2.64\times10^{-3} \mp 1.9\times10^{-4} \Rightarrow \Lambda_0 = 379\ \text{S}\cdot\text{cm}^2$$

Fig.97 Linear plot based on eq(602').

As is calculated on the right of Fig.97, from the slope and the intercept of the plot, the values for the d issociation constant K_a and the limiting molar conductivity Λ_0 of acetic acid are found.

$$\hbar$$

$$\hat{l}_x + i\hat{l}_y = e^{i\varphi}\left(\frac{\partial}{\partial\vartheta} + i\cot\vartheta\frac{\partial}{\partial\varphi}\right)$$

$$k(t) = 4\pi R_{AB} D_{AB} N_A\left\{1 + \frac{R_{AB}}{(\pi D_{AB} t)^{1/2}}\right\}$$

Further Readings

(Books alone. Any articles cited in this book are referred to at relevant places in the book.)

Quantum Mechanics

1. *L.D.Landau and E.M.Lifhitz, Quantum Mechanics-non-relativistic theory、Translated from Russian into English by J.B.Skyes and J.S.Bell.*
2. *D.ter Haar, Selected Problems in Quantum Mechanics,* Academic Press Inc., New York, 1964.
3. *H.Eyring, L.J.Walt, and G.E.Kimball, Quantum Chemistry,* J.Wiley Sons, Inc., New York, 1944, ch.8.
4. *L.I.Shiff, Quantum Mechanics,* Mcgraw-Hill Book Comp, Inc., New York 1955. Translated into Japanese by *T.Inoue* and published from Yoshioka Shotenn.
5. P.A.M.*Dirac, The Principles of Quantum Mechanics,* Oxford University Press, 1947. Translated into Japanese and published from Iwanami Shotenn.
6. *Shinnichirou Tomonaga, Quantum Mechanics* vol.I and II, published in Japanese from Misuzu Publishing Company in1953.
7. *J. E. Straub, Mathematical Methods for Molecular Science: Theory and Applications, Visualizations and Narrative* University Science Books, 2022.
8. *D. A. McQuarrie, Quantum Chemistry,* 2nd ed., 2007. *ISBN 978-1-891389-50-4.*
9. *N. Zettili, Quantum Mechanics: Concepts and Applications.* Wiley,2009. *ISBN 978-0470026793.*

Statistical and Classical Mechanics

1. *Terrell.L.Hills、An Introduction to Statistical Mechanics,* Addison-Wesley Publishing Comp.Inc, 1960.
2. *L.D.Landau and E.M.Lifschitz, Course of Theoretical Physics* vol.5, *Statistical Physics,* chap.5. Translated from the Russian and published in Japanese from Iwanami Shotenn in 1958.
3. *R.Fowler and E.A.Guggenheim , Statistical Thermodynamics* ,Cambridge University Press , 1960.
4. *G.S.Rushbrooke, Introduction to Statistical Mechanics,* Oxford at the Clarendon Press, 1951.

S. Hirayama and M. Yamamoto, *Physical Chemistry Problems and Solutions*,
https://doi.org/10.1007/978-981-97-0165-0

Translated into Japanese by *S.Kubo and T.Kinoshita* and published from Hakusuisha.

5. *ter Haar, Elements of Statistical Mechanics,* Rinehart & Company, NC., 1956.
6. *R.C.Tolman, The Principle of Statistical Mechanics,* Oxford University Press. First printed in 1938 and last printed in 1967.
7. .*E.A.Guggenheim, Applications of Statistical Mechanics,* Clarendon Press, Oxford, 1966.
8. *J.H.Konx, Molecular Thermodynamics, An Introduction to Statistical Mechanics for Chemists,* Wiley-Interscience, New York, 1971.
9. *H.Goldstein, Classical Mechanics,* Addison-Wesley Publishing Company, INC, 1957.
10. *Hideki Yukawa and Matsuhei Tamura, An Introduction to Physics*, published in Japanese in 1961 from Daimeidou.
11. *D. A. McQuarrie, Statistical Mechanics, 2000. ISBN 978-1-891389-15-3.*
12. *Fumiko Yonezawa, Brown Motion,* Kyouritu Publishing Company (27 in One Point Physics Series)、1986.

Thermodynamics

1. *K.S.Pitzer and L.Brewer, Thermodynamics, 2nd Ed.,McGraw-Hill Book Company,* 1961.
2. *H.B.Callen, Thermodynamics, An Introduction to the Physical Theories of Equilibrium Thermostatics and Irreversible Thermodynamics,* John and Wiley & Sons , Inc., 1960.
3. .*I.Prigogenue and R.Defay ,Thermodynamique Chemique, Chemical Themodynamics* I (Japanese translation,1966).
4. *H.Margenau and G.M.Murphy, The Mathematics of Physics and Chemistry,* D.van Nostrand Company, Inc., 1945. Translated into Japanese by *T.Sato and M.Kunisada* and published by Kyoritsu Zensho.
5. *S.Tomonaga, What is Physics* vol.1 and 2, 1979. Published in Japanese from Iwanami Shoten

Chemical Kinetics

1. *S.Glasstone, K.J.Laidler and H.Eyring, The Theory of Rate Processes*, McGraw-Hill Book Comp., Inc., New York, 1941.
2. *P.Erdi and J.Toth, "Mathematical Models of Chemical Reactions —Theory and Applications of Deterministic and Stochastic Models",* Manchester University Press, 1989.
3. *P.J.Robinson and K.A.Holbrook, Unimolecular Reactions,* Wiley-Interscience, 1972.
4. *S.W.Benson, The Foundations of Chemical Kinetics,* McGraw-Hill Book Comp., Inc., New York, 1960.
5. *A.A.Frost and R.G.Peason*、 *Kinetics and Mechanisms,* Wiley, New York, 1961.

6. P*rogress in Reaction Kinetics ed. by G.Porter,* vol.3, Pergamon Press, Oxford, 1965.

7. *P.Erdi and J.Toth, "Mathematical Models of Chemical Reactions —Theory and Applications of Deterministic and Stochastic Models",* Manchester University Press, 1989.

8. *P.J.Robinson and K.A.Holbrook, Unimolecular Reactions*、Wiley-Interscience, London, 1972.

9. *Stephen A.Rice, Comprehensive Kinetics vol.25, Diffusion-limited Reactions, ed. by C.H.Bamford,* Elsevier, 1985, ch.9.

Physical Chemistry

1. *T.Engel and P.Reid, Physical Chemistry,* Person Benjamin Cummings, 2004

2. *R.S.Berry, S.A.Rice and J.Ross*、*Physical Chemistry,* Oxford University Press, 2000.

3. *D.A.McQuarrie and J.D.Simon, Physical Chemistry A Molecular Approach,* University Science Books, 1997.

4. *Walter J. Moore, Basic Physical Chemistry,* Prentice-Hall Inc., 1983.

5. *Walter J. Moore, Physical Chemistry* 4th ed., Prentice-Hall Inc.,1972.

6. *P.Atkins, Physical Chemistry* 6th ed., Oxford University Press, 1998. 8th ed., 2006.

7. *E.A.Moelwyn-Hughes, Physical Chemistry,* 2nd revised ed., Ipswich Litho Press Lit., 1965.

8. *Clyde R. Metz, Theory and Problems of Physical Chemistry, Schaum's OutLine Series,* McGraw-Hill

9. *Koshiyama, Exercises and Solutions in Physical Chemistry,* 1960 (in Japanese).

10. *J.Yamaguchi, S.Hirayama, H.Taniguchi and N.Azuma*、*Basic Physical Chemistry* (in Japanese), Asakura Shoten, 1989.

11.*K.Kano and M.Yamamoto, Joyful Physical Chemistry1-Chemical Thermodynamics and Reaction Kinetics,* Kodannsha Sientific, 2016. Published in Japanese.

12.*M.Yamamoto, Sikeda and K.Kano, Joyful Physical Chemistry2—Quantum Chemistry,* to be published in Japanese in 2023 from Kodannsha Sientific.

13. *D.A. McQuarrie, Mathematics for Physical Chemistry: Opening Doors*, 2008. *ISBN 978-1-891389-56-6.*

14. *D.A. McQuarrie, Mathematical Methods for Scientists and Engineers*, 2003. *ISBN 978-1-891389-29-0*

15. *H. Cox, Problems and Solutions to Accompany McQuarrie and Simon's Physical Chemistry*,1997. *ISBN 978-0-935702-43-9.*

Photochemistry and Molecular Spectroscopy

1. *J.B.Birks, Photophysics of Aromatic Molecules,* Wiley-Interscience, London, 1970.

2. *H.H.Jaffe and Milton Orchin, Theory and Applications of Ultraviolet Spectroscopy, ch.12 ,* John Wiley & Sons, Inc., New York, 1964.

3. *I.B.Berlman、Handbook of Fluorescence Spectra of Aromatic Molecules,* Academic Press, New York, 1966.

4. *G.Hergberg "Molecular Spectra and Molecular Structure 1. Spectra of Diatomic Molecules",* D.van Nostrand Company, Inc., 1950.

5.*J.G.Calvert and J.N.Pitts,Jr., Photochemistry,* John Wiley & Sons, Inc., 1966.

6. *N.J.Turro, Molecular Photochemistry,* Benjamin, 1965.

7. S.P.*McGlynn, T.Azumi and M.Kinoshita, Molecular Spectroscopy of the Triplet State, Prentice Hall, Inc.,* 1969.

8. *J.R.Lakowicz, Principles of Fluorescence Spectroscopy,* 2nd ed., Kluwer Academic/Plenum Publishers 1999。

9. *D.V.O'Conner and D.Phillips, Time-Correlated Single Photon Counting,* Academic Press, 1984. Translated into Japanese by *S.Hirayama and K.Hara* and published from Gakkai Publishing Company.

10. *J.H.D.Eland, "Photoelectron Spectroscopy: An Introduction to Ultraviolet Photoelectron Spectroscopy* (1974). *Japanese translation* (Koudenshi Bunnkouhou) *by K.Kimura and S.Katumata.*

11. *F.Alber Cotton, Chemical Applications of Group Theory*, Interscience Publishers, New York, 1964.

12. *C.N.Banwell and E.M.McCash, Fundamentals of Molecular Spectroscopy*, 4th ed., McGraw-Hill Book Company Europe,1994.

13. Shuuichi Kinoshita et.al., Encyclopedia of Emission, published from Asakura Shotenn in Japanese.

14.*S.Hirayama, Progress in Photochemistry and Photophysics,* vol.V1, ch.1, ed. by *J.F.Rabek,* CRC Press.

15.*Prem Bisht, An Introduction to Photonics and Laser Physics with Applications,* 2022. https://iopscience.iop.org/book/978-0-7503-5226-0

16.*Ch.Zander, J.Enderlein and R.A.Keller(ed.), Single Molecule Detection in Solution, Methods and Applications*, 2002, Wiley-VCH.

Data Analysis

1. *Philip R.Bevington, Data Reduction and Error Analysis for the Physical Science,* McGraw-Hill Book Company, 1969.

2.*K.Maeda, M.Yamamoto and K.Kano, How to Treat Experimental Data Appropriately,* 1997. Published in Japanese from Kagakudoujinn..

3.*D. S. Sivia and S. G. Rawlings,* Foundations of Science Mathematics (Oxford Chemistry Primers, 77 and Worked Problems) 2018. Translated into Japanese by M.Yamamoto and K.Kano. Kagakudoujinn

4. *Shigeru Hannba, Introduction to Gnuplot,* available on internet.

General Chemistry and Others

1. *Linus Pauling , The Nature of the Chemical Bond*、3rd ed., Cornell University Press, 1960.

2. *P.Atkins, T.Overton, J.Rourke, M.Weller and F.Armstrong, Inorganic Chemistry,* 5th ed., Oxford Uiversity Press, 2010.

3. *M.Jones Jr., Organic Chemistry* 3rd ed., W.W.Norton, 2000.

4. *B.H.Maian, University Chemistry* 2nd ed. Addison Wesley Publishing company, Inc.,1968.

5. *J.McMurry and R.C.Fay, General Chemistry; Atoms First,* Prentice Hall, 2010. Translated into Japanese by *H.Hagino, G.Yamamoto and K.Ohno*, and published from Tokyo Kagaku Doujinn.

6. *J.O'M.Bockris and A.K.N.Reddy, Modern Electrochemistry,* vol.1 and 2, Plenum Press, 1970.

7. *P.J.Flory, Principles of Polymer Chemistry,* Cornell University Press, New York, 1953.

8. *C.A.Coulson,"Valence"*,Oxford University Press, 1961

9. *J.O.Hirschfelder, C.F.Curtiss, R.B.Bird, Molecular Theory of Gases and Liquids,* John Wiley & Sons, Inc., New York, 1954.

10. *S.D.Hammann, Physico-Chemical Effects of Pressure,* Butterworth Scientific Publications, 1957.

11. *H. van Vleck, The Theory of Electric and Magnetic Susceptibilities,* Oxford University Press. Japanese translation by *M.Kotani and K.Kanda* and published by Yoshioka Shotenn.

12. *P.Debye, Polar Molecules*, Dover Publications Inc.,1929.

13. *H.Fröhlich, Theory of Dielectrics – Dielectric Constant and Dielectric Loss.*

14. *J.N.Murrell and A.D.Jenkins, Properties of Liquids and Solutions,* 2nd ed., John Wiley $ Sons, 1994.

15. *Techniques of Chemistry,* vol.II, *Organic Solvents —Physical Properties and Methods of Purification,* 3rd ed.

16. *R.W.Gurney, Ionic Processes in Solution,* Dover Publication Inc.,1953. Translated into Japanese by *S.Kikuchi and S.Suzuki* and published by Sangyou Tosho Pub. Co.., Ltd.

General Interest

1. *Frank Close "Too Hot to Handle--The Race for the Cold Fusion", Penguin Books.*
2. *Marcus du Sautoy, The Music of the Primes--Why an Unsolved Problem in Mathematics Matters,* Haper Perennial 2004.
3. *Peter Mason, The Light Fantastic,* Penguin Books Australia, 1981.
4. *V.Pieribone and D.F.Gruber, A Glow in the Dark,* The Belknap Press of Harvard University Press, 2005.
5. *Theodore Gray, Elements A Visual Exploration of Every Known Atom in the Universe,* Black Dog & Leventhal , 2009.
6. *Al Gore, An Inconvenient Truth,* Rodale, 2006.
7. *H.Aldersey-Williams, Periodic Tales, The Curious Lives of the Elements,* Penguin Books, 2011.
8. *Manjit Kumar, Quantum: Einstein, Bohr and the Great Debate About the Nature of Reality,* 2009. *ISBN 9781848310353*

(Access HP of *MITOCW* and click *Audio/Video Course* and select *Physics* in which the 21st lecture on *Electricity and Magnetism* delivered in 2002 is found. About 43 minutes later you will see the relevant experiment in the lecture).

Index

The key words and phrases listed here are rather extensive. So that, the authors do hope, they are greatly helpful in finding the problems or solutions you are concerned with. The gothic letters **P** and **S** put before numerals such as 2-11, 7-38, etc., signify "Problem" and "Solution", respectively. Where advanced problems and exercises are found is indicated at the end of the Index.

[A, a]

S. Hirayama and M. Yamamoto, *Physical Chemistry Problems and Solutions*,
https://doi.org/10.1007/978-981-97-0165-0

[G, g]

[H, h]

[N, n]

[W, w]